Newcastle College Library
02602288

I/S 11/10
KU-574-233
WITHDRAWN

To
f you

INTRODUCTION TO MODERN LIQUID CHROMATOGRAPHY

Third Edition

INTRODUCTION TO MODERN LIQUID CHROMATOGRAPHY

Third Edition

LLOYD R. SNYDER
LC Resources, Inc.
Orinda, CA

JOSEPH J. KIRKLAND
Advanced Materials Technology
Wilmington, DE

JOHN W. DOLAN
LC Resources, Inc.
Amity, OR

WILEY

A John Wiley & Sons, Inc., Publication

Published by John Wiley & Sons, Inc., Hoboken, New Jersey.
Published simultaneously in Canada.

***Library of Congress Cataloging-in-Publication Data*:**

Snyder, Lloyd R.
Introduction to modern liquid chromatography / Lloyd R. Snyder, Joseph J. Kirkland. – 3rd ed. / John W. Dolan.
p. cm.
Includes index.
ISBN 978-0-470-16754-0 (cloth)
1. Liquid chromatography. I. Kirkland, J. J. (Joseph Jack), 1925- II. Dolan, John W. III. Title.
QD79.C454S58 2009
543′.84–dc22

2009005626

Printed in the United States of America.

10 9 8 7 6 5 4 3 2 1

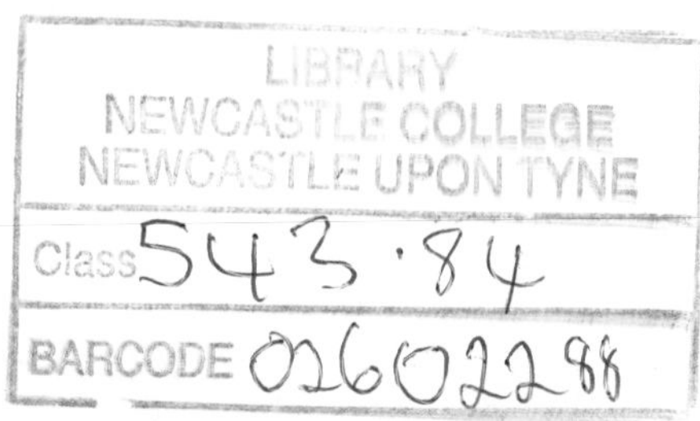

CONTENTS

PREFACE

High-performance liquid chromatography (HPLC) is today the premier technique for chemical analysis and related applications, with an ability to separate, analyze, and/or purify virtually any sample. The second edition of this book appeared in 1979, and for tens of thousands of readers it eventually became their choice of an HPLC reference book. The remarkable staying power of the second edition (with significant sales into the first decade of the present century) can be attributed to certain features which continue to be true for the present book. First, all three editions have been closely tied to short courses presented by the three authors over the past four decades, to an audience of more than 10,000 industrial, governmental, and academic chromatographers. Teaching allows different approaches to a subject to be tried and evaluated, and a pragmatic emphasis is essential when dealing with practicing chromatographers as students. Second, all three editions have tried to combine practical suggestions ("how to?") with a theoretical background ("why?"). Both theory and practice continue to be emphasized so that the reader can better understand and evaluate the various recommendations presented here. Finally, each of the three authors has been an active participant in HPLC research, development, and/or routine application throughout most of their careers.

Since the preparation of the second edition in 1979, there have been major improvements in columns and equipment, as well as numerous advances in (1) our understanding of HPLC separation, (2) our ability to solve problems that were troublesome in the past, and (3) the application of HPLC for new kinds of samples. Whereas six different HPLC procedures received comparable attention in the second edition, today reversed-phase chromatography (RPC) accounts for about 80% of all HPLC applications—and therefore receives major (but not exclusive) attention in the present edition. Over the past three decades the use of HPLC for biological samples, enantiomeric (chiral) separations, and sample purification has expanded enormously, accompanied by a much better understanding of these and other HPLC applications.

Commercial HPLC columns continue to be improved, and many new kinds of columns have been introduced for specific applications, as well as for faster, trouble-free operation. Prior to 1990, HPLC method development was an uncertain process—often requiring several months for the acceptable separation of a sample.

Since then it has become possible to greatly accelerate method development, especially with the help of appropriate software. At the same time HPLC practice is increasingly carried out in a regulatory environment that can slow the release of a final method. These various advances and changes in the way HPLC is carried out have mandated major changes in the present edition.

The organization of the present book, while similar to that of the second edition, has been significantly modified in light of subsequent research and experience. Chapter 1 provides a general background for HPLC, with a summary of how its use compares with other modern separation techniques. Chapter 1 also reviews some of the history of HPLC. Chapter 2 develops the basis of HPLC separation and the general effects of different experimental conditions. Chapters 3 and 4 deal with equipment and detection, respectively. In 1979 the detector was still the weak link in the use of HPLC, but today the widespread use of diode-array UV and mass-spectrometric detection—as well as the availability of several special-purpose detectors—has largely addressed this problem. Chapter 5 deals with the column: the "heart" of the HPLC system. In 1979, numerous problems were associated with the column: peak tailing—especially for basic samples, column instability at elevated temperatures or extremes in mobile-phase pH, and batch-to-batch column variability; today these problems are *much* less common. We also now know a good deal about how performance varies among different columns, allowing a better choice of column for specific applications. Finally, improvements in the column are largely responsible for our current ability to carry out ultra-fast separations (run times of a few minutes or less) and to better separate mixtures that contain hundreds or even thousands of components.

Chapter 6, which deals with the reversed-phase separation of non-ionic samples, extends the discussion of Chapter 2 for these important HPLC applications. A similar treatment for normal-phase chromatography (NPC) is given in Chapter 8, including special attention to hydrophilic interaction liquid chromatography (HILIC). In Chapter 7 the separation of ionized or ionizable samples is treated, whether by RPC, ion-pair chromatography, or ion-exchange chromatography. Gradient elution is introduced in Chapter 9 for small-molecule samples, and as an essential prerequisite for the separation of large biomolecules in Chapter 13; two-dimensional separation—another technique of growing importance—is also discussed. Chapter 10 covers the use of computer-facilitated method development (computer simulation). Other important, general topics are covered in Chapters 11 (Qualitative and Quantitative Analysis) and 12 (Method Validation).

Chapter 13 introduces the separation of large molecules, including both biological and synthetic polymers. HPLC procedures that are uniquely useful for these separations are emphasized: reversed-phase, ion-exchange, and size-exclusion, as well as related two-dimensional separations. Chapter 14 (Enantiomer Separations) marks a decisive shift in approach, as the resolution of enantiomers requires columns and conditions that are sample-specific—unlike most of the HPLC applications described in earlier chapters.

Chapter 15 deals with preparative separations ("prep-LC"), where much larger sample weights are introduced to the column. The big change since 1979 for prep-LC is that we now have a much better understanding of how such separations vary with conditions, in turn making method development much more systematic and efficient. Chapter 16 (Sample Preparation) provides a comprehensive coverage of this important supplement to HPLC separation. As in the case of other HPLC-related topics, the past 30 years have seen numerous developments that today make sample preparation a routine addition to many HPLC procedures. Finally, Chapter 17 deals with HPLC troubleshooting. Despite all our advances in equipment, columns, materials, technique, and understanding, trouble-free HPLC operation is still not guaranteed. Fortunately, our ability to anticipate, diagnose, and solve HPLC problems is now more informed and systematic. One of our three authors (JWD) has been especially active in this area.

Different readers will use this book in different ways. An experienced worker may wish to explore topics of his or her choice, or find an answer to specific problems. For this audience, the Index may be the best starting place. Beginning readers might first skim Chapters 1 through 7, followed by 9 through 10, all of which emphasize reversed-phase HPLC. The latter sequence is similar to the core of the basic HPLC short courses developed by the authors. After this introduction, the reader can jump to chapters or sections of special interest. Other readers may wish to begin with topics of interest from the Contents pages at the front of the book or at the beginning of individual chapters. The present book has been organized with these various options in mind.

This third edition is highly cross-referenced, so as to allow the reader to follow up on topics of special interest, or to clarify questions that may arise during reading. Because extensive cross-referencing represents a potential distraction, in most cases it is recommended that the reader simply ignore (or defer) these invitations to jump to other parts of the book. Some chapters include sections that are more advanced, detailed, and of less immediate interest; these sections are in each case clearly identified by an introductory advisory in *italics*, so that they can be bypassed at the option of the reader. We have also taken pains to provide definitions for all symbols used in this book (Glossary section), along with a comprehensive and detailed index. Finally, attention should be drawn to a "best practices" entry in the Index, which summarizes various recommendations for both method development and routine use.

We very much appreciate the participation of eight collaborators in the preparation of the present book: Peter Schoenmakers (Sections 9.3.10, 13.10), Mike Swartz (Chapter 12), Tim Wehr (Sections 13.1–13.8), Carl Scandella (Section 13.9), Wolfgang Lindner, Michael Lämmerhofer, and Norbert Maier (Chapter 14), Geoff Cox (Chapter 15), and Ron Majors (Chapter 16). Their affiliations are as follows:

Peter Schoenmakers	University of Amsterdam
Mike Swartz	Synomics Pharma
Tim Wehr	BioRad Corp.
Carl Scandella	Carl Scandella Consulting (4404 91st Avenue NE Bellevue, WA 98004)
Wolfgang Lindner, Michael Lämmerhofer, and Norbert Maier	University of Vienna
Geoff Cox	Chiral Technologies
Ron Majors	Agilent Technologies

We also are indebted to the following reviewers of various parts of the book: Peter Carr, Tom Chambers, Geoff Cox, Roy Eksteen, John Fetzer, Dick Henry, Vladimir Ioffe, Pavel Jandera, Peter Johnson, Tom Jupille, Ron Majors, Dan Marchand, David McCalley, Imre Molnar, Tom Mourey, Uwe Neue, Ravi Ravichandran, Karen Russo, Carl Scandella, Peter Schoenmakers, and Loren Wrisley. However, the authors accept responsibility for any errors or other shortcomings in this book.

LLOYD R. SNYDER
J. J. (JACK) KIRKLAND
JOHN W. DOLAN

Orinda, CA
Wilmington, DE
Amity, OR

GLOSSARY OF SYMBOLS AND ABBREVIATIONS

This section is divided into "frequently used" and "less-frequently used" symbols." Most symbols of interest will be included in "frequently used symbols". Equations that define a particular symbol are listed with that symbol; for example, "Equation 2.18" refers to Equation (2.18) in Chapter 2. The units for all symbols used in this book are indicated. Where IUPAC definitions or symbols differ from those used in this book, we have indicated the corresponding IUPAC term (from ASDLID 009921), for example, t_M instead of t_0.

FREQUENTLY USED SYMBOLS AND ABBREVIATIONS

A	the "weak" component in a binary-solvent mobile phase (A/B); in RPC, the A-solvent is water or aqueous buffer; also, "type-A" silica (older, more acidic silica)
ACN	acetonitrile
B (%B)	the "strong" component (and its %-volume) in a binary-solvent mobile phase (A/B); in RPC, the B-solvent is an organic, such as acetonitrile; also, "type-B" silica (newer, less acidic silica; Section 5.2.2.2)
CSP	chiral stationary-phase
CV	coefficient of variation (equivalent to %-relative standard deviation); also, column volumes (Section 13.9)
C_8, C_{18}	Reversed-phase column-packing designations, indicating length of alkyl ligand bonded to the particle
d_c	column inner diameter (mm)
d_p	column-packing particle-diameter (μm)
F	mobile-phase flow rate (mL/min)

H	column plate height (equal to L/N); see also "less-frequently used symbols" below
HIC	hydrophobic interaction chromatography
HILIC	hydrophilic interaction chromatography
i.d.	column or tubing inner diameter (mm)
IEC	ion-exchange chromatography
IPC	ion-pair chromatography
k	retention factor (same as capacity factor k'); equal to $(t_R/t_0) - 1$
k^*	gradient retention factor; Equation (9.5)
L	column length (mm)
LC-MS	liquid chromatography–mass spectrometry
LC-MS/MS	LC-MS with a triple-quadrupole mass spectrometer
M	molecular weight (Da)
MeOH	methanol
MS	mass spectrometry
N	column plate number; Equation (2.9)
n_c	"equivalent" peak capacity, usually referred to as "conditional" or "sample" peak capacity
NPC	normal-phase chromatography
P	pressure drop across the column (psi); bar or atmospheres = 14.7 psi; megaPascal (MPa) = 10 bar = 147 psi; also, partition coefficient (Section 6.2)
PC	peak capacity; Equation (2.30), Figure 2.26*a* (isocratic); Equation 9.20, Figure 9.20 (gradient)
pK_a	logarithm of the acidity constant for an acid or base; Equations (7.2), (7.2a)
R_F	solute fractional migration in TLC; Equation (8.6), Figure 8.8
RI	refractive index
RPC	reversed-phase chromatography
R_s	resolution; Equation (2.23)
S	slope of plots of log k versus $\phi (d \log k/d\phi)$; Equation (2.26)
SEC	size-exclusion chromatography
SPE	solid-phase extraction
T	temperature (°C)
t_D	dwell time (min); equal V_D/F
TFA	trifluoroacetic acid
t_G	gradient time (min); Figure 9.10

t_0	column dead-time (min); also the retention time of a non-retained solute; equal to V_m/F; Equations (2.4a), (2.7)
T-P	touching-peak; Figure 15.9*b*
t_R	retention time (min); Equation (2.5)
type-A	older, more acidic silica (Section 5.2.2.2)
type-B	newer, less acidic silica (Section 5.2.2.2)
UV	ultraviolet absorption
V_D	equipment dwell volume; Section 9.2.2.4
V_m	column "dead-volume"; volume of the mobile phase within a column (mL); Equation (2.7a)
W	baseline peak width *W*; Figure 2.10*a*
w_s	column saturation capacity (g)
w_x	weight of solute injected (g)

α	separation factor; Equation (2.24a)
$\Delta\phi$	change in ϕ during a gradient; Figure 9.2*g*
ε	mobile-phase solvent strength in NPC; Equations (8.2), (8.5); also, dielectric constant
ε^0	value of ε (in NPC) for a pure solvent
ϕ	volume-fraction of the B-solvent (equal to $0.01 \times$ %B)
ϕ^*	value of ϕ during gradient elution for a solute, when the band reaches the column midpoint
ν	reduced velocity; Equation (2.18a)
η	mobile-phase viscosity (cP)

LESS-FREQUENTLY USED (OR LESS-COMMONLY UNDERSTOOD) SYMBOLS AND ABBREVIATIONS

A	absorbance
A	column hydrogen-bond acidity; Equation (5.3)
AAPS	American Society of Pharmaceutical Scientists
AIQ	analytical instrument qualification (or validation)
AMT	analytical method transfer
AOAC	Association of Official Analytical Chemists
APCI	atmospheric pressure chemical ionization
API	active pharmaceutical ingredient (also atmospheric pressure ionization)

A_s	peak asymmetry factor; Figure 2.16*a*
AU	absorbance units (UV detection)
b	fundamental gradient steepness parameter; Equation (9.4)
$\boldsymbol{B}$	column hydrogen-bond basicity; Equation (5.3)
$\boldsymbol{C}$	column ion-exchange capacity or electrostatic interaction; Equation (5.3)
CCD	chemical-composition distribution
CD	cyclodextrin
CDR	chiral derivatizing reagent
CE	capillary electrophoresis
CEC	capillary electrochromatography
CCC	countercurrent chromatography
CLND	chemiluminescent nitrogen detector
C_m	solute concentration in mobile phase
CMPA	chiral mobile-phase additive
CS	chiral selector
C-S	column switching
C_s	solute concentration in stationary phase
Da	Dalton (molecular weight)
DAD	diode-array detector
D_m	solute diffusion coefficient (cm^2/ sec); Equation (2.19)
DMSO	dimethylsulfoxide
EC	electrochemical
EDTA	1,2-ethylenediamine-N, N, N', N'-tetraacetic acid
ELSD	evaporative light scattering detector
EPA	US Environmental Protection Agency
FDA	US Food and Drug Association
F_{opt}	optimum mobile-phase flow rate (mL/min) (Section 2.4.1)
F_s	column-comparison function; Equation (5.4)
$F_s(-\boldsymbol{C})$	value of F_s for non-ionized samples; Equation (6.3)
G	gradient compression factor; Equation (9.15a)
$\boldsymbol{H}$	column hydrophobicity; Equation (5.3)
H-B	hydrogen bond
HFBA	heptaflurobutyric acid
h	reduced plate height; Equation (2.18)
h_p	peak height
HP-TLC	high-performance thin-layer chromatography
HVAC	heating, ventilation, and air-conditioning system

ICH	International Conference on Harmonization
ILE	immobilized liquid extraction
IMAC	immobilized metal affinity chromatography
IPA	isopropanol
IQ	installation qualification
ISO	International Organization for Standardization
IS	internal standard
K	equal to (C_s/C_m)
K_D	SEC distribution coefficient; Figure 13.39; also, Nernst Distribution Law coefficient; Equation (16.1)
k_{EB}	value of k for ethylbenzene (different columns, standard conditions); Equation (5.3)
k_w	extrapolated value of k for solute X with water as mobile phase; Equation (2.26)
k_0	value of k for a solute at the start of gradient elution
LC × LC	comprehensive two-dimensional liquid chromatography
LLE	liquid–liquid extraction
LOD	limit of detection (sometimes called lower limit of detection LLOD)
LOQ	limit of quantification (sometimes called lower limit of quantification or limit of quantitation, LLOQ)
mAU	milli-absorbance units (UV)
MIP	molecular imprinted polymers
MTBE	methyl-*t*-butyl ether
m/z	mass-to-charge ratio
NARP	nonaqueous reversed-phase chromatography
NP	normal-phase (used only with respect to CSP separations)
MWD	molecular-weight distribution
N^*	effective column plate number in gradient elution
o.d.	column or tubing outer diameter (in.)
OQ	operational qualification
P'	overall solvent polarity (Section 2.3.2)
PAH	polycyclic aromatic hydrocarbon
PDA	photodiode-array (detector); also DAD
PEEK	polyetheretherketone (used for fittings and tubing)
PFE	pressurized fluid extraction
PTFE	polytetrafluoroethylene
PVC	polyvinylchloride
PO	polar-organic (used only with respect to CSP separations, Section 14.6.1)

PQ performance qualification
QA quality assurance
QC quality control
QuEChERS Quick, Easy, Cheap, Effective, Rugged, and Safe; Section 16.6.7.5
R fraction of solute molecules in the mobile phase
R^+ a cationic IPC reagent, or a cationic group in an anion-exchange column
R^- an anionic IPC reagent, or an anionic group in a cation-exchange column
$R^\pm$ refers to either R^+ or R^-
RAM restricted access media
RP reversed-phase (used only with respect to CSP separations)
RSD relative standard deviation
S^* column steric interaction; Equation (5.3) (resistance by the stationary phase to penetration by bulky solutes)
SAX strong anion-exchange chromatography
SCX strong cation-exchange chromatography
SD standard deviation
SDME single-drop microextraction
SE standard error
SFC supercritical fluid chromatography
SLE solid-supported liquid-liquid extraction
S/N signal-to-noise ratio
SOP standard operating procedure
TF peak-tailing factor TF; Figure 2.16*a*
THF tetrahydrofuran
TLC thin-layer chromatography
TOF time of flight
T_K temperature (K); Equation (2.8)
USP United States Pharmacopeia
u_x solute migration rate or velocity (mm/min)
U-HPLC ultra-high-pressure liquid chromatography
ULOQ upper limit of quantification (or just upper limit)
V_G gradient volume (gradient time × flow rate) (mL)
V_M gradient mixing volume (mL); Section 9.2.2.4
V_p peak volume (mL)
V_R solute retention volume (mL); equal to $t_R F$
V_s sample volume; Equation (2.29*a*); also, volume of the stationary phase within a column (mL)

WAX	weak anion-exchange chromatography
WCX	weak cation-exchange chromatography
W_0	value of W in the absence of extra-column peak-broadening (Section 2.4.1)
$W_{1/2}$	peak width at half-height; Figure 2.10*a*
X	mole fraction
α^*	separation factor in gradient elution
α'	solute hydrogen-bond acidity; Equation (5.3)
α^H	mobile phase hydrogen-bond acidity; Equation (2.36)
β_2	mobile-phase hydrogen-bond basicity (Section 2.3.1); Equation 2.36)
β'	solute hydrogen-bond basicity; Equation (5.3)
$\Delta \boldsymbol{t}_R$	difference in gradient retention times for a solute (min); Figure 9.15
ε	dielectric constant ε; also molar extinction coefficient
ε_e	inter-particle porosity ε_e
ε_i	intra-particle porosity
ε_T	total column porosity
ϕ_f	final value of ϕ in a gradient separation; Equation (9.2a)
ϕ_0	initial value of ϕ in a gradient separation; Equation (9.2a)
η'	solute hydrophobicity; Equation (5.3)
κ'	solute-effective ionic charge; Equation (5.3)
π	mobile-phase dipolarity; Section 2.3.1
σ	standard deviation of a Gaussian curve; Equation (2.9b)
σ'	solute "bulkiness"; Equation (5.3)
Σ	sum of α, β, and π values for a mobile phase (Section 2.3.1)
Ψ	phase ratio; equal to V_s/V_m
u	mobile-phase velocity (mm/min)
u_e	mobile-phase interstitial velocity (mm/min); $u_e > u$

CHAPTER ONE

INTRODUCTION

High-performance liquid chromatography (HPLC) is one of several chromatographic methods for the separation and analysis of chemical mixtures (Section 1.3). Compared to these other separation procedures, HPLC is exceptional in terms of the following characteristics:

- almost universal applicability; few samples are excluded from the possibility of HPLC separation
- remarkable assay precision (±0.5% or better in many cases)
- a wide range of equipment, columns, and other materials is commercially available, allowing the use of HPLC for almost every application
- most laboratories that deal with a need for analyzing chemical mixtures are equipped for HPLC; it is often the first choice of technique

Introduction to Modern Liquid Chromatography, Third Edition, by Lloyd R. Snyder, Joseph J. Kirkland, and John W. Dolan

As a result, HPLC is today one of the most useful and widely applied analytical techniques. Mass spectrometry rivals and complements HPLC in many respects; the use of these two techniques in combination (LC-MS) is already substantial (Section 4.14), and will continue to grow in importance.

In the present chapter we will:

- examine some general features of HPLC
- summarize the history of HPLC
- very briefly consider some alternatives to HPLC, with their preferred use for certain applications
- list other sources of information about HPLC

1.1 BACKGROUND INFORMATION

1.1.1 What Is HPLC?

Liquid chromatography began in the early 1900s, in the form illustrated in Figure 1.1*a–e*, known as "classical column chromatography". A glass cylinder was packed with a finely divided powder such as chalk (Fig. 1.1*a*), a sample was applied to the top of the column (Fig. 1.1*b*), and a solvent was poured onto the column (Fig. 1.1*c*). As the solvent flows down the column by gravity (Fig. 1.1*d*), the components of the sample (A, B, and C in this example) begins to move through the column at different speeds and became separated. In its initial form, colored samples were investigated so that the separation within the column could be observed visually. Then portions of the solvent leaving the column were collected, the solvent was evaporated, and the separated compounds were recovered for quantitative analysis or other use (Fig. 1.1*e*). In those days a new column was required for each sample, and the entire process was carried out manually (no automation). Consequently the effort required for each separation could be tedious and time-consuming. Still, even at this stage of development, chromatography provided a unique capability compared to other methods for the analysis of chemical mixtures.

A simpler form of liquid chromatography was introduced in the 1940s, called *paper chromatography* (Fig. 1.1*f*). A strip of paper replaced the column of Figure 1.1*a*; after the sample was spotted near the bottom of the paper strip, the paper was placed in a container with solvent at the bottom. As the solvent migrated up the paper by capillary action, a similar separation as seen in Figure 1.1*d* took place, but in the opposite direction. This "open bed" form of chromatography was later modified by coating a thin layer of powdered silica onto a glass plate—as a replacement for the paper strip used in paper chromatography. The resulting procedure is referred to as *thin-layer chromatography* (TLC). The advantages of either paper or thin-layer chromatography included (1) greater convenience, (2) the ability to simultaneously separate several samples on the same paper strip or plate, and (3) easy detection of small amounts of separated compounds by the application of colorimetric reagents to the plate, after the separation was completed.

HPLC (Fig. 1.1*g*, *h*) represents the modern culmination of the development of liquid chromatography. The user begins by placing samples on a tray for automatic injection into the column (Fig. 1.1g). Solvent is continually pumped

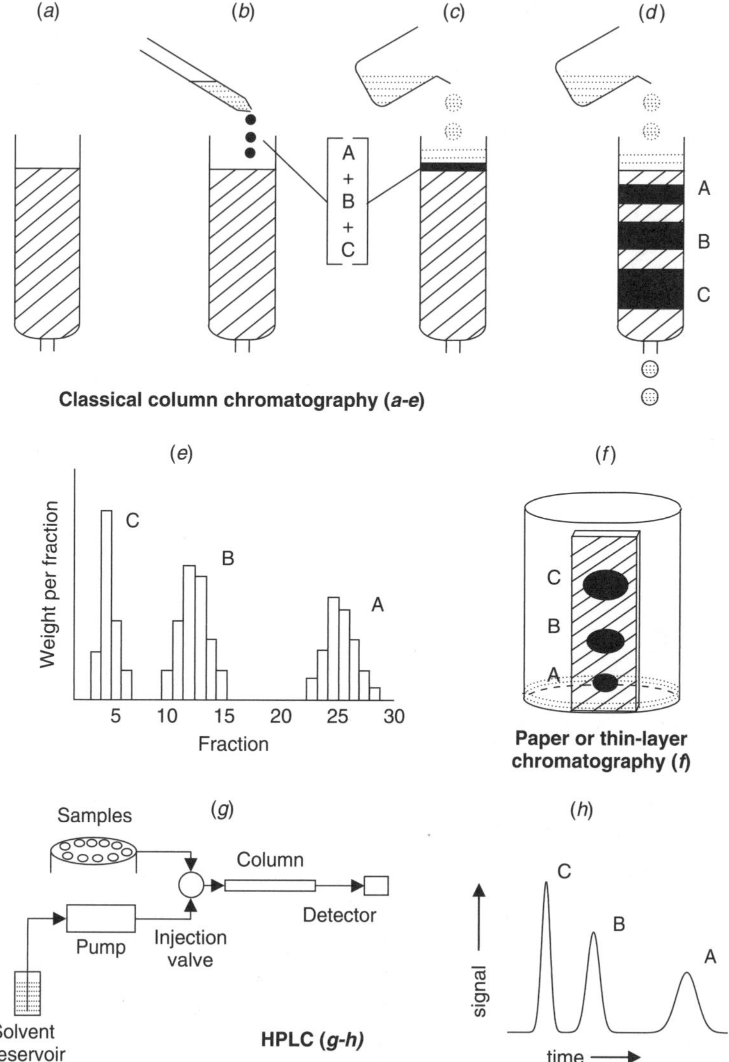

Figure 1.1 Different stages in the development of chromatography.

through the column, and the separated compounds are continuously sensed by a detector as they leave the column. The resulting detector signal plotted against time is the *chromatogram* of Figure 1.1*h*, which can be compared with the result of Figure 1.1*e*—provided that the sample A + B + C and experimental conditions are the same. A computer controls the entire operation, so the only manual intervention required is the placement of samples on the tray. The computer can also generate a final analysis report for the sample. Apart from this automation of the entire process, HPLC is characterized by the use of high-pressure pumps for faster separation, re-usable and more effective columns for enhanced separation, and a better control of the overall process for more precise and reproducible results. More discussion of the history of HPLC can be found in Section 1.2.

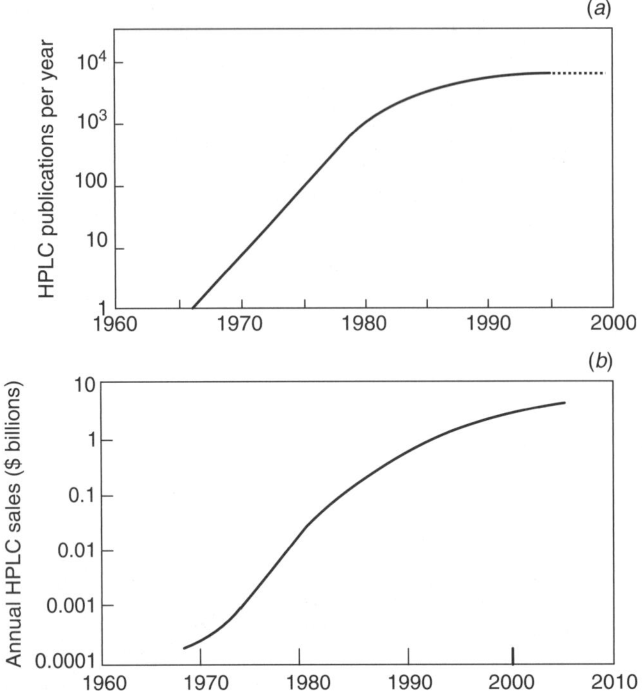

Figure 1.2 The expanding importance of HPLC research and application since 1966. (*a*) Number of HPLC-related publications per year [1]; (*b*) total sales of HPLC equipment and supplies per year (approximate data compiled from various sources).

The growth of HPLC, following its introduction in the late 1960s (Section 1.2), is illustrated in Figure 1.2. In (Fig. 1.2*a*) the annual number of HPLC publications is plotted against time. The first HPLC paper appeared in 1966 [2], and the number of publications grew each year exponentially, leveling off only after 1980. By 1990 the primary requirements of HPLC had largely been satisfied in terms of an understanding of the separation process, and the availability of suitable equipment and columns. At this time HPLC could be considered to have become a mature technique—one that is today practiced in every part of the world. While new, specialized applications of HPLC continued to emerge after 1990, and remaining gaps in our understanding receive ongoing attention, major future changes to our present understanding of HPLC seem unlikely.

As the pace of HPLC research reached a plateau by 1990, a comparable flattening of the HPLC economy took a bit longer—as suggested by the plot in Figure 1.2*b* of annual expenditures against time for all HPLC products (not adjusted for inflation). The money spent annually on HPLC at the present time exceeds that for any other analytical technique.

1.1.2 What Can HPLC Do?

When the second edition of this book appeared in 1979, some examples of HPLC capability were presented, two of which are reproduced in Figure 1.3. Figure 1.3*a*

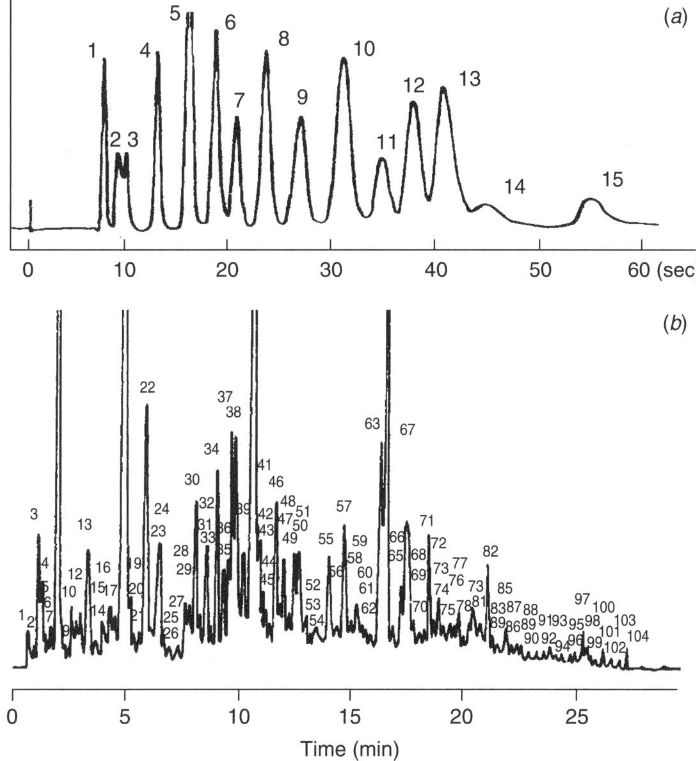

Figure 1.3 Examples of HPLC capability during the mid-1970s. (*a*) Fast separation of a mixture of small molecules [3]; (*b*) high-resolution separation of a urine sample [4]. (*a*) is adapted from [3], and (*b*) is adapted from [4].

shows a fast HPLC separation where 15 compounds are separated in just one minute. Figure 1.3*b* shows the separation power of HPLC by the partial separation of more than 100 recognizable peaks in just 30 minutes. In Figure 1.4 are illustrated comparable separations that were carried out 25 years later. Notice that in Figure 1.4*a*, six proteins are separated in 7 seconds, while in Figure 1.4*b*, *c*, about 1000 peptides plus proteins are separated in a total time of 1.5 hours. The improvement in Figure 1.4*a* compared with Figure 1.3*a* can be ascribed to several factors, some of which are discussed in Section 1.2. The separation of 1000 compounds in Figure 1.4*b*, *c* is the result of so-called two-dimensional separation (Section 9.3.10): a first column (Fig. 1.4*b*) provides fractions for further separation by a second column (Fig. 1.4*c*). In this example 4-minute fractions were collected from the first column and further separated with the second column; Figure 1.4*c* shows the separation of fraction 7. The total number of (recognizable) peaks in the sample is then obtained by adding the unique peaks present in each of the fractions. The enormous progress made in HPLC performance (Fig. 1.4 vs. Fig. 1.3) suggests that comparable major improvements in speed or separation power in the coming years are not so likely.

Some other improvements in HPLC since 1979 have been equally significant. Beginning in the 1980s, the introduction of suitable columns for the separation of proteins and other large biomolecules [7, 8] has opened up an entirely new

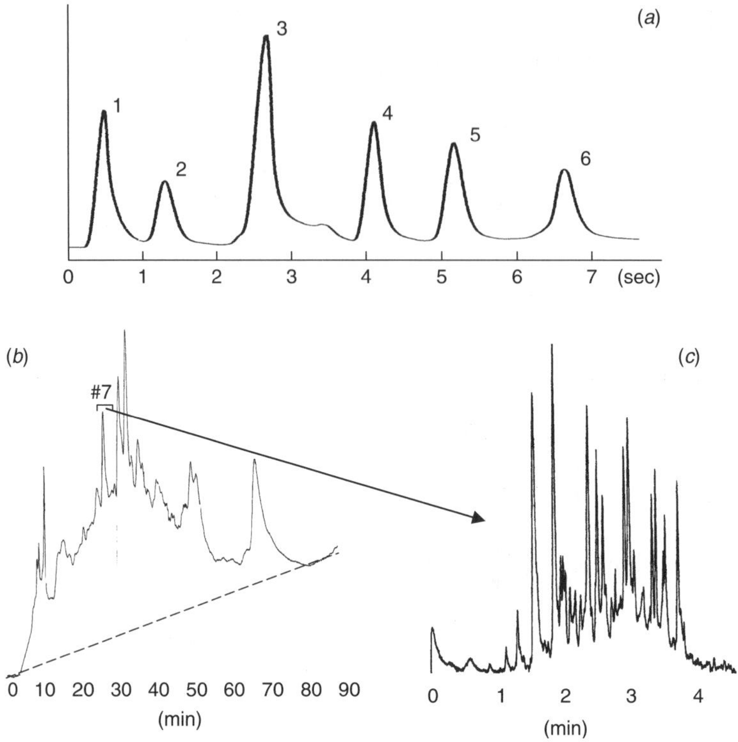

Figure 1.4 Recent examples of HPLC capability. (*a*) Fast separation of six proteins, using gradient elution with a 150 × 4.6-mm column packed with 1.5-μm-diameter pellicular particles [5]; (*b*) initial separation of peptides and proteins from human fetal fibroblast cell by gradient cation-exchange chromatography; (*c*) further separation of fraction 7 (collected between 24–28 min) on a second column by gradient reversed-phase chromatography [6]. Figures adapted from original publications [5, 6].

field of application and facilitated major advances in biochemistry. Similarly the development of chiral columns for the separation of enantiomeric mixtures by Pirkle [9] and others enabled comparable advances in the areas of pharmaceuticals and related life sciences. The use of HPLC for large-scale purification is also increasing, as a result of the availability of appropriate equipment, an increase in our understanding of how such separations should best be carried out, and regulatory pressures for higher purity pharmaceutical products.

1.2 A SHORT HISTORY OF HPLC

We have noted the development of liquid chromatography prior to the advent of HPLC (Section 1.1). For a more complete account of this pre-1965 period, several review articles have been written by Leslie Ettre, our "historian of chromatography":

- precursors to chromatography; developments prior to 1900 [10, 11]
- invention of chromatography by M. S. Tswett in the early 1900s [12]

- rediscovery of chromatography in the early 1930s [13]
- A. J. P. Martin's invention of partition and paper chromatography in the early 1940s [14]
- development of the amino-acid analyzer by S. Moore and W. S. Stein in the late 1950s [15]
- development of the gel-permeation chromatograph by Waters Associates in the early 1960s [16]

Carl Runge, a German dye-chemist born in 1856, first reported crude dye separations by means of a technique similar to paper chromatography [10], but neither he nor others pursued the practical possibilities of this work. In the late 1890s David Day at the US Geological survey carried out separations of petroleum by a technique that resembles classical column chromatography [11]; however, his goal was not the development of a separation technique, but rather the demonstration that petroleum deposits of different quality result from their separation during migration through the ground. As in the case of Runge's work, Day's investigations did not proceed further. In the early 1900s, Mikhail Tswett invented classical column chromatography and demonstrated its ability to separate different plant extracts [12]. This was certainly the beginning of chromatography, but the value of his work was not appreciated for another two decades. In the early 1930s, Tswett's work was rediscovered [13], leading to an explosive subsequent growth of chromatography. The invention of paper chromatography by A.J.P. Martin followed in 1943 [14], accompanied by the development of thin-layer chromatography between the late 1930s and the mid-1950s [17]. This short summary necessarily omits numerous other contributions to the development of chromatography before 1955.

The amino-acid analyzer, introduced in the late 1950s [15], was an important precursor to HPLC; it was an automated means for analyzing mixtures of amino acids by use of ion-exchange chromatography (Section 7.5). This was followed by the invention of gel permeation chromatography (Section 13.7) by Moore [18] and the introduction in the early 1960s of a gel-permeation chromatograph by Waters Associates [16]. Each of these latter techniques was close in concept to what later became HPLC, differing little from the schematic of Figure 1.1*g*. In each case the solvent was pumped at high pressure through a reusable, small-particle column, the column effluent was continuously monitored by a detector, and the output of the device was a chromatogram as in Figure 1.1*h*. What each of these two systems lacked, however, was an ability to separate and analyze other kinds of samples. The amino-acid analyzer was restricted to the analysis of mixtures of amino acids, while the gel-permeation chromatograph was used exclusively for determining the molecular weight distribution of synthetic polymers. In neither case were these devices readily adaptable for the separation of other samples.

During the early 1960s, two different groups embarked on the development of a general-purpose HPLC system, under the leadership of Csaba Horváth in the United States and Josef Huber in Europe. Each of these two men have described their early work on HPLC in a collection of personal recollections [19], and Ettre has provided additional detail on early work in Horváth's laboratory [20]. The immediate results of these two groups, plus related work by others that was carried out a few years later, are described in publications that appeared in 1966 to 1968 [2, 21–24]. The introduction of commercial equipment for HPLC followed in the late 1960s, with systems from Waters Associates and DuPont initially dominating the market. Other companies soon offered competing equipment, and research on HPLC began to accelerate (as seen from Fig. 1.2*a*). By 1971, the first HPLC book had been published [25], and an HPLC short course was offered by the American Chemical Society (Modern Liquid Chromatography), with J. J. Kirkland and L. R. Snyder as course instructors).

Progressive improvements in HPLC from 1960 to 2010 are illustrated by the representative separations of Figure 1.5*a–f*, which show separation times decreasing by several orders of magnitude during this 50-year interval. Figure 1.5*g* shows how this reduction in separation time (◦,—) was related to increases in the pressure drop across the column (- - -) and a reduction in the size of particles (•) that were used to pack the column. In the early days of HPLC the technique was sometimes referred to as "high-pressure liquid chromatography" or "high-speed liquid chromatography," for reasons suggested by Figure 1.5*g*. Figure 1.5*h* shows corresponding changes in column length (•) and flow rate (◦) for the separations of Figure 1.5*a–e*.

A theoretical foundation for the eventual development of HPLC was established well before the 1960s. In 1941, Martin reported [27] that "the most efficient columns ... should be obtainable by using very small particles and high-pressure differences across the length of the column;" this summarized the requirements for HPLC separation in a nutshell (as demonstrated by Fig. 1.5*g*). In the early 1950s, the related technique of gas chromatography was invented by Martin [28]; its rapid acceptance by the world [29] led to a number of theoretical studies that would prove relevant to the later development of HPLC. Giddings summarized and extended this work for specific application to HPLC in the early 1960s [30], work that was later to prove important for both column design and the selection of preferred experimental conditions.

For a further background on the early days of HPLC, see [19, 31–33]. Additional historical details on the progress of HPLC after 1980 are provided by the collected biographies of several HPLC practitioners [34].

1.3 SOME ALTERNATIVES TO HPLC

Two, still-important techniques, each of which can substitute for HPLC in certain applications, existed prior to 1965: gas chromatography (GC) and thin-layer chromatography (TLC). Countercurrent chromatography (CCC) is another pre-1965 technique that, in principle, might compete with HPLC in many applications but falls considerably short of the speed and separation power of HPLC. Several additional, potentially competitive, techniques were introduced after HPLC: supercritical fluid chromatography (SFC) in the 1970s, capillary electrophoresis (CE) in the 1980s, and capillary electrochromatography (CEC) in the 1990s.

1.3.1 Gas Chromatography (GC)

Because GC [35] is limited to samples that are volatile below 300°C, this technique is not applicable for very-high-boiling or nonvolatile materials. Thus about 75% of all known compounds cannot be separated by GC. On the other hand, GC is considerably more efficient than HPLC (higher values of the plate number N),

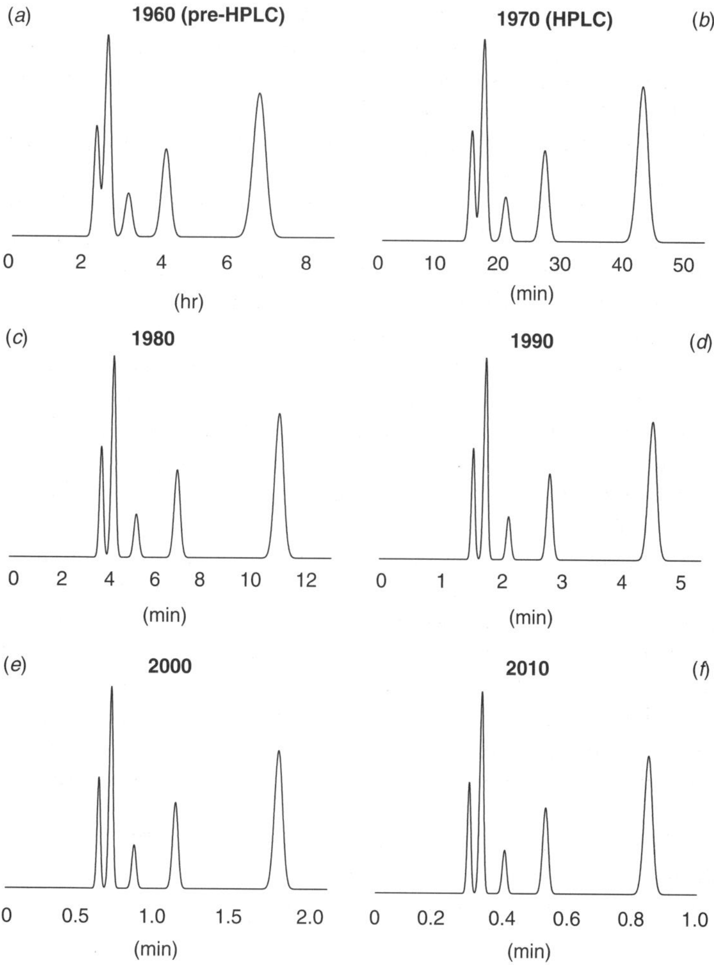

Figure 1.5 Representative chromatograms that illustrate the improvement in HPLC performance over time. Sample: five herbicides. Conditions: 50% methanol-water, ambient temperature. Chromatograms *a–f* are DryLab^R computer simulations (Section 10.2), based on data of [26]; *g* and *h* provide details for the separations of *a–f*. Column-packings of identical selectivity and 4.6-mm-diameter columns are assumed.

which means faster and/or better separations are possible. GC is therefore preferred to HPLC for gases, most low-boiling samples, and many higher boiling samples that are thermally stable under the conditions of separation. GC also has available several very sensitive and/or element-specific detectors that permit considerably lower detection limits.

1.3.2 Thin-Layer Chromatography (TLC)

The strong points of TLC [36] are its ability to separate several samples simultaneously on a single plate, combined with the fact that every component in the sample is visible on the final plate (strongly retained compounds may be missed in

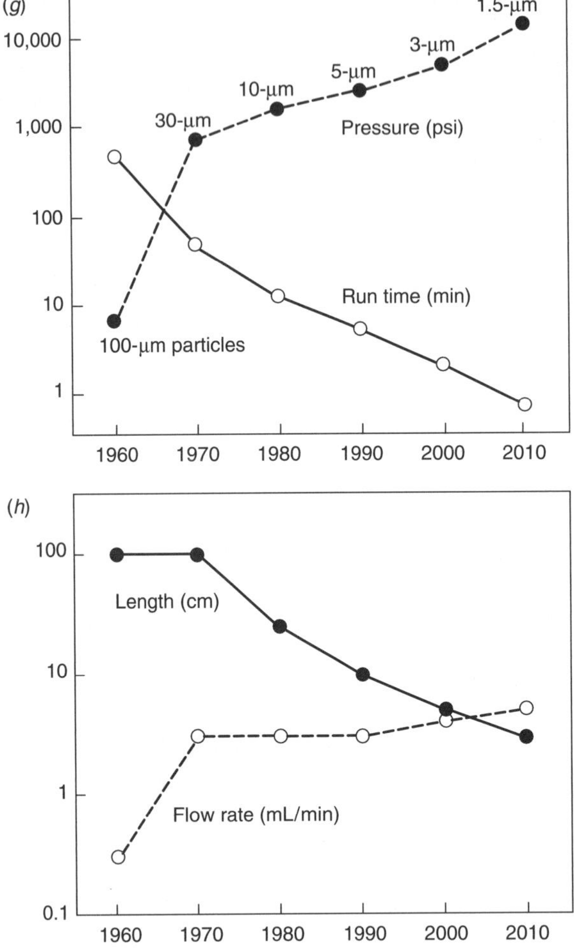

Figure 1.5 (*Continued*)

HPLC). With the advent of specialized equipment for the pressurized flow of solvent across the plate, so-called high-performance TLC (HP-TLC) has become possible. Regardless of how it is carried out, however, TLC lacks the separation efficiency of HPLC (as measured by values of N), and quantitation is less convenient and less precise. At the time of publication of the present book, TLC was used relatively infrequently in the United States for quantitative analysis, although it is a convenient means for semi-quantitative analysis and for the detection of sample impurities. It is widely used for screening large numbers of samples, with little need for sample cleanup (e.g., plasma drug screening). In Europe HP-TLC is more popular than in the United States but much less popular than HPLC.

1.3.3 Supercritical Fluid Chromatography (SFC)

SFC [37] is carried out with equipment and columns that are similar to HPLC. The solvent is, by definition, a supercritical fluid, usually a gas such as CO_2, under conditions of elevated pressure and temperature. SFC can be regarded as an extension of GC, in that supercritical fluids can dissolve and separate samples that are normally considered to be nonvolatile. SFC may be considered as a hybrid of GC and HPLC, as it is characterized by greater separation efficiency than for HPLC (higher N) but lower efficiency than GC. Similarly the solvent in SFC plays a greater role in determining separation than in GC, but less so than in HPLC. Detection sensitivity is also intermediate between what is possible with HPLC compared to GC. A major application of SFC is for the analysis of natural or synthetic polymeric mixtures, for example, the separation of polyphenols as described in [38]. Whereas HPLC may be unable to resolve individual polymeric species with molecular weights above some maximum value, SFC can usually extend this upper molecular-weight limit considerably. SFC has also been used for separating enantiomers, whose very similar retention may require greater separation efficiency (larger value of N).

1.3.4 Capillary Electrophoresis (CE)

CE [1, 39] is not a form of chromatography, but it competes effectively with HPLC for the separation of certain classes of compounds. The principle of separation is the differential migration of sample compounds in a capillary, under the influence of an electric field, with the result that compounds are separated on the basis of their mass-to-charge ratio (m/z); compounds with smaller m/z migrate faster. Consequently compounds that are to be separated by CE must carry an ionic charge. CE is characterized by a greater separation efficiency than for HPLC (higher value of N), especially for the separation of compounds of high molecular weight. However, detection sensitivity is usually much poorer than for HPLC. CE is heavily used for the genomic analysis of various species, based on the fractionation of DNA fragments. CE has also proved popular for analytical separations of enantiomeric samples, where its performance may exceed that of HPLC for two reasons. First, these separations are often difficult and therefore are facilitated by the larger values of N available from CE. Second, HPLC separations of enantiomers usually rely on *chiral columns*. The separation of a particular enantiomeric sample may require the trial-and-error testing of several different (and expensive) columns before a successful separation is achieved. CE allows the use of small amounts of different chiral complexing agents—instead of different columns, allowing for a faster, cheaper, and more versatile alternative to HPLC. The required flow rates for HPLC compared with CE (e.g., mL/min vs. μL/min) make the use of costly chiral complexing reagents impractical for HPLC. Several variations of CE exist, which allow its extension to other sample types; for example, non-ionized compounds can be separated by micellar electrokinetic chromatography [40].

1.3.5 Countercurrent Chromatography

CCC [41, 42] is an older form of liquid–liquid partition chromatography that was later improved in various ways. HPLC with a liquid stationary phase was since replaced by bonded-phase HPLC, the use of CCC as an alternative to HPLC has become relatively less frequent. An often-cited feature of CCC is its freedom from problems caused by irreversible attachment of the sample to the large internal surface present in HPLC columns. However, the improved HPLC columns used today are largely free from this problem. CCC may possess certain advantages for the preparative separation of enantiomers [43]; otherwise, the technique is used mainly for the isolation of labile natural products.

1.3.6 Special Forms of HPLC

The five separation techniques mentioned above (Sections 1.3.1–1.3.5) differ in essential ways from HPLC. Four other procedures, which will not be discussed in this book, can be regarded as HPLC variants. However, much of the information in following chapters can be adapted for use with the following procedures.

Capillary electrochromatography [44, 45] (CEC) is generally similar to HPLC, except that the flow of solvent is achieved by means of an electrical potential across the column (endoosmotic flow), rather than by use of a pump. Because solvent flow is not affected by the size of particles within the column (and column efficiency can be greater for small particles), much larger values of N are, in principle, possible by means of CEC. Higher values of N also result from endoosmotic flow per se. Because of these potentially greater values of N in CEC than in HPLC, considerable effort has been invested since 1995 into making this technique practical. However, major technical problems remain to be solved, and CEC had not become a routine alternative to HPLC at the time this book went to press.

HPLC on a chip [46] is a recently introduced technology for the convenient separation of very small samples. A micro-column (e.g., 43 × 0.06 mm) forms part of the chip, which can be interfaced between a micro pump and a mass spectrometer. The principles of separation are the same as for HPLC with conventional columns and equipment, but a chip offers advantages in terms of separation power and convenience for very small samples.

Ion chromatography [47, 48] is widely used for the analysis of mixtures that contain inorganic anions and cations; for example, Cl^- and Na^+, respectively. While the principles of separation are the same as for ion-exchange HPLC (Section 7.5), ion chromatography involves special equipment and is used mainly for inorganic analysis.

Micellar liquid chromatography is a variant of reversed-phase chromatography in which the usual aqueous-organic solvent is replaced by an aqueous surfactant solution [49]. It is little used at present because of the lower efficiency of these separations.

1.4 OTHER SOURCES OF HPLC INFORMATION

A wide variety of resources is available that can be consulted to supplement the use of the present book. These include various other publications (Sections 1.4.1–1.4.3), short courses (Section 1.4.4), and the Internet (Section 1.4.5).

1.4.1 Books

Literally hundreds of books on chromatography have now been published, as reference to Amazon.com and other internet sources can readily verify. Books on HPLC can be divided into two groups: (1) specialized texts that address the HPLC separation of a certain kind of sample (e.g., proteins, carbohydrates, enantiomers), or by means of special detection (e.g., mass spectrometer, chemical derivatization), and (2) more general books, such as the present book, that cover all aspects of HPLC. Specialized HPLC books are referenced in later chapters that address different HPLC topics. Table 1.1 provides a partial listing of more general HPLC books published after 1995 that might serve as useful supplements to the present book.

1.4.2 Journals

Technical articles that involve HPLC can appear in most journals that deal with the chemical or biochemical sciences. However, the journals below are of special value to those readers wishing to keep abreast of new developments in the field.

- *Analytical Chemistry*, American Chemical Society
- *Chromatographia*, Springer
- *Journal of Chromatographic Science*, Preston
- *Journal of Chromatography A*, Elsevier
- *Journal of Chromatography B*, Elsevier
- *Journal of Liquid Chromatography*, Wiley
- *Journal of Separation Science*, Wiley
- *LCGC*, Advanstar (separate issues for North America and Europe)

1.4.3 Reviews

Review articles that deal with HPLC can be found in the journals listed above and in other journals. Additionally there are series of publications that are devoted in part to HPLC, either as collections of review articles

- *Advances in Chromatography*, Dekker
- *High-Performance Liquid Chromatography. Advances and Perspectives*, Academic Press (published only between 1980 and 1986)

or as individual books:

- *Journal of Chromatography Library*, Elsevier

1.4.4 Short Courses

There are numerous short courses offered either "live" or on the Internet (see Section 1.4.5). For a current listing of short courses, see the back pages of *LCGC* magazine or search the Internet for "HPLC training."

1.4.5 The Internet

The dynamic nature of the Internet ensures that any listing in a book will soon be obsolete. A number of sites are links to other sites and, as such, presumably will be continuously updated:

http://www.lcresources.com

http://matematicas.udea.edu.co/~carlopez/index7.html

Table 1.1

Some HPLC Books of General Interest Published since 1995

Title	Author(s)	Publication Date	Publisher
General texts			
Handbook of HPLC	E. Katz, R. Eksteen, P. Schoenmakers, and N. Miller, eds.	1998	Dekker
High Performance Liquid Chromatography	S. Lindsay	2000	Wiley
High Performance Liquid Chromatography	E. Prichard	2003	Royal Society of Chemistry
HPLC, 2nd ed.	M.C. McMaster	2006	Wiley-Interscience
Modern HPLC for Practicing Scientists	M. W. Dong	2006	Wiley-Interscience
Practical High-Performance Liquid Chromatography, 4th ed.	V. R. Meyer	2006	Wiley-Interscience
Method development			
Practical HPLC Method Development, 2nd ed.	L. R. Snyder, J. L. Glajch, and J. J. Kirkland	1997	Wiley-Interscience
HPLC Made to Measure: A Practical Handbook for Optimization	S. Kromidas	2006	Wiley
Troubleshooting			
LC Troubleshooting	J.W. Dolan	1983–present	Monthly column in *LCGC Magazine*; past columns available at www.chromatographyonline.com
Troubleshooting HPLC Systems: A Bench Manual	P. C. Sadek	1999	Wiley
More Practical Problem Solving in HPLC	S. Kromidas	2005	Wiley
Pitfalls and Errors of HPLC in Pictures 2nd ed.	V. R. Meyer	2006	Wiley

Table 1.1

(*Continued*)

Title	Author(s)	Publication Date	Publisher
Preparative HPLC			
Practical Handbook of Preparative HPLC	D. A. Wellings	2006	Elsevier
HPLC columns			
HPLC Columns: Theory, Technology and Practice	U. D. Neue	1997	Wiley-VCH
HPLC solvents			
The HPLC Solvent Guide	P. C. Sadek	1996	Wiley
Gradient elution			
High-Performance Gradient Elution	L. R. Snyder and J. W. Dolan	2007	Wiley

http://lchromatography.com/hplcfind/index.html

thttp://tech.groups.yahoo.com/group/chrom-L/links

http://userpages.umbc.edu/~dfrey1/Freylink

http://www.infochembio.ethz.ch/links/en/analytchem_chromat.html

http://www.chromatographyonline.com

REFERENCES

1. R. L. Cunico, K. M. Gooding, and T. Wehr, *Basic HPLC and CE of Biomolecules*, Bay Bioanalytical Laboratory, Richmond, CA, 1998, p. 4.
2. C. G. Horváth and S. R. Lipsky, *Nature*, 211 (1966) 748.
3. I. Halasz, R. Endele, and J. Asshauer, *J. Chromatogr.*, 112 (1975) 37.
4. I. Molnar and C. Horváth, *J. Chromatogr. (Biomed App.)*, 143 (1977) 391.
5. T. Issaeva, A. Kourganov, and K. Unger, *J. Chromatogr. A*, 846 (1999) 13.
6. K. Wagner, T. Miliotis, G. Marko-Varga, R. Biscoff, and K. Unger, *Anal. Chem.*, 74 (2002) 809.
7. S. H. Chang, K. M. Gooding, and F. E. Regnier, *J. Chromatogr.*, 125 (1976) 103.
8. W. W. Hancock, C. A. Bishop, and M. T. W. Hearn, *Science*, 153 (1978) 1168.
9. W. H. Pirkle, D. W. House, and J. M. Finn, *J. Chromatogr.*, 192 (1980) 143.
10. H. H. Bussemas and L. S. Ettre, *LCGC*, 22 (2004) 262.
11. L. S. Ettre, *LCGC*, 23 (2005) 1274.
12. L. S. Ettre, *LCGC*, 21 (2003) 458.

13. L. S. Ettre, *LCGC*, 25 (2007) 640.
14. L. S. Ettre, *LCGC*, 19 (2001) 506.
15. L. S. Ettre, *LCGC*, 243 (2006) 390.
16. L. S. Ettre, *LCGC*, 23 (2005) 752.
17. J. G. Kirchner, *Thin-layer Chromatography*, Wiley-Interscience, New York, 1978, pp. 5–8.
18. J. C. Moore, *J. Polymer Sci. Part A*, 2 (1964) 835.
19. L. S. Ettre and A. Zlatkis, eds., *75 years of Chromatography—A Historical Dialog*, Elsevier, Amsterdam, 1979.
20. L. S. Ettre, *LCGC*, 23 (2005) 486.
21. J. F. K. Huber and J. A. R. Hulsman, *J. Anal. Chim. Acta*, 38 (1967) 305.
22. J. J. Kirkland, *Anal. Chem.*, 40 (1968) 218.
23. L. R. Snyder, *Anal. Chem.*, 39 (1967) 698, 705.
24. R. P. W. Scott, W. J. Blackburn, and T. J. Wilkens, *J. Gas Chrommatogr.*, 5 (1967) 183.
25. J. J. Kirkland, ed., *Modern Practice of Liquid Chromatography*, Wiley-Interscience, New York, 1971.
26. T. Braumann, G. Weber, and L. H. Grimme, *J. Chromatogr.*, 261 (1983) 329.
27. A. J. P. Martin and R. L. M. Synge, *Biochem. J.*, 35 (1941) 1358.
28. A. T. James and A. J. P. Martin, *Biochem. J.*, 50 (1952) 679.
29. L. S. Ettre, *LCGC*, 19 (2001) 120.
30. J. C. Giddings, *Dynamics of Chromatography. Principles and Theory*, Dekker, New York, 1965.
31. L. R. Snyder, *J. Chem. Ed.*, 74 (1997) 37.
32. L. S. Ettre, *LCGC Europe*, 1 (2001) 314.
33. L. R. Snyder, *Anal. Chem.*, 72 (2000) 412A.
34. C. W. Gehrke, ed., *Chromatography—A Century of Discovery 1900–2000*, Elsevier, Amsterdam, 2001.
35. R. L. Grob and E. F. Barry, *Modern Practice of Gas Chromatography*, 4th ed., Wiley-Interscience, NewYork, 2004.
36. B. Fried and J. Sherma, *Thin-Layer Chromatography (Chromatographic Science, Vol. 81)*, Dekker, New York, 1999.
37. R. M. Smith and S. M. Hawthorne, eds., *Supercritical Fluids in Chromatography and Extraction*, Elsevier, Amsterdam, 1997.
38. T. Bamba, E. Fukusaki, Y. Nakazawa, H. Sato, K. Ute, T. Kitayama, and A. Kobayashi, *J. Chromatogr. A*, 995 (2003) 203.
39. K. D. Altria and D. Elder, *J. Chromatogr. A*, 1023 (2004) 1.
40. A. Berthod and C. Gárcia-Alvarez-Coque, *Micellar Liquid Chromatography*, Dekker, New York, 2000.
41. Y. Ito and W. D. Conway, eds., *High-Speed Countercurrent Chromatography*, Wiley, New York, 1996.
42. J.-M. Menet and D. Thiebaut, eds., *Countercurrent Chromatography*, Dekker, New York, 1999.
43. E. Gavioli, N. M. Maier, C. Minguillón, and W. Lindner, *Anal. Chem.*, 76 (2004) 5837.
44. K. D. Bartle and P. Meyers, *Capillary Electrochromatography (Chromatography Monographs)*, 2001.
45. F. Svec, ed., *J. Chromatogr. A*, 1044 (2004).

46. H. Yin and K. Killeen, *J. Sep. Sci.*, 30 (2007) 1427.
47. J. S. Fritz and D. T. Gjerde, *Ion Chromatography*, 3rd ed., Wiley-VCH, Weinheim, 2000.
48. J. Weiss, *Handbook of Ion Chromatography*, 3rd ed., Wiley, 2005.
49. A. Berthod and M. C. Gárcia-Alvarez-Coque, *Micellar Liquid Chromatography*, Dekker, New York, 2000.

CHAPTER TWO

BASIC CONCEPTS AND THE CONTROL OF SEPARATION

Introduction to Modern Liquid Chromatography, Third Edition, by Lloyd R. Snyder, Joseph J. Kirkland, and John W. Dolan

2.1 INTRODUCTION

The successful use of HPLC requires an understanding of how separation is affected by experimental conditions: the column, solvent, temperature, flow rate and so forth. In this chapter we review some general features of HPLC for use in the laboratory, in order to develop an adequate separation (method development), to carry out a routine HPLC procedure for sample analysis, or to solve problems as they arise. A descriptive or qualitative approach is usually best suited for understanding both method development and the routine application of HPLC. For this reason *the reader may wish to skim or skip any of the following derivations*—at least initially. Important equations that are useful in practice are enclosed within a box; for example, Equation (2.5).

2.2 THE CHROMATOGRAPHIC PROCESS

A schematic of an HPLC system is shown in Figure 2.1, with emphasis on the flow path of the solvent (solid arrows) as it proceeds from the solvent reservoir to the detector (the solvent is usually referred to as the *mobile phase* or *eluent*). A detailed discussion of each part of the system (HPLC equipment) is given in Chapter 3. After injection of the sample, a separation takes place within the column, and separated sample components leave (are *eluted* or washed from) the column—with detection in most cases by either ultraviolet absorption (UV) or mass spectrometry (MS); see Chapter 4 for details on the use of these and other HPLC detectors. The fundamental nature or "mode" of the separation is determined mainly by the choice of column, as summarized in Table 2.1. For sample analysis, the predominant HPLC mode in use today is *reversed-phase chromatography* (RPC), which features a nonpolar column in combination with a (polar) mixture of water plus an organic solvent as mobile phase. Unless noted otherwise, RPC separation will be assumed in this book. Other HPLC modes are described in later sections of the book, as noted in Table 2.1. In Chapters 2 through 8 we will assume that the composition of the solvent remains the same throughout separation, which is called *isocratic elution*, as opposed to

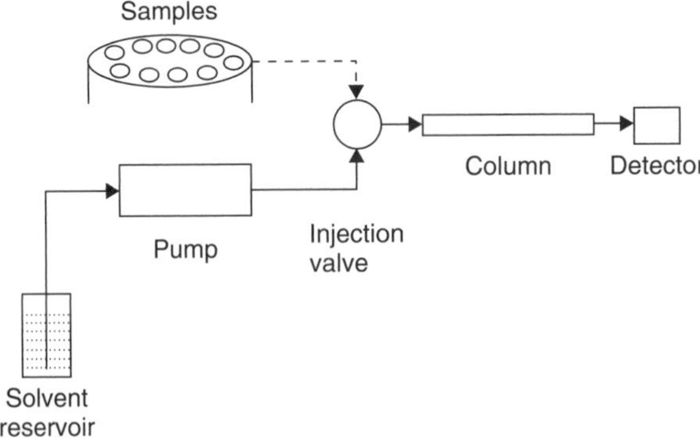

Figure 2.1 Schematic of an HPLC system.

gradient elution where the solvent composition is deliberately changed during the separation (Section 2.7.2, Chapter 9).

The column consists of a cylindrical tube that is typically filled with small (usually 1.5- to 5-μm diameter) spherical particles (Fig. 2.2*a*). These particles are in most cases porous silica, with an individual pore portrayed in Figure 2.2*b* as a cylinder of some specified diameter (typically about 10 nm for use with "small-molecule" samples i.e., molecular weights <1000 Da). The inside of each pore is covered with the *stationary phase*—in this example, C_{18} groups that are attached to the silica particle. Figure 2.2*c* shows a more realistic representation of present-day porous particles for HPLC. The particle is formed by aggregating small, spherical, subparticles as shown. The actual pores are formed by the spaces between the subparticles. Because almost all of the surface of the particle is contained within these pores, most sample molecules are held *inside* the particle rather than on the surface of the particle. That is, the internal surfaces of the pores account for $\gg$99% of the total surface area of the particle; the external surface area (and its effect on separation) is in most cases negligible. The mobile phase surrounds each particle as

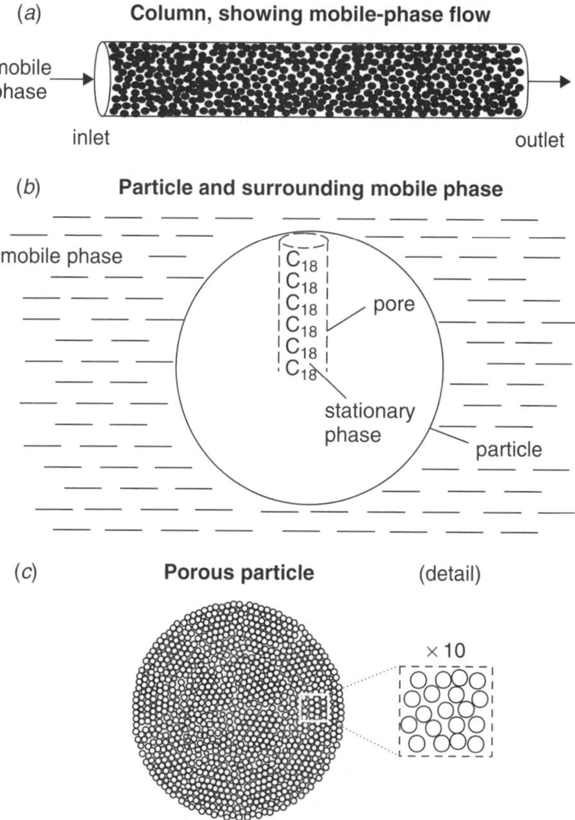

Figure 2.2 The HPLC column. (*a*) Column packed with spherical particles; (*b*) schematic of an individual particle, showing an idealized pore with attached C_{18} groups; (*c*) more realistic picture of a spherical, porous particle, showing detail (10× expansion).

Table 2.1

HPLC Separation Modes

Chromatographic Mode	Comment	Details In
Reversed-phase chromatography (RPC)	The column is nonpolar (e.g., C_{18}), and the mobile phase is a polar mixture of water plus organic solvent (e.g., acetonitrile); RPC is the most widely used mode, especially for water-soluble samples.	Chapter 6, Section 7.3
Normal-phase chromatography (NPC)	The column is polar (e.g., unbonded silica), and the mobile phase is a mixture of less-polar organic solvents (e.g., hexane plus methylene chloride); NPC is used mainly for water-insoluble samples, preparative HPLC, and the separation of isomers.	Chapter 8
Non-aqueous reversed-phase chromatography (NARP)	The column is nonpolar (e.g., C_{18}), and the mobile phase is a mixture of organic solvents (e.g., acetonitrile plus methylene chloride); NARP is used for very hydrophobic, water-insoluble samples.	Section 6.5
Hydrophilic interaction chromatography (HILIC)	The column is polar (e.g., silica or amide-bonded phase), and the mobile phase is a mixture of water plus organic (e.g., acetonitrile); HILIC is useful for samples that are highly polar and therefore poorly retained in RPC.	Section 8.6
Ion-exchange chromatography (IEC)	The column contains charged groups that can bind sample ions of opposite charge, and the mobile phase is usually an aqueous solution of a salt plus buffer; IEC is useful for separating ionizable samples such as acids or bases, and especially for the separation of large biomolecules (e.g., proteins and nucleic acids).	Sections 7.5, 13.4.2
Ion-pair chromatography (IPC)	RPC conditions are used, except that an ion-pair reagent is added to the mobile phase for interaction with sample ions of opposite charge; IPC is useful for the separation of acids or bases that are weakly retained in RPC.	Section 7.4
Size-exclusion chromatography (SEC)	An inert column is used with either an aqueous or organic mobile phase; SEC provides separation on the basis of molecular weight and is used mainly for large biomolecules or synthetic polymers.	Section 13.8

it flows through the column, and sample molecules can enter the particle pores by diffusion (there is normally no significant flow of mobile phase *through* the particle).

Figure 2.3 illustrates a hypothetical separation of a sample that contains three sample compounds (or *solutes*), with individual sample molecules represented by • for solute *X*, □ for solute *Y*, and ▲ for solute *Z*. For clarity, molecules of the mobile phase are not shown, and molecules of the solvent that the sample is dissolved in are portrayed by +. The sample is applied to the column in (Fig. 2.3*a*) is carried through the column by the flowing mobile phase in successive stages (Fig. 2.3*b–d*), and eventually the sample leaves the column (Fig. 2.3*e*) to provide a plot of detector response versus time (a *chromatogram*, or record of the separation). As the separation proceeds in Figure 2.3*a–d*, molecules of sample components *X*, *Y*, and *Z* exhibit two characteristic behaviors: *differential migration* and *molecular spreading*. By Figure 2.3*d*, solutes *X*, *Y*, and *Z* have become separated from each other within the column.

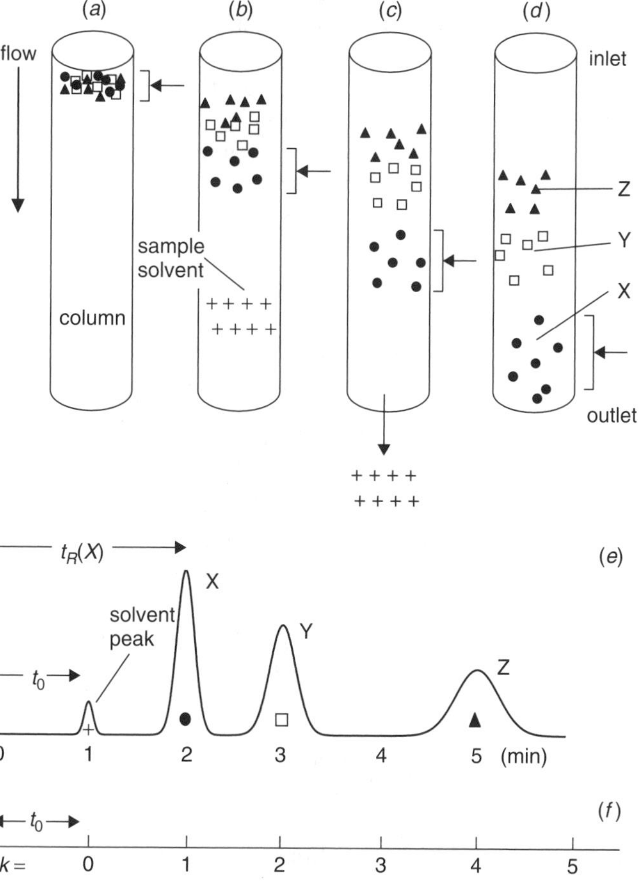

Figure 2.3 Illustration of the separation process in HPLC. (*a–d*) Sequential separation within the column (i.e., as a function of time); (*e*) the final chromatogram; (*f*) estimating values of *k* from the chromatogram (*e*). Solute molecules *X*, *Y*, and *Z* are represented by ●, □, and ▲, respectively; sample solvent molecules are shown by +.

Differential migration (different average speeds at which solute molecules of X, Y, and Z move, or migrate, through the column) forms the basis of chromatographic separation. Without a difference in migration rates for two compounds, their separation cannot occur. In this example molecules of X (•) move fastest, and molecules of Z (▲) move slowest; molecules of the sample solvent or mobile phase are not retained by the column-packing, pass through the column quickest of all, and leave the column first. Solvent molecules that form part of the injected sample are represented in Figure 2.3*b–e* by +.

As a given solute moves through the column, its molecules become increasingly spread out, so as to occupy a larger volume within the column. The volume that encompasses the molecules of a given solute within the column defines what is called a *band*. The width of this solute-volume is measured in the direction of flow, and is defined as the *band width*, as indicated in Figure 2.3*a–d* for solute X by the arrow and bracket alongside molecules of X (•). When a band leaves the column and is recorded in the chromatogram (Fig. 2.3*e*), it is then referred to as a *peak*. The identity of each peak can be determined from the time at which it leaves the column (the *retention time* t_R), while the concentration of each solute in the sample is proportional to peak size (measured either as area or height; see Section 11.2.3). For sufficiently small samples (low-ng to μg injections, as typically used in HPLC assay procedures), peak retention times do not change as sample concentration (and resulting peak size) is varied. In the rest of this chapter we will examine separation further as a function of experimental conditions.

2.3 RETENTION

The retention time t_R for each solute is the time from sample injection to the appearance of the top of the peak in the chromatogram; in Figure 2.3*e* the retention times for solutes X, Y, and Z are, respectively, 2, 3, and 5 minutes. The retention time of the solvent peak at one minute is referred to as the column *dead-time* t_0 (Section 2.3.1) (sometimes t_m is used instead of t_0 to represent column dead-time). The migration rate or velocity u_x at which solute X moves through the column is determined by the fraction R of its molecules that are present in the flowing mobile phase at any time. On average, u_x will be equal to R times the migration rate or velocity u of solvent molecules:

$$u_x = Ru \tag{2.1}$$

For example, if half of the molecules of X are in the mobile phase ($R = 0.5$) and half are in the stationary phase, only half of the molecules are moving at any given time, so the *average* migration rate of X will be one half as fast as that of the solvent.

As illustrated in Figure 2.4, the fraction R of molecules X in the mobile (moving) phase is determined by an equilibrium process:

$$\text{X (mobile phase)} \Leftrightarrow \text{X (stationary phase)} \tag{2.2}$$

Molecules of X in Figure 2.4 are found equally in the mobile and stationary phase at any time, while molecules of Z predominate in the stationary phase; that is, Z

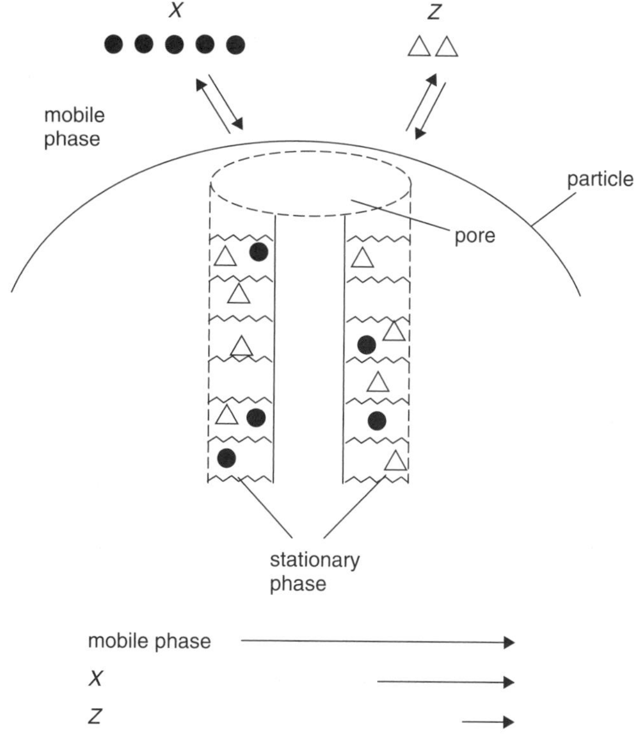

Figure 2.4 Equilibrium distribution of solvent and sample molecules between the mobile and stationary phases, and the resulting effect on solute migration rate. The values of k for solutes X and Z are 1 and 4, respectively. Equal amounts of X and Z in the sample are assumed.

is more retained than X and therefore migrates more slowly (indicated at the base of Fig. 2.4 by arrows, whose lengths denote migration rate). This equilibrium and the migration rate of a given solute are affected by the molecular structure of the solute, the chemical composition of the mobile and stationary phases (the solvent and column), and the temperature. The average pressure within the column can have a small effect on sample retention [1] (see also Section 2.5.3.1), but usually this can be ignored for moderate pressures (e.g., <5000 psi).

2.3.1 Retention Factor *k* and Column Dead-Time t_0

For a given solute, the *retention factor k* (this is still sometimes referred to as the *capacity factor k'*) is defined as the quantity of solute in the stationary phase (s), divided by the quantity in the mobile phase (m). The quantity of solute in each phase is equal to its concentration (C_s or C_m, resp.) times the volume of the phase (V_s or V_m, resp.), which then gives

$$\begin{aligned} k &= \frac{C_s V_s}{C_m V_m} = \frac{C_s/C_m}{V_s/V_m} \\ &= K\Psi \end{aligned} \tag{2.3}$$

where $K = (C_s/C_m)$ is the equilibrium constant for Equation (2.2), and $\Psi = (V_s/V_m)$ is the *phase ratio*—the ratio of stationary and mobile-phase volumes within the column. We will see that k is a very important property of each peak in the chromatogram; values of k can help us interpret and improve the quality of a separation. A solute molecule must be present in either the mobile or stationary phase so that, if the fraction of molecules in the mobile phase is R, the fraction in the stationary phase must be $1 - R$; therefore from Equation (2.3) we have

$$k = \frac{1 - R}{R} \qquad (2.3a)$$

or

$$R = \frac{1}{1 + k} \qquad (2.3b)$$

The retention time (t_R) of X can be defined as distance divided by speed (or band velocity), where the distance is the column length L and the band velocity is u_x:

$$t_R = \frac{L}{u_x} \qquad (2.4)$$

Similarly the retention time of the solvent peak is

$$t_0 = \frac{L}{u} \qquad (2.4a)$$

where u is the average mobile-phase velocity. Eliminating L between Equations (2.4) and (2.4*a*) gives

$$t_R = \frac{t_0 u}{u_x} \qquad (2.4b)$$

which, with $R = u_x/u_0$ (Eq. 2.1) and Equation (2.3b), then gives

$$\boxed{t_R = t_0(1 + k)} \qquad (2.5)$$

Equation (2.5) can also be expressed in terms of retention volume $V_R = t_R F$, where F is the mobile-phase flow rate (mL/min):

$$V_R = V_m(1 + k) \qquad (2.5a)$$

Here V_m is the column *dead-volume*, equal to $t_0 F$ (see the further discussion of V_m and Eq. 2.5a below).

Equation (2.5) can be rearranged to give

$$\boxed{k = \frac{t_R - t_0}{t_0}} \qquad (2.6)$$

which allows the calculation of values of k for each peak in the chromatogram. Visual estimates of k from the chromatogram (based on Eq. 2.6) are often used in practice, because exact values of k are seldom needed for developing a separation (*method development*) or during routine analysis. Thus k is equal to the corrected retention time ($t_R - t_0$), measured in units of t_0, or

$$k = \left(\frac{t_R}{t_0}\right) - 1 \tag{2.6a}$$

As illustrated in Figure 2.3*f* (which corresponds to the chromatogram of Fig. 2.3*e*), the distance t_0 can be used to mark off approximate values of k, beginning at time t_0; thus k equals 1, 2, and 4, respectively, for compounds X, Y, and Z.

We will see in Section 2.4.1 that values of k between about 1 and 10 are preferred for various reasons. Therefore it is important to be able to estimate (or calculate) values of k for the different peaks in a chromatogram, which in turn requires a value of the column dead-time t_0. A value of t_0 can often be obtained from a visual inspection of the initial portion of the chromatogram, as illustrated in Figure 2.5*a*–b. Sometimes the first baseline disturbance assumes the characteristic shape illustrated in Figure 2.5*a*, which is a clear indication of the unretained solvent peak. This t_0-disturbance is usually the result of a change in refractive index (RI) of the mobile phase (due to differences in RI for the sample solvent vs. the mobile phase), which in turn affects the amount of light that passes through the flow cell of the detector. If the sample is dissolved in the mobile phase (usually the preferred choice), a t_0 peak as in Figure 2.5*a* may not be observed.

At other times, especially for the injection of a reaction product, environmental sample, or plant or animal extract, a very large ("excipient" or "junk") peak may be observed at the beginning of the chromatogram (Fig. 2.5*b*). In this case t_0 corresponds to the initial rise of the peak. Sometimes no obvious solvent peak is observed (Fig. 2.5*c*), in which case a value of t_0 can either be measured or estimated.

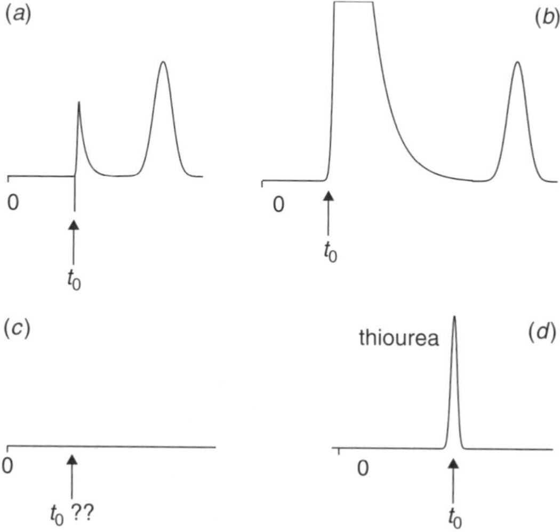

Figure 2.5 Determining the column dead-time t_0.

The most direct procedure for determining t_0 is to inject a solute (dissolved in water or the mobile phase) that is unretained ($k = 0$) and readily detected, as in Figure 2.5*d*. When UV detection below 220 nm is used, thiourea as test solute fulfills both requirements, and is therefore a good choice for the measurement of t_0. Other test solutes have also been used for measuring t_0, for example, uracil or concentrated solutions of a UV-absorbing salt such as sodium nitrate [2–4]. Observed values of t_0 for a given column can vary with mobile-phase composition by as much as ±10–15% for 0–100% B (%B refers to the percent by volume of organic solvent in the mobile phase), but usually t_0 varies by <5% for 20–80% B [5]. For approximate estimates of k as in Figure 2.3*e*, a value of t_0 measured for one value of %B can be assumed to be the same for all values of %B (when only %B is changed).

Alternatively, a value of t_0 can be estimated from the column dimensions and flow rate (for columns packed with fully porous particles):

$$t_0 \approx 5 \times 10^{-4} \frac{Ld_c^2}{F} \tag{2.7}$$

Here L is the column length in mm, d_c is the column inner diameter in mm, F is the flow rate in mL/min, and t_0 is in minutes. For several hundred different RPC columns, it was found that Equation (2.7) agrees with experimental values of t_0 with an average error of only ±10% (1 SD) [6], which again is accurate enough for practical purposes. The column *dead-volume* V_m is related to t_0 as

$$V_m = t_0 F \approx 5 \times 10^{-4} L d_c^2 \tag{2.7a}$$

with L and d_c in mm. The dead-volume V_m represents the total volume of mobile phase inside the column, both inside and outside of the column particles. For example, if $V_m = 2$ mL, and $F = 0.5$ mL/min, then $t_0 = V_m/F = 2/0.5 = 4$ min; t_0 can be regarded as the time required to empty the column of the mobile phase that was originally present in the column.

For the common case where the column inner diameter ≈ 4.6 mm, we can conveniently estimate values of V_m (by combining Eqs. 2.7 and 2.7a):

$$V_m(\text{mL}) \approx 0.01L \qquad \text{(for 4 to 5 mm i.d. columns, with } L \text{ in mm)} \tag{2.7b}$$

Values of t_0 can then be obtained from Equation (2.7b), with $t_0 = V_m/F$. For a further discussion of the measurement, accuracy and significance of column dead-time or dead-volume, see [2–4].

2.3.2 Role of Separation Conditions and Sample Composition

The relative effect of different separation conditions on sample retention k is summarized in the second column of Table 2.2. Table 2.2 is applicable for different HPLC modes, but the following discussion will assume reversed-phase chromatography (RPC). The mobile phase for RPC is usually a mixture of water or aqueous buffer (A-solvent) and an organic solvent (B-solvent) such as acetonitrile or methanol. As the volume-percent of organic solvent (%B) is increased, the retention of all sample compounds decreases. A mobile phase that provides smaller values of k is referred to as a "stronger" mobile phase; similarly water is referred to as a "weak" solvent, and

Table 2.2

Effect of Different Separation Conditions on Retention (k), Selectivity (α), and Plate Number (N)

Condition	k	α	N
%B	++	+	–
B-solvent (acetonitrile, methanol, etc.)	+	++	–
Temperature	+	+	+
Column type (C_{18}, phenyl, cyano, etc.)	+	++	–
Mobile phase pH[a]	++	++	+
Buffer concentration[a]	+	+	–
Ion-pair-reagent concentration[a]	++	++	+
Column length	0	0	++
Particle size	0	0	++
Flow rate	0	0	+
Pressure	–	–	+[b]

Note: ++, major effect; +, minor effect; -, relatively small effect; o, no effect; bolded quantities denote conditions that are primarily used (and recommended) to control k, α, or N, respectively (e.g., %B is varied to control k or α, column length is varied to control N).

[a]For ionizable solutes (acids or bases).

[b]Higher pressures allow larger values of N by a proper choice of other conditions; pressure per se, however, has little direct effect on N (see Sections 2.4.1.1 and 2.5.3.1).

organic solvents are "strong." Typically values of k decrease by a factor of 2 to 3 for a change of +10% B; an example of the effect of %B on sample retention is shown in Figure 2.6, for the separation of a mixture of five herbicides. A mobile phase of 80% B in Figure 2.6*a* results in rapid elution of the sample, with small values of k (0.3–0.8) and poor separation. When %B is decreased (50% B, Fig. 2.6*b*), separation improves, separation or "run time" increases (16 min vs. 1.5 min in Fig. 2.6*a*), and peak heights are reduced because the peaks are wider. Retention normally is controlled within a desired range of k by the choice of %B. The conditions of Table 2.2 can also be varied in order to control separation selectivity (α) or column efficiency (N); see Section 2.5 for details.

Reversed-phase chromatography involves a nonpolar stationary phase or column (e.g., C_{18}) and a polar, water-containing mobile phase. Polar solutes will prefer the polar mobile phase ("like attracts like") and be less retained (larger R, smaller k), while nonpolar solutes will interact preferentially with the nonpolar stationary phase and be more retained (smaller R, larger k). The preferential interaction of a nonpolar solute (*n*-hexane) with the nonpolar stationary phase is illustrated in Figure 2.7*a*, while Figure 2.7*b* shows the preferential interaction of a polar solute (1,3-propanediol) with the polar mobile phase. Figure 2.7*c* is a chromatogram of several mono-substituted benzenes that vary in polarity or "hydrophobicity" because of the nature of the substituent group. Polar (less hydrophobic) groups such as $-NHCHO$, $-CH_2OH$, or $-OH$ reduce retention relative to the unsubstituted solute benzene (shaded peak), while less polar (more hydrophobic) groups such as chloro, methyl, bromo, iodo, and ethyl increase retention.

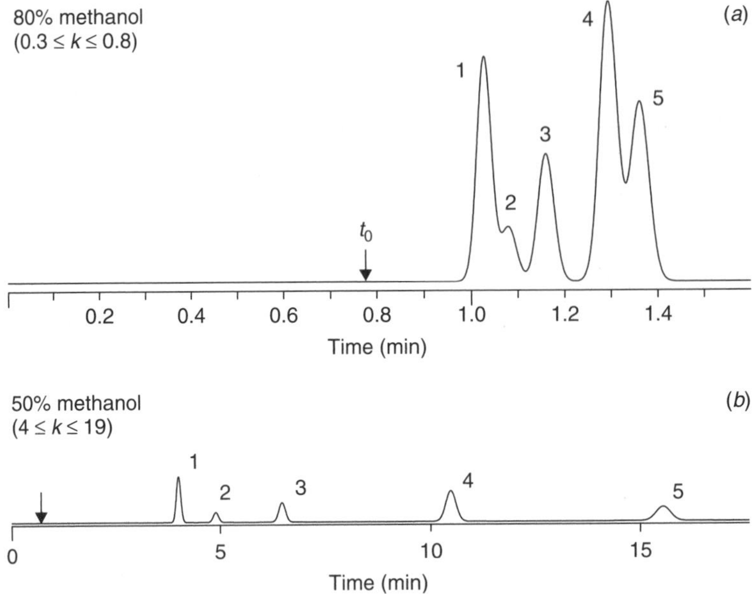

Figure 2.6 Separation as a function of mobile phase %B (%v methanol). Herbicide sample: 1, monolinuron; 2, metobromuron; 3, diuron; 4, propazine; 5, chloroxuron. Conditions, 150 × 4.6-mm, 5-μm C_{18} column; methanol/water mixtures as mobile phase; 2.0 mL/min; ambient temperature. Recreated chromatograms from data of [7].

Ionized acids and bases are much more "polar" and therefore less retained than their neutral counterparts. A change in mobile phase pH that results in increased solute ionization will therefore lead to a decrease in retention time (Section 7.2).

2.3.2.1 Intermolecular Interactions

This section provides additional insight into sample retention as a function of the solute, column, and mobile phase; it also represents more information than is usually required in practice. The reader may therefore prefer to skip to following Section 2.3.2.2, *and return to this section as needed.*

The attraction between adjacent molecules of a solute and solvent is the result of several different intermolecular interactions, as illustrated in Figure 2.8. In principle, a quantitative understanding of these interactions should allow estimates—or even predictions—of retention as a function of molecular structure. While this is usually not possible at the present time (see Section 2.7.7), an understanding of these interactions can prove useful in other ways; for example, when selecting a different column for a change in separation (Section 5.4).

Dispersion interactions (Fig. 2.8*a*) result from the random, instantaneous positions of electrons around adjacent atoms of either the solvent (*S*) or the solute (*X*). Typically the arrangement of electrons around the nucleus of atom *S* will be unsymmetrical at any instant of time (as in Fig 2.8*a*), and this will cause the electrons in adjacent atom *X* to move as shown (due to coulombic repulsion). The result is an instantaneous dipole moment for both *S* and *X* that favors electrostatic attraction. The strength of dispersion interactions increases with the polarizability

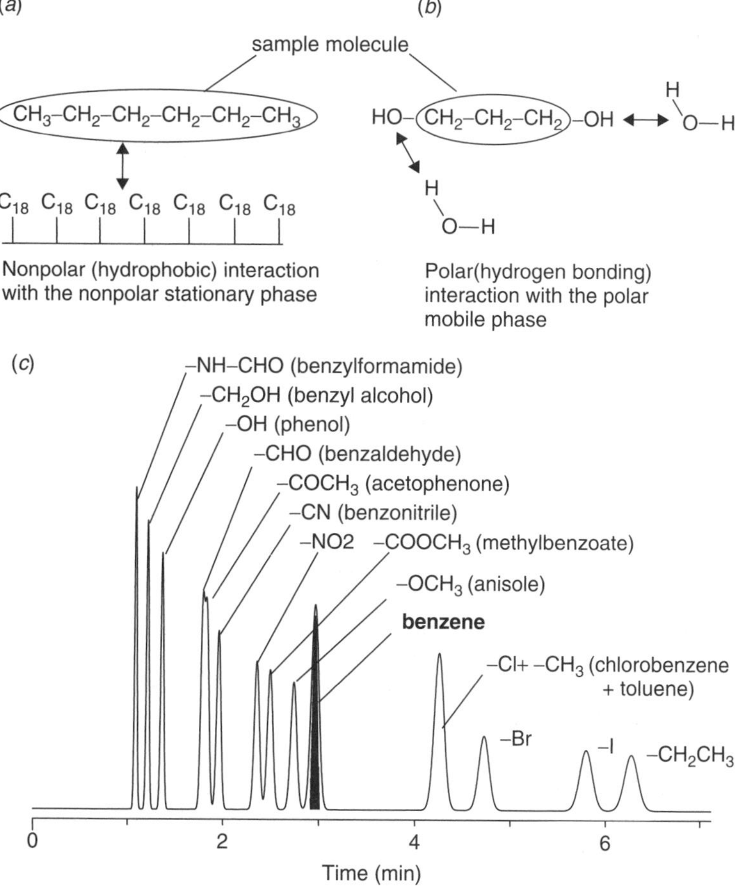

Figure 2.7 Sample polarity and retention. Illustration of the interaction of a nonpolar sample solute with the stationary phase (*a*) and of a polar solute with the mobile phase (*b*); (*c*) effect of different substituents on the retention of monosubstituted benzenes; 150 × 4.6-mm Hypersil C_{18} column, 50% acetonitrile/water as mobile phase, 25°C, 2 mL/min; recreated chromatogram from data of [8].

of each of the two adjacent atoms. Solute polarizability increases with the size of the molecule (number of atoms or molecular weight) and with refractive index [9]; dispersion interactions are therefore stronger for aromatic compounds and for molecules substituted by atoms of higher atomic weight (sulfur, chlorine, bromine, etc.)—provided that molecules are of similar size.

Dispersion interactions exist between every adjacent pair of atoms, and this interaction largely accounts for the physical attraction between molecules of all kinds (especially for less polar molecules). Because of the nonspecific and universal nature of dispersion interactions, they are significant in both the mobile and stationary phases. Dispersion interactions therefore tend to cancel, and they generally play only a minor role in determining *selective* interactions of the kind that result in changes in relative retention when the mobile phase or column is changed. Dispersion interactions contribute to *hydrophobic* interactions, so called because

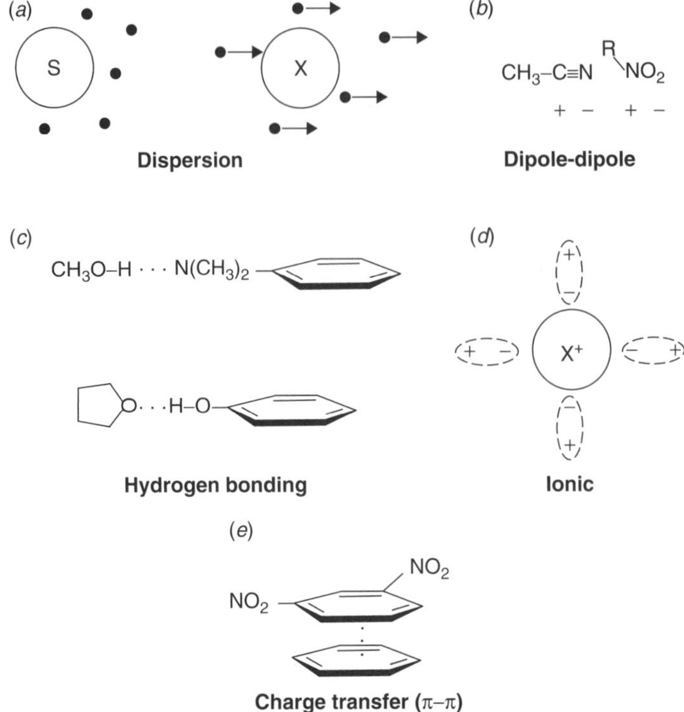

Figure 2.8 Intermolecular interactions that can contribute to sample retention and selectivity.

of the attraction of less polar solutes to nonpolar RPC stationary phases (or their "water-fearing" rejection from the polar aqueous phase). As the strength of dispersion interactions increases (for larger, less polar solute molecules), the solute is increasingly retained.

Dipole–dipole interaction is illustrated in Figure 2.8*b* for the case of dipolar molecules of solvent (acetonitrile, $CH_3C{\equiv}N$) and solute (a nitroalkane, R–NO_2). The functional groups (–C≡N and –NO_2) in these two molecules each have a large, permanent, dipole moment, causing the two molecules to align for maximum electrostatic interaction (positive end of one molecule adjacent to the negative end of the other). The strength of dipole interaction is proportional to the dipole moments of each of the two interacting groups (*not* the dipole moment of an entire, multi-substituted molecule), because dipole interactions are only effective at very close range (i.e., adjacent atoms or groups).

Hydrogen bonding interactions are shown in Figure 2.8*c*, for two cases: an acidic (or proton-donor) solvent (methanol) interacting with a basic (proton-acceptor) solute (*N*,*N*-dimethylaniline), or an acidic solute (phenol) interacting with a basic solvent tetrahydrofuran (THF). The strength of hydrogen bonding increases with increasing hydrogen-bond acidity and basicity of the two interacting species (Table 2.3).

Ionic (coulombic) interaction is illustrated in Figure 2.8*d* for a positively charged sample ion (X^+) interacting with surrounding molecules of a polarizable

Table 2.3

Solvent Selectivity Characteristics

Solvent	Normalized Selectivity[a] H-B Acidity α/Σ	H-B Basicity β/Σ	Dipolarity π^*/Σ	P'[b]	ε[c]
Acetic acid	0.54	0.15	0.31	6.0	6.2
Acetonitrile	0.15	0.25	0.60	5.8	37.5
Alkanes	0.00	0.00	0.00	0.1	1.9
Chloroform	0.43	0.00	0.57	4.1	4.8
Dimethylsulfoxide	0.00	0.43	0.57	7.2	4.7
Ethanol	0.39	0.36	0.25	4.3	24.6
Ethylacetate	0.00	0.45	0.55	4.4	6.0
Ethylene chloride	0.00	0.00	1.00	3.5	10.4
Methanol	0.43	0.29	0.28	5.1	32.7
Methylene chloride	0.27	0.00	0.73	3.1	8.9
Methyl-*t*-butylether	0.00	≈0.6	≈0.4	≈2.4	≈4
Nitromethane	0.17	0.19	0.64	6.0	35.9
Propanol (*n*- or *iso*)	0.36	0.40	0.24	3.9	6.0
Tetrahydrofuran	0.00	0.49	0.51	4.0	7.6
Triethylamine	0.00	0.84	0.16	1.9	2.4
Water	0.43	0.18	0.45	10.2	80

Note: see Appendix I (Table I.4) for additional solvent information.
[a]Values from [11], where Σ refers to the sum of values of α, β, and π^* for each solvent.
[b]Polarity index; values from [12].
[c]Dielectric constant; values from [13].

solvent. The positive charge on the solute ion causes a displacement of charge in the solvent molecules, for maximum electrostatic interaction. The strength of ionic interaction increases for solvents with a larger dielectric constant ε (Table 2.3). Ionic interaction can also occur between a charged sample ion and ions in either the mobile or stationary phases; see the discussion of ion-pair chromatography (Section 7.4.1) and ion-exchange chromatography (Section 7.5.1).

Charge transfer or π–π interaction is illustrated in Figure 2.8*e* for the π-acid (electron-poor) solute 1,3-dinitrobenzene and the π-base (electron-rich) solvent benzene. Interactions of this kind can occur between any two aromatic (or unsaturated) species, with the strength of the interaction increasing for stronger π-bases such as polycyclic aromatics (e.g., naphthalene and anthracene), and for stronger π-acids (e.g., aromatics substituted by electron withdrawing nitro groups). The solvent acetonitrile (a π-acid) can also interact with aromatic solutes by π–π interaction [10].

The polar interactions of various nonionic aliphatic solvents used in HPLC can be described by the *solvent-selectivity triangle* (Fig. 2.9, [11]). The position of each solvent in this plot indicates its *relative* hydrogen-bond acidity α/Σ, hydrogen-bond basicity β/Σ, and dipolarity π^*/Σ. Thus amines are relatively strong hydrogen-bond bases, as indicated by their position near the top of the triangle (large β). Similarly

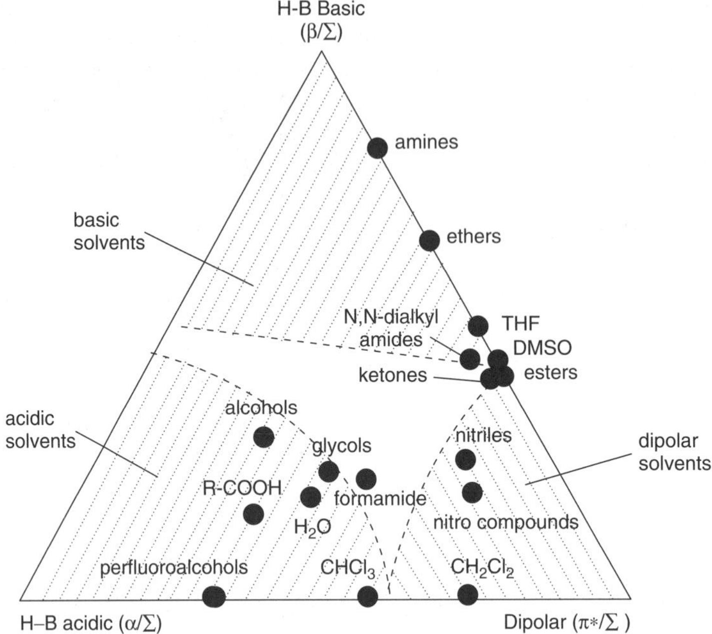

Figure 2.9 Solvent-selectivity triangle for aliphatic solvents of various kinds. See Table 2.3 for values of the solvent properties plotted. Adapted from [11].

nitroalkanes, aliphatic nitriles, and CH_2Cl_2 all have groups with large dipole moments, and they are situated near the lower right-hand side of the triangle. Perfluoroalcohols are especially strong hydrogen-bond donors (and simultaneously very weak acceptors); they and carboxylic acids (R–COOH) are found near the lower left of the triangle (large α). Table 2.3 lists (1) relative contributions to solvent polarity from dipolarity and hydrogen-bond acidity or basicity, (2) a measure of overall solvent polarity (P'), and (3) values of the dielectric constant ε. Larger values of ε for the mobile phase indicate increasing ionic interaction with solute molecules as in Figure 2.8*d*, increasing solubility in the mobile phase for ionic solutes, and smaller values of k for ionic solutes. For a comprehensive review of intermolecular interactions in chromatography, see [14].

More will be said about the solvent-selectivity triangle of Figure 2.9 and solvent selectivity in Chapters 6 and 8. Section 5.4 on *column selectivity* provides a similar treatment of interactions between the solute and the stationary phase.

2.3.2.2 Temperature

Temperature is an important variable in HPLC, as it has a significant effect on values of k. For most solute molecules and customary separation conditions, solute retention varies with temperature according to the Van't Hoff equation, which can be expressed in HPLC as

$$\log k = A + \frac{B}{T_K} \tag{2.8}$$

For a given solute and other conditions unchanged, A and B are temperature-independent constants, and T_K is the temperature (K). Values of k usually decrease with increasing temperature (positive value of B) by 1–2% per °C; thus a 50°C increase will cause about a 2-fold decrease in k. As temperature increases, separation often worsens, while peak heights increase (similar to an increase in %B, as in Fig. 2.6). It should be noted that deviations from Equation (2.8) are not uncommon, sometimes resulting in curved plots of log k against $1/T_K$. In a few cases, retention is observed to *increase* with an increase in temperature. These exceptions to Equation (2.8) can arise for various reasons, including changes with temperature of (1) the ionization of a solute [15, 16], (2) solute molecular conformation [17], and (3) the stationary phase [18].

Temperature also affects the column plate number N and pressure drop (see Section 2.4). The practical use of most current HPLC equipment is limited to temperatures of <80°C (Section 3.7.2), and HPLC column lifetimes often are shorter at temperatures >60°C (Section 5.8). For a further discussion of the role of temperature in HPLC, see Section 2.5.3.1 and [19–20a].

2.4 PEAK WIDTH AND THE COLUMN PLATE NUMBER *N*

As illustrated in Figure 2.3, solute molecules spread out to enclose a larger volume (or form a wider band) during their migration through the column. When the band leaves the column to become a peak in the chromatogram, it will have a width that can be defined in various ways. The *baseline peak width* W is illustrated for the first peak i of Figure 2.10*a*. Tangents are drawn to each side of the peak (through the inflection points), and their intersection with the baseline determines the value of W. When referring to peak width in this book, we will assume values of baseline peak width W. The relative ability of a column to furnish narrow peaks is described as *column efficiency*, and is defined by the *plate number* N:

$$\boxed{N = 16\left(\frac{t_R}{W}\right)^2} \tag{2.9}$$

For example, W for peak i in Figure 2.10*a* is equal to $(4.00 - 3.85) = 0.15$ min, and $t_R = 3.93$ min. Therefore $N = 16 \times (3.93/0.15)^2 = 10{,}980$. Values of N can vary for different samples, separation conditions, and columns (Section 2.4.1). The larger the value of N, the narrower are the peaks in the chromatogram, and the better is the separation.

Peak width can be measured more conveniently (and precisely) by the *half-height peak width* $W_{1/2}$, as illustrated for peak j in Figure 2.10*a*; values of $W_{1/2} \equiv 0.588W$ are reported by many data systems. When the peak width at half height is used to calculate N,

$$N = 5.54\left(\frac{t_R}{W_{1/2}}\right)^2 \tag{2.9a}$$

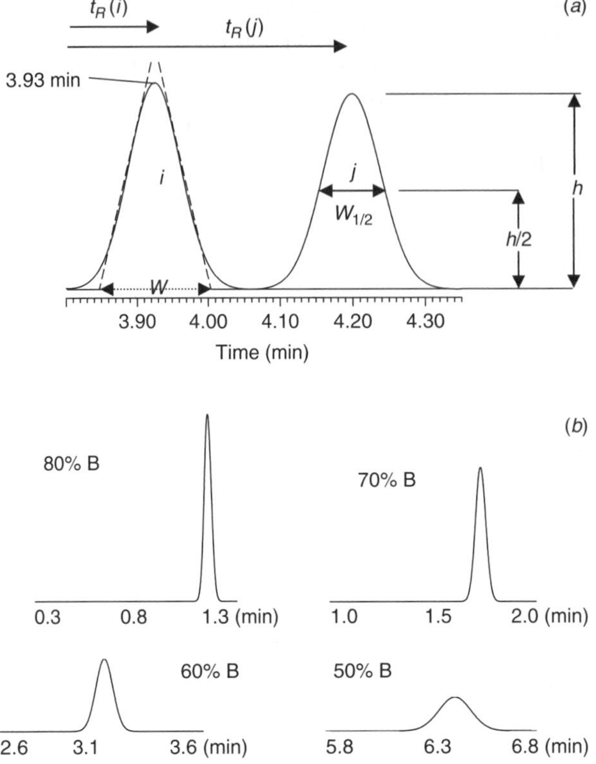

Figure 2.10 Origin and measurement of peak width. (*a*) Measurement of peak width; (*b*) peak 3 of Figure 2.6 as a function of %B, shown with the same time scale for each peak.

Because an ideal chromatographic peak has the shape of a Gaussian curve (Section 2.4.2), peak width is sometimes described in terms of the standard deviation σ of the Gaussian curve, where

$$W = 4\sigma \tag{2.9b}$$

and therefore

$$N = \left(\frac{t_R}{\sigma}\right)^2 \tag{2.9c}$$

Equation (2.9c) can be expressed in other forms, for example, $N = 25(t_R/W_{5\sigma})^2$, where $W_{5\sigma} = 1.25W$ is the so-called 5σ peak width.

Equation (2.9) can be rearranged to give

$$W = 4N^{-0.5}t_R \tag{2.10}$$

or (replacing t_R by Eq. 2.5)

$$W = 4N^{-0.5}t_0(1 + k) \tag{2.10a}$$

Because values of N are approximately constant for the different peaks in a chromatogram, Equation (2.10) tells us that peak width W will increase in proportion to retention time. A continual increase in peak width from the beginning to the end of the chromatogram is therefore observed; for example, see the chromatogram of Figure 2.7*c*.

The area for a given solute peak normally remains approximately constant when retention time is varied by a change in %B, temperature, or the column—so peak height h_p times peak width W will also be constant. For this situation

$$h_p \approx \frac{\text{(constant)}}{W} \approx \frac{\text{(constant)}}{t_R} \tag{2.11}$$

That is, as t_R increases, peak height decreases. An example is shown in Figure 2.10*b* for peak 3 of Figure 2.6 as a function of %B. A reciprocal change is seen in peak height and width as %B is varied, as predicted by Equation (2.11).

2.4.1 Dependence of *N* on Separation Conditions

We will begin by summarizing some practical conclusions about how the column plate number N varies with the column, the sample, and other separation conditions. In following Section 2.4.1.1, we will examine the theory on which these conclusions are based. N can also be described by

$$N = \left(\frac{1}{H}\right) L \tag{2.12}$$

where $H = L/N$ is the column *plate height*. H is a measure of column efficiency per unit length of column; increasing column length (as by replacing a 150-mm long column with a 250-mm column) is therefore a convenient way of increasing N and improving separation (since H is constant for columns that differ only in dimensions).

Consider next the log-log plot of Figure 2.11*a*, which shows how N varies with flow rate F and particle diameter d_p (= 2, 5, or 10 μm), while other separation conditions are held constant. As the mobile-phase flow rate F increases from a starting value of 0.1 mL/min, N first increases, then decreases. For the present conditions, maximum values of N (indicated by •) are found for flow rates of 0.2 to 1.0 mL/min, depending on particle size d_p. A 3-fold increase in F, relative to the "optimum" value F_{opt} for maximum N, has only a minor effect on separation (a decrease in N of ≈20%) but reduces separation time by 3-fold. Therefore flow rates greater than F_{opt} are usually chosen in practice. Flow rates $< F_{opt}$ are highly undesirable, as this means both lower values of N and longer separation times.

A decrease in particle size generally leads to an increase in N, as seen in Figure 2.11*a*. The occurrence of maximum N is seen to occur at higher flow rates for smaller particles, which allows faster separations for columns packed with smaller particles (Section 2.4.1.1). The pressure drop across the column (which we will refer to simply as "pressure" P) increases for smaller particles and higher flow rates (see the dashed lines in Fig. 2.11*a* for P = 2000, 5000, and 15,000 psi). The pressure (in

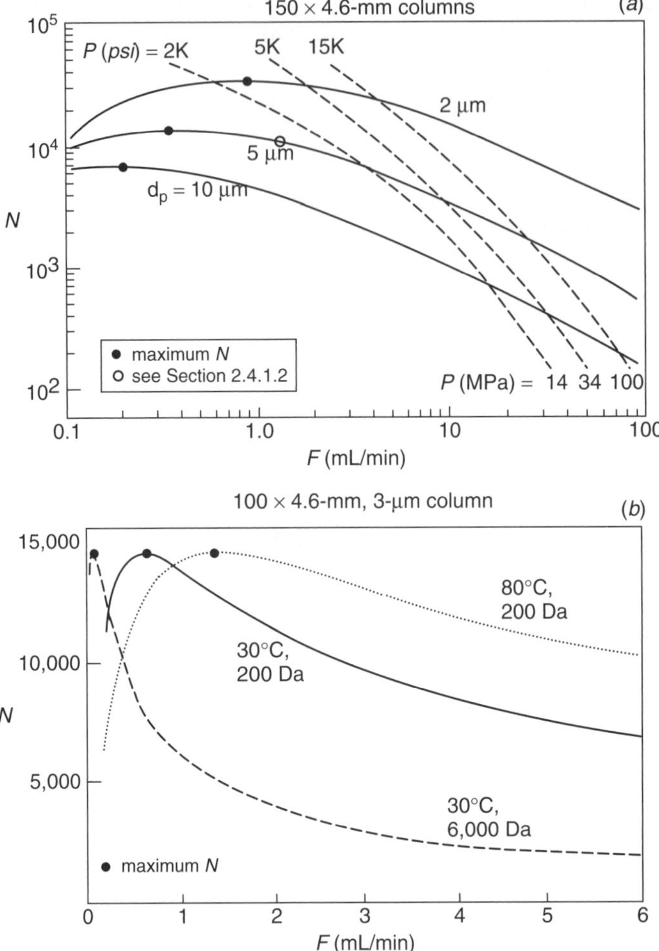

Figure 2.11 Variation of column plate number N with flow rate F, particle diameter d_p, and different conditions. Assumes 50% acetonitrile/water mobile phase. (*a*) Conditions: 150 × 4.6-mm column, 30°C, and a sample molecular weight of 200 Da; (- - -) connects points on curves of N versus F for pressure $P = 2000, 5000$, and 15,000 psi, respectively. (*b*) Conditions: 100 × 4.6-mm column (5-μm particles); other conditions shown in figure. All plots based on Equation (2.18a) with $A = 1, B = 2$, and $C = 0.05$.

psi) for a packed column can be estimated by

$$P \approx \frac{1.25L^2\eta}{t_0 {d_p}^2} \tag{2.13}$$

or from Equations (2.7) and (2.13),

$$P \approx \frac{2500L\eta F}{{d_p}^2 {d_c}^2} \tag{2.13a}$$

Here L is the column length (mm), η is the viscosity of the mobile phase in cP (see Appendix I for values of η as a function of mobile phase composition and temperature), t_0 is in minutes, F is the flow rate (mL/min), d_c is the column internal diameter (mm), and d_p is the particle diameter (μm). Other units of pressure (besides psi) are sometimes used in HPLC: bar ≡ atmospheres = 14.7 psi; megaPascal (MPa) = 10 bar = 147 psi.

Values of P are also affected somewhat by the nature of the particles within the column, and how well the column is packed (Section 5.6). Flow restrictions outside the column (tubing between the pump and detector, sample valve, detector flow cell) add to the total pressure measured at the pump outlet, but the sum of these contributions is usually minor (10–20%) compared to values of P from Equation (2.13)—the exception to this is HPLC systems designed for operation >6000 psi with <3-μm particles, which often exhibit a significant pressure in the absence of a column (e.g., ≥ 1000 psi). Variations in equipment and column permeability can cause Equations (2.13) and (2.13a) to be in error by ±20% or more. Despite the ability of most HPLC systems to operate at 5000 to 6000 psi, it may be desirable to limit the column pressure drop to no more than 3000 psi (Section 3.5). As the pressure typically increases when a column ages, this suggests that the pressure for a routine assay with a new column should not exceed 2000 psi. The latter recommendation is conservative, however, and HPLC systems are now commercially available for routine use at pressures of 10,000 to 15,000 psi or higher (Section 3.5.4.3; [21]).

Next consider Figure 2.11*b*, for a 100 × 4.6-mm column of 5-μm particles, with a mobile phase of 50% acetonitrile/water (note the linear–linear scale and the narrower, more typical range in values of F). Plots of N versus flow rate are shown for three different conditions: 30°C and a 200-Da sample, 30°C and a 6000-Da sample (e.g., a large peptide), or 80°C and a 200-Da sample. Similar plots of N versus F are observed for each of these three examples, except that the flow rate for maximum or optimum N (F_{opt}) is shifted to higher values for higher temperatures, and lower values for larger sample molecules. One conclusion from Figure 2.11*b* is that higher flow rates can be used with separations carried out at higher temperatures, which can in turn be used for faster separations (Section 2.5.3.1). Similarly separations of higher molecular-weight samples will generally require lower flow rates (and longer run times) for comparable values of N (e.g., Table 13.4). The dependence of N on the column, sample molecular weight, and other conditions is summarized in Table 2.4.

2.4.1.1 Band-Broadening Processes That Determine Values of N

The width W and retention time t_R of a peak determine the value of N for the column (Eq. 2.9). Various processes within and outside the column contribute to the final peak volume or width W, as illustrated in Figure 2.12 (note the increases in band width that result for each process [Fig. 2.12*a–e*], indicated by a bracket plus arrow alongside the band). Following sample injection, but before the sample enters the column, molecules of a solute will occupy a volume that is usually small (Section 2.6.1). This is illustrated in Figure 2.12*a*, where individual solute molecules are represented by •. Often the *extra-column* contribution to band width can be ignored (Section 3.9), but that depends on the characteristics of the equipment

Table 2.4

Effect of Different Experimental Conditions on Values of the Plate Number *N*

Condition	Effect on F_{opt}, the Value of F for Maximum N	Effect on N and P of an *Increase* in Specified Condition[a]
Column length L	None	N increases proportionately P increases proportionately
Column diameter d_c	$F_{opt} \propto d_c^2$	None, if flow rate increased in proportion to d_c^2 (recommended)
Column particle size d_p	$F_{opt} \propto 1/d_p$	N decreases (Fig. 2.12) P decreases
Mobile-phase flow rate F	None	N decreases (Fig. 2.12) P increases proportionately
Mobile-phase viscosity η	F_{opt} decreases as η increases	N decreases (Eqn. 2.19) P increases proportionately
Temperature $T(K)$	F_{opt} increases as T increases[b]	N increases (Fig. 2.13*c*) P decreases
Sample molecular weight M	$F_{opt} \propto M^{-0.6}$	N decreases (Fig. 2.13*c*) no effect on P

Note: See discussion of Figures 2.12 and 2.13.
[a]Assumes flow rates equal or exceed the optimum value for maximum N (F_{opt}), as is often the case.
[b]Due to an increase in both T and mobile phase viscosity (Eq. 2.19).

and the size of the column (small-volume columns packed with small particles are especially prone to extra-column band broadening). Consider next the *longitudinal diffusion* of solute molecules along the column, as illustrated in Figure 2.12*b*. This process causes band width to increase with time, and it occurs whether or not the mobile phase is flowing. The time spent by the band during its passage through the column varies inversely with the flow rate, so the contribution to band width from longitudinal diffusion decreases for faster flow.

Eddy diffusion represents another contribution to band broadening (Fig. 2.12*c*). As molecules of the sample are carried through the column in different flow streams (arrows) between particles, molecules in slow-moving (constricted or narrow) streams lag behind, while molecules in fast-moving (wide) streams are carried ahead. This contribution to band broadening is approximately independent of flow rate, and depends only on the arrangement and sizes of particles within the column; band broadening due to eddy diffusion increases for poorly packed columns. *Mobile-phase mass transfer* (Fig. 2.12*d*) is the result of a faster flow of the stream center (much like the middle of a river). As flow rate increases, the center of the stream moves relatively faster, and band broadening increases.

A final contribution to band broadening within the column is *stationary-phase mass transfer* (Fig. 2.12*e*). Some sample molecules will penetrate further into a particle pore (by diffusion) and spend a longer time before leaving the particle (e.g., molecule *i* in Fig. 2.12*e*). During this time other molecules (e.g., *j*) will have moved a shorter distance into the particle and spent less time before leaving the particle. Molecules (e.g., *j*) that spend less time in the particle will move further

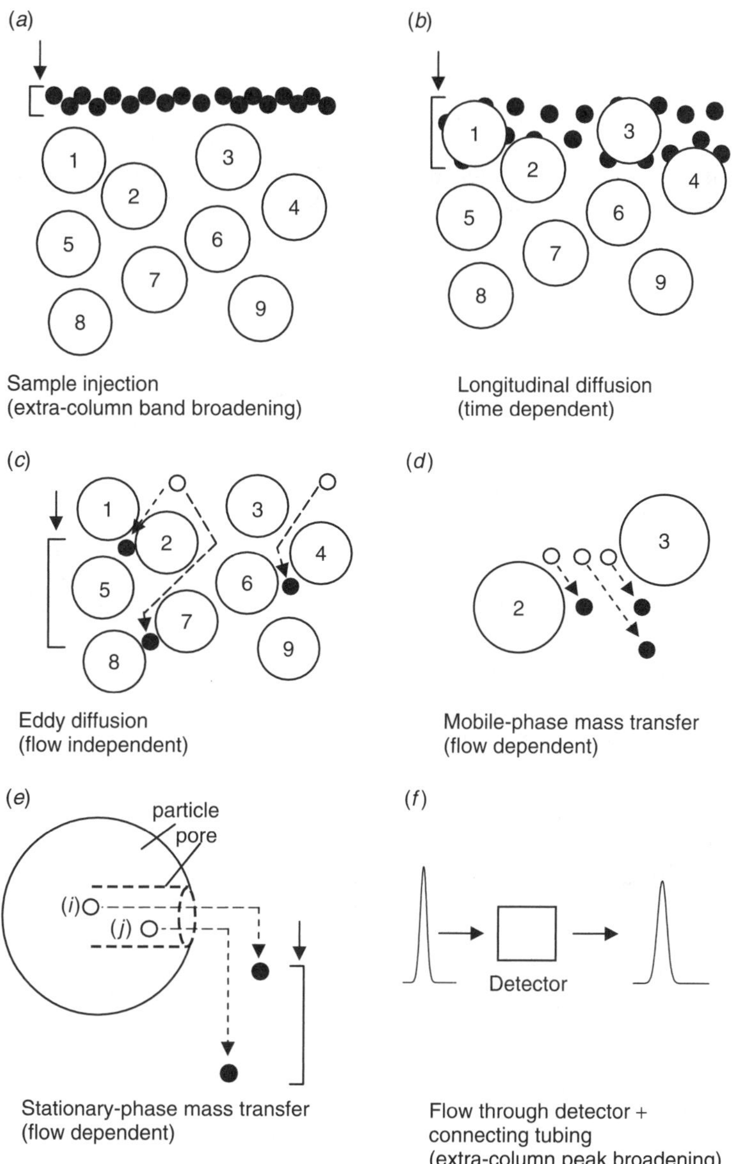

Figure 2.12 Illustration of various contributions to band broadening during HPLC separation. Molecules of a solute represented by ○ (before migration) and ● (after migration); - - -> indicates movement of a solute molecule.

along the column, with a consequent increase in band width. This contribution to band broadening increases as the flow rate increases. Eventually the band leaves the column and passes through the detector (Fig. 2.12*f*), resulting in some additional extra-column peak broadening—as during introduction of the sample to the column (Fig. 2.12*a*).

Band-broadening processes as in Figure 2.12 contribute to the final peak width W as

$$W^2 = \sum W_i^2 \tag{2.14}$$

where W_i is the contribution of each (independent) process i to the final peak width. We can distinguish peak-width contributions that arise either inside or outside of the column. Let W_{EC} represent the sum of extra-column contributions (as in Fig. 2.12*a* plus *f*), and let W_0 indicate the sum of intra-column contributions so that

$$W^2 = W_0^2 + W_{EC}^2 \tag{2.14a}$$

The extra-column peak broadening W_{EC} should be relatively minor in a well-designed HPLC system (Section 3.9), so it will be ignored during the following discussion. Because W_{EC} does not depend on values of k, while W_0 (defined by Eq. 2.10a) increases with k, extra-column band broadening has its largest effect on early peaks in the chromatogram.

The remainder of this section and Section 2.4.1.2 *can be useful for insight into the dependence of N on experimental conditions. This discussion also provides a basis for achieving very fast separations (Section* 2.5.3.2) *and for otherwise optimizing column efficiency and separation. This material is less essential for the everyday use of HPLC separation, however, and can be somewhat challenging. The reader may therefore wish to skip to Section* 2.4.2, *and return to Section* 2.4.1.1 *and* 2.4.1.2 *at a later time. Nevertheless, the material beginning with Equation* (2.17) *and especially Section* 2.4.1.2 *can have great practical value and is very much worth the reader's attention. See also the expanded discussion of band-broadening theory in* [22–25].

The quantity W_0 will henceforth be considered equivalent to the peak width W (Eq. 2.10), which can be expressed in terms of Equation (2.14) as

$W^2 =$	W_L^2	$+$	W_E^2	$+$	W_{MP}^2	$+$	W_{SP}^2
	longitudinal diffusion		eddy diffusion		mobile-phase mass transfer		stationary-phase mass transfer

A combination of Equations (2.10) and (2.12) yields

$$W^2 = \left[\left(\frac{16}{L}\right) t_R^2\right] H \tag{2.15a}$$

where values of $H = L/N$ for different solutes are approximately independent of retention time t_R for a given column of length L and the same experimental conditions. Therefore

$$W^2 \approx (\text{constant})\, H \tag{2.15b}$$

Equation (2.15) can be more directly related to values of N by replacing values of W^2 with values of H (Eq. 2.15b):

$$H = H_L + H_E + H_{MP} + H_{SP} \tag{2.16}$$

The quantities H_L, H_E, H_{MP}, and H_{SP} have the same significance as corresponding values of W in Equation (2.15); H_L is the contribution to H by longitudinal diffusion, and so forth. Recalling our discussion above of Figure 2.12, and noting that values of W^2 are proportional to values of H (Eq. 2.15b), the following expression can be derived from theoretical equations for each of the four contributions to W:

$$H = \underbrace{A}_{\text{eddy diffusion}} + \underbrace{\frac{B}{F}}_{\text{longitudinal mass transfer}} + \underbrace{CF}_{\text{mobile-phase plus stationary-phase mass transfer}}$$

where the coefficients A, B, and C are each constant for a particular solute, column, and set of experimental conditions. If values of F in Equation (2.16a) are replaced by the mobile-phase velocity u, the so-called van Deemter equation results:

$$H = A + \frac{B}{u} + Cu \qquad (2.16b)$$

where A, B, and C represent a different set of constants for a particular solute, column, and set of experimental conditions.

Equation (2.16a) is not quite correct, however, because it assumes that all four contributions to W are independent of each other. This is not the case for eddy diffusion and mobile-phase mass transfer; whenever two inter-particle flow streams combine, remixing occurs, with loss of the velocity profile created by mobile-phase mass transfer (so-called coupling). We must therefore treat these two processes (eddy diffusion and mobile-phase mass transfer) as a single band-broadening event. Because eddy diffusion does not vary with flow rate ($\propto F^0$), while mobile-phase mass transfer does ($\propto F^1$), the combination of the two contributions to band width will vary with some fractional power of $F(F^n)$. Experimental studies suggest a dependence of the combined value of H for eddy diffusion plus mobile-phase mass transfer to the $^1/_3$ flow rate power, which leads to an equation of the form

$$H = \underbrace{\frac{B}{F}}_{\text{longitudinal diffusion}} + \underbrace{AF^{1/3}}_{\text{eddy diffusion + mobile-phase mass transfer}} + \underbrace{CF}_{\text{stationary-phase mass transfer}}$$

(A, B, and C are still another set of constants). A final, generalized relationship between peak width and experimental conditions can be achieved as follows: A, B, and C of Equation (2.16c) are variously functions of the solute diffusion coefficient D_m and/or particle diameter d_p, such that Equation (2.16) can be restated as the so-called Knox equation [25]:

$$\boxed{h = A\nu^{0.33} + \frac{B}{\nu} + C\nu} \qquad (2.17)$$

Values of the coefficients of Equation (2.17) can be assumed for an "average" separation: $A = 1, B = 2$, and $C = 0.05$ (these are very approximate values that vary somewhat with the nature of the column—and how well it is packed—and with values of k). Here we define a *reduced plate height h*,

$$h = \frac{H}{d_p} \tag{2.18}$$

and a *reduced velocity ν*,

$$\nu = \frac{u_e d_p}{D_m} \tag{2.18a}$$

where u_e is the interstitial velocity of the mobile phase, as contrasted with the average mobile-phase velocity u; cgs units are assumed in Equations (2.18) and (2.18a). The total porosity of the column (as a fraction of the column volume) is defined as ε_T, and is composed of the intra-particle porosity ε_i, plus the inter-particle porosity ε_e. The quantity u_e is then equal to $(\varepsilon_T/\varepsilon_e)u$, where the quantity $(\varepsilon_T/\varepsilon_e) \approx 1.6$.

The solute diffusion coefficient D_m (cm^2/sec) can be approximated by a function of solute molecular volume (V_A, in mL), temperature (T, in K), and mobile phase viscosity (η, in cP) by the *Wilke–Chang equation* [26]:

$$D_m = 7.4 \times 10^{-8} \frac{(\psi_B M_B)^{0.5} T}{\eta {V_A}^{0.6}} \qquad (\mathrm{cm^2/sec}) \tag{2.19}$$

Here M_B is the molecular weight of the solvent; the association factor ψ_B is unity for most solvents, and greater than one for strongly hydrogen-bonding solvents. For example, ψ_B equals 2.6 for water and 1.9 for methanol (a value of $\psi_B \approx 2$ can be assumed for typical RPC conditions). Equation (2.19) is reasonably accurate for solutes with molecular weights <500 Da, and it can represent a useful approximation for larger molecules.

The Knox equation (Eq. 2.17) provides a conceptual basis for optimizing conditions, so as to provide maximum values of N in the shortest possible time; it also forms the basis of the predictions of Table 2.4 and the calculated plots of Figure 2.11. That is, *it is possible to predict (approximately) how values of N will change when any experimental condition is changed*. The practical application of Equation (2.17) requires the conversion of values of ν into flow rate. Values of F can be obtained from values of ν as follows: From Equations (2.4a) and (2.7),

$$F \approx 0.0005 {d_c}^2 u \tag{2.20}$$

where flow rate F is in mL/min, column diameter d_c is in mm, and mobile-phase velocity u is in mm/min. Equation (2.18a) for u in mm/min, d_p in μm, D_m in cm^2/sec, and $u_e = 1.6u$ becomes

$$\nu = \frac{2.7 \times 10^{-7} u d_p}{D_m} \tag{2.20a}$$

Combining Equations (2.20) and (2.20a), we have

$$\nu \approx \frac{5.4 \times 10^{-4} F d_p}{{d_c}^2 D_m} \tag{2.20b}$$

and

$$F = \frac{1850 {d_c}^2 \nu D_m}{d_p} \tag{2.20c}$$

Figure 2.13*a* shows a plot of h versus ν(—) based on Equation (2.17) with values of $A = 1, B = 2$, and $C = 0.05$; this plot is independent of the experimental conditions listed in Table 2.4, and therefore applies (approximately) for *all* conditions. A minimum value of $h \equiv h_{opt} \approx 2$ is found in Figure 2.13*a*, corresponding to a value of $\nu \equiv \nu_{opt} \approx 3$ (these "optimum" values of h and ν correspond to maximum values of N). The solid curve of Figure 2.13*a* is repeated as the solid curve of Figure 2.13*b*, except that reduced velocity ν(the x-axis) has now been replaced by the flow rate F (for a particular set of conditions: 50% acetonitrile-water, 30°C, 200-Da solute, 4.6-mm diameter column). When any of the conditions of Table 2.4 change for the example of Figure 2.13*b*, the solid-line plot for 30°C and a 200-Da solute will be shifted right or left, depending on the effect of the change in condition on the flow rate for maximum N (F_{opt} corresponding to $\nu_{opt} \approx 3$). Table 2.4 predicts that F_{opt} will increase for an increase in temperature, which is seen (dotted curve) in Figure 2.13*b*. Similarly an increase in sample molecular weight will decrease F_{opt} and shift the h–ν plot to the left (also seen in Fig. 2.13*b* as the dashed curve). Figure 2.11*b* can be obtained from Figure 2.13*b* by changing values of h to N (Eqs. 2.12, 2.18).

The reason for changes in F_{opt} with conditions (summarized in Table 2.4) is as follows: Any change in condition that involves either particle size d_p or the diffusion coefficient D_m will change the value of ν, provided that mobile-phase velocity u (and flow rate F) are held constant. To maintain the same value of $\nu_{opt} \approx 3$ for minimum h and maximum N, the flow rate must then be changed so as to compensate for the effect of conditions on ν. For example, an increase in temperature will increase the diffusion coefficient D_m (Eq. 2.19) by some factor x, which then results in a decrease in ν by the same factor (Eq. 2.18a). The flow rate corresponding to F_{opt} must accordingly be increased by the factor x, in order to compensate for the temperature increase and hold $\nu = \nu_{opt}$ constant.

Also shown in Figure 2.13*a* are plots of each of the three terms that contribute to Equation (2.17), with the value of ν for minimum h (and maximum N) shown by the arrow ($h \approx 2, \nu \approx 3$). The original Knox equation (Eq. 2.17) has subsequently been refined, leading to modified relationships that are claimed to offer greater accuracy, expanded applicability, and/or greater insight into the basis of column efficiency [27–32]. Additionally values of both B and C depend on the value of k for the solute [33, 34] when stationary-phase diffusion is taken into account [35]. Consequently Equation (2.17) is mainly useful for practical, semi-quantitative application; it has even been described as a "merely empirical expression" [36] (we do not agree!). Nevertheless, its simplicity, convenience, and fundamental basis continue to recommend it as a conceptual tool for everyday practice.

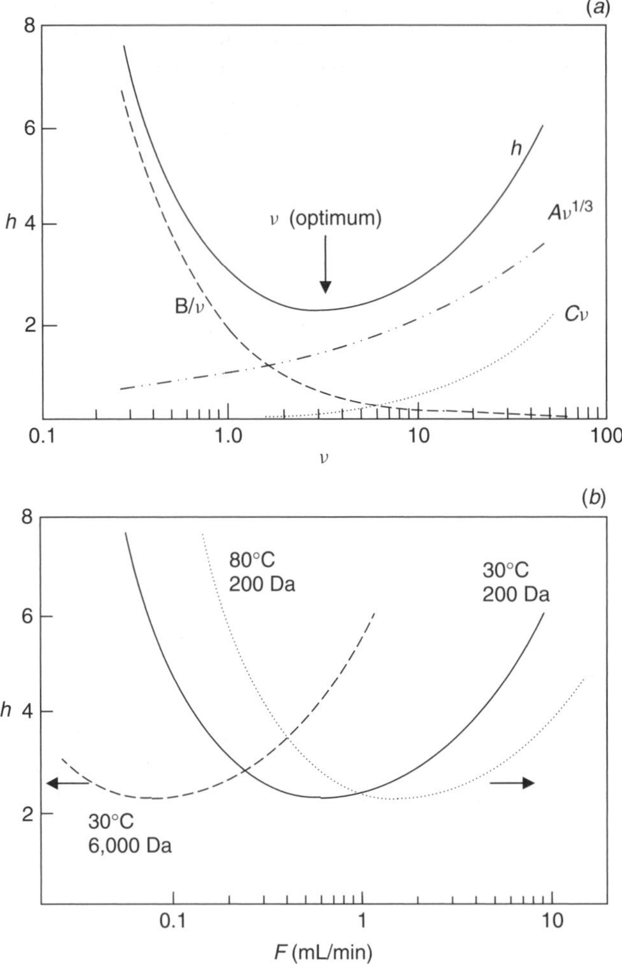

Figure 2.13 Knox equation. (*a*) Plot of reduced plate height h versus reduced velocity ν with $A = 1, B = 2$, and $C = 0.05$. The total plate height h is shown as a solid curve; the three contributions of Equation (2.17a) to h are shown as dashed or dotted curves. (*b*) Effect of a change in solute molecular weight (6000 vs. 200 Da) or temperature (80°C vs. 30°C) on h versus flow rate. All plots based on Equation (2.17) with $A = 1, B = 2$, and $C = 0.05$.

2.4.1.2 Some Guidelines for Selecting Column Conditions

Column conditions can be defined as column length and diameter, particle size, and flow rate. Together these experimental choices by the user determine the value of N, run time, and column pressure P. In general, larger values of N require longer run times, but an increase in the available column pressure can be used to improve the trade-off between N and run time: either larger values of N for the same run time or shorter run times for the same value of N. Column diameters can be varied for different purposes; larger diameters for preparative separations (Chapter 15), or smaller diameters for (1) increased sensitivity when the amount of sample is limited, (2) reduced flow rates for LC-MS (Section 4.14), (3) reduced solvent consumption, or (4) the use of very high-pressure operation (to minimize problems caused by the generation of heat within the column; Section 3.5.4.3). Column diameter d_c does not directly affect the relationship between N, run time, and P, but smaller column diameters mean smaller volume peaks, which are more affected by extra-column peak broadening; small-diameter columns are therefore more demanding in terms of equipment. Similarly pressures higher than about 5000 psi also require special equipment (Section 3.3.5.4) and columns.

Apart from the issue of column diameter, the choice of column conditions can be guided by the theory of Section 2.4.1.1. The following discussion will assume a sample molecular weight <1000 Da, but the separation of larger molecules follows from the discussion of Figure 2.11*b* and Section 13.3.1. Consider the following three possibilities:

A. only a single column configuration is available (e.g., a 150 × 4.6-mm column of 5-μm particles)

B. a single column configuration is available in different column lengths so that length can be varied

C. there is a wide choice of column lengths and particle size

For case A, only the flow rate can be varied (as in Fig. 2.11*a*), and there is a maximum possible flow rate F that is determined by the pressure limit of the system; however, it is often desirable to operate at a lower pressure. In this situation the largest value of F should be selected that results in the desired value of N but does not exceed some maximum pressure. For example, for the 5-μm, 150 × 4.6-mm column of Figure 2.11*a*, assume that a value of $N \geq 10{,}000$ is required, while not exceeding a pressure of 2000 psi. It can be seen(o) that a flow rate of ≈ 2 mL/min will yield a value of $N = 10{,}000$, at a pressure <2000 psi.

For case B, both the flow rate and column length L can be varied. Maximum values of N in minimum time result when the combined choice of values of F and L results in a maximum column pressure. From Equation (2.13*a*) this is equivalent to holding the product FL constant. Figure 2.14*a* illustrates the resulting dependence of N on run time (values of t_0) for three different particle sizes: 2, 5, and 10 μm. A conclusion from Figure 2.14*a* is that the smallest particles (2 micron) are preferred for the shortest separation times (and correspondingly for smaller values of the required N). Thus for a value of $t_0 = 1$ min, values of N are largest for a 2-μm particle column, while values of N are largest for a 10-μm particle column when $t_0 = 1000$ min. Thus, when large values of N are required, larger particles are best; when small values of N are adequate, smaller particles represent a better choice because of shorter run times. At intermediate analysis times (e.g., $t_0 > 5$ min), intermediate-size particles (5 micron) can give the required plate count in less time than the smallest particles.

The data of Figure 2.14*a* are recast into the form of a *kinetic* or *Poppe plot* [37, 38] in Figure 2.14*b*, where the time required to generate one theoretical plate (t_0/N) is plotted against the value of N. This "time per plate" is seen to increase with the value of N; Figure 2.14*b* also emphasizes the advantage of small particles when low values of N are required, and vice versa for large values of N. The dashed lines labeled "10 sec," "100 sec," and "1000 sec" correspond to conditions of constant t_0; other values of t_0 are defined by lines parallel to those shown. Poppe plots

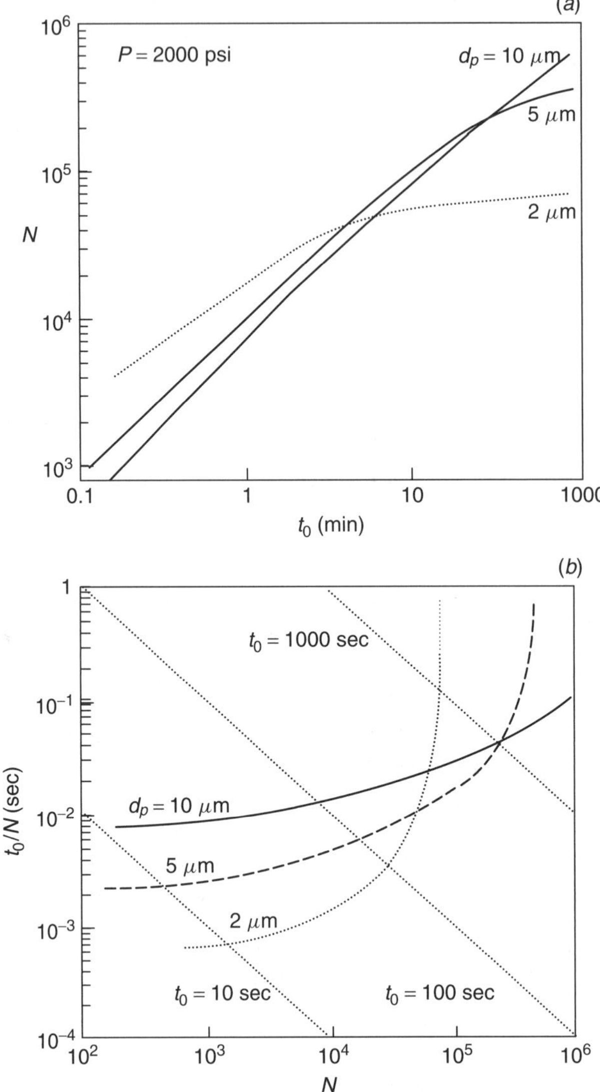

Figure 2.14 Variation of N as a function of column length, holding column pressure constant by varying flow rate. (*a*) N versus t_0 for different particle sizes; (*b*) "Poppe plot" for data of (*a*). Calculated values from Equation (2.17), assuming $A = 1$, $B = 2$, and $C = 0.05$, a viscosity of 0.6 cP (e.g., 60% acetonitrile-water, 35°C), $D_m = 10^{-5}$ cm^2/ sec (corresponds to a sample molecular weight of 300 Da for the latter conditions), and a column pressure $P = 2000$ psi.

are now widely used to compare column performance, as they take both column efficiency and permeability into account.

For case C, we are permitted to vary column length and particle size continuously. While this will be only approximately true in practice, there is not much penalty for using column lengths and particle sizes that are not too different from the optimum. Case C should be considered when the goal is either very fast separation or very large values of N (extreme separation conditions). The past decade has witnessed an increasing demand for faster HPLC separation, sometimes involving run times of a minute or less (Section 2.5.3.2). Similarly biochemists are now faced with the need to separate samples that contain thousands of components (Section 13.4.5); in such cases large values of N become more important. Whether the goal for difficult separations is minimum run time or maximum N, experimental conditions should be selected that favor maximum N per unit time—as in Figure 2.14*b*. It can be shown that the best choice of column length, particle size, and flow rate will always correspond to a minimum value of $h \approx 2$ (or $\nu \approx 3$). Because commercial columns are limited to certain lengths and particle sizes, we can only approximate conditions for $\nu \approx 3$ while at the same time achieving a required value of N in the shortest run time. Fortunately, an increase in ν by 2- to 3-fold has only a small effect on h or N, which gives us considerable flexibility in the selection of a preferred column length or particle size for a given sample. See also [38a].

Figure 2.15, which results from the application of Equation (2.17) with $\nu = 3$, can serve as an example for the selection of conditions that provide maximum N for a given separation time with some maximum pressure. Assume a maximum pressure $P = 2000$ psi, and a value of $k = 3$ for the last peak in the chromatogram (for other values of k, adjust the times in Figure 2.15 in accordance with Eq. 2.5). A value of $N = 5000$ can be achieved in about 15 sec (marked by • and arrow in the figure). Conversely, $N = 300{,}000$ will require about 17 hr (60,000 sec, marked by ◦ and arrow in the figure). In the first case ($N = 5000$), ≈1-μm particles will be required. In the second case ($N = 300{,}000$), 9 μm particles are recommended. That is, fast separations require smaller particles, while large N separations require larger particles (plus much longer separation times)—as in our preceding discussion of Figure 2.14. An increase in the allowable pressure results in either shorter times for the same plate number, or larger values of N for the same time—as the reader can verify from Figure 2.15 for $P = 20{,}000$ compared with 1000 psi. Higher pressures also favor the use of smaller particles, for either faster separation or larger values of N. Very large values of N or very short run times require extreme conditions, as can be inferred from the plot of Figure 2.15.

For various reasons the plots of Figures 2.11 and 2.13 through 2.15 must be regarded as approximate, as well as varying with other conditions according to Table 2.4 (e.g., mobile-phase viscosity, sample molecular weight, and temperature). So these figures should be used primarily as qualitative guides for selecting experimental conditions rather than quantitative data for a given separation.

Flow rates and column lengths corresponding to Figure 2.15 can be determined as follows: From Equation (2.18) with $H = L/N$, and $h = 2$,

$$\begin{aligned} L = NH &= Nhd_p \\ &= 2Nd_p \end{aligned} \tag{2.21}$$

Here cgs units apply: L and d_p in cm. From Equation (2.18a) for $\nu \approx 3$,

$$u = \frac{3D_m}{d_p} \tag{2.21a}$$

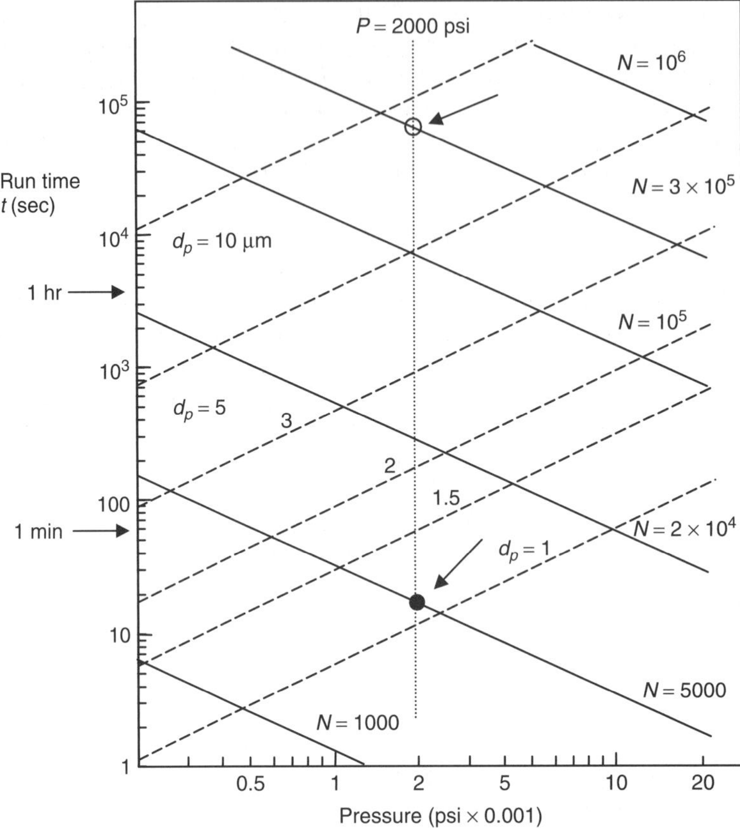

Figure 2.15 Variation of column plate number N with separation time t, column pressure P, and particle diameter d_p. (___) lines representing different values of N; (---) lines representing different particle sizes; (....) line representing a column pressure of 2000 psi. All values in figure are for minimum plate height, $h = 2$ and $k = 3$. Adapted from [23] for a viscosity of 0.6 cP (e.g., 60% acetonitrile-water, 35°C) and $D_m = 10^{-5}$ cm²/sec (corresponds to a sample molecular weight of 300 for the latter conditions).

with F given by

$$F = \frac{3V_m D_m}{L d_p} \qquad (2.21b)$$

where V_m is the column dead-volume (equal to $t_0 F$, Eq. 2.7a). The units of Equation (2.21b) are mL/sec (F), mL (V_m), cm²/ sec (D_m), cm (L), and cm (d_p).

2.4.2 Peak Shape

So far we have assumed symmetrical peaks in the final chromatogram. Under ideal conditions a peak will have a Gaussian shape given by

$$y = (2\pi)^{-0.5} e^{-x^2/2} \qquad (2.22)$$

Here x is $(t - t_R)/\sigma$, where t is time and $\sigma = W/4$; y is the peak height. Actual peaks in a chromatogram will usually depart slightly from a symmetrical, Gaussian shape, typically showing more or less *tailing* as in Figure 2.16*a*. Peak tailing can be characterized in either of two ways (Fig. 2.16*a*): by the *asymmetry factor* A_s (left-hand side of figure) or by the *tailing factor TF* (right-hand side). Values of A_s and TF are related approximately as

$$A_s \approx 1 + 1.5(TF - 1) \tag{2.22a}$$

so that values of A_s are typically somewhat larger than values of TF (Table 2.5).

Figure 2.16*b* shows how peak shape varies for different values of A_s or TF; a value of A_s or $TF = 1.00$ corresponds to a perfectly symmetrical peak. The effect of peak tailing on the *separation* of two adjacent peaks in the chromatogram is illustrated in Figure 2.16*c* (where only peak tailing varies, as measured by the value

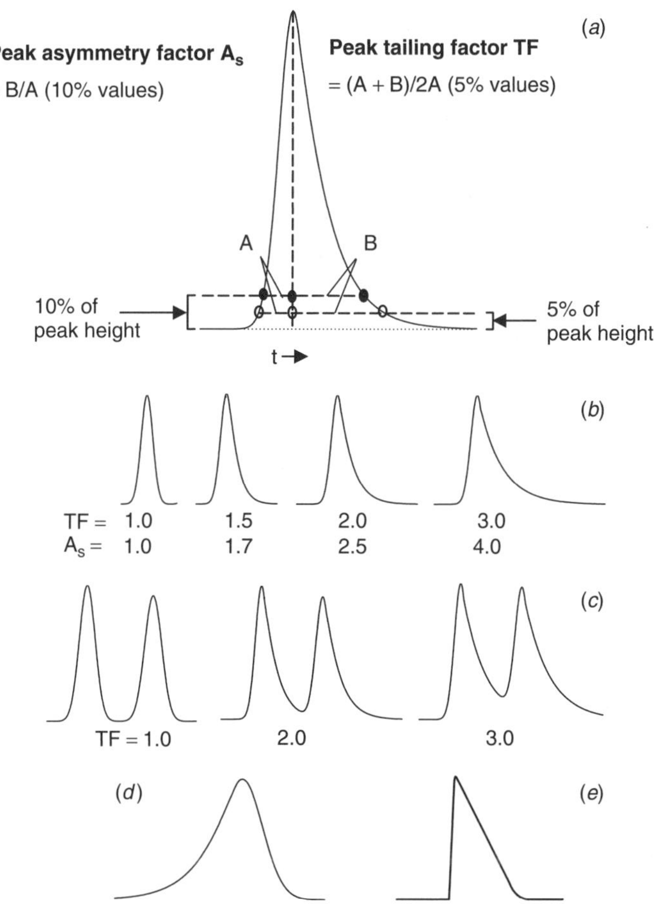

Figure 2.16 Measurement and effects of peak tailing. (*a*) Definitions of peak asymmetry and tailing factors (A_S and TF); (*b*) peak shape as a function of asymmetry (A_S) and tailing (TF) factors; (*c*) effect of peak tailing on separation; (*d*) peak fronting; (*e*) "overloaded" tailing, in contrast to "exponential" tailing in (*a*).

Table 2.5

Relation of Peak-Tailing and Peak-Asymmetry Factors

Peak-Asymmetry Factor (at 10%)	Peak-Tailing Factor (at 5%)
1.0	1.0
1.3	1.2
1.6	1.4
1.9	1.6
2.2	1.8
2.5	2.0

of *TF*). The two peaks with $TF = A_s = 1.0$ are well separated from each other. As peak tailing increases, however, their separation progressively worsens. Many times the extent of peak tailing is quite minor ($TF < 1.2$) and may not be noticed. Tailing of this magnitude ($TF \leq 1.2$) will have a negligible effect on separation, except for the case of a large peak followed by a very small peak (compare Fig. 2.17*d*,*e*). Less often, peak *fronting* will be observed ($TF < 1.0$; Fig. 2.16*d*), with a similar, potentially adverse effect on separation. Major peak tailing ($TF > 2$) can be highly detrimental to both separation and quantitative analysis; a common requirement for routine separations is that $TF < 2$ for all peaks. Quantitation based on peak area (Chapter 11) requires integration of the entire peak; this in turn depends on setting proper integration limits, which can become difficult for the case of severely tailing peaks.

If peak tailing with $TF \geq 2$ *is observed, steps should be taken immediately to correct the problem—whether for a routine assay procedure or during the development of an HPLC method.* The remainder of this book will assume symmetric peaks, unless noted otherwise. Several possible causes of peak distortion or tailing exist:

- "bad" column (plugged frit or void; Section 5.8)
- contaminated column (buildup of strongly retained sample contaminants; Section 5.8)
- column overload (too large a sample; Section 2.6.2)
- a sample solvent that is too strong (Section 2.6.1)
- extra-column peak broadening (Sections 2.4.1, 3.9)
- silanol interactions (including contamination of the column packing by trace metals; Sections 5.4.4.1, 7.3.4.2)
- inadequate or inappropriate buffering (Section 7.2.1.1)
- use of a higher column temperature with cold (inadequately thermostatted) mobile phase (Section 2.5.3.1)

See also Section 17.4.5.3. Before attempting to reduce tailing, two kinds of peak tailing should be distinguished: *exponential* versus *column-overload* tailing. Exponential tailing is characterized by a gradual return of the peak to baseline, as in

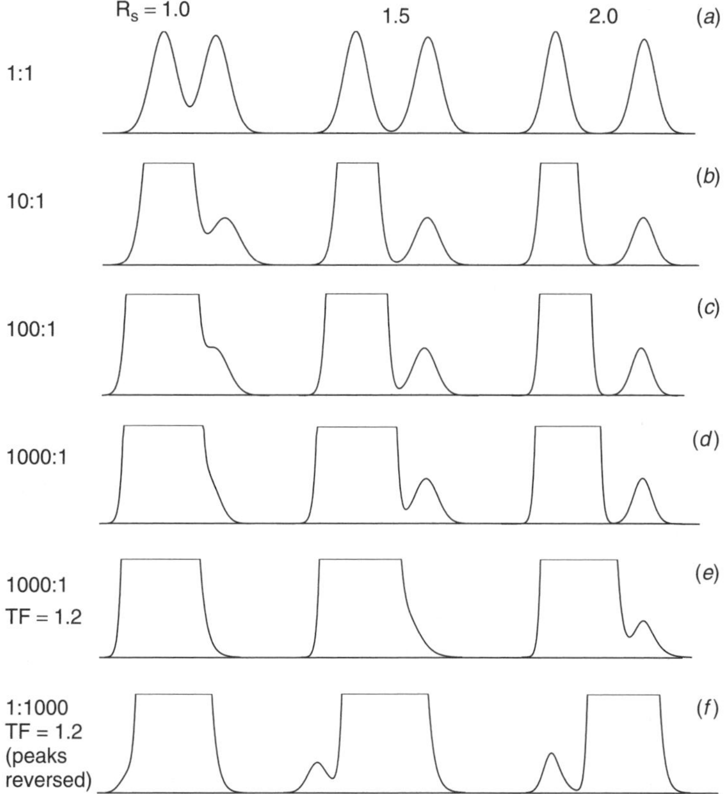

Figure 2.17 Separation as a function of resolution, relative peak size (1:1. 10:1, etc.), and (*e*, *f*) peak tailing.

Figure 2.16*a*; column-overload tailing creates a peak with a "right-triangle" appearance, as in Figure 2.16*e*. A reduction in sample weight will usually reduce the tailing of an overloaded peak as in Figure 2.16*e*, while exponential tailing may not change (or may even *increase*) when sample size is reduced [39, 40]. Usually it is feasible (and desirable) to reduce peak tailing so that $TF \leq 1.5$.

Various attempts have been made to describe the shape of tailing peaks in mathematical terms [41, 42], and to define a plate number N that includes the adverse effect of peak tailing on separation. Values of N for tailing peaks that are calculated by means of Equation (2.9) (and especially Eq. 2.9a) tend to overstate column efficiency; actual column performance as measured by the separation of peaks within the chromatogram usually suggests a lower value of N. The *Dorsey–Foley equation* is widely used, in an attempt to correct values of N for peak tailing [43]:

$$N = \frac{41.7(t_R/W_{0.1})^2}{(B/A) + 1.25} \qquad (2.22b)$$

Here the quantities A and B are defined in Figure 2.16*a* (at 10% of peak height), and $W_{0.1}$ is the value of W measured at 10% of peak height.

Still another attempt at a more realistic value of N (taking peak asymmetry into account) is the so-called 5-σ value:

$$N = 25\left(\frac{t_R}{W_{5\sigma}}\right)^2 \tag{2.22c}$$

Here $W_{5\sigma}$ is the peak width measured at 4.5% of peak maximum, similar to the measurement of N from the width at half height (Eq. 2.9a). The effect of peak tailing on N becomes more pronounced for peak widths that are measured closer to the baseline; that is, $W_{1/2}$ (least effect) $< W < W_{5\sigma}$ (most effect). Equations (2.22b) and (2.22c), as well as other expedients [41], provide at best only approximate measures of column performance for tailing peaks. Peaks that tail cannot be described by a single generic equation (comparable to Eq. 2.22 for symmetrical peaks), as witnessed by comparing the two peaks of Figures 2.16*a* and *e*; consequently no single value of N can accurately describe peak width for peaks with $TF > 1.2$. For the characterization of tailing peaks, we recommend a value of N from Equation (2.9a) *together with* a value of either A_s or TF. In many cases peak tailing is tracked as part of system suitability measurements to assess deterioration of the column; for such applications nearly any measure of peak tailing will serve to successfully track changes in peak shape over time.

2.5 RESOLUTION AND METHOD DEVELOPMENT

HPLC *method development* refers to the selection of separation conditions that provide an acceptable separation of a given sample; for a more detailed (but less up-to-date) account than provided by the present book; see [44]. The separation of two peaks i and j as in Figure 2.10*a* is usually described in terms of their *resolution* R_s = (difference in retention times)/(average peak width):

$$\boxed{R_s = \frac{2[t_{R(j)} - t_{R(i)}]}{W_i + W_j}} \tag{2.23}$$

W_i and W_j are the baseline widths W for peaks i and j, respectively. Better separation (increased resolution) results from a larger difference in peak retention times and/or narrower peaks. Accurate quantitative analysis based on separations as in Figure 2.17*a* is favored by *baseline resolution*, where the valley between the two peaks returns to the baseline. For two peaks of similar size (as in Fig. 2.17*a*), baseline resolution corresponds to $R_s > 1.5$. For preparative separations (Section 15.3.2), baseline resolution also allows a complete recovery of each peak with a purity of 100%.

Examples of resolution are shown in Figure 2.17 for various values of R_s (1.0, 1.5, 2.0) and different peak-size ratios ratios (1:1, 10:1, 100:1, 1000:1). As long as $R_s > 1.5$ and the two peaks are symmetrical ($TF = 1.0$), it is seen that there is little overlap of the two peaks—regardless of relative peak size. Minor peak tailing as in Figure 2.17*e* ($TF = 1.2$) can lead to a significant loss of resolution when a small peak follows a large peak, and in such cases a value of $R_s > 2$ may be necessary.

From a practical standpoint, it is nearly impossible to obtain peaks with $TF \approx 1.0$. Consequently, if a sample component is present in very low concentration, relative to a closely adjacent major component, it is preferable—if possible—to place the impurity peak *before* the main peak, as a means of minimizing the effects of peak tailing on resolution (compare Fig. 2.17*f*,*e*). Changes in relative peak position, so as to place a smaller peak in front of a larger peak, can sometimes be achieved during method development by a change in conditions that affect selectivity (Section 2.5.2).

A common goal of HPLC method development is the separation of every peak of interest from adjacent peaks (with $R_s \geq 2$), normally corresponding to a one-third safety factor (vs. baseline separation with $R_s \approx 1.5$, for peaks of comparable size). The goal of $R_s \geq 2$ takes into account minor peak tailing, peaks of moderately dissimilar size, and the usual slow deterioration of the column over time—with an increase in both peak width and tailing. Even larger values of R_s may be required in some cases. The resolution of two peaks can be improved as described in Sections 2.5.1 through 2.5.3.

When more than two peaks are to be separated, the goal is usually $R_s \geq 2$ for the least well-separated peak-pair. This peak-pair is referred to as the *critical* peak-pair, and its resolution is referred to as the *critical resolution* of the separation; for example, adjacent peaks 1 and 2 in Figure 2.18*a*, for which $R_s = 0.8$, or in Figure 2.18*d*, for which $R_s = 3.9$. Method development usually strives for an acceptable resolution of the critical peak-pair, which then means an adequate separation ($R_s \geq 2$) for all other peaks as well. In this book, *when we refer to the resolution R_s of a chromatogram, we mean the critical resolution* (unless stated otherwise).

For method-development purposes, it is convenient to derive an alternative, approximate expression for resolution from Equations (2.5a) and (2.10a) (assuming equal widths for the two peaks) [45]:

$$\boxed{R_s = \left(\frac{1}{4}\right) \left[\frac{k}{(1+k)}\right] \quad (\alpha - 1) \quad N^{0.5}} \tag{2.24}$$

$$\qquad\qquad (a) \qquad\qquad (b) \qquad\qquad (c)$$

Here resolution is expressed as a function of the retention factor k for the first peak i (term a), the separation factor α (term b), and column efficiency or the plate number N (term c). The separation factor α (a measure of so-called separation *selectivity* or relative retention) is defined as

$$\boxed{\alpha = \frac{k_j}{k_i}} \tag{2.24a}$$

The quantities k_i and k_j are the values of k for adjacent peaks i and j, as in Figure 2.10*a*. For this separation $k \equiv k_1 = 1.55$, $\alpha = 1.12$, and $N = 11{,}000$; resolution is therefore given by $R_s = (1/4)(1.55/2.55)(1.12 - 1)(11{,}000^{0.5}) = 1.9$. Equation (2.24) states that resolution can be improved in any of three different ways (by varying k, α, or N), *but improving separation selectivity (values of α) is by far the most powerful option.*

Equation (2.24) is slightly less accurate than Equation (2.23) but more useful for interpreting chromatograms during method development. Numerical values of

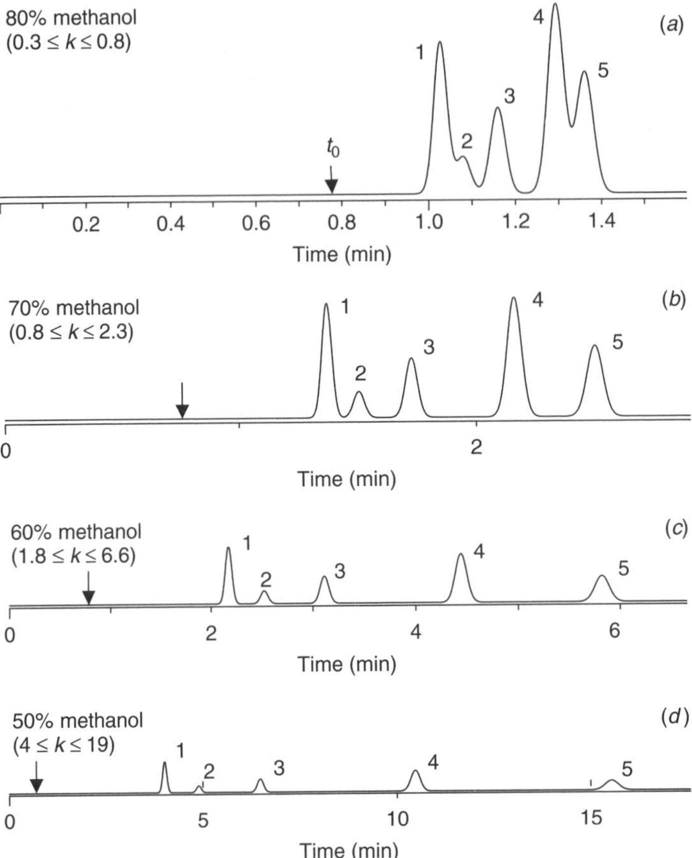

Figure 2.18 Separation as a function of mobile phase %B. Herbicide sample: 1, monolinuron; 2, metobromuron; 3, diuron; 4, propazine; 5, chloroxuron. Conditions, 150 × 4.6-mm, 5-μm C_{18} column; methanol-water mixtures as mobile phase; 2.0 mL/min; ambient temperature. Recreated chromatograms from data of [7].

R_s reported in this book are always calculated from Equation (2.23). Equation (2.24) will be used mainly for an understanding of how resolution depends on various experimental conditions, and as a guide for systematic method development. An alternative expression for R_s in Equation (2.24) is

$$R_s = \frac{1}{4}\left[\frac{k}{1+k}\right]\left[\frac{\alpha-1}{\alpha}\right]N^{0.5} \tag{2.25}$$

The derivations of Equations (2.24) and (2.25) are based on different approximations concerning the widths of the two peaks, and each equation has a similar accuracy. Equation (2.24) has the advantage of greater simplicity for use in guiding method development.

The development of an isocratic HPLC method proceeds by systematically adjusting ("optimizing") experimental conditions until adequate separation is achieved, preferably with a critical resolution $R_s \geq 2$. Equation (2.24) provides a useful guide for isocratic method development, as will be explored in Sections 2.5.1 through 2.5.3. Each of terms *a–c* of Equation (2.24) can be controlled by varying certain separation conditions (Table 2.2). Usually the first step is to choose a column with a sufficient plate number that is likely to separate a sample of the required complexity. In many cases, $N \approx 10,000$ is a good starting point, and this can be achieved either with a 150-mm long column packed with 5-μm particles, or a 100-mm, 3-μm column. The solvent strength (%B) is next varied to achieve an appropriate range in values of k (e.g., $1 \leq k \leq 10$), followed by optimizing selectivity (α). Finally, the column plate number N can be adjusted for a best compromise between the conflicting goals of increased resolution vs. a shorter run time.

The final separation conditions we select are hardly ever truly optimum (the best possible). However, the term "optimized" is often used in the literature to indicate a relatively improved or preferred separation rather than an absolute best separation. We should also note the difference between "local" and "global" optimizations. Local optimization refers to obtaining best values of one or two (seldom more) separation conditions over a limited range in values of each condition, while other (usually nonoptimal) conditions are held constant. Global optimization refers to best-possible values for all conditions that can affect separation or resolution. When chromatographers report an "optimized" procedure, they almost always are using this term to describe an improved separation or local optimum. We will continue this usage in the present book; that is, "optimum" will not be the same as a global best value of resolution.

2.5.1 Optimizing the Retention Factor *k* (Term *a* of Eq. 2.24)

Sample retention k in isocratic elution is usually controlled by varying the mobile-phase composition (%B). The first step is to achieve values of k for the sample that are neither too small nor too large. Relative values of resolution R_s (Eq. 2.24) and peak height h_p (Eq. 2.11) are plotted against k in Figure 2.19, as a way of showing how these two quantities vary with k (assuming no change in α with k). Two peaks at the bottom of Figure 2.19 illustrate how resolution and peak height change when %B is decreased so as to change the average value of k from 1 to 10.

The usual separation goal is $k \leq 10$ for all peaks because this corresponds to narrower, taller peaks for improved detection, as well as short run times so that more samples can be analyzed each day. Values of $k << 1$ can result in poor resolution, especially from the possible overlap of analytes with matrix interferences that typically accumulate near t_0 (as in the "junk" peak of Fig. 2.5*b*). *Therefore* $1 \leq \text{k} \leq 10$ *is usually a goal for all peaks in the final separation* (when method development has been completed). However, at the option of the chromatographer, it is possible to expand this preferred retention range somewhat, for example, to $0.5 \leq k \leq 20$. Alternatively, regulatory agencies may recommend $k \geq 2$ for all peaks of interest in the chromatogram [46], in order to minimize possible interference from sample excipients or other non-assayed ("junk") peaks that elute near t_0. Similarly, for separation with mass spectrometric detection (LC-MS), it is recommended that $k \geq 3$ because of the possibility of ion-suppression effects. When it is found that all peaks of interest cannot be accommodated within some maximum range in k (e.g., $0.5 \leq k \leq 20$), it is then necessary to use gradient elution (Section 2.7.2, Chapter 9).

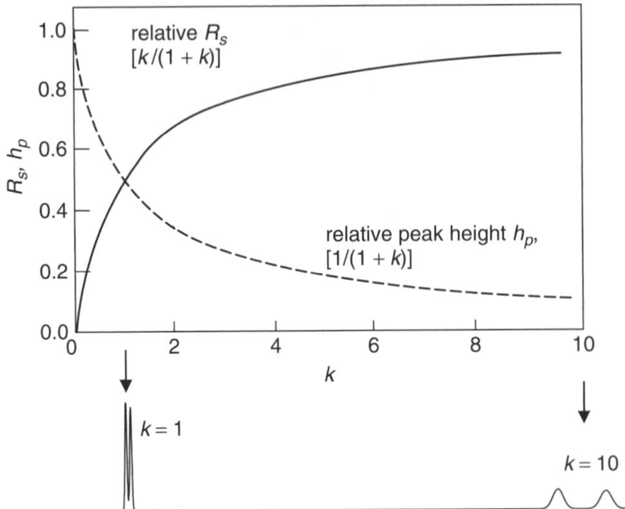

Figure 2.19 Relative resolution R_s and peak height h_p as a function of k or run time. Run time is proportional to the value of $1 + k$ for the last peak, and α is assumed constant.

The effect of a change in %B on separation is illustrated in Figure 2.18 for RPC. With 80% methanol/water (80% B) as the mobile phase (Fig. 2.18*a*), values of k are small ($0.3 \leq k \leq 0.8$), and as a result the sample is poorly resolved ($R_s = 0.8$). With 50% B (Fig. 2.18*d*; $4 \leq k \leq 19$), the sample is very well resolved ($R_s = 3.9$), but the run time is longer than necessary, and later peaks are wide (and short)—with reduced detection sensitivity. An intermediate mobile phase (60% B; Fig. 2.18*c*) provides an acceptable range in k ($1.8 \leq k \leq 6.6$) with a reasonable compromise among resolution ($R_s = 2.6$), detection sensitivity, and run time (6 min). A mobile phase between 60% and 70% B might be an even better choice, offering $R_s > 2.0$ with a shorter run time and increased detection sensitivity.

Changes in RPC retention with change in %B (as in Fig. 2.18) can be described by the empirical relationship [47]

$$\boxed{\log k = \log k_w - S\phi} \tag{2.26}$$

where ϕ is the volume fraction of the B-solvent (equal to $0.01 \times$ %B), k_w is the extrapolated value of k for solute X with water as the mobile phase (for $\phi = 0$), and S is a constant for a given compound when only ϕ is varied. For a review of the historical development of Equation 2.26, see [48] and references therein. For "small" molecules with molecular weights of 100 to 500 Da, $S \approx 4$ [49]. An increase in ϕ by 0.1 unit (e.g., a change in the mobile phase by +10% B) will therefore result in an average decrease in k for all peaks in the sample by a factor of approximately $10^{(0.1\times4)}$, or about 2.5-fold. The latter relationship suggests a systematic procedure for arriving at a satisfactory value of %B so that $1 \leq k \leq 10$ (or any other desired range in k) is achieved.

By this procedure, an initial separation or "run" can be carried out with 80% or 90% B, which usually ensures that the entire sample will be eluted from the column in a short time. In the separations of Figure 2.18, for example, the initial run (Fig. 2.18*a*) with 80% B gives a retention range for the sample of $0.3 \leq k \leq 0.8$. Applying the rule of a 2.5-fold increase in k for each 10%B decrease, we can estimate values of k for lower %B as follows: for 70% B, $2.5 \times 0.3 \approx 0.8$ for peak 1 (first peak) and $2.5 \times 0.8 \approx 2.0$ for peak 5 (last peak), or $0.8 \leq k \leq 2.0$. For 60% B, a similar calculation based on the run with 70% B gives $2.0 \leq k \leq 5$. The latter predicted range in k is well within the desired range of $1 \leq k \leq 10$ (as is the observed range of $1.8 \leq k \leq 6.6$); therefore the second method development experiment should use 60% B as mobile phase (other conditions being the same).

For samples that are much less strongly retained than this sample, the initial experiment with 80% B might yield $k \approx 0$ for all peaks in the chromatogram (i.e., a single peak that comprises all of the sample components). In this case a decrease in %B of at least 30% B will be required for acceptable values of k; the second experiment might therefore substitute (80–30% B) = 50% B as mobile phase, followed by further changes in %B for $1 \leq k \leq 10$—as in the example above based on Figure 2.18. Because of the approximate nature of the "rule of 2.5," it is best not to change %B by >30% B between experiments, or late-eluting peaks may be missed for values of %B that are too low. An alternative (usually preferred) approach for selecting a suitable value of %B makes use of gradient elution for the initial experiment in method development (Section 9.3.1).

2.5.2 Optimizing Selectivity α (Term *b* of Eq. 2.24)

When a further improvement of the separation is needed, relative retention (peak spacing, selectivity, or separation factor α) is next adjusted by varying any of the first seven conditions of Table 2.2: %B, the choice of B-solvent (usually methanol or acetonitrile), temperature, column type, or—for samples that contain acids or bases—mobile phase pH, buffer concentration, and ion-pair reagent concentration. Note in Table 2.2 that changes in each of these conditions can also result in changes in k, so the mobile phase %B may require simultaneous, further adjustment in order to maintain values of k within an acceptable range in k (usually no greater than $0.5 \leq k \leq 20$).

Figure 2.20 provides an illustration of varying selectivity for the separation of a sample that contains six components. In this example, %B was initially varied to achieve $2 \leq k \leq 4$ for the separation of Figure 2.20*a*. We will further vary %B and temperature as means for improving selectivity and resolution. For a mobile phase of 45% B and a temperature of 30°C (Fig. 2.20*a*), peak-pairs 1–3 and 5/6 are poorly resolved ($R_s = 0.3$). A decrease in %B (30% B, Fig. 2.20*b*) results in a better separation of all six peaks, an acceptable retention range ($4 \leq k \leq 12$), but with marginal resolution of peaks 3 and 4 (which were well resolved in Fig. 2.20*a*). A reversal of the positions of peaks 5 and 6 also occurs in Figure 2.20*b*. We can see from these two chromatograms that an intermediate value of %B is likely to improve resolution, by moving peaks 3 and 4 apart—before peak-pairs 1/2 or 5/6 come together. In fact a mobile phase of 33% B (and 30°C) for this sample results in a significant increase in resolution ($R_s = 2.1$, not shown).

If temperature is increased from 30 to 45°C (Fig. 2.20*c*, same %B as Fig. 2.20*a*), further changes in relative retention or selectivity result, but the overall separation is poor ($R_s = 0.4$). If the resolution of this sample is explored further, by trial-and-error

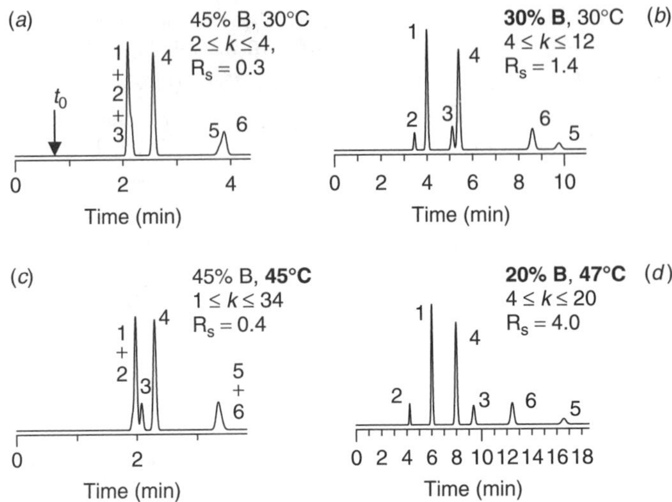

Figure 2.20 Effect of mobile phase %B and/or temperature on the isocratic separation of a six-component sample. Sample: *1*, 3-phenylpropanol; *2*, 1-nitropropane; *3*, oxazepam; *4*, *p*-chlorophenol; *5*, eugenol; *6*, methylbenzoate. Conditions: mobile phase is acetonitrile-water; 150 × 4.6-mm C_{18} column (5-μm particles); 2.0 mL/min; see figure for values of %B and temperature (changed conditions bolded in Figs. 2.20b–d). Note that peak heights are normalized to 100% for tallest peak in each chromatogram. Simulated chromatograms based on data of [50, 51].

changes in *both* %B and temperature, it is possible to achieve a maximum resolution of $R_s = 4.0$ for 20% B and 47°C (Fig. 2.20*d*), provided that we allow a k-value as high as 20. The goal of improving selectivity (as in the examples of Fig. 2.20) can be an increase in resolution, a decrease in run time, or (usually) both. *The selection of conditions for acceptable separation (i.e., method development) should emphasize changes in selectivity*, which can be used for *simultaneous* improvements in both resolution and run time. This will be especially true for samples that are more difficult to separate—those with a large number of components (and a crowded chromatogram), or peaks with very similar retention (e.g., isomeric compounds).

2.5.2.1 "Regular" and "Irregular" Samples

The present section can be useful as an aid in interpreting separation as a function of %B (especially in gradient elution). However, this topic is not essential for using or developing RPC methods. For that reason some readers may prefer to skip to Section 2.5.3, *and return to this section at a later time.*

Samples for reversed-phase separation can be classified as either "regular" or "irregular" [52]. When only %B is varied for a "regular" sample, the chromatogram appears to expand and contract like an accordion, with little, if any, change in the spacing of peaks within the chromatogram. The separations of Figure 2.18 provide a good example for this "regular" sample (a mixture of herbicides). The sample of Figure 2.20, on the other hand, shows a reversal of retention for peaks 5 and 6 when %B is changed (Fig. 2.20*b* vs. Fig. 2.20*a*), as well as less pronounced changes in the relative retention of peaks 1 to 4. This sample can therefore be described as "irregular," in contrast to the "regular" sample of Figure 2.18. A change in %B affects relative retention (or selectivity) for "irregular" samples but not for "regular" samples. Because of the possibility of peak reversals and peak misidentification for "irregular" samples, peak tracking (Section 2.6.4) becomes more difficult for such samples.

"Regular" samples are usually mixtures of structurally related compounds, whereas "irregular" samples are mixtures of more dissimilar molecules—as in Figure 2.20 for this mixture of several compounds of unrelated structure. Nevertheless, predictions of whether a sample is regular or irregular are often uncertain; experiments where %B is varied as in Figure 2.18 are required for reliable answers to this question. Regular and irregular samples can also be defined by means of Equation (2.26). Regular samples will show a strong correlation of values of S and $\log k_w$ for a given sample (i.e., diverging, near parallel plots), whereas irregular samples will show a poor correlation (i.e., intersecting plots). Many samples show a behavior that is intermediate between the examples of Figures 2.18 and 2.20, with less pronounced changes in relative retention as %B changes. For further information concerning regular and irregular retention behavior, see Section 6.3.1 (isocratic elution) or Section 9.2.3 (gradient elution).

2.5.3 Optimizing the Column Plate Number *N* (Term *c* of Eq. 2.24)

2.5.3.1 Effects of Column Conditions on Separation

When selectivity has been adjusted for optimum peak spacing and maximum sample resolution (Section 2.5.2), an adequate separation will often result. Yet a further improvement in separation may be possible by varying *column conditions* (column length, flow rate, particle size), so as to improve the column plate number N (term c of Eq. 2.24). Note that relative retention and peak spacing (values of k and α) will remain the same when only column conditions are changed for isocratic separation; as a result the optimized peak spacing achieved previously by varying α (term b of Eq. 2.24) will not be compromised.

An increase in N leads to an increase in resolution (Eq. 2.24), and usually a longer run time. Conversely, a decrease in N can provide a shorter run time—which may be of interest when $R_s \gg 2$ after optimizing selectivity (see below). Other factors equal, values of N are proportional to column length (Eq. 2.12) and generally increase for a decrease in flow rate or particle size. Run time is proportional to t_0 (Eq. 2.5) when k does not change, and t_0 is proportional to L/F (Eq. 2.7). Therefore run time increases proportionately for an increase in column length or a decrease in flow rate. Similarly the pressure drop P increases for an increase in column length or flow rate, or a decrease in particle size (Eq. 2.13a). Consequently we need to balance run time, resolution, and pressure when we vary column conditions in order to improve separation (Section 2.4.1.2).

As an example of an increase in sample resolution by a change in column conditions, consider the separation of Figure 2.20*b*, where $R_s = 1.4$. In the absence of an improvement in selectivity (as in Fig. 2.20*d*)—which may not be readily possible for some samples—an increase in column length can always be used to increase resolution. Figure 2.21*a* shows the result of an increase in column length from 150 mm in Figure 2.20*b* to 300 mm (e.g., by using two 150-mm columns connected in series). Baseline separation is now achieved, with $R_s = 1.9$.

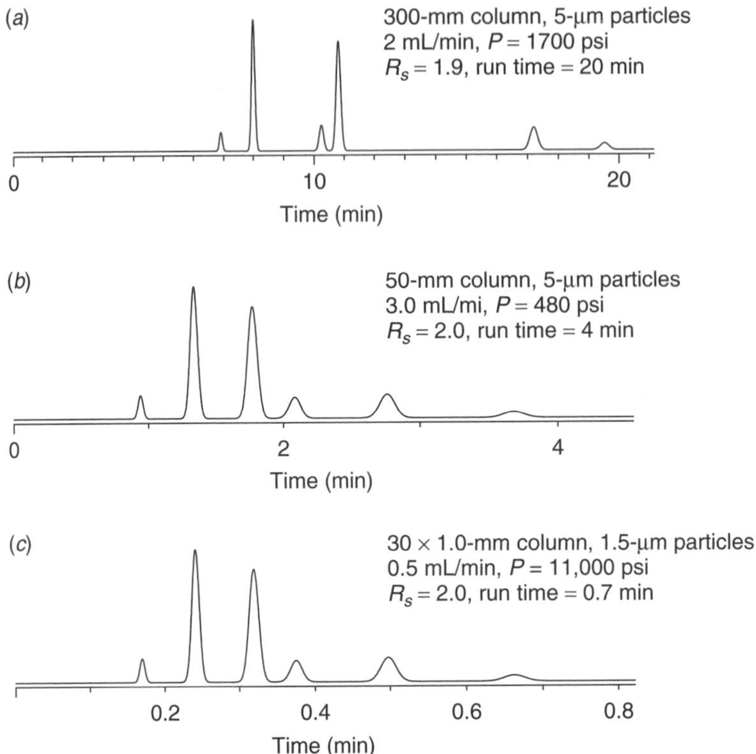

Figure 2.21 Use of a change in column length or flow rate to either increase resolution or decrease run time. (*a*) Separation of Figure 2.20*b* with an increase in column length from 150 to 300 mm, other conditions the same; (*b*) separation of Figure 2.20*d* with a decrease in column length from 150 to 50 mm, and an increase in flow rate from 2.0 to 3.0 mL/min, other conditions the same; (*c*) same as (*b*) except high-pressure operation and a 30 × 1.0-mm column with a flow rate of 0.5 mL/min; column conditions are noted in the figure. Simulated chromatograms based on data of [50, 51].

Although the latter resolution is marginally less than our recommended minimum value ($R_s \geq 2.0$), it should prove acceptable for most applications. The cost of this increased resolution is a doubling of both run time (to 20 min) and pressure (to 1700 psi). In this example the increase in pressure is acceptable.

When optimizing α as in Section 2.5.2, it is often advisable to strive for excess resolution, since this can later be traded for a shorter run time, by using a shorter column and/or a higher flow rate. An example of a change in column conditions that can reduce run time is provided by the separation of Figure 2.20*d*, where $R_s = 4.0$. By shortening the column 3-fold (from 150 to 50 mm) and increasing the flow rate 1.5-fold to 3 mL/min, the run time is decreased 4.5-fold to 4 minutes (Fig. 2.21*b*). At the same time the resolution is acceptable ($R_s = 2.0$), and so is the pressure ($P = 480$ psi). Many samples will have fewer, more widely separated peaks than in this example, allowing their separation in less than a minute by a suitable choice of column conditions.

In the past many laboratories have preferred columns packed with -5μm particles. Such columns are less demanding in terms of equipment (Section 3.9), and are less likely to be plugged by particulates. Today, however, there is increasing use of columns packed with 3-μm particles or smaller (see following Section 2.5.3.2). If a change in column conditions is made for either higher resolution or a faster run time, *the same column packing (e.g., Symmetry C18) is strongly recommended*, in order to avoid any change in column selectivity (Section 5.4).

For further details on the choice of column conditions, see Section 2.4.1.2.

2.5.3.2 Fast HPLC

There is increasing emphasis on very fast separations; for instance, with run times of less than a minute for relatively simple samples (<10–15 components), or a few minutes for more complex mixtures. Assuming the availability of suitable equipment and optimized column conditions, the time required by a separation depends on the value of k for the last peak and the value of α for the least resolved ("critical") peak-pair. Once "best" values of k and α have been established (optimization of selectivity), resolution and run time depend only on N. Conditions that favor fast separation include small particles, short columns, and high flow rates.

Further decreases in run time (with no loss in N) can be achieved by one or more of the following options:

- ultra-high pressure (>6000 psi)
- higher temperature
- particles of special design

High-Pressure Operation. It should be clear from Figure 2.15 and the discussion of Section 2.4.1.1 that a higher pressure can be used to decrease run time, with no loss in N (or resolution). As in the case of separations at lower pressures, small particles should be combined with (relatively) short columns and higher flow rates for fast separation. The use of higher pressures than can be achieved by conventional HPLC systems (with maximums of 6000 psi or 400 bar) is referred to as *ultra–high-pressure liquid chromatography*, or U-HPLC. U-HPLC, with pressures >6000 psi can be used for either better resolution (higher values of N) or reduced run time, as first reported by Rogers [53] and more fully developed by Jorgenson [54]. Commercial HPLC equipment was later introduced that allows operation at pressures of 15,000 psi or more.

As an example of how run time can be reduced with the help of high-pressure operation, Figure 2.21*c* shows the separation of the sample of Figure 2.20*d*, using a 30 × 1.0-mm column, packed with 1.5-μm particles, and a flow rate of 0.5 mL/min; the resulting pressure is 11,000 psi. A run time of only 0.7 min is achieved (vs. 17 min originally), while maintaining a resolution of $R_s = 2.0$. When this result is compared with the separations of Figs. 2.21*a*,*b*, the potential value of higher pressure operation should be apparent. (Note that some commercial U-HPLC systems are limited to flow rates of $\leq$ 5 mL/min, so smaller diameter columns generally are used to enable high mobile-phase linear velocities for fast separations.) Several similar examples have been reported [21], where reductions in run time of 2- to 6-fold were achieved by the use of ultra–high-pressure operation.

However, it should be noted that certain assumed relationships begin to fail significantly as the column pressure increases beyond 5000 psi [55]. Mobile-phase viscosity increases with pressure, so pressure no longer increases proportionately with flow rate. Values of k and α become pressure dependent [1], and therefore dependent on column conditions; this is less noticeable at lower pressures. Finally, heat is generated when a liquid flows through a packed column, and this heat is proportional to the pressure drop across the column. Changes in temperature within the column can have adverse consequences on peak shape and plate number [56], as well as further change values of k and α (undesirable!).

While higher pressure operation has very definite potential advantages, it requires special equipment (Section 3.5.4.3) and columns (Section 5.6.2). High-pressure operation has also been claimed to complicate method development [1, 55]. This is because values of k, diffusion coefficients D_m, mobile-phase viscosity, and other properties that affect separation, are more dependent on pressure, which usually increases during column use (at lower pressures, these properties can be regarded as essentially independent of pressure). For these and other reasons (safety, regulatory, cost, and special problems associated with the use of very high pressures), the extent to which U-HPLC is likely to replace HPLC in the routine laboratory was not clear at the time this book was published.

High-Temperature Operation. HPLC separation is usually carried out at temperatures between ambient and 50°C. The selection of a specific temperature within this range is often made on the basis of optimum selectivity (as in the example of Fig. 2.20*d*). The use of higher separation temperatures (e.g., >100°C) has been suggested as a means of shortening run time and improving resolution, as a result of an increase in N per unit time for a given pressure [57–59]. If we consider Figure 2.15 again, we recall that run time, the plate number N, and column pressure are all interrelated. Provided that we select a particle size, column length, and flow rate for maximum plates per unit time (i.e., "optimum" use of the column, with $\nu \approx 3$), the value of N increases for higher temperatures. An increase in temperature results in both a decrease in mobile-phase viscosity (Appendix I) and an increase of the solute diffusion coefficient D_m. A lowering of mobile-phase viscosity allows a higher flow rate for the same pressure (Eq. 2.13*a*), which is equivalent to an increase in pressure (as in U-HPLC). An increase in D_m results in the same value of N at a higher flow rate and shorter run time (Fig. 2.11*b*). Consequently an increase in temperature can, *in principle*, be used to shorten run time while maintaining the same value of N—or increase N while maintaining run time the same.

The advantages of high-temperature operation are offset by some corresponding disadvantages. First, HPLC at near-ambient temperatures was often selected in the past because of a concern that sample degradation might occur during separation at higher temperature. Although this is a potential complication which many chromatographers might prefer to avoid, the probability of such sample degradation is undoubtedly low for most samples [60–62]. A second problem in the use of higher separation temperatures is the possibility of radial temperature gradients within the column. Without careful thermostating of both the column and the entering mobile phase, severe peak distortion can result (Section 3.7.1). Radial temperature gradients represent a more serious problem when column temperature is increased, although the problem can be minimized by the use of narrower diameter columns that allow faster equilibration of the column temperature. A third problem is column instability at higher temperatures, especially for a mobile phase pH outside the range of 2 to 8. Finally, selectivity generally decreases for higher temperatures, although this may not be true for selected peak-pairs in the chromatogram. The "best" temperature will often represent a compromise between maximum N and maximum α.

Particles of Special Design. A number of column configurations exist, apart from the commonly used fully porous particles: columns packed with either *pellicular* or *shell* (superficially porous) particles (Section 5.2.1.1), and *monolithic* columns (Section 5.2.4). The relative advantages of each of these different column types will be discussed (Section 5.2). Pellicular and shell particles can be especially advantageous for large-molecule separations because of a reduced contribution of the Cv term of Equation (2.17). Pellicular columns have a thin coating of porous packing material on a solid bead and are easily overloaded, which restricts their use to very small samples. Shell columns have a thicker coating of porous packing than pellicular columns and can be used with sample loadings that are almost as large as those for fully porous columns. Monolithic columns are much more permeable than particulate columns, which allows higher flow rates and faster separation, other factors being equal. The relative merits of these and various conventional columns can be evaluated by means of Poppe plots as in Figure 2.14*b*.

2.5.4 Method Development

The preceding discussion of Sections 2.5.1–2.5.3 deals with the selection of experimental conditions for an acceptable HPLC separation, primarily one with baseline resolution and a reasonable run time. Adequate separation by itself, however, is not the complete story; other steps are often involved in method development [44]:

- assessment of sample composition and separation goals
- sample pretreatment
- selection of chromatographic mode
- detector selection
- choice of separation conditions
- anticipation, identification, and solution of potential problems
- method validation and the determination of system suitability criteria

Some of these steps may not be required, or they may need only minimal attention. "Difficult" samples and/or demanding assay procedures may involve more than simply following the steps outlined above and detailed in other chapters of this book. That is, special problems may arise during method development that require a logical response, based on the chromatographer's prior experience and familiarity with or access to the literature. For some such examples of method development, see [62, 63].

2.5.4.1 Assessment of Sample Composition and Separation Goals

At the start of method development, available information about the chemical composition of the sample should be reviewed. If acidic or basic compounds are present, it will be necessary to add a buffer to the mobile phase so as to control its pH; pK_a values for sample compounds, if available, can be useful during method development (Chapter 7). The molecular weight of the sample may also affect the choice of separation conditions for initial experiments (Sections 9.1.1, 13.4.1.4). Samples that contain enantiomers will require the development of a special method for their separation (Chapter 14).

The separation goals for a sample may not be limited to adequate resolution and minimum run time. Depending on the equipment that is available for the routine application of the final method, gradient elution may not be possible (an isocratic method will therefore be required). Trace analysis, as in the determination of compound impurities, may impose additional requirements on both sample preparation and the detector. Quantitative analysis calls for some minimum precision (e.g., ±1–2% for major components and ±10–20% for trace constituents). However, often only some of the sample components will require separation; for example, drugs present in blood or urine, or pesticides in water or soil samples.

2.5.4.2 Sample Pretreatment

Prior to injection the sample may require some processing in order to remove components that can damage the column or interfere with the separation of compounds of interest. Sample pretreatment procedures (Chapter 16) involve multiple steps and use of a wide range of separation media. For this reason these procedures can be more difficult to develop than the subsequent HPLC separation. When a pretreatment procedure for a similar combination of analyte and sample matrix is available (either as developed in the same laboratory or as reported in the literature), its use or adaptation for the sample is often preferred, so complete re-development of a sample-pretreatment procedure can be avoided (this approach is less likely to be applicable for HPLC separation per se; see Section 2.5.4.5).

2.5.4.3 Selection of Chromatographic Mode

Reversed-phase chromatography is the default choice for HPLC method development. However, depending on the sample, other chromatographic modes may be preferable. Often times this does not become apparent until initial experiments with RPC prove unsuccessful. Similarly, if isomeric compounds are present in the sample and prove difficult to separate by RPC, the use of NPC with unbonded silica as column packing will often prove more successful. NPC with unbonded silica is also preferred for preparative-scale separations (Chapter 15).

2.5.4.4 Detector Selection

The variable-wavelength UV detector is usually a first choice when sample components have an adequate chromophore. The molecular structures of suspected sample components may suggest the use of one detector in preference to another (Chapter 4). Nonspecific detectors based on evaporative light scattering or (less often) refractive index may be necessary when a sample contains components of unknown structure that can be missed with UV detection. Mass spectrometric detection (LC-MS) can be used as a supplement to UV detection because of its versatility in dealing with many kinds of samples and separation goals. The future use of LC-MS is expected to increase greatly, and mass spectrometers may one day be the detector of choice for most HLPC applications.

2.5.4.5 Choice of Separation Conditions

The selection of conditions for HPLC separation is discussed in Sections 2.5.1–2.5.3, and also in later chapters that deal with individual separation modes or special samples. For most samples, a systematic, trial-and-error approach can be followed, based on three successive steps. First, mobile-phase strength (%B) is varied until the right retention range is achieved, for example, $1 \leq k < 10$. Second, different separation conditions are explored for acceptable selectivity (values of α) and resolution. The first conditions that should be explored for improved selectivity are changes in %B (e.g., ±10% B) and temperature (e.g., 30–50°C). If some peaks are still overlapped and poorly separated, other conditions can be varied to improve selectivity (as described in later chapters). The third step is to vary column conditions: column length, particle size, and/or flow rate. A change in column conditions can provide a moderate increase in the plate number N and resolution, usually at the expense of a longer separation time (run time). When the sample resolution is better than necessary ($R_s \gg 2$) after optimizing selectivity, a reduction in column length and/or an increase in flow rate can result in a much shorter run time. *In many cases, adequate separation can be achieved within a day or two, based on a small number of experiments as outlined above.*

For methods involving a large number of samples, and where adequate resolution must be combined with run times that are as short as possible, it can be profitable to spend more time initially on "scouting" experiments. The experimentation may be with different columns, different B-solvents, and variations in mobile-phase pH and temperature. Use of gradient elution during the experiments can help avoid the need to separately optimize values of %B for each variable studied.

Still another approach is to search the literature for a separation of the same or similar sample. Trial-and-error modifications of conditions are then followed until an acceptable separation is achieved. *We do not recommend this approach* because possible deficiencies in literature methods can delay subsequent attempts at achieving a final, acceptable separation. A systematic approach based on starting conditions suggested in this book will usually require fewer experiments and result in a better final method. Nevertheless, apart from the selection of separation conditions, literature separations can be useful for selecting a detector and detection conditions (Chapter 4) and/or a sample preparation procedure (Chapter 16) for a specific analyte or sample.

2.5.4.6 Anticipation, Identification, and Solution of Potential Problems

Different problems may be encountered during the development and subsequent routine use of an HPLC procedure. Most problems (poor peak shape, drifting baselines, etc.) are immediately obvious and reflect deficiencies in materials, equipment, or laboratory technique. Chapter 17 (on troubleshooting) provides information on the likely causes of such problems, as well as means for their solution. Some problems can be anticipated in advance, allowing experiments to be carried out that will minimize the likelihood of their occurrence:

- poor retention of very polar samples
- overlooked peaks
- poor batch-to-batch reproducibility of the column
- non-robust separation conditions
- variations in equipment

A common problem in RPC is *poor retention of very polar samples* (Sections 6.6.1, 7.3.4.3). For non-ionized solutes, it may be necessary to switch to normal-phase chromatography (Chapter 8), which retains polar solutes strongly. For weakly retained solutes that are ionized, the use of ion-pair chromatography (Section 7.4) or ion-exchange chromatography (Section 7.5) may be indicated.

Overlooked peaks can arise for two reasons: (1) poor detection sensitivity, or (2) failure of the chromatographic procedure to separate two adjacent peaks (overlapping peaks). Poor detection sensitivity often can be dealt with by the complementary use of a nonspecific detector (Sections 4.11–4.13), which is advisable when using UV detection for samples whose composition is not fully known at the start of method development. Overlapped peaks are more likely to be missed when one peak is much larger than the other. The problem of missing peaks can also be addressed in part by the use of mass spectrometric detection (Section 4.14), which is able to deconvolute overlapping peaks. An alternative approach, following the apparent separation of all peaks in the sample by a "primary" procedure, is the development of an *orthogonal* separation (Section 6.3.6.2), whose selectivity is very different from that of the primary method. An orthogonal separation should be able to move a missing peak to another part of the chromatogram where it is more noticeable. Hidden peaks can arise during method development or during later routine use if an unsuspected component in the new samples under analysis is overlapped by another peak in the chromatogram.

Poor batch-to-batch reproducibility of the column is today an infrequent problem. It is more likely to arise for complex samples where the chromatogram is crowded and many peaks have marginal or barely adequate resolution. Small, unintended changes in column selectivity (Section 5.4.2) can result in a decrease in resolution for one or more peaks. A commonly used approach for avoiding problems due to batch-to-batch variability in column selectivity is as follows: After the conditions for the final method are selected, several different manufacturing lots of the column are tested to confirm equivalent performance. Usually all the tested column lots will provide adequate separation, and this helps eliminate concern about column reproducibility in the future. Additional means for dealing with the possibility of varying column selectivity are discussed in Sections 5.4.2 and 6.3.6.1.

Non-robust separation conditions can result in a loss in resolution from small, inadvertent changes in one or more separation conditions. For example, small variations in mobile-phase pH are difficult to avoid during normal laboratory operation, yet they can result in significant changes in resolution when ionizable compounds are present in the sample (Section 7.3.4). To confirm that the final method is robust, the effect on resolution of small changes in each separation condition should be determined (Section 12.2.6). It is usually possible to modify separation conditions so as to improve method robustness.

Possible *variations in equipment* and their effect on the separation should also be addressed during method development. The most important requirement is the development of standard test procedures that will guarantee satisfactory performance of the equipment (Sections 3.10.1, 3.1.0.2). The holdup or *dwell volume* of equipment used for gradient elution often varies from system to system, and this can lead to failure of the method as a result of consequent changes in relative retention. Various means for dealing with dwell volume variability are presented in Section 9.3.8.

2.5.4.7 Method Validation and System Suitability

When the development of an HPLC method is complete, the method is usually tested to ensure its suitability for the intended purpose, often according to guidelines issued by a regulatory agency (e.g., FDA and ICH). The precision and accuracy of the method can be determined by analyzing suitable samples in replicate, based on specifications for precision, resolution, peak shape, and other factors pertaining to system suitability.

For methods under the oversight of regulatory agencies, a formal validation is required, with extensive documentation. Because a failed validation attempt can require extensive documentation (deviations, investigations, etc.), it is wise to perform "pre-validation" experiments. Pre-validation is simply a test of some subset of the validation (e.g., precision, accuracy, linearity, capability of analyzing a full batch of samples) that demonstrates that the method will successfully pass a formal validation. Our experience has shown that the extra day or so invested in pre-validation will result in a much higher percentage of passing validations for an overall savings in time and money. For a detailed discussion of method validation and system suitability, see Chapter 12.

2.6 SAMPLE SIZE EFFECTS

As long as the weight and volume of the injected sample are sufficiently small, a change in sample weight or volume should affect peak height and area, but not retention times, peak widths, or resolution. For sufficiently large samples, however, *column overload* results; peak widths increase and resolution decreases. HPLC assay procedures are normally carried out with samples whose size is sufficiently small that retention times and resolution do not vary. If it is suspected that the weight or volume of the sample may be too large—so as to degrade the separation, the sample volume can be reduced by half and the separation repeated. If there is no change in retention or resolution, the original sample size was not too large. When the purpose of HPLC separation is the purification of a crude product, it is customary to use a much larger sample (nonlinear separation), so as to maximize the amount of recovered material (Chapter 15).

For typical separations on columns with lengths of 50 to 250 mm, and an internal diameter of 4 to 5 mm, the weight of individual compounds in the sample should be limited to ≤ 50 μg, with a sample volume ≤ 25 μL (when the mobile phase is used as the injection solvent). For smaller diameter columns, sample size should be reduced in proportion to the square of column diameter. However, if the sample contains ionized solutes, column overload may occur for sample weights $> 1\ \mu g$ (Sections 7.3.4.2, 15.3.2.1). An understanding of the effects of sample size on HPLC separation is important for the following reasons:

- to avoid an undesirable change in separation due to a sample size that is too large
- to increase detection sensitivity for trace analysis, by using the largest possible sample size
- to maximize the recovered weight of purified product in preparative HPLC (Chapter 15)

A change in resolution and/or retention that results from the injection of a sample whose volume or weight is too large is referred to, respectively, as *volume overload* and *mass overload*.

2.6.1 Volume Overload: Effect of Sample Volume on Separation

If the sample (dissolved in the mobile phase) is introduced to the column in a volume V_s, and if the baseline volume of a peak for a very small-volume sample is V_{p0}, the peak volume V_p for a larger sample volume will be [64]

$$V_p = \left(\left[\frac{4}{3}\right] V_s^2 + V_{p0}^2\right)^{0.5} \qquad (2.27)$$

Assuming that the concentration of solute in the sample is constant, the effect of the sample volume V_s on peak size and shape is illustrated in Figure 2.22 (assumes no *mass* overload). As sample volume increases from injections 1 to 4, the peak begins to widen and then develops a flat top. For Equation (2.27) and the examples of Figure 2.22, we assume the delivery of an undistorted (i.e., cylindrical) sample plug to the head of the column; however, the sample volume V_s typically is increased by about 50% in the process of being washed from the sample valve.

The peak volume for a small volume of injected sample can be obtained from $V_{p0} = WF$ and Equations (2.7) and(2.10a):

$$V_{p0} \approx 0.002 L d_c^2 N^{-0.5}(1 + k) \qquad (2.28)$$

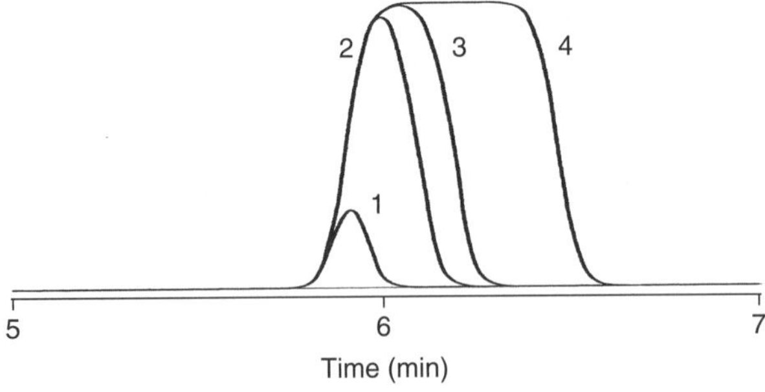

Figure 2.22 Effect of sample volume V_s on peak width and shape. $V_s/V_c = 0.3$ (peak 1); 3 (peak 2); 5 (peak 3); 15 (peak 4). Computer simulations, courtesy of Geoff Cox, Prochrom R&D.

As long as the sample volume V_s is $< 0.4V_{p0}$, the increase in peak width and loss in resolution will be <10% (Eq. 2.27), which is usually acceptable. (Alternatively, for a loss of resolution < 1%, use $V_s < 0.15V_{p0}$.) If we assume a flow rate that is approximately optimum (for $h \approx 2$), then $H = 2d_p, N = L/H = L/2d_p$. For a value of k no smaller than 1, then with Equation (2.28) and $k \geq 1$ we have the allowable sample volume $V_s = 0.4V_{p0}$ given by

$$V_s < 0.14L^{0.5}d_c{}^2d_p{}^{0.5} \qquad (V_s \text{ in } \mu\text{L}, L \text{ and } d_c \text{ in mm } d_p \text{ in } \mu\text{m}) \qquad (2.29)$$

Maximum sample volumes for several different column configurations are as follows: (150 × 4.6-mm, 5-μm particles) $V_s \leq 80$ μL; (100 × 4.6-mm, 3-μm particles) $V_s \leq 50$ μL; (30 × 4.6-mm, 3-μm particles) $V_s < 30$ μL; (30 × 2.1-mm, 1.5-μm particles) $V_s < 4$ μL. If the resolution of early peaks in the chromatogram is not critical, larger sample volumes can be injected.

A sample may be provided as a solution in a solvent other than the mobile phase. When the sample solvent is weaker than the mobile phase, larger sample volumes can be injected without adverse effect on peak width or resolution. Conversely, injection of the sample dissolved in a solvent stronger than the mobile phase often leads to broadening and/or distortion of early peaks in the chromatogram [65–67]; the use of sample solvents that are stronger than the mobile phase is a common mistake and should be avoided if possible. If it is inconvenient to change the sample solvent, smaller injection volumes of sample dissolved in a strong solvent can sometimes be tolerated. A 1:1 dilution of the sample with the weaker A-solvent (e.g., water) followed by injection of a 2-fold larger sample volume may also prove effective for minimizing sample-solvent problems while maintaining the same weight of injected sample (especially for a sample solvent that is >50% B). Larger sample volumes and stronger sample solvents can be used in gradient elution (Chapter 9), because the sample mixes with the weaker mobile phase (lower %B) at the start of the gradient. If in doubt, it is always a good idea to inject half and double the desired sample volume, observe the effect on resolution, and then make adjustments accordingly.

An interesting exception to the conclusions above on the use of a strong sample solvent has been reported [68, 69], for solutes that elute with moderate values of k when pure water (or buffer) is used as mobile phase. Strong solvents such as propanol, tetrahydrofuran, isopropyl acetate, and 4-methyl-2-pentanone can be used to dissolve the sample in the latter case, without adverse effects on peak shape or width for solutes that elute before the sample solvent. Occasionally this observation can be useful for solutes that are not sufficiently soluble in water. For further information on sample volume and separation, see [70, 71].

2.6.2 Mass Overload: Effect of Sample Weight on Separation

Even when a small volume of the sample is injected, it is possible for the weight of dissolved sample to overload the column, causing sample peaks to broaden and change shape. This occurs because the column has a limited capacity to retain sample molecules; that is, the stationary phase adjacent to a band can become saturated with the sample. A representation of peak broadening due to column overload is shown in Figure 2.23. Here a small volume of a sample compound has been injected repeatedly, varying only the sample concentration (and weight); the

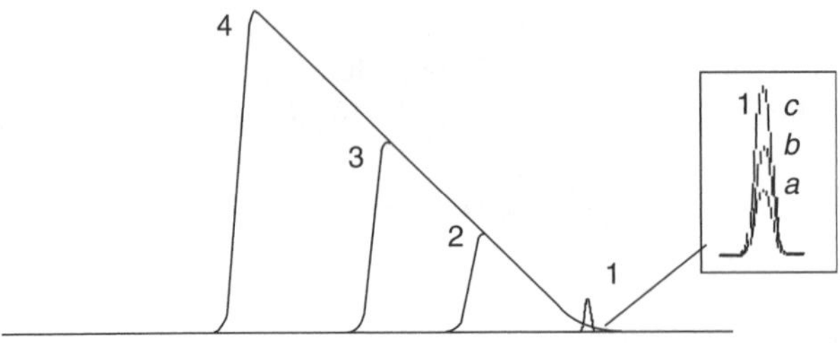

Figure 2.23 Effect of sample weight on peak width and shape. Superimposed solute peaks (1–4) for injections of increasing sample mass; peaks 1 (*a*–*c*) are for small sample weights where peak width is not affected by sample weight.

resulting chromatograms have been superimposed. Injections 1a–c involve small sample weights, so there is no peak broadening or distortion, and peak height is proportional to sample weight. The injection of successively larger sample weights (2, 3, and 4), however, results in the formation of broader peaks with a right-triangle shape. Sample weights increase in Figure 2.23 in the sequence 1a (smallest weight) $< 1b < 1c \ll 2 < 3 < 4$ (largest weight).

As long as the weight of an individual sample component in the injected sample is not excessive (typically <50 μg for 4.6-mm i.d. columns, less for smaller i.d. columns), each band moves through the column without being affected by the presence of other bands in the sample. Consequently it is the weight of *individual* compounds in the sample that normally determines column overload and changes in peak shape, not the total weight of sample. This behavior is illustrated in Figure 2.24. In Figure 2.24*a* the sample size is 2.5 mg for the (overloaded) RPC separation of each of these two solutes (xanthines); this is about 50 times larger than the maximum solute weight for nonoverloaded separation (and the peaks therefore tail). The same weight (2.5 mg each) of the two compounds was injected separately (solid line, *A*′ and *B*′) and as a mixture (dashed line, *A* and *B*), and the three chromatograms

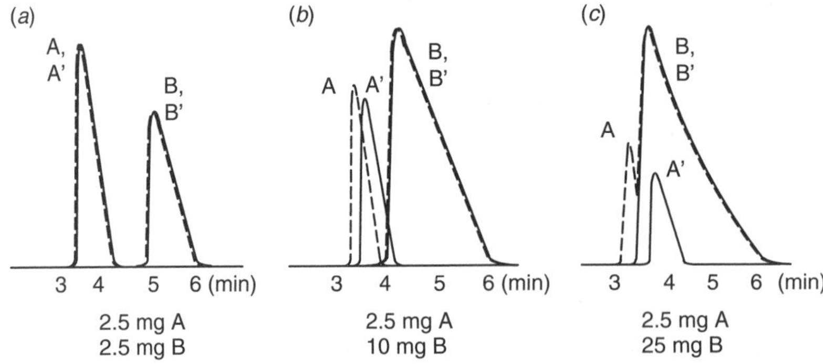

Figure 2.24 Effect on separation of severe mass overload. Sample consists of β-hydroxyethyltheophylline *A* and 7β-hydroxypropyltheophylline *B*, either alone or as a mixture. Sample weights are shown in figure. Conditions: 150 × 4.6-mm column, 5-μm particles; other (isocratic) conditions given in [70]. Peaks labeled *A*′ and *B*′ are for the injection of samples of pure of *A* or *B*; peaks labeled *A* and *B* are for the separation of mixtures of *A* and *B*. Adapted from [72].

were superimposed. There is little difference in the resulting peaks for *A* or *B* in this moderately overloaded separation, whether the compounds are injected alone or in mixture with each other; that is, overloading of the column by one compound does not affect the separation of other peaks in the sample (as long as they are baseline-resolved). A similar behavior is seen in Figure 2.24*b* for separations where the weight of *B* was increased to 10 mg, although here the retention of peak *A* is decreased slightly as a result of the presence of peak *B* in the combined sample. When the weight of compound B is increased sufficiently (25 mg in Fig. 2.24*c*), however, the separation of peak *A is* affected. One message of Figure 2.24 is that column overload with resulting peak tailing will not occur until the weight of an *individual* compound becomes too large (e.g., >50 μg for a column with an internal diameter of 4.6 mm)

2.6.3 Avoiding Problems due to Too Large a Sample

When carrying out routine HPLC separations for sample analysis, it is desirable that values of k, N, and R_s remain constant for different compounds being analyzed by the same procedure. This requirement simplifies both quantitation and peak identification based on retention time. Constant values of k, N, and R_s in turn require sample sizes small enough so that column overload is avoided and separation is not a function of sample size—usually requiring sample weights and volumes that do not exceed some limit (as discussed above). Sample volume is normally held constant for HPLC analysis, so the main concern is then a sample with too large a concentration of the analyte(s). Again, the value of w_{max} is for *each* compound in the sample, *not* for the total sample weight. For example, if no component of the sample comprises more than 10% of the sample weight, the maximum sample weight will be 10-fold greater than for a sample that contains only a single solute.

2.6.3.1 Higher Than Expected Sample Concentrations

If the concentration of an analyte changes from sample to sample, mass overload may result for higher concentration samples, causing loss of resolution and change in retention time. Such an effect on the separation of the analyte concentration or weight should be considered for the final HPLC procedure (following method development), and a maximum analyte concentration or sample weight w_{max} should be established that will avoid problems due to mass overload—including possible overload of the detector causing nonlinear detection. Samples exceeding this concentration should be diluted and re-assayed.

2.6.3.2 Trace Analysis

In trace analysis it is desirable to maximize peak height so as to increase signal to noise. Usually the quantity of a trace analyte injected is too small to overload the column, but other components of the sample may cause column overload, with potentially adverse effects on the separation of the analyte. That is, a large enough injected weight of one compound can affect the separation of a second, adjacent band. This is illustrated in Figure 2.24. Note that in Figure 2.24*a* the presence of compound *B* in the sample does not affect the separation of compound *A*. In Figure 2.24*c*, on the other hand, where the amount injected of compound *B* is increased 10-fold (to 25 mg), the separation and quantitation of compound *A* is markedly affected. For injection of 2.5 mg of *A* alone, the retention time of *A* is 3.6 min; for injection of 2.5 mg of *A* in the presence of 25 mg of *B*, the retention time shifts to 3.1 min, the band becomes narrower, and resolution is poor.

When a sample contains excessive amounts of interfering compounds, the best approach prior to HPLC separation is a sample cleanup to remove the interfering compounds (Chapter 16). In trace analysis it is advantageous to inject the largest possible sample volume. When the peak of interest is well resolved from adjacent peaks and if sufficient sample is available, larger sample volumes can raise peak height further. If the sample is dissolved in a solvent that is much weaker than the mobile phase, still larger volumes can be injected with a proportionate increase in peak size and no additional peak broadening (this approach is especially useful when gradient elution is used). The latter approach for increasing detection sensitivity assumes that large volumes of sample are available.

2.7 RELATED TOPICS

Some additional topics of varying importance are as follows:

- column equilibration
- gradient elution
- peak capacity and two-dimensional separation
- peak tracking
- secondary equilibria
- column switching
- retention predictions based on solute structure

These topics are introduced in this section to provide an adequate background for the more detailed accounts of later chapters.

2.7.1 Column Equilibration

When an HPLC system is turned on and mobile-phase flow has begun, 30 to 60 minutes may be required before the system is ready for use and sample injections can begin. This equilibration time can be shortened by storing the column in the mobile phase for subsequent use, but often it is desirable to flush the system and column at the end of a working day with water, followed by 100% organic solvent (Section 3.10.2.1). Other changes in mobile phase may be necessary during the day, such as for switching from one HPLC assay procedure to another, between repetitive gradient runs, or during method development. After each change in mobile phase, equilibration of the column may require the passage of 10 or more column volumes of the new mobile phase before sample injections can resume. The use of a new column may require an even larger equilibration volume. Whenever there is a change in mobile phase or column, it should be confirmed that the column has been equilibrated before resuming sample analysis or method development. Column equilibration can be checked by injecting replicate samples; when no change is observed in successive chromatograms, the column can be assumed to be equilibrated. For further information on column equilibration, see Sections 7.4.3.2 (ion-pair chromatography), 8.5.2 (normal-phase chromatography), and 9.3.7 (gradient elution).

2.7.2 Gradient Elution

Isocratic elution with a fixed mobile-phase composition works well for many samples and is the simplest form of liquid chromatography. For some samples, however, no single value of %B can provide a generally satisfactory separation, as illustrated by the RPC examples of Figure 2.25*a*, *b* for the separation of a nine-component herbicide sample. With 50% acetonitrile/water (Fig. 2.25*a*), later peaks are very wide and have inconveniently long retention times. As a result run time is excessive (140 min), and later peaks are less easily detected (e.g., peak 9 is only 3% as tall as peak 1). The use of 70% acetonitrile (Fig. 2.25*b*) partly addresses the latter two difficulties, but at the same time it introduces another problem: the poor resolution of peaks 1 to 3. This example illustrates the *general elution problem*: the inability of a single isocratic separation to provide adequate separation within a reasonable run time for samples with a wide range in retention (peaks with very different values of k).

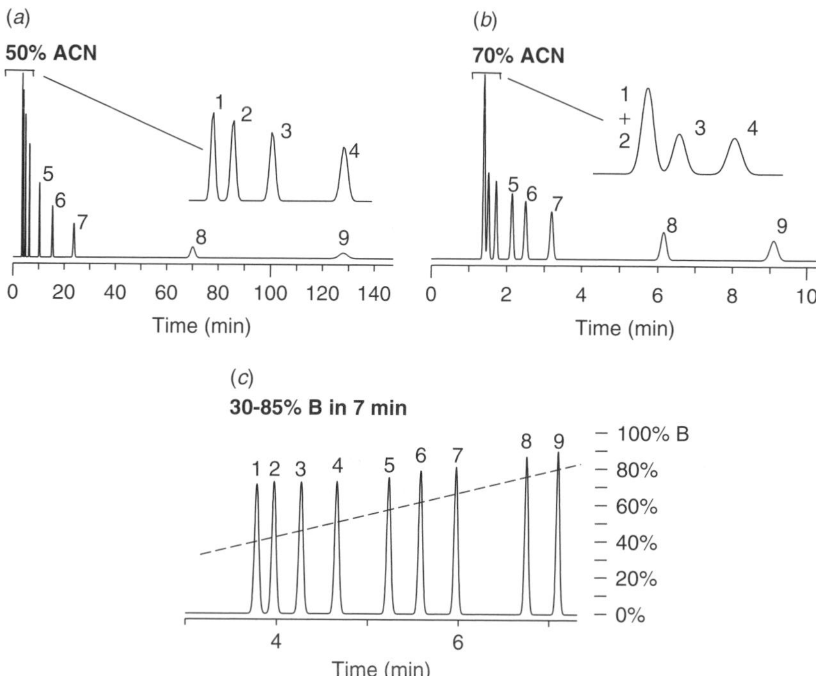

Figure 2.25 Illustration of the general elution problem and the need for gradient elution. The sample is a mixture of herbicides. (*a*) Isocratic elution using 50% acetonitrile (ACN)-water as mobile phase; 150 × 4.6-mm C_{18} column (5-μm particles), 2.0 mL/min, ambient temperature; (*b*) same as (*a*), except 70% ACN-water; (*c*) same as (*a*), except gradient elution: 30–85% ACN in 7 minutes. Computer simulations based on data of [5].

Gradient elution refers to a *continuous* change in the mobile phase during separation, such that the retention of later peaks is continually reduced; that is, the mobile phase becomes steadily stronger (%B increases) as the separation proceeds. An illustration of the power of gradient elution is shown in Figure 2.25*c*, where all peaks for the sample of Figure 2.25*a*, *b* are separated to baseline in a total run time of slightly more than 7 minutes, with approximately constant peak widths and comparable detection sensitivity for each peak (assuming a similar detector response for each solute). The advantages of gradient elution for this sample are obvious. Gradient elution also can be used to deal with several other separation problems, as discussed in Sections 9.1.1 and 13.4.1.4. For a further discussion of gradient elution, see Chapter 9.

2.7.3 Peak Capacity and Two-dimensional Separation

So far we have used critical resolution R_s as the measure of a given separation. This criterion is appropriate when the peaks of interest in a chromatogram can all be resolved to some extent, and our goal is some minimum resolution for all peaks. Some samples contain so many components, however, that it is impractical to achieve a significant resolution for all peaks of interest. Then we need a different measure of "separation power" for various combinations of experimental conditions. The *peak capacity* of a separation refers to the total number of peaks that can be fit into a chromatogram, when every peak is separated from adjacent peaks with $R_s = 1$. An example is shown in Figure 2.26*a*, for a retention range of $0 < k \leq 20$ and $N = 100$. For isocratic separation, peak capacity is given by [73]

$$\begin{aligned} PC &= 1 + \left(\frac{N^{0.5}}{4}\right) \ln\left(\frac{t_{R,z}}{t_0}\right) \\ &= 1 + 0.575N^{0.5} \log\left(\frac{t_{R,z}}{t_0}\right) \end{aligned} \tag{2.30}$$

where $t_{R,z}$ refers to the retention time of the last peak in the chromatogram. For typical separations, with $k \leq 20$ for the last peak and values of N as large as 20,000, $PC = 108$. If we exclude peaks with $k < 0.5$ so that $0.5 \leq k \leq 20$, the peak capacity drops to $PC = 93$; if we require $R_s = 2$ the number of peaks that fit between $k = 0.5$ and 20 drops to 47.

Peak capacity is of much greater importance for separations of complex samples—those containing a very large number of components. It is seldom possible to separate such samples with an acceptable resolution of all peaks, so peak capacity becomes a better measure of overall separation than values of R_s. Separations of complex samples are usually carried out by gradient elution, for which the concept of peak capacity is more relevant (Section 9.3.9.1). Peak capacity is of special interest for so-called two-dimensional (2D-LC) separation (Section 9.3.10), where fractions from a first separation are further resolved in a second separation, as illustrated in the example of Figure 1.4*b*,*c*. There it is seen that a group of overlapping peaks from the first separation (fraction 7) is spread out over the entire chromatogram of the second separation (*orthogonal* separation). Under these circumstances the combined peak capacity for the two separations will be equal to the *product* of peak capacities for each separation. For the example above of an isocratic peak capacity of PC

≈ 100, the 2D-LC peak capacity would be $PC = 100 \times 100 = 10{,}000$. Thus 2D-LC separation provides a lot more room in the combined chromatograms for sample peaks, so it is a powerful technique for separating complex mixtures that contain hundreds or thousands of individual components.

The peak capacity of a separation should not be confused with the number of compounds separated at $R_s = 1$, since it is rarely possible to achieve a regular spacing of peaks as in Figure 2.26*a* [73]. Figure 2.26*b* illustrates the required peak capacity PC_{req} for the separation (where $R_s \geq 1$ for all peaks) of a sample with n components. Prior to the optimization of selectivity as in Section 2.5.2, a random arrangement of peaks within the chromatogram can be assumed. As seen in Figure 2.26*b*, a sample containing 10 components ("random" curve, $n = 10$) would require a peak capacity of about 80 to achieve $R_s \geq 1$ for every peak. However, if separation selectivity has been optimized, critical peak-pairs will be separated to a greater extent, and the required peak capacity would decrease to about 40 ("optimized" curve of Fig. 2.26*b*). See [74] for further details.

2.7.4 Peak Tracking

The interpretation of separations obtained during method development requires *peak tracking* or *peak matching*. For each compound X in the sample, peaks in

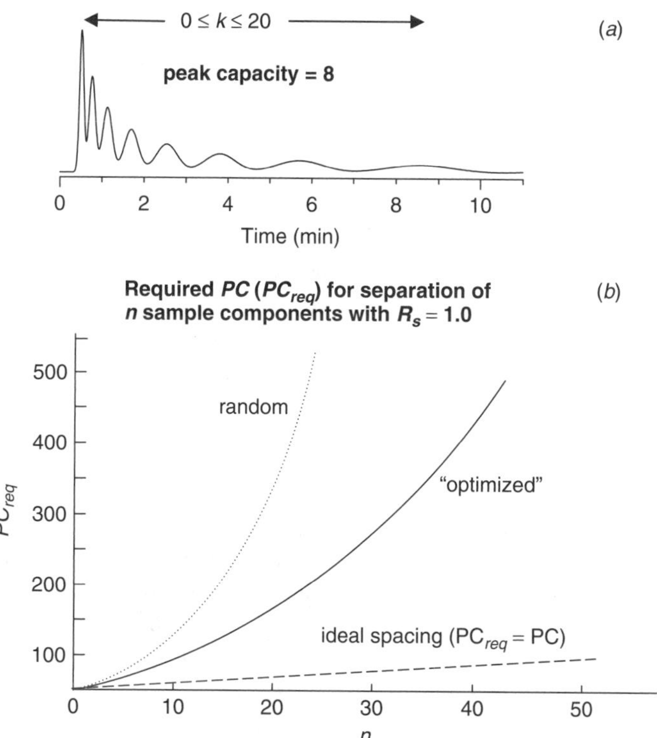

Figure 2.26 Peak capacity. (*a*) Example of peak capacity (PC) for a separation where $PC = 8$; $N = 100$, and $R_s = 1$ for every peak; (*b*) peak capacity required for the separation of a sample that contains n components [74]; "ideal spacing" is from Equation 2.30.

the various method development chromatograms that correspond to X must be characterized or numbered (as in Fig. 2.20). Thus, if peak 1 in run 1 corresponds to compound A (whose chemical structure may or may not be known), it is necessary to know which peak in run 2 also corresponds to A. For many samples this may not be difficult. For example, in Figure 2.20*b*, *d*, the six peaks in each run can be matched on the basis of peak area and relative retention (which usually do not change drastically when separation conditions are varied). Peaks 3 and 4 change places in these two chromatograms, but the areas of these and other peaks are sufficiently different to allow unambiguous peak tracking between the two runs. Manual peak tracking can take advantage of peak area, peak shape, and the observation that retention order changes (when they occur) are usually minor (i.e., a peak for a given compound usually appears in the same region of the chromatogram).

Peak tracking can be much more difficult in other cases, however, for example, when several peaks overlap as in the two separations of Figure 2.20*a*,*c*. While several workers have suggested ways to improve peak tracking with UV detection [75–79], no procedure has proved adequate for all samples. Method development is increasingly making use of mass spectrometer detection (LC-MS), which largely eliminates problems in peak tracking because of the ability of MS detection to (1) recognize each of two overlapped peaks and (2) assign a (usually unique) molecular mass to each peak in the chromatogram [75].

2.7.5 Secondary Equilibria

Chromatographic retention is based on a (primary) equilibrium between a solute molecule X in the mobile and stationary phases (as in Fig. 2.4 and Eq. 2.2):

$$X \text{ (mobile phase)} \Leftrightarrow X \text{ (stationary phase)} \tag{2.2}$$

Solute molecules may undergo further (secondary) equilibria that involve the ionization of acids and bases, ion pairing, complex formation, or isomer interconversion. As a result it is possible for two forms of the solute to be in equilibrium during their migration through the column. A common example is the separation of a partially ionized carboxylic acid, which involves an equilibrium between the ionized and non-ionized forms:

$$\text{R–COOH} \Leftrightarrow \text{R–COO}^- + \text{H}^+ \tag{2.31}$$

The relative concentrations of each form of the molecule will be determined by compound acidity (its pK_a value) and the pH of the mobile phase (Section 7.2), leading to some fraction F^- of the molecules being in the ionized form and some fraction $(1 - F^-)$ being in the neutral form. If the value of k for the ionized form is k^-, and if k_0 refers to k for the non-ionized acid, then a single peak will be observed for the two species, with its retention given by

$$k = F^- k^- + (1 - F^-)k_0 \tag{2.32}$$

As mobile-phase pH is varied, the ionization of an acidic solute and the value of F^- will change, as will the value of k (Section 7.2).

For acid-base equilibria as in Equation (2.31) (for either acidic or basic solutes), it can be assumed that the ionization process will be quite fast, much faster than the time required for a solute molecule to move through the column. As a result each solute molecule will pass back and forth between the ionized and non-ionized states many times during its migration through the column, and its retention will be an average value as described by Equation (2.32). Peak width and shape are not adversely affected by secondary equilibria, despite frequent comments to the contrary. As noted by McCalley [80], "the popular assumption that a mixed-mode mechanism leads inevitably to (peak) tailing is shown to be unfounded." On the other hand, peak tailing for both acids and bases is sometimes observed, primarily because of the properties of the column (Section 5.4.4.1) or inadequate buffering of the mobile phase (Section 7.2.1.1).

When the rate of equilibration between two species is fast, only a single peak will be observed. This is the case for a partially ionized acid, where the two forms R–COOH and R–COO$^-$ rapidly equilibrate during their migration through the column. When the rate of equilibration between two species is slow, peak broadening, distortion, and/or the apperance of separate peaks can result. An example is the interconversion of *cis* and *trans* peptide isomers [81]. At higher temperatures, the interconversion is rapid, and a single, sharp peak is observed for the peptide where isomerization is possible. At lower temperatures, where the interconversion is much slower, two distinct peaks are observed. For intermediate temperatures, a single wide, distorted peak is seen.

2.7.6 Column Switching

Column switching involves the use of two columns connected in a series via a switching valve (Section 3.6.4.1). A sample is injected into the first column, and one or more leaving fractions are transferred sequentially to the second column for further separation. Column switching can be used in each of the following applications:

- sample preparation (Sections 3.6.4.1, 16.9)
- two-dimensional liquid chromatography (2D-LC) (Sections 9.3.10, 13.4.5, 13.10.4)
- increased sampling rate

The use of column switching for sample preparation or 2D-LC usually involves the separation of one or more analytes from a complex sample where compounds of interest are completely overlapped in the first separation (with $\alpha \approx 1.00$). To achieve the separation of compounds with very similar retention, a change in selectivity for the second separation is usually employed—this is generally achieved by the use of both a different column and a different mobile phase. An example of such an application of column switching was illustrated in Figure 1.4*b*,*c*.

Another application of column switching for routine analysis can provide an increase in sampling rate, after conditions have been optimized for the fastest possible separation. A hypothetical example is illustrated in Figure 2.27*a* for the routine assay of peak *c* or *d* (or both peaks). The overall run time is 52 minutes, meaning an assay rate of only slightly more than one sample an hour. Sample pretreatment in this example might be able to remove late-eluting compounds *e* and *f*, in which case the separation time could be reduced to about 25 minutes (a sampling rate of 2.4/hr). If a large number of samples are to be analyzed on a given day, however, it is possible to significantly increase sampling rate for assays such as this by means of a column-switching technique called *boxcar chromatography* [82].

Because the two peaks *c* and *d* in Figure 2.27*a* are well separated from other peaks in the chromatogram, these two peaks can be segregated from other sample components with a shorter column and a faster flow rate—as illustrated in Figure 2.27*b* for a total run time of < 2 minutes (and a potential assay rate of >30 samples/hr). If samples are injected every 2 minutes, a fraction that contains peaks *c* and *d* can be diverted via a switching valve to the column of Figure 2.27*a*. For this way of column switching (Fig. 2.27*c*), a separate pump would deliver the same mobile phase to the second column at 0.5 mL/min, so as to achieve an equivalent separation of peaks *c* and *d* as in Figure 2.27*a* (i.e., with adequate resolution). Because bands *c* and *d* occupy only a small fraction of the second column during their migration through the column, it is possible to simultaneously separate several samples at the same time, as illustrated in Figure 2.27*d*. Here 12 fractions from the first separation can be separated simultaneously, as illustrated by an inside view of column 2 for fractions 1, 6, 10, and 12 at the beginning of this column-switching separation (other peaks not shown).

The final separation by the second column is shown in Figure 2.27*e*; after a delay of about 25 minutes, separated peaks *c* and *d* begin to leave the second column at a rate of 30 samples per hour. Boxcar chromatography relies on the simultaneous separation of different samples within column 2, which requires that two successive samples not overlap during their movement through column 2. To avoid such sample overlap, the rate of sample injections into column 1 must be coordinated with the time required for the peaks of interest (e.g., *c* and *d*) in a given sample to leave column 2.

The use of boxcar chromatography has rarely been reported in the literature [83], and today the availability of mass spectrometric detection might seem to further reduce the potential advantage of this technique for most samples. Where extremely large values of N are required for resolution—as in the preparative separation of compounds differing only in isotopic substitution—boxcar chromatography offers the possibility of achieving a much higher throughput rate than by any other technique.

2.7.7 Retention Predictions Based on Solute Structure

Obviously predictions of retention times from experimental conditions and the molecular structures of sample compounds would be useful for selecting the best conditions for a separation. Unfortunately, sufficiently accurate predictions of this kind were generally not possible at the time this book was published. Where predictions of retention may be useful, however, is for confirmation of the identity of an unknown peak in the chromatogram. The retention k of a compound is determined by its molecular structure and separation conditions. For a given set of conditions, log k can be approximated by

$$\log k = A + \Sigma \Delta R_{M(i)} \tag{2.33}$$

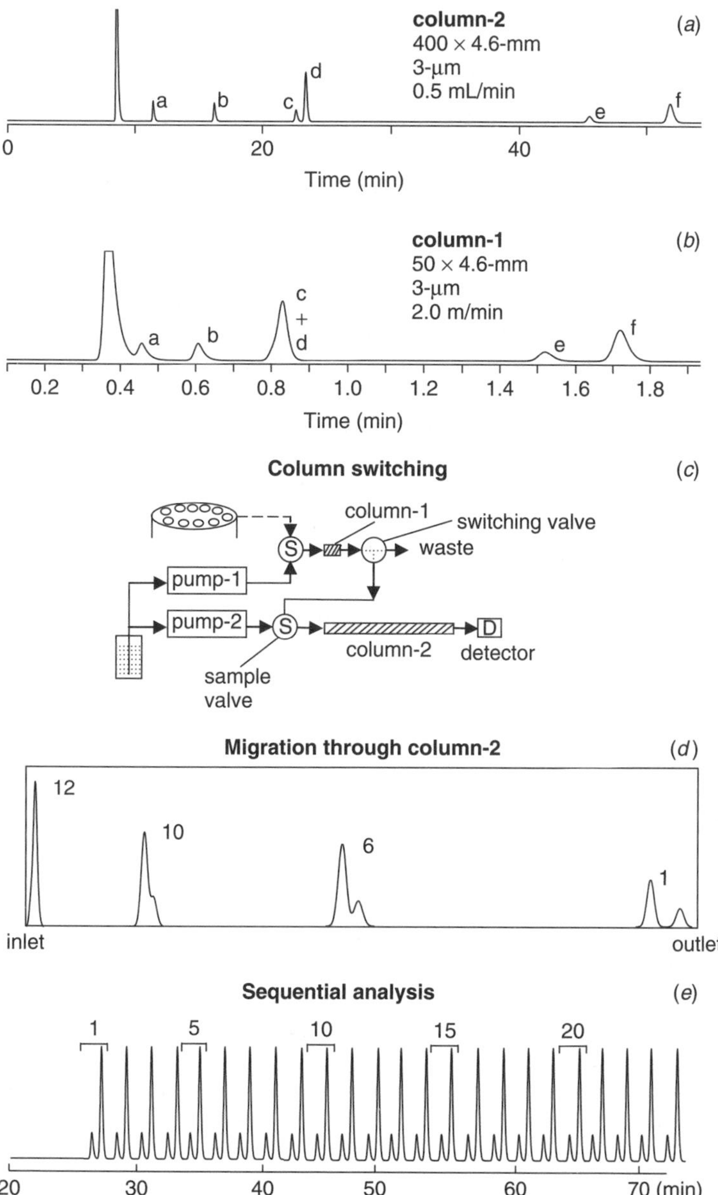

Figure 2.27 Illustration of boxcar chromatography for a hypothetical sample. (*a*) Optimized separation of the sample for acceptable resolution (column 2); (*b*) fast separation of the sample with a shorter column and faster flow rate (column 1); (*c*) equipment setup for separations of the present sample by boxcar chromatography; (*d*) migration of selected sample fractions (1, 6, 10, 12) within column 2, viewed just prior to elution of the fraction for sample 1; (*e*) early part of the chromatogram from column 2.

Here A refers to log k for a parent molecule (e.g., benzene) and $\Delta R_{M(i)}$ is the increase in log k that results from the substitution of group i into the molecule (e.g., insertion of a nitro group i into benzene to form nitrobenzene). Smith [84] has reported values of $\Delta R_{M(i)}$ for a number of common substituent groups and different RPC mobile-phase conditions, allowing estimates of retention as a function of solute molecular composition (for a very limited number of possible solutes and separation conditions).

For the case of a homologous series, Equation (2.33) assumes the form

$$\log k = A + n\alpha_{CH2} \tag{2.34}$$

Here n is the number of methylene groups ($-CH_2-$) within the molecule, and α_{CH2} is the increase in log k due to the addition of one $-CH_2-$ group to the molecule. As a consequence of Equation (2.34), plots of log k for a homologous series versus n are generally observed to be linear (but note the exception of Section 6.2.2 and Fig. 6.5). Relationships similar to Equation (2.34) apply for other compound series based on the presence of some number n equivalent groups in the molecule (e.g., oligomers of polyvinylalcohol [$-CH_2CH_2O-$ repeating groups], polystyrene [$-CH_2(C_6H_5)CH_2-$ repeating groups], etc.) Equations (2.33) and (2.34) are each referred to as the *Martin equation*, in recognition of A. J. P. Martin's first use of these relationships.

In the case of gradient elution, Equation (2.33) becomes

$$t_R \approx A + \Sigma \Delta t_{R(i)} \tag{2.35}$$

where A is the retention time of the parent compound, and $\Delta t_{R(i))}$ is a constant for a given group i that is substituted into the parent compound. Equation (2.35) has been used for the prediction of gradient retention times for a wide variety of solute molecules; for example, triacylglycerols [85], peptides [86], and polysacchrides [87]. In each case these predictions apply only for a specific set of separation conditions.

While Equation (2.33) or (2.35) can prove occasionally useful in estimating where a compound peak should be found within a chromatogram, other factors than the number and kind of substituent groups can have a significant effect on retention, especially for the complex polar molecules that are commonly present in samples for HPLC separation. Since the 1950s a large number of workers have investigated the relationship of sample retention to structure, with the hope of eventually being able to predict retention and separation in the absence of experiments (the "Holy Grail" of chromatography). In general, it has not proved possible to predict chromatographic retention in HPLC with an accuracy that is anywhere near sufficient to support method development (see [88] for a failed example). An interesting exception to these past failures of predictions of retention as a function of solute molecular structure was reported in 2007 [89], where mass spectrometric detection was combined with retention predictions to permit the identification of individual peptides in protein-digest mixtures.

2.7.7.1 Solvation-Parameter Model

A well-documented and widely applied *solvent-parameter* approach has been used to rationalize RPC retention as a function of the sample, column, and separation conditions (see [90, 91] and especially [14]). A non-ionized sample is assumed, in which case retention can be approximated as a result of hydrophobic and hydrogen-bonding interactions among sample, mobile phase, and column. The solvent-parameter model takes the form

$$\log k = C_1 + \underset{(i)}{vV_x} + \underset{(ii)}{rR_2} + \underset{(iii)}{s\pi_2^H} + \underset{(iv)}{a\Sigma\alpha_2^H} + \underset{(v)}{b\Sigma\beta_2} \tag{2.36}$$

A solute retention factor k is related to (1) a constant C_1 that is a function of column and conditions, (2) solute-dependent quantities v, r, s, a, and b, and (3) solute-independent quantities $V_x, R_2, \pi_2^H, \Sigma\alpha_2^H$, and $\Sigma\beta_2$. Terms *i* to *iii* of Equation (2.36) together account for hydrophobic interactions, while terms *iv* and *v* are the result of hydrogen bonding between solute and either the column or the mobile phase. Values of v, r, s, a, and b for a large number of different solutes have been tabulated, and values of $C_i, V_x, R_2, \pi_2^H, \Sigma\alpha_2^H$, and $\Sigma\beta_2$ can be determined for a column and given conditions by the use of appropriate tests solutes.

Equation (2.36) can provide insight into the factors that determine RPC separation, but the errors in predictions of values of k (about ±20%) are too large to be useful for method development. Equation (2.36) is further limited by the fact that it cannot be applied to ionized solutes, and it neglects a number of additional interactions that can affect retention (see the related discussion of Section 5.4).

REFERENCES

1. M. M. Fallas, U. D. Neue, M. R. Hadley, and D. V. McCalley, *J. Chromatogr. A*, 1209 (2008) 195.
2. C. A. Rimmer, C. R. Simmons, and J. G. Dorsey, *J. Chromatogr. A*, 965 (2002) 219.
3. J. M. Bermúdez-Saldaña, L. Escuder-Gilabert, R. M. Villanueva-Camañas, M. J. Medina-Hernández, and S. Sagrado, *J. Chromatogr. A*, 1094 (2005) 24.
4. M. Shibukama, Y. Takazawa, and K. Saitoh, *Anal. Chem.*, 79 (2007) 6279.
5. M. A. Quarry, R. L. Grob, and L. R. Snyder, *J. Chromatogr.*, 285 (1984) 19.
6. L. R. Snyder, unreported data. For a single mobile phase (50% acetonitrile/buffer) and several hundered different columns.
7. T. Braumann, G. Weber, and L. H. Grimme, *J. Chromatogr.*, 261 (1983) 329.
8. L. C. Tan, P. W. Carr, and M. H. Abraham, *J. Chromatogr. A*, 752 (1996) 1.
9. R. A. Keller, B. L. Karger, and L. R. Snyder, in *Gas Chromatography. 1970*, R. Stark and S. G. Perry, eds., Institute of Petroleum, London, 1971, p. 125.
10. K. Croes, A. Steffens, D. H Marchand, and L. R. Snyder, *J. Chromatogr. A*, 1098 (2005) 123.
11. L. R. Snyder, P. W. Carr, and S. C. Rutan, *J. Chromatogr.*, 656 (1993) 537.
12. L. R. Snyder, *J. Chromatogr. Sci.*, 16 (1978) 223.
13. J. A. Riddick and W. B. Bunger, *Organic Solvents*, Wiley-Interscience, New York, 1970.
14. M. Vitha and P. W. Carr, *J. Chromatogr. A*, 1126 (2006) 143.
15. S. M. C. Buckenmaier, D. V. McCalley, and M. R. Euerby, *J. Chromatogr. A*, 1060 (2004) 117.

16. S. Heinisch, G. Puy, M.-P. Barrioulet, and J.-L. Rocca, *J. Chromatogr.*, 1118 (2006) 234.
17. W. R. Melander, A. Nahum and Cs. Horváth, *J. Chromatogr.*, 185 (1979) 129.
18. D. Guillarme and S. Heinisch, *J. Chromatogr. A*, 1052 (2004) 39.
19. J. W. Dolan, *J. Chromatogr. A*, 965 (2002) 195.
20. G. Vanhoenacker and P. Sandra, *J. Sep. Sci.*, 29 (2006) 1822.
20a. S. Heinisch and J. -L. Rocca, *J. Chromatogr. A*, 1216 (2009) 642.
21. S. A. C. Wren and P. Tchelitcheff, *J. Chromatogr. A*, 1119 (2006) 140.
22. J. C. Giddings, Dynamics of Chromatography, *Part 1, Principles and Theory*, Dekker, New York, 1965.
23. G. Guiochon, in *High-Performance Liquid Chromatography. Advances and Perspectives*, Vol. 2, C. Horváth, ed., Academic Press, New York, 1980, p. 1.
24. S. G. Weber and P. W. Carr, in *High Performance Liquid Chromatography*, P. R. Brown and R. A. Hartwick, eds., Wiley-Interscience, New York, 1989, p. 1.
25. J. H. Knox, in *Advances in Chromatography*, Vol. 38, P. R. Brown and E. Grushka, eds., Dekker, New York, 1998, p. 1.
26. C. R. Wilke and P. Chang, *Am. Inst. Chem. Eng. J.*, 1 (1955) 264.
27. J. H. Knox, *J. Chromatogr. A*, 960 (2002) 7.
28. L. Kirkup, M. Foot, and M. Mulholland, *J. Chromatogr. A*, 1030 (2004) 25.
29. F. Gritti and G. Guiochon, *Anal. Chem.*, 78 (2006) 5329.
30. F. Gritti and G. Guiochon, *J. Chromatogr. A*, 1169 (2007) 125.
31. J. G. Dorsey and P. W. Carr, eds., *J. Chromatogr. A*, 1126 (2006) 1–128.
32. K. M. Usher, C. R. Simmons, and J. G. Dorsey, *J. Chromatogr. A*, 1200 (2008) 122.
33. R. W. Stout, J. J. Destefano, and L. R. Snyder, *J. Chromatogr.*, 282 (1983) 263.
34. J. H. Knox and H. P. Scott, *J. Chromatogr.*, 282 (1983) 297.
35. K. Miyabe, *J. Chromatogr. A*, 1167 (2007) 161.
36. G. Desmet, K. Broekhoven, J. De Smet, S. Deridder, G. V. Baron, and P. Gzil, *J. Chromatogr. A*, 1188 (2008) 171.
37. H. Poppe, *J. Chromatogr. A*, 778 (1997) 3.
38. G. Desmet, D. Clicq, and P. Gzil, *Anal.Chem.*, 77 (2005) 4058.
38a. P. W. Carr, X. Wang, and D. R. Stoll, *Anal.Chem.*, 81 (2009) 5342.
39. D. V. McCalley, *J. Chromatogr. A*, 793 (1998) 31.
40. D. M. Marchand, L. R. Snyder, and J. W. Dolan, *J. Chromatogr. A*, 1191 (2008) 2.
41. J. J. Kirkland, W. W. Yau, H. J. Stoklosa, and C. H. Dilks, Jr., *J. Chromatogr. Sci.*, 15 (1977) 303.
42. K. Lan and J. W. Jorgenson, *J. Chromatogr. A*, 915 (2001) 1.
43. J. P. Foley and J. G. Dorsey, *Anal. Chem.*, 55 (1983) 730.
44. L. R. Snyder, J. J. Kirkland, and J. L. Glajch, *Practical HPLC Method Development*, 2nd ed., Wiley-Interscience, New York, 1997.
45. L. R. Snyder and J. J. Kirkland, *Introduction to Modern Liquid Chromatography*, 2nd ed., Wiley-Interscience, New York, 1979, pp. 34–36.
46. *Reviewer Guidance. Validation of Chromatographic Methods*, Center for Drug Evaluation and Research, Food and Drug Administration (Nov. 1994).
47. K. Valkó, L. R. Snyder and J. L. Glajch, *J. Chromatogr.*, 656 (1993) 501.
48. L. R. Snyder, J. W. Dolan, and J. R. Gant, *J. Chromatogr.*, 165 (1979) 3.
49. L. R. Snyder and J. W. Dolan, *J. Chromatogr. A*, 721 (1996) 3.

50. N. S. Wilson, M. D. Nelson, J. W. Dolan, L. R. Snyder, R. G. Wolcott, and P. W. Carr, *J. Chromatogr. A*, 961 (2002) 171.
51. N. S. Wilson, M. D. Nelson, J. W. Dolan, L. R. Snyder, and P. W. Carr, *J. Chromatogr. A*, 961 (2002) 195.
52. L. R. Snyder and J. W. Dolan, *High-Performance Gradient Elution*, Wiley-Interscience, New York, 2007, ch. 1.
53. B. A. Bidlingmeyer, R. P. Hooker, C. H. Lochmuller, and L. B. Rogers, *Sep. Sci.*, 4 (1969) 439.
54. J. E. MacNair, K. E. Lewis, and J. W. Jorgenson, *Anal. Chem.*, 69 (1997) 983.
55. F. Gritti and G. Guiochon, *J. Chromatogr. A*, 1187 (2008) 165.
56. R. G. Wolcott, J. W. Dolan, L. R. Snyder, S. R. Bakalyar, M. A. Arnold, and J. A. Nichols, *J. Chromatogr. A*, 869 (2000) 211.
57. T. Greibrokk and T. Andersen, *J. Chromatogr. A*, 1000 (2003) 743.
58. X. Yang, L. Ma and P. W. Carr, *J. Chromatogr. A*, 1079 (2005) 213.
59. F. Lestremau, A. Cooper, R. Szucs, F. David, and P. Sandra, *J. Chromatogr. A*, 1109 (2006) 191.
60. F. D. Antia and C. Horváth, *J. Chromatogr.*, 435 (1988) 1.
61. J. D. Thompson and P. W. Carr, *Anal. Chem.*, 74 (2002) 1017.
62. K. P. Xiao, Y. Xiong, F. Z. Liu, and A. M. Rustum, *J. Chromatogr. A*, 1163 (2007) 145.
63. L. Deng, H. Nakano, and Y. Wwasaki, *J. Chromatogr. A*, 1165 (2007) 93.
64. J. C. Sternberg, in *Adv. Chromatogr.*, 2 (1966) 205.
65. M. Tsimidou and R. Macrae, *J. Chromatogr.*, 285 (1984) 178.
66. N. E. Hoffman, S.-L. Pan, and A. M. Rustum, *J. Chromatogr.*, 465 (1989) 189.
67. N. E. Hoffman and A. Rahman, *J. Chromatogr.*, 473 (1989) 260.
68. E. Loeser and P. Drumm, *J. Sep. Sci.*, 29 (2006) 2847.
69. E. Loeser, S. Babiak, and P. Drumm, *J. Chromatogr. A*, 1216 (2009) 3409.
70. H. Colin, M. Martin, and G. Guiochon, *J. Chromatogr.*, 185 (1979) 79.
71. S. R. Bakalyar, C. Phipps, B. Spruce, and K. Olsen, *J. Chromatogr. A*, 762 (1997) 167.
72. J. E. Eble, R. L. Grob, P. E. Antle, and L. R. Snyder, *J. Chromatogr.*, 405 (1987) 51.
73. J. M. Davis and J. C. Giddings, *Anal. Chem.* 55 (1983) 418.
74. J. W. Dolan, L. R. Snyder, N, M. Djordjevic, D. W. Hill, L. Van Heukelem, and T. J. Waeghe, *J. Chromatogr. A*, 857 (1999) 1.
75. G. Xue, A. D. Bendick, R. Chen, and S. S. Sekulic, *J. Chromatogr. A*, 1050 (2004) 159.
76. J. L. Glajch, M. A. Quarry, J. F. Vasta, and L. R. Snyder, *Anal. Chem.*, 58 (1986) 280.
77. H. J. Issaq and K. L. McNitt, *J. Liq. Chromatogr.*, 5 (1982) 1771.
78. J. K. Strasters, H. A. H. Billiet, L. de Galan, and B. G. M. Vandeginste, *J. Chromatogr.*, 499 (1990) 499.
79. E. P. Lankmayr, W. Wegscheider, J. Daniel-Ivad, I. Kolossváry, G. Csonka, and M. Otto, *J. Chromatgr.*, 485 (1989) 557.
80. D. V. McCalley, *Adv. Chromatogr.*, 46 (2007) 305.
81. W. R. Melander, J. Jacobson and Cs. Horváth, *J. Chromatogr.*, 234 (1982) 269.
82. L. R. Snyder, J. W. Dolan, and Sj. van der Wal, *J. Chromatogr.*, 203 (1981) 3.
83. A. Nazareth, L. Jaramillo, R. W. Giese, B. L. Karger, and L. R. Snyder, *J. Chromatogr.*, 309 (1984) 357.
84. R. M. Smith, *J. Chromatogr. A*, 656 (1993) 381.

85. J.-T. Lin, L. R. Snyder, and T. A. McKeon, *J. Chromatogr. A*, 808 (1998) 43.
86. C. T. Mant and R. S. Hodges, in *HPLC of Proteins, Peptides and Polynucleotides*, M. T. W. Hearn, ed., VCH, New York, 1991, p. 277.
87. K. Yanagida, H. Ogawa, K. Omichi, and S. Hase, *J. Chromatogr. A*, 800 (1998) 187.
88. T. Baczek, R. Kaliszan, H. A. Claessens, and M. A. van Straten, *LCGC Europe*, 13 (2001) 304.
89. V. Spicer, A. Yamchuk, J. Cortens, S. Sousa, W. Ens, K. G. Standing, J. Q. Wilkens, and O. V. Korkhin, *Anal. Chem.*, 79 (2007) 8762.
90. P. C. Sadek, P. W. Carr, R. M. Doherty, M. J. Kamlet, R. W. Taft, and M. H. Abraham, *Anal. Chem.*, 57 (1985) 2971.
91. C. F. Poole and S. K. Poole, *J. Chromatogr. A*, 965 (2002) 263.

CHAPTER THREE

EQUIPMENT

Introduction to Modern Liquid Chromatography, Third Edition, by Lloyd R. Snyder, Joseph J. Kirkland, and John W. Dolan

3.1 INTRODUCTION

Equipment design for modern HPLC is in a mature state. With certain exceptions (e.g., high-pressure applications, Section 3.5.4.3), major changes in equipment design and features are not often encountered. While small changes from one model to its replacement continue to improve the reliability of HPLC equipment, the rapid obsolescence of HPLC equipment that was once a concern is no longer an issue for most applications.

Analysts beginning their use of HPLC often ask which system or manufacturer is "best." Today there is less distinction between HPLC systems than in the past, and it can be safely said that there are no "bad" HPLC systems currently on the market. This means that a features-and-benefits approach to equipment selection often gives way to choices based on local service and support provided by the equipment vendor. Users in the past often would select specific equipment modules from different vendors and, in a mix-and-match approach, would design their own "ideal" HPLC system. Today this is not common, partly because of the equivalent performance of components between manufacturers, and partly because of the interdependence of the various modules. Usually components chosen from a single manufacturer will work together better than will modules from several manufacturers combined into a single system. Thus the pump, autosampler, and column oven usually are obtained as a unit or as compatible components from a single manufacturer.

The detector may be obtained from a second manufacturer, especially for specialty detectors, such as MS/MS (Section 4.10). Because the major data-system manufacturers often include the ability to control equipment from other vendors, the data system may be chosen from another manufacturer than the pumping components. However, when maintenance, training, repair, and equipment compatibility are considered, most laboratories purchase as many components of the HPLC system as possible from a single vendor and stay with a single manufacturer if multiple HPLC systems are operated in a single facility. An alternative practice is used in some large laboratories, especially those that transfer methods to other sites (Section 12.7). In such cases equipment is selected from several manufacturers in order to allow comparison of method performance on different instruments. This approach helps highlight potential equipment-dependent method-transfer problems that can be addressed prior to transfer of the method to a second laboratory.

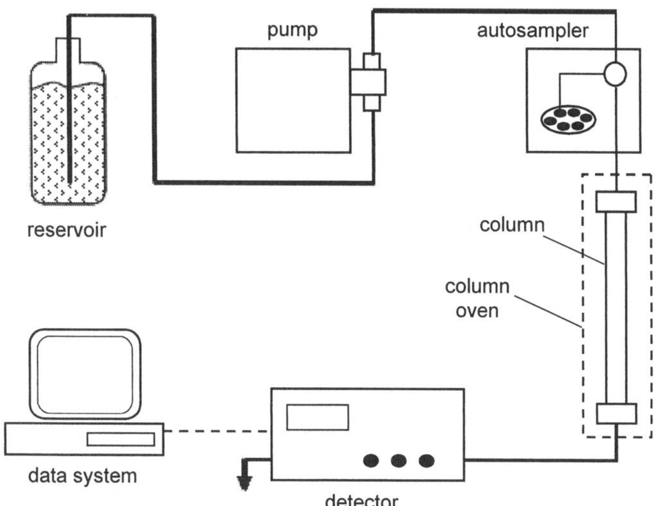

Figure 3.1 HPLC system diagram.

The essential components of an HPLC system are shown in Figure 3.1. Mobile phase is drawn from a reservoir into a pump, which controls the flow rate and generates sufficient pressure to drive the mobile phase through the column. An injector or autosampler is used to place the sample on the column without stopping the pump flow. The separation takes place in the column, which generally resides inside a column oven. The detector responds to changes in analyte concentration during the run. A data system monitors the detector output and provides data processing for both graphic and tabular output of data. A system controller (often combined with the data system) directs the functions of the various modules.

The HPLC system may comprise a group of individual components (often referred to as a "modular" system), or the components may be combined within a single cabinet as an "integrated" system. Because of the precious nature of laboratory bench space, modular systems usually are designed to enable stacking of components for a small footprint, similar to that of an integrated system. In addition to systems designed for analytical applications, HPLC systems may be specially designed for low-flow (micro), high-flow (preparative), or high-pressure applications (Sections 3.5.4 and 15.2). The majority of analytical methods rely on UV detection, but many other detectors are available for specialized applications (Chapter 4).

In this chapter the various components of the HPLC system are discussed, with the exceptions of the detector (Chapter 4) and application of the data system (Chapter 11). Unless stated otherwise, commercially available equipment is assumed in every case.

3.2 RESERVOIRS AND SOLVENT FILTRATION

Mobile-phase reservoirs (Fig. 3.2) are simple yet essential parts of the HPLC system. For isocratic applications using premixed mobile phase, only a single reservoir is

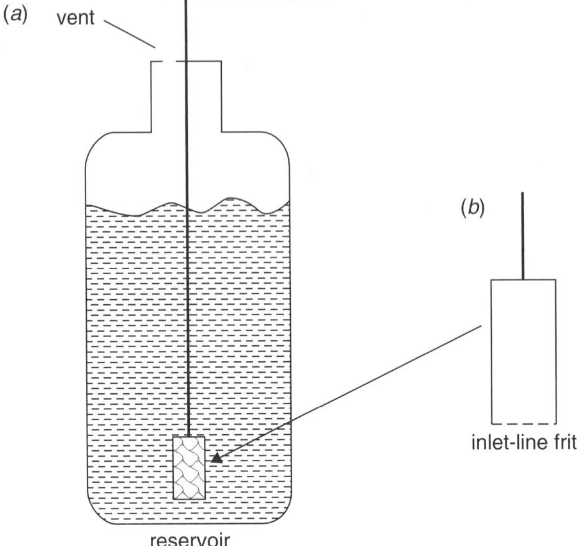

Figure 3.2 Mobile-phase reservoir.

needed. When isocratic mobile phases are blended online or for gradient applications, more than one reservoir is used. Mobile phases must be free of particulate matter, so mobile-phase filtration may be required prior to filling the reservoir.

3.2.1 Reservoir Design and Use

Most reservoir containers (Fig. 3.2*a*) are made of glass, although some applications, such as the determination of Na^+ ions by ion chromatography, require a glass-free system. Laboratory glassware (e.g., Erlenmeyer flasks), heavy-walled glass bottles, or the glass bottles in which the solvents are shipped are the common reservoirs. Some equipment manufacturers supply reservoirs specifically designed for their equipment.

Besides inertness to the mobile phase, cleanliness is the most important reservoir requirement. Glassware should be washed on a regular basis (e.g., weekly), using standard laboratory dishwashing techniques. A cover of some sort should be used to keep dust from entering the reservoir and to minimize evaporation of the mobile phase, but the reservoir should not be so tightly capped that a vacuum forms when mobile phase is pumped out. A threaded cap with an oversized hole (Fig. 3.2*a*) for the mobile-phase inlet line or a piece of aluminum foil crimped around the top of the reservoir are the most popular closure techniques and allow rapid pressure equalization when mobile phase is pumped out. The use of polymeric laboratory film products (e.g., Parafilm®) to cover the reservoir should be avoided, since some mobile phases may extract components that can contaminate the system.

An inlet-line frit (Fig. 3.2*a,b*) is used at the inlet end of the tubing that connects the reservoir and the pump. The frit acts as a weight to keep the inlet tubing at the bottom of the reservoir, but its primary function is to provide backup filtration to remove particulate matter, such as dust, that might enter the reservoir. Since it is not the primary solvent filter, it should not restrict solvent flow to the pump. A frit porosity of $\geq$10 μm is recommended so that solvent can flow freely through the inlet-line frit. This can be confirmed with a siphon test. Disconnect the tube fitting at the pump inlet (high-pressure mixing systems) or solvent proportioning module (low-pressure mixing); if solvent is not flowing freely, start a siphon flowing with a pipette bulb. A good rule of thumb is that the flow through the siphon should be $\geq$10× the required flow rate when the solvent reservoirs are located >50 cm above the point of measurement. For example, if flow rates of 1 to 2 mL/min are typically used, the siphon test should supply >20 mL/min of solvent. Generally, flow rates of >50 mL/min are observed under these conditions. If the siphon delivery is too slow, replace the frit and/or clear any blockage in the tubing. In use, the reservoir should be located higher than the pump inlet (e.g., >50 cm) so as to provide a positive-pressure feed of solvent to the pump for more reliable pump operation.

There are many designs of inlet-line frits available, and these are made of stainless-steel, ceramic, PEEK, and other materials that are inert to the mobile phase. One popular design is sketched in Figure 3.2*b*, in which the intake portion of the frit is on the bottom rather than the sides. This "last drop" design enables the use of more mobile phase in the reservoir before it must be replenished.

3.2.2 Mobile-Phase Filtration

The operation of several parts of the HPLC system can be compromised if particulate matter is present. These parts include proportioning valves, check valves, tubing, and column frits. For this reason it is important to use a particulate-free mobile phase. If prefiltered (i.e., HPLC-grade) solvents are not available, the mobile-phase components should be filtered prior to adding them to the reservoir. For laboratories that work in a regulated environment, a standard operating procedure (SOP) should be written to describe when additional mobile-phase filtration is required and when it is not.

Use of prefiltered solvents is the simplest way to avoid introducing particulate matter into the mobile phase. Commercial HPLC-grade solvents are filtered through submicron filters (generally 0.2 μm) prior to packaging. HPLC-grade water prepared in the laboratory (e.g., Milli-Q water purification) is passed through a final 0.2-μm filter as the last step in purification. If only HPLC-grade liquids are used in the mobile phase, it is common practice not to perform any additional filtration prior to use. However, if non–HPLC-grade reagents or any solid reagents are added to the mobile phase (e.g., phosphate buffer), it is wise to filter all mobile-phase mixtures prior to use.

Mobile phases can be easily filtered with a vacuum-filter apparatus, such as that shown in Figure 3.3. A membrane filter (typically ≈0.5-μm porosity) is mounted on a support frit between the funnel and the vacuum flask. Solvent is poured into the funnel and collected under vacuum-assisted (e.g., water aspirator) filtration. Filter manufacturers provide guides to the selection of the proper filter material for each application. For example, PTFE filters are hydrophobic and work well with pure organic solvents, such as MeOH or ACN, but are too nonpolar to allow rapid filtration of water. The seal between the vacuum flask is made with a ground-glass fitting or an "inert" stopper (e.g., silicone), but it is best not to allow mobile phase to contact the stopper.

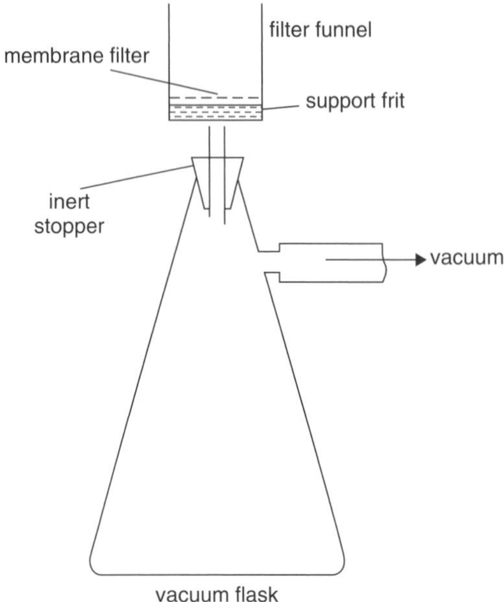

Figure 3.3 Vacuum apparatus for mobile phase filtration.

3.3 MOBILE-PHASE DEGASSING

The presence of air bubbles in the mobile phase is a common problem in the operation of an HPLC system. These bubbles can lead to pump-delivery problems and spurious peaks in the detector output. Most often concern about bubbles can be eliminated by degassing the mobile phase prior to use.

3.3.1 Degassing Requirements

As long as air remains dissolved in the mobile phase, bubble problems are seldom encountered. In principle, hand-mixed isocratic mobile phases should be suitable for use without degassing, but an air-saturated solution may outgas with only a minor drop in pressure, such as when the mobile phase is pulled through the solvent inlet-line filter or when it enters the relatively low-pressure region in the detector cell. For this reason, and for general HPLC operational reliability, degassing of all solvents used for reversed-phase applications is strongly recommended. Outgassing is less of a problem with normal-phase HPLC, so degassing may be considered as optional in such applications. The amount of dissolved gas that must be removed from the mobile phase will vary with the design of the HPLC pump—some pumps are very tolerant to dissolved gas, whereas others require thorough degassing for reliable operation.

Bubble formation can be especially problematic in the case of mobile phases for reversed-phase chromatography (RPC), as illustrated by the data of Figure 3.4. For example, assume that pure water and pure ethanol are each saturated with oxygen, as might be the case if the solvents are exposed to air. When the solvents are blended, the mixture contains an amount of oxygen and solvent that is proportional

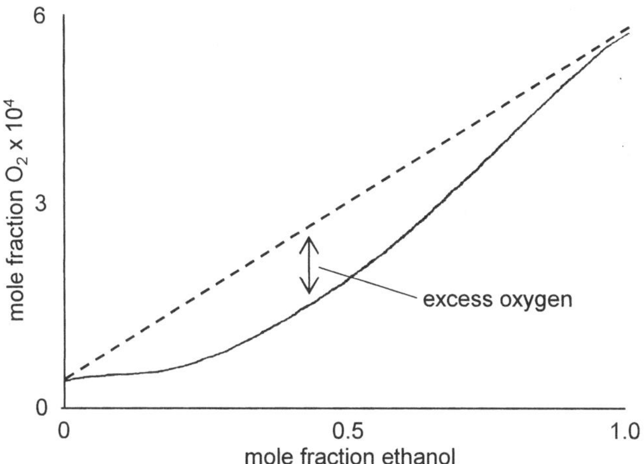

Figure 3.4 Solubility of oxygen in ethanol. (- - -), Oxygen concentration following mixing of air-saturated water and ethanol (before release of excess oxygen); (—), saturation concentration of oxygen in mixture. Adapted from [1].

to the relative concentrations of each solvent (represented by the dashed line in Fig. 3.4). However, oxygen is seen to be less soluble in a solvent mixture (solid line in Fig. 3.4), so the mixture is now supersaturated with oxygen. In such cases oxygen either bubbles out immediately or when it contacts a nucleation site, such as the rough surface of the solvent inlet-line filter. Although Figure 3.4 shows data for oxygen, water, and ethanol, the same principle holds for air (comprising primarily nitrogen and oxygen), buffered water, and other organic solvents, such as acetonitrile or methanol [1]. These data also suggest that it is not necessary to remove all of the dissolved air from solution—just enough that the amount of dissolved air in the mixtures is below the (solid) saturation curve of Figure 3.4.

For most applications, degassing is important primarily to improve pump operation. However, in some cases the presence of dissolved oxygen can degrade detector performance. It has been reported [2] that UV detection (Section 4.3) as low as 185 nm is possible if the detector (and acetonitrile/water mobile phase) is purged with helium to remove oxygen from the optical path of the detector. Under these conditions the apparent detector-lamp response increased and the baseline noise was reduced. Even at higher wavelengths, dissolved oxygen in the mobile phase can elevate the detector background signal, as can be seen in Figure 3.5*a*. At 254 nm, the mobile phase sparged with air gave an increased baseline signal compared to the mobile phase sparged with helium, presumably because of a change in refractive index of the air-sparged mobile phase. Under the same conditions, but with fluorescence detection (Section 4.5), $\approx$75 % of the signal intensity for naphthalene was lost (Fig. 3.5*b*) when the mobile phase was sparged with air instead of helium [1]. When the electrochemical detector (Section 4.6) is operated in the reductive mode, dissolved oxygen creates an unacceptable background signal, so oxygen must be removed from the mobile phase, as by helium sparging (Section 3.3.2). Finally, it is conceivable that dissolved oxygen might react with some samples during separation. So it is important to select a degassing technique that addresses both chemical

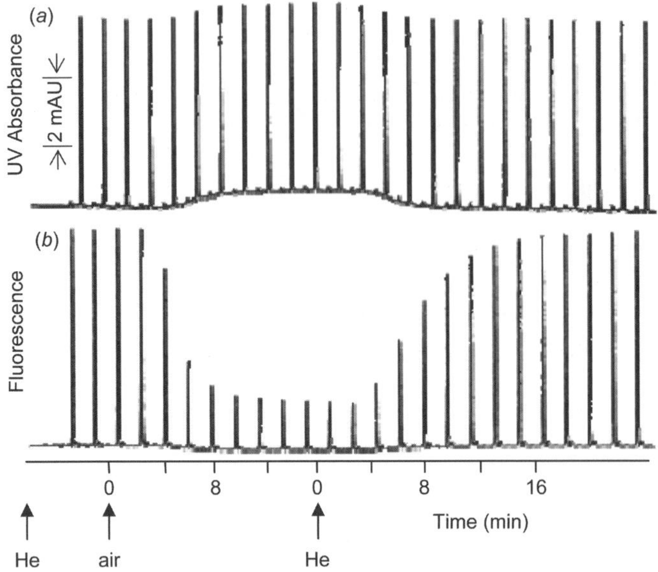

Figure 3.5 Effect of helium sparging on detector response to naphthalene. (*a*) UV detection at 254 nm, (*b*) fluorescence detection at 250-nm excitation and 340-nm emission. He, mobile-phase sparging with helium begins; air, sparging with air begins. Adapted from [1].

problems (e.g., detector response) and physical problems (e.g., bubbles in the pump) that may result from the presence of dissolved gas in the mobile phase.

When off-line degassing is used, such as stand-alone helium sparging or vacuum degassing, the solvent will begin to re-equilibrate with air as soon as the degassing treatment is stopped. For HPLC systems that are highly susceptible to dissolved gas in the mobile phase, off-line degassing may not be sufficient. In such cases continuous helium sparging (Section 3.3.2) or on-line vacuum degassing (Section 3.3.3) are better choices.

3.3.2 Helium Sparging

Helium sparging is the most effective technique for removing dissolved gas from the mobile phase [3] (with the exception of refluxing or distillation), and removes 80–90% of the dissolved air. Typically a frit is used to disperse helium (e.g., at ≈5 psi through a sparging frit) in the reservoir. Under these conditions it takes only one volume of helium to degas an equal volume of mobile phase [4]. This means that just a few minutes of a vigorous sparging stream will adequately degas the mobile phase. Helium itself has such a low solubility in HPLC solvents that a helium-sparged solution is nearly gas free. Excessive sparging of the mobile phase is undesirable, since it can change the composition of the mobile phase through evaporation of the more volatile component(s); however, vigorously sparging a RPC mobile phase for a few minutes is unlikely to cause problems Normal-phase solvents are much more volatile, so helium sparging of a (blended) mobile phase should be used cautiously—if at all. Sparging pure solvents prior to on-line mixing poses no problem, however.

3.3.3 Vacuum and In-line Degassing

For most HPLC systems, the application of a partial vacuum to the mobile phase will remove a sufficient amount of dissolved gas to avoid outgassing problems. Vacuum degassing for 10 to 15 minutes will remove 60–70% of the dissolved gas [3]. In its simplest form, some vacuum degassing takes place during solvent filtration, as in Figure 3.3. This can be enhanced after filtration is complete by replacing the filter funnel with an inert stopper and applying the vacuum for a few more minutes. Some users find that placement of the vacuum flask in an ultrasonic cleaning bath during this process further enhances degassing.

Today in-line (or on-line) degassing is the most popular degassing technique; most HPLC equipment manufacturers include an in-line degasser as either standard or optional equipment with new systems. The operation of the in-line degasser is illustrated in Figure 3.6 for two solvents (A and B; degassers for 1–4 solvents are available), and it is based on the selective permeability of certain polymeric tubing to gas. The degasser is mounted before the pump(s) (high-pressure mixing, Section 3.5.2.1; or hybrid systems, Section 3.5.2.3) or proportioning valves (low-pressure mixing, Section 3.5.2.2). Solvent is passed through a piece of polymeric tubing inside a vacuum chamber; the vacuum pulls the dissolved gas passes through the walls of the tubing; the liquid mobile phase stays inside the tubing (detail

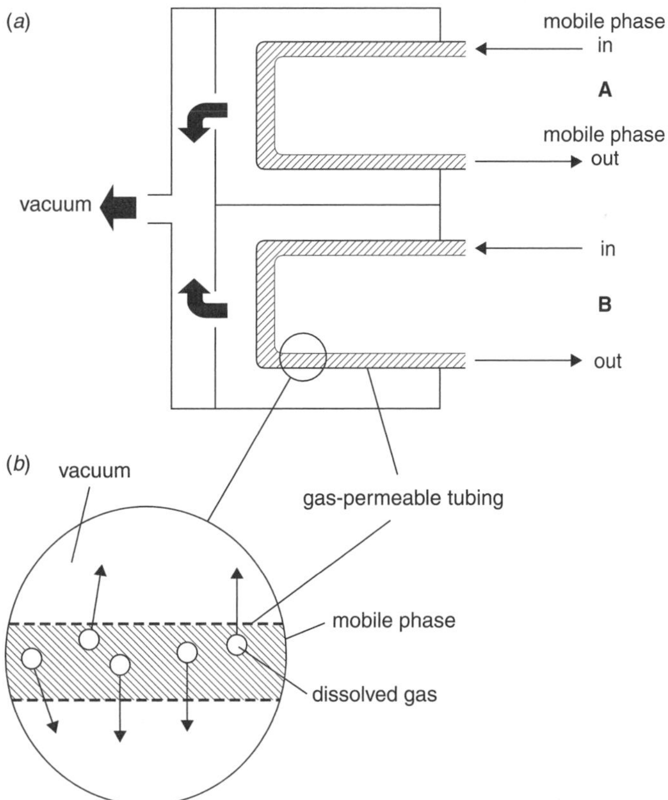

Figure 3.6 Diagram of a membrane degassing apparatus for two solvents, A and B.

of tubing wall in Fig. 3.6*b*). With properly designed systems, sub-milliliter volumes of tubing can be used for each solvent, with degassing effectiveness equivalent to extended vacuum degassing under static conditions for most HPLC operations [5]. Although not quite as effective as helium sparging, for most users the convenience of in-line vacuum degassing and the cost of helium have made this the preferred degassing technique in place of helium sparging. For applications where dissolved-oxygen concentrations are critical (e.g., some fluorescence and electrochemical detection methods) or where extremely low and constant concentrations of dissolved gases must be maintained (e.g., maximum sensitivity refractive-index detection), continuous helium sparging followed by in-line vacuum degassing is the best choice [5].

3.4 TUBING AND FITTINGS

Tubing and the fittings used to connect tubing to various HPLC system components are required for transporting the mobile phase and sample through the chromatograph. If reasonable care is taken in the selection and use of tubing and fittings, problems will seldom be encountered. However, improper selection and use can generate unwanted extra-column volume (Section 3.9), which can compromise the separation—especially for small peak volumes. Additional technical information on tubing and fittings can be found on the Internet [6–7] and in manufacturer's literature, such as [8]. Note that the following discussion refers to the dimensions and thread sizes of tubing and fittings in English units (usually fractional or decimal inches), because this is the way they are commonly supplied in the United States. These products are available in metric sizes in many other markets. As a note of caution, if both English and metric versions of similar products are used, be sure to label them clearly—even if they may *seem* to fit together in some cases, since damage to the part or a leak may result.

3.4.1 Tubing

3.4.1.1 Low-Pressure Tubing

For pressures less than $\approx$100 psi, polymeric tubing generally is suitable. The two primary applications are transport of the mobile phase from the reservoir to the pump, and waste from the detector to the waste container.

On the inlet side of the pump, tubing generally is 1/8-in o.d. and 1/16-in or smaller i.d. The inertness of fluorocarbon tubing (e.g., Teflon®) makes it the first choice for transport of solvents to the pump. Other polymers (e.g., polypropylene and polyethylene) may be suitable as well, but these products are best purchased from an HPLC supplies vendor to ensure that they are of sufficient purity for the application. Teflon tubing is somewhat permeable to gas, so air can diffuse through the inlet tubing into the mobile phase. Generally, air is not a problem, but in some applications (e.g., reductive electrochemical applications) dissolved oxygen in the mobile phase can cause problems. PEEK (polyetheretherketone) tubing is another suitable inlet-line tubing; it is not gas-permeable but it is opaque, so it is not possible to see bubbles inside the tubing. PEEK tubing also has some limitations in terms of chemical compatibility (Section 3.4.1.2).

On the waste side of the detector, Teflon, polypropylene, polyethylene, PEEK, or other relatively inert tubing can be used. For ease of connection with the detector outlet, 1/16-in o.d. tubing generally is used as a waste line. The use of tubing with an internal diameter of 0.010-in i.d. or less will create sufficient back-pressure to keep bubbles in solution until they exit the detector cell—this will help reduce bubble problems in the detector. However, with this technique, one should be careful not to overpressure the detector cell, especially if the flow rate is increased dramatically. It usually is more prudent to use larger i.d. tubing (e.g., ≥0.20-in) and a back-pressure restrictor (available from many HPLC fittings suppliers) at the outlet end of the tubing to maintain 50 to 75 psi back-pressure at any flow rate.

The length of tubing is not critical in low-pressure applications, but it is a good idea to keep these lengths to a convenient minimum. Excessive tubing lengths can result in longer washout times and delayed equilibration. Most low-pressure tubing is cut with a razor blade or, for PEEK, a cutter supplied by the tubing vendor. A flat cut, perpendicular to the tube axis is desired.

3.4.1.2 High-Pressure Tubing

Whereas low-pressure tubing is used primarily to transport mobile phase to the pump or waste from the detector, high-pressure tubing is required elsewhere in the system. Conventional HPLC systems are designed for use with pressures up to 6000 psi between the pump and detector, so the tubing must be able to withstand such pressures. Also tubing used to transport the sample from the autosampler to the column and from the column to the detector must be sufficiently inert that sample adsorption or degradation does not take place, and the tubing length and diameter should be selected so that it does not contribute significantly to peak broadening (Section 3.9). Both stainless steel and PEEK tubing will satisfy these requirements for most applications.

Type 316 stainless-steel tubing is most commonly used for HPLC applications. It is inert to nearly all solvents and has a very high burst strength. It is less convenient to use than PEEK tubing because of its stiffness. PEEK tubing ≤0.030-in i.d. will work up to pressures of ≈7000 psi without rupturing; for higher pressures, stainless-steel tubing is required. PEEK tubing is compatible with most HPLC mobile phases, but it will swell and become brittle when exposed to THF, chlorinated solvents, or DMSO (see [6 or 7] for a full listing of solvent compatibility). It is best to avoid PEEK when these solvents are present. The convenience and flexibility of PEEK tubing make it a good choice for use when connections are changed regularly, such as between the autosampler and column, and column and detector. Whenever the mobile phase contains components corrosive to stainless steel, the use of PEEK tubing will help minimize problems. When connections are made once and then forgotten, such as between the pump and autosampler, stainless-steel tubing is a more trouble-free choice.

Tubing with an outside diameter of 1/16-in is used in most HPLC equipment; for some applications, 1/32-in o.d. tubing is preferred, but it is more fragile. Typical internal diameters for 1/16-in o.d. tubing are listed in Table 3.1. Stainless-steel tubing is readily available in sizes from 0.005-in to 0.046-in i.d., whereas PEEK covers the 0.0025-in to 0.040-in range. The 0.010-in and 0.020-in i.d. sizes are used to transport solvent from the pump to the autosampler, where extra volume

Table 3.1

Internal Diameters for Common High-Pressure Tubing (1/16-in o.d)

Internal Diameter (in)	Internal Diameter (mm)[a]	Volume (μL/cm)
0.0025[b]	0.06	0.03
0.005	0.13	0.13
0.007	0.18	0.25
0.010	0.25	0.51
0.020	0.50	2.03
0.030	0.75	4.56
0.040	1.00	8.11
0.046[c]	1.20	10.72

[a]Nominal value (calculated from inch dimensions).
[b]Available only in PEEK.
[c]Available only in stainless steel.

is not a concern, and these diameters are unlikely to become blocked. In the tubing where sample is present—between the autosampler and the column, and between the column and detector—smaller i.d. tubing (e.g., ≤0.007-in i.d.) is needed to avoid excessive peak broadening. Larger diameter tubing (≥0.030-in i.d.) is used primarily for construction of injector loops, because of the relatively large volume per unit length (Table 3.1).

Care should be taken to select the tubing length and diameter so as to minimize peak-broadening contributions. Table 3.2 shows the impact of various combinations of tubing length and diameter on peak broadening for several representative column configurations. It can be seen that although 0.007-in i.d. tubing is satisfactory for conventional columns of ≥100 × 4.6 mm with ≥3-μm particles, smaller columns require smaller diameter tubing. Applications that use sub-2-μm particles and/or long tubing runs (e.g., LC-MS) will require the use of 0.0025-in i.d. tubing. The smaller the tubing, the more prone it is to blockage from particulates that originate from the mobile phase, pump seal, valve-rotor wear, or sample. Consequently special care must be taken to avoid blockage when using tubing of ≤0.005-in i.d.. Remember that the tubing lengths shown in Table 3.2 are the total of the autosampler-to-column plus column-to-detector connections. Sometimes it is advantageous to use a short piece of 0.007-in i.d. tubing between the autosampler and column, so as to minimize blockage problems, and to use ≤0.005-in tubing between the column and detector, to minimize peak broadening. See Section 3.9 for an additional discussion of extra-column peak broadening.

Because it is difficult to duplicate the quality of factory-cut tubing, it is best to buy precut lengths of stainless-steel tubing. Precut tubing has the added advantage of having been thoroughly cleaned and passivated, so it can be used without further treatment. Bulk stainless-steel (type 316 is recommended) can be purchased for cutting to lengths that cannot be purchased precut. Stainless-steel tubing can be cut easily with a tubing cutter purchased from the tubing supplier. The tubing should be flushed with several milliliters of solvent prior to use (connect the up-stream end of

Table 3.2

Guide to Tubing Length

Column Characteristics				Maximum Length (cm) for 5% Increase in Bandwidth[a]		
L (mm)	d_c (mm)	d_p (μm)	N ($h \approx 3$)	0.0025-in	0.005-in	0.007-in
150	4.6	5.0	10,000	1450	90	25
150	2.1	5.0	10,000	300	20	*
100	4.6	3.0	11,100	580	35	10
100	2.1	3.0	11,100	120	*	*
100	2.1	1.8	18,500	70	*	*
100	1.0	1.8	18,500	15	*	*
50	4.6	3.0	5,500	290	20	*
50	2.1	3.0	5,500	60	*	*
50	1.0	3.0	5,500	15	*	*
50	4.6	1.8	9,200	170	10	*
50	2.1	1.8	9,200	35	*	*
50	1.0	1.8	9,200	*	*	*

$* \leq 10$ cm.
[a]Conditions for linear velocity = 2.5 mm/sec within the column (e.g., 2 mL/min for a 4.6-mm i.d. column), $k = 1$.

the tubing to the HPLC system and direct the outlet to waste), so as to remove any residual oils or particulate matter. PEEK tubing can be cut easily in the laboratory, so it usually is purchased in bulk. For best results, a PEEK tubing cutter is used to score a line around the tubing, then the tubing is flexed to snap it; this gives a higher quality tube end than cutting all the way through the tubing.

3.4.2 Fittings

3.4.2.1 Low-Pressure Fittings

These fittings are used to connect tubing when the pressure will not exceed ≈100 psi. The two most common fitting designs are shown in Figure 3.7. The flared-tubing fitting (e.g., Cheminert®, Fig. 3.7*a*) requires a special tool to flare the tube end. A washer is used between the nut and the flared end to help secure the tubing in the fitting port. These fittings require some skill to flare but are inexpensive and reliable, so they are popular with instrument manufacturers to reduce manufacturing costs. An alternate design uses a ferrule to secure the tube end (e.g., Fingertight®, Fig. 3.7*b*). The ferrule is reversed from the normal orientation in high-pressure fittings (as in Fig. 3.8*a*) so that the flat end contacts the bottom of the fitting port. The nut contains an internal taper that matches the ferrule so that the ferrule is swaged onto the tubing when the nut is tightened. This type of fitting is easy to assemble (the knurled nut is tightened with finger pressure), which has made it a popular alternative to the flared-tubing fitting. The industry standard is to use $\frac{1}{4}$-in nuts with 28 threads per inch (1/4-28); this way low-pressure fittings from different manufacturers are interchangeable.

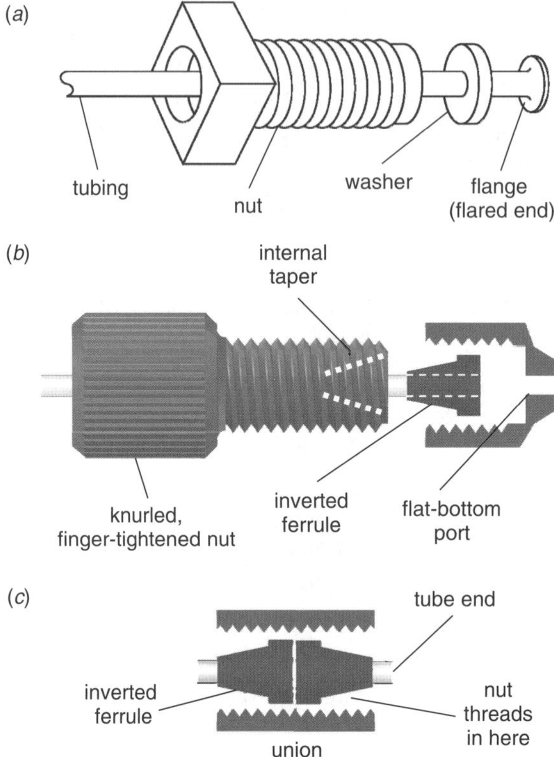

Figure 3.7 Fittings for low-pressure applications. (*a*) Cheminert® fitting, using a washer and flange for the seal; (*b*) Fingertight® fitting, using an inverted ferrule for the seal. (*c*) Low-pressure connection of two tube ends with inverted ferrule of (*b*) in a union (nuts not shown for clarity). (*a*) Courtesy of VICI Valco Instruments Co. Inc.; (*b*) courtesy of Upchurch Scientific, Inc., a unit of IDEX Corporation in the IDEX Health & Science Group.

Tubing connections for low-pressure fittings are made as shown in Figure 3.7*c*. Here two tube ends with inverted-ferrule fittings are connected in a union; the two ferrules butt against each other to make the connection (compare this to the high-pressure union of Fig. 3.8*b*,*c*). In other applications (e.g., a solvent-proportioning manifold, Section 3.5.2.2) a flat-bottomed fitting port is formed in the mating piece, as shown in the partially assembled fitting of Figure 3.7*b*.

Finger pressure is all that is needed to tighten low-pressure fittings, so it is recommended that all such fittings be finger tightened, even if the nut is designed for use with a wrench. When a wrench or pair of pliers is used to tighten the fitting, it is easy to overtighten the fitting, and distort it or damage the threads. For applications where the fitting might vibrate loose (e.g., on the solvent-proportioning manifold in low-pressure mixing systems), some manufacturers offer a lock nut to provide extra security for low-pressure fittings. Remember, in a low-pressure application sometimes a loose fitting will allow air to leak into the liquid stream without liquid leaking out, so a loose fitting does not necessarily create a puddle of mobile phase.

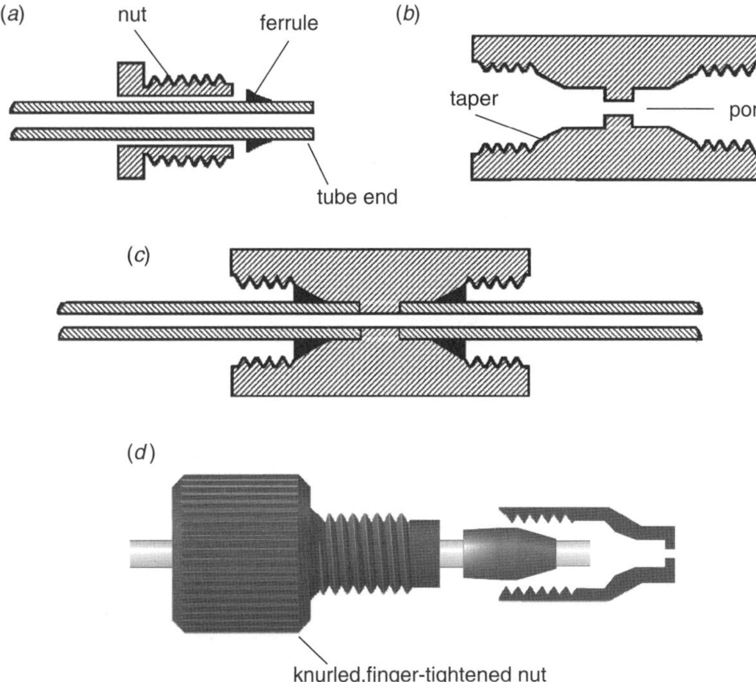

Figure 3.8 Compression fittings for high-pressure tubing connections. (*a*) Conventional stainless-steel nut, ferrule, and tube end; (*b*) union body; (*c*) properly assembled union. (*d*) Finger-tightened PEEK nut, ferrule, and fitting port. Courtesy of Upchurch Scientific, Inc., a unit of IDEX Corporation in the IDEX Health & Science Group.

3.4.2.2 High-Pressure Fittings

All high-pressure tubing connections are made with fittings that use a ferrule to secure the tube end (Fig. 3.8*a*) in the fitting port (Fig. 3.8*b*). The fitting body, whether it is a union, check valve, or other fitting, contains a threaded portion for the nut, a taper for the ferrule, and a cylindrical port where the tube end contacts the fitting. To assemble the fitting, the nut and ferrule are slipped over the tube end, the tube end is inserted into the fitting until it bottoms out in the port, and then the nut is tightened to secure the fitting. When the nut is tightened, compression between the nut and the taper in the fitting body swages (crimps) the ferrule onto the tubing and provides a secure connection. When properly assembled, high-pressure fittings should have no gaps and little or no diameter change between the tubing and the fitting body (Fig. 3.8*c*). Such connections are referred to as *zero-volume* or *zero-dead-volume* connections.

Nearly all manufacturers standardize on nuts with 10-32 threads and use the same ferrule taper for fittings used with 1/16-in o.d. tubing (different standards are used for metric sizes). This means that fittings from different manufacturers are nominally interchangeable. However, different manufacturers' fittings may have different port depths, which means that for stainless-steel fittings, where the ferrule is tightly swaged onto the tubing, the ferrule setback from the end of the tube (Fig. 3.9*a*) may vary between manufacturers. For example, Rheodyne injection

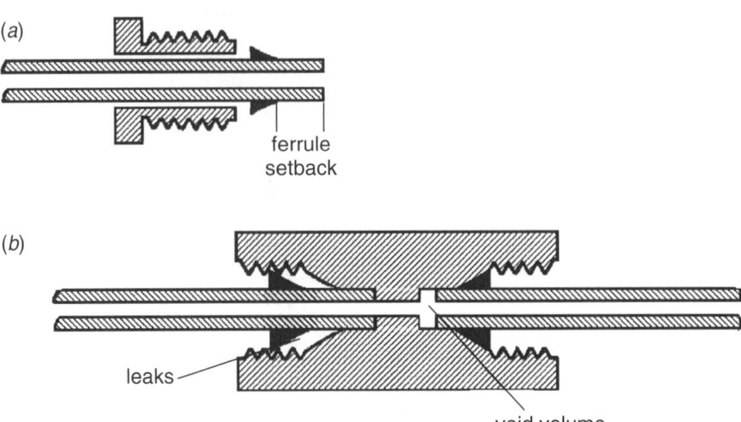

Figure 3.9 Effect of ferrule setback on compression fitting assembly. (*a*) Tube end showing ferrule setback; (*b*) assembly with (left) too large of a ferrule setback, causing leaks, or (right) too small of a ferrule setback, creating a void volume.

valves have noticeably deeper ports than most other fitting designs. Mismatching of the tube end and the fitting body *after* the fitting has been initially assembled can lead to problems, as illustrated in Figure 3.9*b*. For the example with the large ferrule setback, the tube end will reach the bottom of the fitting port before the ferrule contacts its mating taper (left-hand side of Fig. 3.9*b*). This will result in a fitting that leaks. Usually the ferrule can be forced to slide down the tubing as the nut is tightened; in this case the leak may stop and the fitting will be properly assembled. When the tube end is moved back to its original fitting body, however, the ferrule will contact the fitting taper before the tube end bottoms out in the fitting port (right-hand side of Fig. 3.9*b*). This will result in a leak-free connection that has a small void volume at the tube end, which can cause band spreading. To avoid these problems, it is a good idea to stay with a single manufacturer's fittings when stainless-steel fittings are used.

High-pressure fittings are made of PEEK as well as stainless steel. When the ferrule is made of PEEK (or some other polymer), it is not permanently swaged onto the tube end so that the ferrule can be slid easily into the proper position when a tube end is moved from one fitting to another. An added convenience that is popular in this fitting design is the use of a knurled PEEK nut (Fig. 3.8*d*) that can be finger-tightened instead of requiring a wrench. Finger-tightened PEEK fittings can be used for high-pressure fittings in conventional HPLC systems where pressures of 6000 psi are not exceeded. For higher pressure applications, stainless-steel fittings are required. If a PEEK fitting leaks, it is a good idea to turn off the pump, loosen the fitting, push the tubing to the bottom of the fitting port, and retighten the nut before turning the pump on again. This will reduce the risk of having the tubing slip during tightening, leaving a gap, as in Figure 3.9*b*.

Nuts, ferrules, and fitting bodies can be made of PEEK or stainless steel. It is common to use PEEK nuts and ferrules in stainless stainless-steel fitting bodies, but stainless-steel nuts are rarely used with PEEK parts.

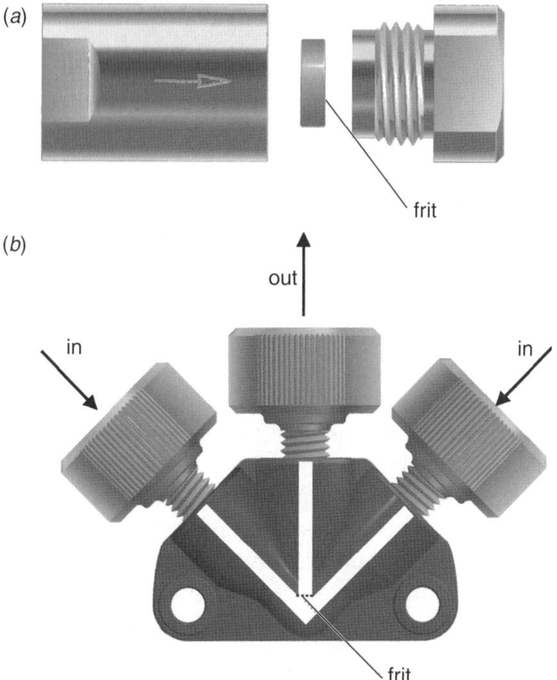

Figure 3.10 Specialty fittings. (*a*) In-line filter showing filter body, nut, and replaceable frit. (*b*) PEEK low-volume tee-mixer. Images courtesy of Upchurch Scientific, Inc., a unit of IDEX Corporation in the IDEX Health & Science Group.

3.4.2.3 Specialty Fittings

Two modifications of standard high-pressure fittings provide special benefits:

- in-line filters
- low-volume mixers

The *in-line filter* (Fig. 3.10*a*) is a modification of the standard union that is used to connect two pieces of tubing. A small-porosity frit (typically 0.5 μm porosity) is used in the in-line filter to remove unwanted debris from the fluid stream. It is a good practice to use an in-line filter just downstream from the autosampler on every HPLC system. This adds an insignificant amount of extra-column peak broadening, yet prevents particulate matter from the mobile phase, pump seals, valve rotors, or samples from reaching the column inlet frit. (Some pumps have built-in small-porosity filters at the pump outlet. These serve to trap particulate matter from pump seals or other upstream sources. In the absence of such filters, some users install an in-line filter between the pump and autosampler to prevent particulate matter from causing problems in the autosampler. The column frit can seldom be changed without damage to the column, whereas the in-line frit is designed for easy replacement. When the in-line filter becomes blocked, as signaled by an increase in system pressure, the frit is replaced and the HPLC system is back in service with minimal downtime. In-line filters are available from many vendors in either stainless-steel or PEEK construction, and with various frit porosities.

A *low-volume mixer* can be used to convert high-pressure-mixing pumps (Section 3.5.2) used for conventional applications to LC-MS-compatible pumps. Typical high-pressure mixers have volumes of 1 mL or more, whereas low-volume static mixers often contain <10 μL of volume. The example shown in Figure 3.10*b* is made of PEEK and contains a 10-μm porosity frit to aid mixing, yet has only 2.2 μL of swept volume. Other designs of low-volume mixers are also available.

3.5 PUMPING SYSTEMS

Nearly all HPLC pumping systems in service today use some variation of the reciprocating-piston pump. Hydraulic-amplifier pumps are used primarily for column packing and also may be used for preparative applications (Section 15.2). Syringe pumps are used as infusion pumps for tuning LC-MS systems, but not for high-pressure solvent delivery. Piston-diaphragm pumps that were once used for HPLC systems are no longer popular. Reciprocating-piston pumps have evolved over the years into reliable units that can provide hundreds of hours of trouble-free operation without maintenance. The precision and accuracy of the pump and its associated mobile-phase mixing system are keys to the success of HPLC as an analytical tool.

Most HPLC pumping systems sold for routine analytical work are designed to work at pressures of up to 6000 psi (400 bar), but most workers operate such systems in the 2000 to 3000 psi (150–200 bar) region. HPLC systems that are promoted for fast analyses or high-pressure work, especially with sub-2-μm particle columns, have higher upper-pressure limits and may allow routine operation in the 8000 to 15,000 psi (550–1000 bar) range or higher. It should be noted that conventional systems operated in the 5000 to 6000 psi region are able to provide some of the benefits of faster runs with smaller particle columns, without the need to purchase specialized equipment. However, when conventional HPLC systems are used at higher pressures (e.g., >3000 psi), care must be taken to prevent leaks. For example, injection-valve rotors may need to be tensioned for higher pressures and fittings may need to be tightened more. Also, the mechanical wear rate of pump seals, injection rotor-seals, and other moving parts usually increases as the system operating pressure is increased.

3.5.1 Reciprocating-Piston Pumps

The single-piston pump shown in Figure 3.11 is the core of all other HPLC pumping systems. The main components are:

- motor
- piston
- pump seal
- check valves

The rotation of the pump motor (driving cam) drives the piston back and forth in the pump head. The piston usually is made of sapphire, although some pumps

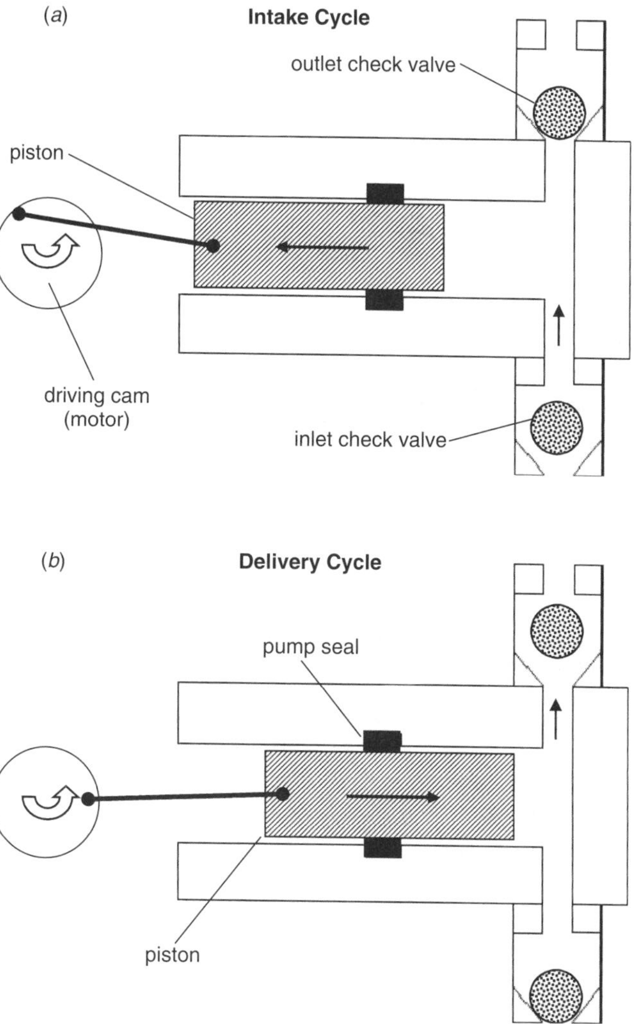

Figure 3.11 Single-piston reciprocating pump. (*a*) Intake (fill) cycle, (*b*) delivery cycle.

use ceramic pistons. A polymeric pump seal is used to prevent mobile phase from leaking out of the pump head. An inlet and outlet check valve control the direction of flow of the mobile phase. During the fill cycle (Fig. 3.11*a*), the piston is pulled out of the pump head, which creates a low-pressure region in the pump head. The outlet check valve closes and the inlet check valve opens, which allows the mobile phase to enter the pump. During the delivery cycle (Fig. 3.11*b*), the piston moves into the pump head, the inlet check valve closes, and the outlet check valve opens, which allows mobile phase to flow to the column.

A dependable check valve is important for reliable pump operation. Ball-type check valves are commonly used, as illustrated in Figure 3.12*a*. The valve comprises a ruby ball and sapphire seat. This combination of materials gives a reliable seal, usually with no assistance other than gravity plus the pressure differential in the

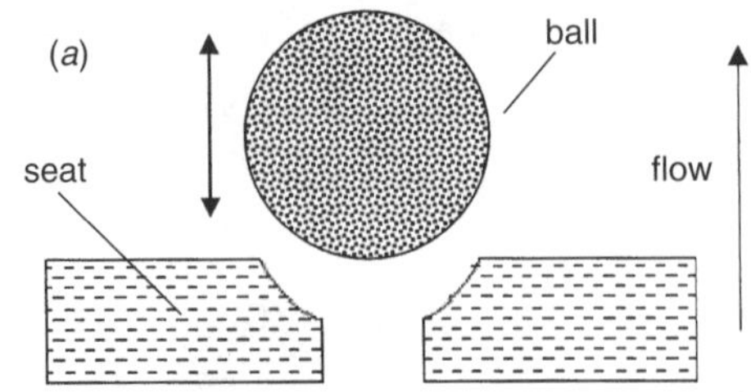

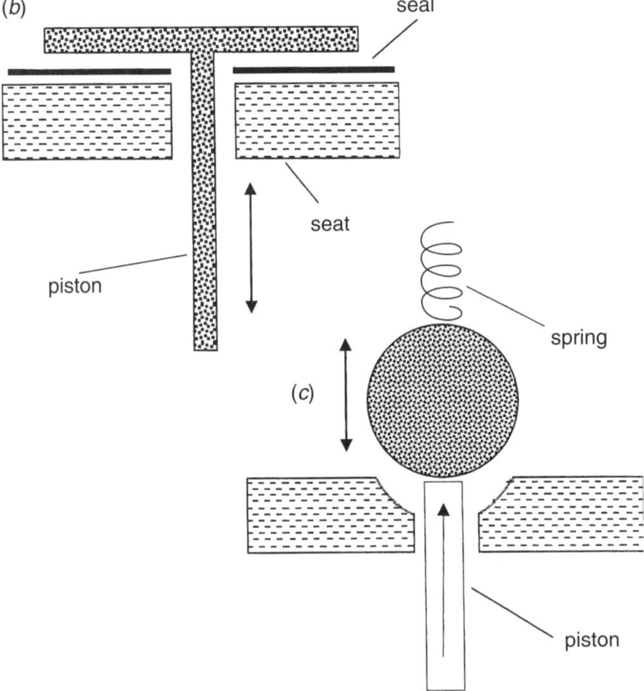

Figure 3.12 Check-valve designs. (*a*) Simple ball and seat; (*b*, *c*) active check valves.

pump head. Ceramic check valves may be used in some pumps, and sometimes a small spring is used to assist check-valve closure.

Operation of the ball-type check valve is quite simple. When the pressure on the inlet side of the valve (below the ball in Fig. 3.12*a*) is higher than that on the outlet side, the ball will be displaced from the seat and solvent will flow through the valve. When the pressure is higher on the outlet side (above the ball in Fig. 3.12*a*), the ball will be pushed against the seat, which creates a seal so that no liquid passes through the valve. As described above, and illustrated in Figure 3.11, the inlet and outlet check valves allow the pump head to alternately fill with mobile phase from the reservoir and deliver mobile phase to the column. An alternative to the ball-type check valve is the active check valve, shown in Figure 3.12*b* and described in Section 3.5.1.3.

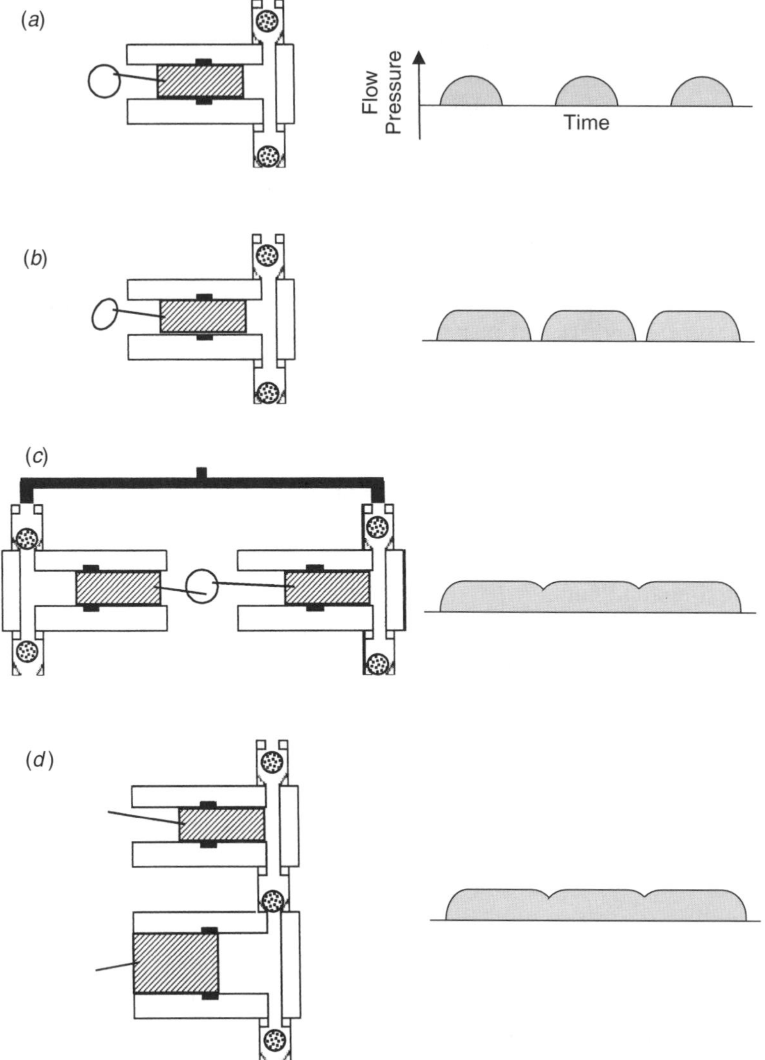

Figure 3.13 HPLC pump designs with corresponding flow and pressure profiles. (*a*) Single-piston reciprocating pump; (*b*) effect of a shaped driving cam; (*c*) dual-piston pump; (*d*) accumulator or tandem-piston design.

The single-piston pump with a constant-speed, linear drive will spend half of the time filling and half of the time delivering solvent. The resulting flow (and pressure) profile is shown in Figure 3.13*a*. These extreme pulses in flow and pressure are undesirable for HPLC applications.

An alternative single-piston design uses a specially shaped (e.g., elliptical) driving cam on the motor, or a variable-speed, stepper-driven motor, to move the piston. With this modification, the piston speed can be varied within each pump cycle so that more than half of the time is spent in the delivery cycle and a small fraction of the time is spent in the fill cycle (Fig. 3.13*b*). Some pumps refine this process further so that the pump controller remembers the pressure on the prior stroke and moves the piston forward rapidly at the beginning of the delivery stroke until the previous pressure is reached. Then the piston slows to a constant speed until the end of the delivery stroke. Even with variable-speed design and extensive pulse dampening, the single-piston pump does not produce a sufficiently stable flow and pressure for analytical HPLC. Two refinements of the single-piston pump are the dual-piston pump (Fig. 3.13*c*, Section 3.5.1.1) and the accumulator-piston pump (Fig. 3.13*d*, Section 3.5.1.2).

In addition to the pump components shown in Figure 3.11, most pumps have a purge valve on the outlet side of the pump that directs the pump output to waste during pump priming, solvent changeover, and bubble removal. Many pumps also have a fitting on the inlet side of the pump that allows for manually priming the pump with the aid of a syringe. In the past, mechanical pulse dampers (e.g., Bourdon tubes) were used to compensate for pump pulsations; today, the need for mechanical pulse damping is due to low-pulse pump designs (e.g., dual-piston or accumulator-piston pumps, Sections 3.5.1.1, 3.5.1.2). Pulse dampers also add unwanted dwell volume for gradient applications (Section 3.5.3).

3.5.1.1 Dual-Piston Pumps

One way to minimize the pulsations of the single-piston pump is to use two pump heads in parallel so that, when one is filling, the other is delivering solvent. This is illustrated in Figure 3.13*c* with opposing pistons driven off the same cam. Although most designs use two cams driven off the same motor, the pistons are mounted in parallel beside each other for operational convenience. With the use of a specially shaped driving-cam with the dual-piston pump, the pump output can be quite smooth, requiring little, if any additional pulse dampening. Dual-piston pumps are one of the two most widely used pump designs for analytical HPLC today.

3.5.1.2 Accumulator-Piston Pumps

An alternative design for the dual-piston pump is the accumulator-piston, or tandem-piston design shown in Figure 3.13*d*. In this case, one piston feeds into the other piston at twice the flow rate. For example, if a flow rate of 1 mL/min was desired, the top piston (Fig. 3.13*d*) would pump at 1 mL/min. While the top piston delivered solvent, the top (outlet) check valve would be open and the intermediate check valve (inlet for the top piston) would be closed. Meanwhile the lower piston would fill at a rate of 2 mL/min, with its inlet check valve open. Next, the lower piston would deliver solvent at 2 mL/min (while the upper piston filled), which would cause its inlet check valve to close and the intermediate check valve to open. Half of this 2 mL/min flow (1 mL/min) would be used to fill the top piston and half would be pumped directly to the column.

The accumulator-piston design, at least in theory, has several advantages over the dual-piston design. Flow never stops to the column, so flow and pressure pulsations should be minimized. Because the check valves can be the most problematic components in the entire HPLC system, the reduction of the number of check valves from four (dual-piston) to three (accumulator-piston) can reduce check valve problems. Furthermore, because solvent is always flowing to the column, the outlet check valve on the top pump is not necessary and it can be eliminated so that just two check valves remain. Although the simplicity of the accumulator-piston design makes it appear to be more reliable, many other factors go into the final pump performance (software, materials, assembly, mechanical tolerances, etc.). Most workers obtain comparable performance with either the dual-piston or accumulator-piston design.

3.5.1.3 Active Check Valve

A final refinement in design can be applied to both the dual-piston and accumulator-piston pump designs. The ball-type check valve (Fig. 3.12*a*) is susceptible to leakage if a tiny bit of debris is lodged between the ball and the seat. The surfaces are very hard (commonly sapphire and ruby), and even with the pressure of the pump pushing them closed, the ball-type valve can leak when particulates are present.

The active check valve is an alternative design that works well as an inlet check valve on the low-pressure side of the pump. As illustrated in Figure 3.12*b*, the active check valve depends on a polymeric seal and a mechanically driven piston to provide the sealing action. During the fill stroke, the piston is lifted off the seal and solvent flows into the pump. During the delivery stroke, the piston is pulled against the seal. The increased surface area relative to the ball-type valve and the soft seal allows the active check valve to seal effectively, even when a small amount of particulate matter is present. In an alternative design of the active check valve (Fig. 3.12*c*), a ball-type check valve is used with a strong spring to close the check valve; a piston below the ball pushes it from the seat to open the valve.

In the active check-valve system with dual-piston pumps (Fig. 3.13*c*), only two ball-type valves are used (the outlet check valves). With the accumulator-piston design, only one ball-type valve remains (the check valve between the upper and lower pump chambers in Fig. 3.13*d*; the outlet check valve is not needed). In both cases a reduction in the number of ball-type check valves improves pump reliability.

3.5.2 On-line Mixing

For isocratic methods the mobile phase can be hand-mixed; no additional mixing is then required within the HPLC system. On the other hand, most users take advantage of the convenience of on-line mixing and use the HPLC system to blend the mobile phase for isocratic methods, as well as for gradient methods where on-line mixing is required. Even when on-line mixing is available, some premixing of the mobile phase often gives quieter baselines, and in some cases, such as use with refractive index detection (Section 4.11), on-line mixing is unsuitable for good baselines at high sensitivity. On-line mixing takes place by one of three techniques:

- high-pressure mixing
- low-pressure mixing
- hybrid systems

3.5.2.1 High-Pressure Mixing

With high-pressure-mixing systems (Fig. 3.14), each solvent is delivered to the mixer by a dedicated pump. The ratio of solvents in the mobile phase is controlled by the

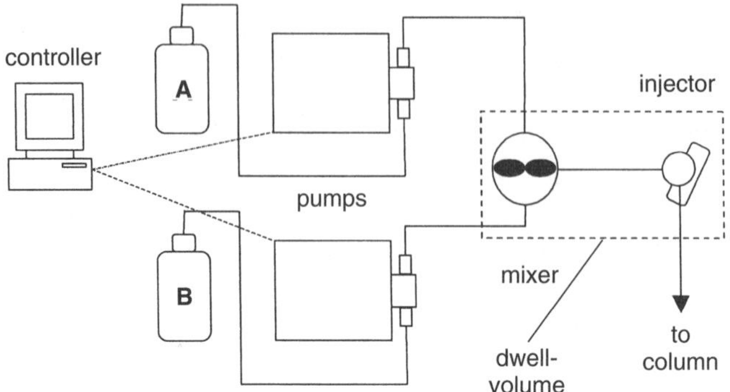

Figure 3.14 HPLC system with high-pressure mixing. Dwell-volume located inside dashed line.

relative flow rates of the pumps. For example, if 1 mL/min of a 60/40 MeOH/water mobile phase (60%B) is selected, the MeOH pump delivers 0.6 mL/min and the water pump delivers 0.4 mL/min. During gradient operation—where %B changes during the separation—the relative pump speeds change with the gradient program. Because the solvents are blended under high pressure, outgassing (Section 3.3.1) is less of a problem than with low-pressure mixing, so degassing problems usually are minimal. High-pressure-mixing systems are generally limited to the simultaneous use of two solvents, since another pump is required for each additional solvent. Some HPLC systems have a solvent-selector valve on one or both pumps that allows the nonsimultaneous use of two or more solvents by the same pump. This can be useful for method development (e.g., ACN and MeOH in separate runs) or for automated system flushing (e.g., flushing with water to remove buffer from the system). High-pressure mixing systems have an advantage over low-pressure mixing in that the standard mixer can be replaced with a micro-mixer (Section 3.4.2.3) for applications, such as LC-MS (Section 4.14), that require minimum dwell-volumes. Parts of the flow-stream that contribute to system dwell-volume are segregated within a dashed box in Figs. 3.14 and 3.15 (see also Section 3.5.3).

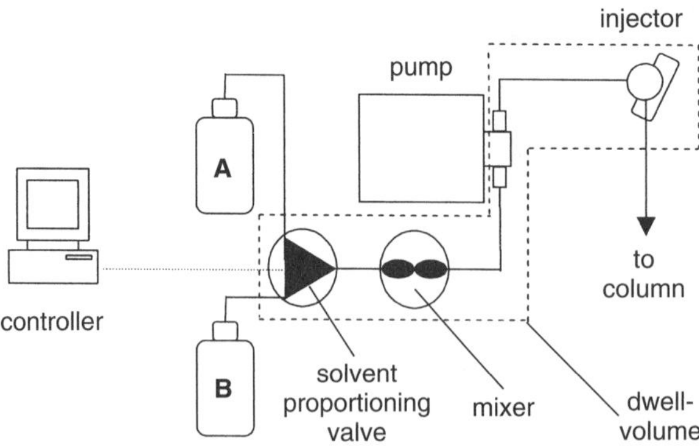

Figure 3.15 HPLC system with low-pressure mixing. Dwell-volume located inside dashed line.

3.5.2.2 Low-Pressure Mixing

With low-pressure-mixing systems (Fig. 3.15), the mobile phase components are blended before they reach the pump. Consequently these systems only require a single pump to deliver the mobile phase to the column. Solvent blending takes place in a proportioning manifold that usually has a capacity for blending up to four different solvents. The pump is operated at a constant flow rate, and the proportioning valve for each solvent is opened momentarily for a time (usually <1 sec) that is proportional to its mobile-phase concentration. Thus, for 1 mL/min of a 75/25 ACN/buffer mobile phase, the pump would deliver at a constant 1 mL/min, the ACN proportioning valve would be open 75% of the time, and the buffer valve 25%. Gradients are formed by a continuous variation of the proportioning-valve–open-time ratios. Because the solvents are mixed on the low-pressure side of the pump at atmospheric pressure, outgassing from the mixed solvent can generate bubbles that will cause pumping problems. This means that mobile-phase degassing is required for all low-pressure-mixing systems. It is also especially important, when low-pressure mixing is used, to make sure that the reservoir inlet-line frits and transfer tubing are not restricted. A restriction in one of the frits (most common with the aqueous phase) can reduce the proportion of that solvent delivered to the mixer and create solvent proportioning errors. To avoid this problem, it is wise to check the frits for free flow by use of the siphon test (Section 3.2.1) on a regular basis (e.g., monthly).

3.5.2.3 Hybrid Systems

Although high-pressure- and low-pressure-mixing systems are popular, at least two manufacturers (Thermo and Varian) produce a pumping system that relies on a hybrid of the two. In these hybrid systems the proportioning valves are mounted directly on the inlet to the pump (Fig. 3.16), with an active check valve (Section 3.5.1.3) used for each solvent. With solvents proportioned into the pump head one at a time in very small volumes, mixing takes place within the pump head under high pressure. This way outgassing problems are minimized (any potential bubbles resulting from mixing stay in solution under pressure), and additional mixer volume is not needed. By mounting the proportioning valves directly on the pump head, the extra dwell-volume normally associated with low-pressure mixing is eliminated. When combined with a small piston-volume (24 μL), one implementation of this pump (Accela from Thermo) lists 65 μL as the dwell-volume (Section 3.5.3), making it very attractive for use with gradient elution with small-peak-volume applications, such as fast HPLC and LC-MS.

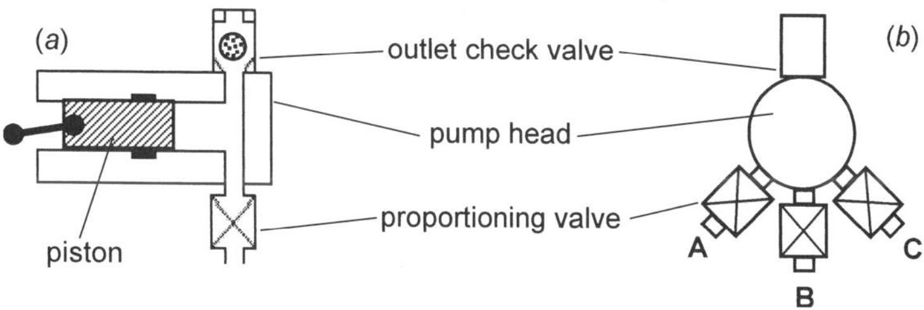

Figure 3.16 Hybrid mixing system with proportioning valves for solvents A, B, and C mounted on pump head. (*a*) Cross-sectional view; (*b*) frontal view.

3.5.3 Gradient Systems

When gradient elution is used (Chapter 9), the mobile-phase composition must be changed during the gradient, so on-line mixing (Section 3.5.2) is required. Both high-pressure- and low-pressure-mixing systems are used widely for gradient applications. The system dwell-volume is a major concern for gradient operations, one that is of little importance for isocratic separations. A difference in dwell-volume between two HPLC systems can have a dramatic effect on the resulting chromatograms (Section 9.2.2.4), and this is one of the primary reasons why gradient methods are hard to transfer (Section 9.3.8.2).

The dwell-volume comprises the system volume from the point at which the solvents are mixed until they reach the column inlet. It can be seen in Figure 3.14 that the primary contributions to dwell-volume in a high-pressure-mixing system are the mixer, the autosampler (injector), and the connecting tubing. For a low-pressure-mixing system (Figure 3.15) additional connecting tubing on the low-pressure side of the pump and the pump head volume are included in the dwell-volume. The dwell-volume should be measured (Section 3.10.1.2) for every gradient HPLC system. In some cases the system can be modified to reduce the dwell-volume, such as by replacement of the mixer in high-pressure-mixing systems (Section 3.4.2.3).

3.5.4 Special Applications

Conventional pumping systems that can generate gradients at flow rates of 0.1 to 10 mL/min and pressures up to 6000 psi are sufficient for the majority of HPLC applications. There are three application areas in which conventional HPLC systems may fall short of the analytical requirements:

- low flow
- high flow
- high pressure

These areas, particularly the first two, are sufficiently specialized to support books of their own, and are described only briefly here.

3.5.4.1 Low-Flow (Micro and Nano) Applications

Separations in the proteomics and other "omics" fields often operate on a scale that is an order of magnitude or more smaller than conventional HPLC separations (for reviews of these applications see [9–10]). Roughly classified by column internal diameter, such separations often are called *micro-LC* (100–1000 μm i.d.) and *nano-LC* (75–300 μm), with no clear distinction between the two. Such applications require flow rates that are well below the lower limit of ≈0.1 mL/min that is available from the pumps described earlier. Specially designed pumps are capable of delivering flow rates as small as 1 μL/min directly—or 50 nL/min with split flow at pressures to 6000 psi (and higher for some instruments designed for high-pressure use, Section 3.5.4.3). Obviously tubing, fittings, autosamplers, and detectors must be scaled accordingly, or extra-column effects will be unacceptable. Special capillary cells for UV detectors or direct introduction of the column effluent into an MS or MS/MS detector are common.

3.5.4.2 High-Flow (Prep) Applications

Preparative applications of HPLC use larger columns and require higher flow rates than conventional HPLC pumps can provide (Section 15.2). For semi-prep applications the flow rates of 10 mL/min that are available from most HPLC pumps may be sufficient. Several manufacturers offer modifications of their conventional pumps that increase the maximum flow rate from $\approx$10 mL/min to between 20 and 50 mL/min. Pressures often are below those encountered for analytical applications, so pressure limits generally are not a concern. Pumping systems that deliver 300 to 2000 mL/min at pressures up to 1800 psi are available. Since many preparative applications are isocratic and flow-rate control is not as critical, pneumatically amplified (constant pressure) pumps can be used for some high-flow applications. Because of the large volumes of solvent used, solvent recycling or recovery systems often are necessary with high-flow applications.

3.5.4.3 High-Pressure Applications

With the current availability of columns packed with sub-2-μm particles (Section 5.2.1.2), the pressure limits of conventional HPLC systems (typically $\leq$6000 psi) may restrict taking full advantage of these small particles (e.g., very fast separations; Sections 2.5.3.1, 9.3.9.2). Several manufacturers offer HPLC systems capable of operation at >6000 psi. High-pressure applications often emphasize high sample-throughput, so run time can be reduced by the use of higher flow rates and shorter columns (typically 50- or 100-mm long). To help increase sensitivity (peak height), small-diameter columns (e.g., 1.0- or 2.1-mm i.d.) are used as well. These separation characteristics reduce peak volumes, and often require modification of HPLC pumps, fittings, and other system components to minimize extra-column volume. Most high-pressure equipment is based on the same design as conventional HPLC equipment, but with added high-pressure and low-volume capabilities, which may limit the range of some system settings. For example, one system (Waters' Acuity UPLC) specifies a maximum flow rate of 2 mL/min at pressures <9000 psi, and 1 mL/min at higher pressures (up to 15,000 psi maximum); however, with small-diameter columns (e.g., 1–2.1-mm i.d.), these flow rates are adequate.

3.6 AUTOSAMPLERS

The introduction of the sample into the column requires that a measured quantity of sample must be added to the flowing, pressurized mobile phase. For open-column, stopped-flow, or some preparative separations, manual sample injection may be satisfactory. But for automated, unattended analysis, which often involves hundreds of samples per day, sample injection must be precise, accurate, and automatic. For such applications, an autosampler is used. Manual injection, popular in the past, is seldom used today except during operator training or in very low throughput environments. Descriptions of the sample-injection process and equipment are presented here in terms of autosamplers, but the same principles apply for manual injection.

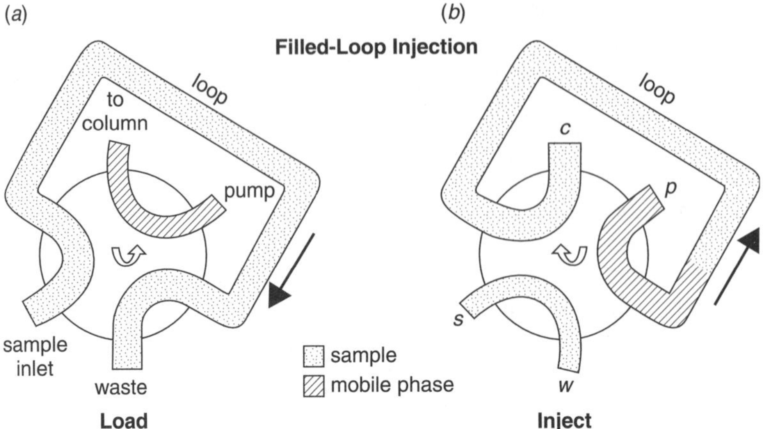

Figure 3.17 Six-port sample injection valve operated in filled-loop mode. (*a*) Load position; (*b*) inject position. Arrow shows direction of flow; *s*, sample inlet; *w*, to waste; *p*, mobile phase from pump; *c*, to column.

3.6.1 Six-Port Injection Valves

The sample-injection valve, originally designed for manual use, is the core component of an autosampler. Although other designs exist, the six-port, rotary valve is the most commonly used. A block of stainless steel comprises the valve body, as shown schematically in Figure 3.17; connections for the sample inlet (s), waste outlet (w), sample loop, pump (p), and column (c) are shown for two positions of the valve: load and inject. In the *load* position (Fig. 3.17*a*), the sample and waste ports are connected to opposite ends of the loop, and the pump is connected to the column. To inject, the rotor is moved to the *inject* position (3.17*b*), and the contents of the loop are swept onto the column; simultaneously, the sample inlet is connected to the waste outlet for flushing, if desired.

A polymeric rotor seal serves to connect three pairs of the connections (see also Fig. 17.3). Valve rotor-seals are usually made of a fluoropolymer or PEEK, and include additional materials to enhance structural integrity. The PEEK used in rotor seals is a different blend than that used for extruded PEEK tubing, so solvent compatibility with tetrahydrofuran and chlorinated solvents does not appear to be a problem [11].

3.6.1.1 Filled-Loop Injection

In the *filled-loop* injection mode, the volume of sample injected is controlled by the volume of the sample loop. For example, when a 20-μL sample loop is used, sample is introduced in the inject position until excess sample exits the waste port. When the valve is moved to the inject position, the 20-μL volume of sample trapped in the loop is pumped onto the column.

Filled-loop injection can be very precise and accurate *if* the sample loop is calibrated and overfilled. It is inconvenient to change sample loops, so if the injection volume must be changed regularly, filled-loop injection generally is not used.

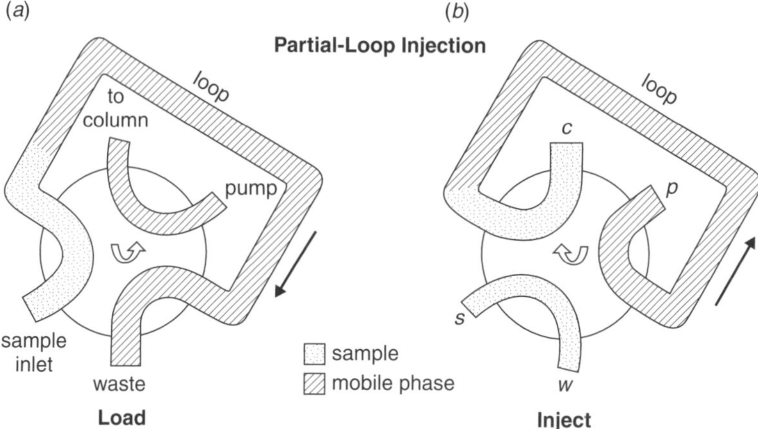

Figure 3.18 Six-port sample injection valve operated in partially filled loop mode. (*a*) Load position; (*b*) inject position. Arrow shows direction of flow; *s*, sample inlet; *w*, to waste; *p*, mobile phase from pump; *c*, to column.

3.6.1.2 Partial-Loop Injection

An alternative to the use of the injection valve in the filled-loop mode is to partially fill the injection loop, often referred to as *partial-loop* injection. Operation is identical to filled-loop injection, except that the loop volume is larger than the injection volume and a measured amount of sample must be placed in the loop. For example, a 100-μL loop might be mounted on the valve, and a 20-μL sample would be measured with a calibrated syringe and pushed into the loop in the load position (Fig. 3.18*a*). (Usually the remainder of the loop contains mobile phase.) The valve-rotor is then moved to the inject position (Fig. 3.18*b*), and the loop contents are pumped onto the column. Note that the plumbing connections are such that the flow direction through the sample loop is reversed in the load and injection positions. This helps ensure the integrity of the injected plug of sample—if the sample were to flow through a large volume of sample loop prior to entering the column, unwanted peak broadening would result.

Partial-loop injection can be precise and accurate *if* the sample aspiration and loop filling are precise and accurate, and if less than half of the loop volume is used. One potential problem related to injection accuracy is illustrated in Figure 3.19*a* [12]. In this case, a 20-μL loop was mounted on the injection valve and different volumes of sample were injected. When less than 10 μL or more than 40 μL were dispensed into the loop, the detector response accurately reflected the injected volume. However, in the region of 10 to 40 μL, the detector response was less than the expected amount. This problem is related to laminar flow (Fig. 3.19*b*) prior to and in the sample loop. When fluids pass through tubing, the molecules at the walls of the tubing are slowed due to friction, resulting in a bullet-shaped flow profile characteristic of laminar flow. The molecules at the center of the stream travel at approximately twice the velocity of those near the walls. Thus it can be seen that if 20 μL of sample is introduced into a 20-μL loop (volume defined by the dashed lines in Fig. 3.19*b*), some of the sample at the beginning of the injection plug will exit the loop to waste, whereas some of the sample at the end might not have entered the loop yet. The result is an injection volume smaller than intended. This

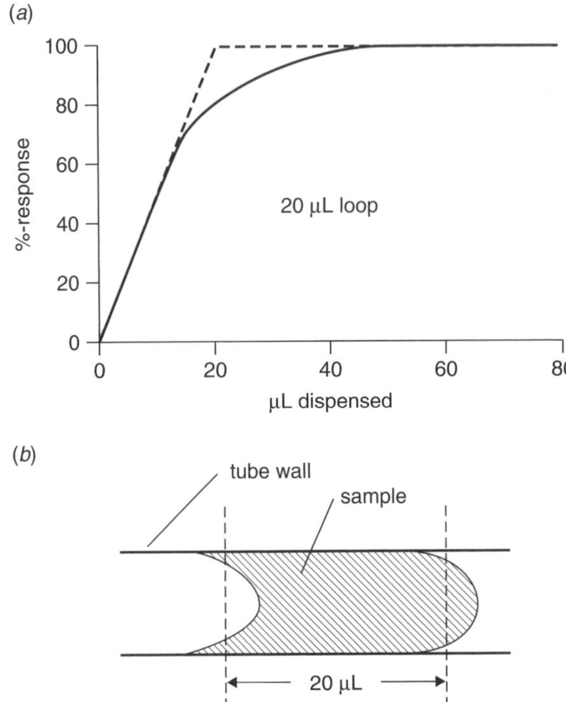

Figure 3.19 Effect of laminar flow on injection accuracy. (*a*) Comparison of detector response when loading different volumes into a 20 µL loop; (*b*) laminar flow profile of sample. Adapted from [12].

simplified description is further complicated by back-flushing and by mixing that takes place when changes in tube diameter or other flow disruptions are encountered. For manual injection, it is best to keep the injected volume constant and ≤50% or ≥200% of the loop volume for maximum accuracy.

3.6.2 Autosampler Designs

Autosamplers have largely replaced manual injectors—primarily for convenience, but autosamplers also provide levels of precision that may not be possible with manual injection. Many of today's autosamplers are capable of <0.5% RSD of peak area for injection volumes of ≥5 µL. Injection-volume *accuracy* may fall short of this level because of errors in calibration of the sample syringe or injection loop. Incomplete loop filling due to laminar flow (Section 3.6.1.2) can also introduce volumetric errors. Usually the precision of injection is more important than accuracy because of the compensating use of standards or calibrators. Autosamplers are best used in a constant-volume injection mode for each method, so the same exact (though not necessarily accurate) sample volume is injected for both calibration standards and samples. When this practice is followed, injection accuracy is less important, and the excellent precision will provide satisfactory analytical results.

Carryover is indicated by the presence of a small peak(s) in a blank chromatogram (no sample injection) that follows a separation where a sample was injected. Carryover results from part of a sample being retained in the system, after a separation is complete; it is especially a problem when the injection of a low-concentration sample follows that of a high-concentration sample (Section 17.2.5.10). Many autosamplers have features that provide for a wash of the needle and sample-contacting passages between injections (to avoid carryover). These range from static wash-vials to active-flush capabilities When a wash solvent is used, the composition is chosen to readily dissolve the sample and to be compatible with the mobile phase.

Samples generally are placed in individual sample vials or well-plates containing 96 or 384 sample wells. Vials most commonly are made of glass, sometimes specially treated to reduce adsorptive losses of sample. Sizes are 1 to 1.5-mL capacity for standard vials and 100 to 300 μL (or less) for microvials or inserts in standard vials. Vial closures use a cap and septum, usually made of silicon rubber and/or Teflon film. Sample plates generally are plastic with volumes of $\leq$1 mL per well. Plate closures are usually a press-on septum mat or iron-on metalized polymer film.

Sample access most commonly is via movement of the sample needle in *xyz* axies to the sample container. Some autosamplers use a rotating tray to bring the sample vials to the needle, while others pick up an individual vial and move it to the needle.

The cycle time is the amount of time it takes an autosampler to complete an injection from the time it is given the initial start signal. At a minimum, the cycle time includes the time it takes to pick up a sample and inject it onto the column. The addition of wash steps or other procedures can increase the cycle time. As long as the autosampler cycle time does not have a significant impact on sample throughput, it is not important. Thus a 1-min cycle time for a 30-minute gradient run is of little consequence, but the same cycle time would greatly reduce throughput for a 4-minute run. Cycle time that is $\leq$5% of the run time, generally is considered acceptable; as of this writing, few autosamplers have cycle times $\leq$15 sec with acceptable levels of carryover and injection precision. One way of reducing the negative impact on the autosampler cycle time is to use a "load-ahead" feature offered by some systems. In this implementation, the autosampler is programmed to perform its wash cycle(s) and pick up the sample while the previous sample is being eluted. As soon as the run is completed, the injection can be made, this reduces the effective cycle time to just a few seconds.

Three common autosampler designs are in common use:

- pull-to-fill
- push-to-fill
- needle-in-loop

3.6.2.1 Pull-to-Fill Autosamplers

The pull-to-fill autosampler design is illustrated in Figure 3.20. A syringe is mounted on a mechanical drive and connected to the injection valve as shown. A sample loop corresponding to the desired injection volume is mounted on the valve. The needle is mounted on a piece of connecting tubing attached to the valve. The needle may be moved to the sample vial, or the vial may be moved to a stationary needle. In the load position (Fig. 3.20*a*) the needle penetrates the septum on the vial and the syringe plunger is withdrawn to pull sample through the needle and connecting

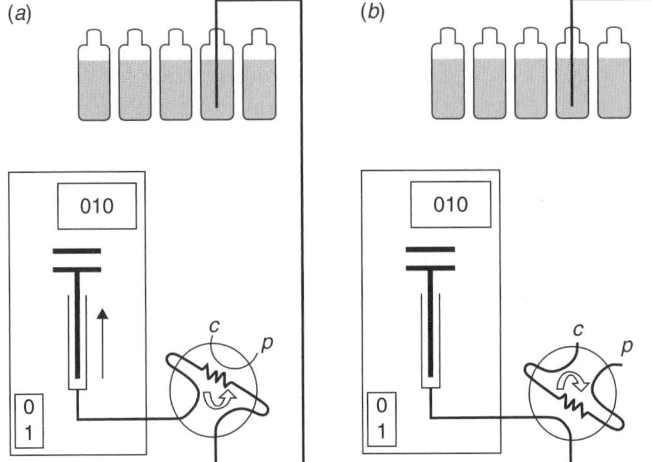

Figure 3.20 Pull-to-fill autosampler design. (*a*) Transfer of sample from sample vial to injector loop; (*b*) injection. *p*, flow from pump; *c*, flow to column.

tubing until excess sample exits the sample loop. The valve rotor then is moved to the inject position (Fig. 3.20*b*), and the sample is pumped onto the column. Note that no needle seal is used in this type of autosampler.

The pull-to-fill autosampler wastes sample because of the relatively large diameter of connecting tubing required to avoid blockage, and the need to flush excess sample through the loop (Section 3.6.1.1). It is therefore best used when the amount of sample is not limited, such as applications used to monitor production processes. The design is simple and reliable, and because it uses an overfilled, fixed-volume loop, it can have very good precision and accuracy.

3.6.2.2 Push-to-Fill Autosamplers

The push-to-fill autosampler is an automated version of the manual injector. In the load mode (Fig. 3.21*a*) the needle draws sample from the sample vial into a connecting tube attached to a mechanically operated syringe. The needle then is withdrawn from the sample vial and pushed into the low-pressure needle-seal in the injection port (Fig. 3.21*b*); sample is then dispensed into the sample loop. Next the valve rotor is moved to the inject position (Fig. 3.21*c*), and the sample is pumped onto the column.

The push-to-fill autosampler can be used in the filled-loop or partially filled loop injection mode (Sections 3.6.1.1 and 3.6.1.2). In the partially filled mode, the precision depends on the precision of the syringe controller. Because it is possible to inject nearly all of the sample, the push-to-fill autosampler does not waste much sample—perhaps 10 μL of sample is left in the needle and connecting passages. The push-to-fill autosampler design uses a low-pressure needle-seal that generally is trouble free. These autosamplers are very popular and are used for a wide range of applications.

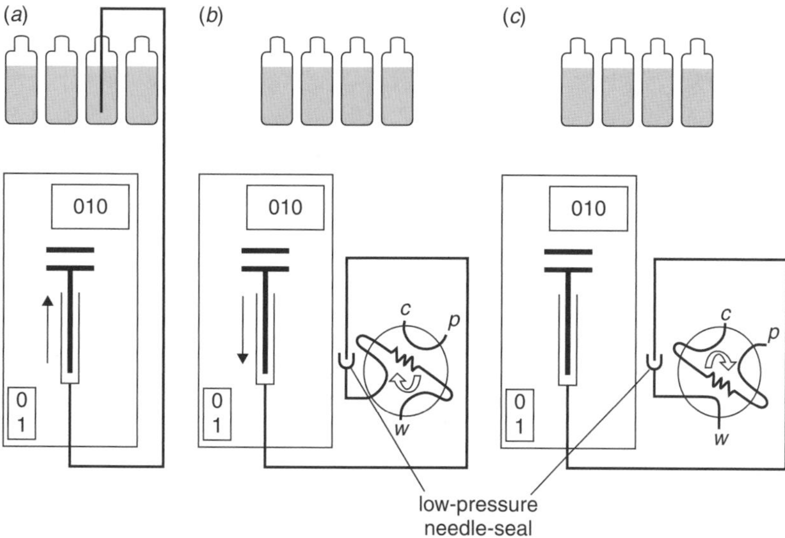

Figure 3.21 Push-to-fill autosampler design. (*a*) Transfer of sample from sample vial to injection needle and syringe; (*b*) filling of sample loop; (*c*) injection. *p*, flow from pump; *c*, flow to column; *w*, to waste.

3.6.2.3 Needle-in-Loop Autosamplers

The needle-in-loop autosampler uses a needle and loop that are one piece (needle-loop in Fig. 3.22*a*). In the load position, as shown in Figure 3.22*a*, the needle picks up the sample from the vial. The needle is then moved to a high-pressure needle-seal in the injection port, and the rotor is turned to inject the sample (Fig. 3.22*b*).

Because the tip of the needle is in the flow stream, the needle-in-loop autosampler injects all of the sample that is withdrawn from the sample vial, so there is no wasted sample with this injection technique. This makes the needle-in-loop autosampler a favorite for methods in which the sample volume is very small. These autosamplers typically use a 100-μL sample needle-loop, which will accommodate most analytical requirements. If a larger injection volume is needed, the needle-loop must be replaced with a larger one; this type of needle-loop is much more expensive than a conventional sample loop. This autosampler also depends on a high-pressure seal between the sample needle and the injection valve, which is a weak point with some implementations of this design. These autosamplers are a popular design because of the low sample waste and generally minimal carryover.

3.6.3 Sample-Size Effects

The amount of sample that is injected can influence the appearance of the chromatogram, not only in peak height or area but also in retention time and peak shape. Peak broadening can be influenced by the volume of sample injected, V_s, and the injection-solvent strength relative to the mobile phase, as discussed below. Sample-mass effects and overload are discussed in Section 2.6.2.

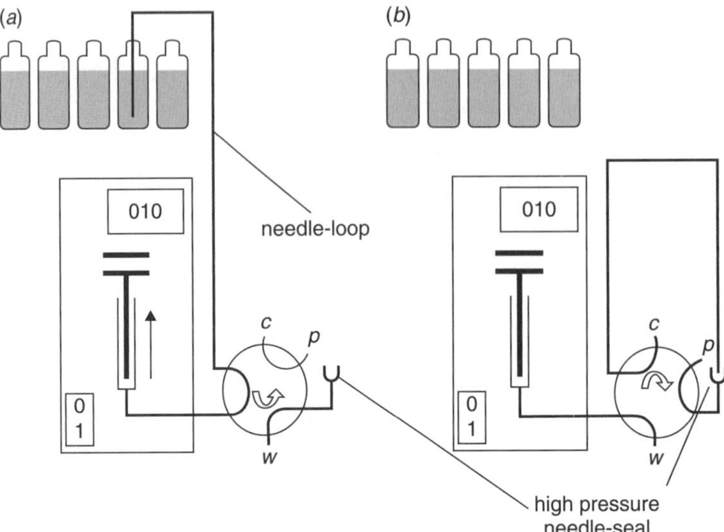

Figure 3.22 Needle-in-loop autosampler design. (*a*) Transfer of sample from sample vial to injector needle-loop; (*b*) injection. *p*, flow from pump; *c*, flow to column; *w*, to waste.

3.6.3.1 Injection Volume

The influence of the injection volume on the peak width was discussed in Section 2.6.1, and is summarized in Table 3.3 and Equation (2.27) as

$$V_p = \left(\left[\frac{4}{3}\right] V_s^{\,2} + V_{p0}^{\,2}\right)^{1/2} \qquad (2.27)$$

where V_p is the observed peak volume, V_{p0} is the volume of the peak due to broadening within the column, and V_s is the injection volume when the mobile phase is used as the injection solvent for isocratic separations. Equation (2.27) can be used to determine how large an injection volume can be made for a given increase in peak volume; if a 5% loss in resolution is acceptable, a 5% increase in peak width can be tolerated. The allowed injection volumes listed in Table 3.3 are calculated for a 5% increase in peak width for various columns and retention factors. (Note that the allowed injection volumes in Table 3.3 are smaller than those calculated by Eq. 2.27, because of the typical increase of V_s by ≈50% as the sample leaves the sample loop; see Section 2.6.1.) It is obvious that smaller injection volumes are required for smaller volume columns, columns that generate larger plate numbers, and/or early-eluted peaks—all of which result in narrower peaks. On the other hand, 4.6-mm i.d. columns generate fairly broad peaks, even when packed with sub-2-μm particles and used in short, 50-mm lengths. For example, a 50 × 4.6-mm, 1.8-μ*m* column gives V_{p0}≈30 μL for $k = 0.5$, with an allowable sample volume of $V_s \approx 5$ μL. Most autosamplers yield an imprecision of ≤0.5% RSD for injection volumes ≥5 μL.

Some autosamplers are able to maintain a similar precision for sample volumes of 1 to 2 μL or smaller. For columns ≤2.1-mm i.d. and packed with ≤3-μm particles, the autosampler must be capable of precisely injecting very small sample

Table 3.3

Allowed Injection Volumes for 5% Band Broadening

Column Characteristics				Peak Volume (Injection Volume)[a] (μL)		
L (mm)	d_c (mm)	d_p (μm)	N ($h \approx 3$)	$k = 0.5$	$k = 2$	$k = 20$
150	4.6	5.0	10,000	90[b] (10)[c]	180 (20)	1,270 (130)
150	2.1	5.0	10,000	20 (2)	40 (4)	265 (25)
100	4.6	3.0	11,100	60 (6)	115 (15)	800 (90)
100	2.1	3.0	11,100	12 (2)	25 (15)	170 (20)
100	2.1	1.8	18,500	9 (1)	20 (2)	130 (15)
100	1.0	1.8	18,500	2 (<1)	5 (<1)	30 (3)
50	4.6	3.0	5,500	40 (4)	80 (10)	565 (60)
50	2.1	3.0	5,500	8 (1)	20 (2)	120 (80)
50	1.0	3.0	5,500	2 (<1)	5 (<1)	25 (3)
50	4.6	1.8	9,200	30 (4)	65 (7)	440 (50)
50	2.1	1.8	9,200	7 (1)	15 (2)	90 (10)
50	1.0	1.8	9,200	2 (<1)	3 (<1)	20 (2)

[a]Equation (2.27), $V_S/V_{po} = 0.16$; note that the latter (theoretical) value of the injection volume V_s has been divided by 1.5, to take into account the spreading of the sample plug as it leaves the loop.
[b]Peak volume (μL)
[c]Injection volume (μL)

volumes. In all cases for isocratic separation, the peak width increases with the retention time, so longer retained peaks can tolerate larger injection volumes. Modification of a method so as to increase retention is one (seldom used) approach for minimizing extra-column peak broadening, because the resulting reduction in peak height and increase in run time is usually a poor trade-off.

3.6.3.2 Injection Solvent

As mentioned in Section 2.6.1, when the injection solvent is not matched to the mobile phase, Equation (2.27) no longer holds. If the injection solvent is sufficiently weaker than the mobile phase, the sample will be concentrated at the head of the column. Based on a change in k of about 2.5-fold for a change in the mobile phase of 10% B (Section 6.2.1), an injection solvent 10% B weaker than the mobile phase should significantly retard the sample as it enters the column; larger solvent-strength differences will be even more effective. When large-volume sample injections are desired, dilution of the sample with water may allow the injection of a larger sample weight (Section 2.6.1). The use of an injection solvent stronger than the mobile phase will adversely affect early-eluted peaks more than more strongly retained ones (Section 17.4.5.3). Also it is important to match the injection solvent for standards and samples.

Injection in solvents stronger than the mobile phase tends to "wash" the sample down the column until it becomes fully diluted in the mobile phase. As with the use of dilute injection solvents, the observed effects are a function of both the injection volume and the difference in solvent strength between the injection solvent and the mobile phase. With 150 × 4.6-mm columns, injection volumes of ≤10 μL in a strong solvent (e.g., 100% B) generally can be tolerated. When larger injection volumes and/or smaller volume columns are used, it is wise to compare retention and peak shape with a small-volume injection of the sample dissolved in mobile phase, in order to see if the results are acceptable. For mobile phases of <50% B, diluting a sample dissolved in 100% B ratio 1:1 with water or buffer will often allow a larger volume and weight of injected sample without adverse consequences.

3.6.4 Other Valve Applications

Automated injection valves are most widely used in autosamplers, but the same valves also are used for other applications. These include:

- column switching
- fraction collection
- waste diversion

3.6.4.1 Column Switching

High-pressure switching valves are available in many configurations other than the simple six-port valve illustrated in Figure 3.17. These may be purchased as motorized valves with switching controlled through the external-events outputs of the HPLC system controller. Two general configurations are popular: two-position valves, such as shown in Figure 3.17, and multi-position valves, which allow a single input tube to be connected to one of many output tubes (see later the discussion of Fig. 3.25). Three applications are discussed here, and another in Section 2.7.6; many additional applications are available on the valve manufacturer's websites (e.g., [6, 13]).

One popular application of the two-position valve, *sample enrichment*, is shown in Figure 3.23. The objective is to concentrate a dilute sample and then inject the concentrated fraction onto the analytical column. An example of this is the concentration of a nonpolar analyte from a water sample (e.g., an environmental monitoring application). In the enrichment phase the valve is set as shown in Figure 3.23a, where the first pump pushes a dilute sample through an enrichment column, while the previous sample is separated on the analytical column, using a second pump. In this example the aqueous sample might be passed through a C_{18} column in a weak mobile phase to trap the nonpolar materials. Once the entire sample is concentrated on the enrichment column, the valve is switched (Fig. 3.23*b*) and the sample is back-flushed onto the analytical column. Because of the reversed direction of flow and the sudden increase in mobile-phase strength, the sample is released from the enrichment column onto the analytical column in a narrow band for analysis. Other applications of column switching as in Figure 3.23 are discussed in Section 16.9.

Figure 3.24 shows a valving configuration that allows *regeneration* of one column while a second column is eluting the sample to the detector. This can increase throughput and be advantageous for gradient applications, while at the same time increasing the utilization rate of an expensive mass spectrometric detector, such as for the analysis of drugs in plasma samples. In the configuration shown in Figure 3.24*a*, the sample is injected and analyzed on column 2 in the normal manner using gradient elution. Meanwhile column 1 is regenerated by the mobile phase delivered by a second pumping system. As soon as the sample is

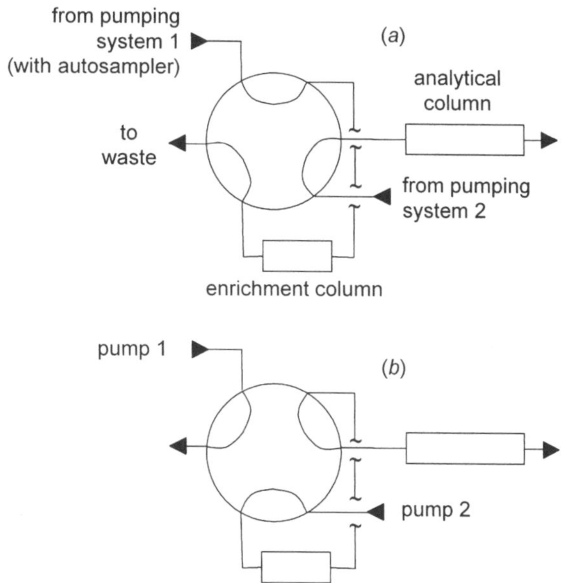

Figure 3.23 Column switching for sample enrichment. (*a*) Valve set for loading sample enrichment column; (*b*) back-flushing enrichment column to analytical column.

eluted from column 2 to the detector, the valves are switched to the configuration of Figure 3.24*b*, and a new sample is injected onto column 1 while column 2 is regenerated. This application increases throughput by the elimination of the time normally spent waiting for the column to be re-equilibrated to the starting conditions after a gradient run. However, it should be noted that other means exist to minimize the time required for column equilibration after gradient elution (Section 9.3.7).

In Figure 3.25, a multi-position valve is used to *select* from one of three columns. With this setup, three separate columns can be evaluated automatically, for use in method development. One application of this technique involves a setup similar to Figure 3.25, but with as many as 32 different chiral columns installed on two 32-port valves (see Section 14.6.1). In a Gatling-gun approach, a sample is sequentially injected on each column in an unattended series of runs. The chromatograms are later inspected to determine which column provided the best separation.

3.6.4.2 Fraction Collectors

Preparative chromatography (Section 15.2.4) requires fraction collection for either (1) the retrieval of individual peaks from a chromatogram or (2) the collection of fractions from an overlapped peak (as part of the purification of some compound). A fraction collector, which is commonly used for this purpose, resembles an autosampler that is used in a reversed mode. A single sample stream from the column is distributed into multiple vials through use of a mechanism that moves the outlet tube to the desired vial. In the simplest implementation a fraction collector is operated on a timed-collection basis. At some selected time after injection, collection starts and the sample is collected for a fixed time in each collection tube. For

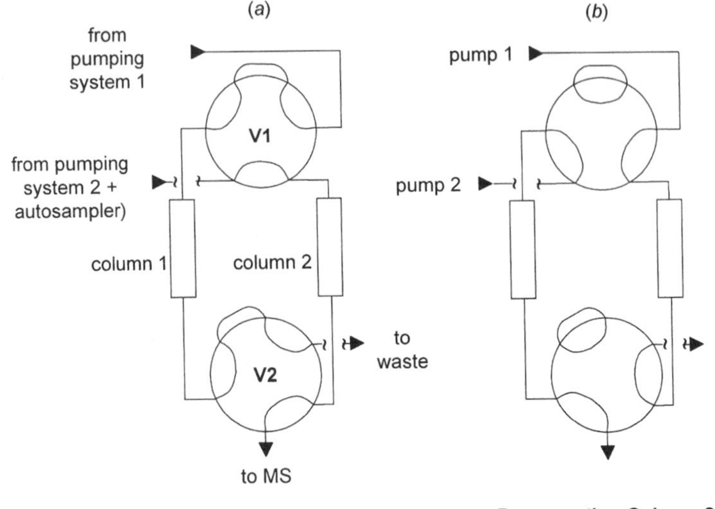

Figure 3.24 Column switching for column regeneration. (*a*) Sample is injected on column 2 and directed to detector (MS) while column 1 is regenerated by pump 1; (*b*) sample is separated on column 1 while column 2 is regenerated. V1, V2, six-port switching valves.

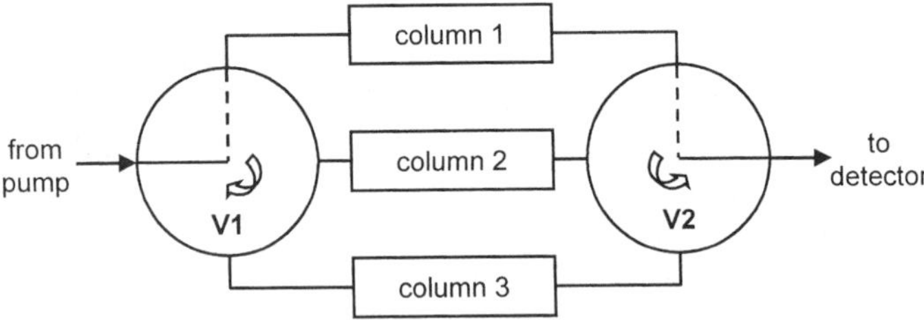

Figure 3.25 Use of multi-port switching valves for selection of one of several columns. As shown valve, V1, directs flow from pump, to column-1, through valve V2, to detector. Connecting passage (- - -) moves to desired column under software control.

example, if the peak of interest was eluted at 7 minutes, the fraction collector might be programmed to start collecting 20-second fractions starting at 6 minutes and ending at 8 minutes. This way several "cuts" across the peak would be collected. Another popular implementation of the fraction collector is to use an electronic circuit to monitor the detector output so that fractions are collected only when a peak is eluted. A delay coil of tubing can be mounted between the detector and the fraction collector to allow for peak detection just prior to the capture of a fraction.

3.6.4.3 Waste Diversion

The possible applications of switching valves in HPLC are practically limitless. Two additional uses of switching valves are discussed here: the diversion of a waste stream to protect the detector, and the recycle of mobile phase for isocratic applications.

One method of sample cleanup for the analysis of drugs in plasma is plasma precipitation (Table 16.11). Although quick and inexpensive, plasma precipitation often leaves a significant nonvolatile protein burden in the sample, which can accumulate in the interface of an MS detector. One way to avoid this is to use a switching valve to divert to waste the portion of the column effluent that contains most of the protein—usually the early portion of the chromatogram near the column dead-volume. When actuated by the system controller as part of the method program, automated waste diversion will minimize interface contamination due to nonvolatile materials in the injected sample.

Although the cost of mobile-phase solvents is not a large fraction of the total expense of sample analysis, it can be significant when the costs of disposal are considered. To reduce this expense, as well as for environmental reasons, some users attempt to reduce the volume of solvents used. One approach is to reuse the mobile phase. The simplest procedure is to direct the waste stream back to the reservoir (mounted on a stir-plate). As the waste stream is mixed with the remaining mobile phase, impurities are diluted and pumped back into the column at a steady state, so no interfering peaks will appear. Over time, however, the contaminant concentration in the reservoir will increase. A simple way to minimize this is to recycle only the portion of mobile phase that does not contain sample peaks. There are several commercial units (e.g., Axxiom's SolventTrak) that include a switching valve and a sensor that monitors the detector output. When a peak is detected, the valve switches the detector effluent to waste; when no peaks are present, the valve directs the effluent to the mobile-phase reservoir. Thus only the "clean" mobile phase is recycled. A quantitative evaluation of mobile-phase recycling can be found in [14].

An alternative way to reduce solvent consumption is to decrease the column internal diameter. Solvent consumption is proportional to the cross-sectional area of the column, so replacement of a 4.6-mm i.d. column by a 2.1-mm i.d. column will reduce solvent usage by $(4.6)^2/(2.1)^2 \approx 5$-fold (the flow rate should be simultaneously reduced by the same amount to maintain a constant linear velocity). The greater importance of extra-column peak broadening for small-diameter columns should be kept in mind, however.

3.7 COLUMN OVENS

It has long been known that column temperature plays an important role in HPLC retention and selectivity. For additional information, see Section 6.3.3 and [15–17] (and associated references).

3.7.1 Temperature-Control Requirements

A rule of thumb for reversed-phase isocratic separation is that a 1°C increase in column temperature will decrease values of k by about 2%. Temperature can also affect chromatographic selectivity (Sections 6.3.3, 7.3.2.2, 8.33, [15]), so close control of column temperature can be important—especially for separations with marginal resolution (e.g., $R_s \leq 2$). If the mobile phase entering the column is not preheated to the temperature of the column, distorted peaks can result. To avoid peak distortion, the temperature of the mobile phase as it enters the column should be within ±6°C of the column temperature.

The influence of temperature on chromatographic separation suggests that in most cases some means of column temperature control is necessary. Even though the laboratory as a whole may have adequate climate control, an individual location in the laboratory may experience temperature fluctuations of several degrees, as the heating, ventilating, and air-conditioning (HVAC) system cycles. The use of a column oven on every HPLC system is strongly recommend to maintain constant temperature. If the column oven does not preheat the mobile phase (generally required only for methods operating at >40°C), a simple preheater can be fabricated from 0.5 m of 0.005-in i.d. stainless-steel tubing. This tubing should be connected to the column inlet and mounted inside the column oven, preferably in contact with a heated surface.

3.7.2 Oven Designs

Three column-oven designs are popular:

- block heater
- air bath
- Peltier heater

Water-bath heaters are a simple alternative, but these are inconvenient and seldom used today. If, for some reason, a column oven is not used, temperature fluctuations should be minimized by wrapping the column in an insulating material (e.g., foam pipe-wrap).

3.7.2.1 Block Heater

The block heater relies on direct contact of the column with a heat source. Commonly the heat is transferred from a grooved aluminum block into which the column is clamped, with heat provided by a cartridge-type heater. The column and heating block are contained in an insulated compartment. In another format, a flexible blanket of heating tape is wrapped around the column. The direct contact of the column with the heater provides efficient heating [12]. If additional solvent preheating is necessary, a preheater can be used as described above.

3.7.2.2 Air Bath

An air-bath heater is constructed the same way as a gas chromatography oven; this design is effective at controlling the column temperature. Air is a less efficient conductor of heat than metal, so column-temperature equilibration may take longer than with a block heater

3.7.2.3 Peltier Heater

Peltier heaters are popular for HPLC work. In addition to heating, they can control the column oven at ambient temperatures or below—without an auxiliary cooling mechanism. However, most Peltier-heated column ovens are not efficient heat conductors because the column seldom is in intimate contact with the heated surface. Most Peltier oven designs include a preheater, which is a piece of capillary tubing embedded in the aluminum heater block of the oven. Often this preheater provides most of the heating for the column [15]; if a preheater is not used with a Peltier oven, it is unlikely that the set temperature will match the true column temperature. A well-designed Peltier oven will include a grooved block that clamps the column to the heated surface (or some other mechanism to ensure column contact) and an embedded preheater enclosed in an insulated compartment.

3.8 DATA SYSTEMS

In the second edition of this book, a total of three paragraphs were devoted to chromatographic data systems. Perhaps no other factor in HPLC practice has been impacted more in the interceding years than data handling. In 1979, dedicated data integrators were available from a few suppliers, but the personal computer (PC) and its ripple-down effect of software to support HPLC were not even on the horizon. At one point HPLC systems comprised separate modules, each controlled by manual settings. Gradually control settings changed from switches and rheostats to microprocessors, and today's modules each contain many microprocessors. These developments allowed the operational control of all the HPLC hardware functions (flow rate, detector wavelength, autosampler control, etc.) by a dedicated *system controller*. At the same time the system controller directed the operation of the HPLC, and chromatographic data were collected, processed, and displayed by a separate *data system* (or data processor). Today, system control and data processing have merged to the point that the device (and its associated software) performs all these functions is simply called a data system; we will use this terminology unless there is a need for clarification.

Today's chromatographic data systems serve many functions In the HPLC operation. Most or all of the following capabilities are available in today's data systems:

- experimental aids (Section 3.8.1)
- system control (Section 3.8.2)
- data collection (Section 3.8.3)
- data processing (Section 3.8.4)
- report generation (Section 3.8.5)
- regulatory functions (Section 3.8.6)

In general, to access all its control capabilities, a data system must be used with HPLC equipment from the same manufacturer as the data system. However, several data-system manufacturers offer control capabilities for instrumentation from the major HPLC equipment suppliers. Data collection and processing are functions that are more universal—many laboratories standardize on one brand of data collection and processing system, and use a central system to collect and process data from a variety of brands and types of laboratory equipment.

3.8.1 Experimental Aids

During method development, a systematic approach will reduce the amount of work involved and increase the likelihood of obtaining an acceptable separation.

Chapter 10 discusses the use of computer simulation as a tool to guide the method development process. Software for method development is available as standalone software that can be used with any HPLC system (e.g., DryLab®), or the software may be specifically designed for use with one brand and model of equipment.

When a computer is connected to the HPLC system, other useful software may be available for tasks that may not be directly related to the separation. Wizards, help files, and other information can assist the user in isolating and solving system problems quickly. Electronic laboratory notebooks and databases can simplify recordkeeping and system maintenance records. Some systems have reminders that can be set to indicate when pump-seal changes or other maintenance should be undertaken. Several manufacturers supply audiovideo files that can be used to guide certain maintenance tasks.

The Laboratory Information Management System (LIMS) is another category of software that allows users to track the flow of samples and sample results through the laboratory. Some of these functions will help with regulatory compliance (Section 3.8.6 and Chapter 12), some are used to help coordinate work activities, and some may be used to support the financial aspects of the laboratory business (e.g., initiate the billing process when sample analysis is complete). For example, consider the application of a LIMS system in a bioanalytical service laboratory that analyzes drugs in plasma. When the samples arrive at the lab, the samples are assigned a number, and information for the sample is added to a sample record in a database. For example, the date and time of receipt, sample condition, sample tracking number, and patient identification (ID) might be entered. The sample tube would have a bar-code label added (if it did not already have one) and would be transferred to a freezer (freezer location, date, and time logged) for holding until analysis. When it is time to analyze the sample, a sample-analysis table is created from the database by the analyst, and samples are pulled from the freezer and moved to the sample preparation lab (date and time recorded). After the sample preparation takes place, the bar code from the original sample container is correlated to the bar code on the sample vial for injection.

Additional data might be added to the database during sample preparation, such as the lot numbers of various reagents. Any remaining raw sample would be returned to the freezer (date, time, and location recorded) A sample-analysis sequence table is created that correlates the autosampler-tray position with the sample ID. After analysis, the data are processed to generate a report of sample concentration for each sample; these data would be automatically entered into data tables in a report template. The analyst would review the data and transfer the report to the quality unit for further review before he sends it to the client (each approval or review would be recorded).

When the report was ready to send, another report would be sent to the accounting department to bill the client for the analyzed samples. At any future point, customized reports could be created for special purposes, such as chain of custody, identification of problems (e.g., correlation of a batch of reagents with aberrant results), number of samples run per instrument per month, and so forth. The capabilities of LIMS systems are often customized for a particular laboratory or application, so the possibilities are practically limitless. In some cases the boundary between a LIMS and a data system gets blurred; LIMS software can duplicate, replace, or take advantage of separate data-system software. For more information, consult one of the LIMS vendors; a Google search of "LIMS vendors" identified one listing of >200 companies that supply LIMS and related software or hardware.

3.8.2 System Control

One of the most widespread uses of the computer in support of HPLC is to control the system. The system-control function provides a single point of control for all operational settings; for example, flow rates, mobile phase composition, column temperature, and detector wavelength. All the settings for a specific method can be stored for easy retrieval and setup when the method is next used, as well as for archival purposes, or for transfer to another HPLC system. Method modification and the development of new methods are simplified, because an existing method can be used as a template, so only the necessary items have to be changed. Most controller software contains features that make it easy to make a permanent electronic record of the specific instrument settings used for each sample run within a laboratory, information that may help when troubleshooting system problems arise and may simplify regulatory audits.

3.8.3 Data Collection

HPLC detectors convert a sample peak into a stream of x- (time) and y- (intensity) data, and sometimes an additional z-variable (e.g., wavelength). These data are generated in an analog or digital format that is then sent to the data system for collection and recording. Most data systems can accept either analog or digital signal inputs and most detectors have both analog and digital outputs. For systems in which the detector and data system are from the same manufacturer, the connection may be as simple as a fiber-optic cable. When the data system and detector are not from the same manufacturer, generally an electrical connection of some kind is required.

Besides the detector output, the data system may record other system settings, such as the pump flow rate, detector wavelength, system pressure, and other settings. In its most sophisticated form the data system can reconstruct the exact conditions used to analyze each sample, including the column and equipment serial numbers, mobile-phase batch numbers, and all of the chromatographic results.

The sampling rate (data rate) for the data system must be sufficiently high so that the collected data accurately represent the peaks; if the data rate is too fast, excessive noise will be collected. A good compromise is to collect $\approx$20 data points across a peak. For narrow peaks, such as those generated by short, small-particle columns, the data system must be capable of a high enough sampling rate to gather 20 data points across a peak. For example, in Table 3.3, the 50 $\times$ 4.6-mm column packed with 1.8-μm particles generates a 30-μL peak for $k = 0.5$. When operated at 1-mL/min, this means that the peak is $\approx$2-sec wide, so a data rate of $\approx$10 Hz is required. However, when operated in the high-throughput mode, the flow rate may be much higher. At 5 mL/min, the same peak would be just 0.4-sec wide, requiring a data rate of $\approx$50 Hz for adequate sampling. Larger retention times and lower flow rates will generate broader peaks, so the sampling rate does not have to be as fast. Many data systems will adjust the sampling rate during a run so that approximately the same number of data points is collected across each peak. If in doubt, *it is always better to collect data at too high a data rate*, since data averaging (bunching) to simplify the data set (and reduce noise) can be done during data processing; data processing will be unable to create additional data points when the original sampling rate is too slow. See Section 11.2.1 for additional information.

3.8.4 Data Processing

Once data are stored by the data system, they can be processed at any future time. Most analysts use a graphical chromatogram for a visual inspection of the quality of the data. As laboratories move to paperless records, the chromatogram may be viewed only on a computer monitor. For qualitative and quantitative analysis (Chapter 11), the collected data are processed to create a simplified output table of retention time and peak area (or height). Additional processing may take advantage of the results of calibration runs to convert the time-and-area results into concentration data (Section 11.4.1).

Further processing of the data may provide additional information to the operator, such as UV spectra, MS, or MS/MS data, and other information from specialty detectors, such as light-scattering detectors.

3.8.5 Report Generation

Most data systems are PC-based, and as such are capable of running any PC-compatible software. This makes it easy to transfer data from the data processing portion of the data system into Excel® and other report-generation software. Thus available tables, chromatograms, graphs, statistical results, and other processed data can be incorporated into a formal report. Some software packages may be able to transfer results in an automatic or semi-automatic fashion so that reports are generated automatically. Besides the convenience and time-saving nature of this process, it can reduce transcription and other operator-related errors. If the software programs have been validated, the amount of time spent checking reports for errors can be greatly reduced.

3.8.6 Regulatory Functions

As more laboratories become subject to regulatory guidelines by government or other regulatory agencies, the integrity of the chromatographic results will become more important. Specific requirements for electronic records, such as 21 CFR Part 11 [18], require electronic audit trails for data when they are stored electronically—often without paper copies. Data systems that are "Part 11 compliant" have built-in functions that remove the burden of record keeping from the individual operator. For example, such systems require that any manual adjustment of baselines during data processing include a record of the reason for the change, the name of the operator, and a time stamp. Data systems that track all the system settings for each injection can simplify the process of proving the validity of a specific result, when a regulatory auditor reviews the data. It is expected that as additional regulatory requirements are placed on the HPLC laboratory, software manufacturers will continue to provide products that help users comply with the regulations.

3.9 EXTRA-COLUMN EFFECTS

The observed peak volume V_p in a chromatogram is influenced by band broadening inside the column V_{p0}, as well as by other system-related factors,

$$V_p = (V_{p0}^2 + 4V_s^2 + V_{pl}^2 + V_{det}^2 + V_{ds}^2)^{0.5} \qquad (3.1)$$

including the volume contributions of the injection process V_s, the tubing and fittings used to plumb the system V_{pl}, the detector cell V_{det}, and the data-system time-constant V_{ds}. The various contributions add as in Equation (3.1) to give the overall peak volume V_p and the related peak width W. Peak broadening that results from factors other than the separation (V_{p0}) arises from *extra-column effects*. For a given method setup, extra-column effects will be constant, but the column contribution will vary with k for the solute. In isocratic separation extra-column peak broadening will therefore be more pronounced for early peaks in the chromatogram with smaller values of V_{p0}. The use of column conditions that generate smaller peak volumes V_{p0}—short, narrow-diameter, small-particle columns (see Table 3.3)—will make extra-column effects more important. The large peak volumes generated by 150 × 4.6-mm columns packed with 5-μm particles tolerate relatively large sample injections, larger tubing diameters, and larger detector flow cells. However, HPLC systems that rely on sub-2-micron particles in small-volume columns must be specially designed to minimize extra-column effects, or the system will be unable to provide satisfactory results. For more information on the influence of the various contributions to peak broadening, see V_p (Section 2.4), V_s (Sections 2.6, 3.6.3), V_{pl} (Sections 3.4.1.2, 3.4.2.2), V_{det} (Section 4.2.4), and V_{ds} (Sections 3.8.3, 11.2.3).

3.10 MAINTENANCE

The quality of the analytical results obtained from an HPLC system and method depends heavily on the ability of the HPLC system to perform reliably and according to specifications. Three main areas need to be addressed for reliable system operation:

- ensure the HPLC system works properly (Section 3.10.1)
- prevent as many problems as possible Section 3.10.2)
- make efficient and effective repairs (Section 3.10.3)

These tasks are covered in the present section. Although no HPLC system is ever problem free, the user can take an active role to minimize problems through the use of some of the techniques described here.

3.10.1 System-Performance Tests

Quantitative measurements of HPLC system performance will allow the user to compare performance over time. This can ensure that the instrument works when it is new, help anticipate future problems, and show that a repair was effective.

3.10.1.1 Installation Qualification, Operational Qualification, and Performance Qualification

One way to demonstrate that a new HPLC system functions properly is to follow the practice of the pharmaceutical industry and perform installation qualification (IQ), operational qualification (OQ), and performance qualification (PQ) tests prior to the release of the system for routine work. The IQ test ensures that the instrument is installed according to the manufacturer's procedures. IQ often is done by the vendor, if installation is included with the purchase of the system. The documentation accompanying the system will outline the IQ test.

OQ demonstrates that the instrument meets the manufacturer's specifications, or some subset of them. The OQ test results also may be included in the documentation. Alternatively, the performance tests outlined in Sections 3.10.1.2 and 3.10.1.3 may be used to compare the test results to the manufacturer's specifications (generally found in the back of one or more of the operator's manuals). The OQ or PQ test often includes a column-performance test (Section 3.10.1.3) to confirm that all the system hardware is working well as a unit.

The PQ test generally is user-designed and may range from extensive tests, such as the performance tests described below, to the analysis of a few mock samples to show that the expected results can be obtained for a specific HPLC method. Once these three tests (IQ, OQ, and PQ) have been performed, the system should be ready for routine use.

3.10.1.2 Gradient Performance Test

The most important part of the test suite for the evaluation of system performance is a pair of experiments to determine the linearity and accuracy of gradient formation, as well as measure the system dwell-volume. These tests apply to system operation whenever on-line mixing is to be used, for both isocratic or gradient methods. The column is removed, and replaced with a piece of narrow-bore connecting tubing. For example, ≈1 m of 0.005-in (≈0.13-mm) i.d. tubing can be used to connect the injector and detector. This provides sufficient pressure to enable reliable operation of the pump check-valves and results in insignificant dead-volume (≈12 μL) or dispersion of the gradient for most systems. Next water is placed in the A-reservoir, and water that contains 0.1% acetone is placed in the B-reservoir. (An alternative is to use methanol in A and 0.1% acetone in methanol in B, but the same base solvent must be used in both reservoirs.) A UV detector is used, with the wavelength set to 265 nm.

Gradient Linearity. The system is programmed first to run a full-range gradient (0–100% B); a 20-minute gradient time is recommended. The flow rate should be set such that the system generates sufficient pressure for reliable check-valve operation; generally, 1 to 3 mL/min will be satisfactory. The autosampler should be in the inject mode, so that mobile phase is pumped through the loop. Because the injector loop normally is in the flow stream during a run (Fig. 3.17*b*), the loop-volume contributes to the dwell-volume. (If the system is usually run with the loop out of the flow stream, this test should be run with the injection valve in the load mode [Fig. 3.17*b*] rather than the inject mode.) The test gradient is then carried out. A plot of the baseline should appear as an S-shaped curve, as illustrated by the solid

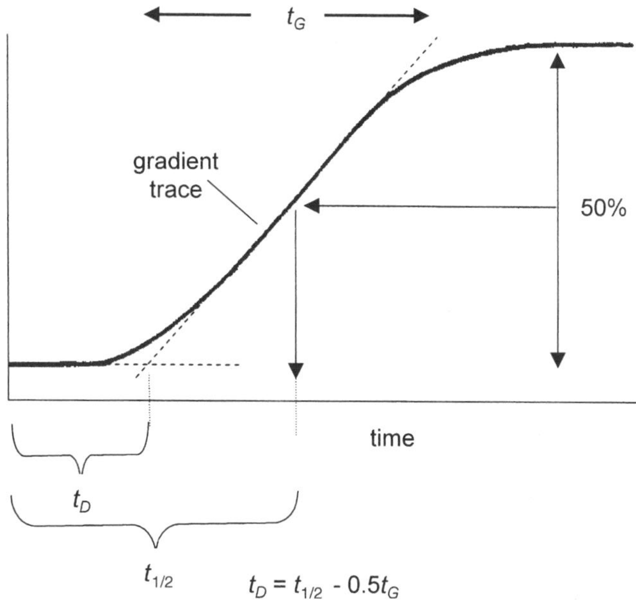

Figure 3.26 Gradient profile for water to water-acetone gradient.

curve of Figure 3.26. This blank gradient can be used as a rough check of gradient linearity, and also measures the system dwell-volume (see below).

Gross deviations from gradient linearity can be checked by comparison to a straight line that fits the middle of the gradient profile (dashed line in Fig. 3.26). The actual gradient (solid curve) should be smooth and not deviate from the line except for the slight "gradient rounding" at the beginning and end of the gradient. If deviations are seen, extra attention should be paid to the gradient step-test (see below). For low-pressure-mixing systems, the gradient-proportioning valve test (see below) also will help isolate gradient-linearity problems [19]. If visible deviations from linearity are observed, these often will appear as a distinct shift or angle in the gradient plot (e.g., Fig. 17.23 and Section 5.5.3.3 of [20]). It is wise to carry out additional step-test measurements in the region where deviations are observed.

Most HPLC systems can be programmed to generate gradient rates of 10%/min or higher. If the system is to be used with steep gradients—such as for high-throughput applications—or a very different flow rate, the gradient linearity can be checked by running the blank gradient test under these conditions. For example, instead of the 20-minute, 2-mL/min gradient of Figure 3.26, it may be more appropriate to test the system with a 5-minute gradient run at 0.5 mL/min if these are more representative operating conditions. Gradient rounding, which usually is not a significant problem for most gradient conditions, tends to be more serious for very steep gradients (or smaller gradient-volumes $t_G F$, where t_G is the gradient time and F is the flow rate; also see Sections 8.1.6.2 and 9.2.2 of [20]). Usually gradient rounding can be reduced by decreasing the system dwell volume (see discussion of Fig. 17.28) while increasing the gradient volume $t_G F$.

Dwell-Volume Determination. The gradient profile (as in Fig. 3.26) can be used to determine the system dwell-volume. Dwell-volume can be measured by means of one of two techniques. The first method is to extend the linearity test line (dashed line in Fig. 3.26) until it intersects the extended baseline. The time between this intersection and the start of the gradient is the dwell-time, as shown in Figure 3.26. Dwell-time t_D can be converted to dwell-volume V_D by multiplying by the flow rate $F : V_D = t_D F$. This method to determine dwell-volume is simple, but it is subject to any errors that can result from inaccuracy in drawing the linearity-test line through the gradient. It also may be inconvenient to make this measurement directly on a computer monitor from a data system output.

A second method for measuring the dwell-volume is less error-prone and more convenient to perform on the computer monitor. This is shown graphically in Figure 3.26. Determine the detector response at the initial baseline (0% B) and at the end of the gradient (100% B). From these two values, locate the point on the plot that the response has reached 50% B and note the time $t_{1/2}$ it took to reach this point. The dwell time t_D equals $t_{1/2}$ minus half the gradient time (e.g., 10 minutes for a 20-minute gradient). The dwell-volume equals $t_D F$.

Gradient Step-Test. This test uses the same system setup and A- and B-solvents. The gradient step-test determines the accuracy of solvent proportioning for selected solvent (or mobile-phase) mixtures. If the system also is used for isocratic methods, this test will check the accuracy of on-line mixing. The system controller is set to deliver a series of solvent mixtures in a stair-step design. A good choice for this test is to use a 10% step size, so that mixtures of 0, 10, 20, ..., 80, 90, 100% B are formed—each for an interval of 3 minutes. Problems are encountered most commonly near 50% B, so an additional step at 45% B and 55% B should be added for a total of 13 steps. The remaining conditions are the same as those described for the gradient linearity test (above). The results for the 40–60% B portion of this test for a well-behaved system are shown in Figure 3.27.

Next calculate actual %B for each step (as in Fig. 3.27) by measuring its height from the baseline (0% B) and dividing by the distance between the 100% B step and the 0% B step. The %B for each step should compare favorably to the programmed

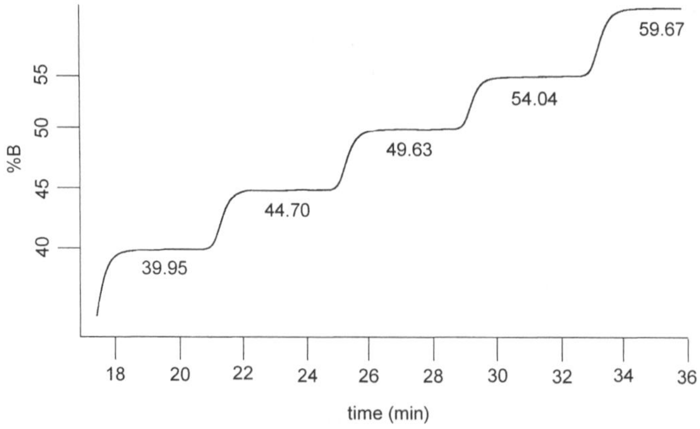

Figure 3.27 Results of a gradient step-test. Measured response (%B) shown below each step.

value for the step. For example, the 40% B setting in Figure 3.27 actually delivered 39.95% B. Typically manufacturers specify accuracy of ±0.5–1% B throughout the mixing range. For example, in Figure 3.27, the 55% B step actually delivered 54.04% B—(barely) within a ±1% criterion. For applications with gradient rates of ≥1%/min, accuracy of ±1% usually is sufficient. When shallower gradients are used, smaller deviations may be required. It may also be possible to improve proportioning accuracy by premixing solvents. For example, the proportioning accuracy of a 15–25% B gradient can be improved 10-fold from ±1% to ±0.1% by the replacement of 100% aqueous solvent in reservoir A with hand-mixed 15/85 organic/aqueous, and 100% organic in B with 25/75 organic/aqueous, plus revision of the program to generate a 0–100% B gradient instead of 15–25% B. The accuracy of on-line mixing for isocratic elution can be similarly improved with this technique.

Gradient Proportioning Valve (GPV) Test. A third test with a similar system setup as above is useful for low-pressure-mixing systems but does not apply to high-pressure mixing. This test checks the accuracy of the proportioning-valve system and its associated control software. For example, consider a four-solvent system (A, B, C, and D). The A and B inlet lines are placed in a reservoir that contains water and the C and D lines are placed in a reservoir that contains 0.1% acetone in water. The baseline is generated by delivery of a 50/50 mixture of A and B. The various combinations of solvents are checked by generation of blends of 90% of A or B with 10% of C or D. For example, the test results shown in Figure 3.28*a* (for an acceptable test result) are for the sequence shown in the caption. The height above baseline of each 90/10 plateau is measured. The difference between the height of the highest and lowest plateaus is divided by the average plateau height to determine the %-range for the various proportioning valve combinations. A plateau range of ≤2% is usually acceptable, although ranges of ≤1% are common for well-behaved systems (Fig. 3.28*b*; see Fig. 17.25 for an example of a failed GPV test).

3.10.1.3 Additional System Checks

In addition to the accuracy and linearity of gradient formation, other factors affect LC-system reliability. The tests listed below can be used on a periodic basis (e.g., annually or semiannually) to help ensure that the system is operating properly. Typical performance values for these tests are listed in Table 3.4.

Flow-Rate Check. Although a small change in the flow rate F usually results in only minor changes in separation, large errors in F can be more serious. Consequently a check of flow-rate accuracy on a periodic basis is recommended. The second-by-second flow-rate accuracy during a run is difficult to measure without specialized equipment (and is not important), but a longer term volumetric check of flow rate can be made easily by carrying out a timed collection of mobile phase in a 10-mL volumetric flask at a flow rate of 1 mL/min under isocratic conditions. For high-pressure-mixing systems the flow rate can depend on solvent compressibility, so it is best to check the flow for representative solvents. For example, check the flow of 100% A with water and 100% B with acetonitrile or methanol. Typical manufacturer's system specifications are ±1% for flow-rate accuracy. A measured flow-rate accuracy should fall within this range for routine operation. In Table 3.4

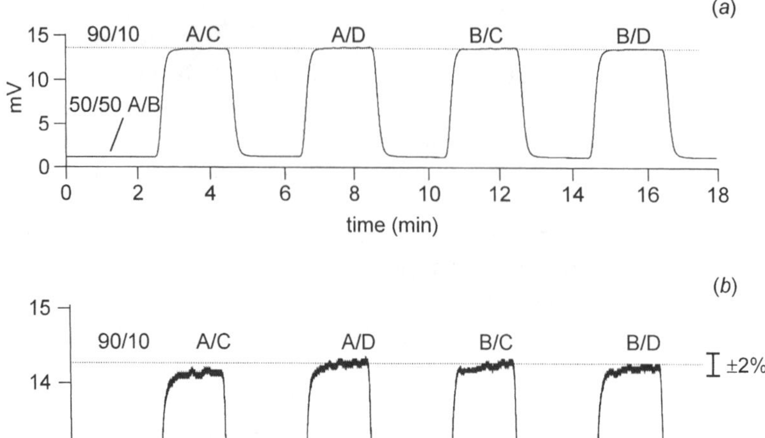

Figure 3.28 Results of a gradient proportioning valve (GPV) test. (*a*) Gradient proportioning valve test results; (*b*) detail showing ±2% allowed difference between plateaus. Baseline is generated by 50:50 A:B; the remaining plateaus are 90:10 A:C, A:D, B:C, and B:D from left to right. Solvents: A = B = water; C = D = 0.1% acetone in water.

Table 3.4

Typical System Performance Parameters for Tests of Section 3.10.1

Parameter/Test	Typical Specification	Typical Acceptance Criteria
Linearity	±1%	Visual linearity 5–95% B
Dwell-volume	Varies	As measured
Step-test (accuracy)	±1%	±1%
GPV[a] test	Not specified	≤2% plateau range[b]
Flow rate (collect 10 mL at 1 mL/min)	±1%	±2% (±12 sec on 10 mL at 1 mL/min)
Pressure bleed-down from 4000 psi	Not specified	15% in 10 min
Retention reproducibility	Not specified	±0.05 min
Area reproducibility (10 μL injection)	±0.3%	±1%

[a] Gradient proportioning valve.
[b] Difference between highest plateau and lowest plateau when C or D solvent is blended with A- or B-solvent, as in Figure 3.28.

the acceptance criterion is set to ±2%, since the combination of measurement errors (volumetric glassware) and the timing start/stop errors will add somewhat to the overall measured error.

Pressure Bleed-Down. Malfunctioning check valves and worn pump seals often show up as deviations in expected values of mixing accuracy, gradient linearity, and flow rate. An additional test of the outlet check valves can be made with a pressure bleed-down test. For this test the outlet tubing from the pump is blocked (by a union and a plug). The high-pressure shutoff limit for the pump is set near its maximum value, for example, 5000 psi (350 bar) for a system capable of 6000 psi (400 bar). The pump is then turned on and allowed to shut off at the upper pressure limit. The maximum pressure is recorded, and 10 minutes later the pressure is recorded again. A pressure drop of $\leq$15% indicates that the check valves are working properly. A larger drop suggests that the outlet check valve(s) should be cleaned or replaced or that the pump seal(s) should be replaced if no outlet check valve is used (e.g., accumulator-piston pump, Section 3.5.1.2). A pressure drop to atmospheric pressure over the 10-minute test is more indicative of a leaky fitting.

Retention Reproducibility. A check of retention reproducibility is an overall check of on-line mixing and pump performance. Although this check can be done with any sample, it is wise to use a sample that can be formulated easily under conditions that can be reproduced at any time. This allows an independent check of system performance to be made for a specific method, and it is a good tool for troubleshooting. Example test conditions are listed in Table 3.5. It is important to use a sample concentration such that the peak is within the detector's linear range and is sufficiently large that baseline noise does not affect the precision of the measurements. For example, a sample that generates a peak height of 0.1 to 0.8 AU would be a good choice for UV detection. Retention-time variation of no more than $\approx$0.05 min (1 S.D.) is acceptable for six replicate injections. Larger variations are an indication of problems. If the other performance tests prove to be okay, but retention reproducibility is poor, leaks or air bubbles are the most likely problem sources. An alternative test, especially if the system is used only for isocratic methods, is to run the retention test under the isocratic conditions listed in Table 3.5 (80% B).

Table 3.5

Retention-Time and Peak-Area Test Conditions

Parameter	Value
Flow rate	1.5 mL/min
Column	C_{18}, 150 × 4.6 mm, 5 μm
Temperature	35°C
Detection	280 nm UV
Mobile phase A	Water
Mobile phase B	Methanol
Gradient/isocratic	5–95% B in 20 min[a]/80% B[b]
Equilibration time	10 min
Sample	5 μg/mL 1-chloro-4-nitrobenzene[c]
Injection volume	10 μL
Retention (typical) gradient/isocratic	14 min[a]/3 min[b] (with $V_D = 2.2$ mL[d])

[a]Gradient test conditions.
[b]Isocratic test conditions; adjust as necessary for $2 < k < 10$.
[c]Other nonpolar aromatic compounds (e.g., toluene, methyl benzoate) can be used.
[d]For other values of the dwell-volume V_D; gradient retention times will change by the difference in dwell-volumes divided by the flow rate (1.5 mL/min).

Peak-Area Reproducibility. The same chromatograms run for retention reproducibility can be used to determine peak-area reproducibility, which is primarily a measure of the autosampler's performance. The variation in peak areas should be <1% RSD based on six injections. Most modern autosamplers will generate values of ≤0.5% RSD for injection volumes of ≥5 μL when operated under standardized conditions (e.g., those of Table 3.5). Poor area reproducibility usually can be traced to an autosampler problem. Peak-area reproducibility should be checked for both the isocratic and gradient operation; similar results should be obtained under both conditions.

New-Column Test. Repeating the column manufacturer's test procedure on a new column is a quick procedure to show that the HPLC hardware is working properly. This test should be done as part of the operational qualification (OQ) or performance qualification (PQ) testing (Section 3.10.1.1) to show that the system is functioning well before its first use. The new-column test also serves as a tool in the "divide-and-conquer" strategy (Section 17.3.1) of problem isolation. Just install a new column and repeat the manufacturer's performance test. The test conditions should be similar to the isocratic conditions listed in Table 3.5, and should be listed in the column test sheet included with the column literature. If the test results are within ≈10% of the column manufacturer's tests (retention times, plate number), the system is working well—the flow rate, column temperature, and mobile-phase composition are correct, and extra-column effects are minimal. (It is rare to obtain a plate number as large as that reported in the manufacturer's test; their column testing systems are optimized for column performance, not routine operation as in a user's laboratory.)

3.10.2 Preventive Maintenance

A good way to improve the reliability of an HPLC system is to anticipate and prevent problems before they occur. Problems with HPLC systems can result from normal wear of components or from the way the system is used. Many problems can therefore be minimized by performing periodic maintenance so that normal-wear items are serviced or repaired before they fail. In addition there are a number of simple laboratory practices that can be undertaken to minimize user-caused system failure.

3.10.2.1 Periodic Maintenance

Regular cleaning and/or component replacement will help make the HPLC system work more reliably. Some of these periodic maintenance tasks are listed in Table 3.6. The recommended frequency of the various actions is typical for HPLC systems used several times a week; the experience in each laboratory may vary from these recommended intervals. The important concept is that a disciplined approach be taken so that every instrument receives regular maintenance. For further details on the parts and modules mentioned below, consult the descriptions in the first part of this chapter.

Reservoir. The reservoir that contains the aqueous solvent (A-solvent) should be cleaned and replenished, so that microbial growth does not occur. Replacement

Table 3.6

Recommended Maintenance Intervals

Item	Action	Frequency[a]
Reservoir	Replace buffer, wash reservoir	Weekly
	Replace organic, wash reservoir	Monthly
	Filter mobile phase	Every batch
	Replace inlet-line frit	6 months
	Degas mobile phase	Daily
Tubing and fittings	Inspect for leaks	Weekly
Pump(s)	Sonicate or replace check valves	As needed
	Replace pump seal(s)	6 months
	Flush to remove buffers	Daily
	Inspect for leaks	Daily
Autosampler	Replace wash solvent and clean Reservoir	Weekly
	Replace valve rotor seal	See manufacturer's recommendations
	Replace needle seal	6 months
	Inspect for leaks	Daily
Column oven	Calibration check	Annually
	Inspect for leaks	Daily
Detector	Inspect for leaks	Weekly
	Additional detector checks	See detector manual
Waste container	Check capacity; empty	Daily
Data system	Per experience	
LC system	System check	6 months
LC method	System suitability	Daily

[a]May vary based on frequency of use, performance history, or SOP.

of the A-reservoir and its contents on a weekly basis should avoid this, although some laboratories stretch this to two weeks. It is a good idea never to "top off" the aqueous solvent, but instead to use a clean reservoir for each new batch of aqueous solvent. The organic reservoir (B-solvent) has a longer cleaning and replacement cycle, because the probability of microbial growth in >30% organic is considerably diminished compared with the aqueous phases. If the A- or B-solvent comprises only HPLC-grade solvents, mobile-phase filtration is not required; otherwise, all mobile phases should be filtered through a membrane filter of $\leq$0.5-μm porosity. The inlet-line frit can be an inadvertent source of re-inoculation of the reservoir contents once it becomes contaminated. The frit is inexpensive, so it should be replaced every six months as a precautionary measure. Mobile-phase degassing is one of the easiest ways to improve system reliability, so all solvents used for reversed-phase HPLC should be degassed (Section 3.3).

Tubing and Fittings. If tubing and fittings are properly assembled, they should work indefinitely without problems. However, leaks do occur occasionally, so it is a good idea to check all tubing and fittings for leaks once a week. A drop of moisture or a small bit of white buffer residue may be the only sign of a leak. If a leak is found, the pump should be shut off, the fitting tightened, and the pump turned back on again. When PEEK fittings and/or tubing are used, the tube end should be reseated in the fitting port prior to tightening the fitting. Sometimes PEEK tubing will slip in the fitting when it is tightened under pressure, so the pump should be turned off during this operation.

Pumps. Sticky or leaky check valves are the most common pump problem in most systems. This is indicated most commonly by system-pressure fluctuations. Many times a leaky check valve can be reconditioned by sonication in MeOH for a few minutes. Prior to sonication, it is best to inspect the check valve to ensure that the components will not fall out and get mixed up. If the valve requires reassembly, dust-free gloves should be used, so as to avoid recontamination of the check valve. (See Section 17.2.5.4 for additional information on check-valve sonication.) Pump seals may last for a year or more, but they are inexpensive and when pump-seal failure occurs, other problems may arise, such as blocked frits. For this reason it is prudent to replace the pump seals twice a year, that is, before significant wear occurs. Buffers left in the pump when it is not used can precipitate and leave abrasive deposits that increase pump-seal wear or cause check-valve problems. It is a good practice to flush all buffers out of the HPLC system before shutting it off for more than an hour or two. If the system must be ready to run at a moment's notice ("stat" conditions), the pump can be set to its minimum flow rate (e.g., 0.1 mL/min), so the mobile phase is always flowing. As with tubing and fittings, pump leaks occur periodically, so it is a good idea to inspect the pump for leaks each day during startup.

Autosampler. Just as the mobile phase and the reservoirs need to be cleaned or replaced to avoid microbial contamination, the autosampler wash-solvent and its reservoir should be replaced periodically. If an aqueous wash-solvent is used, it should be replaced and the reservoir cleaned each week; if organic solvent is used, monthly service should suffice. The injector rotor seal has a long life—100,000 cycles or more if treated properly. Check the manufacturer's recommended service interval in the preventive maintenance section of the operator's manual. The needle seal will wear with continued use. If there are no other guidelines in the operator's manual, replacement of the needle seal every six months will help to avoid problems. The autosampler is the component of the HPLC system that is most likely to leak; it is wise to inspect it daily for leaks.

Column Oven. The column oven should be a trouble-free component. The oven calibration should be checked when the oven is first installed, then on an annual basis. A flow-through thermocouple can be used to check the temperature of the mobile phase exiting the column as a check of the oven temperature and temperature equilibration within the column. During system startup each day or when a new method is started, the column endfittings should be checked for leaks.

Detector. HPLC detectors operate at pressures significantly lower than the column, so leaks at the detector are less common than elsewhere. However, a weekly check is wise. Other detector checks will vary depending on the detector; consult the detector manual for more information.

Waste Container. The waste container is a system component that is easy to overlook. It is wise to check it daily to ensure that it has sufficient capacity for the day's runs. Placement of the waste container in a plastic dishpan or other safety container will help protect against spills when the inevitable overflow occurs.

HPLC System-Check. The system performance checks described in Sections 3.10.1.2 and 3.10.1.3 should be performed once or twice a year. A semiannual check will often catch pending problems that might occur if the check is done on an annual basis. Check the system on a yearly basis if a semiannual check cannot be justified.

HPLC Method-Check. Before samples are run with an analytical method, a system suitability check should be made. This may be the most important test that one can perform for an HPLC system because it shows that both the system and method are suitable for carrying out the desired analysis.

3.10.2.2 Suggestions for Routine Applications

To obtain high-quality data, the HPLC system must perform in a reliable and reproducible manner. This section lists and reviews some additional tips and techniques that will help ensure high quality results.

Reagent Quality. Gradient elution tends to concentrate nonpolar impurities in the A- and B-solvents at the head of the column, followed by their release as the gradient progresses. These impurities can show up as peaks in both blank and sample runs (e.g., Fig. 17.12 and Sections 17.2.5.9, 9.6.2). For this reason it is essential to use HPLC-grade reagents for gradient work. Lower quality reagents may be suitable for isocratic applications, but even the most minor impurities can cause problems with gradient elution. For best results one should use only HPLC-grade reagents for all HPLC work. Aqueous reagents and buffers should be discarded frequently (e.g., weekly) to avoid contamination by microbial growth. Water impurities can be especially problematic.

System Cleanliness. As important as reagent quality in minimizing artifactual peaks is maintaining a clean instrument. The recommendations of Section 3.10.2.1 (reservoirs and pumps) should be followed. Spills, leaks, and other potential sources of contamination should be cleaned up promptly.

Degassing. Although some HPLC systems will operate without degassing the mobile phase, every system will operate more reliably with degassed solvents (Section 3.3). Trapped air bubbles and solvent outgassing are common problems that can be largely avoided by solvent degassing. It is a good idea to purge the pump(s) and solvent inlet lines daily by opening the purge valve(s) and operating the pump at an elevated flow rate (e.g., 5 mL/min) for a few minutes to remove any air bubbles.

Dedicated Columns. Each analytical method should have a column dedicated to that method. It is not a good idea to share columns between methods, since peaks that are not of concern in one method (e.g., late-eluting peaks) may cause interferences in a second method. Sample components may change the selectivity or degrade the column for one method but not another.

Equilibration. Prior to each run, the column should be equilibrated to the same extent as the other runs in the run sequence. Complete equilibration may or may not be necessary for gradient methods (Section 9.3.7).

Priming Injections. Some methods give better results if several "priming" injections are made before the first sample is injected. These injections of standards or mock samples can help saturate slowly equilibrating active sites on the column so that more reproducible separations can be obtained. Priming injections are more often useful for separations of biological samples (Section 13.3.1.4). Sometimes the system suitability injections serve as priming injections.

Ignore the First Injection. Because some methods require a priming process and the first injection may be equilibrated differently than subsequent injections, it is best to avoid use of the first injection for quantitative analysis. The second and subsequent runs will be more reliable than the first injection; a preceding system suitability test (see following) also serves this purpose.

System Suitability. Many methods that run under the oversight of regulatory agencies (FDA, EPA, OECD, USP, etc.) will require a system suitability test prior to sample analysis. System suitability serves as a confirmation that the equipment and analytical method are operating in a fashion that will produce reliable results. Requirements for system suitability tests vary, so the regulatory guidelines should be consulted to help select appropriate tests (Section 12.3). Many workers use retention time and area reproducibility, peak response (detection sensitivity), peak width, peak tailing, resolution, and column pressure, either alone or in combination, as part of the system suitability test. The system suitability sample may be a diluted pure standard, a mock sample in extracted matrix, or some other sample selected to demonstrate system performance. The important concept is to select system suitability samples that test the ability of the method to perform its desired function. Whether or not a system suitability test is required, it is wise to run such a test prior to routine analysis, even if it is just an injection of a standard to see if the retention and peak size are as expected.

Standards and Calibrators. For quantitative analysis (Chapter 11), the response of unknown samples is compared to the response for standards of known concentration. The range of standard concentrations, number of replicates, and sequence of injection may depend on the specific application. Either external or internal standardization may be used. In any event, at least one standard should be injected prior to the analysis of unknown samples, to ensure that the analytical method is working properly before potentially valuable samples are injected.

Table 3.7

Repair Recommendations

User Level	Typical Repairs
Novice/operator	Replace mobile phase, inlet-line filter, in-line filter, guard column, column
Experienced	Check valves, pump seals, proportioning-manifold replacement, injector rotor, detector lamp, tubing and fittings
Factory-trained	Proportioning-manifold repair, electronics, repairs involving module subassemblies, detector flow cell

3.10.3 Repairs

Sooner or later, no matter how much preventive maintenance is practiced, problems will be encountered on every HPLC system. Chapter 17 is dedicated to troubleshooting. In this section, some of the more philosophical aspects of instrument repair are considered.

3.10.3.1 Personnel

Some laboratories rely completely on a service contract from the equipment manufacturer for all maintenance and repairs, whereas other laboratories perform all these tasks themselves. Most laboratories, however, rely on something between these two extremes. Even in the case of a 100% service-contract arrangement, there are repairs that make sense to accomplish with laboratory staff—for time savings, if nothing else. It is a good idea to have a plan in place that determines what repairs are to be made by which personnel. Some recommendations are listed in Table 3.7 and discussed in Section 3.10.3.3.

3.10.3.2 Record Keeping

Laboratories working under the oversight of regulatory agencies may be required to keep maintenance and repair records, but even laboratories for which such records are not required should maintain some level of equipment records. One of the simplest record-keeping techniques is to prepare a three-ring binder with sections for different maintenance activities and records. A separate binder is kept with each HPLC system where it can be referenced easily and is most likely to be used during a maintenance session. It is convenient to organize this record book in three sections:

System Configuration. A record should be made of all the components in a given system configuration, including the model and serial numbers of each component. As changes are made, a fresh copy can be completed. This will facilitate communications with the instrument vendors when service is needed, and will help confirm for regulatory auditors the specific system configuration used with a specific sample.

Maintenance Records. Repair records can be simplified if a template is made that contains blanks for the who, what, why, when, and how questions of a maintenance session. Several blank copies of these pages can be kept in the notebook so that one is available to complete for each maintenance session. Once a year, or after a number of sessions, these pages can be sorted into failure categories (e.g., pump-seals, check valves) and can serve to help develop a system-specific or laboratorywide preventive maintenance program.

System Checks. A third section can contain the records from the periodic instrument system checks (Sections 3.10.1.2, 3.10.1.3). With a readily available historic record, trends from one system check to another can be tracked. These also can provide data that help establish preventive maintenance programs.

Record-keeping technologies could also be used. Computer databases or electronic notebooks are two good options. But, it is important that they are sufficiently convenient, so that data can be easily entered at the time of maintenance or repair; otherwise, valuable records will be lost.

3.10.3.3 Specific Repair Recommendations

Table 3.7 summarizes various repairs that are appropriate for users of different skill levels:

Novice/Operator. Personnel just learning how to run an HPLC system, or workers in a routine environment that "just" run samples, may have limited skills or authority to make repairs. However, little training is needed to replace mobile phases, filters, guard columns, and columns. A system of double-checking or sign-off by a second person will help avoid errors in such activities.

Experienced. As the user gains experience, additional troubleshooting and maintenance skills will be developed. Some people have a natural tendency to become more skilled than others, but most workers can learn how to change check valves, pump seals, and tubing and fittings. More complex problems, such as replacement of detector lamps or low-pressure gradient-proportioning manifolds, can be reserved for more skilled workers.

Factory-Trained. Certain activities should be reserved for specially trained personnel. These may be normal laboratory workers with an excellent troubleshooting and maintenance aptitude, a service group within the company, or a technician contracted from the instrument manufacturer. Electronics problems generally require advanced skills and/or equipment to fix. Other problems that involve disassembly of modules or repair of components that are delicately tuned—such as proportioning valves—may also call for special training.

REFERENCES

1. S. R. Bakalyar, M. P. T. Bradley, and R. Honganen, *J. Chromatogr.*, 158 (1978) 277.
2. S. van der Wal and L. R. Snyder, *J. Chromatogr.*, 255 (1983) 463.

3. J. N. Brown, M. Hewins, J. H. M. van der Linden, and J. H. M Lynch, *J. Chromatogr.*, 204 (1981) 115.
4. L. R. Snyder, *J. Chromatogr. Sci.*, 21 (1983) 65.
5. J. Thompson, personal communication, 2008.
6. http://www.vici.com (Valco Instruments).
7. http://www.idex-hs.com/products/Brand.aspx?BrandID=1 (Upchurch Scientific).
8. J. W. Batts, IV, *All About Fittings*, Upchurch Scientific, Oak Harbor, WA, 2003.
9. Y. Ishihama, *J. Chromatogr. A.*, 1067 (2005) 73.
10. J. Hernandez-Borges, Z. Aturki, A. Rocco, and S. Fanali, *J. Sep. Sci.*, 30 (2007) 1589.
11. S. Stearns, personal communication, 2008.
12. *Technical Notes 5*, Rheodyne, Dec. 1983.
13. http://www.idex-hs.com/products/Brand.aspx?BrandID=8 (Rheodyne).
14. O. Abreu and G. D. Lawrence, *Anal. Chem.*, 72 (2000) 1749.
15. R. G. Wolcott, J. W. Dolan, L. R. Snyder, S. R. Bakalyar, M. A. Arnold, and J. A. Nichols, *J. Chromatogr. A*, 869 (2000) 211.
16. J. W. Dolan, *J. Chromatogr. A*, 965 (2002) 195.
17. P.-L. Zhu, L. R. Snyder, J. W. Dolan, N. M. Djordjeveic, D. W. Hill, L. R. Sander, and T. J. Waeghe, *J. Chromatogr. A*, 756 (1996) 21.
18. *Guidance for Industry: Part 11, Electronic Signatures—Scope and Application*, http://www.fda.gov/cder/guidance/index.htm (2003).
19. J. J. Gilroy and J. W. Dolan, *LCGC*, 22 (2004) 982.
20. L. R. Snyder and J. W. Dolan, *High-Performance Gradient Elution*, Wiley, Hoboken, NJ, 2007.

CHAPTER FOUR

DETECTION

Introduction to Modern Liquid Chromatography, Third Edition, by Lloyd R. Snyder, Joseph J. Kirkland, and John W. Dolan

4.1 INTRODUCTION

The detector for the first liquid chromatographic separations was the human eye, used by Tswett in his classic experiments [1, 2]. For many years quantitative and qualitative analyses were accomplished by the collection of fractions eluted from the column (e.g., Fig. 4.1*a*), followed by off-line analysis using wet chemical, gravimetric, optical, or other analysis techniques. The concentration of analyte in each fraction could be plotted against fraction number, as in Figure 4.1*b*, resulting in a crude chromatogram. Collection and analysis of fractions is time-consuming, usually degrades chromatographic resolution, and is generally inconvenient, so on-line detection has many benefits. Some of the first applications of on-line liquid chromatographic detection included the refractive index detector reported by Tsielius [3] and the conductivity detector by Martin and Randall [4]. However, it was the introduction of the UV detector in 1966 by Horváth and Lipsky [5] that gave HPLC its most widely used detector.

The UV detector was further improved in 1968 by Kirkland [6]; a flow-through cell, plus use of a mercury lamp, resulted in a great improvement in sensitivity (1×10^{-5} AU noise). Single-wavelength detectors (254 nm) based on this principle were by far the most popular through the 1970s—until they were subsequently displaced by the variable-wavelength and diode-array UV detectors.

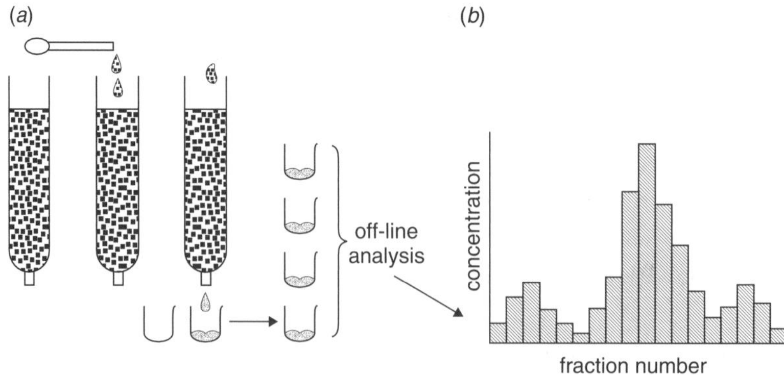

Figure 4.1 (*a*) Open-column LC; (*b*) plot of concentration of analyte in each fraction from (*a*).

The chromatographic detector is a transducer that converts a physical or chemical property of an eluted analyte into an electrical signal that can be related to analyte concentration. Gas chromatography (GC) was well into its second decade of popular use when HPLC began its rise to popularity. This gave GC a head start in detector development as well. Much effort was given in the 1960s and 1970s to adapt GC detectors for HPLC use. A number of such detectors were described, including flame ionization [7, 8], flame photometric [9], and electrolytic conductivity [10]. However, adaptation of GC detectors required elimination of the mobile phase through evaporation, and until the introduction of the electrospray interface for LC-MS, the liquid-to-gas conversion process was very unreliable. As a result GC-based detectors for HPLC were doomed by operational problems. It was only after the development of an efficient nebulizer that detectors relying on mobile-phase-elimination (LC-MS, evaporative light scattering, corona discharge, FTIR, etc.) became practical.

Despite the lack of a universal, sensitive detector, such as the GC flame ionization detector, there are many HPLC detectors that have been successful for general or specific applications. This chapter describes the principles of operation of the most popular HPLC detectors, presents example applications, and—where appropriate—compares the advantages and disadvantages of specific detectors.

4.2 DETECTOR CHARACTERISTICS

The second edition of this book [11] listed nine characteristics of an ideal HPLC detector:

- have high sensitivity and predictable response
- respond to all solutes, or else have predictable specificity
- be unaffected by changes in temperature and carrier flow
- respond independently of the mobile phase
- not contribute to extra-column peak broadening
- be reliable and convenient to use
- have a response that increases linearly with the amount of solute
- be nondestructive of the solute
- provide qualitative information on the detected peak

Of course, no detector possesses all of these characteristics, nor is likely to in the foreseeable future. However, those HPLC detectors that have been the most successful have many of the properties listed above.

4.2.1 General Layout

The HPLC detector is positioned directly after the column (Fig. 4.2) so as to minimize post-column peak broadening (Section 3.9). The column effluent is directed into a detector flow cell, where detection takes place. For most detectors the mobile phase is maintained in a liquid state, detection occurs, and the mobile phase passes out of the detector cell to a waste container, fraction collector, or another detector.

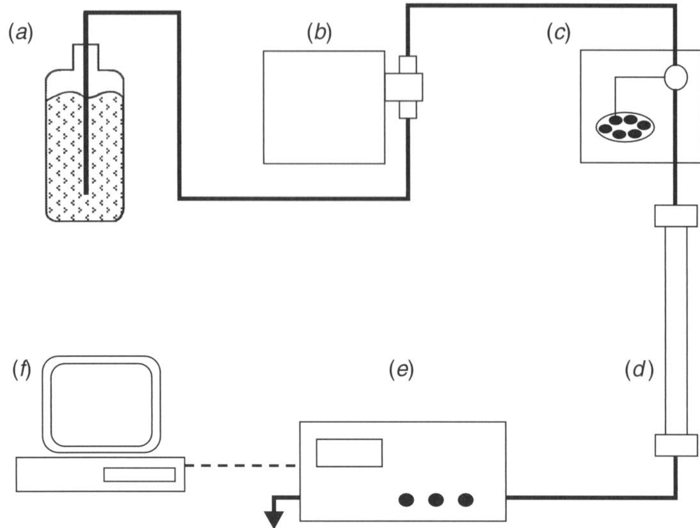

Figure 4.2 HPLC system diagram. (*a*) Mobile phase reservoir; (*b*) pump; (*c*) autosampler or injector; (*d*) column; (*e*) detector; (*f*) data system.

The UV flow cell shown in Figure 4.3*a* has evolved into its present configuration, in which the mobile phase flows through in a Z-shaped path, and UV light enters the quartz window at one end of the cell. The differential absorbance of light is monitored at the other end of the cell by a photosensitive diode, is converted to an electrical signal, and eventually is presented by the data system as a chromatogram. Different detectors have different flow-cell designs, as described in the following detector-specific sections.

Optical detectors, such as UV and especially refractive index detectors, often are sensitive to small temperature changes in the mobile phase, which cause changes in the refractive index, and thus the amount of light that is transmitted through the flow cell. To stabilize the temperature, the detector cell is mounted in a draft free, and sometimes temperature-controlled, location in the detector case. The detector commonly includes a capillary heat-exchanger to stabilize the incoming mobile-phase temperature. One popular configuration is to wrap the capillary tubing around the detector cell body and embed it in a thermally conductive sealant.

Detectors that make their measurements in the liquid state can be susceptible to optical or electrical disturbances when bubbles are present. Thorough degassing of the mobile phase (Section 3.3) often is sufficient to prevent bubble formation within the detector. As an additional precaution a small back-pressure can be applied to the detector outlet tube to prevent bubble formation. We recommend that a spring-loaded check valve be purchased from an HPLC parts-supplier for this purpose. If such a device is used, be careful that it does not exceed the pressure limit of the cell. For example, use a back-pressure regulator that produces 50 to 75 psi of back-pressure with a detector cell with a 150 psi pressure limit. Some workers rely on the use of a piece of capillary tubing (e.g., 1 m × 0.005-in. i.d.) as a back-pressure regulator. But the pressure created by such devices is proportional to the flow rate,

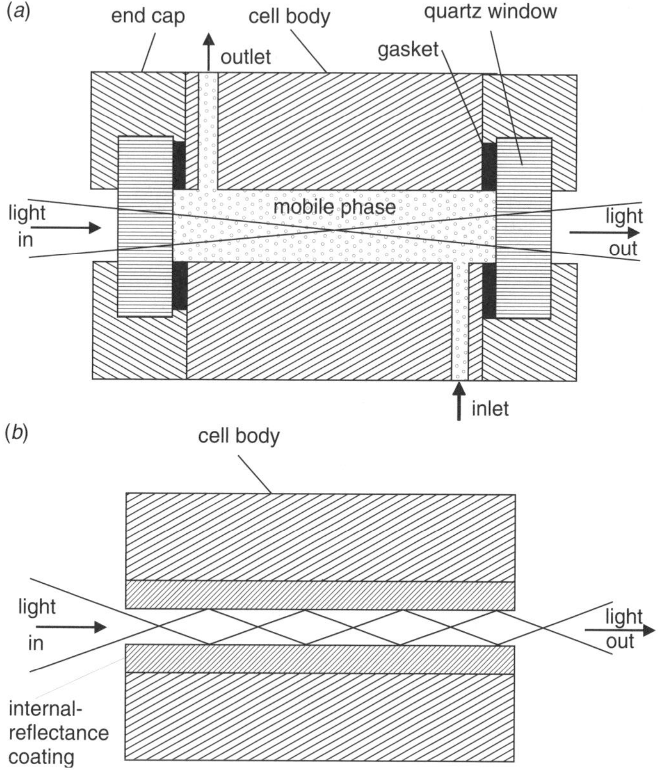

Figure 4.3 UV-detector flow cells. (*a*) Typical construction of a Z-path flow cell; (*b*) light-pipe flow cell lined with internally reflective surface.

so an increase in the pump flow rate can inadvertently create excessive back pressure and cause cell leakage or damage.

4.2.2 Detection Techniques

There are four general techniques that are used for HPLC detection:

- bulk property or differential measurement
- sample specific
- mobile-phase modification
- hyphenated techniques

4.2.2.1 Bulk Property Detectors

A bulk property detector can be considered a universal detector as it measures a property that is common to all compounds. The detector measures a change in this property as a differential measurement between the mobile phase containing the sample and that without the sample. The most familiar of the bulk property detectors is the refractive index detector (Section 4.11). Bulk property detectors have the advantage that they detect all compounds. The universal nature of bulk property detectors can be a disadvantage, as well, because all components of the sample that are eluted from the column will generate a detector signal. This means that additional chromatographic selectivity may be needed to make up for the lack of detection selectivity. As a general rule, universal detectors are inherently less sensitive, since they rely on the difference between two large measurements (solvent vs. solvent + solute).

4.2.2.2 Sample-Specific Detectors

Some characteristic of a sample is unique to that sample, or at least is not common to all analytes, and the sample-specific detector responds to that characterstic. The UV detector is the most commonly used sample-specific detector. It responds to compounds that absorb UV light at a specific wavelength. The distinction between bulk property detectors and sample-specific detectors is somewhat vague; for example, at low wavelength (<210 nm), organic compounds all absorb in the UV to some extent, so the UV detector becomes less selective and more universal at low detection wavelengths. Other common sample-specific detectors rely on the ability of an analyte to fluoresce (fluorescence, Section 4.5), conduct electricity (conductivity, Section 4.8), or react under specific conditions (electrochemical, Section 4.6).

4.2.2.3 Mobile-Phase Modification Detectors

These detectors change the mobile phase after the column to produce a change in the properties of the analyte. Such changes include a specific liquid-phase chemical reaction with the analyte (reaction detectors, Section 4.16), a gas-phase reaction (e.g., corona discharge, Section 4.13; mass spectrometric detectors, Section 4.14), or creation of analyte particles suspended in a gas phase (e.g., evaporative light-scattering detectors, Section 4.12.1).

4.2.2.4 Hyphenated Techniques

Hyphenated techniques refer to the coupling of an independent analytical instrument to the HPLC system to provide detection, and often are abbreviated with a hyphen as LC-(plus the technique). The most common hyphenated technique is LC-MS (Section 4.14), where a mass spectrometer is coupled with an HPLC system. Other less widely used techniques are LC-IR or LC-FTIR (Section 4.15.1) and LC-NMR (Section 4.15.2). One can speculate at what point a detector is no longer considered a hyphenated technique. Certainly the first UV detectors were UV spectrophotometers fitted with flow-through cells, but now UV detectors are just another HPLC detector. Generally, when the detector becomes widely used and the price of the detector is in the same general range as the rest of the HPLC system, it can be considered a detector. Thus it is reasonable to predict that mass spectrometers—which presently cost 2- to 5-fold more than an HPLC system—will eventually be regarded as just another HPLC detector.

4.2.3 Signal, Noise, Drift, and Assay Precision

The ability of a detector to provide precise and accurate quantitative data is a function of the signal size generated by the analyte, background noise, and—to a certain extent—baseline drift. Additional discussion of these topics is deferred to Section 11.2.4.

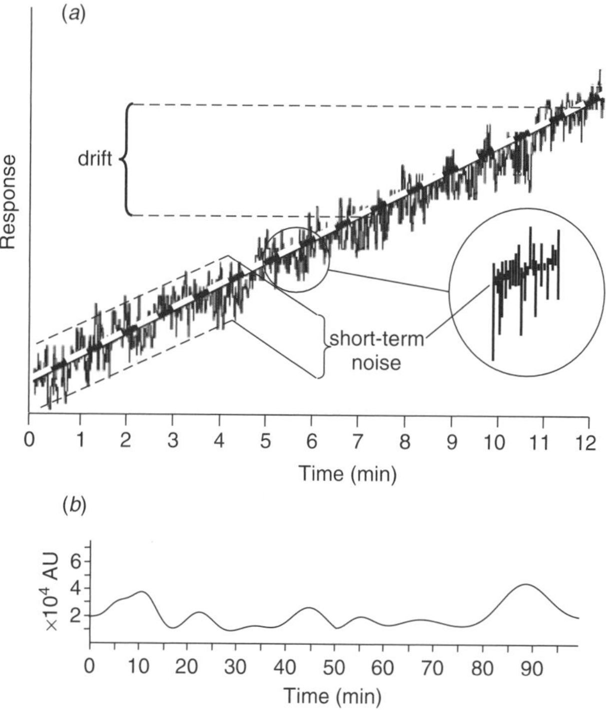

Figure 4.4 Detector noise and drift. (*a*) Baseline, showing noise and drift; inset expanded to show short-term noise; (*b*) long-term noise on baseline.

4.2.3.1 Noise and Drift

As defined in Figure 4.4, *short-term noise* is the baseline disturbance that occurs with a cycle time of $\ll$1 min (inset, Fig. 4.4*a*). Short-term noise is measured manually as the peak-to-peak noise bracketing the extremes of the baseline (between dashed lines of Fig. 4.4*a*; also shown in Fig. 4.7) in appropriate units (e.g., absorbance units for a UV detector). Alternatively, short-term noise can be determined by the data-system (usually with a root-mean-square calculation), if this feature is available. A common contribution to short-term noise is the "buzz" on the baseline resulting from noise contributed by motors or other appliances operated on the same electrical circuit (as in Fig. 4.5). The frequency of this noise corresponds to the 60 Hz electrical frequency, so it is referred to as 60-cycle noise (or 50-cycle in many countries).

Short-term noise often can be reduced or eliminated through the use of electronic noise filtration. Most detectors have user-selectable noise filtration built into their electronics. Alternatively, a resistance-capacitance (*RC*) filter can be built using inexpensive components purchased from Radio Shack plus directions from an introductory physics book. An example of the effectiveness of electronic noise filtration is shown in Figure 4.5, where the noise on the raw signal is reduced by $\approx$300-fold through the application of an *RC* filter with a 1-sec time constant [12]. If the time constant is too large, it will also reduce the signal. As a rule of thumb, if the time constant is less than $\approx$10% of the peak width, the peak signal will not be compromised. For example, a 10,000 plate peak at $k = 2$ generated by a

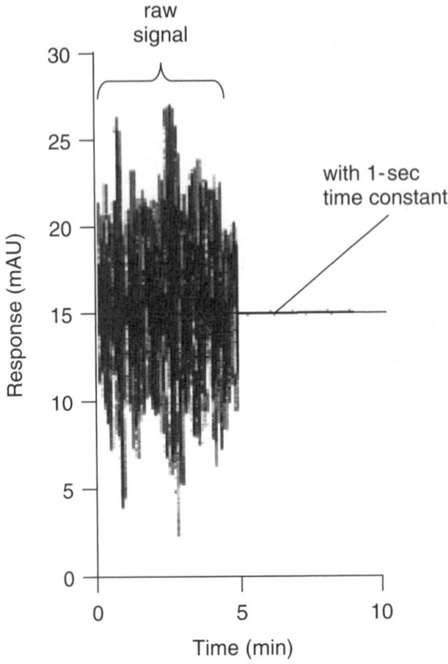

Figure 4.5 Effect of a resistance-capacitance noise filter on detector noise. Raw detector signal with ≈16 mAU short-term noise; 1-sec time constant reduces noise to <1 mAU.

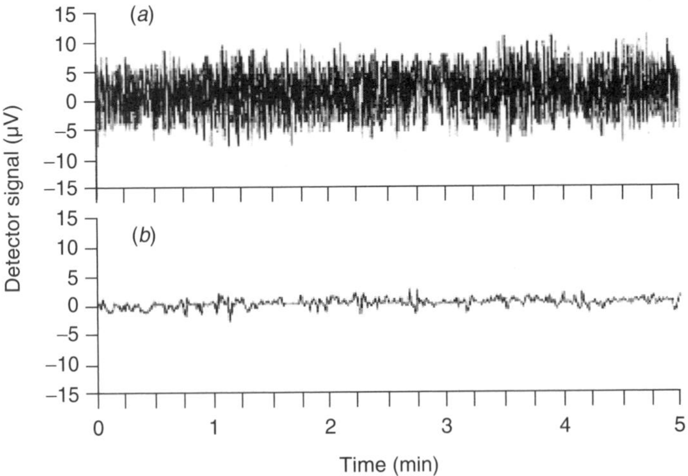

Figure 4.6 Effect of data rate on detector noise. Baseline plots of data generated with (*a*) data collection rate of 15 Hz and (*b*) 1 Hz (with data system input leads shorted). Adapted from [12].

150 × 4.6-mm, 5-μm column operated at 1.5 mL/min would have a peak width of ≈7 sec, which would translate into a suitable time constant of ≤0.7 sec.

The data-system sampling rate (Section 11.2.1.1) also can affect short-term noise. This is illustrated in Figure 4.6, where data sampling rates of 15 Hz (points per sec) and 1 Hz are compared [12]. The noise should be reduced by the square root of the data rate change; therefore the change in the sampling rate from 15 Hz (Fig. 4.6*a*) to 1 Hz (b) should reduce the noise ≈4-fold, as seen in Figure 4.6. The noise in Figure 4.6 is reduced from ≈12 to ≈3 μV, although it looks like much more because the high-frequency noise of Figure 4.6*a* is mostly eliminated in Figure 4.6*b*. A rule of thumb for setting the data sampling rate is to select a rate that collects ≥ 20 points across a peak. For the $N = 10{,}000$, $k = 2$ example above (peak width ≈7 sec), a data rate <7/20 sec or ≥ 20/7 $sec^{-1} \approx 3$ Hz would be suitable. *Remember that it is better to err on the side of oversampling when the data are initially recorded; the data rate can be reduced by post–run processing, but never increased (Section 11.2.1.1).*

Although the detector time-constant (noise filtration) and the data-system sampling rate (or data collection rate for processed data) both can be used to reduce baseline noise, they accomplish this in a different manner. It is therefore wise to evaluate both techniques (individually or in combination) to determine whether one or a combination is more effective at noise reduction.

Long-term noise is the fluctuation in the baseline that occurs with a periodicity in the same range as chromatographic peaks, as in Figure 4.4*b*. The longer period of this type of noise means that electronic noise filtration will not be effective. Long-term noise often is the result of elution of strongly retained materials from previous runs. As retention is increased, peak width is increased, eventually to the point where the peak deteriorates into baseline wander. A column flush with strong solvent (e.g., 100% ACN or MeOH) will often reduce long-term noise.

Baseline *drift* is the gradual shift in the baseline over the course of one or more runs, as illustrated by the heavy dashed line in Figure 4.4*a*. Baseline drift is common with gradient elution because of the difference in detector response (the concentration range over which the detector output is proportional to analyte concentration e.g., UV absorbance) between the starting and ending mobile phase (Section 17.4.5.1). Baseline drift with UV detection can often be compensated by adding a nonretained, UV-absorbing compound to the less-absorbant mobile-phase component (p. 178 of [13]). A change in the temperature of the mobile phase also can cause baseline drift, especially for refractive-index detectors and other detectors that are sensitive to refractive-index changes. Careful temperature control of the column, connecting lines, and detector, as well as isolation of the entire HPLC system from drafts can minimize drift. As long as the data system is able to adequately integrate peaks, a moderately drifting baseline will not have any detrimental effect on data quality.

Short-term and long-term noise are superimposed on baseline drift, as seen in Figure 4.4*a*. A good approach to minimize noise and drift problems is to (1) use appropriate detector time-constants and data-system sampling rates, (2) regularly flush the column with strong solvent, and (3) minimize temperature fluctuations within and near the HPLC system. Remember that some baseline drift is common in gradient elution, because of the difference in detector response to the A- and B-solvents, and may not be changed by the practices above.

4.2.3.2 Signal-to-Noise Ratio (S/N)

This ratio is more important than either the signal or noise alone as an indicator of detector performance for a particular peak. Noise is measured as described in Section 4.3.2.1 for short-term noise, as shown in Figure 4.7. The signal is measured

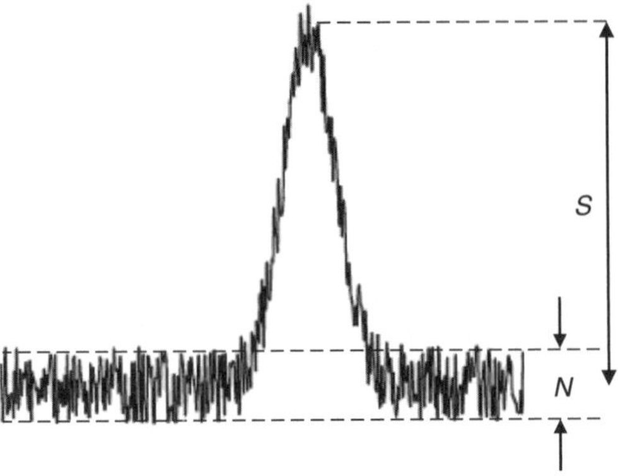

Figure 4.7 Measurement of chromatographic signal (S) and noise (N).

from the middle of the baseline noise to the top of the peak (Fig. 4.7). The contribution of S/N to precision can be estimated as

$$CV = \frac{50}{S/N} \tag{4.1}$$

where CV is the coefficient of variation (equivalent to %-relative standard deviation, %-RSD). The lower limit of detection ($LLOD$) often is described as $S/N = 3$, which would give $CV \approx 16\%$, whereas the lower limit of quantification ($LLOQ$) is $S/N = 10$, for $CV \approx 5\%$ (see also Section 11.2.4). These values of CV are the contribution of S/N to the overall imprecision of the method, so the overall method precision is expected to be worse than the S/N contribution. As long as the imprecision attributable to S/N (or any other single contributor to error) is less than half of the desired method imprecision, S/N will have a minor (<15%) influence on the overall method precision (see the discussion of Eq. 11.2). For example, if the overall method requires imprecision of no more than 2%, a contribution of S/N of <1% should be satisfactory. This suggests that a S/N value of ratio 50:1 or more is required for an overall method imprecision of <2%.

The signal-to-noise ratio can be improved by increasing the signal, reducing the noise, or both, as summarized in Table 4.1. An increase in signal for a given peak or sample may be available from a change in detector setting; for example, the use of a UV wavelength that corresponds to maximum sample absorptivity. A more sensitive detector also may be available, of either the same or different type.

Derivatization (Section 16.12) or other modification of the analyte may make it more responsive to the chosen detector. A more common means of increasing the signal is to inject a larger weight of sample (either a larger sample volume or a sample concentrate; Section 3.6.3). However, column or detector overload will eventually limit the possible increase in signal in this way. Any reduction in peak width should translate into a proportional increase in peak height (area is assumed to be constant); smaller k-values (increase in %B; see examples of Fig. 2.10*b*), narrow-diameter columns, or more efficient small-particle columns can each be used for this purpose.

Table 4.1

Improvement of Signal-to-Noise Ratio

Increase Signal	Decrease Noise
Better wavelength (or other detector adjustment)	Increase detector time constant
More sensitive detector	More data-system signal averaging
Analyte derivatization	Better temperature control
Inject larger weight of sample	Higher reagent/solvent purity
Reduce peak width (volumetric)	Better sample cleanup
Smaller *k*	Constant, pulse-free flow
Smaller column volume	Column switching
Larger plate number	

Any reduction of baseline noise also can improve S/N, for example, signal smoothing by an increase in the detector time-constant or data-system sampling rate (Section 4.2.3.1). Excessive smoothing, however, can reduce the signal intensity. Better temperature control of the column, detector, and general instrument environment also can reduce noise, especially for detectors sensitive to refractive index changes. Purer solvents (e.g., HPLC grade) and better sample cleanup can reduce the introduction of noise-generating contaminants. For gradient applications, changes in the system are sometimes attempted in order to reduce the dwell-volume (Section 9.2.2.4) and the gradient delay time. The mixer-volume comprises a major fraction of the dwell-volume in many systems, but reduction of the mixer-volume can increase baseline noise. Some HPLC systems have optional mixers that can be added to smooth the baseline and reduce noise—these devices can be especially advantageous for isocratic methods run at maximum detector sensitivity. Column switching (Section 16.9) can be used to transfer a desired fraction from a cleanup column to the analytical column, thereby diverting unwanted contaminants to waste, so as to reduce baseline noise.

4.2.4 Detection Limits

Although the signal-to-noise ratio is a measure of the inherent quality of the detector signal, the minimum detectable mass or concentration often is the limiting factor in the usefulness of a detector for a particular application. The term *sensitivity* often is used interchangeably with *detection limit* when describing an HPLC detector. However, in proper use, sensitivity is the slope of a calibration plot, that is, the change in signal per unit change in concentration (or mass) of analyte, whereas detection limit refers to the minimum concentration (or mass) that can be measured. HPLC detectors respond either to the concentration of the sample in the detector cell (e.g., UV detection) or the mass of sample in the detector (e.g., LC-MS).

Detection limits, discussed more thoroughly in Section 11.2.5, are defined as follows: The *limit of detection LOD* is the smallest signal that can be discerned from the noise—with confidence that a peak really is present. Often a S/N of 3 is equated to the LOD. The *lower limit of quantification LLOQ* (sometimes called *limit of quantification* or *limit of quantitation, LOQ*) is the smallest signal that can be measured with the required precision for the method. The $LLOQ$ often is defined as $S/N \geq 10$, but a value of $S/N \geq 50$ may be chosen for high-precision methods. There is a never-ending need for lower and lower detection limits for trace analysis, and assays for which on-column injections of <1 ng are becoming more and more common.

The LOD and $LLOQ$ are directly related to the concentration (or mass) of sample in the detector cell. Thus a longer path-length cell for UV detection is favored in terms of signal intensity. However, the detector cell should be designed with a minimum volume that is compatible with other requirements of the detector. Excess cell volume will result in additional extra-column peak broadening (Section 3.9). This is especially true for small-volume columns, columns packed with small particles, and peaks with $k<2$. For example, with a 50 × 4.6-mm column packed with 3-μm particles and $k<3$, significant peak broadening was observed for an 8-μL UV-detector cell when compared with a 1-μL cell [14]. To minimize the broadening of early-eluted peaks, the detector cell volume V_{det} should be less than approximately one-tenth of the final volume of the peak of interest V_p ($V_p = WF$, where W is the baseline width of the peak [min], and F is flow rate [mL/min]) [15]:

$$V_{det} < 0.1V_p \tag{4.2}$$

(For other peak-broadening contributions to V_p, see Eq. 3.1 in Section 3.9.)

Some examples of the column contribution to peak volume V_{p0} for early-eluted peaks ($k = 2$) for some popular column configurations are shown in Table 4.2. (In a well-behaved system, according to Eq. 2.27 and 3.1, the observed peak volume V_p should not be much larger than V_{p0}.) Table 4.3 lists the detector cell volumes for several UV-detector configurations. For UV detectors (Section 4.4), signal intensity is proportional to path length, so longer path flow cells will have lower detection limits. However, for detector cell diameters <1 mm, signal loss due to light scattering in the cell can be a problem, so special cell designs (e.g., total internal reflectance) are necessary for smaller cell diameters (see the discussion of Section 4.4). The data of Tables 4.2 and 4.3 show that column lengths $L \geq 100$ mm with a diameter $d_c = 4.6$ mm, packed with 5- or 3-μm d_p particles, will work well with the standard 10 × 1.0-mm UV cell (see (Eq. 4.2), but any combination of smaller column dimensions or smaller particles requires smaller cell volumes to avoid unnecessary extra-column peak broadening. (Note that Eq. 3.1 is an approximation, so peak-broadening calculations based on Eq. 3.1, and therefore conclusions based on Tables 4.2 and 4.3, also are approximations.)

4.2.5 Linearity

For quantitative analysis by HPLC (Section 11.4), the detector response must be related to the amount of analyte present. If analyte response y is plotted against analyte concentration x, the simplest, most convenient, and most reliable relationship is $y = mx$, where the slope m is a constant defined as the *sensitivity*. Such a relationship between analyte response and analyte amount is termed *linear*.

Table 4.2

Typical Peak Volumes V_{po}

L (mm)	d_c (mm)	d_p (μm)	V_{p0} (μL)[a]
250	4.6	5	212
150	4.6	5	164
		3	127
	2.1	3	26
100	4.6	3	104
	2.1	3	22
	1.0	3	5
		2	4
50	4.6	3	73
	2.1	3	15
	2.1	2	12
	1.0	2	3

[a]Assumes $k = 2$; reduced plate height $h = 2.5$ (see Eqs. 2.27 and 3.1).

Table 4.3

UV-Detector Cell Volumes

Path Length (mm)	Inner Diameter (mm)	Volume (μL)
10	1.0	8
	0.5	2
	0.25	0.5
5	1	4
	0.5	1
1	1	0.8
	0.5	0.2

For best use over a wide range of sample concentrations, a wide *linear dynamic range* (the concentration range over which the detector output is proportional to analyte concentration, e.g., 10^5 for UV detection) is desired, so that both major and trace components can be determined in a single analysis over a wide concentration range. For example, with a stability-indicating method, peaks $\geq 0.1\%$ of the response of the active ingredient ($= 100\%$) must be reported, which would require a linear range of at least $100/0.1 = 10^3$. Some detectors (e.g., evaporative light scattering) have a narrow linear range of 1 to 2 orders of magnitude. Although less convenient and reliable, a nonlinear calibration curve (e.g., quadratic) can be used—as long as the detector response changes in a predictable manner with sample concentration (or mass).

Table 4.4

HPLC Detectors

Sample-Specific (Sections 4.4–4.10)	Bulk Property (Sections 4.11–4.13)	Hyphenated (Sections 4.14–4.15)	Reaction (Section 4.16)
UV-visible	Refractive index	Mass spectrometric	Reaction
Fluorescence	Light scattering	Infrared	
Electrochemical	Corona discharge	Nuclear magnetic resonance	
Radioactivity			
Conductivity			
Chemiluminescent nitrogen			
Chiral			

4.3 INTRODUCTION TO INDIVIDUAL DETECTORS

The remainder of this chapter (Sections 4.4–4.16) provides a discussion of most HPLC detectors in use today. In Table 4.4, detectors are grouped by technique (sample specific, bulk property, etc.) in approximate order of popularity within each group. Sample-specific detectors will be treated first, and reaction detectors last—with only limited discussion of less-used detectors. Within each section, principles of detector operation are discussed first, followed by one or more example applications. Where appropriate, a comparison with other detectors is included.

A detailed discussion of every detector is beyond the scope of this book. In addition to the references cited in each section, a more general discussion of HPLC detectors can be found in [16, 17].

4.4 UV-VISIBLE DETECTORS

The most widely used detectors in modern HPLC are photometers based on ultraviolet (UV) and visible light absorption. These detectors have a high sensitivity for many solutes, but samples must absorb in the UV (or visible) region (e.g., 190–600 nm). Sample concentration in the flow cell is related to the fraction of light transmitted through the cell by *Beer's law*:

$$\log\left(\frac{I_o}{I}\right) = \varepsilon bc \qquad (4.3)$$

where I_o is the incident light intensity, I is the intensity of the transmitted light, ε is the molar absorptivity (or molar extinction coefficient) of the sample, b is the cell path-length in cm, and c is the sample concentration in moles/L. Light-absorption

HPLC detectors usually are designed to provide an output in *absorbance* that is linearly proportional to sample concentration in the flow cell,

$$A = \log\left(\frac{I_o}{I}\right) = \varepsilon bc \tag{4.4}$$

where A is the absorbance.

Properly designed UV detectors are relatively insensitive to flow and temperature changes. UV photometers that are linear to >2 absorbance units full scale (AUFS) with $<1 \times 10^{-5}$ AU noise are commercially available. With this performance, solutes with relatively low absorptivities can be monitored by UV, and it is possible to detect a few nanograms of a solute having only moderate UV absorbance. The wide linear range of UV detectors ($\approx 10^5$) makes it possible to measure both trace and major components in the same chromatogram.

UV detectors commonly use flow cells of the Z-path design of Figure 4.3*a*, with a 1-mm diameter and 10-mm path length (for a volume of ≈8 μL). This cell volume is adequate for ≥100 × ≥4.6-mm columns packed with ≥3-μm particles (Section 4.2.4), but smaller volume and/or smaller particle columns may experience unacceptable extra-column peak broadening in an 8-μL cell. Shorter path-length cells will reduce the cell volume, but the signal is proportional to the path length (Eq. 4.4)—so sensitivity must be balanced against extra-column peak broadening in choosing the flow cell dimensions. If the refractive index (RI) within the cell changes (e.g., during gradient elution), the amount of energy reaching the photodetector can change; when a light ray hits the side of the flow cell, the ratio of reflected to absorbed light depends on the refractive-index ratio of the mobile phase and cell wall (and the angle of the light ray hitting the cell wall). The latter refractive-index effect plus imperfections in optical alignment make it difficult to successfully use cell diameters smaller than ≈1 mm. One innovation that can minimize this problem is a flow cell design as in Figure 4.3*b*, where the internal surface of the flow cell is coated with a reflective coating—light that strikes the sides of the flow cell is reflected so as to still reach the photodetector [18, 19]. The use of this light-pipe technique allows the cell diameter to be reduced for smaller cell volumes (e.g., 0.25 mm × 10 mm ≈0.5 μL), and thus less peak spreading for use with very small-volume, small-particle columns. Alternatively, a longer, narrower diameter flow cell can be used (increasing b in Eqs. 4.3 and 4.4) for more absorbance in a smaller volume cell (e.g., 0.25 mm i.d. × 50 mm long, with a volume of ≈2.5 μL).

It is not necessary to operate a UV detector at the absorption maximum of the analyte. A hypothetical example of wavelength selection is shown in Figure 4.8. The spectra for two analytes, X and Y, are shown in Figure 4.8*a*, with UV maxima at ≈250 nm and ≈270 nm, respectively. At 280 nm, Y has much stronger absorbance, so it has a much larger peak (Fig. 4.8*b*, same mass on column). At 260 nm, the absorbances of X and Y are approximately equal, so the peaks are of approximately equal size (Fig. 4.8*c*). At 210 nm, both compounds have even stronger absorbance and generate much larger peaks (Fig. 4.8*d*). Notice also the appearance of a new peak Z, which was not observed at higher wavelengths. This general increase in sensitivity at lower wavelengths is one reason for the widespread use of ≤220 nm for detection (near-universal detection). The corresponding loss of *detector selectivity* at lower wavelengths can be a disadvantage for other separations, where it might

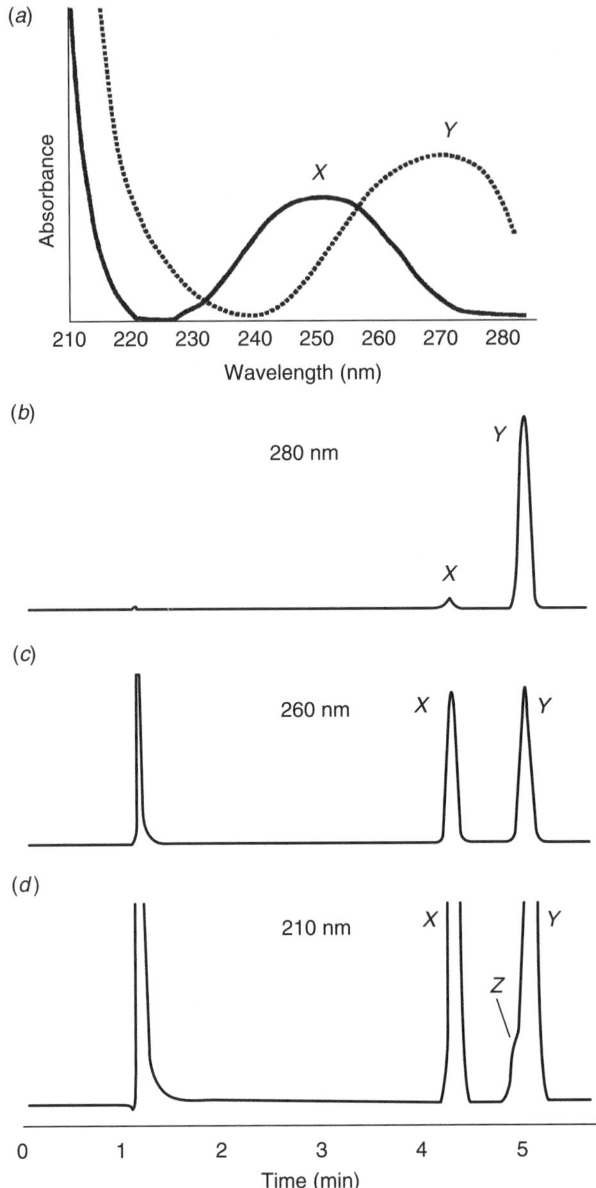

Figure 4.8 Wavelength selectivity for UV detection. (*a*) Absorbance spectra for two hypothetical compounds *X* and *Y*. Chromatograms at (*b*) 280 nm, (*c*) 260 nm, and (*d*) 210 nm.

be undesirable to "see" certain sample constituents (e.g., arising from the sample matrix).

UV-visible spectrophotometric detectors can respond throughout a wide wavelength range (e.g., 190–600 nm), which enables the detection of a broad spectrum of compound types. Almost all aromatic compounds absorb strongly below 260 nm;

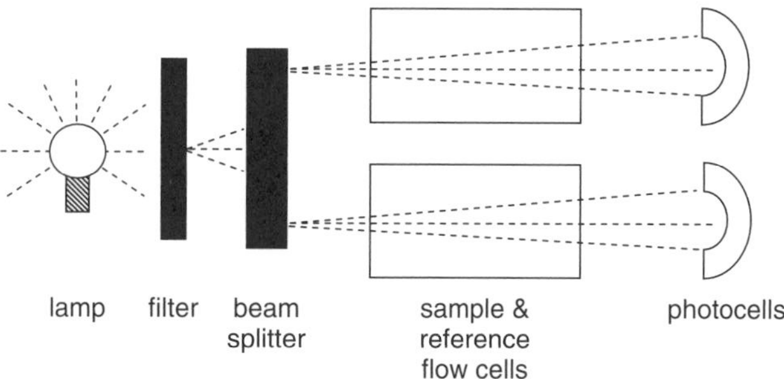

Figure 4.9 Schematic of a fixed-wavelength UV detector. Dashed lines show optical path.

compounds with one or more double bonds (e.g., carbonyls, olefins) can be detected at wavelengths of <215 nm, while the preponderance of aliphatic compounds possess significant absorbance at ≤205 nm. Reversed-phase mobile phases of acetonitrile plus water or phosphate buffer can be used routinely for detection at 200 nm, whereas methanol-containing mobile phases cannot be used below ≈210 to 220 nm, depending on methanol concentration (see Appendix I, Table I.2). The proper selection of the mobile phase makes it possible to operate UV detectors in a near-universal detection mode in the 200- to 215-nm region, where most organic compounds exhibit some UV absorbance. Because of the relatively small absorbance differential between water (or phosphate buffer) and acetonitrile at >200 nm or methanol at >220 nm, UV detectors are also quite useful for gradient elution. Mobile phases with large differences in UV absorbance, such as tetrahydrofuran and water at <240 nm, may not be amenable for use with gradients and UV detection.

UV detectors come in three common configurations. *Fixed-wavelength detectors* (Section 4.4.1) rely on distinct wavelengths of light generated from the lamp, whereas *variable-wavelength* (Section 4.4.2) and *diode-array* (Section 4.4.3) detectors select one or more wavelengths generated from a broad-spectrum lamp.

4.4.1 Fixed-Wavelength Detectors

Figure 4.9 is a generic schematic of a fixed-wavelength UV detector. These detectors were the mainstay of UV detection prior to the introduction of the variable- and diode-array detectors, but they are not widely used today. Their current appeal is low price and simple construction, and they tend to be more popular in the educational environment or other budget-limited settings.

UV radiation at 254 nm from a low-pressure mercury lamp passes through a band-pass filter and beam splitter, and shines on the entrance of the flow cell. Light transmitted through the flow cell strikes the photodetector (usually a photodiode) and is converted to an electronic signal. Most UV detectors operate in a differential absorbance mode, where light also passes through a reference cell, and the difference between the light passing through the sample and reference cells is converted to absorbance, according to Equation (4.4). Although some detectors enable the reference cell to be filled with mobile phase, an air reference is most commonly used, which allows for correction of variations in light intensity from the source lamp, but not for changes in the mobile-phase absorbance.

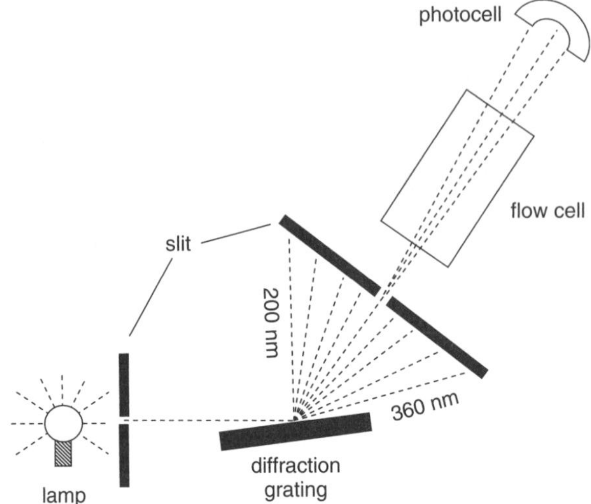

Figure 4.10 Schematic of a variable-wavelength UV detector; reference flow cell not shown. Dashed lines show optical path.

The 254-nm line from the low-pressure mercury lamp is the most popular wavelength for use with the fixed-wavelength UV detector. For historical reasons this wavelength is still popular for applications that use variable- and diode-array detectors, although there is no real reason to use this particular wavelength. Through the use of other lamps (e.g., zinc), phosphors, and other lines in the mercury lamp output, detection at 214, 220, 280, 313, 334, and 365 nm can be accomplished with the fixed-wavelength detector.

4.4.2 Variable-Wavelength Detectors

UV spectrophotometers (variable-wavelength and diode-array detectors) offer a wide selection of UV and visible wavelengths. Such devices have the versatility and convenience of operation at the absorbance maximum of a solute or at a wavelength that provides maximum selectivity, as well as the ability to change wavelengths during a chromatographic run.

The most widely used detector in HPLC today is the variable-wavelength UV detector shown schematically in Figure 4.10. A broad-spectrum UV lamp (typically deuterium) is directed through a slit and onto a diffraction grating. The grating spreads the light out into its component wavelengths, and the grating is then rotated to direct a single wavelength (or narrow range of wavelengths) of light through the slit and detector cell and onto a photodetector. These detectors usually use a sample and reference cell configuration (Section 4.4) for differential detection. For detection in the visible region, a tungsten lamp is used instead of deuterium.

The use of a variable-wavelength detector allows one to program a change in the detection wavelength during a chromatogram. Thus one peak can be detected at 280 nm and another at 220 nm. Although it is possible for many detector models to change the wavelength quickly, so as to generate a UV spectrum for a peak, the results are complicated by the change in analyte concentration during the spectral scan and may be of limited value.

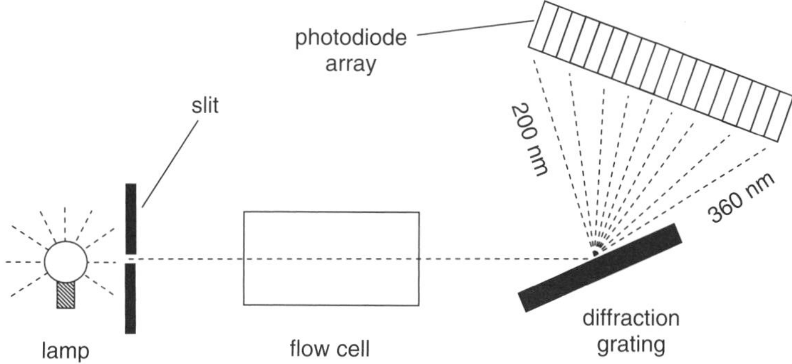

Figure 4.11 Schematic of a diode-array UV detector. Dashed lines show optical path.

4.4.3 Diode-Array Detectors

A schematic of the diode-array detector (DAD, also called photodiode-array, PDA) is shown in Figure 4.11; it has a similar optical path to the variable-wavelength detector, except that the white light from the lamp passes through the flow cell prior to striking the diffraction grating. This allows the grating to spread the spectrum across an array of photodiodes, hence the name *photodiode array* (PDA). The number of photodiodes varies with the specific brand and model of detector, but detectors with 512 or 1024 diodes are common. The signals from the individual photodiodes are processed to generate a spectrum of the analyte. Because the spectra are generated at the same time (vs. single-wavelength monitoring with the variable-wavelength detector), the DAD can contribute to peak identification. The DAD can be operated to collect data at one or more wavelengths across a chromatogram, or to collect full spectra on one or more analytes in a run. Of course, the data-file size is much larger for full-spectra runs, but data compression techniques and inexpensive data storage make this less of a concern than it was in the past.

If two closely eluted peaks have sufficiently different spectra, it may be possible to distinguish the two peaks spectrally. The utility of the DAD to distinguish between two peaks can be understood in conjunction with Figure 4.12, where a partial chromatogram for a closely eluted peak pair X and Y is shown (Fig. 4.12*a*). If the solutes have spectra as shown in Figure 4.12*b* and are monitored at a wavelength where both have significant absorbance, such as 260 nm, the resulting chromatogram will look like a single peak ($X + Y$ in Fig. 4.12*a*); the corresponding peaks are shown for the solutes injected individually. Even though the peaks appear to overlap completely at 260 nm, if other wavelengths are monitored, it may be possible to distinguish between the peaks. For example, if 240 nm is used, only X will respond, whereas only Y will respond at 280 nm. The added selectivity of the detector can be used to compensate, at least in part, for inadequate chromatographic separation. Thus the DAD could simultaneously collect data at 240 and 280 nm during the chromatographic run, and individual chromatograms plotted at 240 or 280 nm would allow quantification of X and Y, even under the partially overlapped conditions of Figure 4.12.

Another common application of the DAD is for peak-purity determination. The software accompanying the DAD accomplishes this by calculating an absorbance

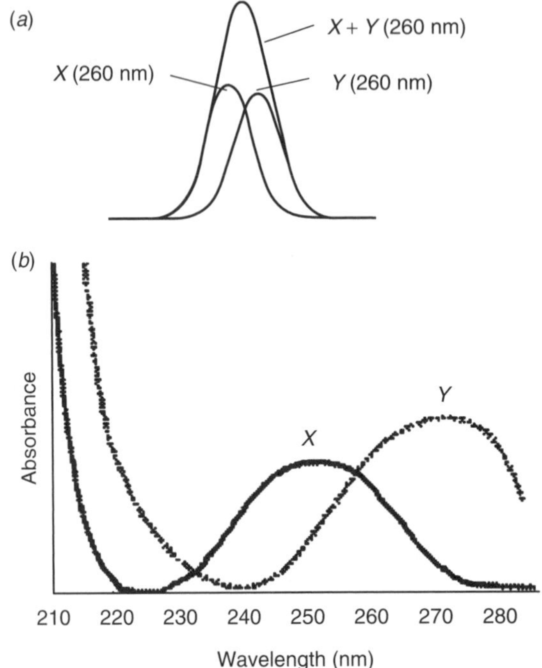

Figure 4.12 Illustration of spectral deconvolution of analytes. (*a*) Hypothetical chromatograms for individual injections of *X* and *Y* at 260 nm shown with combined response $X + Y$ at 260 nm; (*b*) spectra for *X* and *Y*.

ratio across the peak. The same dataset collected at 240 and 280 nm could be used to determine peak purity by calculation of the 240/280 ratio at every point across the peak. If the peak were pure *X* or *Y*, the ratio would be constant, whereas if the mixture of Figure 4.12*a* were present, the ratio would be >1 when *X* was predominant and <1 when *Y* was the major compound. The nonconstant nature of the ratio would indicate the presence of a peak mixture, even though the peaks overlapped chromatographically and appeared as a single peak at 260 nm. Peak-purity algorithms compare the consistency of the spectrum across the entire peak and in some cases can identify the presence of minor impurities (e.g., <1%) that are eluted under the tail of the major peak. For additional examples of the determination of peak purity by DAD, see [20, 21] or the detector manufacturer's literature (e.g., [22]).

4.4.4 General UV-Detector Characteristics

Table 4.5 summarizes the general characteristics of UV detectors. UV detectors are ideal for use with gradient elution; many common, UV-transmitting solvents are available in HPLC grade for use as mobile phases (Tables I.2 and I.3 of Appendix I). The UV detector is very useful for the trace analysis of UV-absorbing solutes, but its widely varying response for different solutes can be a disadvantage if the compound of interest does not absorb in the UV (or visible) region. UV detectors are reliable and easy to operate, and are particularly suitable for use by less-skilled operators.

Table 4.5

UV-Detector Characteristics

Capable of very high sensitivity (for samples that absorb in the UV)
Good linear range (>10^5)
Can be made with small cell volumes to minimize extra-column band broadening
Relatively insensitive to mobile-phase flow and temperature changes
Very reliable
Easy to operate
Nondestructive of sample
Widely varying response for different solutes
Compatible with gradient elution
Detection wavelength can be selected
Internal troubleshooting and calibration checks are common

Built-in test procedures that can be carried out at detector startup identify many potential detector problems and can provide automatic wavelength calibration.

The background, or baseline absorbance, of UV detectors can increase with continued use. This usually indicates that the cell windows have become dirty and need cleaning or replacement. Regular detector-cell flushing (as when the column is flushed) and sample cleanup can make more thorough cell cleaning a rarity. Lamp life, a concern in the past, is seldom an issue today. Useful lifetimes of >2000 hr are common, and internal circuitry monitors lamp performance and can alert the user when the lamp output has deteriorated. Although the linear response range of UV detectors may be >2 AU, according to manufacturer's specifications, most analysts try to operate the detectors at <1 AU for best results. Stabilizing the flow-cell temperature through thermostatting or use of a capillary-tubing heat exchanger helps to reduce noise and drift from flow rate or temperature changes.

Figure 4.13*a* shows an example chromatogram for the determination of derivatized roxithromycin (ROX) in human plasma by UV detection at 220 nm [23]. An internal standard, erythromycin (IS), was added to 50 μL of plasma followed by solid-phase-extraction sample cleanup and derivatization with 9-fluorenylmethyl chloroformate (FMOC-Cl). With UV detection at 220 nm, the method could monitor plasma concentrations of ROX but was unable to reach the LLOQ of <1 μg/mL necessary for pharmacokinetic studies. (See discussion of Section 4.5 for comparison of the UV response of Fig. 4.13*a* for this sample to the fluorescence response of Fig. 4.13*b*.)

4.5 FLUORESCENCE DETECTORS

Fluorescence detectors are very sensitive and selective for solutes that fluoresce when excited by UV radiation. Sample components that do not fluoresce do not produce a detector signal, so sample cleanup may be simplified. For example, a simple acetonitrile/buffer extraction allowed detection of as little as 30 pg of (naturally

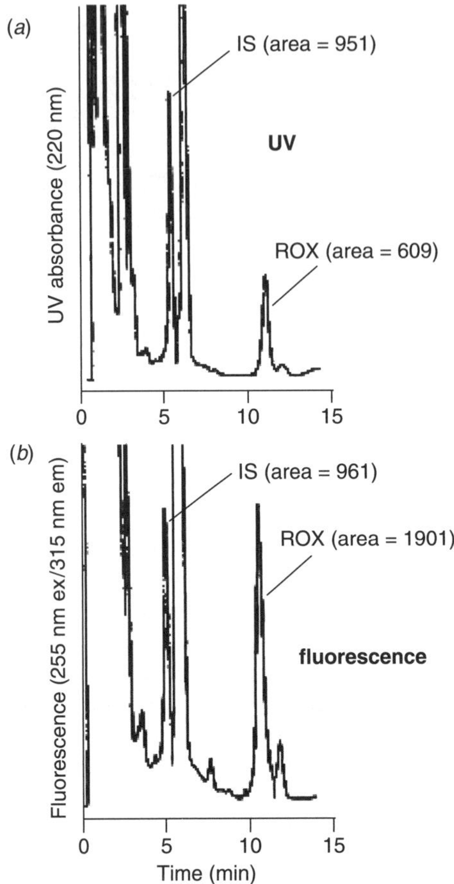

Figure 4.13 Chromatogram for the determination of roxithromycin (ROX) in human plasma by (*a*) UV detection at 220 nm, and (*b*) fluorescence detection (excitation 255 nm, emission 315 nm). Retention: ROX (10.7 min), internal standard erythromycin (5.1 min), both cleaned up by solid-phase extraction and derivatized with 9-fluorenylmethyl chloroformate (FMOC-Cl). Adapted from data of [23].

fluorescing) riboflavin in food products by HPLC with fluorescence detection [24]. Fluorescent derivatives of many nonfluorescing analytes can also be prepared (e.g., [25]), and this approach can be attractive for the selective detection of compounds for which sensitive or selective detection methods are otherwise not available.

A schematic of a fluorescence detector is shown in Figure 4.14. The light source usually is a broad-spectrum UV lamp, such as the deuterium lamp used in UV detectors, or a xenon flash lamp. The excitation wavelength is selected by a filter or monochromator, and it illuminates the sample as it passes through the flow cell. When a compound fluoresces, the desired emission wavelength is isolated with a filter or monochromator and directed to a photodetector, where it is monitored and converted to an electronic signal for data processing. Because fluorescence is emitted in all directions, it is common to monitor the emitted light at right angles to the incident light—this simplifies the optics and reduces background noise. The least

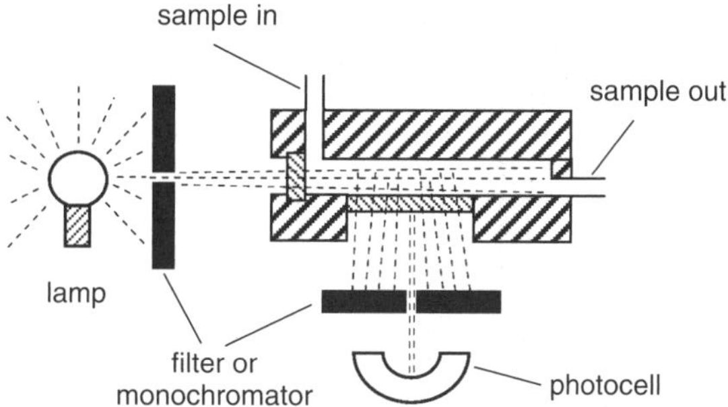

Figure 4.14 Schematic of a fluorescence detector. Dashed lines show optical path.

expensive fluorometers use filters to select both excitation and emission wavelengths, whereas the most expensive use two monochromators (allowing a wide choice for both excitation and emission wavelengths). Remember, the fluorescence process is not 100% efficient, so energy is lost. This means that the emission wavelength always must be at lower energy (higher wavelength) than the excitation wavelength.

For many samples, the fluorescence detector is 100-fold more sensitive than UV absorption—and is one of the most sensitive HPLC detectors. In other cases the sensitivity advantage of fluorescence over UV detection may be smaller but adequate for the task at hand. A comparison of the detector response to roxithromycin (ROX) by fluorescence and UV is shown in the RPC separations of Figure 4.13 [23]. ROX does not fluoresce naturally, so derivatization (9-fluorenylmethyl chloroformate [FMOC-Cl]) of the sample and internal standard (IS) was used to enable detection by fluorescence. When comparing the UV response (Fig. 4.13*a*) to fluorescence (Fig. 4.13*b*), the fluorescence response for the derivatized IS is approximately the same as the UV response, but the derivatized ROX peak response tripled with fluorescence detection. The baseline noise was approximately the same for both UV and fluorescence. This increase in response by the fluorescence method was adequate to reduce the LLOQ to <1 μg/mL of ROX in human plasma, which was required for pharmacokinetic studies.

Because of its high sensitivity the fluorescence detector is particularly useful for trace analysis, or when either the sample size is small or the solute concentration is extremely low. The linear dynamic range of the fluorescence detector usually is smaller than that of UV detectors, but it is more than adequate for most trace analysis applications. While the dynamic range (the range over which a change in sample concentration produces a change in the detector output) of fluorescence detectors can be fairly large (e.g., 10^4), the *linear* dynamic range may be restricted for certain solutes to relatively narrow concentration ranges (as low as 10-fold). For all quantitative analyses using the fluorescence detector (or any other detector, for that matter), the linear range should be determined through the use of appropriate calibration (Section 11.4.1).

In comparison to other detection techniques, fluorescence generally offers greater sensitivity and fewer problems with instrument instability (e.g., from temperature and flow changes). If solvents and mobile-phase additives free of fluorescing materials are used, the fluorescence detector can be used with gradient elution. The major disadvantage of the fluorescence detector is that not all compounds fluoresce.

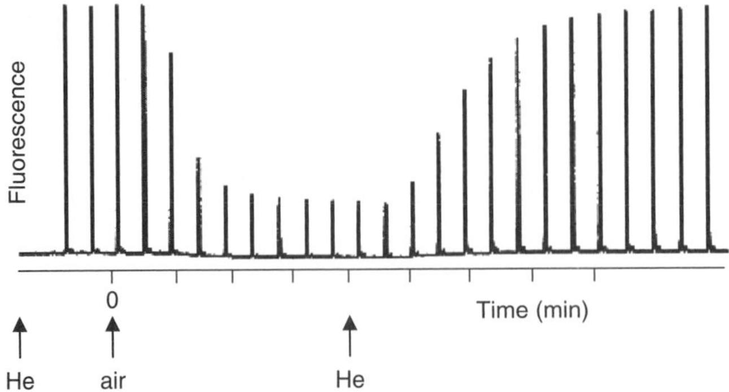

Figure 4.15 Fluorescence quenching of naphthalene by dissolved oxygen in the mobile phase. Mobile phase sparged with helium (He) or air, as shown. Adapted from data of [25].

As with other fluorescence techniques, fluorescence detection can be compromised by background fluorescence of the mobile phase or sample matrix, and by quenching effects. An example of fluorescence quenching is shown in Figure 4.15 [25]. When the mobile phase is sparged with helium, a consistent signal is observed, but when air is bubbled through the mobile phase, the signal drops because oxygen quenches the fluorescence of the naphthalene peak (250-nm excitation, 340-nm emission). Sparging the oxygenated mobile phase with helium then displaces the oxygen and the signal returns to normal. The presence of oxygen in the mobile phase also shifts the baseline slightly, but this is of minor concern.

The use of a laser (laser-induced fluorescence, LIF) as the excitation source is available in the LIF detector. The higher energy of the laser over the conventional deuterium or xenon lamp gives added sensitivity to this detector, but the excitation wavelength range is more limited (300–700 nm vs. 200–700 nm for conventional fluorescence). LIF detection is not widely used with conventional HPLC systems, but is more common with micro applications (micro-LC, capillary LC, capillary electrophoresis, etc.) where a small diameter (e.g., 100-μm i.d.) flow cell is required to limit dispersion.

4.6 ELECTROCHEMICAL (AMPEROMETRIC) DETECTORS

Many compounds that can be oxidized or reduced in the presence of an electric potential can be detected at very low concentrations by selective electrochemical (EC) measurements. By this approach the current between polarizable and reference electrodes is measured as a function of applied voltage. Because a constant voltage normally is imposed between the electrodes, and only the current varies as a result of solute reaction, EC detectors are more accurately termed *amperometric* devices. EC detectors can be made sensitive to a relatively wide variety of compound types, as illustrated in Table 4.6. EC detection is common for the determination of catecholamine and other neurotransmitters. Many of the compounds in Table 4.6 also can be detected by UV absorption, but some compound types (e.g., aliphatic mercaptans, hydroperoxides) sensed by EC detection cannot be detected at all by UV absorption, or only with difficulty and low sensitivity at low wavelengths.

Table 4.6

Some Compound Types Sensed by the EC Detector

Oxidation	Reduction
Phenolics	Ketones
Oximes	Aldehydes
Mercaptans	Oximes
Peroxides	Conjugated acids
Hydroperoxides	Conjugated esters
Aromatic amines, diamines	Conjugated nitriles
Purines	Conjugated unsaturation
Heterocyclic rings[a]	Activated halogens
	Aromatic halogens
	Nitro compounds
	Heterocyclic rings

Note: Compound types generally not sensed include ethers, aliphatic hydrocarbons, alcohols, and carboxylic acids.
[a]Detected depending on structure.

EC detectors can be used only under the condition that the mobile phase is electrically conductive, but this is a minor limitation, since most HPLC separations are done by reversed-phase with water or buffer in the mobile phase. By fine-tuning the detector potential, one can achieve great selectivity for electroactive compounds. The EC detector's sensitivity makes it one of the most sensitive of all HPLC detectors, for example with detection limits to 50 fg on-column of dopamine. However, to operate under high sensitivity, extra care must be taken to use highly purified mobile phases to reduce background noise. In order to reduce the background noise, in some applications the mobile phase is routed through a high-potential pretreatment cell so as to oxidize or reduce background interferences before the mobile phase reaches the autosampler.

A glassy carbon electrode is most commonly used in the electrochemical cell. In the configuration shown in Figure 4.16, the column effluent flows across a glassy carbon electrode, whereas in another popular configuration, the sample flows through a porous graphite electrode. Several electrode styles are available, for example, Figure 4.16*c* shows a dual-electrode configuration. The high susceptibility of the EC detector to background noise and electrode contamination has earned it a reputation as a difficult detector to use. However, newer units are much more trouble free and can provide excellent and reliable results in the hands of a reasonably careful operator.

Figure 4.17 shows the electrochemical detection of acteoside, an active ingredient in many Chinese medicinal plants. Following intravenous administration of acteoside at 10 mg/kg, the analyte was detected in rat brain microdialysate at a concentration of ≈25 ng/mL (≈0.4 ng on-column) by reversed-phase HPLC [26]. More information about electrochemical detectors can be found in [27].

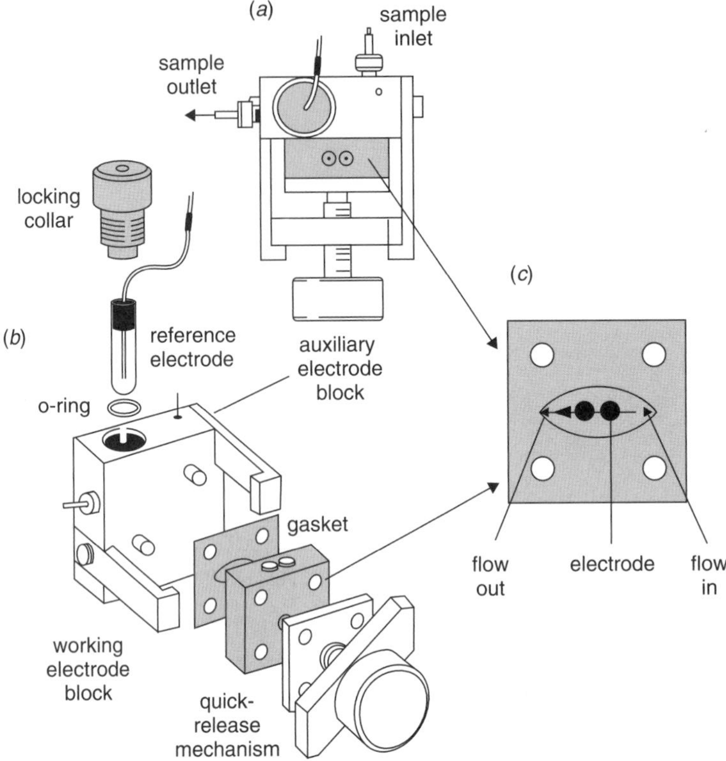

Figure 4.16 Schematic of an electrochemical detector. (*a*) Top view of assembled flow cell; (*b*) exploded diagram of cell; (*c*) detail of dual electrode cell. Courtesy of Bioanalytical Systems, Inc.

4.7 RADIOACTIVITY DETECTORS

Radioactivity detectors are used to monitor radio-labeled solutes as they elute from the HPLC column. Detection is based on the emission of light in the flow cell as a result of radioactive decay of the solute, followed by emission of α-, β-, or γ-radiation. The continuous-flow monitoring of β-radiation in the eluent ordinarily involves the use of a scintillation technique, where the original radiation is converted to light. Depending on the method of combining the eluent and the scintillator, this can be classified as either a homogeneous or heterogeneous system. In *homogeneous* operation, a liquid-scintillation cocktail is mixed with the column effluent prior to entering the detection cell, where emitted light is monitored. Under *heterogeneous* conditions, the column outlet is routed directly into the detector cell, which is packed with beads of a solid scintillant. When adsorption of the analyte on the beads is a problem, the scintillant may be coated onto the walls of the detector cell.

Homogeneous detectors are best used with analytical procedures where recovery of the sample is unimportant. The technique also can be applied to preparative HPLC, when a portion of the sample stream is split off to the detector. Heterogeneous detectors are less sensitive, and therefore better suited for samples with

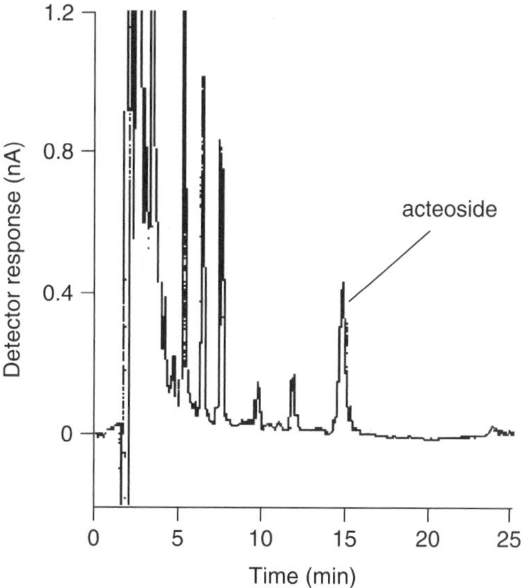

Figure 4.17 Determination of acteoside ($t_R \approx$ 15 min) in rat brain microdialysate with electrochemical detection. Adapted from data of [26].

higher levels of radioactivity (or for larger solute concentrations, as in preparative separations). Heterogeneous systems also are relatively free of chemical quenching effects, and solutes can be recovered easily. However, this detector exhibits relatively low counting efficiency for low-energy β-emitters, such as ^{35}S, ^{14}C, ^{3}H, and ^{32}P, and is better suited for stronger α-, β-, and γ-emitters (e.g., ^{131}I, ^{210}Po, and ^{125}Sb). One application of the radioactivity monitor is to determine the complete distribution and mass balance of a radio-labeled pharmaceutical dosed in an experimental animal. Such determinations are difficult, if not impossible, without the aid of radio-labeled drugs.

Radiochemical detectors have a wide response range and are insensitive to solvent change, making them useful with gradient elution. With radioactivity detectors, it may be necessary to compromise sensitivity to improve chromatographic resolution and speed of analysis. Detection sensitivity is proportional to the number of radioactive decays that are detected, and this number is proportional to the volume of the flow cell and inversely proportional to the flow rate (proportional to residence time, which allows more atoms to decay during passage of a peak through the flow cell). Larger flow-cell volumes increase extra-column peak broadening and can diminish resolution, while slower flow rates mean an increase in separation time. Because detection sensitivity is often marginal, larger flow cells are generally preferred for radioactivity detection.

In practice, peak tailing and peak broadening in a radiometric flow cell can be minimized by working with columns of larger volume (assuming that sufficient sample is available for larger mass injections to compensate for sample dilution). With radioactivity detection, a compromise between chromatographic resolution and detector sensitivity must be reached, the exact nature of which depends on the analytical requirements.

4.8 CONDUCTIVITY DETECTORS

Conductivity detectors use low-volume detector cells to measure a change in the conductivity of the column effluent as it passes through the cell. Conductivity detectors are most popular for ion chromatography and ion exchange applications in which the analyte does not have a UV chromophore. Analysis of inorganic ions (e.g., lithium, sodium, ammonium, potassium) in water samples, plating baths, power plant cooling fluids, and the like, is an ideal use of the conductivity detector. Organic acids, such as acetate, formate, and citrate are also conveniently detected by conductivity.

Conductivity detection can be compromised by the presence of a conductive mobile phase; for example, the mobile-phase buffer. Thus the presence of the buffer greatly increases the conductance of the mobile phase, which is only slightly increased by the presence of the solute. One way to minimize this problem is to use a suitable buffer in combination with a suppressor column (ion exchanger), in order to reduce the background conductivity of the mobile phase. For example, consider the need to detect one or more anionic solutes. The use of a Na_2CO_3-$NaHCO_3$ buffer with a cation-exchange suppressor column (termed an *anion suppressor* in ion chromatography terms) in the H^+ form will eliminate Na^+ and other cations from the mobile phase, and convert carbonate to weakly acidic carbonic acid. This reduces the conductivity of the mobile phase and allows an easier detection or small concentrations of anionic solutes. The application of a suppressor column is illustrated in Figure 4.18 for the dramatic improvement in conductivity detector response to F^-, Cl^-, and SO_4^{2-}.

4.9 CHEMILUMINESCENT NITROGEN DETECTOR

One advantage that gas chromatography has over HPLC is the availability of several element-specific detectors, allowing selective detection of compounds containing nitrogen, sulfur, or phosphorus. In the 1970s much effort was given to developing element-specific detectors for HPLC, but for the most part the results have been discouraging. One exception is the chemiluminescent nitrogen detector (CLND), which was reported as early as 1975 [28]. Several commercial implementations and refinements have resulted in today's CLND.

The HPLC column effluent is nebulized with oxygen and a carrier gas of argon or helium and pyrolyzed at 1050°C. Nitrogen-containing compounds (except N_2) are oxidized to nitric oxide (NO), which is then mixed with ozone to form nitrogen dioxide in the excited state (NO_2*). NO_2* decays to the ground state releasing a photon, which is detected by a photometer. The signal is directly proportional to the amount of nitrogen in the original sample, so calibrants of known nitrogen content can be used to quantify the nitrogen content of unknown analytes. This is illustrated in Figure 4.19*a* [29], where the injection of 50-ng nitrogen equivalents of 7 different compounds give detector responses that are constant within ±10%. Care must be taken to maintain a nitrogen-free mobile phase, so the use of acetonitrile is ruled out. Many solvents are compatible with the CLND, as is shown in Figure 4.19*b* for the response of the injection of 1 mg each of 6 nitrogen-free solvents, compared to an injection of 1 ng nitrogen-equivalent of a standard.

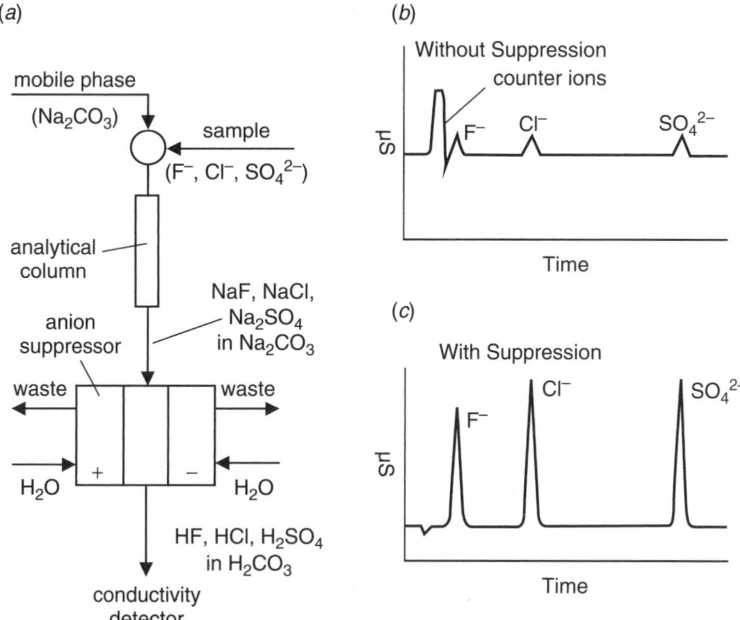

Figure 4.18 Use of an anion suppressor column to enhance conductivity detector response to anionic analytes. (*a*) Schematic of instrumentation; (*b*) conductivity detector output without suppressor column; (*c*) chromatogram with suppressor column in use. Courtesy of Dionex.

One detector manufacturer claims detection limits equivalent to 0.1 ng of nitrogen. A practical example is seen in Figure 4.20*a* [30] for the detection of 13 underivatized amino acids by ion-pair chromatography and CLND. The response per nitrogen atom is within 6% RSD, with detection limits of ≈0.3 to 0.5 μg/mL for the amino acids. Figure 4.20*b* shows the chromatogram for an injection of 10 μL of wine filtered through a 1000-Da filter (note overloaded proline peak shows shorter retention and strong tailing compared to *a*; see Section 2.6 for further discussion of overload).

4.10 CHIRAL DETECTORS

Chiral drug candidates often are encountered in the development of new pharmaceutical compounds. Different enantiomers can possess different efficacy, toxicology, or other pharmacological characteristics, and the final product generally is a single enantiomer or a known mixture of enantiomeric forms. Chromatographic separation of the enantiomers (Chapter 14) is vital to the analysis of such mixtures. Detection and identification can be further aided by the use of detectors that respond selectively to specific chiral forms.

Chiral detectors come in three different formats; each of these uses the same principles as stand-alone instrumentation, but in a flow-cell format. *Polarimeters* (PL) measure the degree of rotation of polarized light (typically in the 400–700 nm range) as it passes through the sample. The degree of rotation is dependent on

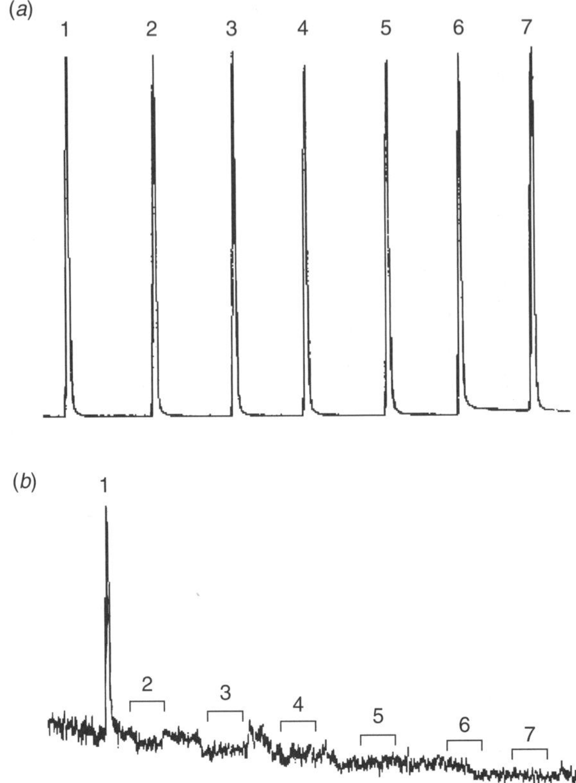

Figure 4.19 Response of chemiluminescent nitrogen detector (CLND) for different compound types. (*a*) Response of 50 ng nitrogen-equivalent of *1*, *N*,*N*-dimethyl aniline; *2*, nitrobenzene; *3*, miconazole nitrate; *4*, nicotinamide; *5*, 4-acetamidophenol; *6*, glycine; *7*, caffeine. (*b*) Response of *1*, 1 ng nitrogen-equivalent of standard to 1 mg injected solvents: *2*, acetone; *3*, ethyl acetate; *4*, hexane; *5*, isopropanol; *6*, methanol; *7*, water. Adapted from [29].

both the concentration of the chiral compound and its molecular structure. *Optical rotary dispersion* (ORD) detectors operate on a similar principle to polarimeters, but use lower wavelengths (e.g., the 365-nm mercury emission line), which in theory should give stronger signals. *Circular dichroism* (CD) detectors are based on measuring the difference in absorption of right and left circularly polarized light when an analyte passes through the flow cell. For strong CD signals, it is desired that the analyte have a chromophore with absorption in the 200 to 420 nm range.

In the example of Figure 4.21 [31], the response of CD, ORD, and UV detection is compared for the chiral chromatographic separation of ibuprofen enantiomers. The CD detector generates peaks with signal to noise (S/N) about 5-fold larger than the ORD detector, but only half that of the UV detector. Note that the two chiral detectors produce both negative and positive peaks. Another study [32] compared the response of PL, ORD, and CD detection for 6 pharmaceutical compounds. For naproxen, CD was about 6-fold more sensitive than PL, and 24-fold more than

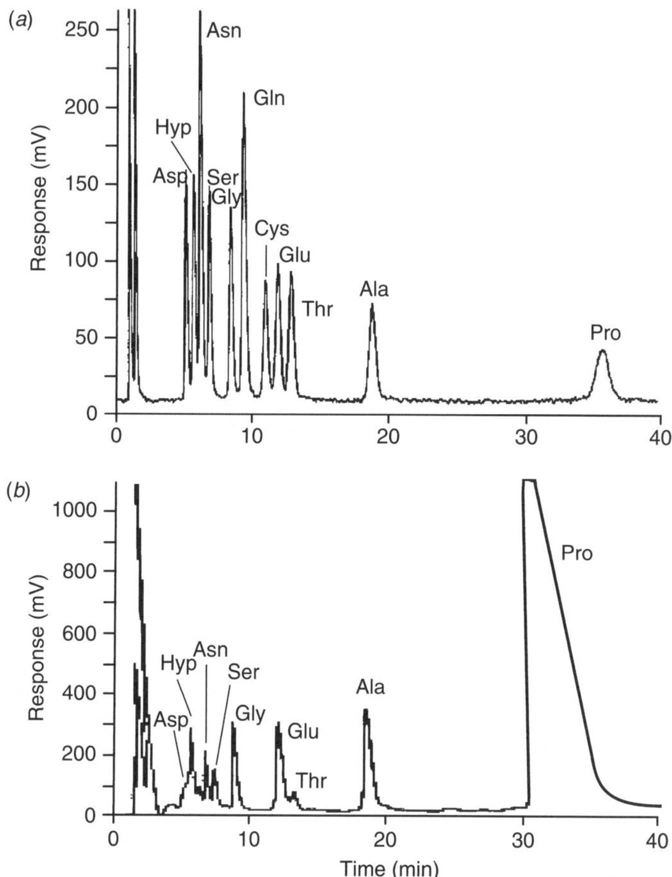

Figure 4.20 Response of chemiluminescent nitrogen detector (CLND) for amino acids. (*a*) 10 μL injection of a 0.1 mM standard solution of 13 underivatized amino acids; (*b*) 10 μL injection of wine filtered through a 1000 Da filter. Adapted from data of [30].

ORD. The relative response for the various test compounds varied, but CD was superior in all cases.

4.11 REFRACTIVE INDEX DETECTORS

The differential refractometer, or refractive index (RI) detector, responds to a difference in the refractive index of the column effluent as it passes through the detector flow cell. The RI detector is a bulk-property detector that responds to all solutes, if the refractive index of the solute is sufficiently different from that of the mobile phase. The most popular RI detector design is the deflection refractometer illustrated in Figure 4.22*a*. Light from the source lamp (typically tungsten) is directed through a pair of wedge-shaped flow cells. One cell is the reference cell, typically containing a trapped (static) sample of mobile phase; column effluent is directed through the sample cell. As the light passes through the detector cells,

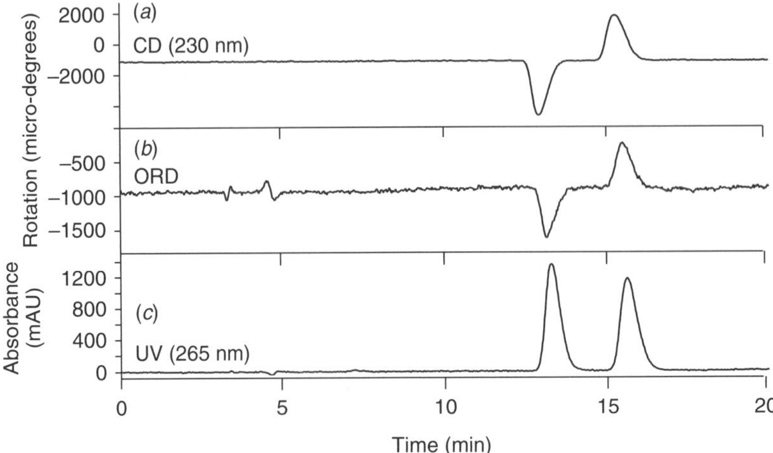

Figure 4.21 Comparison of response of circular dichroism (CD), optical rotation (ORD), and UV detectors for a 10 μg injection of ibuprofen. (*a*) CD at 230 nm, $S/N = 49.6$; (*b*) ORD, $S/N = 10.9$; (*c*) UV at 265 nm, $S/N = 113.4$. Adapted from [31].

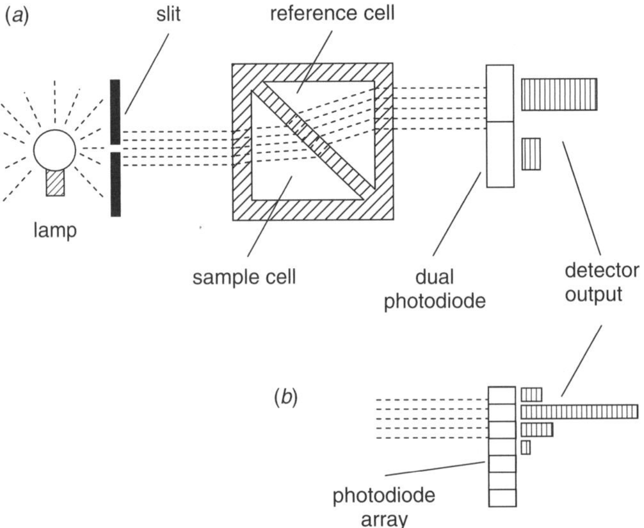

Figure 4.22 Schematic of a deflection refractive index (RI) detector. (*a*) Dual-photodiode detector; (*b*) photodiode array detector (lamp and flow cell not shown). Dashed lines show optical path.

it is refracted differently, depending on the instantaneous conditions in the cell. A pair of photodiodes measures the change in refraction (position of the beam) of the light passing through the flow cell and converts this to an output voltage. The conventional RI detector uses two photodiodes, as shown in Figure 4.22*a*. As the refractive index of the sample solution changes, the light is deflected so that the amount of light reaching each photodiode changes. More recent application

Table 4.7

Characteristics of Refractive Index Detectors

Excellent versatility; all solutes can be detected
Moderate sensitivity
Generally not useful for trace analyses
Not useful for gradient elution
Efficient heat-exchanger required
Sensitive to temperature changes
Reliable, fairly easy to operate
Nondestructive

of photodiode-array technology to the RI detector allows multiple photodiodes to be used for detection, as shown in Figure 4.22*b*. This configuration is claimed to improve the dynamic range of the RI detector and increase detector sensitivity.

Table 4.7 summarizes the characteristics of RI detectors. Because they respond to all solutes, these devices have excellent versatility if the mobile phase is properly selected. For maximum RI detector sensitivity, the mobile phase should have a refractive index as different from the solute as possible (Table I.3 of Appendix I). However, even under optimum conditions RI detectors possess only modest sensitivity. Although this detector generally is not useful for trace analysis, it is possible under optimum conditions to quantify peaks at the 0.1% concentration level. A severe limitation of RI detectors is that they are unsuitable for use with gradient elution, since it would be exceedingly difficult to match the refractive indexes of the reference and sample streams (see exception in the discussion of Fig. 4.25*a*, Section 4.12.1). Even isocratic mobile-phase composition changes that are insignificant with UV detection can show up as baseline noise or ripple. For best results hand-mixed mobile phases will give quieter baselines than those prepared by on-line mixing. Despite the sensitivity limitation and impracticality in gradient elution, the differential refractometer is widely used, particularly in size-exclusion chromatography, where sensitivity is not as critical.

The sensitivity of RI detectors to temperature change also represents a severe limitation. Current models of RI detectors have been carefully designed to minimize temperature fluctuations through the use of constant-temperature detection environments and efficient heat exchangers to thermally equilibrate the mobile phase stream with the detector. For best performance, the RI detector should be turned on at all times, or allowed to warm up for at least two hours prior to use. Another good tip is to insulate the tubing connecting the column to the detector so as to minimize temperature fluctuations. Refractometers are convenient and reliable, although generally not as trouble free and easy to operate as UV detectors.

RI baseline drift can result when changing from one bottle of "pure" solvent or "identical" hand-mixed mobile phase to another. Baseline drift can be severe when different solvents are involved, until the first solvent is completely flushed from the HPLC equipment and column. To maintain a homogeneous composition of the mobile phase during a series of runs, a sufficiently large volume of mobile

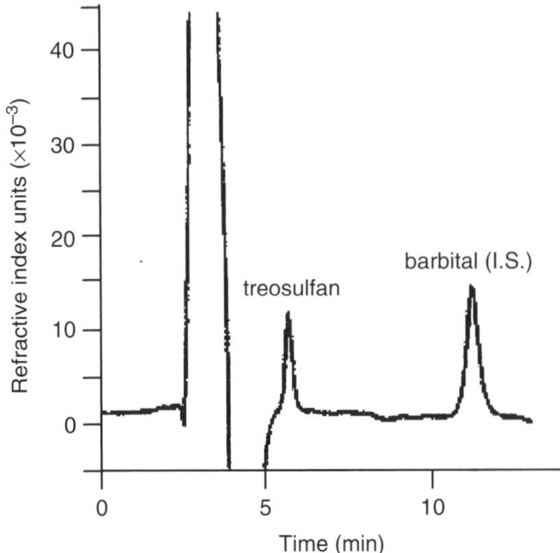

Figure 4.23 Refractive index detector response for 560 μg/mL treosulfan 1 hr after onset of intravenous infusion; barbital is used as an internal standard. Adapted from data of [33].

phase should be formulated, with continuous stirring within the reservoir. For acceptable baseline stability, any change in the solvent composition (due to degassing, evaporation, water vapor pickup, etc.) normally should be avoided.

In the past RI detectors based on a Fresnel design or interferometric detection were available, but the deflection refractometer is most popular today. For many years, the RI detector was the only option for "universal" detection with HPLC. Today, light-scattering detectors (Section 4.12) are replacing RI detectors for many applications. Low-wavelength UV detection (<210 nm) also provides better sensitivity than RI for many compounds that have very weak UV absorbance at higher wavelengths (see Fig. 4.25 for some comparisons of UV, RI, and ELSD responses).

The sensitivity of RI detection usually precludes its use in routine drug monitoring, but in some cases it has proved useful for the determination of drug concentrations in biological samples, for example, when high drug concentrations are present, and other detection techniques have failed. In the example of Figure 4.23 the RI detector is used to measure treosulfan (L-threitol-1,4-methanesulfonate) levels in pediatric plasma [33]. Treosulfan is an antitumor drug that is toxic to stem cells, and is administered intravenously prior to a stem cell transplant to kill all the native stem cells. Figure 4.23 shows a chromatogram for 560-μg/mL treosulfan in pediatric plasma following infusion of the drug; adjusting for sample preparation, this is equivalent to an injection of 83 μL of plasma.

4.12 LIGHT-SCATTERING DETECTORS

In recent years improvements in light-scattering detectors have led to their replacing the refractive index (RI) detectors for many applications. One reason for this transformation (which has also boosted the practicality of mass spectrometric detectors)

Table 4.8

Comparison of Refractive Index and Light-Scattering Detectors

Property	RI[a]	ELSD[a]	CNLSD[a]	LLSD[a]
Universal response	+	+	+	+
Sensitivity	–	0	+	na
Gradient compatibility	–	+	+	+
Volatile mobile phase required	No	Yes	Yes	No
Temperature sensitivity	–	+	+	+
Provides qualitative data to assist structural determination	–	–	–	+

Note: +, good; o, intermediate; –, very poor; na, does not apply (detector designed for qualitative information, not sensitivity).
[a]RI, refractive index; ELSD, evaporative light-scattering detector; CNLSD, condensation nucleation light-scattering detector; LLSD, laser light-scattering detector.

is the ability of the light-scattering detector to efficiently nebulize the column effluent and evaporate the mobile phase. The most popular is the *evaporative light-scattering detector* (ELSD). The *condensation nucleation light-scattering detector* (CNLSD) is a modification of the ELSD that can provide increased performance. On the high-end of the price range are *laser light-scattering detectors* (LLSD), which occupy more of a specialty application niche than the ELSD and CNLSD. A comparison of some of the properties of the refractive index and light scattering detectors is presented in Table 4.8.

4.12.1 Evaporative Light-Scattering Detector (ELSD)

Evaporative light-scattering detectors (ELSD) are based on evaporation of the mobile phase, followed by measurement of light scattered by particles of nonvolatile analyte. The ELSD principle is illustrated in Figure 4.24. Column effluent is nebulized in a stream of nitrogen or air and evaporated in a heated drift tube, leaving nonvolatile particles suspended in the carrier gas stream. Light scattered by the particles is detected by a photodetector mounted at a fixed angle from the incident beam. The ELSD should respond to most compounds that are analyzed by HPLC, but sensitivity may decrease for more volatile analytes. Detector response is related to the absolute quantity of analyte present, not its spectral properties.

The ELSD, like the refractive index (RI) detector, is considered universal, so it has potential to be used for "any" sample. ELSD has the advantage over RI of having a response independent of the solvent, so it can be used with gradient elution and is insensitive to temperature or flow-rate fluctuations. The selection of the mobile phase for ELSD has similar restrictions as mass spectral detectors (Section 4.14) in that the mobile phase must be volatile and free of nonvolatile additives (e.g., phosphate buffer). Once the ELSD is adjusted (e.g., carrier flow rate, drift-tube temperature) for the mobile-phase conditions, it should provide acceptably stable operation. Linearity is somewhat limited (10- to 100-fold), but with the selection of appropriate calibration standard concentrations, ELSD can be useful for quantitative work over a wider range in analyte concentration.

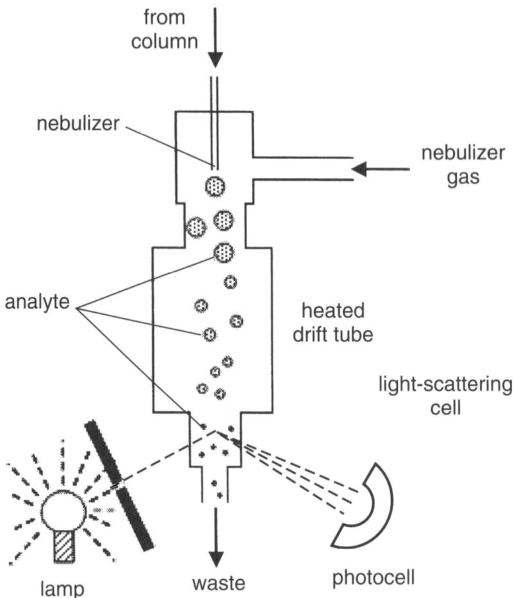

Figure 4.24 Schematic of an evaporative light scattering detector (ELSD).

In general, ELSD provides a 10- to 100-fold improvement in sensitivity over the RI detector, with detection limits of 1- to 100-ng on-column. For some samples the sensitivity gain can be much greater, as is seen in Figure 4.25*a* for the separation of a triglyceride sample with detection by ELSD, whereas the UV detector at 205 nm and the RI detector do not respond to the triglycerides. Note that this separation is via gradient elution in the nonaqueous reversed-phase (NARP) mode (Section 6.5). Whereas water/organic gradients are not suitable for RI detection, acetonitrile and dichloromethane are sufficiently similar in refractive index that a changing mixture can be tolerated by the RI detector. The chromatograms of Figure 4.25*b* illustrate the superiority of the ELSD over the RI detector for a polyethylene sample analyzed by high-temperature (160°C) GPC.

4.12.2 Condensation Nucleation Light-Scattering Detector (CNLSD)

The condensation nucleation light-scattering detector (CNLSD) is an enhancement of the standard ELSD for improved sensitivity and linear range. Following evaporation of the mobile phase, a saturated stream of solvent is added to the particles in the carrier gas. The particles act as condensation nuclei and the solvent condenses onto the particles, causing them to grow to a size where they are more easily detected by light-scattering detection [34]. Early work in this field [34] used butanol vapor, but current instrumentation uses water as the condensing solvent. The applications of the CNLSD are the same as those for the ELSD. In general, the CNLSD gives 10- to 100-fold improvement in sensitivity over the classic ELSD configuration. Manufacturer's applications literature [35] shows detection of inorganic ions (Li^+, Na^+, K^+) at 0.5-ng on-column, linearity for sucrose of three orders of magnitude, and five orders of magnitude of dynamic range.

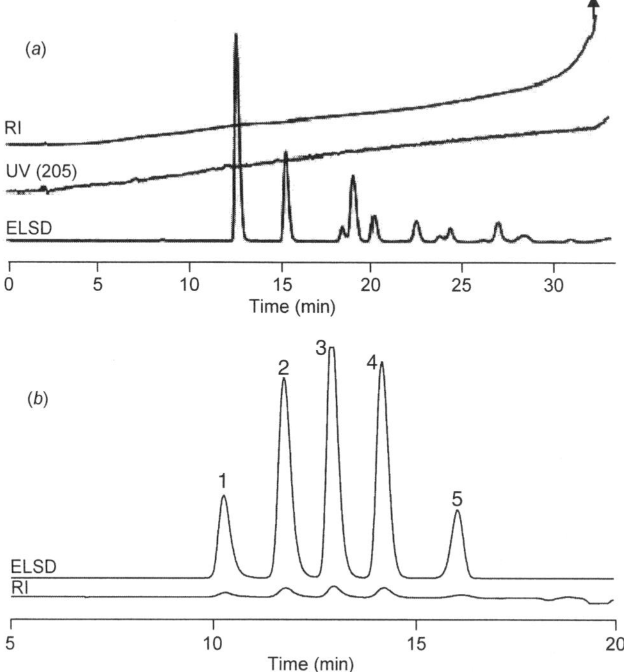

Figure 4.25 Comparison of ELSD detector response. (*a*) ELSD versus refractive index (RI) and UV at 205 nm for triglyceride sample. Shimadzu Premier C18 column; acetonitrile/dichloromethane gradient; 1 mL/min; 30°C. (*b*) ELSD versus RI for the analysis of polystyrene standards by high-temperature (160°C) GPC; 200 μg sample on PL-Gel Mixed B column. Sample molecular weights: *1*, 2,560,000 Da; *2*, 320,000 Da; *3*, 59,500 Da; *4*, 10,850 Da; *5*, 580 Da. (*a*) Courtesy of Shimadzu Corporation; (*b*) courtesy of Varian Polymer Laboratories.

4.12.3 Laser Light-Scattering Detectors (LLSD)

Laser light-scattering detectors (LLSD; also called *multi-angle light-scattering*, MALS) generally refer to HPLC detectors that make light-scattering measurements in solution, as opposed to the ELSD or CNLSD systems that measure light scattered by particles suspended in a gas. LLSD use a laser light source directed on the flow cell as the sample passes through in the mobile phase. Scattered light is measured at multiple angles (e.g., 3–18 different angles) and can be used, with the proper mathematical transformations, to determine the mass of the analyte in the absence of reference standards. These detectors are useful in conjunction with size-exclusion chromatography (see Chapter 13) for the determination of molecular weights of synthetic polymers and biological molecules in the range of 10^3 to 10^6 Da. Figure 4.26 shows superimposed UV chromatograms (280 nm) for a protein kinase fragment and three protein standards (ADH trimer, BSA and ADH monomer). Also shown are the LLSD-determined molecular weights (*y*-axis; 3 separate runs). The kinase has a theoretical mass of 53.5 kDa, whereas the molecular weight of the kinase peak by LLSD is about 108,000, indicating that this is a dimer peak. The expected molecular weights of the standards are 141,000 (ADH), 67,000

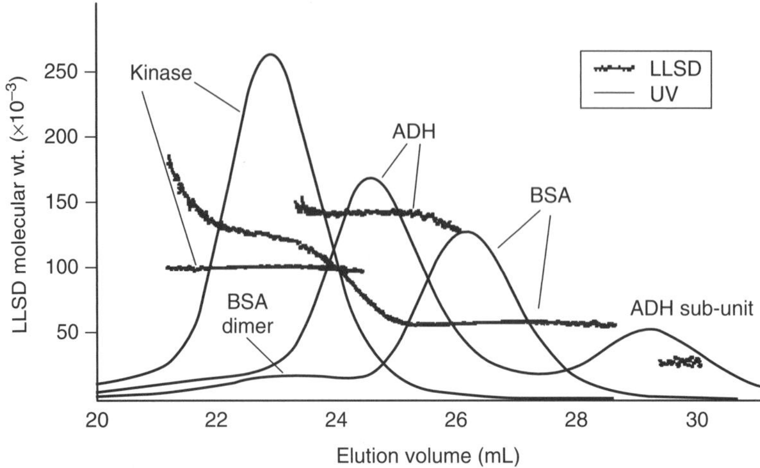

Figure 4.26 Size-exclusion separation of several proteins, with detection by laser (multi-angle) light-scattering detector (LLSD) and UV at 280 nm. Molecular weights by LLSD are plotted on the *y*-axis. Kinase, BSA, and ADH each run separately. Adapted from Wyatt Technology Corporation.

(BSA), and 35,000 Da (ADH sub-unit), which closely match values by LLSD in Figure 4.26. The BSA dimer (135,000 Da) is observed to elute earlier (23.3 mL) than ADH (24.7 mL) despite its lower molecular weight. This demonstrates the greater accuracy of LLSD for molecular-weight determinations, compared to values from size-exclusion measurements (Sections 13.8, 13.10.3.1).

4.13 CORONA-DISCHARGE DETECTOR (CAD)

The corona-discharge detector, also called the *charged-aerosol detector* (CAD) is classified as a universal HPLC detector because it responds to most analytes. The function of the CAD is illustrated in the schematic diagram of Figure 4.27. Column effluent is nebulized and the mobile phase is evaporated, the same as by the evaporative light-scattering detector (Section 4.12.1) or the mass spectrometer (Section 4.14). Analytes in the gas phase are then mixed with a stream of nitrogen gas that has been positively charged by a corona-discharge device. The charge is transferred to the analyte particles, and high-mobility charged species are removed in an ion trap to improve signal quality. The remaining charged analyte ions generate a signal that is read by an electrometer.

The CAD is sensitive to nearly any compound that is sufficiently less volatile than the mobile phase so that remains in the gas phase after the mobile phase is evaporated. As with other evaporative detectors, the mobile phase is restricted to volatile components (e.g., no phosphate buffer); it also requires particles that can be charged in the detector. CAD has been applied to sugars and other carbohydrates as an alternative detector to RI or ELSD, with detection limits ($S/N = 3$) for oligosaccharides of 5-ng on-column and a dynamic range of $>10^4$ [36]. The example of Figure 4.28 shows that the CAD can be applied to impurities analysis at

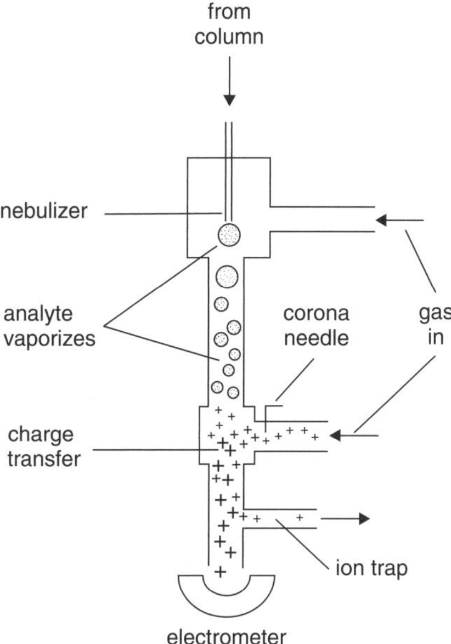

Figure 4.27 Schematic of the corona discharge detector.

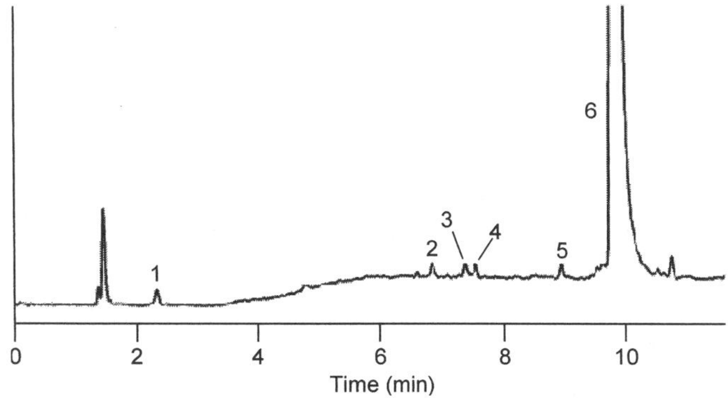

Figure 4.28 Response of corona-discharge detector to 10 μg on-column of sulfadimethoxine (6) plus 5 ng on-column each of related substances: *1*, sulfaguanidine; *2*, sulfamerazine; *3*, sulfamethazine; *4*, sulfamethizole; *5*, sulfamethoxazole; and *6*, sulfadimethoxin. Adapted from data of [39].

the 0.05% level relative to the active pharmaceutical ingredient (API) [37]. In this case 5-ng on-column of 5 related sulfonamide drugs (Fig. 4.28, peaks 1–5) are easily detected in the presence of 10-μg on-column of sulfadimethoxine (6).

4.14 MASS SPECTRAL DETECTORS (MS)

Hyphenated HPLC detectors refer to the coupling of an independent analytical instrument (e.g., MS, NMR, FTIR) to the HPLC system to provide detection. The mass spectrometric (MS) detector is the most popular hyphenated HPLC detector in use today (other hyphenated detectors are discussed in Section 4.15). MS detection has become the standard detector system for *bioanalytical* methods—the analysis of pharmaceutical compounds in biological systems (e.g., plasma or urine). MS detectors are also widely used in the R&D setting to provide structural information or confirmation of unknowns, although it does not have the mass-resolution capability of traditional stand-alone mass spectrometers. MS detectors come in two popular configurations. The *single-stage detector*, sometimes called an MSD (*mass selective detector*), is used to measure a single ionic species for each analyte, often the protonated molecular ion ($M + H$). (Within a given run, more than one analyte ion can be monitored by switching back and forth between different m/z values or scanning between ions.) Instruments using this type of detection are referred to as LC-MS. A more complex detector design isolates the primary ionic species (parent or precursor ion), fragments it into additional ions (daughter or product ions), and monitors one or more of these product ions. This process, sometimes called *multiple reaction monitoring* (MRM), gives added selectivity when the transition from precursor to product ion is used as a "signature" of a specific analyte. Such systems are referred to as LC-MS/MS. We will refer to LC-MS/MS when this specific technique is used, and LC-MS for the single-stage methodology or when it is not important whether the system is LC-MS or LC-MS/MS. A more detailed discussion of MS detection as applied to gradient elution can be found in Section 8.1 of [13], much of which is equally valid for isocratic separation. Additional information about LC-MS and LC-MS/MS detection can be found in books dedicated to the subject (e.g., [38–40]).

4.14.1 Interfaces

The development of the MS detector interface is perhaps the most important factor in the successful application of mass spectrometry as an HPLC detection technique. MS detectors manipulate and detect ions in the gaseous phase, so for the MS to be useful as an HPLC detector, the mobile phase must be evaporated and sample ions must be generated. This is the function of the MS detector interface. The mobile phase is converted from liquid to gas phase, with an expansion in volume of $\approx$1000-fold; at the same time the pressure must be reduced from atmospheric pressure (760 torr) to 10^{-5} to 10^{-6} torr within the 5 to 10 cm flow path of the interface. Pressure is reduced by pumping most of the vaporized sample and mobile phase to waste (no concentration takes place); only a tiny fraction of the sample is drawn into the MS itself. The two most popular interfaces are *electrospray ionization* (ESI) and *atmospheric pressure chemical ionization* (APCI).

4.14.1.1 Electrospray Interface (ESI)

The electrospray interface (Fig. 4.29) adds a charge to analytes in the mobile phase by placing a potential (e.g., 3–5 kV) on the stainless-steel nebulizer spray-tip ("capillary" in Fig. 4.29). The mobile phase is sprayed into the heated interface, where the solvent evaporates, leaving ions in the gaseous state. ESI is the most commonly used interface for bioanalytical applications because it is a "softer" ionization technique and is less likely to cause undesirable analyte degradation.

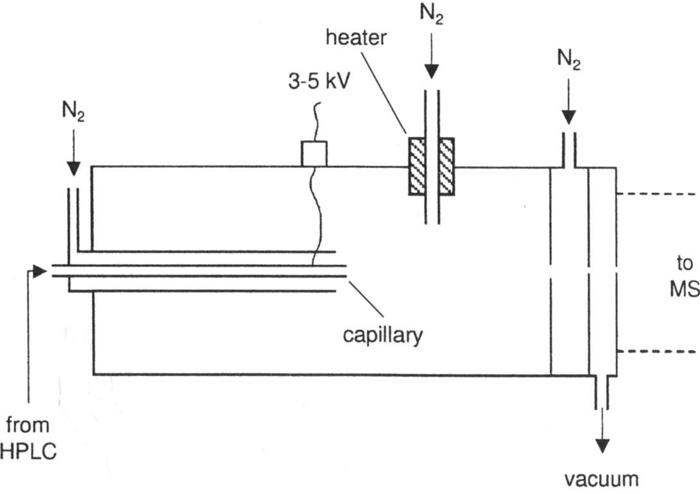

Figure 4.29 Schematic of the electrospray interface (ESI) for the LC-MS detector.

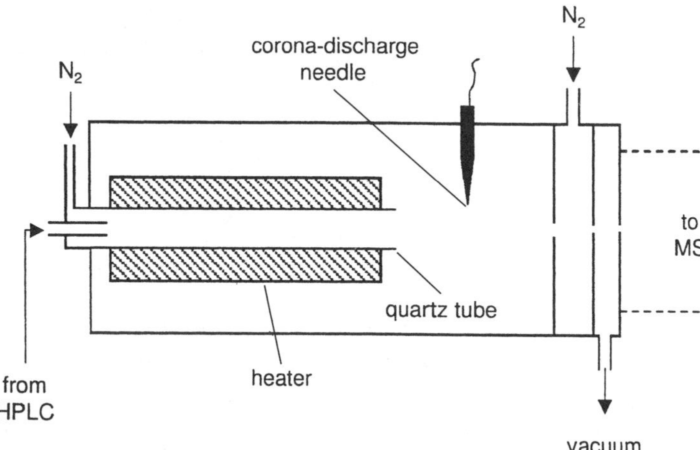

Figure 4.30 Schematic of the atmospheric pressure chemical ionization interface (APCI) for the LC-MS detector.

4.14.1.2 Atmospheric-Pressure Chemical-Ionization Interface (APCI)

This interface (Fig. 4.30) vaporizes the mobile phase first, and then uses a corona discharge to add a charge to the analyte in the gas phase. The APCI technique is used for compounds that do not ionize well with ESI (often more stable, smaller molecular-weight compounds and some nonpolar compounds), but under harsher conditions, so it is more likely than ESI to cause sample degradation, especially with heat-labile compounds. APCI has been shown to have fewer matrix-ionization problems than ESI. APCI and ESI have different ionization mechanisms, so the response and selectivity may vary significantly between the two interfaces. Either interface can be operated in the positive- or negative-ion mode, resulting in the generation of positively or negatively charged sample ions (most often achieved by adding or removing a proton from the analyte molecule).

4.14.1.3 Other Interface Designs

Other interfaces are available for LC-MS besides ESI and APCI. For example, instead of the corona discharge of the APCI, the *atmospheric-pressure photoionization interface* (APPI) uses a UV lamp to generate photons, which in turn ionize the analyte. At least one interface splits the column effluent to allow simultaneous generation of ions using ESI and APCI, which can be useful for screening applications where the sample ionization properties are not known.

4.14.1.4 Flow-Rate Considerations

To minimize the work required by the interface, a smaller column i.d. is selected than is used for more traditional HPLC separations, so as to reduce the mobile-phase flow rate. Although LC-MS interfaces can operate with a flow rate of 1 mL/min, they are more reliable with lower flow rates. The use of 2.1-mm i.d. columns allows the use of flow rates of 0.2 to 0.5 mL/min, with linear velocities (and separation) comparable to flow rates used with conventional 4.6-mm i.d. columns (1.0–2.5 mL/min). Short, 30- to 50-mm-long columns packed with 3- to 5-μm particles provide fast separations of the usual (simple) mixtures encountered in bioanalytical applications, namely an analyte, an internal standard, and one or two metabolites. For more complex mixtures, longer column lengths (100–150 mm) may be required, in order to obtain larger column plate numbers (with longer run times). Capillary columns (e.g., <1 mm) often are used with proteomics applications and a "nanospray" interface designed for low- to submicroliter flow rates (Section 13.4.1.6).

4.14.2 Quadrupoles and Ion Traps

Two designs of mass filters are predominant for LC-MS (single-stage) applications: quadrupoles and ion traps. Time-of-flight (TOF) designs (Section 4.14.3) also are growing in popularity. Quadrupoles use a set of four rods and a carefully controlled electric field to isolate selected ions from the sample. Ions of a selected mass-to-charge ratio (m/z) are then passed to an electron multiplier for detection, providing a selective response for the desired analyte. Ion traps use a ring electrode in combination with end-cap electrodes to accomplish the same isolation of desired ions, followed by detection. Both quadrupoles and ion traps can be set up to change rapidly from monitoring one mass to another, and thus generate a spectrum (scan) across a range of masses. An alternate mode of operation allows the detector to "simultaneously" detect co-eluting compounds, such as an analyte and internal standard, by switching back and forth between data collection channels for each mass during the elution of the peaks. As discussed below, quadrupole MS detectors are favored for quantitative analysis, whereas ion traps have advantages for qualitative (structural) applications.

Single-stage MS detectors of the above-mentioned kind are used in less expensive LC-MS units; however, additional structural discrimination is needed for more

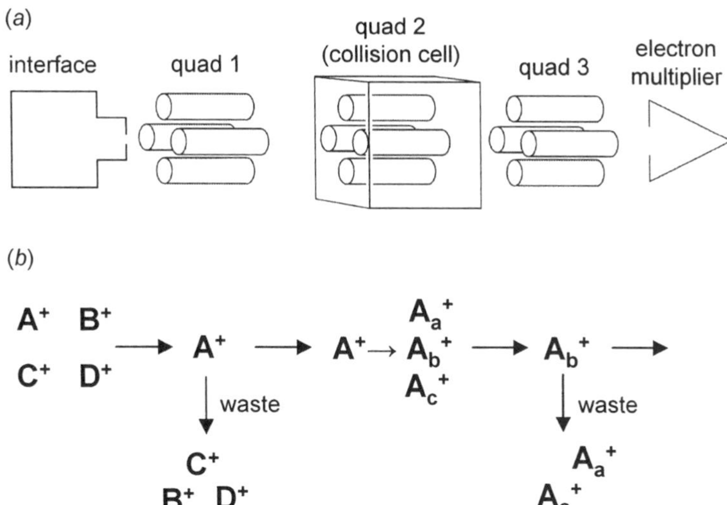

Figure 4.31 Triple quadrupole mass spectrometer. (*a*) Schematic; (*b*) MS/MS experiment for $A^+ > A_b^+$ precursor > product ion transition.

selective detection. The *triple-quadrupole* (Fig. 4.31*a*), or *tandem*, MS detector can provide additional selectivity compared to that obtained with a single-quadrupole unit. Sample ions generated in the interface (A^+, B^+, C^+, D^+ in Fig 4.31*b*) enter the first quadrupole. The ions of a given m/z (A^+) are isolated in the first quadrupole and sent to a second quadrupole (collision cell), which is filled with an inert gas (nitrogen or argon). The ions are fragmented ($A^+ \rightarrow A_a^+$, A_b^+, A_c^+) in the collision cell and passed to a third quadrupole. The third quadrupole then isolates specific ion fragments (e.g., A_b^+) and passes them to the electron multiplier for measurement. The transition from the initial ion (*precursor* or *parent*) to the fragment ion (*product* or *daughter* ion) provides a unique "signature" ($A^+ > A_b^+$ in Fig. 4.31*b*) for an analyte, and greatly increases the selectivity of the triple-quadrupole (MS/MS) over the single-quadrupole detector. (Note that the conventional notation is "$A^+ > A_b^+$" to represent the transition signature of the precursor A^+ to the product ion A_b^+. We will use this shorthand, while using "$A^+ \rightarrow A_a^+$, A_b^+, A_c^+" to represent the fragmentation process itself.)

Ion traps accomplish multiple-stage fragmentation and the isolation of a preferred product ion in the same physical space (vs. in different parts of the detector as in the triple quadrupole of Fig. 4.31). First, ions are generated in the interface and passed into the ion trap (ion accumulation, Fig. 4.32). Ions of a desired m/z are held, while the remaining ions are sent to waste (ion isolation 1). The isolated ions (A^+ in Fig. 4.32) are then fragmented ($A^+ \rightarrow A_a^+$, A_b^+, A_c^+) and the desired fragment m/z is isolated (A_b^+, ion isolation 2). The ions can then be sent to the electron multiplier for detection or the process can be continued (further fragmentation of A_b^+, isolation, etc.). The ion trap is capable of performing this operation over and over, isolating and breaking ion fragments into successively smaller fragments (with a corresponding loss of sensitivity for each fragmentation step). This is useful for structural identification, but historically the ion trap has not been as good for quantitative work as the quadrupole because of space-charge effects (ion interactions

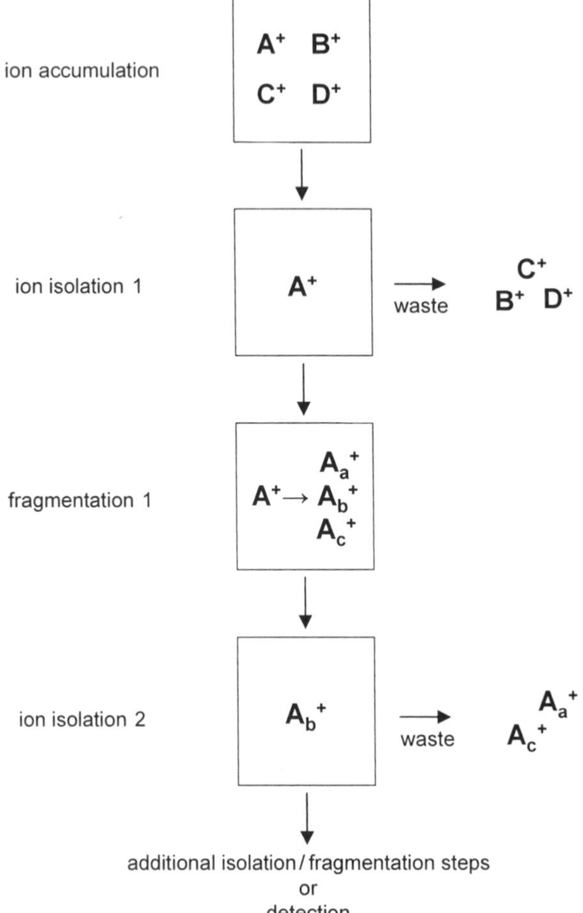

Figure 4.32 Ion-trap mass spectrometer. MS/MS experiment for $A^+ > A_b^+$ precursor > product ion transition. Note that the different boxes represent the same physical part of the system at different times.

within the detector) and variability in the output signal intensity. Thus quadrupoles (single and triple) tend to be more widely used for routine quantitative work, whereas ion traps are preferred when structural identification is needed, such as in metabolite isolation and identification.

4.14.3 Other MS Detectors

In addition to the quadrupole and ion trap, several other LC-MS detector configurations are available. One of the most popular of these is the *time-of-flight* (TOF) MS detector, as illustrated in Figure 4.33. In the TOF, ions are generated in the interface and are accelerated with a specific energy and directed through a drift tube to the detector. The velocity of an ion traveling through the drift tube will be related to the amount of energy applied, so for the same applied energy, lower mass ions will travel more quickly to the detector than will larger ions. The time taken to transit the

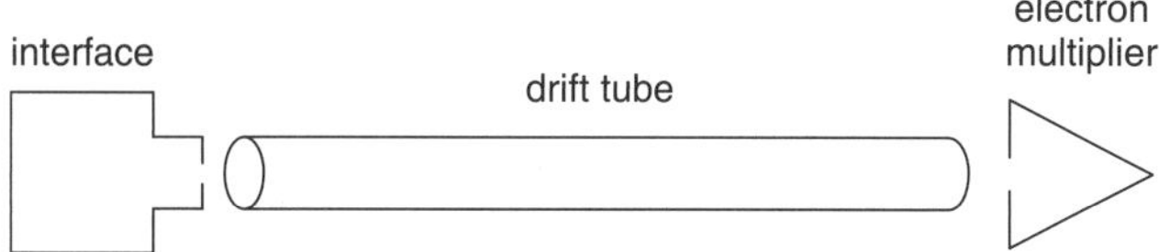

Figure 4.33 Linear time-of-flight (TOF) mass spectrometer.

drift tube is then correlated to the mass (m/z) of the ion. Mass resolution is related to the length of the drift tube. Spatial limitations limit the length of the drift tube, so one popular configuration uses an electrostatic mirror ("reflectron") to increase the effective length of the drift tube, and thus improve mass resolution. With sufficient path length, the TOF can provide higher mass resolution than the quadrupole, so it is useful for structural work. It should be noted, however, that none of the LC-MS detectors have the mass-resolution capability of the more traditional, dedicated mass spectrometers.

Many combinations of MS detector designs are available for MS-MS detection. As mentioned above (Section 4.14.2), the triple quadrupole uses one quadrupole to isolate the precursor ion, a second quadrupole as the collision cell for fragmentation, and a third quadrupole for isolation of fragment ions. Two TOF sections can be connected to a collision cell to provide similar function with a TOF-TOF. Mixing the ion isolation sections is also possible. For example the Q-TOF uses a quadrupole for the first stage, a collision cell, and then a TOF for the final stage. Each design has certain advantages (and disadvantages)—and the instrument manufacturers are glad to explain why their favorite configuration is better than all the others!

4.15 OTHER HYPHENATED DETECTORS

Hyphenated detectors, as defined in Section 4.2.2.4, refer to the combination of an HPLC and another analytical instrument. It seems that nearly every stand-alone analytical instrument has been connected to an HPLC system at some time or place, so the number of possible combinations of hyphenated detectors is large. If each detector provides quantitative and/or qualitative information about the sample, it follows that multiple detectors attached to a single HPLC system could provide even more information. In at least one case, four spectrometers were connected to a single HPLC. This multiple hyphenation has been called "hypernation," and is reviewed in [41]. However, only LC-MS has reached the level of acceptance that it is used widely and applied in nearly every application area of HPLC today. Two other hyphenated detectors, infrared and nuclear magnetic resonance, also are commercially available with HPLC interfaces. These detectors are used primarily as tools to aid structural elucidation of unknowns in mixtures, rather than for routine quantitative analysis.

4.15.1 Infrared (FTIR)

Fourier transform infrared spectroscopy (FTIR) is a popular tool for providing chemical structural information. When coupled to an HPLC system, IR data can

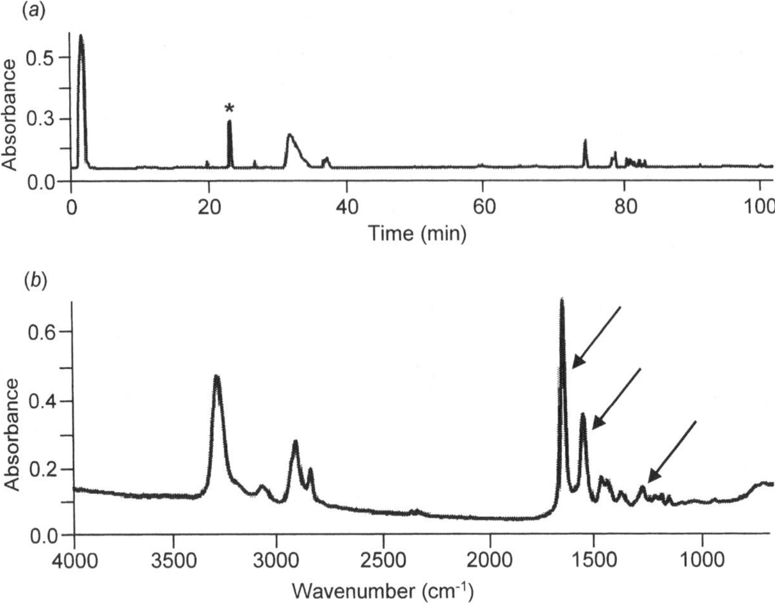

Figure 4.34 LC-FTIR response for *E. coli* extracts. (*a*) Composite chromatogram of largest IR absorbance versus time; (*b*) individual FTIR spectrum for sample at 23 min (* in *a*). Adapted from data of [41].

be obtained for separated components; that is, it makes possible IR analysis of a mixture. The mobile-phase background spectrum can confuse interpretation of IR spectra, so flow-through IR detectors are of limited use. An alternative is to use an interface that evaporates the mobile phase and deposits the sample on a window (e.g., ZnSe) that is transferred automatically into the IR instrument for analysis. In this case, the detection is not continuous but rather in small, discrete samples. (Sensitivity can be increased by increasing the time over which each IR reading takes place, but this slows analysis time.) However, the data are sufficient for applications such as that shown in Figure 4.34 for the LC-FTIR analysis of *E. coli* extracts [42]. The sample was separated on a C18 column using a gradient of 0.1% formic acid and acetonitrile. The composite chromatogram (Fig. 4.34*a*) was obtained by plotting the largest IR absorbance value in every spectrum over time. The individual spectrum at 23 minutes (*) is shown in Figure 4.34*b*. The spectrum is identified as a protein by the various amide vibration bands at 1642, 1550, and 1268 cm^{-1} (arrows) plus other spectral features. See [43] for a review of LC-FTIR.

4.15.2 Nuclear Magnetic Resonance (NMR)

Nuclear magnetic resonance (NMR) can be used as an HPLC detector in three different operating modes [44]. In the simplest, the column effluent is directed through the NMR flow cell and NMR data are gathered on-the-fly. Because of time limitations, the sensitivity of this technique is low, but ^{1}H NMR has proved useful. Recent advances have enabled 2D NMR and ^{13}C NMR to be used in *flow-through* mode for some applications. An alternate technique is *stopped-flow* sampling, usually performed by trapping a segment of the chromatogram in a loop and sending it to the NMR for stopped-flow analysis. Both flow-through and stopped-flow LC-NMR must be used with proton-free mobile phases, so D_2O often is used instead of water (and deuterated ACN or CD_3OD instead of MeOH), along with narrow-bore columns to keep solvent consumption low. A third alternative is *LC-SPE-NMR*, where a sample of column effluent is trapped on an SPE cartridge. The cartridge can then be treated to remove any water, and the sample can be eluted with a small volume of acetonitrile into the NMR cell for analysis. Both of the latter techniques can be operated in short cycles (e.g., 20 sec) to minimize loss of chromatographic resolution. See [45] for a review of LC-NMR.

Figure 4.35 shows an application of stopped-flow LC-NMR to the analysis of organic matter in a water sample [46]. An gradient ion-pairing separation was performed on a C_{18} column using a mobile phase of 0.01M tetrabutylammonium hydrogen sulfate in D_2O (A-solvent) and acetonitrile (B-solvent), with 20-second sampling into a holding loop across the regions of interest. NMR spectra were obtained for selected regions of the UV chromatogram (Fig. 4.35*a*); the spectrum of Figure 4.35*b* represents the aromatic region of the ^{1}H NMR spectrum for the sample at 16 minutes (*) in Figure 4.35*a*.

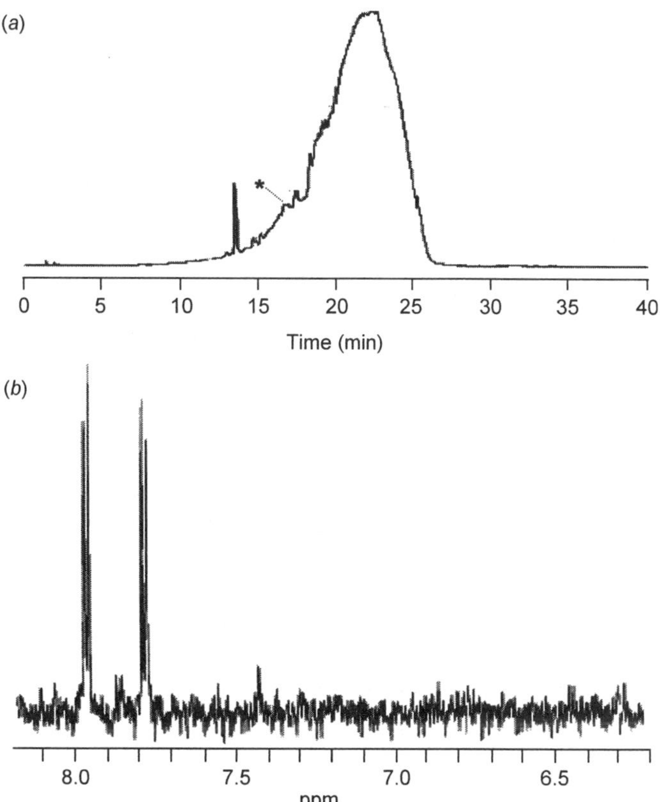

Figure 4.35 Stopped-flow LC-NMR response for organic matter in water. (*a*) UV chromatogram for sample at 280 nm; (*b*) aromatic region of ^{1}H NMR spectrum taken at (*). Adapted from data of [46].

4.16 SAMPLE DERIVATIZATION AND REACTION DETECTORS

Despite the wide variety of HPLC detectors available today, some compounds cannot be detected, or detection limits may not be sufficiently low for practical application. One way around this problem is to derivatize the analyte to a new compound or complex that will have sufficient response to conventional detectors (e.g., UV or fluorescence). The derivatization reaction may be performed before the sample is injected or after the analyte is eluted from the column. New derivatization reactions are regularly reported in the literature, which should be consulted for detection of specific compounds or compound types. Books of derivatization reactions are available (e.g., [47]; also see references of Section 16.12), while a very practical source for information is to consult the applications literature of the manufacturers of reaction detectors (e.g., [48]).

Pre-column derivatization can be used to enhance detection limits, the chromatographic characteristics of the analyte (less often), or both. As a part of sample pretreatment, pre-column derivitization may be done manually or in automated fashion. Some autosamplers are designed to derivatize the sample just prior to injection. Derivatization reactions can be instantaneous or slow, but if done in batchwise fashion, even slow reactions may provide acceptable sample throughput. An example of pre-column derivatization to enhance the fluorometric detection of low-molecular-weight acids is shown in Figure 4.36 [49]. In this case, 9-chloromethylanthracene plus tetrabutylammonium bromide (a catalyst) is reacted with the sample at 80°C for 50 minutes as part of sample pretreatment. Fluorescence detection (excitation 256 nm, emission 412 nm) gave linear response of 1 to 250 ng/mL of monofluoroacetate in serum and a detection limit ($S/N = 3$) of 0.25-ng/mL serum.

Post-column derivatization (reaction detectors) involves reacting the sample as it travels between the column outlet and the detector cell. This can be mediated chemically, photochemically, or in combination. Table 4.9 summarizes some of the desirable characteristics of a post-column reaction detector and the selected chemistry. The detector design usually incorporates additional tubing between the column and detector to provide reagent mixing and sufficient reaction time. However, extra tubing will increase extra-column peak broadening, so the system must be designed to balance reaction requirements with peak-broadening effects. One method

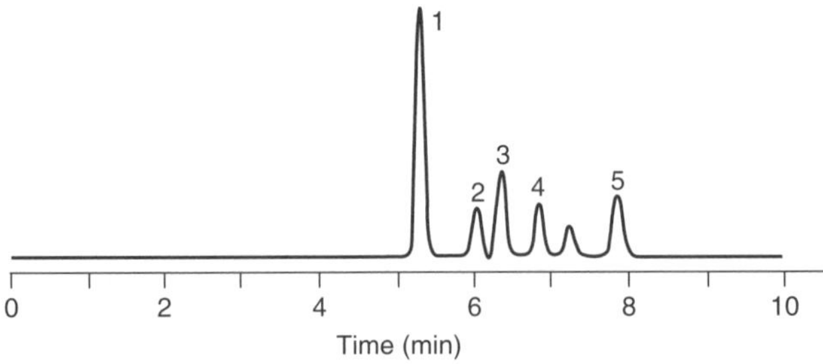

Figure 4.36 Pre-column derivatization of organic acids to provide fluorescence response (excitation at 256 nm, emission at 412 nm). *1*, 9-Chloromethylanthracene derivatization reagent. Derivatives of *2*, monofluoroacetate; *3*, formic acid; *4*, acetic acid; *5*, propionic acid. Adapted from data of [48].

Table 4.9

Requirements of Post-column Reaction Detection

Minimum dispersion
Completeness of reaction in a short time
Reproducibility
Stability of reagents
Solubility of reagents and products
Minimum detector response to reagents

to reduce extra-column peak broadening during reaction is to use a *knitted reactor*, where narrow-bore flexible tubing is knitted into a series of tight radius bends to create turbulent flow and reduce dispersion. Generally, a fast reaction is desired to minimize the reactor volume requirements. Also reactions that go to completion tend to be more reproducible than those that only go partially to completion. The reaction chemistry must be chosen so that the reagents are sufficiently stable to use over the course of a batch of samples, and both the reagents and the reaction products must be soluble in the mobile phase. Of course, the reagents should not interfere with detection of the product, so detector response to the reagents should be minimal. Reagent kits for popular post-column reactions (e.g., orthophthaldehyde, OPA, and ninhydrin for amino acid analysis) are commercially available. The purchase of such kits is convenient and provides standardized concentrations and quality control of reagents that may not be available when performing in-lab formulation of reagents.

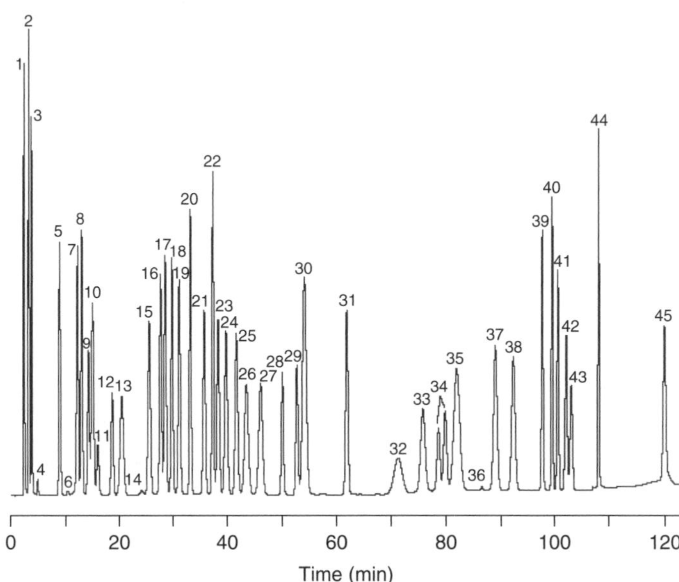

Figure 4.37 UV response at 405 nm after post-column derivatization with ninhydrin of 45 amino acids and related compounds. Courtesy of Pickering Laboratories, Inc.

Figure 4.37 shows an example of the popular post-column derivatization of amino acids with ninhydrin followed by UV detection at 405 nm. In this case 45 amino acids and related substances are separated on an ion exchange column followed by reaction with ninhydrin at 130°C in a 500-μL reactor (residence time ≈45 sec). Sample derivatization and reaction detectors have gradually lost popularity as other HPLC detectors have been introduced and been made more effective.

REFERENCES

1. G. Hesse and H. Weil, eds., *Michael Tswett's First Paper on Chromatography*, M. Woelm, Eschwege, 1954.
2. V. G. Berezkin, ed., *Chromatographic Adsorption Analysis: Selected Works of M. S. Tswett*, Ellis Horwood, New York, 1990.
3. A. Tiselius and D. Claesson, *Arkiv. Kemi Mineral. Geol.*, 15B (18) (1942) 1.
4. A. J. P. Martin and S. S. Randall, *Biochem. J.*, 49 (1951) 293.
5. C. G. Horváth and S. R. Lipsky, *Nature*, 211 (1966) 748.
6. J. J. Kirkland, *Anal. Chem.*, 40 (1968) 391.
7. E. Haahti and T. Nikkari, *Acta Chem. Scand.*, 17 (1963) 2565.
8. R. P. W. Scott and J. G. Lawrence, *J. Chromatogr. Sci.*, 11 (1970) 65.
9. B. G. Julin, H. W. Vandenborn, and J. J. Kirkland, *J. Chromatogr.*, 112 (1975) 443.
10. J. W. Dolan and J. N. Seiber, *Anal. Chem.*, 49 (1977) 326.
11. L. R. Snyder and J. J. Kirkland, *Introduction to Modern Liquid Chromatography*, 1st ed., Wiley, New York, 1979, p. 136.
12. J. W. Dolan, *LCGC*, 14 (1996) 378.
13. L. R. Snyder and J. W. Dolan, *High-Performance Gradient Elution*, Wiley, New York, 2007.
14. J. J. Kirkland, W. W. Yau, H. J. Stoklosa, and C. H. Dilks, Jr., *J. Chromatogr. Sci.*, 15 (1977) 303.
15. J. C. Sternberg, in *Advances in Chromatography*, Vol. 2, J. C. Giddings and R. A. Keller, eds., Dekker, New York, 1966, p. 205.
16. D. Parriott, *A Practical Guide to HPLC Detection*, Academic Press, San Diego, 1993.
17. R. P. W. Scott, in *Handbook of HPLC* (*Chromatographic Science Series*, Vol. 78), E. Katz, R. Eksteen, P. Schoenmakers, and N. Miller, eds., Dekker, New York, 1998, ch. 15.
18. D. T. Bach, *US Patent 4,867,559* (Sept. 19, 1989).
19. M. H. Garrett, *US Patent 6,542,231* (Apr. 1, 2003).
20. G. Cameron, P. E. Jackson, and M. V. Gorenstein, *Chem. Aust.*, 60 (1993) 288.
21. A. G. Frenich, J. R. Torres-Laasio, K. De Braekeleer, D. L. Massart, J. L. Martinez-Vidal, and M. M. Galera, *J. Chromatogr. A*, 855 (1999) 487.
22. *Waters 996 Photodiode Detector: Peak Purity III: Interpretation of Peak Purity Plots* Appl. Note 970983, Waters, 1996.
23. F. K. Glowka and M. Karazniewicz-Lada, *J. Chromatogr. B*, 852 (2007) 669.
24. P. Vinas, N. Balsalobre, C. Lopez-Erroz, and M. Hernandez-Cordoba, *J. Agr. Food Chem.*, 52 (2004) 1789.
25. S. R. Bakalyar, M. P. T. Bradley, and R. Honganen, *J. Chromatogr.*, 158 (1978) 277.
26. Y.-T. Wu, T.-R. Tsai, L.-C. Lin, and T.-H. Tsai, *J. Chromatogr. B*, 853 (2007) 281.

27. I. N. Ackworth, *Coulometric Electrode Array Detectors for HPLC (Progress in HPLC-HPCE*, Vol. 6), Brill Academic, Boston, 1997.
28. R. E. Parks and R. L. Marietta, US Patent 4018562 (1975).
29. *Model 8060 Nitrogen Specific HPLC Detector*, brochure A-8060-99-25M, Antek Instruments, Houston, TX, 1999.
30. K. Petritis, C. Elfakir, and M. Dreux, *LCGC Europe*, 14 (2001) 389.
31. Jasco Incorporated, *LCGC Applications Notebook* (June 2003) 26.
32. L. Kott, W. B. Holzheuer, M. M. Wong, and G. K. Webster, *J. Pharm. Biomed. Anal.*, 43 (2007) 57.
33. F. Glowka, M. K. Lada, G. Grund, and J. Wachowiak, *J. Chromatogr. B*, 850 (2007) 569.
34. L. B. Allen, J. A. Koropchak, and B. Szostek, *Anal. Chem.*, 67 (1995) 659.
35. Quant Technologies, QT-500 datasheet (2/ 2007).
36. D. Asa, *Amer. Lab.*, 38(7) (2006) 16.
37. T. Gorecki, F. Lynen, R. Szucs, and P. Sandra, *Anal. Chem.*, 78 (2006) 3186.
38. W. M. A. Niessen, *Liquid Chromatography-Mass Spectrometry* (*Chromatographic Science Series*, Vol. 97), 3rd. ed., Taylor and Francis, London, 2006.
39. M. C. McMaster, *LC-MS: A Practical User's Guide*, Wiley, Hoboken, NJ, 2005.
40. *Current Practice in Liquid Chromatography-Mass Spectrometry*, W. M. A. Niessen and R. D. Voyksner, eds., Elsevier, Amsterdam, 1998.
41. I. D. Wilson and U. A. Th. Brinkman, *J. Chromatogr. A.*, 1000 (2003) 325.
42. S. W. Huffman, K. Lukasiewicz, S. Geldart, S. Elliott, J. F. Sperry, and C. W. Brown, *Anal. Chem.*, 75 (2003) 4606.
43. G. W. Somsen, G. Gooijer, and U. A. Th. Brinkman, *J. Chromatogr. A.*, 856 (1999) 213.
44. S. Down, *Spectro. Europe*, 16 (2004) 8.
45. K. Albert, *J. Chromatogr. A*, 856 (1999) 199.
46. A. J. Simpson, L.-H. Tseng, M. J. Simpson, M. Spraul, U. Braumann, W. L. Kingery, B. P. Kelleher, and M. H. B. Hayes, *Analyst*, 129 (2004) 1216.
47. K. Blau and J. M. Halket, eds., *Handbook of Derivatives for Chromatography*, 2nd ed., Wiley, New York, 1993.
48. *Application Manual: Amino Acids*, Pickering Laboratories, Mountain View, CA, 2002.
49. Z. Xie, W. Shi, L. Liu, and Q. Deng, *J. Chromatogr. B*, 857 (2007) 53.

CHAPTER FIVE

THE COLUMN

Introduction to Modern Liquid Chromatography, Third Edition, by Lloyd R. Snyder, Joseph J. Kirkland, and John W. Dolan

5.1 INTRODUCTION

The column—the "heart" of the HPLC system—has changed greatly from the beginning of HPLC in the mid-1960s. Columns have been improved for greater separation speed and efficiency, as well as increased stability and reproducibility. New stationary phases have been introduced for the extension of HPLC to a wider range of sample types, or for better separations of compounds that have in the past proved problematic. Today it is rare that a column cannot be found for solving a particular HPLC separation problem. Early columns were made of glass, but the need for higher pressure operation quickly led to the exclusive use of metal columns. With the passage of time, columns became shorter and particles smaller (Fig. 1.5*g*, *h*). Column lengths of 30 to 250 mm are commonly used today, with particles that are 1.5 to 5 μm in diameter (larger particles are used for preparative separations; see Chapter 15).

In this chapter we will examine the HPLC column, including the bare particle or *support*, the added ("bonded") stationary phase, and the column hardware. Porous-silica particles in packed beds (as in Fig. 2.2) are most commonly used. Particles can also be made from solids other than silica; the main value of non-silica particles is their greater stability for use with extremes of temperature or mobile-phase pH. Of special current interest are columns packed with very small particles (<3 μm) for carrying out fast separations (Section 2.5.3.2).

Although very few workers will produce their own bonded-phase particles ("column packings"), we will describe briefly how these materials are made; this information can prove helpful in choosing a column or for troubleshooting. Column characteristics that affect reversed-phase retention and selectivity (Section 5.4) can also influence our selection of a column for a particular application. Column-packing methods (Section 5.6) are described as further background for how "good" columns are prepared. Column testing is covered in Section 5.7. Finally, column specifications and column handling techniques are covered in Sections 5.7 and 5.8, as an aid for good laboratory practice.

5.2 COLUMN SUPPORTS

Column packings consist of a rigid support plus an attached stationary phase (as in Fig. 2.2*b*, which shows a silica particle with attached C_{18} groups). In some cases the support and stationary phase are the same; for example, an unbonded silica particle is often used for normal-phase chromatography (Chapter 8). Silica *monoliths* (Section 5.2.4) represent an alternative column support of more recent vintage. Monoliths refer to columns composed of an interconnected, porous bed, as opposed to columns packed with distinct particles (a monolith may be thought of as one big particle that fills the entire column). In this section we will describe column supports, their characterization, and how their properties affect their final use. Later, we will discuss (1) how these supports can be modified to create column packings for different purposes (Section 5.3) and (2) how the final particles are packed into the column (Section 5.6).

5.2.1 Particle Characterization

Particles can be characterized by their configuration or type, physical dimensions (particle diameter), the nature and size of pores within the particle, and surface area. The importance of particle diameter in influencing separation has been discussed (Fig. 1.5 and Section 2.4). Pore size and surface area are usually related reciprocally; as pore diameter increases, the surface area decreases in roughly the same proportion. The phase ratio Ψ (Section 2.3.1) is roughly proportional to surface area, and retention is proportional to Ψ (Eq. 2.3); therefore retention increases as pore size decreases and column surface area increases. The maximum weight of sample that can be injected is also proportional to surface area (Section 2.6.2), so a greater surface area is usually desirable—which suggests the use of the smallest possible pore diameter (and smallest pores). However, solute molecules must be able to enter the pores without hindrance, and this requires pores that are larger than the solute molecule. For compounds with molecular weights <500 Da, the average pore diameter should preferably be about 9 nm or larger. Larger molecules require larger pores; for example, proteins are usually separated with 30-nm-pore particles (Section 13.3.1.1). The *interstitial volume* of the column is the space between particles; it is usually about 40% of the total column volume.

The physical characteristics of particles are important, and manufacturers monitor these particle properties in various ways. The size of the particles is especially important, as this largely determines the efficiency of the packed column. While optical microscopy and air or liquid elutriation (or classification) methods can provide this information, instruments for the measurement of particle size (e.g., the Coulter counter) are more convenient and quantitative; they can also provide a particle-size distribution. A narrow particle-size distribution is preferred, as this favors larger values of N (Section 5.2.2.1), and provides lower column pressure for comparable efficiency.

Particle surface-area, average pore-diameter, and pore-diameter distribution typically are measured by the adsorption of nitrogen or argon, using the *Brunauer–Emmett–Teller* (BET) procedure. Particles with average pore-diameters >30 nm (which are used less commonly) are preferably characterized by mercury intrusion, which also can be used to measure pores as small as 3 nm. Mercury intrusion does not work well for fragile particles (with a large pore volume) or for soft polymer particles. Reference [1] should be consulted for additional information on the measurement of the physical characteristics of HPLC particles.

5.2.1.1 Particle Type

As illustrated in Figure 5.1, several particle configurations are currently available for HPLC. *Totally porous silica particles* (Fig. 5.1*a*) are the most common because of their greater column capacity (allowing the injection of a larger sample weight) and availability in a wider variety of options (stationary phase, particle and pore size, column dimensions, etc.). The most popular particles have diameters in the 1.5- to 5-μm range. Today these particles are often prepared by the aggregation of much smaller spheres (Section 5.2.2.3).

Pellicular particles (Fig. 5.1*b*) consist of solid, spheres that are covered with a very thin surface layer of stationary phase. These silica- or polymer-based particles (presently 1.5 to 2.5 μm in diameter, but much larger in the past) display larger

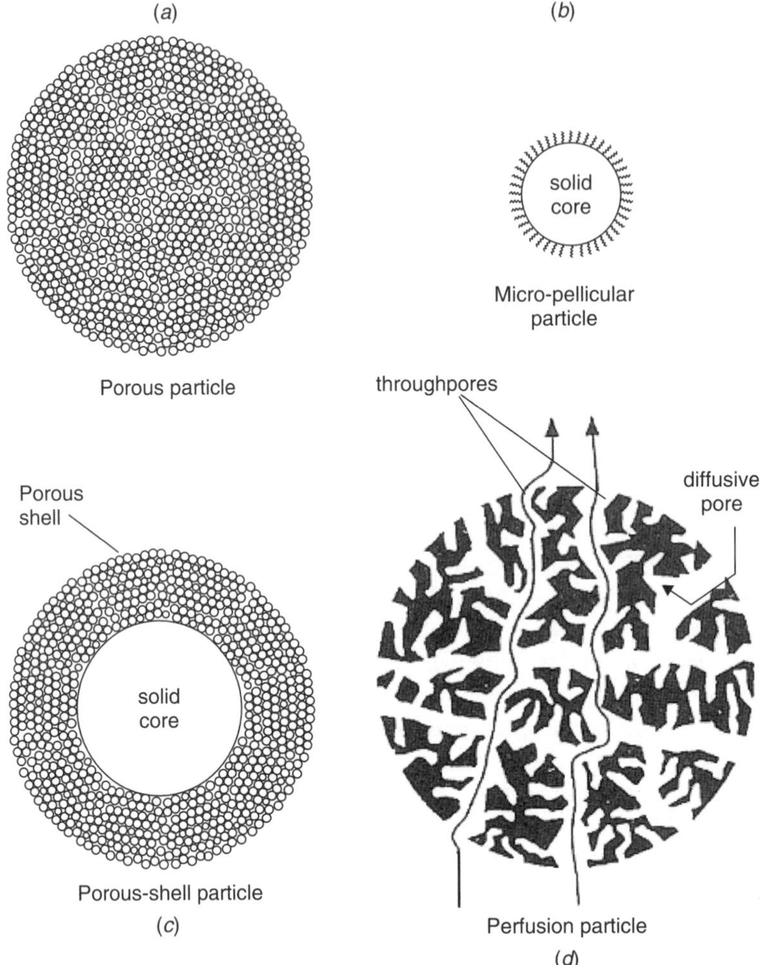

Figure 5.1 Different particle configurations.

values of N for macromolecules, because of better stationary-phase mass-transfer (smaller C-term, Section 2.4.1.1; [2, 3]). Because of their very low surface area, these particles exhibit greatly reduced retention and are limited to much smaller weights of injected sample (with a proportional decrease in detection sensitivity). Pellicular particles are best suited for the analysis of major components (rather than trace impurities), and have been used mainly for the separation of large biomolecules (Chapter 13).

Superficially porous particles (also called *fused-core™ particles, shell particles*, or *controlled-surface-porosity particles*) have a solid core with a porous outer shell (Fig. 5.1*c*). These particles typically have diameters of 2 to 5 μm, with porous shells of 0.25 to 0.5 μm in thickness. Superficially porous particles have much greater surface areas than pellicular particles (~3/4 as large as fully porous particles), providing longer retention and allowing a larger weight of injected sample. The thin, outer porous shell permits faster separations—especially for macromolecules

[4, 5]. These particles also provide somewhat higher values of N than totally porous particles (Section 5.2.2.1).

Perfusion particles (Fig. 5.1*d*) contain very large (e.g., 400- to 800-nm) pores (*through-pores*), connected to a network of smaller (e.g., 30- to 100-nm) pores. By comparison, most other particles have pore diameters of 8 to 30 nm (*diffusive pores*). At high flow rates, solutes are carried into and out of perfusion particles by a combination of diffusion and flow of mobile phase *through* the particle [6, 7]. This additional contribution to stationary-phase mass transfer is claimed to reduce band broadening, especially for large molecules at higher flow rates. Perfusion particles are usually larger (e.g., 10 μm), permitting their use at lower pressures. These particles are best suited for preparative-scale separations of macromolecules such as proteins; they are little used for the analysis of small molecules.

5.2.1.2 Particle Size and Pore Diameter

From 1980 to 2000, particles with diameters of ~5 μm were generally preferred for routine separation. Particles of this size represent a good compromise in terms of efficiency, pressure drop, convenience, equipment requirements, and column lifetime. However, columns with particle diameters of 3 μm and smaller are becoming more popular, mainly as a means for decreasing run time and increasing sample throughput (Sections 2.5.3.2, 9.3.9.2). Because of the very narrow peaks generated by small-particle columns, any extra-column peak broadening must be minimized (Sections 2.4.1.1, 3.9), especially when smaller diameter columns are used. Columns with particle diameters <2 μm are used mainly with equipment that is capable of higher pressure operation (e.g., up to 15,000 psi; Section 3.5.4.3).

Table 5.1 summarizes some physical characteristics of HPLC particles and their effect on separation from favorable ("4") to unfavorable ("1"). Totally porous and superficially porous particles with pore diameters of 8 to 12 nm are used most often for the separation of molecules smaller than 10,000 Da. These column-packings have surface areas of about 125 to 400 m^2/g, which allow injected sample weights of ≈50 μg (for column inner-diameters of 4–5 mm). Because peaks as small as a few nanograms can be measured, both major and trace components can be assayed in a single separation (or "run") with these columns.

Larger molecules (usually biomolecules such as proteins) require larger pores, to avoid restricted diffusion within the pore, with a resulting decrease in column efficiency (Section 13.3.1.1). Column packings with ≈30-nm pores typically are used for separating molecules larger than 10,000 Da. If the pore diameter is at least four times larger than the hydrodynamic diameter of the solute molecule, hindered diffusion and lower values of N will be avoided [8]. Wider pores also mean smaller surface areas and decreased sample weights. C_{18} columns are used most often for reversed-phase chromatography (RPC); the physical properties of some popular columns of this type are shown in Table 5.2.

5.2.2 Silica Supports

Silica in the form of either particles or monoliths is the most commonly used support for the production of HPLC packings. High mechanical strength is a strong advantage for silica particles, allowing the formation of packed beds that are stable for long periods and high operating pressures. Silica-based columns also provide

Table 5.1

Separation Characteristics of Particles for HPLC Columns

Particle Type	Separation Speed	Pressure	Ruggedness	Operator Convenience	Column Stability[a]
5-μm totally porous	1	4	4	4	4
3.5 totally porous	2	3	4	4	4
2- to 3-μm totally porous	3	2	2	3	3
<2-μm totally porous	4	1	2	1	3
5-μm superficially porous	2	4	3	3	3
2- to 3-μm superficially porous	4	2	4	3	4
1.5-μm pellicular (nonporous)	4	1	2	1	1
Pore diameter					
7- to 12-nm pores (150–400 m^2/g)	Small-molecule separations (<10,000 Da)				
15- to 100-nm pores (5–150 m^2/g)	Large-molecule separations (>10,000 Da)				

Note: Ratings in terms of advantage from moderate (1) to high (4).
[a]Ability to tolerate high pressures or a rapid change in pressure.

Table 5.2

Properties of Some Commercial C_{18} Particles

Packing Material	Pore Diameter (nm)	Surface Area (m^2/g)	Carbon Load (%)	Bonded-Phase Coverage ($\mu mol/m^2$)
Ace C_{18}	10	300	15.5	na[c]
Ascentis C_{18}	10	450	25	3.7
Halo C_{18}[a]	9	150	8	3.5
Hypersil Gold C_{18}	17.5	220	10	na
Luna C_{18} (2)	10	400	17.5	3.0
Sunfire C_{18}	10	340	16	na
TSK-GEL ODS-100V	10	450	15	na
XBridge C_{18}[b]	13.5	185	na	3.1
Zorbax XDB-C_{18} Plus	9.5	160	8	na

[a]Superficially porous particles; other particles are totally porous.
[b]Hybrid stationary phase (Section 5.3.2.2).
[c]Data not available.

higher values of N, compared to other support materials (Section 5.2.5). Spherical particles can be synthesized with a wide choice of pore sizes (e.g., 10, 30, 100 nm), particle sizes (e.g., 1.5, 2.7, 3.5, 5 μm), and in different particle configurations (as in Fig. 5.1). Silica also can be used for monolithic columns (Section 5.2.4).

Another advantage of silica is that it can be bonded with different ligands (e.g., C_8, C_{18}, phenyl, and cyano; Section 5.3.1) for use with different samples and to change separation selectivity. Silica-based columns are compatible with all organic solvents and water, and do not swell or shrink with a change of solvent (as can be the case for polymeric particles; Section 5.2.3); silica particles are especially suited for gradient elution, where the mobile-phase composition changes during the separation. Silica begins to dissolve in the mobile phase at pH > 8 [9], which can result in a short lifetime for columns packed with silica particles. However, special bonded-phases are available that stabilize silica particles at higher pH (Section 5.3.2.1). Alternatively, supports other than silica can be used for increased column stability at high pH (Section 5.2.5). In the case of basic solutes, another limitation of silica has been a tendency for tailing peaks. This problem has largely been resolved by the use of higher purity silica (Section 5.2.2.2).

Most silica particles in use today are spherical in shape. Spherical particles are more easily and reproducibly packed to yield efficient columns, and spherical particles tend to be stronger. Large, irregular particles are used mainly for preparative separations because of the lower cost of these materials and less need for high values of N (Section 15.3.1.2). Silica-based particles are available that can withstand pressures of 15,000 psi (especially superficially porous and pellicular particles, because of their solid core). Spherical silica particles can be synthesized by several different methods (Section 5.2.2.3 [1, 4]). Figure 5.2*a* shows the visual appearance (surface topography, shape and size distribution) of some commercially available silica particles, while a closer look at some popular (≈3-μm-diameter) particles is shown in Figure 5.2*b*. Significant deviations from a true spherical shape can be seen in some of these examples.

5.2.2.1 Column Efficiency

Particle size is a primary factor in determining column efficiency as measured by the plate number N. This is illustrated in Figure 5.3 for several columns of varied particle diameter; the plate height H (inversely proportional to N) is plotted versus mobile-phase velocity u (proportional to flow rate). As the diameter d_p of the porous particles decreases from 5 to 1.8 μm, the plate height H decreases—corresponding to an increase in column efficiency per mm of column length. For well-packed columns of totally porous particles, the reduced plate height $h = H/d_p$ for a small molecule is ≈2, which until about 2006 was accepted as a lower limit for well-packed columns filled with totally porous particles (Section 2.4.1).

The type of particle and its particle-size distribution can also affect column efficiency, presumably by influencing the homogeneity of the packed bed. An example is shown in Figure 5.3 for the 2.7-μm superficially porous column. This column (with $h \approx 1.5$) is more efficient than the column packed with 1.8-μm fully porous particles. The superior efficiency of the 2.7-μm column may result from two separate factors. First, the particle-size distribution of these particles is extremely narrow (±5–6%, 1 SD), as shown in Figure 5.4. This can be compared with a

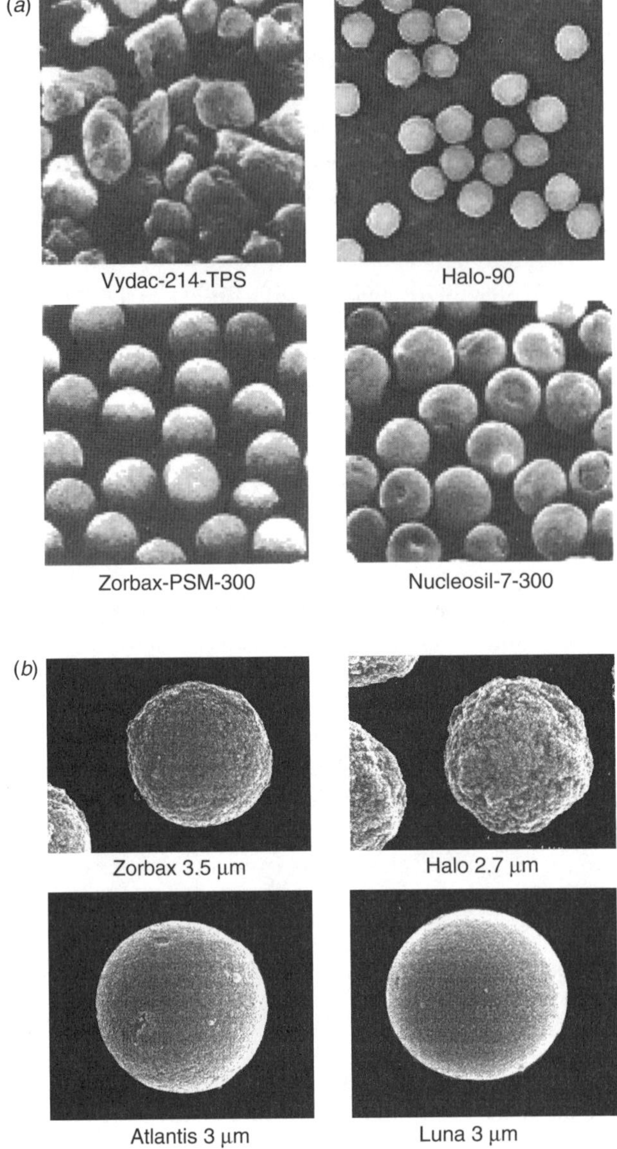

Figure 5.2 Visual appearance of several silica particles for RPC; magnification in (*b*) is 7× greater than in (*a*).

±15–20% (1 SD) particle-size distribution for most commercial packings of similar size [5], as in the example of Figure 5.4. Computer simulations [10] suggest that a narrower particle-size distribution should result in columns with larger values of N, as well as improved permeability (i.e., a lower pressure drop, other factors equal); see also [11, 12]. Columns with a narrower range in particle size are also more stable [13].

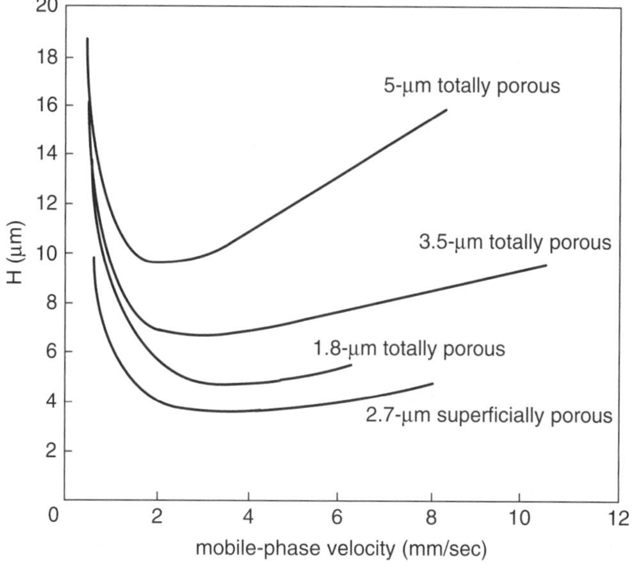

Figure 5.3 Column efficiency as a function of particle size and type. Sample, naphthalene. Conditions: 50 × 4.6-mm, C_{18} columns; mobile phase is 60% acetonitrile-water mobile phase; 23°C. Courtesy of Advanced Materials Technology.

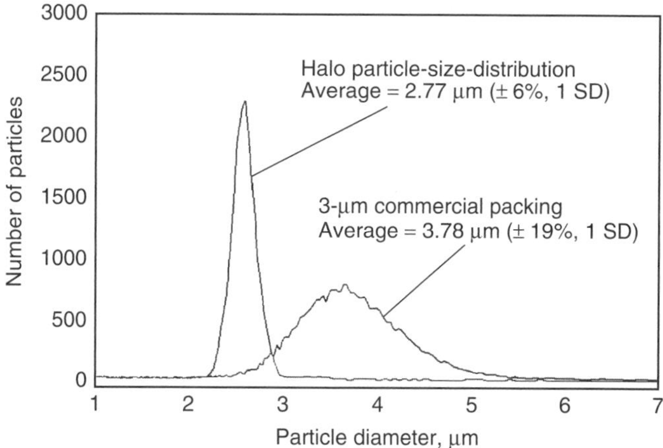

Figure 5.4 Narrower particle-size-distribution for the superficially porous (Halo™) packing of Figure 5.3, compared with that of a commercial totally porous packing. Courtesy of Advanced Materials Technology.

A second possible reason for the exceptional efficiency of the superficially porous column of Figure 5.3 is the higher density of these particles (~1.4 g/cc), which is 30–70% greater than that of totally porous particles. A more dense particle may pack more efficiently, as is the case for larger particles. Figure 5.5 shows a transmission electron micrograph cross section of a superficially porous particle, whose structure has a higher density solid core and a porous shell. Figure 5.6*a* shows

Figure 5.5 Cross-section of superficially porous (Halo™) particles with 9-nm pores (electron micrograph). Courtesy of Advanced Materials Technology.

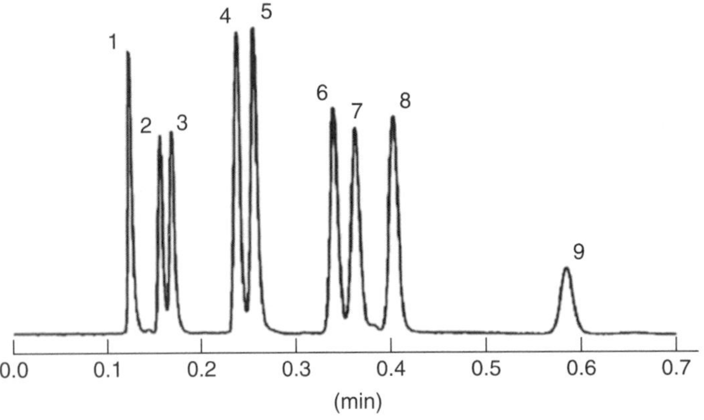

Figure 5.6 Fast separation by means of a superficially porous column. Sample: *1*, acetone; *2*, tebuthiron; *3*, thiazuron; *4*, fluometuron; *5*, diuron; *6*, propanil; *7*, siduron; *8*, linuron; *9*, diflubenzuron. Conditions: 50 × 4.6-mm Halo C_{18} column (2.7-μm particles); mobile phase is 45% acetonitrile/water mobile phase; 4.0 mL/min; 60°C; 3400 psi. Courtesy of Advanced Materials Technology.

an example of a very fast separation of a herbicide sample by means of a 2.7-μm superficially porous column.

5.2.2.2 Nature of the Silica Surface

The chemical nature of the unmodified silica surface (which varies with the manufacturing process) strongly influences its properties [14–17]. A surface layer of silanol groups (–SiOH) with a concentration of ≈8 μmoles/m^2 is a feature of all fully hydrated silicas; these silanol groups must be present for the reaction of silanes with the silica to form a bonded phase (Section 5.3.1). Most of these silanol groups

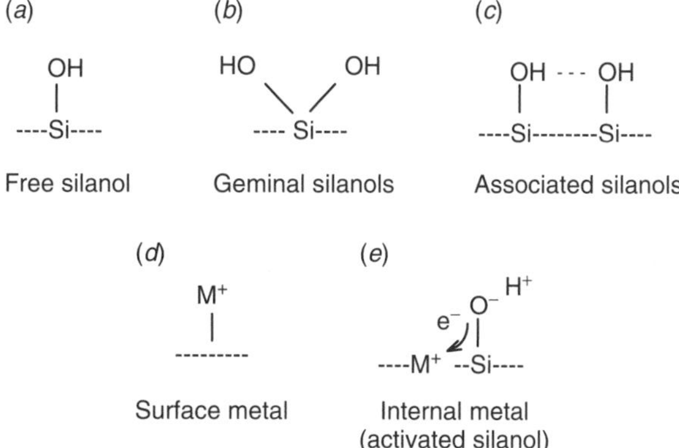

Figure 5.7 Silica surface showing different types of silanols.

are lost when silica is heated above 800°C, rendering the silica useless for HPLC. Three different silanol groups are present on the surface of a hydrated silica [14, 16–20], as illustrated in Figure 5.7*a–c*. Column performance is strongly affected by silanol acidity; free (non–hydrogen-bonded) silanols are relatively more acidic and have been associated with lower values of *N* and increased peak tailing for basic solutes.

The purity of the silica support has an even greater effect on silanol acidity and column performance. Certain metals (especially Al[III] and Fe[III]) are potential contaminants that can increase silica acidity by withdrawing electrons from the oxygen of the silanol (Fig. 5.7*e*), as well as interact directly with chelating solutes (Figs. 5.7*d*, 5.20*i*). These various consequences of metal contamination can result in tailing peaks and poor recovery for some solutes. Present state-of-the art chromatographic silicas are much more pure, as illustrated by the data of Table 5.3 for one such silica (tests also exist for the direct measurement of silica acidity and complexation; Section 5.4.1 and [18, 21]). Silica particles and resulting packings and columns can be classified as type A or B [15], based on their purity or cation-exchange behavior (Section 5.4.1).

Table 5.3

Representative Analysis by ICP-AES/MS for Trace Elements in Zorbax Rx-SIL

Element	Content (ppm)
Na	10
Ca	2
K	<3
Al	1.5
Fe	3
Mg	4
Zn	1

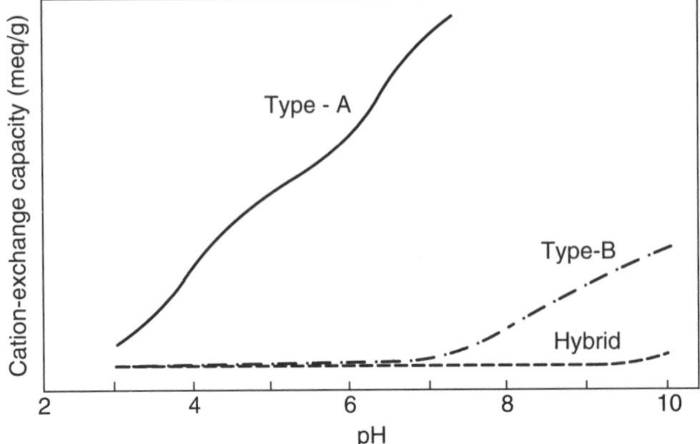

Figure 5.8 Relative acidity (silanol ionization) of different HPLC supports, as measured by the retention of Li^+ ion. Adapted from [22].

Type-A silicas are less pure and more acidic, and were used exclusively for HPLC prior to 1990; they are still useful for some less-demanding applications such as sample preparation (Section 16.6) and preparative chromatography (Chapter 15). Silanol ionization varies with both silica type and mobile-phase pH, and can be measured by the relative retention of an inorganic cation such as Li^+. Figure 5.8 illustrates silica ionization or "cation-exchange capacity" for a type A, a type B, and a "hybrid" C_{18} column (Section 5.3.2.2). The relative acidity of these three C_{18} columns decreases in the order type A $\gg$ type B > hybrid.

Basic solutes and ionic or ionizable compounds are usually best separated with newer type-B columns. Figure 5.9 compares the separation of some basic

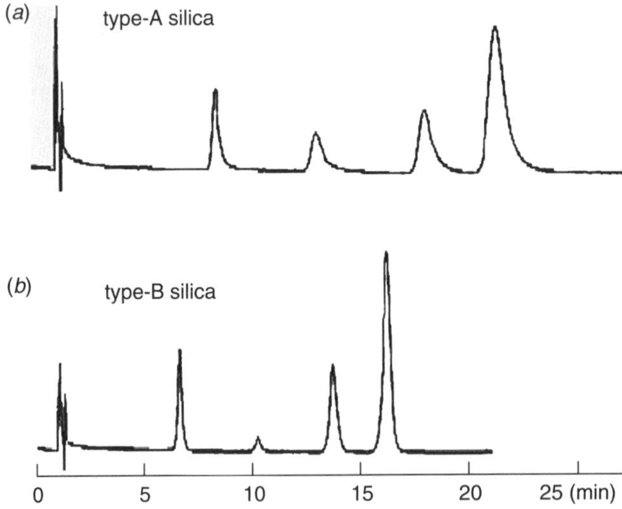

Figure 5.9 Separation of protonated basic compounds on type-A (*a*) compared with type-B (*b*) columns. Sample: four tricyclic antidepressants. Conditions: 150 × 4.6-mm C_{18} columns; mobile phase is 30% acetonitrile-water with pH-2.5 phosphate buffer. Adapted from [19].

drugs with C_{18} columns based on either type-A and -B silica. The type-B column (Fig. 5.9*b*) is seen to provide peaks that are narrower and more symmetrical. The selection of a less-acidic column as a means of minimizing peak tailing is discussed in Section 5.4.4.1. Commercial columns that were introduced after 1990 are mainly type B.

5.2.2.3 Particle Preparation

Several methods are used to prepare particles for HPLC columns. Totally porous silica particles can be made by a sol-gel procedure, which involves the emulsification of a silica sol in an immiscible nonpolar liquid, followed by (1) the formation of droplets and their conversion into spherical beads of silica hydrogel, (2) drying, and (3) classification into a narrow particle-size range. By controlling pH, temperature, and the concentration of the silica sol, particles with the desired size and pore size distribution can be obtained. An alternative approach to emulsification is to spray-dry silica sols or solutions of finely divided silica, so as to form spherical, totally porous particles. The resulting pore structure of these materials can be modified by hydrothermal treatment in an aqueous media (or with steam) at elevated temperatures and pressures.

Still another approach for preparing totally porous silica particles is to aggregate or assemble microparticles, as illustrated in Figure 5.10. One such procedure is to disperse a silica sol of defined size in a polar liquid, add a solution of polymerizable organic material such as formaldehyde and urea or melamine, and then initiate the coacervation of the silica with the polymer to form spherical, uniformly sized aggregates. The resulting material is first heated to eliminate the organic polymer, and then sintered at higher temperature to strengthen the interconnected network of silica sol particles. Small, silica-sol particles are used to prepare particles with smaller pores; larger pores are formed from larger silica sol-particles. This aggregation method can also be used to prepare totally porous particles of zirconia.

So-called hybrid (organic/inorganic) particles are prepared by the co-polymerization of organo-silanes with tetraalkyl-*o*-silicate or other organic silicates. The pore structure of particles made by this approach can be modified by appropriate hydrothermal treatments. After hydrolysis, the surface can be further modified by reaction with other silanes to produce the desired final ligand.

Superficially porous silica particles are prepared by layering silica-sol particles onto a spherical, solid core of silica. The general approach is to alternate layers of oppositely charged organic polymer and negatively charged silica sol until the desired shell thickness is obtained. The organic interlayer is then removed by heating,

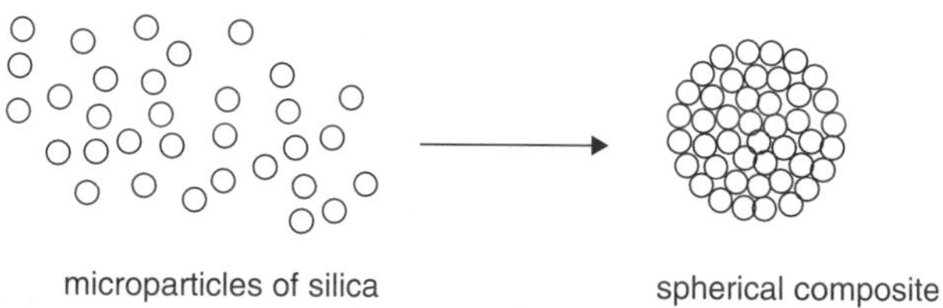

Figure 5.10 Aggregation of microparticles to form totally porous particles.

and the particles are strengthened by sintering. Small particles of the silica sol create smaller pores, while larger silica sols are used for larger pores.

An "inverse" approach has been used to make graphitized-carbon particles (Section 5.2.5.3). For example, a polymerizable organic material is introduced into the pores of silica particles and polymerized within the pores. The resulting silica/polymer particles are treated with hydrofluoric acid to dissolve the silica sol and leave spherical, porous-polymer particles. Finally, these particles are heated to a high temperature in the absence of oxygen for graphitization. For further details of the preparation of particles for HPLC columns, see [1].

5.2.3 Porous Polymers

Porous-polymeric particles (primarily, cross-linked polystyrene) have been used for reversed-phase, ion-exchange, and size-exclusion columns. Other polymers, for example, substituted methacrylates and polyvinyl alcohols, are also commercially available—but are used less often. Polymeric columns can be derivatized with RPC ligands, such as $-C_{18}$, $-NH_2$, and $-C{\equiv}N$ in order to create differences in column selectivity, or with ionic groups for use in ion-exchange chromatography. Compared to the number of commercially available silica-based packings, however, only a small number of different RPC ligands are currently available for polymer-based particles. A main advantage of porous, cross-linked polystyrene particles is that they are pH stable; any mobile-phase pH is acceptable ($0 \leq \text{pH} \leq 14$). Polymeric columns can be used for the RPC separation of strongly basic samples at $\text{pH} \geq 10$, in order to suppress sample ionization and avoid the tailing peaks that are sometimes seen with silica-based packings (Section 7.3.4.2). However, newer, type-B RPC columns that are stable at higher pH have largely replaced the latter application.

Polymer-based particles modified with ionizable functional groups ($-COOH$, $-SO_3H$, $-NH_2$, and NR_3^+) are used mainly for ion-exchange separations (Sections 7.5, 13.4.2). Polymeric columns can be made with narrow pores for the separation of small solute molecules, or wide pores for large molecules such as proteins. Because of their stability at high pH, these polymer-based columns allow strongly retained sample contaminants to be purged from the column with 0.1 M sodium hydroxide (Section 13.4.2.1).

Some limitations exist on the use of porous, polymer-based packings. These columns exhibit lower values of N, compared to silica-based columns with particles of similar size. A potential problem with polymer particles is that they can swell (or shrink) to varying extents with different organic solvents, and thus cause loss of column efficiency or increase in column pressure. Adverse effects due to particle swelling are more noticeable in gradient elution because the B-solvent concentration (%B) changes during the separation. Some polymeric packings have been designed to minimize swelling, however, rendering them suitable for use with gradient elution.

5.2.4 Monoliths

Since 1990 monoliths have received increasing attention. Monolithic columns are cast as a coherent, rigid cylinder by in situ polymerization, and they can be made from either silica or polymer. Figure 5.11 shows electron micrographs of monoliths made from (Fig. 5.11*a*) silica and (Fig. 5.11*b*) polymer. Monoliths are available as either columns or disks; the latter are used primarily for peptide

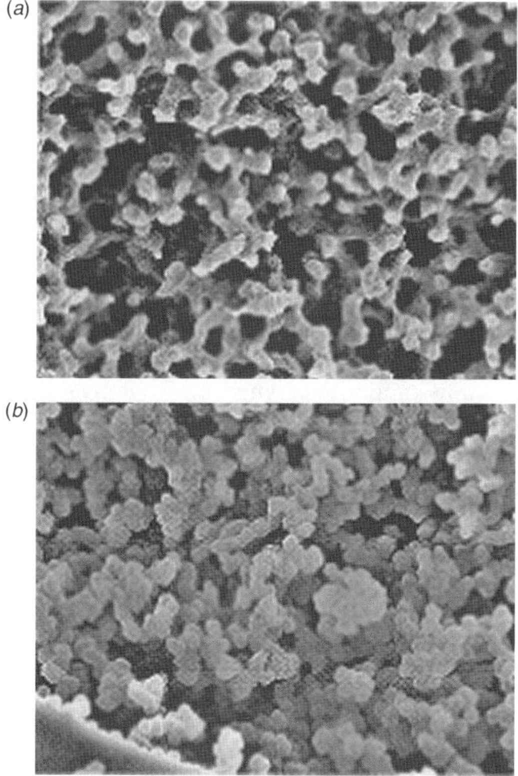

Figure 5.11 Cross section of representative monolith packings (electron micrographs). (*a*) Silica-based; (*b*) polymeric. Reprinted from [23] with permission of EDM Chemicals Inc. (*a*) and F. Svec (*b*).

synthesis and other applications of biological interest, and these will not be discussed further. This section describes both silica- and polymer-based monolithic columns, their advantages and limitations, and some important applications. The history, development, and characteristics of monoliths have been reviewed [24–27].

5.2.4.1 Silica-Based Monoliths

Silica monoliths are prepared as rods from a single piece of porous silica, so packing particles into a length of column tubing (Section 5.6) is no longer required. Commercial monoliths are not produced inside a column blank; rather, the monolith is formed, dried, and then encapsulated with a polymeric coating. The surface modification of the resulting silica monolith (e.g., the bonding of C_{18} ligands) is carried out later. The monolith bed contains pores of two types: large *macropores*, with diameters of ~2 μm, and small *mesopores* with diameters of ~10 nm. The resulting surface area (e.g., 300 m^2/g) is contributed mainly by the mesopores, while flow through the column occurs in the macropores. The interstitial volume of a conventional packed bed is about 40% of the column volume, and this is virtually independent of particle size or particle-size distribution. Monoliths, on the other hand, can be made with interstitial volumes (defined as the volume of the macropores) as high as 80%.

The diameters and total volume of the monolith macropores determine column permeability and the pressure required to achieve a certain flow rate. Because these macropores are quite large, the required pressure for flow through a monolith is much less than for conventional columns of packed particles with "equivalent" particle diameters d_p (as estimated from later Eq. 5.5). Compared to particulate columns, higher flow rates can be used with monoliths, without a corresponding loss in column efficiency. One study with small-molecule solutes suggests that commercial monolith columns have efficiencies that are equivalent to columns packed with 4- to 5-μm particles [28]. An important question is how monoliths compare with conventional packed columns in both pressure drop and values of N. Another study suggests that monolithic columns are *only beneficial* when high-resolution separations are needed (e.g., $N > 100{,}000$) [29], but not all investigators agree [30]. For further details on how monoliths compare with conventional packed columns, see the review of [24].

Some limitations for silica-based monoliths include a restricted range of available column dimensions, a limited variety of stationary phases, and a tendency of available materials to exhibit tailing peaks (e.g., tailing factor ~1.2), even for neutral (non-ionizable) compounds [28]. At the time that the present book was written, monoliths were finding only limited application in routine laboratories, but research at that time was promising [31], especially for capillary monolith columns. Limited studies also suggest that silica-based monoliths can be prepared with a reproducibility that is acceptable for routine analysis [28, 32]. The future impact of these columns remains to be seen.

5.2.4.2 Polymer-Based Monoliths

Monoliths made from organic polymers are available for a wide range of applications. These columns can be made by the copolymerization of monomers such as styrene/divinylbenzene with monovinyl/divinyl methacrylate. The copolymerization of other monomers with different functionalities allows the formation of particles with different ligands and RPC selectivity. These polymeric monoliths are directly formed into column tubes, including capillary columns (0.5-mm diameter and smaller). During the production of a monolith, two sets of characteristics can be controlled simultaneously: the nature of the polymerizable material and the pore structure. Polymer-based monoliths with wider pores appear well-suited for the separation of large biomolecules (peptides, proteins, oligonucleotides, etc.), as well as for capillary columns, where long lengths can be used for separating complex mixtures that require large N [33].

5.2.5 Other Inorganic Particles

Although silica is usually the preferred support material, packings based on other inorganic structures have been found useful for certain applications. This section will review these other inorganic supports and describe their preferred use. Unmodified metal oxides have a different surface chemistry than silica, which has led to three different procedures for attaching the stationary phase: (1) deposition of a polymer or carbon layer onto the surface, (2) covalent attachment of a ligand, and (3) use of a strongly interacting mobile-phase additive. We will describe particles made from zirconia, alumina, titania, and graphitized carbon. Table 5.4 lists some commercially

Table 5.4

Commercially Available Column-Packings Based on Zirconia or Alumina

Column Packing[a]	Particle Type	Particle Diameter (μm)	% Carbon	Pore Size (nm)	Surface Area (m^2/g)
ZirChrom-PBD	Zirconia	3 or 5	na[b]	30	30
ZirChrom-PS	Zirconia	3 or 5	na	30	30
Aluspher RP-select B	Alumina	5	na	10	170
Millpore PBD	Alumina	5	7.2	9	110
Unisphere	Alumina	10	5.1	22	37
GammaBond RP-1 (PBD)	Alumina	5	na	8	na
GammaBond RP-8	Alumina	5	na	8	na

[a]Taken in part from [35].
[b]Not available.

Table 5.5

Comparison of Silica and Metal-Oxide Particles for Chromatography

Property	Silica	Titania	Alumina	Zirconia
Particle monodispersity	3	2	2	2
Pore structure	3	na[b]	na	2
Surface area/pore diameter	4	2	2	2
Surface chemistry	3	na	2	2
Mechanical strength	4	na	na	4
Chemical stability	2	2	3	4
Thermal stability	2	na	na	4
Column efficiency	3	na	2	3
Energetic homogeneity[a]	3	1	1	1

Note: Ratings from low or neutral (1) to high or favorable (4).
[a]Particles other than silica possess strong adsorption sites, which can lead to low values of N and incomplete recovery of the sample.
[b]Not available.

available columns based on zirconia and alumina. Some advantages and limitations of columns made from various inorganic supports are summarized in Table 5.5.

5.2.5.1 Zirconia

The covalent bonding of a ligand onto zirconia particles has been generally unsuccessful, because the ligand-particle bond has proved to be unstable under most conditions used for RPC. Useful zirconia-based packings are usually made by depositing or polymerizing an organic phase onto the particle surface. Polybutadiene or polystyrene is the usual stationary phase, resulting in packings whose selectivity is somewhat similar to that of alkylsilica packings (at least for non-ionizable solutes). These polymer-based zirconia materials show outstanding stability for both

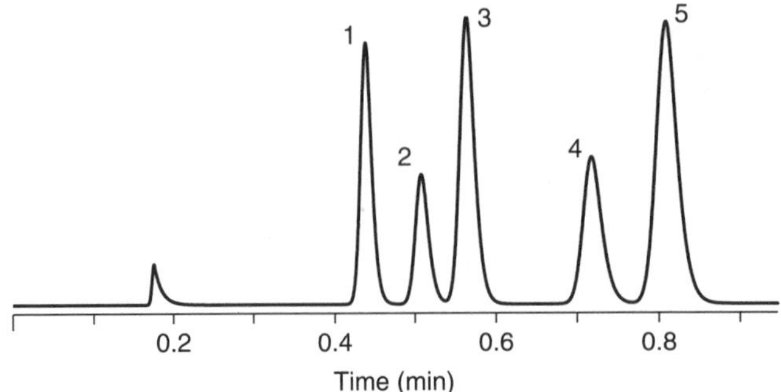

Figure 5.12 High-temperature separation of a pharmaceutical mixture. Sample: *1*, doxylamine; *2*, methapyrilene; *3*, chloropheniramine; *4*, meclizine; *5*, triprolidine. Conditions: 100 × 4.6-mm ZirChrom-PBD® column (zirconia); 20% acetonitrile/water with added tetramethylammonium hydroxide to control pH-13; 4.2 mL/min; 140°C; 2850 psi. Courtesy of ZirChrom Separations, Inc.

low- and high-pH mobile phases ($1 \leq \text{pH} \leq 13$), and they can be used at very high temperatures ($\leq 160°\text{C}$) [34]. Figure 5.12 shows a separation on a zirconia column at pH-13 and 140°C. Early columns packed with polymer-coated zirconia displayed rather poor efficiency, apparently because of poor stationary-phase mass transfer. While their performance continues to be improved, zirconia columns are still waiting (as of the time this book was written) for a critical application where they perform demonstrably better than silica-based RPC columns. However, this may reflect the present limited impact of high-temperature separation (>60°C).

For ionizable solutes and zirconia-based columns, poor peak shapes often result, regardless of how the packing is prepared. Consequently, when zirconia-based packings are used with ionizable solutes, special mobile-phase additives (e.g., phosphate or fluoride) are required for good peak shape and reasonable values of N [34, 35]. For a sufficiently large concentration of the additive, values of N and peak shape for basic solutes are similar to those found for alkylsilica columns (for $\text{pH} \leq 10$). At the time this book was published, somewhat poorer results were obtained for the separation of peptides and proteins, compared to separations with alkylsilica columns.

Zirconia-based ion-exchangers are also available commercially. A weak anion-exchanger can be formed by coating zirconia particles with polyethyleneimine, followed by cross-linking with 1,4-butanedioldiglycidylether. This approach produces columns that are stable from pH range 3 to 9 and can be used to separate organic acids, inorganic anions, and highly polar compounds such as sugars. Chemically and thermally stable, strong anion-exchange columns are also available for separations at $\leq 100°\text{C}$, over the pH range of 1 to 13. These packings are formed by cross-linking zirconia-coated polyethyleneimine with 1,10-diiododecane or a similar compound [34].

Carbon-clad zirconia is a uniquely selective packing, compared to other RPC columns; it is prepared by passing a reduced pressure of organic vapor over porous zirconia at a temperature of ~700°C [34]. Carbon-clad zirconia differs from alkyksilica packings in being more hydrophobic, is better able to separate polar and nonpolar geometrical isomers, and is also capable of π–π interactions

(Section 5.4.1). This packing is stable from pH 0.3 to 14 at 40°C, and is thermally stable to ≥ 200°C at neutral pH. Peak shape and column efficiency tend to be poor at lower temperatures (e.g., 35°C), but these columns have been under development for a much shorter time than silica-based columns—so future improvements seem likely.

5.2.5.2 Alumina and Titania

RPC columns based on an alumina support were reported in the early 1980s [36], and have since been reviewed [37]. The chemistry of the alumina surface is more like that of zirconia than silica, and alumina columns are also stable at higher pH. As covalently bonded alumina is not stable, coating or polymerizing a polymer onto the surface is used in place of silane derivatization. Because of its strongly adsorptive properties, few reversed-phase applications of coated alumina have been reported. Unbonded alumina is not used at present for HPLC separation, but it has found a role for sample preparation (Section 16.6.5.2).

Commercial columns packed with titania particles are also available, but these are used mainly for normal-phase separations. Reversed-phase materials based on titania generally show no advantages over more traditional silica- and zirconia-based particles, so they have not achieved widespread use. On balance, silica remains by far the most popular support, because of a better compromise among important column properties (Table 5.5).

5.2.5.3 Graphitized Carbon

Porous, graphitic carbon (PGC) is a very different column-packing, consisting of fully porous, spherical carbon particles that are formed from flat sheets of hexagonally arranged carbon atoms [38, 39]. The carbon atoms have a fully satisfied valence that results in very different retention and selectivity, compared to other columns. PGC can be used for both reversed- and normal-phase separation, and is stable at $1 \leq pH \leq 14$ and $\leq 200°C$. However, its reduced particle strength limits the maximum pressure that can be used with these columns.

PGC retains polar compounds by a combination of strong hydrophobic, electronic, and dipolar interactions [38], so that polar solutes can be preferentially retained even under RPC conditions. The selectivity of graphitized-carbon columns is difficult to predict, compared to conventional bonded-phase columns, and this can make method development more difficult. Also column efficiency and peak shape can be somewhat poorer than for conventional RPC columns. However, porous carbon shows a special ability to separate stereo- and diastereoisomers, as well as positional isomers for which poor or no separation occurs with conventional packings. Figure 5.13 shows a separation of hippuric acid and its methyl-substituted isomers on a PGC column, using a low-pH mobile phase.

5.3 STATIONARY PHASES

The column stationary phase determines retention and selectivity (Sections 2.3, 5.4), and it must meet certain practical requirements, for example, acceptable stability,

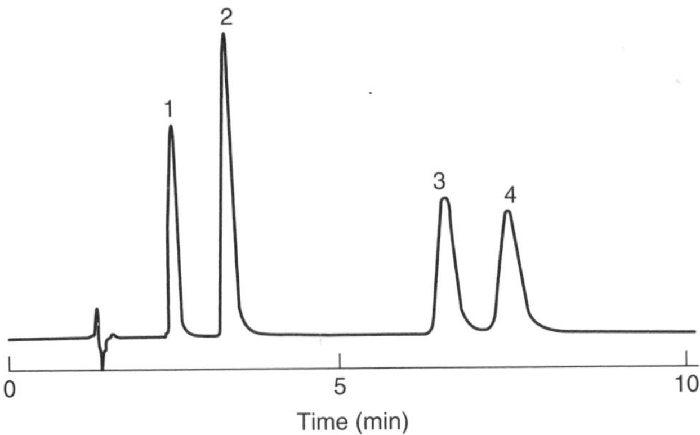

Figure 5.13 Separation on a graphitized-carbon column of hippuric acid and its methyl-substituted isomers. Sample: *1*, 2-methylhippuric acid; *2*, hippuric acid; *3*, 3- methylhippuric acid; *4*, 4- methylhippuric acid. Conditions: 100 × 4.6-mm Hypercarb column; mobile phase is 30% acetonitrile, 30% isopropanol, and 40% water with 0.1% TFA; 1.0 mL/min; 25°C. Courtesy of Thermo Scientific.

reproducibility, peak shape, and column efficiency N. In this section we review the preparation, nature, and properties of different stationary phases—apart from their selectivity, which is discussed in the following Section 5.4. Most stationary phases are organic in nature, either covalently bound to or (rarely) mechanically deposited on the particle. In some cases the surface of the unmodified particle *is* the stationary phase, for example, unmodified silica for use in normal-phase chromatography (including hydrophilic interaction chromatography, HILIC). We will assume that we start with a silica particle, prior to adding the stationary phase. Procedures used for other supports were referred to in the previous Sections 5.2.3–5.2.5.

5.3.1 "Bonded" Stationary Phases

RPC packings usually are made by covalently reacting ("bonding") an organosilane with the silanols on the surface of a silica particle to form the stationary phase or *ligand R*:

$$X_3\text{–}Si\text{–}R + \equiv Si\text{–}OH \rightarrow \equiv Si\text{–}O\text{–}Si(X_2)\text{–}R + HX \quad (5.1)$$

(silane) (silanol) (final phase)

The functional group X is often –Cl or –OEt, and/or $-CH_3$ (Fig. 5.14), in which case the reaction by-product HX is HCl or ethanol. Silanes substituted with other groups X are also used, as will be discussed. Some bonded-phase packings are made via the monofunctional reaction of Figure 5.14*a*. Here a single silane reagent reacts with a single, surface-silanol group, for example, chlorodimethyl-octadecylsilane (where the ligand $R = C_{18}$) reacts to form a *monomeric* C_{18} column. Other commercial packings are formed from a surface reaction with a trifunctional (or difunctional) silane, as illustrated in Figure 5.14*b*, *c* (although two silane-silica bonds are shown here, three such bonds are also possible). Depending on the reaction conditions, polymerization of the stationary phase can result in the latter case (use of a difunctional

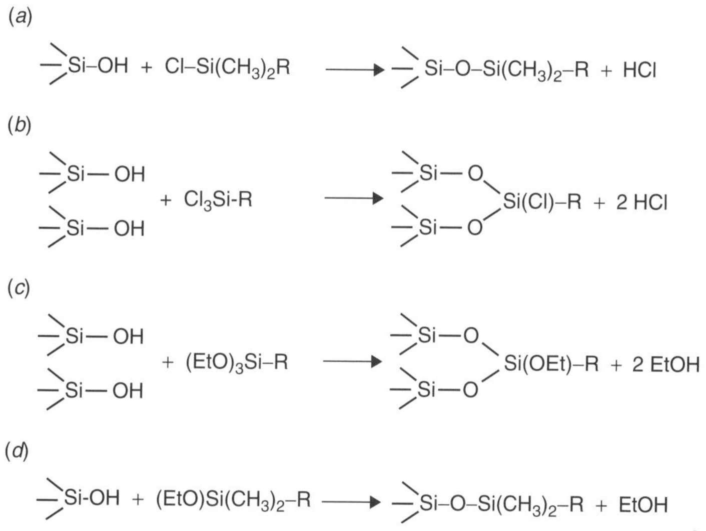

(*a*)

$$\equiv Si\text{-}OH + Cl\text{-}Si(CH_3)_2R \longrightarrow \equiv Si\text{-}O\text{-}Si(CH_3)_2\text{-}R + HCl$$

(*b*)

$$\begin{matrix}\equiv Si\text{—}OH \\ \equiv Si\text{—}OH\end{matrix} + Cl_3Si\text{-}R \longrightarrow \begin{matrix}\equiv Si\text{—}O \\ \equiv Si\text{—}O\end{matrix} Si(Cl)\text{-}R + 2\,HCl$$

(*c*)

$$\begin{matrix}\equiv Si\text{—}OH \\ \equiv Si\text{—}OH\end{matrix} + (EtO)_3Si\text{-}R \longrightarrow \begin{matrix}\equiv Si\text{—}O \\ \equiv Si\text{—}O\end{matrix} Si(OEt)\text{-}R + 2\,EtOH$$

(*d*)

$$\equiv Si\text{-}OH + (EtO)Si(CH_3)_2\text{-}R \longrightarrow \equiv Si\text{-}O\text{-}Si(CH_3)_2\text{-}R + EtOH$$

Figure 5.14 Synthesis of various bonded-phase column packings by the reaction of a silane with silica. (*a*, *d*), Monomeric packings; (*b*, c), potentially polymeric packings.

or trifunctional silane), yielding a *polymeric* stationary phase or column (not to be confused with a "polymer column"; Section 5.2.3). As we will see, the properties of monomeric and polymeric columns are significantly different in important respects.

Several different kinds of silane–silica reactions have been used to prepare HPLC columns, as illustrated in Figure 5.15. Figure 5.15*a* illustrates a "vertical" polymerized phase that results from the reaction of a di- or trifunctional silane (as in Fig. 5.14*b* or Fig. 5.14*c*). In the example of Figure 5.15*a*, the silane that initially reacts with the surface further reacts with one or more additional silane molecules to give a polymeric phase (in the presence of water; see below). These phases tend to be more stable than monomeric phases at both low and high pH, as the "heavier" surface coverage of these packings slows down the attack of the mobile phase on both the silica and the ligand–silica bond. However, packings of this type tend to be less reproducible in terms of retention and selectivity because of variable (inadequately controlled) silane polymerization.

"Horizontal" polymerization with self-assembled silanes (C_3 plus C_{18}) yields the general structure shown in Figure 5.15*b*. Here Si atoms of adjacent silanes are connected to each other through oxygen atoms (siloxane linkages, Si–O–Si), while each silane is connected to the silica via another siloxane bond. Columns prepared in this way have been reported to exhibit superior stability in both low- and high-pH applications [40], but no commercial columns of this type had been announced at the time this book went to press.

The monomeric phase of Figure 5.15*c* is most widely used for RPC columns; packings with several different functional groups (ligands) are commercially available (Section 5.3.3, Table 5.7); the silane side-groups are usually methyl groups, as in Figure 5.15*c*. These packings are commonly prepared by reacting dimethylchloro- or dimethylethoxy-silanes with the silica support (Fig. 5.14*a*, *d*): one silane molecule reacts with one silanol group. The advantage of this one-to-one reaction is that a

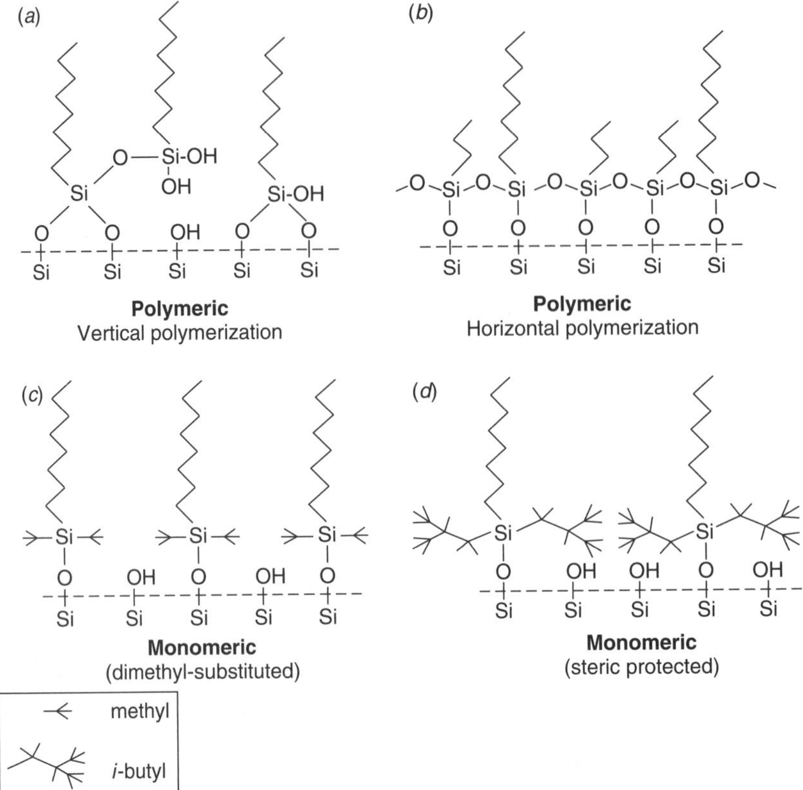

Figure 5.15 Some alternative bonded phases based on different reaction conditions. Adapted from [40].

reproducible, well-defined bonded phase results. Packings made in this way often exhibit the highest column efficiency because of rapid diffusion of the solute into and out of the less-crowded stationary-phase layer. Conversely, packings with multifunctional, highly polymerized stationary phases can exhibit slower solute diffusion and lower values of N, especially at higher flow rates.

The silane reactions of Figure 5.14*c*, *d* are typically carried out with alkoxysilanes that have reactive R-groups (ligands) such as $-C_3-NH_2$ or $-C_3-O-CH(OH)-CH_2OH$ (to give an amino or diol column, respectively). Stationary phases with certain ligands (e.g., those containing reactive amino or hydroxyl groups) cannot be prepared from chlorosilanes because of undesirable secondary reactions of the ligand. These reactive stationary phases are instead made from alkoxysilanes, as in Figure 5.14*c* or *d*. Alkoxysilane reactions are somewhat slower than those with chloro- and dimethylamino-silane, generally requiring longer reaction times and higher silane concentrations for equivalent reaction yields.

The *sterically protected* silane stationary phase of Figure 5.15*d* [41–43] is a variation of the monomeric phase of Figure 5.15*c*, where the methyl groups of the silane in Figure 5.15*c* are replaced by *i*-propyl or *i*-butyl in Figure 5.15*d*. The latter large, bulky side-groups interfere with the hydrolysis of the bonded silane, as illustrated in Figure 5.16*b* and compared with Figure 5.16*a*. Each Si–O–Si bond is

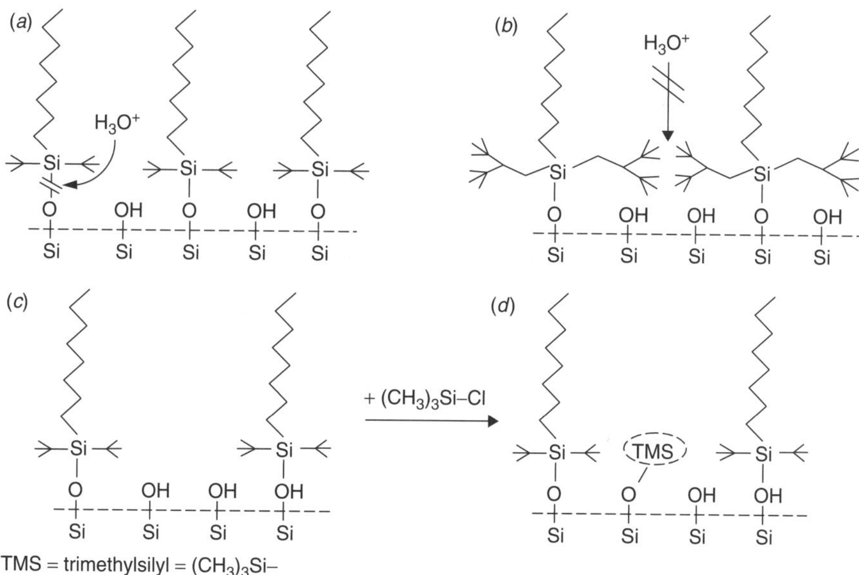

Figure 5.16 Options for increasing the stability of alkylsilica columns. (*a*, *b*), protection of the—Si–O–bond by a steric-protected bonded phase (for low-pH conditions only); (*c*, *d*) protection of the bonded phase by end-capping.

individually protected by the size of the two bulky side-groups in Figure 5.16*b*. Steric protection is useful for separations at low pH (but not at high pH) because low pH catalyzes the breaking of the O–Si bond [41]. Sterically protected stationary phases are available with a variety of ligands (e.g., C_8, C_{18}, cyano, phenyl), each of which show exceptional stability for use with low-pH mobile phases. Because low-pH mobile phases are preferred for the separation of ionic samples (Section 7.3.4.2), sterically protected columns are especially useful for the separation of biological samples such as peptides and proteins. Because of the steric bulk of the protecting silane groups, these packings have a lower surface concentration of the ligand, and exhibit lower retention than comparable dimethyl-substituted phases—as shown by the data of Table 5.6 for monomeric columns. Bonded bidentate-silane stationary phases (as in Fig. 5.14*b*) are more stable for high-pH applications [44].

Prior to the silane reaction, a fully hydroxylated silica will have a surface-silanol concentration of $\approx 8\ \mu mol/m^2$. However, the size of an attached silane results in some overlap of adjacent silanols, which inhibits their further reaction with the silane reagent. The ligand concentration for a fully reacted packing will therefore seldom exceed 4 $\mu mol/m^2$ (leaving half or more of the original silanols unreacted). As shown in Table 5.6, as the chain length or cross section of the silane increases, the percentage of reacted silanol groups (and ligand concentration) decreases. Almost 50% of the silanol groups remain unreacted for the smallest silane (trimethyl). Although unreacted silanols may be inaccessible to a reacting silane, they may still be able to interact with the solute molecule (Section 5.4.1).

Polymeric phases are used (to a limited extent) because of their greater stability and unique selectivity (Section 5.4.1.2). Whereas the preparation of monomeric phases (as in Fig. 5.14*a*) must be carried out in a water-free reaction medium,

Table 5.6

Effect of Silane Bonded-Phase Chain Length and Bulk on Silica Support Coverage

Bonded Phase Ligand	Surface Coverage ($\mu mol/m^2$)	Reacted Silanols (%)
Trimethyl	4.1	51
Dimethyl-3-cyanopropyl	3.8	48
Dimethyl-*n*-butyl	3.8	48
Dimethyl-*n*-octyl	3.5	44
Dimethyl-*n*-octadecyl	3.2	40
Sterically protected columns		
Triisopropyl	2.2	28
Diisopropyl-3-cyanopropyl	2.1	26
Diisopropyl-*n*-octyl	2.0	25
Diisobutyl-*n*-octadecyl	1.9	25

polymeric phases require the presence of water during part of their synthesis. The extent of reaction or polymerization is controlled by varying the amount of water added to the reaction [45]. The reaction of an alkyltrichlorosilane with silica particles is carried out in the presence of water [45].

To minimize unwanted interactions with residual silanol groups (Section 7.3.4.2), column packings for RPC are usually *endcapped*, by a further reaction of the bonded phase with a small silane such as trimethylchlorosilane or dimethyldichlorosilane (Fig. 5.16*d*). This procedure decreases the concentration of unreacted silanols, as well as their interaction with retained solute molecules—but does not totally eliminate silanol-solute interaction (end-capping increases the percentage of *reacted* silanols by only 20–30% [46], corresponding to a somewhat smaller decrease of *unreacted* silanols). Small ligands (e.g., end-capping trimethylsilyl groups) are more susceptible to hydrolysis and loss at low pH, which can lead to changes in retention and selectivity. On the other hand, end-capped columns are more stable at intermediate and higher pH.

Still another kind of bonded phase is based on "type-C" silica. The polar silanols of a typical type-B silica are first reacted to form a nonpolar, silicon-hydride surface (Fig. 5.17*a*). The latter type-C silica can be used without further change for either RPC or normal-phase chromatography. Type-C silica can be modified (for RPC) by the addition of alkyl groups (Fig. 5.17*b*). Good reproducibility and stability are claimed for these packings, even when used in 100% water mobile phases over a pH range of 1.5 to 10 [47]. These packings were relatively new at the time this book was published. Information on their properties and use was quite limited, but they were commercially available (MicroSolve Technology; Eatontown, NJ).

Alkyl and aromatic ligands present in a packing (including end-capping) can be identified following their removal from the silica particle [48]. The packing is first treated with aqueous hydrofluoric acid, which cleaves the ligand–silica bond. The ligand reaction-product can then be characterized by GC, NMR, and/or mass spectrometry.

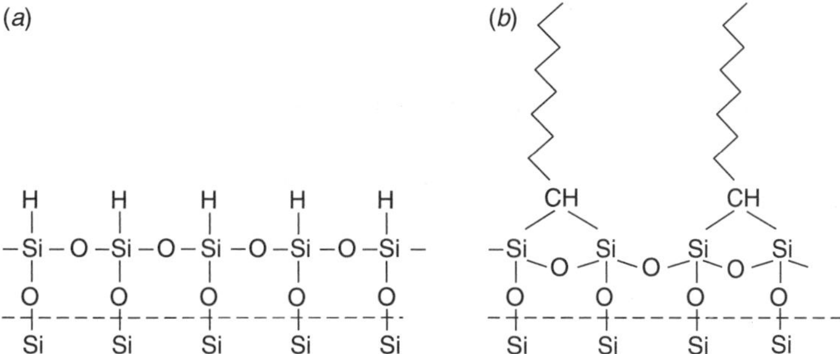

Figure 5.17 Type-C silica (*a*) and the resulting bonded phase (*b*) (bidentate C_{18}).

5.3.2 Other Organic-Based Stationary Phases

5.3.2.1 Mechanically Held Polymers

As in the case of metal-oxide supports other than silica (Section 5.2.5), mechanically held polymers such as cross-linked polystyrene and polybutadiene have been used as stationary phases for silica-based particles [49]. Because of their poorer efficiency and reproducibility, as well as the lack of phases with different functional groups, little use has so far been made of these columns.

5.3.2.2 Hybrid Particles

Hybrid particles are formed by polymerizing two monomers (e.g., tetramethoxysilane and tetraethoxysilane), to form an *organic/inorganic* structure as in Figure 5.18. The chemistry for these materials was first introduced by Unger et al. [50] in

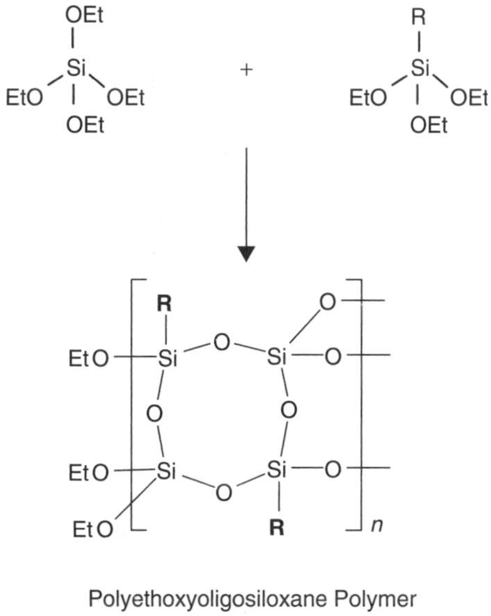

Polyethoxyoligosiloxane Polymer

Figure 5.18 Synthesis of organic/inorganic hybrid particle. Courtesy of Waters Corporation.

1977, and later developed more fully by Waters Corporation. Hybrid particles are formed using a silane that contains hydrolytically stable Si–C bonds. These stable bonds are part of the matrix, which improves the pH stability of such packings relative to silica-based packings. The first such packing was prepared from a 2 to 1 ratio of tetraethoxysilane and methyltriethoxysilane (XTerra). Later a hybrid packing with further improved pH stability was created from tetra-ethoxysilane and ethyl-bis-triethoxysilane (XBridge). These particles possess excellent stability when used with both low- and high-pH mobile phases, as well as a mechanical strength that allows their use at pressures to 15,000 psi. As with silica particles, hybrid particles can be derivatized with various ligands (C_8, C_{18}, etc.) [51]. Hybrid packings are especially useful for the high-pH separation of non-ionized basic solutes—allowing improved peak shapes and larger sample weights (Section 15.3.2.1).

5.3.2.3 Columns for Highly Aqueous Mobile Phases

RPC separations of very polar samples may require small values of %B in order to achieve values of $k \geq 1$, although other means exist for increasing the retention of such samples (Sections 6.6.1, 7.3.4.3, and Chapter 8). When RPC is used with low values of %B, several problems may be encountered: a *decrease* in sample retention with time, decrease in values of N, and long equilibration times when changing from one mobile phase to another [52, 53]. This behavior is the result of stationary-phase *de-wetting* (sometimes incorrectly called phase collapse), with the consequent expulsion of mobile phase from the pores of the particle [54, 55]. Thus the pressure P required to force the mobile phase into a particle pore of diameter d_{pore} is

$$P = \frac{-4\gamma \cos\theta}{d_{pore}} \tag{5.2}$$

where γ is the surface tension of the mobile phase, and θ is the contact angle between the stationary and mobile phases. The value of θ is $>90^\circ$ for a C_{18} stationary phase and water as mobile phase, meaning that pressure is required to force water into the pores. If the pressure P is insufficient to force a highly aqueous mobile phase into all the pores of the particle, solute molecules will be excluded from these pores. The required pressure increases for more hydrophobic columns (C_{30}, more pressure; C_1 less pressure), and for particles with smaller pores. P also increases for smaller values of %B. Dewetling and the loss in retention can be reversed by flushing the column with methanol or another organic solvent [56].

Problems from column de-wetting arise mainly when the column is de-pressured. If the column is initially filled with mobile phase of >50% B, all pores will be filled with mobile phase at pressures normally used in HPLC. This will commonly be the case when beginning a series of RPC separations, as it is recommended to store the column with 100% acetonitrile as fill solvent. If the mobile phase is then changed to a lower value of %B *while maintaining the column pressure*, de-wetting is less likely to occur. However, when planning to carry out RPC separations with <5% B, it is advisable to select a column that is less likely to undergo de-wetting (shorter ligand lengths, more polar ligands, lower ligand concentration; see also Section 5.4.4.2).

In an attempt to solve the problem of de-wetting, special "aqueous reversed-phase" packings have been developed that allow the use of mobile phases that contain >95% water. Examples of this kind include columns with embedded polar groups, polar end-capping (indicated by such terms as "polar," "AQ," "hydrosphere," "aqua," "aquasil"), or a lower concentration of the alkyl ligand. Wide-pore columns or columns with a shorter ligand are also less susceptible to de-wetting, but such columns are also less retentive and therefore less useful for the RPC separation of very polar samples.

5.3.3 Column Functionality (Ligand Type)

Apart from the differences in RPC stationary phases described in Sections 5.3.1 and 5.3.2, the chemical composition of the ligand can vary. Ligands for several, commercially available column types are described in Table 5.7 and illustrated in the

Table 5.7

Functional Groups Found in HPLC Stationary Phases

Functional Group	Mode[a]	Comment
C_3	RPC	Used primarily for separations of proteins
C_4	RPC	
C_5	RPC	
C_8	RPC	Most commonly used columns; similar retention and selectivity
C_{18}	RPC	
C_{30}	RPC	Used mainly for carotene separation
Phenyl	RPC	Commonly used column, mainly for a change in selectivity
Embedded-polar-group (amide, carbamate, urea)	RPC	Commonly used column, mainly for or use with water-rich mobile phases (<5% B), to improve peak shape for basic solutes, or for a change in selectivity
Perfluorophenyl (PFP)	RPC	Less commonly used column, mainly for a change in selectivity
Cyano	RPC, NPC	Less commonly used column
NH_2 (amino)	RPC, NPC, IEC	Less commonly used column
Diol	RP, NP, SEC	Mainly used for SEC
WAX	IEC	Used mainly for separating inorganic ions or large biomolecules (Section 13.4.2)
WCX	IEC	
SAX	IEC	
SCX	IEC	

[a]RPC, used for reversed-phase chromatography; NPC, used for normal-phase chromatography; IEC, used for ion-exchange chromatography; SEC, used for size-exclusion chromatography.

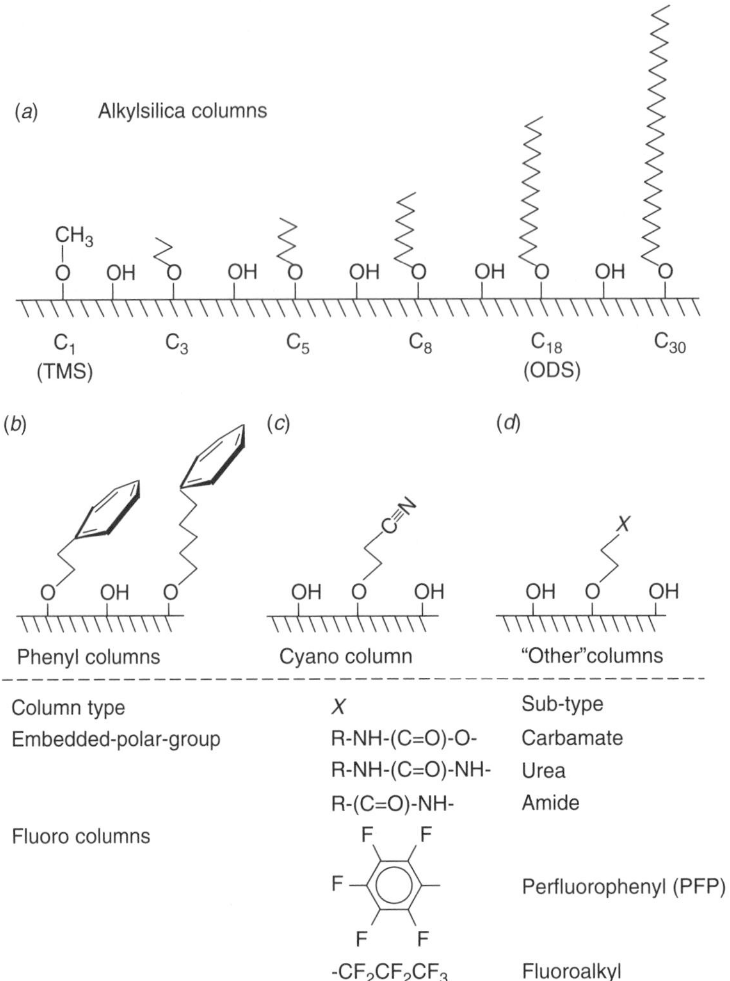

Figure 5.19 RPC columns classified according to the ligand (figures omit the connecting silane group [–$Si(CH_3)_2$–]).

simplified cartoons of Figure 5.19 (the—$Si[CH_3]$ group is omitted in Fig. 5.19*a*–*d*). The ligand of a RPC column is often an alkyl group, for example, C_3, C_8, C_{18} (Fig. 5.19*a*). Alternatively, the ligand may consist of phenylpropyl or phenylhexyl, called *phenyl* columns (Fig. 5.19*b*). If the ligand is –C_3–C≡N (Fig. 5.19*c*), we have a *cyano* column. The alkyl group may also be substituted by other functional groups *X* (Fig. 5.19*d*), and this gives rise to the additional column types listed at the bottom of Figure 5.19. So-called embedded-polar-group (EPG) phases have been growing in popularity, because of their compatibility with low %B mobile phases, their reduced silanol interactions, and unique selectivity (Section 5.4.1); peak shape for basic solutes is usually quite good with these columns. The ligands in these phases contain amide, carbamate, urea (all of which are strong hydrogen-bond bases), or other polar functional groups embedded within the ligand structure. Some EPG packings tend to be less stable than comparable alkyl or aryl columns. The nature of the ligand mainly determines column selectivity, which is the subject of following Section 5.4.

5.4 COLUMN SELECTIVITY

Column selectivity can be important for different reasons. During method development a change of column may be necessary to improve selectivity and increase resolution (Sections 2.5.2, 5.4.3). For the latter application we must be able to identify a second column with quite *different* selectivity. When a routine RPC procedure is used at different times and places, a replacement column from the same source may not be immediately available locally, or too costly, or impractical for other reasons (Sections 5.4.2, 6.3.6.1). In this case we must identify a column of equivalent (or at least similar) selectivity. For either situation, we require a quantitative procedure that allows us to compare column selectivity. Column selectivity is also related to certain problems that can arise during either method development or the routine use of an RPC procedure: peak tailing, the deterioration of a column during use, and "de-wetting" of the column when used with mobile phases that are predominantly aqueous (Section 5.3.2.3). Finally, knowledge concerning column selectivity helps us understand sample retention as a function of the column and solute molecular structure, in turn preparing us to better deal with various separation challenges.

In the remainder of Section 5.4, we will first discuss the *basis* of column selectivity, which can be attributed to different interactions between solute molecules and the column. This will lead to quantitative values of those properties of a column that determine its selectivity. Finally, we will discuss the *use* of these column-selectivity properties for both method development and the routine use of an RPC procedure.

5.4.1 Basis of RPC Column Selectivity

As discussed in Section 2.3.2.1, solute retention is determined by various interactions among the solute, mobile phase, and stationary phase (column). The relative importance of different solute–column interactions—and column selectivity—depends on the composition of the stationary phase and the molecular structure of the solute. Figure 5.20 illustrates eight different interactions that can affect column selectivity:

(a) hydrophobic interaction

(b) steric exclusion of larger solute molecules from the stationary phase (here referred to as "steric interaction")

(c) hydrogen bonding of an acceptor (basic) solute group by a donor (acidic) group within the stationary phase (usually a silanol –SiOH)

(d) hydrogen bonding of a donor (acidic) solute group by an acceptor (basic) group within the stationary phase (represented here by a group "X")

(e) cation-exchange or electrostatic interaction between a cationic solute and an ionized silanol ($-SiO^-$) within the stationary phase; also repulsion of an ionized acid (e.g., $R-COO^-$)

(f) dipole–dipole interaction between a dipolar solute group (a nitro group in this example) and a dipolar group in the stationary phase (a nitrile group for a cyano column)

(g, h) $\pi-\pi$ interaction between an aromatic solute and either a phenyl group (phenyl column) (*g*), or a nitrile group (cyano column) (*h*)

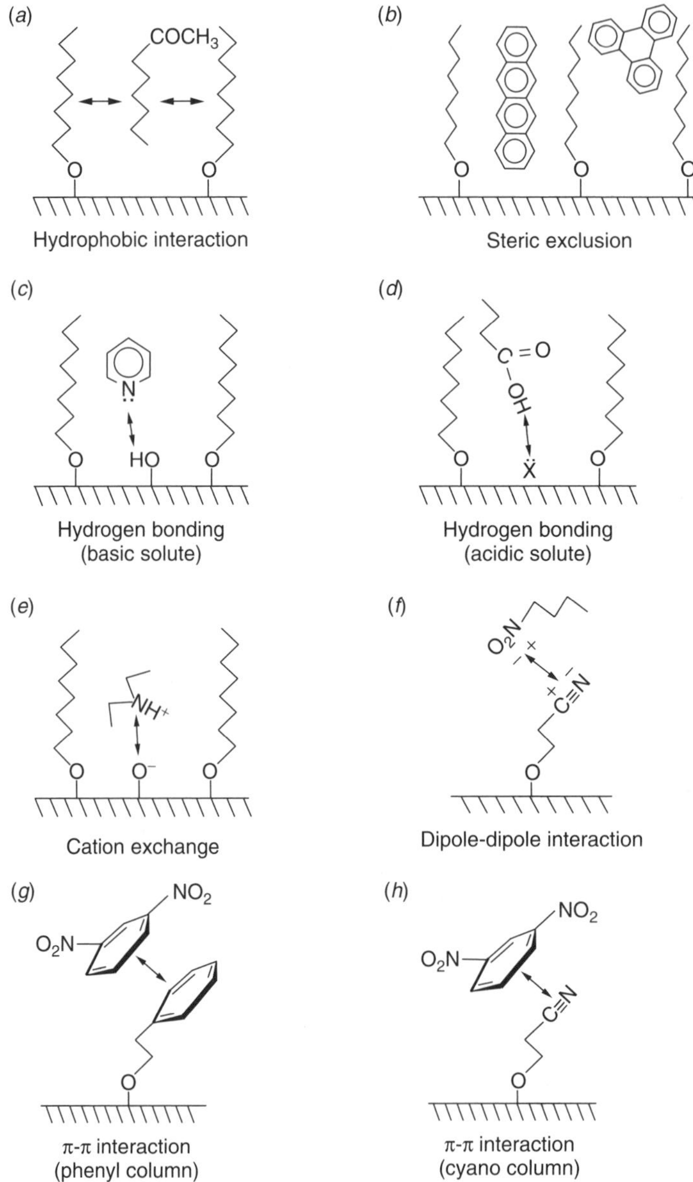

Figure 5.20 Solute-column interactions that determine column selectivity (figures omit the connecting silane group [–Si(CH_3)$_2$–]).

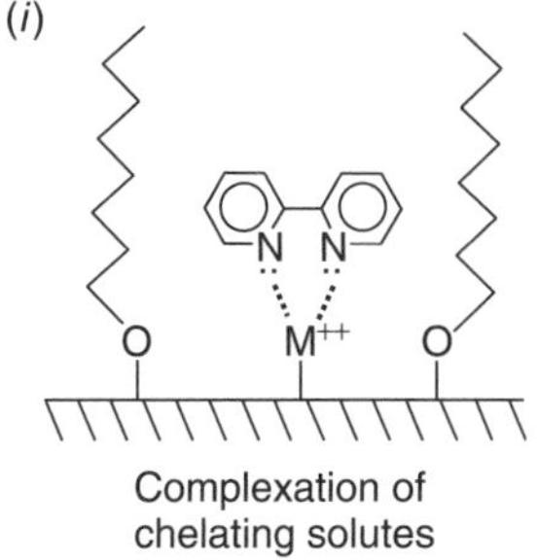

Figure 5.20 (*Continued*)

(i) complexation between a chelating solute and metal contaminants on the particle surface

Interactions (*a–e*) can be significant for every column; dipole interactions (*f*) are only important in the case of cyano columns, and $\pi - \pi$ interactions (*g*, *h*) occur only for phenyl and cyano columns [57]. Both dipole and $\pi - \pi$ interactions are inhibited by the use of acetonitrile as B-solvent, which further minimizes their importance for separations with acetonitrile. Complexation with surface metals (*i*) can result from the use of a less pure, type-A silica, leading to broad, tailing peaks (very undesirable); the chelating solute α,α-bipyridyl has been used to test columns for metal complexation. Because phenyl and cyano columns are used less often, and type-A columns are not recommended, we will emphasize interactions (*a–e*) in this chapter (but see [57, 58]).

5.4.1.1 Solute–Column Interactions

The various solute–column interactions of Figure 5.20, which determine column selectivity, have been understood in general terms since the 1980s; see [59] for a good discussion of recent attempts at characterizing column selectivity. However, only after 2000 did it become possible to reliably characterize RPC column selectivity in terms of these interactions. This was accomplished by the development and application of the *hydrophobic-subtraction model* [60–62], which recognizes that hydrophobic interactions are by far the most important contribution to RPC retention. If only hydrophobic interactions were significant, a plot of values of log *k* for one column against another would give a straight line with no scatter of data around the line. As seen in Figure 5.21, this is approximately the case for these two C_{18} columns (Inertsil ODS-3 and Stablebond C18)—however, values of log *k* for aliphatic amides (□) and protonated strong bases (○) fall below the best fit to these data. These latter deviations are due to interactions of these solute molecules with silanol groups (silanol interactions are more significant for the StableBond C18 column). These and other smaller deviations $\delta \log\, k$ from this plot (see the expanded inset of Fig. 5.21) represent contributions to retention from nonhydrophobic interactions *b–e* of Figure 5.20. It is possible to analyze values of $\delta \log\, k$ for the combination of different solutes and columns so as to separately evaluate the five interactions of Figure 5.20*a–e*. For columns other than phenyl or cyano (i.e., those for which only interactions *a–e* of Fig. 5.20 are significant), values

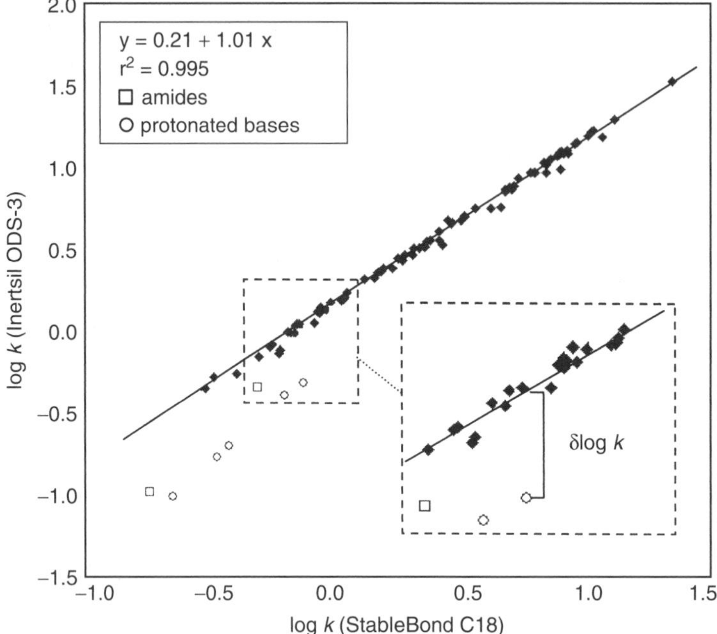

Figure 5.21 Comparison of retention on two different C_{18} columns. Data for 90 different organic compounds. Conditions: 15 × 4.6-mm columns; 50% acetonitrile-water, pH-2.8 phosphate buffer; 2.0 mL/min; 35°C. Adapted from [61].

of k can be related to the interactive properties of the solute and the column:

$$\log\left(\frac{k}{k_{EB}}\right) = \underset{(i)}{\eta' \boldsymbol{H}} - \underset{(ii)}{\sigma' \boldsymbol{S}^*} + \underset{(iii)}{\beta' \boldsymbol{A}} + \underset{(iv)}{\alpha' \boldsymbol{B}} + \underset{(v)}{\kappa' \boldsymbol{C}} \tag{5.3}$$

Here k and k_{EB} are values of the retention factor for a given solute and the reference compound ethylbenzene (EB), respectively. Terms *i–v* of Equation (5.3) correspond, respectively, to the interactions of Figure 5.20*a–e*. Quantities η', σ', β', α', and κ' refer to properties of the solute molecule: hydrophobicity (η'), "bulkiness" (σ'), hydrogen-bond (H-B) basicity (β'), H-B acidity (α'), and effective ionic charge (κ'). Corresponding column parameters are of primary practical interest: $\boldsymbol{H}$, hydrophobicity; $\boldsymbol{S}^*$, steric interaction, or resistance by the stationary phase to penetration by bulky solutes; $\boldsymbol{A}$, H-B acidity; $\boldsymbol{B}$, H-B basicity; and $\boldsymbol{C}$, ion-exchange capacity or electrostatic (coulombic) interaction. Columns with similar values of $\boldsymbol{H}$, $\boldsymbol{S}^*$, etc., will possess similar selectivity and provide a similar retention order for peaks within the chromatogram. Columns with different values of $\boldsymbol{H}$, $\boldsymbol{S}^*$, etc., will differ in selectivity and provide changes in relative retention. The hydrophobic-subtraction model and Equation (5.3) best summarize our present understanding of RPC retention and column selectivity [59].

We will next relate terms *i–v* of Equation (5.3) to the interactions of Figure 5.20*a–e*. *Hydrophobic interaction* is illustrated in Figure 5.20*a*, by the interaction of the solute 2-*n*-octanone ($CH_3COC_6H_{13}$) with the ligand groups of a C_8 column. Values of column hydrophobicity ***H*** increase for longer ligands (e.g., C_{18} vs. C_8), a higher concentration of ligand groups on the silica surface (μmoles/m^2), smaller pore diameters (e.g., 8- vs. 30-nm pores), and the presence of column end-capping. An increase in ***H*** increases the retention of more hydrophobic molecules, those with larger values of η'.

Steric exclusion or "steric interaction" is illustrated in Figure 5.20*b* by the retention of two polycyclic aromatic hydrocarbon (PAH) isomers: the narrow, long naphthacene and the more "bulky" triphenylene. Naphthacene is better able to squeeze between adjacent ligands, but if the spacing of column ligands is increased (lower ligand concentration), it becomes easier for the bulky triphenylene to enter the stationary phase. The column parameter ***S**** measures the "tightness" of the stationary phase or the difficulty that bulky solute molecules experience in squeezing between the ligands; larger values of ***S**** mean a "tighter" stationary phase and relatively less retention of bulky solute molecules. Values of ***S**** increase for longer ligands, a higher concentration of the ligand (ligands closer together), and smaller pore diameters. Solute bulkiness is measured by its value of σ'. Steric exclusion is a somewhat complex phenomenon; see Section 5.4.1.2 below for further insights.

Hydrogen bonding of a non-ionized basic solute (e.g., pyridine) by a column silanol is illustrated in Figure 5.20*c*. The hydrogen-bond acidity ***A*** of the column is due to the presence of surface silanols, and therefore decreases when the column is end-capped (due to the removal of some silanols and blocking of others; see the example of Fig. 5.16*d*). The silanols of type-A columns are usually more acidic than those present in type-B columns; therefore values of ***A*** tend to be larger for type-A columns. The H-B basicity of the solute is measured by its value of α'; unprotonated amines and amides are more basic and have larger values of α', while nitriles and nitro compounds are much less basic and have smaller values of α'. Most other polar compounds have intermediate H-B basicities and values of α'.

Hydrogen bonding of a solute that is a H-B acid (e.g., butyric acid) is illustrated in Figure 5.20*d*. The H-B basic group "*X*:" in the stationary phase is not specified because the nature of *X* differs for different kinds of RPC columns. For type-B alkylsilica columns, there is so far no compelling explanation for what groups "*X*:" consist of. It has been suggested that water dissolved into the stationary phase corresponds to these groups "*X*:", on the basis of an inverse correlation of values of ***A*** and ***H***—as well as other arguments. In the case of some type-A alkylsilica columns with large values of ***B***, contaminating metals in the silica appear to comprise the "*X*:" groups. Alternatively, for embedded-polar-group columns, the polar group (which is usually a H-B base) very likely corresponds to the "*X*:" group. An increase in ***B*** leads to increased retention of carboxylic acids, which have large values of α'; the retention of other H-B acids, such as alcohols and phenols, is less dependent on ***B*** (smaller values of α')—although phenols are preferentially retained on embedded-polar-group columns (i.e., phenols appear as stronger H-B acids when these columns are used).

Column *ion-exchange capacity* **C** is a measure of the ionization of the silica and the accessibility of ionized silanols. Silanol ionization and/or accessibility (and values of **C**) increase (1) as mobile-phase pH increases, (2) for non–end-capped columns, and (3) for type-A versus type-B columns. While the main effect of larger values of C is to increase the retention of protonated bases, it also results in a *decrease* in retention for ionized acids (because of electrostatic repulsion); the value of κ' for a solute is approximately equal to its molecular charge (e.g., +1 for fully protonated bases, −1 for fully ionized acids). The main difference in selectivity for type-A versus type-B columns is determined by their low-pH values of **C**; type-B columns have values of **C** < 0.25 at pH 2.8, while type-A columns have **C** > 0.25. For columns with values of **C** < 0.00 at low pH, it is believed that these columns carry a net *positive* charge [63], presumably the result of protonated amine groups that are introduced during the manufacturing process for some columns. Values of ***H, S*, A***, and ***B*** are assumed not to change with the pH of the mobile phase.

Values of the column-selectivity parameters ***H, S****, etc., have been measured for over 400 different columns; see [64] for a partial listing, or for a current list of values contact one of the authors (or http://www.USP.org/USPNF/columnsDB.html). Average values of these column parameters are summarized at the top of Table 5.8 for several different kinds of RPC column. Within a given column type, there is also a significant variation in values of ***H, S****, etc., as illustrated at the bottom of Table 5.8 for several type-B C_{18} columns. Consequently not all columns of a given kind can be regarded as equivalent in terms of selectivity. Apart from values of ***H*** and ***S****, for example, average retention as measured by values of k for ethylbenzene (last column of Table 5.8) increases with the surface area of the particle.

5.4.1.2 Shape Selectivity

The following, minor digression examines two distinct forms of steric exclusion; for now, the reader may prefer to skip to Section 5.4.2.

Two separate manifestations of steric exclusion have been described: *steric interaction*, as measured by term *ii* of Equation (5.3), and *shape selectivity* [65]. Differences between these two phenomena are illustrated in Figure 5.22 for the separation of two isomeric hydrocarbons on a polymeric column (Fig. 5.22*a*) and a monomeric column (Fig. 5.22*b*). The basis of shape selectivity is illustrated in Figure 5.22*a* for a "narrow" molecule *i*, a "wide" molecule *j*, and a polymeric alkylsilica column. The "wide" molecule *j* is excluded from part of the stationary phase because its minimum cross-section (double-headed arrow) exceeds the spacing between ligands (molecule *j* cannot "squeeze" between the ligands). In a monomeric column (Fig. 5.22*b*), the ligands are further apart, so as to allow access of both narrow and wide molecules (*i* and *j*) to the stationary phase. When a molecule has access to the stationary phase, steric exclusion affects retention in a different way; now the hydrodynamic diameter of the molecule becomes important, rather than its minimum cross-section. When the hydrodynamic diameter of the solute molecule is comparable in size to the spacing between ligands (as for molecule *i*), the retained molecule is restricted in its possible orientations within the stationary phase. This restriction of the solute molecule reduces its retention, in a similar way as for size-exclusion chromatography (Section 13.8.1). Note that shape selectivity and steric interaction lead to dissimilar effects on retention as a function of molecular shape; thus they clearly represent two different contributions to retention [61]. Because of the "either-or" nature of shape selectivity, it can result in relatively large changes in relative retention—whereas steric interaction has a smaller effect on RPC selectivity.

Figures 5.22*c*, *d* illustrates the potential advantage of shape selectivity for the separation of polycyclic aromatic hydrocarbons (PAHs). The polymeric column

Table 5.8

Characterization of Column Selectivity by means of the Hydrophobic-Subtraction Model (Eq. 5.3)

Column type	*H*	*S**	*A*	*B*	C (pH-2.8)	C (pH-7.0)	k_{EB}
Different column types							
C_1 (type-B)	0.41	−0.08	−0.08	0.02	0.04	0.66	1.2
C_3 (type-B)	0.60	−0.12	−0.08	0.04	−0.08	0.81	2.8
C_8 (type-B)	0.84	0.00	−0.12	0.02	−0.03	0.25	5.4
C_{18} (type-B)	0.99	0.01	−0.01	0.00	0.00	0.24	8.8
C_{18} (type-B, wide-pore)	0.95	0.01	−0.05	0.01	0.22	0.31	3.2
C_{18} (type-B, monolith)	1.01	0.02	0.12	−0.02	0.11	0.31	3.2
C_{18} (type-B, hybrid)	0.98	0.01	−0.14	−0.01	0.13	0.05	6.3
C_{18} (polar end-capped)	0.90	−0.04	−0.02	0.02	−0.02	0.40	7.4
C_{18} (type-A)	0.94	−0.05	0.14	0.01	0.79	1.18	6.4
C_{30} (type-B)	1.05	−0.01	0.09	−0.02	−0.08	0.45	13.0
Embedded-polar-group	0.74	0.00	−0.22	0.12	−0.27	0.53	5.9
Phenyl (type-B)	0.63	−0.12	−0.20	0.02	0.13	0.68	2.7
Cyano (type-B)	0.43	−0.09	−0.49	0.00	0.02	0.72	1.0
Perfluorophenyl (PFP)	0.65	−0.11	−0.25	0.01	0.40	0.96	4.3
Fluoroalkyl	0.66	−0.07	−0.11	0.03	0.87	1.18	3.7
Zirconia-base	0.97	0.01	−0.62	0.00	2.01	2.01	0.8
Different narrow-pore, type-B C_{18} columns[a]							
Halo-C18[b]	1.11	0.05	0.01	−0.05	0.06	0.04	6.1
Zorbax StableBond C18[c]	1.00	−0.03	0.26	0.00	0.14	1.04	7.6
Zorbax Eclipse XDB-C18[c]	1.08	0.02	−0.06	−0.03	0.05	0.09	9.1
Kromasil 100–5C18[d]	1.05	0.04	−0.07	−0.02	0.04	−0.06	12.5
ProntoSIL 120–5 C18 SH[e]	1.03	0.02	−0.11	−0.02	0.11	0.40	8.7
Inertsil ODS-3[f]	0.99	0.02	−0.15	−0.02	−0.47	−0.33	10.9
Alltima C18[g]	0.99	−0.01	0.04	−0.01	0.09	0.39	11.5
Nucleodur C18 Gravity[h]	1.06	0.04	−0.10	−0.02	−0.08	0.32	11.0
ACE 5 C18[i]	1.00	0.03	−0.10	−0.01	0.14	0.10	7.9
Chromolith[j]	1.00	0.03	0.01	−0.01	0.10	0.19	3.1
Luna C18(2)[k]	1.00	0.02	−0.12	−0.01	−0.27	−0.17	9.6
Gemini C18 110A[k]	0.97	−0.01	0.03	0.01	−0.09	0.19	8.0
Discovery C18[l]	0.98	0.03	−0.13	0.00	0.18	0.15	4.8
Hypurity C18[m]	0.98	0.03	−0.09	0.00	0.19	0.17	5.6
Hypersil GOLD[m]	0.88	0.00	−0.02	0.04	0.16	0.48	3.9
Symmetry C18[n]	1.05	0.06	0.02	−0.02	−0.30	0.12	9.8
Xterra MS C18[n]	0.98	0.01	−0.14	−0.01	0.13	0.05	6.3
Sunfire C18[n]	1.03	0.03	0.04	−0.01	−0.19	−0.10	9.9
TSKgel ODS-100Z[o]	1.03	0.02	−0.13	−0.03	−0.06	−0.16	11.6

[a] All columns 5-μm particles; data of [73].
[b] Advanced Materials Technology; [c] Agilent; [d] Akzo Nobel; [e] Bischoff; [f] GL Science; [g] Grace-Alltech; [h] Macherey Nagel; [i] ACT; [j] Merck; [k] Phenomenex; [l] Supelco; [m] Thermo/Hypersil; [n] Waters; [o] Tosoh Bioscience.

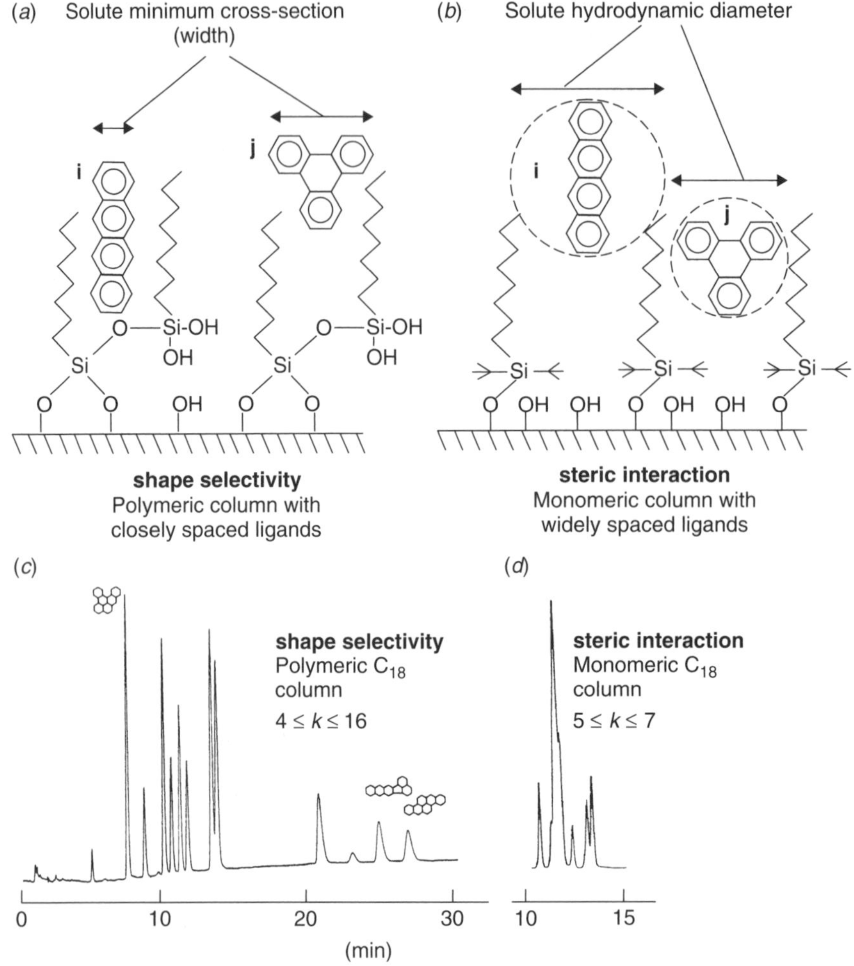

Figure 5.22 Different manifestations of steric exclusion. Shape selectivity (*a*) compared with steric interaction (*b*). (*c*) Separation of a mixture of 13 polycyclic aromatic hydrocarbons on a polymeric column. (*d*) Separation of same sample with same conditions on a monomeric column. (*c*) and (*d*) are adapted from [65].

for the separation of Figure 5.22*c* exhibits greater shape selectivity and therefore provides a much greater differentiation (and better separation) of these different isomeric C_{22} PAHs, versus the corresponding separation in Figure 5.22*d* with a monomeric column (where shape selectivity is minimal). Long, narrow molecules (compared to those that are short and wide) are preferentially retained when shape selectivity is more important, while short, wide solute molecules (of similar molecular weight) are more retained when steric interaction is dominant. As a rule, we can say that shape selectivity is more important when C_{30} or polymeric columns are used, and sample molecules are both large and have very different ratios of length to width. Most RPC separations are carried out with monomeric columns other than C_{30}, in which case steric interaction and values of S^* largely define the effect of steric exclusion on column selectivity. For further details on the practical utility of shape selectivity, see Section 6.3.5.2.

5.4.2 Column Reproducibility and "Equivalent" Columns

Column manufacturers try to ensure that each column (e.g., Waters Symmetry C18) has similar properties and will perform satisfactorily and reproducibly in a routine RPC assay. Consequently the plate number N and column pressure drop for each column usually is measured prior to its sale (Section 5.7); columns whose values of N fall below some minimum value are discarded. Similarly other tests are carried out by the manufacturer (Section 5.7) to ensure that column selectivity stays the same from one batch to the next of the column packing (similar to the measurements of values of ***H***, ***S****, etc.). An example is shown in Figure 5.23 for several successive batches of Zorbax^R Rx-C_{18}, where the retention times for dimethylaniline and toluene are plotted against the batch number. Values of k for the two solutes vary by ±4% (1 SD), mainly as a result of small, unimportant differences in the surface areas of the silica particles. The ratio of these two k-values (α) is a more direct measure of column selectivity (primarily the important column-selectivity parameter C of Eq. 5.3); values of α vary by only ±0.5%. Consequently it appears that the selectivity of different batches of this column packing should be similar, especially for separations that involve protonated basic compounds.

During the 1970s and 1980s HPLC column manufacturing had not yet developed to its present advanced state, and column selectivity often varied significantly among different batches of a particular column. More recently column reproducibility has improved (e.g., see [67]), but it is still possible that one column batch will differ enough from another in terms of selectivity to result in a failed separation—especially for demanding separations. Various means exist for dealing with the latter problem (Section 6.3.6.1), one of which is to select an "equivalent" column from a different source. Using values of ***H***, ***S****, etc., as in Table 5.8*b*, it is possible to identify one or more columns with similar values of ***H***, ***S****, etc., by means

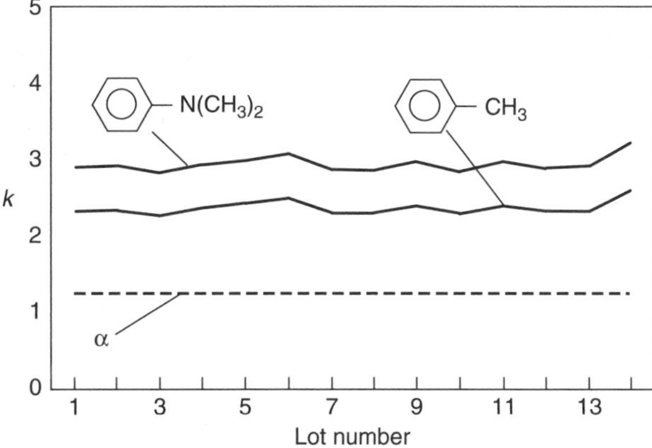

Figure 5.23 Monitoring different batches of column packing for possible changes in selectivity. Sample: dimethylaniline and toluene. Conditions: 150 × 4.6-mm Zorbax Rx-C_{18} columns; 50% acetonitrile-water plus pH-7 phosphate buffer; 1.6 mL/min; 22°C. Adapted from [66].

of a column-comparison function F_s [61]:

$$F_s = \{[12.5(H_2 - H_1)]^2 + [100(S^*_2 - S^*_1)]^2 + [30(A_2 - A_1)]^2 + [143(B_2 - B_1)]^2 + [83(C_2 - C_1)]^2\}^{0.5} \quad (5.4)$$

Values of H_1 and H_2 refer to values of H for columns *1* and 2, and similarly for the remaining column parameters S^*, A, etc. A value of $F_s \leq 3$ indicates that the two columns are similar in selectivity and can be substituted for each other in any RPC separation (for less challenging separations, larger values of F_s can be tolerated). The ability of Equation (5.4) to identify columns of equivalent selectivity has been demonstrated for a dozen different routine RPC separations that were developed and used in several different laboratories [68] (an example is given in Fig. 6.19). Software for the comparison of column selectivity by means of Equation (5.4), and values of H, S^*, etc., for more than 400 RPC columns can be accessed at the US Pharmacopeia website (http://www.usp.org/USPNF/columnsDB.html). Alternatively, contact one of the authors for a current database of values of H, S^*, etc., for different RPC columns.

5.4.3 Orthogonal Separation

Just as it is sometimes necessary to identify two different columns of similar selectivity, at other times we need a second column of different selectivity (usually combined with a change in mobile phase). Several different reasons for a change in selectivity can be identified. First, when developing a reversed-phase separation, a large change in separation selectivity may be needed in order to improve the resolution of certain peaks in the chromatogram. Second, during method development for samples of initially unknown composition, there may be a concern that a minor component might be overlapped by a larger peak in the chromatogram—and therefore missed in the final analysis. Third, a similar situation may arise in the use of a routine procedure for future samples that might contain additional, unanticipated impurities. In either of the latter two cases it is desirable to have available an "orthogonal" separation, for which selectivity is different from that provided by the original assay procedure; if two peaks overlap in the routine separation, they are then more likely to be separated (and observable) in the orthogonal separation. The use of Equation (5.4) for identifying columns of different selectivity in the latter two applications has been demonstrated for a dozen different routine assay procedures [69] (see the example of Fig. 6.21).

Fourth, orthogonal columns are required for the technique of *thermally tuned tandem-column optimization* [70], in which two columns connected in series are operated at different temperatures in order to better control selectivity. Finally, orthogonal columns may be required for two-dimensional separation, where two different columns are used sequentially for the separation of a given sample (Section 9.3.10). In all these applications two RPC columns of very different selectivity correspond to two columns with a large value of F_s in Equation (5.4).

It has been suggested [71] that the selection of an orthogonal column can be further improved by emphasizing the separation of neutral solutes (for which values of C are unimportant in determining column selectivity). For separations of neutral solutes, the corresponding column selectivity function will be

$$F_s(-C) = \{[12.5(H_2 - H_1)]^2 + [100(S^*{}_2 - S^*{}_1)]^2 + [30(A_2 - A_1)]^2 + [143(B_2 - B_1)]^2 \tag{5.4a}$$

For maximum column orthogonality, the simultaneous conditions $F_s(-C) \geq 50$ and $F_s \geq 100$ have been recommended [71].

5.4.4 Other Applications of Column Selectivity

Values of ***H***, ***S****, etc., reflect the nature and properties of the stationary phase, which are also related to certain common problems associated with the column:

- peak tailing
- stationary-phase "de-wetting"
- column degradation during routine use

5.4.4.1 Peak Tailing

Tailing peaks were discussed in Section 2.3.2, primarily from the standpoint of their effect on resolution. Peak tailing in RPC occurs mainly for protonated basic solutes [72] and type-A alkylsilica columns [73]. Thus the use of type-B columns (values of **C**[2.8] ≤ 0.25) largely solves this problem. Additional means for reducing peak tailing for basic solutes are discussed in Section 7.3.4.2. Even for type-B columns, however, fully ionized compounds tend to tail whenever the sample size exceeds about 1 μg for a 4.6-mm-diameter column. The reason is the mutual repulsion of ionized molecules (of the same charge) when concentrated into the stationary phase (Section 15.3.2.1 and [72]). Non-ionized solutes do not tail until a 50-fold larger sample weight is injected.

Tailing peaks may also occur (less frequently) when carboxylic acids are separated with low-pH mobile phases [73]. Such peak tailing is more likely for type-A columns but can also occur for type-B columns that are more basic (i.e., have larger values of ***B***). Opposite to the case of protonated bases, non-ionized acids tend to tail for sample weights <1 μg but give symmetrical peaks for sample weights of 1 to 50 μg. For further details, see [73].

5.4.4.2 Stationary-Phase De-Wetting

Column de-wetting (Section 5.3.2.3) is more likely for narrow-pore, more hydrophobic columns; that is, columns with larger values of ***H***. As seen in Table 5.8, the average value of $H = 0.99$ for a narrow-pore (≤ 12 nm), type-B C_{18} column (the most popular column). Columns with lower values of ***H*** will be less likely to experience column de-wetting. Polar-end-capped columns (average $H = 0.90$) are specifically designed to minimize column de-wetting but are otherwise similar to type-B C_{18} columns in terms of selectivity (because the effect of ***H*** on selectivity is less important, as can be seen from the weighting factors in Eq. 5.4).

5.4.4.3 Column Degradation

When a column is used for routine analysis, the stationary phase is gradually lost due to attack by the mobile phase on either the silane–silica bond (at low pH) or the silica itself (at high pH). Either process results in an increase in the number of silanols (therefore increased values of the column-selectivity parameters A and C) and a decrease in column hydrophobicity (values of H) [73]. As a result column selectivity can be expected to gradually change with further use of the column. Usually a column is discarded and replaced by a new column when selectivity changes (or the plate number drops) to the point of peak overlap.

5.5 COLUMN HARDWARE

5.5.1 Column Fittings

Stainless-steel columns with standard end fittings (Section 3.4.2) are available with a wide choice of column dimensions and different packings. The greatest efficiency, reproducibility, and ruggedness are found for columns of this type. Less-costly stainless-steel *cartridge columns* are also available with a wide range of packings, but these are mostly used for less-demanding routine assays. Cartridge columns are not supplied with end fittings, so they require reusable holders.

Pressures as high as 15,000 psi ($\approx$1000 bar or $\approx$100 MPa) are now achievable with some HPLC instruments. Columns should therefore be constructed to (1) withstand such pressures and (2) resist chemical attack by the mobile phase. To meet these goals, stainless-steel tubing is used for most columns. In rare cases where stainless steel might be attacked by the mobile phase or react with the sample, titanium or glass-lined steel columns are available. Fused-silica capillaries can be used with mass spectrometric detection, and heavy-wall glass tubing for pressures as high as 600 psi have been used for preparative separations. Polymer-based PEEK columns with very heavy walls are available for use with pressures up to 6000 psi. In all cases the inside walls of the column tubing blank should be smooth with a mirror finish; the condition of the column wall greatly influences the homogeneity of the packed bed and therefore the ultimate efficiency of the column.

End fittings and connectors for the column must be designed to have a minimum dead-volume, in order not to contribute significantly to extra-column peak broadening (Sections 2.4.1.1, 3.9). Hardware such as compression fittings, tubing, connectors, detector cells, and so forth, should be assembled so as to minimize unswept corners or stagnant pools that can act as mixing vessels and broaden peaks. Metal-to-metal inlet seals are used for columns where input pressures exceed 6000 psi; some seals allow inlet pressures as high as 15,000 psi. For lower inlet pressures (e.g., <6000 psi), inlet compression fittings with seals composed of organic polymers such as PEEK can be used successfully. See Section 3.4 for further details.

Introduction of the sample to the column as a sharp, minimum-volume "plug" is required for best results. Therefore porous frits or screens used at the column ends to retain the packing must minimize extra-column peak broadening. Porous stainless-steel, titanium, or Hastelloy frits about 0.2 mm thick are used most often. Thin, stainless-steel screens provide the least peak broadening, but they are more difficult to use—especially for higher inlet pressures. The porosity of porous frits or screens must be substantially smaller than the size of the packing particles. This is especially true for a wider particle-size-distribution, where the smallest particles may be able to leak through or plug the frit. For example, 2-μm porosity frits are generally adequate for columns of 5-μm particles, or for 3-μm particles that have a narrow particle-size distribution. Most columns of 3-μm particles use 0.5-μm frits, whereas sub-2-μm-particle columns use 0.2-μm outlet frits to retain the packing material in the column. Because 0.2- and 0.5-μm frits are susceptible to plugging by particulates, samples and mobile phases often require careful filtration when used with small-particle columns.

Straight (rather than curved or coiled) columns are almost always used. Experience has shown that glass columns are rarely needed, even for separating sensitive biological samples such as some peptides and proteins. While columns of glass-lined stainless-steel and aluminum-clad rigid polymer (PEEK) columns are available, it is rare that such materials are required.

Radial-compression columns (pp. 67–69 of [12]) consist of particles that are loosely packed into a soft, polymeric cylinder. Prior to use, hydraulic pressure is applied to the exterior of the column, so as to squeeze the particles together and form a compact bed. These columns are available for both analytical and preparative applications, with the advantages of lower cost and metal-free construction. Today these columns are not widely used because of awkward temperature control and other inconveniences, as well as added cost when connecting columns in series (for an increase in N). *Axial-compression columns* are packed loosely and then compressed by means of a close-fitting piston that enters one end of the column. These columns are used exclusively for large-scale preparative separations [74, 75].

5.5.2 Column Configurations

A wide range of column sizes are available, depending on the intended application. Table 5.9 summarizes some column dimensions that are commercially available for various types of column packings. Routine methods usually are developed with 3- to 4.6-mm i.d. columns, equipped with compression fitting and packed with 2.5- to 5-μm particles. Such columns represent a good compromise among convenience, efficient performance and adequate column lifetime. Columns of 2.1-mm i.d. often

Table 5.9

Typical Column Configurations

Type	Inner Diameter (mm)	Length (mm)	Particle Size (μm)
Analytical	1–4.6	30–250	1.5–10
Cartridge	3–4.6	50–100	3–10
Microbore	1, 2.1	50–250	2–8
Semi-preparative	8–10	100–250	5–200
Preparative	20–50	100–250	50–200

Note: Stainless-steel columns; glass, glass-lined, plastic and PEEK are also available in some configurations.

are used when interfacing with mass spectrometry, as the mobile-phase flow rates used with these columns are more compatible with the requirements of this detector (Section 4.14). Columns of $\leq$2.1-mm i.d. often exhibit 15–25% smaller plate numbers, because of (1) equipment extra-column peak broadening, and (2) difficulty in packing narrow-diameter columns. Automatic sample injectors can be a significant source of extra-column peak broadening when using narrow-bore columns, because of the required very small peak volumes. Columns of 1-mm i.d. and capillary columns as small as 50-μm i.d. sometimes are used with mass spectrometric detection, but they are generally not suited for routine application. Narrow-bore columns are more difficult to pack efficiently, and special equipment is needed when using these very small internal-diameter columns in order to minimize extra-column peak broadening.

5.6 COLUMN-PACKING METHODS

For most readers, the following section will have little application to laboratory practice, and can therefore be bypassed. Today most chromatographers obtain columns from a commercial source, and have no need to pack their own columns.

Commercial columns have been under development for several decades. Duplicating their efficiency, reliability, and reproducibility is beyond the capability of most laboratories that are responsible for HPLC method development and analysis. This chapter is *not* intended as a manual for the preparation of HPLC columns. Rather, we will review some principles for the packing of HPLC columns, so that the reader can at least appreciate how good columns are made. For those who might wish to pack their own columns for special applications, see [76, 77] for further details.

5.6.1 Dry-Packing

Early columns for HPLC were packed with irregular particles in the 45- to 50-μm size range, using vibration procedures that had been used previously for gas chromatography. With the introduction in the late 1960s of dense, ≈25-μm spherical particles with solid cores (both pellicular and superficially porous packings), a "tap-fill" dry-pack method could be used to produce efficient, stable columns [77]. The "tap-fill" method is still useful for the preparation of columns with larger particles, especially for preparative chromatography (Chapter 15). With the subsequent introduction of much smaller ($\leq$10-μm) particles for HPLC, dry-packing procedures were found to yield inefficient, unstable columns. Other column-packing approaches were therefore required.

5.6.2 Slurry-Packing of Rigid Particles

Early HPLC particles had a relatively wide particle-size distribution, and in some cases, an irregular shape. Such particles apparently tend to segregate across the column cross-section according to size, leading to an inefficient and unstable column bed. Methods were therefore required to minimize this problem. Packing columns with small, rigid particles is best done by filling the column blank under pressure with a slurry of particles in some liquid [26, 76, 77]. Three different variations of this approach can be used, in each case making use of conditions that minimize the size-separation of particles during column-packing.

- balanced-density liquids
- high-viscosity liquids
- low-viscosity unbalanced slurry

To suspend the packing, the *balanced-density* procedure uses a slurry liquid (e.g., mixtures of tetrabromoethane with a less-dense solvent) whose density is similar to that of the liquid-filled particle. The *high-viscosity* procedure makes use of a high-viscosity slurrying liquid (e.g., glycerin). Both of these approaches impede the settling and sizing of particles during the packing process. However, both methods require more time and generally do not produce the higher column efficiency of the following method.

The subsequent availability of porous silica microspheres (≤ 5 μm) with a relatively narrow particle-size distribution led to the *low-viscosity unbalanced-slurry* method. Because particle sizing during column packing is not so critical for very small particles, a balanced-density or high-viscosity slurry is no longer needed. Low-viscosity liquids such as tetrahydrofuran, chloroform, or mixtures of these with other low-viscosity liquids can be used, allowing the rapid production of packed columns that are both efficient and stable.

5.6.2.1 Selection of Slurry Liquid

Packing columns by means of any of the slurry procedures is relatively simple: slurry the particles with a liquid, place the slurry in a reservoir, attach this reservoir to a high pressure pump, then force the slurry into an empty column blank (with the outlet frit and end-fitting in place) by starting the pump (Fig. 5.24). During this

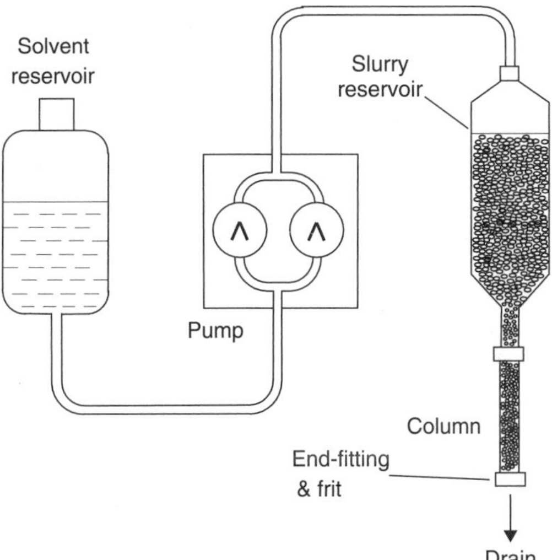

Figure 5.24 Schematic of equipment for packing columns by the slurry procedure. Adapted from [76].

process particles are held at the column outlet by means of a porous frit or screen, while the slurry solvent passes through. When the column blank is full, the column is removed from the packing apparatus and a second porous frit or screen is placed at the inlet of the column. Despite the simplicity of this description, many important issues must be addressed for the production of efficient, stable, and reproducible columns [26, 76].

A critical factor is the need for suspension of the particles in the slurry liquid without aggregation. Thus the selection of the slurry liquid is critical. The suitability of a liquid for avoiding particle aggregation usually can be determined by means of the following, simple test. Particles are added to the slurry liquid, followed by inserting the mixture into an ultrasonic bath. The mixture is then examined with an ordinary microscope. If the particles are freely dispersed and nonaggregated, as illustrated by the cartoon of Figure 5.25*a*, the liquid may be acceptable. However, if the particles tend to aggregate as in Figure 5.25*b*, the liquid is likely a poor choice. Even when the particles are properly dispersed, a liquid may not be optimum for a particular column packing; the only real test is the performance of the resulting column.

The selection of a low-viscosity slurry liquid depends on the nature of the packing particles. To prevent particle aggregation, the interaction of the particle with the slurry liquid should be stronger than interactions between particles. For example, C_8- or C_{18}-modified particles (which contain polar, unreacted silanol groups) should be packed with a slightly polar liquid such as tetrahydrofuran, methyl-t-butyl ether—or mixtures such as acetonitrile/chloroform, chloroform/methanol, or chloroform/acetone (the use of hydrocarbons such as hexane usually is not successful). More polar column packings (unmodified silica or silica that is bonded with polar ligands such as amino or diol) require methanol or some other polar liquid for the slurry. When packing capillary columns with small particles, it is not clear whether particle nonaggregation in the slurry is required for good columns; further study is needed. However, for conventional columns with internal diameters $\geq$1 mm, a nonaggregating slurry liquid is strongly recommended.

Once a nonaggregating slurry liquid has been chosen, some further aspects of slurry packing should be considered.

Packing particle-size. Particles $\geq$3 μm in diameter are relatively easy to pack into efficient, stable beds. Resulting columns should exhibit minimum reduced plate heights h of 2 to 2.5, and values of the tailing factor $<$1.2 (for small, nonpolar

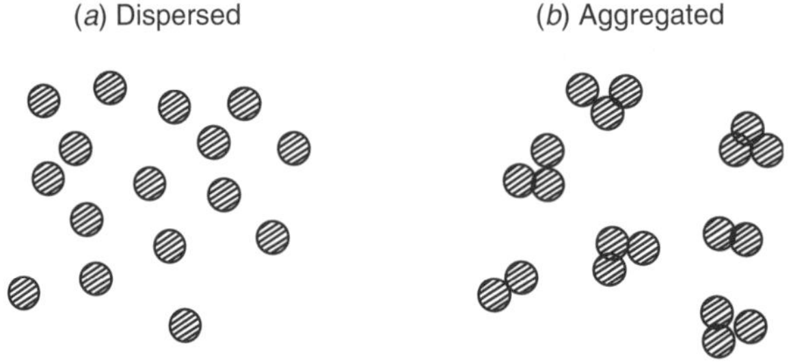

Figure 5.25 Comparison of dispersed and aggregated particles (different slurry-packing solvents).

solutes). Values of $h \leq 1.5$ have been reported for superficially porous particles of 2.7-μm diameter [5, 13]. Particles with sizes of ≤2 μm are more difficult to pack, and less efficient columns may result.

Particle-size distribution. Many commercial packings possess relatively narrow particle-size distributions, with standard deviations of 15–20% from the average. Superficially porous, 2.7-μm particles have been described with a standard deviation of only 5%, and this may be in part responsible for the higher efficiency of these columns (Section 5.2.2.1).

Slurry concentration. A particle concentration of 7–15% usually works best. The exact concentration of particles in the slurry liquid (for best results) depends on particle type, the slurry liquid, and the packing apparatus—and must be determined empirically.

Slurry-apparatus design. The design and configuration of the reservoir to deliver the packing into the column blank can be important. Best results are obtained when the packing is delivered though tubes that maintain the same internal diameter from the bottom outlet of the reservoir into the column blank inlet (as in Fig. 5.24). Constant-pressure pumps (e.g., Haskell pumps) are favored over constant-flow pumps for this operation.

Packing pressure. The pressure required for a dense column bed depends on the size of the particle, column length, and internal diameter. For stable columns, the packing pressure should be ≥50% higher than the maximum pressure at which the column will be used, and the delivery of slurry into the column blank must be a fast as possible.

Surface finish of the column-blank wall. The walls of the column expand or "balloon" slightly during the pressure loading process, and the particles of packing are also compressed [76]. After the column has been filled and the pressure released, the particles decompress and some are forced to move along the column wall. Consequently the inside wall of the column should have a smooth, mirror finish, in order to avoid abrading particles and creating fines that can result in a lower column efficiency.

Particle strength. Particles are compressed during the pressurization process and deform elastically; some of this compression will compensate for internal stresses encountered during use of the column at a higher pressure or temperature. If particles do not have sufficient strength, they will eventually fracture or crush, resulting in an inefficient column with a higher than normal pressure. Some columns are rated at a maximum pressure of 9000 to 15,000 psi; the particles in such columns must be strong enough to maintain their structure during slurry packing at even higher pressures.

5.6.2.2 Rigid Polymeric Particles

The balanced-density slurry-packing procedure is also used for preparing columns of hard, polymeric particles ("gels") such as cross-linked polystyrene. The same hardware and general procedure is used as for rigid particles. In contrast to rigid particles, hard gels must first be allowed to swell in the liquid in which they are to be packed. As organic gels have a lower density than rigid solids, lower density slurring liquids are used (e.g., acetone/perchlororethylene mixtures). Because hard-gel particles are not as strong as rigid particles, the packing pressures must be lower. Polymeric particles ("resins") for ion-exchange also can be slurry-packed by the balanced-density method, using aqueous liquids. The recommended procedure is as follows: A thick slurry of the swollen, ion-exchange resin is prepared in a salt solution such as calcium chloride, whose density can be matched to that of the resin. The slurry is then forced by high pressure into the column blank, using the same apparatus and technique as for rigid particles. The packing pressure depends on the strength of the resin particles, which is generally a function of the degree of cross-linking of the resin. A pressure of $\sim$5000 psi can be used for the strongest ion-exchange resins. Packed columns must be carefully flushed with the mobile phase to ensure removal of the salt before use.

5.6.3 Soft Gels

Soft-gel particles such as those used for gel filtration (size exclusion, Section 13.8) cannot be dry-packed, nor can the high-pressure slurry-packing method be used (soft column packings compress and deform at relatively low pressures). Therefore columns of soft gels usually are packed using a gravity-based, slurry-sedimentation method. It is best to follow the procedure recommended by the manufacturer.

5.7 COLUMN SPECIFICATIONS

5.7.1 Manufacturing Standards

As a means of ensuring reproducible column performance, manufacturers set standards or specifications for each column (defined by model number) that they produce—however, there is little uniformity in these standards from one manufacturer to another. Many manufacturers supply a written specification for each column, including target values for various column characteristics and actual performance data. Other manufacturers assume that data for one or more columns from a manufacturing lot or batch will be representative for all columns in that lot. Data reported for the column can vary from one manufacturer to the next, or even between product lines for a single manufacturer.

It is important to discriminate between data that are specific to an individual column, and data that depend on properties of the bulk packing (a batch test). Column-specific data are the plate number N, peak asymmetry A_s (or tailing factor), and column pressure P. A test chromatogram for each individual column should report values of N, A_s, and P, together with the detailed test conditions, so that users can confirm these results on their own equipment. However, values of N, A_s, and P determined by the user can differ somewhat because of differences in HPLC equipment.

Information from the batch test includes physicochemical data (e.g., surface area, pore size, pore volume, particle size, carbon content, μmoles/m^2 of ligand) and/or a batch-test chromatogram. Physicochemical data are not of immediate use to most chromatographers, but these data may be helpful for future troubleshooting, and should therefore be kept on file. The test solutes used for the batch-test chromatogram should include compounds of varied functionality (e.g., acids, bases, neutrals) whose retention can indicate undesirable changes in batch-to-batch column selectivity (Section 5.4.2). Retention data for the test solutes (values of k and α) should fall within a narrow range of values (usually specified by the manufacturer). It is important to keep these data and chromatograms on file, for use if problems arise with a routine procedure.

Not all manufacturers provide all of the data for each column. Sometimes, only average values for each column model number are reported, and this information may be found in different places: the column insert, the manufacturer's website, company literature, or scientific publications. Some suppliers warrant their columns against defects for a period of use so that the user can be assured of a certain column lifetime. Additional useful data can also be found in care-and-use manuals, brochures, or websites; e.g., recommended operating conditions such as pH and temperature ranges, as well as solvents that are not recommended for a particular column.

5.7.2 Column Plate Number

The value of N reported is usually for separation conditions that are close to "ideal" (low-viscosity mobile phase, a small, neutral solute molecule, near-optimum flow rate). This value of N will often differ from that found for other solutes and/or operating conditions, for reasons described in Section 2.4.1. For columns of totally porous particles, the following equation can be used to estimate the plate number for a well-packed column and conditions that have been optimized for maximum N:

$$N \approx \frac{500L}{d_p} \tag{5.5}$$

Here the column length L is in mm, and the particle diameter d_p is in μm. Table 5.10 shows typical plate-number values (neutral solute molecules with molecular weights of ≈200 Da) for well-packed HPLC columns of various lengths, particle sizes and types. The values in Table 5.10 assume a column diameter of 4.6 mm; a column diameter ≤2 mm can result in values of N that are lower, possibly because of the less-efficient packing of small-diameter columns, but mainly because of extra-column peak broadening.

The conditions used to measure N for a given column are usually specified by the manufacturer so that the user can confirm this value of N when poor column performance is suspected. If the measured plate number is less than 80% of the reported value, if the performance of the HPLC system has been verified, *and if the column has not been mistreated or used extensively*, the column should be returned to the supplier. However, it is unusual for a column not to meet the manufacturer's specifications. Lower plate numbers and higher asymmetry values measured by users for a new column are almost always the result of a system that has excessive extra-column peak broadening—which will be larger for small-diameter and/or short columns, especially those with particles smaller than 5 μm.

When carrying out a routine HPLC analysis over a period of time, it is desirable to dedicate one or more columns to this assay. It can also be useful to keep track of values of N and peak asymmetry (A_s or TF) for one or more of the sample compounds, during routine analysis. These values of N and peak asymmetry can then be compared with values originally determined for the method, when using a

Table 5.10

Approximate Plate Number for Well-Packed Columns under Optimized Test Conditions

Particle Type[a]	Particle Diameter (μm)	Column Length (mm)[a]	Plate Number N
Totally porous	5	30	2,500–3,000
Totally porous	5	50	4,500–5,000
Totally porous	5	100	8,000–10,000
Totally porous	5	150	12,000–15,000
Totally porous	5	250	20,000–25,000
Totally porous	3.5	30	3,000–4,000
Totally porous	3.5	50	5,500–7,000
Totally porous	3.5	100	10,500–14,000
Totally porous	3.5	150	17,000–21,000
Fused-core	2.7	30	7,000–8,000
Fused-core	2.7	50	9,000–11,000
Fused-core	2.7	100	18,000–22,000
Fused-core	2.7	150	28,000–34,000
Totally porous	1.8[b]	30	6,000–7,000
Totally porous	1.8	50	10,000–12,000
Totally porous	1.8	100	20,000–25,000

Note: Small, neutral test-solute, low viscosity mobile phase, ambient temperature, measured at the plate-height minimum.

[a]Estimated values for 4.6 mm i.d. columns.

[b]An average for commercially available, sub-2-μm particles.

new column that met the manufacturer specifications. If problems are encountered with a routine method that might be caused by the column, values of N and either A_s and TF can be compared with prior values for a "good" column. In this way a "bad" column can be confirmed as the cause of the problem.

Every column used for a routine method has a finite lifetime (number of injected samples before column failure) that depends on separation conditions—especially mobile-phase pH and temperature. For "clean" samples, 1000 to 2000 analyses should be possible for must silica-based columns, particularly reversed phase columns. However, for other samples (with minimal sample preparation), such as extracts of blood, plant or animal tissue, or soil, 200–500 analyses is a more typical column lifetime.

5.8 COLUMN HANDLING

The performance and life of the column depend on how it is used and handled. During heavy use with "dirty" samples (especially samples from biological sources), columns can develop severe peak tailing (Fig. 5.26*a*) or double peaks for each component (Fig. 5.26*b*)—usually the result of a partly blocked frit, a contaminated

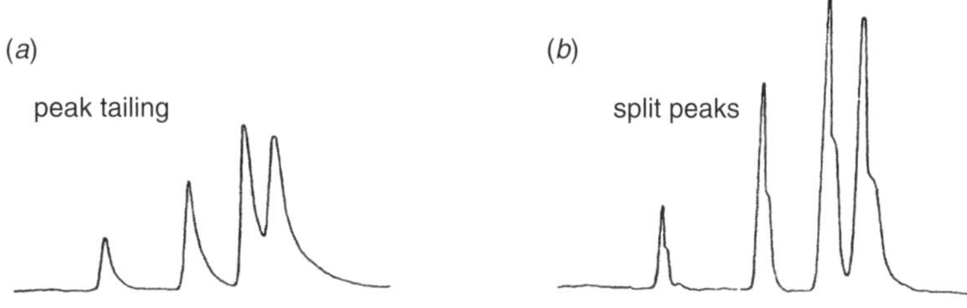

Figure 5.26 Examples of peak tailing (*a*) and split peaks (*b*).

column, or deterioration of the column packing. The following restorative measures are sometimes effective, but the time and effort involved are often not cost-effective. Usually it is more economical to simply replace the column.

A blocked frit or contaminated column can sometimes be restored by periodically purging the column with a strong solvent. A 20-column-volume purge (about 30 mL for a 150 × 4.6-mm column) with a mixture of 96% dichloromethane and 0.1% ammonium hydroxide in 4% methanol is often effective for RPC columns. Pure methanol or isopropanol can be used for normal-phase columns. Flushing a RPC column (at least) daily with a strong solvent, such as methanol or acetonitrile, can enhance column performance and lifetime for isocratic separations. This approach removes strongly retained sample components that can build up at the column inlet. *Back-flushing* a column with a strong solvent at 0.2 to 0.5 mL/min may be more effective, so as to avoid driving the column contaminants into the column. However, some manufactures recommend against back-flushing because their columns are fitted with an inlet frit that has larger pores that might allow particles to be swept out (consult the column care-and-use instructions for a specific column). In gradient elution, clearing the column of strongly retained components can be accomplished by using a steep or step gradient (Section 9.2.2.5). The use of a 0.5-μm porosity in-line filter (Section 3.4.2.3) can be highly effective in preventing blockage of the column inlet frit, and is highly recommended.

To reduce the possible impact of "dirty" samples on column lifetime, sample pretreatment is commonly used (Chapter 16). A *guard column* can also be used to protect the column, and it is recommended for routine analysis. A guard column is short (e.g., 10–20 mm) and preferably contains a packing that is the same as or similar to that in the main column (Section 5.4.2). The guard column captures strongly retained sample components (and particulates), and prevents these from fouling the analytical column. Guard columns must be replaced at regular intervals, before the column becomes saturated with strongly retained sample components that then pass into the analytical column. However, because of their added expense and inconvenience, some users prefer to avoid guard columns and replace the main column more frequently. The use of guard columns with low-volume, high-efficiency columns (e.g., sub-2-μm columns) requires special care, because of the greater importance of extra-column peak broadening when small columns are used.

For columns that are not well packed, a sudden pressure surge (as during sample injection) can cause a void at the column inlet, with a decrease in column performance. Fortunately, this problem is no longer common for columns from established manufacturers. While pressure-related problems are uncommon for silica-based columns, other types of particles (e.g., organic gels, graphitized carbon) are more fragile and less able to withstand sudden changes in flow, pressure, or temperature.

Column performance can be reduced significantly by a loss of the bonded phase during use, leading to a short column lifetime. The recommendations of the manufacturer, especially with regard to mobile-phase pH and temperature, should be observed. A low-pH mobile phase (e.g., pH < 2.5) can cause some hydrolysis of Si–O bonds, with loss of bonded silane (see Fig. 5.16*a*). Short-chain ligands (e.g., $-C_3$, $-C_3{-}C{\equiv}N$) are least stable in a low-pH environment, while longer chain alkyl groups (C_{18}, C_8) are usually adequately stable for a mobile-phase pH between 2.5 and 7, and a temperature $\leq 40^\circ$C. As previously noted, sterically protected stationary phases (Fig. 5.16*b*) provide additional stability at low pH. High-pH mobile phases (e.g., pH > 8) can slowly dissolve silica-based packings, again resulting in a degradation of column performance. Columns of hybrid silica-silane particles (Section 5.3.2.2) and those based on zirconia (Section 5.2.5.1) are especially resistant to degradation by high-pH mobile phases; some other special stationary phases (e.g., bonded bidentate silanes, Fig. 5.14*b* or *c*) are also more stable for high-pH applications [44].

Stationary-phase loss from silica-based columns at low pH is accelerated at higher temperatures. Therefore higher temperatures as a means of improving column performance or separation selectivity should be used carefully. Many workers have reported good results for conventional C_{18} and C_8 columns when operated at 40–60°C. Sterically protected stationary phases can be used routinely with temperatures up to about 90°C, and polymer-coated zirconia can be used satisfactorily at temperatures up to at least 150°C [34].

Degradation of silica-based columns at higher pH occurs via dissolution of the silica, and its degradation is also accelerated by higher temperatures. Figure 5.27 provides an example where the cumulative loss of silica at 60°C is almost 20-fold greater than at 40°C. When using a column at intermediate and high pH with both phosphate and carbonate buffers, temperatures above 40°C with silica-based

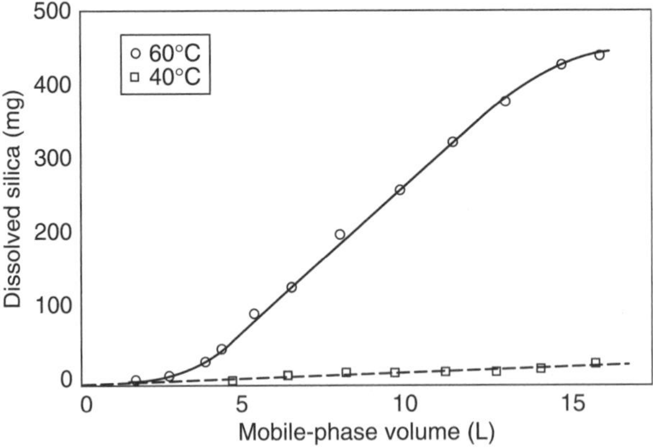

Figure 5.27 Effect of temperature on silica-support dissolution. Column: Zorbax Rx-C18, 15 × 0.46 cm; continuous nonrecycled 20% acetonitrile/80% sodium phosphate buffer, 0.25 M, pH 7.0. Flow rate: 1.0 mL/min. Adapted from [78].

columns should be avoided because of rapid dissolution of the silica support [78]. On the other hand, use of organic buffers (e.g., TRIS, HEPES, citrate) may increase column lifetime over that with phosphate and carbonate buffers when operating at intermediate and higher pH [79]. One speculation is that these basic, partially hydrophobic, organic buffer compounds tend to bind to unreacted silanol groups on the packing so as to create an additional barrier to the dissolution of the silica support [79]. However, the advantage of the organic buffers may be misleading, as the actual pH of the buffer/organic solvent mobile phase may be somewhat lower than that measured for the buffer solution itself (the pH of phosphate and carbonate buffers in organic-containing solvents is somewhat higher that the aqueous buffer [80]); see the further discussion of Section 7.2.3.

The performance and lifetime of a column can also be affected by improper storage of the mobile phase. Aqueous buffers (especially with pH $\approx$7) encourage microbial growth if they are stored for more than a day at room temperature. The resulting particulates can in turn block the column inlet, reducing N and increasing the column pressure. Therefore it is good practice to formulate buffers daily. However, the presence of $\geq$20% organic solvent in the mobile phase, or an absence of oxygen due to helium sparging, can inhibit bacterial growth and prolong buffer life.

When removed from the system, the column is best stored in a nonprotic solvent such as acetonitrile (100% B). For short-term applications (overnight or a few days) it is convenient (and acceptable) to leave the mobile phase in place. However, prolonged storage with buffered solutions, particularly those with high concentrations of water or alcohols, should be avoided. Prior to storage, the column should be flushed with 5 to 10 column volumes of the same aqueous-organic mobile phase but *without buffer* before an additional 5-column-volume flush with 100% organic phase (this avoids precipitation of the buffer within the column). Flushing columns with pure water for long periods should be avoided because of stationary-phase de-wetting (especially more hydrophobic columns, e.g., C_{18}). To prevent columns from drying out, they should be tightly capped for storage.

Special handling is required for columns of <2-μm particles. Such columns are fitted with inlet and outlet frits that have very narrow pores (e.g., 0.2 μm) in order to retain these very small particles. Therefore both the sample and mobile phase should be passed through 0.2-μm filters to ensure that particulates do not block the frits and degrade column performance. Small-particle columns are often used for fast separations (run times of <1 min), in which case resulting peaks have very narrow widths—measured either in time or volume. This requires instrumental conditions that minimize potential extra-column effects that artificially broaden peaks:

- short, low-volume tubing that connects the sampling valve, column, and detector
- detector microcells of low volume (e.g., 1 μL)
- small sample volumes (1–2 μL preferred)
- fast detector response (e.g., $\leq$0.1 sec)
- high data-capturing rate (at least 20 points/sec or 20 Hz)

For further details, see [81, 82].

REFERENCES

1. K. K. Unger, *Porous Silica*, Elsevier, Amsterdam, 1979, p. 294.
2. K. Kalghatgi and C. Horváth, *J. Chromatogr.*, 443 (1988) 343.
3. K. K. Unger and H. Giesche, *Ger. Pat.* DE-3543 143.2 (1985).
4. J. J. Kirkland, F. A. Truszkowski, C. H. Dilks, Jr., and G. S. Engel, *J. Chromatogr. A*, 890 (2000) 3.
5. J. J. Kirkland, T. J. Langlois, and J. J. DeStefano, *Amer. Lab.*, 39 (2007) 18.
6. N. B. Afeyan, N. F. Gordon, I. Mazsaroff, L. Varady, S. P. Fulton, Y. B. Yang, and F. E. Regnier, *J. Chromatogr.*, 519 (1990) 1.
7. D. Whitney, M. McCoy, N. Gordon, and N. Afeyan, *J. Chromatogr. A*, 807 (1998) 165.
8. L. R. Snyder and M. A. Stadalius, in *High-performance Liquid Chromatography: Advances and Perspectives*, Vol. 4, C. Horvath, ed., Academic Press, San Diego, 1986.
9. R. K. Iler, *The Chemistry of Silica*, Wiley, New York, 1979, p. 639.
10. J. Billen, P. Gzil, F. Lynen, P. Sandra, P Van der Meeren, and G. Desmet, Poster, HPLC 2006, San Francisco, June 2006.
11. K. K. Unger and E. Weber, *A Guide to Practical HPLC*, Git Verlag, Darmstadt, 1999, p. 45.
12. U. D. Neue, *HPLC Columns*, Wiley-VCH, New York, 1997, p. 82.
13. J. J. DeStefano, T. J. Langlois and J. J. Kirkland, *J. Chromatogr. Sci.*, 46 (2008) 1.
14. J. Köhler, D. B. Chase, R. D. Farlee, A. J. Vega, and J. J. Kirkland, *J. Chromatogr.*, 352 (1986) 275.
15. J. Köhler and J. J. Kirkland, *J. Chromatogr.*, 385 (1987) 125.
16. J. Nawrocki, *Chromatographia*, 31 (1991) 177.
17. J. Nawrocki, *Chromatographia*, 31 (1991) 193.
18. H. Engelhardt, H. Low, and W. Götzinger, *J. Chromatogr.*, 544 (1991) 371.
19. J. J. Kirkland, B. E. Boyes, and J. J. DeStefano, *Amer. Lab.*, 26 (1994) 36.
20. D. W. Sindorf and G. E. Maciel, *J. Am. Chem. Soc.*, 105 (1983) 1487.
21. P. G. Dietrich, K.-H. Lerche, J. Reusch, and R. Nitzsche, *Chromatographia*, 44 (1997) 362.
22. A. Méndez, E. Bosch, M. Rosés, and U. D. Neue, *J. Chromatogr. A*, 986 (2003) 33.
23. M. Jacoby, Chemical and Engineering News, Dec. 11, 2006.
24. F. Svec and C. G. Huber, *Anal. Chem.*, 78 (2006) 2100.
25. G. Guiochon, *J. Chromatogr. A*, 1168 (2007) 101.
26. K. K. Unger, R. Skudas, and M. M. Schulte, *J. Chromatogr. A*, 1184 (2008) 393.
27. F. Svec, T. Tennikova, and Z. Deyl (eds.), *Monolithic Materials: Preparation, Properties and Applications*, J. Chromatogr. Library, Vol 67, Elsevier, Amsterdam, 2004.
28. M. Kele and G. Guiochon, *J. Chromatogr. A*, 960 (2002) 19.
29. X. Wang, D. R. Stoll, P. W. Carr, and P. J. Schoenmakers, *J. Chromatogr. A*, 1125 (2006) 177.
30. S. Eeltink, G. Desmet, G. Vivó-Truyols, G. P. Rozing, P. J. Schoemnmakers, and W. Th. Kok, *J. Chromatogr. A*, 1104 (2006) 256.
31. T. Hara, H. Kobayashi, T. Ikegami, K. Nakanishi, and N. Tanaka, *Anal. Chem.*, 78 (2006) 7632.
32. B. Bidlingmeyer, K. K. Unger, and N. von Doehren, *J. Chromatogr. A*, 832 (1999) 11.

33. L. Geiser, S. Eeltink, F. Svec, and J. M. J. Fréchet, *J. Chromatogr. A,* 1140 (2007) 140.
34. J. Nawrocki, C. Dunlap, J. Li, J. Zhao, C. V. McNeff, A. McCormick, and P. W. Carr, *J. Chromatogr. A,* 1028 (2004) 31.
35. J. Nawrocki, C. Dunlap, A. McCormick, and P. W. Carr, *J. Chromatogr. A,* 1028 (2004) 1.
36. U. Bien-Vogelsang, A. Deege, H. Figge, J. Köhler, and G. Schomburg, *Chromatographia,* 19 (1984) 170.
37. J. J. Pesek and M. T. Matyska, *J. Chromatogr. A,* 952 (2001) 1.
38. P. Ross, *LCGC,* 18 (2000) 14.
39. J. H. Knox, B. Kaur, and G. R. Millward, *J. Chromatogr.,* 352 (1986) 3.
40. M. J. Wirth and H. O. Fatunmbi, *Anal. Chem,.* 65 (1993) 822.
41. J. J. Kirkland, J. L. Glajch, and R. D. Farlee, *Anal. Chem.,* 61 (1988) 2.
42. J. L. Glajch and J. J. Kirkland, US Pat. 4,705,725, 1987
43. J. L. Glajch and J. J. Kirkland, US Pat. 4,847,159, 1989.
44. J. J. Kirkland, J. B. Adams, M. A. Van Straten, and H. A. Claessens, *Anal. Chem.,* 70 (1998) 4344.
45. L. C. Sanders and S. A. Wise, *Anal. Chem., 56* (1984) 504.
46. N. S. Wilson, J. Gilroy, J. W. Dolan, and L. R. Snyder *J. Chromatogr. A,* 1026 (2004) 91.
47. L. Brown, B, Ciccone, J. J. Pesek, and M. T. Matyska, *Amer. Lab.*, Dec. (2003) 23.
48. K. Miyabe and N. Orita, *Talanta,* 36 (1989) 897.
49. M. R. Buchmeiser, *J. Chromatogr. A,* 918 (2001) 233.
50. K. K. Unger and J. Schick-Kalb, US Pat. 4,017,528, 1977.
51. K. D. Wyndham, J. E. O'Gara, T. H. Walter, K. H. Glose, N. L. Lawrence, B. A. Alden, G. S. Izzo, C. J. Hudalla, and P. C. Iraneta, *Anal. Chem.*, 75 (2003) 6781.
52. R. E. Majors, *LCGC,* 20 (2002) 516.
53. B. A. Bidlingmeyer and A. D. Broske, *J. Chromatogr. Sci.*, 42 (2004) 100.
54. T. H. Walter, P. Iranetaund, and M. Capparella, *J. Chromatogr. A*, 1075 (2005) 177.
55. R. Eksteen and J. Thoma, in: *Chromatography, the State of the Art*, Vol. 1, Akademia Kiado, Budapest, 1985.
56. L. Zhang, L. Sun, J. I. Siepmann, and M. R. Shure, *J. Chromatogr. A,* 1079 (2005) 127.
57. K. Croes, A. Steffens, D. H. Marchand, and L. R. Snyder, *J. Chromatogr. A,* 1098 (2005) 123.
58. M. R. Euerby, P. Petersson, W. Campbell, and W. Roe, *J. Chromatogr. A,* 1154 (2007) 138.
59. U. D. Neue, *J. Sep. Sci.,* 30 (2007) 1611.
60. N. S. Wilson, M. D. Nelson, J. W. Dolan, L. R. Snyder, R. G. Wolcott, and P. W. Carr, *J. Chromatogr. A,* 961 (2002) 171.
61. L. R. Snyder, J. W. Dolan, and P. W. Carr, *J. Chromatogr. A,* 1060 (2004) 77.
62. L. R. Snyder, J. W. Dolan, and P. W. Carr, *Anal. Chem.,* 79 (2007) 3255.
63. D. H. Marchand and L. R. Snyder, *J. Chromatogr. A*, 1209 (2008) 104.
64. L. R. Snyder and J. W, Dolan, *High Performance Gradient Elution*, Wiley-Interscience, Hoboken, NJ, 2007, p. 420.
65. L. C. Sander and S. A. Wise, *J. Chromatogr. A,* 656 (1993) 335.
66. J. J. Kirkland, *Amer. Lab.*, 26 (1994) 28K.

67. U. D. Neue, E. Serowik, P. Iraneta, B. A. Alden, and T. H. Walter, *J. Chromatogr. A*, 849 (1999) 87.
68. L. R. Snyder, A. Maule, A. Heebsch, R. Cuellar, S. Paulson, J. Carrano, L. Wrisley, C. C. Chan, N. Pearson, J. W. Dolan, and J. Gilroy, *J. Chromatogr. A*, 1057 (2004) 49.
69. J. Pellett, P. Lukulay, Y. Mao, W. Bowen, R. Reed, M. Ma, R. C. Munger, J. W. Dolan, L. Wrisley, K. Medwid, N. P. Toltl, C. C. Chan, M. Skibic, K. Biswas, K. A. Wells, and L. R. Snyder, *J. Chromatogr. A*, 1101 (2006) 122.
70. Y. Mao and P. W. Carr, *LCGC*, 21 (2003) 150.
71. J. W. Dolan and L. R. Snyder, *J. Chromatogr. A*, 1216 (2009) 3467.
72. D. V. McCalley, *Adv. Chromatogr.*, 46 (2008) 305.
73. D. M. Marchand, L. R. Snyder, and J. W. Dolan, *J. Chromatogr. A*, 1191 (2008) 2.
74. H. Colin. P. Hilaireau, and J. De Tournemire, *LCGC*, 8 (1990) 302.
75. F. Couillard, Chromatography Apparatus, US Patent 4,597,866, 1986-07-01.
76. J. J. Kirkland and J. J. DeStefano, *J. Chromatogr. A*, 1126 (2006) 50.
77. L. R. Snyder and J. J. Kirkland, *Introduction to Modern Liquid Chromatography*, 2nd ed., Wiley, New York, 1979, ch. 5.
78. H. A. Claessens, M. A. Van Straten, and J. J. Kirkland, *J. Chromatogr. A*, 728 (1996) 259.
79. J. J. Kirkland, M. A. Van Straten, and H. A. Claessens, *J. Chromatogr. A*, 797 (1998) 111.
80. G. W. Tindall and R. L. Perry, *J. Chromatogr. A*, 988 (2003) 309.
81. J. J. Kirkland, *J. Chromatogr. Sci.*, 38 (2000) 535.
82. F. Gerber, M. Krummen, H. Potgeter, A. Roth, C. Siffrin, and C. Spoendlin, *J. Chromatogr. A*, 1036 (2004) 127.

CHAPTER SIX

REVERSED-PHASE CHROMATOGRAPHY FOR NEUTRAL SAMPLES

Introduction to Modern Liquid Chromatography, Third Edition, by Lloyd R. Snyder, Joseph J. Kirkland, and John W. Dolan

6.1 INTRODUCTION

This chapter describes the separation of neutral samples by means of reversed-phase chromatography (RPC). By a "neutral" sample, we mean one that contains no molecules that carry a positive or negative charge—usually as the result of the ionization of an acid or a base. Although a neutral sample implies an absence of acidic and basic solutes, this is not necessarily the case. Depending on mobile phase pH, any acids or bases in the sample may be present largely (e.g., 90%+) in the neutral (non-ionized) form—in which case their chromatographic behavior is similar to that of non-ionizable compounds. The separation of "ionic" samples (which contain one or more ionized compounds) by RPC is covered in Chapter 7.

RPC is usually a first choice for the separation of both neutral and ionic samples, using a column packing that contains a less polar bonded phase such as C_8 or C_{18}. The mobile phase is in most cases a mixture of water and either acetonitrile (ACN) or methanol (MeOH); other organic solvents (e.g., isopropanol [IPA], tetrahydrofuran [THF]) are used less often. A preferred organic solvent for an RPC mobile phase will be water-miscible, relatively nonviscous, stable under the conditions of use, transparent at the lowest possible wavelength for UV detection, and readily available at moderate cost. Commonly used B-solvents can be ranked in terms of these properties as follows:

$$\text{ACN (preferred)} > \text{MeOH} > \text{IPA} \gg \text{THF (less useful)}$$

In a few countries, ACN is considered sufficiently toxic to limit its general use, but this is not true elsewhere. See Appendix I for further information concerning solvent properties and the choice of B-solvent for a given application. Samples that contain acids or bases normally require a buffered mobile phase, in order to maintain a constant pH throughout the separation (Chapter 7). Strongly retained, very hydrophobic samples may require a water-free mobile phase (nonaqueous reversed-phase chromatography [NARP], Section 6.5). Normal-phase chromatography (Chapter 8) can also provide acceptable separations of very hydrophobic samples, as sample hydrophobicity contributes little to retention for this HPLC mode. Preferred conditions for the isocratic separation of neutral samples by RPC are listed in Table 6.1.

Compared to other forms of HPLC (normal-phase, ion-exchange chromatography, etc.; Table 2.1), separations by RPC are usually more convenient, robust, and versatile. RPC columns also tend to be more efficient and reproducible, and are available in a wider range of choices that include column dimensions, particle size, and stationary-phase type (C_1–C_{30}, phenyl, cyano, etc.; Section 5.3.3). The solvents used for RPC tend to be less flammable or toxic, and are more compatible with UV detection at wavelengths below 230 nm for increased detection sensitivity (Table I.2 of Appendix I). An additional advantage of RPC is generally fast equilibration of the column after a change in the mobile phase—or between runs when using gradient elution (Section 9.3.7). Finally, because RPC has been the dominant form of HPLC since the late 1970s, a better practical understanding of this technique has evolved. This usually means an easier development of better separations. All of the foregoing reasons have contributed to the present popularity of RPC.

Table 6.1

Preferred Conditions for the Separation of Neutral Samples by Reversed-Phase Chromatography

Condition	Comment
Column[a]	Type: C_8 or C_{18} (type-B)
	Dimensions: 100 × 4.6-mm
	particle size: 3 μm
	Pore diameter: 8–12 nm
Mobile phase	Acetonitrile/water
Flow rate	2.0 mL/min[b]
Temperature	30 or 35°C[b]
%B	To be determined[c]
Sample	Volume ≤ 25 μL
	Weight ≤ 50 μg
k	$1 \leq k \leq 10$

[a]Alternatively, use a 150 × 4.6-mm column of 5-μm particles; flow rate, column dimensions, and particle size can be varied, depending on the anticipated difficulty of the separation and the maximum allowable column pressure (Section 2.4.1)

[b]Initial values, which may be changed during method development (Section 2.5); a temperature 5–10°C above ambient is suggested in most cases. Also consult the column manufacturer's recommendations for a maximum column temperature.

[c]Varies with the sample; start with 80%B and adjust further as described in Section 2.5.1.

Many organic compounds have limited solubility in either water or the water-organic mobile phases used for RPC, but this is rarely a practical concern. Thus very small weights (nanograms or low micrograms) of individual solutes are usually injected, so the required sample concentration is usually only a few micrograms/mL or less. In those cases where sample solubility in water or water-organic mixtures is exceptionally poor (very hydrophobic samples), the use of normal-phase chromatography with nonaqueous mobile phases may be preferred (Section 8.4.1).

Some samples are less well separated by RPC. For example, very polar molecules may be retained weakly in RPC ($k \ll 1$), even with 100% water as mobile phase; these samples may require a different approach (Section 6.6.1). Similarly enantiomers require separation conditions that exhibit chiral selectivity (Chapter 14). While many achiral isomers can be separated by RPC (Section 6.3.5), these compounds are often better separated by normal-phase chromatography using an unbonded silica column (Section 8.3.4.1). Finally, normal-phase chromatography is often a better choice for preparative HPLC (Chapter 15).

6.1.1 Abbreviated History of Reversed-Phase Chromatography

Prior to the invention of RPC in 1950 by A. J. P. Martin [1], the chromatographic separation of neutral samples was carried out with a polar column (or stationary phase) and a less polar mobile phase; such separations are now referred to as "normal-phase" chromatography (NPC). As RPC involves a *less* polar column and a *more* polar mobile phase, the two phases can be regarded as interchanged or "reversed." The first RPC packings were made from silica particles that had been reacted with $(CH_3)_2SiCl_2$, so as to render the surface nonpolar. A nonpolar stationary phase was then used to coat the particles, allowing their use for reversed-phase liquid–liquid partition chromatography [2], a HPLC technique that is rarely used today.

Bonded-phase columns represented the next major advance in "high-performance" RPC. Early C_{18} columns of this type were prepared by the covalent attachment of polyoctadecylsiloxane to silica particles [3, 4]. This was followed a few years later by the introduction of columns where individual C_{18} groups (rather than a C_{18} polymer) were attached to the particle (Section 5.3.1). Beginning in the early 1970s, RPC columns of the latter type were used increasingly because of their many advantages. In the mid-1970s, the use of RPC underwent an explosive growth in popularity; several hundred separations are cited in [5] for the period 1976 to 1979. Whereas earlier applications of RPC emphasized the separation of more hydrophobic ("lipophilic") samples, the classic paper by Horváth [6] demonstrated that RPC could also be used for the separation of relatively polar compounds, especially water-soluble samples of biochemical interest. A little later, RPC was further adapted for the separation of ionizable compounds (including enantiomers) by the addition of ion-pairing compounds to the mobile phase [7] (*ion-pair chromatography*, Section 7.4). In the late 1970s, several groups reported the first RPC separations of large peptides and proteins, using short-chain, wide-pore column packings. Today, RPC is the dominant HPLC mode and accounts for a substantial majority of all HPLC separations.

6.2 RETENTION

Retention in RPC was discussed briefly in Section 2.2. Because very polar molecules interact more strongly with the polar mobile phase, these compounds are less retained and leave the column first. Similarly less polar compounds prefer the nonpolar stationary phase and leave the column last. Thus molecules of similar size are eluted in RPC approximately in order of decreasing polarity. An example is provided by Figure 6.1*a*, where it is seen that the more-polar benzonitrile (1) appears in the chromatogram first, followed by the increasingly less polar anisole (2), and finally toluene (3). A more detailed example of RPC retention as a function of solute polarity is provided by Figure 2.7*c*; see also the related discussion of RPC column selectivity in Section 5.4.

Retention in RPC is largely the result of interactions between a solute molecule and either the mobile phase or the column (Section 2.3.2.1). An increase in %B (volume-% of organic solvent in the mobile phase) makes the mobile phase less polar ("stronger") and increases the strength of interactions between solute and solvent molecules. The result is decreased retention for all solute molecules when %B is increased. This is illustrated by the separation of Figure 6.1*b* (with a mobile phase of 60% B) compared to that of Figure 6.1*a* (40% B). An increase in temperature weakens the interaction of the solute with both the mobile phase and column, and

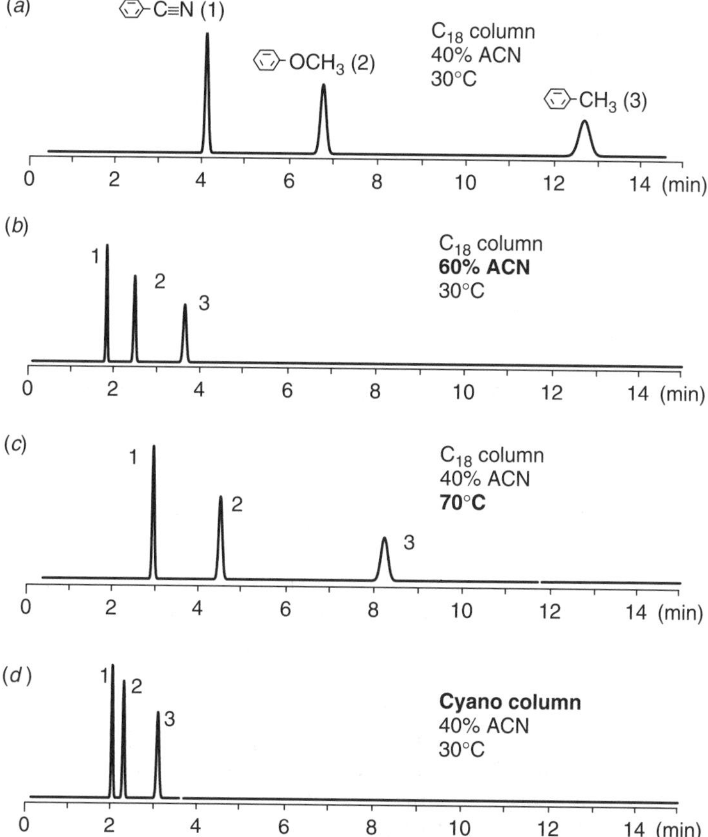

Figure 6.1 Retention in RPC as a function of temperature and the polarity of the solute, mobile phase and column. Sample: as indicated in figure. Conditions: (*a-c*) 150 × 4.0-mm 5-μm) Symmetry C_{18} column, and (*d*) 150 × 4.6-mm (5-μm) Zorbax StableBond cyano column; 2.0 mL/min; mobile phase is acetonitrile/water, with mobile-phase composition (%B) and temperature indicated in figure (bolded values represent changes from [*a*]). Chromatograms recreated from data of [8, 9].

decreases retention; compare Figure 6.1*c* (70°C) and Figure 6.1*a* (30°C). Finally, a decrease in column hydrophobicity weakens the interaction between the solute and column, and reduces retention; compare Figure 6.1*d* (more-polar cyano column) and Figure 6.1*a* (less-polar C_{18} column).

6.2.1 Solvent Strength

As noted in Section 2.4.1, retention in RPC varies with mobile phase %B as

$$\boxed{\log k = \log k_w - S\phi} \tag{6.1}$$

where k_w refers to the (extrapolated) value of k for 0% B (water as mobile phase), S is a constant for a given solute when only %B is varied, and ϕ is the volume-fraction of organic solvent B in the mobile phase ($\phi \equiv 0.01\%$ B). The value of %B selected

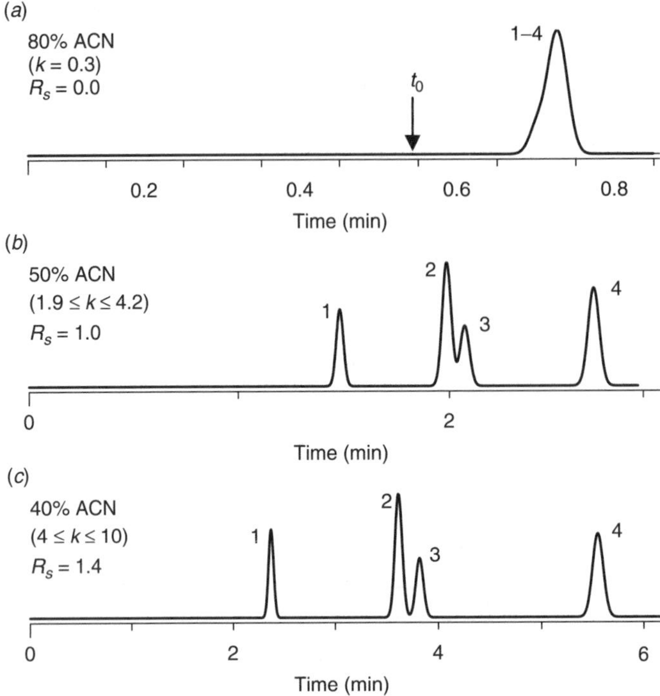

Figure 6.2 Separation of a mixture of four nitro-substituted benzenes as a function of solvent strength (%B). Sample: *1*, nitrobenzene; *2*, 4-nitrotoluene; *3*, 3-nitrotoluene; *4*, 2-nitro-1,3-xylene. Conditions: 100 × 4.6-mm (3-μm) Zorbax C8 column; mobile phase consists of acetonitrile-water mixtures (varying %B); 35°C; 2 mL/min. Chromatograms recreated from data of [10].

for the final separation should provide values of k for the sample that are within a desired range (e.g., $1 \leq k \leq 10$), while at the same time maximizing solvent-strength selectivity (Section 6.3.1).

A suitable value of %B can be obtained as described in Section 2.5.1: start with 80% B, and then reduce %B in steps until the desired retention range of $1 \leq k \leq 10$ is obtained. Figure 6.2 provides an example of this approach. The first separation with 80% B (Fig. 6.2*a*) provides very little retention ($k = 0.3$), so a change to 50% B is tried for the next experiment (Fig. 6.2*b*). The retention range for the sample is now reasonable ($1.9 \leq k \leq 4.2$), but the resolution is inadequate ($R_s = 1.0$). A further decrease in %B will usually (but not always) increase resolution; for samples with molecular weights < 500 Da, a 10%B decrease will increase values of k by a factor of about 2.5, which suggests a mobile phase of 40% B for the next experiment (Fig. 6.2*c*). Resolution is increased moderately ($R_s = 1.4$) but is still inadequate. As a further decrease in %B will result in values of $k > 10$, some other means of further increasing resolution may be necessary: a change in selectivity (Section 6.3) or a change in column conditions (Section 2.5.3).

It should be noted that Equation (6.1) is not an exact relationship but an approximation. For example, values of log k for a representative solute (4-nitrotoluene; compound-2 in Fig. 6.2) are plotted against %B in Figure 6.3

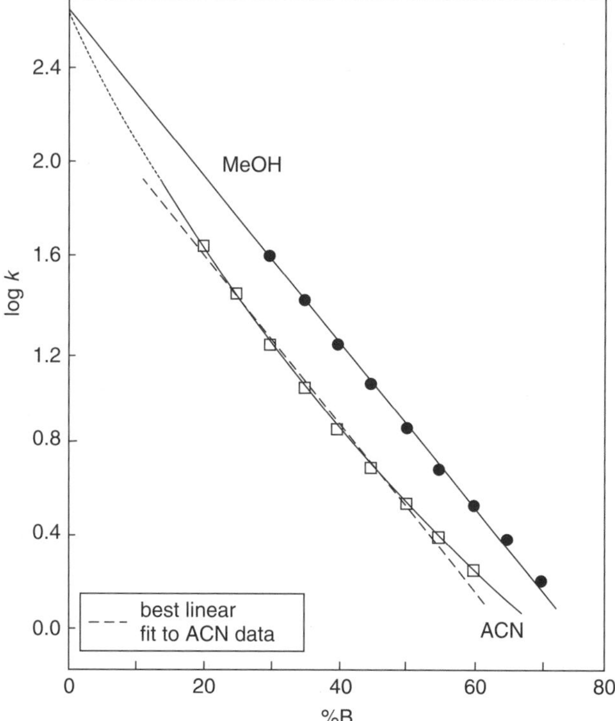

Figure 6.3 Variation of log k with %B. Sample is 4-nitrotoluene. Conditions: 250 × 4.6-mm (5-μm) Zorbax C8 column; mobile phase consists of organic/water mixtures; 35°C; 2 mL/min. Created from data taken from [10].

for both acetonitrile (ACN, □) and methanol (MeOH, •) as the B-solvent. Whereas Equation (6.1) predicts a linear plot, a slightly curved plot results for ACN as B-solvent. The data for MeOH fall closer to the linear curve in Figure 6.3 that is fitted to these data. This behavior is typical of other samples and experimental conditions [11]; more linear plots are usually obtained for MeOH, compared to the use of ACN or other organics as B-solvent. However, over the usual range in k that is of interest (e.g., $1 \leq k \leq 10$), *Equation* (6.1) *is adequately accurate for either MeOH or ACN as the B-solvent.*

6.2.2 Reversed-Phase Retention Process

The following discussion provides further insight into the basis of RPC retention. However, it is mainly of academic interest, with little practical value for the application of RPC. A results-oriented reader may prefer to skip to following Section 6.3.

The nature of RPC retention (the retention "mechanism") has been the subject of a large number of research studies, as summarized in several reviews and research publications [5, 12–16]. Some of the questions that this work has addressed include:

- positioning of the solute molecule within the stationary phase (adsorption or partition?)

- dependence of solute retention on mobile-phase composition (k vs. ϕ)
- conformation of the alkyl ligands that form the stationary phase ("extended" vs. "collapsed")

The positioning of the solute molecule within the stationary phase might occur in any of the ways pictured in Figures 6.4*a–c* for a C_8 column. *Solvophobic interaction* (Fig. 6.4*a*) assumes that the solute molecule aligns with and is attached to a ligand group (C_8 in this example). *Adsorption* (Fig. 6.4*b*) implies that the solute molecule does not penetrate into the stationary phase, but is retained at the interface between the stationary and mobile phases. *Partition* (Fig. 6.4*c*) considers the stationary phase to be similar to a liquid phase, into which the solute molecule dissolves. Notice that the stationary phase consists of alkyl ligands plus an organic solvent that is preferentially extracted from the mobile phase by the C_8 groups of the stationary phase. In both solvophobic interaction and partition, the solute molecule lies within the stationary phase.

When discussing the mechanism of RPC retention, Horváth's solvophobic-interaction model [5] is commonly cited: (relatively) hydrophobic solute molecules prefer to adhere to the hydrophobic alkyl ligands—so-called hydrophobic retention. Soon after the introduction of RPC for HPLC, it was observed that RPC retention (values of k) correlates with partition coefficients P for the distribution of the solute between octanol and water (Fig. 6.4*d*; [17]); this suggests that a partition process best describes RPC retention. However, later studies showed that correlations of *log P* versus *log k*, as in Figure 6.4*d* (for amino acids) are less pronounced when the sample consists of molecules with more diverse structures, which makes the latter argument on behalf of partition less compelling.

In the early 1980s [18] a surprising observation was made for the RPC retention of various homologous series ($CH_3–[CH_2]_{n-1}–X$), where X represents a functional group such as $–OH$ or $–CO_2CH_3$. Plots of log k versus carbon-number n were found to exhibit a discontinuity for a value of n that is approximately equal to the carbon-number n' for the stationary-phase ligand (e.g., $n' = 8$ for $C_8 \equiv CH_3–[CH_2]_7–$). This anomalous behavior is illustrated in Figure 6.5*a* by a hypothetical plot of log k versus n for a homologous series and a C_8 column; a discontinuity in the expected linear plot (dashed line) is observed (arrow) when n equals 8 for the solute ($CH_3–[CH_2]_7–X$). It was concluded from this observation that the contribution to retention for successive $–CH_2$-groups in the solute becomes slightly smaller when the length of the solute molecule just exceeds the length of the alkyl ligand. The reason for this discontinuity in the plot of Figure 6.5*a* is visualized in Figure 6.5*b*, where $n = 12$ for solute *ii* exceeds the value of $n' = 8$ for the column ligand (the situation for $n = n' = 8$ is illustrated by solute *i*). In the example of Figure 6.5*b* with $n = 12$, the end of the molecule likely folds back onto or into the stationary phase—rather than extending into the mobile phase as shown.

Presumably there is a decreased interaction with the column for solute molecules that are too long to penetrate fully into the stationary phase (or attach to a single column ligand), with a corresponding decrease in the retention of $–CH_2$-groups that "stick out of" the stationary phase. The contribution to retention of each $–CH_2$-group can be defined as α_{CH2} = ratio of k values for successive homologs, and this value is normally assumed constant (see discussion of Eq. 2.34). However, for

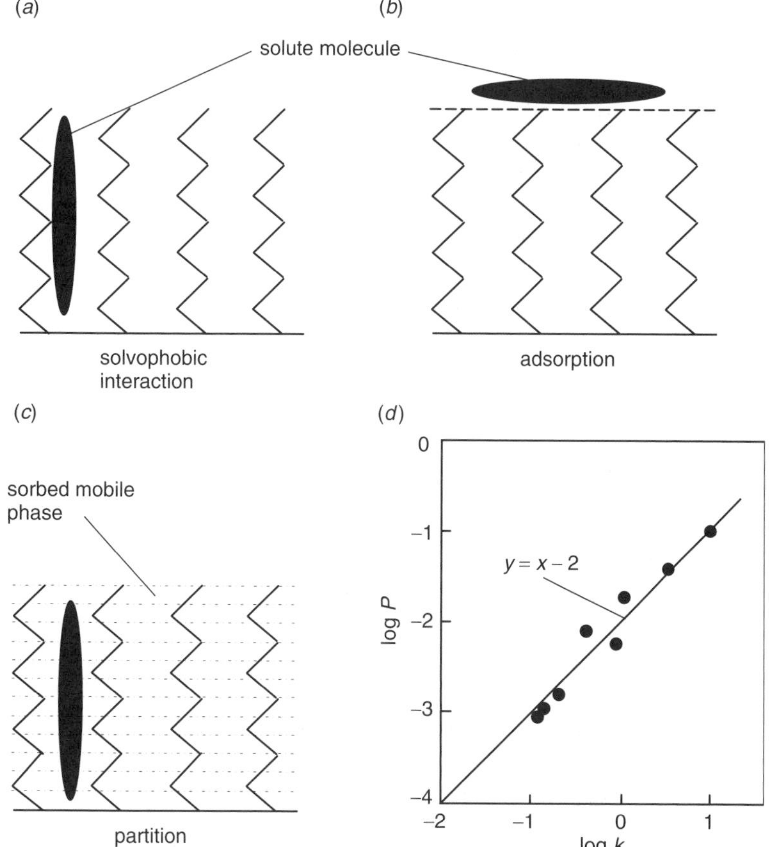

Figure 6.4 Different possibilities for the retention of a solute molecule in reversed-phase chromatography. (*a*) Solvophobic interaction; (*b*) adsorption; (*c*) partition; (*d*) comparison of RPC retention (k) with octanol-water partition P [17]; sample; eight amino acids; column: C_8; mobile phase: aqueous buffer (pH-6.7); 70°C.

excluded $-CH_2$-groups, the value of α_{CH2} should be somewhat smaller. This "exclusion" effect is demonstrated experimentally in Figure 6.5*c* for a C_{18} column, where α_{CH2} is plotted versus n. There is a distinct break in the plot in the vicinity of $n = n' = 18$ (dashed vertical line). Similar breaks are shown in the plots of Figures 6.5*d*,*e* for the C_8 and C_6 columns, respectively, but no significant penetration of the solute is possible for the C_1 column of Figure 6.5*f*. The data of Figures 6.5*c–e* suggest that small solute molecules can penetrate the stationary phase of a C_6, C_8, or C_{18} column, so this would rule out a purely adsorption process (as in Fig. 6.4*b*) for other small solutes. Figure 6.5 also supports the solvophobic-interaction model of Figure 6.4*a*.

Figure 6.5 is suggestive of possible conclusions concerning retention in RPC, but more complicated arguments have been offered concerning the adsorption and partition processes [14, 15]. It should be noted that the nature of the stationary phase (and presumably the retention process) varies with experimental conditions. Increasing amounts of the B-solvent (e.g., acetonitrile) are taken up by the stationary

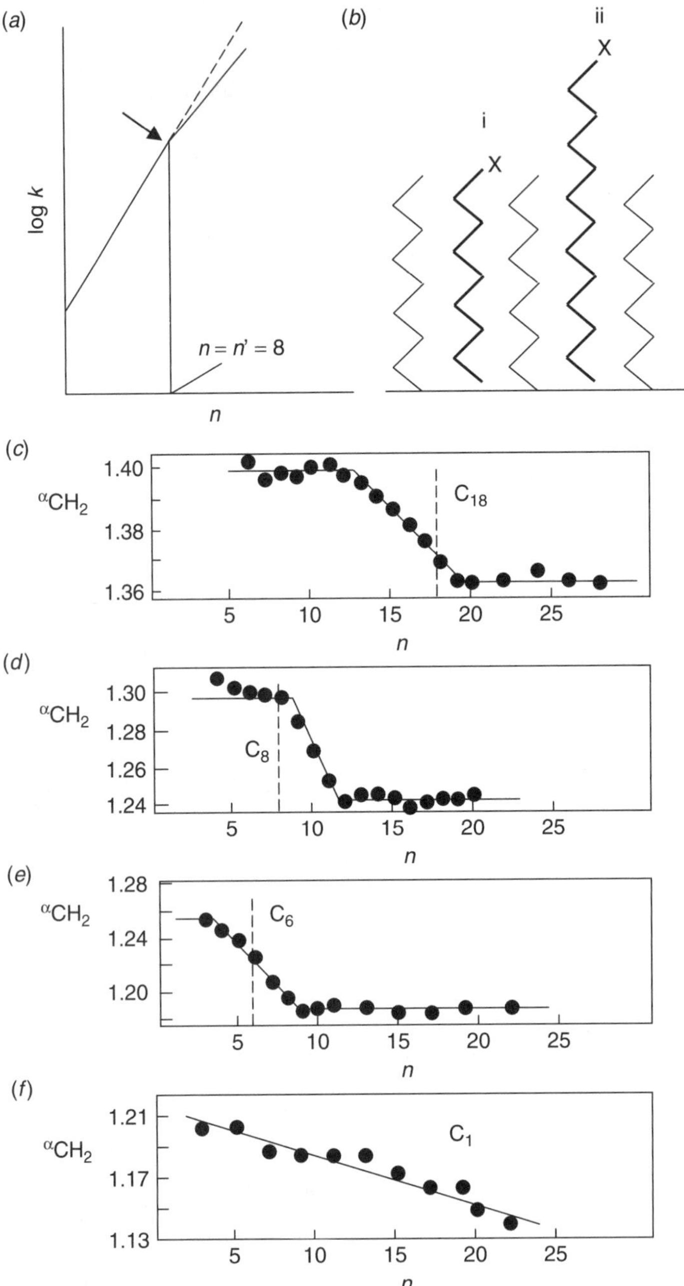

Figure 6.5 Retention as a function of alkyl chain length (for both the solute and column). (*a*) Illustrative plot of log *k* versus number of $-CH_2$-groups *n* for a homologous series $CH_3-(CH_2)_{n-1}-X$;C_8 column; (*b*) illustration of the "overlapping" of alkyl chains in the solute and column; (*c–f*) plots of experimental methylene selectivity α_{CH2} versus carbon number n_c for indicated columns of differing ligand length. Average data for several homologous series; 90% methanol-water as mobile phase; 25°C. Figures are adapted from [18].

phase as %B increases. Likewise some solutes may interact with underivatized silanols present on the particle surface (Section 5.4.1). For these and other reasons the precise nature of the retention process is likely to vary with the column, the solute, and experimental conditions. Horváth anticipated this situation early on [17], in noting that "a clear distinction between partition and adsorption in RPC of nonpolar [solutes] [and] with no apparent thermodynamic or practical significance ... [so that] this issue may not be worth further investigation." The authors of this book find it difficult to argue with this conclusion.

Speculation concerning the nature of RPC retention has also been based on retention as a function of mobile-phase composition (%B). While Equation (6.1) is a purely empirical relationship, several theory-based equations for k as a function of ϕ have been derived [11]. The resulting expressions for k versus ϕ are in some cases slightly more reliable than Equation (6.1)[19], largely because of additional fitting parameters. However, Equation (6.1) is generally adequate for practical application, and is much more convenient to use. The major assumptions involved in all previous theoretical derivations of k versus ϕ negate any value in their use for interpreting the nature of RPC retention.

Stationary phase ligand conformation has been claimed to play a role in the "mechanism" of RPC retention. The use of mobile phases that are predominantly aqueous ($\phi \approx 0$) can lead to greatly reduced sample retention—the opposite of that predicted by Equation (6.1). When first observed, this reduced retention was attributed to "phase collapse," whereby alkyl ligands clump together and tend toward a horizontal rather than vertical alignment with the particle surface. This retention behavior was subsequently shown to arise not from phase collapse but rather from exclusion of mobile phase and sample molecules from particle pores as a result of surface-tension effects ("de-wetting," Section 5.3.2.3 and [2, 21]). More recent studies [22, 22a] suggest that ligand conformation does not change as a function of either mobile-phase composition or the relative coverage of the particle surface by ligands.

6.3 SELECTIVITY

The most effective way to improve the resolution (or speed) of a chromatographic separation is to initiate a change in relative retention (selectivity). For the separation of non-ionic samples by RPC, changes in selectivity can be achieved by a change in solvent strength (%B), temperature, solvent type (e.g., ACN vs. MeOH as the organic solvent), or column type (e.g., C_{18} vs. cyano). The relative effectiveness of a change in conditions for a change in selectivity varies roughly as

temperature (least effective) < %B < solvent type ≈ column type (most effective)

However, each of the four conditions above for changing selectivity can be useful for different samples or separation goals, as discussed next.

6.3.1 Solvent-Strength Selectivity

In the examples of Figures 6.1 and 6.2, relative retention does not change when solvent strength (%B) is varied. As %B decreases (and k increases), the resolution

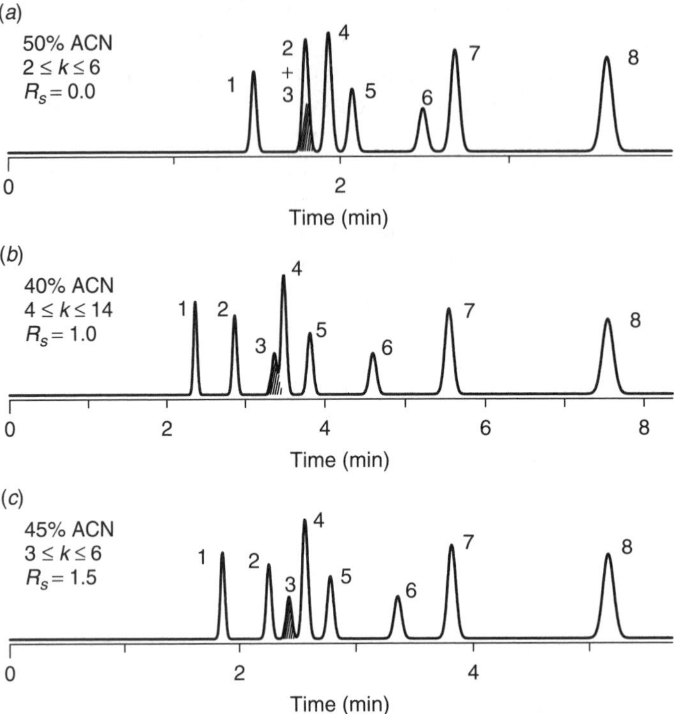

Figure 6.6 Separation of a moderately irregular sample (mixture of eight nitro-aromatic compounds) as a function of solvent strength (%B). Sample: *1*, nitrobenzene; *2*, 2,6-dinitrobenzene; *3*, benzene (shaded peak); *4*, 2-nitrotoluene; *5*, 3-nitrotoluene; *6*, toluene; *7*, 2-nitro-1,3-xylene; *8*, 1,3-xylene. Conditions: 100 × 4.6-mm (3-μm) Zorbax C8 column; mobile phase consists of acetonitrile/water mixtures; 35°C; 2 mL/min. Chromatograms recreated from data of [10].

of all peaks improves, but their relative spacing stays essentially the same. In Section 2.5.2.1 we defined samples as in Figures 6.1 and 6.2 as *regular*. Figure 6.6 shows the separation of a sample where %B is varied, but relative retention does *not* remain the same. An initial separation with 50% B (Fig. 6.6*a*) shows a complete overlap of peaks 2 and 3 (shaded). When the mobile phase is changed to 40% B (Fig. 6.6*b*), peak 3 moves toward peak 4 and partially overlaps it. From these two experiments it can be seen that an intermediate value of %B should result in an improved separation, which is observed for the separation of Figure 6.6*c* (45% B). Samples such as this, where relative retention changes with solvent strength, are referred to as *irregular*. Regular samples are often composed of structurally similar molecules; for example, in the separation of Figure 6.2 the sample is a mixture of mono-nitro alkylbenzenes. The sample of Figure 6.6, on the other hand, exhibits a somewhat greater molecular diversity: it is a mixture of alkylbenzenes that contain 0, 1, or 2 nitro-substituents.

Changes in %B often lead to significant changes in relative retention, with maximum resolution occurring for an intermediate value of %B. Despite this obvious fact, practical workers often overlook solvent-strength selectivity as a useful tool for optimizing relative retention and resolution. To take maximum advantage of solvent-strength selectivity, the allowable retention range can be expanded from $1 \leq k \leq 10$ to $0.5 \leq k \leq 20$. When conditions are varied so as to change selectivity, it is important to keep track of which peak is which. The numbering of each peak in the chromatogram (as in Fig. 6.6) may not be obvious; in such cases *peak tracking* will be required (Section 2.6.4).

The remainder of this section, which expands on the discussion above of regular and irregular samples, is somewhat detailed; the reader may prefer to skip (or skim) this discussion and go on to Section 6.3.2.

The dependence of retention on %B for regular as opposed to irregular samples is further illustrated in Figure 6.7. Figure 6.7*a* shows plots of log k against %B for a regular sample; a mixture of nine herbicides of similar molecular structure (see Fig. 2.6 for the separation of several of these compounds as a function of %B). The slope of each plot increases for more retained solutes (in the order $1 < 2 < 3\ldots$). For plots of log k versus %B for regular samples, this results in a characteristic, "fan-like" appearance—with no intersection of one plot by another (no change in relative retention). Values of α and resolution increase continuously for regular samples as %B is decreased. Another way to describe the behavior of regular samples is in terms of Equation (6.1). For regular samples, the slopes S of plots as in Figure 6.7*a* are highly correlated with extrapolated values of log k for water as mobile phase (log k_w). Figure 6.7*b* shows such a plot for the data of Figure 6.7*a*; an excellent correlation is noted ($r^2 = 1.00$). Corresponding correlations of log k_w versus S for the regular samples of Figures 6.1*a*,*b* and 6.2 give $r^2 = 0.99$ for each.

A similar treatment as in Figures 6.7*a*,*b* for a regular sample is shown in Figures 6.7*c*,*d* for the irregular sample of Figure 6.6. In Figure 6.7*c*, plots of log k versus %B frequently intersect (marked by •), unlike the behavior of the regular sample in Figure 6.7*a*. As a result several peak reversals occur over this range in %B (e.g., peaks 3–4). Similarly a plot of S versus log k_w for the irregular sample in Figure 6.7*d* shows a somewhat poorer correlation ($r^2 = 0.87$). Samples that contain acidic and/or basic compounds (unlike the examples of Fig. 6.7) tend to be more irregular; However, most samples—whether "ionic" or "neutral"—exhibit some degree of irregularity and solvent-strength selectivity can be a useful tool for improving the resolution of such samples.

6.3.2 Solvent-Type Selectivity

Changes in %B may fail to achieve adequate resolution of a given sample. An illustration is shown in Figure 6.8 for the separation of a mixture of substituted benzenes. In this example, a change in mobile phase from 46 to 34% ACN results in some change in relative retention (e.g., peaks 2–3, 8–9), but that has no significant effect on the separation of overlapped peaks 3 and 4. Consequently some other change in conditions that affect selectivity will be necessary—for example, a change from ACN to MeOH as the B-solvent. When 61% MeOH is used in place of 46% ACN (Fig. 6.9*a*), solvent strength is about the same (similar run times), but several further changes in relative retention are seen due to solvent-type selectivity. Although previously unresolved peaks 3 and 4 are now well separated, peaks 1 and 2 (which were well resolved with ACN as B-solvent) are overlapped. With results such as this for two mobile phases with different B-solvents, a *mixture* of the two mobile phases will often provide a better separation. This proved to be the case for the present sample. A 1:1 blend of the mobile phases of Figure 6.8*a* (46% ACN) and Figure 6.9*a* (61% MeOH) was prepared, and used to obtain the separation of Figure 6.9*b*. The new mobile phase (containing 23% ACN + 30% MeOH) provides baseline resolution ($R_s = 1.8$).

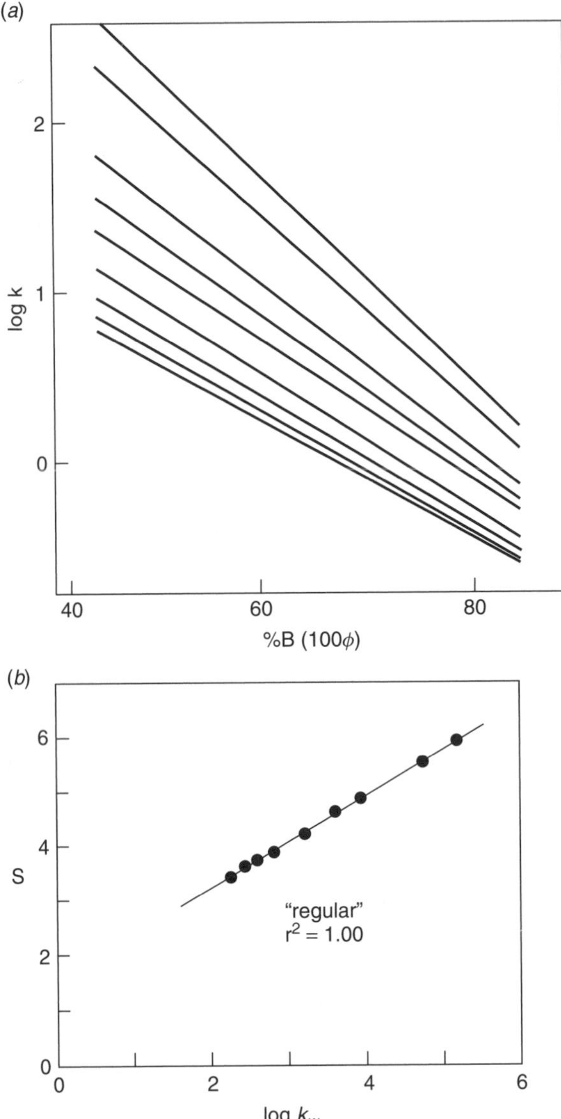

Figure 6.7 Variation of log *k* with %B for regular and irregular samples. (*a*) Regular sample (a mixture of herbicides [24]), separated on a C_{18} column with methanol-water as mobile phase (see Fig. 2.6 for other conditions); (*b*) plot of values of *S* versus log k_w for data of (*a*); (*c*) irregular sample of Figure 6.6; conditions as in Figure 6.6; (*d*) plot of values of *S* versus log k_w for data of (*c*). (*a*, *b*) Adapted from [25].

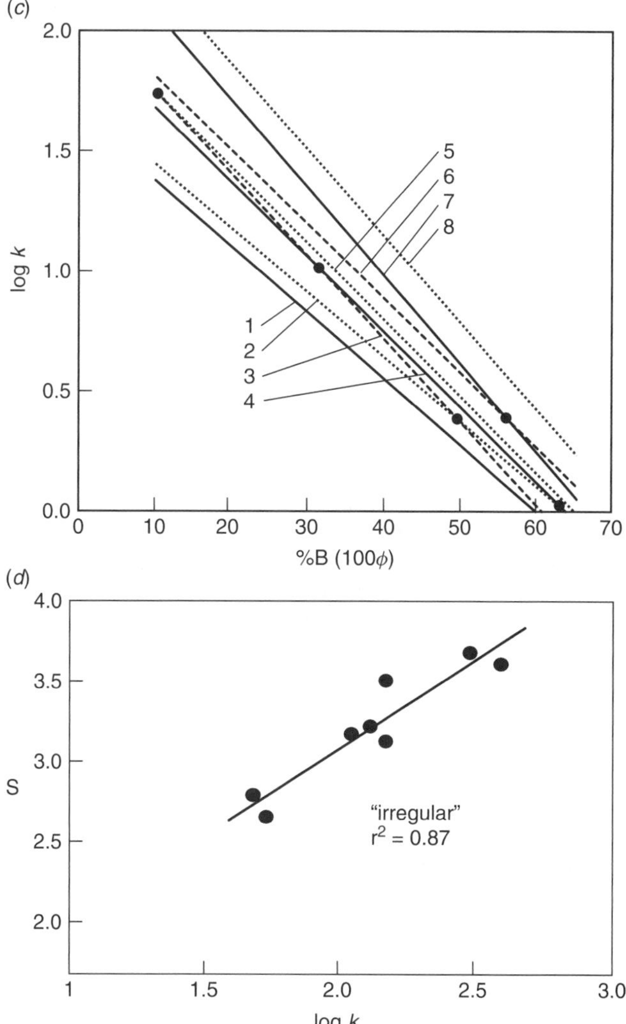

Figure 6.7 (*Continued*)

While the separation of Figure 6.9*b* is much improved, further variation in the proportions of mobile phase from Figure 6.8*a* (46% ACN) and Figure 6.9*a* (61% MeOH) may provide some more resolution. Peaks 1 to 4 in Figure 6.9*b* are less resolved than remaining peaks 5 to 10 (i.e., peaks 1–4 are "critical"), so we will limit our discussion to these peaks. Figures 6.10*a–c* replicate earlier chromatograms for peaks 1 to 4 from Figures 6.8 and 6.9, for changes in the relative proportions of ACN and MeOH. As MeOH replaces ACN in going from Figure 6.10*a–c*, it is seen that peaks 2 and 3 (shaded) move toward the front of the chromatogram. In Figure 6.10*b* (1:1 blend of mobile phases from Fig. 6.10*a*,*c*), the critical (least resolved) peak-pair is 2/3, and we can improve its resolution by a further movement

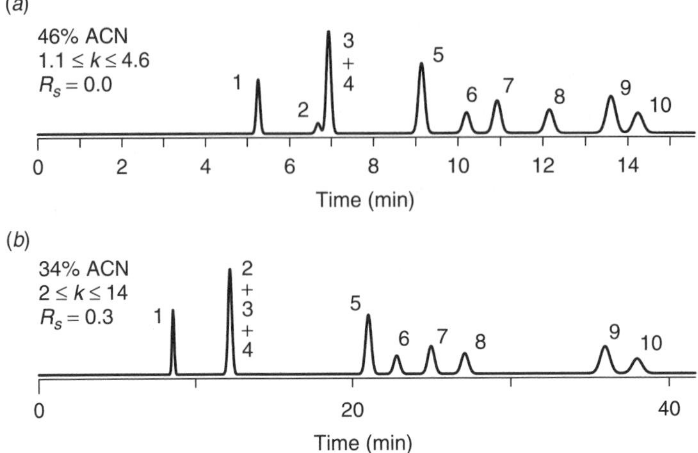

Figure 6.8 Separation of a mixture of substituted benzenes as a function of solvent strength (%B). Sample: *1*, *p*-cresol; *2*, benzonitrile; *3*, 2-chloroaniline; *4*, 2-ethylaniline; *5*, 3,4-dichloroaniline; *6*, 2-nitrotoluene; *7*, 3-nitrotoluene; *8*, toluene; *9*, 3-nitro-*o*-xylene; *10*, 4-nitro-*m*-xylene. Conditions: 250 × 4.6-mm (5-μm) C_8 column; acetonitrile–pH-6.5 buffer mobile phase; 35°C; 1.0 mL/min. Chromatograms recreated from data of [26]. Note that all compounds are non-ionized under these condition, so the sample is effectively neutral.

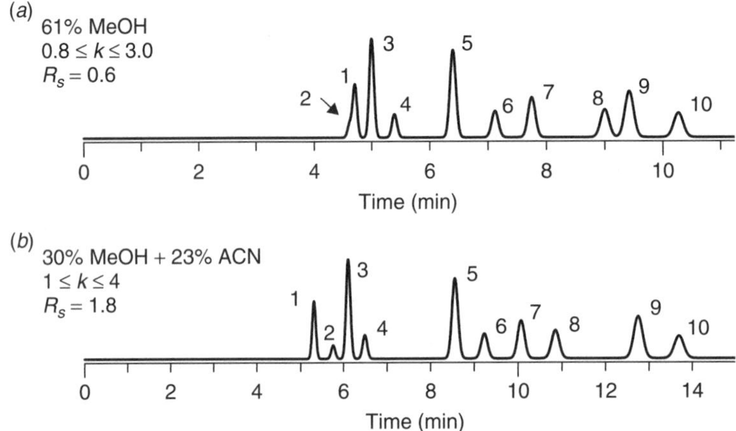

Figure 6.9 Solvent-type selectivity. Separation of a mixture of substituted benzenes with methanol or mixtures of methanol-acetonitrile as mobile phase. Same sample and conditions as in Figure 6.8, except as noted in figure. Chromatograms recreated from data of [26].

of peak 2 away from peak 3 and toward peak 1. This can be achieved by increasing the proportion of MeOH in the final mobile phase (with respect to the mobile phase of Fig. 6.10*b*). As seen in Figure 6.10*d*, a mixture of 57% Figure 6.10*c* (61% MeOH) and 43% Figure 6.10*a* (46% ACN) positions peak 2 midway between peaks 1 and 3 for maximum resolution ($R_s = 2.0$ vs. $R_s = 1.8$ in Fig. 6.10*b*). Achieving an optimum final separation in this example involves simple interpolation between

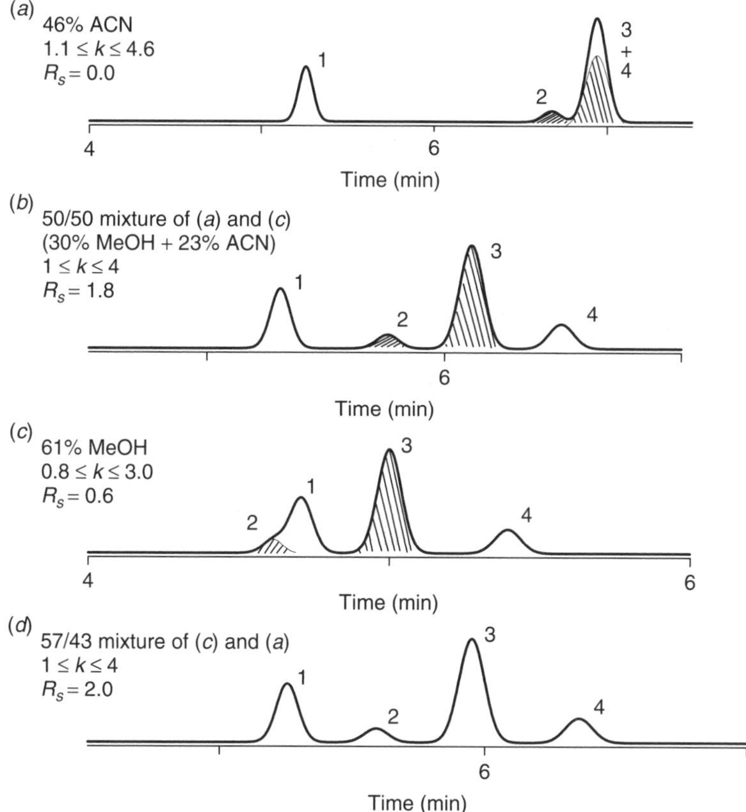

Figure 6.10 Solvent-type selectivity: fine-tuning the B-solvent. Same sample and conditions as in Figures 6.8 and 6.9 (peaks 1–4 only), plus added figure (*d*); (*b*) is 30% MeOH + 23% ACN, and (*d*) is 35% MeOH + 20% ACN. Chromatograms recreated from data of [26].

the two preceding experiments (Fig. 6.10*b*,*c*), much like the example of Figure 6.6 where %B was optimized. Similar optimizations of selectivity can be carried out more conveniently by the use of computer simulation (Chapter 10).

Although improvements in resolution may be explored by changing the B-solvent (solvent-type selectivity), the new mobile phase must have a similar solvent strength in order to maintain comparable values of k and run time. In Figure 6.9*a*, a higher %-MeOH was used (61% vs. 46% ACN) because methanol is a weaker (more retentive) B-solvent than acetonitrile. When changing solvent type, we can estimate the necessary change in %B for the new B-solvent by means of the *solvent nomograph* of Figure 6.11. Here similar %B values for different B-solvents fall on vertical lines. Recall that in the previous example we needed to replace 46% ACN with a similar strength MeOH-water mobile phase. From the diagram of Figure 6.11, we see that 46% ACN should be about equivalent in strength to 57% MeOH (each of these two mobile phases is marked by • in Fig. 6.11); this is close to the mobile phase actually selected in Figure 6.9*a* (61% MeOH). Note that the run time in Figure 6.9*a* (11 min) is somewhat shorter than that in Figure 6.8*a* (15 min),

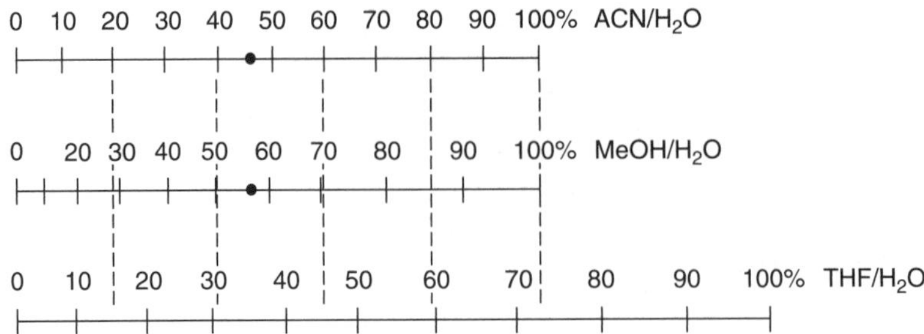

Figure 6.11 Solvent-strength nomograph for reversed-phase HPLC (adapted from [28]). Two mobile phases of equal strength (46% ACN and 57% MeOH) marked by •, as an example.

as could be expected from Figure 6.11; that is, 61% MeOH is stronger than the 57% MeOH recommended by Figure 6.11.

To summarize, when using a change in solvent type to change selectivity and increase resolution, *proceed as follows*: If ACN was used in the initial separation, and MeOH is to be substituted, Figure 6.11 shows how to estimate the best %-MeOH for use in the second separation. Next examine the ACN and MeOH chromatograms to determine if a mixture of ACN and MeOH is likely to give a better separation than either B-solvent alone (as in Fig. 6.9*b*). Conditions can be improved as illustrated by Figure 6.10 and discussed further in Section 6.4.1. If MeOH was used initially as the B-solvent, use Figure 6.11 to estimate the equivalent % ACN for a change in B-solvent without affecting the solvent strength.

Solvent-type selectivity arises from interactions among solute, B-solvent, and column, as described in Sections 2.3.2.1 and 5.4.1. The preferential retention of the B-solvent in the stationary phase means that a stronger interaction of the B-solvent and solute will normally lead to *increased* retention (this may at first seem counterintuitive, in terms of the discussion of Fig. 2.7*a*,*b*). However, even qualitative predictions of the effect of a given B-solvent on the relative retention of different solute classes (e.g., phenols, ethers) are difficult at best. For further discussion, see [23].

6.3.3 Temperature Selectivity

Retention as a function of temperature can usually be described by

$$\log k = A + \frac{B}{T_K} \tag{6.2}$$

where A and B are constants for a given compound when only the absolute temperature T_K (K) is varied. Plots of log k versus $1/T_K$ should therefore yield straight lines, as in Figure 6.12*a* for a number of polycyclic aromatic hydrocarbons (PAHs). An increase in column temperature by 1°C usually results in a decrease in values of k by 1–2% for each peak in the chromatogram [29]. The effect of a change in temperature on selectivity is usually minor for the RPC separation of neutral samples. However, exceptions have been noted, as in the case of PAHs [30] and plant pigments [31]. Figure 6.12*b* is an extension of Figure 6.12*a*, with data for three additional PAHs (dashed curves A–C); compounds 1 to 6 are flat, fused-ring

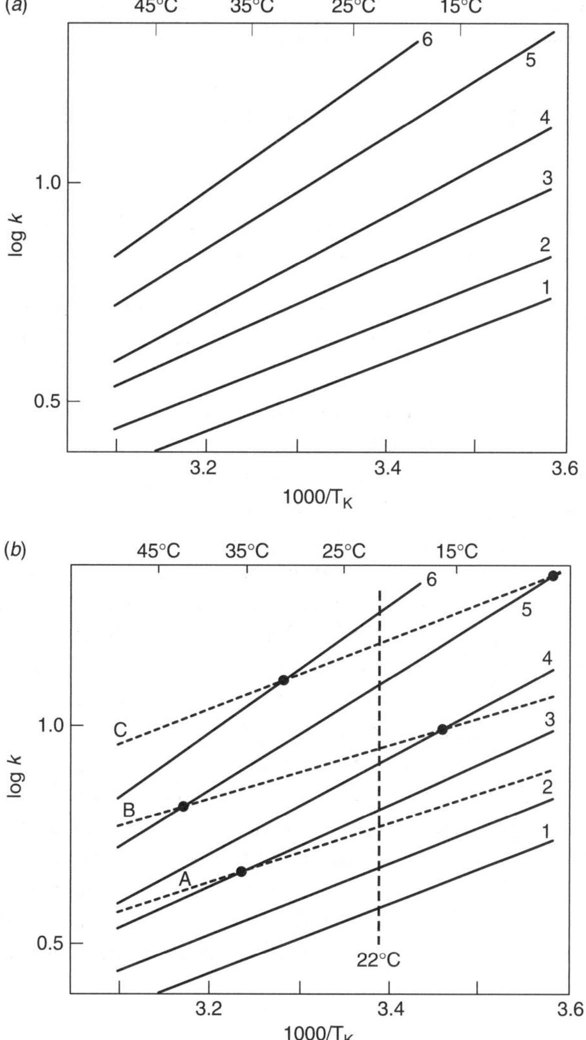

Figure 6.12 Retention of polycyclic aromatic hydrocarbons as a function of separation temperature. Conditions: 250 × 4.6-mm (5-μm) Chromegabond-C_{18} column; 80% ACN-water; 1.0 mL/min. (*a*) Sample (fused-ring aromatic hydrocarbons): *1*, anthracene; *2*, fluoranthene; *3*, triphenylene; *4*, chrysene; *5*, 3,4-benzofluoranthene; *6*, 1,2,5,6-dibenzoanthracene. (*b*) Sample same as (*a*), plus added poly-aryls: *A*, 1,1′,-dinaphthyl; *B*, 1,3,5-triphenylbenzene; *C*, 9,10-diphenylanthracene. Adapted from [30].

PAHs such as anthracene (peak 1), while compounds A to C are three-dimensional (nonflat) poly-aryls such as 1,1′,-dinaphthyl (peak A). It is seen that retention reversals (marked by •) occur in Figure 6.12*b* for five adjacent peak-pairs as temperature is varied between 15 and 45°C. (Note the °C scale at top of figure): B/5 at 42°C, A/3 at 36°C, C/6 at 31°C, B/4 at 16°C, and C/5 at 6°C. In this example there are major changes in selectivity as temperature is varied. As a result several temperatures exist between 10 and 70°C where baseline resolution ($R_s \geq 1.5$) is possible, for example, a temperature of 22°C, as shown in Figure 6.12*b* by a vertical dotted line.

Changes in resolution with temperature for this sample are better seen in the *resolution map* of Figure 6.13*a*: a plot of critical resolution as a function of temperature. The resolution map of Figure 6.13*a* is enhanced in Figure 6.13*b*, to provide additional insight into this very useful tool. Thus potentially critical peak-pairs (e.g., B/4, C/6, A/3) that correspond to different line segments in Figure 6.13*a* are identified in Figure 6.13*b*. Similarly arrows in Figure 6.13*b* indicate four temperatures where baseline resolution ($R_s \geq 1.5$) is possible. Resulting separations for the latter four preferred temperatures are shown in Figures 6.13*c*–*f* (note the changed retention order in each of these examples). The three poly-aryls A–C (shaded peaks) are seen to be retained more strongly at higher temperatures relative to the fused-ring PAHs 1–6 (see the further discussion of following Section 6.3.3.1). The best separation ($R_s = 2.1$) is provided by a temperature of 22°C. While changes in temperature sometimes can be effective in altering separation selectivity for neutral samples, changes in selectivity with temperature are much more likely for partly ionized acids or bases (Section 7.3.2.2). For a further discussion of temperature selectivity in RPC, see [32, 32a].

6.3.3.1 Further Observations

The following treatment provides a more complete picture of temperature selectivity for neutral samples. However, the reader may prefer to skip this discussion and continue with Section 6.3.4.

While Equation (6.2) is generally an adequate representation of RPC retention as a function of temperature, more complex changes in k with temperature are sometimes observed [7, 33–35]. Sometimes k *increases* when the temperature is increased, or plots of log k versus $1/T$ may be curved instead of linear. These exceptions to Equation (6.2) are often associated with acidic or basic solutes and a mobile phase pH that is close to the pK_a value of the solute (see Section 7.2). In such cases a change in temperature may result in a change in the relative ionization of an acid or base (Section 7.2.3), with a large resulting change in solute retention—in addition to (and sometimes opposing) the normal effect of temperature that is described by Equation (6.2).

The dependence of log k on temperature is determined by the value of B in Equation (6.2), which is proportional to the enthalpy of retention, ΔH. Values of B usually increase for larger values of k; as a result plots of log k versus $1/T$ for different solutes (other conditions the same) often resemble the fan-shaped plots of Figure 6.12*a* for several fused-ring PAHs. The latter behavior can be compared with the similar example of Figure 6.7*a* for regular samples as %B is varied; in both cases (change in %B or T) a change in condition has the largest effect on k for the most retained solute (with the largest values of k). For most neutral samples, changes in retention order with temperature are not expected (as observed in Fig. 6.12*a*), but maximum resolution may still be observed at an intermediate temperature. However, a change in temperature is generally less useful for improving the resolution of neutral samples (similar to the case of "regular" samples and solvent-strength selectivity).

As noted in Figure 6.12*b*, values of *B* tend to be relatively smaller for more compact molecules such as phenylnaphthalene (a poly-aryl), compared to less compact molecules such as anthracene (a fused-ring PAH). Similar temperature-selectivity effects have been observed for gas chromatography [36], where values of *B* in Equation (6.2) decrease in the order *n*-alkynes > *n*-alkanes > branched alkanes > cycloalkanes. That is, values of *B* in GC also decrease for less extended, more compact molecule—possibly for similar reasons.

6.3.4 Column Selectivity

During the early days of HPLC, a change of column was often used as a means of varying selectivity and improving resolution. Indeed column selectivity represents a powerful means for altering relative retention and improving the separation of neutral samples. However, the use of column selectivity alone for the purpose of systematically improving separation has a serious limitation, compared to changes in %B, solvent type, or temperature. When any of the latter conditions are varied,

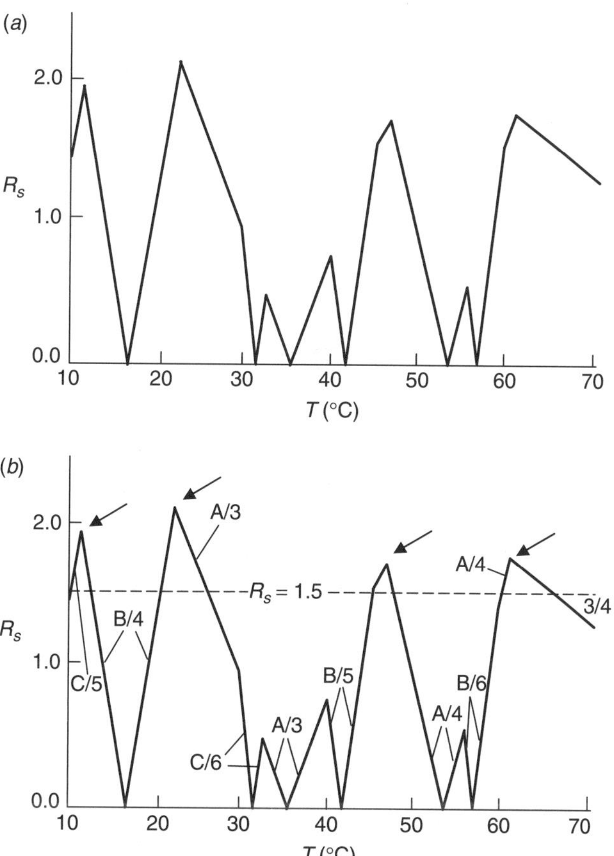

Figure 6.13 Separation of polycyclic aromatic hydrocarbons as a function of temperature. Sample and conditions of Figure 6.12. (*a*, *b*) resolution map; (*c*–*f*) chromatograms for different optimum temperatures. Adapted from [30].

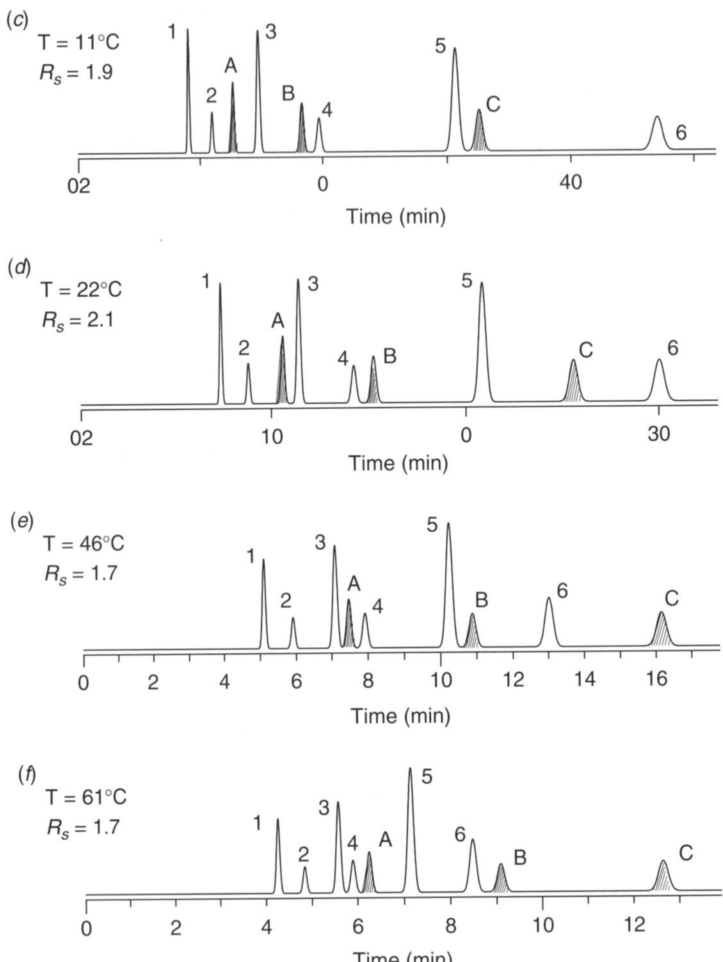

Figure 6.13 (*Continued*)

continuous changes in selectivity are possible, and there will often be an *intermediate* condition that corresponds to maximum resolution. In the examples of Figure 6.13, this greatly increased the likelihood of an acceptable separation, compared to the use of only one or two arbitrary temperatures. When the column is changed, the selection of an intermediate separation (between that provided by either column) is inconvenient. As a result a change in just the column is less likely to result in an improved separation (but see the further discussion of Section 6.4.1.4). Changes in the column are also less convenient than changes in continuous variables.

Examples of a change in selectivity when only the column is changed are shown in Figure 6.14 for a 10-component neutral sample and four columns of varying selectivity. Despite significant changes in selectivity for this sample when the column is changed, no single column provides a resolution of $R_s > 0.8$, as long as other conditions remain fixed (as in this example). If two different columns are connected together, additional separation possibilities are created, but this procedure

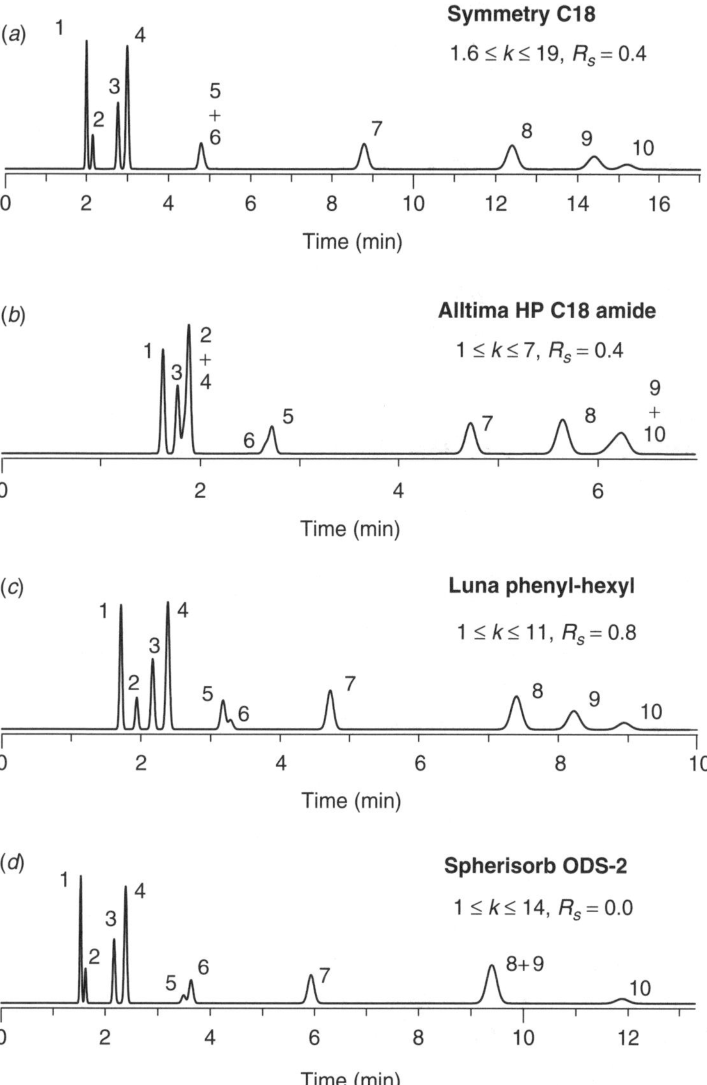

Figure 6.14 Separation of a mixture of 10 organic compounds of diverse structure on four different columns. Sample: *1*, 4-nitrophenol; *2*, 5,5-diphenylhydantoin; *3*, acetophenone; *4*, benzonitrile; *5*, 5-phenylpentanol; *6*, anisole; *7*, toluene; *8*, *cis*-chalcone; *9*, ethylbenzene; *10*, *trans*-chalcone. Conditions: 150 × 4.6-mm (5-μm) columns; 45% acetonitrile-water; 35°C; 2.0 mL/min. Chromatograms recreated from data of [8, 9].

is inconvenient and only marginally better than the use of a single column (in the example of Fig. 6.14, no combination of these four columns can improve the separation of Fig 6.14*c*). The real advantage of a change in column selectivity is only achieved when a change in column is accompanied by a simultaneous change in one or more other conditions, for example, %B, solvent type, and/or temperature for the case of neutral samples (Section 6.4.1.4).

We presently have a more complete understanding of the basis of column selectivity than for other kinds of selectivity (solvent strength, solvent type, temperature, etc.). As discussed in Section 5.4.1, column selectivity can be quantitatively defined by five different characteristics:

- column hydrophobicity ***H***
- column steric resistance S^*
- column hydrogen-bond acidity ***A***
- column hydrogen-bond basicity ***B***
- column cation-exchange capacity **C**

For the case of neutral samples, only the first four column parameters (***H, S*, A, B***) are important. The difference in selectivity of two columns for a neutral sample can be characterized by a selectivity function $F_s(-C)$ that is defined by Equation (6.3) below (Section 6.3.6.1; see also Section 5.4.3). For the columns of Figure 6.14, compared to the Symmetry column, values of $F_s(-C)$ are equal to 26 for the Alltima column, 12 for the Luna column, and 16 for the Spherisorb column. A value of $F_s(-C) > 5$ can lead to potentially significant changes in relative retention, as observed for this example. Larger changes in relative retention as a result of a change in column are observed for ionic samples because the column parameter C has a much larger effect on column selectivity for such samples.

6.3.5 Isomer Separations

Compounds that resemble each other structurally tend to have similar values of k and are often more difficult to separate. The RPC separation of isomeric compounds is therefore commonly regarded as challenging. While this may be somewhat true in general, one review of the RPC separation of 137 different isomer-pairs with alkylsilica columns [37] found that 90% of these samples could be separated with $R_s > 1.0$ by simply optimizing %B (k varied up to by 4-fold) and temperature (T varied by as much as 25°C). An example is shown in Figure 6.15 for the separation of three hydroxytestosterone isomers on a C_{18} column as a function of temperature. At 40°, all three isomers co-elute; at 28°C, the three isomers are separated with $R_s = 0.8$. Lower temperatures generally favor a better separation of isomers.

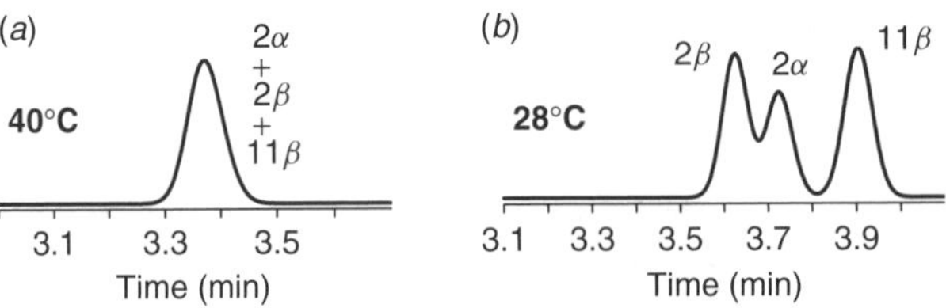

Figure 6.15 Separation of three hydroxytestosterone isomers (2α, 2β, 11β) as a function of temperature. Conditions: 250 × 4.6-mm (5-μm) C_{18} column; 40% acetonitrile-water; 2 mL/min. Adapted from [38].

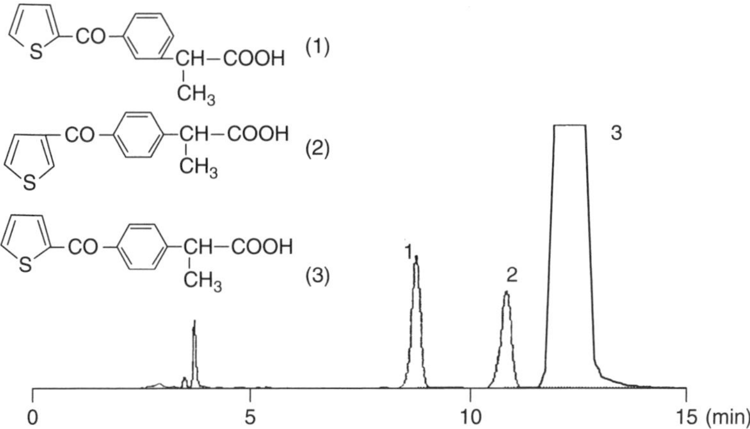

Figure 6.16 Separation of isomers with a cyclodextrin-bonded column. Conditions: 250 × 4.6-mm (5-μm) Cyclobond I column; 30% acetonitrile–pH-4.5 buffer; 35°C; 2.0 mL/min. Adapted from [40].

6.3.5.1 Enhanced Isomer Selectivity

While it is possible to achieve the baseline separation of some isomers by RPC with alkylsilica columns, the use of a cyclodextrin column may be a better choice [39–41], as illustrated by the example of Figure 6.16. The enhanced isomer selectivity provided by cyclodextrin columns may be due to the presence of –OH groups in the cyclodextrin molecule (allowing for hydrogen bonding between solute and column). The greater rigidity of the three-dimensional cyclodextrin molecule, in contrast to flexible alkylsilica ligands, may also provide a more demanding steric fit for one isomer over another. However, even better isomer selectivity is generally possible by the use of normal-phase chromatography with a silica column (Section 8.3.5).

Silver-ion complexation of olefins has been found to be a useful means for enhancing the RPC separation of *cis*- from *trans*-olefin isomers. The addition of Ag^+ to the mobile phase results in a preferential interaction with *cis* olefins. The resulting Ag^+-olefin complex is more polar and hence prefers the mobile phase (Section 2.3.2), resulting in its decreased retention compared to the *trans* isomer. Compounds with varying degrees of *cis*-unsaturation can be separated according to the number of *cis* double bonds in the molecule. Silver-ion complexation has been used for the separation of unsaturated fatty acids or their derivatives [42], but is also able to create changes in selectivity for nitrogen- or sulfur-containing heterocyclic compounds [43]. Today silver-ion complexation for the enhanced separation of *cis*- from *trans*-olefins is in most cases carried out with silver-impregnated ion-exchange columns [42].

6.3.5.2 Shape Selectivity

This special form of column selectivity was discussed in Section 5.4.1.1. Although it has received considerable attention in the literature, shape selectivity has limited practical application. The reader may therefore wish to skip to following Section 6.3.6.

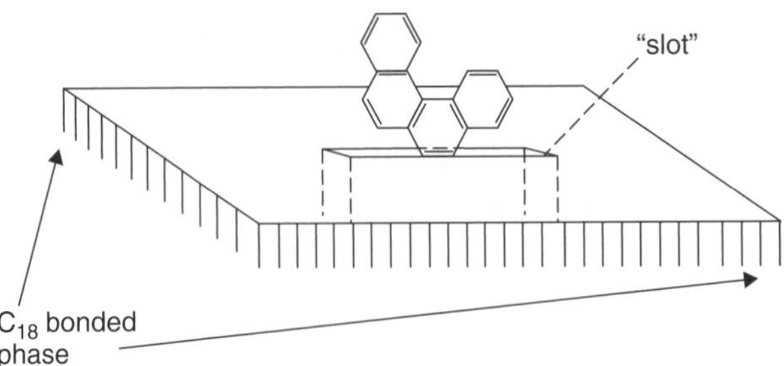

Figure 6.17 "Slot" model of "shape selectivity."

Shape selectivity refers to the preferential retention of planar polycyclic aromatic hydrocarbons (PAHs) and certain other compounds (carotenes, steroids) on *polymeric* alkylsilica columns [44–46]. As a result excellent separations of these isomeric solutes are often observed with polymeric columns. Two solute characteristics favor the increased retention of a PAH isomer on a polymeric column: molecular planarity (or reduced molecular thickness), and a larger ratio of molecular length to breadth (L/B). It is assumed that polymeric stationary phases are only accessible via narrow openings or "slots" so that planar and/or narrower molecules are better able to enter the stationary phase, and are therefore better retained.

This "slot" model of retention on polymeric columns is illustrated in Figure 6.17, where a nonplanar PAH is shown being inserted into a narrow "slot" in the stationary phase of a polymeric column. Because of the greater "thickness" of this nonplanar molecule, it will not be able to completely enter the narrow slot. A more planar, less thick PAH would be better able to access the stationary phase—and therefore be retained more strongly. For separation on polymeric columns (which have a higher coverage of alkyl ligands), it is found that isomeric PAHs are much better resolved than is the case for monomeric columns [45]. This was illustrated in Chapter 5 by the separation of a mixture of 12 C_{22} PAH isomers on a polymeric column (Fig. 5.23*c*), compared to separation on a monomeric column (Fig. 5.23*d*). However, monomeric columns generally perform better than polymeric columns in other respects (Section 5.2.3) and are much more widely used. Because only a few classes of compounds are better separated on polymeric columns, shape selectivity is rarely tried as a means of improving separation during method development.

6.3.6 Other Selectivity Considerations

So far our discussion of separation selectivity has emphasized its use during method development—the systematic variation of conditions to maximize resolution and/or shorten run time. Once a satisfactory separation has been achieved in this way, certain adverse possibilities require consideration. Thus after a routine method has been developed, it is likely to be used over an extended period and in different laboratories. Columns degrade during routine use and must therefore be replaced from time to time. As discussed in Section 5.4.2, column selectivity may vary slightly from one manufacturing batch to another, in which case a replacement column of the same kind from the same source may no longer provide an acceptable separation. This is more likely for separations with > 10 peaks, especially when combined with marginal resolution ($R_s < 2$). When a replacement column exhibits changed selectivity and fails to provide adequate resolution, it is necessary to locate a different column that is equivalent to the original column, or to vary separation conditions so as to minimize the effect of a change in column selectivity (Section 6.3.6.1). An equivalent replacement column may also be required if the original column has been discontinued and is no longer available, or if the original column is unavailable at another site where the separation is to be carried out.

Another problem unrelated to column variability is the possibility that a sample component may be overlapped by another peak and therefore missed in the final RPC analysis. The discovery of such hidden peaks can be aided by the use of *orthogonal* separation (Section 6.3.6.2).

6.3.6.1 Equivalent Separation

Two approaches can be used to restore an original separation when column selectivity changes: (1) a change in column source (i.e., part number), or (2) a change in separation conditions ("method adjustment"). Which procedure is applicable may depend on the recommendations of a regulatory body (Section 12.8) when an HPLC method falls under governmental regulation.

A *change in column source* requires the identification of an alternative, "equivalent" column. Most RPC columns can be characterized by five selectivity parameters (**H**, **S***, **A**, **B**, and **C**), as discussed above and in Section 5.4.1. Equivalent columns will have similar values of each of these column parameters. A column comparison function F_s has been defined for two columns *1* and *2* (Eq. 5.4) and any sample (ionic as well as neutral). For the case of neutral samples, the column parameter **C** is not relevant, so Equation (5.4) simplifies to [47]

$$\text{neutral samples only: } F_s(-\mathbf{C}) = \{[12.5(\mathbf{H}_2 - \mathbf{H}_1)]^2 + [100(\mathbf{S^*}_2 - \mathbf{S^*}_1)]^2 + [30(\mathbf{A}_2 - \mathbf{A}_1)]^2 + [143(\mathbf{B}_2 - \mathbf{B}_1)]^2\}^{0.5} \quad (6.3)$$

where subscripts 1 or 2 refer to values of a specific parameter (**H**, **S***, etc.) for each column. For two columns that are exactly equivalent, F_s and $F_s(-\mathbf{C}) = 0$. When F_s or $F_s(-\mathbf{C}) \leq 3$ for two columns, it is highly likely that they will provide equivalent separation (i.e., critical R_s values that differ by < 0.5 units).

An example of the application of Equation (6.3) for a gradient method and an ionic sample is shown in Figure 6.18. An original RPC procedure was developed, giving the separation shown in Figure 6.18*a* for a drug product and 10 impurities (total of 11 peaks, each marked by an *). Two possible replacement columns were selected for which $F_s \leq 3$ (vs. the column of Fig. 6.18*a*), giving the separations shown in Figures 6.18*b* and *c*. These three separations (*a*–*c*) are seen to be essentially equivalent, with very similar resolution for each compound. A fourth column (with $F_s = 10$) gave the result of Figure 6.18*d*, where the last two peaks are seen to overlap (arrow). Consequently this last column is not equivalent to the starting column of Figure 6.18*a*, as suggested by its larger value of F_s.

Method adjustment aims at correcting for differences in column selectivity, when locating an equivalent column proves to be impractical or impossible. Instead of changing the column, other separation conditions are varied as a means of counteracting the change in column selectivity. An example of this approach is shown

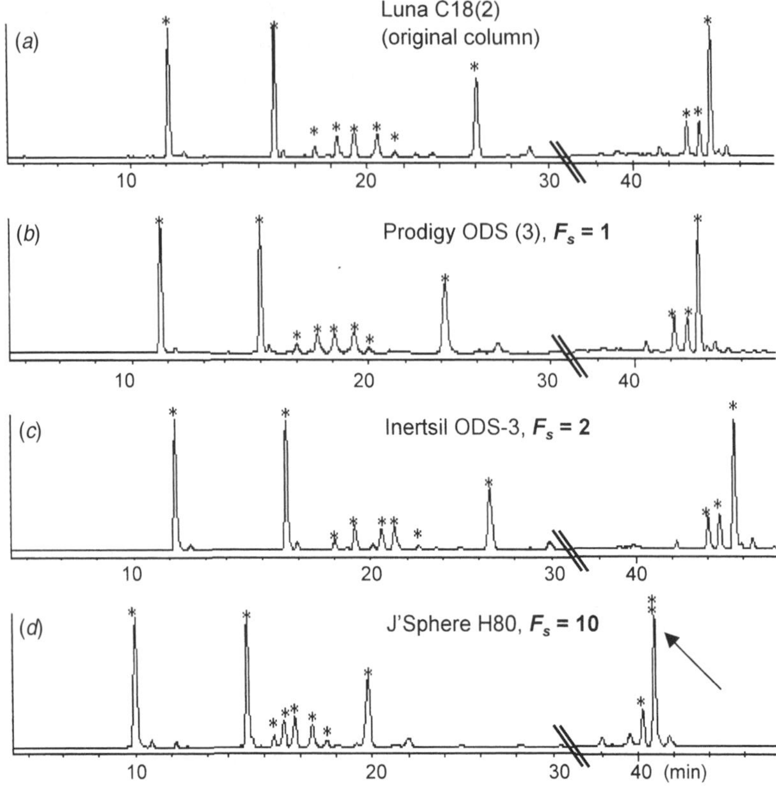

Figure 6.18 Example of the use of values of F_s to select columns of similar selectivity for possible replacement in a routine HPLC assay. Gradient separations where only the column is changed for the separations of *a–d*. Asterisks mark peaks of interest, values of F_s calculated from Equation (5.5) (ionic [not neutral] sample). Reproduced with permission from [48].

in Figure 6.19. The original separation with column *A* is shown in Figure 6.19*a*, with baseline resolution ($R_s = 1.7$). At a later time a new column *B* (same type, different production lot) was used and the separation of Figure 6.19*b* resulted; the observed resolution is unacceptable, due to the increased overlap of peak-pairs 2 and 3 ($R_s = 1.3$) and 6/7 ($R_s = 1.2$)—the result of a change in column selectivity. When carrying out method adjustment, the first step is to determine how separation changes when one or more experimental conditions are varied. For neutral samples, possible choices in conditions include %B and temperature (preferred), or the variation of the proportions of two organic solvents that together comprise the B-solvent (e.g., mixtures of ACN and MeOH).

The effect of a change in temperature or %B for column *B* is shown in Figure 6.19*c*,*d*. We see that an increase in temperature (Fig. 6.19*c*) reduces the resolution of peak-pair 2 and 3 (a change in R_s of −0.8 units) but has an opposite effect on peak-pair 6 and 7 (a change in R_s of +0.2 units). A decrease in temperature could therefore restore baseline resolution for peaks 2 and 3 but would further *decrease* the resolution of peaks 6 and 7 (which are already poorly resolved).

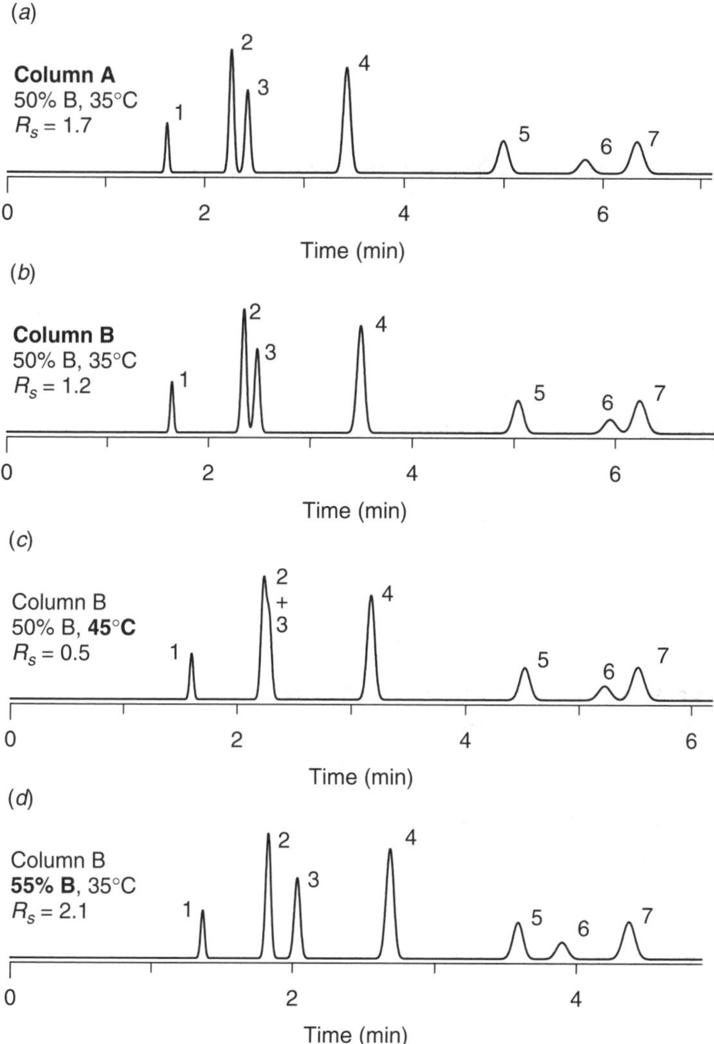

Figure 6.19 Example of method adjustment for a seven-component mixture of neutral compounds. Sample: *1*, oxazepam; *2*, flunitrazepam; *3*, nitrobenzene; *4*, 4-nitrotoluene; *5*, benzophenone; *6*, *cis*-4-nitrochalcone; *7*, naphthalene. Conditions: 150 × 4.6-mm C_{18} column (*B* differs from *A* only in a 10% lower ligand coverage); 2.0 mL/min; acetonitrile-water mobile phases; other conditions shown in figure. Separations *a–d* recreated from data of [8, 9].

An increase in %B, (Fig. 6.19*d*), on the other hand, results in improved resolution for each peak-pair, and acceptable resolution ($R_s = 2.1$).

The approach of Figure 6.19 is straightforward for this sample, in that a change in only one condition (%B) was able to correct for the change in selectivity of column *B*. Often this is not the case, and then a simultaneous change in two (or more) conditions may be necessary to adjust for a change in column selectivity. While a selection of two or more new conditions can be made by trial and error, a more efficient approach has been described [50]. Retention data are required for four experimental runs, as in Figure 6.19*a–d*, following which a simple mathematic procedure can predict conditions for the closest possible match to the original separation of Figure 6.19*a*, using the replacement column of Figure 6.19*b*. For allowed changes in conditions during method adjustment, see Section 12.8.

The problem of restoring an equivalent separation (when column selectivity changes between different production lots) is best anticipated during method development, rather than addressed after the problem arises. Three different options exist during method development:

1. check the reproducibility of different production lots of the column used for the final method
2. confirm the identity of one or more equivalent replacement columns
3. carry out method adjustment for one or more nonequivalent replacement columns

Option 1 should be our first choice. Several different lots of the selected column can be evaluated for equivalent separation. Usually it will be found that all of the column lots tested provide adequate separation. If this is not the case, it may be necessary to replace the original column with a different (more reproducible) column.

Option 2 is a useful supplement to option 1, even when option 1 confirms that different production lots of the original column are equivalent (later production lots may *not* be equivalent!). By identifying one or more equivalent (replacement) columns during method development (e.g., Fig. 6.18*b*,*c*), any of these columns can serve as a replacement in the event that future lots of the original column exhibit changed selectivity and are no longer suitable [48].

Option 3 can be used whenever option 2 fails (no alternative columns are equivalent to the original column). If the required changes in conditions are minor, the use of an alternative column with method adjustment may be considered as an equivalent method, *not* requiring complete re-validation of the method (Section 12.8). Another version of method adjustment that avoids the use of *changed* conditions for a replacement column is to select separation conditions *during* method development that provide equivalent separations for both the original and one or more nonequivalent replacement columns [51].

6.3.6.2 Orthogonal Separation

The problem of missing or "hidden" peaks can arise during method development when two compounds remain unseparated despite changes in separation conditions. If this situation is overlooked, the final method will be unacceptable because of the missing peak. Even when all the compounds of interest have been separated during method development, later samples may contain additional (unexpected) components that might be missed if overlapped by another peak. Missing peaks are most likely when the overlapped peak is small compared to other peaks in the sample (e.g., a sample impurity), or when the number of possible compounds in a sample is initially unknown.

The problem of missing peaks can be addressed by the use of selective detection (mainly mass spectrometry) and/or the development of an *orthogonal* separation: a separation with very different selectivity that is therefore likely to separate two peaks that were overlapped in the original (primary) method. For method development where mass spectrometric detection (LC-MS) is often employed, several sets of orthogonal conditions have been proposed that employ changes in the column, B-solvents, and/or mobile phase pH [52–53]. Two or more of these procedures can be used with a given sample to minimize the possibility that two solutes will be overlapped and therefore missed during method development. The specificity of LC-MS, combined with very different separation selectivity, makes it highly unlikely that any sample component will be overlooked.

Once method development is completed for a given (hopefully representative) sample, there is always the possibility that future samples may be found to contain additional compounds not present in the original sample. These extra compounds might arise because of changes in a manufacturing process, contamination of the sample during processing or handling, or for other reasons. Routine HPLC assay procedures usually involve UV rather than MS detection; therefore an orthogonal procedure based on UV detection needs to be available for laboratories where the routine assay is performed. A further requirement is the complete separation of known sample components by the orthogonal method.

A general procedure for the development of an orthogonal separation that meets the requirements above for routine use has been described and proved useful in several different laboratories [49]. Starting with the primary method used for routine application, a column of very different selectivity is selected, based on a large value of $F_s(-C)$ (Eq. 6.3) for the orthogonal versus primary columns. In addition to a change in the column, the B-solvent is changed; if ACN was used for the primary method, MeOH is used for the orthogonal method, and vice versa. Because the compounds separated by the routine procedure may not be fully resolved at this point, separation temperature and %B are next optimized for the maximum resolution of the sample (Section 6.4.1.2). Finally, the orthogonality of the latter method is evaluated from a plot of log *k* for the orthogonal method against log *k* for the routine method. If insufficient orthogonality has been achieved at this point, further changes in separation conditions can be explored (use of a different column, change in pH, etc.). When an ionic sample is involved (Chapter 7), a large value is sought of F_s (Eq. 5.4) rather than $F_s(-C)$. The latter general procedure, which can be extended to gradient elution and used for either neutral or ionic samples, can be further improved [47, 54] for greater orthogonality of the two separations.

An example of an orthogonal method that was developed as above is shown in Figure 6.20 for a primary method based on gradient elution. In Figure 6.20*a* the separation of a sample by the routine method is shown. A major component 3 and four impurities 1, 2, 4, and 5 are separated to baseline. When the routine method was first developed, it was established that only these five components were present in samples manufactured at that time. When the orthogonal method of Figure 6.20*b* was developed at a later time and applied to a new batch of this active ingredient, another sample impurity 6 was discovered that is overlapped in the routine method by the major component 3 (and therefore would be missed in the primary method of Fig. 6.20*a*).

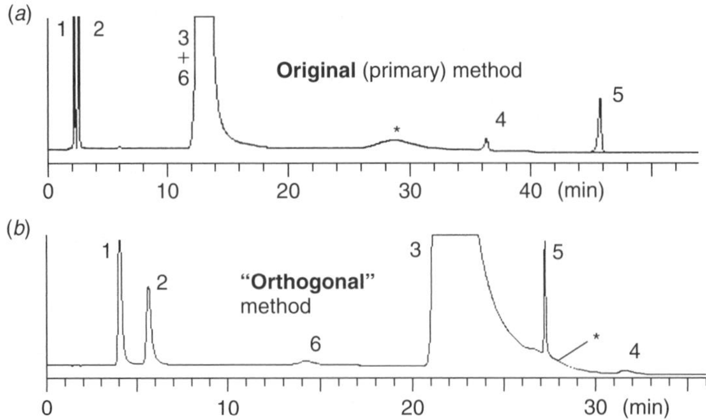

Figure 6.20 Comparison of separation by an original versus "orthogonal" method. Gradient separations where the column and organic solvent are changed (mobile-phase pH = 6.5 for both *a* and *b*). Asterisks mark gradient artifacts (not solute peaks). Reproduced with permission from [49].

6.4 METHOD DEVELOPMENT AND STRATEGIES FOR OPTIMIZING SELECTIVITY

A general approach to method development was presented in Section 2.5.4 and is summarized in Figure 6.21. Seven method-development steps are listed in Figure 6.21*a*. Step 1 consists of a review of initial information about the sample and a preliminary set of goals for the final separation (required resolution, e.g., $R_s \geq 2$; desired separation time, e.g., $\leq$10 min; etc.). A specialized approach to method development may be indicated for some samples (large biochemical or polymeric molecules, Chapter 13; enantiomeric isomers, Chapter 14; inorganic ions, not discussed in present book), or if preparative separation is intended (Chapter 15). Step 2 (sample pretreatment) is required for some samples, those that cannot be injected without damaging the column or that contain interfering substances that are likely to overlap peaks of interest. Step 3 involves an initial choice of chromatographic mode; RPC will be selected in most cases, but this decision can be modified after initial experiments where separation conditions are varied (step 3 of Fig. 6.21*b*). The choice of detector (usually UV and/or MS) is made in step 4. The selection of separation conditions (step 5) is usually considered the main part of method development, is detailed in Figure 6.21*b*, and discussed further below. Step 6 deals with some common problems that can arise after a method is developed, when it is used for routine application. Step 7, which deals with method validation and system suitability, represents another critical part of method development.

The selection of chromatographic conditions is examined further in Figure 6.21*b*. Based on what is known about the sample, initial conditions are selected for the separation (e.g., as in Table 6.1) and an initial run is carried out (step 1 of Fig. 6.21*b*). For an initial isocratic separation, %B is then varied for adequate retention: $1 \leq k \leq 10$ (step 1a). Alternatively, a single gradient separation can be used (step 1b; see Section 9.3.1). The initial separation(s) may exhibit various problems (step 2): tailing peaks, poor retention of the sample (even for an aqueous mobile phase), excessive retention of the sample (even for 100% B), or too many peaks (a "complex" sample). Whatever problem might be encountered in the initial separation(s), the problem should be resolved before proceeding further (see Section 6.6). In some cases a change in separation conditions or chromatographic mode may be indicated (step 3); in this case, return to step 1 and start over.

Before beginning experiments for the optimization of selectivity (step 5), the presence in the sample of acidic, basic, and/or neutral compounds should be confirmed (step 4). When the composition of the sample is known before method development starts, step 4 can be omitted. For other samples, individual peaks can be identified as acids, bases, or neutrals by a 0.5-unit change in mobile-phase pH (see Section 7.2 and examples of Figs. 7.2 and 7.3). The all-important selection of conditions for optimized selectivity (step 5) will vary for different samples and chromatographic modes; this topic is addressed below and in individual chapters that focus on sample type and/or chromatographic mode (Chapters 6–9, 13, and 14). Finally, when the optimization of selectivity is completed, column conditions

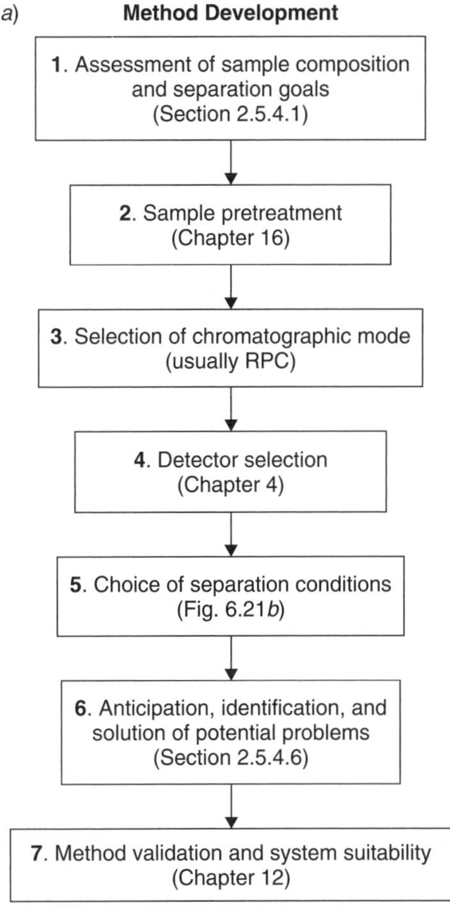
(a)
Method Development
1. Assessment of sample composition and separation goals (Section 2.5.4.1)
2. Sample pretreatment (Chapter 16)
3. Selection of chromatographic mode (usually RPC)
4. Detector selection (Chapter 4)
5. Choice of separation conditions (Fig. 6.21b)
6. Anticipation, identification, and solution of potential problems (Section 2.5.4.6)
7. Method validation and system suitability (Chapter 12)

Figure 6.21 General method-development approach for use in this and following chapters.

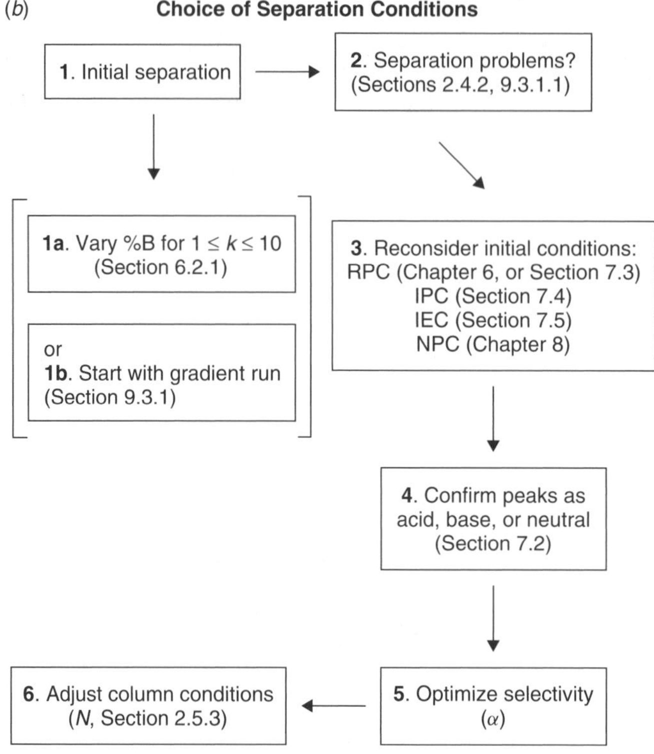

Figure 6.21 (*Continued*)

can be varied (step 6) for the purpose of either increasing resolution or reducing run time.

Samples that are relatively easy to separate may require no more than the selection of a final value of %B, which involves only a few experiments in which %B is varied. If gradient elution is used during initial method development experiments, only a single experiment is needed in order to select a value of %B for $1 \leq k \leq 10$ (Section 9.3.1). Many samples will require a further improvement of separation selectivity (preceding Section 6.3); for some samples this may involve the simultaneous change of two or more separation conditions. Various procedures for such multiple-variable optimization will be described next.

6.4.1 Multiple-Variable Optimization

Multiple-variable optimization in each case relies on an *experimental design*: a plan for the required experiments, as illustrated in Figure 6.22 for certain combinations of conditions that affect selectivity for neutral samples. In each case it is assumed that %B has been varied initially, so as to define a range in %B that provides adequate retention of the sample, for example, 40–50%B, so that $0.5 \leq k \leq 20$ for every peak (when varying %B for a change in selectivity, a wider k-range than the usual $1 \leq k \leq 10$ is recommended). By way of illustration, first consider Figure 6.22*a*. Experiments 1 and 2 are carried out first (i.e., a change in %B only). These two

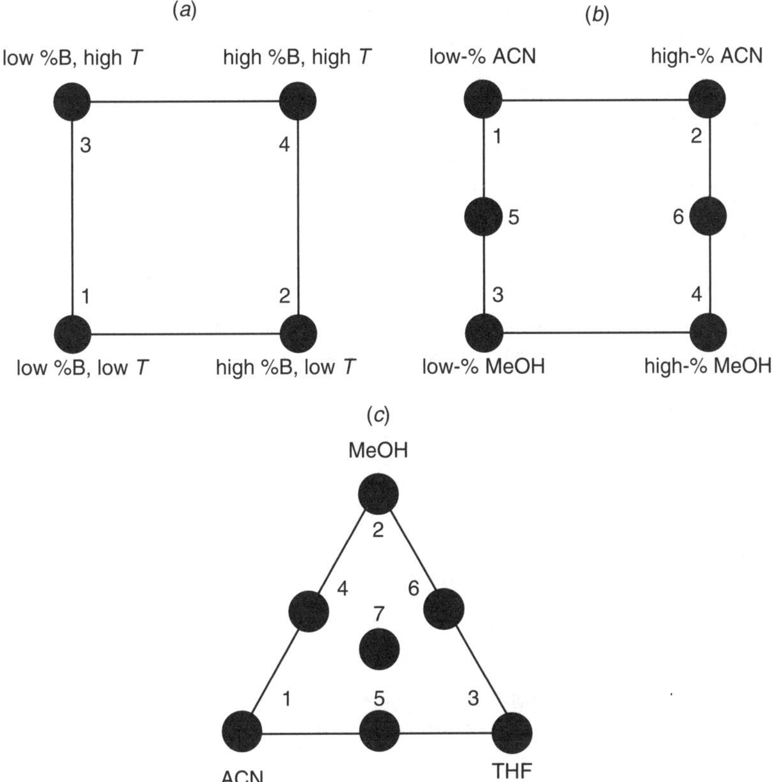

Figure 6.22 Experimental designs for the simultaneous optimization of various separation conditions for optimum selectivity. (*a*) Solvent strength (%B) and temperature (*T*); (*b*) solvent strength and solvent type (MeOH and ACN); (c) solvent type (MeOH, ACN, and THF).

experiments may suggest successful separation for some final value of %B, where both $0.5 \leq k \leq 20$ and resolution is adequate. If acceptable separation cannot be attained in this way, experiments 3 and 4 are carried out next (repeat experiments 1 and 2 at a higher temperature *T*). These four experiments can be interpolated (or extrapolated) to estimate values of *T* and %B that provide optimum selectivity and maximum resolution. A final experiment with these promising conditions is then carried out to confirm the predicted separation.

Experimental-design experiments, because of the simultaneous variation of two (or occasionally more) different conditions, can be difficult to interpret (especially for samples that contain a large number *n* of components; e.g., for $n > 10$). For this reason experimental design is often used in combination with computer simulation (Chapter 10). A reliable peak-tracking procedure will also be necessary (Section 2.7.4).

6.4.1.1 Mixtures of Different Organic Solvents

Two organic solvents (B-solvents) have been noted previously as especially suitable for RPC: acetonitrile (ACN) and methanol (MeOH). Tetrahydrofuran (THF) is

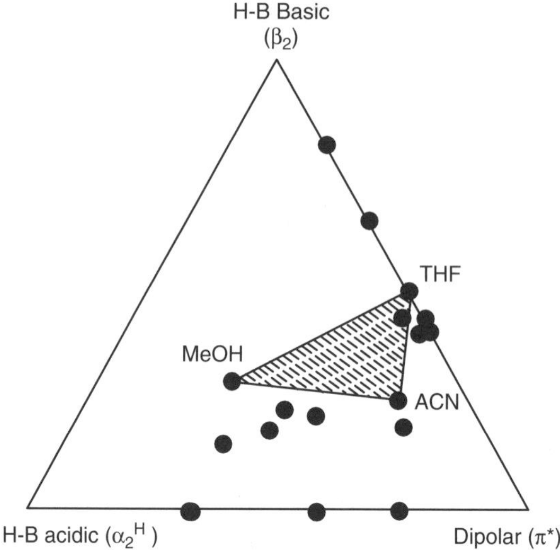

Figure 6.23 Preferred solvents for maximum change in solvent-type selectivity.

used less often because of its higher UV cutoff, susceptibility to oxidation, slower column equilibration when changing the mobile phase (e.g., from THF/water to ACN/water), and incompatibility with PEEK tubing (Section 3.4.1.2). However, the oxidation of THF during use can be minimized by the addition of water [55], and many samples do not require detection below 230 nm, thus making the use of THF practical in some cases. Appendix I contains additional information on the properties of these three B-solvents. *The remainder of Section* 6.4.1.1 *describes a procedure that is no longer commonly used; the reader may therefore wish to skip this discussion.*

The solvent-selectivity triangle of Figure 2.9 has been used to compare the solvent-type selectivity of different B-solvents. As suggested by Figure 6.23, significant differences in solvent selectivity exist among different mixtures of ACN, MeOH, and THF with water [56]. These three B-solvents can be used to optimize solvent-type selectivity by varying the proportions of each solvent in the mobile phase. Initial experiments with ACN as B-solvent will have identified a value of %B such that $0.5 \leq k \leq 20$ for the sample. Corresponding values of %-MeOH and %-THF (for equal solvent strength or a similar range in k) can then be obtained from the nomograph of Figure 6.11. The resulting three binary-solvent mobile phases can be blended next in various proportions, as illustrated in Figure 6.22*c*. An example of the application of the experimental design of Figure 6.22*c* is shown in Figure 6.24 for the separation of a 9-component mixture of substituted naphthalenes. In this example, run 1 is 52% ACN/water; run 2 is 63% MeOH/water; and run 3 is 39% THF/water. While selectivity varies among runs 1 to 3 of Figure 6.24, two or more peaks are poorly resolved in each separation.

The next step is to carry out further experiments in which the mobile phase is varied by blending equal portions of mobile phases 1 to 3. Thus a 1:1 (by volume) blend of mobile phases 1 and 2 results in mobile phase 4: a 26/32/42% mixture of ACN, MeOH, and water. Similarly mobile phases 5 to 7 are prepared as follows:

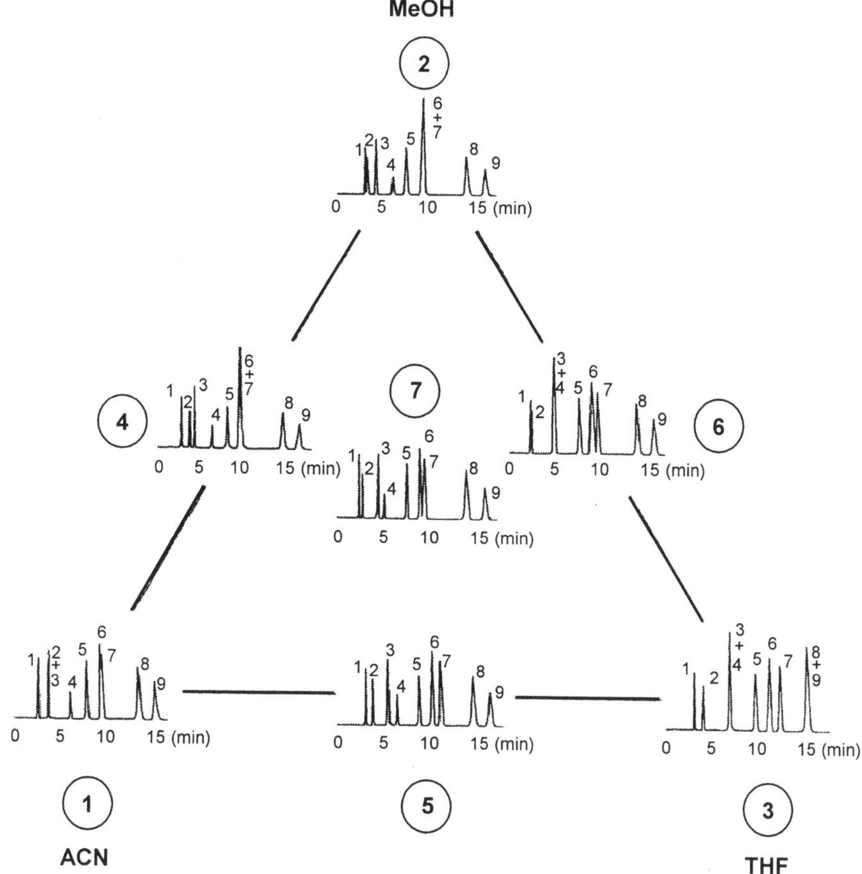

Figure 6.24 Use of seven solvent-type-selectivity experiments for the separation of a mixture of nine substituted naphthalenes. Sample substituents are: *1*, 1-$NHCOCH_3$; *2*, 2-SO_2CH_3; *3*, 2-OH; *4*, 1-$COCH_3$; *5*,1-NO_2; 6, 2-OCH_3; 7, -H (naphthalene); *8*, 1-SCH_3; *9*, 1-Cl. Conditions: 150 × 4.6-mm C_8 column; 40°C; 2.0 mL/min. Mobile phases (circled): *1*, exchange: 1, ACN; 2, MeOH; *2* exchange: 1, ACN; 2, MeOH; *3*, 39% tetrahydrofuran/water; *4*, 1:1 mixture of *1* and 2; *5*, 1:1 mixture of *2* and 3; 6, 1:1 mixture of *1* and 3; 7, 1:1:1 mixture of *1*, *2*, and *3*. Recreated from data of [57].

phase 5, a 1:1 blend of mobile phases 1 and 3 (26/20/54% ACN/THF/water); phase 6, a 1:1 blend of mobile phases 2 and 3 (32/20/48% MeOH/THF/water); and phase 7, a 1:1:1 blend of mobile phases 1, 2, and 3 (17/21/13/49% ACN/MeOH/THF/water). An examination of the latter four chromatograms in Figure 6.24 shows that baseline resolution is achieved with mobile phase 5 (ACN/THF). In most cases further improvements in selectivity and resolution are possible by blending mobile phases 1 to 7 to obtain intermediate mixtures. This approach can be simplified by eliminating unpromising mixtures. For example, runs 1 and 2 in Figure 6.24 show that peaks 6 and 7 overlap in each run; this suggests that no mixture of mobile phases 1 and 2 (e.g., mobile phase 4) is likely to improve the separation of peaks 6 and 7 (as seen in the separation of run 4).

The experimental design of Figures 6.22*c* and 6.24 can be further improved by using computer simulation to predict separation as a function of mobile phase composition [57]. Other work suggests expanding the number of experimental mobile phases from seven in Figure 6.22*c* to 15 [58], for a more reliable prediction of retention as a function of the mobile phase (when using computer simulation). However, the seven mobile phases of Figure 6.22*c* are usually sufficient for a visual (non–computer-assisted) selection of the final optimum mobile phase. By carrying out the experiments of Figure 6.24 sequentially ($1 \rightarrow 2, 2 \rightarrow 3$, etc.) and observing the result of each separation when completed, an acceptable separation may result without completing all seven experiments. For example, with this approach for the sample of Figure 6.24, only six experiments would be required (runs 1–6)—or five experiments if run 4 is bypassed for the reasons discussed above.

When using this (or similar) multiple-solvent optimization approach, the reader is cautioned that more complex mobile phases are more prone to problems than simpler ones. For this reason an "adequate" ternary separation, as in run 5 of Figure 6.24, may be preferable (because of reliability) over a quaternary separation that uses all four solvents to achieve a slightly improved separation.

6.4.1.2 Simultaneous Variation of Solvent Strength and Type

If the potential disadvantages of THF as B-solvent are considered unacceptable for a given sample, an alternative approach is the variation of the proportions of ACN, MeOH, and water in the mobile phase [26]. The experimental design for the latter procedure is shown in Figure 6.22*b*. Beginning with results for two mobile phases with varying %-ACN that cover a range in k of as much as 0.5 to 20 (runs 1 and 2), MeOH/water mobile phases of equivalent strength (runs 3 and 4) are selected with the help of Figure 6.11. Finally, mobile phases 5 and 6 are selected by blending mobile phases 1 and 3 (1:1) and 2 and 4 (1:1), respectively. The use of data collected according to the experimental design of Figure 6.22*b* then allows an optimum mobile phase to be selected by interpolation. A partial example of this approach for optimizing selectivity was described in Figures 6.8 to 6.10 (data were not reported there for runs 3 and 5 of Fig. 6.22*b*, which were poorly resolved). The use of all six runs of Figure 6.22*b* would not change the final optimum conditions of Figure 6.10*d*.

A simpler, but less effective procedure for optimizing %B and solvent type is to successively vary %B for mobile phases that contain first one then another B-solvent (no blending of different B-solvents). An example is shown in Figure 6.25 for the separation of six steroids. In the initial two experiments (Fig. 6.26*a*,*b*) %-ACN is varied. As peaks 2 and 3 are unseparated in each run, ACN alone is unable to resolve this sample.

The next two experiments (Fig. 6.25*c*,*d*) vary %-MeOH, using the nomograph of Figure 6.11 to choose values of %-MeOH that provide similar retention as for the separations of Figure 6.25*a*,*b*. Now all peaks can be separated with near-baseline resolution ($R_s = 1.4$ for critical peak-pair 4/5 in Fig. 6.25*d* [35% B]). As the resolution of peak-pair 4/5 improves with increasing %-MeOH (while that of peak-pair 1/2 decreases), a slight increase in %-MeOH is suggested. For 37%-MeOH a resolution of $R_s = 1.5$ (baseline resolution) was found, with a small reduction in separation time compared to 35%-MeOH (separation not shown).

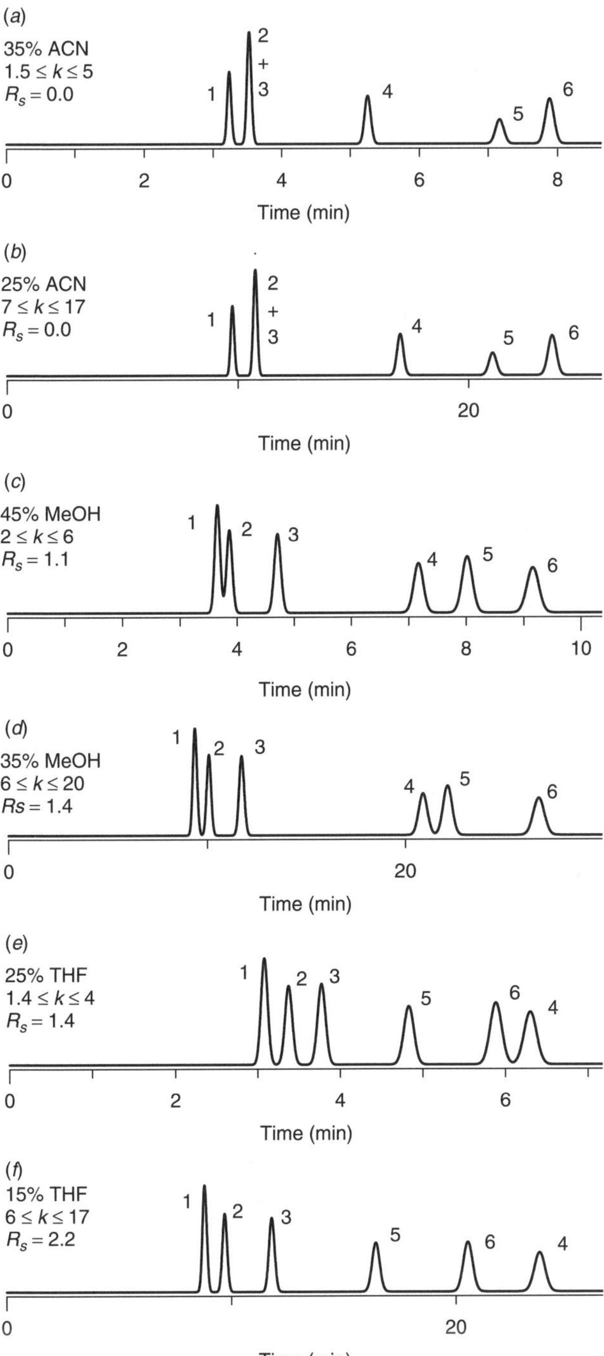

Figure 6.25 Separation of six steroids by changes in solvent strength (%B) and type. Sample: *1*, prednisone; *2*, hydrocortisone; *3*, cortisone; *4*, dexamethasone; *5*, corticosterone; *6*, cortexolone. Conditions: 250 × 4.6-mm C_8 (5-μm) column; different mobile phases are organic/water mixtures, as indicated in figure; 35°C; 2.0 mL/min. Chromatograms recreated from data of [59].

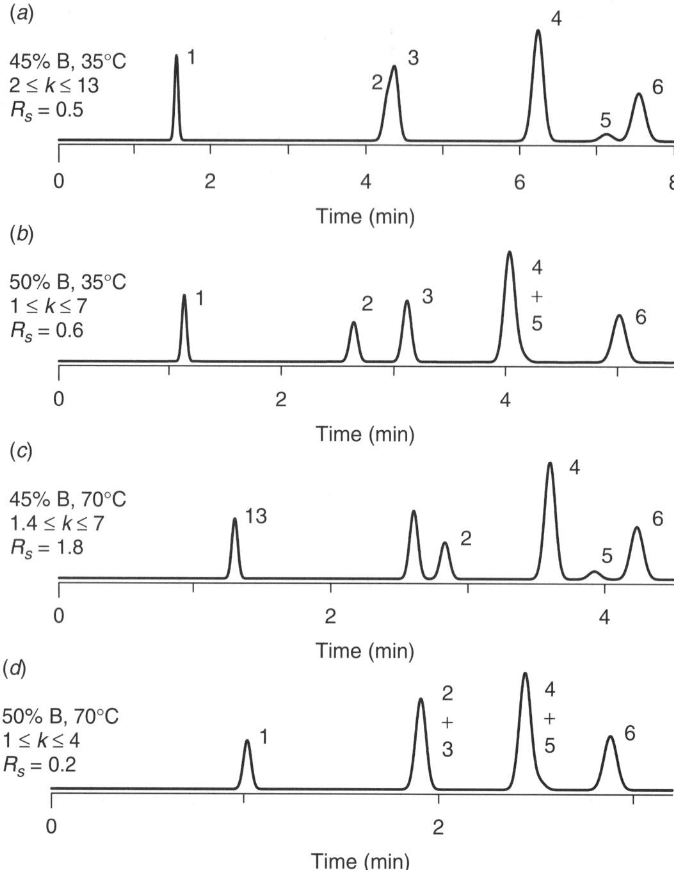

Figure 6.26 Separation of a mixture of 6 organic compounds of diverse structure by changes in solvent strength (%B) and temperature. Sample: *1*, methylbenzoate; *2*, benzophenone; *3*, toluene; *4*, naphthalene; *5*, phenothiazine; *6*, 1,4-dichlorobenzene. Conditions: 125 × 3.0-mm C_{18} column; mobile phase acetonitrile/water mixtures; 1.0 mL/min. Chromatograms recreated from data of [66].

Because a separation that is barely baseline resolved is considered marginal for a final method (assuming the use of a new, nondegraded column), THF as the B-solvent can be considered next (Fig. 6.25*e*,*f*). Even better resolution is obtained for 15%-THF (Fig. 6.25*f*, $R_s = 2.2$), and this cannot be improved by further changes in %-THF without exceeding $k = 20$. The procedure of Figure 6.25 for optimizing solvent type and strength can be terminated at any step if an adequate resolution and run time are achieved. Thus all six experiments of Figure 6.25 will be unnecessary for some samples but may be insufficient for others.

6.4.1.3 Simultaneous Variation of Solvent Strength and Temperature

It was noted in Table 2.2 that changes in solvent strength (%B) and/or temperature (T) are less effective than a change in solvent type. However, the *simultaneous* optimization of both %B and T is often adequate for achieving baseline resolution [60–65], while being more convenient and less tedious than alternative multiple-variable optimization schemes. The experimental design for the simultaneous variation of %B and T is described by Figure 6.22*a*, while an example of its application is shown in Figure 6.26. In Figure 6.26*a,b*, %-B is varied for $T = 35^{\circ}$C. There is considerable change in selectivity as %-B is varied, and a resolution of $R_s = 1.4$ is possible for 41% B, with a k-range of $3 \leq k \leq 20$ (separation not shown). The temperature is next changed to 70°C, with the results of Figure 6.26*c,d*. A reasonable separation ($R_s = 1.8$) is obtained for 45% B and 70°C. Further experiments varying %B and T did not increase resolution, as long as retention was maintained within the range $0.5 \leq k \leq 20$.

The use of a simultaneous variation of %B and T has become increasingly popular for the following reasons: (1) only four initial experiments are required, once a value of %B for a reasonable range in k has been established; (2) with on-line mixing of the A- and B-solvents, all four experiments can be carried out automatically, without operator intervention (assuming temperature control by the system controller); (3) there are none of the experimental problems associated with other means of optimizing selectivity [26, 67]; (4) this procedure is often adequate for achieving the desired selectivity and resolution of a sample; and (5) peak matching tends to be easier than for other experimental designs. For more details about this approach for optimizing selectivity and resolution, see [63]. For a variation on this technique, see also [64].

6.4.1.4 Change of the Column with Variation of One or More Other Conditions

It was seen in Figure 6.14 that the use of these four columns (with other conditions the same) resulted in a maximum resolution of $R_s = 0.8$. Such a result for this particular sample is not unusual because a change in column does not allow the convenient use of "intermediate" conditions, as is the case for a change of other conditions in the examples of Figure 6.6 (%B), Figure 6.10 (blended B-solvents), or Figures 6.13 and 6.14 (temperature). A better approach, when changing the column, is to *combine* a change in column with a further change in one or more other conditions that affect selectivity. A procedure that we recommend (and that many laboratories now use) is a change in the column combined with simultaneous changes in %B and temperature (as in Fig. 6.22*a*). When this procedure was applied to the four columns of Figure 6.14, the optimized separations of Figure 6.27*a–d* resulted. Now the range in resolution for the three columns has been increased from $0.0 \leq R_s \leq 0.8$ to $1.2 \leq R_s \leq 1.5$. If N is increased for the Symmetry C18 column (Fig. 6.27*a*) by increasing column length (from 150 to 250 mm) and decreasing flow rate (from 2.0 to 1.0 mL/min), a resolution of $R_s = 2.0$ is achieved, with no increase in pressure (but an increase in run time from 7.5 to 25 min).

In many cases, simply optimizing %B and temperature for columns of different selectivity (large values of F_s) will lead to a satisfactory separation. For other, more demanding samples, a greater change in selectivity may be needed. For such cases solvent-type optimization as in Figure 6.22*b* might be combined with a change in column [68].

It was noted in Section 6.3.4 that it is inconvenient to vary column selectivity continuously, as by mixing two different column packings in different proportions to

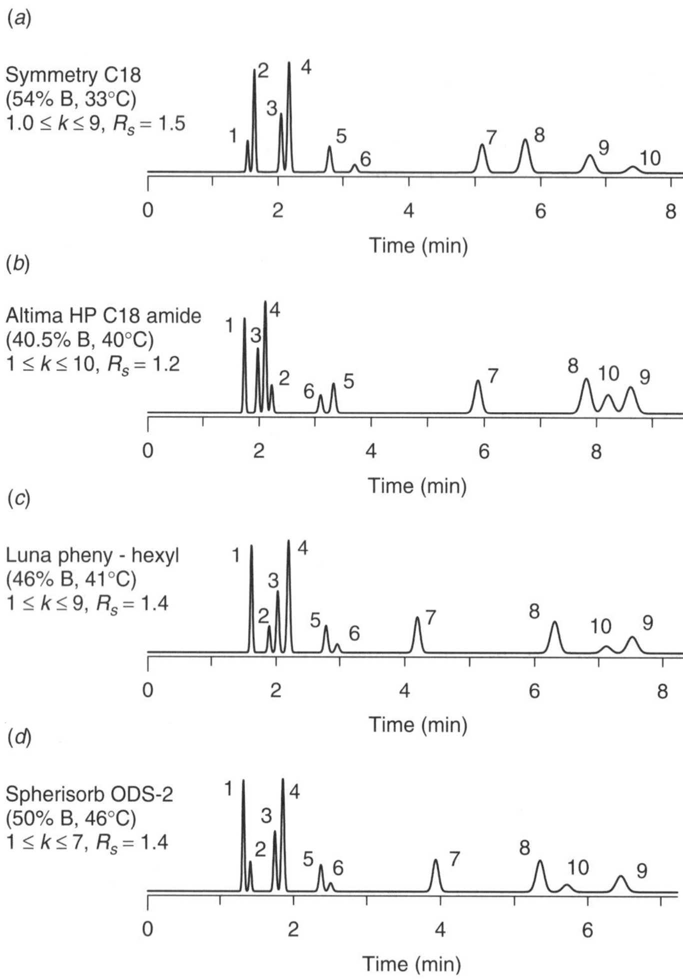

Figure 6.27 Optimized separation of a mixture of 10 organic compounds of diverse structure on four different columns by varying solvent strength (%B) and temperature. Sample and conditions as in Figure 6.14, except as indicated in figure. Chromatograms recreated from data of [8, 9, 69].

form a single column. Two alternatives to the latter procedure have been suggested, neither of which has so far found much practical application. In one approach, small lengths of different columns are connected in series. By varying both column type and length, column selectivity can be varied in discontinuous fashion (called "phase-optimized liquid chromatography" or POPLC® from Bischoff). Experiments with individual columns define sample retention for each column type, and it is then possible to predict retention for different combinations of columns and column length [71]. A second approach uses two different columns in series, with separate control of the temperature for each column (called "thermally tuned tandem-column approach" [72]). Because sample retention decreases at higher temperatures, the relative contribution of either of the two columns to overall selectivity can be increased by lowering its temperature relative to that of the other column. Each of these two procedures appears somewhat complicated for routine application.

6.4.2 Optimizing Column Conditions

Column conditions (column length L and diameter d_c, particle size d_p, flow rate F) are preferably optimized *after* other conditions are varied for optimized selectivity and maximum resolution (step 6 of Fig. 6.21*b*). Particle size, column length and diameter, and flow rate are usually selected prior to the start of method development (e.g., as recommended in Table 6.1) to provide a sufficient plate number for the separation of the sample under study. Following the selection of other conditions (column, mobile phase, temperature), column length and flow rate can be varied to either increase resolution (at the cost of increased run time) or decrease run time (when the initial resolution $\gg 2$). Figure 6.28 shows examples of each of these cases. The separation of Figure 6.28*a* has marginal resolution ($R_s = 1.1$), which must be increased. This is most effectively done by an increase in column length, as illustrated by Figure 6.28*b* for an increase in column length from 150 to 250 mm. Baseline resolution is now (barely) achieved ($R_s = 1.5$), at the cost of a 2/3 increase in run time and pressure. An even greater increase in resolution is desirable; however, the latter example illustrates that an increase in column length comes at a cost of increased run time and pressure, which effectively limits the possible increase in resolution. A decrease in flow rate is an alternative option for increasing resolution, but this is generally not worthwhile. For the example of Figure 6.28*a* a reduction in flow rate from 2.0 to 1.0 mL/min (not shown) results in an insignificant increase in resolution (to $R_s = 1.2$), while run time is doubled to 10 min. Alternatively, the particle size can be reduced, for example, with a 150 × 4.6-mm column of 3-μm particles. In this case run time remains the same as in Figure 6.28*a*, pressure increases to 2900 psi, and $R_s = 1.6$. In each of these examples, resolution, run time, and pressure can be varied to meet some final goal. A reasonable overall compromise might be achieved with a 250 × 4.6-mm, 3-μm column and a flow rate of 1.0 mL/min: $R_s = 2.1$, 2400 psi, with a run time of 15 minutes. See also Section 2.4.1.2.

Figure 6.28*c*,*d* provides an example of a decrease in run time when the initial resolution is more than adequate. In Figure 6.28*c* the run time is 30 minutes and $R_s = 4.8$. By decreasing column length and increasing flow rate, run time can be shortened drastically, while maintaining $R_s > 2.0$ and a pressure < 2000 psi. In Figure 6.28*d* these combined changes in column length and flow rate result in a 10-fold decrease in run time, with $R_s = 2.1$. By decreasing column length in the identical proportion as flow rate is increased, the pressure can be maintained constant (not done in the example of Fig. 6.28*c*,*d*).

6.5 NONAQUEOUS REVERSED-PHASE CHROMATOGRAPHY (NARP)

Separation by nonaqueous reversed-phase chromatography (NARP) is reserved for very hydrophobic samples that are retained strongly and not eluted by 100% ACN as mobile phase (lipids, synthetic polymers, etc. [74–77]). The mobile phase for NARP separations will therefore consist of a mixture of more polar (A-solvent) and

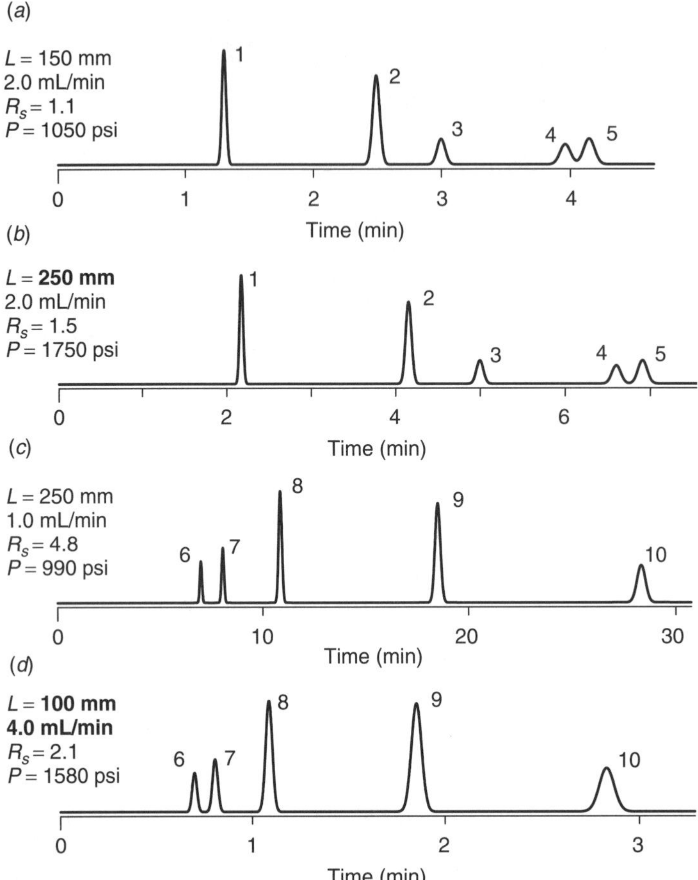

Figure 6.28 Illustrations of a change in column conditions to either improve resolution or decrease run time. Sample components (non-ionized for these conditions; pH-2.6): *1*, phthalic acid; *2*, 2-nitrobenzoic acid; *3*, 2-fluorobenzoic acid; *4*, 3-nitrobenzoic acid; *5*; 2-chlorobenzic acid; *6*, 4-chloroaniline; *7*, 3-fluorobenzic acid; *8*, 2,6-dimethylbenzoic acid; *9*, 2-chloroaniline; *10*, 3,4-dichloroaniline. Conditions: 4.6-mm C_{18} columns (5-μm) with indicated lengths *L*; mobile phase is 30% ACN-buffer for (*a*) and (*b*); 40% ACN-buffer for (*c*) and (*d*); 40°C in (*a*) and (*b*), 30°C in (*c*) and (*d*); flow rates indicated in figure. Chromatograms recreated from data of [73].

less polar (B-solvent) organic solvents. Often the A-solvent will be ACN or MeOH, while the B-solvent can be THF, methylene chloride, chloroform, methyl-*t*-butyl ether (MTBE), or other less polar organic solvents. Sample retention is controlled by varying %B and/or the polarity of the B-solvent, which can be approximated by the value of P' in Table I.4 of Appendix I.

Figure 6.29 shows an example of NARP for the separation of various carotenes (Fig. 6.29*a*) in a mixture of standards (Fig. 6.29*b*) and in an extract from tomato (Fig. 6.29*c*). Very hydrophobic samples are often insoluble in aqueous solutions,

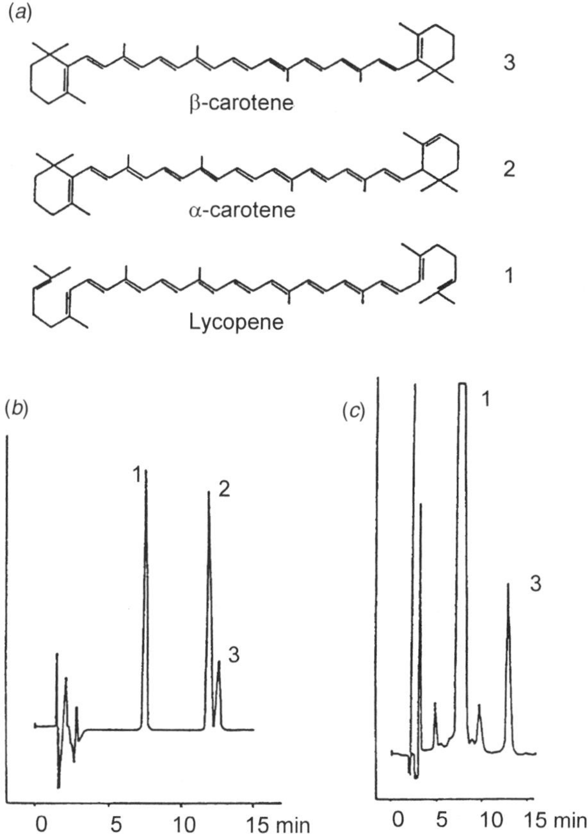

Figure 6.29 Non-aqueous reversed-phase (NARP) separations of carotenes. Conditions: 250 × 4.6-mm C_{18} column; 8% chloroform-ACN mobile phase; 2.0 mL/min; ambient temperature. Adapted from [75].

which can be another reason to use NARP for such samples. From a practical standpoint, if NARP is chosen for a separation, all water must be washed from the HPLC system and column prior to switching to the nonaqueous mobile phase. Generally a 30-minute flush with ACN or MeOH is sufficient.

6.6 SPECIAL PROBLEMS

One reason why RPC is more popular than other HPLC separations is that there are fewer problems in its use. Two possible problems with RPC that require attention are (1) poor retention for very polar samples and (2) peak tailing.

6.6.1 Poor Retention of Very Polar Samples

This problem was noted in Section 6.1. Solutes that are very polar may not be retained with $k \geq 1$, even when pure water (0% B) is used as mobile phase. This problem is more often encountered in the case of ionized solutes, which are much less retained than their non-ionized counterparts (e.g., R–COO$^-$ vs. R–COOH). For ionized solutes their RPC retention can usually be increased by a change in mobile-phase pH (so as to decrease solute ionization; Section 7.2), or the addition of an ion-pair reagent to the mobile phase (Section 7.4).

When attempting the separation of very polar, non-ionic samples by RPC, some columns exhibit a *decrease* in retention when mobile phases with < 5%B are used ("stationary-phase de-wetting"). Some columns are designed to avoid this problem, while the problem can be further minimized by following certain procedures (Section 5.3.2.3). When sample retention must be increased, even with the use of water as a mobile phase, the choice of column can provide some further control over sample retention. For example, columns with a higher surface area (smaller pore diameter) provide generally larger values of k. Graphitized-carbon columns (Section 5.2.5.3) are known to retain some very polar non-ionized solutes preferentially, although the use of these columns is constrained by their high cost and limited stability.

When the sample is poorly retained by RPC, the preferred approach is often the use of normal-phase chromatography—because polar solutes are preferentially retained by the more polar stationary phase. Hydrophilic interaction chromatography (HILIC; Section 8.6), which is a variation of NPC, is especially useful in this connection; it can be used with aqueous mobile phases, and has other advantages when used in combination with mass spectrometric detection (LC-MS).

6.6.2 Peak Tailing

Tailing peaks can arise for a number of different reasons (Section 17.4.5.3), often for acids or bases as solutes (Sections 5.4.4.1, 7.3.4.2). Whenever markedly tailing peaks are observed (e.g., with asymmetry factors A_s >2), steps should be taken to correct the problem. When peak tailing is observed during routine analysis, usually a replacement of the column or guard column will solve the problem. If peak tailing is encountered during method development, it is important restore good peak shape by a change in conditions, before carrying out further experiments. For further information on peak tailing, see Sections 7.3.4.2, 7.4.3.3, and 17.4.5.3.

REFERENCES

1. G. A. Howard and A. J. P. Martin, *Biochem. J.*, 46 (1950) 532.
2. L. R. Snyder and J. J. Kirkland, *Introduction to Modern Liquid Chromatography*, Wiley-Interscience, New York, 1974, ch. 8.
3. J. J. Kirkland and J. J. DeStefano, *J. Chromatogr. Sci.*, 8 (1970) 309.
4. J. A. Schmit, R. A. Henry, R. C. Williams, and J. F. Dieckmann, *J. Chromatogr. Sci.*, 9 (1971) 645.
5. W. R. Melander and C. Horváth, in *High-Performance Liquid Chromatography. Advances and Perspectives*, Vol. 2, C. Horváth, ed., Academic Press, New York, 1980, pp. 113–319.
6. C. Horváth, W. Melander, and I. Molnar, *J. Chromatogr.*, 125 (1976) 129.
7. M. T. W. Hearn, in *Ion-pair Chromatography*, M. T. W. Hearn, ed., Dekker, New York, 1985.

8. N. S. Wilson, M. D. Nelson, J. W. Dolan, L. R. Snyder, R. G. Wolcott, and P. W. Carr, *J. Chromatogr. A*, 961 (2002) 171.
9. N. S. Wilson, M. D. Nelson, J. W. Dolan, L. R. Snyder, and P. W. Carr, *J. Chromatogr. A*, 961 (2002) 195.
10. L. R. Snyder and M. A. Quarry, *J. Liq. Chromatogr.*, 10 (1987) 1789.
11. K. Valkó, L. R. Snyder, and J. L. Glajch, *J. Chromatogr.*, 656 (1993) 501.
12. P. W. Carr, D. E. Martire, and L. R. Snyder, eds., *Retention in Reversed-Phase HPLC (J. Chromatogr.*, Vol. 656, 1993).
13. C. F. Poole, *The Essence of Chromatography*, Elsevier, Amsterdam, 2003.
14. L. C. Tan and P. W. Carr, *J. Chromatogr. A*, 775 (1997) 1.
15. A. Ailaya and C. Horváth, *J. Chromatogr. A*, 829 (1998) 1.
16. P. Nikitas, A. Pappa-Louisi, and P. Agrafiotou, *J. Chromatogr. A*, 1034 (2004) 41.
17. I. Molnar and Cs. Horváth, *J. Chromatogr.*, 142 (1977) 623.
18. A. Tchapla, H. Colin, and G. Guiochon, *Anal. Chem.*, 56 (1984) 621.
19. J. Ko and J. C. Ford, *J. Chromatogr. A*, 913 (2001) 3.
20. B. A. Bidlingmeyer and A. D. Broske, *J. Chromatog,. Sci.* 42 (2004) 100.
21. T. H. Walter, P. Iraneta, and M. Capparella, *J. Chromatogr. A*, 1075 (2005) 177.
22. L. C. Sander, K. A. Lippa, and S. A. Wise, *Anal. Bioanal. Chem.*, 382 (2005) 646.
22a. J. L. Rafferty, J. I. Siepmann, and M. R. Schure, *J. Chromatogr. A*, 1204 (2008) 11.
23. A. Klimek-Turek, T. H. Dzido, and H. Engelhardt, *LCGC Europe*, 21 (2008) 33.
24. T. Braumann, G. Weber, and L. H. Grimme, *J. Chromatogr.*, 261 (1983) 329.
25. L. R. Snyder and J. W. Dolan, *High-Performance Gradient Elution*, Wiley, New York, 2007, pp. 19–21.
26. L. R. Snyder, *Today's Chemist at Work*, 5 (1996) 29.
27. P. J. Schoenmakers, H. A. H. Billiet, and L. de Galan, *J. Chromatogr.*, 185 (1979) 179.
28. P. J. Schoenmakers, H. A. H. Billiet, and L. de Galan, *J. Chromatogr.*, 218 (1981) 259.
29. R. G. Wolcott, J. W. Dolan, and L. R. Snyder, *J. Chromatogr. A*, 869 (2000) 3.
30. J. Chmielowiec and H. Sawatzky, *J. Chromatogr. Sci.*, 17 (9790) 245.
31. J. W. Dolan, L. R. Snyder, N. M. Djordjevic, D. W. Hill, D. L. Saunders, L. Van Heukelem, and T. J. Waeghe, *J. Chromatogr. A*, 803 (1998) 1.
32. J. W. Dolan, *J. Chromatogr. A*, 965 (2002) 195.
32a. S. Heinisch, and J. L. Rocca, *J. Chromatogr. A*, 1216 (2009) 642.
33. S. Heinisch, G. Puy, M.-P. Barrioulet, and J. L. Rocca, *J. Chromatogr. A*, 1118 (2006) 234.
34. J. W. Coym and J. G. Dorsey, *J. Chromatogr. A*, 1035 (2004) 23.
35. J. W. Dolan, L. R. Snyder, N. M. Djordjevic, D. W. Hill, D. L. Saunders, L. Van Heukelem, and T. J. Waeghe, *J. Chromatogr. A*, 803 (1998) 1.
36. L. R. Snyder, *J. Chromatogr.*, 179 (1979) 167.
37. L. R. Snyder and J. W. Dolan, *J. Chromatogr. A*, 892 (2000) 107.
38. P. L. Zhu, J. W. Dolan, L. R. Snyder, N. M. Djordjevic, D. W. Hill, J.-T. Lin, L. C. Sander, and L. Van Heukelem, *J. Chromatogr. A*, 756 (1996) 63.
39. D. W. Armstrong, W. Demond, A. Alak, W. L. Hinze, T. E. Riehl, and K. H. Bui, *Anal. Chem.*, 57 (1985) 234.
40. F. C. Marziani and W. R. Cisco, *J. Chromatogr.*, 465 (1989) 422.
41. M. Paleologou, S. Li, and W. C. Purdy, *J. Chromatogr. Sci.*, 28 (1990) 311.

42. B. Nikolova-Damyanov, in *HPLC of Acyl Lipids*, J.-T. Lin and T. A. McKeon, eds., HNB Publishing, New York, 2005, p. 221.
43. B. Voach and G. Schomburg, *J. Chromatogr.* 149 (1978) 417.
44. L. C. Sander and S. A. Wise, *J. Chromatogr. A* 656 (1993) 335.
45. L. C. Sander, M. Pursch, and S. A. Wise, *Anal. Chem.*, 71 (1999) 4821.
46. L. C. Sander, K. A. Lippa, and S. A. Wise, *Anal. Bioanal. Chem.*, 382 (2005). 646.
47. J. W. Dolan and L. R. Snyder, *J. Chromatogr. A*, 1216 (2009) 3467.
48. J. W. Dolan, A. Maule, L. Wrisley, C. C. Chan, M. Angod, C. Lunte, R. Krisko, J. Winston, B. Homeierand, D. M. McCalley, and L. R. Snyder, *J. Chromatogr. A*, 1057 (2004) 59.
49. J. Pellett, P. Lukulay, Y. Mao, W. Bowen, R. Reed, M. Ma, R. C. Munger, J. W. Dolan, L. Wrisley, K. Medwid, N. P. Toltl, C. C. Chan, M. Skibic, K. Biswas, K. A. Wells, and L. R. Snyder, *J. Chromatogr. A* 1101 (2006) 122.
50. J. W. Dolan, L. R. Snyder, T. H. Jupille, and N. S. Wilson, *J. Chromatogr. A* 960 (2002) 51.
51. J. W. Dolan, L. R. Snyder, and T. Blanc, *J. Chromatogr. A*, 897 (2000) 51.
52. G. Xue, A. D. Bendick, R. Chen, and S. S. Sekulic, *J. Chromatogr. A*, 1050 (2004) 159.
53. E. Van Gyseghem, M. Jimidar, R. Sneyers, D. Redlich, E. Verhoeven, D. L. Massart, and Y, Vander Heyden, *J. Chromatogr. A*, 1074 (2005) 117.
54. D. H. Marchand, L. R. Snyder, and J. W. Dolan, *J. Chromatogr. A*, 1191 (2008) 2.
55. J. Zhao and P. W. Carr, *LCGC*, 17 (1999) 346.
56. L. R. Snyder, *J. Chromatogr. B*, 689 (1997) 105.
57. J. L. Glajch, J. J. Kirkland, K. M. Squire, and J. M. Minor, *J. Chromatogr.*, 199 (1980) 57.
58. A. C. J. H. Drouen, H. A. H. Billiet, P. J. Schoenmakers, and L. de Galan, *Chromatographia*, 10 (1982) 48.
59. L. R. Snyder, M. A. Quarry, and J. L. Glajch, *Chromatographia*, 24 (1987) 33.
60. P. L. Zhu, J. W. Dolan, L. R. Snyder, N. M. Djordjevic, D. W. Hill, J.-T. Lin, L. C. Sander, and L. Van Heukelem, *J. Chromatogr. A*, 756 (1996) 63.
61. J. W. Dolan, L. R. Snyder, N. M. Djordjevic, D. W. Hill, D. L. Saunders, L. Van Heukelem, and T. J. Waeghe, *J. Chromatogr. A*, 803 (1998) 1.
62. J. W. Dolan, L. R. Snyder, N. M. Djordjevic, D. W. Hill, L. Van Heukelem, and T. J. Waeghe, *J. Chromatogr. A*, 857 (1999) 1.
63. R. G. Wolcott, J. W. Dolan, and L. R. Snyder, *J. Chromatogr. A*, 869 (2000) 3.
64. A. Gonzalez, K. L. Foster, and G. Hanrahan, *J. Chromatogr. A*, 1167 (2007) 135.
65. J. W. Dolan, L. R. Snyder, T. Blanc, and L. Van Heukelem, *J. Chromatogr. A*, 897 (2000) 37.
66. J. W. Dolan, L. R. Snyder, N. M. Djordjevic, D. W. Hill, D. L. Saunders, L. Van Heukelem, and T. J. Waeghe, *J. Chromatogr. A*, 803 (1998) 1.
67. P. L. Zhu, L. R. Snyder, J. W. Dolan, N. M. Djordjevic, D. W. Hill, L. C. Sander, and T. J. Waeghe, *J. Chromatogr. A*, 756 (1996) 21.
68. J. J. DeStefano, J. A. Lewis, and L. R. Snyder, *LCGC*, 10 (1992) 130.
69. L. R. Snyder, J. W. Dolan, and P. W. Carr, *J. Chromatogr. A*, 1060 (2004) 77.
70. K. Valkó, S. Espinosa, C. M. Du, E. Bosch, M. Rosés, C. Bevan, and M. H. Abraham, *J. Chromatogr. A*, 933 (2001) 73.
71. Sz. Nyiredy, A. Szucs, and L. Szepesy, *J. Chromatogr. A*, 1157 (2007) 122.
72. Y. Mao, and P. W. Carr, *Anal. Chem.*, 73 (2001) 1821.

73. P. L. Zhu, J. W. Dolan, and L. R. Snyder, D. W. Hill, L. Van Heukelem, and T. J. Waeghe, *J. Chromatogr. A*, 756 (1996) 51.

74. N. A. Parris, *J. Chromatogr.*, 157 (1978) 161.

75. M. Zakaria, K. Simpson, P. R. Brown, and A. Krstulovic, *J. Chromatogr.*, 176 (1979) 109.

76. N. E. Craft, S. A. Wise, and J. H. Soares, *J. Chromatogr.*, 589 (1992) 171.

77. H. J. A. Philipsen, *J. Chromatogr. A*, 1037 (2004) 329.

CHAPTER SEVEN

IONIC SAMPLES: REVERSED-PHASE, ION-PAIR, AND ION-EXCHANGE CHROMATOGRAPHY

Introduction to Modern Liquid Chromatography, Third Edition, by Lloyd R. Snyder, Joseph J. Kirkland, and John W. Dolan

7.1 INTRODUCTION

Chapter 6 dealt with the separation of neutral (non-ionized) molecules by means of reversed-phase chromatography (RPC). The present chapter extends this treatment to the HPLC separation of "ionic" samples; these are mainly mixtures that contain acids and/or bases (with or without neutral compounds), but they can include compounds that are totally ionized between pH-2 and pH-12 (e.g., tetralkylammonium salts, sulfonic acids). In the early days of HPLC, ionic samples often presented special problems—partly the result of less suitable column packings that were available at that time but also because of a limited understanding of how such separations are best carried out. Although these past limitations have been largely overcome, the separation of ionic samples remains somewhat more demanding when compared with separations of neutral samples. Before 1980, ion-exchange chromatography (IEC, Section 7.5) was commonly selected for the separation of acids and bases, but today RPC (Section 7.3) and—to a lesser extent—ion-pair chromatography (Section 7.4) have become preferred procedures for the separation of "small," ionizable molecules (<1000 Da). However, IEC is still used heavily for the separation of large biomolecules such as proteins (Chapter 13); for additional details on IEC separation, see Sections 13.4.2, 13.5.1, and 13.6.3.

7.2 ACID–BASE EQUILIBRIA AND REVERSED-PHASE RETENTION

The RPC retention of neutral samples decreases for less hydrophobic (more polar) molecules (Sections 2.3.2.1, 6.2). When an acid (HA) or base (B) undergoes ionization (i.e., is converted from an uncharged to a charged species), the compound becomes much more polar or hydrophilic. As a result its retention factor k in RPC can be reduced 10-fold or more:

uncharged molecule   ionized molecule

$$\text{(acids)}\quad HA \Leftrightarrow A^- + H^+ \tag{7.1}$$

$$\text{(bases)}\quad B + H^+ \Leftrightarrow BH^+ \tag{7.1a}$$

hydrophobic (more retained in RPC)   hydrophilic (less retained in RPC)

Acids lose a proton and become ionized when the mobile-phase pH is increased; bases gain a proton and become ionized when mobile-phase pH decreases. The ionization of an acid (HA) or base (B) can be related to its acidity constant K_a:

$$\text{(acids)} \quad K_a = \frac{[A^-][H^+]}{[HA]} \tag{7.2}$$

or

$$\text{(bases)} \quad K_a = \frac{[B][H^+]}{[BH^+]} \tag{7.2a}$$

Here [HA] and $[A^-]$ are the concentrations of the free and ionized acidic solute HA; [B] and $[BH^+]$ refer to the concentrations of the free and protonated basic solute B. The pK_a value ($= -\log K_a$) of an acid or base is given by the Henderson–Hasselbalch equation:

$$\text{(acids)} \quad pK_a = pH - \log\left(\frac{[A^-]}{[HA]}\right) \tag{7.3}$$

or

$$\text{(bases)} \quad pK_a = pH - \log\left(\frac{[B]}{[BH^+]}\right) \tag{7.3a}$$

For example, the pK_a value in water of a (weakly basic) substituted aniline will fall within a range of about $4 \leq pK_a \leq 6$, while the pK_a of a (strongly basic) aliphatic amine will usually lie between 9 and 11. Values of pK_a in the literature for different acids or bases usually refer to solutions in buffered-water at near-ambient temperatures. If the mobile phase contains organic solute, or if the temperature is much different from ambient, values of both pH and pK_a can change significantly (Section 7.2.3).

Retention as a function of pH and sample ionization is illustrated in Figure 7.1 for the separation of a hypothetical sample composed of carboxylic acid HA (solid curve in Fig. 7.1*a*) and aliphatic-amine B (dashed curve in Fig. 7.1*a*). In Figure 7.1*a*, solute ionization (left-hand scale) is plotted against mobile-phase pH for each solute; the dark circles mark the pH where each compound is half ionized ($pH \equiv pK_a = 5.0$ for HA, and 9.0 for B). Values of k (right-hand scale in Fig. 7.1*a*) decrease with increasing solute ionization and are given as a function of pH and pK_a by

$$\text{(acids, bases)} \quad k = k^0(1 - F^{\pm}) + k^{\pm}F^{\pm} \tag{7.4}$$

Here k^0 is the value of k for the non-ionized molecule (HA or B), $k^{\pm}$ is the value of k for the fully ionized molecule (A^- or BH^+), and $F^{\pm}$ is the fractional ionization of the molecule ($0 \leq F^{\pm} \leq 1$).

$$\text{(acids)} \quad F^{\pm} = \frac{1}{1 + [H^+]/K_a} \tag{7.4a}$$

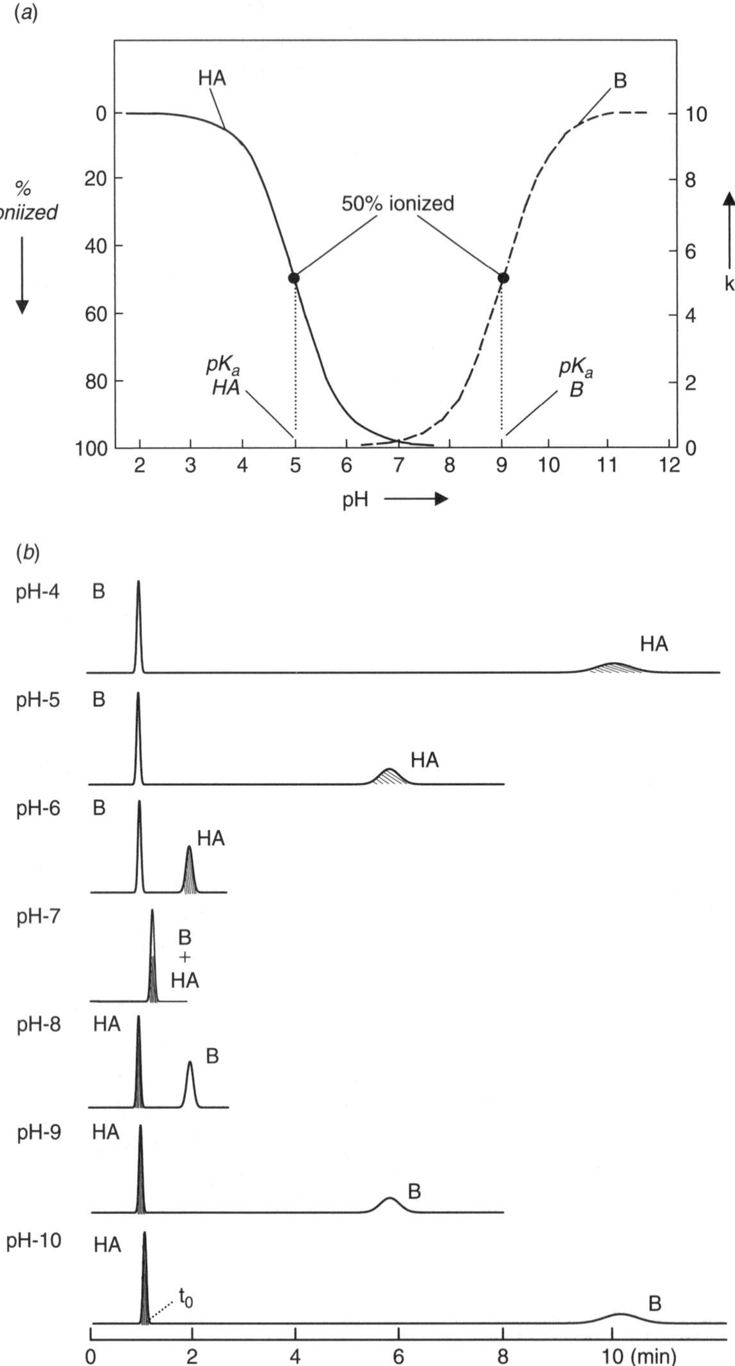

Figure 7.1 Hypothetical illustration of the RPC separation of an acidic compound HA from a basic compound B as a function of pH. (*a*) Ionization of HA and B as a function of mobile-phase pH and effect on k; (*b*) sample separation as a function of mobile-phase pH; values of k^0 for HA and B are assumed equal, $k^{\pm} = 0$, and $t_0 = 1.0$ min.

and

$$\text{(bases)} \quad F^{\pm} = \frac{1}{\{1 + (K_a/[H^+])\}} \tag{7.4b}$$

When pH = pK_a (or $[H^+] = K_a$), a compound is half ionized ($F^{\pm} = 0.5$).

Figure 7.1*b* shows chromatograms for the separation of HA and B as a function of mobile-phase pH. As the pH of the mobile phase increases from 4 to 10, the acid HA (shaded peak) becomes more ionized and less retained, while the base B eventually becomes less ionized and more retained (for pH = 7, $k = 0.1$ for each peak). It can be appreciated from this example that a change in mobile-phase pH can be a powerful means of controlling relative retention (selectivity) and separation for samples that contain acids and/or bases. The relative retention of two acids (or bases) can vary with pH when their values of pK_a, k^0, and/or $k^{\pm}$ differ (usually the case). Some authors have claimed that the separation of partly ionized solutes (e.g., where significant amounts of both HA and A^- are present at the same time) will necessarily lead to poor peak shape, but *there is no empirical or theoretical basis for this belief*.

A similar representation of RPC retention behavior as a function of pH is shown in Figure 7.2, for the variation of retention time t_R as a function of mobile-phase pH for a hypothetical, weakly basic solute with $pK_a = 5.0$ (e.g., a substituted aniline or pyridine). When pH is varied over a sufficiently wide range, solute retention exhibits a characteristic S-shaped plot as shown; this retention plot mirrors the ionization of the sample as in Figure 7.1*a*. At the midpoint of this retention versus pH curve (solid circle in Fig. 7.2), the mobile-phase pH is equal to the pK_a value of the solute. The mobile-phase pH is often chosen in order to control selectivity and resolution. When the mobile-phase pH $\approx pK_a$ for a critical compound or compounds, a change in pH will provide a maximum change in retention and separation. Thus mobile-phase pH should fall within region "II" of Figure 7.2 (pH = $pK_a \pm 1.0$), if we want to change selectivity and resolution by varying pH. However, as discussed below, a mobile-phase pH that allows a greater control over selectivity (i.e., region II) can mean a less reproducible separation—one of many necessary compromises in HPLC method development.

When an acid or base is half-ionized, a change in pH of 0.1 unit will result in a change of k by about 10%. For typical separation conditions, a 10% change in k for a solute can result in a change in resolution of as much as $\pm 2.5 R_s$ units, meaning a possible change in separation from baseline resolution ($R_s > 1.5$) to complete overlap ($R_s = 0$). Thus, if a solute is half-ionized, a change in mobile-phase pH by 0.1 unit can cause a complete loss of resolution. This suggests that mobile-phase pH may need to be controlled within about 0.02 units for such a separation, which could prove difficult for many laboratories (see Section 7.3.4.1). In order to avoid pH-related variations in retention, the mobile-phase pH can be selected to be different from the pK_a values of all sample components, by at least ±1.5 pH-units (regions I and III of Fig. 7.2). As the majority of compounds have pK_a values > 4, low-pH separations ($2 \leq pH \leq 3$) are more likely to be less sensitive to small changes in pH—which is one reason for beginning method development with a low-pH mobile phase (as recommended in this chapter). Separations at high pH (≥ 10) can also be used for this purpose, although special columns are required which are stable under these conditions (Sections 5.2.5, 5.3.2).

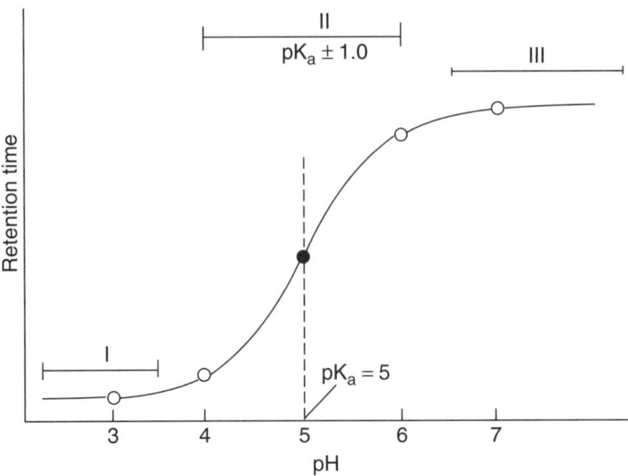

Figure 7.2 RPC retention as a function of pH. A basic solute with $pK_a = 5.0$ is assumed.

A further illustration of the dependence of separation on pH is provided by Figure 7.3, for several compounds of varying acidity or basicity. Figure 7.3*a* maps retention time versus mobile-phase pH for five solutes: compound 1 is salicylic acid (a relatively strong carboxylic acid), compound 2 is phenobarbital (a weak acid), compound 3 is phenacetin (a neutral compound in this pH range), compound 4 is nicotine (a weak base), and compound 5 is methylamphetamine (a strong base). Figure 7.3*b*–e shows the corresponding chromatograms for the separation of this sample as a function of mobile-phase pH. Note, for example, the relative (and absolute) change in retention for strongly basic compound 5 (shaded peak); as pH increases, compound 5 becomes less ionized, and more retained.

Several points are worth making about the example of Figure 7.3. First, for this mixture of acids, bases, and neutrals, a change in pH is a powerful means of varying relative retention and thereby optimizing resolution. A maximum resolution of $R_s = 7.2$ can be obtained for this sample at pH-8.3 (Fig. 7.3*f*) in a time of 28 minutes. Alternatively, baseline separation ($R_s = 2.0$) can be obtained in the shortest time (11 min) at pH-5 (Fig. 7.3*c*). However, by reducing column length from 300 to 50 mm for the separation at pH-8.3, and increasing flow rate from 2.0 to 5.0 mL/min, run time can be shortened to 2 minutes (Fig. 7.3*g*), while maintaining $R_s \geq 2.0$.

Second, this sample contains acids and bases with a wide range in pK_a values (see following discussion) and therefore exhibits sizable changes in retention for small changes in pH throughout the range $3 < pH < 9$. Consequently, either a careful control of mobile pH will be required for the separation of this sample (Section 7.3.4.1) or conditions must be selected that provide excess resolution ($R_s \gg 2$). For example, the separation of Figure 7.3*g* with $pH = 8.3$ and $R_s = 2.0$ could be made more robust by using a 10-cm column (for $R_s = 2.8$ in a run time of 4 min), holding other conditions the same.

Finally, the shape of a plot of retention versus pH for a peak allows a determination of its sample type (acid, base, or neutral), and a rough estimate of its pK_a value. Thus compounds in Figure 7.3*a* whose retention increases significantly as

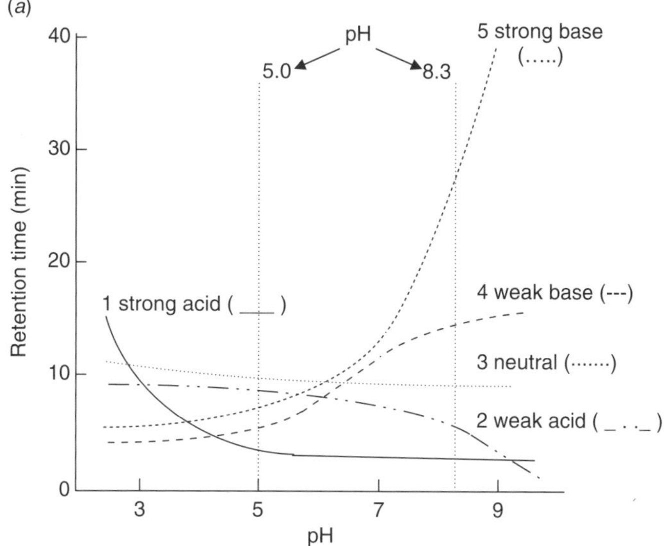

Figure 7.3 Effect of mobile-phase pH on RPC retention as a function of solute type. Sample: *1*, salicylic acid; *2*, phenobartitone; *3*, phenacetin; *4*, nicotine; *5*, methylamphetamine (shaded peak). Conditions for separations (*a–f*): 300 × 4.0-mm C_{18} column (10-μm particles); 40% methanol-phosphate buffer; ambient temperature; 2.0 mL/min. Flow rate is 5.0 mL/min and column length is 50-mm in (*g*). Adapted from [1], with chromatograms (*b–g*) recreated by computer.

pH increases are bases (4 and 5), those whose retention decreases with an increase in pH are acids (1 and 2), and compounds that show little change in retention with pH (3) are either neutral or are fully ionized over the pH range studied. Compounds 2 and 4 are seen to have pK_a values of about 8 and 6.5, respectively. While the pK_a values of compounds 1 and 5 cannot be estimated accurately (a complete retention vs. pH curve is required), it is safe to say that $pK_a \geq 9$ for compound 5, and $pK_a \leq 3$ for compound 1.

The relationship between RPC retention and mobile-phase pH is more complicated for amphoteric compounds that contain both acidic and basic groups. This is illustrated in Figure 7.4 for the retention of two amino acids as a function of pH. A molecule of each compound contains both an acidic –COOH group and a basic $-NH_2$ group. As a result minimum retention is observed at intermediate pH values, because for $4 < pH < 8$ both the carboxyl and amine groups are ionized. More precisely, the molecule is maximally ionized in this pH range, even though the *net* charge is zero (different ionized groups within a molecule—even of different sign—can each prefer the more polar mobile phase).

7.2.1 Choice of Buffers

Whenever acids or bases are separated, it is necessary to buffer the mobile phase in order to maintain a constant pH and reproducible retention during the separation. The use of a pH meter to measure (and control) pH will be less precise when the mobile phase contains organic solvent because the electrode response tends to drift

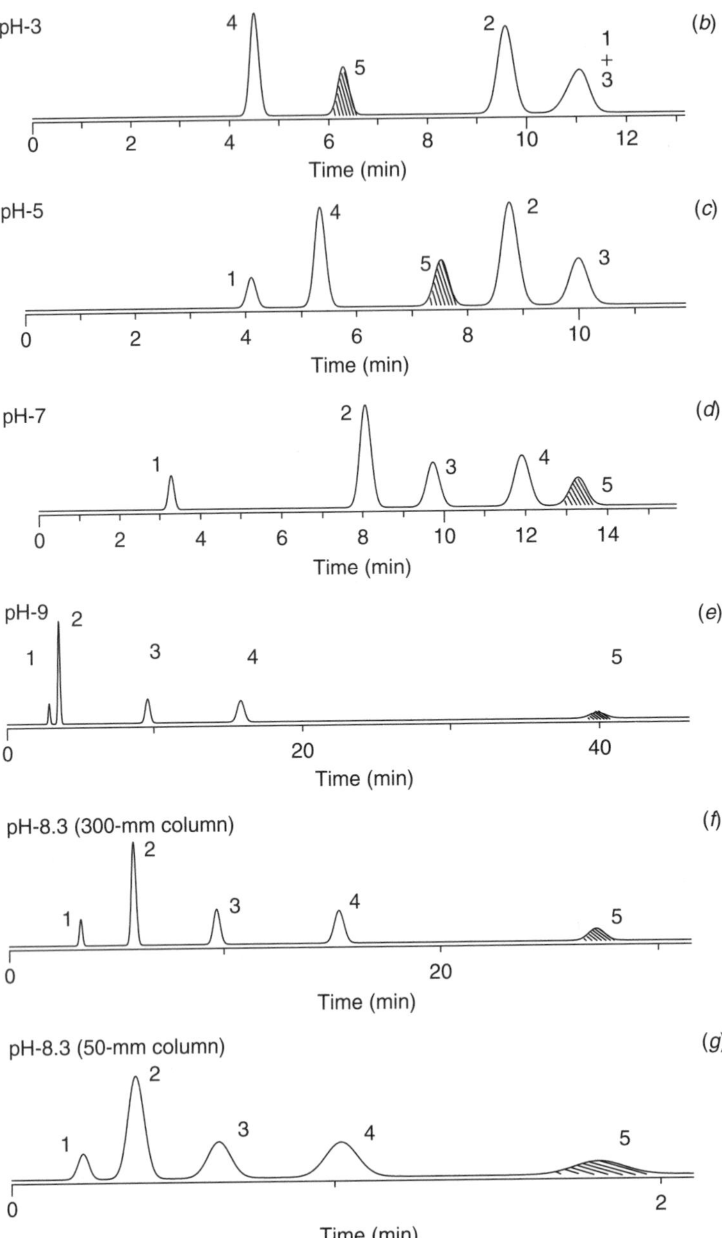

Figure 7.3 (*Continued*)

for organic-water solutions. Consequently *we recommend that pH measurements be carried out for the A-solvent (aqueous buffer) prior to the addition of organic to form the final mobile phase*. The pH of the final mobile phase (including organic solvent) can then be equated to (or labeled as) that of the A-solvent, although the actual mobile-phase pH will be somewhat different (Section 7.2.3). This uncertainty

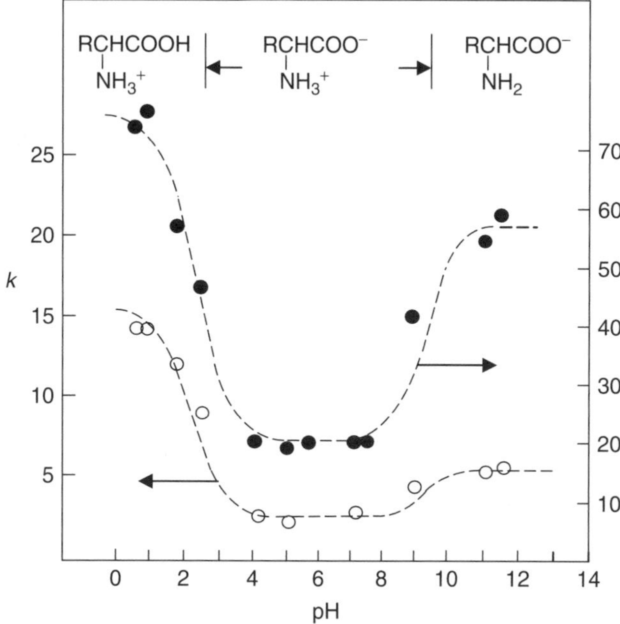

Figure 7.4 Dependence of RPC retention on mobile-phase pH for amphoteric compounds. Sample: phenylalanine (•), leucine (o). Conditions: polystyrene column, 40-mM phosphate buffer as mobile phase. Adapted from [2].

concerning the final mobile-phase pH is unimportant for the routine application of RPC. When the A-solvent is prepared in this way, different laboratories should be able to obtain the same final mobile-phase pH within ±0.04 to 0.05 units [3]. If a closer control of mobile-phase pH is required, see Section 7.3.4.1. *When we refer to mobile-phase pH in this book, we will generally mean the pH of the A-solvent.* Directions for the preparation of buffer solutions of varied pH and buffer-type are given in Appendix II.

In selecting a buffer for RPC separation, several buffer properties may prove relevant:

- pK_a and buffer capacity
- solubility
- UV absorbance (when UV detection is used)
- volatility (when mass-spectrometric or evaporative light-scattering detection is used)
- ion-pairing properties
- stability and compatibility with the equipment

The first four buffer properties are usually the most important.

7.2.1.1 Buffer pK_a and Capacity

"Buffer capacity," or the ability of the buffer to maintain a constant pH, depends on

- pK_a value of the buffer
- buffer concentration
- pH of the mobile phase

Just as for the ionization of a sample component in Figure 7.1*a*, the fractional ionization of the buffer as a function of pH can be expressed by Equations (7.2) or (7.2a); that is, buffer and solute ionization are identical functions of mobile-phase pH and pK_a. Maximum buffering occurs when the concentrations of the two forms of the buffer (e.g., HA and A^-) are equal; that is, when the buffer pK_a equals the mobile-phase pH. Buffer capacity decreases as values for the buffer pK_a and mobile-phase pH become more different. Consequently the first requirement of the buffer is that its pK_a value should be within ± 1.0 units of the selected mobile-phase pH (this requirement can be relaxed to ± 1.5 unit for higher concentrations of the buffer). The buffer capacity of the mobile phase is proportional to buffer concentration, which typically falls within a range of 5 to 25 mM. To minimize the possibility of inadequate buffering of the sample during RPC separation, *it is generally desirable for the sample to be dissolved in the mobile phase* (or buffered to the same pH as the mobile phase); this practice becomes especially important for lower concentrations of the mobile-phase buffer, and/or for larger volumes of injected sample.

Table 7.1 provides a list of buffers that can be used in RPC, along with pertinent properties such as buffer pK_a and the mobile-phase pH range in which the buffer is effective. For separations with UV detection, and a mobile-phase pH ≤ 8, popular buffers include phosphate, trifluoroacetate, acetate, and formate. In addition ammonium bicarbonate can also be considered. The pK_a values of ammonia (9.2) and bicarbonate (10.3) overlap, hence somewhat extending the buffering range of ammonium bicarbonate ($8.2 \leq$ pH ≤ 11.3). This buffer is volatile and therefore compatible with LC-MS; however, when the buffer pH < 8.5, loss of CO_2 (e.g., from excessive degassing) may lead to an unintended increase in pH. Because of this instability it is recommended to prepare fresh ammonium bicarbonate buffers daily.

The remaining discussion of this section (7.2.1.1) *is more detailed. The reader may wish to skip this digression, proceed to Section* 7.2.1.2, *and return to the present section as needed.*

We can define the "effective buffer capacity" of the mobile phase to mean that an increase in this quantity will result in fewer problems due to insufficient buffering. The effective buffer capacity of the mobile phase increases for:

1. a smaller difference between values of the buffer pK_a and the pH of the mobile phase (change either the buffer or pH)
2. a greater difference between the mobile-phase pH and the pK_a of the solute (for large differences, the solute is either non-ionized or completely ionized; buffering is then much less important)
3. increased buffer concentration
4. smaller volumes of injected sample
5. a sample whose pH is adjusted to that of the mobile phase

An example of inadequate buffering is provided by the chromatograms of Figure 7.5 for the solute 3,5-dimethylaniline as a function of mobile-phase pH, using a 25 mM potassium phosphate buffer. Despite this sizable buffer concentration,

Table 7.1

Buffers for Use in HPLC Separation

Buffer Acid[a]	pK_a (25°C)	Approximate Buffer Range	UV Cutoff[b]	Comments
Trifluoroacetic acid	>2	1.5–2.5	210 nm (0.1 %)	Ion-pairing, volatile
Phosphoric acid	2.1	1.5–3.5	<200 nm (10 mM)	Limited solubility
Monophosphate	7.2	6.0–8.5	<200 nm (10 mM)	Limited solubility
Diphosphate	12.3	11.0–13.5	<200 nm (10 mM)	Limited solubility
Citric acid	3.1	2.0–4.5	230 nm (10 mM)	Equipment problems[c]
Monocitrate	4.7	3.5–6.0	230 nm (10 mM)	Equipment problems[c]
Dicitrate	5.4	4.0–7.5	230 nm (10 mM)	Equipment problems[c]
Formic acid	3.8	2.5–5.0	210 nm (10 mM)	Volatile
Acetic acid	4.8	3.5–6.0	210 nm (10 mM)	Volatile
Carbonic acid	6.4	5.0–7.5	<200 nm (10 mM)	Volatile[d]
Monocarbonate	10.3	9.0–11.5	<200 nm (10 mM)	Volatile[e]
Bis-tris propane.HCl[e]	6.8	75.5–8.0	215 nm (10 mM)	Possibly unstable[f]
Bis-tris propane	9.0	7.5–10.5	225 nm (10 mM)	Possibly unstable[f]
Tris.HCl[g]	8.0	7.0–9.5	205 nm (10 mM)	Possibly unstable[f]
Ammonia.HCl	9.2	8.0–10.5	200 nm (10 mM)	
o-Boric acid	9.1	8.0–10.5	200 nm (10 mM)	
mono-*o*-borate	12.7	11.5–14.0	200 nm (10 mM)	
1-methylpiperidine.HCl	10.1	9.0–11.5	215 (10 mM)	Possibly unstable[f]
Triethylamine.HCl	11.0	9.5–12.5	<200 (10 mM)	Possibly unstable[f]

[a] Buffer composed of buffer acid plus ionized acid, such as for phosphoric acid, H_3PO_4 and $H_2PO_4^-$.
[b] Aqueous solutions; absorbance <0.5 AU at wavelengths above cutoff
[c] Claimed to attack stainless steel; we have experienced equipment problems associated with the long-term use of citrate buffer.
[d] Use of this buffer is impractical, because of loss of CO_2 from the reservoir.
[e] 1,3-Bis[tris(hydroxymethyl)methylamino] propane.
[f] Ammonium carbonate is unstable at pH-7.0 (loss of CO_2 to atmosphere), but stable at pH-8.5 [11]; amine buffers tend to oxidize, with a large increase in UV absorbance.
[g] Tris(hydroxymethyl)aminomethane.

a slight tailing of the peak is seen in Figure 7.5 for a mobile-phase pH = 3.0. As the pH increases to 3.5 (Fig. 7.5*b*) and 4.0 (Fig. 7.5*c*), the peak progressively broadens and becomes more distorted (peak fronting)—the result of decreased buffer capacity. The pK_a values of solute and buffer are 3.8 and 2.1, respectively, so as pH increases above 3.0 in this example, the difference between the buffer pK_a value and mobile-phase pH (factor 1 above) increases, while the difference between mobile-phase pH and the pK_a of the solute decreases (factor 2 above). Together, this results in a decrease of the effective buffer capacity as the mobile-phase pH increases from 3 to 4—and a progressive deterioration of peak shape. Poor peak shape in each of the examples of Figure 7.5 could be improved by any of the experimental factors 3 to 5 on p. 312 (increased buffer concentration, etc.)

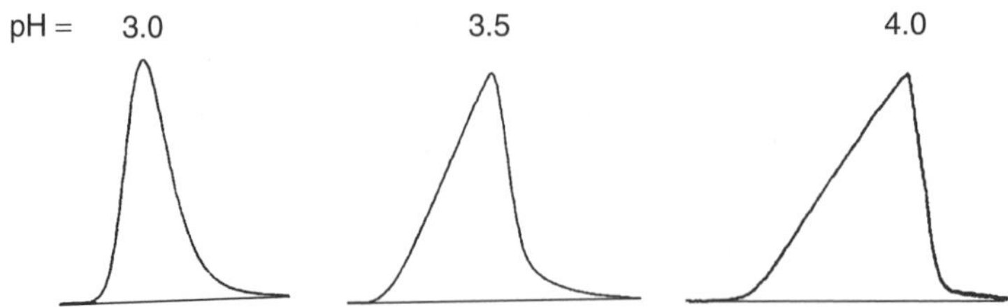

Figure 7.5 Effect of insufficient buffer capacity on peak shape for 3,5-dimethylaniline as solute. Conditions: column, 250 × 4.6-mm cyano column (5-μm particles); 25% methanol-buffer, buffer is 50-mM potassium monophosphate; 35°C; 1 mL/min. Adapted from [4].

When we wish to reduce buffer concentration for any reason (limited buffer solubility, increased UV absorbance, etc.), we need to optimize other contributions to buffer capacity (factors 1, 2, 4, or 5 on p. 312); for example, choose a mobile-phase pH that provides either minimal or maximal ionization of the sample (factor 2). For further discussion of buffer capacity in RPC, see [5].

7.2.1.2 Other Buffer Properties

The remaining buffer properties listed at the end of Section 7.2.1 should also be considered when developing a RPC separation for an ionic sample.

Buffer Solubility. Organic buffers are usually adequately soluble in all organic-water mobile phases (0–100% B), but many inorganic buffers have limited solubility in mobile phases which are predominantly organic (high %B). Consequently there is a danger that combining the A- and B-solvents may result in buffer precipitation, which could lead to blockage of the column or HPLC equipment. If there is any doubt as to whether a mobile phase might precipitate, the complete solubility of the buffer in the final mobile phase should be confirmed first (over the intended pH range), especially when the A- and B-solvents are mixed by the HPLC pumping system. Thus varying proportions of the A- and B-solvents can be combined manually in a container and observed over a period of 30 minutes or so. If any cloudiness develops, or a precipitate is observed for a given mobile-phase composition (%B), mobile phases of that %B or higher should be excluded or the buffer concentration should be reduced (see the discussion of [5, 6] for further details). Buffer solubility is of special concern for separations by gradient elution (Section 9.3.1).

Buffer solubility is affected by several separation conditions [6]. The buffer counter-ion is one such factor; for acidic buffers such as phosphate, buffer solubility usually increases as

sodium salt (least soluble) < potassium salt < ammonium salt (most soluble)

Similarly buffer solubility varies with the organic solvent (B-solvent), provided that comparisons are made for the same %B and pH:

tetrahydrofuran (least soluble) < acetonitrile < methanol (most soluble)

Buffer solubility is also affected by the relative ionization of the buffer; as the charge on the buffer ion increases (e.g., $HPO_4^{=}$ vs. $H_2PO_4^{-}$), the buffer becomes less soluble in high-%B mobile phases. Thus the choice of buffer and other separation conditions permits a considerable control over buffer solubility. Because most isocratic RPC separations of ionic samples are carried out with mobile phases of <60% B, however, buffer solubility is usually not a problem. To a lesser extent this is also true for gradient elution because, when buffer is added only to the A-solvent (the usual practice for inorganic buffers), buffer concentration in the mobile phase becomes inversely proportional to %B (also true for isocratic elution method development).

Buffer solubility is usually only an issue for inorganic buffers, especially phosphate. One study [6] has reported that potassium phosphate has an ambient-temperature solubility at pH-7 of 10 mM for either 85% MeOH-water or 75% ACN-water (with higher solubilities at lower %B, and vice versa for higher %B). At pH-3, a solubility of 10 mM can be achieved with 85% MeOH or 85% ACN. A phosphate-buffer concentration of 1 to 2 mM, combined with other favorable choices from the list "effective buffer capacity" of Section 7.2.1.1, should allow %B values as high as 90% for either methanol or acetonitrile as B-solvent. Fortunately, %B-values this high are rarely required for ionized compounds, in which case buffer solubilty may no longer be an issue.

Detector Requirements. The absorbance of the buffer is proportional to buffer concentration and adds to the absorbance of the water-organic mixture used for the mobile phase. Table 7.1 provides a rough guide for assessing whether a given buffer will result in a significant increase in the UV absorbance of the mobile phase. The influence of buffers on gradient baseline drift is illustrated in Figures 17.7 to 17.10. Additional information on buffer absorbance versus wavelength is provided in Table I.2 of Appendix I.

Mass spectrometric detection (LC-MS) requires a volatile buffer. Common choices include trifluoroacetic acid (TFA), acetic acid, formic acid, and their ammonium salts. For the separation of basic compounds at low pH, with volatile buffers such as formic or acetic acid, it is preferable to select a mobile phase with a higher ionic strength; that is, a higher buffer concentration and a pH where the buffer is significantly ionized (i.e., choose a mobile-phase pH that is fairly close to the pK_a value of the buffer [7]). Otherwise, even small weights of injected sample can result in column overload and peak tailing because of the ionic repulsion of retained molecules of protonated bases BH^+ (Section 15.3.2.1). For example, ammonium formate can be used as buffer at a pH of 3.5 to 4.0, whereas the use of formic or acetic acid alone (at a lower pH) provides much less ionization of the buffer. With reduced buffer ionization, column overload and peak tailing become more likely. For additional information on the best choice of buffer for MS detection, see Section 4.14.

Ion-Pairing by the Buffer. So far we have assumed that the sole effect of the buffer on sample retention and separation is to control mobile-phase pH and sample ionization. Additionally an ionized buffer X can interact with an ionized solute

A^- or BH^+ by ion-pairing:

$$\begin{array}{rcl} & \text{ionized solute} & \text{ion pair} \\ (\text{acids}) & A^- + X^+ \Leftrightarrow & A^-X^+ \quad (7.5) \\ (\text{bases}) & BH^+ + X^- \Leftrightarrow & BH^+X^- \quad (7.5a) \\ & \text{hydrophilic (less retained in RPC)} & \text{hydrophobic (more retained in RPC)} \end{array}$$

Buffers that ion-pair are usually organic ions, and they tend to be more hydrophobic than inorganic ions. For example, trifluoroacetate is a significant ion-pairing agent (Section 13.4.1.2), whereas phosphate is not. Ion-pairing by inorganic buffers is usually not significant, although some studies suggest that even phosphate may undergo weak ion-pairing with protonated bases [8]. For further information on ion-pairing by the buffer, see Section 7.4.2.1.

Buffer Stability and Equipment Compatibility. Inorganic buffers and carboxylic acids are usually stable, but amines are prone to oxidation with a consequent increase in their UV absorbance at wavelengths <250 nm. For example, although pure triethylamine (TEA) should be transparent at wavelengths $\geq$ 200 nm, one study [9] reported that a 1% aqueous solution of TEA (90 mM) has an absorbance >2 AU at 220 nm.

Citrate buffers have been claimed to attack stainless steel, whereas other reports contradict this claim as applied to HPLC equipment [10]. One laboratory has experienced occasional equipment problems that appeared associated with the use of citrate buffer [11], but this behavior has not been confirmed. While it appears likely that citrate concentrations <10 mM can be used without concern, and its use is convenient for applications where pH is to be varied continuously over the range $2 < \text{pH} < 7$, caution is nevertheless advised. Overall, the (real or imagined) problems in the use of citrate, its relatively high UV cutoff of 230 nm (Table 7.1), and the availability of favorable alternative buffers (e.g., phosphate plus acetate) make citrate a seldom-used buffer by most workers.

7.2.1.3 Preferred Buffers

UV detection in the 200 to 220 nm region is often used for samples with low concentrations of compounds that absorb poorly at higher wavelengths. Phosphate buffers have been preferred for this application, as long as the mobile-phase pH can be accommodated within the ranges specified in Table 7.1 ($\text{pH} \leq 3.5, 6.0 \leq \text{pH} \leq 8.5$, or $\text{pH} \geq 11.0$). Formate and acetate cover the pH range of 2.5 to 6.0, can be used at wavelengths of 210 nm or higher, and are volatile for LC-MS detection; for LC-MS detection at low pH with basic solutes, ammonium acetate or formate buffers provide an increase in ionic strength that allows higher sample concentrations without peak tailing [12]. Phosphate, acetate, and formate are the most commonly used buffers for separations with a mobile-phase $\text{pH} < 8.0$.

There is increasing use of higher pH mobile phases ($\text{pH} > 8$) as a result of the development of RPC columns that are stable at high pH (Section 5.2.5). Borate and ammonia have been used to some extent as buffers for high-pH operation, but note the further discussion of Section 5.8 concerning column stability at high pH. See also Appendix II for details on the more convenient preparation of some common buffers of required pH.

7.2.2 pK_a as a Function of Compound Structure

When selecting a range of mobile-phase pH values within which to carry out method development (i.e., optimization of pH), it can be useful to know the approximate pK_a values of the various sample components. This information allows the mobile phase to be restricted within an appropriate range of pH values. For example, a mobile-phase pH range that varies from pK_a − 1 to pK_a + 1 (for the sample) is useful for controlling retention and selectivity by changes in mobile-phase pH. Similarly a mobile-phase pH outside this range will result in a more robust method that is less sensitive to small (unintended) changes in pH. Values of pK_a vary widely for different organic compounds, but a large number of pK_a values have been determined experimentally [13], and additional pK_a values can be estimated on the basis of solute molecular structure. One very extensive source of experimental plus predicted pK_a values (ACD/pK_a DB: pK_a prediction) is available from Advanced Chemistry Development (Toronto, Canada).

Exact pK_a values for sample components are not required in RPC method development, and in many cases the chemical structures of sample components may not even be known. Values of pK_a are determined by ionizable acid or base groups attached to the solute molecule, for example, –COOH, $-NH_2$. This means that estimates of pK_a can be obtained from literature pK_a values for compounds of similar functionality (e.g., benzoic acid, as a representative of aromatic carboxylic acids). Table 7.2 summarizes pK_a values in water for some common acid- or base-substituent groups present in typical sample molecules. It is also possible to infer values of pK_a from experimental plots of retention against pH, as for peaks 2 and 4 in Figure 7.3*a*.

7.2.3 Effects of Organic Solvents and Temperature on Mobile-Phase pH and Sample pK_a Values

This detailed section is less essential in everyday operation, so the reader may wish to skip to Section 7.3. Nevertheless, the conclusions of this section are potentially useful for an accurate interpretation of the relationship between sample retention and mobile-phase pH.

Literature values of pK_a for different compounds (as in Table 7.2) are usually reported for aqueous solutions at near-ambient temperatures. Quite often, however, RPC separations of ionic samples are carried out at higher temperatures, with mobile phases that contain varying amounts of organic solvent. Both mobile-phase pH and values of pK_a for the sample can be affected by added organic (specifically, the value of %B) and by temperature. However, a knowledge of the true (mobile-phase) pH and solute pK_a values as a function of %B and temperature has little practical importance, so far as *routine* RPC assays are concerned; it is only important that %B and temperature are maintained constant for all runs so that solute ionization and retention remain unchanged from run to run. In the case of *method development*, however, approximate values of pK_a can be useful for selecting the pH of the mobile phase (as discussed above). It is useful in this connection to define an "effective" pK_a value for the solute, which takes into account the effects of %B

Table 7.2

Approximate pK_a Values for Acidic or Basic Functional Groups (aqueous solutions)

	pK_a			
	Acid		Base	
Group	Aliphatic	Aromatic	Aliphatic	Aromatic
Sulfonic acid, $-SO_3H$	1	1		
Amino acid, $-C(NH_2)-COOH$[a]	3		10	
Carboxylic acid, $-COOH$	5	4		
Thiol, $-SH$	10	7		
Purine		3		9
Phenol, $-OH$		11		
Pyrazine			1	
Sulfoxide, $-SO-$			2	
Thiazole			2	
Amine, $-NH_2$, $-NR_2$			10	5
Pyridine				5
Imidazole				5
Piperazine			10	

Note: Values can vary by 1 to 2 pK_a units or more as a result of adjacent groups in the molecule.
[a]See Figure 13.1 for values of individual amino acids.

and temperature on the pK_a values of both solute and buffer (Tables 7.1 and 7.2). Effective pK_a values for the solute can be used with the pH of the buffer (*not* the buffer-organic mobile phase) to estimate solute ionization and retention as a function of pH. An effective pK_a value is equivalent to the value that can be inferred from an experimental plot of retention against mobile-phase pH (i.e., *buffer* pH, as in Fig. 7.2).

7.2.3.1 Effect of %B on Values of Effective pK_a for the Solute

The pH of the mobile phase depends on the pK_a value for the buffer, while a change in %B will affect pK_a values of both the buffer and the solute [14–16]. If the buffer and solute are each acidic (e.g., phosphate buffer and a carboxylic acid solute), changes in pK_a with %B will be similar for both solute and buffer—and hence cancel approximately. For the latter case, effective pK_a values for the solute can be assumed equal to the values of Table 7.2. The same will be true when both the buffer and solute are basic (i.e., effective solute pK_a values that are similar to values measured in water at room temperature). However, when the buffer is acidic and the sample is basic, and vice versa, the apparent change in pK_a with %B can be substantial, so that effective pK_a values for the solute will no longer be the same as literature values measured in water.

Commonly used RPC buffers are more often acidic than basic (e.g., phosphate, acetate, formate). For the case of basic solutes and acidic buffers (with methanol as B-solvent), a *decrease* in the apparent pK_a by about 0.03 units can be expected for each 1% increase in %B (experimental data of [17, 18]). For example, consider the separation of several substituted anilines, using 25% methanol-phosphate buffer, as reported in [17]. Effective pK_a values for these solutes would be expected to be $25 \times 0.03 = 0.75$ units lower than literature values. For eight different solutes with literature pK_a values of 2.7 to 5.3, experimental pK_a values (as in Fig. 7.2) for this sample were lower by an average 0.7 ± 0.1 units (1 SD). Similarly, for six substituted benzoic acids separated with 40% methanol and acetate buffer [17], effective pK_a values were the same as literature values (± 0.1 unit, 1 SD). The effect of added acetonitrile on effective pK_a values is similar to that for the addition of methanol.

The effect on relative retention of changes in effective pK_a values with %B is equivalent to a change in mobile-phase pH (the "effective" pH of the mobile phase), which suggests that a change in %B can have a larger effect on relative retention and selectivity for ionic samples than for neutral samples. This has been observed for gradient elution as a function of gradient steepness [19, 20], which is equivalent to a change in %B for isocratic elution (Section 9.1.3).

In the discussion above, the mobile-phase pH is equated to that of the measured pH of the aqueous buffer, a procedure that we recommend for reasons given in Section 7.2.1. The pH of the final, water-organic mobile phase could be measured instead (less conveniently, and less reproducibly), as suggested by some workers [14–16]. However, possible changes in solute pK_a values with %B would still require correction; that is, the use of pH values measured in the mobile phase does not solve the problem of solute pK_a values that vary with %B. Again, *we strongly recommend measuring the pH of the buffer, not the final mobile phase.*

7.2.3.2 Effect of Temperature on Values of pK_a

Values of pK_a for both the buffer and sample can vary with temperature [21–25]. For protonated basic solutes, a lowering of effective pK_a values with increasing temperature [24] can lead to a decrease in solute ionization and an atypical *increase* in retention at higher temperatures. Because of such changes in pK_a with temperature, significant changes in relative retention with temperature can also result for ionic samples—more so than for neutral samples [19, 20]. That is, temperature selectivity will generally be greater for ionic samples. While the correction of pK_a values for a change in temperature should be possible (as above for a change in %B), at present there are no simple guidelines for this purpose. As temperature is seldom varied over a very wide range (e.g., 30–50°C is typical), the effect of temperature on solute pK_a values will usually be small and can therefore be ignored when using estimated values of pK_a for method development.

7.3 SEPARATION OF IONIC SAMPLES BY REVERSED-PHASE CHROMATOGRAPHY (RPC)

Reversed-phase chromatography (RPC) should be a first choice for the separation of mixtures of ionizable organic compounds. Method development for the RPC separation of ionic samples (Section 7.3.3) proceeds in similar fashion as for neutral samples, with some important differences that are developed in the remainder of this section. The following information on RPC separation (Sections 7.3.1, 7.3.2) should be useful for both method development and in troubleshooting routine separations.

7.3.1 Controlling Retention

Following an initial experiment, mobile-phase strength (%B) can be varied to obtain a desirable retention range (e.g., $1 \leq k \leq 10$), the same way as for neutral samples (Sections 2.5.1, 6.2.1). Alternatively (and generally preferable), initial experiments can be carried out using gradient elution, as discussed in Section 9.3.1. Once a value of %B has been selected for the reasonable retention of the sample, the next step is the adjustment of separation selectivity for optimal relative retention and maximum resolution.

7.3.2 Controlling Selectivity

For the separation of neutral samples, selectivity can be varied by changing solvent strength (%B), temperature, the B-solvent, or the column. These variables also affect the separation of ionic samples, usually to a greater extent than for neutral samples. In addition separation selectivity for ionic samples is strongly affected by mobile-phase pH, and to a lesser extent by the kind and concentration of the buffer. In the past, mobile-phase additives for the suppression of silanol activity (alkylamines, quaternary ammonium compounds) have been added to the mobile phase, and these additives can further change selectivity. Today, however, the widespread preference for type-B columns (Section 5.2.2.2) has virtually eliminated the need for silanol suppression by such additives. Depending on the nature of the sample, all of these separation conditions may exert a significant effect on relative retention and resolution, as discussed below. One study [25] ranked the relative importance of different conditions in affecting the selectivity of RPC separations of ionic samples as follows:

$$\text{pH (most important)} > \text{solvent type} \approx \text{column type} > \%\text{B} > \text{temperature}$$
$$\gg \text{buffer concentration and type (least important).}$$

7.3.2.1 Mobile-Phase pH

The ability of a change in mobile-phase pH to affect the relative retention of ionizable samples is apparent from the example of Figure 7.3. We have noted that solute retention changes with pH only when the pH of the mobile phase is within $\approx \pm 1.5$ units of the pK_a value of the solute (Fig. 7.2). Consequently, if mobile-phase pH is to have an effect on separation selectivity, the pH must be similar to pK_a values of the sample constituents. Carboxylic acids and amines are the most commonly encountered examples of ionic solutes; reference to Table 7.2 suggests for a mobile-phase pH between 2 and 3 that bases ($pK_a \approx 5-10$) will be fully ionized, and acids ($pK_a \approx 5$) will be in the neutral form. This is only approximately true, since it overlooks the effects of %B and temperature on values of pK_a, as well as changes in pK_a that can result from the presence of different substituents in the solute molecule. Consequently, while pH selectivity is usually reduced at low pH, it can still be significant—depending on the sample and the value of %B.

Another example of a change in relative retention with pH is shown in Figure 7.6 for a mixture of substituted benzoic acids (peaks 1–4) and anilines (peaks 5–7). As mobile-phase pH is increased from 3.2 to 4.3 (Fig. 7.6*a–c*), the retention of acids 1 to 4 decreases, while the retention of bases 5 to 7 (shaded peaks) increases. For a mobile-phase pH of 4.3 or higher, the acidic compounds 1 to 4 are mainly in the ionized form and therefore retained weakly; similarly, at higher pH the basic compounds 5 to 7 are largely non-ionized and more strongly retained. As a result for a mobile-phase pH > 4 there is a separation of these acids and bases into two groups of peaks. An optimum mobile-phase pH = 3.4 (Fig. 7.6*d*) provides acceptable resolution of the sample. However, even at this relatively low pH, the separation of Figure 7.6*d* is seen to be somewhat sensitive to small changes in pH;

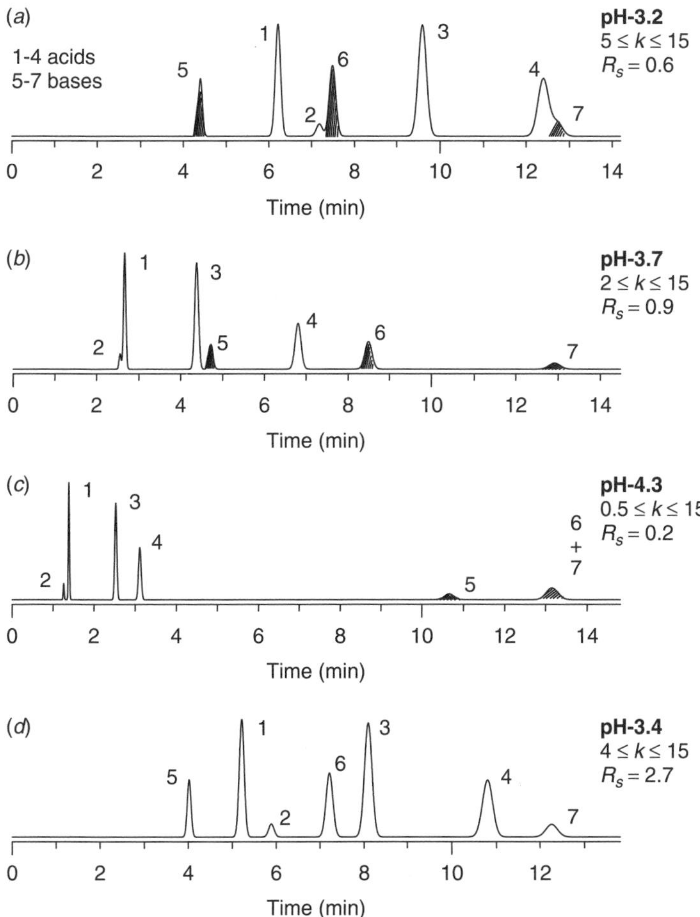

Figure 7.6 Effect of mobile-phase pH on the RPC separation of a mixture of acids and bases. Sample: *1*, 2-fluorobenzoic acid; *2*, 3-chlorobenzoic acid; *3*, 3-nitrobenzoic acid; *4*, 3-fluorobenzoic acid; *5*, 3,5-dimethylaniline; *6*, 4-chloroaniline; *7*, 3-chloroaniline. Conditions: 150 × 4.6-mm C_{18} column (5-μm particles); mobile phase, 13% acetonitrile-buffer (buffer is citrate plus phosphate); 2.0 mL/min; 35°C. Peaks for basic compounds 5 to 7 are shaded. Chromatograms based on data of [19].

the buffer pH should therefore be maintained within ±0.1 pH units of the specified value for reproducible separation. The presence of partly ionized acids and bases in the same sample for $3 < \text{pH} < 4$ is the reason for both changes in relative retention with pH and the marginally robust nature of the separation of Figure 7.6*d*.

7.3.2.2 Solvent Strength (%B) and Temperature

In Figure 7.7 the effects of a change in %B and temperature on relative retention are illustrated for the sample of Figure 7.6, in each case for the same mobile-phase pH = 3.2. It is apparent that significant changes in relative retention occur as either temperature or %B is changed. These changes in relative retention can be attributed to the same factors that are operative for the separation of neutral compounds (e.g., Fig. 6.26), *plus* more important changes in the "effective" mobile-phase pH as a result of change in either %B or temperature (Section 7.2.3). Consequently a change in %B or temperature usually has a larger effect on the relative retention (and resolution) of ionic samples than neutral samples, as noted above. In the examples of Figure 7.7 we see an increase in the *relative* retention of bases 5 to 7 (shaded peaks) compared with acids 1–4 as either %B or temperature increase (while the *absolute* retention of all compounds decreases). This implies that the pK_a values of the sample bases have decreased as a result of an increase in either %B or temperature, which is equivalent to an increase in mobile-phase pH for these basic compounds

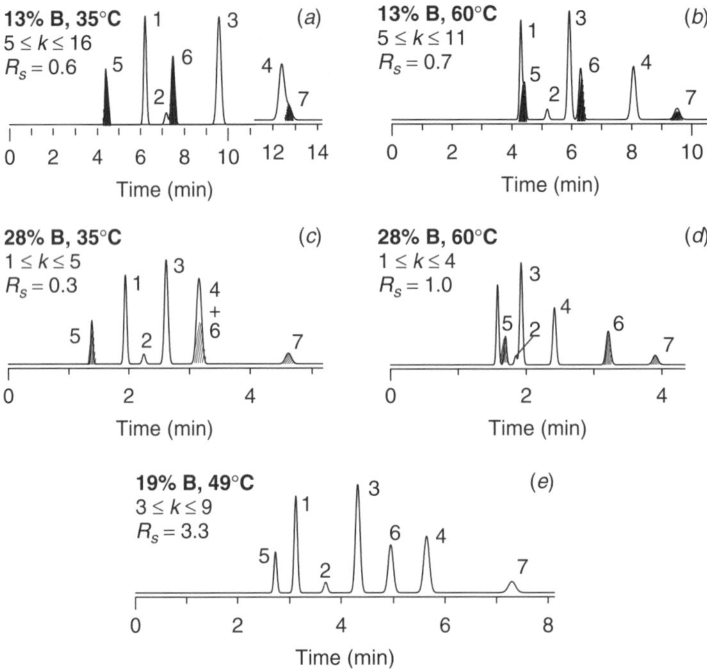

Figure 7.7 Effect of mobile-phase strength (%B) and temperature on the separation of a mixture of acids and bases. Sample: same as in Figure 7.6; conditions also the same, except pH = 3.2, and values of %B and temperature are noted in the figure. Peaks for basic compounds 5 to 7 are shaded. Chromatograms based on data of [19].

(see the discussion of Section 7.2.3 for the case of an acidic buffer and basic solutes). A maximum resolution of $R_s = 3.3$ is observed in Figure 7.7*e* for 19% B and 49°C. The latter optimized separation can be found by trial-and-error changes in %B or temperature, but a more efficient procedure is the use of computer simulation (Section 10.2.2).

The remainder of this section represents an alternative way of interpreting the changes in retention of Figure 7.7. Since it is not essential to an understanding of the effects of %B and temperature on retention, the reader may prefer to skip to the following Section 7.3.2.3.

The effects of changes in %B, temperature, or other conditions on relative retention can be further interpreted in terms of

$$\text{(acids, bases)} \quad k = k^0(1 - F^{\pm}) + k^{\pm}F^{\pm} \tag{7.4}$$

Because $k^0 \gg k^{\pm}$, this relationship can be simplified to

$$k \approx k^0(1 - F^{\pm}), \tag{7.6}$$

where k^0 is the value of k for the neutral (non-ionized) molecule, and $F^{\pm}$ is the fractional ionization of the solute for a given mobile-phase pH. An increase in either temperature or %B will lead to a decrease in values of k^0 for the solute, regardless of whether it is ionic or neutral. Additionally a change in conditions that also changes the "effective" mobile-phase pH (and therefore values of $F^{\pm}$) can have a further effect on the separation of an ionic sample. Thus in Figure 7.7 an increase in either %B or temperature appears to increase mobile-phase pH slightly (equivalent to a decrease in pK_a values for these solutes)—with a preferential retention of basic solutes 5 to 7.

7.3.2.3 Solvent Type

A change in solvent type (e.g., methanol replacing acetonitrile) is expected to have a comparable effect on the relative retention of both ionic and neutral samples. In addition any change in "effective" pK_a values as a result of this change in B-solvent can further affect the relative retention of ionic samples—similar to the case of a change in %B or temperature. The latter effect should lead to larger changes in relative retention for ionic as opposed to neutral samples when the B-solvent is changed. This was observed to be the case in one study [26], where the average change in values of α for 45 neutral solutes was ± 0.04 units for a change in B-solvent from 50% ACN to 45% ACN + 5% MeOH. The corresponding change in α for 22 ionic compounds was ± 0.09 units (twice as large as for neutral solutes). See Sections 6.3.2 and 6.4.1 for a related discussion of the effect of solvent type on the separation of non-ionic samples.

7.3.2.4 Column Type

Separations of a neutral sample with four columns of different type (different ligands) were illustrated in Figure 6.14. Similar separations of an ionic sample are shown in Figure 7.8 for three of the same columns (note that an "ionic" sample may also contain neutral solutes, as in this example). A comparison of results for these two

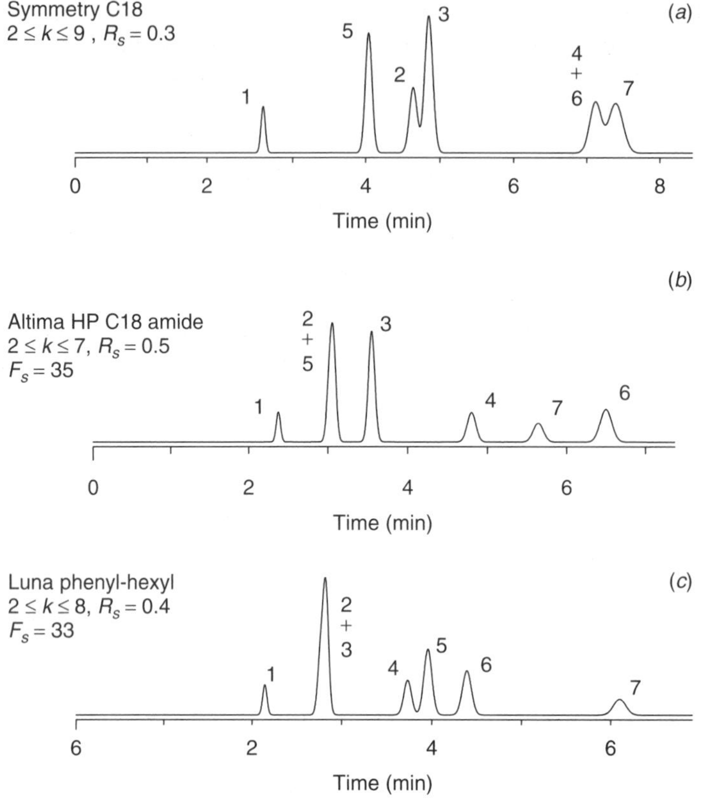

Figure 7.8 Separation of an ionic sample as a function of column type. Sample: *1*, 5-phenylpentanol; *2*, 4-*n*-hexylaniline; *3*, toluene; *4*, ethylbenzene; *5*, 4-*n*-butylbenzoic acid; *6*, *trans*-chalcone; *7*, mefenamic acid. Conditions: 150 × 4.6-mm columns (5-μm particles); 50% acetonitrile-buffer (buffer is pH-2.8 phosphate buffer); 35°C, 2.0 mL/min. F_s values versus Symmetry C18. Recreated separations based on data of [27, 28].

samples (Fig. 6.14 vs. Fig. 7.8) shows somewhat greater changes in relative retention for the ionic sample of Figure 7.8; this is expected, as discussed below. We also see marginal resolution ($R_s < 1$ for the least-resolved peak-pair) for each column in Figure 7.8, similar to the result of Figure 6.14 for the separation of a neutral sample with these same columns. That is, a change in just the column is not likely to provide a significant improvement in overall ("critical") resolution. However, a change in column *combined* with optimized values of other conditions (e.g., %B and temperature) is much more promising, as shown in Figure 6.27 for the separation of a neutral sample on these same columns. Differences in selectivity among these three columns can be described by the column-comparison function F_s (Section 5.4.2): Symmetry/Altima, $F_s = 35$; Symmetry/Luna, $F_s = 33$. In each case these F_s values suggest significant differences in column selectivity, although much larger differences can be achieved with other pairs of columns.

Another example of the effect of the column on selectivity is shown in Figure 7.9 for the separation of a mixture of five, fully protonated strong bases (1–5), five partly ionized weak acids (6–10), and a neutral reference compound (11; shaded peak). In Figure 7.9*a*, mixtures of *either* the strong bases or weak acids plus neutral compound 11 are separated on each of these three columns. In Figure 7.9*b* corresponding separations of samples containing all 11 compounds are shown. The relative retention of the fully protonated strong bases (1–5) of Figure 7.9 is most affected by values of the ion-exchange capacity C for the column (Section 5.4.1); larger values of C mean a greater retention of protonated bases. Values of C at pH-2.8 for these columns are, respectively, −0.47 (Inertsil), −0.30 (Symmetry), and 0.18 (Discovery). As expected, the relative retention of basic solutes 1 to 5 increases in proceeding from the Inertsil to the Symmetry to the Discovery column (note the retention ranges for peaks 1–5 in Fig. 7.9*b*, indicated at the top of each chromatogram by arrows). The relative retention of the weak acids 6 to 10 and neutral compound 11 are quite similar on the three columns because their retention is not affected by values of C.

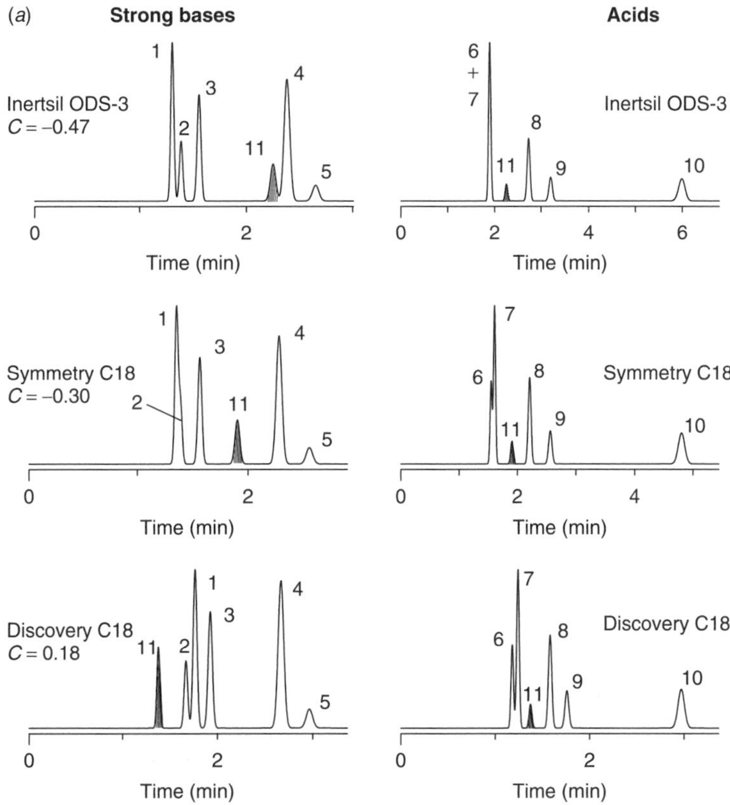

Figure 7.9 RPC separation of an ionic sample as a function of column type. Sample: (bases) *1*, propranolol; *2*, prolintane; *3*, diphenhydramine; *4*, nortriptyline; *5*, amitriptyline; (acids) *6*, 2-fluorobenzoic acid; *7*, 3-cyanobenzoic acid; *8*, 2-nitrobenzoic acid; *9*, 3-nitrobenzoic acid; *10*, 2,6-dimethylbenzoic acid; (neutral compound, shaded) *11*, benzylalcohol. Conditions: 150 × 4.6-mm columns (5-μm particles); 30% acetonitrile-buffer (buffer is pH-2.8 phosphate buffer); 35°C; 2.0 mL/min. (*a*) Separation of acids 6 to 10 or bases 1 to 5, in each case with neutral-compound 11 added; (*b*) separation of all 11 compounds. Recreated separations based on data of [26, 29].

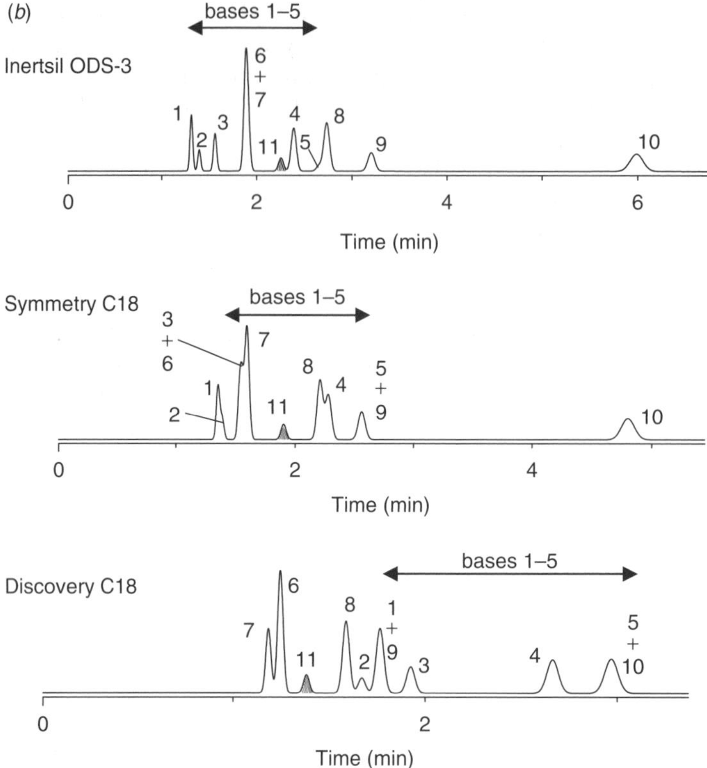

Figure 7.9 (*Continued*)

7.3.2.5 Other Conditions That Can Affect Selectivity

Conditions that are less used today for the control of RPC selectivity include:

- buffer type (e.g., phosphate, acetate, ammonium)
- buffer concentration
- amine modifiers

Buffer type is not commonly considered as a means of controlling selectivity for the separation of ionic samples. As discussed in Section 7.2.3, however, buffer type can affect the "effective" pK_a of a solute, which is equivalent to a change in pH. The largest changes in relative retention will occur when a basic buffer (e.g. ammonium, triethylamine, etc.) replaces an acidic buffer such as phosphate or acetate, and vice versa. Another way in which buffer type can contribute to selectivity is by ion pairing (Section 7.2.1.2). More hydrophobic buffers such as trifluoroacetate TFA or (especially) heptafluorobutyrate HFBA can ion-pair with protonated bases BH^+ and selectively increase their retention [30]. The increase in retention for protonated bases increases with the positive charge on the solute molecule, as in the case of peptides which contain multiple, basic amino-acid residues (Fig. 13.8; Section 13.4.1.2). The use of TFA and HFBA as buffers is not subject to problems that are common for other ion-pair separations (Section 7.4.3).

Buffer concentration usually has only a minor effect on relative retention for separations at low pH on modern (type-B) alkylsilica columns [26]. However, for separations at pH > 6 and/or older, type-A columns (Section 5.2.2.2), protonated bases can be retained by ion exchange as a result of interaction with ionized silanols of silica-based column packings. Ion-exchange retention decreases as mobile-phase ionic strength increases (Section 7.5.1), with the result that an increase in buffer concentration will tend to decrease the retention of protonated bases [31].

Amine modifiers, such as triethylamine or tetrabutylammonium salts, have been added to the mobile phase in the past, primarily as a means of suppressing unwanted silanol interactions (Section 7.3.4.2). By interacting with stationary-phase silanol groups, amine modifiers can suppress ion exchange by the sample, thereby resulting in decreased retention for protonated bases. These modifiers are little used today because (1) modern RPC columns (type-B) are largely free of unwanted silanol interactions and (2) the use of amine modifiers can be inconvenient, requiring long column-equilibration times in some cases.

7.3.3 Method Development

Method development is similar for the RPC separation of either ionic or neutral samples, as summarized in Figure 6.21*a*. Seven method-development steps are defined there, of which only one (step 3: choosing separation conditions) differs significantly for the separation of ionic samples. The choice of separation conditions for either ionic or neutral samples is summarized in Figure 6.21*b* and includes the following steps:

1. choose starting conditions
2. select %B for $1 \leq k \leq 10$
3. adjust conditions for improved selectivity and resolution
4. vary column conditions for a best compromise between resolution and run time

Method development for ionic samples differs from that for neutral samples mainly with respect to steps 1 and 3 above. *Method development should always start with a new (unused) column*, as exposure of a column to previous samples and conditions can change its selectivity so as to make it impossible to replicate the column at a later time.

7.3.3.1 Starting Conditions (Step 1)

Table 7.3 suggests conditions for the initial separation of a mixture of acids and/or bases, conditions that are similar to those recommended for the initial separation of neutral samples (Table 6.1). The main difference for ionizable samples is the need for a buffered mobile phase. Because ionic samples are usually less strongly retained in RPC, the value of %B for the initial mobile phase is likely to be a bit lower than for neutral samples. However, it is best to start development at 80% B, so as to reduce the risk of missing a late-eluted solute with a mobile phase that is too weak. Alternatively (and preferably), an initial gradient-elution run can be used to determine the best value of %B for isocratic separation (Section 9.3.1).

Table 7.3

Representative Conditions for the Separation of Ionic Samples by Means of Reversed-Phase Chromatography

Condition	Comment
Column[a]	Type: C_8 or C_{18} (type-B)
	Dimensions: 100 × 4.6-mm
	Particle size: 3 μm
	Pore diameter: 8–12 nm
Mobile phase	80 % acetonitrile-buffer; buffer is 10 mM potassium phosphate, adjusted to pH–2.5[b]
Flow rate	2.0 mL/min[b]
Temperature	30 or 35°C[b]
%B	Determined by trial and error[c]
Sample	Volume ≤ 50 μL
	weight ≤ 10 μg
k	$1 \le k \le 10$

[a]Alternatively, use a 150 × 4.6-mm column of 5-μm particles; *note that a new (unused) column should always be selected at the start of method development*.
[b]Initial values will be varied during method development (Section 2.5); the starting pH of the mobile phase can also be varied.
[c]Start with 80%B and adjust further as described in Section 2.5.1.

The choice of starting mobile-phase pH and buffer depends on (1) the separation goals and (2) what the chromatographer knows about the sample. We recommend carrying out initial separations with a mobile-phase pH of 2.5 to 3.0, using phosphate for UV detection, or ammonium formate for LC-MS. Problems arising from peak tailing (Section 7.3.4.2), column instability (Section 5.3.1), or a lack of method robustness (Section 12.2.6) are somewhat less likely for a mobile-phase pH of 2.5 to 3.0. For samples that contain strong bases, there is increasing use of high-pH mobile phases (pH > 8), in order to minimize the ionization of basic solutes during separation. Decreased sample ionization results in stronger retention, and can favor symmetrical peaks and more robust RPC methods. However, when the mobile phase pH is > 8, special columns are required to avoid the dissolution of the silica particles with resulting failure of the column (Sections 5.2.5, 5.3). The main advantage of a mobile-phase pH > 8 for strongly basic samples is that a larger sample weight can be injected (Section 15.3.2.1), with a resulting increase in detection sensitivity for an assay procedure, or increase in yield for preparative separations. Whatever mobile-phase pH is used, care should be taken to ensure adequate buffering capacity (Section 7.2.1.1).

7.3.3.2 Optimizing Selectivity (Step 3)

Any of the separation conditions described in Section 7.3.2 can be varied in order to improve relative retention and maximize resolution. The *simultaneous* variation of two different separation conditions will generally prove more effective; the same two-variable procedures described for neutral samples in Section 6.4.1 can be applied for ionic samples. We recommend that temperature and %B should be varied first over a range that results in solute k-values within the range $0.5 \leq k \leq 20$; see the example of Figure 7.7 and the discussions of Section 6.4.1.3 and [32]. If further changes in selectivity are needed, simultaneous changes in pH and solvent strength can be used for ionizable samples [33]. When varying two conditions simultaneously, simulation software (Section 10.2) is especially helpful for determining conditions that correspond to maximum resolution.

More than two conditions can be simultaneously optimized for the control of selectivity; for example, varying %B and temperature for different columns (Section 6.4.1.4) is a popular and effective strategy. A few other examples have been reported of the simultaneous optimization of three different variables [34, 35], each of which can be varied continuously (i.e., excluding column type as a variable). However, this approach can require a formidable number of experiments, for example, 32 experiments for the simultaneous optimization of %B, temperature, and pH in one example [34].

7.3.4 Special Problems

RPC separations of ionic samples are subject to two problems that do not occur for the separation of neutral samples.

7.3.4.1 pH Sensitivity

As noted in Section 7.2 for the RPC separation of ionizable samples, relative retention can be quite sensitive to small (unintended) variations in mobile-phase pH. The ability of most laboratories to replicate the pH of the buffer by means of a pH meter is typically no better than ±0.05 to 0.10 units; variations in mobile-phase pH of this magnitude may be unacceptable for some separations. *For this reason the robustness of the final method in terms of pH should be a major concern during method development for ionic samples.*

There are several ways in which the problem of pH sensitivity can be minimized. First, determine the pH sensitivity of the method. If the mobile-phase pH must be held within narrow limits (±0.1 unit or less), precise pH control can be achieved by accurately measuring the buffer ingredients (either by weight or volume), rather than by using a pH meter to adjust the buffer to a desired pH (see Appendix II for some examples). Second, as an alternative to the precise adjustment of pH in this way, carry out separations with mobile phases that are, respectively, 0.2 pH units higher and lower than the required pH. The inclusion of these chromatograms in the method procedure can be used by an operator to guide the correction of mobile-phase pH when needed (Section 12.8).

Finally, the best approach for a method that proves to be too pH sensitive is to re-optimize conditions so as to obtain a method that is more robust. This will sometimes require a change in pH to a value that differs by more than ±1 pH unit from the pK_a values of critical solutes (those whose resolution can be compromised by small changes in pH). Minor changes in other conditions can also result in a more robust separation. See the further discussion of Section 12.2.2.6 and [36].

7.3.4.2 Silanol Effects

Protonated basic solutes BH^+ can interact with stationary-phase silanols by ion exchange (a buffer in the potassium form is assumed):

$$BH^+ + SiO^-K^+ \Leftrightarrow BH^+SiO^- + K^+ \tag{7.7}$$

This interaction can lead to increased retention, peak tailing, and column-to-column irreproducibility. Problems of this kind are most pronounced when older, type-A columns (Section 5.2.2.2) are used, because type-A silica is contaminated by Al^{3+}, Fe^{2+}, and other heavy metals. Metal contamination increases silanol acidity, results in a higher concentration of SiO^- groups for all mobile-phase pH values, and likely contributes to poor column reproducibility. Newer columns made from purer, type-B silica are largely free of metal contamination, and fewer associated problems are encountered in their use.

Even when newer, type-B columns are used, the separation of basic compounds can lead to peak tailing [37]. The origin of peak tailing with type-B columns appears to differ for separations with mobile phases of high or low pH. For a mobile-phase pH < 5, tailing peaks usually resemble rounded right triangles, as in the example of Figure 2.15*e*. For a pH ≥ 6, exponential peak tailing as in Figure 2.15*a* is more often seen. Low-pH tailing is now believed due to charge repulsion between retained ionized molecules (Section 15.3.2.1; [38]). As a result the column overloads more quickly for basic samples than for neutral samples, and peak tailing can become noticeable for injections of more than 0.5 μg of a basic compound (assumes a column diameter of 4–5 mm). Low-pH peak tailing can be reduced somewhat by an increase in mobile-phase ionic strength. For example, the use of buffers at a mobile-phase pH that favors buffer ionization results in a higher ionic strength, even when buffer molarity is unchanged; this approach for reduced peak tailing of bases has been recommended when volatile formate buffers are used for LC-MS [39].

The tailing of basic samples on type-B columns at pH-7 and above is less well understood, but may be the result of slow sorption-desorption of molecules of BH^+ [37]. The extent of tailing is affected by the nature of the B-solvent [40, 41], with

$$\text{acetonitrile (worst)} > \text{methanol} \approx \text{tetrahydrofuran (best)}$$

Peak tailing is generally decreased by the use of higher column temperatures or higher %B, conditions that also favor lower values of k. The use of columns with smaller values of **C** (cation-exchange capacity values; Section 5.4.1) is likely to minimize peak tailing for a mobile-phase pH < 5. "Hybrid" particles (Section 5.3.2.2) do not exhibit silanol ionization below pH-8, and the peak shape of protonated bases is good for pH < 8 [16]. For a good review of peak tailing for basic solutes, see [16].

Small weights of undissociated carboxylic acids can exhibit tailing peaks for some columns (Section 5.4.4.1), but peak shape *improves* for larger samples. The origin of such peak tailing is as yet unknown. It is possible nevertheless to identify columns that are less likely to exhibit this problem (Section 5.4.4.1; [42]).

In the case of type-A columns, various means have been employed in the past to reduce ion exchange (Eq. 7.7) and associated deleterious effects [43, 44]:

- suppress silanol ionization (use low-pH mobile phases)
- suppress the ionization of basic solutes B (use high-pH mobile phases)
- suppress ion exchange (use high–ionic-strength mobile phases)
- block ionized silanols (add amine modifiers to the mobile phase)
- use end-capped columns

Silica-based RPC columns can degrade more rapidly when the mobile phase pH is <2.5 or >8.0, which limits the use of extreme pH to control silanol or solute ionization. Some type-B columns are now available for operation outside these pH limits (see the discussion of Section 5.3). Ion-exchange and related adverse silanol effects can also be minimized by the use of higher buffer concentrations; the buffer cation competes with the solute in the equilibrium of Equation (7.7). Buffer cations such as K^+, Li^+ and NH_4^+ are more effective than Na^+ in suppressing silanols and minimizing peak tailing. The addition to the mobile phase of amine modifiers such as triethylamine and dimethyloctylamine was popular at one time for improving the separation of basic samples, but today the predominant use of type-B columns has rendered these (inconvenient) additives unnecessary. End-capping the column (Section 5.3.1) tends to shield silanols from the solute and typically improves peak shape.

7.3.4.3 Poor Retention of the Sample

Very polar samples are poorly retained in RPC, as noted for neutral samples in Section 6.6.1. The same problem is even more common for ionic samples—because of the greater polarity of ionized molecules. However, there are additional means for increasing sample retention in this case. Poor retention of an ionic solute is usually due to its ionization, which can result in more than a 10-fold decrease in values of k. The simplest approach for solutes that are acidic or basic is a change in mobile-phase pH that results in decreased solute ionization. Alternatively, ion-pair chromatography (IPC, Section 7.4) can be used to similar effect, especially for permanently ionized solutes such as quaternary-ammonium compounds. Hydrophilic interaction chromatography (HILIC) is also effective for very polar samples (Section 8.6) and is usually a better choice than IPC for this purpose.

7.3.4.4 Temperature Sensitivity

The relative retention of ionized solutes tends to be more dependent on temperature than is the case for neutral samples. Therefore the need for accurate column thermostatting (Section 3.7) can be more important for ionic samples. The use of (unthermostatted) separations at ambient temperature is not recommended for any sample, and it is especially problematic for the separation of ionic samples.

7.4 ION-PAIR CHROMATOGRAPHY (IPC)

Ion-pair chromatography (IPC) can be regarded as a modification of RPC for the separation of ionic samples. The only difference in conditions for IPC is the addition of an ion-pairing reagent R^+ or R^- to the mobile phase, which can then interact with ionized acids A^- or bases BH^+ in an equilibrium process:

$$\begin{array}{rcl} & \text{ionized solute} & \text{ion pair} \\ \text{(acids)} & A^- + R^+ \Leftrightarrow & A^-R^+ \quad (7.8) \\ \text{(bases)} & BH^+ + R^- \Leftrightarrow & BH^+R^- \quad (7.8a) \\ & \text{hydrophilic solute (less retained in RPC)} & \text{hydrophobic ion-pair (more retained in RPC)} \end{array}$$

The use of IPC can thus create similar changes in sample retention as by a change in mobile-phase pH (Section 7.2), but with greater control over the retention of either acidic or basic solutes, and without the need for extreme values of mobile-phase pH (e.g., pH < 2.5 or > 8). Typical ion-pairing reagents include alkylsulfonates $R{-}SO_3^-$ (R^-) and tetraalkylammonium salts R_4N^+ (R^+), as well as strong (normally ionized) carboxylic acids (trifluoroacetic acid, TFA; heptafluorobutyric acid, HFBA [R^-]), and so-called chaotropes (BF_4^-, ClO_4^-, PF_6^-).

When first introduced in the 1970s, high-performance IPC was found to reduce peak tailing for basic solutes. This and its ability to increase the retention of weakly retained ionized acids and bases for acceptable values of k were primary reasons for its use at that time. Additionally IPC provides further options for the control of selectivity in the separation of ionic samples. Today the predominant use of type-B columns has reduced the importance of peak tailing, and we now have a better understanding of how best to control RPC selectivity. The poor RPC retention of very hydrophilic acids and bases (especially strong bases that remain ionized for pH < 8) can also be addressed in other ways, for example, (1) by the use of high- or low-pH mobile phases in order to minimize solute ionization and increase retention (combined with the use of columns that are stable at pH extremes) and (2) by the use of hydrophilic interaction chromatography (HILIC, Sect. 8.6). Consequently there is much less need for IPC today because of its greater complexity and other problems (Section 7.4.3).

When developing an HPLC separation, we recommend starting with RPC, followed by the addition of an ion-pairing reagent *only when necessary*. When, or for what applications, might IPC be recommended? IPC separation involves two additional variables (type and concentration of the IPC reagent) that can be used for further control of selectivity. As will be seen below, the effects of an added IPC reagent on solute retention are reasonably predictable, when we know whether a particular peak corresponds to an acid, base, or neutral. Consequently the retention of both acidic and basic solutes can be varied continuously so as to optimize their separation, when other changes in RPC conditions fail to achieve acceptable resolution.

IPC can also be used to narrow the retention range of a sample, so samples that might otherwise require gradient elution can be separated isocratically. An example is shown in Figure 7.10, for a proprietary sample that includes a drug-product X plus several preservatives and degradants. In Figure 7.10*a*, RPC separation is shown with a mobile phase of 30% methanol-buffer (pH-3.5). The neutral preservative, propylparaben PP, is strongly retained, while the basic drug X and its degradants

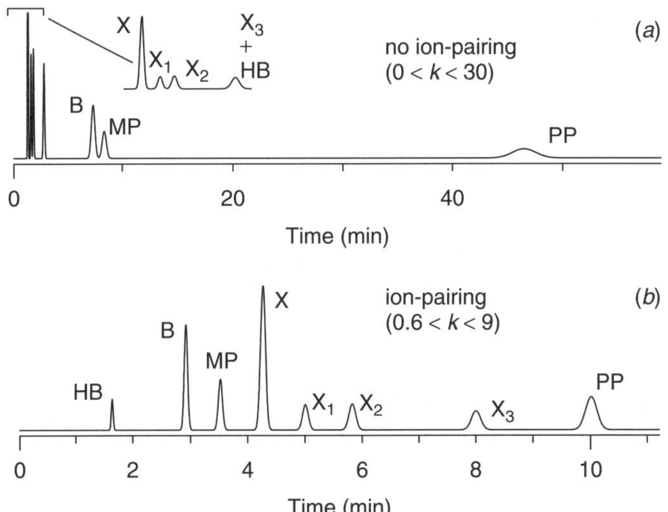

Figure 7.10 RPC separation of a proprietary mixture of acids, strong bases and neutrals. Sample: X, strongly basic (proprietary) drug substance; X1, X2, X3, strongly basic degradants of X; MP and PP, methyl and propyl paraben preservatives; B, benzoic acid; HB, hydroxybenzoic acid (degradant of MP and PP). Conditions: (*a*) 150 × 4.6-mm column (5-μm particles); 30% methanol-buffer mobile phases (buffer is pH-3.5 acetate); 30°C; 2.0 mL/min. (*b*) Same as (*a*), except mobile phase is 45% methanol-buffer plus 65-mM octane sulfonate; 1.5 mL/min. Adapted from [46].

X_1–X_3 are weakly retained (because they are in the protonated form as BH^+). Two other sample compounds, benzoic acid B (another preservative) and hydroxybenzoic acid HB (a paraben degradant), are acidic, while methyl paraben MP is also a neutral preservative. The separation of Figure 7.10*a* exhibits an excessive retention range ($0 \leq k \leq 30$), which would normally suggest gradient elution as an alternative (Section 9.1). Because compounds X–X_3 are strongly basic, an increase in their isocratic retention (relative to the rest of the sample) by an increase in pH was deemed impractical. Thus a mobile-phase pH > 8 would be necessary (requiring a column that is stable at high pH), but this would lead to $k \ll 1$ for acidic compounds B and HB, while having no effect on the retention of neutral compounds MP and PP. Thus no practical change in pH is able to narrow the retention range of this sample so as to provide *k*-values for all compounds in an acceptable range of values (e.g., $0.5 \leq k \leq 20$).

Isocratic elution was preferred for the sample of Figure 7.10, so the use of IPC was investigated as an alternative to gradient elution. The addition of a sulfonate IPC reagent would be predicted to lead to strongly increased retention for the sample cations (X–X_3), accompanied by a modest decrease in the retention of both sample acids (B and HB) and neutral compounds (MP and PP); see Section 7.4.1.2 below. The separation of Figure 7.10*b* was therefore carried out with octane sulfonate as IPC reagent (for a preferential increase in the retention of X–X_3), plus a stronger mobile phase (45% B vs. 30% B in Fig. 7.10*a*) for a reduction in *k* for the neutral solute PP. The resulting decrease in retention range ($0.6 \leq k \leq 9$) now allows the baseline separation of this sample within a reasonable time (11 min).

The remainder of this section provides a short description of the basis of ion-pair separation, followed by a discussion of how separation depends on various conditions. For further details, see [45] and Chapter 7 of [46].

7.4.1 Basis of Retention

Two possible retention processes or "mechanisms" exist for separation by IPC. As an example, we will use the ion-pairing of an ionized acidic solute A^- by a tetraalkylammonium IPC reagent R^+. The ion-pairing of a protonated basic solute B^+ by an alkylsulfonate IPC reagent R^- can be described similarly.

One hypothesis for IPC retention assumes that an ion-pair forms in solution, as described by Equation (7.8a). The resulting ion-pair A^-R^+ is retained by the column; that is, the solute retention equilibrium as described by Equation (2.2) in Section 2.2 is replaced by

$$A^-R^+(\text{mobile phase}) \Leftrightarrow A^-R^+(\text{stationary phase}) \tag{7.9}$$

According to this hypothesis, retention is governed by (1) the fraction of solute molecules A in the mobile phase that are ionized (determined by mobile-phase pH and the solute pK_a value), (2) the concentration of the IPC reagent and its tendency to form an ion pair (the equilibrium constant for Eqs. 7.8 or 7.8a), and (3) the value of k for the ion-pair complex A^-R^+ (which will be greater for more hydrophobic IPC reagents).

An alternative picture of IPC retention assumes that the IPC reagent is retained by the stationary phase, with retention then occurring by an ion-exchange process (Section 7.5.1), for example, for an ionized acid A^- and IPC reagent R^+X^-:

$$\begin{aligned} &A^-(\text{mobile phase}) + R^+X^-(\text{stationary phase}) \Leftrightarrow \\ &A^-R^+(\text{stationary phase}) + X^-(\text{mobile phase}) \end{aligned} \tag{7.9a}$$

That is, the ion-pair reagent R^+X^- first attaches to the stationary phase, and then the sample ion A^- replaces the counter-ion X^- in the stationary phase. Either of these two IPC retention processes (Eqs. 7.9 or 7.9a) might predominate for a given separation, but which mechanism plays the more important role is neither easy to determine nor important in practice. It has been shown that these two retention mechanisms are virtually equivalent [47], and both provide similar predictions of retention as a function of experimental conditions. Consequently either process can be assumed in practice. We will use the ion-exchange process of Equation (7.9a) in the following (simplified) discussion, because this retention mechanism appears to us to be easier to understand and to apply in practice.

7.4.1.1 pH and Ion Pairing

Further insight into IPC retention as a function of mobile-phase pH is provided by Figure 7.11 for the case of an acidic sample (a carboxylic acid RCOOH) and a positively charged IPC reagent (tetrabutylammonium, TBA^+). In Figure 7.11*a*, no IPC reagent is added to the mobile phase (i.e., RPC separation), so the non-ionized acid RCOOH is preferentially retained by the C_8 stationary phase (shown as a

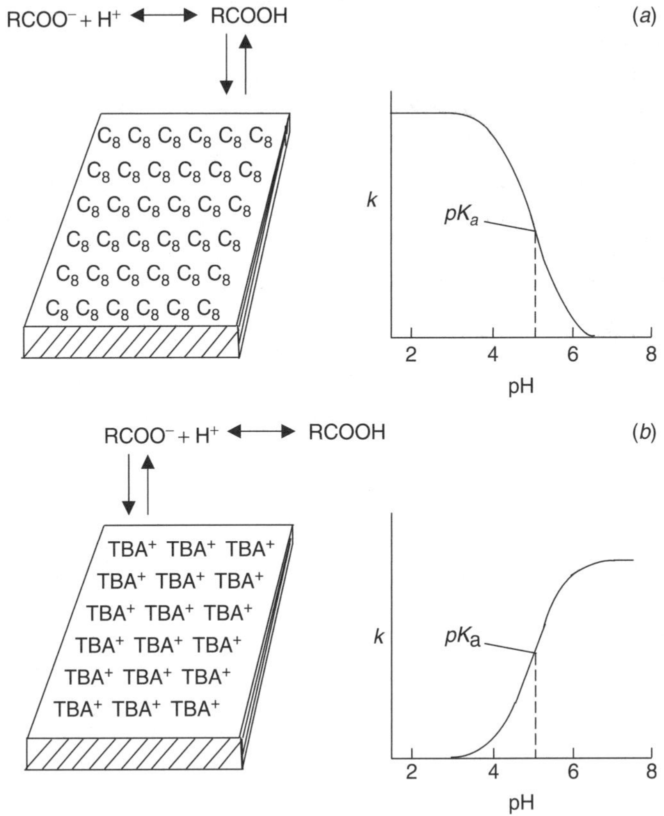

Figure 7.11 Representation of retention for an acidic solute RCOOH as a function of mobile-phase pH; (*a*) RPC and (*b*) IPC with tetrabutylammonium ion (TBA^+) as IPC reagent in high concentration.

surface with attached C_8 groups). Also shown on the right side of Figure 7.11*a* is a plot of retention k as a function of mobile-phase pH; retention decreases with increased pH, due to the greater ionization of RCOOH (i.e., characteristic RPC retention for an acidic solute as pH is varied).

In Figure 7.11*b*, the retention of the same compound RCOOH is shown, except that the IPC reagent TBA^+ has been added to the mobile phase in sufficient concentration to cover the entire stationary phase surface—hence blocking RPC interaction of the sample (non-ionized RCOOH) with the column C_8 groups. Now the ionized acid $RCOO^-$ is preferentially retained by ion exchange with TBA^+ in the stationary phase (Eq. 7.9a). The dependence of k on mobile-phase pH is seen to be the reverse of that in Figure 7.11*a* for RPC (no ion-pairing); retention in Figure 7.11*b* increases with increasing mobile-phase pH and the consequent increasing ionization of the solute, due to ion-pairing of the ionized solute. If fully protonated bases BH^+ are present in the sample (e.g., strong bases), their retention will decrease for increasing concentrations of the IPC reagent (TBA^+), due to the repulsion of the positively charged BH^+ ions by the positively charged TBA^+ ions in the stationary phase, as well as by the decreased availability of C_8 groups—which are covered by sorbed TBA^+. For examples of these generalizations, see the discussion of Figure 7.12 in the following section.

7.4.1.2 Ion-Pair Reagent: Concentration and Type

It is possible to continuously vary the nature of retention, from RPC retention as in Figure 7.11*a* to IPC retention as in Figure 7.11*b*, by varying the concentration of the IPC reagent in the stationary phase. The concentration of the reagent in the stationary phase can be varied by changing its concentration in the mobile phase. We will use the example of Figure 7.12 to illustrate the effect of IPC reagent concentration on solute retention. In this example a tetrabutylammonium phosphate IPC reagent (TBA^+) is assumed initially, with a C_8 column. Consider first the equilibrium uptake of $TBA^+ \equiv R^+$ by the stationary phase (solid curve for TBA^+ at the bottom of Fig. 7.12*a*). The concentration of R^+ in the stationary phase $[R^+]_s$ is plotted against its concentration in the mobile phase $[R^+]_m$, showing a continued increase in stationary phase concentration as $[R^+]_m$ increases, until the stationary phase becomes saturated with R^+ (with no further change in concentration of R^+ in the stationary phase for further increase in $[R^+]_m$). The form of this plot is typical of the uptake of sample or other molecules by a RPC column, when the sample concentration in the mobile phase is increased (so-called Langmuir adsorption; Section 15.3.1.1 and Eq. 15.1).

Two different plots of $[R^+]_s$ versus $[R^+]_m$ are shown in Figure 7.12*a*: a solid curve for TBA^+, and a dashed curve for tetraethylammonium (TEA^+). Because TBA^+ is the more hydrophobic of the two IPC reagents, it is retained by the stationary phase more strongly and saturates the column at a lower concentration of R^+ in the mobile phase. The extent of ion-pairing will depend on the fractional saturation of the stationary phase, and this is seen to depend on (1) the IPC reagent concentration in the mobile phase and (2) the hydrophobicity or retention of the IPC reagent. IPC reagent hydrophobicity and retention will increase for an increase in the carbon number of the reagent (number of CH_3- plus CH_2- groups in the reagent molecule), making TBA^+ (with 16 carbons) more hydrophobic and more retained than TEA^+ (with 8 carbons). We also see in Figure 7.12*a* (dotted lines) that a larger concentration y of a less hydrophobic reagent (TEA^+) can result in the same uptake $[R^+]_s$ of reagent by the column as a lower concentration x of a more hydrophobic IPC reagent (TBA^+), consequently resulting in similar ion pairing and retention of the solute A^- (e.g., see later Fig. 7.15 and the accompanying discussion).

Consider next the change in sample retention k as the IPC reagent concentration increases (Fig. 7.12*b*). Assume a fully ionized acidic solute $RCOO^-$, whose RPC retention is essentially zero for some value of %B. When an IPC reagent is added to the mobile phase, k will increase initially as a result of the interaction of $RCOO^-$ with R^+ in the stationary phase. Once the stationary phase is saturated with R^+, no further increase in retention of $RCOO^-$ can occur; however, a further increase in the mobile-phase concentration of R^+ is accompanied by an increase in the concentration of its counter-ion X^- (e.g., $H_2PO_4^-$). An increase in the $H_2PO_4^-$ concentration $[H_2PO_4^-]_m$ then competes with $RCOO^-$ for ion-exchange retention (Eq. 7.9a), leading to a gradual decrease in k as shown in Figure 7.12*b*. Thus preferred concentrations of the IPC reagent in the mobile phase should not exceed a value that begins to saturate the column—so that solute retention then declines.

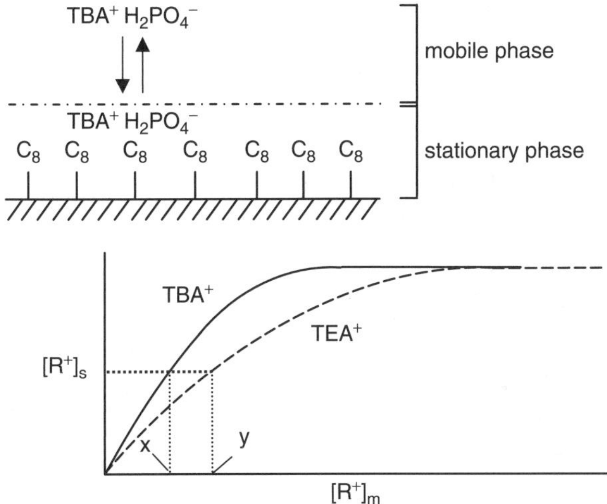

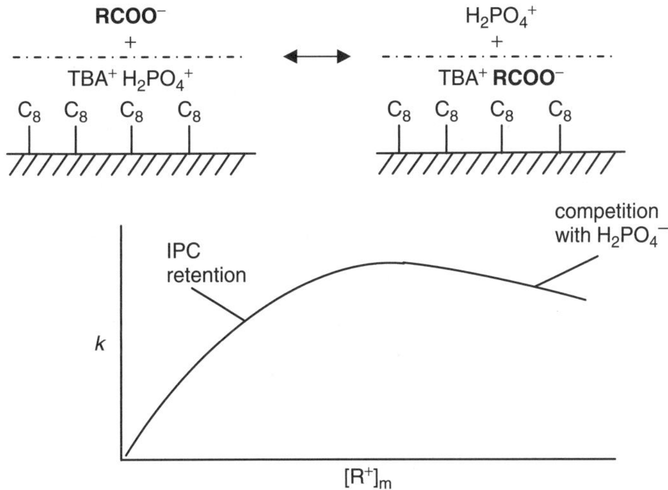

Figure 7.12 Representation of the uptake of IPC reagent by the column, and its effect on solute retention.

7.4.1.3 Simultaneous Changes in pH and Ion Pairing

When mobile-phase pH and the concentration of the IPC reagent are varied simultaneously, a remarkable control is possible over retention range and relative retention for ionic samples (as anticipated by Fig. 7.10). This can be visualized for an individual solute from the plots of k against either pH or IPC reagent concentration in Figures 7.11*b* and 7.12*b*. Now consider that similar plots will result for other solutes, but with the possibility of different dependencies of k on pH and IPC reagent concentration. An example of separation as a function of change in both

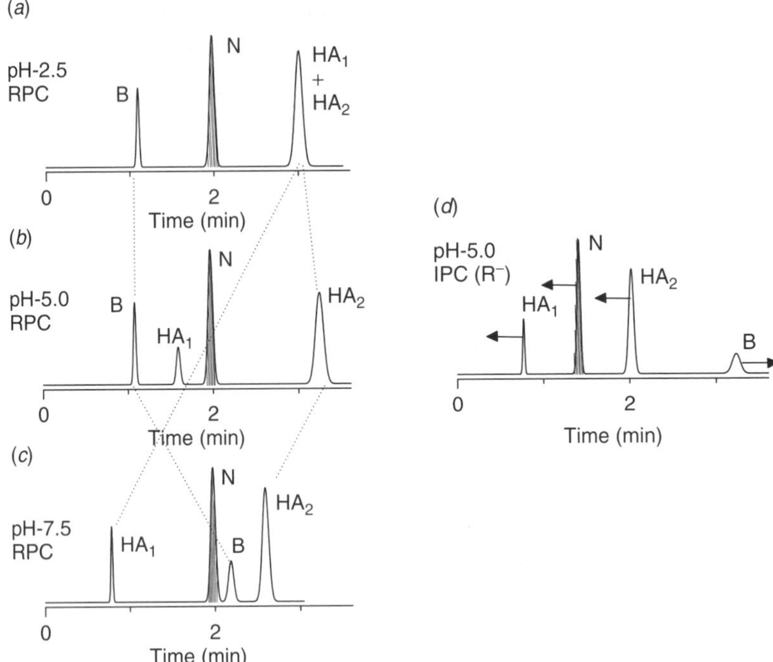

Figure 7.13 Example of the separation of an ionic sample where both mobile-phase pH and IPC reagent concentration are varied. Sample: B, pseudoephedrine; N, glycerol guaicolate (a neutral compound, shaded); HA_1, sodium benzoate; HA_2, methylparaben (a phenol). Conditions: 150×4.6-mm C_8 column (5-μm particles); 30% methanol-citrate buffer with 130-mM hexane sulfonate for the IPC mobile phase; 50°C; 3.0 mL/min. Adapted from [51].

mobile-phase pH and IPC reagent concentration is illustrated in Figure 7.13. The sample consists of a neutral compound (N; shaded peak), a weakly basic compound (B), an acidic compound (HA_1) and a weakly acidic compound (HA_2). Three separations were carried out (Fig. 7.13*a–c*), with pH varying but without any IPC reagent in the mobile phase. The dotted lines track changes in relative retention for each peak. A fourth separation (Fig. 7.13*d*) uses a mobile phase with intermediate pH plus added hexane sulfonate (R^-) as the IPC reagent.

Consider first the retention of the neutral compound N (shaded peak). As pH is varied in the separations of Figure 7.13*a–c*, there is little change in its retention—as expected. When the IPC reagent (R^-) is added in Figure 7.13*d*, the retention of N is reduced as a result of partial blockage of the stationary-phase surface by sorbed R^-. Next consider the retention of weakly basic compound B. Its retention increases for pH > 5, and the addition of IPC reagent in Figure 7.13*d* (pH = 5.0), increases retention even more (the arrows in Fig. 7.13*d* indicate changes in retention vs. the separation of Fig. 7.13*b* [same pH]). The various changes in the retention of B are the result of a decreasing ionization of this basic solute ($BH^+ \rightarrow B$) as pH increases, while addition of the IPC reagent R^- confers a negative charge on the column that attracts the positively charged BH^+. Finally, acidic compounds HA_1 and HA_2 show decreased retention with increase in pH, and their retention is further decreased by the addition of the IPC reagent R^- (because of ionic repulsion between A^- and R^- in the stationary phase, plus partial coverage of the stationary phase by sorbed molecules of R^-).

Finally, we can see that as a result of change in both mobile-phase pH and IPC reagent concentration, major changes in relative retention result for each of the conditions of Figure 7.13:

Low pH, no ion pairing (*a*):

$$B < N < AH_1 = AH_2$$

Intermediate pH, no ion pairing (*b*):

$$B < HA_1 < N < HA_2$$

High pH, no ion pairing (*c*):

$$HA_1 < N < B < HA_2$$

Intermediate pH, ion pairing, *d*:

$$HA_1 < N < HA_2 < B$$

The best separation for this sample is seen in Figure 7.13*d*, using an IPC reagent at pH-5.0.

To summarize, the concentration and type of the IPC reagent can be varied for systematic changes in ion pairing, with predictable effects on retention range and relative retention. When an anionic IPC reagent (e.g., an alkylsulfonate) is added to the mobile phase, the retention of ionized basic compounds will be increased, and the retention of neutral and (especially) acidic compounds will be decreased. When a cationic IPC reagent (e.g., a tetraalkylammonium salt) is used, the retention of ionized acidic compounds will be increased, while that of neutral and (especially) basic solutes will be decreased. These changes in retention will be greater for larger concentrations of the IPC reagent in the mobile phase. Changes in mobile-phase pH that increase the ionization of a compound will increase the effect of the IPC reagent on separation.

7.4.2 Method Development

IPC method development is similar to that for the RPC separation of ionic samples (Section 7.3.3). The same seven method-development steps listed in Figure 6.21*a* for neutral samples still apply, with only step 3 ("choosing separation condition") differing for IPC separation. As in the case of RPC method development, the choice of separation conditions for IPC includes the following steps:

1. choose starting conditions
2. select %B for $1 \leq k \leq 10$ (with IPC reagent in the mobile phase)
3. adjust conditions for improved selectivity and resolution

4. vary column conditions for best compromise between resolution and run time

In most cases, an RPC separation (without IPC reagent present) will have been attempted initially, including a study of changes in mobile-phase pH (*our recommendation*). Consequently it is likely that different peaks in the chromatogram can be assigned as neutrals, acids, or bases (as in the example of Fig. 7.3). This approach also explores the possibility of a non-IPC, RPC separation—with a simpler mobile phase and one less likely to have IPC-related problems (Section 7.4.3). Even when IPC separation is anticipated at the beginning of method development, initial experiments should proceed in similar fashion as described in Section 7.3.3 (development of an RPC method for ionic samples)—without addition of the IPC reagent to the mobile phase. The latter experiments will define the approximate %B required for an acceptable retention range (e.g., $1 < k < 10$) or at least a %B value for an average value of k that falls within this range. As in the case of RPC method development for ionic samples, only steps 1 and 3 above differ for IPC.

7.4.2.1 Choice of Initial Conditions (Step 1)

The requirement of both a buffer and an IPC reagent in the mobile phase may favor the use of methanol as B-solvent, because of the greater solubility of these additives in methanol (TFA and HFBA, however, have adequate solubility in acetonitrile but are weaker IPC reagents). If acetonitrile is used in preliminary RPC experiments (our recommendation), and if solubility problems are subsequently encountered with this solvent, methanol can be substituted (guided by the solvent strength nomograph of Fig. 6.11). In most cases an alkylsulfonate will be chosen as IPC reagent for samples that contain basic compounds, while a tetraalkylammonium salt will be used for acidic samples. When both acids and bases are present in the sample, either type of IPC reagent may prove useful, but it is *not* recommended to add both reagents to the mobile phase. The reagents tend to ion-pair with each other, with cancellation of their net effect on separation; the use of two IPC reagents would also complicate method development. An alkylsulfonate is preferred when it is necessary to selectively increase the retention of basic solutes, while a tetraalkylammonium salt can increase the retention of acidic solutes.

For mixtures of acids and bases, a low-pH mobile phase plus an alkylsulfonate IPC reagent is a good starting point because the pH suppresses ionization of acids and the IPC reagent retains the bases. The final choice of one or the other of these reagents can be determined from information acquired during preliminary RPC experiments; specifically, retention as a function of mobile-phase pH. Both sulfonates and quaternary ammonium salts can be used with UV detection at a wavelength of 210 nm or higher.

The discussion of Section 7.4.1.2 and Figure 7.12*a* suggest that similar separations can be obtained with different concentrations of two different alkyl-sulfonates (or quaternary ammonium salts), for example, a lower concentration of a C_8-sulfonate, or a higher concentration of a C_6-sulfonate. This is generally correct [47, 48], but the practical question is then: Which IPC reagent and which concentration should be used for an initial IPC experiment? One study [49] has provided an approximate answer to this question, as summarized in Figure 7.14.

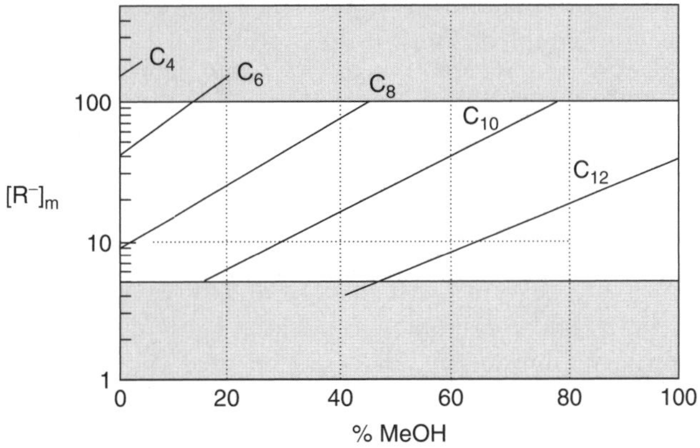

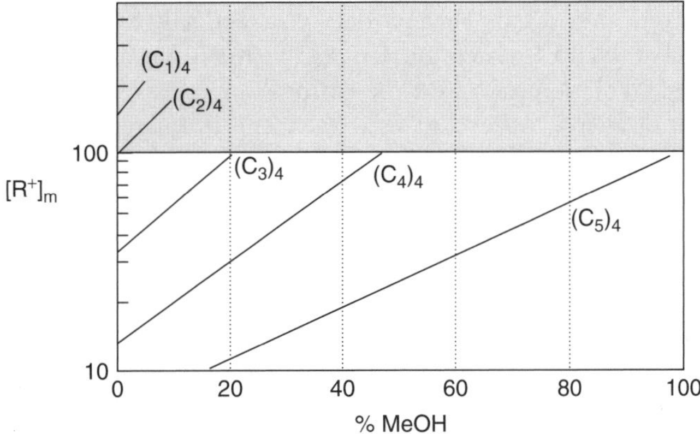

Figure 7.14 Recommended IPC reagents and concentrations as a function of mobile-phase %B: (*a*) sulfonate IPC reagent; (*b*) quaternary ammonium IPC reagent. Adapted from [49]. Conditions corresponding to shaded regions are not recommended.

The recommended starting concentrations of different sulfonates (Fig. 7.14*a*) or quaternary ammonium salts (Fig. 7.14*b*) are plotted against %-MeOH in the mobile phase. For example, given a mobile phase of 40%B, and the planned addition of an alkylsulfonate as IPC reagent, Figure 7.14*a* suggests that either 75 mM of octanesulfonate (C_8) or 15 mM of decanesulfonate (C_{10}) would be a suitable starting concentration. Alternatively, for the use of a tetraalkylammonium salt (Fig. 7.14*b*) with a mobile phase of 40% B, either 70 mM of tetrabutylammonium (C_4) or 20 mM of tetrapentylammonium (C_5) is recommended. It is desirable to select an initial concentration of the IPC reagent between 5 and 100 mM, as indicated by the unshaded regions of Figure 7.14*a, b*.

The (initial) concentrations recommended in Figure 7.14 will provide about $^1/_3$ surface coverage by the reagent. Varying the concentration up or down from this initial value then allows a significant change in reagent uptake by the column, with predictable changes in relative retention (Section 7.4.1.2). If acetonitrile is used as B-solvent for IPC separation, Figure 7.14 can still be used for estimates of IPC reagent type and concentration, but the equivalent value of %B for MeOH should be (very approximately) doubled. Thus, if the mobile phase consists of 20% ACN, a value of $2 \times 20 \approx 40\%$ MeOH should be used in Figure 7.14 [50] for the purpose of selecting an IPC reagent and its initial concentration. That is, acetonitrile is a stronger solvent than methanol, so a lower value of %-ACN is equivalent to a higher value of %-MeOH.

An example that illustrates some of the principles above is provided by Figure 7.15, where the separation of a mixture of water-soluble vitamins is examined as a function of IPC reagent concentration and type. As the sample includes compounds with both acidic (anionic) and basic (cationic) character, either a sulfonate or quaternary ammonium salt could be used for ion pairing. If a sulfonate is selected, Figure 7.14 suggests for this mobile phase (15% methanol-buffer [pH-3.2]) the use of hexane, heptane, or octane sulfonate as IPC reagent. Separations of the sample with varying concentrations of hexane sulfonate are shown in Figure 7.15*a–c*. Peaks 1 to 3 exhibit little change in retention with changing reagent concentration and can be regarded as effectively neutral (neither anionic or cationic). Peaks 4 and 6 show an increase in retention as the reagent concentration increases, so these peaks must be cationic (protonated bases or quaternary

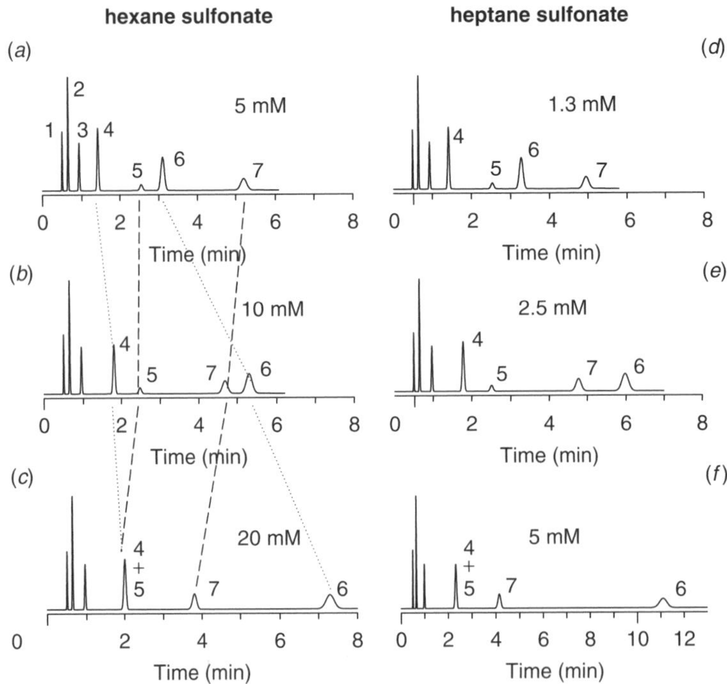

Figure 7.15 IPC separation of a sample of water-soluble vitamins as a function of IPC reagent concentration and type. Sample: *1*, ascorbic acid; *2*, niacin; *3*, niacinamide; *4*, pyridoxine; *5*, folic acid; *6*, thiamine; *7*, riboflavin. Conditions; 83 × 4.6-mm C_8 column (3-μm particles); 15% methanol- buffer (pH-3.2); 35°C; 2.0 mL/min. Chromatograms recreated from data in [48].

ammonium compounds). Similarly peaks 5 and 7 exhibit decreased retention as the reagent concentration increases and are therefore anionic (ionized acids). From these initial experiments (Fig. 7.15*a–c*), it appears that a hexane sulfonate concentration of 6 to 7 mM would provide maximum resolution of peaks 4 to 7 (placing peak 5 midway between peaks 4 and 6, but without moving peak 7 too close to peak 6).

Figure 7.15*d–f* shows corresponding separations with heptane sulfonate as IPC reagent; in each case, the reagent concentration is reduced fourfold compared to separations with hexane sulfonate as IPC reagent. The resulting separations for Figure 7.15*a* and *d*,*b* and *e*, or *c* and *f*, are each quite similar (but not identical). That is, *essentially* the same separation can be achieved for this sample with a lower concentration of a more hydrophobic IPC reagent, but the selectivity may not be exactly the same. *The arbitrary substitution of one IPC reagent for another in a previously developed method is therefore not recommended.*

Inorganic reagents (or "chaotropes") such as ClO_4^-, BF_4^-, and PF_6^- have also been used in IPC [52, 53], in place of the usual alkane sufonates. Because of the lesser retention of inorganic reagents, it is likely that the retention mechanism is based on Equation (7.9)—ion-pairing in the mobile phase—rather than Equation (7.9a)—sorption of the IPC reagent. Chaotropes are advantageous in being better suited for gradient elution (less baseline noise and drift) and are more soluble in mobile phases with larger values of %B. The relative ion-pairing strength of various anions (including both buffers and IPC reagents) increases in the following order:

$$H_2PO_4^- < HCOO^- < CH_3SO_3^- < Cl^- < NO_3^- \ll$$
$$CF_3COO^- < BF_4^- < ClO_4^- < PF_6^-$$

Only the last four anions are useful for IPC. Because inorganic IPC reagents (chaotropes) are less strongly retained by the stationary phase, this can mean faster equilibration of the column when changing the mobile phase (Section 7.4.3.2). The effect of chaotropes in altering the retention of protonated bases appears much more pronounced for acetonitrile as B-solvent, compared to methanol or tetrahydrofuran [54].

7.4.2.2 Control of Selectivity (Step 3)

The separation conditions available for the control of selectivity in IPC include:

- pH
- IPC reagent type (sulfonate, quaternary ammonium salt, chaotrope)
- IPC reagent concentration
- solvent strength (%B)
- solvent type (ACN, MeOH, etc.)
- temperature
- column type
- buffer type and concentration

Despite the large number of variables that can affect selectivity in IPC, usually only a few of these conditions need to be investigated during method development. Furthermore the effects on retention of several of the conditions above are interrelated. Thus a change in mobile-phase pH will, in some cases, give similar results as a change in IPC reagent concentration; for example, an increase in pH or an increase in the concentration of an alkylsulfonate IPC reagent will in each case result in an increase in the retention of basic solutes (and a decrease in retention for acids). Also we have seen that the primary effect of a change in %B or temperature may be the result of associated changes in the "effective" pH of the mobile phase—hence providing similar changes in relative retention as for a change in mobile-phase pH. Other examples of this kind are noted below.

Mobile Phase pH and IPC Reagent Type or Concentration. The combined effects of these conditions on IPC separation were discussed in detail above (Section 7.4.1), and are best investigated first during IPC method development. It is more convenient to vary the concentration of the IPC reagent (as in Fig. 7.15*a–c*), than to change the IPC reagent (as in Fig. 7.15*f* vs. *a*).

Solvent Strength. When %B is varied for the RPC separation of ionic samples (Section 7.3.2.2; Fig. 7.7), changes in both absolute and relative retention can be expected. In some cases these change in retention can be related to corresponding changes in the "apparent" pH of the mobile phase (or values of pK_a for the solute; Section 7.2.3). Corresponding changes in relative retention with %B should also occur for IPC separation, but with an added feature. Thus, if %B is increased, the uptake of the IPC reagent by the column will decrease, just as for the case of sample molecules. Consequently a change in %B should lead to predictable changes in relative retention for peaks that are strongly affected by ion pairing. An example is presented in Figure 7.16 for the same sample of Figure 7.10. An initial separation with 40% MeOH and octanesulfonate as IPC reagent is shown in Figure 7.16*a*, with the four protonated bases (X, X_1–X_3) distinguished as shaded peaks (remaining peaks correspond to either neutral or acidic solutes). When the mobile phase is changed to 45% MeOH in Figure 7.16*b*, and 50% MeOH in Figure 7.16*c*, the retention of all peaks decreases (solvent strength effect), but the four bases become even less retained *relative* to the remaining neutral and acidic peak (they move toward the front of the chromatogram). This behavior is predictable, as the increase in %MeOH will result in a decrease in the retention of the IPC reagent (R^-) by the stationary phase. The reduced concentration of R^- in the stationary phase means a reduction in ion-pairing for the cationic species X, X_1–X_3, and therefore their reduced retention—apart from the *general* decrease in retention for all peaks when %MeOH is increased. The variation of %MeOH in this example shows the exceptional power of a change in %B in IPC to affect band spacing and resolution (note the optimized separation for 45% MeOH in Fig. 7.16*b*).

Solvent Type. A change in solvent type usually leads to changes in IPC selectivity, for either neutral samples (Section 6.3.2; Fig. 6.9) or (especially) ionic samples (Section 7.3.2.3). Because of the added effect of the B-solvent on the uptake of IPC reagent by the column, a change of solvent type in IPC can result in even larger changes in selectivity than in RPC. Figure 7.17 provides a striking

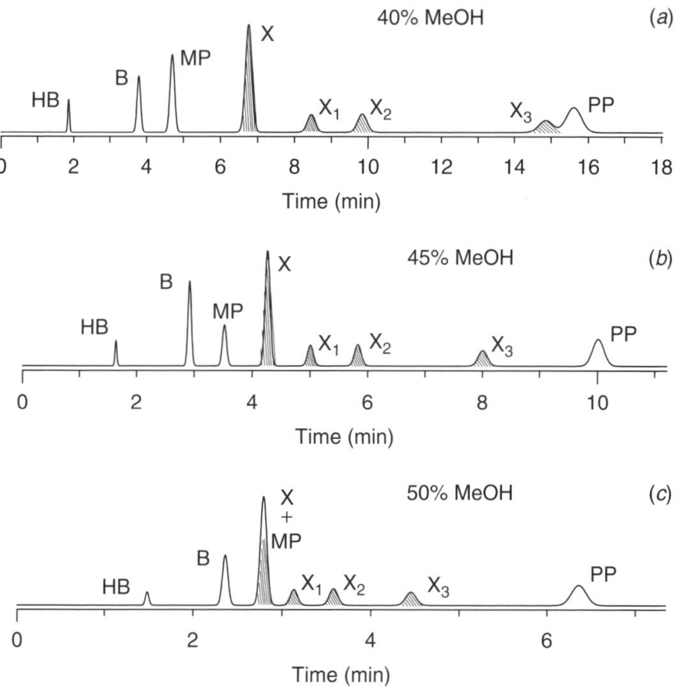

Figure 7.16 Solvent-strength selectivity in IPC separation. Sample and conditions as in Figure 7.10*b*, except for indicated % methanol. Peaks for protonated bases X, X1–X3 shaded. Adapted from [45].

example of solvent-type selectivity in IPC, for a change of B-solvent from MeOH in Figure 7.17*a* to ACN in Figure 7.17*b*, with use of the solvent nomograph of Figure 6.11. A combined variation of mobile-phase pH, IPC reagent concentration, and solvent type should prove especially effective for the separation of challenging ionic samples [48, 50].

Temperature. A change in temperature for IPC should also have a pronounced effect on relative retention. Temperature will alter the amount of IPC reagent held by the column. For this reason *temperature control during IPC separation is especially important.*

Column Type and Buffer. We have seen that column type can have a major effect on selectivity in RPC separations of ionic samples, and it seems likely that this will also be true for IPC separation. However, the partial coverage of the stationary phase surface by IPC reagent may tend to mask the contribution of the column per se to sample retention. In view of the many other ways in which selectivity can be controlled in IPC, the use of column type for this purpose should not be a first choice, nor is it likely to be especially promising. Likewise the effect of buffer type and concentration on IPC separation can be significant, but larger, more predictable changes in selectivity can be obtained by varying pH, %B, temperature, and/or the type and concentration of the IPC reagent.

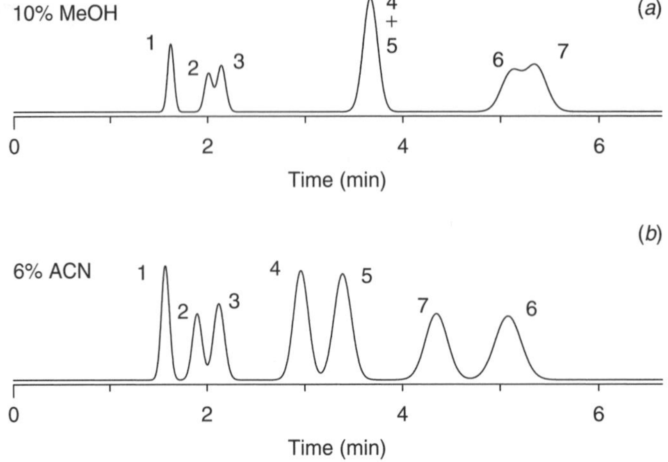

Figure 7.17 Solvent-type selectivity in the IPC separation of a catechol amine sample. Sample: *1*, noradrenaline; *2*, adrenaline; *3*, octopamine; *4*, 3,4-dihydroxyphenylalanine; *5*, dopamine; *6*, isoprenol; *7*, tyrosine. Conditions: 150 × 4.6-mm C_{18} column (5-μm particles); pH-2.5 phosphate buffer plus 2-mM octane sulfonate IPC reagent; 25°C; 1 mL/min. Adapted from [50].

7.4.2.3 Summary

Developing an IPC separation can proceed as follows:

1. select initial conditions for RPC separation (Section 7.3.3.1)
2. vary %B as in RPC, in order to determine a value for an appropriate retention range (e.g., $1 \leq k \leq 10$)
3. vary pH in order to tentatively identify various peaks in the chromatogram as acidic, basic, or neutral (unless peak identities are known by injecting standards)
4. at some stage of further RPC method development, consider the possible value of or need for IPC separation (Section 7.4)
5. if IPC separation is chosen, choose an IPC reagent and its initial concentration as described in Section 7.4.2.1.
   A prior knowledge of the composition of the sample, or previous experiments where pH is varied (step 3) should indicate the choice of either a sulfonate IPC reagent (for increasing the retention of acidic solutes) or a quaternary ammonium salt (for basic or cationic solutes). Alternatively, for basic solutes, a chaotrope can be used as IPC reagent to increase their retention.
6. optimize relative retention (selectivity)
   Simultaneous changes in %B and temperature are expected to be highly effective. For further changes in selectivity, mobile-phase pH and the IPC reagent concentration can be varied (e.g., Figs. 7.13, 7.15). If an additional change in selectivity is needed (unlikely), vary other separation conditions listed at the beginning of Section 7.4.2.2.

7. vary column conditions for further improvements in either resolution or run time (Section 2.5.3)

7.4.3 Special Problems

The separation of ionic samples by IPC is subject to some of the same requirements as for RPC:

- a need for the close control of mobile-phase pH in some cases (e.g., ±0.10 units or better)
- a need for reproducible temperature control (more so than for RPC)

In addition, there are certain problems in IPC that are either absent from RPC separation or differ in some respect for IPC:

- greater complexity of operation and more challenging interpretation of results
- artifact peaks
- slow column equilibration after changing the mobile phase
- poor peak shape for poorly understood reasons

The *greater complexity* of IPC compared to RPC has been noted. There are more variables to choose from in method development or to control during routine operation. While this greater complexity can be manageable, it is nevertheless a distraction that tends to make IPC less attractive. On the other hand, tailing peaks for protonated basic solutes are less likely in IPC (less ionic repulsion of adjacent molecules BH^+ in the stationary phase), and IPC is a more powerful means (when needed) for optimizing the relative retention of ionic samples.

7.4.3.1 Artifact Peaks

Both positive and negative peaks are sometimes observed when the sample solvent (without sample) is injected in IPC (blank run). These artifact peaks can interfere with the development of an IPC method or its routine use. For this reason blank runs should be carried out both during method development and subsequent routine applications, in order to avoid any confusion due to artifact peaks.

Problems with artifact peaks are usually the result of differences in composition of the mobile phase and sample solvent. Such problems can be magnified by the use of impure IPC reagents, buffers, or other mobile-phase additives. A good general rule in IPC is to match the compositions of the sample solvent and mobile phase as closely as possible (including, if necessary, the IPC reagent concentration). Smaller volume sample injections are also recommended (e.g., <25 μL if possible). If problems with artifact peaks persist, a different lot or source of the IPC reagent should be tried. For a general discussion of artifact peaks, see Section 17.4.5.2 and the discussion of [55, 56].

7.4.3.2 Slow Column Equilibration

When a new mobile phase is used, the column must be flushed with a sufficient volume to equilibrate the column (Section 2.7.1). In IPC, both the uptake and release of the IPC reagent by the column can be slow under some circumstances, leading to incomplete equilibration of the column by the new mobile phase. For this reason it is essential to confirm that sample retention is reproducible after a change in the mobile phase, when either the old or new mobile phases contain an IPC reagent (several hours of flow of the new mobile phase may be required to confirm complete column equilibration; see the example below). Column equilibration can be especially slow when the IPC reagent is more hydrophobic (e.g., decane sulfonate vs. octane sulfonate), as well as when quaternary ammonium salts are used with type-A columns [47]. When an IPC reagent is to be replaced, it may be necessary to first remove the previous IPC reagent from the column with a special wash solvent (see below), followed by equilibration of the column with the new mobile phase.

Anionic reagents (e.g., alkane sulfonates) can be removed with a wash solvent composed of 50–80% methanol-water. Quaternary ammonium salts and type-A columns require the use of 50% methanol-buffer (e.g., 100 mM potassium phosphate at pH 4–5; the added potassium phosphate serves to reduce the interaction of the quaternary ammonium group with ionized silanols in the stationary phase). In either case a minimum of 20 column volumes of wash solvent should be used before checking for retention reproducibility (that is, column equilibration) with the new mobile phase.

The initial equilibration of the column with a mobile phase that contains an IPC reagent may prove to be unexpectedly slow. The IPC separation of Figure 7.10*b* was at first believed to equilibrate after washing the column with 20 to 30 column volumes of mobile phase [46], since the replicate injections indicated no significant change in retention times. When samples were subsequently run for an extended period, however, it was found that a very slow decrease in retention for basic compounds X–X_3 occurred over a period of 11 hours, suggesting a very slow approach to column equilibrium. To avoid the need for a 12-hour equilibration at the beginning of every new series of routine runs, it was necessary to store the column filled with mobile phase (containing the IPC reagent) upon completion of each series of runs. This expedient allowed much more rapid column equilibration during startup for assays by IPC, and this is a procedure that we recommend when a separation is to be repeated every day or two. Column lifetime may be reduced, however, when the column is stored in this way (Section 5.8).

The slow equilibration of the column with more hydrophobic IPC reagents can create problems if gradient elution is used. Retention may be less reproducible, baselines can be erratic, and other separation problems may arise. For this reason gradient elution with an IPC reagent added to the mobile phase is usually not recommended, especially for more hydrophobic IPC reagents. An exception can be made for the weakly ion-pairing buffer trifluoroacetic acid (TFA), and for chaotropes such as ClO_4^-, BF_4^-, and PF_6^-, since all of these are less susceptible to slow column equilibration. Passage through a column of 10 to 20 column volumes of the mobile phase is usually adequate for mobile phases that contain TFA or chaotropic reagents. For this and other reasons the use of the latter ion-pair reagents is finding increasing application.

Finally, because of the slow equilibration of the IPC reagent with the column, it is possible that not all of the IPC reagent will be washed from the column, even with aggressive washing procedures. For this reason *we recommend that columns that have been used with IPC not be used subsequently for RPC separations without*

IPC reagents (TFA and chaotropes represent an exception to this warning). A trace of IPC reagent remaining on such a column could cause differences in selectivity that would not be reproduced upon replacement with a new column. Changing from one IPC reagent to another, however, should be less problematic.

7.4.3.3 Poor Peak Shape

Peak tailing of bases usually does not arise in IPC because type-B columns are generally used, and the alkylsulfonate IPC reagent can further minimize the effect of column silanols (the IPC reagent competes with ionized silanols for interaction with protonated bases). Some studies have found peak fronting in IPC to be corrected by operating at a higher column temperature [57]. In one case, conversely, IPC provided better peak shape at a *lower* temperature [45]. The separations were in each case carried out with type-A columns; it is reasonable to expect that peak shape in IPC will be less problematic, as when type-B columns are used. If poor peak shape and/or low values of the column plate number N are observed in IPC, a *change* in temperature (either lower or higher) should be explored.

7.5 ION-EXCHANGE CHROMATOGRAPHY (IEC)

Ion-exchange chromatography (IEC) is an important separation technique, but for the most part, today, with limited applications:

- mixtures of inorganic ions (ion chromatography)
- biomolecules, including amino acids, peptides, proteins, and especially oligonucleotides (Sections 13.4.2, 13.5.1)
- carbohydrates (Section 13.6.3)
- carboxylic acids
- sample preparation (Chapter 16)
- two-dimensional separation (Sections 9.3.10, 13.4.5)

In addition inadvertent ion-exchange interactions can occur and contribute to the separations of ionic samples by RPC (Eq. 7.7, Section 5.4.1).

When HPLC became available in the late 1960s, IEC was a strong candidate for the analysis of any mixture that contained organic acids or bases. Together with liquid–liquid partition and adsorption chromatography, IEC then accounted for most of the reported separations by HPLC. Since that time, however, RPC has taken over most of these applications for small-molecule samples (molecular weights <1000 Da). The reasons for this decline in the use of IEC include (1) lower plate numbers N compared to RPC, (2) greater user-familiarity with RPC separation, and (3) the increased complexity of IEC separations (compared to RPC). Additionally some HPLC equipment is less well suited for typical IEC conditions (high concentrations of salt in the mobile phase, salts such as halides, which are corrosive to stainless steel, etc.). We will first summarize the applications of IEC noted above and then discuss the general principles of IEC separation.

Ion chromatography has represented a major application of IEC since its introduction by Small in 1974 [58]. Prior to that time the analysis of mixtures of inorganic

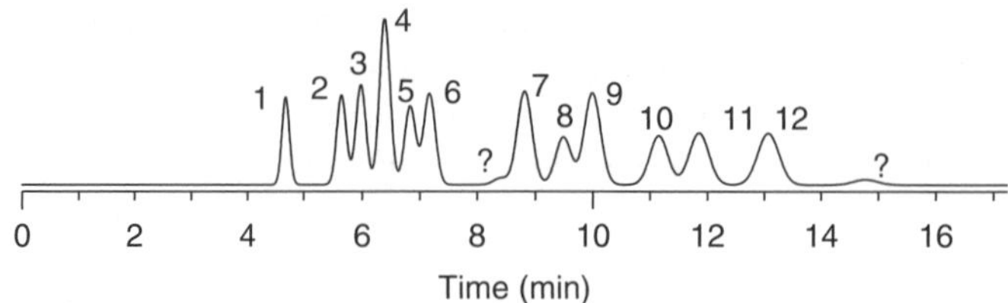

Figure 7.18 Separation of carboxylic acids mixture by ion-exclusion chromatography. Sample: *1*, oxalic; *2*, maleic; *3*, citric; *4*, tartaric; *5*, gluconic; *6*, maliic; *7*, succinic; *8*, lactic; *9*, glutaric; *10*, acetic; *11*, levulinic; *12*, propionic. Conditions: 300 × 7.8-mm cation-exchange column (9-μm particles); 0.006 N sulfuric acid-water mobile phase; 65°C; 0.8 mL/min. Chromatogram redrawn from [57].

ions by other means was tedious and required specialized equipment of limited general applicability. The attempted use of HPLC for separating inorganic solutes was constrained mainly by the lack of a suitably sensitive detector. Ion chromatography overcame this difficulty by the use of ion suppression with conductivity detection. In this book we will not provide a further discussion of ion chromatography; the reader is instead referred to several books on the technique [59–61].

The separation of *biomolecules* by IEC predates the introduction of HPLC by about a decade. During the 1970s HPLC columns were introduced that permitted fast, high-resolution separations of amino acids, peptides, proteins, nucleotides, oligonucleotides, and nucleic acids by means of IEC. These important applications of IEC are discussed in Chapter 13.

Separations of *carbohydrates* by HPLC are possible by means of either hydrophilic interaction chromatography (HILIC, Section 8.6) or by anion exchange chromatography (Section 7.5.7). As discussed below, the latter technique, with amperometric detection, is generally preferred, especially when a greater detection sensitivity is required. See also the analysis of carbohydrate fragments from the digestion of glycosylated proteins (Section 13.6.3).

Carboxylic acids are often separated on IEC columns by ion-exclusion, using acidified water as mobile phase in order to suppress solute ionization. These separations do not involve ion exchange but instead are based on a partition process similar to that involved in RPC. An example of such a separation is provided in Figure 7.18. Relatively hydrophilic samples of this kind are retained weakly on most RPC columns, but more-polar IEC columns are able to provide stronger retention and acceptable k values. Ion-exclusion chromatography (by means of ion-exchange columns) continues to be popular for the assay of samples that contain carboxylic acids, as illustrated by several examples described in [62]. For a further discussion of how experimental conditions affect ion-exclusion separations of carboxylic acids, see [63].

Sample preparation (Section 16.6.5.1) remains a very important application of IEC, albeit as a low-efficiency (non-HPLC) supplement to analysis by RPC or other procedures. The effective use of IEC for sample preparation requires a basic understanding of how retention depends on separation conditions, as briefly reviewed in this section.

Two-dimensional separation is mentioned in Section 2.7.3, and further discussed in Sections 9.3.10 and 13.4.5. This technique is reserved for very complex samples that contain too many components to be separated in a single HPLC run. The principle of operation is the use of a first HPLC separation to achieve partial separation of the sample, followed by injection of fractions into a second column for the further resolution of individual solutes. If IEC with an aqueous mobile phase is used for the first separation, the resulting aqueous fractions can be injected directly onto a second RPC column.

7.5.1 Basis of Retention

IEC separations are carried out on columns with ionized or ionizable groups attached to the stationary-phase surface (Section 7.5.4). For example, cation-exchange columns for the IEC retention of protonated bases (BH^+) might contain sulfonate groups $-SO_3^-$ of opposite charge. Similarly anion-exchange columns for the retention of ionized acids (A^-) might be substituted with quaternary ammonium groups such as $-N(CH_3)_3^+$. Retention in IEC is governed by a competition between sample ions and mobile-phase counter-ions for interaction with stationary-phase ionic groups of opposite charge. IEC retention can be illustrated by the cation-exchange retention of a protonated basic solute BH^+ with K^+ as the counter-ion:

$$BH^+ + R^-K^+ \Leftrightarrow K^+ + R^-BH^+ \tag{7.10}$$

Here R^- refers to an anionic group attached to the column packing (e.g., $-SO_3^-$), which can bind either the sample ion BH^+ or a mobile-phase counter-ion K^+ by coulombic attraction. Equation (7.10) can be generalized for both acidic and basic sample ions that have an absolute charge $|z| \equiv m$ (e.g., fully ionized oxalic acid as solute, $^-OOC{-}COO^-$, with $z = -2, m = 2$). For a cationic solute X^{+m}, and a counter-ion Y^+, retention is described by

$$X^{+m} + m(R^-Y^+) \Leftrightarrow X^{+m}R^-{}_m + mY^+ \tag{7.11}$$

Here the stationary-phase group R^- refers to an anion-exchange group R^- (e.g., $-SO_3^-$). For an anionic solute X^{-m}, Equation (7.11) becomes

$$X^{-m} + m(R^+Y^-) \Leftrightarrow X^{-m}R^+{}_m + mY^- \tag{7.12}$$

Values of the retention factor k in IEC for a univalent counter-ion Y^+ or Y^- in cation- or anion-exchange respectively can be derived from the equilibrium of either Equation (7.11) or (7.12):

$$\log k = \mathrm{a} - m \log \mathrm{C} \tag{7.13}$$

where C is the molar concentration of the counter-ion Y^+ or Y^- in the mobile phase, a is a constant (equal to log k for C = 1M), and m is the absolute value of the charge z on the solute molecule X; a and m are constants for a given sample compound, column, salt, buffer, mobile phase pH, and temperature. An illustration of Equation (7.13b) is shown in Figure 7.19 for the anion-exchange separation of four polyphosphates with z equal −3, −4, −6, and −8 (tri-, tetra-, hexa-, and octa-phosphates, respectively). Numerous examples of the validity of Equation (7.13b) for isocratic IEC have been reported (e.g., [64, 65]).

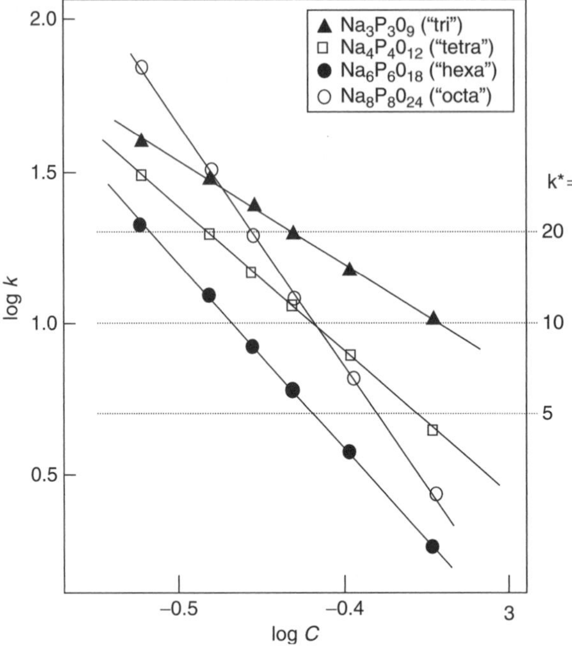

Figure 7.19 Illustration of the dependence of log k on counter-ion concentration (log C) in isocratic IEC. Sample: four polyphosphates described in figure; conditions: 500 × 4.0-mm TSKgel SAX anion exchange column; aqueous KCl salt solutions (buffered at pH-10.2 with EDTA) as mobile phase; 30°C. Adapted from [64].

In Equations (7.10) to (7.13) we treat ion exchange as involving a stoichiometric process, where one molecule of retained solute displaces a certain number of counter ions that are tightly held by the individual charges on the stationary phase surface. While these relationships are adequately reliable in practice, they nevertheless represent a simplification of the actual ion-exchange process. For details concerning the fundamental nature and theory of ion-exchange retention, see [66].

7.5.2 Role of the Counter-Ion

Mobile phases for IEC usually consist of water, a buffer to control pH, and a salt (or counter-ion) to adjust sample retention (solvent-strength control). Because solute retention with IEC columns is usually the result of both IEC and RPC interactions with the column, the addition of methanol or acetonitrile to the mobile phase can further increase solvent strength. However, the primary control of retention is usually accomplished by changing the concentration of the counter-ion (C); an increase in C results in a decrease in retention for solutes that are ionized and retained (Eq. 7.13). For univalent solutes where $m = 1$, an increase in C leads to a proportional decrease in values of k. When the charge m on the solute is larger, a faster decrease in k results as C is increased. Thus, when two solutes have different values of m, a change in C will result in changes in relative retention (with possible peak reversals). This is illustrated in Figure 7.20 for the sample described in Figure 7.19. As the concentration of Cl^- changes from 0.32 to 0.35 to 0.40 M,

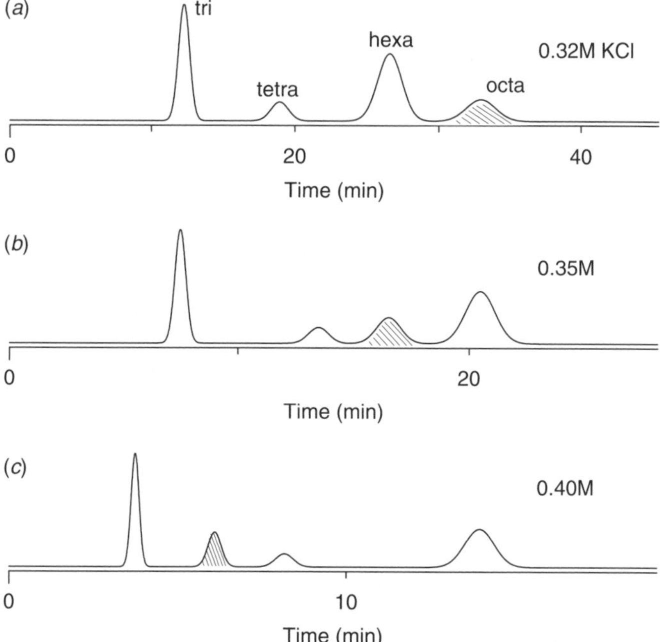

Figure 7.20 Examples of a change in counter-ion (KCl) concentration for the separations of Figure 7.19. Recreated separations for data of [64], assuming a 150 × 4.6-mm column (5-μm particles), 2.0 mL/min, and $N = 1000$.

a decrease in the retention time results for each peak, but at the same time the "octa" peak (shaded) moves toward the front of the chromatogram—with a change in retention order (compare Figs. 7.20 and 7.19).

Different mobile-phase counter-ions are retained more or less strongly by ion exchange, so that a change in the counter-ion can also be used to increase or decrease solvent strength and overall sample retention. Generally, counter-ions with a higher charge will be more effective at reducing sample retention. The relative ability of an ion to bind more strongly, suppress sample retention, and provide smaller values of k increases in the following order:

(anion exchange) F^- (larger values of k for solutes) $< OH^- <$ acetate$^- < Cl^- <$ $SCN^- < Br^- < NO_3^- < I^- <$ oxalate$^{-2} < SO_2^{-2} <$ citrate^{-3} (smaller values of k)

(cation exchange) Li^+ (larger values of k for solutes) $< H^+ < Na^+ < NH_4^+ <$ $K^+ < Rb^+ < Cs^+ < Ag^+ < Mg^{+2} < Zn^{2+} < Co^{2+} < Cu^{2+} <$ $Cd^{2+} < Ni^{2+} < Ca^{2+} < Pb^{2+} < Ba^{2+}$ (smaller values of k)

A change in counter-ion (e.g., from Cl^- to NO_3^-) can also affect values of a in Equation (7.13b), in turn leading to possible changes in selectivity.

7.5.3 Mobile-Phase pH

IEC is typically used for acidic or basic samples. As only the charged (ionized) molecule is retained by ion exchange, values of k for a monovalent solute ($m = 1$) will be proportional to the ionization of the solute; for example, as mobile-phase pH is decreased so that an acid goes from fully ionized to half ionized, the value of k will decrease by half. Similarly the ionization and retention of bases will be decreased as mobile-phase pH increases. This behavior is the opposite of retention changes with pH in RPC (Fig. 7.11*a*) but is the same as in IPC (Fig. 7.11*b*).

7.5.4 IEC Columns

Based on the kind of ionic group $R^{\pm}$ that forms part of the stationary phase, four general kinds of IEC columns are available: strong and weak anion exchangers (SAX, WAX), and strong and weak cation exchangers (SCX, WCX). Strong IEC columns contain groups $R^{\pm}$ that are completely ionized over the usual pH range of interest ($2 \leq pH \leq 13$). For strong anion-exchange columns, the most commonly used group R^{+} is $-N(CH_3)_3{}^{+}$; for strong cation-exchange columns, the most common group R^{-} is $-SO_3{}^{-}$. Weak IEC columns contain groups $R^{\pm}$ with pK_a values in an intermediate range (e.g., $4 \leq pK_a \leq 10$); consequently the ionization of these groups (and the ion-exchange capacity of the column) can change with mobile-phase pH (see Fig. 13.16).

Because the mobile-phase pH for IEC is chosen for solute ionization, and this pH may be outside the $2 < pH < 8$ region recommended for the operation of most silica-based columns, most IEC columns use a polymeric support, such as methacrylate or styrene-divinylbenzene polymers.

Weak anion-exchange columns are commonly substituted with amine groups (e.g., $-NH_2$); these columns begin to lose their charge and ion-exchange capacity when the mobile-phase pH increases much above the pK_a value of the amine group ($5 \leq pK_a \leq 10$; Table 7.2). Weak anion-exchangers are therefore used primarily with acidic mobile phases ($pH \leq 6$) that can significantly protonate the amine groups. Weak cation-exchangers are commonly substituted with carboxyl groups ($-COOH$), with $pK_a \approx 5$, so their ionization begins to decrease when the mobile-phase pH drops below 7. Weak cation-exchangers are used primarily with basic mobile phases (e.g., $pH \geq 8$). Because the ionization and ion-exchange capacity of weak IEC columns can be reduced by the use of an appropriate mobile-phase pH, solutes that are strongly retained and might be difficult to remove from a strong IEC column can be eluted more easily from weak IEC columns by a change in mobile-phase pH. Column selectivity will also differ for weak versus strong ion exchangers, which can be another reason for their use with a particular sample. See the further discussion of Section 13.4.2.1.

7.5.5 Role of Other Conditions

Other IEC conditions are varied primarily for a change in relative retention or selectivity:

- salt or buffer type
- addition of organic solvent to the mobile phase
- temperature

For the IEC separation of small organic solutes, few generalizations have been offered for the effect of the above conditions on selectivity. However, it is known that changes in each condition can affect selectivity.

7.5.6 Method Development

The usual goal of IEC separation is the retention and resolution of a mixture of either anions (ionized acids) or cations (protonated bases). When IEC is used for sample pretreatment, conditions usually are selected for the selective capture of acids or bases for further analysis by HPLC. If the goal is the high-performance separation and analysis of the sample by IEC, then either acids or bases can be resolved and analyzed—but not both simultaneously. If separation involves retention of acids, an anion-exchange column should be selected. For bases, a cation-exchange column will be used. Usually a strong ion-exchange column is preferred (at least initially). As in the development of all HPLC methods, experiments for the determination of optimum IEC conditions should start with a new (unused) column.

The general approach used for RPC method development (Sections 6.4, 7.3.3) also can be followed for IEC separation. An aqueous mobile phase will be used initially, with addition of 1 to 5 mM of a suitable buffer plus a variable concentration of some salt (e.g., NaCl). The concentration of the salt (or counter-ion) is then varied by trial and error (or by gradient elution) in order to achieve a desirable retention range (e.g., $1 \leq k \leq 10$). Changes in selectivity can then be investigated as discussed above.

7.5.7 Separations of Carbohydrates

Carbohydrate mixtures can be separated by either hydrophilic interaction chromatography (HILIC, Section 8.6) or by IEC. Carbohydrates have pK_a values of about 12, which means that high-pH mobile phases can effect their ionization and allow their separation by anion-exchange chromatography (AEC). AEC separations of carbohydrates is now generally preferred because of the greater sensitivity of amperometric detection (detection limits < 1 nanomole [67]), combined with the possibility of influencing selectivity by small changes in mobile-phase pH. An example of such a separation is shown in Figure 7.21 for the separation of a sample that contains 11 different sugars. Other studies have demonstrated a significant role for temperature in affecting peak spacing and resolution for these separations [69]. See also Section 13.6.

7.5.8 Mixed-Mode Separations

Mixed-mode columns can be thought of as hydrophobic ion-exchangers—in contrast to the more hydrophilic columns used for conventional IEC. As a result these columns exhibit both RPC and IEC behaviors. The original use of these columns was suggested by their different selectivity, compared to either RPC or IEC columns, and this feature continues as a reason for their use. An example is shown in Figure 7.22, where separations by RPC (Fig. 7.22*a*) and mixed-mode cation-exchange (Fig. 7.22*b*) are compared, for a nitrogen-mustard mixture (small, hydrophilic amines). The better retention (and resolution) of early-eluting peaks 1 to 6 in Figure 7.22*b* is obvious,

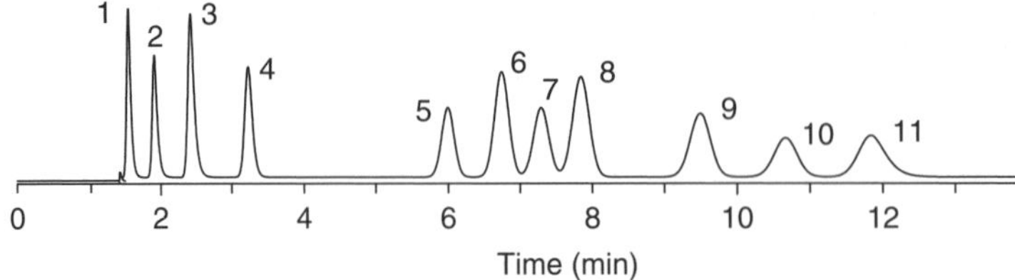

Figure 7.21 Separation of a mixture of carbohydrate standards by anion-exchange chromatography with amperometric detection. Sample: *1*, *myo*-inositol; *2*, D-sorbitol; *3*, lactitol; *4*, L-fructose; *5*, rhamnose; *6*, D-galactose; *7*, D-glucosamine; *8*, D-glugose; *9*, D-mannose; *10*, D-fructose; *11*, D-ribose. Conditions: 300 × 4-mm anion-exchange column (5-μm particles); mobile phase, aqueous 5-mM NaOH + 1-mM $Ba(OAC)_2$; ambient temperature; 1 mL/min. Adapted from [68].

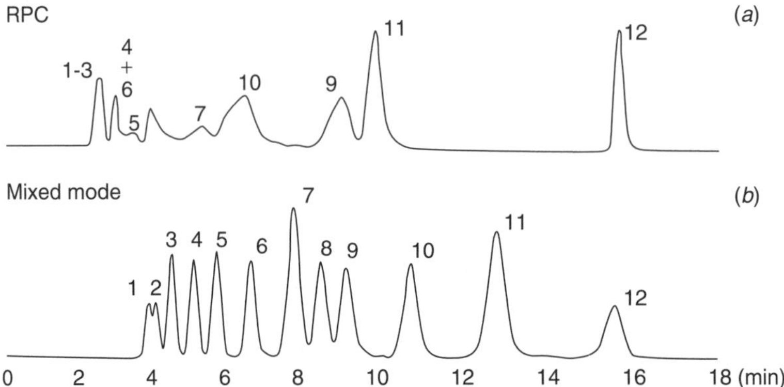

Figure 7.22 Separation of a nitrogen-mustard mixture by RPC (*a*) versus mixed-mode IPC (*b*). Sample: a mixture of small, hydrophilic amines. Conditions in (*b*): 150 × 2.1-mm Primesep 100 column (5-μm particles) (SIELC Technologies, USA); 40% acetonitrile/buffer (0.1% TFA); 0.2 mL/min. Adapted from [70].

while better peak shapes for compounds 9 and 10 are also observed. Another example of mixed-mode separation with a cation-exchange column (PrimeSep SIELC) has been reported to give a "unique anthocyanin elution pattern" for the analysis of grape juice [71], a pattern that facilitates peak identification and quantitation. Retention in mixed-mode separations can be controlled by varying %B, pH, and buffer or salt concentration. Separation conditions that affect selectivity in RPC or IEC can be used to vary relative retention.

Mixed-mode separation also offers an answer to the problem of poor retention in RPC of certain strong bases, as the latter compounds can be more strongly retained by interaction with a negatively charged column [72]. An additional advantage of mixed-mode columns in this respect is their higher loadability for ionized samples. Finally, mixed-mode columns are virtually unique in being able to simultaneously separate mixtures of anions, cations, zwitterions, and neutrals [73]. A mixed-mode cation-exchange column which is especially stable at low pH, while maintaining exceptional efficiency, was reported shortly before the present book was sent to the publisher [74]. Comparisons of separations by a conventional C_{18}

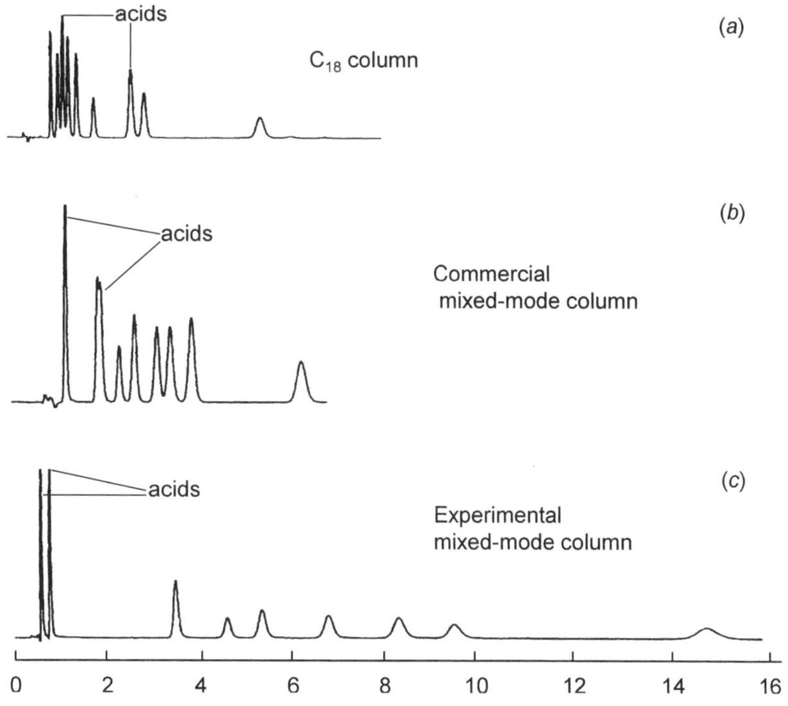

Figure 7.23 Separations of a mixture of acids, bases, and three amino acids by means of different columns. Conditions: 50 × 4.6-mm columns (5-μm particles), 1.0 mL/min; (*a*) C_{18} column, 10% ACN/aqueous 0.1% TFA, 40°C; (*b*) commercial mixed-mode column (PrimeSep 200), 10% ACN/aqueous 0.01% TFA, 40°C; (*c*) experimental mixed-mode column, 24% ACN/0.02% TFA, 65°C. Adapted from [75].

column, a commercial mixed-mode column (hydrophobic cation-exchanger), and the latter column are shown in Figure 7.23. Note the stronger retention of bases in Figure 7.23*c*, despite the higher temperature and stronger mobile phase (24%B), as well as their better resolution—possibly the result of an expanded retention range. Mixed-mode phases for solid phase extraction (SPE) have also been found useful for sample preparation (Section 16.6.7.1).

REFERENCES

1. P. J. Twitchett and A. C. Moffat, *J. Chromatogr.*, 111 (1975) 149.
2. E. P. Kroef and D. J. Pietrzyk, *Anal. Chem.*, 50 (1978) 502.
3. L. R. Snyder, A. Maule, A. Heebsch, R. Cuellar, S. Paulson, J. Carrano, L. Wrisley, C. C. Chan, N. Pearson, J. W. Dolan, and J. Gilroy, *J. Chromatogr. A*, 1057 (2004) 49.
4. J. A. Lewis, J. W. Dolan, L. R. Snyder, and I. Molnar, *J. Chromatogr.*, 592 (1992) 197.
5. U. D. Neue, A. Méndez, K. Van Tran, and D. M. Diehl, in *HPLC Made to Measure*, S. Kromidas, ed., Wiley-VCH, Hoboken, NJ, 2006, pp. 71–87.
6. A. P. Schellinger and P. W. Carr, *LCGC*, 22 (2004) 544.
7. D. V. McCalley, *Anal. Chem.*, 75 (2003) 3404.

8. F. Gritti and G. Guiochon, *J. Chromatogr. A*, 1033 (2004) 43–; 1132 (2006) 51.
9. J. B. Li, *LCGC*, 10 (1992) 856.
10. J. W. Dolan *LCGC*, 8 (1990) 212.
11. L. R. Snyder, unreported data.
12. D. V. McCalley, *J. Chromatogr. A*, 987 (2003) 17.
13. R. F. Doerge, ed., *Wilson and Gisvold's Textbook of Organic Medicinal and Pharmaceutical Chemistry*, 8th ed., J. B. Lippincott, Philadepha, app. B.
14. M. Rosés, *J. Chromatogr. A*, 1037 (2004) 283.
15. X. Subirats, E. Bosch, and M. Rosés, *J. Chromatogr. A*, 1121 (2006) 170.
16. D. V. McCalley, *Adv. Chromatogr.*, 46 (2008) 305.
17. J. A. Lewis, D. C. Lommen, W. D. Raddatz, J. W. Dolan, L. R. Snyder, and I. Molnar, *J. Chromatogr.*, 592 (1992) 183.
18. I. Canals, K. Valkó, E. Bosch, and M. Rosés, *Anal. Chem.*, 73 (2001) 4937.
19. P. L. Zhu, J. W. Dolan, L. R. Snyder, D. W. Hill, L. Van Heukelem, and T. J. Waeghe, *J. Chromatogr. A*, 756 (1996) 51.
20. P. L. Zhu, J. W. Dolan, L. R. Snyder, N. M. Djordjevic, D. W. Hill, J.-T. Lin, L. C. Sander, and L. Van Heukelem, *J. Chromatogr. A*, 756 (1996) 63.
21. C. B. Castells, C. Ràfols, M. Rosés and E. Bosch, *J. Chromatogr. A*, 1002 (2003) 41.
22. C. B. Castells, L. G. Gagliardi, C. Ràfols, M. Rosés, and E. Bosch, *J. Chromatogr. A*, 1042 (2004) 23.
23. L. G. Gagliardi, C. B. Castells, C. Ràfols, M. Rosés, and E. Bosch, *J. Chromatogr. A*, 1077 (2005) 159.
24. S. M. C. Buckenmaier, D. V. McCalley, and M. R. Euerby, *J. Chromatogr. A*, 1060 (2004) 117.
25. J. Pellett, P. Lukulay, Y. Mao, W. Bowen, R. Reed, M. Ma, R. C. Munger, J. W. Dolan, L. Wrisley, K. Medwid, N. P. Toltl, C. C. Chan, M. Skibic, K. Biswas, K. Wells, and L. R. Snyder, *J. Chromatogr. A*, 1101 (2006) 122.
26. N. S. Wilson, M. D. Nelson, J. W. Dolan, L. R. Snyder, and P. W. Carr, *J. Chromatogr. A*, 961 (2002) 195.
27. J. Gilroy, J. W. Dolan and L. R. Snyder, *J. Chromatogr. A*, 1000 (2003) 757.
28. J. J. Gilroy, J. W. Dolan, P. W. Carr, and L. R. Snyder *J. Chromatogr. A*, 1026 (2004) 77.
29. N. S. Wilson, M. D. Nelson, J. W. Dolan, L. R. Snyder, R. G. Wolcott, and P. W. Carr, *J. Chromatogr. A*, 961 (2002) 171.
30. D. Guo, C. T. Mant, and R. S. Hodges, *J. Chromatogr.*, 386 (1987) 205.
31. J. L. Glajch, J. C. Gluckman, J. G. Charikofsky, J. M. Minor, and J. J. Kirkland, *J. Chromatogr.*, 318 (1985) 23.
32. R. G. Wolcott, J. W. Dolan, and L. R. Snyder, *J. Chromatogr. A*, 869 (2000) 3.
33. P. Haber, T. Baczek, R. Kaliszan, L. R. Snyder, J. W. Dolan, and C. T. Wehr., *J. Chromatogr. Sci.*, 38 (2000) 386.
34. S. Pous-Torres, J. R. Torres-Lapasió, J. J. Baeza-Baeza, and M. C. García-Álvarez-Coque, *J, Chromatogr. A*, 1163 (2007) 49.
35. S. Pous-Torres, J. R. Torres- Lapasió, J. J. Baeza-Baeza, and M. C. García-Álvarez-Coque, *J, Chromatogr. A*, 1193 (2008) 117.
36. F. Vanbel, B. L. Tilquin, and P. J. Schoenmakers, *J. Chromatogr. A* 697 (1995) 3.

37. D. V. McCalley, *LCGC*, 17 (1999) 440.
38. S. M. C. Buckenmaier, D. V. McCalley, and M. R. Euerby, *Anal. Chem.*, 74 (2002) 4672.
39. D. V. McCalley, *J. Chromatogr. A*, 1075 (2005) 57.
40. D. V. McCalley, *J. Chromatogr. A*, 708 (1995) 185.
41. D. V. McCalley, *J. Chromatogr. A*, 738 (1996) 169.
42. D. M. Marchand, L. R. Snyder, and J. W. Dolan, *J. Chromatogr. A*, 1191 (2008) 2.
43. M. A. Stadalius, J. S. Berus, and L. R. Snyder, *LCGC*, 6 (1988) 494.
44. J. Nawrocki, *J. Chromatogr. A*, 779 (1997) 29.
45. M. T. W. Hearn, ed., *Ion-pair Chromatography: Theory and Biological and Pharmaceutical Applications*, Dekker, New York, 1985.
46. L. R. Snyder, J. J. Kirkland, and J. L. Glajch, *Practical HPLC Method Development*, 2nd ed., Wiley-Interscience, New York, 1997, ch. 7.
47. J. H. Knox and R. A. Hartwick, *J. Chromatogr.*, 204 (1981) 3.
48. M. W. Dong, J. Lepore, and T. Tarumoto, *J. Chromatogr.*, 442 (1988) 81.
49. A. Bartha, G. Vigh, and Z. Varga-Puchony, *J. Chromatogr.*, 499 (1990) 423.
50. A. Bartha, G. Vigh and J. Ståhlberg, *J. Chromatogr.*, 485 (1989) 403.
51. A. P. Goldberg, E. Nowakowska, P. E. Antle, and L. R. Snyder, *J. Chromatogr.*, 316 (1984) 241.
52. J. M. Roberts, A. R. Diaz, D. T. Fortin, J. M. Friedle, and Stanley D. Piper, *Anal. Chem.*, 74 (2002) 4927.
53. J. Flieger, *J. Chromatogr. A*, 1113 (2006) 37.
54. J. Flieger, *J. Chromatogr. A*, 1175 (2007) 207.
55. J. W. Dolan and L. R. Snyder, *Troubleshooting HPLC Systems*, Humana Press, Clifton, NJ, 1989, pp. 429–435.
56. S. Levin and E. Grushka, *Anal. Chem.*, 58 (1986) 1602.
57. E. Rajaklä, *J. Chromatogr.*, 218 (1981) 695.
58. H. Small, T. S. Stevens, and W. C. Bauman, *Anal. Chem.*, 47 (1975) 1801.
59. P. R. Haddad and P. E. Jackson, *Ion Chromatography*, Elsevier, Amsterdam, 1990.
60. J. S. Fritz and D. T. Gjerde, *Ion Chromatography*, 3rd ed., Wiley-VCH, Weinheim, 2009.
61. J. Weiss, *Handbook of Ion Chromatography*, 3rd ed., Wiley, 2005.
62. M. More, M. I. H. Helaleh, Q. Xu, W. Hu, M. Ikedo, M.-Y. Ding, H. Taoda, and K. Tanaka, *J. Chromatogr. A*, 1039 (2004) 129.
63. J. Qiu, *J. Chromatogr. A*, 859 (1999) 153.
64. Y. Baba and G. Kura, *J. Chromatogr. A*, 550 (1991) 5.
65. J. E. Madden and P. R. Haddad, *J. Chromatogr. A*, 850 (1999) 29.
66. J. Ståhlberg, *J. Chromatogr. A*, 855 (1999) 3.
67. P. J. Andralojc, A. J. Keys, A. Adam, and M. A. J. Parry, *J. Chromatogr. A*, 814 (1998) 105.
68. T. R. Cataldi, C. Campa, M. Angelotti, and S. A. Bufo, *J.Chromatogr. A*, 855 (1999) 539.
69. E. Landberg, A. Lundblad, and P. Påhlsson, *J. Chromatogr. A*, 814 (1998) 97.
70. H.-C. Chua, H.-S. Lee, and M.-T. Sng, *J. Chromatogr. A*, 1102 (2006) 214.

71. J. L. McCallum, R. Yang, J. C. Young, J. N. Strommer, and R. Tsao, *J. Chromatogr. A*, 1148, (2007) 38.
72. N. H. Davies, M. R. Euerby, and D. V. McCalley *J. Chromatogr. A*, 1138 (2007) 65.
73. J. Li, S. Shao, M. S. Jaworsky, and P. T. Kurtulik, *J. Chromatogr. A*, 1185 (2008) 185.
74. H. Luo, L. Ma, Y. Zhang, and P. W. Carr, *J. Chromatogr. A*, 1182 (2008) 41.
75. H. Luo, L. Ma, C. Paek, and P. W. Carr, *J. Chromatogr. A*, 1202 (2008) 8.

CHAPTER EIGHT

NORMAL-PHASE CHROMATOGRAPHY

Introduction to Modern Liquid Chromatography, Third Edition, by Lloyd R. Snyder, Joseph J. Kirkland, and John W. Dolan

8.1 INTRODUCTION

In the early 1900s, when chromatography was first developed (Section 1.2), columns were packed with polar, inorganic particles such as calcium carbonate or alumina. The mobile phase used in these experiments was a less-polar (water-free) solvent such as ligroin (a saturated hydrocarbon fraction from petroleum). For the next 60 years, this procedure continued to be the most common ("normal") way in which chromatography was carried out. For this reason the use of a polar stationary phase (with a less-polar mobile phase) is today referred to as *normal-phase chromatography* (NPC). Another term used to describe NPC is *adsorption chromatography*, in recognition of the fact that retained solute molecules are attached to (or adsorbed onto) the surface of particles within the column (Section 8.2).

After the introduction of high-performance reversed-phase chromatography (RPC) in the 1970s, the use of NPC for HPLC analysis became increasingly less common. This was in part the result of the greater convenience of RPC, as well as its advantages for the separation of many samples of biological origin and/or medical interest. Some problems that are common to NPC (Section 8.5) have also played a role in its declining popularity compared with RPC.

Today NPC is useful mainly for (1) analytical separations by thin-layer chromatography (TLC, Section 1.3.2), (2) the purification of crude samples (preparative chromatography and sample preparation, Chapters 15, 16), (3) the separation of very polar samples that are poorly retained and separated by RPC, or (4) the resolution of achiral isomers (Section 8.35). NPC may also prove beneficial occasionally for other samples, by virtue of its unique characteristics; for example, samples that contain very nonpolar compounds that are of no interest to the analyst. The latter sample constituents would be strongly retained by RPC, necessitating either long run times, sample preparation, or the use of gradient elution; with NPC, very nonpolar compounds elute near t_0, and do not create a problem for isocratic separation (e.g., see Section 8.4.3). In any case, *it is often best to postpone the use of NPC until after RPC has been tried and found wanting*.

Prior to 1970 a wide variety of inorganic packings were used for NPC: alumina, magnesia, magnesium silicate (Florisil), and diatomaceous earth (Celite, kieselguhr), to name a few examples. By the advent of HPLC, however, synthetic (unbonded) silica had become the column packing of choice for both column chromatography and TLC. The advantages of silica for NPC include:

- a more neutral, less active surface, with less likelihood of undesirable sample reactions during separation
- strong particles of controlled size and porosity that can withstand the high pressures required in HPLC
- a generally higher surface area, allowing larger weights of injected sample for either increased detection sensitivity or increased yields in preparative chromatography
- greater purity and reproducibility, permitting more repeatable separations
- reasonable cost and availability

While a preference for silica has continued to the present day, other column options for NPC have emerged over time. Three polar-bonded-phase packings

(Section 5.3.3), chemically similar to those used in RPC, were introduced for NPC during the 1970s: (1) cyano columns, where $-(CH_2)_3-C{\equiv}N$ groups are bonded to silica particles, (2) diol columns bonded with $-(CH_2)_3-O-CH_2-CHOH-CH_2OH$ groups, and (3) amino columns with $-(CH_2)_3-NH_2$ ligands. The differing properties of these bonded-phase columns for NPC are discussed in Section 8.3.4, and some reasons for their use in place of unbonded silica can be inferred from the discussion of Section 8.5, which deals with problems associated with the use of silica columns in NPC.

During the 1990s silicas of higher purity (type-B; Section 5.2.2.2) became commercially available, and these materials gradually displaced the less pure type-A silica used previously for analytical NPC separations. Some advantages of type-B silica for NPC are discussed in Section 8.5. The latest version of NPC is so-called hydrophilic interaction chromatography (HILIC; Section 8.6), also called aqueous NPC. HILIC column-packings consist of either (a) silica particles that are bonded with polar hydrophilic groups such as amides or (b) bare silica. For either kind of HILIC column, the mobile phase is a mixture of water and organic solvent—as opposed to the water-free mobile phases that have traditionally been used for NPC. HILIC provides some of the convenience that is characteristic of RPC, while minimizing other problems associated with the use of silica columns and nonaqueous mobile phases (Section 8.5).

In the present chapter, unless noted otherwise, we will assume the use of unbonded, type-B silica columns. The surface of a silica particle is covered with silanol groups $\equiv$Si–OH (Section 5.2.2.2) which are mainly responsible for its chromatographic properties. These silanol groups are relatively strong proton donors that can interact with and retain solute molecules that contain hydrogen-bond acceptor groups (any molecule with available electrons or a dipole moment). The silica surface also strongly attracts small polar molecules such as water, which can lead to certain problems discussed in Section 8.5. For further details on the role of the column in NPC separation, see Section 8.3.4 (column selectivity).

8.2 RETENTION

Because the column in NPC is more polar than the mobile phase, more-polar solutes will be preferentially retained or adsorbed—the opposite of RPC. This is illustrated in Figure 8.1*a* for the separation of several mono-substituted benzenes, using a silica column with 20% $CHCl_3$-hexane as mobile phase; the more-polar solvent $CHCl_3$ is the B-solvent and the less-polar hexane is the A-solvent. Here the less-polar solutes benzene (–H) and chlorobenzene (–Cl) leave the column first, while the more-polar aniline ($-NH_2$), benzoic acid (–COOH), and benzamide ($-CONH_2$) leave the column last. This retention behavior can be contrasted with RPC retention (Fig. 2.7*c*), where retention *decreases* with increasing solute polarity. Figure 8.1*b* compares retention (log *k*) in NPC and in RPC for several mono-substituted benzenes. As expected, there is a negative correlation of retention for NPC over RPC—corresponding *approximately* to a reversal of retention order. While the correlation of Figure 8.1*b* is moderately strong ($r^2 = 0.76$), there is also significant scatter of the data. That is, NPC separation cannot be regarded as the *exact* opposite of RPC retention. Keep in mind that relative retention in both NPC and RPC can

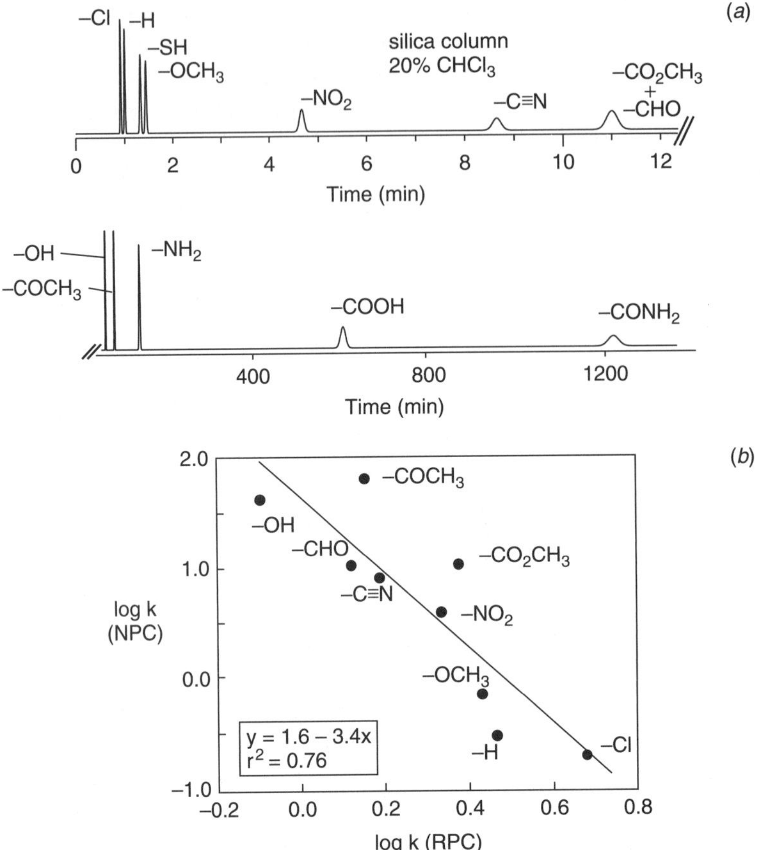

Figure 8.1 Example of normal-phase retention as a function of solute polarity. Sample: mono-substituted benzenes (substituents indicated for each peak; e.g., –H is benzene, –Cl is chlorobenzene). Conditions: 150 × 4.6-mm silica (5-μm particles); 20% $CHCl_3$-hexane mobile phase; ambient temperature; 2.0 mL/min. (*a*) Chromatogram is recreated from data of [1]; (*b*) retention of (a) compared with RPC retention from Figure 2.7*c* for benzenes substituted by the same functional group (50% acetonitrile-water as RPC mobile phase).

also vary significantly with changes in the column, mobile phase, or temperature, all factors that contribute to the scatter of retention plots as in Figure 8.1*b*.

Apart from the approximately inverted retention order for the sample in NPC as opposed to RPC, there are two additional differences in NPC retention that are related to (1) the number *n* of alkyl carbons in the solute molecule (its carbon number C_n), and (2) isomeric solutes. These two general characteristics of NPC versus RPC are illustrated in the separations of Figure 8.2. Figure 8.2*a* shows the RPC separation of 17 alkyl-substituted anilines with a C_8 column and 60% MeOH as mobile phase. As the value of C_n increases, retention increases for RPC (but not for NPC). Isomeric solutes of identical alkyl-carbon number (e.g., C_1, consisting of *o*-, *m*-, and *p*-methylanliline) are seen to be bunched together, while solutes of differing carbon number (e.g., C_1 vs. C_2) are well separated. As summarized in Figure 8.2*e*, average retention times in RPC increase regularly as the carbon number increases (by an average 1.4-fold per additional carbon in this example).

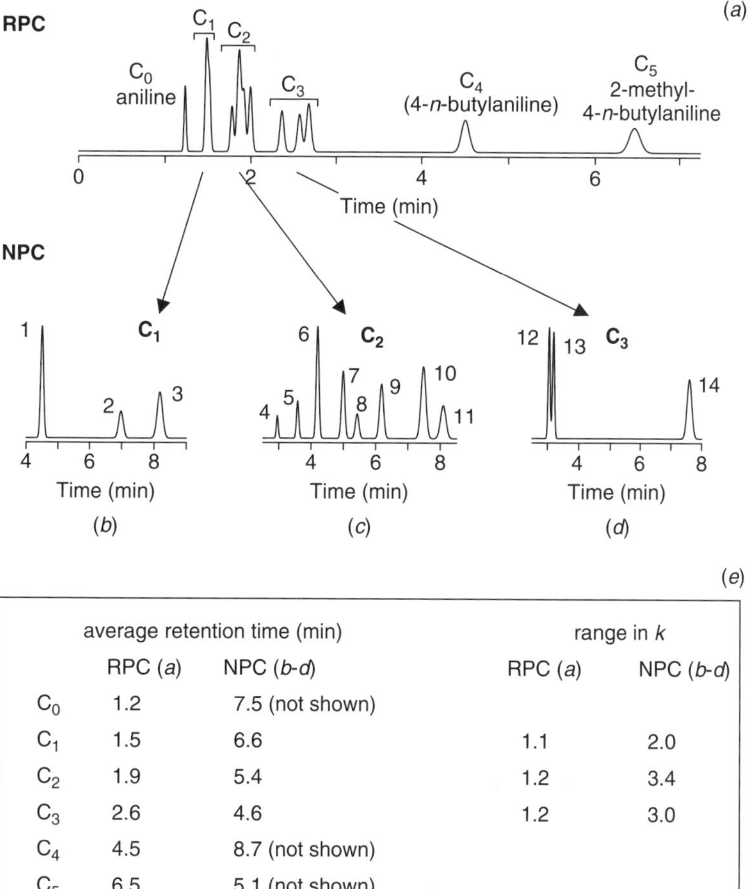

	average retention time (min)		range in k	
	RPC (*a*)	NPC (*b-d*)	RPC (*a*)	NPC (*b-d*)
C_0	1.2	7.5 (not shown)		
C_1	1.5	6.6	1.1	2.0
C_2	1.9	5.4	1.2	3.4
C_3	2.6	4.6	1.2	3.0
C_4	4.5	8.7 (not shown)		
C_5	6.5	5.1 (not shown)		

Figure 8.2 Comparison of NPC separation (*a*) with RPC separation (*b–d*) for a mixture of alkyl-substituted anilines. Conditions: 150 × 4.6-mm C_8 column (5-μm particles) in (*a*), 150 × 4.6-mm cyano column (5-μm particles) in (*b–d*); mobile phase is 60% methanol–pH-7.0 buffer in (*a*), and 0.2% isopropanol-hexane in (*b*); ambient temperature and 2.0 mL/min in (*a*) and (*b*). Sample (peak numbers): *1–3*, 2-, 3- and 4-methylaniline; *4*, 2,6-dimethylaniline; *5*, 2-ethylaniline; *6*, 2,5-dimethylaniline; *7*, 2,3-dimethylaniline; *8*, 2,4-dimethylaniline; *9*, 3-ethylaniline; *10*, 4-ethylaniline; *11*, 3,4-dimethylanilne; *12*, 2,4,6-trimethylaniline; *13*, 2-*i*-propylaniline; *14*, 4-*i*-propylaniline. Chromatograms reconstructed from data of [2].

Figure 8.2*b–d* illustrates the further separation of fractions C_1, C_2, and C_3 from Figure 8.2*a* by NPC (using a cyano column with a mobile phase of 0.2% isopropanol/hexane). It is seen that there is no consistent change in retention time for NPC as the number of alkyl carbons increases (see the summary of Fig. 8.2*e*). That is, NPC can separate solutes of differing functionality (as in Fig. 8.1*a*), but differences in solute carbon number have much less effect on retention. For this reason NPC has been used in the past for *compound-class separations* of petroleum-related materials [3] and lipid samples [4]. NPC permits the group-separation of petroleum samples into saturated hydrocarbons, olefins, benzenes, and various polycyclic aromatic hydrocarbons—according to the number of double bonds in the molecule, but with little effect of differences in alkyl substitution or solute molecular weight. Similarly lipid samples can be resolved into mono-, di-, and tri-glycerides (as well as other compound classes).

Isomeric solutes are usually much better separated by NPC than by RPC, as seen in Figure 8.2*b* (C_1 or methyl-substituted anilines), Figure 8.2*c* (C_2 or dimethyl- plus ethyl-substituted anilines), and Figure 8.2*d* (C_3 or trimethyl-, methylethyl-, and *n*-propyl-substituted anilines). The RPC retention of isomers (e.g., C_1 anilines) is generally similar (with marginal separation), while the reverse is true for NPC (more varied values of k for different isomers, and better separation). The greater isomer-selectivity of NPC versus RPC is also shown by the range in values of k for each set of isomers; that is, the ratio of k-values for the most retained and least retained isomers ("range in k," Fig. 8.2*e*). The spread in k values for a group of isomers (C_1, C_2, or C_3) ranges from 1.1- to 1.2-fold for RPC, versus 2.0- to 3.4-fold for NPC, which is many times larger for NPC.

When two isomeric compounds prove difficult to separate by RPC (Section 6.3.5), NPC will usually prove more effective. Thus, for preparative separations (Chapter 15), where the largest possible values of α are desirable, NPC with a silica column is strongly recommended for the separation of achiral isomers (for chiral isomers, see Chapter 14). For a review of the RPC separation of isomers, see Section 6.3.5; for a further discussion of isomer separation by NPC, see Section 8.3.5.

If in doubt as to whether a separation is based on NPC or RPC retention, a simple test is to vary the polarity of the mobile phase. If retention increases *with increased mobile-phase polarity, RPC is involved; if retention* decreases *with increased mobile-phase polarity, NPC can be assumed. This test holds whether the stationary phase is unbonded or bonded silica, and whether the mobile phase is aqueous or nonaqueous (note that water is the most polar of all solvents).*

8.2.1 Theory

Retention in NPC is best described by a *displacement* process, based on the fact that the silica surface is covered by a monolayer of solvent molecules that are adsorbed from the mobile phase [1, 5, 6]. Consequently, for a solute molecule to be retained in NPC, one or more previously adsorbed solvent molecules must be displaced from (leave) the silica surface in order to make room for the adsorbing solute. Displacement in NPC is illustrated in Figure 8.3*a*, *b* for a relatively nonpolar solute (chlorobenzene) and a weaker, less-polar mobile-phase solvent methylene chloride. When a molecule of chlorobenzene moves from the mobile phase in Figure 8.3*a* to the stationary phase in Figure 8.3*b*, one or more pre-adsorbed solvent molecules CH_2Cl_2 must be displaced from the stationary phase and return to the mobile phase. In this example the adsorbed solute molecule is assumed to lie flat on the surface of the silica, and to occupy an area that is indicated in Figure 8.3*b* by the dotted rectangle that surrounds the molecule of retained chlorobenzene. By reference to Figure 8.3*a*, it is seen that this same area was originally occupied by (approximately) two retained molecules of CH_2Cl_2. Consequently the resulting retention equilibrium can be written as

$$Y_{(m)} + nM_{(s)} \Leftrightarrow Y_{(s)} + nM_{(m)} \tag{8.1}$$

where n is the number of solvent molecules M displaced by a retained solute molecule Y ($n = 2$ in the example of Fig. 8.3*a,b*). $Y_{(m)}$ and $Y_{(s)}$ refer to a molecule of solute Y in the mobile and stationary phases, respectively, while $M_{(m)}$ and $M_{(s)}$ refer to a molecule of solvent M in the mobile and stationary phases, respectively. The quantity n in Equation (8.1) is thus the ratio of molecular areas for the solute with relation to the mobile phase.

Retention differs for a more-polar mobile-phase solvent such as tetrahydrofuran (THF) and a more polar solute such as phenol (Fig.8.3*c,d*). Here the interaction of solvent and solute molecules with surface silanols will be stronger, as indicated by the arrows that connect the two interacting species—in contrast to the weaker

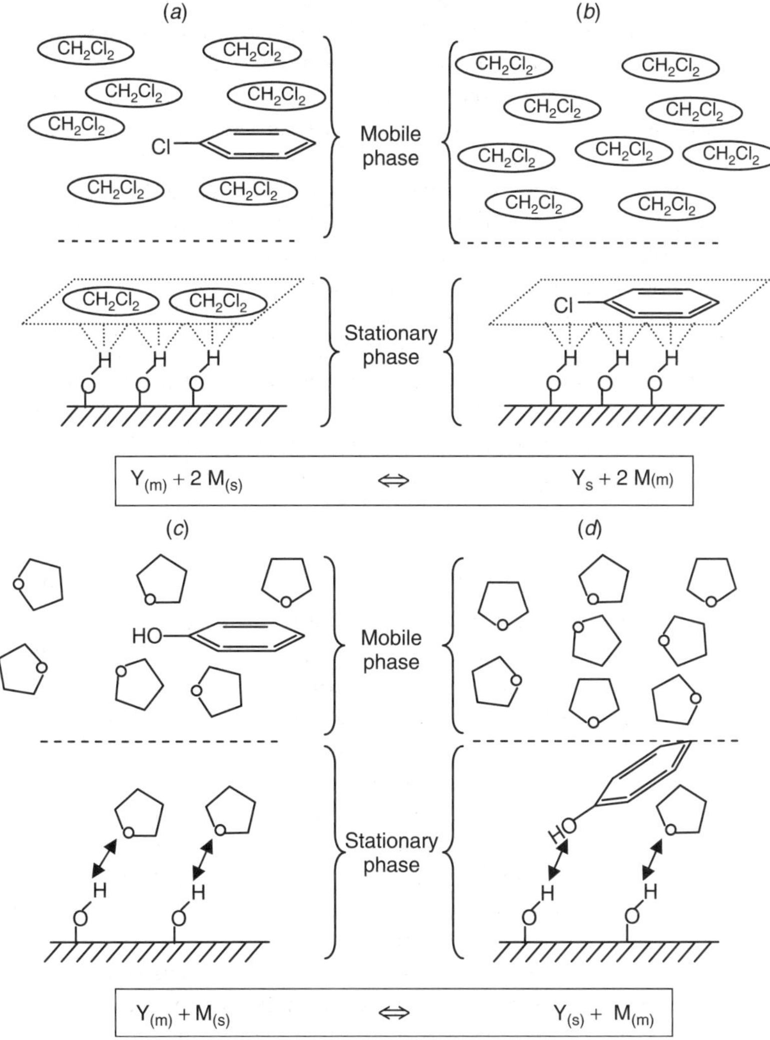

Figure 8.3 Hypothetical examples of solute retention on silica for chlorobenzene (*a,b* non-localized) and phenol (*c,d* localized). Mobile phase in (*a,b*) is a less-polar solvent (CH_2Cl_2); mobile phase in (*c,d*) is a more-polar solvent (tetrahydrofuran, THF).

and less specific interactions shown in Figure 8.3*a,b*. As a result there is a ratio 1 : 1 interaction of a surface silanol with a polar group in a molecule of either solute or mobile phase—called *localized* adsorption. Under these conditions adsorbed molecules can assume a vertical rather than flat configuration, as illustrated by the retention of phenol in Figure 8.3*d*. In the example of Figure 8.3*a,b*, the entire solute molecule (chlorobenzene) is attracted more strongly to the silica surface than are molecules of the less-polar solvent CH_2Cl_2. In Figure 8.3*c,d*, the very polar –OH group of the phenol solute interacts strongly with a silanol (–OH) group on the silica surface, while the less-polar phenyl group is attracted less strongly than are molecules of the strong (very polar) solvent THF. As a result the phenyl group cannot compete with molecules of THF for a place on the silica surface; the phenyl group therefore dangles out into the mobile phase, tethered to the surface by the interaction of a silanol with solute-hydroxyl groups. For solutes with some number n of more strongly retained (very polar) substituent groups (e.g., –OH, $-NH_2$), Equation (8.1) will still apply, with n referring now to the number of polar groups in the solute molecule that can simultaneously interact with silica silanols. However, for less-polar molecules of the mobile phase as in Figure 8.3*a,b*, n refers to the ratio of areas required by molecules of the solute and mobile-phase, respectively.

The competition of solute and solvent molecules for a place on the silica surface will be affected by the interactions of each molecule with the mobile and stationary phases. Because polar interactions predominate in NPC, polar molecules of solute or solvent will interact much more strongly with the more-polar silica surface than with the less-polar mobile phase. As a first approximation we can ignore interactions between solute and solvent molecules and consider only interactions of the solute and solvent with the stationary phase. This allows the derivation of a simple equation for k as a function of the concentration of the B-solvent in a binary mobile phase A–B [1, 6]:

$$\log k = \log k_A - A_s \varepsilon \tag{8.2}$$

where k_A is the value of k for a nonpolar A-solvent (for which ε is zero), A_s refers to the molecular area of the solute molecule, and the mobile-phase solvent strength ε can be calculated as a function of (1) the solvent strengths ε_A and ε_B of the pure A- and B-solvents, respectively, (2) the molecular area of the B-solvent, and (3) the concentration (%B) of the B-solvent in the mobile phase (see Eq. 8.5 in Section 8.2 following).

The remainder of this section provides a quantitative, somewhat detailed, discussion of retention as a function of the solute and mobile phase. The reader may prefer to skip to Section 8.2.2 *(which provides a practical summary of the discussion below) and return to this section as appropriate.*

Equation (8.2) applies approximately for all solvents and solutes. For more polar solutes and B-solvents, as in Figure 8.3*c,d*, polar solute groups can attach to individual silanols. So Equation (8.1) now applies with n equal to the number of polar groups in the solute molecule (rather than the area of the solute molecule). If %B is large enough, and the B-solvent is strong enough, the silica surface will be covered almost entirely by the B-solvent (to the exclusion of the A-solvent); under these conditions the concentration of *adsorbed* B-solvent will not vary much when

%B is changed. For this case the *Soczewinski equation* can be derived from Equation (8.1) [7, 8]. Thus the equilibrium for Equation (8.1) can be written as

$$K_{eq} = \frac{X_{Y,s} {X_{M,m}}^n}{X_{Y,m} {X_{M,s}}^n} \tag{8.3}$$

where $X_{Y,s}$ and $X_{Y,m}$ refer to the concentrations (mole-fractions) of solute Y in the stationary (s) and mobile (m) phases, respectively. Similarly $X_{M,s}$ and $X_{M,m}$ are the mole-fractions of the B-solvent M in the stationary (s) and mobile (m) phases. The retention factor k can be written as

$$k = \psi \left(\frac{X_{Y,s}}{X_{Y,m}} \right) \tag{8.3a}$$

where ψ is the *phase ratio*: the volume-ratio of stationary phase to mobile phase within the column. If the silica surface is covered almost entirely by the B-solvent, $X_{M,s} \approx 1$, Equations (8.3) and (8.3a) then yield

$$\log k = \log k_B - n \log X_B \qquad \text{(Soczewinski equation)} \tag{8.4}$$

where n is the number of B-solvent molecules displaced by the solute (approximately equal to the number of polar substituent groups in the solute molecule); k_B refers to the value of k for pure B-solvent as the mobile phase (100% B), and X_B is the mole-fraction of the B-solvent in the mobile phase. Equation (8.4) can be regarded as a special case of Equation (8.2), whenever $X_{M,s} \approx 1$ (i.e., for larger concentrations of the B-solvent).

Equation (8.4) is more conveniently (and approximately) expressed as

$$\log k \approx \log k_B - n \log \phi \tag{8.4a}$$

where ϕ is the volume-fraction of the (polar) B-solvent in the mobile phase ($= 0.01 \times$ %B). An example of the application of Equation (8.4a) is shown in Figure 8.4 for two different (polar) solutes, and a mobile phase composed of hexane (A) plus tetrahydrofuran (B), with %B varying between 8 and 27%. In each case an approximately linear plot of log k against ϕ is observed. Values of n from Equation (8.4a) for the two plots of Figure 8.4 are each equal to about 1.4, which is somewhat greater than the expected value of $n = 1.0$ (as there is just one polar –OH group in each solute molecule); this is a consequence of the approximate nature of Equation (8.4*a*). Equations (8.4) and (8.4a) are reliable for more-polar B-solvents ($\varepsilon^0_B > 0.4$) and higher concentrations of the B-solvent (>10% B)—conditions that assure that the surface of the silica will be covered almost entirely by molecules of the B-solvent (instead of the A-solvent). Typically values of n in Equation (8.4a) fall between one and two for representative small-molecule solutes (molecular weights <500 Da), so that on average a change in %B by a factor of two (e.g., a change in mobile phase from 100% B to 50% B) will change values of k by a factor of 2 to 4.

Similar plots of log k against log %B as in Figure 8.4 are shown in Figure 8.5*a* for two less-polar solute molecules, using the weaker B-solvent chloroform ($\varepsilon^0_B = 0.26$ with lower values of %B. Because these conditions fail to meet the above

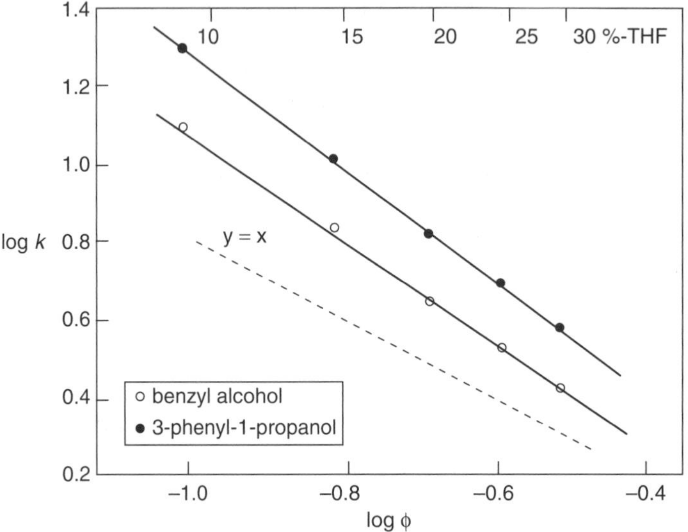

Figure 8.4 Dependence of log k on log %B for solutions of a polar (localizing) B-solvent. Sample: benzyl alcohol and 3-phenyl-1-propanol. Conditions: silica column; tetrahydrofuran (THF)-hexane mobile phases; 25°C. Data from [11].

requirements for the application of Equation (8.4a), the resulting plots are curved rather than straight (as required by the theory upon which Eqn. 8.4a is based), and the best-fit values of n are $\ll 1$. That is, a change in %B by a factor of 2 for this example produces a change in k by much less than 2- to 4-fold. However, the use of the more accurate Equation (8.2) for these data (Fig. 8.5*b*) yields a near-linear fit of values of log k versus ε^0 as expected.

8.2.2 Solvent Strength as a Function of the B-Solvent and %B

Solvent strength in NPC depends on the polarity of the solvent; more polar solvents are stronger, resulting in smaller values of k for a sample. The strength of a pure solvent for unbonded silica as column packing can be expressed by the solvent-strength parameter ε of Equation (8.2); for pure solvents, ε will be referred to as ε^0. As the value of ε increases, the solvent becomes stronger, and solute k-values decrease. Values of ε^0 for some commonly used NPC solvents are listed in Table 8.1; for additional values of ε^0 for other pure B-solvents, see Appendix I. The value of ε for a mixture of A- and B-solvents is given by [1, 6]

$$\varepsilon = \varepsilon_A^O + \frac{\log[N_B 10^{n_B(\varepsilon_B^0 - \varepsilon_A^0)} + 1 - N_B]}{n_B} \tag{8.5}$$

Here, ε_A^0 and ε_B^0 refer to the ε^0 values of pure solvents A and B, N_B is the mole-fraction of solvent B in the mobile phase, and n_B represents the relative size (area) of solvent B (relative to a value of $n_B = 6$ for benzene as B-solvent). Equation (8.5) assumes a fully active (non–water-deactivated) adsorbent such as silica.

When a weaker A-solvent (e.g., hexane) is mixed with a stronger B-solvent (e.g., CH_2Cl_2), the resulting mobile phase will have an intermediate strength and value of ε that is given by Equation (8.5). Figure 8.6 is a solvent nomograph for NPC with a silica column (similar to that for RPC in Fig. 6.11), which compares the strengths of different mobile-phase mixtures in terms of their values of ε (see values at top of Fig. 8.6, calculated from Eq. 8.5). A change in ε by 0.05 units will change values of k by roughly a factor of 2. For example, a mobile phase of 50% methylene chloride-hexane has $\varepsilon = 0.24$ (dotted vertical line of Fig. 8.6). If a change

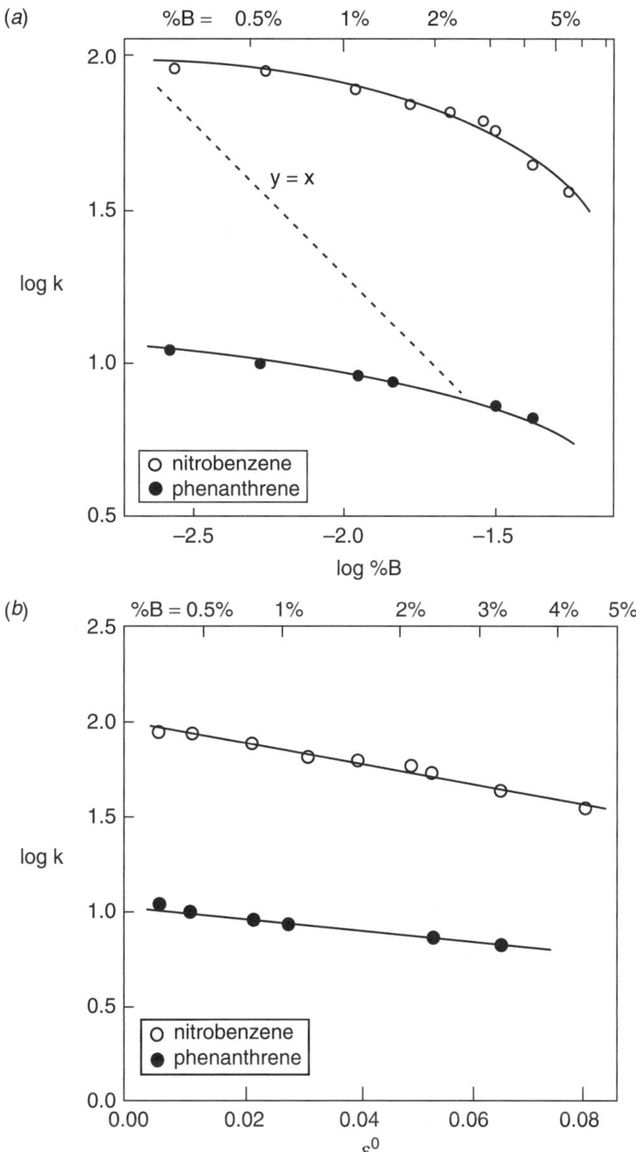

Figure 8.5 Dependence of log k on log %B for dilute solutions of a lesspolar (nonlocalizing) B-solvent. Sample: nitrobenzene and phenanthrene. Conditions: silica column; $CHCl_3$-hexane mobile phases; 25°C. (*a*) unsatisfactory correlations of log k with log ϕ (Eq. 8.4); (*b*) improved correlations of log k with ε (Eq. 8.2). Data from [11].

Table 8.1

Solvent Strength (ε^o), Molecular Area (n_B), and UV Absorbance as a Function of Wavelength for Normal-Phase Solvents

Solvent	ε^{0g}	n_B	Absorbance (AU) at Indicated Wavelength (nm)						
			200	210	220	230	240	250	260
Acetonitrile[a,b]	0.52	3.1	0.03	0.02	0.00	0.00	0.00	0.00	0.00
Ethoxynonafluorobutane	0.01	[h]	Presumed similar to ethyl ether						
Chloroform[c]	0.26	5.0	>1.0	>1.0	>1.0	>1.0	>1.0	0.25	0.08
Ethyl acetate[b]	0.48	5.2	>1.0	>1.0	>1.0	>1.0	>1.0	>1.0	0.10
Ethyl ether[d,e]	0.43	4.4	>1.0	>1.0	0.46	0.27	0.18	0.10	0.05
Hexane, heptane	0.00	[h]	0.35	0.20	0.07	0.03	0.02	0.01	0.00
Methanol[a]	0.70	3.7	>1.0	0.53	0.23	0.10	0.04	0.02	0.01
Methylene chloride[c]	0.30	4.1	>1.0	>1.0	>1.0	>1.0	0.09	0.00	0.00
Methyl-*t*-butyl ether[d]	0.48	4.1	>1.0	0.70	0.54	0.45	0.28	0.10	0.05
n-Propanol[f]	0.60	4.4	>1.0	0.65	0.35	0.15	0.07	0.03	0.01
i- Propanol[f]	0.60	4.4	>1.0	0.44	0.20	0.11	0.05	0.03	0.02
Tetrahydrofuran[d,e]	0.53	5.0	>1.0	>1.0	0.60	0.40	0.21	0.18	0.09

Sources: Data from [6, 13].
[a]immiscible with hexane.
[b]Nonbasic localizing.
[c]Nonlocalizing.
[d]Basic localizing.
[e]Easily oxidized and therefore less useful in practice.
[f]Very strong (localizing), proton-donor solvent; classification as "basic" or "nonbasic" may not be relevant.
[g]Values from [6], derived as described in [1].
[h]Values of n_B for A-solvents are not required in Equation (8.5).

in values of k by 2-fold is needed, a mobile phase of 30% methylene chloride-hexane ($\varepsilon = 0.19$) will increase k by about 2-fold, while 95% methylene chloride-hexane ($\varepsilon = 0.29$) will *decrease* k by about 2-fold.

Mobile phases with the same values of ε in Figure 8.6 should provide similar values of k for a given sample; for example, suppose that 50% CH_2Cl_2-hexane ($\varepsilon = 0.24$) has been found to provide $1 \leq k \leq 10$ for given sample and a silica column. From Figure 8.6 we can predict that 3.5% MTBE-hexane, 6% THF-hexane, or 4% ethyl acetate-hexane will each have a similar solvent strength ($\varepsilon = 0.24$); each of these four mobile phases should therefore provide a retention range of about $1 \leq k \leq 10$ for the same sample. Figure 8.6 is thus useful for selecting a different B-solvent in order to change selectivity (Section 8.3.2), while maintaining the same solvent strength and similar values of k. Because of potentially large changes in solvent-type selectivity for NPC (Section 8.3.2), the NPC solvent-nomograph of Figure 8.6 is somewhat more approximate than the corresponding nomograph of Figure 6.11 for RPC.

An example of NPC separation as a function of %B is shown in Figure 8.7 for an arbitrary mixture of organic compounds, using a silica column with mixtures of

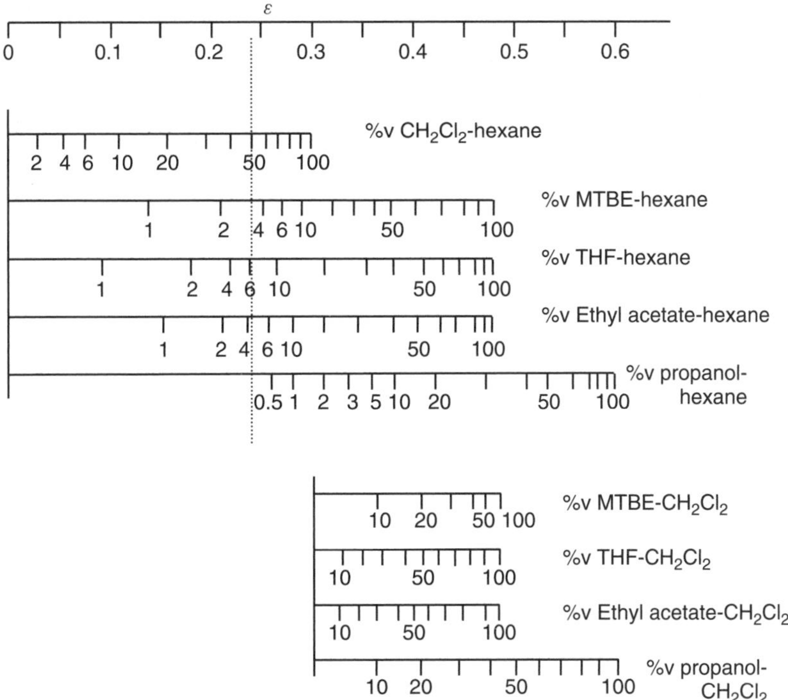

Figure 8.6 Solvent nomograph for normal-phase chromatography and silica columns. Adapted from [12].

ethyl acetate (B) and cyclohexane (A) as mobile phase. An increase in %B by a factor of two (40% B in Fig. 8.7*b* vs. 20% B in Fig. 8.7*a*) leads to a decrease in values of k by a factor of 2 to 3 in this example. Changes in relative retention with %B are also seen in Figure 8.7, as discussed in following Section 8.3.1 (solvent-strength selectivity). Because of these changes in relative retention with %B, an intermediate mobile-phase composition (Fig. 8.7*c*) provides the best resolution for this sample.

8.2.3 Use of TLC Data for Predicting NPC Retention

Thin-layer chromatography (TLC) can be a useful complement to the use of HPLC with a silica column. *Corresponding* separations by TLC and column chromatography (involving the same sample, mobile phase, temperature, and especially the same silica as stationary phase) should yield similar values of k for each compound in the sample. As illustrated in Figure 8.8, the positions of separated bands (spots) on a TLC plate can be used to estimate values of k for a corresponding column separation. The R_F value of a solute in TLC is defined as its fractional migration from the original sample spot (point at which the sample is applied) toward the solvent front (end of solvent migration during TLC). Thus a sample band that migrates half as far as the solvent front (e.g., spot D in Fig. 8.8) would have $R_F = 0.5$; the R_F values of solutes A through D in Figure 8.8 are 0.15, 0.25, 0.38, and 0.50, respectively.

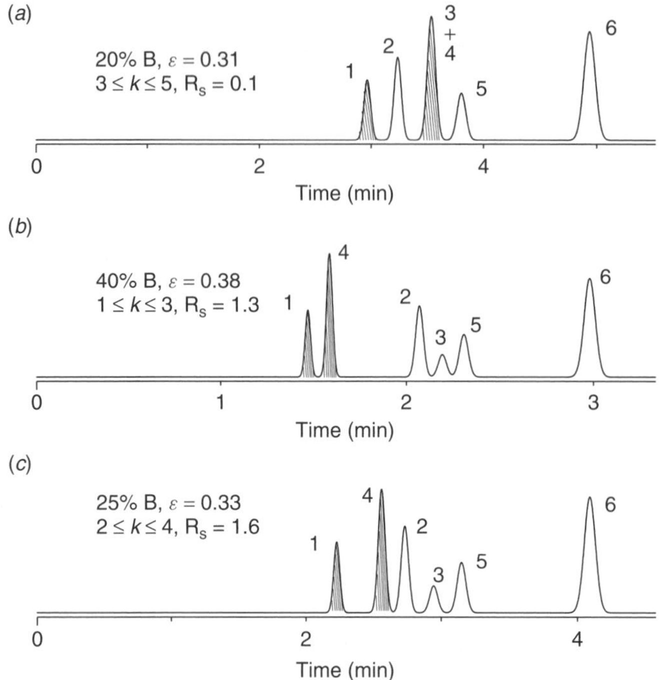

Figure 8.7 Solvent-strength selectivity in normal-phase chromatography. Sample: *1*, 2-aminonaphthalene; *2*, 2,6-dimethylquinoline; *3*, 2,4-dimethylquinoline; *4*, 4-nitrophenol; *5*, quinoline; *6*, isoquinoline. Conditions: 150 × 4.6-mm silica column (5-μm particles); ethylacetate (B)-cyclohexane (A) mixtures as mobile phase; ambient temperature; 2.0 mL/min. Peaks 1 and 4 are shaded to emphasize their change in relative retention as %B is varied. Chromatograms recreated from data of [14].

Corresponding values of k can be obtained from the relationship

$$k = \frac{1 - R_F}{R_F} \tag{8.6}$$

In the example of Figure 8.8, the R_F values of the spots vary from 0.15 to 0.50, corresponding to $1 \leq k \leq 5.7$. Similar k-values are expected for a corresponding HPLC separation (use of the same mobile phase with a silica column), hence providing an acceptable retention range ($1 \leq k \leq 10$) for this sample.

TLC is sometimes used prior to NPC separation with a column, as a convenient way of anticipating possible problems and/or exploring different mobile phases. Sample components (bands) that do not move ($R_F \approx 0.0$) will be visible in TLC, whereas any strongly retained solutes in column NPC will remain in the column and hence be undetected and perhaps missed. A related advantage of a preliminary TLC separation is that samples that might otherwise contaminate a column—and therefore require pretreatment for column chromatography (Chapter 16)—can usually be applied directly in TLC because a TLC plate is used only once and then discarded. By trial-and-error changes in %B (e.g., for methylene chloride-hexane mixtures) a value of %B that can provide values of $1 \leq k \leq 10$ for HPLC separation

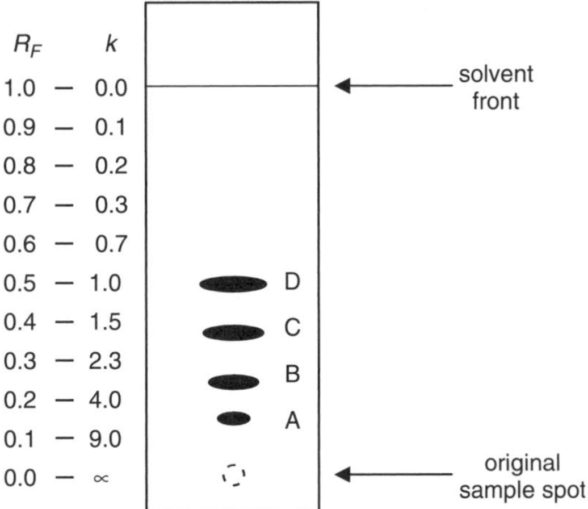

Figure 8.8 Use of thin-layer chromatography (TLC) for selecting a mobile phase to be used for HPLC with a silica column (hypothetical sample).

can be obtained in a short time by means of TLC—provided that isocratic elution is possible. If $1 \leq k \leq 10$ is the goal for all peaks in isocratic column chromatography, R_F values in a corresponding TLC separation should fall between 0.1 and 0.5. TLC experiments may show that the sample cannot be separated isocratically by column NPC (no value of %B results in sample bands with $0.1 \leq R_F \leq 0.5$). Changes in relative retention as a result of solvent-strength selectivity (Section 8.3.1) or solvent-type selectivity (Section 8.3.2) can also be explored by TLC—often more conveniently than by HPLC (see the example of Section 8.4.3).

While the use of TLC as described above is reasonably reliable, it should be noted that *solvent demixing* (Section 8.5.2) can lead to misleading results in some cases. When dilute solutions (<10% B) of strong B-solvents ($\varepsilon^0 > 0.4$) are used as the mobile phase in TLC, the B-solvent can be strongly retained by the silica, leading to its removal from the mobile phase during separation. As a result the concentration of B-solvent in the mobile phase will be lowered, the mobile-phase strength will be reduced, and observed values of R_F will be too small; thus values of k that are estimated from R_F values may be too large when solvent demixing occurs. This problem can be minimized by first equilibrating the TLC plate with the mobile phase. The plate should be placed in the developing chamber for 15 minutes—but without allowing the mobile phase to touch the plate and initiate sample migration [9]. This way the vapor above the solvent will equilibrate the plate and minimize any solvent demixing, after which the plate is lowered to contact the mobile phase and begin the separation.

TLC can be especially useful for monitoring preparative separations. Thus several fractions from a separation by column chromatography can be conveniently analyzed at one time, by spotting a single (suitably large) plate with multiple samples—each in a different lane of the plate. Fractions that are seen to contain the desired product, uncontaminated by adjacent impurity peaks, can be combined for further processing.

The detection of sample bands in TLC as in Figure 8.8 normally requires the addition of a visualization agent to the plate. For different means of band detection in TLC, see [10].

8.3 SELECTIVITY

Changes in retention order as a result of a change in separation conditions can be quite pronounced when NPC is used with a silica column—often much more so than in RPC. This ability to exert a greater control over relative retention with NPC is especially useful for preparative separations, where large α values for a compound to be purified allow much larger sample weights and a corresponding reduction in the effort and cost of sample purification (Section 15.3.2).

8.3.1 Solvent-Strength Selectivity

An example of solvent-strength selectivity can be seen in Figure 8.7. Although the relative retention of peaks 2, 3, 5, and 6 does not change much as %B is varied, peaks 1 and 4 (shaded) move toward the front of the chromatogram when %B is increased. It is significant that molecules 2, 3, 5, and 6 are each substituted with the same polar group (–N=), whereas peak 1 is 2-aminonaphthalene ($-NH_2$), and peak 4 is 4-nitrophenol ($-NO_2$, –OH); that is, molecules of solutes 1 and 4 are substituted by different polar groups. The change in k with %B for peak 4 is greater than for other compounds in this sample, and this can reasonably be attributed to the presence of two polar groups in the molecule ($-NO_2$, –OH) compared with only a single polar group for the other compounds (–N= or $-NH_2$); that is, n in Equation (8.4a) should equal 2 for peak 4 (vs. $n = 1$ for the remaining peaks), and therefore its retention should change more for a given change in %B. The somewhat different behavior of peak 1 compared with peaks 2, 3, 5, and 6 in this respect may be due to its different functionality: a $-NH_2$ group for peak-1 as opposed to a –N= group for peaks 2, 3, 5, and 6.

To summarize, pronounced changes in relative retention with a change in %B (solvent-strength selectivity) are more likely for a sample that contains compounds substituted by different polar groups, and especially for compounds with differing numbers of polar groups in the solute molecule—and therefore different values of n in Equation (8.4a).

8.3.2 Solvent-Type Selectivity

Before changing the B-solvent, %B should be varied so as to achieve $1 \leq k \leq 10$, while at the same time maximizing resolution (as in Fig. 8.7*c*). If further changes in selectivity (α) and resolution are needed, the B-solvent can be changed while maintaining the same solvent strength (same approximate value of ε in Fig. 8.6) for $1 \leq k \leq 10$. A change of the (more-polar) B-solvent is usually the most effective means for changing relative retention in NPC. An example of solvent-type selectivity is shown in Figure 8.9, for the separation of 12 naphthalene solutes that are substituted with different polar groups. In each case the mobile-phase strength is approximately the same ($\varepsilon = 0.25$; see Fig. 8.6 for the mobile phases of Figs. 8.9*a*,*b*),

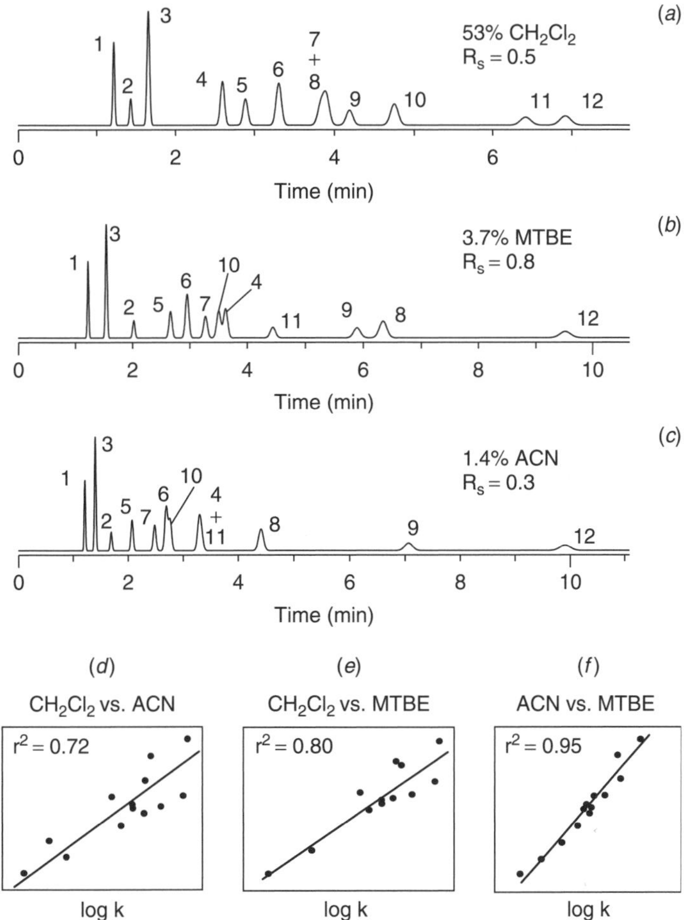

Figure 8.9 Example of solvent-type selectivity for normal-phase chromatography. Sample: *1*, 2-methoxynapthalene; *2*, 1-nitronapthalene; *3*, 1,2-dimethoxynapthalene; *4*, 1,5-dinitronapthalene; *5*, 1-naphthaldehyde; *6*, methyl-1-naphthoate; *7*, 2-naphthaldehyde; *8*, 1-naphthylnitrile; *9*, 1-hydroxynaphthalene; *10*, 1-acetylnapthalene; *11*, 2- acetylnapthalene; *12*, 2-hydroxynaphthalene. Conditions: 150 × 4.6-mm silica column (5-μm particles); mobile phases (%v) indicated in figure (50% water-saturated), except that (*c*) contains 6% added CH_2Cl_2 to achieve miscibility of ACN (hexane is the A-solvent in each case) 35°C; 2 mL/min. (*a–c*) Separations with indicated mobile phases; (*d–f*) correlations of retention data from (*a–c*). Chromatograms recreated from data of [15].

as are the run times (7–10 min), so solvent-strength selectivity should be about the same. Numerous changes in relative retention can be seen among these three separations, as a result of differences in solvent-type selectivity:

$$(53\%CH_2Cl_2) \quad 1 < 2 < 3 < 4 < 5 < 6 < 7 = 8 < 9 < 10 < 11 < 12$$

$$(3.7\%MTBE) \quad 1 < 3 < 2 < 5 < 6 < 7 < 10 < 4 < 11 < 9 < 8 < 12$$

$$(1.4\%ACN) \quad 1 < 3 < 2 < 5 < 7 < 6 \approx 10 < 4 = 11 < 8 < 9 < 12$$

Experimental studies [16] have shown that solvent-type selectivity in NPC depends mainly on the strength of the B-solvent (ε_B^0). As ε_B^0 increases, the B-solvent becomes more strongly attached to a specific silanol, resulting in *localized* adsorption of the B-solvent. Solvent fixation in this way is illustrated in Figure 8.3*c* by the arrow that connects surface silanols with molecules of the B-solvent (THF). The latter (localized) interaction of silanol and solvent can be contrasted with the weaker, more diffuse interaction of silanols with a less-polar (nonlocalizing) solvent in Figure 8.3*a*. In the separations of Figure 8.9, the B-solvents are CH_2Cl_2 with $\varepsilon_B^0 = 0.30$ (Fig. 8.9*a*), methyl-*t*-butyl ether (MTBE) with $\varepsilon_B^0 = 0.48$ (Fig. 8.9*b*), and acetonitrile (ACN) with $\varepsilon_B^0 = 0.52$ (Fig. 8.9*c*); MTBE and ACN are localizing solvents (as are mobile phases that contain these two solvents), while CH_2Cl_2 is not.

As the polarity of the solute increases, it too will become more strongly attached to (or *localized* onto) one or more silanols, as in the example of phenol in Figure 8.3*d*. Because the B-solvent and solute compete with each other for a place on the silica surface, an increase in B-solvent localization will result in a relatively greater reduction in *k* for solutes that are substituted by more-polar functional groups (and are therefore more localized), compared to less-polar, less-localized groups. That is, localized molecules of solute and B-solvent compete for the same positions (adsorption sites) on the silica surface. Nonlocalizing solutes will be less affected by the localization of the mobile phase on individual silanols, resulting in changes in relative retention (as in Fig. 8.9) that are determined by the relative localization of different solute molecules. See [6, 17] for a more detailed account of solvent and solute localization, and its effects on relative retention and solvent-type selectivity.

Localizing B-solvents, such as ACN and MTBE, can exhibit smaller, but significant differences in selectivity. For nine different localizing B-solvents [15–17] it was found that four of these solvents (nitromethane, ACN, acetone, and ethyl acetate) were very similar in terms of selectivity. The remaining solvents (dimethylsulfoxide, triethylamine, THF, ethyl ether, and pyridine) were significantly different in this respect. When these solvents are compared in terms of the solvent-selectivity triangle of Figure 2.9, it is seen that the latter five solvents fall within the group labeled "basic solvents," while the remaining four solvents are in the "dipolar solvents" (or nonbasic) group. It therefore appears that solvents can be characterized in terms of NPC selectivity as nonlocalizing (e.g., methylene chloride), basic localizing (e.g., MTBE), and nonbasic localizing (e.g., ethylacetate or ACN). Figure 2.9 allows other B-solvents to be classified as basic localizing or nonbasic localizing. For example, methyl-*t*-butylether is a commonly used (localizing) solvent in NPC; as seen in Figure 2.9, ethers such as MTBE are classified as basic.

Relative retention is compared among the three B-solvents of Figure 8.9*a–c* in Figure 8.9*d–f*. It is seen that the largest differences in relative retention (or change in selectivity) occur for the nonlocalizing B-solvent methylene chloride rather than for either the non-basic localizing ACN (Fig. 8.9*d*) or the basic localizing MTBE (Fig. 8.9*e*). That is, correlations for these plots are not very strong ($0.72 \leq r^2 \leq 0.80$). When retention for the two localizing B-solvents is compared (Fig. 8.9*f*), relative retention is more similar for these two solvents ($r^2 = 0.95$), but still sufficiently different to result in useful differences in selectivity (as seen in Fig. 8.9*b* vs. Fig. 8.9*c*). Thus solvent localization is the main source of solvent-type selectivity in NPC, but localizing solvents can be further differentiated as either basic or nonbasic. The commonly used solvents for NPC in Table 8.1 are characterized as nonlocalizing, basic localizing, or nonbasic localizing. When exploring solvent-type selectivity, a B-solvent of each type should be tried—as in Figure 8.9. Alcohols are very strong, proton-donor solvents; while they can be classified as localizing and basic, they should provide a moderate, further change in selectivity—especially for samples that contain strong proton acceptors.

Once separations have been carried out with nonlocalizing, basic-localizing, and nonbasic-localizing B-solvents as in Figure 8.9 (with %B adjusted to give $1 \leq k \leq 10$ for each mobile phase), two or more of these mobile phases can be blended to achieve an intermediate selectivity and further increase resolution; Figure 8.10 shows the best achievable resolution for the sample of Figure 8.9, as a result of blending the mobile phases from the separations of Figure 8.9*b,c* (to give 0.03% ACN/0.1% CH_2Cl_2/3.7% MTBE/96% hexane). A systematic selection of the best mobile-phase mixture can be carried out in a similar way as was described for optimizing solvent-type selectivity in RPC (Fig. 6.24); see Section 8.4.2 and Figure 8.15 for details.

A best choice of A- and B-solvents for varying solvent-type selectivity depends on several factors:

- B-solvent type: nonlocalizing, basic localizing, or nonbasic localizing
- solvent miscibility
- solvent UV cutoff (solutes that are weakly UV-absorbing may require detection at lower wavelengths)

Potentially useful solvents for NPC are listed with their properties in Table 8.1. For a full exploration of solvent-type selectivity, three different B-solvents will be required: nonlocalizing, basic localizing, and nonbasic localizing. For a nonlocalizing B-solvent, methylene chloride is preferable to chloroform by virtue of its lower UV absorbance and lower toxicity. However, detection can only be carried out at wavelengths >230 nm. Methylene chloride is miscible with all of the other solvents in Table 8.1.

There are five possible basic-localizing solvents listed in Table 8.1. Ethyl ether and tetrahydrofuran are susceptible to oxidation by air, and for this reason are not recommended. Both *n*- and *i*-propanol are quite strong ($\varepsilon^0_B = 0.60$), which means that mobile phases with $\varepsilon^0 < 0.25$ require propanol concentrations of <0.5% (as seen in Fig. 8.6); the use of very low concentrations of the B-solvent can create problems (Section 8.5) and should therefore be avoided if possible—especially when

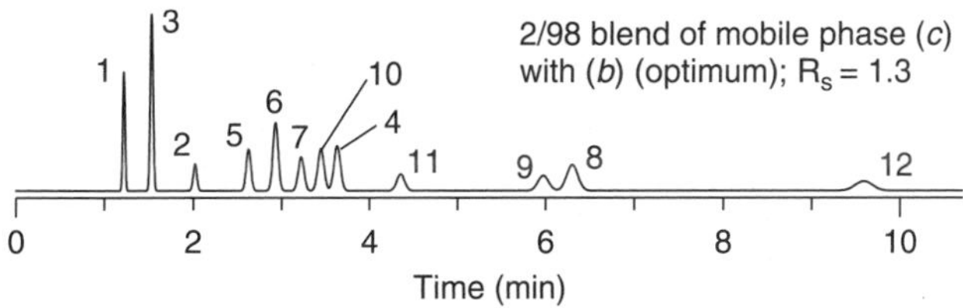

Figure 8.10 Optimized selectivity and resolution for sample of Figure 8.9. Conditions as in Figure 8.9, except that mobile phase is a 98/2 blend of mobile phases (*b*) and (*c*).

using bare-silica columns. When choosing a basic-localizing B-solvent, it is suggested that MTBE be used for mobile phases with $\varepsilon^0 < 0.48$, and either *n*- and *i*-propanol for $\varepsilon^0 > 0.48$. Neither MTBE nor propanol present miscibility or detection problems.

Ethyl acetate and ACN are each candidates for a nonbasic-localizing B-solvent. Ethyl acetate suffers in terms of UV absorbance (detection with this solvent is only possible at >250 nm), while ACN and hexane are not fully miscible. The addition of a co-solvent (e.g., methylene chloride, as in Fig. 8.9*c*) allows the use of hexane with ACN. But determining the required addition of the co-solvent is somewhat tedious, and the estimation of ε values for mixtures of these three solvents is not straightforward [18]. However, a computer program (LSChrom) is available for the calculation of ε-values [19, 20]. An alternative to the need for a co-solvent is the use of ethoxynonafluorobutane as the A-solvent [21] (available from 3M, and miscible with ACN or MeOH—but expensive) with ACN as the B-solvent—this allows UV detection at wavelengths $\geq$ 220 nm. Hexane and ethoxynonafluorobutane can be considered as interchangeable in terms of solvent strength ($\varepsilon^0 \approx 0.00$).

8.3.3 Temperature Selectivity

The effect of temperature on NPC selectivity has received only limited attention; one of a few reported examples is shown in Figure 8.11. In Figure 8.11*a,b*, separation is shown for 22 and 55°C, using methylene chloride as the (nonlocalizing) mobile phase. It is seen that selectivity does not change appreciably with temperature, and this may be generally true for less-polar samples and nonlocalizing mobile

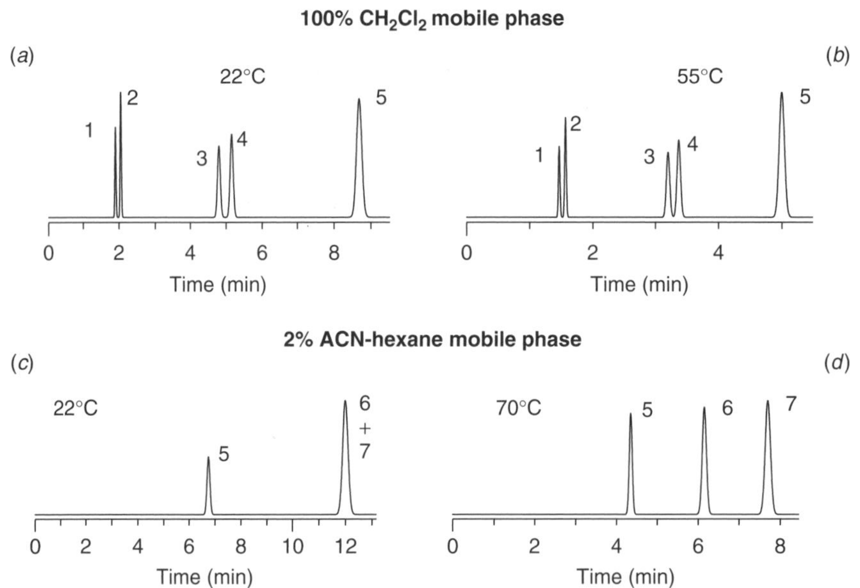

Figure 8.11 Effect of temperature on relative retention in NPC. Sample: *1*, nitrobenzene; *2*, methyl benzoate; *3*, benzaldehyde; *4*, acetophenone; *5*, α-methyl benzyl alcohol; *6*, benzyl alcohol; *7*, 3-phenyl-1-propanol. Conditions: 250 × 4.6-mm silica column (5-μm particles); 2 mL/min; mobile phase and temperature shown in figure. Chromatograms recreated from data of [22].

phases. For the separations of Figure 8.11*c,d* with a localizing mobile phase (2% ACN-hexane), a change in temperature leads to a marked change in the relative retention of localized-solute peaks 6 and 7. Changes in selectivity with temperature such as that of Figure 8.11*c, d* may be the result of the relatively reduced retention of localizing solvents at higher temperatures, in which case similar changes in relative retention may result for a change from a localizing to a nonlocalizing B-solvent. A change in temperature as a means of changing NPC selectivity is limited, in practice, by the relatively low boiling points of the more useful solvents, and may have only marginal utility. Alternatively, large changes in selectivity can be achieved by changing the B-solvent—minimizing the need for further changes in selectivity as by a change of temperature.

8.3.4 Column Selectivity

Because silica is most often used in NPC, and because a change of B-solvent is an effective means for varying selectivity, a change of column is not often used for the sole purpose of changing NPC selectivity. A more common reason for using a polar-bonded-phase column in preference to unbonded silica (primarily for assay methods) is to avoid the problems described in Section 8.5: poor separation reproducibility from run to run, slow column equilibration, or difficulty in using gradient elution for samples with a wide retention range. Several studies have been reported [23–27] of retention for different test solutes, using each of the three types of polar-bonded-phase NPC columns (cyano, diol, amino). It appears that selectivity is a complex function of both column type and the choice of B-solvent [27], suggesting that trial-and-error changes in both column type and B-solvent can result in significant variations in relative retention.

Figure 8.12 compares the separation of a mixture of aromatic compounds with hexane as mobile phase and (1) a cyano column, (2) a diol column, and (3) an amino column. The separation of the same sample on a silica column (Fig. 8.12*d*) can be estimated (very approximately) from other published data [1]. A comparison of these four separations suggests that run time (or "column strength") varies as

$$\text{silica} \gg \text{amino} > \text{diol} > \text{cyano}$$

Note the logarithmic time scale for the silica separation of Figure 8.12*d*, which correctly implies that the range in retention (and selectivity) for a sample with a silica column can be far wider than for a polar-bonded-phase column.

Separation selectivity or relative retention varies moderately from column to column in Figure 8.12, as expected; proton-donor solutes are retained more strongly on amino columns relative to other solutes, and less strongly on cyano columns [26, 27]. It should also be noted that the sample of Figure 8.12 can be separated isocratically with any of the polar-bonded-phase columns (Fig. 8.12*a–c*), whereas separation of this sample with a silica column (Fig. 8.12*d*) would require gradient elution (if gradient elution is even feasible; see Section 8.5.2). This observation should be true for any sample: *isocratic separation is more likely to be possible with the use of a polar-bonded-phase column than with silica.*

Finally, large values of α (and the possibility of baseline resolution) are more likely when a silica column is used. Similarly changes in selectivity with change

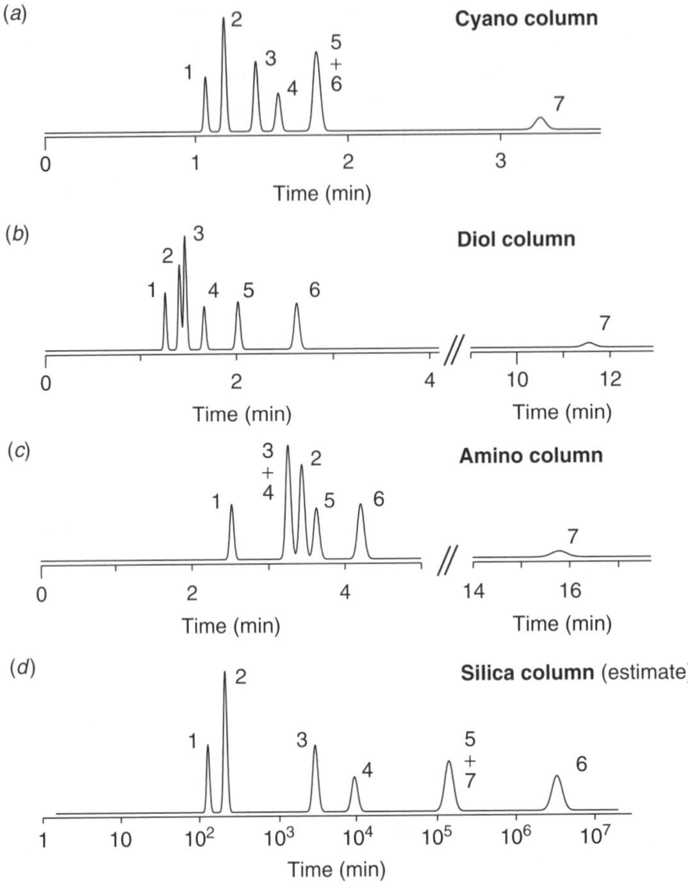

Figure 8.12 Comparison of retention and selectivity among different NPC columns. Sample: *1*, chrysene; *2*, perylene; *3*, 1-nitronaphthalene; *4*, 1-cyanonaphthalene; *5*, 2-acetonaphthalene; *6*, naphthalene-2,7-dimethylcarboxylate; *7*, benzyl alcohol. Conditions: 150 × 4.6-mm columns (column type indicated in figure); hexane mobile phase; 35°C; 2.0 mL/min. Chromatograms (*a* – *c*) reconstructed from data of [26]; (*d*) estimated from data of [1] (note extreme change in retention range for silica column *d* vs. polar-bonded columns *a*–*c*).

in the B-solvent (solvent-type selectivity) are much more pronounced for silica columns—compared to polar-bonded-phase columns. Consequently a change from silica to a polar-bonded-phase column, with further variation of the mobile phase (change in %B and/or B-solvent), is unlikely to lead to a better resolution of a peak-pair that has not been separated after varying solvent-type selectivity with a silica column.

8.3.5 Isomer Separations

We have noted that isomers are generally better separated by NPC than by RPC. Similarly limited data [15, 26] suggest that isomer separation is generally more pronounced on silica columns, compared to polar-bonded-phase columns. However, the separations of Figure 8.2*b–d* with a cyano column make clear that polar-bonded-phase columns *are* able to separate some isomers. Isomer-selectivity, using NPC with a silica column, can be attributed to (at least) three possible characteristics of isomeric molecules:

- steric hindrance of a polar substituent by an adjacent nonpolar substituent
- electron donation or withdrawal from a polar group by a second substituent in the solute molecule
- the relative positions of different polar groups within the molecule, and the planarity of the solute molecule

Figure 8.13 illustrates each of these three effects. In Figure 8.13*a,b* it is seen that for the separation of two methylaniline isomers the interaction of the polar $-NH_2$ group with a surface silanol will be interfered with by an *o*-methyl group (Fig. 8.13*b*), because of steric hindrance, but not by a *p*-methyl group (Fig. 8.13*a*). The steric hindrance created by the *o*-methyl group will be further enhanced by the adjacent silica surface. Consequently a molecule with a sterically hindered polar substituent should be less retained than an isomer in which steric hindrance is absent or less pronounced. Figure 8.2*b* provides an experimental example of the contribution of steric resistance to isomer selectivity, where *p*-methylaniline (peak 3) is more strongly retained than *o*-methylaniline (peak 1) in this NPC separation.

In Figure 8.13*c,d*, steric hindrance of the $-NH_2$ group by the methyl substituent does not occur for either the *m*- or *p*-isomer. However, the methyl group in the *para* position is more effective at transferring electrons to the $-NH_2$ group, in turn increasing its hydrogen-bond basicity and retention; that is, a *p*-CH_3 group has a more negative value of the Hammett σ-parameter [28] than does a *m*–CH_3 group. Therefore the *p*-methyl isomer should be more retained than the *m*-methyl isomer in this example. This is confirmed in Figure 8.2*b* where *p*-methylaniline (peak 3) is more retained than *m*-methylaniline (peak 2). Similar examples of the effects of methyl substitution on retention are provided by the separations of Figure 8.2*c,d*.

In Figure 8.13*e,f* the relative retention of these two dimethoxyethylene isomers will be affected by which isomer can better position itself adjacent to the surface, so as to allow *each* polar methoxy group in the molecule to interact with an adjacent silanol group on the surface. In this example it appears that *cis*-1, 2-dimethoxyethylene will be more strongly retained, but any such prediction must be regarded as tentative; silanol groups are distributed randomly about the silica surface, and the solute molecule is free to adapt various positions on the silica surface. Consequently there is usually little basis for predicting which of two positional isomers will have its polar substituents more closely matched to the positions of neighboring silanols. In some cases adjacent polar groups within a solute molecule (as in *cis*-1,2-dihydroxyethylene) can interact intra-molecularly (Fig. 8.13*g*), possibly competing with and reducing inter-molecular interactions that increase retention. These and other contributions to isomer separation on silica (as in Fig. 8.13*a–f*) are examined in detail in Chapter 11 of [1].

Finally, approximately planar molecules are more easily matched to the (roughly) planar silica surface so that more planar isomers are preferentially retained. An example is provided in Figure 8.14 for the separation of five isomers of the compound retinol. In the RPC separation of Figure 8.14*a*, there is little separation of these isomers, whereas in the NPC separation of Figure 8.14*b* every peak is at least partly resolved. Peak-5 (the all-*trans*) isomer is more nearly planar and is preferentially retained. Solutes with an increasing number of *cis*-linkages tend to be increasingly less planar and less retained.

Let us now compare all of the contributions above to isomer separation for a silica column versus separation on (1) polar-bonded-phase NPC columns or (2) RPC columns. In the case of polar-bonded-phase NPC columns, steric hindrance effects (Figs. 8.13*a*,*b*) will be less important because the silica surface is further removed from the polar cyano, diol, or amino group of the stationary phase—hence contributing less to steric hindrance between the solute and the stationary phase. Similarly the matching of polar groups in the solute molecule with polar groups in the stationary phase (Figs. 8.13*e*,*f*) will be easier for a polar-bonded-phase column (with less effect on isomer selectivity) because the cyano, diol, or amino groups are not rigidly positioned on the surface but are connected to the silica surface by a flexible $-CH_2-CH_2-CH_2-$linkage. Finally, the attraction of polar groups in the solute molecule to the polar stationary phase is weaker for polar-bonded-phase columns than for silica, which in turn reduces the effect of each of the contributions to isomer separation in Figure 8.13. Consequently *isomer separations*

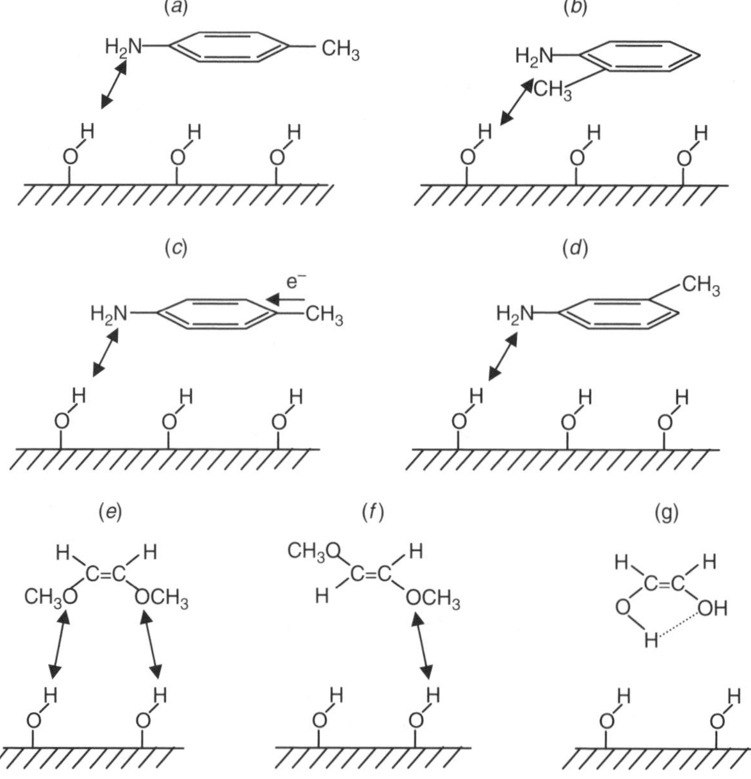

Figure 8.13 Factors that contribute to isomer selectivity for NPC separation on silica columns. (*a*, *b*) Steric hindrance; (*c*, *d*) electron donation; (*e*, *f*) relative positions of polar groups within the solute molecule; (*g*) intramolecular hydrogen bonding of two polar groups.

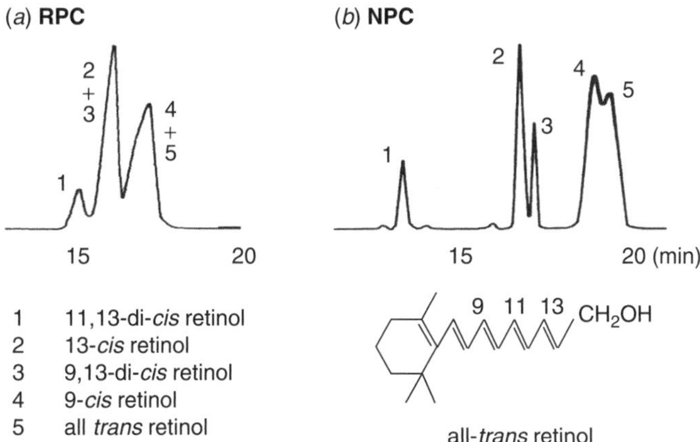

Figure 8.14 Comparison of isomer selectivity for separation by RPC (*a*) and NPC (*b*). Sample shown in figure (retinal isomers). Conditions: (*a*) 200 × 4.4-mm C_{18} column; 80% methanol-water; 1.0 mL/min; 40°C. (*b*) 250 × 4.0-mm silica column; 8% dioxane-hexane; 1.0 mL/min. Figure is adapted from [29].

on polar-bonded-phase columns will usually be less pronounced, compared to separation on silica.

For the case of RPC separation, the interaction of polar solute groups with the stationary phase is much weaker than in NPC, which minimizes each of the effects of Figure 8.13 (as in the case of polar-bonded-phase columns), and reduces isomer selectivity. While corresponding interactions with the polar mobile phase are possible, the latter interactions are generally weaker than corresponding interactions of solute and mobile phase with silica silanols, and less subject to steric effects. There is one exception to this conclusion for RPC, however, in the case of separations on cyclodextrin columns (Section 6.3.5). Isomer resolution is more pronounced for the latter columns, and this may be interpreted as follows: First, the cyclodextrin molecule possesses a cavity into which a solute molecule can enter (see Figs. 14.17, 14.18), and some solute molecules may fit this cavity better than others. Second, the cyclodextrin molecule possesses multiple –OH substituents with fixed positions within the molecule. So far as their effect on isomer separation, these cyclodextrin—OH groups may play a similar role in RPC as for silanols in NPC (as in Fig. 8.13*e*,*f*).

8.4 METHOD-DEVELOPMENT SUMMARY

The first step in NPC method development should consist of a review of the goals of separation, including reasons why NPC is being considered. Unless some problem can be anticipated for the use of RPC—or has been experienced in prior RPC separations of the sample—RPC is normally a best first choice at the beginning of method development. Some applications for which NPC might be considered initially include:

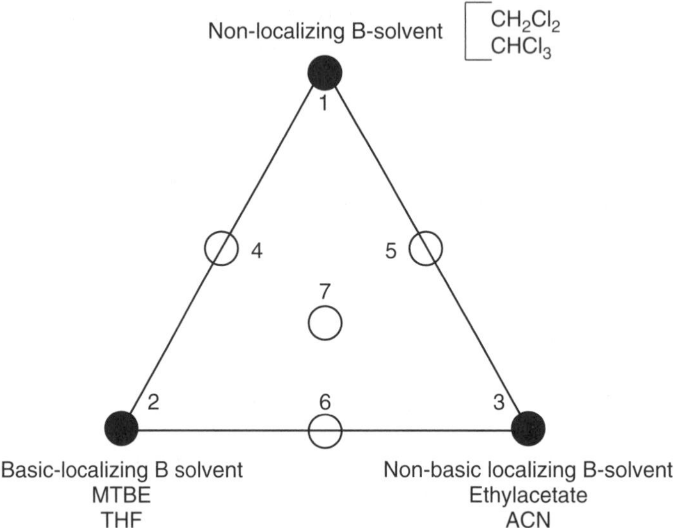

Figure 8.15 Use of seven experiments for the optimization of solvent-type selectivity in NPC. A-solvent: hexane, heptane, or ethoxynonafluorobutane, or (for $\varepsilon^0 > 0.30$) CH_2Cl_2. For a very strong localizing B-solvent use *n*- or *i*-propanol.

- the purification of crude samples
- the separation of isomers
- orthogonal separation
- samples that contain hydrophobic interferences
- samples that contain very polar analytes (e.g., unretained in near-100% water by RPC)

The *purification* (preparative separation) of organic-soluble samples is usually best carried out with NPC and a silica column. The removal of solvent from separated fractions is easier for the organic solvents used in NPC, compared with the higher boiling water used in RPC. Larger values of α for the solute to be purified are usually possible with NPC and silica columns, meaning that larger sample weights can be injected (other conditions the same). Finally, the use of TLC can be convenient for the preliminary assay of fractions from preparative NPC.

As noted above (Section 8.3.5), *isomeric compounds* are usually much better separated by means of NPC with silica columns. However, many isomers are easily separated by RPC, so it may be preferable to try RPC first for such samples—unless prior experience with related samples suggests otherwise.

Orthogonal separations (Section 6.3.6.2) can be used to check whether a proposed method has separated all of the compounds in a sample. In principle, the selectivity of an orthogonal method should be as different as possible from that of the original method. In the past some workers have assumed that separations based on a different principle (e.g., NPC vs. RPC) are more likely to provide a large enough difference in selectivity so that any pair of peaks overlapped in one method will be separated by an orthogonal method. While this suggests the use of NPC for orthogonal separations, it is also possible to design two orthogonal RPC methods (Section 6.3.6.2), especially for the case of ionic samples whose selectivity is more dependent on separation conditions (Section 7.3.2).

Hydrophobic interferences in a sample are a less common reason for the use of NPC. Many samples contain hydrophilic additives or impurities that are much less retained in RPC than sample peaks of interest; these non-analytes then appear as "junk" or "garbage" peaks near t_0 (as in the example of Fig. 2.5*b*). If these hydrophilic non-analytes do not overlap peaks of interest, they may not need to be removed from the sample prior to RPC separation. When sample non-analytes are more hydrophobic than peaks of interest, they will elute much later in RPC, resulting in a very long run time (as in the example of Section 8.4.3 below). Either sample pretreatment will be required for their removal (Chapter 16) or gradient elution can be used to elute hydrophobic impurities within a reasonable time. Either of the latter two approaches may be inconvenient or ineffective for some samples. In such cases NPC separation may be the best solution, inasmuch as hydrophobic non-analytes will be retained less strongly than peaks of interest, and leave the column near t_0.

Very polar samples are often insufficiently retained by RPC (Section 6.6.1), whereas they are likely to be well retained by NPC. For this application, HILIC (Section 8.6) is often a first choice, especially for ionic samples.

NPC method development proceeds in similar fashion as for RPC method development. The same seven steps described in Section 6.4 also apply for NPC:

1. define the goals of separation: resolution, run time, detection limits, and so on
2. carry out sample preparation (Chapter 16)
3. choose separation conditions
4. verify column reproducibility, choose alternative columns (Section 6.3.6.1)
5. develop a routine orthogonal separation (Section 6.3.6.2)
6. carry out method validation (Section 12.5)
7. develop a system suitability test (Section 12.3)

For the case of preparative separations (Chapter 15), the choice of separation conditions is of primary importance, and method-development steps 1 and 4 through 7 can often be ignored. For all applications the primary difference in the case of NPC versus RPC is step 3, the choice of separation conditions. The following four steps apply for the development of any isocratic HPLC separation (see Section 9.3.10 for gradient methods):

1. select starting conditions
2. adjust %B for an acceptable retention range (e.g., $1 \leq k \leq 10$)
3. optimize selectivity
4. optimize column length, flow rate, and (possibly) particle size for the best compromise between resolution and run time

Table 8.2

Method Development for Normal-Phase Chromatography

Step	Approach	Comments
1. Starting conditions	Choose column and A- and B-solvents from Table 8.1	A silica column is usually preferred; diol-silica is an alternative; hexane and methylene chloride are good first choices for the A- and B-solvents, respectively.
2. %B for $1 \leq k \leq 10$	Vary % CH_2Cl_2-hexane	Can use either TLC or HPLC; start with 100% CH_2Cl_2; a 2-fold change in %B will change k by about 3-fold; if a stronger B-solvent is needed, try MTBE, then *i*-propanol.
3. Optimize α	Change B-solvent	If CH_2Cl_2/hexane can be used as mobile phase, use Figure 8.6 to choose a MTBE/hexane and/or ethyl acetate/hexane mobile phases of same ε value.[a]
	Blend B-solvents	If further changes in selectivity are needed, blend above mobile phases as in Figure 8.15.

[a]If CH_2Cl_2-hexane mixtures are too weak ($k > 10$), then try other mobile phases from Figure 8.6.

Steps 1 to 3 for NPC are discussed below and summarized in Table 8.2. Step 4, the choice of final column conditions, is carried out in the same way as for RPC separation (Section 2.4.3).

8.4.1 Starting Conditions for NPC Method Development: Choice of Mobile-Phase Strength and Column Type (Steps 1, 2; Table 8.2)

In RPC method development, it was recommended to begin with a 100 × 4.6-mm C_{18} column (3-μm particles) and 80% ACN as mobile phase (Table 6.1), followed by the adjustment of % ACN so as to achieve $1 \leq k \leq 10$. For NPC, either a silica or polar-bonded-phase column may represent a reasonable starting point. Silica columns are preferred for preparative separations and the resolution of isomers, while polar-bonded-phase columns present fewer problems for assay procedures, and are advantageous for wide-range samples that might require gradient elution if a silica column is used.

The selection of an initial mobile phase involves a number of considerations, as reviewed in Section 8.3.2.1. A choice of mobile-phase strength (%B) is conveniently made on the basis of exploratory TLC separations with silica plates (Section 8.2.3). Alternatively, gradient elution with a polar-bonded-phase column can also be used (Section 9.3.2), especially if the latter column will be use for the final separation (instead of silica). In either case, the goal should be the estimation of a mobile-phase composition that will provide $1 \leq k \leq 10$ in a subsequent column separation (equivalent to $0.1 \leq R_F \leq 0.5$ in a TLC separation). We recommend an initial TLC separation with a silica plate and 100% methylene chloride (B-solvent) as mobile phase. If values of R_F are too high, then the mobile phase is too strong, and lower values of % methylene chloride (with hexane as A-solvent) can be investigated—using Section 8.2.2 as a guide. Values of R_F that are too low with 100% methylene chloride as mobile phase will require a stronger B-solvent; in this case mixtures of MTBE and hexane are a good choice for subsequent TLC separations. If a still stronger B-solvent is required, try mixtures of *n*- or *iso*-propanol with hexane as A-solvent. In each of the latter cases the solvent nomograph of Figure 8.6 can be useful in selecting values of %B for successive experiments.

During the adjustment of %B for $1 \leq k \leq 10$, a *decrease* in values of k by a factor of two requires an increase in ε by $\approx$0.05 units (a decrease in ε by $\approx$0.05 units will result in a 2-fold *increase* in k). For example, assume that a TLC separation with 100% methylene chloride as mobile phase gives $0.5 \leq R_F \leq 0.8$ for the sample bands, corresponding to $0.2 \leq k \leq 1$ (Fig. 8.8 or Eq. 8.5). In this case we need an increase in k by 5- to 10-fold. A decrease in ε by 0.05, 0.10, or 0.15 units should increase values of k by a factor of about 2, $2^2 = 4$, or $2^3 = 8$, respectively. For this example, a decrease in ε^0 by 0.15 units is suggested. Referring to Figure 8.6, we see that ε for 100% methylene-chloride is 0.30, which means that the recommended mobile phase will have $\varepsilon^0 = 0.15$; a mobile phase of 18% methylene chloride-hexane would therefore be suggested by Figure 8.6. Because estimates of k as a function of %B can be less accurate in NPC, further adjustment of %B will usually be required to achieve the desired retention range of $1 \leq k \leq 10$, as well as to take advantage of solvent-strength selectivity.

At the conclusion of exploratory TLC studies with silica plates, we need to choose a NPC column (unless TLC can be used for routine analysis). Silica columns provide larger values of α for preparative separations and the resolution of isomers, while polar-bonded-phase columns are less subject to the various problems described in Section 8.5, and are more suitable for samples whose retention range exceeds $0.5 \leq k \leq 20$ (as estimated from preliminary TLC separations). Retention with polar-bonded-phase columns is much weaker than for silica (smaller values of k); if the mobile phase selected by means of TLC as above (with a silica plate) has $\varepsilon < 0.15$, retention on a polar-bonded-column may be too weak to provide $1 \leq k \leq 10$—even with pure hexane as mobile phase ($\varepsilon^0 = 0.00$). In this case the use of a polar-bonded-phase column is not an option.

Once a mobile-phase composition has been selected from initial TLC studies (for $0.1 \leq R_F \leq 0.5$, or $1 \leq k \leq 10$), this mobile phase can be used with a silica column. If a polar-bonded-phase column is used instead, then mobile-phase strength should be lowered by about 0.15 ε-units. For example, if the recommended mobile phase were to consist of 40% methylene chloride-hexane (for a silica column), the corresponding value of $\varepsilon^0 = 0.22$; for a polar-bonded-phase column, the value of ε^0 in Figure 8.6 should then be $0.22 - 0.15 \approx 0.07$. From Figure 8.6, a value of $\varepsilon^0 = 0.07$ corresponds to 5% methylene chloride-hexane, which can be used as mobile phase for an initial separation on a polar-bonded-phase column. Keep in mind that estimates of mobile-phase strength from Figure 8.6 for silica (based on a 2-fold change in k for a change in ε^0 by 0.05 units) are approximate, and the extension of this rule to polar-bonded-phase columns is even less reliable.

8.4.2 Strategies for Optimizing Selectivity (Step 3; Table 8.2)

The determination of a suitable value of %B for $1 \leq k \leq 10$ will usually involve experiments where %B is varied; solvent-strength selectivity (Section 8.3.1) can be explored at the same time. If further changes in selectivity are required, solvent-type selectivity (Section 8.3.2) should be investigated next—because of its very large effect on relative retention and resolution. Three different B-solvents will be needed for a full exploration of solvent-type selectivity: nonlocalizing, basic localizing, and nonbasic localizing, as illustrated in Figure 8.9. These three binary-solvent mixtures can then be mixed with each other in various proportions to provide any intermediate selectivity (as in the example of Fig. 8.10). A general plan or "experimental design" for the latter approach is illustrated in Figure 8.15, with a list of possible B-solvents of each type.

Initial experiments with CH_2Cl_2-hexane can be used to determine a value of %B (and ε) for acceptable retention ($1 \leq k \leq 10$). The latter separation corresponds to experiment 1 of Figure 8.15. If a further improvement in selectivity and resolution are desired, experiments 2 and 3 of Figure 8.15 are carried out (e.g., mixtures of MTBE-hexane and ethylacetate-hexane that have the same value of ε as in experiment 1; see Fig. 8.6). An examination of the latter three chromatograms will indicate whether further blending of these three mobile phases (experiments 4 to 7) can provide any improvement in selectivity and resolution. This procedure for optimizing solvent-type selectivity can be quite powerful, especially when silica columns are used. The plan of Figure 8.15 can be compared with the similar optimization of solvent-type selectivity for RPC in Figure 6.24 (see related text for details). When the mobile phases of experiments 1 to 3 of Figure 8.15 are blended, the resulting values of ε may not remain the same, requiring a change in the concentration of the A-solvent. A computer program for the more accurate prediction of values of ε as a function of mobile-phase composition has been reported [30], based on the procedure of [31]; a demo copy of the software is available [19].

8.4.3 Example of NPC Method Development

A method was required for samples containing the polar drug paclitaxel in mixture with a more hydrophobic polymer (poly[sebacic-recinoleic ester-anhydride]) [32]. Structures of these two entities are shown at the bottom of Figure 8.16. Because the drug is more polar than the polymer, an assay by RPC would have required either prior separation of polymer from the drug (because of very late elution of the polymer in RPC) or gradient elution. For this reason NPC separation was explored, with the objective that the polymer would leave the column *before* the drug (in the vicinity of t_0) and thereby preclude a need for sample pretreatment. Initial studies were carried out by means of TLC with silica plates. The use of 100% methylene chloride yielded $R_F = 0.00$ for the drug, so the stronger B-solvents tetrahydrofuran (THF, $\varepsilon^0 = 0.53$) and methanol (MeOH, $\varepsilon^0 = 0.70$) were investigated next, in mixture with methylene chloride. Mobile phases composed of 2–5% MeOH–CH_2Cl_2 appeared promising from the TLC results presented in Table 8.3 (THF would be a less desirable choice; Section 8.3.2.1); 1.5% MeOH–CH_2Cl_2 with a silica column provided the satisfactory separation of Figure 8.16*a*, with UV detection at 240 nm. Note that $k \approx 3$ for the paclitaxel peak in the separation of Figure 8.16*a*, whereas TLC separation predicts $k \approx 6$. Somewhat approximate predictions of NPC retention from TLC are expected, but such predictions can still be useful—as in the present example.

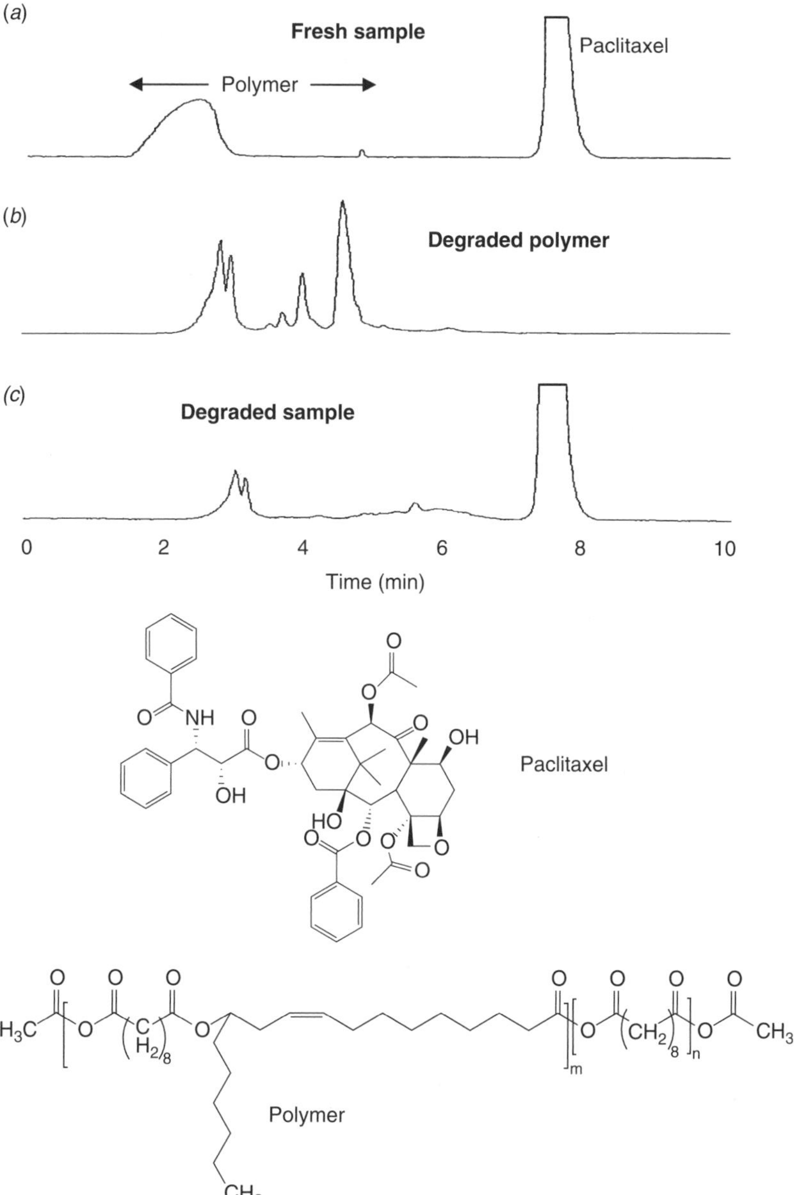

Figure 8.16 NPC assay of paclitaxel in the presence of a polymeric additive. Conditions: 250 × 4.0-mm silica column (5-μm particles); 1.5% methanol-methylene chloride; 25°C; 1 mL/min. (*a*) Fresh sample; (*b*) degraded sample of polymer (stored at pH-7.4 and 37°C for 60 days); (*c*) degraded sample of paclitaxel plus polymer. Reprinted with permission from [32].

Table 8.3

Exploratory TLC Separations of Paclitaxel-Polymer Samples

MeOH–CH_2Cl_2			THF–CH_2Cl_2		
%-MeOH	R_F Paclitaxel	R_F Polymer	%-THF	R_F Paclitaxel	R_F Polymer
1	0.04	0.05	2	0.00	0.56
2	0.22	1.00	4	0.00	0.64
3	0.29	1.00	9	0.12	1.00
4	0.35	1.00	20	0.52	1.00
5	0.46	1.00	30	0.88	1.00

Source: Data from [32].

The method of Figure 8.16*a* was also intended for use with thermally stressed samples, as carried out in the experiments of Figure 8.16*b,c*. It appears that thermal degradation of the polymer (Fig. 8.16*b*) does not result in the formation of peaks that overlap the paclitaxel peak and thereby compromise its assay. For other details of this NPC method development, see [32].

8.5 PROBLEMS IN THE USE OF NPC

Several interrelated problems can occur during NPC separation with silica as column packing:

- poor separation reproducibility (including extreme sensitivity to mobile-phase water content)
- solvent demixing
- slow column equilibration when changing the mobile phase
- tailing peaks

The first three problems arise from the very strong interaction of small, polar molecules with surface silanols; this in turn can have a dramatic effect on sample retention.

8.5.1 Poor Separation Reproducibility

Sample retention times in NPC can vary from day to day, or even within the same day, as a result of significant variations in room humidity, and consequent changes in the water content of nominally "dry" mobile-phase solvents. This effect is illustrated in Figure 8.17*a* for the elution of benzanilide as the solute with methylene chloride as the mobile phase and a silica column. In this example, the mobile phase was prepared by blending different volumes of methylene chloride that were either water-free or water-saturated, so as to achieve different concentrations of water in the final mobile phase (see top of Fig. 8.17*a*). For example, blending equal volumes of the two solvents would result in 50% water saturation of the final mobile phase (or 0.08% water). As the water content of the mobile phase is increased, the retention of the solute decreases sharply (from $k = 9$ to $k = 2$) because of the increasing coverage of the silica surface by adsorbed water (water interacts very strongly with silica). It can be appreciated from this example that small changes in room humidity can lead to significant changes in %-water saturation, which in turn can lead to variable sample retention. When a silica column is used, this variability of mobile-phase water-content is the most common cause of variable NPC retention—provided that the column has been properly equilibrated before samples are injected (Section 8.5.2).

It is possible to minimize variations in mobile-phase water content and sample retention by controlling the water content of the mobile phase, as described above (blending the water-free with the water-saturated mobile phase). An alternative, more convenient approach is the addition of small amounts of a very polar solvent to the mobile phase. As seen in Figure 8.17*b*, the addition of small amounts of methanol brings on a reduction in sample retention similar to that of added water—caused by a comparable deactivation of the silica surface. Presumably the addition of methanol renders the column less susceptible to variations in water content. The amount of methanol required for silica deactivation will vary with mobile-phase composition, and be less for weaker mobile phases (with smaller values of ε). The limited miscibility of methanol and hexane suggests that propanol should be substituted when necessary. Note that as the concentrations of either water or methanol in the mobile phase increase, sample retention decreases, and may call for adding hexane to the mobile phase to decrease ε. The ability of silica to separate isomers and other solutes will be compromised by excessive deactivation of the column.

Variable retention due to changes in mobile-phase water content, as in Figure 8.17*a*, should be less pronounced for (1) more-polar mobile phases with larger values of ε or (2) less-polar, bonded-phase columns (because water is less tightly bound to such columns). Consequently polar-bonded-phase columns are less likely to be affected by problem than silica columns. Retention variability with *silica* columns may not even be a problem if the sample, mobile phase, the relative constancy of room humidity are controlled, and if solvents are transferred from their original bottles to the reservoir with minimum exposure to the atmosphere.

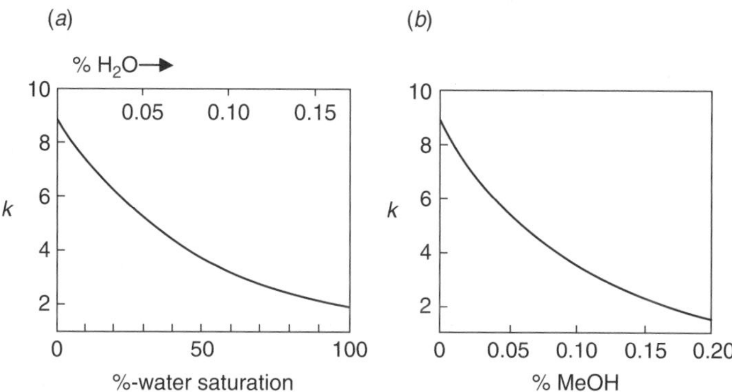

Figure 8.17 Effect of polar deactivators on NPC retention. Sample: benzanilide; conditions: type-B silica column; mobile phase, methylene chloride with varying concentrations of water (*a*) or methanol (*b*); 35°C. Adapted from [37].

It is therefore prudent to wait until retention variability becomes an issue before adjusting the %-water-saturation of the mobile phase or adding a polar solvent such as methanol.

A second possible cause of retention variability in NPC can arise from changes in mobile-phase composition as a result of helium sparging. The evaporation of the mobile phase that results can lead to a preferential loss of one solvent over the other [33], as well as changes in the water content of the mobile phase. Such changes in mobile-phase composition can be reduced by the use of on-line degassing of the mobile phase (Section 3.3.3).

8.5.2 Solvent Demixing and Slow Column Equilibration

When dilute solutions of a polar solvent first contact a silica column or TLC plate, the polar solvent will be selectively taken up by the silica. This leaves a mobile phase that is depleted in the polar solvent, generally resulting in a lowering of observed R_F values and larger values of k. Solvent demixing is mainly a problem for TLC or gradient elution, since isocratic separations are usually preceded by column equilibration (Section 2.7.1). Solvent demixing in TLC reduces the mobile-phase strength, so separations by TLC and column chromatography may no longer be equivalent. Consequently values of k estimated from TLC experiments may be too high, as in the example of Section 8.4.3 (which may or may not be due to solvent demixing). In gradient elution with a silica column, where the concentration of a polar B-solvent increases during the separation, solvent demixing can lead to an interruption of the gradient and a related deterioration of separation. For this reason gradient elution in NPC with a silica column is often avoided. Nevertheless, Meyer has shown [34] that gradients of hexane and methyl-*t*-butylether ($\varepsilon^0 = 0.48$) do not present problems of this kind. The same may be true of other gradients with silica columns, as long as the difference in ε^0 values of the A- and B-solvents is no greater than 0.5. See also the discussion of NPC gradient elution in [35].

For isocratic NPC separation, solvent demixing is not ordinarily a problem, but it may require a longer pre-equilibration of the column. Because of solvent demixing, when changing mobile phases the equilibration of the column will often be slower in NPC than in RPC; this will require a larger volume of mobile phase (for column equilibration) before injecting samples. Column equilibration in NPC can be especially slow for the case of less-polar mobile phases, if the %-water saturation of the new mobile phase is very different from that of the previous mobile phase. The reason is that the capacity of the column for water (weight water/weight stationary phase) is much greater than that of the mobile phase, so a large volume of mobile phase must pass through the column in order to transfer enough water between column and mobile phase to reach equilibrium. Polar-bonded-phase columns will be less subject to solvent demixing and slow column equilibration because polar solvents are less strongly retained by these columns (compared to silica columns); polar-bonded-phase columns are therefore more amenable for use with gradient elution.

8.5.3 Tailing Peaks

Peak tailing has traditionally been a more important problem for NPC than for RPC. Similarly the column plate number N is often smaller in NPC than in RPC, possibly because of slower diffusion on the silica surface (stationary-phase diffusion [36]). Just as in the case of RPC (Sections 5.2.2.2, 7.3.4.2), column performance for NPC is much improved by the use of type-B silica in place of the older type-A silica [37]; peak shapes are generally better, and plate numbers higher. For this reason the use of type-B silica is strongly recommended for analytical separations by means of NPC. For preparative separations, the higher cost of type-B silica may not be justified, especially as these separations often tolerate lower values of N (Section 15.4.1.1).

8.6 HYDROPHILIC INTERACTION CHROMATOGRAPHY (HILIC)

Hydrophilic interaction chromatography (HILIC) can be regarded as normal-phase chromatography with an aqueous-organic mobile phase [38–40]; for this reason it is sometimes referred to as "aqueous normal-phase chromatography." HILIC columns are more polar than RPC columns, and the more-polar water serves as the stronger B-solvent in HILIC—so that an increase in %-water results in a *decrease* in sample retention (the opposite of RPC behavior). An example is provided in Figure 8.18 for the separation of a mixture of neutral oligosaccharides by HILIC, using mobile phases of water and acetonitrile; as the %-water increases, retention decreases (see also the similar example of Fig. 8.19 for the HILIC separation of several peptides). More-polar and/or ionized solutes tend to be more strongly retained in HILIC, other factors equal (again, the opposite of RPC, but typical of NPC). In most HILIC separations the mobile phase is varied from 3 to 40% water. There is usually little retention ($k \approx 0$) for water concentrations>40%, although occasionally—for some solutes and columns—retention can begin to increase as the water concentration increases beyond 40% (i.e., onset of RPC behavior) [42].

The preferential retention of polar solutes in HILIC means that many samples that exhibit poor retention in RPC ($k \approx 0$) can be better separated by HILIC. HILIC is also characterized by several other potentially advantageous features [43]:

- good peak shape for basic solutes
- enhanced mass-spectrometer sensitivity
- possibility of direct injection of samples that are dissolved in a primarily organic solvent (which would be unsuitable for RPC; Section 2.6.1)
- higher flow rates (or lower column pressures) possible, because of the lower viscosity of the mobile phase (Table I.4 in Appendix I)

While relative retention for NPC with a silica column tends to be the reverse of RPC retention (Fig. 8.1*b*), HILIC retention is often intermediate between these two extremes. The latter observation may reflect the fact that solute "polarity" is a complex function of (1) molecular structure and (2) the kinds of sample-column interactions that are important for retention and especially selectivity. Many (perhaps most) HILIC separations are carried out by means of gradient elution. However, the following discussion for isocratic HILIC separations is equally applicable for gradient elution with HILIC; see the further discussion of Section 9.5.3.

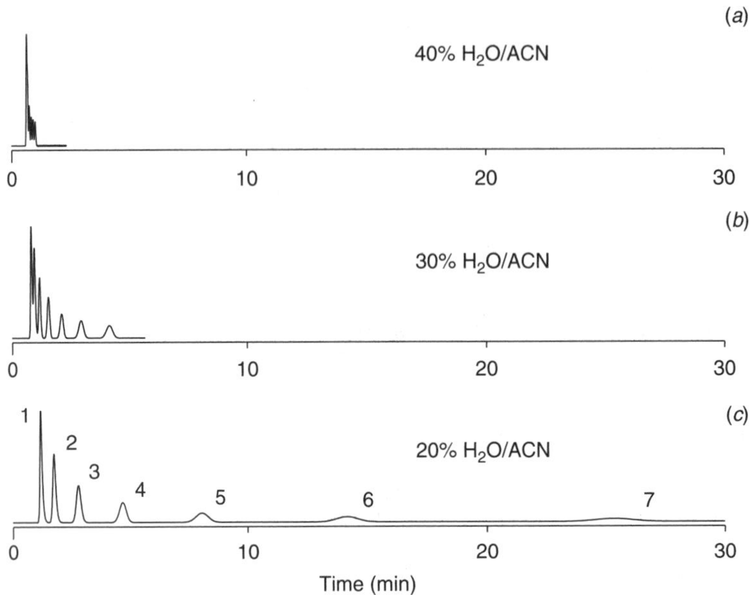

Figure 8.18 Separation of a mixture of derivatized oligosaccharides by HILIC with mobile phases of varying %-water. Conditions: 200 × 4.6-mm PolyHydroxyethyl A column (5-μm particles); mobile phases are water-acetonitrile as indicated in the figure; 2 mL/min. Values of n in the figure for each peak refer to the number of saccharide units in the corresponding oligosaccharide. The chromatograms are recreated from the data of [38].

8.6.1 Retention Mechanism

Although the "retention mechanism" for HILIC has received considerable attention in the literature, this topic is mainly of academic interest. Inasmuch as the subject has very limited practical application, the reader may wish to skip this section.

Retention in HILIC is believed to involve a partitioning of the solute into a water layer that is formed on the surface of the column packing [38, 39, 42, 44]. For silica as column packing and HILIC conditions, it has been shown [42] that the column dead volume V_m continues to decrease as the mobile-phase water increases from 0 to 30%; this has been attributed to the buildup of a layer of water on the silica surface. The observation that HILIC separations require at least 2–3% water in the mobile phase confirms the importance in HILIC separation of this water layer, which presumably comprises the stationary phase. It is also possible that the stationary phase includes some organic solvent (acetonitrile) from the mobile phase, with further contributions to solute retention from both the silica surface (silanols) and any column ligands that might interact with solute molecules.

A distinction can be made between *adsorption* on and *partition* into the stationary phase, as in the case of RPC (Section 6.2.2.1). It is believed by some investigators that these two processes can be distinguished by a comparison of values of k as a function of the volume-fraction (ϕ) of the B-solvent (water in this case). In RPC, which is believed to involve a partition process (in at least some cases), approximately linear plots of log k versus ϕ are observed (Fig. 6.3). In NPC with silica as column packing, where adsorption is believed to prevail, more linear plots

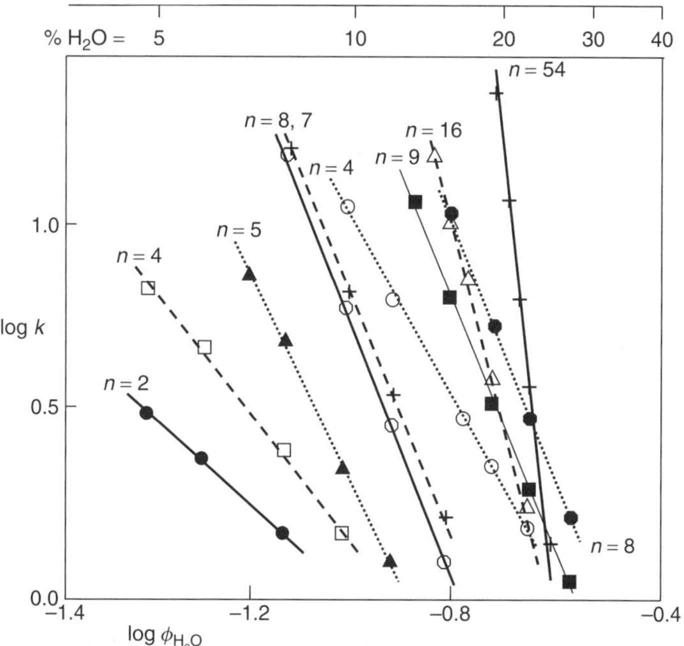

Figure 8.19 Changes in HILIC retention of various peptides with mobile-phase strength (varying %-water). Conditions: 250 × 4.6-mm TSK gel Amide-80 column; 1.0 mL/min; 40°C. The sample peptides can each be characterized by the number *n* of amino-acid sub-units in the molecule (values of *n* shown). Adapted from [41].

of log k versus log ϕ are found (Fig. 8.4). When the relationship of k to ϕ for HILIC separations is examined, no clear-cut relationship is found for all systems, but generally plots of log k versus log ϕ tend to be more linear; see the example of Figure 8.19 for several peptides (values of n in Fig. 8.19 indicate the number of amino-acid residues in the solute molecule). While linear log k–log ϕ plots suggest that HILIC retention involves adsorption rather than partition, the HILIC stationary phase—and the various possible interactions of the solute with the water layer, silica surface, and any ligands attached to the silica—is surely quite complex. Any conclusions as to whether "adsorption" or "partition" predominates in HILIC are therefore premature, as well as of little practical value.

8.6.2 Columns

Almost all present HILIC columns are created from silica particles. Aminopropyl column packings ("amino" columns) were used initially, mainly for carbohydrate separations; these columns are well suited for this application, as they prevent the formation of double peaks for each solute as a result of anomer resolution. Subsequently a variety of different bonded-silica packings have been employed for HILIC [39, 40, 45], which can be categorized as follows: bare silica, polar neutral (e.g., cyanopropyl), diol-bonded, amide-bonded, polypeptide-bonded, positively charged amine-bonded (anion-exchange), negatively charged (cation-exchange), and zwitterionic phases. While the differing selectivity of these column packings [46] can

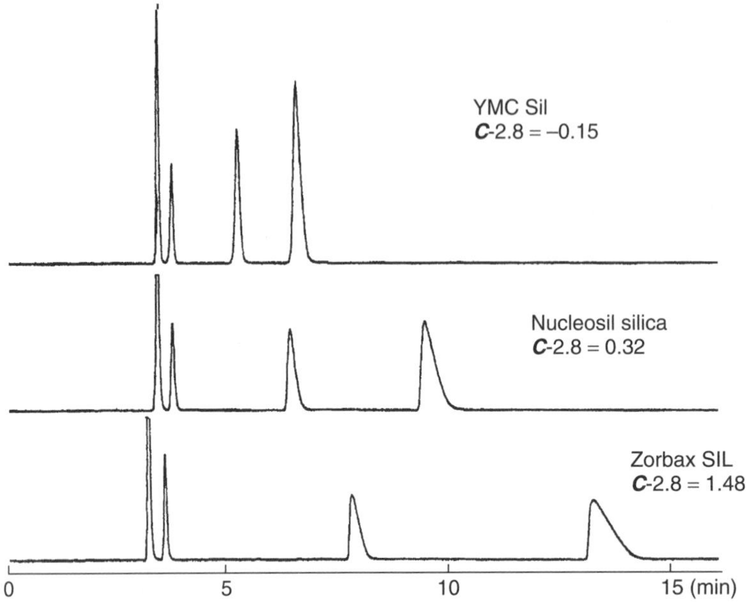

Figure 8.20 Tailing of pyrimidines on some HILIC columns. Conditions: 250 × 4.6-mm silica columns (YMC SIL, Nucleosil silica, and Zorbax SIL); mobile phase: 75% acetonitrile/buffer (5-mM phosphoric acid); 1.0 mL/min; ambient. Values of **C**-2.8 were measured for C_{18} RPC columns from same source. Adapted from [48].

prove useful during method development, *unbonded silica* appears to be the packing of choice for many HILIC separations when mass spectrometric detection (LC-MS) is used [39, 43, 47]—because of an absence of stationary-phase bleed. However, irreversible retention of some compounds on silica has been reported, in which case an amide-bonded column—a commonly used alternative [40]—can be used. For an extensive list and discussion of columns for HILIC, with manufacturer websites, see [39].

An early publication [48] noted significant peak tailing for the HILIC separation of basic pyrimidines at low pH on bare silica (Fig. 8.20). It is seen in Figure 8.20 that peak tailing increases for columns with larger values of the column parameter **C**-2.8 for the corresponding C_{18}-bonded silica (Section 5.4.1). Values of **C**-2.8 >0.25 for C_{18} columns are indicative of a less-pure, more acidic, type-A silica (Section 5.2.2.2), suggesting that peak tailing as in Figure 8.20 is associated with the use of type-A silica. As most silica columns used today are type-B, tailing as in Figure 8.20*b,c* should be less likely for HILIC separations with type-B silica columns. Irreversible binding of the solute should also be less likely when using type-B silica columns.

8.6.3 HILIC Method Development

Several publications have reviewed the effects of different experimental conditions on HILIC separation [38, 39, 45, 46, 44]. Method development for the HILIC separation of a small-molecule sample can be carried out in similar fashion as for RPC or NPC. A 100–150 × 4.6-mm HILIC column (3–μm particles) is first selected; as noted above, bare silica is often preferred as column packing, with amide packings as a good alternative. An initial separation can be carried out with a strong mobile phase; for example, 60% acetonitrile buffer (acetonitrile is by far the most commonly used organic solvent for HILIC). The %-acetonitrile can then be increased in 10% increments, until the desired retention range is obtained (e.g., $1 \leq k \leq 10$); see the similar procedure of Section 2.4.1 for RPC, as well as the HILIC example of Fig. 8.18. Alternatively, an initial gradient run may be more effective (Section 9.3.1). Formic acid, ammonium acetate or ammonium formate are commonly used as HILIC buffers for the separation of ionizable compounds, in concentrations of 2 to 20 mM; mobile-phase pH usually falls within a range of 3 to 7. The latter buffers are volatile (for MS detection) and soluble in all mixtures of organic and water.

Once a value of %-water has been selected for acceptable retention, conditions can be varied to optimize selectivity. Changes in relative retention may occur as a result of a change in %B, as can be inferred from the plots of Figure 8.19. Thus the retention sequence for the last five peptides in the chromatogram is

$$n = 54 > 16 > 8 > 9 > 4 \qquad \text{for } 15\%\ H_2O$$

and

$$n = 8 > 9 > 4 > 54 = 16 \qquad \text{for } 25\%\ H_2O$$

Changes in selectivity as %-water is varied have also been observed by others [46, 44, 49, 50]. While such changes in relative retention are often minor and therefore less useful, they are easily recognized during the adjustment of %-H_2O for the purpose of achieving acceptable retention (e.g., $1 \leq k \leq 10$). A few studies have demonstrated changes in HILIC selectivity as a function of temperature [44, 50], suggesting that the simultaneous optimization of both %B and temperature may be a promising approach (similar to the case for RPC; Section 7.3.2.2).

The effect of mobile-phase pH on solute ionization is predictable, as in the example of Figure 7.1a; as pH increases, ionization increases for acids and decreases for bases. Since HILIC retention *increases* for more ionized solutes, the retention of acids will increase with an increase in pH, while that of bases will decrease. When the mobile-phase pH is within ±1 unit of solute pK_a values, and solute pK_a values vary (frequently the case for ionizable solutes), large changes in selectivity should result from a change in pH. A change in column type is also expected to result in significant changes in selectivity, especially considering the wide range in stationary phase functionality that is commercially available; for some examples, see [40, 51]. A change in the organic A-solvent (from acetonitrile) is also possible, but infrequently used for a change in selectivity; other solvents are generally stronger (Fig. 8.21), which can result in insufficient retention of the sample; acetonitrile usually also results in higher values of N [44].

The usual changes in column length and flow rate are also available as a means for increasing resolution or decreasing run time (Section 2.5.3). The flow rate will be limited for acceptable pressure (Section 2.4.1), and smaller diameter columns are normally used for mass spectrometric detection (Section 4.14). The efficiencies of HILIC columns appear comparable to or even better than those of RPC columns [40]; a nice example is shown in the two separations of Figure 8.22. The average plate number for the separation of Figure 8.22*a* with a 150-mm column of 2.7-μm, superficially porous silica particles is $N = 36{,}000$, while that for the 450-mm column

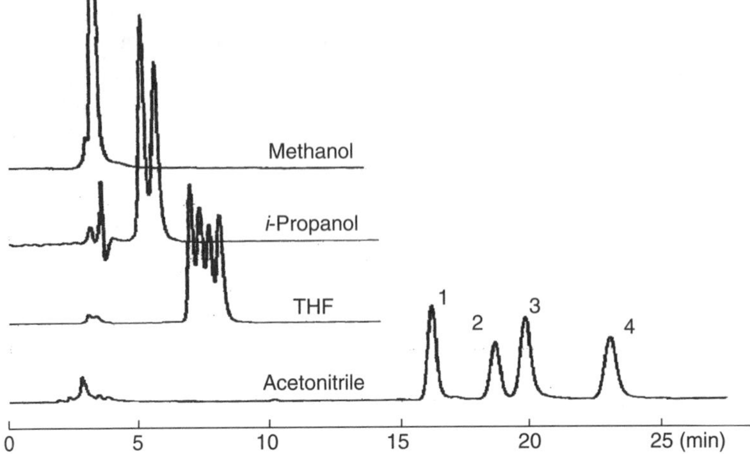

Figure 8.21 Effect of B-solvent on HILIC separation. Sample: epirubicin and its analogues. Conditions: 250 × 4.6 Kromasil KR100-5SIL (silica) column. Mobile phase: 90% organic/pH-2.9 buffer; 1.0 mL/min; ambient. Adapted from [52].

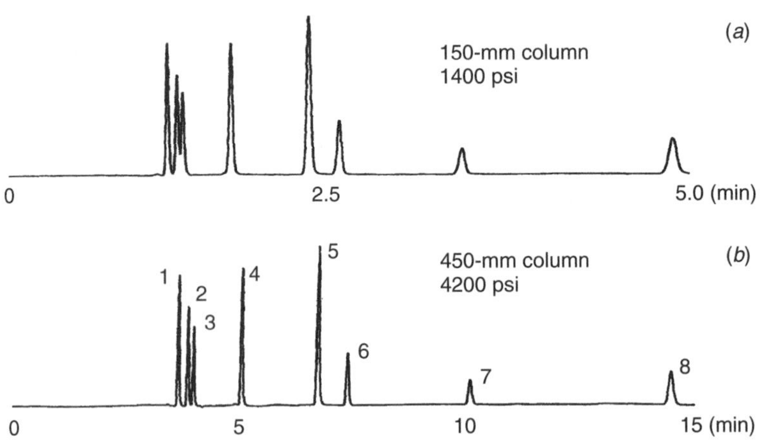

Figure 8.22 High-efficiency HILIC separation. Sample: *1*, phenol; *2*, 2-naphthalenesulfonic acid; *3*, *p*-xylenesulfonic acid; *4*, caffeine; *5*, nortriptyline; *6*, diphenhydramine; *7*, benzylamine; *8*, procainamide. Conditions:150 × 4.6-mm Halo silica column (Advanced Materials Technology, Wilmington, DE) (*a*); three columns as in (*a*) connected in series (*b*); 75% acetonitrile–pH-3.0 buffer; 1.0 mL/min; 30°C. Adapted from [51].

in Figure 8.22*b* is $N = 103{,}000$. This seems rather remarkable, considering the short run times.

A review of the recent literature indicates that reported HILIC separations are often carried out with gradient elution and mass spectrometric detection. The comments above for isocratic separation generally apply to gradient elution as well; see Section 9.5.3 for details on HILIC separations based on gradient elution.

8.6.4 HILIC Problems

Problems with *peak shape* (both fronting and tailing) seem to be somewhat more common when using HILIC, than for RPC [46, 50, 54]; this may reflect the fact that HILIC was introduced more recently and has not been used as extensively as RPC, as well as the possible use of type-A silica for some column packings—especially columns introduced before 1995 (e.g., see Fig. 8.20 and the related discussion). Further improvements in HILIC columns combined with a better understanding of how to manage these peak-shape problems seem likely.

When peak-shape problems are encountered, an increase in mobile-phase buffer concentration should be tried first. Some samples may require a buffer concentration as high as 100 mM in order to achieve acceptable peak shapes. The sample should be dissolved in the mobile phase, but in some cases an increase in %-organic of the sample solvent can be beneficial. If peak-shape problems persist, a change in column or mobile-phase pH may be advisable. The mobile phase should always contain some water, preferably at least 3–5%.

Column bleed has been observed for some bonded-phase columns [41] when using mass-spectrometric detection. A change to a silica column or a different bonded-phase column usually solves the problem. *Irreversible sorption* of some sample components has been reported for silica columns when using HILIC. If the problem arises, a change to a bonded-phase HILIC column may be advisable. *Slow equilibration of the column* with mobile phases that contain a different buffer or buffer concentration may also occur, but problems of this kind are much less likely for HILIC than for NPC with a silica column.

REFERENCES

1. L. R. Snyder, *Principles of Adsorption Chromatography. The Separation of Nonionic Organic Compounds*, Ml Dekker, New York, 1968, chs. 8 and 10.
2. L.-AA. Truedsson and B. E. F. Smith, *J. Chromatogr.*, 214 (1981) 291.
3. L. R. Snyder and D. L. Saunders, in *Chromatography in Petroleum Analysis*, K. H. Altgelt and T. H. Gouw, eds., Dekker, New York, 1979, ch. 10.
4. A. Kuksis, in *Chromatography*, E. Heftmann, ed., 5th ed., Elsevier, Amsterdam, 1993, p. B171.
5. L. R. Snyder and H. Poppe, *J. Chromatogr.*, 184 (1980) 363.
6. L. R. Snyder, in *High-Performance Liquid Chromatography: Advances and Perspectives*, Vol. 3. C. Horváth, ed., Academic Press, New York, 1983, p. 157. tab. 1
7. E. Soczewinski, *Anal. Chem.*, 41 (1969) 179.
8. E. Soczewinski, *J. Chromatogr. A*, 965 (2002) 109.
9. L. R. Snyder, *J. Chromatogr.*, 63 (1971) 15.
10. B. Fried and J. Sherma. *Thin-Layer Chromatography (Chromatographic Science*, Vol. 81), Ml Dekker, New York, 1999.
11. R. P. W. Scott and P. Kucera, *J. Chromatogr.*, 112 (1975) 425.
12. V. R. Meyer and M. D. Palamareva, *J. Chromatogr.*, 641 (1993) 391.
13. *Solvent Guide*, Burdick and Jackson Labs., Muskegon, MI, 1980.
14. E. Soczewinski, T. Dzido, and W. Golkiewicz, *J. Chromatogr.*, 131 (1977) 408.
15. J. L. Glajch, J. J. Kirkland, and L. R. Snyder, *J. Chromatogr.*, 238 (1982) 269.

16. L. R. Snyder, *J. Chromatogr.*, 63 (1971) 15.
17. L. R. Snyder, *J. Planar Chromatogr.*, 21 (2008) 315.
18. L. R. Snyder, J. L. Glajch, and J. J. Kirkland, *J. Chromatogr.*, 218 (1981) 299.
19. M. D. Palamareva and H. E. Palamareva, *J. Chromatogr.*, 477 (1989) 235.
20. Ch. E. Palamareva and M. D. Palamareva, *LSChrom, Ver. 2.1, Demo version*, 1999, http://www.lschrom.com.
21. M. Z. Kagan, *J. Chromatogr. A*, 918 (2001) 293.
22. L. R. Snyder and J. J. Kirkland, *Introduction to Modern Liquid Chromatography*, 2nd edn., Wiley-Interscience, New York, 1979, pp. 391–392.
23. L. R. Snyder and T. C. Schunk, *Anal. Chem.*, 54 (1982) 1764.
24. E. L. Weiser, A. W. Salotto, S. M. Flach, and L. R. Snyder, *J. Chromatogr.*, 303 (1984) 1.
25. W. T. Cooper and P. L. Smith, *J. Chromatogr.*, 355 (1986) 57.
26. A. W. Salotto, E. L. Weiser, K. P. Caffey, R. L. Carty, S. C. Racine, and L. R. Snyder, *J. Chromatogr.*, 498 (1990) 55.
27. P. L. Smith and W. T. Cooper, *J. Chromatogr.*, 410 (1987) 249.
28. L. P. Hammett, *Physical Organic Chemistry*, McGraw-Hill, New York, 1940, ch. 7.
29. B. Stancher and F. Zonta, *J. Chromatogr.*, 234 (1982) 244.
30. M. D. Palamareva and H. E. Palamareva, *J. Chromatogr.*, 477 (1989) 225.
31. J. L. Glajch and L. R. Snyder, *J. Chromatogr.*, 214 (1981) 21.
32. B. Vaisman, A. Shikanov, and A. J. Domb, *J. Chromatogr. A*, 1064 (2005) 85.
33. L. R. Snyder, *J. Chromatogr. Sci.*, 21 (1983) 65.
34. V. R. Meyer, *J. Chromatogr. A*, 768 (1997) 315.
35. P. Jandera, *J. Chromatogr. A*, 965 (2002) 239.
36. R. W. Stout, J. J. DeStefano, and L. R. Snyder, *J. Chromatogr.*, 282 (1983) 263.
37. J. J. Kirkland, C. H. Dilks, and J. J. DeStefano, *J. Chromatogr.*, 635 (1993) 19.
38. A. Alpert, *J. Chromatogr.*, 499 (1990) 177.
39. P. Hemstrom and K. Irgum, *J. Sep. Sci.*, 29 (2006) 1784.
40. T. Ikegami, K. Tomomatsu, J. Takubo, K. Horie, and N. Tanaka, *J. Chromatogr. A*, 1184 (2008) 474.
41. T. Yoshida, *J.Chromatogr. A*, 811 (1998) 61.
42. D. V. McCalley and U. D. Neue, *J.Chromatogr. A*, 1192 (2008) 225.
43. D. V. McCalley, *J. Chromatogr. A*, 1171 (2007) 46.
44. Z. Hao, B. Xiao, and N. Weng, *J. Sep. Sci.* 31 (2008) 1449.
45. B. Dejaegher, D. Mangelings, and Y. Vander Heyden, *J. Sep. Sci.*, 31 (2008) 1438.
46. Y. Guo and S. Gaiki, *J. Chromatogr. A*, 1074 (2005) 71.
47. W. Naidong, *J. Chromatogr. B*, 796 (2003) 209.
48. B. A. Olsen, *J. Chromatogr. A*, 913 (2001) 113.
49. X. Wang, W. Li, and H. T. Rasmussen, *J. Chromatogr. A*, 1083 (2005) 58.
50. C. Dell'Aversano, P. Hess, and M. A. Quilliam, *J. Chromatogr. A*, 1081 (2005) 190.
51. D. V. McCalley, *J. Chromatogr. A*, 1171 (2007) 46.
52. R. Li and J. Huang, *J. Chromatogr. A*, 1041 (2004) 163.
53. D. V. McCalley, *J. Chromatogr. A*, 1193 (2008) 85.
54. C. Dell'Aversan, G. K. Eaglesham, and M. A. Quilliam, *J. Chromatogr. A*, 1028 (2004) 155.

CHAPTER NINE

GRADIENT ELUTION

Introduction to Modern Liquid Chromatography, Third Edition, by Lloyd R. Snyder, Joseph J. Kirkland, and John W. Dolan

9.1 INTRODUCTION

Gradient elution was introduced in Section 2.7.2 as a means for dealing with samples that are unsuitable for isocratic elution. The most common reason for the use of gradient elution is a sample whose retention range exceeds the preferred goal for isocratic separation ($1 \leq k \leq 10$). As discussed in Section 2.5.1, it is possible to expand this retention range somewhat, for example, to $0.5 \leq k \leq 20$. However, many samples cover a much wider range in k-values, making gradient elution essential for their separation. An example of a sample that cannot be separated successfully by isocratic elution is shown in Figure 9.1*a*. Here a mixture of 14 toxicology standards is injected into a C_{18} column, using a mobile phase of 50% acetonitrile buffer. The excessive retention range for this sample (k-values that range 0–50) results in the poor resolution of early peaks 1 to 6, and excessive retention times for later peaks 13 and 14. Later peaks are also very broad and therefore not very tall—consequently their measurement may be compromised (poor signal/noise ratio; Section 4.2.3). No single change in %-acetonitrile (%B) would result in the adequate separation of the entire sample; larger values of %B would mean smaller values of k and even poorer resolution of early peaks in the chromatogram, while smaller values of %B would further increase values of k and run time—and make the measurement of later peaks still more difficult. Figure 9.1*a* is a good example of the *general elution problem* for samples with a wide range of k-values, which is the main reason for gradient elution.

If adjacent groups of peaks from Figure 9.1*a* could be processed separately (with different values of %B), the improved isocratic separations of Figure 9.1*b*–*f* would result. Thus the use of 10% B as mobile phase for peaks 1 to 3 (Fig. 9.1*b*) results in an average value of $k \approx 3$ for these peaks, and their baseline resolution. Similarly the separation of peaks 4 to 6 with a mobile phase of 25% B (Fig. 9.1*c*) also provides an average value of $k \approx 3$ with good resolution. Likewise the separations of peaks 7 to 8, 9 to 11, and 12 to 14 with 45, 62, and 75% B, respectively (Figs. 9.1*d*–*f*), result in $k \approx 3$ for each group of peaks. Thus each of these sample-fractions can be separated isocratically with reasonable resolution ($R_s \geq 2$) and separation time (6–8 min), as well as providing narrower, taller peaks for improved detection. The only requirement is a mobile phase that provides $k \approx 3$ for each group of peaks; however, as different values of %B are required for each set of peaks, isocratic separation of the total sample with $1 \leq k \leq 10$ is not possible.

Gradient elution (Fig. 9.1g) is a means of realizing the benefits shown in the isocratic separations of Figures 9.1*b*–*f* by means of a single run. Thus at the

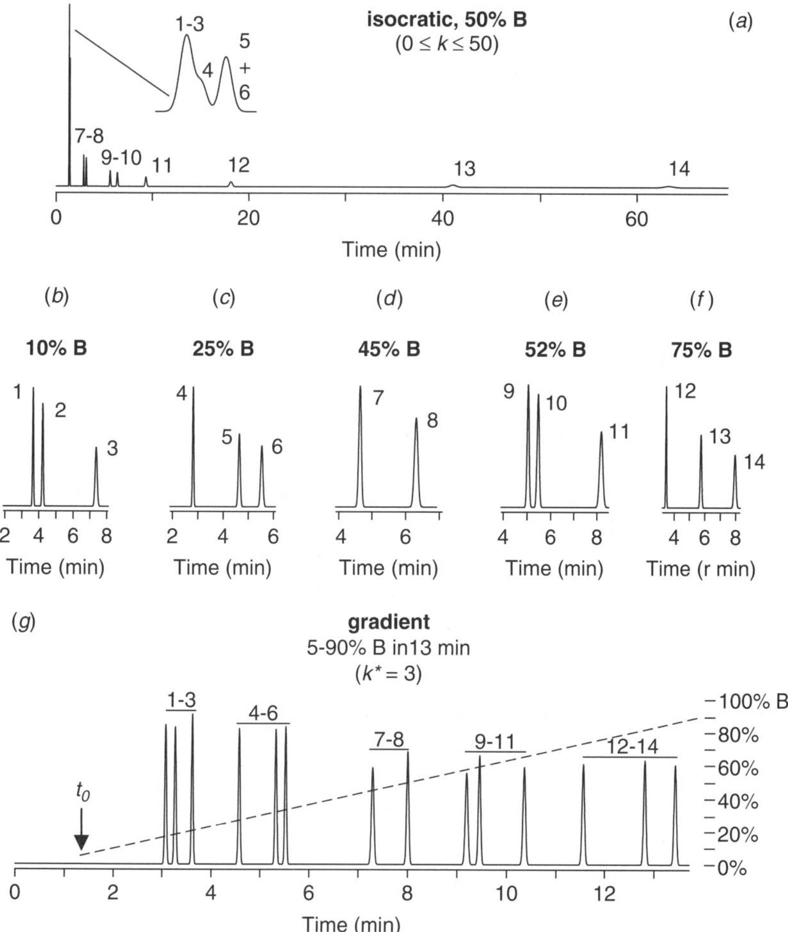

Figure 9.1 Example of the general elution problem. Sample: 14 toxicology standards. Conditions: 250 × 4.6-mm (5-μm) C_{18} column; mobile phase is ACN (B) and pH-2.5 phosphate buffer (A); 65°C; 2.0 mL/min. (a) isocratic separation with 50% B; ($b - f$), isocratic separation of indicated compounds (peaks) with 10%, 25%, 45%, 52%, and 75% B, respectively ($k \approx 3$); (c) gradient elution as indicated. Chromatograms are computer simulations based on the experimental data of [1].

beginning of the gradient, peaks 1 to 3 move through the column with an average value of $k \approx 3$, while peaks 4 to 14 lag behind near the column inlet. As %B continues to increase during the gradient (indicated by dashed line, marking %B at column outlet vs. time), later peaks become less strongly retained and then also move through the column, again, with average values of $k \approx 3$. As we will see in Section 9.1.3.2, values of k in gradient elution tend to be similar for all peaks in the chromatogram, and can be easily controlled by the choice of gradient time and flow rate. Unless stated otherwise, the present chapter refers to RPC separation; however, the same *general* principles and conclusions apply for other HPLC separations (ion-exchange, normal-phase, etc.). For a more comprehensive and detailed account of gradient elution than is presented in this book, see [2].

9.1.1 Other Reasons for the Use of Gradient Elution

Apart from the need for gradient elution in the case of wide-polarity-range samples like that of Figure 9.1, there are a number of other situations that favor or require the use of gradient elution:

- high-molecular-weight samples
- generic separations
- efficient HPLC method development
- sample preparation
- peak tailing

High-molecular-weight compounds, such as peptides, proteins, and oligonucleotides, are usually poor candidates for isocratic separation, because their retention can be extremely sensitive to small changes in mobile-phase composition (%B). For example, the retention factor k of a 50,000-Da protein can change by 3-fold as a result of a change in the mobile phase by only 1% B. This behavior can make it extremely difficult to obtain reproducible isocratic separations of macromolecules in different laboratories, or even within the same laboratory. Furthermore the isocratic separation of a mixture of macromolecules usually results in the immediate elution of some sample components (with $k \approx 0$ and no separation), and such slow elution of other components (with $k \gg 100$) that it appears that the sample never leaves the column; that is, the retention range of such samples is often extremely wide (isocratic k-values for different sample components that vary by several orders of magnitude). With gradient elution, on the other hand, irreproducible retention times for large molecules are seldom a problem, and resulting separations can be fast, effective, and convenient (Chapter 13).

Generic separations are used for a series of samples, each of which is made up of different components; for example, compounds A, B, and C in sample 1, compounds D, E, and F in sample 2, and so forth. Typically each sample will be separated just once within a fixed separation time (run time), with no further method development for each new sample. In this way hundreds or thousands of related samples—each with a unique composition—can be processed in minimum time and with minimum cost. Generic separations by RPC (with fixed run times, for automated analysis) are only practical by means of gradient elution and are commonly used to assay combinatorial libraries [3], as well as other samples [4]. Generic separation is often combined with mass spectrometric detection [5], which allows both the separation and identification of the components of samples of previously unknown composition—without requiring the baseline resolution of peaks of interest.

Efficient HPLC method development [6] is best begun with one or more gradient experiments (Section 9.3.1). A single gradient run at the start of method development can replace several trial-and-error isocratic runs as a means for establishing the best solvent strength (value of %B) for isocratic separation. An initial gradient run can also establish whether isocratic or gradient elution is the best choice for a given sample.

Sample preparation (Chapter 16) is required in many cases because some samples are unsuitable for direct injection followed by isocratic elution. Interfering peaks, strongly retained components, and particulates must first be removed. In some cases, however, gradient elution can minimize (or even eliminate) the need for sample preparation. For example, by spreading out peaks near the beginning of a gradient chromatogram (as in Fig. 9.1*g* vs. Fig. 9.1*a*), interfering peaks (non-analytes) that commonly elute near t_0 can be separated from peaks of interest. Similarly strongly retained non-analytes at the end of an isocratic separation can result in excessive run times, because these peaks must clear the column before injection of the next sample. Gradient elution can usually remove these late-eluting compounds within a reasonable run time (Section 9.2.2.5).

Peak tailing was a common problem in the early days of chromatography, and the reduction of tailing was an early goal of gradient elution [7]. Because of the increase in mobile-phase strength during the time a band moves through the column in gradient elution, the tail of the band moves faster than the peak front, with a resulting reduction in peak tailing and peak width (Section 9.2.4.3). However, peak tailing is today much less common, and other means are a better choice for addressing this problem when it occurs (Section 17.4.5.3).

9.1.2 Gradient Shape

By *gradient shape*, we mean the way in which mobile-phase composition (%B) changes with time during a gradient run. Gradient elution can be carried out with different gradient shapes, as illustrated in Figure 9.2*a–f*. Most gradient separations use linear gradients (Fig. 9.2*a*), *which are strongly recommended during the initial stages of method development*. Curved gradients (Fig. 9.2*b,c*) have been used in the past for certain kinds of samples, but for various reasons such gradients have been largely replaced by segmented gradients (Fig. 9.2*d*). Segmented gradients can provide most of the advantages of curved gradients, are easier to design for different samples, and can be replicated by most gradient systems. The use of segmented gradients for various purposes is examined in Section 9.2.2.5. Gradient delay or "isocratic hold" (Section 9.2.2.3) is illustrated by Figure 9.2*e*; an isocratic hold can also be used at the end of the gradient. Step gradients (Fig. 9.2*f*), where an instantaneous change in %B is made during the separation, are a special kind of segmented gradient. They are used infrequently—except at the end of a gradient separation for cleaning late-eluting compounds from the column; a sudden increase in %B (as in *i* of Fig. 9.2*f*) achieves this purpose. A step gradient that provides a sudden decrease in %B (as in *ii* of Fig. 9.2*f*) can return the gradient to its starting value for the next separation. In the past, step gradients were sometimes avoided because of a concern for column stability; with today's well-packed silica-base columns, however, step gradients can be used without worry about column damage.

A linear gradient can be described (Fig. 9.2*g*) by the initial and final mobile-phase compositions, and gradient time t_G (the time from start-to-finish for the gradient). We can define the initial and final mobile-phase compositions in terms of %B, or we can use the volume-fraction ϕ of solvent B in the mobile phase (equal to 0.01%B): values ϕ_o and ϕ_f, respectively. The change in %B or ϕ during the gradient is defined as the *gradient range* and is designated by $\Delta\phi = \phi_f - \phi_0$ (or the equivalent Δ%B = [final%B] − [initial%B]). In the present book, values of %B and ϕ will be used interchangeably; that is, ϕ always equals 0.01%B, and 100% B ($\phi = 1.00$) signifies pure organic solvent in RPC. For reasons discussed in Section

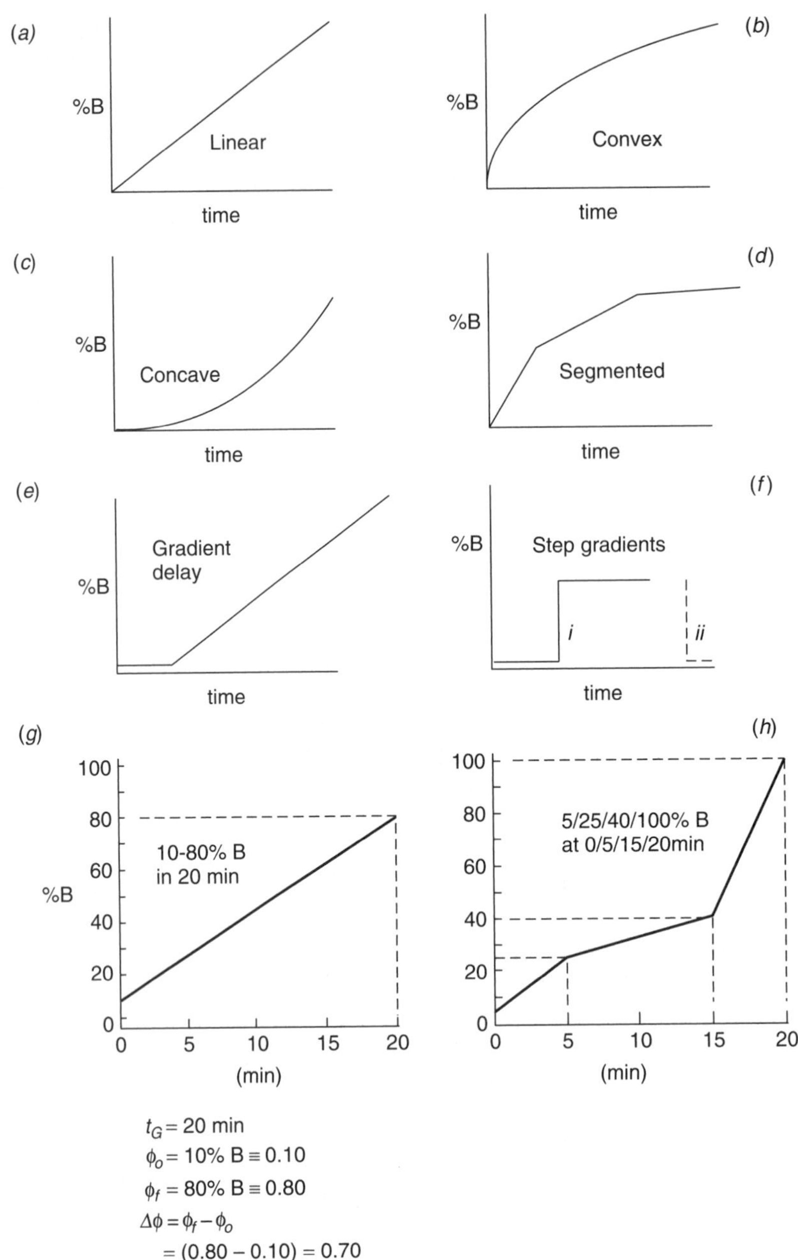

Figure 9.2 Illustration of different gradient shapes (plots of %B vs. time).

17.2.5.3, it is sometimes desirable for the A- and/or B-solvent reservoirs to contain *mixtures* of the A- and B-solvents, rather than pure water and organic, respectively; for example, 5% acetonitrile/water in the A-reservoir and 95% acetonitrile/water in the B-reservoir. For the latter example, a (nominal) 0–100% B gradient would then correspond to 5–95% acetonitrile, with $\Delta\phi = (0.95 - 0.05) = 0.90$.

By a *gradient program*, we refer to a description of how the mobile-phase composition changes with time during a gradient. Linear gradients represent the simplest program, for example, a gradient from 10–80% B in 20 minutes (Fig. 9.2*g*), which can also be described as 10/80% B in 0/20 min (10% B at 0 min to 80% B at 20 min). Segmented programs are usually represented by values of %B and time for each linear segment in the gradient, for example, 5/25/40/100%B at 0/5/15/20 min for Figure 9.2*h*.

9.1.3 Similarity of Isocratic and Gradient Elution

A peak moves through the column during gradient elution in a series of small steps, in each of which there is a small change in mobile-phase composition (%B). That is, gradient separation can be regarded as the result of a large number of small, isocratic steps. Separations by isocratic and gradient elution can be designed to give similar results. The resolution achieved for selected peaks in either isocratic or gradient elution will be about the same, when *average* values of k in gradient elution (during migration of each peak through the column) are similar to values of k in isocratic elution, and other conditions (column, temperature, A- and B-solvents, etc.) are the same. Isocratic and gradient separations where the latter conditions apply are referred to as *corresponding*. Thus the isocratic examples of Figure 9.1*b*–*f* can be compared with the "corresponding" gradient separation of Figure 9.1*g*. The isocratic separations of individual groups of peaks in Figure 9.1*b*–*f* each occur with $k \approx 3$, while in the gradient separation of Figure 9.1*g* the equivalent value of k for *each* peak is also ≈ 3. We see in this example that the peak resolutions of Figure 9.1*b*–*f* are similar to those of Figure 9.1*g*.

In isocratic elution we can change values of k by varying the mobile-phase strength (%B). In gradient elution, average values of k can be varied by changing other experimental conditions—as described below in Section 9.2.

9.1.3.1 The Linear-Solvent-Strength (LSS) Model

This section provides a quantitative basis for the treatment of gradient elution in this chapter. However, the derivations presented here are of limited practical utility per se (although necessary for a quantitative treatment of gradient elution). The reader may wish to skip to Section 9.1.3.2 *and return to this section as needed.*

Isocratic retention in RPC is given as a function of %B (Section 2.5.1) by

$$\log k = \log k_w - S\phi \tag{9.1}$$

For a given solute, the quantity k_w is the (extrapolated) value of k for $\phi = 0$ (water or buffer as mobile phase), and $S \approx 4$ for small molecules (<500 Da). A linear gradient can be described by

$$\%B = (\%B)_0 + \left(\frac{t}{t_G}\right)[(\%B)_f - (\%B)_0] \tag{9.2}$$

Here %B refers to the mobile-phase composition at the column inlet, $(\%B)_0$ is the value of %B at the start of the gradient (time zero), $(\%B)_f$ is the value of %B at the finish of the gradient, t is any time during the gradient, and t_G is the gradient time. We can restate Equation (9.2) in terms of ϕ, the volume-fraction of B:

$$\begin{aligned}\phi &= \phi_0 + \left(\frac{t}{t_G}\right)(\phi_f - \phi_0) \\ &= \phi_0 + \left(\frac{\Delta\phi}{t_G}\right)t \end{aligned} \tag{9.2a}$$

where ϕ_0 is the value of ϕ at the start of the gradient, ϕ_f is the value of ϕ at the end of the gradient, and $\Delta\phi = (\phi_f - \phi_0)$ is the change in ϕ during the gradient (the *gradient range*); see Figure 9.2*g*. The quantity ϕ refers to values at the column inlet, measured at different times t during the gradient. Thus the mobile-phase composition at time $t = 0$ (the start of the gradient) is $\phi = \phi_0$, provided that no delay occurs between the gradient mixer and the column inlet (Section 9.2.2.3).

Equations (9.1) and (9.2*a*) can be combined to give

$$\begin{aligned}\log k &= \log k_w - S\phi_0 - \left(\frac{\Delta\phi S}{t_G}\right)t \\ &= \qquad C_1 \qquad - \qquad C_2 t\end{aligned} \tag{9.3}$$

For a linear gradient, a given solute, and specified experimental conditions C_1 and C_2 are constants, so log k varies linearly with time t during the gradient (the value of k in Eq. 9.3 refers to the value of k measured at the column inlet at any given time t). Gradients for which Equation (9.3) applies are called *linear-solvent-strength* (LSS) gradients; linear RPC gradients are therefore (approximately) LSS gradients. Exact equations for retention and peak width can be derived for LSS gradients (Section 9.2.4). LSS separations are much easier to understand and to control, compared to the use of other gradient shapes. Finally, LSS gradients provide a better separation of most samples that require gradient elution.

A fundamental definition of gradient steepness b for a given solute is

$$b = \frac{V_m \Delta\phi S}{t_G F} \tag{9.4}$$

or as $t_0 = V_m/F$,

$$b = \frac{t_0 \Delta\phi S}{t_G} \tag{9.4a}$$

This definition of gradient steepness follows from Equation (9.3), which can be written as

$$\log k = \log k_w - S\phi_0 - \left(\frac{t_0 \Delta\phi S}{t_G}\right)\left(\frac{t}{t_0}\right)$$

or

$$\log k = \log k_w - S\phi_0 - b\left(\frac{t}{t_0}\right) \tag{9.4b}$$

where $(\log k_w - S\phi_0)$ for a given gradient and solute is equal to log k at the start of the gradient (and therefore varies with ϕ_0; see later Eq. 9.7). A larger value of b corresponds to a faster decrease in k with time, or a steeper gradient. Retention times and peak widths in gradient elution can be derived from the relationships above (see Section 9.2.4).

9.1.3.2 Band Migration in Gradient Elution

Consider next how individual solute bands move through the column during gradient elution (Fig. 9.3). For an initially eluted compound i in Figure 9.3*a*, the solid curve ($x[i]$) marks the fractional migration x of band i through the column as a function of time (note that $y = 1$ on the y-axis represents elution of the band from the column; $y = 0$ represents the band at the column inlet). Band migration is seen to accelerate with time, resulting in an upward-curved plot of x versus t. Also plotted in Figure 9.3*a* is the *instantaneous* value of k for band-i (dashed curve, $k[i]$) as it migrates through the column. The quantity $k(i)$ is the value of k at time t for an isocratic mobile phase whose composition (%B) is the same as that of the mobile phase in contact with the band at time t. Peak width and resolution in gradient elution depend on the *median* value of k: the instantaneous value of k when the

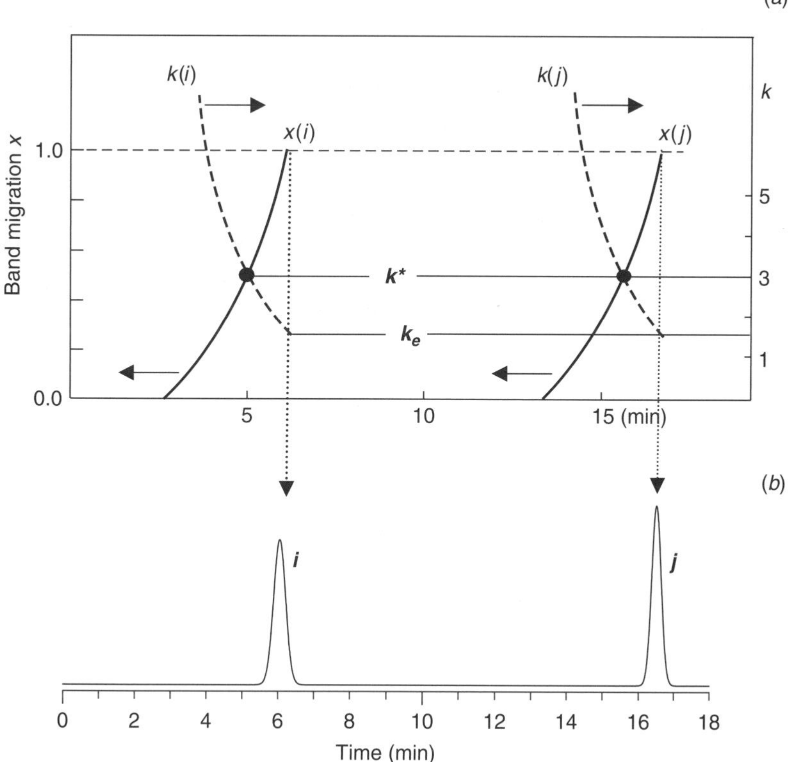

Figure 9.3 Peak migration during gradient elution. (*a*) Band-migration x and instantaneous values of k related to time, showing average (k^*) and final values of k (at elution, k_e); (*b*) resulting chromatogram.

band has migrated halfway through the column. This median value of k in gradient elution is defined as the *gradient retention factor* k^*. Peak width is determined by the value of k when the peak leaves the column (defined as k_e, equal to $k^*/2$). A similar plot for a second band j (with values of $x = x[j]$, and $k = k[j]$) is also shown in Figure 9.3*a*. The resulting chromatogram for the separation of Figure 9.3*a* is shown in Figure 9.3*b*.

A comparison of band migration in Figure 9.3*a* for the two compounds i and j shows a generally similar behavior, apart from a delayed start in the migration of band j because of its stronger initial retention (larger value of k_w). Specifically, values of k^* and k_e for both early and late peaks in the chromatogram are approximately the same for solutes i and j, suggesting that resolution and peak spacing need not decrease for earlier peaks, as in isocratic elution for small values of k (compare the gradient separation of peaks 1–6 in Fig. 9.1*e* with their isocratic separation in Fig. 9.1*a*). Values of k_e are also usually similar for early and late peaks in gradient elution, meaning that peak widths (and heights) will be similar for both early and late peaks in the chromatogram (contrast the peak heights for peaks 1–14 in the gradient separation of Fig. 9.1*e* with these same peaks in the isocratic separation of Fig. 9.1*a*). *The relative constancy of values of k^* and k_e for a linear-gradient separation are responsible for the pronounced advantages of gradient over isocratic elution for the separation of wide-range samples such as that of Figure* 9.1.

9.2 EXPERIMENTAL CONDITIONS AND THEIR EFFECTS ON SEPARATION

The gradient retention factor k^* of Figure 9.3 has a similar significance in gradient elution as the retention factor k has in isocratic elution. Values of k in isocratic elution are important for the understanding and control of separation, and we will see that values of k^* play the same role in gradient elution. The value of k^* depends on the solute (its value of S in Eq. 9.1) and experimental conditions: gradient time t_G, flow rate F, column dimensions, and the gradient range $\Delta\phi$ [2]:

$$\boxed{k^* = \frac{0.87 t_G F}{V_m \Delta\phi S}} \qquad (9.5)$$

Here V_m is the column dead-volume (mL), which can be determined from an experimental value of t_0 and the flow rate F (Section 2.3.1; $V_m = t_0 F$). Values of S for different samples with molecular weights in the 100 to 500 Da range can be assumed equal to about 4. *This means that values of k^* for different solutes in a given linear-gradient separation (with constant values of t_G, F, V_m, and $\Delta\phi$) will all be about the same.*

Let us next compare isocratic and gradient separation in terms of values of k and k^*, for the same sample and similar conditions (same A- and B-solvents, column, flow rate, and temperature). The isocratic separations of Figure 9.4*a*–*c* illustrate the effect of a change in %B (and k), for mobile phases of 70, 55, and 40% B. Similar values of k^* in gradient elution can be achieved by varying gradient time t_G (Eq. 9.5 with $S = 4$); see Figure 9.4*d*–*f*, where $k^* = 1$, 3, and 9 for $t_G = 3$, 10, and 30 minutes, respectively. Isocratic and gradient separations will be referred to as *"corresponding"* when the average value of k in the isocratic separation equals the value of k^* for the gradient separation (for the same sample and experimental conditions, except that %B and k^* are allowed to vary). In the example of Figure 9.4, separations (*a*) and (*d*) are "corresponding," as are separations (*b*) and (*e*), and (*c*) and (*f*). "Corresponding" separations as in these examples should be similar in terms of resolution and average peak heights—except that peaks in the gradient separation can be taller by as much as 2-fold.

For either isocratic or gradient elution, an increase in k or k^* corresponds to an increase in run time (other conditions the same). In isocratic elution, resolution increases for larger values of k (Eq. 2.24), as observed in Figure 9.4*a–c* ($R_s = 0.4$, 1.7, and 3.4). For similar values of k^* in gradient elution (Fig. 9.4*d–f*), the observed resolution is seen to be about the same for each "corresponding" separation ($R_s = 0.4$, 1.7, and 3.6). Finally, peak widths in isocratic elution increase with k (decrease in %B), resulting in decreased peak heights. Again, similar changes in peak width and height are observed in gradient elution as k^* is varied in Figure 9.4*d–f*. Thus changes in %B for isocratic elution, or gradient time in gradient elution, lead to similar changes in run time, resolution, and peak heights.

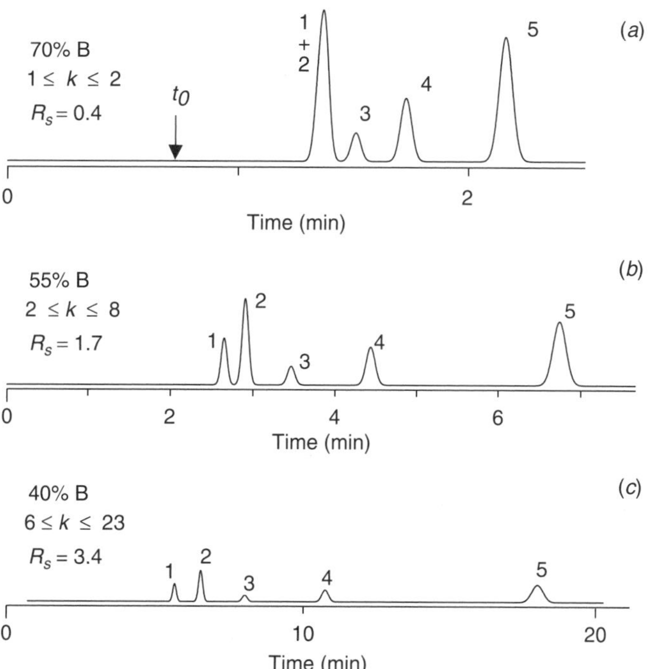

Figure 9.4 Isocratic (*a–c*) and gradient (*d–f*) separations compared for a regular sample and change in either %B or gradient time. Sample: *1*, simazine, *2*, monolinuron; *3*, metobromuron; *4*, diuron; *5*, propazine. Conditions: 150 × 4.6-mm C_{18} column (5-μm particles); methanol-water mobile phase (%B or gradient conditions indicated in figure); ambient temperature; 2.0 mL/min. Note that actual peak heights are shown (*not* normalized to 100% for tallest peak). Chromatograms recreated from data of [8].

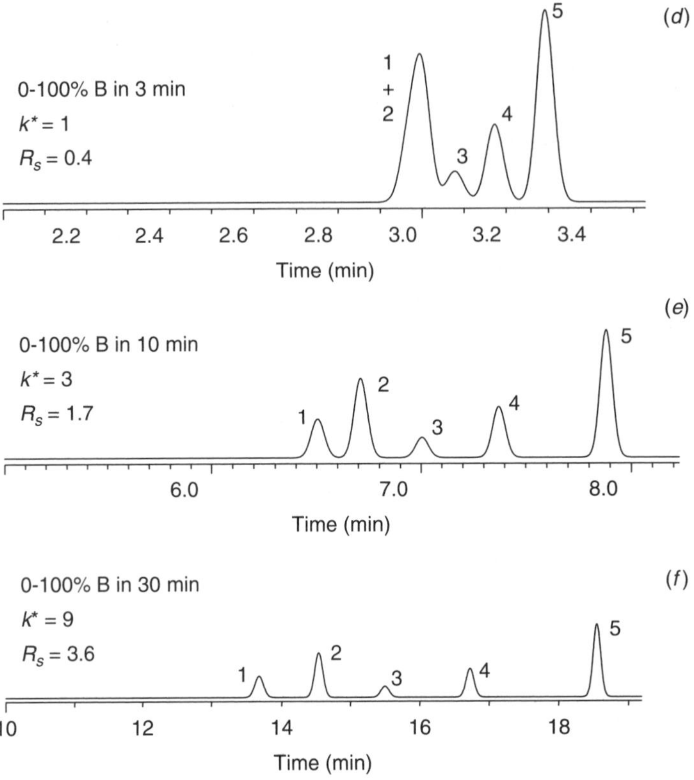

Figure 9.4 (*Continued*)

The sample of Figure 9.4 can be described as "regular" (Section 2.5.2.1) because there are no changes in relative retention when k or k^* are varied by varying isocratic %B or gradient time, respectively (holding other conditions constant). Consequently critical resolution increases continuously in Figure 9.4*d*–*f* as gradient time (and k^*) is increased. A similar series of experiments are shown in Figure 9.5 for an "irregular" sample (Section 2.5.2.1), composed of a mixture of substituted anilines and benzoic acids. Relative retention for an "irregular" sample changes as either isocratic %B or gradient time is varied. As in Figure 9.4, the same trends in *average* resolution, peak heights, and run time result in Figure 9.5 when gradient time is increased. However, changes in relative retention also occur for the sample of Figure 9.5 when gradient time is changed (note the changes in relative retention of shaded peak 3 and—to a lesser extent—peaks 7–10). As a result of these changes in relative retention with t_G, maximum ("critical") resolution for this sample occurs for an intermediate gradient time of 10 minutes (Fig. 9.5*b*; $R_s = 0.9$), whereas the resolution of the "regular" sample in Figure 9.4 continues to increase as gradient time (and k^*) increases. For "irregular" samples a change in either k (isocratic) or k^* (gradient) will result in similar changes in relative retention; consequently maximum sample resolution may not correspond to the largest possible value of k or k^* for such samples.

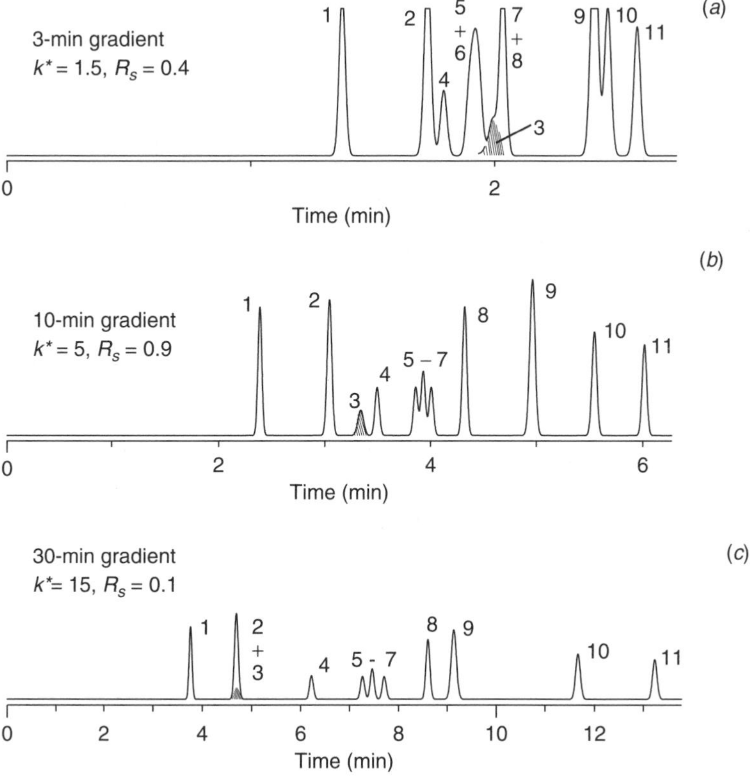

Figure 9.5 Separations of an irregular sample as a function of gradient time t_G. Sample: a mixture of substituted anilines and benzoic acids. Conditions: 100 × 4.6-mm C_{18} column (3-μm particles), 2.0 mL/min, 42°C, 5–100% acetonitrile-pH-2.6 phosphate buffer in (*a*) 5 minutes, (*b*) 15 minutes, and (*c*) 30 minutes. Peak 3 is cross-hatched to better illustrate changes in relative retention for this sample as gradient time is varied. Note that actual peak heights are shown (*not* normalized to 100% for tallest peak). Chromatograms recreated from data of [9].

9.2.1 Effects of a Change in Column Conditions

Column conditions—column length and diameter, flow rate, and particle size—affect the column plate number N (Section 2.4.1) and run time. Column conditions are chosen at the start of method development, then sometimes changed after other separation conditions have been selected—in order to either improve resolution or reduce run time (Section 2.5.3). In isocratic elution, a change in column conditions has no effect on values of k or relative retention. Resolution and run time usually increase for an increase in column length or a decrease in flow rate, while peak heights decrease for longer columns and faster flow. These changes in isocratic separation, when only column length or flow rate is changed, are illustrated in Figure 9.6*a*–*c* for the "regular" sample of Figure 9.4. Figure 9.6*a* represents a starting separation, while Figures 9.6*b* and 9.6*c* show the results of an increase in either column length or flow rate, respectively. Note the resulting changes in run time, resolution, and peak heights for these isocratic separations as column conditions are varied.

When changing experimental conditions during method development for isocratic elution, it is desirable to first vary conditions that affect values of k and α, so as to optimize selectivity and resolution. If a further improvement in separation is desired, by varying column conditions, the previously optimized values of k and α will not change for isocratic separation. With constant values of k and α, the interpretation of subsequent experiments is also simplified—as only N and run time can change. For gradient elution, the situation is more complicated—as values of k^* vary with column length and flow rate (Eq. 9.5). For values of k^* and α to remain constant while varying column conditions for gradient elution, it is necessary to hold values of $(t_G F/L)$ constant (Eq. 9.5; V_m is proportional to column length L, provided that the column diameter is not changed). For changes in column length L or flow rate F, a concomitant change in gradient time t_G is the most convenient way of maintaining constant values of k^* and α. For an x-fold change in L, gradient time should be changed by the same factor x. For an x-fold change in F, gradient time should be changed by $1/x$-fold. Just as a change in isocratic values of L or F results in a change in run time, changes in gradient values of L or F result in the same relative change in run time—as long as constant values of k^* are maintained by changing gradient time.

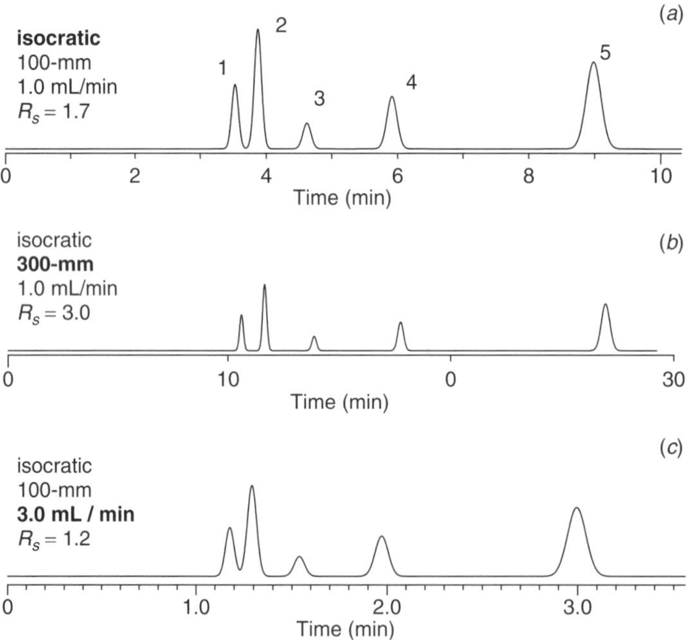

Figure 9.6 Isocratic and gradient elution compared for a regular sample and change in column length or flow rate. Sample and conditions as in Figure 9.4, except for varying column length and flow rate (as indicated in figure); 55% B for isocratic runs (a–c), 0–100% B for gradient runs (d–h). Note that actual peak heights are shown (*not* normalized to 100% for tallest peak). Chromatograms recreated from data of [8].

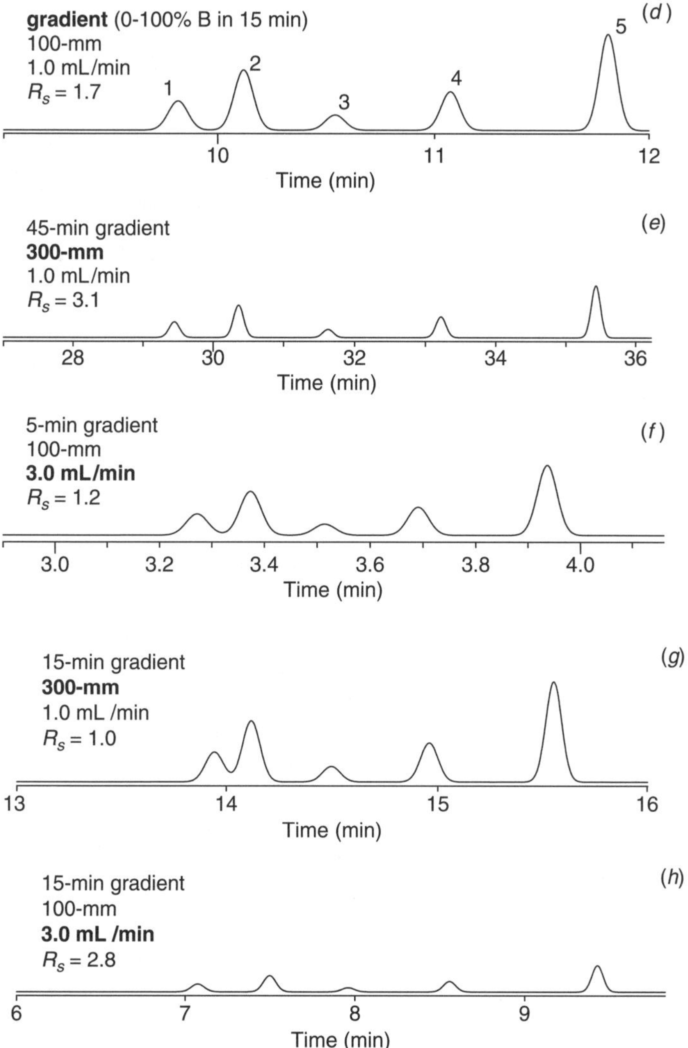

Figure 9.6 (*Continued*)

The gradient separations of Figure 9.6*d–f* illustrate the effects of the same changes in column length and flow rate as in the isocratic separations of Figure 9.6*a–c*, while holding k^* constant by varying gradient time t_G. For the "corresponding" separations of Figure 9.6*b*,*e*, where column length is increased from 100 to 300 mm (and gradient time in *e* is increased from 15 to 45 min), there is a similar increase in run time (by a factor of 3) and resolution ($R_s = 3.0$ [isocratic] and 3.1 [gradient]). Peak heights are decreased in each run, as a result of an increase in peak width. Likewise for the corresponding separations of Figure 9.6*c*,*f* where flow rate is increased from 1.0 to 3.0 mL/min (and gradient time in *f* is decreased from 15 to 5 min), there is a similar decrease in run time (by a factor of 3) and resolution ($R_s = 1.2$ [isocratic] and 1.2 [gradient]). Peak heights are increased in

Table 9.1

Contrasting Changes in Separation as Flow Rate *F* or Column Length *L* is Changed for Isocratic versus Gradient Elution (Examples of Fig. 9.6)

Elution Mode	Original Separation		Increase L by 3-fold		Increase F by 3-fold	
	R_s	Average Peak Height[d]	R_s	Average Peak Height[d]	R_s	Average Peak Height[d]
1. Isocratic[a]	1.7	(1.0)	3.0	0.6	1.2	0.8
2. Gradient (t_G varies, k^* constant)[b]	1.7	(1.0)	3.1	0.6	1.2	0.7
3. Gradient (t_G constant, k^* varies)[c]	1.7	(1.0)	1.0	1.0	2.8	0.3

[a]Figure 9.6*a–c*.
[b]Figure 9.6*d–f*.
[c]Figure 9.6*g–h*.
[d]Relative values, versus original separation.

the separations of Figure 9.6*c*,*f*, as a result of narrower peaks. The examples of Figure 9.6*a–f* confirm the similarity of gradient and isocratic elution for changes in column conditions, when values of k or k^* are held constant. Details of the separations of Figure 9.6 are summarized in Table 9.1.

When *only* column dimensions or flow rate are changed in gradient elution (i.e., gradient time unchanged), changes in k^* will also occur (Eq. 9.5; see also Eq. 9.5*c* on p. 431). Resulting separations may then appear surprising to workers who expect similar results as in isocratic elution (as in Figs. 9.6*a–c*). This is illustrated in Figure 9.6*g*,*h*, for the same changes in column length or flow rate as in Figure 9.6*e*,*f*, *while holding gradient time constant at 15 min so that k^* is no longer constant.* For the latter conditions, resolution decreases when column length is increased (Fig. 9.6*g*, $R_s = 1.0$), and increases when flow rate is increased (Fig. 9.6*h*, $R_s = 2.8$). In the latter case (Fig. 9.6*e*,f), the *opposite* behavior is found for gradient elution when k^* is allowed to vary. *For this reason, when changing column length or flow rate in gradient elution, gradient time should be changed at the same time so as to maintain values of k^* constant and—more important—retain the same relative retention or selectivity.*

To conclude, "corresponding" separations by isocratic or gradient elution (i.e., with similar values of k and k^*) will generally exhibit similar values of resolution and peak heights. Run times will change to the same extent, when any column condition (or combination of column conditions) is changed for both isocratic and gradient runs, as long as k^* (or k) is held constant.

9.2.2 Effects of Changes in the Gradient

Changes in the gradient can be made intentionally—or unintentionally as a result of a change in equipment. These changes in the gradient can be summarized as follows:

- a change in %B at the start of the gradient (initial-%B; Section 9.2.2.1)
- a change in %B at the end of the gradient (final-%B; Section 9.2.2.2)
- gradient delay (Section 9.2.2.3)

- a change in equipment (dwell-volume, Section 9.2.2.4)
- segmented gradients (Section 9.2.2.5)

9.2.2.1 Initial-%B

The usual goal of a change in initial-%B is to shorten run time, by removing empty space in the early part of a gradient chromatogram, as illustrated in Figure 9.7. A change in initial-%B (and therefore a change of the gradient range $\Delta\phi$), *without* a change in gradient time, would also change values of k^* (Eq. 9.5)—which can be undesirable. In the present section we will examine the effects of a change in initial-%B while holding k^* constant (by varying gradient time t_G in proportion

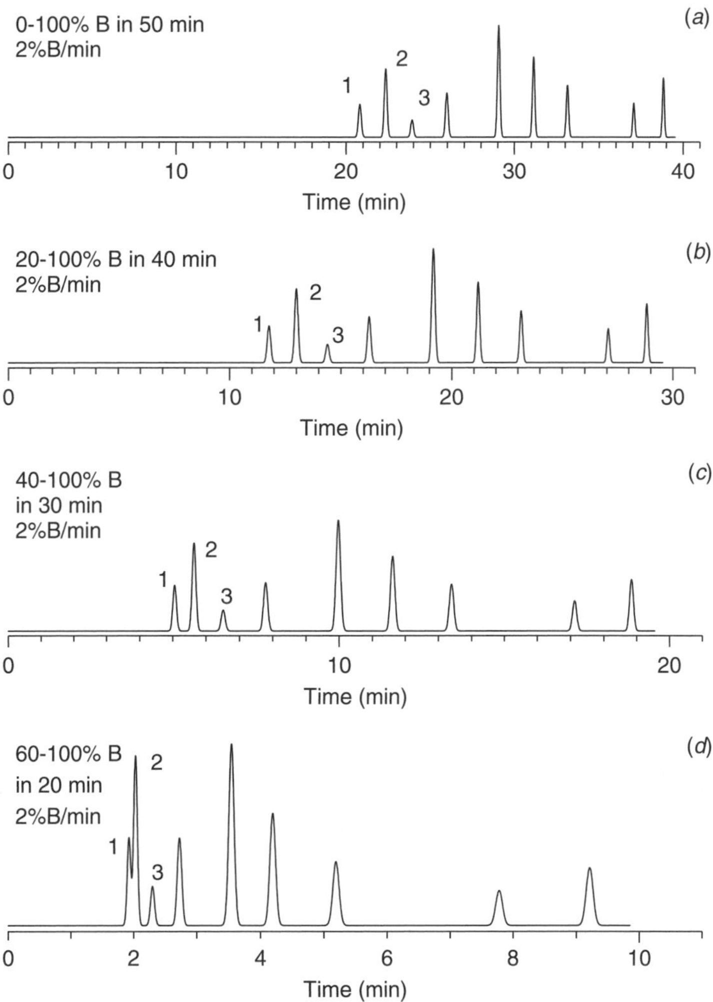

Figure 9.7 Effect of a change in initial %B for the gradient separation of a "regular" sample. Sample: a mixture of herbicides. Conditions; 150 × 4.6-mm (5-μm) C_{18} column; ambient temperature; 2.0 mL/min; methanol-water mobile phase; gradient time adjusted to maintain $k^* = 4$. Other conditions indicated in the figure.

to $\Delta\phi$), thus holding $(\Delta\phi/t_G)$ and k^* constant. *Keep in mind that if only %B is changed, while holding other conditions constant, resulting changes in separation will represent the combined effect of change in k^* and the value of initial-%B.* It is much easier to interpret and optimize separation, if k^* is held constant when initial-%B (or some other condition) is varied (as in the preceding example of changes in column length or flow rate).

Figure 9.7 illustrates the effects of a change in initial-%B for the separation of a "regular" sample. In successive separations, Figure 9.7*a*–*d*, the value of %B at the start of the gradient is increased (resulting in a reduction of the gradient range $\Delta\phi$), while simultaneously shortening gradient time t_G so as to keep $\Delta\phi/t_G$ and k^* constant. For an increase in initial-%B from 0 to 20% (Fig. 9.7*b*), $\Delta\phi$ is shortened by 20%, so a similar 20% shortening of gradient time is required (from 50 to 40 min), in order to maintain k^* constant (Eq. 9.5). The separation of Figure 9.7*b* remains essentially the same as in Figure 9.7*a*, except that all peaks leave the column 10 minutes earlier—and run time is reduced by 20%. When initial-%B is increased further to 40%B (Fig. 9.7*c*), a slight change in peaks 1 and 2 is observed: the heights of these peaks have increased a bit, and their resolution has decreased a bit, too ($R_s = 2.7$ vs. $R_s = 4.0$ in Fig. 9.7*a*). However, separation is still acceptable, and run time has been shortened by another 10 minutes. Finally, in Figure 9.7*d*, the initial-%B is increased to 60%, with a considerable increase in the heights of early peaks, as well as markedly lower resolution for peaks 1 and 2 ($R_s = 0.9$). In this case the shortest run time with acceptable resolution occurs for approximately 40% B at the start of the gradient (Fig. 9.7*c*).

Because early peaks elute fairly late in the 0–100% B gradient of Figure 9.7*a*, these peaks are strongly retained initially at the column inlet. As a result their values of k^* are given by Equation (9.5) (average $k^* \approx 3.7$). When the initial-%B of the gradient is increased to 20% B (Fig. 9.7*b*), the initial peaks are still well retained, and k^* still equals 3.7. When initial-%B is increased further in Figures 9.7*c* (40% B) and 9.7*d* (60% B), peaks at the beginning of the chromatogram leave the column in a still stronger mobile phase, but now with lower values of k^* (Eq. 9.5 is strictly applicable only for peaks that are strongly retained at the start of the gradient; for weakly retained peaks, see Eq. 9.5f in following Section 9.2.4.1). This decrease in values of k^* for early peaks, when initial%-B is increased sufficiently, results in narrower, higher peaks—usually with reduced resolution.

Because values of k^* decrease for early peaks when initial-%B is increased enough, changes in *relative* retention can also result for "irregular" samples. As a result resolution has been observed in some cases to *increase* when the initial-%B is increased [10], despite the corresponding decrease in k^*. See the further discussion of Section 9.2.3 for the gradient separation of "irregular" samples.

9.2.2.2 Final-%B

Figure 9.8 illustrates the effect of changing the final-%B for the "regular" sample and separation of Figure 9.7*a*, with the goal of a reduction in run time. The separation in Figure 9.8*a* is for a gradient of 0–100% B in 50 minutes. Subsequent changes in the final-%B value are accompanied by changes in gradient time so as to keep $(\Delta\phi/t_G)$ and k^* constant (as in Fig. 9.7 for changes in initial-%B). Thus, for a 20% shortening of $\Delta\phi$ to a final-%B of 80% in Figure 9.8*b*, the gradient time is also

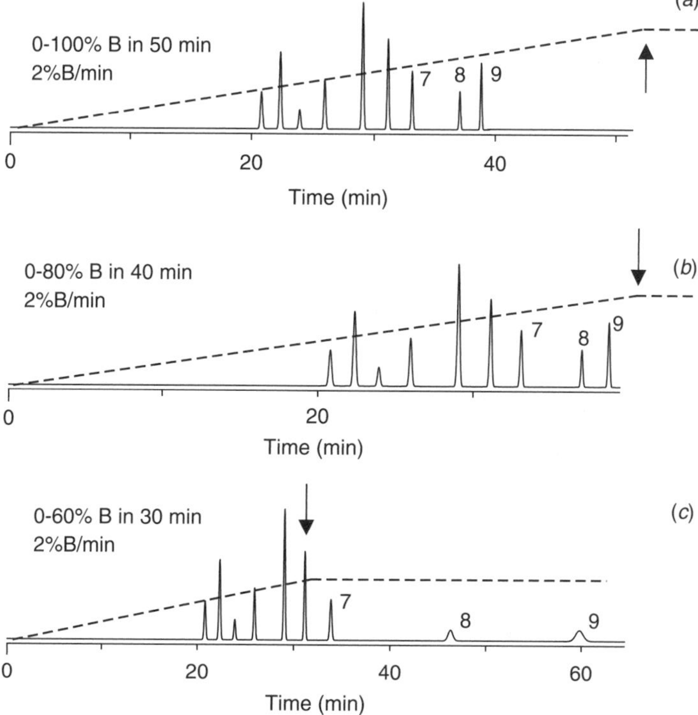

Figure 9.8 Effect of a change in final %B for the gradient separation of the regular sample of Figure 9.7. Conditions as in Figure 9.7; gradient time adjusted to maintain $k^* = 4$. Dashed lines indicate the gradient: values of %B at the column outlet (so as to correspond to peaks in the chromatogram). Arrows mark end of gradient as it leaves the column. Other conditions indicated in the figure.

shortened by 20% (from 50 to 40 min). For the separation of Figure 9.8*b*, there is no change in separation because the last peak in the sample leaves the column before the gradient has ended (see arrow). A further shortening of the gradient to 0–60% B in 30 minutes (Fig. 9.8*c*), however, results in elution of peaks 7 through 9 *after* the end of the gradient, so these peaks leave the column under isocratic conditions. As a result peak width and resolution increase for peaks 7 through 9, as does run time, because of larger values of k^* for these peaks (Note that Eq. 9.5 only applies for peaks that are eluted *during* the gradient; peaks eluting after the gradient will have larger values of k^*). Figure 9.8*c*, where the value of final-%B is reduced too much, can be compared with Figure 9.7*d*, where initial-%B is increased too much; in each case the resulting separation is unsatisfactory—either resolution is too low or run time is too long.

As long as the last peak leaves the column before the end of the gradient, there is no effect of a change in final-%B on separation (provided that $t_G/\Delta\phi$ is held constant), other than to decrease run time for smaller values of final-%B. In most cases it will be advisable to end the gradient as soon as the last peak leaves the column, but not before. The elution of peaks after the gradient wastes run time and leads to undesirable peak broadening (Fig. 9.8*c*). The effect of final-%B on separation is similar for both "regular" and "irregular" samples (no change in relative retention or elution order), as long as late elution of peaks is avoided and $(t_G/\Delta\phi)$ is held constant. For some samples the use of a very steep gradient can lead to elution of the last peaks after the gradient, even when the gradient ends with 100% B (and less steep gradients do not result in late elution). However, this situation does not present any special problem; it is only necessary to wait for the last peak to leave the column (by adding an isocratic hold at the end of the gradient; for example, 0/60/60% B in 0/30/60 min for the separation of Fig. 9.8*c*) before starting the next gradient (although the gradient of Fig. 9.8*b* is obviously a better choice).

From the combined examples of Figures 9.7 and 9.8, it can be concluded that a gradient of 40–80% B in 20 minutes represents a suitable shortening of the original gradient (vs. Fig. 9.7*a*; 0–100%B in 50 min). This separation is shown in Figure 9.9*a*; sample resolution is acceptable, with a 60% decrease in run time compared to the separation of Figure 9.7*a*, and no unacceptable loss in resolution or other problems.

9.2.2.3 Gradient Delay

Gradient delay (also referred to as an *isocratic hold*) refers to isocratic elution for some period of time prior to the start of the gradient. The effect of a gradient delay is illustrated in Figure 9.9 for the "regular" sample of Figure 9.7. Figure 9.9*a* shows a chromatogram for a 40–80% B gradient without a gradient delay, where the first peak in the chromatogram does not leave the column until well after the arrival of the gradient at the *outlet* of the column (the column dead-time t_0 is indicated by the arrow). When a 5-minute gradient delay is added (Fig. 9.9*b*), the effect is to increase retention times by 2 to 5 minutes, but the two chromatograms of Figures 9.9*a* and *b* are otherwise quite similar (there is also a typical, modest increase in resolution for early peaks in Fig. 9.9*b*).

When initial peaks leave the column close to the start of the gradient, a gradient delay can have a more noticeable effect on the separation—especially if early peaks are not well resolved. This is illustrated in the similar examples of Figure 9.9*c* (no delay) and Figure 9.9*d* (with delay), for the same sample but different starting gradient conditions. In the separation of Figure 9.9*d*, peaks 1 through 3 leave the column isocratically *during* the gradient delay (note the arrow in Fig. 9.9*d* that marks the arrival of the gradient at the column outlet). As can be seen in these latter two examples, peaks 1 and 2 are poorly separated in Figure 9.9*c* ($R_s = 1.1$), whereas in Figure 9.9*d* their separation is much improved ($R_s = 2.3$). The better resolution of early peaks in Figure 9.9*d* as a result of the gradient delay can be attributed to larger values of k^* for these peaks compared to the separation of Figure 9.9*c* (see later Eq. 9.5g). Peaks 1 through 3 for Figure 9.9*d* show the expected increase in peak width characteristic of isocratic separation, whereas later peaks, eluted under gradient conditions, exhibit narrower peak widths—typical of gradient separation.

When peaks elute near the end of the gradient, the effect of an initial gradient delay is to increase retention time by the same amount as the delay, with no change in relative retention. For example, the last two peaks in Figure 9.9*b*,*d* are delayed by 5 minutes relative to Figures 9.9*a*,*c*—exactly the amount of the gradient delay. This behavior holds for both regular and irregular samples.

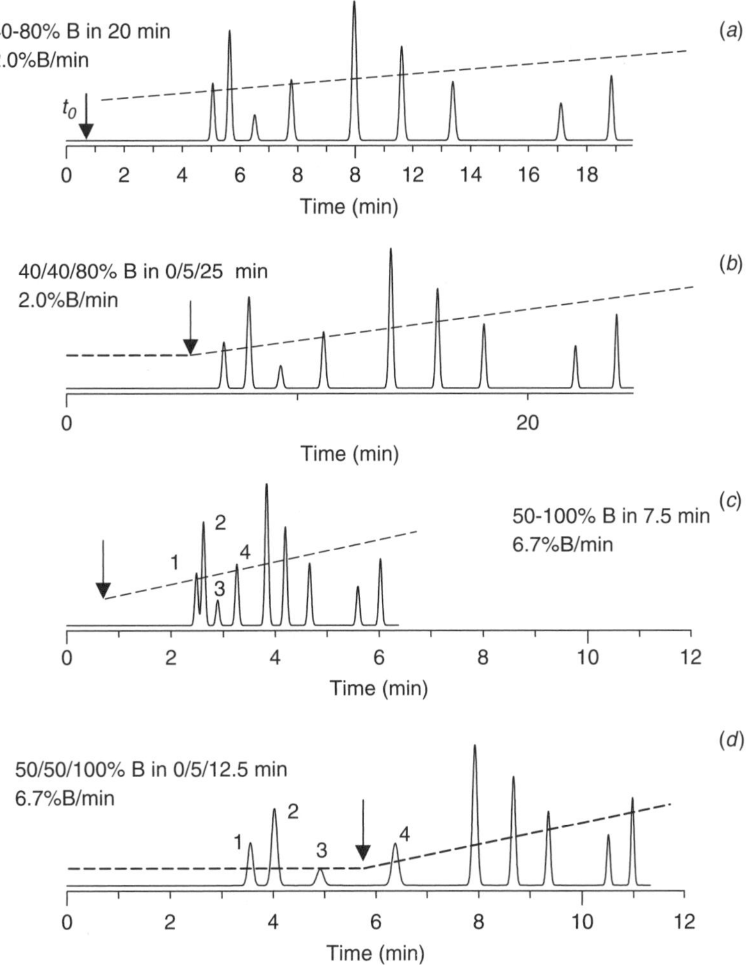

Figure 9.9 Effect of gradient delay on the gradient separation of the herbicide sample of Figure 9.4. Conditions: 150 × 4.6-mm (5-μm) C_{18} column; 30°C; 2.0 mL/min; methanol-water mobile phase; gradient time adjusted to maintain $k^* = 4$. Peak heights *not* normalized to 100%; gradient indicated by (- - -), and arrows mark start of the gradient (measured at the column outlet). Other conditions indicated in the figure.

A gradient delay is sometimes used to increase the resolution of early peaks in the chromatogram, as in the example of Figure 9.9*d* compared to that of Figure 9.9*c*. For separations that start at a higher %B (e.g., Fig. 9.9*c*), however, resolution can best be improved by simply reducing the initial value of %B in the gradient (compare separations in Fig. 9.7*d* vs. Fig. 9.7*c*). On the other hand, when the initial-%B of the gradient is close to zero (and a significant reduction in initial-%B is therefore not feasible), a gradient delay may be the most convenient alternative; still there are other means for increasing *k* in this situation (Section 6.6.1). Note that relative retention does not change when a gradient delay is used for a "regular" sample, as in Figure 9.9. However, because a gradient delay can affect values of k^* for early peaks in the chromatogram, changes in relative retention can occur for "irregular" samples (see Section 9.2.2.4, and later Fig. 9.13*f* vs. Fig. 9.13*a*).

9.2.2.4 Dwell-Volume

Every instrument used for gradient elution will have a certain holdup volume (called the *dwell-volume* V_D) equal to the volume of the gradient mixer plus that of the mobile-phase flow path between the mixer and the column inlet (Section 3.5.3; Figs. 3.13 and 3.14). Values of V_D can vary for different gradient equipment, from a fraction of a mL for modern equipment to several mL for older equipment. The existence of a dwell-volume is equivalent to the intentional use of a gradient delay, so the effects on separation of varying dwell-time $t_D = V_D/F$ can therefore be inferred from the examples of Figure 9.9 for a gradient delay. The actual gradient entering the column is delayed by a time t_D, while the gradient leaving the column is delayed further by the column dead time t_0 (Fig. 9.10). Values of V_D for a particular gradient system can be determined as described in Section 3.10.1.2.

When a gradient method is transferred from one HPLC system to another, differences in the dwell-volume V_D of the two systems can result in changes in separation. Often an HPLC method will be developed on a newer system in an R&D laboratory, while routine assays will be carried out on an older system in a production laboratory. As a result the dwell-volume may be greater for a method in routine operation, compared to the method procedure issued by the R&D laboratory. For a "regular" sample, as in the examples of Figure 9.9, an increase in dwell-volume will cause an increase in retention times for all peaks, possibly with some reduction in peak height and increase in resolution for early peaks in the chromatogram (as in the example of Fig. 9.9*d*). Relative retention will remain unchanged for different values of V_D. When the dwell-volume is changed for "irregular" samples, however, changes in relative retention *can* occur for early

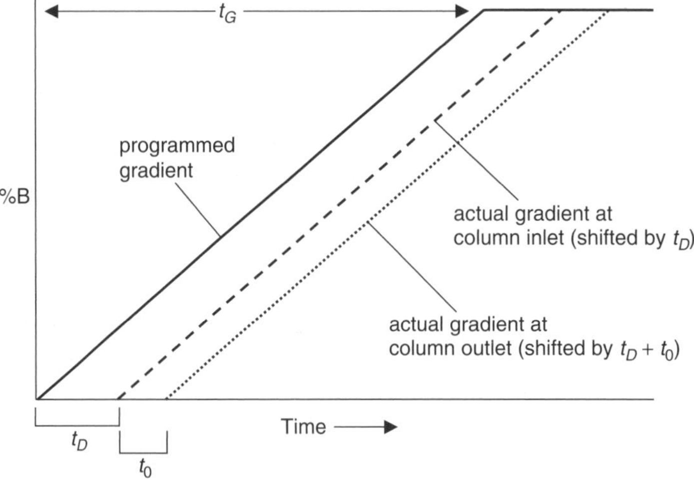

Figure 9.10 Effect of dwell-volume on the gradient. (____), Programmed gradient selected by the user; (- - -) actual gradient at the column inlet, taking the dwell-volume of the system into account; (......) actual gradient at the column outlet, assuming a dwell time t_D.

peaks, and this can lead to a change in the resolution of early peaks (see the later example of Fig. 9.13*f* vs. Fig. 9.13*a*)—sometimes unacceptably. These and other problems relating to equipment dwell-volume are discussed in Section 9.3.8.2.

A similar situation can arise when the column size (and dead-volume V_m) is changed because the effect of the dwell-volume on relative retention for early peaks is determined by the ratio V_D/V_m. When changes are made in the column-volume, it may be necessary to adjust the dwell-volume in proportion to column volume, in order to maintain the same relative retention and resolution for early peaks in the chromatogram. For example, if column diameter d_c is reduced for use with LC-MS, the dwell volume should be reduced in proportion to d_c^2. (A reduction in dwell volume by the user usually is possible with high-pressure-mixing systems, but not with low-pressure-mixing systems.) If column diameter is increased for scaling up a preparative separation, a similar increase in dwell-volume may be necessary—although this can be duplicated more conveniently by the addition of an isocratic hold at the start of the gradient. See [11, 12] and Section 3.5.3 of [2] for further details.

When a test gradient is carried out as in Section 3.10.1.2, some distortion is normally observed at each end of the gradient (Fig. 3.26). This *gradient rounding* results from dispersion of the A- and B-solvents as the mobile phase flows into the gradient mixer and on to the column inlet; gradient rounding is more pronounced for low-pressure-mixing gradient systems. The extent of gradient rounding increases for larger values of V_D and can be described quantitatively in terms of the equipment *mixing volume* $V_M (V_M \approx V_D)$. Gradient rounding has little effect on separation, unless the value of V_M becomes comparable to that of the *gradient volume* $V_G = t_G F$. For a further discussion of the effect of mixing volume on gradient shape and separation, see Section 17.4.6.1 and pp. 394–396 of [2].

9.2.2.5 Segmented Gradients

Segmented gradients, as in Figure 9.2*d*, are used for different purposes:

- to clean the column between sample injections
- to shorten run time
- to increase resolution by adjusting selectivity for different parts of the chromatogram (for "irregular" samples only)

Segmented or step gradients for *cleaning the column* are often employed when separating environmental or biological samples because the presence of extraneous, strongly retained sample components (non-analytes) can foul the column. When separating samples of this kind, *and* where the gradient required to elute all peaks of interest ends short of 100% B, it is customary to follow the initial gradient with a steep gradient segment or step that ends at or near 100% B. Figure 9.11*a* shows the linear gradient separation of a mixture of peptides from a tryptic digest of recombinant human growth hormone (*rh*-GH). Nineteen peptides are baseline-separated in 50 minutes. In Figure 9.11*b* the separation of Figure 9.11*a* is followed by a gradient step from 40% B to 100% B in one minute, in order to purge the column of any sample components that are not eluted by the gradient of Figure 9.11*a*. This increase in steepness at the end of the gradient is usually followed

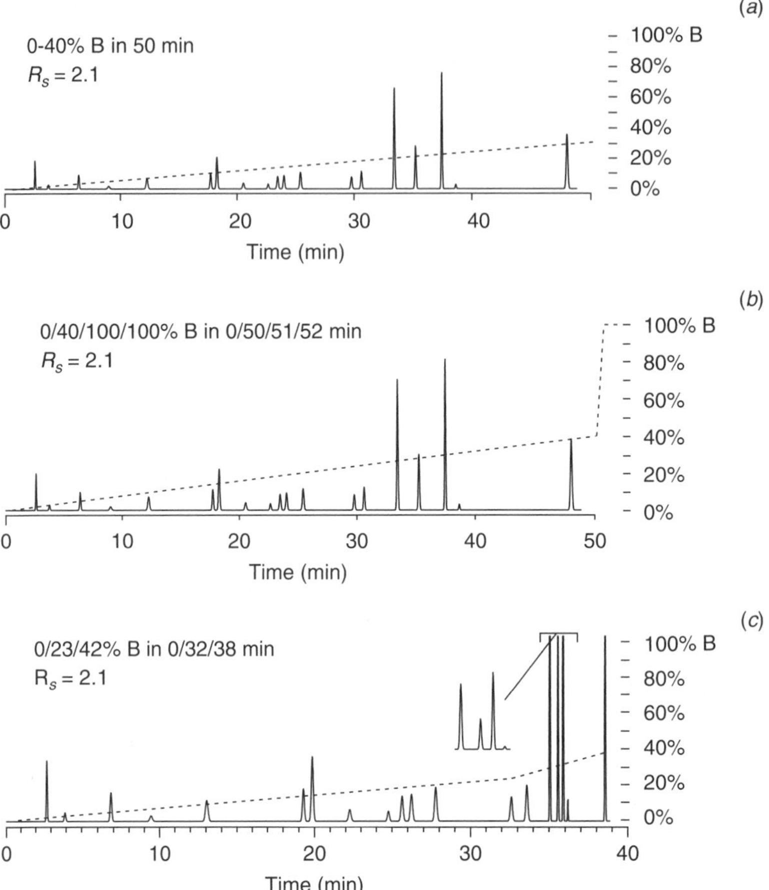

Figure 9.11 Gradient separations of a peptide digest of recombinant human growth hormone. Conditions: 150 × 4.6-mm C_{18} column (5-μm); 45°C; 2.0 mL/min. (*a*) 0–40% B in 50 min; (*b*) same as in (*a*), except a steep gradient segment is added in order to remove strongly retentive "junk" from the column; (*c*) same as in (*a*), except a second gradient segment is added in order to accelerate elution of the last two peaks in the chromatogram. Gradient indicated by (- - -). Chromatograms recreated from data of [13].

by a short isocratic hold. Thus the final gradient in Figure 9.11*b* is 0/40/100/100%B in 0/50/51/52 min.

Shortening run time is illustrated in Figure 9.11*c* for the sample of Figure 9.11*a*, without a final column-cleaning gradient step (which could be added, if needed). Because the last five peaks in the chromatogram are resolved with $R_s \gg 2$, it is possible to increase gradient steepness for these peaks, so as to reduce their retention times while maintaining $R_s \geq 2$ for all peaks. This way run time is shortened from 50 minutes in Figure 9.11*a* to 40 minutes in Figure 9.11*c*.

Increasing resolution by adjusting selectivity for different parts of the chromatogram can sometimes be achieved with a segmented gradient; gradient steepness (and values of k^*) for different segments are optimized for different critical

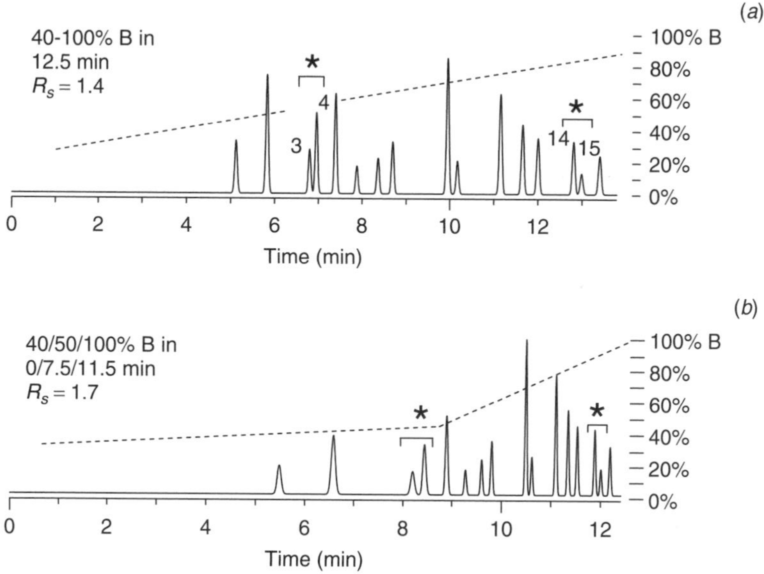

Figure 9.12 Separation of a mixture of 16 polycyclic aromatic hydrocarbons, adapted from Figure 6.4 of [6]. Conditions: 150 × 4.6-mm (5-μm) C_{18} column; 35°C; 2.0 mL/min. (*a*) Separation with an optimized linear gradient; (*b*) separation with an optimized two-segment gradient. Gradient indicated by (- - -). See [6] for further details.

peak-pairs. An example is shown in Figure 9.12 for the separation of a mixture of polycyclic aromatic hydrocarbons; peak-pairs 3–4 and 14–15 (marked by *) are critical. Whereas peak-pair 3–4 is better separated with a flatter gradient (larger values of k^*), the separation of peaks 14 and 15 improves for a steeper gradient (smaller k^*). In Figure 9.12*a*, the slope of a linear gradient has been selected for maximum critical resolution of the sample. Maximum critical resolution corresponds to equal resolution for each of these two peak-pairs because a change in gradient steepness will increase resolution for one peak-pair while decreasing resolution for the other. However, the resolution of each peak-pair can be improved by the segmented gradient of Figure 9.12*b*, which combines a flatter gradient for peaks 3 and 4 with a steeper gradient for peaks 14 and 15. The small increase in R_s shown in Figure 9.12*b* (+0.3R_s-units vs. Fig. 9.12*a*) is typical of the effect of segmented gradients. It is rare to achieve an increase in resolution of more than ≈ 0.5 units with segmented gradients. In the absence of computer simulation (Section 10.2.3.4), the time required to develop such separations may not be worthwhile.

Segmented gradients are not often used for improving resolution as in Figure 9.12 because their ability to enhance resolution without increasing run time is usually limited [14]. An increase in critical resolution as a result of the use of segmented gradients requires at least two critical pair-pairs that elute, respectively, early and late in the chromatogram (as in Fig. 9.12). Otherwise, the partial migration of the second peak-pair under the influence of the initial gradient segment will result in little or no overall advantage from the use of the second gradient segment. However, this limitation of segmented gradients for an increase in sample resolution becomes less important for high-molecular-weight samples such as proteins [15, 16], since there is less migration of later peaks during an earlier gradient segment, and therefore less effect of the earlier segment on the resolution of later peaks. The use of segmented gradients for the purpose of increasing critical resolution is therefore somewhat more practical for the separation of mixtures of large biomolecules. However, there are other—generally more useful—means for optimizing resolution by changing selectivity and relative retention (Section 9.3.3). Also separations that use segmented gradients to improve resolution are likely to be less reproducible when transferred to another piece of equipment.

A more detailed examination of the use of segmented gradients in this way is offered in [17, 18]. Computer programs have also been reported for the automated development of optimized segmented gradients [14, 19, 20]. Stepwise elution involving step gradients can be regarded as a simple (if less generally effective) kind of segmented gradient; a theory of such separations has been described [21].

9.2.3 "Irregular Samples"

The following section discusses gradient separations where relative retention changes for an "irregular" sample as a result of a change in some condition that affects k^* (gradient time, flow rate, etc.). These examples are intended to supplement preceding examples in Figures 9.4 and 9.6 to 9.9 for "regular" samples, by illustrating changes in relative retention for "irregular" samples as a function of changes in conditions that affect k^*. The reader may choose to skip to Section 9.3, and return to this section at a later time—or as needed. However, this treatment can add to the reader's intuitive understanding of gradient elution, as well as find occasional practical application.

Changes in k^* can result from a change in any of the experimental conditions included in Equation (9.5) (t_G, F, V_m or column length L, $\Delta\phi$), as well as from a change in initial-%B, the introduction of a gradient delay, or a change in dwell volume. An increase in k^* will result in an *average* increase in retention time, resolution, and peak width for all samples, as illustrated by Figures 9.4 and 9.5 for changes in gradient time. In the case of "irregular" samples (Fig. 9.5) a change in k^* will also cause relative retention to change, which can result in a change in resolution for certain peaks. *Any* change in k^* for a given "irregular" sample will result in similar changes in relative retention and resolution, regardless of how k^* is caused to vary. This is illustrated in the remainder of this section for various changes in gradient or column conditions, using the examples of Figure 9.13 for selected peak-pairs (2–3 and 8–9) from the irregular sample of Figure 9.5. Because many real samples fall in the "irregular" sample category, the following discussion is expected to reflect the kind of changes most users will observe with changes in gradient elution conditions.

A starting separation of peak-pairs 2–3 and 8–9 of the "irregular" sample of Figure 9.5 is shown in Figure 9.13*a*. These two peak-pairs have been chosen because their resolution responds in opposite fashion to a change in k^* (as a consequence of difference in *S*-values for these four solutes: $S_3 > S_2$; $S_8 < S_9$; see the similar examples of Fig. 6.7*c*). Consider first an *increase in gradient time* from 5 to 20 minutes (Fig. 9.13*b*), corresponding to an increase in average k^* from 5 to 20. As a result the retention of peak 2 *relative* to that of peak 3 increases, and the resolution of peak-pair 2–3 therefore increases. At the same time the relative retention of peak 9 relative to peak 8 *decreases* when gradient time is increased, and the resolution

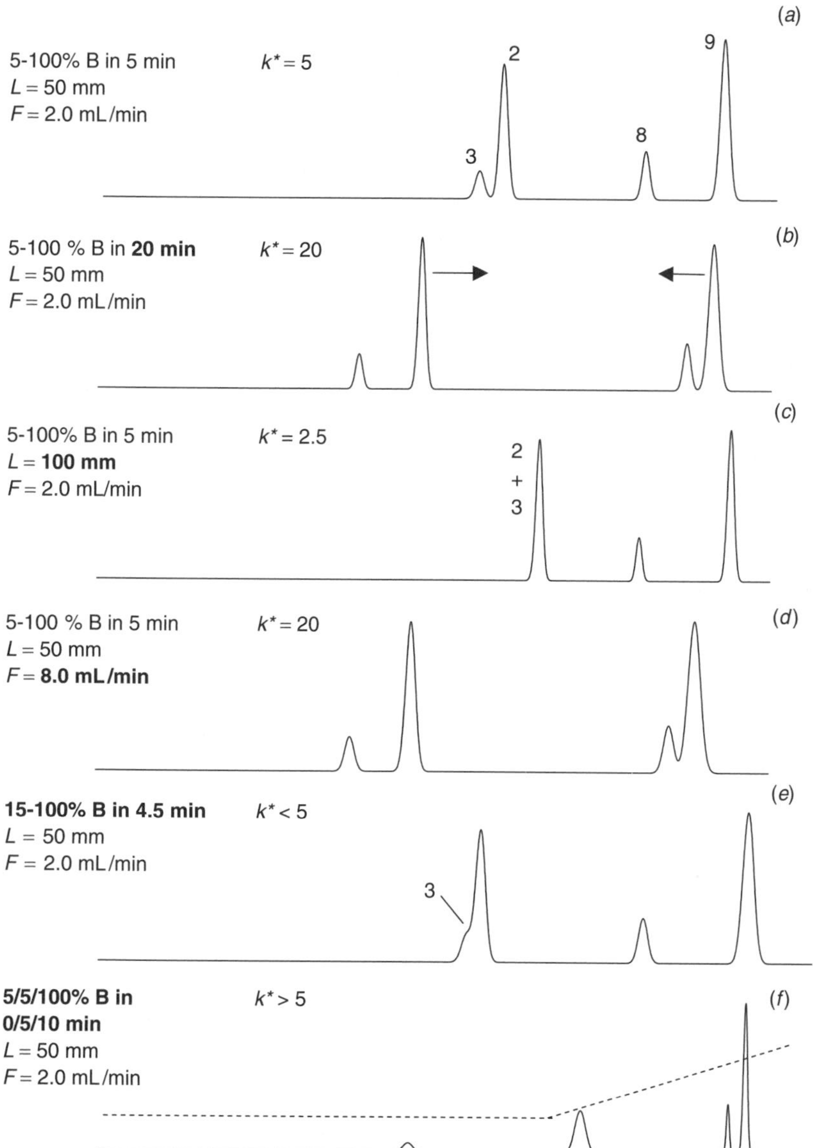

Figure 9.13 Changes in peak spacing with changes in gradient conditions. Sample consists of peaks 2, 3, 8, and 9 of the irregular sample of Figure 9.5. Conditions: 28°C. The arrows in (*b*) indicate the relative movement of peaks 2 and 9 as a result of an increase in gradient time and k^*. Gradient indicated by (- - -).

of this peak-pair decreases. Similar changes in relative retention and resolution for these two peak-pairs can be expected for changes in any other condition, which results in an increase in k^*. Opposite changes in relative retention will occur when k^* is decreased.

In Figure 9.13*c*, *column length L is increased* from 50 to 100 mm, while other conditions remain the same as in Figure 9.13*a*; the value of k^* decreases by a factor of 2 to $k^* = 2.5$ (Eq. 9.5c below). As expected from this decrease in k^* (relative to the separation of Fig. 9.13*a*), the changes in relative retention seen in Figure 9.13*b* compared to Figure 9.13*a* are reversed in Figure 9.13*c*: peak 2 now moves *toward* peak 3 with a decrease in resolution, while peak 9 has moved *away* from peak 8, with an increase in resolution.

The effect of an *increase in flow rate* (from 2.0 to 8.0 mL/min) is seen in Figure 9.13*d*. Because k^* has increased from 5 to 20 (Eq. 9.3), a similar change in relative retention is expected as for an increase in gradient time (Fig. 9.13*b*): again, peak 2 moves away from peak 3 with an increase in resolution, and peak 9 moves toward peak 8, with a decrease in resolution.

When *%B at the start of the gradient* (ϕ_o) is increased while holding $\Delta\phi/t_G$ constant (Fig. 9.13*e*), values of k^* calculated from Equation (9.5) remain the same. However, *actual* values of k^* for early-eluting peaks are decreased (Eq. 9.11), despite holding ($t_G/\Delta\phi$) constant. Thus Equation (9.5) no longer applies for early peaks in the chromatogram, resulting in the movement of peak 2 toward peak 3. The value of k^* for later peaks 8 and 9 is somewhat less affected by the increase in initial %B, so the relative retention and resolution of peaks 8 and 9 are less affected (compared to the separation of Fig. 9.13*a*).

Finally, in Figure 9.13*f*, a *gradient delay* (or increase in dwell time t_D) of 5 minutes is introduced into the separation of Figure 9.13*a* (other conditions the same). As in the preceding example (Fig. 9.13*e*), the value of k^* calculated from Equation (9.5) is unchanged ($k^* = 5$), but the effect of a gradient delay is to reduce the effect of the gradient on initial peaks in the chromatogram. This in turn means *effectively* higher values of k^* for these early peaks (Eq. 9.12). As a result a similar change in relative retention and resolution results as in Figure 9.13*b*, for an increase in gradient time—but to a somewhat lesser extent for later peaks 8 and 9 (whose values of k^* are less affected by either a gradient delay or a change in initial %B). A change in dwell-volume and dwell-time (due to a change in gradient system) would give the same result as this change in gradient delay in Figure 9.13*f*.

Resolution is also affected by changes in k^* and N^* (see Eq. 9.15c below), *apart* from changes in relative retention. The former contributions to resolution may occasionally confuse the dependence of resolution on relative retention.

9.2.4 Quantitative Relationships

The LSS model allows the derivation of a number of exact relationships for retention and peak width; these equations form the basis of computer simulation for gradient elution (Section 10.2). Apart from computer simulation and the dependence of k^* on experimental conditions (Eq. 9.4), following Equations (9.5a) to (9.15) have somewhat limited practical application. *For this reason the reader may wish to skip to Equation* (9.16) *at the end of this section, and return to the remainder of this section as needed.* For the derivation of the various equations contained in this section, and for details on their application, see Chapter 9 of [2].

Linear RPC gradients are assumed for each of the following equations. Values of k^* can be described by a relationship that corresponds to Equation (9.1) for isocratic elution:

$$\log k^* = \log k_w - S\phi^* \tag{9.5a}$$

where ϕ^* refers to the value of ϕ for mobile phase in contact with the solute band when it has reached the column midpoint. Values of k_w and S are the same for either isocratic or gradient elution.

V_m can also be estimated (Eq. 2.7a, which assumes a total column porosity $\varepsilon_T = 0.65$) from column length L and internal diameter d_c:

$$V_m \approx 5 \times 10^{-4} L d_c^{\,2} \qquad \text{(units of } L \text{ and } d_c \text{ in mm)} \tag{9.5b}$$

For the usual column diameter of 4.6 mm, it is convenient to approximate V_m by 0.01 times the column length in mm; for example, $V_m \approx 1.5$ for a 150×4.6-mm column. Combining Equations (9.5) and (9.5b), we have

$$k^* = \frac{1740 t_G F}{L d_c^{\,2} \Delta\phi S} \tag{9.5c}$$

or for $S \approx 4$ for small solute molecules,

$$k^* \approx \frac{450 t_G F}{L d_c^{\,2} \Delta\phi} \qquad \text{(for solutes} < 500 \text{ Da}, S \approx 4) \tag{9.5c}$$

Thus k^* will increase for larger values of t_G and F or smaller values of column length L, column diameter d_c or gradient range $\Delta\phi$. From Equations (9.4) and (9.5), we see also that k^* is related to the gradient-steepness parameter b:

$$k^* = \frac{0.87}{b} \tag{9.6}$$

That is, the value of k^* decreases for steeper gradients with larger values of b.

9.2.4.1 Retention Time

The calculation of retention time t_R of a solute in gradient elution takes different forms, depending on (1) whether a significant dwell volume is assumed ($V_D > 0$) and (2) whether the initial value of k at the start of the gradient (k_0) is small. The value of k_0 is given by

$$\log k_0 = \log k_w - S\phi_0 \tag{9.7}$$

If k_0 is large, and if $V_D = 0$,

$$t_R = \left(\frac{t_0}{b}\right) \log(2.3 k_0 b + 1) + t_0 \tag{9.8}$$

$$\approx \left(\frac{t_0}{b}\right) \log(2.3 k_0) + t_0 \tag{9.8a}$$

If k_0 is large, and if $V_D > 0$,

$$t_R = \left(\frac{t_0}{b}\right) \log(2.3 k_0 b + 1) + t_0 + t_D \tag{9.9}$$

$$\approx \left(\frac{t_0}{b}\right) \log(2.3k_0 b) + t_0 + t_D \qquad (9.9a)$$

Here $t_D = V_D/F$ is the column *dwell-time*.

If k_0 is small, and if $V_D > 0$,

$$\boxed{t_R = \left(\frac{t_0}{b}\right) \log\{2.3k_0 b[1 - \left(\frac{t_D}{t_0 k_0}\right)] + 1\} + t_0 + t_D} \qquad (9.10)$$

Equation (9.10) is valid, regardless of the values of k_0 or V_D. Equations (9.8) to (9.10) assume that the peak does not elute before or after the gradient. For equations that cover the latter cases, see [22]. Equation (9.9) is often a reasonable approximation for gradient separations and is frequently cited in the literature (although different symbols are sometimes used; see pp. *xxv–xxvi* of [2]).

Values of the gradient retention factor k^* can also vary with values of V_D and k_0. For small values of k_0, and $V_D = 0$,

$$k^* = \frac{1}{1.15b + (1/k_0)} \qquad (9.11)$$

For small k_0 and $V_D > 0$ (or any values of k_0 and V_D),

$$\begin{aligned} k^* &= \frac{k_0}{2.3b[(k_0/2) - (V_D/V_m)] + 1} \\ &= \frac{k_0}{2.3b[(k_0/2) - (t_D/t_0)] + 1} \end{aligned} \qquad (9.12)$$

Thus a small value of k_0 leads to smaller values of k^*, compared to values from Equation (9.4) or (9.6). Likewise, for larger values of t_D (or a gradient-delay time t_{delay}), the value of k^* will be larger, compared to values from Equation (9.4) or (9.6).

9.2.4.2 Measurement of Values of S and k_w

Values of S and k_w can be obtained from isocratic values of k as a function of ϕ from Equation (9.7), or from two gradient runs where only gradient time is varied. When values of k_0 are large for gradient elution, Equation (9.9a) accurately describes linear-gradient retention in RP-LC. For this case it is possible to calculate values of $\log k_w$ and S for each compound in any sample, based on two experimental gradient runs where only gradient time is varied. Thus suppose gradient times for the two experiments of t_{G1} and t_{G2} ($t_{G1} < t_{G2}$), with a ratio $\beta = t_{G2}/t_{G1}$. Given values of t_R for a given solute in run-1 (t_{R1}) and run-2 (t_{R2}), a value of b_1 can be calculated as

$$b_1 = \frac{t_0 \log \beta}{t_{R1} - (t_{R2}/\beta) - (t_0 + t_D)(\beta - 1)/\beta} \qquad (9.13)$$

Similarly

$$\log k_0 = \left[\frac{b_1(t_{R1} - t_0 - t_D)}{t_0}\right] - \log(2.3b_1) \tag{9.13a}$$

Insertion of b_1 into Equation (9.4a) allows the calculation of a value of S, while log k_w is then calculable as $\log k_0 + S\phi_0$.

9.2.4.3 Peak Width

Peak width W in gradient elution is defined in the same way as for isocratic separation (Section 2.3) and is given by any of the following equivalent equations:

$$W = (4N^{*-0.5})Gt_0\left(1 + \frac{1}{2.3b}\right) \tag{9.14}$$

$$\equiv (4N^{*-0.5})Gt_0\left(1 + \frac{k^*}{2}\right) \tag{9.14a}$$

$$\equiv (4N^{*-0.5})Gt_0(1 + k_e) \tag{9.14b}$$

That is, W can be related to gradient steepness b, a value of k^*, or the value of k when the peak leaves the column (k_e); as noted in Section 9.1.3.2, $k_e = k^*/2$. The peak compression factor G describes the narrowing of a peak in gradient elution, due to the faster migration of the band tail (in a higher%B mobile phase) compared to the band front (in a weaker%B mobile phase) [23, 24]. G can be related to gradient steepness b [25]. First define the quantity p as

$$p = \frac{2.3k_0b}{k_0 + 1} \approx 2.3b \tag{9.15}$$

for large k_0. G is then given in terms of p as

$$G = \left\{\frac{(1 + p + [p^2/3])}{(1 + p)^2}\right\}^{0.5} \tag{9.15a}$$

Values of G vary with gradient steepness b as follows: for $0.05 < b < 2$ (corresponding to $17 > k^* > 0.4$), $1 > G > 0.6$; that is, large b or small k^* corresponds to smaller G. Thus the value of G varies from 0.6 for very steep gradients to 1.0 for very flat gradients. A more convenient equation for G can be derived from the similarity of equations for isocratic and gradient elution (Eqs. 2.24 and 9.15 below):

$$G \approx \frac{1 + k^*}{1 + 2k^*} \tag{9.15b}$$

For values of $k^* \geq 1$, Equation (9.15b) is accurate within a few percent.

The theory of peak compression in gradient elution was well developed by 1981, but subsequent experimental studies failed to confirm this phenomenon until

2006 [24]. It is now believed that this past uncertainty concerning peak compression was mainly the result of a moderate failure of Equation (9.1), combined with the use of Equation (9.14) instead of Equation (9.14*b*); the latter relationship is more accurate when plots of log k against ϕ are slightly curved (i.e., failure of Eq. 9.1).

9.2.4.4 Resolution

An equation analogous to Equation (2.24) for isocratic elution can be derived for gradient elution [26]. Starting with Equation (2.23), Equation (9.8a) can be substituted for values of $t_{R(j)}$ and $t_{R(i)}$. Values of W_i and W_j can be replaced by a single peak width W (Eq. 9.14), and the quantity G can be approximated by Equation (9.15b). to give

$$R_s = \left(\frac{2.3}{4}\right) N^{*1/2} \log\alpha \left[\frac{k^*}{1+k^*}\right] \tag{9.15c}$$

With the final approximation 2.3 log(α) ≈ (α − 1), for small values of α, we have

$$\boxed{R_s = \left(\frac{1}{4}\right)\left(\frac{k^*}{1+k^*}\right)(\alpha^* - 1)N^{*0.5}} \tag{9.16}$$

Here α^* is the value of the separation factor α when the band-pair reaches the middle of the column (at which time $k \equiv k^*$), and N^* is the value of N when the band reaches the middle of the column. Values of N^* in gradient elution are the same as N in isocratic elution, when $k = k^*$. Equation (9.16) is primarily of conceptual value; it describes how resolution depends on k^*, the separation factor or selectivity, and the column plate number. *We will find this relationship useful in our following discussion of gradient method development (Section* 9.3). Equation (2.23), which defines resolution for both isocratic and gradient elution, is more accurate than Equation (9.16) and is used in this book for all calculations of resolution—but Equation (2.23) is of little use as a guide for method development.

9.3 METHOD DEVELOPMENT

Method development for a gradient separation (Table 9.2, Fig. 9.14) is conceptually similar to the development of an isocratic procedure (Section 6.4, Fig. 6.21). The composition of the sample must first be considered (step 1 of Table 9.2 and Fig. 9.14), in order to establish appropriate starting conditions. Defining the goals of separation comes next (step 2), for example, as baseline resolution ($R_s \geq 2.0$), the shortest possible run time, and conditions that favor (or do not hinder) the detection and measurement of individual peaks of interest. Other aspects of method development that are similar for isocratic or gradient separation include:

- a possible need for sample pretreatment prior to injection (Chapter 16)
- checking that all experiments are reproducible (replicate runs)
- verifying column reproducibility (two or more columns from different lots; Section 9.3.8)

Table 9.2

Outline for the Development of a Routine Gradient Separation (Compare with Fig. 9.14)

Step	Comment
1. Review information on sample	a. Molecular weight >5,000 Da? (see Chapter 13) b. Mobile phase buffering required? c. Sample pretreatment required?
2. Define separation goals	Section 6.4
3. Carry out initial separation (run 1)	a. Conditions of Table 9.3; 10-min gradient b. Any problems? (Section 9.3.1.1, Fig. 9.17) c. Isocratic separation possible? (Fig. 9.15)
4. Optimize gradient retention k^*	Conditions of Table 9.3 should yield an acceptable value of $k^* \approx 5$
5. Optimize separation selectivity α^*	Increase gradient time by 3-fold (run 2, 30 min); increase temperature by 20°C (runs 3 and 4); see examples of Figure 9.18
5a. If best resolution from step 5 is $R_s \ll 2$, or if very short run times are required, vary conditions further in order to optimize peak spacing (for maximum R_s or minimum run time)	a. Replace acetonitrile by methanol and repeat runs 1–4 b. Replace column and repeat runs 1–4 c. Change pH and repeat runs 1–4 d. Consider use of segmented-gradients (Section 9.3.5; least promising)
6. Adjust gradient range and shape	a. Select best initial and final values of %B for minimum run time with acceptable R_s b. Add a steep gradient segment to 100%B for "dirty" samples (e.g., Fig. 9.11*b*) c. Add a steep gradient segment to speed up separation of later, widely spaced peaks (Fig. 9.11*c*) d. Add an isocratic hold to improve separation of peaks eluting at start of gradient (Fig. 9.9*d*)
7. With best separation from step 5 or 6, choose best compromise between resolution and run time	Vary column conditions (Section 9.3.6)
8. Determine necessary column equilibration between successive sample injections	Using the conditions selected above, carry out successive, identical separations while varying the equilibration time between runs; select a minimum equilibration time that provides acceptable separation (Section 9.3.7)

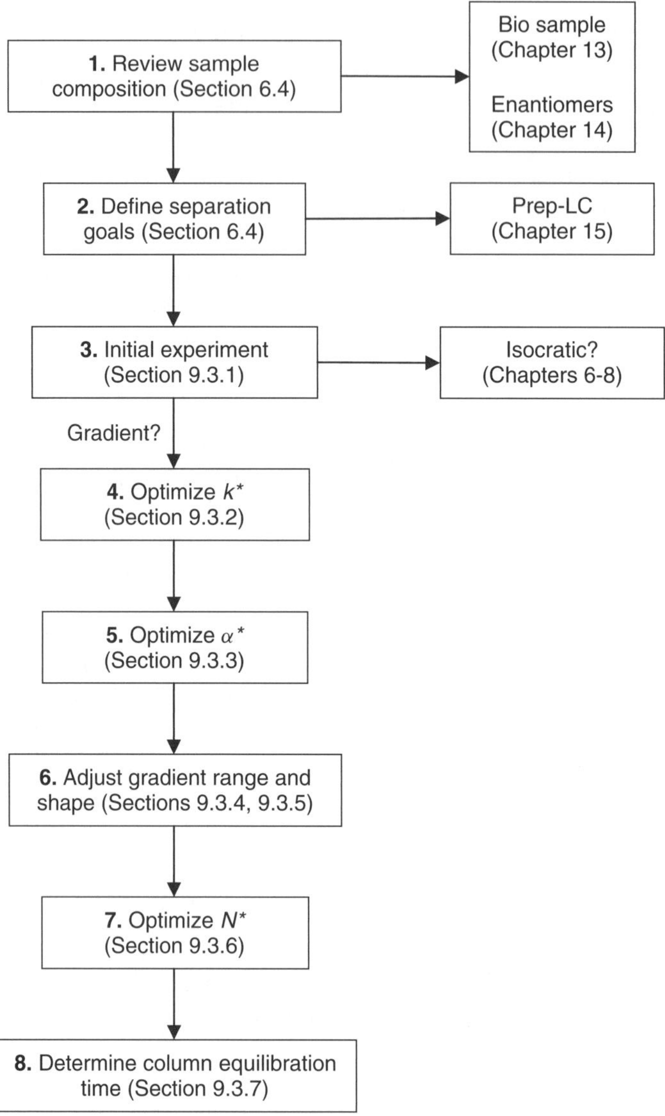

Figure 9.14 Plan for gradient method development.

- carrying out method validation and developing a system suitability test at the end of method development (Chapter 12)
- developing an orthogonal method to ensure that all peaks of interest have been included in the primary assay procedure (Section 6.3.6.2; [27])

The method-development requirements above are the same as discussed in Section 6.4 for isocratic separation, so we will not consider them further in this chapter. The main differences between gradient and isocratic method development are in the experiments used to arrive at a final separation (steps 3–8 of Table 9.2 and Fig. 9.14).

9.3.1 Initial Gradient Separation (Step 3 of Table 9.2)

Ideally a full-range gradient (0–100% B) is preferred for the initial experiment; if the sample contain acids or bases, the A-solvent should contain a buffer (Section 7.2). However, initiating the gradient at 0% B can create problems for some columns, due to nonwetting of the stationary phase by organic-free water (Section 5.3.2.3; see also [28, 29]). For this reason it is better to initiate the gradient at 5% B or higher, unless it is known that the column can tolerate a totally aqueous mobile phase (0% B), or column pressure can be continuously maintained after an initial wetting of the column with 100% B. Problems with stationary-phase wetting are more likely for heavily bonded C_{18} columns (larger H; Section 5.4.1) than for columns that are lightly bonded, contain embedded polar groups, or are end-capped with polar groups (stationary-phase de-wetting can be avoided for most columns by following the protocol of Section 5.3.2.3). If buffer solubility is limited for 100% ACN (at the end of a 5–100% ACN gradient), the final-%B of the gradient may need to be lowered or the buffer concentration reduced. However, when the buffer is added only to the A-solvent (the usual case), buffer precipitation may not be a problem. See Section 7.2.1.2 for further details on buffer solubility.

9.3.1.1 Choosing between Isocratic and Gradient Elution

The first gradient run is important as a means of (1) assessing the likely difficulty of method development and (2) planning further experiments. Table 9.3 recommends specific starting conditions: 5–100% ACN/buffer (or water) in 10 min, a 100 × 4.6-mm, 3-μm C_8 or C_{18} column, 30 or 35°C, and 2.0 mL/min. These conditions should result in an average value of $k^* \approx 5$ for the separation (Eq. 9.5) that is large enough to provide acceptable average resolution, while restricting the pressure drop to ≈ 2500 psi. Other column configurations and flow rates are also acceptable (e.g., 150 × 4.6-mm, 5-μm column, 1–2 mL/min), as long as acceptable values of k^* and pressure are maintained. Equation (2.13*a*) can be used to estimate the column pressure; in gradient elution, the maximum pressure during the run is determined by the maximum mobile-phase viscosity—see Table I.5 of Appendix I. The gradient time t_G can be varied to maintain a value of $k^* \approx 5$,

$$t_G = \frac{1.15k^* V_m \Delta\phi S}{F} \tag{9.17}$$

or for $k^* = 5$ and $S = 4$,

$$t_G = \frac{23 V_m \Delta\phi}{F} \tag{9.17a}$$

For example, a 100 × 4.6-mm, 3-μm column ($V_m \approx 1.0$ mL) with acetonitrile as the B-solvent, a temperature of 30°C, and a flow rate of 2.0 mL/min should have a pressure of ≈2500 psi; a value of $k^* \approx 5$ would require a gradient time of (23 × 1.5 × 0.95/2), or 11 minutes. Smaller diameter columns with flow rates reduced in proportion to the square of column-diameter are another option.

Table 9.3

Preferred Conditions for the Initial Experiment in Gradient Method Development (Small-molecule sample, 100–1000 Da, Assumed)

Column	Type	C_8 or C_{18} (type-B)
	Dimensions	100 × 4.6-mm[a]
	Particle size	3 μm[a]
	Pore diameter	8–12 nm
Mobile phase	Sample contains no acids or bases	Acetonitrile/water
	Sample contains acids and/or bases	Acetonitrile/aqueous buffer (pH 2.5–3.0)[b]
Flow rate		2.0 mL/min
Temperature		30 or 35°C
Gradient		5–100% B in 10 min
Sample		
	Volume	≤50 μL
	Weight	≤10 μg
k^*		≈5

[a]Other column dimensions and particle sizes can be used, as discussed in Section 9.3.1.
[b]10–25 mM buffer in A-solvent only; see Section 7.2.1 for further details on buffer composition, including concentration.

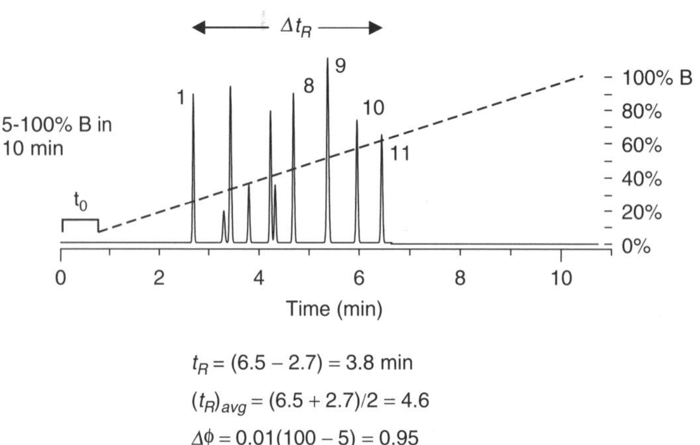

$t_R = (6.5 - 2.7) = 3.8$ min

$(t_R)_{avg} = (6.5 + 2.7)/2 = 4.6$

$\Delta\phi = 0.01(100 - 5) = 0.95$

Figure 9.15 Use of a standard gradient run to determine whether isocratic or gradient elution is best for the sample. In this example the "irregular" sample of Figure 9.4 was separated with the recommended initial conditions of Table 9.3: 5–100% acetonitrile in 10 min, 100 × 4.6-mm (3-μm) C_{18} column, 2.0 mL/min, 30°C. Gradient indicated by (- - -).

A representative, initial gradient run is shown in Figure 9.15, based on the "irregular" sample of Figure 9.5 and the gradient conditions of Table 9.3. This first experiment can be used to answer two questions [2,30–33]: (1) will gradient elution be required for the sample, and (2) if isocratic separation is possible, what isocratic mobile phase should be tried first in order to achieve $1 \leq k \leq 10$ for all peaks? Before using this initial gradient run to draw any conclusions (as in following paragraphs), it is important to establish that the column has been adequately equilibrated—see the discussion of Section 9.3.8.1.

In the initial separation of Figure 9.15, retention times for the first and last peaks (1 and 11) are equal to 2.7 and 6.5 minutes, respectively. The latter retention times determine whether isocratic separation is feasible. First, calculate the *difference* in retention times (Δt_R) for peaks 1 and 11 (6.5–2.7), or $\Delta t_R = 3.8$ min. Also, calculate the *average* retention time for the first and last peaks, $(t_R)_{avg} = (6.5 + 2.7)/2 = 4.6$ minutes. Samples that have small values of Δt_R can be separated isocratically with $1 \leq k \leq 10$, while samples with larger values of Δt_R may require gradient elution. An approximate rule for deciding whether to use isocratic or gradient elution is as follows: if $\Delta t_R/t_G \leq 0.25$, use isocratic elution; if $\Delta t_R/t_G \geq 0.40$, use gradient elution. For intermediate values of $\Delta t_R/t_G$, either isocratic or gradient elution may prove best. For the example of Figure 9.15, $\Delta t_R/t_G = 3.8/10 = 0.38$, so isocratic elution is (barely) an option while gradient elution seems preferable.

When isocratic separation is feasible, the recommended %B for the isocratic mobile phase can be estimated from the value of $(t_R)_{avg}$ [2]. For the conditions of Table 9.3

$$\text{isocratic } \%B \approx 9.5[(t_R)_{avg} - t_D] - 2 \tag{9.18}$$

Here t_D is the dwell-time of the gradient system, equal to V_D/F. As an example of the application of Equation (9.18), consider the separation of Figure 9.15 with peaks 9 to 11 omitted (so as to make this a better candidate for isocratic separation). The dwell-volume V_D and dwell-time t_D for this separation are approximately zero, while values of t_R for the first and last peaks are 2.7 and 4.7 min, respectively. Values of Δt_R and $(t_R)_{avg}$ are then 2.0 and 3.7 min, respectively. The resulting value of $\Delta t_R/t_G = 2.0/10 = 0.20$, so isocratic elution is preferred (as discussed above).

For the preceding example with $(t_R)_{avg} = 3.7$ and $t_D \approx 0$, the best mobile phase for an isocratic separation of this sample is 33% B (Eq. 9.18). This separation is shown in Figure 9.16*a* (same column, temperature, and flow rate), where $0.6 \leq k \leq 7$. The retention range of the latter separation is just outside the target range of $1 \leq k \leq 10$ (Eq. 9.18 is only an estimate!), but retention can be improved by a small decrease in %B. Thus values of k for the separation of Figure 9.16*a* should be increased by a factor of about 1.5, in order to achieve $k > 1$ for the first peak. Section 2.4.1 suggests that a decrease of 10% B in the mobile phase will increase log k by 0.4-units ("rule of 2.5"), whereas we wish an increase in log k by a factor of log 1.5 = 0.18 units. This suggests a decrease in %B of (0.18/0.40) × 10% = 4.5% B, to a final value of 28.5% B. The latter separation is shown in Figure 9.16*b*, where $1 \geq k \geq 10$ as desired. There are two overlapping peak-pairs (2 + 3 and 5 + 6), which likely can be separated by varying other conditions (see Section 7.3.2 for ways to improve isocratic selectivity for this ionic sample).

The recommendations above assume that if a sample can be separated isocratically with $1 < k < 10$, then isocratic elution is the preferred option. This assumption has been challenged [34], on the basis that gradient elution is usually faster and equally satisfactory in other respects, when optimized conditions are used for both isocratic and gradient runs, and when column equilibration between successive gradient runs has been reduced as much as possible (Section 9.3.7). On

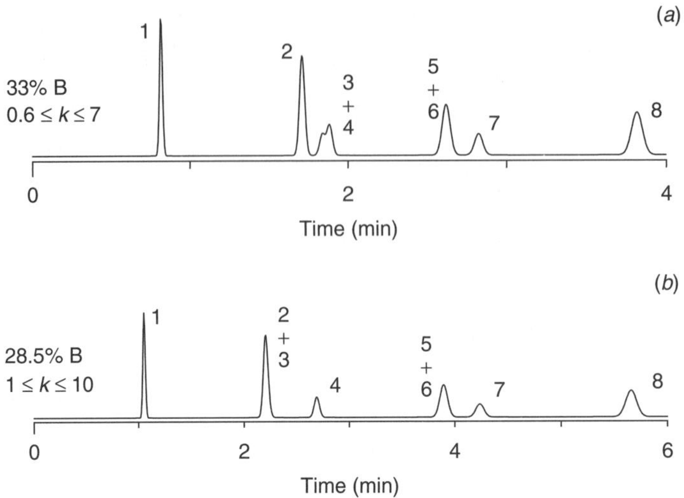

Figure 9.16 Isocratic separation of the sample of Figure 9.15 (peaks 1–8 only). (*a*) Separation with 33% B as predicted from the initial gradient separation in Figure 9.15. Conditions as in Figure 9.15, except isocratic; (*b*) separation for 28.5% B, as described in the text.

the basis of results for a single sample [34], gradient elution was recommended whenever $\Delta t_R/t_G \geq 0.10$. At present, however, many laboratories have a strong preference for isocratic elution—regardless of somewhat longer run times—because gradient elution is still considered more susceptible to problems than isocratic elution, and less easy to transfer between laboratories. In time this bias against gradient elution may diminish, and gradient equipment may be improved so as to make very short equilibration times convenient. The proposal of [34] to use gradient elution whenever $\Delta t_R/t_G \geq 0.10$ may then prove more popular.

9.3.1.2 Possible Problems

The initial gradient run may also be used to highlight some potential problems with the separation:

- tailing peaks
- early elution
- late elution
- complex samples
- artifact peaks

Tailing peaks (Section 2.4.2) may be encountered in the initial gradient separation. In such cases it is important to correct the problem before proceeding further (Section 17.4.5.3). If the correction of peak tailing (by a change of separation conditions) is delayed until a later time, the resulting changes in selectivity with possible loss in resolution may require additional method-development experiments that could otherwise have been avoided.

Figure 9.17 illustrates three additional problems that may be apparent from an initial gradient run. *Early elution* of peaks in RPC, as in Figure 9.17*a*, is not

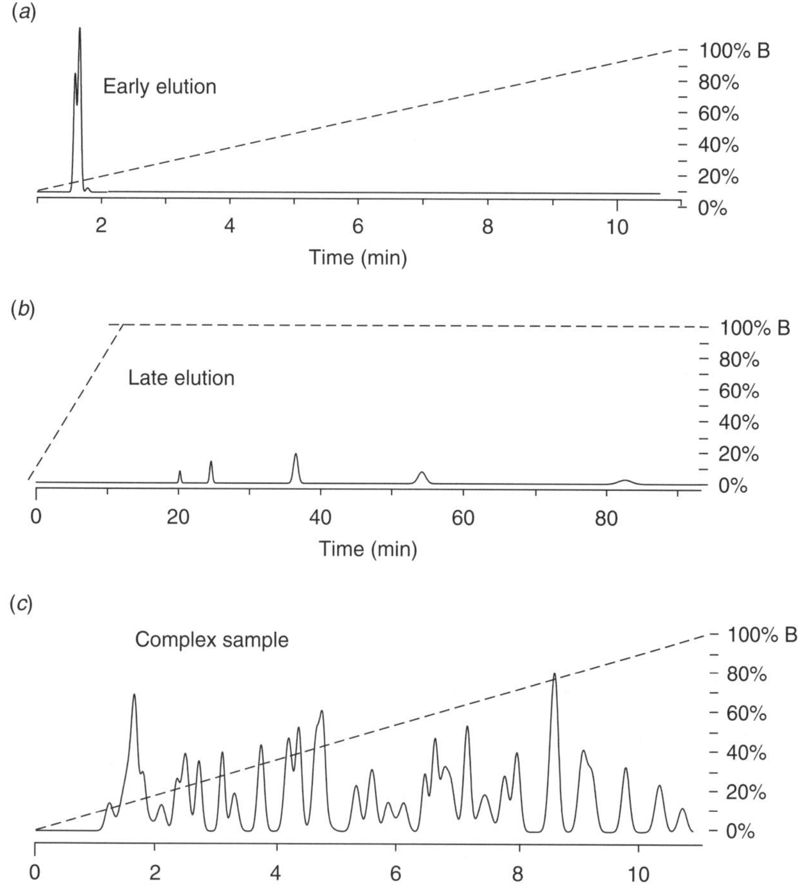

Figure 9.17 Potential problems in gradient elution. (*a*) Non-retentive sample; (*b*) excessively retentive sample; (*c*) sample contains too many components. Gradient indicated by (- - -).

uncommon for small, polar molecules, especially ionized acids or bases. Some improvement in separations such as that of Figure 9.17*a* can be obtained by a reduction in initial %B for the gradient (if feasible), or by the use of an initial isocratic hold as in Figure 9.9*d*. For other, more effective, means of dealing with early elution, see Section 6.6.1, or try normal-phase chromatography (Chapter 8)—especially HILIC (Section 8.6), which is especially well suited for use with gradient elution.

Late elution as in Figure 9.17*b* (or an absence of peaks during the gradient) suggests that the sample may be too nonpolar for separation with the usual RPC conditions. In such cases an acetonitrile/buffer gradient can be replaced by a gradient from acetonitrile to a less-polar solvent such as tetrahydrofuran or (better) methyl-*t*-butyl ether, either of which is a stronger RPC solvent than acetonitrile (buffer solubility should be checked for either of the latter two gradients, although a buffer is often not required for very nonpolar samples). Alternatively, a less hydrophobic column (lower value of *H*; Section 5.4.1) or normal-phase chromatography (Chapter 8) can be tried.

Complex samples with >15 components can result in crowded chromatograms, as in Figure 9.17*c*. For such samples it is unlikely that a single reversed-phase separation will be able to separate all peaks to baseline. If every sample component is of interest, it may be necessary to develop a more powerful separation scheme. Two-dimensional (2D) chromatography (Sections 9.3.10, 13.4.5) is the most commonly used option for dealing with complex samples; fractions from an initial run are further resolved in a second, "orthogonal" separation. If only a few sample components are of interest, however, a better choice is sample preparation (Chapter 16), followed by a conventional isocratic or gradient separation.

Another problem that is sometimes encountered in gradient elution is the appearance of *artifact* peaks that do not correspond to sample components. Artifact peaks usually arise from impurities in either the A- or B-solvents used to form the gradient, but occasionally dissolved air in the sample can result in an "air peak." This problem can be anticipated by carrying out a "blank" gradient (without injection of the sample) at the very beginning of each day. A blank gradient is also useful for recognizing (and correcting) baseline drift during the gradient (Section 17.4.5.1). See the related discussion of Section 7.4.3.1 for further details.

9.3.2 Optimize k^* (Step 4 of Table 9.2)

Further improvements in separation can be guided by Equation (9.16), that is, the optimization of k^*, α^*, and N^*. This approach for gradient elution is exactly analogous to the similar use of Equation (2.24) for isocratic method development, as described here and in following Sections 9.3.3 to 9.3.6.

The initial gradient conditions recommended in Table 9.3 will result in an average value of $k^* \approx 5$ for most small-molecule samples, those with molecular weights <1000 Da (for higher-molecular-weight samples, see Chapter 13). Thus, unlike isocratic method development, the first gradient-elution experiment can be carried out in a way that guarantees $1 \leq k^* \leq 10$. The initial separation of the irregular sample of Figure 9.5 with these conditions is shown in Figure 9.15 and repeated in Figure 9.18*a*. The latter separation is reasonably promising, with only one overlapping peak-pair (5–6, indicated by the arrow). The next step is to vary separation conditions so as to improve peak spacing (selectivity) and resolution.

9.3.3 Optimize Gradient Selectivity α^* (Step 5 of Table 9.2)

Changes in values of α^* can be achieved by varying any of the first seven isocratic conditions of Table 2.2: solvent strength (a change in t_G is equivalent to a change in %B in isocratic elution), B-solvent (e.g., methanol replaces acetonitrile), temperature, column type, mobile-phase pH, buffer concentration, or ion-pair-reagent concentration. Each of these seven variables has a comparable effect on relative retention and selectivity for both gradient and isocratic elution. A growing body of evidence [35–42] suggests that gradient time and temperature should be changed first, as a preferred means for adjusting values of α^* during initial method-development experiments (while maintaining $0.5 \leq k^* \leq 20$). Therefore we recommend an increase in gradient time by a factor of 2 to 3 for the second method-development experiment. Starting with the separation of Figure 9.18*a*, gradient time was increased from 10 to 30 minutes, other conditions held constant; the resulting separation is shown in Figure 9.18*b*, with $k^* \approx 15$. While there are significant changes in relative retention,

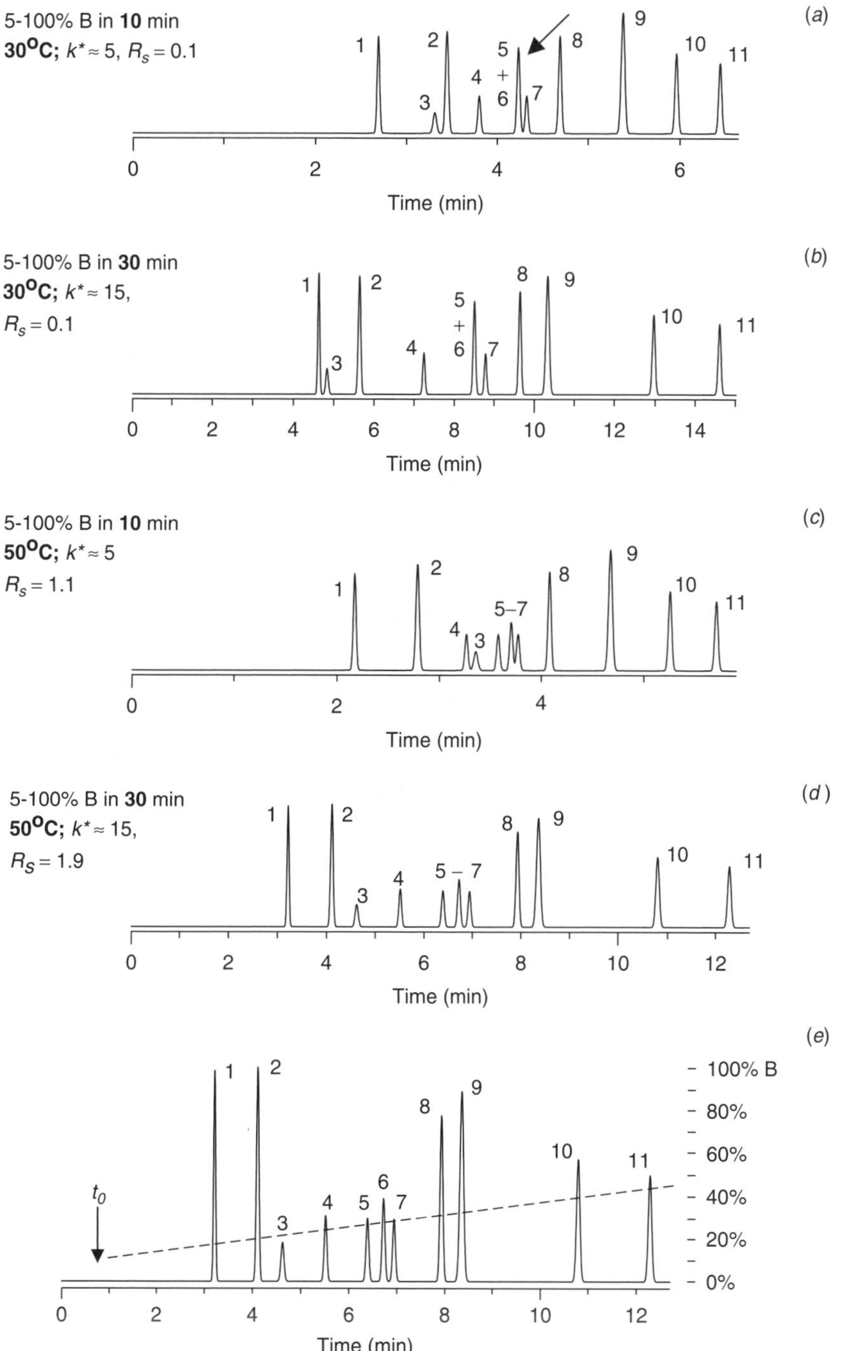

Figure 9.18 Gradient separations of the "irregular" sample of Figure 9.15 as a function of gradient time and temperature ($a - d$). Conditions: 100 × 4.6-mm (3-μm) C_{18} column, 5–100% acetonitrile–pH-2.6 phosphate buffer; 2.0 mL/min; gradient times and temperatures indicated in figure; (e) shows gradient details for (d).

there is no change in the separation of critical peak-pair 5–6. If a 3-fold change in gradient time does not significantly change the resolution of an overlapping peak pair, further changes in gradient time are unlikely to provide much additional benefit—as long as other conditions are held constant.

The next step is a change in temperature. The third and fourth method-development runs are illustrated in Figure 9.18*c*,*d*, where the runs of Figures 9.18*a* and *b* are each repeated with a change in temperature from 30° to 50°C. Because peak-pair 5–6 was unresolved in the first two runs, the primary question is whether peaks 5 and 6 can be separated at the higher temperature. A large increase in resolution for peaks 5 and 6 is seen in Figure 9.18*c* ($R_s = 2.1$), but peaks 6 and 7 are now critical ($R_s = 1.1$). An increase in gradient time (Fig. 9.18*d*) results in better resolution of peaks 6 and 7 ($R_s = 1.9$)—and of the entire sample. These results suggest that a further increase in gradient time might provide better overall resolution, but no significant increase in resolution resulted when t_G was increased for this sample—due to the increasing overlap of peaks 2 and 3.

The resolution of Figure 9.18*d* might be improved by a true optimization of gradient time and temperature (Section 10.2.2), but the conditions of Figure 9.18*d* will be regarded as adequate for the moment.

9.3.4 Optimizing Gradient Range (Step 6 of Table 9.2)

The next step in gradient method development is to consider (1) whether the gradient range $\Delta\phi$ can be shortened (with a decrease in run time), and (2) whether the use of a segmented gradient (Section 9.3.5) might lead to either a faster separation or better resolution. The approximately optimized separation of Figure 9.18*d* is repeated in Figure 9.18*e*, overlaid by the gradient as it leaves the column (delayed by a time t_0). The first peak (1) leaves the column at 3.2 min, at which time its accompanying mobile phase is 14% B. Similarly, the last peak (11) leaves at 12.3 minutes in a mobile phase of 42% B.

It is recommended to terminate the gradient just after the elution of the last peak. In the example of Figure 9.18*e* the retention time of the last peak is 12.3 min. If the gradient time is shortened in this way, the final %B in the gradient must be reduced proportionately in order to maintain k^* constant (so as to preserve the optimum peak spacing of Fig. 9.18*e*). That is, $t_G/\Delta\phi$ in Equation (9.5) must be held constant; for the present example, $t_G/\Delta\phi = 30/0.95 = 31.6$. The value of ϕ at the time a peak elutes from the column (ϕ_e) can also be calculated by

$$\phi_e = \phi_0 + \frac{\Delta\phi(t_R - t_o - t_D)}{t_G} \tag{9.19}$$

where ϕ_o is the value of ϕ at the start of the gradient, and t_R is the retention time of the peak. For the last peak in Figure 9.18*e*, $\phi_e = 0.05 + 0.95(12.3–0.5–0.0)/30 = 0.42$ (note that $t_0 = 0.5$ and $t_D \approx 0.0$ in this example). That is, the new (shortened) gradient should end at 42% B. The new value of $\Delta\phi$ is then 42–5% = 37% or 0.37. As $t_G/\Delta\phi = 31.6$ should remain constant (to avoid changes in relative retention, the new value of t_G is $31.6 \times 0.37 = 11.7$ minutes (i.e., a final gradient of 5–42% B in 11.7 min, with other conditions kept the same as in Fig. 9.18*e*). This new gradient will result in the same chromatogram but end at 12.2 minutes (equal $t_G + t_0$).

It is advisable to extend the gradient somewhat beyond the time that the last peak leaves the column because of gradient rounding (Section 3.10.1.2). We might therefore increase the gradient time to 13 minutes, which then requires an increase in final-%B to maintain k^* constant. As $t_G/\Delta\phi = 31.6$ for the "optimized" separation of Figure 9.18*e*, the new value of $\Delta\phi = 13/31.6 = 0.41$, and the final%-B equals $41 + 5\% = 46\%$; that is, a gradient of 5–46% B in 13 minutes.

In principle, the gradient run time could be shortened further by increasing initial-%B (while decreasing t_G so as to hold k^* constant). In this example, however, resolution became smaller for any increase in initial-%B (due to changes in relative retention for this irregular sample, similar to the example of Fig. 9.13*e*). Consequently the value of initial-%B was left unchanged at 5% B. For other samples, it may be possible to increase initial-%B in order to reduce run time, with no loss in resolution.

9.3.5 Segmented (Nonlinear) Gradients (Step-6 of Table 9.2 continued)

The preceding discussion of gradient elution assumes that we are dealing with linear gradients. Various reasons for the possible use of a segmented gradient in place of a linear gradient were summarized in Section 9.2.2.5: (1) to clean the column between sample injections, (2) to shorten run time, or (3) to improve separation by adjusting selectivity for different parts of the chromatogram. Because of the excess resolution between peaks that follow peak 9 in the separation of Figure 9.18*e*, run time could be shortened by an increase in gradient steepness after peak 9 leaves the column. See the similar example of Figure 9.11*c*. Keep in mind, however, that gradient rounding may vary between different equipment, which can make segmented gradients less reproducible—as well as require an increase in final %B.

Cleaning the column is a common reason for the use of segmented gradients, while shortening run time and improving separation by the use of segmented gradients are less often feasible or desirable. For further details, see Section 9.2.2.5.

9.3.6 Optimizing the Column Plate Number N^* (Step 7 of Table 9.2)

The column plate number $N \equiv N^*$ is affected by column dimensions, particle size, and flow rate (called *column conditions*, Section 2.5.3), as well as by sample molecular weight (Section 2.4.1.1). Particle size and column diameter are usually selected prior to the start of method development (e.g., as recommended in Table 9.3). An increase in column length usually results in an increase in N^*, resolution, and run time (as in Figs. 9.6*e* vs. Fig. 9.6*d*). Conversely, run time can be shorted by a decrease in column length and/or an increase in flow rate (as in Fig. 9.6*f* vs. Fig. 9.6*d*). After varying conditions for improved selectivity α^* (step 5 of Table 9.2), and adjusting gradient range and shape (step 6 of Table 9.2), the resulting separation may exhibit a resolution that is either (1) too low ($R_s < 2$) or (2) greater than needed ($R_s \gg 2$). In either case, a change in column conditions can be used to improve separation; any resulting changes in the pressure drop across the column should be kept in mind (Eq. 2.13).

In isocratic elution, changes in column length or flow rate do not affect relative retention or selectivity because values of k and α are not affected when column conditions are varied. When changing column length L or flow rate F in gradient elution, however, a change in either of these two conditions alone will result in a change in k^* (Eq. 9.5c). For 'irregular' samples this can result in changes in selectivity. As selectivity for a gradient method should have been optimized (step 5 of Table 9.2) prior to a change in column conditions, *it is important to maintain the same values of k^* (and α^*) when changing column conditions and* N^*. This can be achieved by maintaining $(t_G F/L)$ constant (Eq. 9.5c); for example, if column length is doubled, gradient time must also be doubled so that t_G/L stays constant; if flow rate is doubled, gradient time must be decreased by half so as to keep $t_G F$ constant. For examples of this approach to optimizing N^*, see Figure 9.6*d*–*f*. As long as values of k^* are maintained constant in this way, a change in column length or flow rate has the same effects on run time and resolution in either isocratic or gradient elution. A minor exception to this rule can occur for the resolution of early peaks in the chromatogram for larger values of V_D—regardless of whether k^* is held constant (Section 9.2.2.4 [11]) [11].

When column conditions are changed for a segmented gradient, the time t_{seg} for each segment must be adjusted so as to maintain $t_{seg}F/L$ constant. For example, consider the separation of Figure 9.11*c*, where the gradient is 0/23/42% B in 0/32/38 min. If column length were doubled, the length of each segment t_{seg} would also require doubling, so that the new gradient would be 0/23/42% B in 0/64/76 min.

9.3.7 Determine Necessary Column-Equilibration Time (Step 8 of Table 9.2)

After method development is complete, in most cases the resulting HPLC procedure will be used for routine sample analysis. During this application of the method the column must be washed between successive gradient runs with a sufficient volume of mobile phase whose composition matches that of the mobile phase at the start of the gradient (e.g., 5% B in the examples of Fig. 9.18). This column-equilibration step is intended to allow for (1) the holdup volume V_D (or dwell-time $t_D = V_D/F$) of the gradient equipment, (2) gradient rounding (Section 3.10.1.2), and (3) slow equilibration of the stationary phase (removal of excess B-solvent) when switching from high %B at the end of one gradient to low %B at the beginning of the next gradient.

Figure 9.19 illustrates the possible consequence and correction of the combined effects of dwell-volume, gradient rounding, and slow column equilibration, when sequential sample injections are made during routine analysis. In Figure 9.19*a* the solid lines describe a series of programmed gradients in terms of time, while the arrows mark the times when samples 1, 2, and so forth, are injected at the beginning of each gradient. These 5–100% B gradients in 10 minutes are followed by a between-run equilibration with 5% B for one minute (the equilibration time $t_{eq} = 1$ min). The complete programmed gradient is therefore 5/100/5/5%B in 0/10/10/11 min. If the system dwell-volume and gradient rounding are negligible, and if column equilibration is fast, the actual gradient should be the same as the programmed gradient—and injection of each sample would then occur one minute after completion of the previous gradient. The same separation of each sample would then result.

Figure 9.19*b* expands on the example of Figure 9.19*a* by introducing some additional features of an actual gradient (which are common in practice): a significant

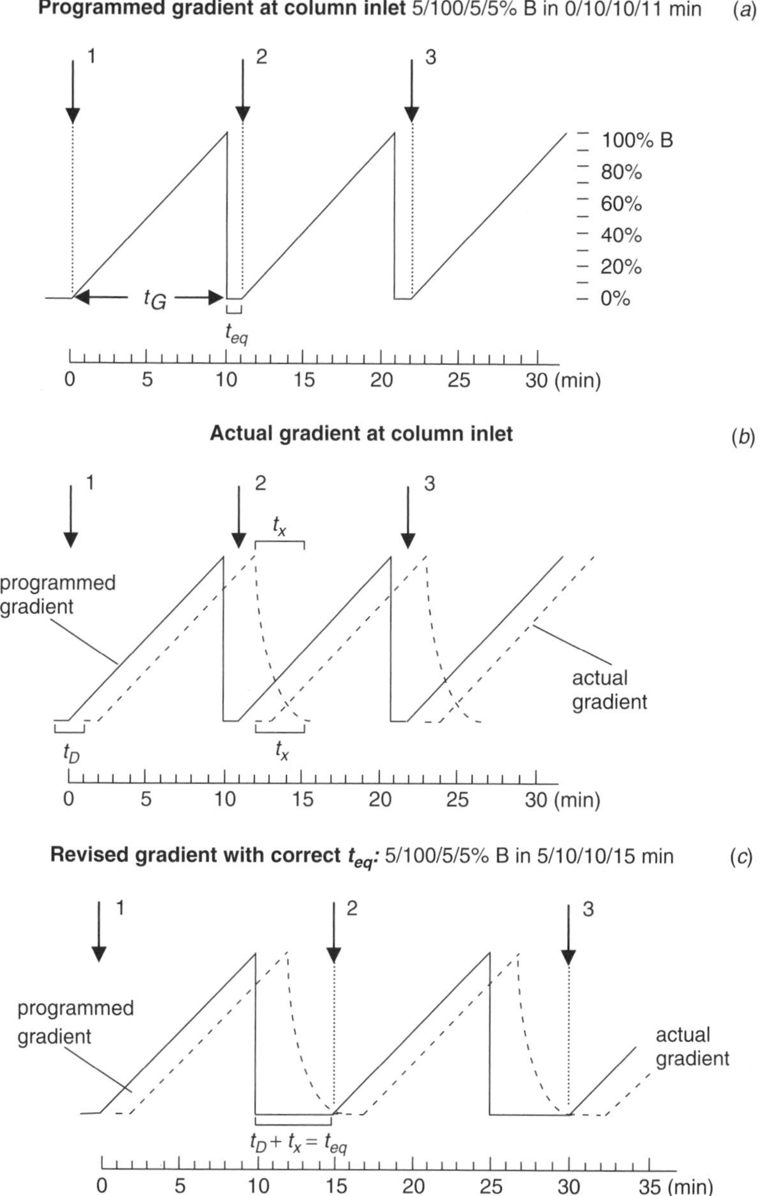

Figure 9.19 Illustration of different contributions to column nonequilibration in gradient elution. (*a*) Ideal gradient; (*b*) more realistic gradient; (*c*) addition of an adequate between-run equilibration time for the gradient of (*b*). (—) programmed gradient; (- - -) actual gradient at the column inlet; 1, 2, etc., refer to injections of sample 1, 2, etc.

equipment dwell volume with gradient rounding and/or slow column equilibration. The dashed curves (- - -) in Figure 9.19*b* describe the *actual* (individual) gradients at the column inlet; there is a significant dwell-volume, resulting in a dwell-time $t_D = 2$ min. Consequently the arrival of the gradient at the column inlet is delayed by 2 minutes. Sample injection at the time the programmed gradient begins (the usual case) is still acceptable for the first sample, but later samples are now injected one minute *before* the previous gradient has been completed. Because samples 2, 3, and so forth, are injected into a mobile phase with much higher %B (just prior to the completion of the gradient), the result would be very small values of k^* for early bands—and an unacceptable loss in resolution for these peaks. Due to gradient rounding and/or slow column equilibration, there is a slow decrease in %B from 100% B to 5% B at the end of each gradient. A time $t_x = 3$ min is required for the return of the gradient to the initial-%B value. Note that adjacent gradients (%B values) add in this example (not shown in Fig. 9.19*b*).

The detrimental effects of dwell-volume plus the slow return of the gradient to baseline can be eliminated by the use of an equilibration time t_{eq} that is made equal to $t_D + t_x$, as illustrated in Figure 9.19*c*. With this change in the gradient (5/100/5/5%B in 0/10/10/15 min), injection of each sample occurs at the start of its programmed gradient (with return of the preceding gradient to the initial-%B value); now the same (acceptable) separation is achieved for all samples. The required value of t_{eq} can be determined by trial and error—where successive sample injections are made for a given value of t_{eq}, then repeated for different values of t_{eq}. The preferred value of t_{eq} is the lowest value that gives an acceptable separation for successive samples. The trial-and-error approach also has the benefit of including any effect of autosampler delay in the equilibration process.

For method-development experiments, 10 column-volumes or more of the starting mobile phase($\phi = \phi_o$) should be passed through the column before starting the next gradient run (corresponding to $10V_m/F$, or a time equal to $10t_0$). Otherwise, any change in the time between successive experiments (often the case in method development) may result in variable column equilibration and resulting changes in retention and separation. For a *routine* gradient assay as in Figure 9.19, however, the time devoted to column equilibration can be reduced to t_{eq}, with a corresponding shortening of run time (compared to $10t_0$) and an increase in the number of samples that can be analyzed each day. Furthermore *partial* equilibration of the column (i.e., incomplete return of the mobile phase to initial-%B) may be acceptable for routine analysis, with a reduction of the equilibration time to a value $< t_{eq}$. Thus, if the resolution of early peaks is not compromised, and if the equilibration time is the same for each gradient run, each sample is treated the same and each separation will be the same—despite incomplete column equilibration.

For HPLC systems that do not permit delayed injection, the preferred partial equilibration time t'_{eq} for routine analysis will be given by $t_D \leq t'_{eq} \leq (t_D + t_x)$, where t'_{eq} must be determined by trial and error (i.e., minimum allowable value of t'_{eq}). Values of t_x are usually comparable to values of t_D, which provides an initial estimate of $t_{eq} \approx 2t_D$ (followed by trial-and-error changes in equilibration time in order to determine t'_{eq}). For systems that allow injection of the sample at any time following the start of the gradient, sample injection can be programmed to occur at a time t_D after the start of each gradient. The corresponding equilibration time t'_{eq} is then $\leq t_x$. However, methods based on sample injection after the gradient begins can only be carried out with gradient systems that allow delayed injection.

Whatever column equilibration time is allowed for a routine assay procedure, it should be confirmed that repetitive sample injections yield the same (acceptable) chromatograms and reproducible data, except possibly for the first run (which can be discarded). The required column equilibration time will be less for systems with a smaller dwell-volume and reduced gradient rounding, and can be reduced further for the case of systems that permit delayed sample injection. Changes in the plumbing of the system can be used to minimize the effects of gradient rounding [43] and further reduce t_{eq}, although most users will not make use of this option (but future equipment may eliminate the need for changes in system plumbing). For additional details on column equilibration in gradient elution, see [2, 44, 45].

9.3.8 Method Reproducibility

Some causes of irreproducible results or poor method precision are the same for both isocratic and gradient separations (Section Section 11.2). Other sources of irreproducibility are either unique to gradient elution, or more likely for this technique:

1. poor control of experimental conditions from run to run
2. malfunctioning or poorly designed equipment
3. insufficient column equilibration between gradient runs
4. differences in equipment dwell volume

The contributions above to separation variability can impact both method development (following Section 9.3.8.1), and the subsequent routine use of a gradient method (Section 9.3.8.2). Additionally the accuracy and precision of gradient assays (and the interpretation of method development experiments) can be compromised by drifting baselines during a gradient run, as well as artifact peaks that are independent of the sample (Section 17.4.5.2). To rule out such problems, *we strongly recommend that every series of gradient runs be preceded by a blank gradient*: a gradient run without sample injection, or (better) with injection of only the sample solvent (Section 3.10.1.2).

9.3.8.1 Method Development

Consider first the need for repeatable data during method development, where it is advisable to replicate each experiment so as to verify that the data obtained are reproducible from run to run; this is especially important for gradient elution experiments. Retention times in duplicate, back-to-back runs should not vary by more than some set amount; for example, ±0.02 min or ±0.1%, whichever is larger.

Poorly controlled experimental conditions and malfunctioning equipment (items 1 and 2) fall largely under the heading of good laboratory technique. For purposes of the present discussion, we will assume that all experimental conditions are controlled within limits necessary for repeatable separation. We will also assume that the equipment is operating properly, and that column performance meets the manufacturer's specifications (Section 3.10.1.2). Apart from operator and equipment issues, however, a major objective of method development should be a final method that can tolerate small, largely unavoidable changes in gradient conditions, temperature, and mobile phase composition (pH, buffer concentration, etc.), from day to day and from system to system. If a method appears not to be robust, efforts should be made to reduce the dependence of the method on experimental conditions, by examining both method robustness and resolution as a function of conditions (Section 12.2.6).

Insufficient column equilibration (item 3) is a major source of variable retention in gradient elution, so a column-equilibration step between each run or experiment is necessary (Section 9.3.7). Retention-time repeatability should be checked initially for two replicate, successive runs that use the selected minimum equilibration time (e.g., $10t_0$) between the two runs, but with a 2-fold longer equilibration time prior to the first run. An equilibration time $>10t_0$ min may be required for some samples and/or separation conditions, and very slow changes in retention may occur over a longer time period [43] (but have little effect on method development). To ensure reproducibility of method-development runs, it is prudent to allow more than the minimum required equilibration time between runs; this can be trimmed to reduce the run time when the method is finalized for routine use.

Differences in equipment dwell volume (item 4) can significantly affect experimental results (Section 9.2.2.4). For this reason *it is strongly recommended to carry out all method-development experiments for a given sample on the same (or equivalent) equipment*, in order to avoid changes in dwell-volume among different experiments.

9.3.8.2 Routine Analysis

During method development it is necessary to anticipate possible changes in the separation that might inadvertently occur when the method is transferred to another laboratory for routine analysis. *Variation in experimental conditions and malfunctioning equipment* (items 1 and 2 above) can be recognized by system suitability tests (Section 12.3.2.9). *Column equilibration* (item 3) should be handled differently in routine analysis than in method development. During method development, a between-run equilibration time of at least $10t_0$ min is usually acceptable. For routine analysis, where the equilibration time between runs is generally fixed, it is desirable to shorten the equilibration time as much as possible, in order to minimize the time between sample injections (Section 9.3.7)—as well as the overall run time.

Differences in equipment dwell volume (item 4) are a common reason for the failure of a gradient method during method transfer or routine application on a different HPLC system. The dwell-volume V_D can vary significantly between different gradient systems; older equipment usually has larger values of V_D. A different gradient system will often be used to carry out routine assays, compared to the system used to develop the method. If the second system has a different dwell-volume (V_D) compared to the original system, unacceptable changes in separation can result (Section 9.2.2.4), especially for irregular samples. When the value of V_D for the second system is smaller, this difference in dwell-volumes can be compensated by adding a gradient-delay time t_{delay} for the separation carried out on the second system, as this is equivalent to an increase in dwell-volume. The length of this gradient delay in minutes should be made equal to the difference in dwell-times t_D for the two systems ($t_D = V_D/F$).

A second gradient system with a larger dwell-volume (the more likely case) presents a more difficult problem. For this reason *an effort should be made in method development to anticipate the maximum dwell-volume likely to be encountered in other laboratories that will use a given procedure*. The original method can be developed with a total gradient delay (equal $t_D + t_{delay}$) that effectively increases dwell-time to the maximum value of t_D expected in other labs to which the method will be transferred; a value of t_{delay} can then be selected in each transfer lab, so as to compensate for differences in dwell-volume relative to the original equipment (so that $t_D + t_{delay}$ remains constant for each system). Alternatively, a delayed injection of the sample (if the system allows this option) can be used to effectively reduce the dwell volume of the second system. See the more detailed discussion of [2]—as well as [46], where other options for dealing with varying dwell-volume are discussed.

9.3.9 Peak Capacity and Fast Separation

The peak capacity (PC) of an isocratic separation was discussed in Section 2.7.3. The definition of peak capacity is the same for both isocratic and gradient elution; PC equals the maximum number of peaks that can be inserted into a given chromatographic space (e.g., a gradient chromatogram) with a resolution $R_s = 1.0$ for all adjacent peaks (a defined run time is assumed). In a gradient separation, where every peak has approximately the same peak width W, peak capacity can be approximated in terms of the gradient time t_G as

$$
\begin{aligned}
PC &= 1 + \left(\frac{t_G}{W}\right) \\
&\approx \frac{t_G}{W}
\end{aligned}
\tag{9.20}
$$

Figure 9.20 illustrates this definition of peak capacity for a (hypothetical) separation in a gradient time of 10 minutes. For the example of Figure 9.20*a*, the average peak width $W = 0.2$ min. Therefore PC for this example equals $t_G/W = 10/0.2 = 50$, as illustrated in Figure 9.20*b*, where 50 peaks, each with $W = 0.2$ min, can be fit into the 10-minute chromatogram with $R_s = 1$ for each adjacent peak-pair. The concept of peak capacity has been used to evaluate the relative performance of separations by gradient elution, in place of measurements of the column plate number N (which are possible, but less convenient; see p. 38 of [2]). As a measure of separation effectiveness, values of PC are especially useful for samples that generate more peaks than can be individually separated to baseline, that is, "complex" samples as in Figure 9.17*c* or Figure 9.20*a* (e.g., peptide digests, plant extracts, etc).

"Peak capacity" is a hypothetical (if measurable) quantity that generally overestimates the separation power of an actual gradient chromatogram. Thus, in the separation of Figure 9.20*a*, peaks appear only between 2 and 9 minutes, so that only a fraction of the gradient chromatogram is actually used: (9–2)/10 = 70%. When sample peaks are confined within part of the chromatogram (as in Fig. 9.20*a*), rather than being distributed over the entire chromatogram, the *effective* peak capacity of the separation is less than the value of PC defined in Figure 9.20*b* for

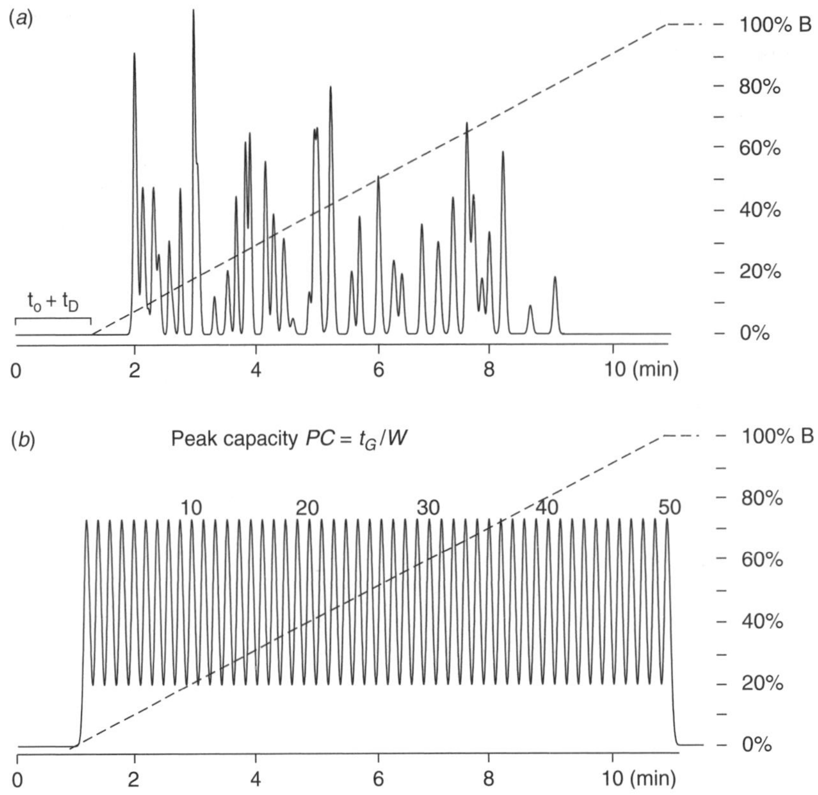

Figure 9.20 Peak capacity in gradient elution. (*a*) Hypothetical separation; (*b*) illustration of peak capacity *PC* for separation of (*a*).

the full gradient. In actual separations as in Figure 9.20*a*, separation performance is better defined by the number of resolved peaks that can be fit between the first and last peaks in the chromatogram (not necessarily the beginning and end of the gradient). The latter quantity will be referred to as the *equivalent* peak capacity $n_c = PC\ (t_Z - t_A)/t_G$; n_c has also been called *conditional peak capacity* or *sample peak capacity* [47]. In practice, equivalent peak capacities will be smaller than values of *PC* given by Equation (9.20) ($n_c = 35$ for the example of Fig. 9.20*a*). Note that the selection of conditions that maximize PC will also tend to maximize n_c.

Because a single gradient run can be inadequate for the separation of complex samples, two-dimensional (2D) separation is often employed (Sections 9.10, 13.4.5, 13.10.4). In 2D separation, fractions from a first column are transferred to a second column for further separation (as in the example of Fig. 1.4*b*,*c*). If the two separations are orthogonal (i.e., uncorrelated retention times; Section 6.3.6.2), the peak capacity of the combined separations equals the product of the peak capacities for each separation. A common goal is to maximize the peak capacity of each separation—but in minimum overall run time [47–58]. In this section we will approach an understanding (and control) of peak capacity on the basis of similar considerations as for isocratic elution (Section 2.4.1). It is also relevant to note that detection by mass spectrometry (MS) and multi-wavelength absorbance effectively add an extra dimension to one- or two-dimensional separations [53], hence further increasing the overall peak capacity.

9.3.9.1 Optimized Peak Capacities

The primary value of the present section is limited to just two applications: very fast gradient separations (Section 9.3.9.2) and two-dimensional HPLC (2D-LC) (Section 9.3.10). Much of the additional detail of this section had not appeared in the literature as of 2009. Unless the reader has an immediate interest in fast or 2D-LC separation, it may be advisable to skip to Section 9.4.

It is useful to express peak capacity in terms of experimental conditions that can be optimized for maximum values of PC and/or minimum run times. For a full-range gradient ($\Delta\phi = 1$), Equation (9.20) can be expressed as

$$PC = \frac{t_Z - t_A}{W} \tag{9.21}$$

where t_A and t_Z refer to the retention times for two peaks that elute, respectively, at the start and finish of the gradient (e.g., peaks 1 and 50 in Fig. 9.20*b*). Values of W can be assumed to be approximately equal for peaks A and Z, so Equation (9.21) is equivalent to the resolution of these two peaks (Eq. 2.23). Resolution R_s in Equation (9.15c) can therefore be replaced by PC to give

$$PC = \left(\frac{2.3}{4}\right) N^{*0.5} \log \alpha^* \left[\frac{k^*}{1 + k^*}\right] \tag{9.21a}$$

If we assume equal values of S for peaks A and Z (although S is usually larger for the later peaks in the chromatogram), then α^* will equal the ratio of k_w-values for peaks Z and A (Eq. 9.5a), which in turn is equal to $\Delta\phi S$. Equation 9.21*a* then takes the form

$$PC = \underset{(i)}{\left(\frac{2.3}{4}\right) S\Delta\phi} \quad \underset{(ii)}{N^{*0.5}} \quad \underset{(iii)}{\left[\frac{k^*}{1 + k^*}\right]} \tag{9.22}$$

This expression for peak capacity PC assumes a full-range gradient ($\Delta\phi = 1$), and the value of S is determined by the molecular weight M of the sample (Section 13.4.1.4):

$$S \approx 0.25 M^{0.5} \tag{9.23}$$

For samples with $M \leq 500$, a value of $S \approx 4$ can be assumed. Term *i* of Equation (9.22) is therefore constant for a sample of defined molecular weight. Term *ii* varies with column length, particle size, flow rate, temperature, and sample molecular weight (Section 2.3.1). Finally, term *iii* varies with gradient time, flow rate, column length, and S or sample molecular weight (Eq. 9.5*c*). The elaboration of Equation (9.22) in terms of the latter experimental variables can lead to fairly complex relationships that are challenging to both interpret and apply.

A simpler, more convenient use of Equation (9.22) is as follows, assuming conditions that provide maximum values of *PC* for a given run time and some maximum allowable column pressure *P*. In Section 2.4.1.1 it was argued that the best use of a column (for some required value of *N* in minimum time) occurs when the value of the reduced plate height *h* is a minimum, corresponding to a value of the reduced mobile-phase velocity $\nu \equiv hd_p/D_m \approx 3$ (case "C" in Section 2.4.1.1, which assumes a wide choice of column lengths and particle sizes). Optimum values of *N* (isocratic elution) are then defined (as in Figure 2.15) as a function of particle size d_p, pressure, and separation time (for given values of mobile-phase viscosity η and solute diffusion coefficient D_m). Specific values of the column length, flow rate, and particle size then result for each separation time and pressure (Eqs. 2.13a and 2.21b), such that $\nu = 3$ (optimum value). As separation time increases, column length and particle size must increase, and flow rate decrease—so as to maintain constant pressure (while also maintaining $\nu = 3$ by varying d_p). From Equation (9.22), it can be shown that values of *PC* are proportional to $k^{*3/4}/(1 + k^*)$, which then results in an optimum value of $k^* = 3$.

A summary of optimal values of *PC* as a function of experimental conditions is shown in Figure 9.21*a* for a small-molecule sample (note the essential similarity of this plot, based on gradient elution, and the corresponding isocratic plot of Figure 2.15 for values of *N*). For these optimized conditions, we see that peak capacity increases with gradient time and pressure, while the required particle diameter also increases with gradient time. For very short runs, especially at higher pressures, column packings that are presently unavailable are required ($d_p < 1.5$ μm). In these cases, various (sub-optimum) expedients are available to maximize peak capacity for very short gradients (e.g., $t_G \leq 1$ min). The data of Figure 9.21*a* also assume ideal, equipment-related conditions, which are difficult to attain for very fast separations (Section 9.3.9.2)

When the optimized conditions of Figure 9.21*a* require particle sizes < 1.5 μm (i.e., for very short gradients), larger particles can be used, with some loss in peak capacity. Also, as long as ν (and N^*) is optimized, the selection of other values of k^* within the range $1 \leq k^* \leq 10$ (i.e., by varying gradient time only) leads to a ≤ 10% reduction in the optimum value of *PC* for a given run time (value of t_G). *This then provides a simple means for varying gradient time (over a 10-fold range) while maintaining near-optimum values of PC over the gradient time.* The achievement of sub-optimum, shorter gradients is illustrated in Figure 9.21*b*, for columns of different (discrete or noncontinuous) lengths packed with 3-μm particles; flow rate is also varied to maintain $P = 6000$ psi. The solid line represents optimal values of *PC* taken from Figure 9.21*a* for $P = 6000$ psi (where particle size and column length were allowed to vary continuously). Each of the other curves shown in this figure correspond to values of k^* that vary from 1 to 10—but with particle diameter fixed at 3 μm, and $P = 6000$ psi. The use of these sub-optimum conditions is seen to result in some loss in peak capacity, compared to optimum values for a given gradient time—but this loss in *PC* is usually less than half. Note that the optimum column length in this example for $d_p = 3$ μm is $L = 1500$ mm. Table 9.4 summarizes conditions for the optimized separations of Figure 9.21*b* (6000 psi).

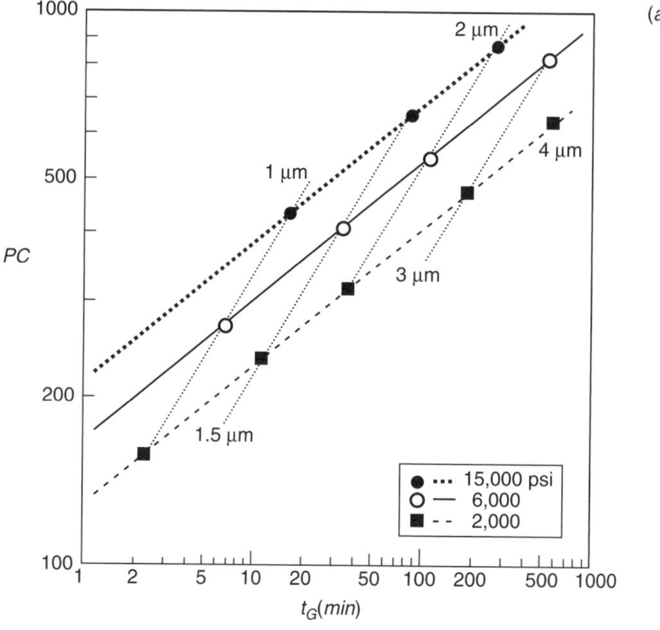

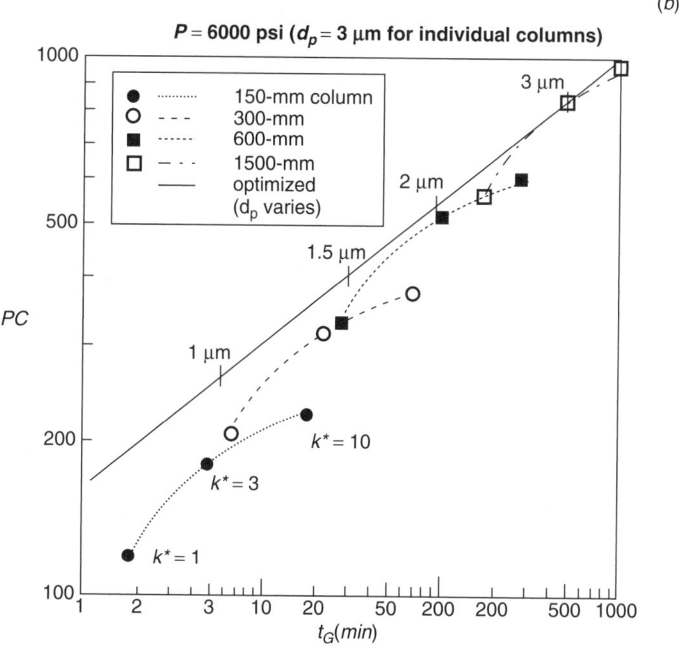

Figure 9.21 Optimized peak capacity as a function of particle diameter, gradient time and column pressure. Data for a small-molecule sample ($S = 4, D_m = 10^{-5}$; 0–100% B acetonitrile-buffer gradient ($\eta = 0.75$); calculations based on Equation (2.17) with $A = 1, B = 2$, and $C = 0.05$. (*a*) Optimized values ($\nu = 3, k^* = 3$) for $P = 2000$, 6000, and 15,000 psi; (*b*) sub-optimized values for a pressure of 6000 psi and 3-µm particles (k^* allowed to vary for each column length).

Table 9.4

Separation Conditions for Optimized Conditions in Figure 9.21*a*

t_G (min)	d_p (μm)	L (mm)	F (mL/min)	k*	PC[a]
1	0.6	10	1.8	3	170
3	0.8	30	1.4	3	230
10	1.1	80	1	3	310
30	1.4	170	0.8	3	400
100	1.9	420	0.6	3	540
300	2.5	960	0.4	3	710
1000	3.4	2400	0.3	3	960

Note: Column diameter of 4.6 mm is assumed, with a pressure of 6000 psi
[a]Calculations based on $h = \nu^{0.33} + 2/\nu + 0.05\nu$.

The examples of Figure 9.21*b* can be more fully appreciated in terms of a relationship for gradient time. From Equation (9.5) we have

$$t_G = \frac{1.15k^* V_m \Delta\phi S}{F} \quad (9.24)$$

so a reduction in k^* results in a decrease in gradient time, and vice versa. A decrease in column length L (proportional to V_m) while holding pressure and k^* constant requires a proportionate increase in flow rate F, resulting in a decrease in gradient time that is proportional to L/F (or to L^2).

In most cases, maximum or optimum values of PC for a given gradient time will require an intermediate particle size (e.g., 2.2 μm) or column length (e.g., 65 mm) that are unavailable, especially for a limited range of columns from a preferred source. However, the use of a moderately different particle size (2- or 3-μm) can be compensated by the use of sub-optimum values of k^* as above, with only a moderate loss in peak capacity. Similar plots as in Figure 9.21*a*,*b* result for the separation of higher molecular-weight samples, but with generally higher peak capacities and a need for still smaller particles.

9.3.9.2 Fast Gradient Separations

Gradient separations with run times of a few min or less (sometimes referred to as "ballistic gradients" [59]) are needed for high-volume testing, where thousands of samples must be analyzed at acceptable cost—and therefore minimum run time. Short run times are also needed in two-dimensional HPLC (Section 9.3.10) for the second-dimension separation, in order to analyze a large number of fractions from the first-dimension separation, during the time required by the initial separation. As run time for a given assay is decreased below a few minutes, the performance of the equipment becomes limiting. Aside from previously discussed requirements of column length, flow rate, and particle size (in connection with Fig. 9.21), fast separations require (1) very small values of the dwell-volume V_D, (2) sample injections that can be performed within a second or two, (3) fast detector response for peaks with $W < 1$ sec, (4) an ability to carry out very steep gradients (e.g., 1–2%B/sec; Fig. 17.28), and (5) rapid or off-line data processing (Section 3.8.4). Fast separations are normally limited to small values of the column plate number N^* or peak capacity *PC*, because possible values of *PC* decrease for shorter gradients (Fig. 9.21). However, small values of N^* or *PC* are still acceptable in many cases:

- samples with fewer, easily separated components
- following the extensive optimization of separation selectivity, especially the use of two or more conditions that affect selectivity (Table 2.2)
- separations with a tolerance for small values of R_s, because of either selective detection (e.g., LC-MS) or an acceptance of reduced accuracy in assay results
- the second separation in 2D-HPLC

Samples with only a few easily separated components can be assayed using smaller values of N^*. This can also be true for samples that contain a larger number of components, when *separation selectivity has been extensively optimized* (resulting in maximum values of α), and for *separations with a tolerance for small values of* R_s. The *second separation in 2D-HPLC* can often be carried out with a smaller value of N^* because the number of sample components will have been drastically reduced, their values of α tend to be larger for the second, orthogonal separation, and MS is often used for detection.

Apart from the equipment needed for fast separation, the choice of column dimensions, particle size, and flow rate and the maximum allowable pressure for the system determine (in theory) the minimum separation time for a required value of *PC*. Thus the smallest available particles and highest possible pressure will (in principle) allow the fastest separation for some required sample resolution; see Figure 9.21*a* and the discussion of [60]. However, practical constraints for a given gradient system will qualify the latter conclusion to some extent (column lengths may be limited to some minimum value, flow rates cannot be greater than some maximum value, and extra-column effects cannot be entirely avoided). Finally, the equilibration time t_{eq} between successive gradient runs must be made as short as possible (Section 9.3.7), which is predominantly a function of the equipment (its dwell volume and gradient rounding).

Several reports [43, 59, 61–65] provide both examples and further experimental details for "fast" gradient elution. Figure 9.22*a* shows the separation of a model sample in 1.6 minute, while Figure 9.22*b* shows the result of successive injections every 1.6 minute (arrows mark the time of each injection). The 5-μm-particle column used in this example is not especially well suited for fast separation, but in this case fast separation is favored by large α-values—as might result from an extensive optimization of selectivity. Separation speed can also be enhanced by the use of higher temperatures (Section 2.4.1).

9.3.10 Comprehensive Two-Dimensional HPLC (with Peter Schoenmakers)

Section 9.3.9 examined conditions (particle size, column length, flow rate) for maximum peak capacity within a given gradient time. Other conditions can be varied further to optimize relative retention and maximize critical resolution, as discussed in Section 9.3.3. However, even these steps will be insufficient for samples that contain hundreds or thousands of individual compounds—as in the case of

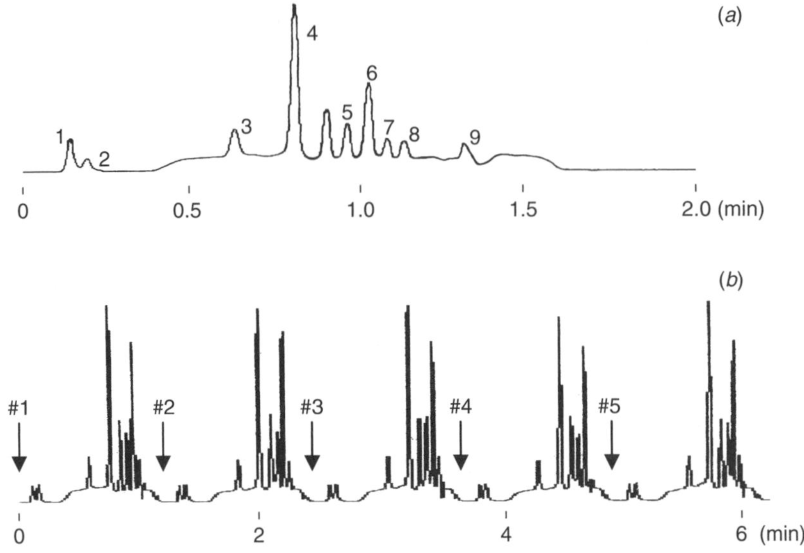

Figure 9.22 Example of fast gradient separation. Sample: *1*, uracil; *2*, acetone; *3*, *N*-benzylformamide; *4–9*, C_2-C_7 alkylphenones. Conditions: 50 × 2.1-mm (5-μm) C_{18} column; 0 – 100% B in 1 min; A-solvent is 3/7/90% *n*-propanol/acetonitrile/water; B-solvent is 3/97/0% *n*-propanol/acetonitrile/water; 1.0 mL/min; 40°C. (*a*) single chromatogram; (*b*) successive injection of five samples at 1.6-min intervals. Arrows mark the time of each sample injection. Figures adapted from [43].

proteolytic digests of the human proteome (Section 13.4.5). In the latter case we require a considerable increase in peak capacity over that which can be achieved by a single separation (as in Figure 9.21). This increase in peak capacity can be achieved by two-dimensional HPLC (2D-LC), in which some or all fractions from an initial (first-dimension) gradient separation are collected and injected into a second (second-dimension) HPLC system with subsequent gradient separation. 2D-LC separation can be carried out off-line (collecting fractions) or on-line (column switching [Section 2.7.6] with one or two switching valves). If all fractions are subjected to the second-dimension separation, and if fractions are taken so frequently that the first-dimension separation is largely maintained, we have *comprehensive two-dimensional liquid chromatography*, conveniently abbreviated to LC × LC [66].

9.3.10.1 Principles of LC × LC

Some aspects of comprehensive 2D-LC are illustrated in Figure 9.23, where a portion (two peaks) from the first-dimension chromatogram is shown. A number of fractions (shaded rectangles) are collected across each peak, and each of these fractions is then injected into the second-dimension column—resulting in the chromatograms shown at the top of Figure 9.23 for each fraction (in this example, the second peak from the first separation is resolved into two peaks [*i* and *j*] by the second separation). The latter chromatograms are obtained using a single detector, which is positioned after the second-dimension column. The series of chromatograms (one for each fraction from the first-dimension separation) is stored in the computer, and the data can then

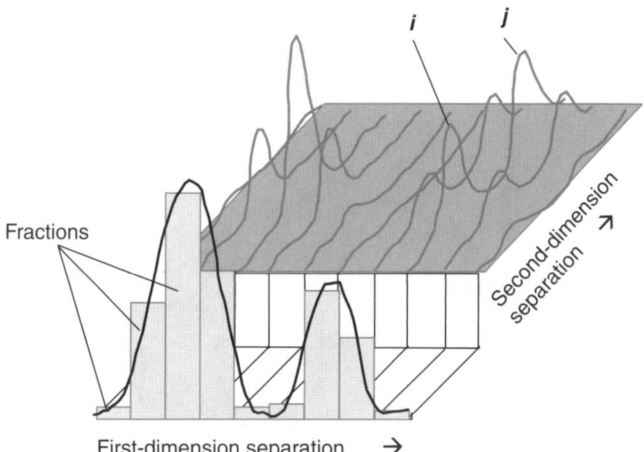

Figure 9.23 Hypothetical example of comprehensive 2D separation of a single analyte. Several fractions are collected in the first-dimension separation for each peak; each fraction then occurs in several second-dimension chromatograms. Peaks *i* and *j* are present in the second peak of the first-dimension separation.

be presented in various ways; for example, (quasi) three-dimensional, contour, or color plots.

An example of a three-dimensional plot is shown in Figure 9.24*a*, where each peak or solute is defined by its retention time in the first- and second-dimension separations; peak absorbance (proportional to concentration) is shown by the height of each peak. In this case some of the data are obscured (small peaks behind larger peaks), but this can be overcome by computer-rotation of the plot. An example of a two-dimensional plot is shown in Figure 9.24*b*, with the data (as in Fig. 9.24*a*, but a different sample) now observed from above. Different shades represent different peak intensities (darker spots correspond to higher peaks, as in Fig. 9.24*a*).

If a comprehensive two-dimensional analysis is performed automatically and in real time, the second-dimension separation must be much faster than the first-dimension separation. For example, if the first separation has a gradient time of 250 minutes, and 500 fractions (not an excessive number) are collected, the second separation can take no longer than 0.5 minute. If the fractions are collected by a fraction collector, they can be analyzed more slowly in the second dimension, but in that case the total analysis time can be very long. As can be seen in Figure 9.21*a*, a faster second-dimension separation means a much lower peak capacity; shorter columns, smaller particles, and higher flow rates are favored for the second separation.

To achieve the full potential of 2D-LC (as measured by peak capacity), any separation accomplished in the first dimension must not be undone when the second-dimension separation is implemented [67]. That is, no remixing of the materials separated in the first dimension is allowed (even though each fraction from the first column can be assumed to be remixed during its transfer to the second column). An initial conclusion [68] was that each first-dimension peak must be sampled at least 4 times, in order to avoid a considerable loss in 2D peak capacity. However, this can lead to a very large number of collected fractions,

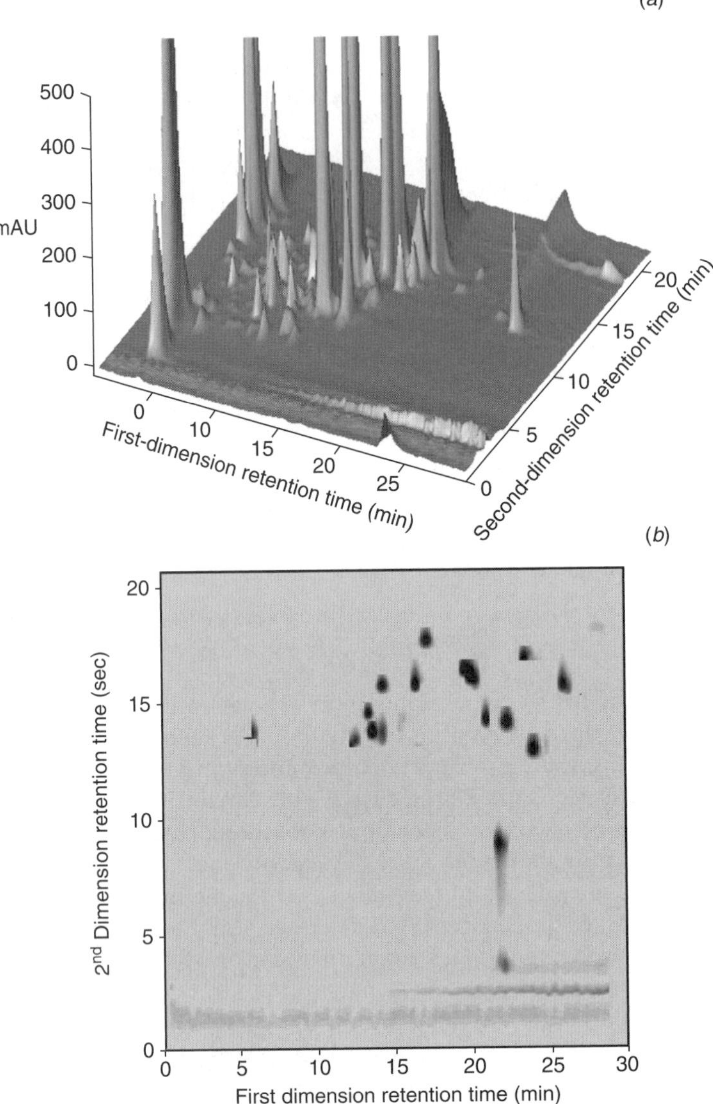

Figure 9.24 Examples of 2D separation. (*a*) Three-dimensional representation of a comprehensive two-dimensional chromatogram of a mutant-maize extract; (*b*) gray-scale 2D plot from a comprehensive LC × LC separation of an indolic-metabolite standard mixture.

with a corresponding increase in effort. A subsequent analysis of the effect of sampling frequency on peak capacity suggests that a better compromise is a sampling rate of two fractions per first-dimension peak [69], which still results in a 2-fold loss in total peak capacity [57] because of the smaller number of collected fractions.

For a well-designed LC × LC separation (Section 9.3.10.4), a given component will be present in several of the serially collected fractions, and will therefore appear in several sequential second-dimension chromatograms (as in Fig. 9.23). Each component must be correctly identified in the adjoining chromatograms, and its peak areas must be added for quantitative analysis [71]. Because (1) the peak is split into several fractions, (2) each fraction is diluted during the second-dimension separation, and (3) fast-gradient baselines are often problematic, detection can be a serious bottleneck for LC × LC. UV detectors have been popular for LC × LC, but detection with a mass spectrometer is much more powerful because of its advantages of peak identification and deconvolution (Section 4.14).

9.3.10.2 Peak Capacity

As a rule of thumb, high-resolution one-dimensional LC allows peak capacities in the hundreds, whereas LC × LC provides peak capacities in the thousands. For example, by reference to Figure 9.21*b*, consider a 200-minute first-dimension separation, followed by 100 separations for 2 minutes in the second dimension. Assume a maximum column-pressure of 6000 psi; then the peak capacity *PC* for the first dimension will be about 650. Similarly, for the second dimension, $PC \approx 130$, from which $PC \approx 650 \times 130 \approx 85,000$. However, the effective peak capacity n_c will be much smaller for several reasons:

- separations in the first and second dimension that are non-orthogonal
- use of only part of the chromatogram
- under-sampling of the first dimension

When peak capacity for a 2D separation is calculated as the product of peak capacities for the two separations, it is assumed that the *two separations are completely orthogonal* (retention times in the first dimension independent of the retention times in the second dimension). In practice, this is never even approximately the case in LC × LC, especially for any non-ionizable compounds that are present in the sample [72]. Because *only part of the chromatogram* is filled with peaks, a further reduction in effective peak capacity results (as discussed in Section 9.3.9). Finally, it is impractical to collect enough fractions in the first dimension to completely avoid *undersampling*.

Little control is possible over the above contributions to a loss of effective peak capacity; as a result values of n_c for LC × LC have seldom exceeded ≈ 2000 at the time the present book was published. The latter number can be effectively increased, however, by the use of selective detection (e.g., LC-MS). The main challenge is to select two separations that are as nearly orthogonal as possible. This normally requires the use of two different separation modes (Section 9.5); for example, ion exchange for the first dimension, followed by RPC for the second dimension. Alternatively, if separation conditions can be made sufficiently orthogonal (as discussed in Chapters 5–7), two RPC separations may represent a comparable choice for maximum orthogonality in LC × LC.

9.3.10.3 Instrumentation for LC × LC

Off-line LC × LC can be carried out with conventional instrumentation, as described in Section 13.4.5. LC × LC with valve-switching is a bit more complicated but more convenient. While one fraction is being analyzed on the second-dimension column,

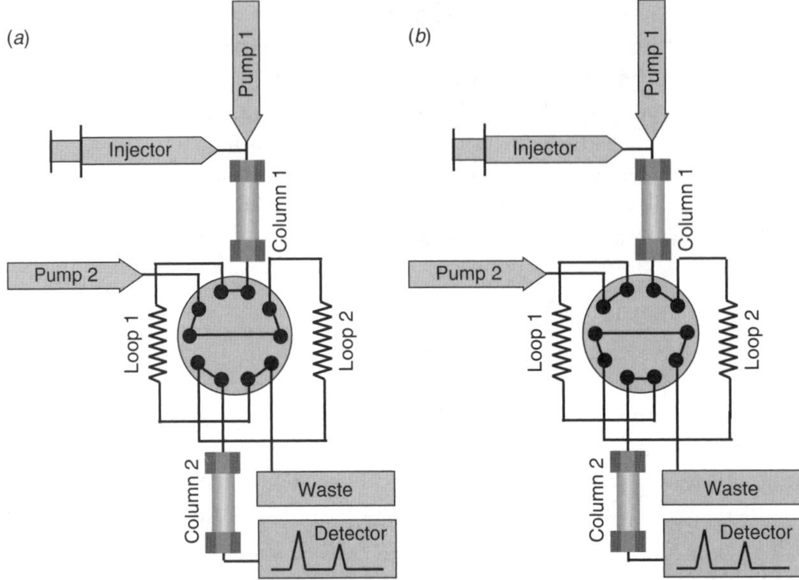

Figure 9.25 Possible valve configuration for LC × LC. In the first stage (*a*), loop 1 is loaded with effluent from the first-dimension column, loop 2 is connected between pump 2 and the second-dimension column. During the next stage (*b*), the contents of loop 1 are injected into the second-dimension system, while loop 2 is filled with effluent from the first-dimension column.

a second fraction from the first separation is being collected, ready for the next injection. One popular arrangement employs a two-way, ten-port switching valve, as illustrated in Figure 9.25.

In Figure 9.25*a*, loop 1 is loaded with effluent from column 1. When the valve is switched, the contents of loop 1 are flushed by pump 2 onto column 2. Thus a (second-dimension) chromatogram is obtained for the contents of loop 1. At the same time, loop 2 is being filled with the next fraction from column 1. When the valve is switched again (Fig. 9.25*b*), a second-dimension chromatogram is obtained for the contents of loop 2. If the loop size is significantly greater than the volume of a fraction, all of the fraction is collected in the loop and the total sample is separated and analyzed. If ${}^2t_{an}$ is the second-dimension analysis time (also called the *cycle time*) and ${}^1F$ the flow rate in the first-dimension separation, the volume of the loop (V_{loop}) should meet the following criterion:

$$V_{loop} > {}^2t_{an}\,{}^1F \qquad (9.25)$$

An LC × LC system consists of two pumping systems, but only one detector is required. The detector must be both fast (small time constant) and sensitive—because of the dilution of the sample that occurs during two consecutive separations.

9.3.10.4 Method Development for LC × LC

The selection of conditions for LC × LC proceeds differently than for 1D separation, where the primary goal is the optimization of separation selectivity for some minimum resolution (e.g., $R_s \geq 2$) for all peaks of interest. For various reasons, this is not be practical for LC × LC. The main goals in developing a fully automated 2D separation include the following:

- orthogonal conditions for the two separations
- compatibility of the two mobile phases
- selection of conditions for maximum peak capacity for each separation, within the constraints that the gradient time for the second separation will be limited to a small fraction of that for the first separation

Orthogonal Conditions. RPC separates mainly on the basis of solute hydrophobicity, while ion exchange chromatography (IEC) separates on the basis of molecular charge or ionization—the two separations can therefore be regarded as roughly orthogonal. For this and other reasons discussed below, these two separation modes are often used together for 2D-LC. Two RPC separations can also be made near-orthogonal, by selecting two orthogonal columns (Sect. 5.4.3) with different B-solvents, temperature, and especially pH [73]. Other possible combinations of separation conditions or modes are possible, but are subject to some of the problems discussed below.

Mobile-Phase Compatibility. The sample solvent for the second dimension is the first-dimension mobile phase. Two important characteristics of the mobile phases for the first- and second-dimension separation are *miscibility* and *elution strength*. The injection of a significant volume of a solvent that is immiscible with the mobile phase—as when combining RPC with nonaqueous normal-phase chromatography—can lead to severe peak broadening and/or distortion. If the mobile phase associated with the fraction from the first dimension is a strong solvent for the second dimension (resulting in a small value of k), peak distortion and broadening can again occur in the second separation (Section 17.4.5.3). The use of IEC followed by RPC largely avoids problems with mobile-phase compatibility because the aqueous IEC mobile phase is (1) miscible with RPC mobile phases and (2) is a very weak RPC mobile phase.

A starting point for the design of an LC × LC system is to select an acceptable first-dimension separation time (t_G) and pressure. From this we can estimate PC for this separation, as well as suitable conditions (particle size, column length, and flow rate; see Fig. 9.21*a*, Table 9.4, and the discussion of Sections 9.3.9.1). The resulting peak capacity allows an estimate of average peak width W (Eq. 9.20), which is of critical importance if we are to take several fractions for the first-dimension peak (as in Fig. 9.23). A good recommendation [69] is to choose two fractions for each peak (i.e., a collection time equal to $w/2$); for further information on the effect of fraction size on peak capacity, see [74]. If we know the time and the pressure available for the second-dimension separation, we can optimize the column and separation conditions in the same way as for the first separation. A possible optimized configuration for LC × LC is described in Table 9.5, which is somewhat more ambitious than contemporary practice at the time this book was published.

Because the first-dimension separation of Table 9.5 is slow, the corresponding optimum column is long; ≈950 mm—or four 250-mm columns in series. In contrast,

Table 9.5

Representative Conditions for an LC × LC Separation

Parameter	First Dimension	Second Dimension[a]
Pressure drop (psi)	6000	6000
Eluent viscosity (cPoise)	0.75	0.75
Analyte diffusivity (cm^2/sec)	10^{-5}	10^{-5}
Gradient time	300 min	30 sec
Column diameter (mm)	1	4.6
Injection volume (μL)	5	20
Flow rate (μL/min)	20	1500
Particle size (μm)	5	1.5
Column length (mm)	950	25
k^*	4	3
Peak capacity	700	100
Peak width (sec)/number of fractions	26 sec/1400	
Dilution factor		
$k = 0$	1	1
$k = 3$	3.5	1
Injection band broadening (%)		
$k = 0$	21	264
$k = 3$	1	33

[a]Alternative choices are possible, with similar gradient times and peak capacities.

the second-dimension column should be short (about 25 mm in the present example). The particle size is conventional (5 μm) for the first dimension, but very small (1.5 μm) for the second dimension, because fast separations are favored by smaller particles (Fig. 9.21). Band broadening in the second dimension can be reduced by minimizing the volume of fractions from the first column—which is favored by a smaller diameter of the first column, relative to the second. However, this implies that peaks will be greatly diluted in the second separation (because the first, narrow-diameter column will have a smaller sample capacity). Thus selecting the column diameters means striking a balance between peak width and detection sensitivity. For further details, see [58].

9.4 LARGE-MOLECULE SEPARATIONS

The gradient separation of large molecules (peptides, proteins, nucleic acids, synthetic polymers, etc.) occurs in essentially the same way as for small molecules (100–1000 Da) [2]. Consequently changes in gradient or column conditions (t_G, ϕ_0, ϕ_f, F, L, etc.) will affect the separation of large-molecules in the same general way as discussed in Sections 9.1 through 9.3 for small molecules. Large molecules *do* have some special characteristics that play a role in their gradient separation

(Sections 13.3, 13.4.1.4). In this section we will examine one of these characteristics: the increase of values of S as solute molecular weight M increases, as described approximately by Equation (9.23) (see also Fig. 13.11 for some examples).

The main consequence of an increase in S with M is its effect on values of k^* (Eq. 9.4). If gradient conditions are selected to give a value of $k^* = 5$ for a small-molecule sample (as in Table 9.3), the same conditions for a large-molecule sample will result in a smaller value of k^* (because of larger values of S) and poorer resolution. To achieve the same value of $k^* = 5$ for a large-molecule sample, the gradient time must be increased, by the ratio of S-values for the two samples. For example, if $M = 10{,}000$ Da for the large-molecule sample, S will be about $0.25 \times (10{,}000)^{0.5} \approx 25$ (Eq. 9.23). As $S \approx 4$ for small molecules, the gradient time should be (25/4) = 6-fold larger for the large-molecule sample. If the gradient time for the small-molecule samples is 10 minutes (Table 9.3), this should be increased to about one hour for the large-molecule sample, in which case, k^* will also equal 5 for the large-molecule sample. Gradient separation of large-molecule samples thus require more time, or larger values of t_G, other factors equal. Chapter 13 provides a detailed discussion of the separation of large-molecule samples (in most cases by gradient elution).

9.5 OTHER SEPARATION MODES

Our discussion of gradient elution in Sections 9.1 through 9.4 has assumed RPC separation. Other separation modes exist, as summarized in Table 9.6. The gradient separation of small molecules by any of the procedures of Table 9.6 takes place in similar fashion as for RPC, so changes in gradient or column conditions will affect separation in approximately the same way as discussed in Sections 9.1 through 9.3 for RPC. For example, an increase in gradient time t_G or flow rate F, or a decrease in column volume V_m or gradient range $\Delta\phi$, will increase k^*—with predictable consequences for average resolution or peak width (Eq. 9.4).

9.5.1 Theory

For each of the separation modes of Table 9.6, isocratic values of k as a function of %B are given either by Equation (9.1) or by

$$\log k = \log k_B - n \log \phi \tag{9.26}$$

where log k_B is defined in Table 9.6, n is a constant for a given solute and experimental conditions (also see Table 9.6), and ϕ refers to the volume fraction of the B-solvent (= 0.01% B; see Table 9.6 for a definition of the B-solvent for different separation modes).

For linear-gradient separations that are described by Equation (9.26), the gradient retention factor k^* is given by [2]

$$k^* = \frac{t_G F}{1.15[V_m n \log(\phi_f/\phi_0)]} \tag{9.27}$$

Table 9.6

Dependence of *k* on %B for Different Separation Modes (Eq. 9.1 or 9.26)

Separation Mode	Dependence of k on %B	Definition of k_w (Eq. 9.1) or k_B (Eq. 9.26)	Dependence of S or n on the solute
Reversed-phase (RPC)	Eq. (9.1)	(k_w) value of isocratic k for A-solvent (water or buffer) as mobile phase	S increases with solute molecular size (Eq. 9.21)
Normal-phase (NPC, Chapter 7)	Eq. (9.24)	(k_B) value of isocratic k for B-solvent (more polar solvent) as mobile phase	n increases with the number of polar groups in the solute molecule (Section 8.2)
Ion-exchange (IPC, Chapter 7)	Eq. (9.24)	(k_B) value of isocratic k for mobile phase with salt concentration (ϕ) = 1.00 M	n = number of charges on solute molecule (assumes mono-valent buffer)
Hydrophobic interaction (HIC, Chapter 13)	Eq. (9.1)	(k_w) value of isocratic k for A-solvent as mobile phase (higher salt concentration)	$S \approx 0.14M^{0.37}$ [2]
Hydrophilic interaction (HILIC, Chapter 8)	Eq. (9.24)	(k_B) value of isocratic k for A-solvent (organic) as mobile phase	n increases with the number of polar groups in the solute molecule (Section 8.6)

Here the value of n varies for different solutes and separation modes (as defined in Table 9.6); ϕ_f and ϕ_0 refer to values of $\phi \equiv \phi_B$ at the beginning (0) and end (f) of the gradient, respectively.

For small-molecule samples, values of n for different separation modes usually vary between 1 and 4; a value of $n \approx 2$ can be used as a starting approximation, prior to method development. Thus, for a NPC separation with a 150 × 4.6-mm column, a flow rate of 2 mL/min, and gradient of 10–100% B in 10 minutes, the value of k^* would be approximately $(10 \times 2)/(1.5 \times 2 \times \log 10) = 6.7$, that is, not much different from the value of $k \approx 4$ for similar RPC conditions. For an IEC separation with a 150 × 4.6-mm column, a flow rate of 2 mL/min, and gradient of 10–100 mM in 10 minutes, the same value of k^* results from Equation (9.26) (note also for IEC that the salt concentration ϕ is the sum of concentrations of salt plus buffer). For further details on the theory of gradient elution for separation modes other than RPC, see Chapter 13 of [2] and [75].

9.5.2 Normal-Phase Chromatography (NPC)

Section 8.5 summarized some disadvantages of isocratic normal-phase chromatography (NPC) with silica columns. With the exception of HILIC (Section 9.5.3), these same problems apply equally for corresponding gradient separations. When the A- and B-solvents are quite different in polarity or strength, solvent demixing (Section 8.5.2) is a potentially serious problem [76] (very few gradient NPC separations with silica columns have been reported in the past 30 years). The use of polar-bonded-phase columns (Section 8.3.4) largely avoids the problem of solvent demixing in gradient elution.

9.5.3 Hydrophilic-Interaction Chromatography (HILIC)

Isocratic hydrophilic-interaction chromatography (HILIC) separations were reviewed in Section 8.6; most conclusions presented there apply equally for HILIC gradient elution. The use of HILIC gradient elution is as convenient and free from problems as are gradient separations by RPC—which in part accounts for the increasing popularity of HILIC. The applicability of HILIC for different kinds of samples can be visualized in the hypothetical separations of Figure 9.26*a*,*b*, where gradient separations by RPC and HILIC are compared. A series of solutes (1–29) of decreasing polarity (or increasing hydrophobicity) is visualized, where RPC retention increases in this order. The shaded peak 20 corresponds approximately to toluene, which provides a reference point for comparing these two separations. Compounds 1 to 6 are unretained by RPC (because of their greater polarity), while compounds 19 to 29 are unretained by HILIC because of their greater hydrophobicity. Compounds 26 to 29 are very hydrophobic, might not be eluted in RPC with this acetonitrile/buffer gradient, and therefore require the use of nonaqueous RPC (NARP; Section 6.5) for their effective separation.

For the corresponding HILIC separation of Figure 9.26*b*, compounds 1 to 6 are strongly retained and well separated—in contrast to their poor retention by RPC. This ability of HILIC to separate polar compounds that are unretained or poorly retained by RPC represents its primary advantage. Compounds 7 to 17 (indicated by double-headed arrows in both chromatograms) are retained on both columns, so that these compounds of intermediate polarity or hydrophobicity can be separated by either RPC or HILIC. The gradient separations of such a sample by both RPC and HILIC are shown in Figure 9.27*a*,*b* for a mixture of peptides.

9.5.3.1 Applications

The hypothetical gradient separations of Figure 9.26 suggest a reversal of relative retention for separations by RPC and HILIC, and an inverse correlation of solute retention (i.e., non-orthogonal separation). This is only roughly the case for actual separations, as shown by the examples of Figure 9.27*a*,*b*—and summarized in the plot of gradient retention times in Figure 9.27*c* (for which $r^2 = 0.00$; i.e., orthogonal separation). Whereas Figure 9.26 suggests that two compounds that are overlapped in a RPC separation will also be overlapped in a HILIC separation, this usually is not the case. Furthermore changes in other conditions (gradient steepness, temperature, column type, etc.) are known to further affect selectivity in both RPC and HILIC.

The analysis of samples composed of compounds with widely varying polarity may require the use of both HILIC and RPC for adequate retention and subsequent separation. Using the example of compounds 1 to 25 in Figure 9.26, RPC could be used for the analysis of compounds 13 to 25, and HILIC could

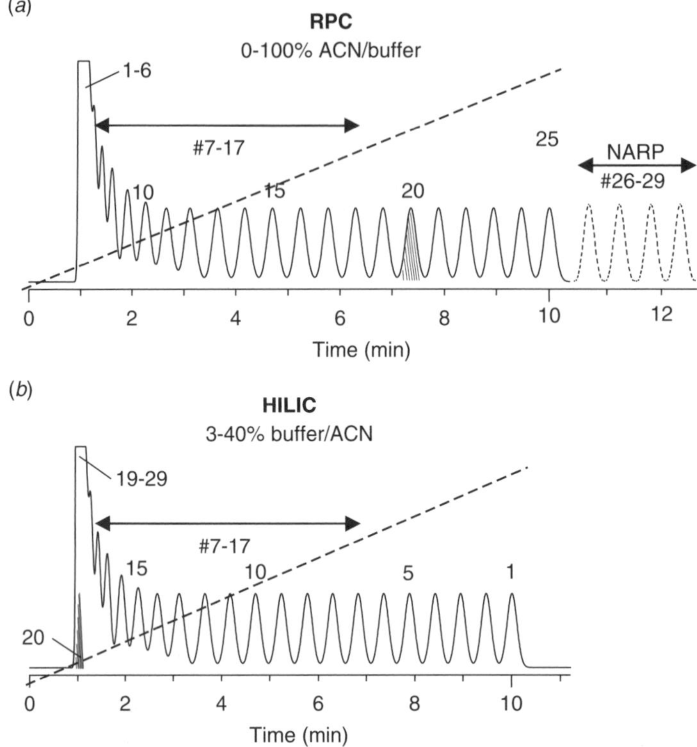

Figure 9.26 Hypothetical example of retention in RPC (*a*) and HILIC (*b*) as a function of solute polarity (solute polarity decreases from compound 1 to 29). Same sample assumed for (*a*) and (*b*).

be used for the balance of the sample (compounds 1 to 12). An example is provided by the HILIC separation of Figure 9.28, which was applied to the unretained fraction from the separation of a wheat gluten hydrolysate by RPC gradient elution.

9.5.3.2 Separation Conditions

Bare silica or bonded-amide columns are often used, with gradients that begin with 3–5% aqueous buffer (B)/acetonitrile (A) and end with 40–60% B, in a time of 20 to 60 minutes; other conditions are similar to those used for RPC (Table 9.3). Gradient HILIC separations are often employed for the same reasons cited in Section 13 for isocratic HILIC (usually for polar samples that are poorly retained in RPC, and especially for use with LC-MS). In addition, for samples that are retained by both RPC and HILIC, the two separation modes are often assumed to be orthogonal—allowing their use for 2D separation (Section 9.3.10). However, fractions from a first- dimension RPC separation will be in a strong solvent for the second-dimension HILIC separation, and vice versa (see Section 9.3.10.4)

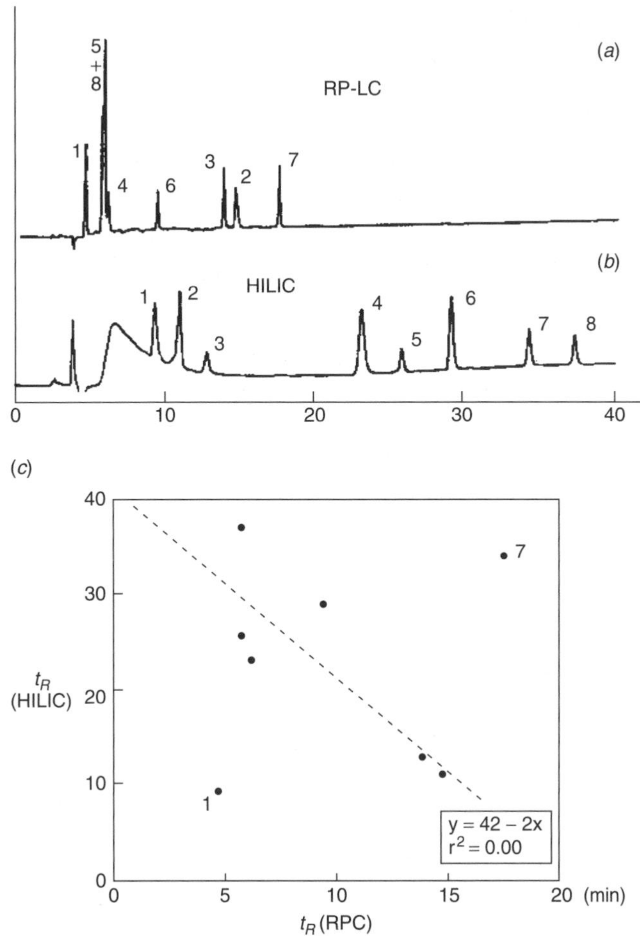

Figure 9.27 Selectivity in RPC versus HILIC for a mixture of peptides. Conditions: (*a*) C_{18} column; 5–55% acetonitrile–aqueous buffer (pH-2) in 83 minutes; (*b*) TSKgel Amide-80 column; 3–45% aqueous buffer (pH-2)–acetonitrile in 70 minutes; (*c*) comparison of retention times from separation of (*b*) versus (*a*). Adapted from [77].

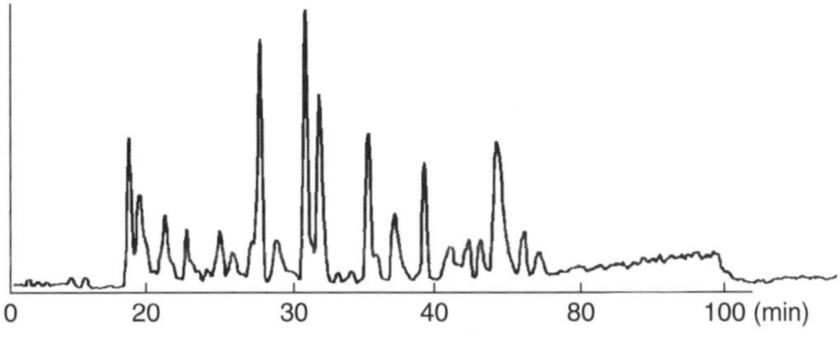

Figure 9.28 Separation by HILIC of a polar, unretained fraction from RPC of a wheat gluten hydrolysate. Conditions: 250 × 1.5-mm TSK Gel Amide 80 column; 10–40% buffer (pH-7.0)–acetonitrile in 50 minutes; 100 μL/min. Reproduced from [78] with permission.

9.5.4 Ion-Exchange Chromatography (IEC)

As noted in Section 7.5, ion-exchange chromatography (IEC) is rarely used today for small-molecule separations, with the exception of carbohydrates, carboxylic acids, and ion chromatography for inorganic ions. IEC in a gradient mode is much more popular for the separation of large biomolecules such as polysaccharides, proteins, nucleic acids, and viruses (Chapter 13).

9.6 PROBLEMS

Various problems can be encountered with gradient elution that are either less common or nonexistent for isocratic elution. These include:

- solvent demixing
- ghost peaks
- baseline drift

9.6.1 Solvent Demixing

In gradient elution, demixing refers to the preferential uptake by the stationary phase of the B-solvent, because of its greater affinity for the stationary phase. In most cases solvent demixing is not sufficiently pronounced in RPC to compromise separation. Its effects are to slightly increase retention for initial peaks in the chromatogram. Solvent demixing is especially a problem in separations by normal-phase chromatography [79] with base silica as column packing. This can result in complete retention of the B-solvent during the early part of the gradient, followed by a sudden breakthrough of B-solvent in the column effluent. The result is to elute all early peaks together as a single peak, with a major loss in sample resolution. For a further discussion of solvent demixing in gradient elution, see [2].

9.6.2 Ghost Peaks

These refer to extraneous peaks in the chromatogram that do not correspond to sample components. These can be the result of impurities in the mobile-phase solvents, buffers, or other additives. Ghost peaks can also arise from carryover of sample from one injection to the next, or as a result of adsorption of sample in the autosampler or on the column inlet frit. For a general discussion of ghost peaks, see [80] and [81]. Means for the elimination of ghost peaks are discussed in Section 17.4.5.2.

9.6.3 Baseline Drift

This is common in gradient elution, as a result of differences in the detector response for the A- and B-solvent. As a result the baseline can drift during the gradient. Baseline drift is common in RPC with UV detection because organic solvents (the B-solvent) generally absorb more strongly than water, especially at low wavelengths. For a further discussion of baseline drift, see Section 17.4.5.1.

REFERENCES

1. P. L. Zhu, L. R. Snyder, J. W. Dolan, N. M. Djordjevic, D. W. Hill, L. C. Sander, and T. J. Waeghe, *J. Chromatogr. A*, 756 (1996) 21.
2. L. R. Snyder and J. W. Dolan, *High-Performance Gradient Elution*, Wiley-Interscience, New York, 2007.
3. F. Leroy, B. Presle, F. Verillon, and E. Verette, *J. Chromatogr. Sci.*, 39 (2001) 487.
4. S. R. Needham and T. Wehr, *LCGC Europe*, 14 (2001) 244.
5. J. Ayrton, G. J. Dear, W. J. Leavens, D. N. Mallet, and R. S. Plumb, *J. Chromatogr. B*, 709 (1998) 243.
6. L. R. Snyder, J. J. Kirkland, and J. L. Glajch, *Practical HPLC Method Development*, 2nd ed., Wiley-Interscience, New York, 1997.
7. L. Hagdahl, R. J. P. Williams, and A. Tiselius, *Arkiv. Kemi*, 4 (1952) 193.
8. T. Braumann, G. Weber, and L. H. Grimme, *J. Chromatogr.*, 261 (1983) 329.
9. P. L. Zhu, L. R. Snyder, J. W. Dolan, N. M. Djordjevic, D. W. Hill, L. C. Sander, and T. J. Waeghe, *J. Chromatogr. A*, 756 (1996) 21.
10. P. Jandera and M. Špaček, *J. Chromatogr.*, 366 (1986) 107.
11. L. R. Snyder and J. W. Dolan, *J. Chromatogr. A*, 799 (1998) 21.
12. A. P. Schellinger and P. W. Carr, *J. Chromatogr. A*, 1077 (2005) 110.
13. W. Hancock, R. C. Chloupek, J. J. Kirkland, and L. R. Snyder, *J. Chromatogr. A*, 686 (1994) 31.
14. T. Jupille, L. Snyder, and I. Molnar, *LCGC Europe*, 15 (2002) 596.
15. B. F. D. Ghrist, B. S. Cooperman, and L. R. Snyder, *J. Chromatogr.*, 459 (1989) 1.
16. B. F. D. Ghrist and L. R. Snyder, *J. Chromatogr.*, 459 (1989) 63.
17. B. F. D. Ghrist and L. R. Snyder, *J. Chromatogr.*, 459 (1989) 25.
18. V. Concha-Herrera, G. Vivó-Truyols, J. R. Torres-Lapasió and M. C. Garciá-Alvarez-Coque, *J. Chromatogr. A*, 1063 (2005) 79
19. S. V. Galushko and A. A. Kamenchuk, *LCGC Intern.*, 8 (1995) 581.
20. H. Xiao, X. Liang, and P. Lu, *J. Sep. Sci.*, 24 (2001) 186.
21. W. Markowski and W. Golkiewicz, *Chromatographia*, 25 (1988) 339.
22. P. Jandera, *Adv. Chromatogr.*, 43 (2004) 1.
23. L. R. Snyder and D. L. Saunders, *J. Chromatogr. Sci.*, 7 (1969) 195.
24. U. D. Neue, D. H. Marchand, and L. R. Snyder, *J. Chromatogr. A*, 1111 (2006) 32.
25. H. Poppe, J. Paanakker, and M. Bronkhorst, *J. Chromatogr.*, 204 (1981) 77.
26. L. R. Snyder and J. W. Dolan, *Adv. Chromatogr.*, 38 (1998) 115.
27. J. Pellett, P. Lukulay, Y. Mao, W. Bowen, R. Reed, M. Ma, R. C. Munger, J. W. Dolan, L. Wrisley, K. Medwid, N. P. Toltl, C. C. Chan, M. Skibic, K. Biswas, K. A. Wells, and L. R. Snyder, *J. Chromatogr. A*, 1101 (2006) 122.
28. Z. Li, S. C. Rutan and S. Dong, *Anal. Chem.*, 68 (1996) 124.
29. R. Majors, *LCGC*, 20 (2002) 516.
30. L. R. Snyder and J. W. Dolan, *J. Chromatogr. A*, 721 (1996) 3.
31. P. J. Schoenmakers, H. A. H. Billiet, and L. De Galan, *J. Chromatogr.*, 205 (1981) 13.
32. P. J. Schoenmakers, A. Bartha, and H. A. H. Billiet, *J. Chromatogr.*, 550 (1991) 425.
33. P. Schoenmakers, in *Handbook of HPLC*, E. Katz, R. Eksteen, P. Schoemnakers, and N. Miller, eds., Dekker, New York, 1998, pp. 218–220.

34. A. P. Schellinger and P. W. Carr, *J. Chromatogr. A*, 1109 (2006) 253.
35. L. R. Snyder and J. W. Dolan, *J. Chromatogr. A*, 892 (2000) 107.
36. J. W. Dolan, L. R. Snyder, N, M. Djordjevic, D. W. Hill, L. Van Heukelem, and T. J. Waeghe, *J. Chromatogr. A*, 857 (1999) 41.
37. P. L. Zhu, J. W. Dolan and L. R. Snyder, D. W. Hill, L. Van Heukelem, and T. J. Waeghe, *J. Chromatogr. A*, 756 (1996) 51.
38. W. Hancock, R. C. Chloupek, J. J. Kirkland, and L. R. Snyder, *J. Chromatogr. A*, 686 (1994) 31, 45.
39. J. W. Dolan, L. R. Snyder, N. M. Djordjevic, D. W. Hill, L. Van Heukelem, and T. J. Waeghe, *J. Chromatogr. A*, 857 (1999) 1.
40. J. W. Dolan, L. R. Snyder, T. Blanc, and L. Van Heukelem, *J. Chromatogr. A*, 897 (2000) 37.
41. J. W. Dolan, *J. Chromatogr. A*, 965 (2002) 195.
42. J. W. Dolan, L. R. Snyder, N. M. Djordjevic, D. W. Hill, D. L. Saunders, L. Van Heukelem, and T. J. Waeghe, *J. Chromatogr.*, 803 (1998) 1.
43. A. P. Schellinger, D. R. Stoll, and P. W. Carr, *J. Chromatogr. A*, 1064 (2005) 143.
44. A. P. Schellinger, D. R. Stoll, and P. W. Carr, *J. Chromatogr. A*, 1192 (2008) 41.
45. A. P. Schellinger, P. Adam, D. R. Stoll, and P. W. Carr, *J. Chromatogr. A*, 1192 (2008) 54.
46. J. García, J. A. Martínez-Pontevedra, M. Lores, and R. Cela, *J. Chromatogr. A*, 1128 (2006) 17.
47. V. R. Meyer, *J. Chromatogr. A*, 1187 (2008) 138.
48. U. D. Neue and J. R. Mazzeo, *J. Sep. Sci.*, 24 (2001) 921.
49. U. D. Neue, *J. Chromatogr. A*, 1079 (2005) 153.
50. D. R. Stoll, X. Wang, and P. W. Carr, *Anal. Chem.*, 78 (2006) 3406.
51. S. E. G. Porter, D. R. Stoll, S. C. Rutan, P. W. Carr, and Cohen, *Anal. Chem.*, 78 (2006) 5559.
52. X. Wang, W. E. Barber, and P. W. Carr *J. Chromatogr. A*, 1107 (2006) 139.
53. D. R. Stoll, X. Li, X. Wang, P. W. Carr, S. E. G. Porter, and S. C. Rutan, *J. Chromatogr. A*, 1168 (2007) 3.
54. M. Gilar and U. D. Neue, *J. Chromatogr. A*, 1169 (2007) 139.
55. U. D. Neue, *J. Chromatogr. A*, 1184 (2008) 107.
56. D. R. Stoll, X. Wang, and P. W. Carr, *Anal. Chem.*, 80 (2008) 268.
57. J. M. Davis, D. R. Stoll, and P. W. Carr, *Anal. Chem.*, 80 (2008) 461, 8122.
58. X. Li, D. R. Stoll, and P. W. Carr, *Anal. Chem.*, 81 (2009) 845.
59. L. A. Romanyshyn and P. R. Tiller, *J. Chromatogr. A*, 928 (2001) 41.
60. K. D. Patel, A. D. Jerkovich, J. C. Link, and J. W. Jorgenson, *Anal. Chem.*, 76 (2004) 5777.
61. H. Chen and C. Horváth, *J. Chromatogr. A*, 705 (1995) 3.
62. S. K. Paliwal, M. de Frutos, and F. E. Regnier, in *Methods in Enzymology*, Vol. 270. W. S. Hancock and B. L. Karger, eds., Academic Press, Orlando, 1996, p. 133.
63. J. J. Kirkland, F. A. Truszkowski, and R. D. Ricker, *J. Chromatogr. A*, 965 (2002) 25.
64. L. Xiong, R. Zhang, and F. E. Regnier, *J. Chromatogr. A.*, 1030 (2004) 187.
65. P. Brown and E. Grushka, eds., *Design of Rapid Gradient Methods for the Analysis of Combinatorial Chemistry Libraries and the Preparation of Pure Compounds*, Dekker, New York, 2001.

66. P. J. Schoenmakers, P. Marriott, and J. Beans, *LCGC Europe*, 16 (2003) 335.
67. J. C. Giddings, *Multidimensional Chromatography: Techniques and Applications*; Dekker, New York, 1990.
68. R. E. Murphy, M. R. Schure, and J. P. Foley, *Anal. Chem.*, 70 (1998) 1585.
69. K. Horie, H. Kimura, T. Ikegama, A. Iwatsuka, N. Saad, O. Fiehn, and N. Tanaka, *Anal. Chem.*, 79 (2007) 3764.
70. D. R. Stoll, J. D. Cohen, and P. W. Carr, *J. Chromatogr. A*, 1122 (2006) 123.
71. S. Peters, G. Vivó-Truyols, P. J. Marriott, and P. J. Schoenmakers, *J. Chromatogr. A*, 1156 (2007) 14.
72. J. W. Dolan and L. R. Snyder, *J. Chromatogr. A*, 1216 (2009) 3467.
73. M. Gilar, P. Olivova, A. E. Daly, and J. C. Gebler, *Anal. Chem.*, 77 (2005) 6426.
74. K. Horie, H. Kimura, T. Ikegami, A. Iwatsuka, N. Saad, O. Fiehn, and N. Tanaka, *Anal. Chem.*, 79 (2007) 3764.
75. P. J. Schoenmakers, G. Vivo-Truyols, and W. M. C. Decrop, *J.Chromatogr. A*, 1120 (2006) 282.
76. P. Jandera, *J. Chromatogr. A*, 965 (2002) 239.
77. P. Jandera, *J. Chromatogr. A*, 1126 (2006) 195.
78. T. Yoshida, *J. Chromatogr. A*, 60 (2004) 265.
79. H. Schlichtherle-Cerny, M. Affolter, and C. Cerny, *Anal. Chem.*, 75 (2003) 2349.
80. P. Jandera, *J. Chromatogr. A*, 965 (2002) 239.
81. S. Williams, *J. Chromatogr. A*, 1052 (2004) 1.

CHAPTER TEN

COMPUTER-ASSISTED METHOD DEVELOPMENT

10.1 INTRODUCTION

Computer-assisted method development is a broad term, one that might be applied to the use of any software that facilitates method development. In this chapter we will

Introduction to Modern Liquid Chromatography, Third Edition, by Lloyd R. Snyder, Joseph J. Kirkland, and John W. Dolan

emphasize commercial *computer-simulation* software that can predict separation as a function of one or more experimental conditions, by means of experimental data from a few preliminary separations. Because of this software's ability to predict separation—as opposed to carrying out "real" experiments—we can reduce the amount of experimental work that is required while ensuring the "best" conditions for the final method.

A simple example for isocratic RPC is the prediction of separation as a function of mobile-phase strength (%B); for this application, two experimental runs are required prior to computer simulation. In the example of Figure 10.1, experiments using a mobile phase of 10 and 20% B (other conditions unchanged) are shown in Figure 10.1*a*,*b*. Resulting data are entered into the computer: retention times for each solute in each run, %B values for the two runs (10, 20% B), and other experimental conditions (A- and B-solvent compositions, column dimensions and particle size, flow rate, temperature; see Section 10.2.1). The computer can now be interrogated for predictions of separation (simulations) as a function of changes in mobile-phase %B, flow rate, column dimensions, and particle size.

The most useful information provided by computer simulation is usually a *resolution map*, as illustrated in Figure 10.1*c* for the separation of this mixture of seven acids and bases. A resolution map is a plot of the critical resolution R_s, for the two least-resolved peaks, as a function of the condition or conditions that were varied in the initial experimental runs (Section 6.3.3 and Fig. 6.13*b* provide additional information on resolution maps). In this example %B was varied, so R_s is plotted as a function of %B. In Figure 10.1*c* we see that maximum resolution occurs for three different conditions: 9, 17, and 25% B, with the largest resolution for 17% B. We can also request a simulated chromatogram for any value of %B, as illustrated in Figure 10.1*d*–*f* for the latter "optimum" %B values. The computer can provide further information for each of these (and other) separations as a function of %B; the range in values of k is especially useful in this regard (indicated for each of the separations in Figs. 10.1*d*–*f*). For the usually recommended range of $1 \leq k \leq 10$, it is seen that the separation of Figure 10.1*f* (25% B, $1 \leq k \leq 6$) comes closest to this goal. However, some laboratories might prefer the separation of 17% B ($3 \leq k \leq 12$) in Figure 10.1*e* for its greater resolution. For the present sample, however, the retention range for any one of these three separations might be regarded as acceptable, depending on the goals of separation.

Once an acceptable %B has been selected, it may be possible to trade excess resolution for a shorter run time, by changing column conditions. For example, the resolution of the separation with 25% B ($R_s = 2.4$) could be reduced without adverse effects by a decrease in column length and/or an increase in flow rate. Computer simulation allows the user to explore the effects of changing column dimensions, particle size, or flow rate. Figure 10.1*g* shows one such simulation (for 25% B): a reduction in column length by half (from 150 to 75 mm), a change in particle size from 5 to 3 μm, and no change in flow rate. Acceptable resolution ($R_s = 2.1$) and pressure ($P = 2200$ psi) are achieved in a run time of only 3 minutes, while retaining a preferred retention range of $1 \geq k \geq 6$.

The final choice of an "optimized" separation may depend on considerations other than resolution, run time, and retention range; for example, an acceptable

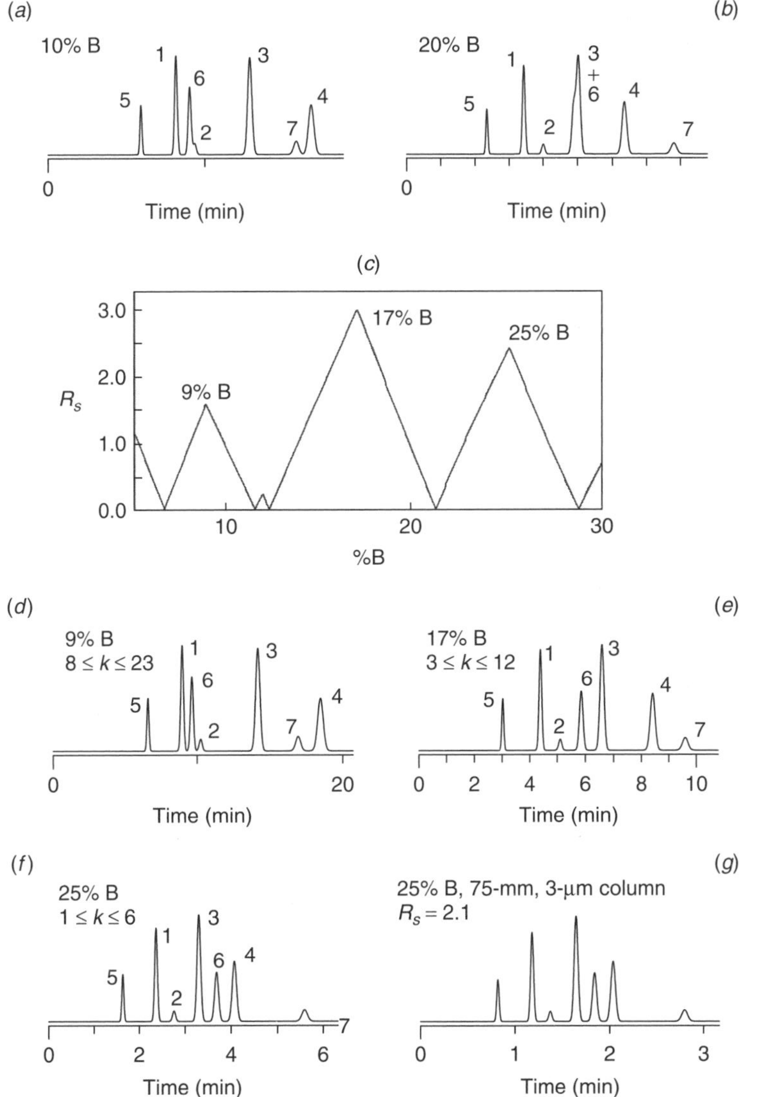

Figure 10.1 Illustration of computer simulation for the separation of a mixture of acids and bases. Sample (a mixture of acids and bases) and conditions are the same as in Figure 7.7. (*a*,*b*) Initial experimental separations; (*c*) the resolution map predicted from the data of (*a*, *b*); (*d*–*f*) predicted chromatograms for various values of %B (corresponding to maximum resolution); (*g*) predicted chromatogram for conditions that favor fast separation (with adequate resolution). Computer simulations based on data of [1].

column pressure, narrow peaks for more sensitive detection, and the relative importance of different peaks in the chromatogram. For this reason, computer-simulation software should be flexible, so as to allow the user to examine the consequences of various changes in separation conditions—rather than just offer a single, "optimized" separation to the user.

10.1.1 Basis and History of Computer Simulation

Computer simulation makes use of various empirical and theoretical relationships; for example, Equations (2.7), (2.10), (2.18), (2.23), and (2.26), and Equations (9.7) to (9.14*a*), as means for predicting retention time and peak width for either isocratic or gradient elution. Often computer simulation begins with a relationship between values of k and %B (or $\phi = 0.01$ %B). For the case of RPC,

$$\log k = \log k_w - S\phi \tag{10.1}$$

Here k_w is the value of k for $\phi = 0$ (i.e., water as mobile phase), and S is a constant for a given solute when all conditions except %B are held fixed. Two actual separations, where either %B (for isocratic elution) or gradient time (for gradient elution; Section 9.2.4.2) is varied, permit the calculations of values of k_w and S for each solute in the sample. For NPC a similar relationship between k and ϕ exists (Eq. 8.4*a*). Once values of k_w and S are known, values of k and retention time t_R can then be predicted for any values of %B (or gradient time), column dimensions, or flow rate (Eqs. 2.5 and 9.10).

Peak widths W can be calculated in various ways: (1) the assumption of some value of the plate number N for the initial separations (e.g., $N = 10{,}000$ for a 100-mm column with 3-μm particles), (2) measurements of peak widths for the initial experimental runs, or (3) calculation of peak widths based on values of N from Equation (2.17). Computer simulation can be extended to the case of segmented gradients, columns of different size, changes in flow rate, and so forth, by various relationships [2]. The accuracy of computer simulation has been confirmed in numerous experimental studies, as summarized in [2]. Retention times are usually reliable within a few percent while, more important, predicted values of resolution R_s are typically accurate to ±10%, which is generally adequate for use in method development.

The first example of computer simulation for HPLC was reported in 1978 by Laub and Purnell [3]; they described the use of a resolution map for isocratic RPC as a function of temperature. A few years later [4], Glajch, Kirkland, et al., reported an isocratic procedure for optimizing solvent type, based on the experimental plan of Figure 10.2*f*, which requires seven experimental runs with different proportions of ACN, MeOH, and THF (see also Section 6.4.1.1; Fig. 6.24). At the same time [5], Deming and coworkers presented a similar scheme for simultaneously optimizing mobile-phase pH and the concentration of an ion-pair reagent. The isocratic approach of Glajch and Kirkland was extended to gradient elution in 1983 [6]. In 1985 DryLab^R software was introduced and subsequently expanded into the most comprehensive and widely used computer-simulation software presently available [7, 8]. The latter software is described in Section 10.2, as an example of various possible applications of computer simulation. Similar computer-simulation programs were developed by others after 1985, as reviewed and/or compared in [9–12] and summarized in Section 10.3.4.

10.1.2 When to Use Computer Simulation

Method development can be pursued with or without the help of computer simulation, so it is important to weigh the potential pros and cons of computer simulation for each application.

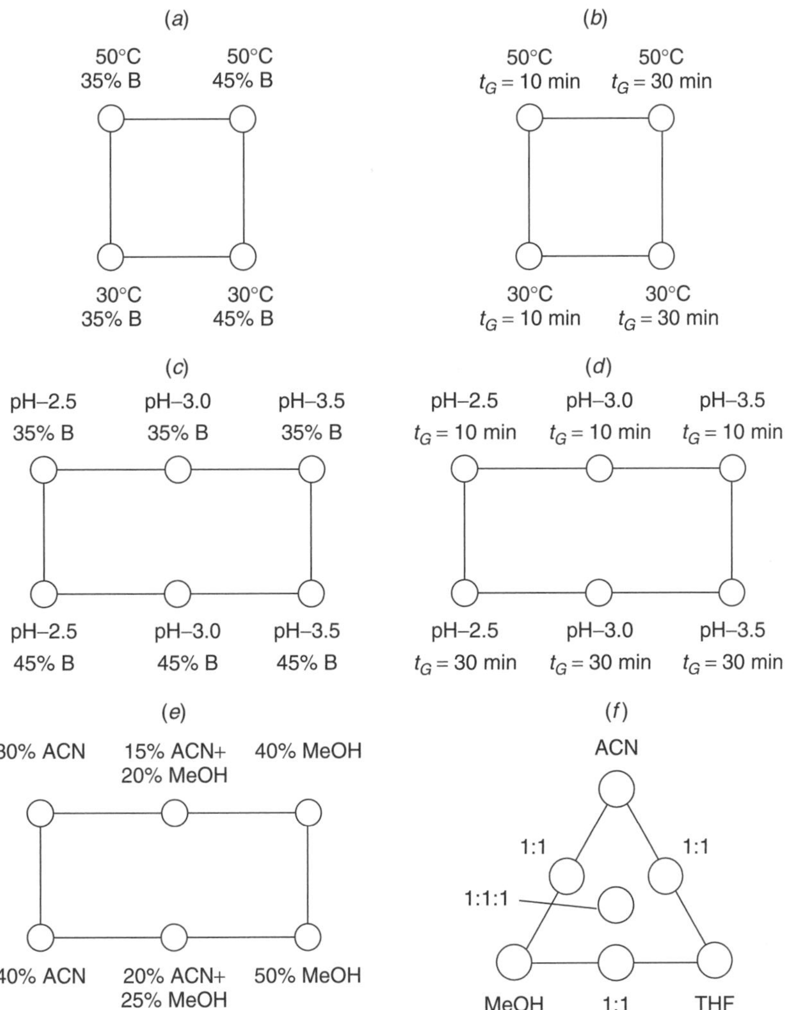

Figure 10.2 Examples of different systematic approaches for optimizing separation selectivity (experimental design); (*a*, *c*, *e*, *f*) isocratic experiments, (*b*, *d*) gradients.

10.1.2.1 Advantages

The value of computer simulation is likely to be greatest when one or more of the following conditions applies:

- separations by gradient elution
- complex samples that contain 5 to 10 components or more
- very short run times are required (e.g., a minute or less)
- method robustness is critical
- minor improvements in resolution or run time are worthwhile

The isocratic separation of a sample that contains only a few components—and where run time is not critical—may be developed quickly and easily by means of a few trial-and-error experiments. In such cases the advantage of computer simulation is often marginal, although this overlooks other possible benefits of computer simulation (ability to fine-tune separation for faster separation, explore method robustness, etc.). *Gradient elution*, on the other hand, involves additional separation variables that are more conveniently explored via computer simulation. Similarly *complex samples* with lots of closely bunched peaks often require a large number of experiments to achieve a final successful separation; many of these experiments can be conveniently replaced by computer simulation. *Very short run times* can present a similar challenge, one also requiring several experiments.

When *method robustness* is critical, computer simulation allows a quick examination of the effects of changes in different conditions on the separation, without a need for additional experiments. *Small improvements* in separation via trial-and-error changes in conditions can be effected easily by means of computer simulation. *Simultaneous variation* in two separation conditions that affect selectivity (Table 2.2) is a powerful tool that is often required for complex samples or other demanding separations. In such cases computer simulation can be used to determine optimal conditions by means of a minimum number of experimental runs (Fig. 10.2). Changes in *relative retention* as conditions are varied can confuse peak identification and the interpretation of experimental runs, especially when a large number of components are present in the sample. With computer simulation individual peaks are automatically matched between runs (following peak tracking for the initial experiments); the interpretation of simulated experiments is thereby greatly simplified. Examples of each of these various advantages of computer simulation will be shown in Section 10.2.

10.1.2.2 Disadvantages

The cost of computer simulation software, and the time required for the chromatographer to become familiar with its usage, represent the main impediments to a wider use of this approach. When method development is carried out infrequently, and samples are usually easy to separate, the potential value of computer simulation may be at best marginal. Another barrier to the use of computer simulation is the belief that the chromatographer has no *need* for computer simulation, because of his or her experience and competence. While computer simulation is seldom essential for carrying out method development, it can frequently reduce cost and improve the quality of the final method, even for experienced chromatographers. On the other hand, computer simulation is not a substitute for competence; chromatographic skills are still important for its effective use.

A less important objection to computer simulation is that the predicted chromatograms are "ideal" rather than "real." Thus baselines are assumed not to drift, peaks may be assumed to be symmetrical (although peak tailing can be taken into account by computer simulation; Section 10.2.3.5), and baseline artifacts or extraneous peaks are usually ignored. However, such chromatographic artifacts often detract from a final method, and are best eliminated before computer simulation is started. In other cases artifacts may not affect the interpretation of the separation or the choice of final conditions, and can be ignored.

Finally, any mistakes in data entry (including mismatched peaks) can result in major errors in predicted separations. However, errors in data entry can be reduced by the automatic transfer of data from the data system to the computer-simulation program. Other errors that might occur from the use of computer simulation are usually obvious—and easily corrected—when predicted separations and (confirming) experimental chromatograms do not agree.

10.2 COMPUTER-SIMULATION SOFTWARE

Computer simulation is best used within an overall strategy of method development, as described in other sections of this book and illustrated in Section 10.4. Thus initial simulations should first examine the "best" retention range (values of k for isocratic elution or k^* for gradient elution) by varying isocratic %B or gradient time. At the same time any changes in selectivity as a function of %B or t_G should be noted; maximum resolution may correspond to an intermediate value of (as in Fig. 10.1*c*). If further changes in relative retention are needed, other variables that affect selectivity should be explored next. The various plans ("experimental designs") of Figure 10.2 summarize some of the more popular approaches. After one of these sets of experiments is carried out (e.g., four actual runs in the plan of Fig. 10.2*a*), the effects of simultaneous changes of two different separation conditions can be simulated. In principle, any two separation conditions that affect selectivity can be modeled as in Figure 10.2. After experiments have been carried out according to any of the choices of Figure 10.2, computer simulation can be used to select the best isocratic or gradient conditions (Section 9.2.2); column conditions (column dimensions, particle size, flow rate) can also be varied as a means of increasing resolution or decreasing run time.

10.2.1 DryLab Operation

This section illustrates some useful features of computer simulation that form part of the DryLab software. Many of these features can also be found in other commercial computer-simulation software.

An experimental design and separation mode are first selected, which defines the number of experimental calibration runs that will be required for computer simulation (as in Fig. 10.2). In this chapter we will limit our discussion to computer simulation for RPC. We will select the experimental design of Figure 10.2*b* as example, for the isocratic separation of a mixture of eight corticosteroids—but based on initial *gradient-elution* experiments. This requires four initial runs at two different values of gradient time t_G (20, 60 min) and temperature T (30, 60°C), in order to simultaneously optimize %B and temperature T. The following experimental data are entered into the computer (see Fig. 10.3*a*): equipment dwell volume (5.5 mL; used only for initial experiments by gradient elution), column dimensions (250 × 4.6 mm), particle size (5 μm), flow rate (2.0 mL/min), mobile phase composition ("elution data;" water, A; acetonitrile, B), gradient range (0–100% B), and retention times plus peak areas for each compound in each of the four experimental runs. Finally, it is necessary to carry out peak matching (Section 10.2.4), where retention data for each peak are matched with a given compound in the sample. Peak matching can be carried out manually, or facilitated by the computer; the peaks of Figure 10.3*a*

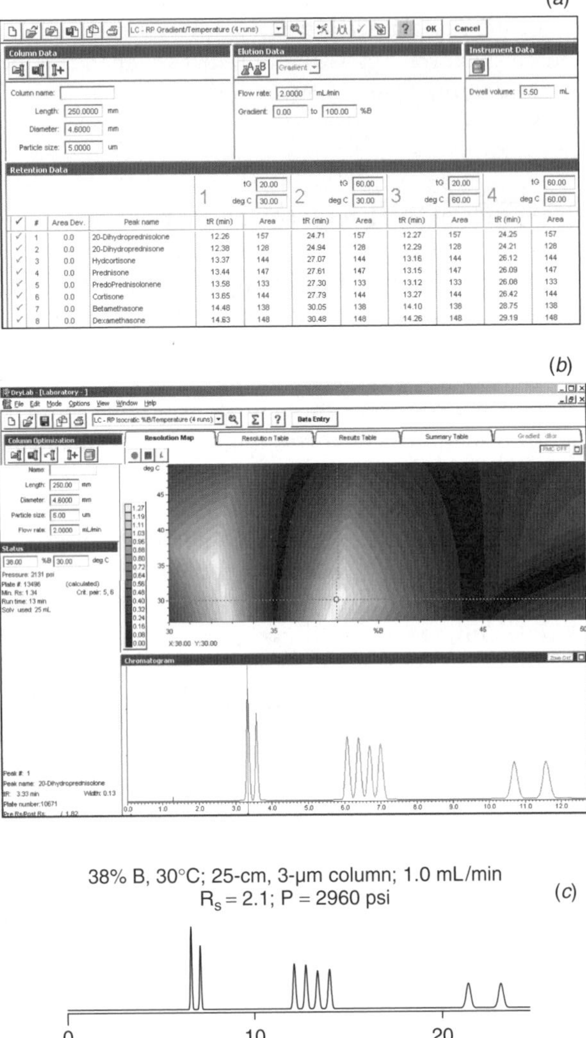

✓	#	Area Dev.	Peak name	1 (tG 20.00, deg C 30.00) tR (min)	Area	2 (tG 60.00, deg C 30.00) tR (min)	Area	3 (tG 20.00, deg C 60.00) tR (min)	Area	4 (tG 60.00, deg C 60.00) tR (min)	Area
✓	1	0.0	20-Dihydroprednisolone	12.26	157	24.71	157	12.27	157	24.25	157
✓	2	0.0	20-Dihydroprednisone	12.38	128	24.94	128	12.29	128	24.21	128
✓	3	0.0	Hydcortisone	13.37	144	27.07	144	13.16	144	26.12	144
✓	4	0.0	Prednisone	13.44	147	27.61	147	13.15	147	26.09	147
✓	5	0.0	PredoPrednisolonene	13.58	133	27.30	133	13.12	133	26.08	133
✓	6	0.0	Cortisone	13.65	144	27.79	144	13.27	144	26.42	144
✓	7	0.0	Betamethasone	14.48	138	30.05	138	14.10	138	28.75	138
✓	8	0.0	Dexamethasone	14.63	148	30.48	148	14.26	148	29.19	148

Figure 10.3 Examples of data entry (*a*) (gradient data) and laboratory (*b*) screens for isocratic computer simulation by means of DryLab. The sample is a mixture of eight corticosteroids [13]. Conditions in (*b*): 250 × 4.6-mm C_{18} column (5-μm particles); mobile phase, acetonitrile-water; 38% B; 30°C; 2.0 mL/min; (*c*) computer simulation for separation of sample with conditions of (*b*), except for a change in particle size (3 μm) and flow rate (1.0 mL/min).

have been matched satisfactorily (as confirmed by the "area dev" column, which measures the %-deviation of predicted from actual peak areas).

With the completion of data entry, a *Laboratory* screen for computer simulations can be selected (Fig. 10.3*b*). Since predictions for an isocratic separation are desired, the %B-°C output mode is selected. Several display options are shown in the tabs above the resolution map: resolution map, resolution table, results table, and so forth. A resolution map has been selected in the example of Figure 10.3*b*;

values of R_s are indicated in shades of gray for various combinations of °C (y-axis) and %B (x-axis). In this case maximum $R_s = 1.34$ occurs for $T = 30°C$ and %B = 38% (note the cross-hairs in Fig. 10.3*b*). When the cursor is placed within the resolution map on the values of %B and T for maximum resolution (or any selected set of values of %B and T), a chromatogram for those conditions is immediately displayed (bottom of Figure 10.3*b*). An alternative choice of conditions (32.6% B, 30°C, when entered into the "status" box) provides a similar resolution ($R_s = 1.30$), but requires a longer run time (22 min vs.12 min for the example of Fig. 10.3*b*). Further considerations that mitigate against the selection of T = 30°C and 32.6% B are as follows: First, the retention range is less favorable ($4 \leq k \leq 16$, compared with $2 \leq k \leq 8$). Second, for the run with 32.6% B and 30°C, small changes in %B can result in a sizable decrease in resolution—which is less true for $T = 30°C$ and %B = 38%, so the latter separation is expected to be more robust. All of the latter considerations are immediately apparent from an inspection of Figure 10.3*b*.

The user can carry out further simulations, such as a change in column conditions, without a need for additional laboratory experiments. Note that the pressure (2131 psi) is displayed, which can be used to avoid conditions that would overpressure the pump. As the resolution of this separation might be unacceptable ($R_s = 1.3$), we can improve resolution by either increasing column length or decreasing particle size. Figure 10.3*c* shows the prediction for a change to a column of the same size (250 × 4.6-mm) packed with 3-μm particles, in place of the previous 5-μm column. Because of the increase in pressure due to the smaller particles, it is necessary to reduce the flow rate from 2.0 mL/min in Figure 10.3*b* to 1.0 mL/min in the separation of Figure 10.3*c*. The resulting resolution ($R_s = 2.1$) and pressure ($P = 2960$ psi) are acceptable, although the run time is doubled to 24 minutes. After exploring the options above, the best-possible separation might be considered inadequate because of an excessive run time. In this case a different column can be selected (or methanol can be substituted for acetonitrile as B-solvent), and the entire procedure repeated (four more experimental runs varying gradient time and temperature).

10.2.2 Gradient Optimization

The four experimental runs of Figure 9.18 correspond to the experimental design of Figure 10.2*b* (varying temperature and gradient time). In this case, however, we will seek an optimized gradient separation (rather than an isocratic separation, as in Fig. 10.3). The resulting resolution map is shown in Figure 10.4*a*. The cross-hairs (and arrow) in the latter resolution map correspond to conditions for maximum resolution ($R_s = 2.1$): a gradient of 5–100% in 35 minutes, and a temperature of 47°C. The resulting chromatogram is shown in Figure 10.4*b*. Because the last peak leaves the column at 14 minutes (before the gradient ends), the gradient can be shortened to 5–43%B in 14 minutes (see Section 9.3.4). Previously the (manual) trial-and-error optimization of gradient time and temperature for this sample yielded "optimized" conditions of 5–46%B gradient in 14 minutes at 50°C temperature, with a resulting resolution of $R_s = 1.9$. In this example (Fig. 10.4) computer simulation provides only a slight improvement in resolution compared to manual, trial-and-error method development. However, computer simulation does confirm that the final conditions selected are truly "best," since the resolution maps summarize predictions for several hundred combinations of conditions (different "virtual experiments").

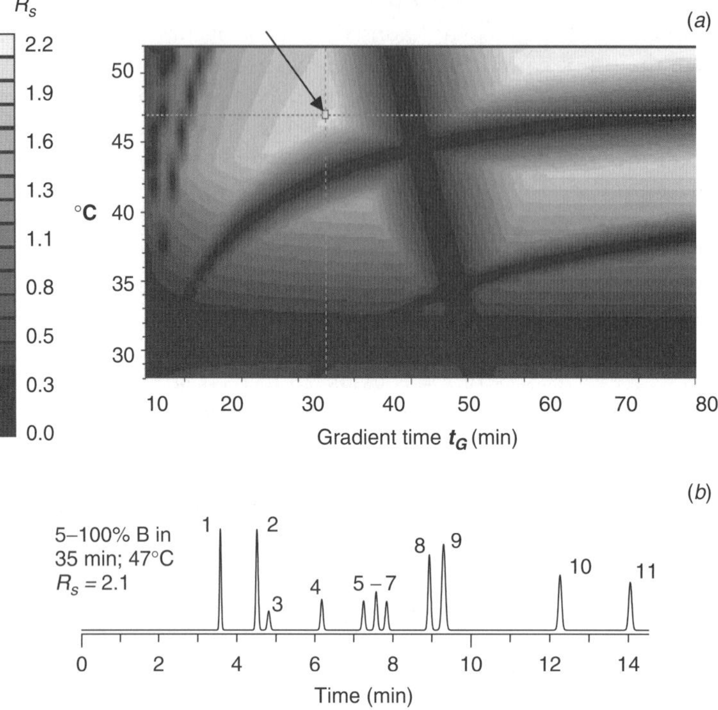

Figure 10.4 Computer simulation for the optimization of gradient time and temperature. Same conditions and 'irregular' sample as for Figure 9.18 (5–100% B calibration gradients). (*a*) Resolution map; (*b*) optimized separation.

A further inspection of the map of Figure 10.4*a* shows that resolution decreases only slightly for a shorter gradient time, especially if temperature is decreased at the same time. Thus, for a 5–46%B gradient in 12 min at 45°C, resolution is hardly changed ($R_s = 2.0$), while run time is shortened by 2 minutes. Computer simulation is thus useful for fine-tuning a separation and so achieves maximum resolution and/or minimum run time by exploring further changes in conditions by trial and error. Minimizing run time in this fashion can be important when a method will be used for hundreds or thousands of samples. Method robustness can also be checked by making small changes in experimental conditions. In the example of Figure 10.4*a* a decrease in temperature by 2°C would result in a serious loss of resolution ($R_s = 1.3$). When a method is carried out with a different HPLC system, uncertainties in the actual temperature of ±1–2°C are not uncommon (Section 3.7). The sensitivity of the present method to unintended changes in temperature can be greatly reduced by small changes in the final values of %B and T for the method, with only a modest sacrifice in resolution. For example, the choice of a gradient time of 13 minutes (5–46% B gradient) and a temperature of 48°C gives a resolution of $R_s = 1.9$, but a change in temperature of ±2°C results in no loss of resolution; that is, the latter method is robust with respect to moderate temperature variations. Trial-and-error simulations of this kind can be carried out within a few seconds each (no new experiments required).

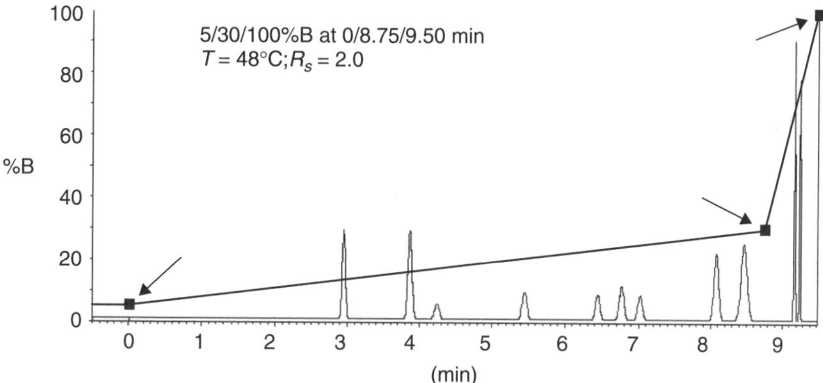

Figure 10.5 Trial-and-error computer simulations for the design of a segmented gradient (for an improved separation of Figure 10.4*b*).

A large number of *segmented gradients* can be explored intuitively in a very short time by means of computer simulation, allowing the easy development of a final method that meets the needs of the user. This is illustrated in Figure 10.5, which shows the use of the DryLab computer screen for the design of such gradient shapes. The preceding example ($t_G = 13$ min and 48°C) is the starting point. During the design of a gradient, each of the three inserted points shown in Figure 10.5 (marked by arrows) can be dragged via a computer mouse to any desired values of time and %B (thereby changing the gradient program), while the resulting chromatogram is simultaneously displayed as in Figure 10.5. The gradient shown in Figure 10.5 shortens the run time to 9.5 minutes, while maintaining a rugged separation with regard to small changes in temperature (see the related discussion of Section 9.2.2.5).

10.2.3 Other Features

Computer-simulation software can offer the user a number of other options, as summarized in Table 10.1 (see items 4–9).

10.2.3.1 Isocratic Predictions from Gradient Data

Once computer simulation has been initiated on the basis of two gradient experiments, where only gradient time is varied, it is also possible to predict isocratic separation as a function of %B. Similarly, when four experiments, as in Figure 10.2*b*, are carried out, isocratic separation as a function of %B and temperature can be predicted, as illustrated in Figure 10.3. This ability to predict isocratic separations on the basis of gradient experiments allows an evaluation of isocratic elution as a possible alternative to gradient separation. As the range in k for the sample decreases, isocratic elution becomes increasingly preferred. Predictions of isocratic separation from gradient data are somewhat less reliable, compared to the use of isocratic experiments for this purpose. Thus, if an initial gradient experiment suggests that isocratic elution is preferable (Section 9.3.1), the latter experiment might be followed by the four isocratic experiments of Figure 10.2*a*, rather than continuing with gradient experiments as in Figure 10.2*b*. (For a further discussion of the prediction of isocratic elution from gradient runs, see [14]). It should be

Table 10.1

Computer-Simulation Options

Option	Comment
1. Simulation of chromatograms for isocratic %B or different gradient conditions, and temperature	Section 10.1
2. Use of resolution maps for easy selection of optimized conditions	Section 10.1
3. Selection of best column conditions	Section 10.1
4. Isocratic predictions from gradient data	Simulation of changes in t_G and T also allow predictions of isocratic elution as a function of %B and T (Sections 10.2.1, 10.2.3.1).
5. Designated-peak selection	R_s and R_s-maps calculated only for peaks of interest; peaks not overlapping the peak of interest are ignored (Section 10.2.3.2).
6. Change in other conditions	Separation can be simulated for any variables that affect selectivity (Table 2.2); a change in column requires experimental data for that column (Section 10.2.3.3)
7. Computer-selection of a multi-segment gradient	Manual or automated searches possible (Section 10.2.3.4)
8. Peak tailing	Can be simulated (Section 10.2.3.5)
9. Two-run procedures for improving resolution	Different gradients for different peaks in same sample (Section 10.2.3.6)

noted that although isocratic separation can be predicted from gradient experiments, the prediction of gradient elution from isocratic experiments is not convenient; for maximum flexibility starting method development with gradient experiments often makes the most sense.

10.2.3.2 Designated-Peak Selection

Many chromatograms contain peaks that are of no direct interest to the chromatographer. For example, the analysis of biological or environmental samples for specific compounds may be complicated by the presence of numerous interfering peaks. Similarly, in preparative separation (Chapter 15), we are often concerned with the purification and recovery of a single compound in the sample. When only some peaks in the chromatogram are of interest, it is important to resolve each of these peaks from *all* remaining peaks—but not the separation of interfering peaks from each other, as this is unnecessary. As an illustration of designated-peak selection, consider the example of Figure 10.4, but assume that it is required to assay only peaks 3, 8, and 10 in the presence of the remaining "interfering" peaks. A final separation is therefore required that will separate these three peaks from each other *and* from any other peaks that might overlap 3, 8, and 10 (this is a much easier task than the separation of all 11 peaks from each other).

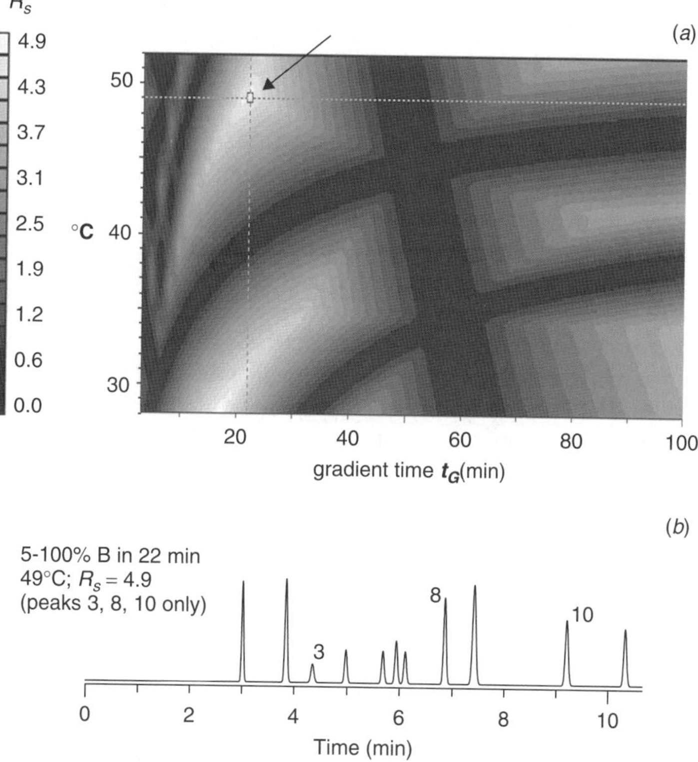

Figure 10.6 Optimizing the separation of selected peaks in the 'irregular' sample of Figure 10.4; use of computer simulation to select optimum values of gradient time and temperature for the separation of peaks 3, 8, and 10 from remaining peaks. (*a*) Resolution map; (*b*) best separation for 5–100% B in 33 minutes and 49°C.

Using computer simulation, we can designate peaks 3, 8, and 10 as peaks of interest. When a resolution map for designated peaks only is next requested (peaks 3, 8, and 10 in this example), values of the critical resolution R_s will be plotted versus temperature and gradient time for *just* peaks 3, 8, and 10. This is illustrated in Figure 10.6*a*, which indicates conditions for the best separation (cross-hairs and arrow in *a*). The corresponding chromatogram is shown in Figure 10.6*b* (numbers mark peaks of interest). As anticipated, the possible critical resolution for the separation of only three of the 11 peaks in this sample is much greater ($R_s = 4.9$) than for the separation of all 11 peaks ($R_s = 2.1$ in Fig. 10.4); a shorter run time is also required (11 vs.14 min). The latter "excess" resolution can be traded for a much shorter run time by a suitable change in column conditions. For example, reducing column length by half and doubling the flow rate reduces run time to 3 minutes, while maintaining a resolution of $R_s = 3.0$ and leaving column pressure unchanged.

10.2.3.3 Change in Other Conditions

Other conditions that affect selectivity can be modeled by computer simulation by varying either one or two different conditions at a time. In addition to the choices of Figure 10.2, any combination of %B or gradient time, temperature, B-solvent mixtures, mobile phase pH, buffer, or ion-pair reagent concentration can be simulated for RPC. Normal-phase and ion-exchange separations can also be modeled for most conditions that affect their selectivity.

10.2.3.4 Computer Selection of the Best Multi-Segment Gradient

Other than for cleaning the column as in Figure 9.11*b*, two-segment gradients are used primarily for one of three general purposes. First, the resolution of bunched peaks at the beginning of the gradient often can be improved with an isocratic hold (e.g., Fig. 9.9*c*, *d*). Second, the separation of critical peak-pairs at the beginning and end of a separation—when the peak-pairs respond differently to changes in gradient steepness—can be improved by the use of segments of differing steepness (e.g., Figs. 9.12*a*,*b*). Third, runs with excessive resolution at the end often can be shortened by using a steeper gradient to compress the end of the chromatogram (e.g., Fig. 10.5). Computer simulation can provide a trial-and-error examination of a large number of two-segment gradients (as in Fig. 10.5), followed by the selection of a gradient that yields the best selectivity for maximum resolution and/or shortest run time. However, the application of this approach to several samples [15] suggests that the advantage of two-segment gradients for further improvement in resolution (as in the example of Fig. 9.12*b*) is often marginal; thus it is rare for a segmented gradient to improve resolution by as much as 0.5 units. Segmented gradients for the purpose of maximizing resolution can be developed automatically by some computer software [15–18].

When compared with the simultaneous optimization of gradient time and separation temperature (a preferred approach), there appears to be little advantage in the sole use of multi-segment gradients for increasing resolution [15]. A minor exception is the use of segmented gradients for the separation of samples that contain large molecules such as proteins (Section 9.2.2.5). A disadvantage of segmented gradients is that they can contribute to problems in method transfer because of gradient rounding (Section 3.10.1.2; [19]).

10.2.3.5 Peak Tailing

Peak tailing should be corrected before experimental method-development separations are carried out. For some samples, however, this may not be possible. Because separation can be strongly affected by peak tailing, computer simulation should take this into account. Some examples of peak tailing were shown in Figures 2.16 and 2.17, all of which were created by computer simulation with the DryLab program. While moderate peak tailing usually has a limited effect on separation and resolution, this is not the case for two compounds whose size is very different (compare Fig. 2.16*e* and Fig. 2.16*d*). In these cases it is important that computer simulation can reliably simulate peak tailing, based on the tailing of peaks in the input runs.

10.2.3.6 Two-Run Procedures for the Improvement of Sample Resolution

When the sample contains 15 or more components, it may be difficult to achieve adequate resolution (e.g., $R_s \geq 2$; see Section 2.3.4) in a single gradient separation. An alternative, occasionally successful, approach for such samples is the use of two *different* gradient procedures (or "runs") for the same sample [20], but with no change in either the column or the A- and B-solvents. For example, the two runs might each use a different gradient time t_G and a different temperature T that allows the two runs to be carried out without any additional operator intervention during the analysis of a set of samples. All samples would be assayed first with one set of conditions, followed by their re-analysis using the second set of conditions (different values of t_G and T). The goal is adequate resolution of every peak of interest in one or the other of the two runs, which then allows an assay of all the peaks in the sample based on a composite of the two runs. By means of computer simulation, it is relatively easy to select best conditions for each of the two runs so that overall critical resolution for the sample is maximized [20]. A similar approach has been described for isocratic separation, which uses two 'complementary' runs [21].

10.2.3.7 Examples of Computer Simulation as Part of Method Development

Numerous examples of computer simulation have been reported, as summarized in [22] and p. 119 of [2]. Many of these reports are intended as illustrations of computer simulation, or as examples of its accuracy. Other publications report the use of computer simulation for the actual development of a routine method. Fifteen examples of the latter, more representative applications of computer simulation are summarized in Table 10.2.

10.2.4 Peak Tracking

When varying conditions during method development that can change relative retention, it is important to keep track of which peak is which in each chromatogram. Means for matching peaks were discussed in Section 2.7.4. Peak tracking is even more important for computer simulation because errors in peak assignment often result in large errors in predicted separations. Automated peak tracking, based on relative retention and peak areas, is included in DryLab software. Because this software's procedure is restricted to relatively simple separations ($\leq$10 peaks, limited changes in relative retention, no more than 10-fold variation in peak area), its accuracy can be improved by the use of supplemental software (Peak Match®; Molnar Institut, Berlin). Mass spectrometric detection may be required for samples containing many peaks whose sizes vary over orders of magnitude. Once peak tracking has been carried out for the experiments used in computer simulation, peaks are automatically matched for all subsequent simulations. Despite the importance of accurate peak tracking when using computer simulation, errors in peak tracking will be apparent when predicted and experimental separations are compared (the predicted retention times for one or more peaks will not match the corresponding experiment).

10.2.5 Sources of Computer-Simulation Software

Computer simulation software is available from several companies or groups, for example, ACD/LC Simulator® (Advanced Chemistry Development, Toronto, Ontario, Canada); ChromSmart® (Agilent, Palo Alto); ChromSword® (Merck KGaA, Darmstadt, Germany); DryLab® (Molnar Institut, Berlin), Turbo Method Development (PerkinElmer, Shelton, CT); Osiris^R [31]; and Preopt-W^R [32]. These software packages are for the most part limited to simulations of RPC.

Table 10.2

Examples of Computer Simulation as Part of Method Development for "Real" Applications

Feature	Results (Comments)
Designated peak selection; optimization of gradient time [8]	a. Measurement of a single impurity in a sample that contained >15 components b. Preparative purification of a single peak c. Measurement of three components in samples containing as many as 10 related compounds
Best multi-segment gradient [8]	Analysis of a 7-component sample
Best multi-segment gradient [23]	Multi-segment gradients were able (for the first time) to separate all 19 of the 30S ribosomal proteins; 33 of 34 of the 50S ribosomal protein could be similarly separated.
Best multi-segment gradient [8, 18, 24]	Separations of 7-, 12-, and 19-component mixtures by means of segmented gradients
Vary T and t_G plus a change in B-solvent[8, 25]	Where the use of an initial B-solvent was unsuccessful, variation of T and t_G was repeated with different B-solvents (ACN, MeOH, or 2-propanol)
Isocratic predictions from gradient data [8]	Computer simulation for varying T and t_G provided excess isocratic resolution, followed by changes in column length and flow rate for reduced run time
Selection of best column and gradient time [26]	16 columns were investigated for the best separation of a 6- and an 8-component mixture
Selection of best column and B-solvent [27]	Two gradient runs ($t_G = 20$ and 50 min) were repeated for three columns and two B-solvents (ACN, MeOH); separation with the best column/solvent choice was further optimized using a Plankett-Burmann design
Selection of best column, temperature and gradient time [28]	Gradient time and temperature were optimized for 9 different columns; this allowed the best conditions to be selected for various separation goals
Variation of gradient time and ratio of ACN and MeOH in B-solvent [29]	B-solvent was varied from 0–100% ACN/MeOH (4 solvent mixtures), each of which was run at two gradient times for the optimum separation of a 8-component mixture
Simple variation of gradient time or isocratic % [30]	Four examples for different samples

In addition Virtual Column (Dionex, Sunnyvale, CA) is available for predicting separations by ion chromatography. See also several reviews and/or comparisons of software of this kind [9–11, 33]. For a detailed review of some technical requirements of computer simulation, see [34]. (New products are introduced with regularity as others are discontinued, so this list is by no means complete and is expected to soon be out of date.)

10.3 OTHER METHOD-DEVELOPMENT SOFTWARE

Besides the software for computer simulation as described above, other software packages are available to support method development. Some features of such software include the following:

- predictions of solute retention from molecular structure
- predictions of solute pK_a-values from molecular structure
- selection of reversed-phase columns with similar or different selectivity
- expert systems for method development

Sections 10.3.1 to 10.3.5 briefly review software for these applications but do not exhaust the possibilities for computer-assisted method development.

10.3.1 Solute Retention and Molecular Structure

Attempts at relating chromatographic retention to solute molecular structure have a long history. In principle, if it were possible to predict solute retention as a function of molecular structure and experimental conditions (mobile phase, stationary phase, temperature), there would be no need for actual experiments during method development. As discussed in Section 2.6.7, the possibility of reliable predictions of this kind appears remote at present. Nevertheless, predictive software based on molecular structure has been offered at various times in the past—and will probably continue to be offered in the future.

ChromSword and the ACD/LC Simulator (cited in Section 10.2.5) provide an alternative option for computer simulation, where one or more experimental runs are replaced by predictions of retention based on molecular structure. An independent evaluation of this feature [11] concluded that "Predictions based on molecular structure alone are not very accurate and are not likely to provide useful separation information." Similar software has been available for more specialized applications (protein digests [35], metabolized drugs [36]), but the reliability of such software is similarly in question. More recent attempts at predicting RPC separation are somewhat more promising [37] but aim at a different goal—namely peak identification when used in combination with mass spectrometric detection. The required predictive accuracy for the latter goal is much less than would be required for optimizing resolution.

10.3.2 Solute pK_a Values and Molecular Structure

Compound pK_a values determine solute retention as a function of mobile-phase pH. Consequently a knowledge of pK_a values for compounds present in the sample can be useful for planning and interpreting method-development experiments (Section 7.2). Software for the retrieval or estimation of pK_a values (water as mobile phase and 25°C) can be obtained from various sources (Advanced Chemistry Development, Toronto, Ontario, Canada; Intertek ASG Laboratory, Manchester, UK). However, as noted in Section 7.2.3, values of pK_a vary with both temperature and mobile-phase composition. The latter values of pK_a may be of limited use in practice, where organic-water mobile phases are the rule and temperature can vary. Experiments where mobile-phase pH is varied will generally provide more reliable estimates of sample pK_a values for method development (see Section 7.2 and the examples of Fig. 7.3). The approximate pK_a values summarized in Table 7.2 for different functional groups or compound classes should suffice in most cases for method development.

10.3.3 Reversed-Phase Column Selectivity

RPC column selectivity can be characterized by values for five column characteristics (Section 5.4.1). Appropriate software (Column Match®, Molnar Institut, Berlin; http://www.usp.org/USPNF/columnsDB.html) allows different columns to be compared in terms of selectivity, in turn allowing the selection of replacement columns of similar selectivity, or columns of very different selectivity when changes in relative retention are needed (Section 5.4).

10.3.4 Expert Systems for Method Development

The possibility of fully automatic method development by means of computer software (*expert systems*) has been under investigation since the mid-1980s [9, 38]. Ideally information on the nature of the sample and separation goals would be entered into the computer, the computer would recommend initial separation conditions, a sample would be injected, the results of this first injection would be interpreted by the computer, and subsequent experiments would be selected by the computer and used for successive computer simulations—until a successful separation is obtained. Any problems encountered during this process would be solved by the computer. Previous attempts in this direction either have failed to achieve commercial success or have been limited to optimizing separation after the user selects initial conditions [39]. However, the appeal of this approach is strong, so improved products are sure to be introduced in the future.

10.4 COMPUTER SIMULATION AND METHOD DEVELOPMENT

Computer simulation is not intended to *replace* the various strategies for method development that have been presented in preceding chapters. Rather, computer simulation should be used to augment real trial-and-error experiments during method development. The selection of conditions for adequate resolution in a minimum separation time is often the main consideration in method development—a goal for which computer simulation can be especially effective. Two examples of the use of computer simulation in this way are presented in Sections 10.4.1 and 10.4.2.

10.4.1 Example 1: Separation of a Pharmaceutical Mixture

Consider the development of an RPC method for a sample that consists of 12 derivatives of lysergic acid diethylamide (LSD). An initial gradient separation with the conditions recommended in Table 9.3 was first carried out (Fig. 10.7*a*). From this initial chromatogram with a 10-minute gradient, it appears that gradient elution is the preferred option (Section 9.3.1, $\Delta t_R/t_G = 0.33$), although isocratic elution is also possible. Therefore the next step is to carry out a separation with a 30-minute

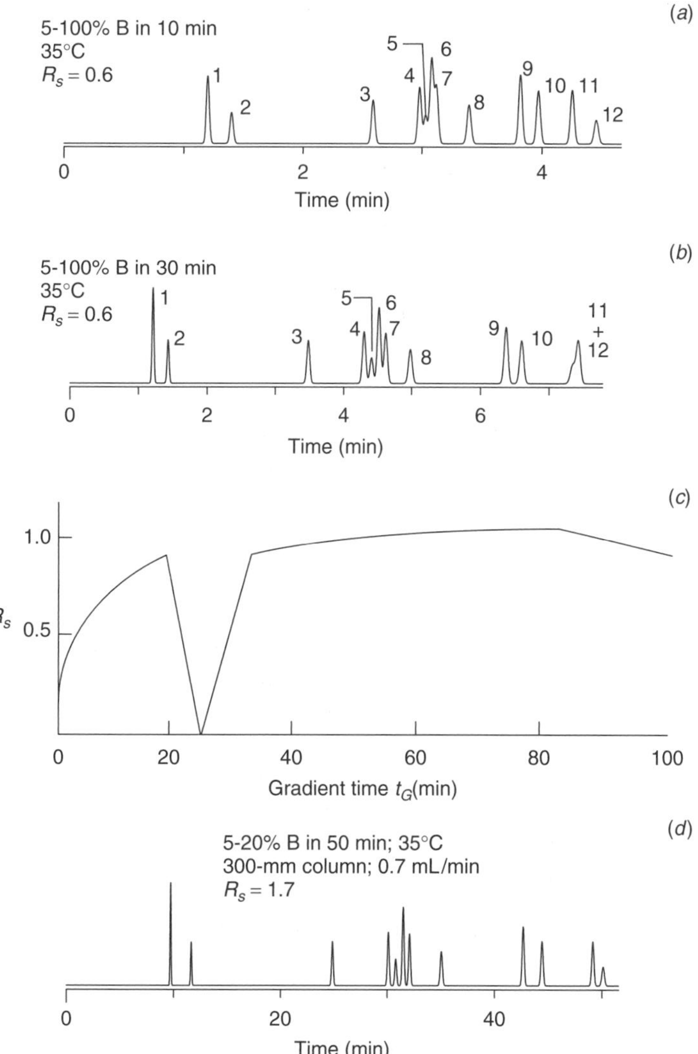

Figure 10.7 Illustration of a strategy for method development based on the use of computer simulation for the selection of final separation conditions. Sample: mixture of 12 derivatives of lysergic acid diethylamide (LSD). Conditions: C_{18} column; acetonitrile/water gradients; other conditions varied. (*a*) 5–100% B in 10 minutes, 100 × 4.6-mm (3-μm) C_{18} column; 35°C; 2.0 mL/min; (*b*), same as in (*a*), except 5–100% B in 30 minutes; (*c*) resolution map; (*d*) final separation with conditions indicated in figure. Simulated separations based on data of [13].

gradient (other conditions the same); this separation is shown in Figure 10.7*b*. Resolution ($R_s = 0.6$) is unacceptable for either run, but computer simulation can be used to determine a gradient time for maximum resolution. The resolution map of Figure 10.7*c* indicates two "optimum" gradient times: 19 minutes ($R_s = 0.9$), and 80 minutes ($R_s = 1.1$). The average value of $k^* = 6$ for $t_G = 19$ min is preferable to $k^* = 25$ for $t_G = 80$ min; a shorter gradient time is also generally preferred.

Because resolution is still far from adequate, the best next step is to explore a further change in selectivity. Our recommendation is to carry out two additional experimental separations where temperature is varied, which with the two separations of Figure 10.7*a*,*b* yield the experimental design shown in Figure 10.2*b* (gradient times of 10 and 30 min, temperatures of 35 and 55°C). Unfortunately, resolution was not improved by this simultaneous optimization of gradient time and temperature.

Because resolution is unacceptable ($R_s = 0.9$) at this stage of method development, conditions for improved resolution need to be explored—by further changes in either selectivity or column conditions. A change in selectivity can proceed by selecting a different B-solvent (e.g., methanol instead of acetonitrile), column (Section 5.4.3), or mobile-phase pH (for samples that contain acids or bases). When computer simulation is used, changes in column conditions and the gradient do not require additional experiments, so this option should be pursued first. The simplest change in column conditions is an increase in column length, accompanied by decrease in flow rate, so as to maintain column pressure the same. For the present sample, a 300-mm column with a flow rate of 0.7 mL/min gives a resolution of $R_s = 1.7$, but with an increase in gradient time to 171 minutes.

We can next reduce the gradient range to save time. Figure 10.7*d* shows the resulting separation for a gradient of 5–20% B in 50 minutes, where $R_s = 1.7$. Other changes in gradient shape were explored in an effort to improve the compromise between resolution and separation time, but these proved unsuccessful. Considering the difficulty of the separation, the observed resolution might be considered adequate. However, if the 50-minute run time is a problem (because of a large number of samples to be assayed), further attempts at optimizing selectivity should be explored—as suggested above. Any increase in resolution as a result of improved selectivity can always be traded for a shorter run time (Sections 6.4.2, 9.3.6).

While the example of Figure 10.7 might be regarded as somewhat disappointing in its outcome, the amount of experimental effort required for an exploration of all these options was minimal (four experimental runs)—as was the time required for all these simulations. At the same time a number of unsuccessful options have been dispensed with, allowing the chromatographer to focus on other, more promising lines of attack.

10.4.2 Example 2: Alternative Method Development Strategy

The preceding example emphasizes the optimization of gradient time and/or temperature for the optimization of selectivity. Other, similar strategies are illustrated in Figure 10.2. A different approach is to change one condition at a time (e.g., column type), while optimizing either %B or gradient time for each set of conditions. An example of this approach is described in [27], for the separation of a 10-component pharmaceutical mixture that contained three active ingredients and seven impurities or degradation products. All solutes were non-ionizable, so buffering the mobile phase was not required. This sample required gradient elution, and the method-development approach followed is outlined in Figure 10.8*a*.

The initial experiments evaluated three different columns for the separation: NovaPak C18, Luna C18, and Discovery RP Amide C16. Gradients from 10 to 90% B were carried out in times of 20 and 50 minutes, using either acetonitrile

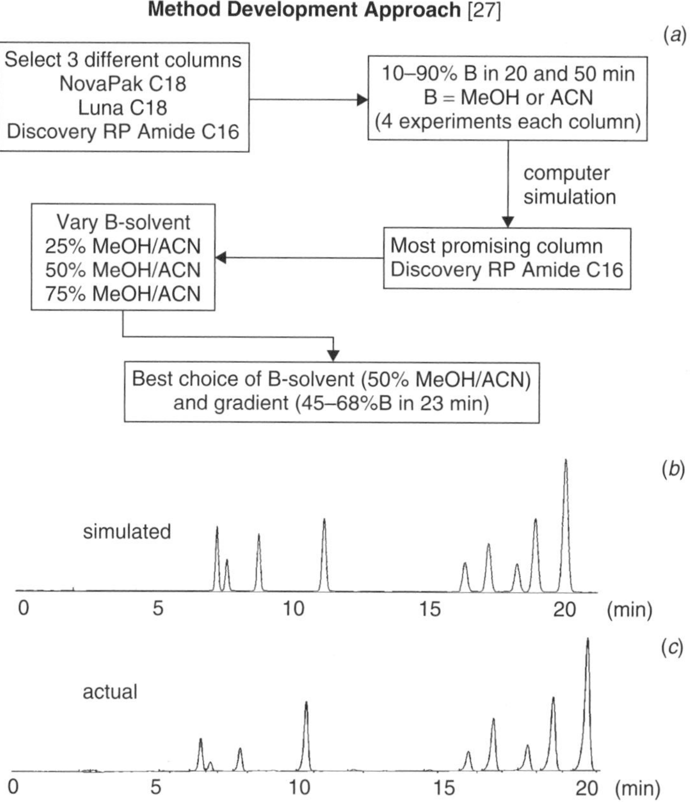

Figure 10.8 Separation of a pharmaceutical mixture of 10 components using computer simulation. (*a*) Outline of experiments; (*b*) simulated final separation (DryLab); (*c*) actual final separation. Conditions for both (*b*) and (*c*): 250 × 4.6-mm Discovery RP Amide C16 column; 45–68% B in 23 minutes (B is 50% MeOH/ACN); 40°C; 1.0 mL/min. Adapted from [27].

(ACN) or methanol (MeOH) as B-solvent (12 initial scouting runs). Each of the six pairs of experiments for a given column and B-solvent was used as input for computer simulation (DryLab), in order to establish the optimum gradient time. The most promising results were obtained for the Discovery RP Amide C16 column with ACN as B-solvent, but one pair of compounds were still overlapped ($R_s \approx 0$). However, comparing results for the two B-solvents with the Discovery RP Amide C16 column suggested that a mixture of ACN and MeOH as B-solvent would enable the separation of all 10 compounds in the sample. Mixtures of 25, 50, and 75% MeOH/ACN as B-solvent were tried. These resulted in an adequate separation for the sample with 50% MeOH/ACN as the B-solvent. Finally, the gradient range was optimized (45–68% B in 23 min) to give the predicted separation of Figure 10.8*b*, with the actual separation shown in Figure 10.8*c*.

A similar method-development approach is described in [40], where 12 different columns were evaluated with changes in temperature and mobile phase, using computer simulation to interpret results and guide further experiments. However, such a large number of method-development experiments will rarely be necessary—especially if a smaller number of columns of differing selectivity are chosen, as described in Sections 5.4.3, 6.3.6.2, and 7.3.2.4 (see also [41]).

10.4.3 Verifying Method Robustness

Following the selection of experimental conditions for a method, its robustness should be verified. Computer simulation can be used to check the effects of unintended variations in various conditions, for example:

- system dwell volume V_D (gradient methods only)
- mobile-phase pH
- temperature
- mobile-phase %B
- flow rate

These five separation conditions are listed in approximate order of decreasing importance, by their effect on method robustness.

Values of V_D vary from one HPLC system to another, leading to possibly significant changes in gradient separation for some samples (Section 9.2.2.4). The effect of a change in V_D on a gradient separation is easily investigated by means of computer simulation. For the LSD method describe above (Fig. 10.7*d*), the dwell-volume was 1.1 mL. For a change in V_D to a value between 0 and 5 mL (a range that should cover most systems in use today), the effect on resolution is no greater than ±0.1 unit in R_s. Consequently the latter method is robust to changes in V_D. Computer simulation is especially useful for designing gradient methods that are insensitive to typical changes in V_D from system to system (Section 9.2.2.4).

Excessive sensitivity to changes in *mobile-phase pH* is a common reason for a lack of method robustness, as discussed in Section 7.3.4.1. If computer simulation is used to optimize mobile-phase pH, it can also determine the robustness of the method with respect to small variations in pH.

The *temperature* of an HPLC system can be controlled in various ways (Section 3.7), but small differences in temperature can occur, based on the value selected. Again, computer simulation can be used to determine the effect of a change in temperature on separation. For the LSD method above (Fig. 10.7*d*), a change in temperature of ±2°C results in a loss in resolution of no more than $0.1R_s$-unit. Consequently this method can tolerate small changes in temperature from system to system.

When *mobile-phase %B* is controlled by on-line mixing, %B can vary by as much as 1–2% relative. For isocratic methods where computer simulation was used to select an optimized value of %B, the effect of a change in %B by ±2% relative can be determined by computer simulation. The similar effect of uncertainty in %B for gradient elution can be determined by computer simulation, by entering different values of the gradient range into the computer (e.g., for a nominal 5–100% B gradient, examine gradients of 4–100%, 6–100%, etc.). For the present LSD example, an uncertainty of ±2% in %B in the gradient results in changes in resolution of <0.1 R_s-unit.

Flow rate is usually controlled within ±1–2% by most HPLC systems. For the LSD method above, the effect of a change in flow rate by ±2% can again be determined by computer simulation: <0.1 R_s-unit.

It should be noted that when computer simulation is used to model method robustness, this does not eliminate the need to demonstrate robustness experimentally (Section 12.2.6). However, simulations can be used during method development to avoid conditions that are not robust, as well as help select the appropriate variation in each parameter for testing. For example, if the simulation suggests robustness in pH only to ±0.2 units, the actual robustness test could be made with ±0.2 pH units of variation, not at ±0.3 units, where failure would be likely.

10.4.4 Summary

Whatever approach is adopted during method development, computer simulation can replace many of the experimental steps that are designed to optimize the final separation, with a reduction in experimental effort and the selection of better final conditions. Computer simulation has advanced considerably since its introduction in 1985, allowing increasingly accurate predictions of separation for a wider range of experimental conditions. The future may see a greater use of computer simulation as an integral part of the HPLC system—so-called automatic method development. However, the best use of computer simulation for more demanding separations will always involve a close coordination of the skills and experience of the chromatographer with the capabilities of computer simulation. *Computers should be kept on tap, not on top* (to paraphrase Winston Churchill's remark during World War II (about scientists ... not computers).

REFERENCES

1. P. L. Zhu, J. W. Dolan, L. R. Snyder, N. M. Djordjevic, D. W. Hill, J.-T. Lin, L. C. Sander, and L. Van Heukelem, *J. Chromatogr. A*, 756 (1996) 63.
2. L. R. Snyder and J. W. Dolan, *High-Performance Gradient Elution*, Wiley-Interscience, Hoboken, NJ, 2007.
3. R. J. Laub and J. H. Purnell, *J. Chromatogr.*, 161 (1978) 49.
4. J. L. Glajch, J. J. Kirkland, K. M. Squire, and J. M. Minor, *J. Chromatogr.*, 199 (1980) 57.
5. B. Sachok, R. C. Kong, and S. N. Deming, *J. Chromatogr.*, 199 (1980) 317.
6. J. J. Kirkland and J. L. Glajch, *J. Chromatogr.*, 255 (1983) 27.
7. I. Molnar, *J. Chromatogr. A*, 965 (2002) 175.
8. L. R. Snyder and L. Wrisley, in *HPLC Made to Measure: A Practical Handbook for Optimization*, S. Kromidas, ed., Wiley-VCH, Weinheim, 2006, 56785.
9. J. L. Glajch and L. R. Snyder, eds., *J. Chromatogr.*, 485 (1989)
10. P. J. Schoenmakers, J. W. Dolan, L. R. Snyder, A. Poile, and A. Drouen, *LCGC*, 9 (1991) 714.
11. T. Baczek, R. Kaliszan, H. A. Claessens, and M. A. van Straten, *LCGC Europe*, 14 (2001) 304.
12. S. Kromidas, ed., *HPLC Made to Measure: A Practical Handbook for Optimization*, Wiley-VCH, Weinheim, 2006, pp. 565–623.
13. J. W. Dolan, L. R. Snyder, N, M. Djordjevic, D. W. Hill, D. L. Saunders, L. Van Heukelem, and T. J. Waeghe, *J. Chromatogr. A*, 803 (1998) 1.
14. R. G. Wolcott, J. W. Dolan, and L. R. Snyder, *J. Chromatogr. A*, 869 (2000) 3.

15. T. Jupille, L. Snyder, and I. Molnar, *LCGC Europe*, 15 (2002) 596.
16. S. V. Galushko and A. A. Kamenchuk, *LCGC Intern.*, 8 (1995) 581.
17. S. V. Galushko and A. A. Kamenchuk, and G. L. Pit, *Amer. Lab.*, 27 (1995) 33G.
18. V. Concha-Herrera, G. Vivo-Trujols, J. R. Torres-Lapasió, and M. C. Garcia-Alvarez-Coque, *J. Chromatogr. A*, 1063 (2005) 79.
19. D. D. Lisi, J. D. Stuart and L. R. Snyder, *J. Chromatogr.*, 555 (1991) 1.
20. J. W. Dolan, L. R. Snyder, N. M. Djordjevic, D. W. Hill, L. Van Heukelem, and T. J. Waeghe, *J. Chromatogr. A*, 857 (1999) 21.
21. G. Vivo-Truyols, J. R. Torres-Lapasió, and M. C. Garcia-Alvarez-Coque, *J. Chromatogr. A*, 876 (2000) 17.
22. L. R. Snyder and J. W. Dolan, *Adv. Chromatogr.*, 38 (1998) 115.
23. B. F. D. Ghrist, B. S. Cooperman, and L. R. Snyder, *J. Chromatogr.*, 459 (1989) 43.
24. R. Bonfichi, *J. Chromatogr. A*, 678 (1994) 213.
25. I. Molnar, *J. Chromatogr. A*, 948 (2002) 51.
26. R. M. Krisko, K. McLaughlin, M. J. Koenigbauer, and C. E. Lunte, *J. Chromatogr. A*, 1122 (2006) 186.
27. W. Li and H. T. Rasmussen, *J. Chromatogr. A*, 1016 (2003) 165.
28. L. Van Heukelem and C. S. Thomas, *J. Chromatogr. A*, 910 (2001) 31.
29. M. R. Eurby, F. Scannapieco, H.-J. Rieger, and I. Molnar, *J. Chromatogr. A*, 1121 (2006) 219.
30. N. G. Mellisch, *LCGC*, 9 (1991) 845.
31. S. Heinisch, E. Lesellier, C. Podevin, J. L. Rocca, and A. Tschapla, *Chromatographia*, 44 (1997) 529.
32. R. Cela and M. Lores, *Comput. Chem.*, 20 (1996) 175.
33. A. Tchapla, *Analusis*, 20(7) (1992) 71.
34. N. Lundell, *J. Chromatogr.*, 639 (1993) 97.
35. C. T. Mant and R. S. Hodges, in *High-Performance Liquid Chromatography of Peptides and Proteins: Separation, Analysis and Confirmation*, C. T. Mant and R. S. Hodges, eds., CRC Press, Boca Raton, 1991, p. 705.
36. K. Valko, G. Szabo, J. Rohricht, K. Jemnitz, and F. Darvas, *J. Chromatogr.*, 485 (1989) 349.
37. V. Spicer, A. Yamchuk, J. Cortens, S. Sousa, W. Ens, K. G. Standing, J. A. Wilkins, and O. V. Krokhin, *Anal. Chem.*, 79 (2007) 8762.
38. T.-P. I, R. Smith, S. Guhan, K. Taksen, M. Vavra, D. Myers, and M. T. W. Hearn, *J. Chromatogr. A*, 972 (2002) 27.
39. E. F. Hewitt, P. Lukulay, and S. Galushko, *J. Chromatogr. A*, 1107 (2006) 79.
40. M. Peffer and H. Windt, Fresenius, *J. Anal. Chem.*, 369 (2001) 36.
41. D. M. Marchand, L. R. Snyder, and J. W. Dolan, *J. Chromatogr. A*, 1191 (2008) 2.

CHAPTER ELEVEN

QUALITATIVE AND QUANTITATIVE ANALYSIS

11.1 INTRODUCTION

The HPLC system can provide both qualitative and quantitative data. Qualitative information serves to identify analytes, while quantitative results define how much of each analyte is present. Three major factors affect the quality of these results. First, the HPLC hardware must operate in a predictable and repeatable fashion, so as to generate data that are sufficiently precise and accurate for the application at hand. Second, the data system and associated software must be able to convert the HPLC

Introduction to Modern Liquid Chromatography, Third Edition, by Lloyd R. Snyder, Joseph J. Kirkland, and John W. Dolan

detector output signal into meaningful qualitative and quantitative information. Third, the results from the routine application of a method also depend on the quality of the chromatogram (resolution, peak shape, baseline drift, etc.). In this chapter, however, we will assume "good" chromatography (no problems with the HPLC-system hardware, data system, or chromatography). With the exception of some specific examples, troubleshooting and correcting system problems are left to Chapter 17.

We will begin (Section 11.2) with how the data system measures the signal from the detector, including some sources of error (Section 11.2.4) and how the limits of a method are established (Section 11.2.5). Next, Section 11.3 (qualitative analysis) will cover some of the techniques used to identify analytes. Finally, Section 11.4 (quantitative analysis) will examine how data are used to answer the question of "how much"? The scope of this chapter is necessarily limited, so for more detailed information on many of the topics, the reader is referred to general texts on analytical chemistry and statistics, as well as the references cited in this chapter. In particular, reference [1] contains more information about chromatographic integration than most readers will need in a lifetime.

11.2 SIGNAL MEASUREMENT

The HPLC detector (Chapter 4) is a transducer that converts the concentration (or mass) of analyte in the column effluent into an electrical signal. The data system then transforms this signal into a plot of intensity against time (a chromatogram). The data initially are in the form of either analog or digital signals; digital data are required for storage and manipulation, so analog signals must be converted to a digital format prior to storage. Software associated with or external to the data system then converts the digital signal into something useful to the chromatographer—a chromatogram, a data table, or some other presentation of the data. The key quantities at any point in the chromatogram are the time (x-value) and intensity (y-value), from which are obtained the retention time (Section 11.2.2) and peak area or height (Section 11.2.3).

11.2.1 Integrator Operation

When the second edition of this book [2] was written, strip-chart recorders were common in many laboratories as primary data-gathering devices for HPLC systems, although some laboratories used dedicated integrators for data collection. Many of the measurements and calculations were made by hand, with little more than a ruler and hand-held calculator to aid the process. With the introduction of the personal computer (PC) in the early 1980s, and subsequent development of PC-based data systems, data collection for HPLC was revolutionized. Today nearly every HPLC system uses computer-based data collection and analysis. Some users refer to *integrators* as small, dedicated data-collection systems that gather chromatographic data from a single HPLC and produce very simple reports (e.g., retention time and area tables). *Computer-based data systems*, on the other hand, usually offer additional features, including instrument control and specialized data processing capabilities for one or more HPLC systems. For the present discussion, we will use the terms "integrator" and "data system" (including its software) interchangeably. Data systems use a special set of terms (language) that describe settings or chromatographic characteristics. Several of these terms are mentioned in the following discussion and are summarized in later Figure 11.2. Terms vary somewhat from one manufacturer to another, but the same functions are common to most systems. *Note: Readers who already know how an integrator works—or don't care—may want to skip the rest of Section* 11.2.1.

11.2.1.1 Data Sampling

The data system measures the signal intensity at a high sampling rate throughout the chromatogram (generally 20–100 Hz [1]), as illustrated in Figure 11.1 ("data slices"). Because the chromatographic baseline rarely is true zero, the baseline is determined and the region below the baseline is subtracted from each data point, resulting in a set of corrected data slices that represent the chromatographic signal at each point in the chromatogram (Fig. 11.1). A high sampling rate will generate a large data file very quickly (e.g., a 20-min run sampled at 100 Hz creates $>10^5$ data points), so the data files can be large—regardless of inexpensive data storage. A peak can be defined at near maximum accuracy with 100 points across the peak, and more points do not improve the peak description [1], so *bunching* (Fig. 11.2*a*) of the raw data can reduce the file size while maintaining peak integrity. For example, a peak with $k = 1$ and $N = 10{,}000$, for a 150 × 4.6-mm column operated at 1 mL/min, will have a 6σ width of ≈11 seconds. This converts to an effective sampling rate of ≈9 Hz for 100 points across the peak, so a data collection rate of 10 Hz would be adequate to fully describe the peak. (Note that to minimize confusion, in this section we will refer to the data *sampling rate* for the original, raw signal and the data *collection rate* for the resultant bunched or simplified data set stored for further processing.) Thus every 10

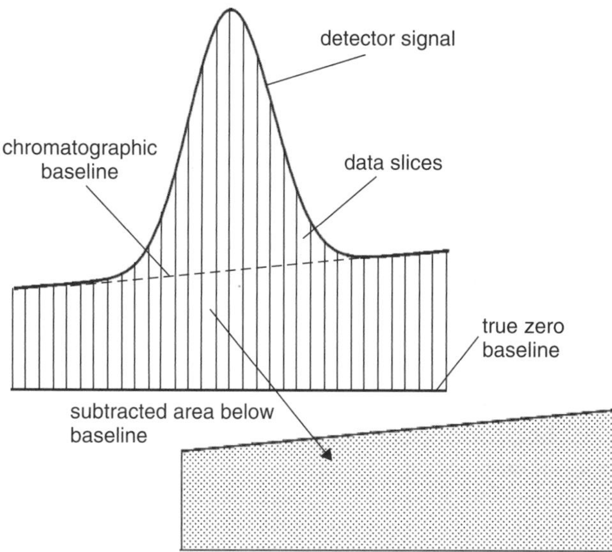

Figure 11.1 Illustration of peak integration by area slices and subtraction of area below the chromatographic baseline.

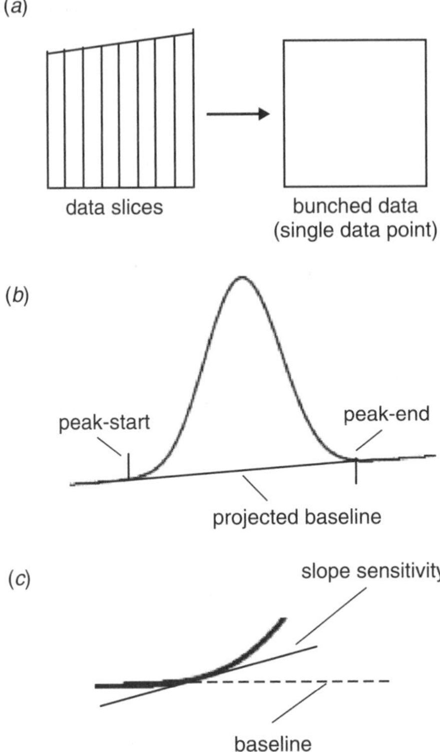

Figure 11.2 Common integrator settings. (*a*) Data bunching; (*b*) peak-start and peak-end markers, baseline extended before and after peak; (*c*) slope sensitivity to detect presence of a peak.

adjacent, original points could be combined to reduce the data-file size and convert the effective data collection rate from 100 Hz to 10 Hz. Software can further reduce the file size by (1) adjusting the bunching rate across the chromatogram as peak widths increase for later peaks, and (2) using other data compression techniques.

Although 100 points across a peak fully defines the peak, for purposes of quantification, 20 (bunched) data points are sufficient to describe a peak. In addition to reduction of the data-file size, this bunching of data reduces the baseline noise as a function of the square root of the number of points that are combined. Thus reducing the data collection rate from 100 points per peak to 20 points per peak reduces the noise by a little more than 2-fold, yet does not noticeably compromise the quantitative information contained in the data. Some examples of a chromatogram at various bunching rates are shown in Figure 11.3. As the number of bunched points increases from the initial raw signal (Fig. 11.3*a*, $\approx$250 points across peak, relative peak height 1.00), the noise is greatly reduced with minor loss in peak height (e.g., Fig. 11.3*b*, $\approx$25 points, 0.98 height). However, if too many points are bunched, the peak gets very "steppy," while the peak heights are lowered and the valley between the peaks may be raised (Fig. 11.3*d*, $\approx$6 points, 0.93 height).

It should be noted, that for maximum data integrity, the sampling rate for the original data collection (e.g., 100 Hz) or an oversampled bunching (e.g., 10 Hz for

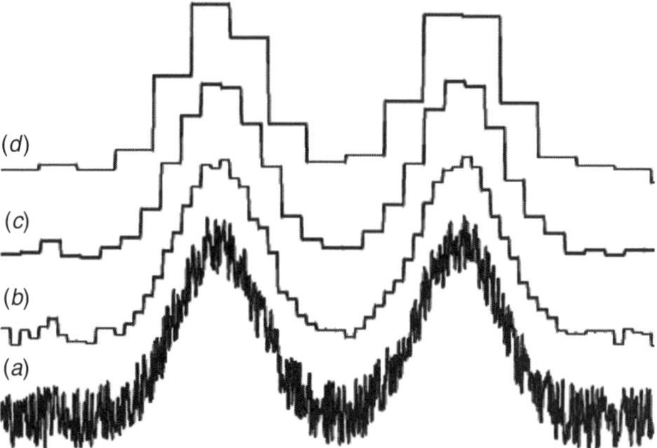

Figure 11.3 Effect of data bunching or data rate on signal and noise. (*a*) Raw signal showing noise, ≈250 points across peak; (*b*) 10 points from (*a*) per bunch; relative peak height 98%; (*c*) 20 points from (a) per bunch; relative height 97%; (*d*) 40 points from (*a*) per bunch; relative height 93%.

the example above) should be stored as "raw data." All subsequent data treatments (e.g., bunching to 20 points per peak) is performed without destroying the original raw data. Thus, if mistakes are made, or if the original data need to be treated in another manner, the raw data are available. *Data sampled at too high a data rate always can be simplified by bunching, but data sampled at too low a data rate cannot be divided to create more data points.* As a safety net, the data rate for stored raw data generally will be higher during method development or early application of the method; as the method is put into routine use, lower data rates will be used that adequately describe the peak(s), yet conserve data storage space.

11.2.1.2 Peak Recognition

One of the main functions of the data system is to recognize the presence of a peak in the chromatogram. It does this by monitoring the value of the detector signal and comparing it to the values of neighboring slices. When the signal increases for several consecutive slices, generally 5 to 10, a peak is recognized and a *peak-start* time (Fig. 11.2*b*) is recorded. The same data evaluation takes place on the tail of the peak so that, when a predetermined number of slices are not smaller than their predecessors, a *peak-end* time (Fig. 11.2*b*) is recorded. The *slope sensitivity* setting (Fig. 11.2*c*) determines how much change is required to identify a peak-start or peak-end, and this may vary based on the amount of baseline noise, the intensity of the peak, user choice, or other factors. Data systems include peak-detect algorithms that facilitate the detection of peaks on a sloping baseline, when peaks are not fully resolved, and in many other non-ideal separation conditions. Once the peak-start and peak-end points have been established, the peak width can be determined. Based on the assumption of a Gaussian curve as a peak shape, this can be reported as the 6σ width, 2.35σ at half the peak height ("width at half-height"), or other values based on the standard Gaussian distribution.

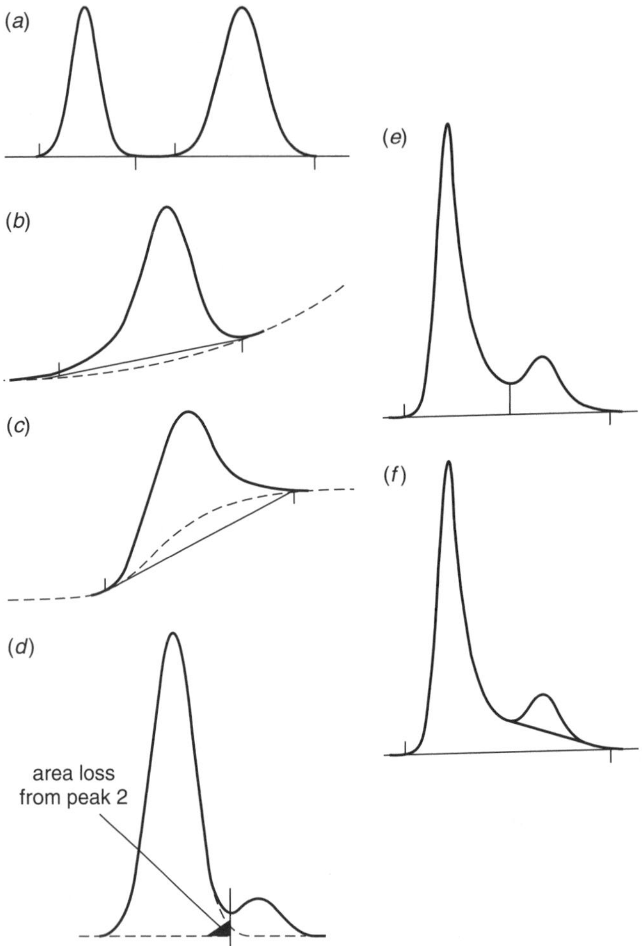

Figure 11.4 Peak integration. (*a*) Integration of two well-resolved peaks on a noise-free baseline without drift. (*b*) Loss of peak area by tangent skim on rising concave baseline. (*c*) Excess peak area by tangent skim on rising convex baseline. (*d*) Proper use of a perpendicular drop for overlapping Gaussian peaks; for peaks of unequal size, the smaller peak is under-reported. (*e*) Use a perpendicular drop to integrate small peak after a tailing larger peak; the smaller peak is over-reported. (*f*) Use of skimming to integrate a small peak after a larger tailing peak. Adapted from [1].

11.2.1.3 Integration of Non-Ideal Chromatograms

Peaks that are well resolved and elute on a flat baseline, as in Figure 11.4*a*, are easy to identify and integrate. When peaks are not fully resolved, or the baseline is noisy or drifts, it is more difficult for the integrator to determine where each peak starts and stops, and therefore how many slices of data to assign to each peak. Peak *skimming*, as in Figure 11.4*b, c*, is the usual means of dealing with drifting baselines, but errors in the measurement of peak area can arise because of uncertainty in the actual position of the baseline below the peak. The measured peak area will be too small in the case of a concave baseline (Fig. 11.4*b*), and too large for a convex baseline (Fig. 11.4*c*). Much effort has been expended in trying to determine the best skimming technique (linear, exponential, etc.), but a linear skim as in Figure 11.1 is usually the best overall compromise [1].

When two peaks are not fully resolved, the most common way to separate them is to use a *perpendicular drop* from the valley between the peaks to the baseline drawn between the baseline before and after the peaks, as in Figure 11.4*d*. For equal sized, symmetric peaks, this technique accurately assigns the peak area to each peak, but for unequal, symmetric peaks the area is underestimated for the smaller peak, and overestimated for the larger peak [3]. If the peaks are of unequal size and one or both peaks tail, as in Figure 11.4*e*, the uncertainty in assignment of peak areas increases. At some point a change from the use of a perpendicular drop (Fig. 11.4*e*) to a skim (Fig. 11.4*f*) will be warranted. As a rule of thumb, if the smaller peak is <10% of the height of the larger peak, a skim should be used; if the smaller peak is >10%, a perpendicular is appropriate [1]. However, whatever skimming technique is used, the accuracy of resulting peak areas will be compromised—especially for smaller and/or tailing peaks. In some cases peak height may be preferred to area as a means of quantitation (Section 11.2.3).

11.2.1.4 Common Integration Errors

No matter how well designed the data-system software, it may not match the skill of the chromatographer for accurate integration. It is desirable to adjust the integration settings so as to do the best possible job of identifying peaks and determining how to assign peak area. This becomes more difficult the smaller the signal-to-noise ratio—potentially resulting in integration errors. Three common examples are shown in Figure 11.5. The baseline under a peak usually is assigned by identifying the baseline before and after a peak, then connecting the two baselines with a straight line. However, negative peaks and other baseline disturbances can confound this process, as illustrated in Figure 11.5*a*. Here the negative peak before the peak of interest results in a baseline that is too low (solid line), artificially increasing the area assigned to the peak. The baseline needs to be redrawn (dashed line). As mentioned in Section 11.2.1.3, it can be difficult to make a decision about whether to use a perpendicular drop or a tangent skim with a pair of poorly resolved peaks. In Figure 11.5*b* a perpendicular drop was incorrectly assigned (solid line), so this needs to be adjusted to a tangent skim (dashed line) for more accurate integration. Curved skimming algorithms (e.g., Gaussian) are available on some data systems, but it must be realized that all skimming techniques are estimates and will never give as accurate or consistent results as those obtained for baseline-resolved peaks.

One of the most difficult tasks for an integrator is to determine when a peak ends. This is complicated when tailing peaks, rising or falling baselines, and/or excessive baseline noise are present. In the case of Figure 11.5*c* the peak-end point was assigned too early and needs to be replaced with a later point (arrow in Fig. 11.5*c*). This is perhaps the most common error encountered with peaks sizes near the limit of detection or lower limit of quantification. One way to minimize this problem is to stop integration when the peak is sure to have left the detector. This can be accomplished by visually determining when the peak has returned to baseline (e.g., at the right end of the dashed baseline in Fig. 11.5*c*). At this time, set a "force peak-end" function in the data system (the name of the function will vary for

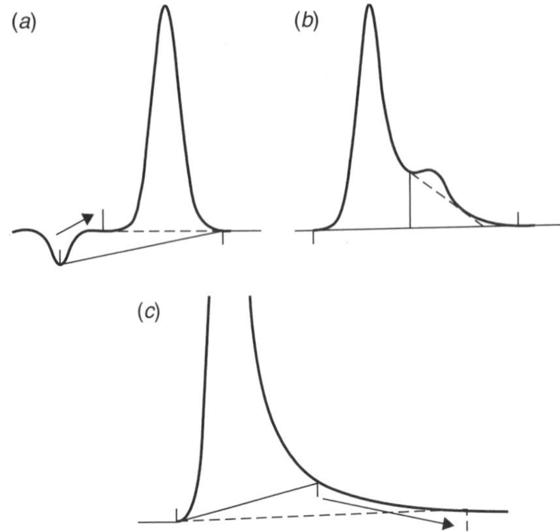

Figure 11.5 Common integration errors. (*a*) Improper peak-start assignment by the presence of a negative peak prior to analyte peak. Arrow shows move to a new peak-start point for proper integration (dashed line). (*b*) Improper integration of overlapping peaks (perpendicular drop); tangent skim (dashed line) gives proper integration. (*c*) Improper location of peak-end point. Arrow shows move to a new peak-end for proper integration (dashed line).

different systems). Rather than trying to find the peak-end point using the normal processes, the integrator will now automatically assign it to the forced time point for all chromatograms. It is a good practice to visually examine every chromatogram to confirm that the integration is correct and to make necessary adjustments if it is not. Some users feel that it is "cheating" to adjust baselines after the initial integration is done, but the simple fact is that a data system cannot properly integrate all peaks all the time. Of course, any such manual baseline changes must be carried out consistently and without reference to a desired answer.

11.2.1.5 Additional Suggestions

The preceding discussion covers only a portion of the factors that contribute to a well-integrated chromatogram. The software in today's data systems is very sophisticated, and is designed to properly identify peaks, adjust baselines, determine bunching rates, and so forth. When starting a new method, it usually is most efficient to run several representative samples and allow the integrator to assign the initial settings. After the results have been reviewed, the pre-assigned settings can be adjusted to correct for consistent errors, such as separation of two peaks by a perpendicular drop or a tangent skim. In other words, first let the integrator operate in the usual way, and then make any necessary changes in settings as needed.

To help protect data integrity, the pharmaceutical industry has a set of regulations called the electronic signatures rules, and referred to by the federal code, 21 CFR Part 11 [5]. One requirement of these rules is that any changes to raw data (e.g., baseline adjustments or peak-start–stop assignments) must be identified with the operator's name, the date and time, and the reason for the change; and the original raw data must be preserved for later examination. Most data-system software is designed to accommodate these requirements through the use of a built-in audit trail. If this feature is turned on, each change requires entry of a comment, and the operator's electronic signature is added along with a date-and-time stamp. For example, in the case of the adjustment made for Figure 11.5*c*, the comment might be "wrong peak end." Even if the use of such audit trails is not required by your specific industry, activation and use of this feature is a good practice—it is an easy way to track changes made to the data in case the need arises to reexamine the decision at a later time.

The last paragraph of an excellent reference book on chromatographic integration [1, p. 191] provides a word of caution to the chromatographer:

As long as integrators use perpendiculars and tangents and draw straight baselines beneath peaks, they are of use only in controlled circumstances, when chromatography is good. Even then, the use of integrators requires vigilance from the operator and skill in assessing and assigning parameters. ***Integrators cannot improve bad chromatography*** *[emphasis added], only the analysts can do that [provide better methods]—and at the end of the day that is what they are paid for.*

11.2.2 Retention

Analyte retention is a primary measurement that is used for the qualitative identification of a compound (Section 11.3.1). Retention most commonly is measured as retention time, t_R, usually in decimal minutes (e.g., 6.54 min) but sometimes as seconds for fast separations (e.g., 36.4 sec). Occasionally retention is measured in volume units (e.g., 4.35 mL), but today this practice is rare. *Relative retention* times are used in many USP monographs and other methods. In such cases retention is reported relative to the retention time of a reference peak (e.g., the ratio of retention times of the two peaks). This method of reporting retention compensates somewhat for changes in absolute retention time, especially when a method is transferred from one HPLC system to another (of course, true peak identity needs to be established when methods are transferred, not just assumed via relative retention; see Section 12.7).

Retention time is measured from the time of injection to the top of the peak for each analyte of interest (Fig. 2.3*e*). The retention time should also correspond to the time at which the highest data slice was gathered. If all the chromatographic conditions are held constant (flow rate, temperature, mobile phase composition, etc.), t_R should be constant (assuming a sample that is sufficiently small; Section 2.6). This also assumes a properly operating HPLC system, such that retention varies <0.02 to 0.05 minutes between injections within a single day's run.

For calculations of retention factor k (Section 2.3.1), we also need to measure the column dead-time t_0. In most cases it will be adequate to identify t_0 as the retention time reported for the unretained "solvent" peak at the beginning of the chromatogram (Figs. 2.3*e*, 2.5*a*). For clean samples, no baseline disturbance at t_0 may be detectible, while some detectors (e.g., LC-MS; Section 4.14) may not report any change in signal at t_0. In the latter cases the determination of t_0 may require a separate measurement (Section 2.3.1).

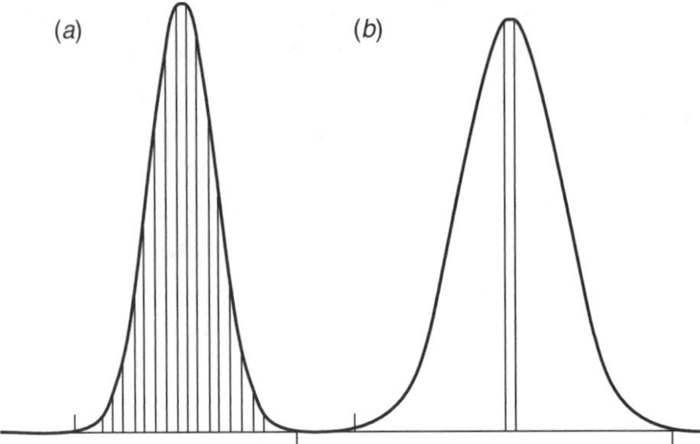

Figure 11.6 (*a*) Peak area measurement by data slices. (*b*) Peak height measurement as the largest data slice.

11.2.3 Peak Size

Chromatographic data are collected as a series of time-voltage values (Section 11.2.1.1), with signal intensity commonly reported in units of μV-sec. *Peak area* is reported as the sum of all corrected data slices between the peak-start and peak-end points, as illustrated in Figure 11.6*a*. *Peak height* is reported as the value of the largest data slice of a peak, and corresponds to the slice at the retention time of the peak (Fig 11.6*b*). In the days when manual integration was used, peak height often was found to give less error, because it only required one measurement (height), whereas peak area required two measurements (height and width). However, with today's integrators, peak area is the most popular way to report peak size. In some cases peak height is expected to give better results, such as when partially resolved peaks do not overlap significantly at each other's peak maximum. In such cases the peak height will still correspond to the "pure" peak, whereas peak area may be compromised by improper assignment of area, as in Figure 11.4*d*. If in doubt, have the integrator report both heights and areas; then compare the results to see which are more precise and accurate (e.g., test the height and area data for precision and accuracy according to Sections 12.2.1 and 12.2.2).

11.2.4 Sources of Error

The *accuracy* of results is defined as the closeness of a measured value of an analyte to its true value. Accuracy is dependent on (1) the calibration of the system with reliable standards and (2) the resolution of adjacent peaks. *Precision* of results is a measure of how close replicate measurements are to the same value—at different times, for different instrumentation, and/or with different operators. The ability to control instrumental operating conditions for a particular method determines the precision of the method. It is desirable to have results that are both precise and accurate. Still it is possible to have accurate, but not precise results, or precise, but not accurate values. Precise *and* accurate quantitative analyses can be obtained only when careful attention is given to all phases of the analysis, from initial sample collection to final report generation. Some of the major sources of error are:

- sampling and cleanup (Section 11.2.4.1)
- chromatography (Section 11.2.4.2)
- detection (Section 11.2.4.3)
- peak measurement (Section 11.2.4.4)
- calibration (Section 11.2.4.5)

11.2.4.1 Sampling and Cleanup

Any analytical result is based on the assumption that the analyzed sample is representative of the population from which it was obtained. So the first challenge is to obtain a sample (water, soil, powder, plasma, tissue, etc.) that meets this requirement (Section 16.1.1). Once the sample is obtained, it must be transported and stored in a manner that ensures sample integrity until the time of analysis. Most samples require sample preparation or cleanup steps (Chapter 16) prior to their introduction to the HPLC system. Each step of sample processing, including all of the ancillary instrumentation and reagents, contributes to the overall method error. Once the processed sample is placed in the autosampler tray, it is subject to additional errors, such as stability, evaporation, and injection variability. Some of these errors can be compensated—at least in part—by the use of internal standards (Section 11.4.1.2); other errors cannot be corrected.

11.2.4.2 Chromatography

Although sample preparation is the most likely source of large errors and imprecision, several different aspects of the chromatographic process contribute to variability of analytical results. The *chromatographic separation* is of primary concern. As discussed in Section 11.2.1, peaks that are baseline-resolved with flat, low-noise baselines are much easier to integrate; such peaks will give more consistent results. It is obvious that known peaks should be separated from each other, but a good method should also ensure that potential interferences (co-administered drugs, metabolites, degradants, byproducts, etc.) are separated from the peaks of interest. As noted above, *peak tailing* can make the measurement of peak area less reliable. In some cases nonlinear *adsorption* or other concentration-dependent interactions with the column will cause further variability in the results, often for small-mass injections or biological molecules. When the signal-to-noise ratio (S/N) is $<\approx 100$, noise can become an important contributor to imprecision (Section 4.2.3); larger peaks generally give more precise and accurate results. Finally, instrument variability (column aging, flow rate, temperature, etc.) can contribute to imprecision. Each of these factors can influence overall method precision and accuracy.

11.2.4.3 Detection

HPLC detectors (Chapter 4) are transducers that convert the concentration or mass of analyte in the mobile phase into an electrical signal. Some detectors, such as the refractive index (RI) detector (Section 4.11) are very sensitive to changes in temperature or mobile-phase composition; this can add significant noise and uncertainty to the measurement. The most common detectors, the variable-wavelength and diode-array UV detectors (Section 4.4), are much less affected by these factors, but dirty flow cells, bubbles, and aging lamps can contribute to error in the detector output. Other detectors are subject to compromised signals due to suppressed ionization (LC-MS, Section 4.14), fluorescence quenching (fluorescence detector, Section 4.5), or other factors specific to the detection technique. All detectors have a limited linear range, above or below which the response per unit mass of analyte (sensitivity) changes; if the response is not compensated by the calibration curve, errors will be introduced. Finally, any electronic or software aberrations that result in a faulty conversion of the amount of analyte into a proportional electric signal will add to method imprecision; usually these contributions are minor and do not affect data quality.

11.2.4.4 Peak Measurement

As discussed in Section 11.2.1, there are many opportunities for errors when converting the HPLC detector output into values of analyte retention time and area. Location of the peak-start and peak-end points, separation of partially resolved peaks by perpendicular drop or tangent skim, location of the peak maxima, the amount of bunching or smoothing of data slices, and estimation of the baseline location under the peak are just some of the possible sources of error. Methods with well-separated peaks on flat, low-noise baselines will help minimize the amount of error contributed during the integration process, and this should be a primary goal during method development.

11.2.4.5 Calibration

Errors due to calibration fall into two general categories: instrument calibration and method calibration. *Instrument calibration* has been formally defined in a series of tests called instrument qualification, operational qualification, and performance qualification (IQ/OQ/PQ), as described in Section 3.10.1.1. These, plus a periodic system performance test (Section 17.2.1), help ensure that the various instrument components (pump, autosampler, detector, etc.) perform in an acceptable manner both individually and combined as an HPLC system. A properly calibrated instrument should make minimal contributions to the overall method error, but there will always be some variability in flow rate, mobile phase composition, temperature, and so forth.

Method calibration (Section 11.4.1) refers to the selection of reference standards, generation of a calibration curve (plot of concentration vs. response; also called a standard curve), and choice of the curve-fitting algorithm. Each of these steps can contribute to method error. Reference standards should be of known purity; if an internal standard (IS) is used, it should properly mimic the analyte in the sample preparation steps, and be accurately measurable in the chromatogram.

One measure of method performance is *linearity*, usually reported as the coefficient of determination r^2 (although r^2 *alone* is not always adequate, as discussed in Section 11.4.1.5). "Linearity" is shown if a plot of response vs. concentration fits a $y = mx + b$ relationship that passes acceptably close to the origin (e.g.,

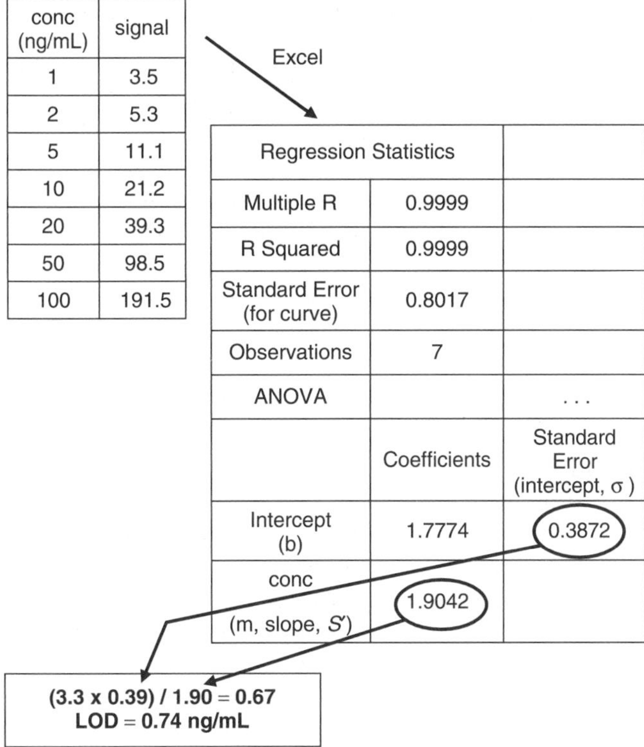

conc (ng/mL)
signal
1
3.5
2
5.3
5
11.1
10
21.2
20
39.3
50
98.5
100
191.5
Excel
Regression Statistics
Multiple R
0.9999
R Squared
0.9999
Standard Error (for curve)
0.8017
Observations
7
ANOVA
. . .
Coefficients
Standard Error (intercept, σ)
Intercept (b)
1.7774
0.3872
conc (m, slope, S′)
1.9042
(3.3 x 0.39) / 1.90 = 0.67
LOD = 0.74 ng/mL

Figure 11.7 Calculation of limit of detection (LOD) from calibration curve data.

$b \leq$ the standard error of the y-intercept), and a linear regression generates some minimum value of r^2 (e.g., $r^2 \geq 0.98$). In such cases a single-point calibration curve may be justified. A *single-point calibration* comprises a single calibration standard (or the average of multiple injections of the same concentration), usually with a concentration near the target concentration for samples (e.g., 100% of a dosage form); the model $y = mx$ is assumed. Many pharmacopeial methods use single-point calibration.

When the calibration plot does not pass through the origin, or the range of possible sample concentrations covers several orders of magnitude, a *multi-point calibration* usually will be more appropriate. An example of this is seen for the data of Figure 11.7, where expected sample concentrations cover two orders of magnitude (1—100 ng/mL), so a multi-point calibration is chosen. For these data the plot is linear, as demonstrated by $r^2 = 0.9999$, but the y-intercept (1.7774 ng/mL) is >4 times the standard error of the y-intercept (0.3872), so a $y = mx + b$ model (with $b \neq 0$) is appropriate: $y = 1.9042x + 1.7774$.

A subset of multi-point calibration—*two-point calibration*—is appropriate and convenient when the calibration plot is linear (e.g., high value of r^2) but does not pass through the origin, and the range of expected sample concentrations is narrow (e.g., less than less than 1 order of magnitude). In such cases two calibration standards are made at two concentrations, preferably near the ends of the range (e.g., 80 and 120% for samples expected to be 100 ± 20%). Any extrapolation beyond the concentration of standards used in any multi-point calibration curve adds uncertainty to the measurement. Generally, it is easiest (and appropriate) to use linear calibration for HPLC methods, but in some cases quadratic or other curve-fitting methods may be required. Error always is introduced when a generalized model of response (i.e., calibration curve) is used to quantify specific samples, since there is error associated with each calibration point that contributes to the overall calibration error. An important part of method validation is to carry out a sufficient number of experiments under controlled conditions with known analyte concentrations so as to determine accuracy and precision throughout the method range (Sections 12.2.1, 12.2.2, 12.2.5).

11.2.5 Limits

A given HPLC method should produce results with acceptable precision and accuracy within a certain range of analyte concentrations. The *limit of detection* (LOD) is the smallest concentration at which an analyte can be confidently determined as being present in a sample, but not necessarily quantified with an exact value. The *lower limit of quantification*, also called the *limit of quantification* or *quantitation* (LLOQ or LOQ), is the smallest concentration of analyte that can be reported as a quantitative value with acceptable precision and accuracy. The *upper limit of quantification* (ULOQ or just *upper limit*) is the largest concentration of analyte that can be reported as a quantitative value with acceptable precision and accuracy. No matter what technique is used to establish these limits, it is strongly recommended [6] to verify the limits with injection of spiked samples at the limit concentrations. These three limits (LOD, LLOQ, ULOQ) are discussed in more detail below. There are several different ways to establish these method limits, and they do not all give equivalent results. For methods that are performed under the oversight of a regulatory agency, it is best to use the limits tests defined by the particular agency. If the method is for nonregulatory use, the decision about which limits tests to use is at the option of the laboratory.

Method limits are strongly linked to the use of the final data. For example, with bioanalytical methods (e.g., drugs in plasma), precision and accuracy are expected to be no worse than ±15% RSD at all concentrations above the LLOQ and ±20% at the LLOQ [6]. On the other hand, a method for release or stability of a drug product is expected to have ≤1% RSD at the 100% concentration [7]. The signal-to-noise ratio (S/N), as shown in Figure 4.7, will influence the overall method error. A simple estimate of the contribution of S/N to the overall method variability is [8]

$$\%\text{-RSD} \approx \frac{50}{S/N} \tag{11.1}$$

This contribution to method variability combines with other sources of error (sampling, sample preparation, etc.) to determine the overall method variability.

Each source of error x in a method accumulates as the sum of the variances x^2:

$$E_T = (E_1^2 + E_2^2 + \ldots + E_n^2)^{0.5} \tag{11.2}$$

where E_T is the total error and $E_1, E_2, \ldots, E_n$ are the contributions of error from each source, 1, 2, ..., n. For example, E_1 might be the error due to sample preparation,

E_2 the error due to the autosampler, or E_3 the error due to signal-to-noise (Eq. 11.1). As a rule, if the RSD of an error source is less than half the total RSD, its contribution to total RSD will be less than 15%. The largest error source in Equation (11.2) will dominate the result, so to reduce the total error, the largest source of error (often S/N at the LOD or LLOQ) should be reduced first. If a single source of error is larger than the acceptable total error, the desired total error will not be reached until this source is reduced.

11.2.5.1 Limit of Detection (LOD)

There are four ways that values of LOD have been estimated:

- visual evaluation
- signal-to-noise ratio
- response standard deviation and slope
- %-RSD-based

These techniques are interrelated, in that they are different approaches to the same goal, but they may give somewhat different results.

Visual evaluation of the chromatogram for the presence or absence of an analyte can be used, but it is highly subjective and susceptible to analyst bias. Therefore it is not recommended.

The *signal-to-noise ratio* can be used to estimate the LOD. A $S/N = 3$ (measured as in Fig. 4.7) is commonly accepted as a definition of LOD. If the LOD can be determined by means of S/N measurements made automatically by the data system for multiple injections, an average value of $S/N = 3$ then corresponds to the LOD. However, if both signal and noise measurements are made manually, the resulting LOD value may be somewhat subjective, due to (unintended) operator bias. In either case several injections at the determined LOD should be made to verify the corresponding LOD analyte concentration. From Equation (11.1) a concentration for which $S/N = 3$ should have its %-RSD $\approx$17%, which should approximately match the %-RSD for multiple injections at the LOD.

Use of the *response standard-deviation and slope* is also a straight-forward method for estimating the LOD [9]:

$$\text{LOD} = \frac{3.3\sigma}{S'} \tag{11.3}$$

where σ is the standard deviation for a calibration curve and S' is its slope (e.g., unit response per ng/mL of concentration). An example of this procedure is illustrated in Figure 11.7 with the aid of an Excel spreadsheet. Data for a calibration curve are collected and entered into a table of analyte concentration vs. detector signal (area or height). A linear regression in Excel is carried out with the data, yielding a table of regression statistics; σ in Equation (11.3) is set equal to the standard error of the y-intercept (0.3872) and S' is set equal to the concentration coefficient (1.9042). Substituted into Equation (11.3), the values in Figure 11.7 give LOD = $(3.3 \times 0.3872)/1.9042 = 0.67$ ng/mL.

Reference [9] allows either the standard error (SE) of the curve or SE of the y-intercept to be used for σ in Eq. (11.3); we disagree and recommend using the

SE of the y-intercept (based on a $y = mx + b$ regression, with b $\neq 0$, even if a forced-zero curve will be used). The SE of the curve is an average of values for the entire curve (including high and low concentrations); for lower concentrations that are more pertinent at the LOD, the SE for an individual concentration decreases, and the %-RSD increases. LOD and LLOQ values should be based on the error at those concentrations, not an average for the curve—the SE of the y-intercept is a better estimate of this than is the SE of the entire curve.

An advantage of Equation (11.3), as opposed to the use of $S/N = 3$, is that the estimated LOD does not have to coincide with one of the calibration standards used to generate the calibration curve. For example, standards were injected at 1 and 2 ng/mL for the regression of Figure 11.7, yet the estimated LOD was 0.67 ng/mL. Equation (11.3) is only an estimate of the LOD, so it is important to confirm this LOD by injecting several samples spiked at 0.67 ng/mL to verify that peaks are indeed detected by the data system (the visual approach) and $S/N \approx 3$—or %-RSD $\approx$17% (Eq. 11.1).

Finally, some users set the LOD based on a specific %-RSD. This *%-RSD-based* value of the LOD might be defined as the minimum concentration below which the %-RSD exceeds 17% (corresponding to $S/N \approx 3$ from Eq. 11.1). The use of this approach requires values of %-RSD as a function of analyte concentration, as are typically carried out during method validation.

It is only by making multiple injections at the proposed LOD that one can have confidence that the correct LOD has been chosen. The LOD estimate from S/N or the method of Equation (11.3) can be confirmed by making 5 to 6 injections at the estimated LOD to give added statistical support—the calculation of the %-RSD at the LOD.

11.2.5.2 Lower Limit of Quantification (LLOQ or LOQ)

Values of LLOQ can be determined by three of the same procedure that are used for determining LOD values:

- signal-to-noise ratio
- response standard deviation and slope
- %-RSD based

If the *signal-to-noise* approach is taken, $S/N = 10$ is commonly accepted as the LLOQ, which can be verified by an RSD of $\leq$5% (Eq. 11.1). The *response standard deviation and slope* procedure is based on the calibration curve, as described in Section 11.2.5.1. For LLOQ, Equation (11.3) is modified to [9]

$$\text{LLOQ} = \frac{10\sigma}{S'} \tag{11.4}$$

With the data of Figure 11.7, LLOQ = 2.0 ng/mL would be calculated from Equation (11.4) by using σ = SE of the y-intercept (see discussion in Section 11.2.5.1).

%-RSD-based values of LLOQ are directly related to the allowed imprecision of a result—which is commonly 5%-RSD (for $S/N = 10$, Eq. 11.1) but can be any value that meets the requirements of the assay. For bioanalytical methods the LLOQ is defined as the smallest concentration for which %-RSD does not exceed 20% [6].

For such methods, sample preparation errors can be a significant contribution to total error, in addition to the contribution of S/N at the LLOQ. Thus $S/N \leq 10$ at the LLOQ (according to the S/N definition of LLOQ) is a reasonable expectation for multiple injections of calibrators; the S/N contribution to total error (Eq. 11.2) should be no more than $(50/10) = 5\%$ (Eq. 11.1)—unlikely the primary source of error. Whatever value of %-RSD is allowed, the LLOQ can then be determined from a plot of %-RSD against the analyte concentration.

No matter what definition is chosen for LLOQ (or LOD), the data at the LLOQ (or LOD) must be sufficiently precise and accurate for the intended application. *LLOQ and LOD should always be verified with samples spiked at the appropriate concentration.*

11.2.5.3 Upper Limits

Whereas techniques to determine the LOD and LLOQ are specified in regulatory guidelines (e.g., [6, 9]), the upper limit of the method is defined in these guidelines only as the upper end of the calibration curve or highest quantifiable amount of analyte within the required precision and accuracy. No techniques are given to determine this amount. For analytical applications of HPLC, usually the lower limits are of most concern. The upper limit is dictated by the highest concentration tested (highest calibration curve concentration) or the point where detector nonlinearity or saturation starts to become a problem.

Assays of a drug substance (pure drug) or drug product (formulated drug), including associated impurities or degradation products, require that the method perform well at both the upper and lower limits of quantification. In such cases the reporting limits often are specified as a percentage of the response for the active pharmaceutical ingredient (API) at the normal dosage level. For example, impurities in new drug substances must be reported at the 0.05% level and quantified at the 0.1% level relative to the API [10]. This requires that the method have a linear response (or other defined curve shape) over a range of $>10^3$. This generally is not a problem with well-behaved detectors, such as UV (Section 4.4), but some detectors, such as the evaporative light-scattering detector (ELSD, Section 4.12.1), may have a much more limited linearity range, precluding them from such applications or requiring innovative techniques to work around the shortcomings of the detector.

11.2.5.4 Samples Outside Limits

Method validation (Section 12.5) is meant to define the performance of a method over the working range, often between the LOD and upper limit. When samples are encountered that exceed the method limits, adjustments in the method process may be required, if valid data are to be obtained. Whether the sample concentration is lower or higher than the method range, extrapolation of the calibration curve is strongly discouraged. Nonlinear behavior, due to sample adsorption at the low end, detector saturation at the high end, or other factors, is common enough that extrapolated data are not to be trusted. It is advisable to choose one of the alternatives listed below.

Samples that are above the upper limit of the calibration curve often can be diluted into the method range, and thus allow useful results to be obtained. Generally, it is best to dilute the sample with the appropriate blank sample matrix (comprising all the constituents of the sample excluding the analyte; for example, blank plasma or the excipients in a pharmaceutical product) or sample diluent (i.e., injection solvent). For example, make a known dilution of an over-limit plasma sample with blank plasma prior to sample pretreatment, or a dilution of a pesticide formulation with additional injection solvent prior to injection. In applications where occasional over-limit samples are likely to be encountered, such as preliminary pharmacokinetic studies, it is wise to include dilution tests as part of the validation process. For example, spiked samples could be prepared to demonstrate that a 10-fold dilution of an over-limit sample gives the same result as a sample prepared at 1/10 the concentration. The method then could be written to allow dilution of any sample, up to 10-fold over-range, into the concentration range approved for analysis.

Samples that have concentrations below the standard curve may require a larger mass of injected sample. In some cases, this can be accomplished simply by injecting a larger sample volume. In other cases, concentration of the sample during sample pretreatment, or less dilution, can provide a solution. In any event, the process should be validated so that one has confidence in the quality of the resulting data. Some methods may be written so that samples with peaks between the LOD and the LLOQ are reported as below limit of quantification (BLQ), indicating that analyte is present but in an insufficient amount to quantify. In every case, the means adopted should be appropriate to the use of the final results.

11.3 QUALITATIVE ANALYSIS

Qualitative measurements are those that identify or help to identify the structure of an analyte. In general, chromatography is a weak tool for qualitative analysis, but a well-behaved HPLC system coupled to an appropriate detector can make it much more suitable. Three approaches are common for qualitative analysis by HPLC:

- retention time
- on-line qualitative analysis
- off-line analysis

Remember that no matter which technique is used, it is much easier to prove that two peaks are *not* the same compound than to prove that they are the same. That is, compounds with closely related structures usually have similar retention times and UV spectral characteristics, so retention time plus UV spectra may not be sufficient for peak identification—it may be necessary to use mass spectrometry, NMR, FTIR, or other techniques to confirm peak identity. Nevertheless, the identification of a peak is generally easier when its behavior and properties can be shown to be identical to those of a specific compound or standard—as opposed to a compound that is not available.

11.3.1 Retention Time

The most common technique for qualitative analysis by HPLC is to compare the retention time t_R of the analyte to that of a reference standard (as discussed in Section 11.2.2, relative retention is often used instead of absolute retention time).

If all HPLC conditions (mobile-phase composition, temperature, flow, etc.) are kept constant, the retention time should be constant. Of course, conditions are never exactly constant, and this results in small variations in retention from run to run (e.g., ±0.02–0.05 min). If an injected analyte falls within the retention range of the standard, this supports the conclusion that the standard and analyte peaks are the same compound. However, retention time is *characteristic, but not unique*; more than one compound can have the same retention time.

In order to minimize the possibility of confusion of one compound for another with the same retention time, efforts should be taken during method development to ensure that adequate resolution is achieved between the analyte of interest and any likely interfering substances. The examples of adjacent peaks in Figure 2.17 can serve as a guide for how much resolution is required for this purpose, which depends considerably on the relative size of two adjacent peaks and how much they tail. Further considerations are (1) the likelihood that the column plate number and peak tailing can change over time, (2) relative retention can vary with inadvertent—usually small—changes in separation condition, and (3) the observed retention time can differ from the true retention time when two peaks overlap. Retention also can be influenced by the sample matrix; for example, the retention time of a pure reference standard may differ slightly from t_R of the same compound in plasma. This is one reason why methods for drugs in biological samples should be calibrated using matrix-based standards [6] (by spiking known concentrations of the reference standard into blank [drug-free] matrix). Because several factors can create uncertainty in the use of retention time for confirming peak identity (qualitative analysis), *it is best to limit this technique to methods where a particular analyte is likely to be present and no other sample components are likely to overlap the analyte peak*.

Another retention-related technique for qualitative analysis is co-injection of a reference standard. In this case the sample is injected; then the reference standard is added to the sample and it is injected again. This technique is related to the method of standard addition (Section 11.4.1.4) used for quantitative analysis. If the peaks in the two injections have the same retention times, peak widths, and peak shapes (within normal variability), there is additional evidence to conclude that the two compounds are identical. On the other hand, if co-injection produces a broader peak, a distorted peak, or two peaks, it is strong evidence that the reference standard and analyte are not the same.

Finally, the use of retention-time predictions, literature values for retention time, or retention times of related substances are never sufficiently accurate to confirm the identity of a compound—although such estimates may prove useful for certain purposes (e.g., the combination of a retention estimate with mass spectral information for peak identification). The use of a retention time alone to identify an analyte should be limited to comparisons with the retention of a known reference standard, where the presence of interfering peaks is unlikely.

11.3.2 On-line Qualitative Analysis

Structural elucidation of unknown analytes can be performed with the aid of HPLC detectors, but rarely is HPLC detection as effective as off-line qualitative procedures. Several HPLC detectors (e.g., UV, NMR, or IR), provide qualitative spectral information about the sample; other detectors, such as the chemiluminescent nitrogen detector, laser light-scattering detector, MS, or chiral detectors, generate more specific and quantitative information about the analyte, such as nitrogen content, approximate mass, analyte molecular weight, or optical rotation, respectively. Detectors usually collect information while the sample passes through the detector flow cell, so the time available for measurement during passage of the peak through the flow cell is similar in magnitude to the peak width—often only a few seconds. A further constraint is that the sample is usually very dilute. Stopped-flow operation is also possible, with the advantage of increasing the time allowed for measurement.

When compared to the same instrumental techniques used in an off-line, stand-alone application—where analysis time is not limited and analyte concentration often is much greater—the information content of on-line techniques is consequently reduced. For this reason on-line data may be most useful for proving that a particular peak is *not* a specific compound, rather than establishing chemical structure. However, when a reference standard is available, the combination of retention time with a single detector response can be sufficient to legally prove the identity of an analyte. An example of this is the use of LC-MS (or LC-MS/MS) in the forensic analysis of drugs of abuse. Finally, in many cases the presumed identity of a peak (e.g., a metabolite of a drug) plus qualitative information from on-line detection may be sufficient for tentative structural confirmation. When data from several HPLC detectors (e.g., FTIR, MS, or chemiluminescent nitrogen) are combined, the structural identity of an analyte can be inferred with greater confidence.

11.3.2.1 UV Detection

The diode-array UV detector (Section 4.4.3), and less commonly the variable-wavelength UV detector (Section 4.4.2) in the stopped-flow scanning mode, can generate UV spectra of chromatographic peaks as they pass through the detector flow cell. UV spectra alone, whether obtained on-line or off-line, rarely have enough information content to assign an analyte structure. The spectra may be sufficient to help confirm the presence of a compound suspected to be in a sample, but the spectral similarity of structurally similar compounds usually prevents any final conclusion about structure. For example, the UV spectra may be sufficient to confirm which peak is the active ingredient in a drug dissolution sample, but it would not be satisfactory to prove the present of a drug of abuse in a forensic situation.

11.3.2.2 LC-MS

The mass spectral detector (Section 4.14), especially in the MS-MS mode, can provide sufficient spectral information to confirm the identity of a peak. The quadrupole LC-MS in the MS-MS mode generates data on precursor-to-product ion transitions that can be used to help elucidate the structure of an unknown, especially when several different transitions can be obtained from the same analyte. The ion-trap LC-MS has the capability to generate additional structural information in the MS^n mode, where product ions may be successively fragmented into smaller product ions (Section 4.14.2). However, with each successive fragmentation, the sample is diluted, reducing the quality of the data. The time-of-flight LC-MS also measures analyte-mass information that can help to provide structural identity. Because the mass resolution (e.g., $m/z \approx 1$) of LC-MS detectors is much lower than stand-alone MS units, fractional mass differences cannot be used for structural elucidation.

11.3.2.3 LC-FTIR

The Fourier transform infrared detector (Section 4.15.1) is used most commonly by trapping and evaporating aliquots of the column effluent, followed by spectroscopic measurements. The LC-FTIR can generate valuable structural information (e.g., Fig. 4.34) that can be used to determine or confirm the chemical structure of a chromatographic peak.

11.3.2.4 LC-NMR

The nuclear-magnetic-resonance LC detector (Section 4.15.2) in the flow-through or stopped-flow mode can provide valuable structural information about a peak (e.g., Fig. 4.35). For ^{1}H NMR, deuterated solvents are required for the mobile phase, or the mobile phase must be evaporated—this can restrict the scope of application of LC-NMR.

11.3.2.5 Chemiluminescence Nitrogen Detector (CLND)

The CLND (Section 4.9) responds to the nitrogen content of the analyte. Because the detector response is proportional to the molar content of nitrogen in the sample, the detector can be calibrated with compounds of known nitrogen content. The molar nitrogen content of an analyte can be determined from the detector response and the mass of analyte injected (e.g., Fig. 4.20). Although this information is not sufficient to determine molecular structure, it can aid in structural analysis.

11.3.2.6 Laser Light-Scattering Detector (LLSD)

The LLSD (Section 4.12.3) can assign an approximate molecular weight to a macromolecular analyte, without the need for a reference standard of the analyte. This capability can be sufficient to distinguish between monomeric and dimeric forms of an analyte (e.g., Fig. 4.26). However, the accuracy of LLSD for determining analyte molecular weight is far below that of LC-MS.

11.3.2.7 Chiral Detectors

Chiral detectors (Section 4.10) can distinguish between enantiomeric forms of an analyte (e.g., Fig. 4.21), and give the sign of rotation. However, no other information is provided about the structure of the analyte.

11.3.2.8 Off-line Analysis

If fractions are collected from the HPLC effluent stream in the semipreparative or preparative mode (Chapter 15), a sufficient amount of pure analyte may be collected to enable off-line analysis for structural determination. Because larger quantities of sample are available and the time-frame restrictions of on-line analysis are removed, off-line structural analysis often can provide conclusive structural identity of a trapped peak. Traditional FTIR, NMR, and mass spectral analysis can be performed; with sufficient sample, wet chemical tests, X-ray crystallography, or other analytical techniques also may be applied.

11.4 QUANTITATIVE ANALYSIS

Whereas the HPLC can provide qualitative information about a sample, its real strength is shown in quantitative analysis. Other than the analytical balance, pH meter, or volumetric pipette, it is likely that HPLC is the most commonly used quantitative tool in the analytical laboratory. For its reliable application, five requirements must be met. First, the HPLC system and its associated method must work in a reproducible manner that provides the requisite precision and accuracy (Sections 11.2.4.2, 11.2.4.3). Second, the data system (Section 11.2) must precisely and accurately convert the detector signal into time and response data. Third, the system must be properly calibrated (Section 11.4.1) to allow measurement of unknown sample concentrations against known quantities of reference standards. Finally, all of the data must be processed in a manner that assures that the overall procedure performs at the required level to comply with appropriate regulatory standards (e.g., Section 12.5) or other end-use requirements for the data. Finally, separation conditions must be such as to enable stable baselines and adequate resolution ($R_s > 1.5$). However, the latter requirement has been dealt with in other chapters and will not be repeated here.

11.4.1 Calibration

Calibration is the process by which the detector response per unit concentration (or mass) of analyte is determined. Some detectors respond to analyte concentration (e.g., UV, Section 4.4), whereas others respond to analyte mass (e.g., evaporative light scattering, Section 4.12.1). In the present discussion we will assume that the detector is concentration sensitive; for the most part the exact same procedures are followed for mass-sensitive detectors. The two most common calibration techniques are *external standardization* and *internal standardization. Area normalization* often is used for purity analyses and other applications where relative concentration is more important than absolute concentration—or where standards for calibration do not exist. The *method of standard addition* is a specialized calibration technique of particular use when a blank sample is not available, and the sample matrix may affect the retention time and/or peak area response for the analyte. For such sample matrices (e.g., plasma), it is also strongly recommended (e.g., [6]) that the calibration standards be prepared in blank matrix (i.e., all the components that are normally found in a sample, excluding the analyte). This, of course, applies to both external and internal standardization (area-normalization and standard addition techniques already have the matrix present). In cases where the sample matrix has little influence on retention or selectivity, such as environmental water samples, or the assay of pure compounds or simple mixtures, matrix-based calibrators may not be required.

In order to obtain accurate results from a method, the calibration curve must adequately represent the concentration-response relationship for the analyte. One way to help improve the accuracy of the calibration curve is to evaluate whether *curve weighting* is appropriate. This topic is discussed in Section 11.4.1.5.

11.4.1.1 External Standardization

A matrix-based set of *calibration standards* is prepared. Usually this is done by accurately weighing a quantity of reference standard of known purity and diluting it in water or buffer to make a primary stock. This stock then is added to the sample matrix (e.g., sample diluent, blank plasma, water, soil, or other matrix appropriate to the sample type), to the concentration corresponding to the highest point on the *calibration curve.* (Some laboratories refer to the standard curve as a *line;* it also is commonly called a *standard curve.* "Curve" is generally used to describe this plot, even though it is most often a linear plot.) Further dilutions are made in matrix to prepare standards that span the method range, including the lowest point on the curve, and sometimes a standard at the limit of detection (LOD). It is customary to include a blank-matrix sample to demonstrate that interferences are not present in the blank.

The two most popular ways to prepare the calibration-curve samples are to use a linear or exponential (sometimes incorrectly called "logarithmic") dilution scheme. For example, with a standard curve covering a method range of 1 to 100 ng/mL, standards might be prepared at 0, 1, 20, 40, 60, 80, and 100 ng/mL for a linear dilution, or 0, 1, 2, 5, 10, 20, 50, and 100 ng/mL for an exponential dilution. At the same time as the calibration standards are prepared, or during preparation of analytical samples, it is a good idea to prepare *quality control standards* (Section 12.3) that will be used to check method performance within a batch of samples. If sample preparation is required, the calibration (and quality control) standards are then processed through the normal sample preparation process, yielding extracted calibration standards for injection.

For external standardization, the same volume of calibration standard at each concentration (level) is injected in sequence from lowest to highest concentration. The low-to-high sequence tends to minimize any carryover-related bias in the curve. The highest concentration standard can be followed by a blank (zero-concentration) standard to check for carryover (Section 17.2.5.10), as well as to avoid carryover bias if the following sample has a low concentration of the analyte. It is best to inject the same volume of different standard concentrations when running the calibration curve, rather than different volumes of the same concentration; the injection volume delivered by most autosamplers is very precise but not necessarily as accurate (Section 3.6.1).

A calibration plot can be constructed manually with the aid of spreadsheet software (e.g., Microsoft Excel), or with the data-system software. It is best to use the data-system software, because in most cases it can be validated, and transfer of data from the raw-data tables into the software is seamless. Excel and similar programs are flexible and work well, but are not considered validated (or validate-able) software by some regulatory guidelines published by authorities such as FDA or ICH, so additional data checking will need to be done to make sure the results are error-free. An example of an external calibration plot is presented in Figure 11.8 using the data from Table 11.1. The slope of the calibration plot, S', is then used to calculate the concentration of unknown samples:

$$\text{ng/mL analyte} = \frac{\text{area analyte}}{S'} \tag{11.5}$$

In the case of Figure 11.8 (with a linear-regression forced through the origin; i.e., $x = 0, y = 0$), $S' = 200.8$ (area units)/(ng/mL analyte). So an analyte peak of 862 area counts would have a concentration of $(862/200.8) = 4.3$ ng/mL. Equation (11.5) assumes that the trend line intercepts the x-axis at $y = 0$, which may or

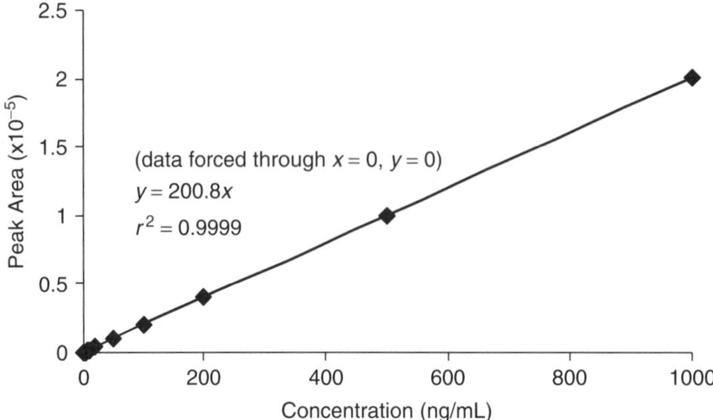

Figure 11.8 Calibration curve based on external standardization data of Table 11.1 (curve forced through zero).

Table 11.1

Calibration Curve Data for Figures 11.8 to 11.10 for Same Analyte and Separation Conditions

	Response		
Concentration (ng/mL)	External Standard[a]	Internal Standard[b]	Standard Addition[c]
0			487
1	215	0.0408	729
2	416	0.0789	911
5	976	0.185	1,435
10	2,056	0.390	2,529
20	4,060	0.770	
50	9,921	1.88	
100	20,140	3.82	
200	40,163	7.62	
500	99,796	18.9	
1000	201,123	38.2	

[a] Area units (Fig. 11.8).
[b] Analyte/IS ratio (Fig. 11.9).
[c] Area with standard added at concentration in column 1 (Fig. 11.10).

may not be the case (Section 11.2.4.5). Alternatively, carry out the linear-regression without forcing the fit through zero (0, 0 point); this is the usual approach taken by data processing software. In the present case, the data of Table 11.1 yield a regression equation (without forcing zero) of $y = 201x - 38$. Solving for x and inserting $y = 862$ gives $x = 4.5$ ng/mL, which adjusts for the (slight) nonzero intercept. The calculated value represents the concentration of analyte in the injected sample; any weighing, dilution, or other sample processing corrections need to be applied to this value before the final sample concentration is reported.

Because the external standard method assumes that the area of the analyte peak accurately represents the concentration of analyte in the original sample, external standardization is best used with methods that involve minimal sample manipulation between the initial sampling process and injection. Therefore solid or liquid samples that undergo weighing, pipetting, dilution, dissolution, and/or filtration processes are good candidates for external standardization. Pharmaceutical dissolution analysis involves placing one or more drug tablets in a dissolution bath of known volume, taking samples at specific time points, filtering the samples, and injecting them. An environmental water sample might be aliquotted by volume, shaken in a measured volume of solvent, filtered, and injected. In both of these cases, it is easy to track the concentration of the injected sample relative to the initial untreated sample, so they would be good candidates for external standardization.

A variation of the external standardization method is *single-point calibration* (Section 11.2.4.5). In this technique experiments during method development and validation are performed to show that analyte response is proportional to its concentration over the method range. Then a single standard is injected, and the analyte concentration in an unknown sample is determined by the ratio of the areas of the standard and the unknown (equivalent to use of Eq. 11.5). Usually the range of the method is narrow, such as ±20% of the target dose of a drug in tablet form. For example, dissolution testing of drug products designed for immediate release can be tested with single-point calibration if supporting data have been gathered to show the validity of this technique [11].

11.4.1.2 Internal Standardization

Internal standardization is superior to external standardization whenever there are sample preparation steps (Chapter 16) in which sample loss can take place. For example, the determination of drugs in plasma often involves solid-phase or liquid–liquid extraction with variable volume recovery, evaporation to dryness, and reconstitution in the injection solvent. At each of these steps the initial and final sample volume seldom are the same for every sample, but the internal standard (IS) tracks such changes, making it possible to obtain precise and accurate results.

The primary difference between internal standardization and external standardization is that an IS is added to samples and calibrators prior to sample pretreatment; calculations of analyte concentration are based on the ratio of areas for the analyte and IS. Calibrators are prepared by weighing and serial dilution, just as for the external standard method (Section 11.4.1.1). Aliquots of the calibration standards (e.g., 200 μL of spiked plasma) are then mixed with an IS solution (e.g., 10 μL), as are all samples—standards and samples are then processed in the same way. The IS solution is prepared in water or buffer at a concentration such that a small volume (e.g., $\leq 5\%$ of the sample volume) will generate a peak of sufficient size (e.g., $S/N > 100$) for measurement with suitable precision and accuracy. As in the case of external standardization, standards are injected in a low-to-high concentration sequence.

The ratio of the area of analyte to area of IS in each of the calibration samples is calculated (e.g., Table 11.1) and a plot of this ratio against the analyte concentration is made (Fig. 11.9). The linear regression for these data gives $y = 0.0381x - 0.0073$, with a standard error of the y-intercept of 0.0172. Since the y-intercept (absolute

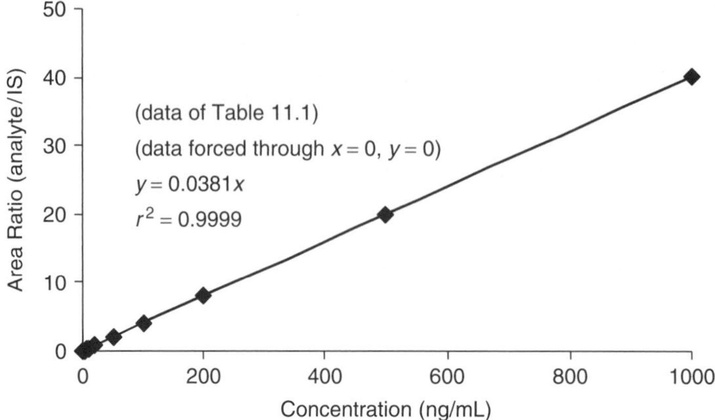

Figure 11.9 Calibration curve based on internal standardization data of Table 11.1.

value) is <SE, it is appropriate to force the curve through the origin (Section 11.2.4.5); that is, $y = 0.0381x$. For this example, $S' = (0.0381$ratio units)/(ng/mL analyte). Calculations of unknown samples are carried out in the same manner as for external standardization, except that the analyte-to-IS ratio is used instead of the absolute analyte area:

$$\text{ng/mL analyte} = \frac{\text{analyte/IS area-ratio}}{S'} \tag{11.6}$$

Thus a sample for which the analyte area is 15,345 and the IS = 4725, the ratio = 3.25. From Equation (11.6), (3.25 ratio units)/0.0381 = 85.2 ng/mL analyte.

For an IS to properly perform its functions, it must have certain properties, several of which are summarized in Table 11.2. The IS or a sample component with the same retention time must never be present in the sample, or an invalid (low) assay result will be obtained. It is desirable to have the IS peak elute near the analyte so that it experiences a similar chromatographic history. If possible, elution just after the analyte is preferable, because if the IS has the correct retention time and area

Table 11.2

Internal Standard Properties

1. Never found in sample
2. Similar k to analyte, preferably eluted after analyte
3. Equivalent sample preparation properties to analyte (pK_a, log $P_{o/w}$, etc.)
4. Well-resolved from analyte (or stable-label for MS)
5. Stable
6. Pure or known purity
7. Compatible detector response (IS peak should have $S/N > 100$)
8. Structurally similar to analyte (desirable, but not essential if property 3 is applicable)

response, it is known that all peaks eluted earlier also were eluted under the proper chromatographic conditions. Peaks eluted after the IS may be subject to bubbles or equipment errors that would not be apparent from an examination of the IS peak.

One of the main roles of the IS is to correct for variations in analyte recovery during sample pretreatment; the IS should therefore possess chemical properties (extraction coefficients, pK_a values for partly ionized analytes, etc.) that are similar to those of the analyte. When several analytes are present in a single sample, it may be desirable to have more than one IS, but this usually is not essential. The IS needs to be well-resolved from the analyte so that peak measurements are not compromised. The exception is for LC-MS applications where a stable, isotopically labeled compound often is used as the IS. Stable-label standards do not require chromatographic resolution if the MS resolution is adequate to distinguish clearly between the analyte and IS (usually the case). Isotopically labeled standards that co-elute with the analyte (typical for ^{13}C standards) will not be subject to changes in separation conditions (e.g., a bubble) or detector conditions (e.g., spray plume irregularity in the MS interface). Thus co-eluted standards more closely mimic the analyte than standards that do not co-elute (often the case for deuterated standards).

The IS needs to be sufficiently stable that it will not change during sample preparation and chromatographic separation. Although it is not necessary for the IS to be highly pure, as long as it is stable, it is important that no IS impurities interfere with the analyte response. The IS also requires an acceptable detector response. Some lists of IS requirements suggest that the IS should be structurally similar to the analyte. There is no basis for this requirement—as long as the IS has the other required properties. However, an IS that is structurally similar to the analyte is more likely to be suitable, so most users choose a structurally similar IS. Good IS candidates are structural analogues of the analyte or related compounds that are not likely to be present in the sample. In some cases it is convenient to make a "flip-flop" method, where two related compounds are used as the IS for each other. For example, in method 1, compound X is used as the IS for analyte Y, and in method 2, compound Y is used as the IS for analyte X; the limitation of this technique, of course, is that both X and Y can never be present in the same sample. This technique would work with two related drugs that were never co-dosed but had closely related extraction and chromatographic properties

11.4.1.3 Area Normalization

A standard feature of data systems is to report *percent peak area*. This is obtained by adding the areas of all peaks in a chromatogram and reporting each peak as a percentage of the total. This report format is convenient for screening a chemical reaction for completion and approximate product purity as well as other applications. A related report is *area normalization*, in which one peak is chosen as a reference peak and all other peaks are reported as a percentage of the reference peak. Area normalization is common for methods used to test drug stability or assay impurities in drug products. In the latter case, any peak $\geq 0.1\%$ of the active pharmaceutical ingredient (API) must be reported and identified, whereas peaks $\geq 0.05\%$ must be reported but not necessarily identified [10]. Both area-% and area normalization are convenient because they do not require standards for each peak. However, both procedures rely on the assumption that the detector response for nonstandardized (e.g., unknown) peaks is the same as for peaks for which standards are available; this assumption may or may not be appropriate, depending on the sample composition and choice of detector. UV detection is notorious for order-of-magnitude differences in sensitivity for different compounds.

11.4.1.4 Standard Addition

The method of standard addition ("spiking") can be useful when a sample blank cannot be obtained, and the sample matrix can affect analyte recovery and/or response. For example, when measuring insulin levels in plasma, it is impossible to obtain plasma without insulin, so standard addition can be used.

Standard addition can be based on a single-point or multiple-point calibration. For single-point calibration, the sample is split into two fractions. One fraction is spiked with a known concentration of the standard, and both fractions are analyzed. The calibration factor is obtained as

$$S' = \frac{\text{area}_\text{s} - \text{area}_\text{ns}}{\text{conc}_\text{s}} \tag{11.7}$$

where area_s and area_{ns} are the areas of the spiked and nonspiked samples, respectively, and conc_s is the concentration of standard *added* to the spiked sample. Using the data of Table 11.1, we see that for $\text{conc}_s = 2$ ng/mL, $\text{area}_s = 911$ and $\text{area}_{ns} = 487$. $S' = (911 - 487)/(2 \text{ ng/mL}) = 212$ area units/(ng/mL). Now the concentration of the non-spiked sample (conc_{ns}; shown as 0 ng/mL in Table 11.1) can be determined as

$$\text{conc}_\text{ns} = \frac{\text{area}_\text{ns}}{S'} \tag{11.8}$$

or $\text{conc}_\text{ns} = 487/212 = 2.3$ ng/mL.

For multiple-point calibration using standard addition, the sample is split into $n + 1$ fractions, where n is the number of standards to be used. Then n samples are spiked, each with a different concentration of standard and all samples are analyzed. A calibration curve is plotted, as shown in Figure 11.10 for the data of Table 11.1. Note that the regression line is extended to the left until it intersects the x-axis (arrow in Fig. 11.10). The value of the intercept x corresponds to $-\text{conc}_\text{ns}$. Linear regression of the data of Table 11.1 gives $y = 201x + 496$. Solving for x and inserting $y = 0$ gives $x = -2.5$, so $\text{conc}_{ns} = 2.5$ ng/mL. The result (<1 SD difference) is the same as that obtained above with Equations (11.7) and (11.8).

It should be stressed that the method of standard addition does not correct for baseline variation or other sample interferences. These problems must be handled in the usual way, before the standard addition procedure is applied. In effect, this approach is a form of in situ calibration, and it can be very useful when the more traditional techniques of external or internal standardization cannot be used. As noted in Section 11.3.1, the method of standard addition also can be useful in confirming peak identity, although its value for this purpose is no greater than the use of a retention time.

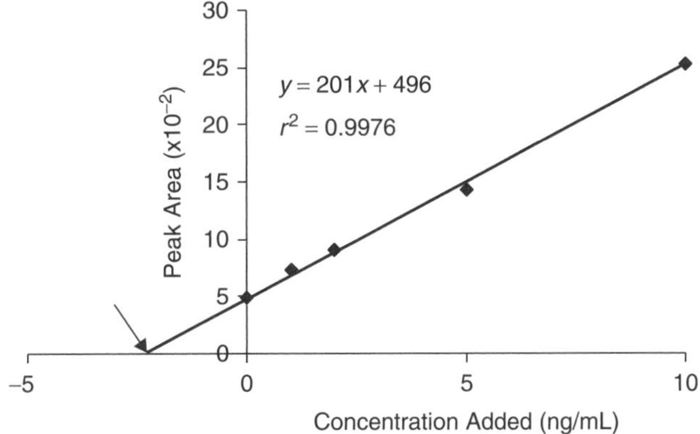

Figure 11.10 Use of the method of standard addition to determine analyte concentration at x-axis intercept (arrow); data of Table 11.1.

11.4.1.5 Evaluating Calibration Curves

Plots for wide concentration-range calibration curves, such as those of Figures 11.8 and 11.9 can be hard to interpret when a linear x-axis is used, because the low concentration points are crowded together. An alternate way to examine the data is to make a plot of %-error against log concentration (a "%-error plot"). %-Error is determined by using the regression equation to calculate the theoretical y-value for each concentration; error is calculated as (experimental value − theoretical value)/theoretical value, and expressed as %-error.

The benefit of the %-error plot is shown in Figure 11.11 for a data from a hypothetical method-validation study. Calibrators were injected at 10 concentrations: 1, 2, 5, 10, 20, 50, 100, 200, 500, and 1000 ng/mL. The plot of %-error versus (linear) concentration of Figure 11.11*a* shows that there is much more relative error at lower concentrations, but this figure is hard to interpret because of the crowding of data points at low concentrations. The %-error versus log concentration plot of Figure 11.11*b* solves the crowding problem, and now the error at each concentration can be easily examined. The data fall into two sets—those above 20 ng/mL with a relatively constant error of $\approx\pm1\%$ (1 SD)—while remaining concentrations show increasing relative error as concentration is reduced (dashed lines of Fig. 11.11*b*). This pattern is expected, as S/N makes a larger contribution to relative error (Eq. 11.1) at low concentrations. Other than the (normal) increase in error at low concentrations, the data of Figure 11.11 have acceptable regression statistics ($r^2 = 0.9999$, y-intercept <SE), so a multiple-point calibration curve with forced zero ($y = mx$; Sections 11.4.1.1, 11.4.1.2) is appropriate.

The %-error plot can highlight problems with calibration curves. Data similar to that of Figure 11.11*a, b* are shown in the example of Figure 11.11*c*, which emphasizes the importance of picking the proper y-intercept. If a multiple-point calibration is chosen, with the curve forced through zero, the %-error plot of Figure 11.11*c* results. Lower concentrations show increasing error, with an average error $\approx50\%$ for 1 ng/mL. For a bioanalytical method, where maximum error allowed at the LLOQ $= \pm20\%$, the lowest concentration with an average error of $\leq20\%$ is 5 ng/mL ($\approx9\%$ error); this limits the application of the method to a

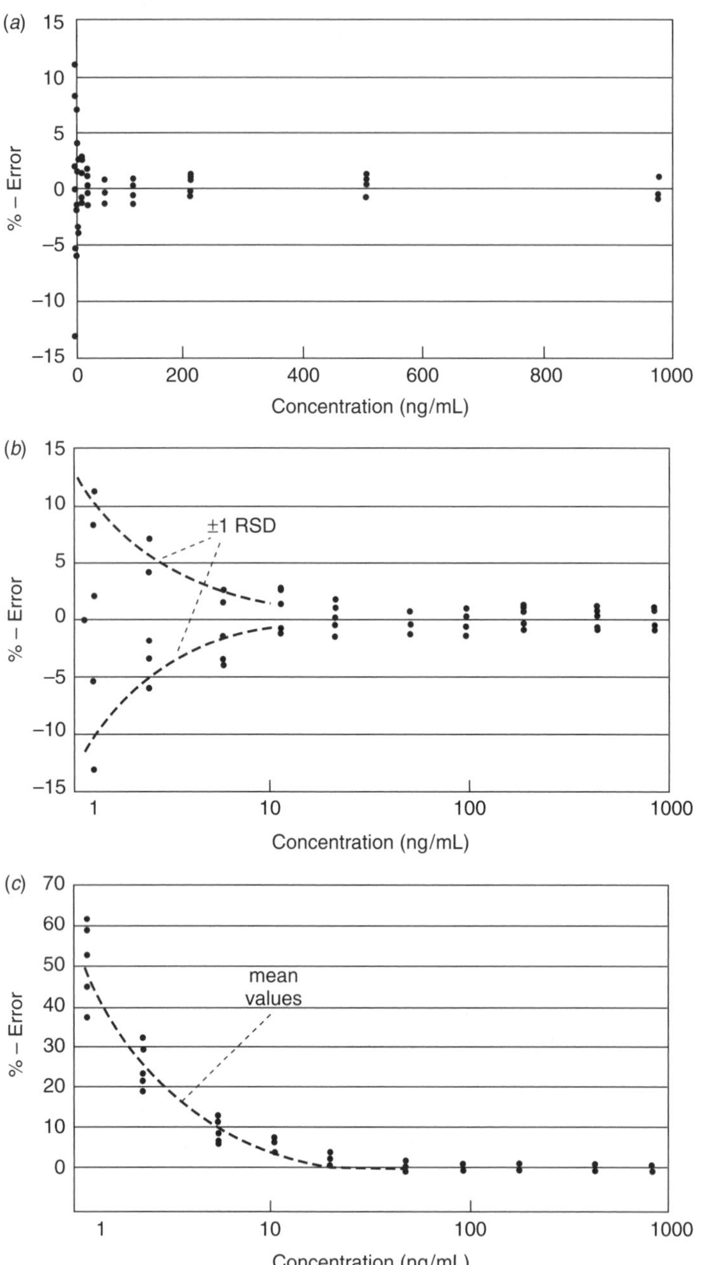

Figure 11.11 Use of %-error plots to examine calibration-curve data. (*a*) %-Error versus (linear) concentration; (*b*) data of (*a*) plotted as %-error versus log concentration (*y*-intercept <SE; curve forced through zero); (*c*) %-error versus log concentration (*y*-intercept >SE; curve improperly forced through zero).

range of 5 to 1000 ng/mL. For this data set the y-intercept >SE, so the proper curve fit is $y = mx + b$ (with $b \neq 0$). The resulting %-error plot (not shown) closely resembles Figure 11.11*b*, with an average error of <6% throughout the curve. Now the calibration curve allows a bioanalytical method to be applied over a range of 1 to 1000 ng/mL. It is interesting to note that for both curve fits (with $b = 0$ or $b \neq 0$), $r^2 > 0.9999$, so r^2 alone is not sufficient to ensure good performance of a multiple-point calibration.

Low r- or r^2-values can indicate that there are problems with a calibration curve, but the converse is not necessarily true—large r^2-values do not guarantee a well-behaved curve. A plot of %-error against log concentration (%-error plot) is a useful way to make a visual examination of data. *We recommend examining all calibration-curve data with a %-error plot as a means of highlighting potential problems with the method.*

11.4.2 Trace Analysis

HPLC is used for the analysis of samples of widely varying concentration. The term *trace analysis* often is used to describe small sample concentrations. One way to define trace analysis is to describe samples for which the precision of measurement is affected by the concentration, often with a transition point to trace analysis when $S/N < \approx 100$ (Section 4.3). Other than the problems associated with dealing with low concentrations and small signals, trace analysis is little different from the analysis of more concentrated samples. With trace analysis, peak height measurements may be preferred over peak area. We recommend evaluating both peak height and peak area; choose the final measurement technique based on the one that gives the best precision and accuracy. For additional information, see Sections 2.6.3.2, 4.2.4, and 11.2.5; for discussion related to specific detectors, consult the appropriate detector discussion in Sections 4.4 through 4.16.

11.5 SUMMARY

Use of the HPLC as a qualitative or quantitative analytical tool requires that the system be operating properly and that the data system be set up to accurately determine peak retention times and peak areas or heights. Consideration has to be taken relative to resolution requirements for the peaks of interest, including the influence of relative peak size and shape. Although the HPLC system is not as useful a tool for qualitative analysis as some dedicated instruments (e.g., FTIR, NMR, or high-resolution MS), with the help of the appropriate detector(s) it can provide valuable qualitative information for many applications. Liquid chromatography shows its strongest assets with quantitative analysis. HPLC can be used for trace analysis of pollutants in river water, drugs and their metabolites in biological systems, or impurities in reagents. It is also useful for determining content uniformity of pharmaceutical products with high precision and accuracy. Quantitative analysis relies on selection of appropriate reference standards and proper calibration so that the results are of high quality and can withstand the scrutiny of review by regulatory agencies.

REFERENCES

1. N. Dyson, *Chromatographic Integration Methods*, 2nd ed., Royal Society of Chemistry, Letchworth, UK, 1998.
2. L. R. Snyder and J. J. Kirkland, *Introduction to Modern Liquid Chromatography*, 2nd ed., Wiley, New York, 1974.
3. V. R. Meyer, *J. Chromatogr. Sci.*, 33 (1995) 26.
4. R. Q. Thompson, *J. Chem. Ed.*, 62 (1985) 866.
5. *Guidance for Industry: Part 11, Electronic Records; Electronic Signatures—Scope and Application*, USFDA-CDER, Aug. 2003, http://www.fda.gov/ora/compliance_ref/part11.
6. *Guidance for Industry: Bioanalytical Method Validation*, USFDA-CDER, May 2001, http://www.fda.gov/cder/guidance/index.htm.
7. *Reviewer Guidance: Validation of Chromatographic Methods*, USFDA-CDER, Nov. 1994, http://www.fda.gov/cder/guidance/index.htm.
8. L. R. Snyder, J. J. Kirkland, and J. L. Glajch, *Practical HPLC Method Development*, 2nd ed., Wiley-Interscience, New York, 1997, p. 71.
9. *Validation of Analytical Procedures: Text and Methodology Q2(R1)*, International Conference on Harmonization, Nov. 2005, http://www.ich.org/LOB/media/MEDIA417.pdf.
10. *Impurities in New Drug Substances Q3A(R2)*, International Conference on Harmonization, Oct. 2006, http://www.ich.org/LOB/media/MEDIA422.pdf.
11. *Specifications: Test Procedures and Acceptance Criteria for New Drug Substances and New Drug Products: Chemical Substances Q6A*, International Conference on Harmonization, Oct. 1999, http://www.ich.org/LOB/media430.pdf.
12. J. C. Miller and J. N. Miller, *Statistics for Analytical Chemistry*, Halsted Press-Wiley, New York, 1984, secs. 4.9–4.10.

CHAPTER TWELVE

METHOD VALIDATION

with Michael Swartz

Introduction to Modern Liquid Chromatography, Third Edition, by Lloyd R. Snyder, Joseph J. Kirkland, and John W. Dolan

12.1 INTRODUCTION

Quality is a commonly used word in the world of analytical chemistry. Quality encompasses many aspects of the laboratory; in this chapter it refers to the development and application of HPLC methods. Our primary focus will be the validation of HPLC methods, a process that underlies the quality of the method and laboratory results. Other aspects of quality in the HPLC laboratory are also discussed in this chapter, especially quality control and quality assurance (Section 12.9).

Method validation establishes, by means of laboratory studies, that the performance characteristics of the test method meet the requirements of the intended analytical application. Method validation provides an assurance of reliability during normal use, and this process is sometimes referred to as providing documented evidence that the method does what it is intended to do. Regulated laboratories must carry out method validation in order to be in compliance with governmental or other regulatory agencies. A well-defined and documented method-validation process not only satisfies regulatory compliance requirements but also provides evidence that the system and method are suitable for their intended use, and aids in method transfer. In 1987, the US Food and Drug Association (FDA) first designated the specifications listed in the current edition of the *United States Pharmacopeia* (USP) as those legally recognized to determine compliance with the Federal Food, Drug, and Cosmetic Act [1–2]. More recently, new information has been published that updates the previous guidelines and provides more detail and harmonization with International Conference on Harmonization (ICH) guidelines [3–4]. The inclusion and/or definition of some terms differs for the FDA, USP, and ICH, but harmonization on a global basis has provided much more detail than was available in the past. So it may be useful to downplay any differences between global regulatory requirements.

An HPLC method may be referred to as an "analytical procedure," "analytical method," "assay procedure," "test method," or just "method." In the present discussion, we generally will use "test method" or (less often) "HPLC method" for any of these methods. The largest number of HPLC methods in use today is carried out in the pharmaceutical industry; for this reason we will describe method validation for pharmaceutical applications. Other regulated industries have well-defined processes in place for method validation as well. For example, environmental monitoring laboratories are under the oversight of the Environmental Protection Agency (EPA) [5], whereas some other organizations rely on directives of the International Organization for Standardization (ISO) [6].

Validation and other laboratory practices are regulated by the FDA, USP, ICH, EPA, and related organizations. Because HPLC methods developed by an industrial analytical laboratory often are used in a manufacturing department, the analytical laboratory may be constrained by manufacturing practices and regulations. Two of the most common references to these practices are cGMP (current Good Manufacturing Practice, e.g., [7–8]) and the ISO 9000 Global Management Standards [9] and related ISO (International Organization on Standardization) documents. These aspects of laboratory regulation are not discussed further in this chapter.

In nonregulated industries and academic laboratories, there also is a need for high-quality test methods that provide reliable data. The use of good scientific practices is often assumed but is not always the case, so method validation is strongly recommended even where it is not required by regulation. The reader should be able to take the information presented here for pharmaceutical applications and use it as a basis for other areas of application.

Method validation can be regarded as just one part of an overall validation process that encompasses at least four distinct steps: (1) software validation, (2) analytical instrument qualification or validation (AIQ; Section 3.10.1), (3) method validation, and (4) system suitability. The overall validation process begins with validated software and a qualified instrument; then a test method is developed and subsequently validated using the qualified system. Finally, the performance of the test method on a given day can be confirmed by means of a system suitability test. Each of these four steps is critical to method performance.

Two guidelines are important for any method validation process: USP Chapter 1225, Validation of Compendial Methods [2], and the International Conference on Harmonization (ICH) Guideline, Validation of Analytical Procedures: Text and Methodology Q2 (R1) [4]. Although the subject of the current discussion is HPLC, both the USP and ICH guidelines apply to any analytical procedure, technique, or technology used in a regulated laboratory. It should be noted that the USP publishes official test methods, often called *compendial* methods, that are accepted by the USP as already validated. The USP also publishes guidelines that should be applied to the validation of test methods not developed by the USP (it is assumed that the USP published methods were subject to the same guidelines). This chapter concentrates on the application of the USP (and other regulatory agency) guidelines to HPLC methods developed by independent laboratories (i.e., not by the agencies themselves).

Even though the USP is the sole legal document in the eyes of the FDA, this chapter draws from both USP and ICH guidelines, as appropriate, for definitions and methodology. For the most part the FDA, USP, and ICH guidelines agree. Where the guidelines disagree, it is up to the user to decide on an appropriate interpretation of the guidelines. Often this is the responsibility of the user's quality assurance unit (Section 12.9), and may be aided by review of the latest regulatory actions (e.g., FDA-issued Form 483 Inspectional Observations). For the present discussion, the regulatory publications will be referred to generically as "guidelines." In addition to these guidelines, sometimes a regulatory body publishes other information that can be useful for interpretation of the guidelines. One of these is a "reviewer guidance" [10] published by the FDA. This document is intended to help FDA auditors determine what comprises a good test method, so many users try to adhere to the suggestions of this document to ensure that their test methods will pass regulatory scrutiny. In addition to the general process of method validation, we will discuss terms, definitions, and related topics, and—where possible—provide examples to illustrate how these general guidelines apply to HPLC.

A major difference between this chapter and other chapters in this book is the regulatory oversight of validated methods in the pharmaceutical, environmental, and certain other industries. Rules pertaining to method validation are described in official documents that originate at different times, are written by different people (with different writing skills), and are released by different agencies (with varying internal policies). This bureaucratic process inevitably results in documents that can be ambiguous, inconsistent, and difficult to interpret. In this chapter we try to impart some unity to the requirements and guidelines contained in these various regulatory pronouncements. We also try to make the discussion practical for the average user. Nevertheless, the bureaucratic language of regulatory recommendations and requirements could not entirely be masked. Consider this limitation as good practice for dealing with the official documents.

Method validation may seem to have a vocabulary of its own. This chapter therefore begins with a discussion of important terms and definitions (Section 12.2). A procedure that ensures that a test method can provide valid data on a given day is the system suitability test, described in Section 12.3. Without documentation, there is no proof of method validity; some aspects of method documentation are described in Section 12.4. Validation of test methods for drug substance (pure drug) and drug product (formulated drug) have requirements (Section 12.5) that are distinctly different from bioanalytical methods that measure drugs in biological matrices (Section 12.6). Once a test method has been validated, it often must be transferred to another laboratory for routine application; some of the principles of analytical method transfer (AMT) are discussed in Section 12.7. Many times when test methods are transferred, they do not work exactly as they did in the original laboratory, and over time, most methods require some adjustment to meet system suitability requirements. Methods can be *adjusted* to meet system suitability, but if they require more substantial changes (are *modified*), they must undergo as least some re-validation; the topic of adjustment vs. modification is covered in Section 12.8. Finally, a good test method and its application require strong quality control and quality assurance programs, as described in Section 12.9.

12.2 TERMS AND DEFINITIONS

Several analytical performance characteristics may be investigated during any method validation protocol:

- accuracy
- precision/ruggedness

- specificity
- limit of detection
- limit of quantitation
- linearity
- range
- robustness

Although most of these terms are familiar and are used daily in any regulated HPLC laboratory, they sometimes mean different things to different people. For example, ruggedness, which forms a part of any well-designed precision study, is often confused with robustness. The following standard definitions (Sections 12.2.1–12.2.6), of applications in the pharmaceutical industry should clarify any confusion. In this context, *drug substance* refers to the pure chemical drug, also called the *active pharmaceutical ingredient* (API). The *drug product* refers to the product that is sold to the consumer; it usually contains one or more drug substances plus *excipients*—(other chemicals, fillers, colors, etc.).

Several types of test methods are used to measure the API and/or impurities, related substances, excipients, and so forth. The major method types discussed in this chapter are assay, impurity (also related substances), dissolution, and bioanalytical methods. A test method used for *assay* is one that measures the active ingredient concentration in a drug product or substance. A *content uniformity* method is similar to an assay method, but it targets the measurement of the variability in drug concentration within a batch of samples. An *impurity* test measures the (generally unintentional) minor components present in the substance or product that originate from raw material manufacturing, product manufacturing, or degradation during storage or processing. A *stability-indicating* method is used to quantify the presence of impurities (degradants) generated through a forced degradation of the API; it is assumed that this test will enable measurement of any impurities generated during normal or accelerated shelf-life testing of a drug substance or product. Any degradants found in this way may be included in the impurity test. A *dissolution* assay measures the concentration of API in a solution designed to simulate release of the drug from a formulation under the conditions of administration of the drug (e.g., in simulated stomach fluids). Whereas the preceding test methods are for drug product or drug substance, a *bioanalytical* method (Section 12.6) is used to determine the concentration of a drug in a biological system, most commonly plasma.

12.2.1 Accuracy

Accuracy is the measure of exactness of an analytical method, or the closeness of agreement between an accepted reference value and the value found in a sample. Established across the range of the method, accuracy is measured as the percentage of analyte recovered by the assay. For the drug substance, accuracy measurements are obtained by comparison of the results to the analysis of a standard reference material, or by comparison to results from a second, well-characterized method. For the assay of the drug product, accuracy is evaluated by the analysis of synthetic mixtures spiked with known quantities of the analytes. For the quantification of impurities, accuracy is determined by the analysis of samples (drug substance or drug

Table 12.1

Determination of Method Accuracy/Recovery and Precision

Sample Concentration (mg)	Accuracy/Recovery		
	Replicate 1	Replicate 2	Replicate 3
1.000	98.91%	98.79%	98.44%
2.000	99.08%	98.54%	98.39%
3.000	98.78%	98.68%	98.01%
Mean	98.62%		
Standard deviation	0.32%		
Relative standard deviation	0.32%		
Acceptance criteria	Accuracy (mean)	Precision (RSD)	
	98–102%	≤2.0%	
Assessment	Pass	Pass	

product) spiked with known amounts of impurities (if impurities are not available, see specificity, Section 12.2.3).

Table 12.1 illustrates a representative accuracy study. To document accuracy, the guidelines recommend that data be collected from a minimum of nine determinations over a minimum of three concentration levels covering the specified range (i.e., three concentrations, three replicates each). The data should be reported as the percent recovery of the known or added amount, or as the difference between the mean and true value with confidence intervals (±1SD). In Table 12.1, data are shown relative to 100%, and the mean recovery for $n = 9$ samples is 98.62% with %RSD = 0.32%. In this example both the accuracy and precision pass the pre-defined acceptance criteria of 98–102% and ≤2%, respectively.

12.2.2 Precision

The precision of an analytical method is defined as the closeness of agreement among individual test results from repeated analyses of a homogeneous sample. Precision is commonly performed as three different measurements: repeatability, intermediate precision, and reproducibility.

12.2.2.1 Repeatability

The ability of the test method to generate the same results over a short time interval under identical conditions (intra-assay precision) should be determined from a minimum of nine determinations. Their repeatability should cover the specified range of the procedure (i.e., three concentrations, three repetitions each) or from a minimum of six determinations at 100% of the test or target concentration. Representative repeatability results are summarized in Table 12.2, where results are summarized for six replicate injections of the same sample. The 0.12% RSD easily passes the ≤2% acceptance criterion.

Table 12.2

Determination of Repeatability by Replicate Injections of the Same Sample

Injection	Response
1	488,450
2	488,155
3	487,986
4	489,247
5	487,557
6	487,923
Mean	488,220
Standard deviation	582.15
RSD	0.12%
Acceptance criteria (RSD)	≤2%
Assessment	Pass

12.2.2.2 Intermediate Precision

Intermediate precision refers to the agreement between the results from within-laboratory variations due to random events that might normally occur during the use of a test method, such as different days, analysts, or equipment. To determine intermediate precision, an experimental design should be employed so that the effects (if any) of the individual variables can be monitored. Typical intermediate precision results are shown in Table 12.3. In this study, analysts from two different laboratories prepared and analyzed six sample preparations from one batch of samples and two preparations each from two additional batches (all samples are assumed to be the same concentration); all data from each analyst were pooled for the summary in Table 12.3. The analysts prepared their own standards and solutions, used a column from a different lot, and used a different HPLC system to evaluate the sample solutions. Each analyst successfully attained the precision requirements of ≤2% RSD, and the %-difference in the mean values between the two analysts was 0.7%, which indicates that there is no difference in the mean values obtained (Student's *t*-test, $P = 0.01$).

12.2.2.3 Reproducibility

Documentation in support of collaborative studies among different laboratories should include the standard deviation, relative standard deviation (or coefficient of variation), and the confidence interval. Table 12.4 lists some results typical of a reproducibility study. To generate the data shown here, analysts from two different laboratories (but not the same analysts involved in the intermediate precision study) prepared and analyzed six sample preparations from one product batch and two preparations each from two additional batches (all samples are assumed to be the same concentration). They prepared their own standards and solutions, used a column from a different lot, and used a different HPLC system to evaluate the

Table 12.3

Measurement of Intermediate Precision

	Amount	
	Analyst One	Analyst Two
Mean	13.9 mg	14.0 mg
Standard deviation	0.05 mg	0.03 mg
% RSD	0.36	0.21
% Difference (means)	0.70%	
Acceptance criteria (RSD)	≤2%	
Assessment	Pass	

Table 12.4

Measurement of Reproducibility

	Amount	
	Lab One	Lab Two
Mean	14.0 mg	13.8 mg
Standard deviation	0.07 mg	0.14 mg
% RSD	0.50	1.01
% Difference (means)	1.43%	
Acceptance criteria (RSD)	≤2%	
Assessment	Pass	

sample solutions. Each analyst successfully attained the precision requirements of ≤2% RSD, and the %-difference in the mean values between the two analysts was 1.4%, indicating that there is no difference in the mean values obtained (Student's t-test, $P = 0.01$).

12.2.2.4 Ruggedness

Ruggedness is defined in past USP guidelines as the degree of reproducibility of test results obtained by the analysis of the same samples under a variety of conditions, such as different laboratories, analysts, instruments, reagent lots, elapsed assay times, assay temperature, or days. Ruggedness is a measure of the reproducibility of test results under the variation in conditions normally expected from laboratory to laboratory and from analyst to analyst. The use of the term ruggedness, however, is falling out of favor; the term is not used by the ICH but is instead addressed in guideline Q2 (R1) [4] under the discussion of intermediate precision (Section 12.2.2.2, within-laboratory variations: different days, analysts, equipment, etc.) and reproducibility (Section 12.2.2.3, between laboratory variations from collaborative studies).

12.2.3 Specificity

Specificity is the ability to measure accurately and specifically the analyte of interest in the presence of other components that may be expected to be present in the sample. Specificity takes into account the degree of interference from other active ingredients, excipients, impurities, degradation products, and so forth. Specificity in a test method ensures that a peak's response is due to a single component (no peak overlaps). Specificity for a given analyte is commonly measured and documented by resolution, plate number (efficiency), and tailing factor.

For identification purposes, specificity is demonstrated by (1) separation from other compounds in the sample, and/or (2) by comparison to known reference materials.

Separation from Other Compounds. For assay and impurity tests, specificity can be shown by the resolution of the two most closely eluted compounds that might be in the sample. These compounds usually are the major component or active ingredient and the closest impurity. If impurities *are* available, it must be demonstrated that the assay is unaffected by the presence of spiked materials (impurities and/or excipients). If the impurities *are not* available, the test results are compared to a second, well-characterized procedure. For assay, the two results are compared. For impurity tests, the impurity profiles are compared. Comparison of test results will vary with the particular test method but may include visual comparison as well as retention times, peak areas (or heights), peak shape, and so forth.

Comparison to Known Reference Materials. Starting with the publication of *USP 24*, and as a direct result of the ICH process, it is now recommended that a peak-purity test based on diodearray (DAD) detection or mass spectrometry (MS) be used to demonstrate specificity in chromatographic analyses. Modern DAD technology (Section 4.4.3) is a powerful tool used to evaluate specificity. DAD detectors can collect spectra across a range of wavelengths for each data point collected across a peak, and through software processes, each spectrum can be compared (to the other spectra collected) to determine peak purity. Used in this manner, DAD detectors can distinguish minute spectral and chromatographic differences not readily observed by simple overlay comparisons.

However, DAD detectors can be limited, on occasion, in the evaluation of peak purity by a lack of UV response, as well as by the noise of the system and the relative concentrations of interfering substances. Also the more similar the spectra are, and the larger the concentration ratio is, the more difficult it is to distinguish co-eluted compounds. Mass spectrometry (MS) detection (Section 4.14) overcomes many of these limitations of the DAD, and in many laboratories it has become the detection method of choice for method validation. MS can provide unequivocal peak purity information, including exact mass, structural and quantitative information. The combination of both DAD and MS on a single HPLC instrument can provide complementary information and ensure that interferences are not overlooked during method validation.

12.2.4 Limit of Detection and Limit of Quantification

The limit of detection (LOD) is defined as the lowest concentration of an analyte in a sample that can be detected but not necessarily quantified. It is a limit test that specifies whether an analyte is above or below a certain value. The limit of quantification (LOQ, also called limit of quantitation) is defined as the lowest concentration of an analyte in a sample that can be quantified with acceptable precision and accuracy by the test method.

Determination of the LOQ is a two-step process. Regardless of the method used to determine the LOQ, the limits should first be estimated from experimental data, such as by signal-to-noise ratio or the slope of a calibration curve (Sections 4.2.4, 11.2.5). Second, the latter value must be confirmed by results for samples formulated at the LOQ. For further details, see Section 11.2.5.

12.2.5 Linearity and Range

Linearity is the ability of the test method to provide results that are directly proportional to analyte concentration within a given range. Linearity generally is reported as the variance of the slope of the regression line (e.g., standard error from an Excel regression analysis, as in Fig. 11.7). Range is the interval between the upper and lower concentrations of analyte (inclusive) that have been demonstrated to be determined with acceptable precision, accuracy, and linearity using the test method as written. The range is normally expressed in the same units as the results obtained by the test method (e.g., ng/mL). Guidelines specify a minimum of five concentration levels for determining range and linearity, along with certain minimum specified ranges that depend on the type of test method [2, 4]. Table 12.5 summarizes typical minimum ranges specified by the guidelines [4]. Data to be reported generally include the equation for the calibration curve line, the coefficient of determination (r^2), and the curve itself, as illustrated in Figures 11.8 and 11.9 based on the data of Table 11.1.

12.2.6 Robustness

The robustness of an analytical procedure is defined as a measure of its capacity to obtain comparable and acceptable results when perturbed by small but deliberate variations in specified experimental conditions. Robustness provides an indication of the test method's suitability and reliability during normal use. During a robustness study, conditions are intentionally varied to see if the method results are affected. The key word in the definition is *deliberate*. Example HPLC variations are illustrated

Table 12.5

Example Minimum Recommended Ranges

Type of Method	Recommended Minimum Range
Assay	80–120% of the target concentration
Impurities[a]	From the reporting level of each impurity, to 120% of the specification
Content uniformity	70–130% of the test or target concentration
Dissolution	±20% over the specified range of the dissolution test

[a]For toxic or more potent impurities, the range should reflect the concentrations at which these must be controlled.

Table 12.6

Typical Variations to Test Robustness in Isocratic Separations

Factor	Limit Range
Organic solvent concentration	±2–3%
Buffer concentration	±1–2%
Buffer pH (if applicable)	±0.1–0.2 pH units
Temperature	±3°C
Flow rate	±0.1–0.2 mL/min
Detector wavelength	±2–3 nm for 5-nm bandwidth
Injection volume	Depends on injection type and size
Column lots	2–3 different lots

Table 12.7

Typical Variations to Test Robustness in Gradient Separations

Factor	Limit Range
Initial gradient hold time	±10–20% of hold time
Slope and length	Slope determined by the gradient range and time; adjust gradient time by ±10–20% and allow the slope to vary
Final hold time	Adjust to allow last-eluted compound to appear in chromatogram
Other variables listed in Table 12.6 (as appropriate)	

in Tables 12.6 and 12.7 for isocratic and gradient methods, respectively. Variations should be chosen symmetrically about the value specified in the test method (e.g., for ±2%, variations of +2 and −2%), to form an interval that slightly exceeds the variations that can be expected when the test method is implemented or transferred. For example, if the buffer pH is adjusted by titration and the use of a pH meter, the typical laboratory has an error of ±0.1 pH units. To test robustness of a test method to variations in a specified pH-2.5 buffer, additional buffer might be prepared and tested at pH-2.3 and pH-2.7 to ensure that acceptable analytical results are obtained. For instrument settings, manufacturers' specifications can be used to determine variability. The range evaluated during the robustness study should not be selected to be so wide that the robustness test will purposely fail, but rather to represent the type of variability routinely encountered in the laboratory. Challenging the test method to the point of failure is not necessary. One practical advantage of robustness tests is that once robustness is demonstrated over a given range of an experimental condition, the value of that condition can be adjusted within that range to meet system suitability without a requirement to revalidate the test method (Section 12.8).

Robustness should be tested late in the development of a test method, and if not, is typically one of the first method characteristics investigated during method validation. However, throughout the method development process attention should be paid to the identification of which chromatographic conditions are most sensitive to small changes so that, when robustness tests are undertaken, the appropriate variables can be tested. Robustness studies also are used to establish the system suitability test to make sure that the validity of the entire system (including both the instrument and the test method) is maintained throughout method implementation and use. In addition, if the results of a test method or other measurements are susceptible to variations in experimental conditions, these conditions should be adequately controlled and a precautionary statement included in the method documentation.

To measure and document robustness, the following characteristics should be monitored:

- critical peak pair resolution R_s
- column plate number N (or peak width in gradient elution)
- retention time t_R
- tailing factor TF
- peak area (and/or height) and concentration

Replicate injections should be made during the robustness study to improve the estimates (e.g., %RSD) of the effect of an experimental-variable change. In many cases multiple peaks are monitored, particularly when some combination of acidic, neutral, or basic compounds is present in the sample. It may be useful to include in the method document a series of compromised chromatograms illustrating the extremes of robustness (see discussion of Fig. 1.5 of [11]). Such examples, plus corrective instructions, can be useful in troubleshooting method problems.

12.3 SYSTEM SUITABILITY

Although not formally a part of method validation according to the USP, system suitability tests are an integral part of chromatographic methods [6]. System suitability tests are used to verify that the resolution and precision of the system are adequate for the analysis to be performed. System suitability tests are based on the concept that the equipment, electronics, analytical operations, and samples comprise an integral system that can be evaluated as a whole.

System-suitability tests check for adequate system performance before or during sample analysis. Characteristics such as plate number, tailing factor, resolution, and precision (repeatability) are measured and compared to the method specifications. System-suitability parameters are measured during the analysis of a "system suitability sample" (a mixture of the main components and expected degradants or impurities, formulated to simulate a representative sample). However, samples consisting of only a single peak (e.g., a drug substance assay where only the API is present) can be used, provided that a column plate number and tailing factor are specified in the test method. Replicate injections of the system suitability sample are compared to determine if requirements for precision are met. Unless otherwise

Table 12.8

FDA System Suitability Recommendations

Parameter	Recommendation	Comments
Retention factor k	$k > 2$	Peak should be well resolved from other peaks and the t_0-peak
Injection repeatability	RSD $\leq$1% for $n \geq 5$	Measured at time samples are analyzed
Resolution R_s	$R_s > 2$	Measured between peak of interest and closest potential interfering peak
Tailing factor TF	$TF \leq 2$	
Column plate number N	$N > 2000$	Column characteristics not specified

Source: Data from [10].

specified by the test method, data from five replicate injections of the analyte are used to calculate the relative standard deviation when the test method requires RSD $\leq$2%; data from six replicate injections are used if the specification is RSD >2%.

In a regulated environment, system suitability tests must be carried out prior to the analysis of any samples. Following blank injections of mobile phase, water, and/or sample diluent, replicate system suitability injections are made, and the results compared to method specifications. If specifications are met, subsequent analyses can continue. If the method's system suitability requirements are not met, any problems with the system or method must be identified and corrected (possibly as part of a formal out-of-specification [OOS] investigation), and passing system suitability results must be obtained before sample analysis is resumed. To provide confidence that the test method runs properly, it is also recommended that additional system suitability samples (quality control samples or check standards) are run at regular intervals (interspersed throughout the sample batch); %-difference specifications should be included for these interspersed samples to make sure the system still performs adequately over the course of the entire sample run. Alternatively, a second set of system suitability samples can be included at the end of the run.

In 1994 the FDA published a reviewer guidance document regarding the validation of chromatographic methods that includes the last-published recommendations for system suitability [10]. These recommendations are summarized in Table 12.8. Many practitioners believe that test methods that satisfy the criteria of Table 12.8 will reduce the risk of criticism during a regulatory audit. These guidelines serve as useful examples; however, actual specifications are set by the user and can vary significantly according to the method.

12.4 DOCUMENTATION

Validation documentation includes the protocol used to carry out the validation, the test method, and the validation report. These documents should be written as controlled documents as part of a quality system (Section 12.9) that ensures compliance with appropriate regulations.

12.4.1 Validation Protocol

The validation protocol specifies the requirements (validation procedures and acceptance criteria) to be satisfied. Where possible, the protocol should reference standard operating procedures (SOPs) for specific work instructions and analytical methods. The protocol must be prepared and approved before the official validation process begins. In addition the validation protocol typically contains the following:

- protocol title
- purpose of the test method to be validated
- description of the test and reference substances
- summary of the test method to be validated, including the equipment, specified range, and description of the test and reference substances; alternatively, the detailed method description may be referenced or appended to the protocol
- validation characteristics to be demonstrated
- establishment and justification of the acceptance criteria for the selected validation characteristics
- dated signature of approval of a designated person and the quality unit

The *protocol title* is a brief description of the work or study to be performed, for example, "Validation of the Test Method for the HPLC Assay of API X in Drug Product Y." The *purpose* should specify the scope and applicability of the test method. The *summary* must adequately describe the actual written test method (which contains enough detail to be easily reproduced by a qualified individual). To reduce repetition, however, the test method often is included by reference or as an appendix to the protocol. The specific *validation characteristics* (accuracy, precision, etc.) to be evaluated are also included in the protocol, because these are dependant upon the type of analytical method (Section 12.5). *Acceptance criteria* for method validation (e.g., allowable error or imprecision) often are established during the final phase of method development (sometimes referred to as "pre-validation" experiments; see Section 12.4.2). The *designated quality unit representative* reviews and approves the protocol, to ensure that the proper regulatory regulations will be met and the proposed work will satisfy its intended purpose (Section 12.9).

Experimental work outlined in the validation protocol can be designed such that several appropriate validation characteristics are measured simultaneously. For example, experiments that measure accuracy and precision can be used as part of linearity studies, LOD, and LLOQ can be determined from the range and linearity data, and the solution stability of the sample and standard can use the same preparations that test accuracy and precision. Executed in this manner, the experimental design makes the most efficient use of time and materials.

12.4.2 Test Method

The test method is the formal document that contains all of the necessary detail to implement the analytical procedure on a routine basis. The test method is a controlled document with revision control (the requirement that any document changes are authorized, and all revisions are available for later comparison), approvals at the appropriate levels (including the quality unit, Section 12.9), and written with enough detail to warrant only one possible interpretation for any and all instructions. A typical test method will include the following:

- descriptive method title
- brief method description or summary
- description of the applicability and specificity, along with any special precautions (e.g., safety, storage, and handling)
- list of reagents, including source and purity/grade
- equipment, including the HPLC and any other equipment necessary (balances, centrifuges, pH meters, etc.)
- detailed instrument operating conditions, including integration settings
- detailed description of the preparation of all solutions (mobile phases, diluents), standards, and samples
- system suitability test description and acceptance criteria
- example chromatograms, spectra, or representative data
- detailed procedures, including an example sample queue (the order in which standards and samples are run)
- representative calculations
- revision history
- approvals

Once drafted, test methods often are subjected to a pre-validation stage, to demonstrate that acceptance criteria will be met when the formal validation takes place. The pre-validation stage typically consists of an evaluation of linearity and accuracy. Sometimes a test of robustness, if it has not already been evaluated during method development, is then carried out. The validation process usually will proceed more smoothly, and with lower risk of failure, if the ability to pass all the key validation criteria is confirmed during the pre-validation stage. A draft method will become an official test method after a full validation of its intended purpose.

12.4.3 Validation Report

The validation report is a summary of the results obtained when the proposed test method is used to conduct the validation protocol. The report includes representative calculations, chromatograms, calibration curves, and other results obtained from the validation process. Tables of data for each step in the protocol, and a pass/fail statement for each of the acceptance criteria are also included. A validation report generally consists of the following sections:

- cover page with the title, author, and affiliations
- signature page dated and signed by appropriate personnel, which may include the analyst, the group leader, a senior manager, and a quality control and/or a quality assurance representative
- an itemized list of the validation characteristics evaluated, often in the form of a table of contents

- an introduction or objective
- method summary including instrument and solution preparation specifics
- validation results in subsections organized by the characteristic studied. Each subsection should include a brief summary of the applicable protocol, and the mean, standard deviation, relative standard deviation, acceptance criteria, and assessment (pass or fail).
- any deviations from the protocol, planned or observed, and the impact (if any) on the validation
- any amendments to the protocol, with explanations and approvals
- conclusion

A properly designed validation protocol can serve as a template for the validation report. For example, in the protocol a test can be described and the acceptance criteria listed. For the validation report, this information is supplemented with supporting results, a reference to the location and identity of the raw data, and a pass/fail statement.

12.5 VALIDATION FOR DIFFERENT PHARMACEUTICAL-METHOD TYPES

The USP recognizes that is it not always necessary to evaluate every analytical performance characteristic for every test method. The type of method and its intended use dictate which performance characteristics need to be investigated, as summarized in Table 12.9 [12]. Both the USP and ICH divide test methods into four separate categories:

- assays for the quantification of major components or active ingredients (category 1 methods)
- determination of impurities or degradation products (category 2 methods)
- determination of product performance characteristics (category 3 methods)
- identification tests (category 4 methods)

These test methods and categories generally apply to drug substances and drug products, as opposed to bioanalytical samples, covered in Section 12.6.

12.5.1 Category 1 Methods

Category 1 tests target the analysis of major components, and include test methods such as content uniformity and potency assay. The latter methods, while quantitative, are not usually concerned with low concentrations of analyte, but only with the amount of the API in the drug product. Because of the simplicity of the separation (the API must be resolved from all interferences, but any other peaks in the chromatogram need not be resolved from each other), emphasis is on speed over resolution. For assays in category 1, LOD and LLOQ evaluations are not necessary because the major component or active ingredient to be measured is normally present at high concentrations. However, since quantitative information is desired, all of the remaining analytical performance characteristics are pertinent.

Table 12.9

Analytical Performance Characteristics to Measure vs. Type of Method

Analytical Performance Parameter	Category 1: Assays	Category 2: Impurities		Category 3: Specific Tests	Category 4: Identification
		Quantitative	Limit Tests		
Accuracy	Yes	Yes	*	*	No
Precision	Yes	Yes	No	Yes	No
Specificity	Yes	Yes	Yes	*	Yes
LOD	No	No	Yes	*	No
LLOQ	No	Yes	No	*	No
Linearity	Yes	Yes	No	*	No
Range	Yes	Yes	No	*	No
Robustness	Yes	Yes	No	Yes	No

Source: Data from [12].
Note: *May be required, depending upon type of test. For example, although dissolution testing falls into category 3, as a quantitative test, measurements typical of category 1 are used (with some exceptions).

12.5.2 Category 2 Methods

Category 2 tests target the analysis of impurities or degradation products (among other applications). These assays usually look at much lower analyte concentrations than category 1 methods, and are divided into two subcategories: quantitative and limit tests. If quantitative information is desired, a determination of LOD is not necessary, but the remaining characteristics are required. Methods used in support of stability studies (referred to as stability-indicating methods) are an example of a quantitative category 2 test. The situation reverses itself for a limit test. Since quantification is not required, it is sufficient to measure the LOD and demonstrate specificity and robustness. For a category 2 limit test, it is only necessary to show that a compound of interest is either present or not—that is, above or below a certain concentration. Methods in support of cleaning validation and environmental EPA methods often fit into this category. Although, as seen in Table 12.9, it is never necessary to measure both LOD and LLOQ for any given category 2 method, it is common during validation to evaluate both characteristics (more out of tradition than necessity).

12.5.3 Category 3 Methods

The characteristics that must be documented for test methods in USP-assay category 3 (specific tests or methods for product performance characteristics) are dependent on the nature of the test. Dissolution testing is an example of a category 3 method. Since it is a quantitative test optimized for the determination of the API in a drug product, the validation characteristics evaluated are similar to a category 1 test for a formulation designed for immediate release. However, for an extended-release formulation, where it might be necessary to confirm that none of the active ingredient has been released from the formulation until after a certain time point, the characteristics to be investigated would be more like a quantitative category 2 test that includes LOQ. Because the analytical goals may differ, the category 3 evaluation characteristics are very dependant on the actual test method, as indicated in Table 12.9.

12.5.4 Category 4 Methods

Category 4 identification tests are qualitative in nature, so only specificity is required. Identification can be performed, for example, by comparing the retention time or a spectrum to that of a known reference standard. Freedom from interferences is all that is necessary in terms of chromatographic separation.

12.6 BIOANALYTICAL METHODS

Bioanalytical methods refer to test methods for the analysis of drugs and their metabolites in biological samples, commonly plasma or urine but can include other animal tissues. Sometimes bioanalytical methods are confused with the analysis of biomolecules, such as proteins, peptides, and oligonucleotides—the latter separation techniques are discussed in Chapter 13. Bioanalytical methods are used in clinical pharmacology, bioavailability, toxicology, bioequivalence, and other studies that require pharmacokinetic evaluation in support of various drug applications to regulatory agencies, such as the FDA. In a regulated laboratory, bioanalytical methods must be validated to demonstrate that they are reliable and reproducible for their intended use (as for any other analytical method).

Test methods for finished product, raw materials, or active pharmaceutical ingredients (APIs) each have their own development and validation challenges. Bioanalytical methods are further complicated by the nature of the sample matrices, the trace concentrations of drug and metabolites encountered, and (potentially) the complexity of the required instrumentation.

The sensitivity and selectivity of bioanalytical methods are critical to the success of preclinical and clinical pharmacology studies. As with any other test method, the performance characteristics of a bioanalytical method must be demonstrated (by documented laboratory data) to be reliable and reproducible for its intended use. Joint industry and regulatory conferences have been held to discuss this topic (e.g., [13–15]). As a result of the first two conferences in May 2001, the FDA issued a guidance document for validating bioanalytical methods [16]. In contrast to performance criteria for drug substance or drug product methods, where specific performance criteria are listed (e.g., precision and accuracy), bioanalytical method regulations are listed as "guidelines." The general interpretation of these guideline documents is that if test methods are developed that adhere to their recommendations, there will be less likelihood of a negative regulatory action. In other words, if the recommendations of the guidelines are *not* followed, one should be sure to develop a logical and scientifically supported statement to show that alternative performance criteria are justified.

Regulated bioanalysis usually involves an HPLC system coupled to a triple-quadrupole mass spectrometer (LC-MS/MS, Section 4.14). The sensitivity and selectivity of the LC-MS/MS allows for the quantification of analytes with acceptable precision and accuracy at concentrations lower than most other HPLC detectors. Typically short, small-particle columns (e.g., 30–50 × 2.1-mm i.d. packed with ≤3 μm particles) are used for the fast separations needed for the large number of samples generated by clinical studies. Either isocratic or gradient separations with run times <5 min are common. Sample preparation (Chapter 16) to remove excess protein and other potential interferences can require as much effort to develop as the HPLC method. Automation of both sample preparation and analysis is common.

The development and use of a bioanalytical method can be divided into three parts:

- reference standard preparation
- method development and validation
- application of the validated test method to routine drug analyses

Each of these processes is discussed in following sections.

12.6.1 Reference Standard Preparation

Reference standards are necessary for quantification of the analyte in a biological matrix. These are used both for calibration (standard) curves and to check method performance (quality control, QC, samples). Reference standards can be one of three types: (1) standards whose purity is certified by a recognized organization (e.g., USP compendial standards), (2) reference standards obtained from another commercial source (e.g., a company in the business of the sales of general or specialty chemicals), and (3) custom-synthesized standards. Whenever possible, the standard should be identical to the analyte, or at least an established chemical form (e.g., free acid or base, or salt). In each case the purity of the standards must be demonstrated by appropriate documentation, usually a certificate of analysis. Supporting documentation such as the lot number, expiration date, certificates of analysis, and evidence of identity and purity should be kept with other method data for regulatory inspection. Compounds used for internal standards (often isotopically labeled drug) must have similar data to support purity.

12.6.2 Bioanalytical Method Development and Validation

The key bioanalytical performance characteristics that must be validated for each analyte of interest in the matrix include accuracy, precision, selectivity, range, reproducibility, and stability. In practice, to develop the test method and validate the method, four areas are investigated:

- selectivity
- accuracy, precision, and recovery
- calibration/standard curve
- stability

From each of these investigations, data are gathered to support the remaining characteristics.

12.6.2.1 Selectivity

The selectivity of a test method shows that the analyte can be accurately measured in the presence of potential interferences from other components in the sample (including the sample matrix). Interferences can take the form of endogenous matrix components (proteins, lipids, etc.), metabolites, degradation products, concomitant medication, or other analytes of interest. The FDA guidelines recommend the analysis of blank samples of the appropriate biological matrix from at least six different sources. For example, plasma from each source should be spiked with known concentrations of analyte at the lower limit of quantification (LOQ or LLOQ) to show that accurate results can be obtained. Similarly a blank extract of each matrix should be analyzed to show the absence of interferences. In cases of rare or difficult to obtain matrix (e.g., plasma from an exotic species or human tissue), the six-matrix requirement is relaxed.

12.6.2.2 Accuracy, Precision, and Recovery

The *accuracy* of a bioanalytical method is defined as the closeness of test results to the true value—as determined by replicate analyses of samples containing known amounts of the analyte of interest; results are reported as deviations of the mean from the true value. The FDA guidelines recommend the use of a minimum of five determinations per concentration, and a minimum of three concentrations over the expected range (a minimum of 15 separately prepared samples). The guidelines further recommend that the mean value be within ±15% of the actual value except at the LLOQ, where ±20% is acceptable.

The *precision* of a bioanalytical method measures agreement among test results when the method is applied repeatedly to multiple samplings of a homogeneous sample. As in recent ICH guidelines, precision can be further divided into repeatability (within-run or intra-batch) determinations, and intermediate (between-run or inter-batch) precision [4]. The FDA guidelines recommend the use of a minimum of five determinations per concentration, and a minimum of three concentrations over the expected range. The imprecision measured at each concentration level should not exceed 15% RSD, except for the LLOQ, which should not exceed 20% RSD. Usually the same data are used to determine both precision and accuracy.

The assay *recovery* relates to the extraction efficiency, and this is determined by a comparison of the response from a sample extracted from the matrix to the reference standard (with appropriate adjustments for dilution, etc.). The recovery of the analyte can be <100%, but it must be quantitative. That is, it should be precise and reproducible. Recovery experiments should be carried out at three concentrations (low, medium, and high), with a comparison of the results for extracted samples vs. unextracted samples (adjusted for dilution). Often it is impractical to analyze unextracted samples (e.g., injection of unextracted plasma will ruin most HPLC columns), so creative ways to show recovery may need to be devised. For example, a liquid–liquid extraction of spiked matrix might be compared to extraction of a matrix-free aqueous solution; or recovery from a solid-phase extraction might be determined by calculation of volumetric recovery and comparison of the response from an extracted sample to a known concentration of reference standard.

12.6.2.3 Calibration/Standard Curve

A calibration curve (also called a standard curve, or sometimes a "line") illustrates the relationship between the instrument response and the known concentration of the analyte, within a given range based on expected values. The simplest model that describes the proportionality should be used (e.g., a linear fit is preferred over a quadratic curve-fitting function). Calibration for bioanalytical methods usually is more complicated than for API assays, which typically have linear calibration plots that pass through the origin and may only require one calibration standard concentration. Because a significant amount of sample manipulation takes place in the typical sample preparation procedure, internal standards (Section 11.4.1.2) are preferred for most bioanalytical methods. At least four out of six nonzero standards (67%) should fall within ±15% of the expected concentration (±20% at the LLOQ). The calibration curve should be generated for every analyte in the sample, and prepared in the same matrix as the samples by addition of known concentrations of the analyte to blank matrix. The FDA guidelines suggest that a calibration curve should be constructed from six to eight nonzero samples that cover the expected range, including the LLOQ. In addition, noninterference is shown by the analysis of a blank sample (nonspiked matrix sample processed without internal standard) and a zero sample (nonspiked matrix processed with internal standard). Two conditions must be met to determine the LLOQ: (1) analyte response at the LLOQ should be >5 times the blank response, and (2) the analyte peak should be identifiable, discreet, and reproducible with an imprecision of $\leq 20\%$ and an accuracy of at least 80–120%.

12.6.2.4 Bioanalytical Sample Stability

Stability tests determine that the analyte (and internal standard) does not break down under typical laboratory conditions, or if degradation occurs, it is known and can be avoided by appropriate sample handling. Many different factors can affect bioanalytical sample stability; these include the chemical properties of the drug, the storage conditions, and the matrix. Studies must be designed to evaluate the stability of the analyte during sample collection and handling, under both long-term (at the intended storage temperature) and short-term (bench top, controlled room temperature) storage conditions, and through any freeze–thaw cycles. The conditions used for any sample-stability studies should reflect the actual conditions the sample (including working and stock solutions) may experience during collection, storage, and routine analysis. Stock solutions should be prepared in an appropriate solvent at known concentrations. The stability of stock solutions should also be ascertained at room temperature over at least six hours, and storage-condition stability (e.g., in a refrigerator) should be evaluated as well. In addition, since samples commonly will be left on a bench top or in an autosampler for some period of time, it is also important to establish the stability of processed samples (e.g., drug and internal standard extracted from sample matrix) over the anticipated run time for the batch of samples to be processed. Working standards should be prepared from freshly made stock solutions of the analyte in the sample matrix. Appropriate standard operating procedures (SOPs) should be followed for the experimental studies as well as the poststudy statistical treatment of the data.

The FDA guidelines recommend a minimum protocol that includes freeze and thaw stability plus short- and long-term temperature stability. For freeze–thaw stability, three spiked-matrix sample aliquots at each of the low and high concentrations should be exposed to three freeze–thaw cycles. The samples should be kept at the storage temperature for 24 hours and then thawed at room temperature (without heating). When completely thawed, the samples should be refrozen for 12 to 24 hours, then thawed again; this procedure is repeated a third time. Analysis of the sample then proceeds after completion of the third freeze–thaw cycle.

For short-term temperature stability, three aliquots (at each of the low and high concentrations) are thawed and kept at room temperature for a time that is equal to the maximum (e.g., 4–6 hr) the samples will be maintained at room temperature prior to their analysis.

The storage time for a long-term stability evaluation should bracket the time between the first sample collection and the analysis of the last sample (often 12 months or more); the sample volume reserved should be sufficient for at least three separate time points. At each time point, at least three aliquots (at each of the low and high concentrations) stored under the same conditions as the study samples (e.g., −20°C or −70°C) should be tested. In a long-term stability study the concentration of the stability samples should be determined using freshly made standards. The mean of resulting concentrations should be reported relative to the mean of the results from the first day of the study.

12.6.3 Routine Application of the Bioanalytical Method

Once the bioanalytical method has been validated for routine use, system suitability and QC samples are used to monitor accuracy and precision, and to determine whether to accept or reject sample batches. QC samples are prepared separately and analyzed with unknowns at intervals according to the number of unknown samples for a sample batch. Duplicate QC samples (prepared from the matrix spiked with the analyte) at three concentrations (low, near the LLOQ, midrange, and high) are normally used. The minimum number of QC samples (in multiples of three—low, midrange, and high concentration) is recommended to be at least 5% of the number of unknown samples, or six, whichever is greater. For example, if 40 unknowns are to be analyzed, $40 \times 5\% = 2$, so 6 QCs are run (2 low, 2 midrange, 2 high); or for 200 samples, $200 \times 5\% = 10$, so 12 QCs are run (4 each, low, midrange, and high). At least four out of every six QC sample results should be within ±15% of their respective nominal value. Data representative of typical results obtained by LC-MS/MS for the analysis of QC samples (at concentrations of 10, 35, 1000, 4400, and 5000 pg/mL of plasma) are listed in Table 12.10. As mentioned previously, for acceptable method validation, both the imprecision at each concentration level (%RSD), and the accuracy (%Bias) must be $\leq$15% ($\leq$20% at the LLOQ). In Table 12.10, the maximum %RSD ($\leq$3.9%) and maximum %Bias ($\leq$11.0%) values at all concentration levels were well within the validation guidelines.

System suitability, sample analysis, acceptance criteria, and guidelines for repeat analysis or data reintegration should all be performed according to an established SOP. The rationale for repeat analyses, data re-integration, and the reporting of results should be clearly documented. Problems from inconsistent replicate analysis, sample processing errors, equipment failure, or poor chromatography are some of the issues that can lead to a need to re-analyze samples. In addition recent interpretations [15] of bioanalytical guidelines indicate that a certain number of samples

Table 12.10

Example Bioanalytical LC-MS/MS QC Results

Measured concentration (pg/mL)					Target (pg/mL)	n	Mean	SD	%RSD	%Bias
QC1	QC2	QC3	QC4	QC5	10.0	10	11.1	0.402	3.6	+11.0
11.8	35.7	1009.8	4670.3	5425.0	35.0	10	34.8	1.37	3.9	−0.6
11.1	37.1	1036.0	4796.4	5334.5	1000.0	10	997.0	35.45	3.6	−0.3
11.4	35.4	1047.2	4684.9	5180.9	4400.0	10	4630.0	160.5	3.5	+5.2
10.4	36.0	975.8	4964.3	5241.6	5000.0	10	5138.3	199.4	3.9	+2.8
10.8	34.6	1047.8	4628.6	5285.6						
10.9	34.9	986.5	4564.3	5049.0						
10.9	33.6	971.8	4491.9	5009.2						
10.8	32.6	960.4	4404.1	4883.7						
11.3	33.2	956.7	4539.5	5170.8						
11.4	34.4	977.8	4558.6	4802.7						

be reanalyzed on a routine basis to ensure method performance (sometimes referred to as "incurred sample reproducibility").

12.6.4 Bioanalytical Method Documentation

As discussed previously in Section 12.4, good record keeping and documented SOPs are an essential part of any validated test method. Once the validity of a bioanalytical method is established and verified by laboratory studies, pertinent information is provided in an assay validation report. Data generated during method development and QC should be available for audit and inspection. Documentation for submission to the FDA should include (1) summary information, (2) method development and validation reports, (3) reports of the application of the test method to routine sample analysis, and (4) other miscellaneous information (e.g., SOPs, abbreviations, and references).

The *summary information* should include a tabular listing of all reports, protocols, and codes. The documentation for *method development and validation* should include a detailed operational description of the experimental procedures and studies, purity and identity evidence, method validation specifics (results of studies to determine accuracy, precision, recovery, etc.), and any protocol deviations with justifications. Documentation of the *application of the test method to routine sample analysis* is usually quite extensive. It should include:

- summary tables describing sample processing and storage
- detailed summary tables of analytical runs of pre-clinical or clinical samples
- calibration curve data
- QC sample summary data including raw data, trend analysis, and summary statistics

- example chromatograms (unknowns, standards, QC samples) for up to 20% of the subjects
- reasons and justification for any missing samples or any deviations from written protocols or SOPs
- documentation for any repeat analyses, or re-integrated data

12.7 ANALYTICAL METHOD TRANSFER (AMT)

In a regulated environment, it is rare for the laboratory that develops and validates a test method to perform all of the routine sample testing. Instead, once developed and validated (in the originating, or "sending" laboratory), test methods are commonly transferred to another laboratory (the "receiving" laboratory) for implementation. However, the receiving laboratory must still be able to get the same results, within experimental error, as the sending laboratory. The objective of a formal method-transfer process is to ensure that the receiving laboratory is well-trained, qualified to run the test method in question, and able to obtain the same results (within experimental error) as the sending laboratory. The development and validation of robust test methods and a strict adherence to well-documented SOPs are the best ways to ensure the ultimate success of the method.

The process that provides documented evidence that the analytical method works as well in the receiving laboratory as in the sending laboratory, is called *analytical method transfer* (AMT). The topic of AMT has been addressed by the American Association of Pharmaceutical Scientist (AAPS, in collaboration with the FDA and EU regulatory authorities), the Pharmaceutical Research and Manufacturers of America (PhRMA), and the International Society for Pharmaceutical Engineering (ISPE) [17–18]. The PhRMA activities resulted in what is referred to as an Acceptable Analytical Practice (AAP) document that serves as a suitable first-step guidance document for AMT [19].

In essence, the AMT process qualifies a laboratory to use a test method; regulators will want documented proof that this process was completed successfully. Only when both of these processes (qualification and documentation) are complete can the receiving laboratory obtain cGMP "reportable data" from their laboratory results. AMT specifically applies to drug product and drug substance methods, but the same principles can apply to bioanalytical methods (Section 12.6). A typical example is when AMT takes place between a research group that develops the test method and a quality-control group responsible for the release of the finished product. Any time information moves from one group to another (e.g., from a pharmaceutical company to a contract analytical laboratory), proper AMT should be observed. Both the sending and the receiving laboratory have certain responsibilities in the AMT process; these are listed in Table 12.11.

Before initiating AMT, several pre-transfer activities should take place to minimize unexpected problems in method transfer. If not previously involved with the test method, the receiving laboratory should have an opportunity to review the method prior to the transfer, and to carry out the method so as to identify any potential issues that may need to be resolved prior to finalization of the transfer protocol. The sending laboratory should provide the receiving laboratory with all

Table 12.11

Analytical Method Transfer: Sending and Receiving Laboratory Responsibilities

Sending Laboratory Responsibilities	Receiving Laboratory Provides
Create the transfer protocol	Qualified instrumentation
Execute training	Personnel
Assist in analysis	Systems
Acceptance criteria	Protocol execution
Final report (with receiving laboratory)	Final report (with sending laboratory)

of the validation results, including robustness study results, as well as documented training.

12.7.1 Analytical Method-Transfer Options

The foundation of a successful AMT is a properly developed and validated method. A good robustness study will also help facilitate method transfer. A well-designed AMT process requires that a sufficient number of samples should be run to support a statistical assessment of method performance, because a single test is no indication of how well a test method will perform over time. A formal AMT is not always necessary, however. In-process tests or research methods do not require a formal transfer; a system suitability test is employed as the basis for the transfer. In all cases sound scientific judgment should guide the AMT requirements.

Several different techniques can be used for AMT. These include:

- comparative testing
- complete or partial method validation or revalidation
- co-validation between the two laboratories
- omission of a formal transfer, sometimes called a transfer waiver

The choice of which option to use depends on the stage of development in which the test method is to be used (early or late stage), the type of method (e.g., compendial vs. noncompendial; simple or complex), and the experience and capabilities of the laboratory personnel.

12.7.1.1 Comparative Testing

The most common AMT option used is to compare test data from two (or more) laboratories. This is accomplished when two or more laboratories perform a pre-approved protocol that details the criteria used to determine whether the receiving laboratory is qualified to use the test method being transferred. The data resulting from the joint exercise are compared to a set of predetermined acceptance criteria. For example, a blinded set of samples and blanks at known concentrations might be provided to both the sending and receiving laboratories; the individual laboratory results would then be compared to the true values to qualify the receiving laboratory. Comparative testing can also be used in other postapproval situations that involve additional manufacturing sites and/or contract laboratories. In general, comparative testing is most often used for late-stage methods and for the transfer of more complex methods.

12.7.1.2 Co-validation between Laboratories

Comparative testing (Section 12.7.1.1) traditionally requires a validated method as a prerequisite to AMT. However, another option for AMT is to involve the receiving laboratory from the beginning in the actual validation of the test method to be transferred. By completing a co-validation study, the receiving laboratory is considered qualified to perform the test method for release testing. To perform this transfer option, the receiving laboratory must be involved in identifying the intermediate precision validation characteristics to be evaluated and the experimental design. By inclusion of data from all laboratories involved in the study, it is possible to have the validation report serve as proof of AMT, without requiring a separate validation study by the receiving laboratory.

12.7.1.3 Method Validation and/or Revalidation

Another technique that can be used for AMT involves the receiving laboratory's repeating some or all of the originating laboratory's validation experiments. As with co-validation (Section 12.7.1.2), by completing part of a validation study, the receiving laboratory is considered qualified to perform routine release testing. With this process, the laboratory staff and quality unit determine how much testing is required to satisfy AMT.

12.7.1.4 Transfer Waiver

A transfer waver is used when a formal AMT is not needed (e.g., compendial methods) and for some other situations that may warrant omission of a formal AMT. A transfer waiver is considered when:

- the receiving laboratory currently tests the product with another method
- a test method is in use for a dosage form comparable to the new product
- the test method (or one very similar) is already in use for another application
- the new method involves changes that do not significantly alter the use of an existing test method
- the personnel accompany the transfer of the test method from one laboratory to another

When a transfer waiver is indicated, the receiving laboratory can use the test method without generation of any comparative data. However, the reasons for the waiver must be documented.

12.7.2 Essentials of AMT

Many interrelated components are necessary to achieve a successful AMT. As in any validation process, documentation is essential both for the process and the results. All steps, from the AMT protocol and ending to the transfer report, must be documented for compliance purposes.

12.7.2.1 Pre-approved Test Plan Protocol

A protocol must be in place that describes the general transfer process and the acceptance criteria, before implementing an AMT. This document usually takes the form of a standard operating procedure (SOP) that describes the details of the AMT protocol or test plan specific to the product and method. This document should clearly define the scope and objective of the AMT, all of the respective laboratories' responsibilities, a list of all the methods that will be transferred (if the AMT comprises more than one method), a rationale for any test methods not included (i.e., the transfer waiver), as well as acceptance criteria. It should also include the selection process for materials and samples to be used in the AMT. The protocol should include certificates of analysis for any samples and reference materials used.

Representative, homogeneous samples should be used that are identical for both laboratories. Selection of proper samples is very important; usually samples that are not "official" production lots (e.g., pre-GMP materials) or a "control lot" are chosen so that an out-of-specification (OOS) investigation is not required if an unexpected result is obtained. Remember, the purpose of the method transfer is to assess method performance, not to identify changes in samples or matrix.

Instrumentation and associated settings should also be described. A best-case scenario would have each laboratory use common instrumentation (e.g., transfer the instrumentation from the sending laboratory to the receiving laboratory); if this is not the case, and it rarely is, then the sending laboratory should consider the use of instrumentation common to the receiving laboratory (e.g., same brand and model) to identify any potential issues prior to a formal AMT. Intermediate precision validation studies also commonly take instrument differences into account.

12.7.2.2 Description of Method/Test Procedures

The documentation should include not just the mechanics of performing the test method but also validation data and any idiosyncrasies in the method. Any precautions that must be taken to ensure successful results should be included in the method description. The test method should be written in a manner that ensures only one possible interpretation (e.g., use *v*/*v* notation if volume measurements are made instead of weighing). Clear equations and calculations, if appropriate, should be specified; example calculations often help eliminate misinterpretation of the instructions.

12.7.2.3 Description and Rationale of Test Requirements

This should include specific information, such as the number of replicates and lots to be tested, as well as the rationale for how each characteristic was chosen. This section should describe any system suitability parameters established for the test method.

12.7.2.4 Acceptance Criteria

Before the transfer takes place, documentation must stipulate how the results will be evaluated. Since statistical evaluations are usually employed, clear instructions on the number of batches and replicates, for example, are needed. It is common for simple statistics to be used for acceptance criteria, such as the mean and standard deviation from repeated use of the test method in the sending and receiving laboratory. More sophisticated statistics, such as the *F*-test, or Student's *t*-test, are also commonly applied. The proper use of statistics can provide an unbiased and objective comparison of results using a predetermined procedure listed in the protocol documentation. Appropriate statistical references (e.g., [20–22]) should be consulted for more detail. Since specifications vary with the test method, instrumentation, sample, and other variables, specific performance criteria are not listed in the PhRMA guidance [19]. A partial summary of the ISPE's list [17] of recommended experimental design and acceptance criteria is presented in Table 12.12.

12.7.2.5 Documentation of Results

The AMT Report summarizes the results of the AMT. The report certifies that the acceptance criteria were met, and that the receiving laboratory is fully trained and qualified to run the test method. In addition to a summary of all of the experiments performed and the results obtained, the report should list all of the instrumentation used in the transfer. It is important to include in the AMT report any observations made while performing the transfer. Observations in the form of feedback can be used to further optimize a test method or to address special concerns that might not have been anticipated by the sending laboratory.

Sometimes, of course, the receiving laboratory may not meet the acceptance criteria in the AMT protocol. When this situation arises, the transfer failure should be addressed by an existing policy that dictates specifically how the situation should be handled. An investigation should be initiated and documented in the summary report, and appropriate corrective action should be taken.

12.7.3 Potential AMT Pitfalls

Many of the common pitfalls encountered during AMT can be prevented with a little up-front work. It cannot be stressed enough that the robustness studies performed during late method development or early method validation play a critical roll in the success of AMT. During the robustness studies many of the critical elements might have been identified and noted as a precautionary statement in the test method. Intermediate precision validation studies can serve to identify potential AMT issues. By anticipating that instruments, experience, training, and procedural interpretations can differ from laboratory to laboratory, many common pitfalls can be avoided.

12.7.3.1 Instrument Considerations

Differences in instrumentation, such as component design and performance, are responsible for many adverse effects encountered during AMT. Injector cycle times, detector wavelength accuracy, on-line mobile-phase accuracy and mixing characteristics, and gradient dwell-volumes are just a few of the differences that can result in different results for the same sample run by the same test method on two different instruments [11, 23]. Methods that require that the HPLC system operate near its limits, such as high-throughput, high-resolution, or trace analysis methods can be particularly problematic. Anticipation of such problems during method development can help simplify the AMT process.

12.7.3.2 HPLC Columns

Columns have represented a significant source of variability in test method results, but there is much less concern with the current generation of columns. The test method should specify the brand and other details of the column to be used, as

Table 12.12

Experimental Design and Acceptance Criteria for Analytical Method Transfer

Type of Method	Number of Analysts	Lots[a] or Units	Acceptance Criteria	Notes
Assay	2	3 lots in triplicate	Two-sample t-test with intersite differences of $\leq 2\%$ at 95% CI are required.[b]	Each analyst should use different instrumentation and columns, if available, and independently prepare all solutions. All applicable system suitability criteria must be met.
Content uniformity	2	1 lot	Include a direct comparison of the mean $\pm 3\%$ and variability of the results (%RSD) as a two-sample t-test with intersite differences of $\leq 3\%$ at 95% CI.	If the method for content uniformity is equivalent (e.g., same standard and sample concentrations, HPLC conditions and system suitability criteria) to the assay method, then a separate AMT is not required.
Impurities, degradation products	2	3 lots in duplicate[c]	For high levels, a two-sample t-test is required, with intersite differences of $\leq 10\%$ at 95% CI; for low levels, criteria are based on the absolute difference of the means $\pm 25\%$.	All applicable system suitability criteria should be met. The LOQ should be confirmed in the receiving laboratory, and chromatograms should be compared for the impurity profile. All samples should be similar with respect to age, homogeneity, packaging, and storage. If samples do not contain impurities above the reporting limit, then spiked samples are recommended.
Dissolution	na	6 units for immediate release, 12 units for extended	Meet dissolution specifications in both laboratories, and the two profiles should be comparable, or based on the absolute difference of the means, $\pm 5\%$.	A statistical comparison of the profiles or the data at the test end time point(s) similar to that used for the assay may be performed.

Table 12.12

(Continued)

Type of Method	Number of Analysts	Lots[a] or Units	Acceptance Criteria	Notes
Identity		1 unit	Chromatography: confirm retention time. Spectral identification and chemical testing can also be used, assuming operators are sufficiently trained and the instrumentation can provide equivalent results.	
Cleaning validation		2 spiked samples, one above, one below specification	Spiked levels should not deviate from the specification by an amount >3× the validated standard deviation of the method, or 10% of the specification, whichever is greater.	Essentially a limit test. Low and high samples to confirm both positive and negative outcomes are required.

Source: Data from [17].

[a]A "lot" or "batch" is test material that has been manufactured at the same time with the same equipment; samples within a lot are considered to be identical; samples between lots may have (minor) variations; "unit" is a single dosage unit (e.g., a single tablet, vial, syringe, or patch).

[b]Confidence interval (CI).

[c]Triplicate if the impurities and assay are determined in the same test procedure.

well as approved alternative columns (if any). Using the blanket statement "or equivalent" can lead to problems, especially for inexperienced workers, and should be avoided. The *United States Pharmacopeia* (USP) is addressing this issue by publishing databases that use chromatographic tests to classify column selectivity [24] (for additional discussion of equivalent columns, see Sections 5.4.2, 6.3.6.1). With the aid of an appropriate database, users can quickly identify columns that are equivalent to the one currently in use. The use of techniques such as this should help reduce problems in locating an alternative column that is truly equivalent when the original (specified) column is either not available or no longer provides the required selectivity.

Column temperature is another source of variability; all test methods should use a column oven with mobile-phase pre-heating (Sections 3.7.1, 6.3.2.1). Retention-time variability and changes in chromatographic selectivity with temperature can be reduced through the use of a temperature-controlled column oven. Column-oven temperature calibration and uniformity can vary between brands and models of ovens, so this should be addressed in the AMT as well.

12.7.3.3 Operator Training

Training can be addressed at any time, but it is wise to train new users of the method before formal AMT. Despite all the upfront work, errors are still made; either honest mistakes, or errors in procedure that result from test method ambiguities. Procedures should be written (and proofread) so that there is only one possible interpretation of how to perform the test method, with enough detail so that nothing is left to chance.

12.8 METHOD ADJUSTMENT OR METHOD MODIFICATION

"Official" or "validated" methods can be found in a number of places, such as the USP, or in the Official Methods of Analysis of the Association of Official Analytical Chemists (AOAC). Methods in both New Drug Applications (NDA) and Abbreviated New Drug Applications (ANDA) are also considered to be standardized, official, validated methods. To use an official method "as is" for the first time, a laboratory must perform a verification to demonstrate that both the instrument and method performance criteria are met [25–27]. However, if the desired results cannot be obtained, an adjustment, or modification (change), to a standard method might be needed.

Although test methods in the USP ("compendial methods") are considered to be validated, adjustments to USP methods have been allowed in order that system suitability requirements are met; such instructions may be included in individual monographs. But method changes usually require some degree of revalidation. So an important question is, At what point does an *adjustment* become a *change or modification*? Historically, if adjustments to the test method are made within the boundaries of any robustness studies that were performed, no further actions are warranted, as long as system suitability criteria are satisfied. However, any adjustment outside of the bounds of the robustness study constitutes a change to the test method and may require a re-validation.

In 1998 Furman et al. proposed a way to classify allowable adjustments [28]. But it was not until 2005 that official guidance appeared on the topic [27, 29–30]. Although USP guidance on this topic was recently included into USP Chapter 621 on chromatography [12], the FDA Office of Regulatory Affairs (ORA) has had guidance in place for a number of years [27]. Table 12.13 summarizes the adjustments allowed for various HPLC variables taken from both the USP and ORA documents. Adjustments outside of the ranges listed in Table 12.13 constitute modifications, or changes, that are subject to additional validation. Sound scientific reasoning should be used when determining whether to make a method adjustment or a method change to a specific test method. For example, if robustness studies have shown that the method conditions allow less change for a variable than that listed in Table 12.13, or when robustness testing have shown that more change is allowed, the robustness results (as summarized in the validation report) should prevail.

Although the criteria in Table 12.13 might seem quite straightforward, many of the criteria do not completely account for recent advances in HPLC technology, especially columns with much smaller particle sizes (e.g., <2 μm) operated at higher flow rates (linear velocities) and pressures (e.g., >6000 psi). Also missing in the guidelines is a discussion of gradient adjustments/modifications. For example,

Table 12.13

Maximum Specifications for Adjustments to HPLC Operating Conditions

Variable	Maximum Specification[a]	Comments
pH	±0.2 units	
Buffer salt concentration	±10%	pH variation must be met.
Concentrations of minor mobile-phase components	Only components specified at 50% or less are considered: ±30% or ±2% absolute, whichever is larger; maximum change of ±10% absolute; no component can be reduced to zero	See Section 12.8.3 for examples and discussion.
UV-detector wavelength	No deviations	A validated procedure must be used to verify that wavelength error is ≤ ± 3 nm.
Column length	±70%	
Column inner diameter	±50% (ORA) ±25% (USP)	For USP, see [12].
Flow rate	±50%	
Injection volume	Reduced as far as consistent with accepted precision and detection limits	Increase to as much as 2× volume is specified as long as there are no adverse chromatographic effects.[b]
Particle size	Reduced by as much as 50%	
Column temperature	±10% USP, ±20% ORA	

Source: Data from [7, 12, 26–30].
[a]See Section 12.8 for examples.
[b]Adverse chromatographic effects include factors such as baseline, peak shape, resolution, linearity, and retention times.

although solvent composition is addressed, compensation for gradient dwell-volume (Section 9.2.2.4) when changing between different column dimensions must be considered for equivalent results. In addition identical column chemistry, while not explicitly stated, is implied.

Adjustments to HPLC systems in order to comply with system suitability requirements should not be made in order to compensate for column failure or system malfunction. To prevent specification "creep," adjustments are only made from the original test method conditions and are therefore not subject to continuous adjustment. Adjustments are permitted only when suitable reference standards are available for all compounds used in the system suitability test and only when those standards are used to show that the adjustments have improved the quality of the chromatography so as to meet system suitability requirements. The suitability of the test method under the new conditions must be verified by assessment of the relevant analytical performance characteristics. Since multiple adjustments can have a cumulative effect in the performance of the system, any adjustments should be considered carefully before implementation. Finally, one word of caution: just because limits for a variable are listed in Table 12.13 does not mean that they can be made with impunity; system suitability is the final test of the appropriateness of any change within the limits of Table 12.13.

12.8.1 pH Adjustments

As shown in Table 12.13, the pH value of the buffer in the mobile phase can be adjusted by as much as ±0.2 pH units. For example, a mobile-phase pH of 2.5 could be adjusted in the range of $2.3 \leq \text{pH} \leq 2.7$. Adjustment of the pH should, however, take into account the pK_a's of the compounds of interest; since for a pH near the pK_a, even a 0.1 unit change in the pH can result in significant (>10%) changes in retention time [31]. Studies show that for many compounds, the ±0.2 unit allowed change makes sense only if the test method is operated well away from the compound pK_a (e.g., pH < 4 for basic compounds; or at pH < 3 or pH > 7 for acidic compounds) [32].

12.8.2 Concentration of Buffer Salts

The concentration of the salts used in the preparation of the aqueous buffer used in the mobile phase can be adjusted to within ±10%, provided the pH variation (Section 12.8.1) is met. For example, a mobile-phase buffer containing a 20-mM phosphate buffer could be adjusted in the range of 18 to 22 mM. See Section 7.2.1 for additional discussion of buffer effects.

12.8.3 Ratio of Components in the Mobile Phase

The adjustment of the ratio of mobile phase components (% buffer or organic solvent) is allowed within certain limitations. Minor components of the mobile phase (≤50%) can be adjusted by ±30% relative. However, the change in any component cannot exceed ±10% absolute (i.e., in relation to the total mobile phase). For example, for a 50:50 water:MeOH mobile phase, a 30% change is 15% absolute, which exceeds the limit of ±10% absolute for any one component. Therefore an adjustment only in the range of 60:40 to 40:60 is allowed. For a mobile phase of 95:5 water:MeOH, 30% of 5% is 1.5% absolute, but since up to ±2% is allowed, a mobile phase of 93:7 to 97:3 could be used.

Adjustment can be made to only one minor component in a ternary mixture. For example, with a mobile phase of 60:35:5 buffer:MeOH:ACN, a 30% change in MeOH (35%) would be 10.5% absolute, exceeding the ±10% absolute allowed, so the MeOH content could be adjusted in the range of 25–45%. For the ACN, 30% of 5% is 1.5%, but ±2% is allowed, so the ACN content could be adjusted from 3 to 7%. In the case of this ternary mixture, either the MeOH or the ACN concentration could be changed, but not both. In each case a sufficient portion of the highest concentration component should be used to give a total of 100%. Additional examples of adjustments for binary and ternary mixtures are outlined in USP Chapter 621 and can be consulted for more detail [12].

Although Table 12.13 may allow an adjustment of the mobile phase, it should be emphasized that this can alter chromatographic selectivity. Therefore proceed with caution when making mobile-phase adjustments to existing methods.

12.8.4 Wavelength of the UV-Visible Detector

Deviations from the wavelengths specified in the test method are not permitted. The procedure specified by the detector manufacturer, or another validated procedure, should be used to verify that error in the detector wavelength is $\leq \pm 3$ nm.

12.8.5 Temperature Adjustments

In the case of HPLC column temperature, a ±10 change (or ±20%, depending on the guideline consulted) is allowed. It should be noted that temperature differences can have significant selectivity and retention effects (Section 6.3.2.1).

12.8.6 Column Length, Diameter, and Particle-Size Adjustments

A few inconsistencies exist in the guidelines regarding flow rate, column internal-diameter and length, and particle-size adjustment criteria. It is possible to reduce flow rate and internal diameter more than that listed in Table 12.13 (up to 50% allowed by ORA, 25% by USP) with no effect on retention (or selectivity) as long as the mobile-phase linear velocity is constant (isocratic separation assumed). In its most recent update [12] the USP has allowed for greater adjustments if the linear velocity is maintained.

Although column length, internal diameter, and particle-size adjustments are listed separately in Table 12.13, these variables really need to be considered together, and when correctly scaled in accordance with well-known theoretical principles (including constant linear velocity), equivalent separations will result even outside the recommended adjustment criteria [33]. For example, if the length-to-particle size ratio (L/d_p) is kept constant, an identical separation well outside the recommended limits can be obtained for a 50-mm, 1.7-μm column as for a 300-mm, 10-μm column ($L/d_p = 3$ for both) as long as an increase in the flow rate inversely proportional to the particle size is maintained (and, of course, the stationary-phase chemistry must be identical, extra-column peak broadening must be minimized, and the pressure must be maintained within acceptable limits). In cases such as this, where the regulatory guidelines are deficient, it is wise to write an addendum to the test method (or a separate SOP) that describes the allowable adjustments to a method, with appropriate evidence to support the adjustment.

12.9 QUALITY CONTROL AND QUALITY ASSURANCE

The terms "quality control" and "quality assurance" often are used interchangeably. However, in a properly designed and managed quality system, the two terms have separate and distinct meanings, and functions. Quality assurance can be thought of as related to *process* quality, whereas quality control is related to the quality of the *product*. In a given organization, it does not matter what the functions are named, but the responsibilities for these two activities should be clearly defined. Both quality assurance and quality control make up the "quality unit," and are essential to the production of analytical results that are of high quality and are compliant with the appropriate regulations.

12.9.1 Quality Control

Quality control (QC) is the activity that determines the acceptability or unacceptability of a product, and is determined by the comparison of a product against the original specifications that were created before the product was manufactured. In some organizations, the QC group is responsible for the use of the test method to perform analysis of a product. Other tasks related to QC may include documented reviews, calibrations, or additional types of measurable testing (sampling, etc.) that reoccur more often than activities associated with quality assurance. QC will usually require the involvement of those directly associated with the research, design, or production of a product. For example, in a laboratory-notebook peer-review process, a QC group would check or monitor the quality of the data, look for transcription errors, check calculations, and verify notebook sign-offs. All of these activities help to ensure an acceptable product.

12.9.2 Quality Assurance

Quality assurance (QA) is determined by senior management policies, procedures, and work instructions—and by governmental regulations. At the beginning of the validation process, QA may provide guidance for the development of or review of validation protocols and other validation documents. During the analytical stage, QA's job is to ensure that the proper method or procedure is in use and that the quality of the work meets company and governmental guidelines and regulations. QA can be thought of as the activity that will determine how quality control tasks will be carried out—and then verify that they were performed properly. As opposed to quality control checks (that focus on the product), quality assurance focuses on the process used to test a product. QA is more likely to be performed by managers, by corporate level administrators, or third-party auditors through the review of the quality system, reports, archiving, training, and qualification of the staff that performs the work.

12.10 SUMMARY

Method validation constantly evolves and is just one part of the overall regulated-environment activities. The validation process starts with instrument qualification (Section 17.2.1.1) before an HPLC instrument is placed online, and continues long after method development, optimization, and transfer—living on with the test method during routine use. A well-defined and documented validation process provides regulatory agencies with evidence that the system and test method are both suitable for their intended use. It also assures that the guidelines established meet method validation requirements and specifications.

The bottom line is that all parties involved should be confident that an HPLC method will give results that are sufficiently accurate, precise, and reproducible for the analysis task at hand. Formal method validation is just a set of tools to use to accomplish this task. Whether or not a formal validation is required, performance of good, justifiable science as part of an established quality system will help to ensure that the resultant test method, and the data that it generates will survive the scrutiny of any reviewer. The contents of this chapter concentrate on pharmaceutical methods, but the same principles can be applied to any HPLC method so as to ensure that it is suitable for its intended use

REFERENCES

1. *Guideline for Submitting Samples and Analytical Data for Methods Validation*, USFDA-CDER (February 1987), http://www.fda.gov/cder/guidance/ameth.htm.
2. *United States Pharmacopeia No. 31-NF 26*, (2008), ch. 1225.
3. *Analytical Procedures and Methods Validation*, USFDA-CDER (Aug. 2000), http://www.fda.gov/cder/guidance/2396dft.htm.
4. *Harmonized Tripartite Guideline, Validation of Analytical Procedures, Text and Methodology, Q2 (R1)*, International Conference on Harmonization, (Nov. 2005), http://www.ich.org/LOB/media/MEDIA417.pdf.
5. *Guidance for Methods Development and Methods Validation for the Resource Conservation and Recovery Act (RCRA) Program*, US EPA, (1995), http://www.epa.gov/epawaste/hazard/testmethods/pdfs/methdev.pdf.
6. *ISO/IEC 17025, General Requirements for the Competence of Testing and Calibration Laboratories*, (2005), http://www.iso.org/iso/iso_catalogue/catalogue_tc/catalogue_detail.htm?csnumber=39883.
7. *Current Good Manufacturing Practice in Manufacturing, Processing, Packing, or Holding Of Drugs*, 21 CFR Part 210, http://www.fda.gov/cder/dmpq/cgmpregs.htm.
8. *Current Good Manufacturing Practice for Finished Pharmaceuticals*, 21 CFR Part 211, http://www.fda.gov/cder/dmpq/cgmpregs.htm.
9. International Organization for Standardization, http://www.iso.org/iso/home.htm.
10. *Reviewer Guidance, Validation of Chromatographic Methods*, USFDA (November 1994) http://www.fda.gov/cder/guidance/cmc3.pdf.
11. L. R. Snyder, J. J. Kirkland, and J. L. Glajch, *Practical HPLC Method Development*, 2nd ed., Wiley-Interscience, New York, 1997.
12. *United States Pharmacopeia No. 31-NF 26*, (2008), ch. 621.
13. V. P. Shah, K. K. Midha, and S. V. Dighe, *Pharm. Res.*, 9 (1992) 588.
14. V. P. Shah, K. K. Midha, J. W. A. Findlay, H. M. Hill, J. D. Hulse, I. J. McGilvaray, G. McKay, K. J. Miller, R. N. Patnaik, M. L. Powell, A. Tonelli, C. T. Viswanathan, and A. Yacobi, *Pharm. Res.*, 17 (2000) 1551.
15. C. T. Viswanathan, S. Bansal, B. Booth, A. J. DeStafano, M. J. Rose, J. Sailstad, V. P. Shah, J. P. Skelly, P. G. Swann, and R. Weiner, *AAPS J.*, 9(1), (2007) E30. See also: www.aapsj.org.
16. *Guidance for Industry, Bioanalytical Method Validation*, USFDA-CDER (May 2001), http://www.fda.gov/cder/guidance/4252fnl.pdf.
17. *ISPE Good Practice Guide: Technology Transfer*, ISPE, Tampa, FL (Mar. 2003), http://www.ispe.org/cs/ispe_good_practice_guides_section/ispe_good_practice_guides.
18. *PhRMA Analytical Research and Development Workshop*, Wilmington DE, 20 Sept. 2000.
19. S. Scypinski, D. Roberts, M. Oates, and J. Etse, *Pharm. Tech.*, (Mar. 2004) 84.
20. J. C. Miller and J. N. Miller, *Statistics for Analytical Chemistry*, Ellis Horwood, Chichester, UK, 1986.

21. *NIST/SEMATECH e-Handbook of Statistical Methods*, http://www.itl.nist.gov/div898/handbook.
22. P. C. Meier and R. E. Zund, *Statistical Methods in Analytical Chemistry*, Wiley, New York, 1993.
23. M. Swartz and R. Plumb, unpublished results.
24. H. Pappa and M. Marques, presentation at USP Annual Scientific Meeting, Denver, 28 September 2006. See also: http://www.usp.org/USPNF/columns.html.
25. M. E. Swartz and I. S. Krull, *LCGC,* 23 (2005) 1100.
26. *Pharmacopeial Forum*, 31(2) (Mar.–Apr. 2005) 555.
27. *FDA ORA Laboratory Procedure, ORA-LAB.5.4.5*, USFDA (09/09/ 2005). See also: http://www.fda.gov/ora/science_ref/lm/vol2/section/5_04_05.pdf.
28. W. B. Furman, J. G. Dorsey, and L. R. Snyder, *Pharm. Technol.*, 22(6) (1998) 58.
29. *Pharmacopeial Forum*, 31(3) (May–Jun. 2005) 825.
30. *Pharmacopeial Forum*, 31(6) (Nov.–Dec. 2005) 1681.
31. M. E. Swartz and I. S. Krull, *LCGC*, 23 (2005) 46.
32. M. E. Swartz, unpublished data on the analysis of tricyclic amines at pH-7.2.
33. M. E. Swartz and I. S. Krull, *LCGC*, 24 (2006) 770.

CHAPTER THIRTEEN

BIOCHEMICAL AND SYNTHETIC POLYMER SEPARATIONS

with Timothy Wehr, Carl Scandella, and Peter Schoenmakers

Introduction to Modern Liquid Chromatography, Third Edition, by Lloyd R. Snyder, Joseph J. Kirkland, and John W. Dolan

13.1 BIOMACROMOLECULES

Since liquid chromatography was first developed, it has been an important tool for the isolation and characterization of biomolecules. However, the extension of HPLC to the successful separation of biopolymers such as polypeptides, nucleic acids, and carbohydrates required the development of column packings that were tailored for these molecules. This chapter will concentrate on the HPLC separation of these three most important classes of biomacromolecules, with an emphasis on analytical and semipreparative applications. We can assume that the general principles of HPLC separation for "small" molecules apply equally to the separation of biopolymers. However, the size and structure of a biomolecule lead to some important differences that will be examined in this chapter. As an introduction to the present chapter, the reader is encouraged to first review relevant earlier chapters, *especially Chapter 2* on basic concepts and the control of separation, and *Chapter 9* on gradient elution.

The primary chromatographic modes for the low-pressure separation of biomacromolecules have been ion exchange, size exclusion, hydrophobic interaction, metal chelate, and affinity chromatography; the HPLC versions of the first four techniques will be discussed here. For a detailed discussion of affinity chromatography, see [1]. In addition reversed-phase HPLC (RPC) has been hugely successful in the separation and characterization of peptides, and it serves as one of the major analytical tools for the development and characterization of protein-based biopharmaceuticals. The RPC separation of peptides and proteins will therefore be a major topic in this chapter. For more general guidelines for the preparative separation of all samples, see Chapter 15.

13.2 MOLECULAR STRUCTURE AND CONFORMATION

Macromolecules found in living cells are polymers consisting of subunits of similar chemical properties, such as amino acids, nucleotides, and sugars. The amino-acid sequence of proteins and the nucleotide sequences of RNA and DNA are precisely specified by the genetic code. In contrast, the carbohydrate sequences in glycoprotein side chains are determined by the specificity of the biosynthetic enzyme systems and the availability of substrates, so they may be more variable with respect to structure and sites of attachment on the polypeptide backbone. The properties of the assembled polymer depend on the properties of the individual subunits, as well as how they are positioned within the molecule. These two aspects of biopolymer organization (subunit properties and three-dimensional structure) influence both biological function and chromatographic behavior. Although it was earlier thought that the chromatography of biopolymers depends on different principles than for small molecules, it has been shown that biopolymers interact chromatographically in the same manner as small molecules, albeit with complexities introduced by polymer size, folding state, and three-dimensional structure [2, 3]. These macromolecules, proteins in particular, show complex behavior in solution with respect to their structure, stability, and aggregation state. This behavior restricts the choice of chromatographic conditions.

13.2.1 Peptides and Proteins (Polypeptides)

The fundamental subunits of polypeptides are amino acids, each of which consists of a carboxylic acid group, an amino group, and a side chain (Fig. 13.1). Amino acids differ in their side chains, which can be neutral and hydrophilic (e.g., serine, threonine), neutral and hydrophobic (e.g., leucine, phenylalanine), acidic (aspartic acid, glutamic acid), or basic (lysine, arginine, histidine). In polypeptide biosynthesis the carboxyl group of one amino acid (or *residue*) is linked to the amino group of the next amino acid with loss of water to form an amide or peptide bond (–CONH–). Of special interest is the amino acid cysteine, whose side-chain –SH group can be linked to that of another cysteine to form a disulfide bond (–SS–). Also noteworthy is the imidazole group of histidine, which can form coordination complexes with metal cations. The structures of the 20 common protein amino acids are shown in Figure 13.1, with their single- and three-letter codes, and the pKa values of the ionogenic side chains.

13.2.1.1 Primary Sequence

This comprises the sequence of amino acids in the molecule (Fig. 13.2*a*). Peptides consist of 40 amino acids or less, with a mass of no more than about 5000 Da. Proteins are larger polypeptide chains that contain up to several hundred amino acids, with masses from 5000 to 250,000 Da or greater. Peptides with fewer than 15

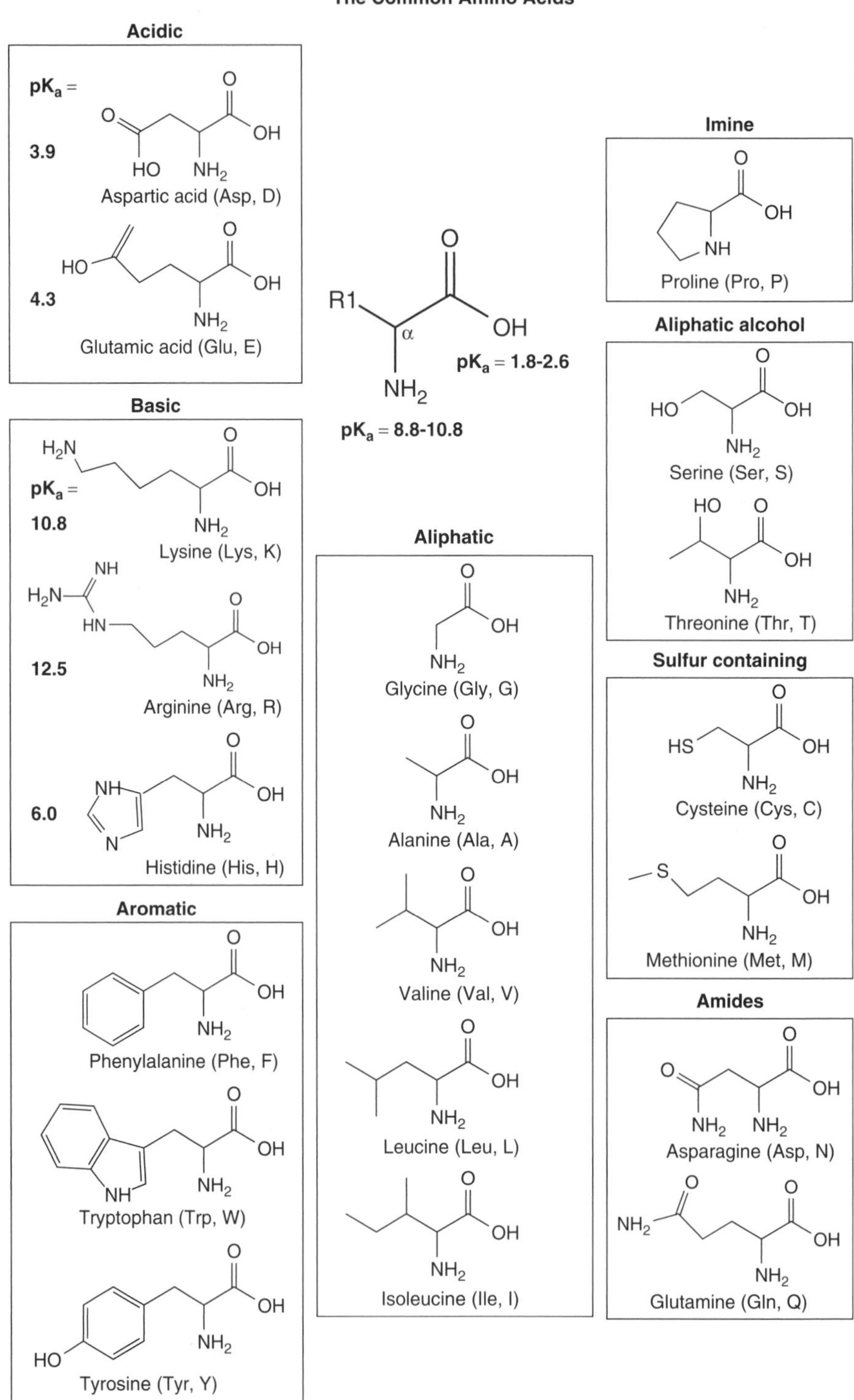

Figure 13.1 Structures of the amino acids commonly found in proteins. The amino acids are divided into groups according to the chemical properties of the side chains. The pK_a values for the ionogenic side chains are shown for acidic and basic amino acids. Adapted from [7].

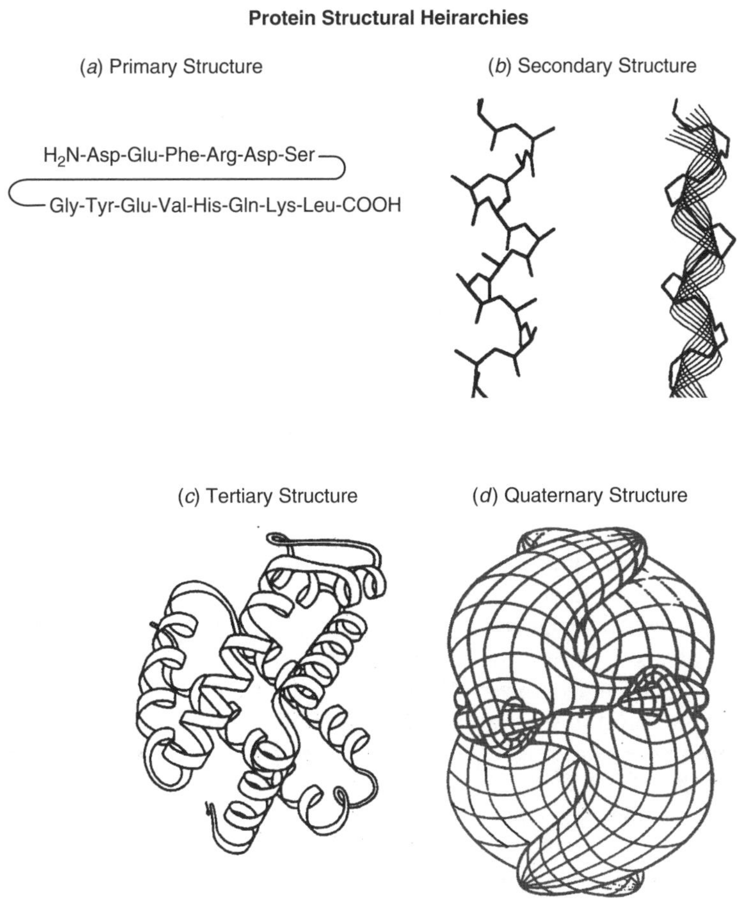

Protein Structural Heirarchies

(*a*) Primary Structure

H_2N-Asp-Glu-Phe-Arg-Asp-Ser-Gly-Tyr-Glu-Val-His-Gln-Lys-Leu-COOH

(*b*) Secondary Structure

(*c*) Tertiary Structure

(*d*) Quaternary Structure

Figure 13.2 Polypeptide structures. (*a*) Linear arrangement of amino acids in a polypeptide determines the primary structure. (*b*) Arrangement of amino acids of a 14-residue alanine homo-oligomer as an α-helical secondary structure, showing representation as a stick figure, and with only the backbone shown, overlain with a ribbon representation of the helix. (*c*) Ribbon diagram of the backbone of the hemoglobin β-subunit. (*d*) Schematic representation of the multi-sub-unit enzyme catalase. Adapted from [7, 8].

amino acid residues exist in solution as random coils, and they behave substantially like small organic molecules in chromatography. As peptide length begins to exceed 15 residues, molecular folding introduces increasing structure, as described below.

13.2.1.2 Secondary Structure

The spontaneous intramolecular interactions of a polypeptide during biosynthesis results in a secondary structure in which the three-dimensional shape of the final molecule is determined. Examples of the secondary structure (Fig. 13.2*b*) include the α-helix, which is stabilized by hydrogen bonds between residues located at intervals of about four amino acids along the primary sequence, and the β-sheet, which forms by hydrogen bonding between adjacent linear segments of primary sequence.

13.2.1.3 Tertiary and Quaternary Structure

The final folded structure of a single polypeptide chain is the *tertiary structure*, which may consist of combinations of helices, β-sheets, turns, and random coil sections (Fig. 13.2*c*). Combinations of secondary-structure elements may exist as *domains*, the fundamental units of tertiary structure; each domain contains an individual hydrophobic core built from secondary structural units. The tertiary structure is stabilized by the summation of a great number of weak interactions, including hydrogen bonding, ionic bonds, and hydrophobic forces. In addition the tertiary structure may depend on disulfide bonds between cysteine residues, which can covalently join remote segments of the primary sequence.

Quaternary structure represents the association of two or more folded protein chains to form a complex (13.2*d*) and depends on the same interactions involved in tertiary structure. The association of protein subunits (and conformational changes within the subunits) often plays a functional role in the regulation of protein activity. Similarly protein aggregation can be altered by the binding of substrates and small-molecule effectors.

Denaturation refers to both a functional and a physical change in the state of the native (bioactive) protein. Functionally, denaturation results in a loss of biological activity. Physically, denaturation occurs when the folding state of protein is altered or abolished, resulting in loss of secondary and higher order structures. Denatured proteins in a random-coil state often form aggregates that precipitate from solution. The environment of the protein molecule (either dissolved in the mobile phase or bound to the stationary phase) is a common cause of denaturation. Denaturation with loss of secondary, tertiary, and quaternary structure commonly occurs during RPC, but is less likely in ion-exchange, hydrophobic interaction, or size-exclusion chromatography.

13.2.1.4 Post-translational Modifications

A protein's primary sequence, which is a direct reflection of the nucleotide sequence in its associated gene, largely determines folding. However, many proteins are modified after translation (the initial creation of the protein) by the addition of one or more groups, and these *post-translational modifications (PTMs)* are not inferable from the gene sequence. The same gene sequence may direct the synthesis of proteins with different PTMs when expressed in different cells. A huge variety of PTMs have been described, but the most frequent are addition of sugar groups to the side chains of serine, threonine, or asparagine residues (glycosylation) and phosphorylation of serine, threonine, or tyrosine groups. Some PTMs are important biologically because they are involved in the regulation of protein function, in signal transduction, and in receptor-ligand interactions, while others result from mistreatment of the protein during isolation and handling. From a separation standpoint, the presence of PTMs may alter the interaction of a protein with a chromatographic surface and its retention.

13.2.2 Nucleic Acids

13.2.2.1 Single-Stranded Nucleic Acids

Single-stranded nucleic acids consist of a linear chain of nucleotides (Fig. 13.3), with each nucleotide consisting of a purine (adenine or guanine) or pyrimidine base

(*a*) Oligonucleotide composition

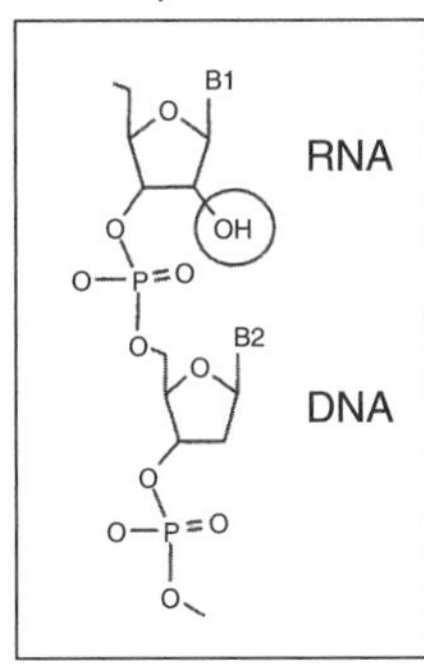

(*b*) Common nucleobases

Adenine Guanine

Thymine Cytosine Uracil

(*c*) Backbone-modified oligonucleotides

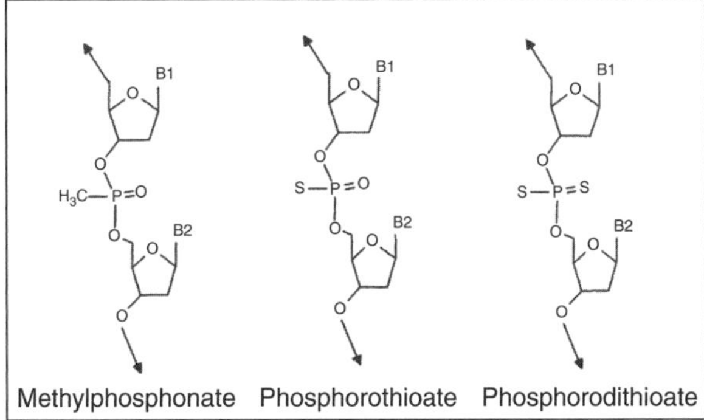

Figure 13.3 Structure of nucleic acids. (*a*) Schematic composition of a single-stranded oligonucleotide; in RNA the 2′ ribose position is hydroxylated (circled), whereas it is not in DNA. B1 and B2 represent the nucleobases, shown in (*b*). Adapted from [7].

(cytosine or thymine for DNA, cytosine or uracil for RNA) (Fig. 13.3*b*) linked to the C-1 carbon of ribose (RNA) or deoxyribose (DNA) (Fig. 13.3*a*). Nucleotide residues are linked through phosphodiester bonds between the 3′ hydroxyl of one nucleotide and the 5′ hydroxyl of the successive nucleotide. *Oligonucleotides* are short (usually single-stranded) nucleic acids, typically 13 to 25 bases in length, although lengths of 100 bases are sometimes referred to as oligonucleotides. Backbone-modified oligonucleotides (Fig. 13.3*c*) are synthetic derivatives used in "antisense" therapy, where the modified compound is able to combine with and deactivate the messenger RNA associated with a pathogen—because of the complementarity of the two molecular entities (as in following Section 13.2.2.2).

13.2.2.2 Double-Stranded Nucleic Acids

These consist of two complementary polynucleotide chains in a helical structure, with both chains coiled around a common axis, and with the two chains oriented in opposite directions (Fig. 13.4). Bases attached to the external sugar-phosphate backbone are situated inside the helix and participate in specific, interchain hydrogen bonds, with adenine (A) pairing with thymine (T) or uracil (U), and guanine (G) pairing with cytosine (C). As with native proteins, the molecular structure

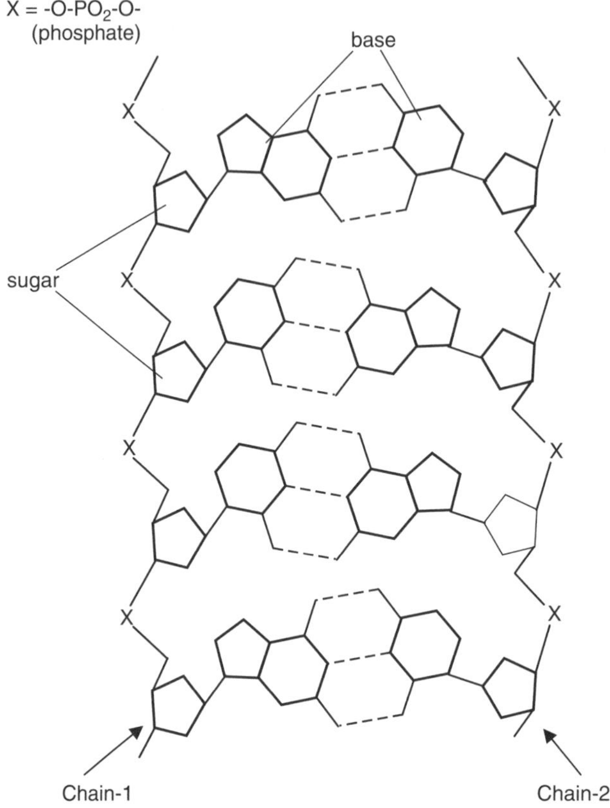

Figure 13.4 Schematic diagram of the helical structure of DNA, showing hydrogen bonding between complementary base pairs.

of a double-stranded nucleic acid depends on the contribution of many individual interactions. Under physiological conditions, hydrogen bonding and stacking interactions between bases stabilize the helical structure. Exposure of the helix to elevated temperature or extremes of pH can induce separation of the two strands. Single-stranded nucleic acids may form intramolecular, base-paired segments if they have regions of internal complementarity; these structures can be dissociated by heat or chemically denaturing conditions.

13.2.3 Carbohydrates

Polysaccharides play a variety of roles throughout the biosphere. Glucose homopolymers provide nutritional storage in both animals and plants. Glycogen, a large branched polymer of glucose residues linked by main-chain α-1,4 glycosidic bonds, is the sugar-storage entity in animals. Starch, the sugar-storage polymer in plants, consists of unbranched (amylase) or branched chains (amylopectin) of glucose with α-1,4 linkages. Both glycogen and starch exist as helical structures. Cellulose, the structural polysaccharide in plants, is a linear polysaccharide with β-1,4 glycosidic linkages.

Oligosaccharides play important functional roles as components of glycoproteins, including integral membrane proteins and many secreted proteins such as antibodies and clotting factors. Oligosaccharides participate in immune cell recognition and cell–cell communication, and also contribute to protein stability and the maintenance of cell structure. Oligosaccharides are added as co-translational or post-translational modifications to proteins, and are covalently linked to the side chains of serine or threonine (*O*-linked oligosaccharides) or the side chain of asparagine (*N*-linked oligosaccharides). The predominant sugars in glycoproteins (Fig. 13.5) are glucose, galactose, mannose, fucose, *N*-acetylgalactosamine (GalNAc), and *N*-acetylglucosamine (GlcNAc). In *O*-linked glycoproteins, the carbohydrate is attached to the protein by a GalNAc residue, while the attachment point in *N*-linked oligosaccharides is via GlcNAc. In *N*-linked glycoproteins (the predominant form in mammals), the oligosaccharide consists of a common core-structure consisting of two GlcNAc residues and three mannose residues. Additional sugars attached to this core form a diverse family of oligosaccharide structures, including high-mannose oligosaccharides and complex oligosaccharides containing GlcNAc,

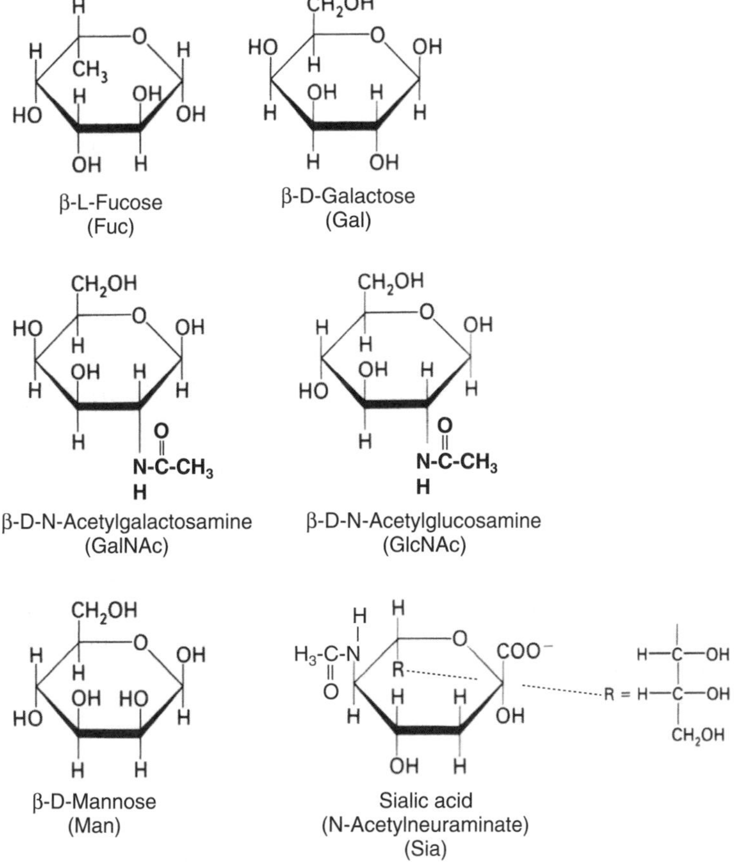

Figure 13.5 Sugar residues commonly found in glycoproteins.

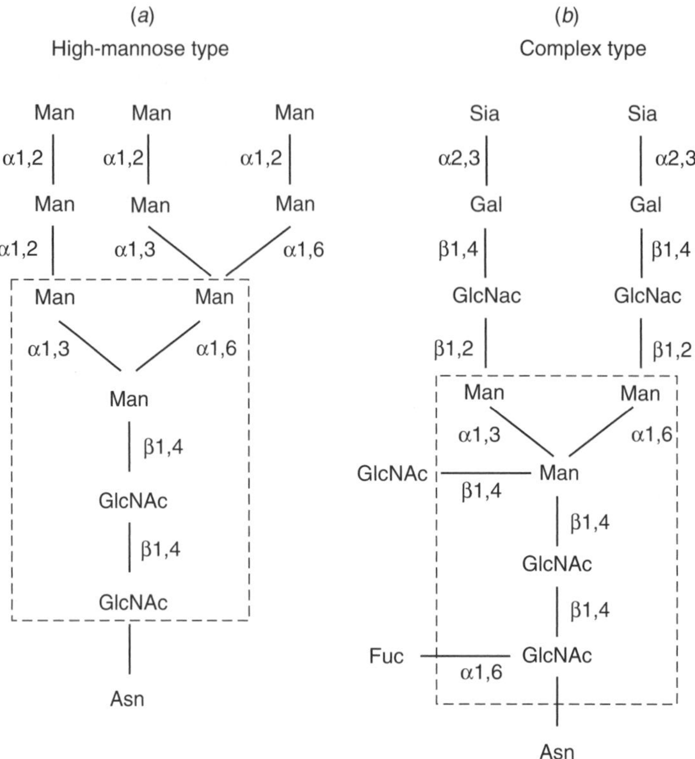

Figure 13.6 *N*-linked oligosaccharides of (*a*) high-mannose type, and (*b*) complex type. The common core structure consisting of three mannose residues and two GlcNAc residues is indicated (dashed enclosure).

galactose, sialic acid, and fucose residues (Fig. 13.6). Glycoproteins vary in the number of glycosylation sites, the occupancy of these sites, and the oligosaccharide structure at each occupied site.

13.2.4 Viruses

In the 1990s the advent of gene therapy generated a need for the purification of recombinant (laboratory-created) viruses; chromatography emerged as the preferred technique for the purification of large quantities of material for gene-therapy trials. Virus purification is also required for the production of some vaccines. Viral purification has traditionally been done by cesium-chloride density-gradient ultracentrifugation [4], but scale-up is not practical. This method results in variable purity and poor yields, and can require removal of CsCl [5].

Recombinant adenoviruses (rAd) are common vectors for gene therapy, and they have served as models for the development of chromatographic methods for the purification and analysis of viruses. The adenovirus particle contains 85% protein and 13% DNA and has a molecular weight of 167×10^9 Da. It consists of an icosahedral protein shell or capsid (70 to 100 nm in diameter) surrounding a protein core that contains the linear, double-stranded DNA genome [11]. The capsid also contains some additional, minor polypeptide elements. The adenovirus genome consists of a linear double-stranded DNA molecule of 35 to 36 kilo base-pairs. One distinguishing feature of viruses, so far as their chromatographic separation, is their enormous relative size—which restricts the penetration of the virus molecule into a porous column-packing.

13.3 SPECIAL CONSIDERATIONS FOR BIOMOLECULE HPLC

The size and shape of biopolymer molecules, as well as the need in preparative applications for maintaining biological activity, require special consideration for the choice of column, mobile phase, and temperature—any of which conditions can affect the recovery of biological activity. Preferred conditions vary for each chromatographic mode and sample type, as discussed in Sections 13.4 through 13.7. However, some general comments can be made with regard to column characteristics and stability as a function of conditions (Section 13.3.1).

General principles of method development are provided in Chapters 6 through 9 for individual chromatographic modes, while other aspects of method development are dealt with in Chapters 11 and 12. The latter material is largely applicable for all solute molecules, both large and small. Additional considerations that are important for biomolecules are addressed in this chapter; for additional information, see [9].

13.3.1 Column Characteristics

The large size of biomacromolecules requires particular attention to the selection of the pore size and particle diameter of the column packing. Analyte stability and mass recovery, a possible need for nondenaturing conditions, and column stability also affect the final choice of column.

13.3.1.1 Pore Size

Column capacity (Section 15.3.2.1) and retention are a function of the amount of stationary phase available for sample interaction, which is in turn proportional to the *accessible* surface area of the packing (Section 5.2.1). For smaller molecules (<1000 Da), particles with pore-diameters of 8 to 12 nm permit free access of solutes into the pore system, such that the solute can sample the entire surface area (typically ≈ 250 m^2/g for a 10-nm-pore particle) and diffuse freely within the pores (so as not to compromise column efficiency N). In contrast, large biopolymers can be excluded partly or entirely from pores of this size so that they interact only with the external surface of the particle (which represents $< 1\%$ of the total surface and column capacity within the pores). In order to achieve adequate column capacity, retention, and column efficiency, particles should be used which have pore diameters large enough to permit an easy entry and exit of the biomolecule.

The relationship between molecular weight M and molecular size for globular and random coil proteins is shown in Table 13.1. To avoid peak broadening due to restricted diffusion of the protein within the pore, the pore diameter should exceed the solute diameter by a factor of 3 or more [2, 6]. However, surface area decreases approximately in proportion to increasing pore size (Table 13.2), so that an optimum pore size allows access of the protein to the pores—without unduly compromising

Table 13.1

Protein Diameter and Molecular Weight Compared

Molecular Weight (kDa)	Hydrodynamic Diameter	
	Random Coil (nm)[a]	Globular (nm)[b]
1	2.6	1.6
10	8.2	3.5
100	25.8	7.6
1000	81.6	16.3

Source: Data from [9].
[a]Applies to separation under denaturing conditions (including RPC).
[b]Usually the native (nondenatured) protein.

Table 13.2

Effect of Pore-Diameter on Surface Area

Pore-Diameter (nm)	Surface Area (m^2/g)[a]
10	250
30	100
100	20
400	5–10

[a]Approximate values that vary with pore-volume.

surface area and column capacity. The combined effects of solute and pore size on retention are illustrated in Figure 13.7. In Figure 13.7*a*, maximum retention of small peptides angiotensin I and II (approximately 1000 Da each) is observed with pores of 10-nm diameter. For smaller pores, exclusion of the two peptides occurs with a decrease in retention; for larger pores, surface area is reduced with a corresponding decrease in both column capacity and retention. For proteins (Fig. 13.7*b*), significant pore penetration is achieved only with the 30-nm-pore packing, and all three proteins are largely excluded from particles with smaller pore-diameters. In practice, columns with pore diameters of about 30 nm are satisfactory for proteins of ≤ 50 kDa, while columns with pore diameters of 100 to 400 nm are preferred for large globular and/or denatured proteins. Note that particles with pore diameters ≥ 100 nm will have reduced surface area, and often exhibit poor mechanical strength.

It should be kept in mind that the hydrodynamic diameter of a protein increases approximately 2- to 3-fold upon denaturation (Table 13.1). Therefore, if proteins are to be separated under denaturing conditions, a larger column-pore size required (particularly important for size-exclusion chromatography [SEC]).

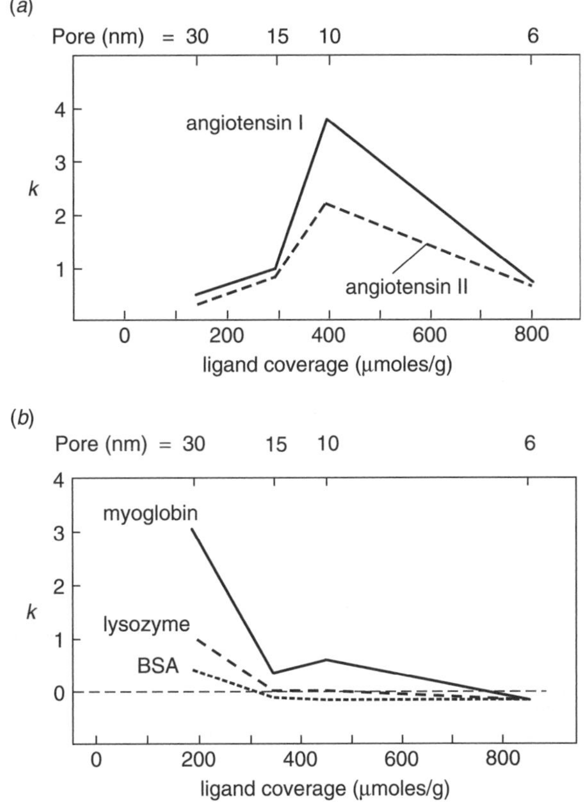

Figure 13.7 Effect of pore and solute-molecule size on retention in RPC with C_{18} columns. (*a*) Small peptides (≈1000Da); 35% acetonitrile/pH-2.3 phosphate buffer; (*b*) proteins (> 10,000 Da); 49% acetonitrile-pH-2.3 buffer. Note the column-packing pore-diameter at top of each figure; "ligand coverage" (*x*-axis) is proportional to surface area. Adapted from [10].

13.3.1.2 Particle Size

Columns packed with fully porous, 3.5- to 5-μm particles are currently preferred for small-molecule analytical separations, and these columns are also widely used for biomolecule applications. However, the slow diffusion of biopolymers results in reduced column efficiency and increased peak widths compared to small molecules (Section 2.4.1). This can be counteracted by the use of much lower mobile-phase flow rates, but with correspondingly longer separation times. Several approaches have been pursued in order to improve column efficiency for large biomolecules. As discussed in Section 2.4.1.1, the effects of slow diffusion on column efficiency can always be mitigated by the use of smaller particles, with particles as small as 1.5-μm finding increasing use for large-molecule separations.

A further improvement in the plate number *N* for large biomolecules can be achieved with small-diameter, nonporous ("pellicular") particles (Section 5.2.1.1). The absence of pores in these packings eliminates slow diffusion within the pores, while the external surface area—although quite small—may provide sufficient column capacity for the analysis of major sample components. So-called superficially porous particles (Section 5.2.1.1) have a solid core, which speeds up the movement of molecules into and out of the particle pores—with an increase in column plate number N, but only a small decrease in column capacity. So-called perfusion chromatography (Section 5.2.1.1 [12]) uses packings that contain very large through-pores that allow flow of mobile phase *through* the particle. In principle, this can also minimize the effects of slow diffusion into and out of the particle, although these columns are used mainly for preparative separations of large biomolecules.

Finally, the replacement of packed beds with a monolith (Section 5.2.4) is still another option. Monolithic columns consist of a continuous, interconnected skeleton with through-pores for transport of mobile phase and solutes through the column. As a result monoliths can be operated at high flow rates with modest pressures, and with little decrease in column efficiency. Both polymer-based and silica monoliths are commercially available. Polymeric monoliths include polymethacrylate and polystyrene-divinylbenzene materials, and they are available in both column and disk formats for analytical and preparative chromatography. These materials also contain a bimodal pore structure of large and small pores.

13.3.1.3 Support Characteristics and Stability

Porous silica has several properties that favor its use as a chromatographic support (Section 5.2.2). Unfortunately, silica has other properties that limit its use for the separation of biomolecules. The separation of peptides by RPC is generally performed under acidic conditions ($pH < 3$), and ion-exchange separations are often carried out under neutral or alkaline conditions ($pH > 7$); some silica-based columns experience reduced stability outside the limits of $2.5 \leq pH \leq 7.5$ (Sections 5.3.1, 5.3.2.1). Separations are sometimes carried out at elevated temperature in order to improve peak shapes or optimize selectivity, but bonded-phase silicas exhibit decreased stability at temperatures above 40°C, especially at extremes of pH. However, the use of suitable columns and other conditions can reduce the adverse effects of mobile-phase pH and temperature on column stability.

Another major potential problem with the use of silica-based columns for biomolecule separations can arise from strong interactions between the silica surface and the solute, resulting in wide, tailing peaks and loss of sample due to irreversible adsorption. Problems of this kind are much more pronounced for older, "type-A" columns; higher-purity, "type-B" columns are therefore strongly recommended (Section 5.2.2.2). End-capped RPC columns (Section 5.3.1) are also more effective in minimizing undesirable sample-column interactions.

Several strategies have been pursued to improve the performance of bonded-phase silicas for biopolymer chromatography, and present-day silica-based RPC columns are the columns of choice for most peptide separations. Because of the limitations of silica-based packings for other modes of chromatography (e.g., ion exchange and hydrophobic interaction), however, polymeric column packings are used mainly for these applications (Section 5.2.3). Polymers such as polystyrene-divinyl benzene and polymethacrylate can be formed into porous particles that can be used directly for chromatography (e.g., PS-DVB for RPC). Alternatively, polymeric columns can be functionalized so as to introduce specific groups (e.g., ionic moieties for ion exchange; Section 7.5.4). These packings are stable over a broad pH-range (including $pH < 2$, $pH > 10$) and can be used at higher temperatures that would destroy a silica-based column. They are also stable to pressures of 4000 to 5000 psi. A major advantage of polymeric materials for preparative and process-scale applications is the ability to clean them with strong bases, in order to remove contaminants such as endotoxins. Endotoxins (typically lipopolysaccharides from host cells used in the production of biopharmaceuticals) cause inflammatory responses, and their introduction into drug products destined for human use must be avoided.

13.3.1.4 Recovery of Mass and Biological Activity

In applications where HPLC is used to isolate material for further characterization or other uses, the analyte must be recovered with good yield. If bioactivity is to be preserved, the biopolymer must also maintain its native conformation. These requirements require careful selection of the chromatographic mode and separation conditions. RPC can be denaturing for proteins, and generally it is not used for the recovery of larger proteins. However, denatured peptides or small proteins from a RPC separation can usually be restored to full bioactivity by exposure to organic-free buffer with appropriate ionic conditions. The use of other chromatographic modes with harsh elution conditions (extremes of pH, elevated temperature) can also compromise the recovery of intact, active species. The mass recovery of a polypeptide can be reduced by its adsorption to active sites on the packing, or by entrapment within the pore system. Sample loss due to adsorption can be minimized by pretreatment of the column with a surrogate biopolymer (e.g., bovine serum albumin for proteinaceous samples), in order to deactivate the column prior to use. Sample loss due to protein unfolding within the pore system can also be minimized by using large-pore supports or less denaturing conditions. The tendency of HPLC conditions to denature a protein can be ordered as follows: RP $\gg$ HIC $\approx$ IEC $>$ SEC. For further details on polypeptide recovery, see [13].

13.3.2 Role of Protein Structure in Chromatographic Behavior

Protein retention in HPLC can be understood as an interplay between protein structure and the chromatographic process. In the case of small solute molecules, most parts of the molecule are in contact with the stationary phase. For large peptides and especially proteins, this may not be possible because only the surface of these three-dimensional molecules (their *contact area*) can be in contact with the stationary phase. Several resulting retention relationships were described in a seminal publication by Regnier [14], and these can be summarized by a series of postulates:

- The weak chemical forces that govern protein conformation and surface recognition (ionic, hydrophobic, hydrogen bonding) are the same as those involved in chromatographic interactions.
- It is not possible for all the amino acids in a protein to simultaneously contact the stationary-phase surface (even more so for the native molecule).
- Only residues located at the protein surface have an impact on chromatographic behavior, and only a fraction of those residues (those within the contact area) are involved in stationary-phase interactions.

- The heterogeneous distribution of residues on the protein surface allows some portions of the surface to dominate chromatographic behavior; these interactive regions may not be the same for different chromatographic modes.
- Structural changes that alter the protein surface can change chromatographic behavior if they occur within the contact area or alter the surface of the contact area.
- Interaction with the stationary or mobile phase can alter protein secondary, tertiary, and quaternary structure.

We will return to these concepts in our following discussion of different chromatographic modes.

13.4 SEPARATION OF PEPTIDES AND PROTEINS

The selection of the appropriate mode for peptide and protein chromatography is dictated by the goals of the separation. If high recovery of protein mass and biological activity is required, relatively gentle chromatographic techniques, such as ion exchange, hydrophobic interaction, and size exclusion are preferred. If the aim is to resolve proteins based on size, or carry out a class separation between large and small molecules, size-exclusion chromatography is the method of choice. RPC is used less often for the purification of larger proteins, because of its tendency to denature—which can degrade separation and compromise both mass recovery and biological activity. For the purification of peptides and smaller proteins, however, RPC has been remarkably successful; it is the universal first choice for separating peptide mixtures. In the area of proteomics, which seeks to characterize the entire protein composition of a cell or tissue, extraordinarily complex mixtures of peptides must be separated prior to MS-MS analysis (Section 4.14); two-dimensional (2-D) HPLC based on ion exchange followed by RPC is often used (Section 13.4.5).

13.4.1 Reversed-Phase Chromatography (RPC)

Several features of reversed-phase chromatography (RPC) are responsible for its wide use in the analysis and purification of peptides and proteins. The high efficiency of RPC columns packed with small particles provides increased resolution and high peak capacities (Section 9.3.9.1) for the separation of complex mixtures. The selectivity of RPC also favors the resolution of peptides with very similar structures. The solvents used in RPC are compatible with UV detection, and their volatility permits solvent removal from recovered fractions. Most important, aqueous-organic mobile phases are compatible with electrospray ionization, so RPC is almost always the technique used with detection by mass spectrometry (Section 4.14.1.1). A large selection of mobile phases and columns is available for solving a given separation problem, although in practice a few generic methods are used for most samples.

An important feature of peptide, and especially protein, behavior in RPC is a strong dependence of solute retention on small changes in solvent strength (Section 13.4.1.4). For example, a 29 kDa protein has been shown to exhibit a 20% change in retention time for a variation of only ±0.1% organic solvent concentration [15]. This usually makes the isocratic separation of even simple polypeptide mixtures impractical; gradient elution is almost always required.

13.4.1.1 Column Selection

Silica-based RPC packings exhibit acceptable stability for commonly used separation conditions, and they are often preferred over polymer-based materials because of their higher column efficiency. Small-pore silicas ($\approx$10 nm diameter) are satisfactory for peptides but large pore silicas (pore diameter$\geq$30 nm) are preferred for protein separations. Proteolytic digests can contain peptides of widely varying sizes (200– $\approx$ 2500 Da for tryptic digests, larger for LysC and AspN digests), so the use of packings with 30-nm pores may be necessary for some peptide samples. Other considerations in column selection are the stationary-phase ligand and its bonding density.

The average selectivity of columns with different ligands is summarized in Table 5.8a and the related text, including values of the column-hydrophobicity parameter $\boldsymbol{H}$. The most popular columns are based on straight-chain alkyl groups (C_4, C_8, C_{18}), which exhibit increased retention and stability with increasing alkyl length. Trimethyl (C_1) columns are the least hydrophobic ($\boldsymbol{H} = 0.41$) and have been used for the separation of proteins that are too strongly retained on longer ligand (more hydrophobic) columns. However, under conditions usually employed for protein separations, C_1 columns are easily hydrolyzed and quite unstable; octyl (C_8, $\boldsymbol{H} = 0.84$) and octadecyl (C_{18}, $\boldsymbol{H} = 0.99$) columns are generally the first choice for separating peptides, because of their greater stability and suitable retention characteristics.

For gradient separations of peptides and proteins, the concentration of the B-solvent rarely exceeds 60–80%B. Under these conditions the selectivities of butyl, octyl, and octadecyl columns for proteins and peptides appear comparable [16]. Phenyl and cyano columns are likely to exhibit different selectivity than straight-chain alkyl columns, but cyano columns are generally less stable when used at either low or high pH. Butyl columns are much less hydrophobic and less retentive; these columns are therefore preferred for very hydrophobic species, such as membrane proteins and the more hydrophobic polypeptides generated by cyanogen bromide cleavage.

Monomeric columns (Section 5.3.1) exhibit higher efficiencies, and they are usually a first choice. Polymeric columns are more stable but may exhibit lot-to-lot variability due to poor reproducibility of the polymerization reaction. The concentration of the ligand (μ moles/m^2) also affects column stability and the effects of residual surface silanol groups.

13.4.1.2 Mobile-Phase Selection

In RPC the mobile phase consists of aqueous buffer (A-solvent) and an organic solvent (B-solvent). For biochemical separations, the "buffer" is often a dilute acid such as phosphoric, trifluoroacetic, formic, or acetic—with a pH of 2 to 3.5. Sample retention in isocratic elution (values of k) can be controlled by varying %B—as in the case of small-molecule samples (Section 2.3.2). For gradient separations, an increase in gradient steepness leads to reduced values of k^* and usually poorer separation (see Section 13.4.1.4 below). The mobile-phase composition can affect separation selectivity, detector compatibility, and (to a lesser extent) column efficiency.

Aqueous Component. The hydrophobicity (and therefore retention) of a polypeptide is heavily dependent on the ionization state of the amino-acid termini (approximate pK_a values of 2.4 and 9.8) and of the ionizable side chains of internal residues (pK_a values in Fig. 13.1). Thus the pH of the mobile phase can have a profound effect on polypeptide retention, and it is necessary to control mobile-phase pH by the addition of a buffer (Section 7.2). In practice, separations of proteins and peptides are most often performed at low pH, where a dilute acid can serve as buffer. Under these conditions silanol ionization is suppressed, reducing undesirable interactions with protonated solutes and resulting peak tailing. Commonly used organic acids (e.g., trifluoroacetic acid) can also ion-pair with protonated solutes, resulting in increased retention for peptides and improved peak shapes for proteins. The ionization of terminal and side-chain carboxyl groups is suppressed at low pH, which further increases retention.

The most widely used acid for the RPC separation of peptides and proteins is trifluoroacetic acid (TFA), with a concentration of 0.05–0.1% (approximately 5–10 mM). The stronger acidity of TFA allows a lower mobile-phase pH ($\approx$2), and lower concentrations are therefore required compared to formic or acetic acid. A lower TFA concentration also reduces background absorbance when UV detection is used at low wavelengths. An added benefit of TFA is its high volatility, which facilitates solvent removal from collected sample fractions in preparative applications; however, purified proteins typically retain significant amounts of bound TFA, which can be removed by dialysis or diafiltration. The higher UV absorbance of organic acids necessitates detection at higher wavelengths—with increased baseline drift and noise (all of which adversely affects sensitive detection). For the trace analysis of peptides, phosphoric acid (which is transparent at 200 nm and above) can be used in place of an organic acid, but proteins often exhibit poor peak shape with phosphoric acid.

The use of TFA as buffer with mass-spectrometric detection (LC-MS) can be problematic when using electrospray ionization. In negative-ion detection, the high concentration of TFA-anion can suppress solute ionization. In positive-ion detection, TFA forms such strong ion-pairs with peptides that ejection of peptide pseudomolecular ions into the gas phase is suppressed. This problem can be alleviated by the postcolumn addition of a weaker, less volatile acid such as propionic acid [17]. This "TFA fix" allows TFA to be used with electrospray sources interfaced with quadrupole MS systems. A more convenient solution to this TFA problem, however, is to simply replace TFA with acetic or formic acid.

Tailing peaks are sometimes observed for peptides, when separated by RPC at low pH. Peak tailing is usually associated with protonated-amine groups within the solute molecule (Section 7.3.4.2). For modern, type-B alkylsilica columns, peak tailing at low pH usually depends on the weight of injected peptide, with a resulting overloading of the column (due to the mutual repulsion of positively charged solute ions in the stationary phase; Section 15.3.2.1). Column overload (and peak tailing) in these cases usually occurs for injections of >1 μg of peptide onto a 4.6-mm ID column (or lower weights for smaller column ID's). It is possible to inject somewhat larger weights of a peptide by increasing the ionic strength of the mobile phase; for example, by the use of fully ionized acids such as TFA or phosphoric acid, or significantly ionized buffers such as ammonium acetate or formate at higher pH [18, 19].

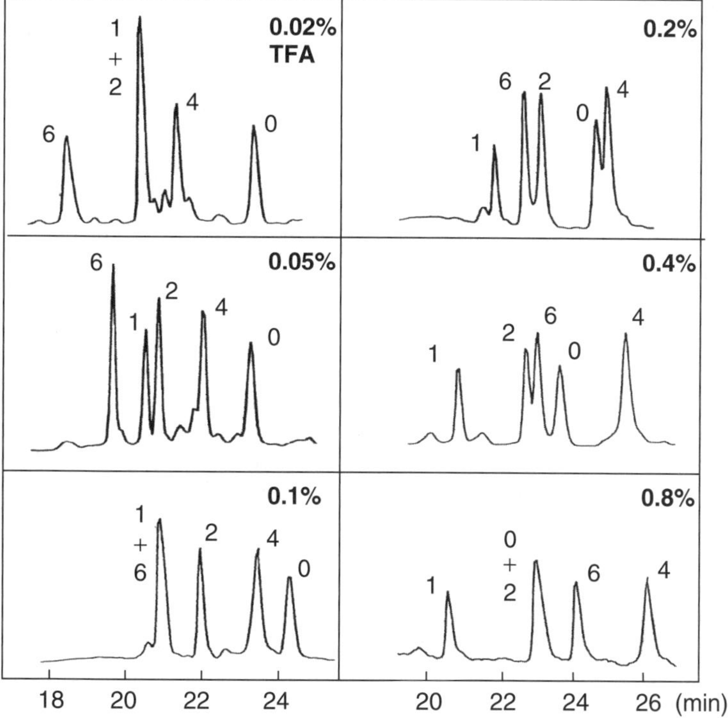

Figure 13.8 Effect of TFA concentration on the RPC retention of basic peptides. Sample: synthetic peptides with varying numbers of basic amino acids in the molecule (the number of these basic groups is indicated for each peak). Conditions: 250 × 4.6-mm C_{18} column; acetonitrile-water gradients with indicated amount of added TFA; 26°C; 1.0 mL/min. Adapted from [20].

TFA is capable of ion-pairing with protonated, basic peptides, as illustrated in Figure 13.8. Here a mixture of synthetic peptides that contain 0, 1, 2, 4, or 6 basic residues was separated with varying concentrations of TFA. As the TFA concentration is increased, the retention of a basic peptide increases because of ion-pairing (Section 7.4.1), and the effect is greater for peptides with a larger number of basic groups. As a result relative retention changes for a change in TFA concentration. The ability to change relative retention based on peptide charge can be useful during method development.

If the separation of proteins that differ in surface modifications is desired, it may be advisable to use conditions that are nondenaturing [21]. The standard, low-pH conditions described above are then inappropriate, and mobile phases buffered near neutrality are required. Buffers based on ammonium acetate, ammonium bicarbonate, and triethylammonium phosphate may also prove more useful in resolving polypeptide variants with differing post-translational modifications, amino-acid substitutions, or oxidation and deamidation products [21].

Triethylamine-phosphate (TEAP) as buffer, with acetonitrile as B-solvent, was recommended initially for the RPC separation of peptides [22]—primarily for the improvement of peak shape and analyte recovery. However, the current use of type-B columns for peptide separations has largely eliminated any need for this buffer in RPC.

Organic Component. The usual B-solvents for the RPC separation of peptides and proteins are acetonitrile, methanol, propanol, and isopropanol—with acetonitrile being the most popular. The major criteria for selecting the B-solvent include low UV absorbance and viscosity, which favor acetonitrile. Solvent selectivity, cost, toxicity, and purity are also considerations. Peptides and proteins can be detected by the UV absorbance of the peptide bond at 205 to 220 nm. Acetonitrile has good optical transparency in this region, and it is compatible with detection at 205 nm. The use of methanol or propanol necessitates detection at >210 nm, to avoid excessive baseline drift in gradient elution. The UV response of polypeptides decreases at longer wavelengths, with a reduction in detection sensitivity. UV detection at 205 to 220 nm is generally employed for peptides and proteins, providing "universal" detection for all polypeptides. Tyrosine and tryptophan residues have local absorbance maxima at 270 and 280 nm, respectively, allowing the selective detection of peptides and proteins that contain these residues. However, the absorbance of the latter polypeptides at 270 to 280 nm is much lower than at 205 nm.

Acetonitrile-water mixtures exhibit lower viscosity than alcohol-water, which favors faster, higher resolution separations (Section 2.3.1). The comparative behavior of acetonitrile, methanol, and isopropanol as B-solvents is illustrated in Figure 13.9 (gradient separations of a series of synthetic peptides). The analytes in this study were octapeptides of identical structure, except for different amino acids in positions 2 and 3 of the peptide [23]. Total analysis time increases in the order isopropanol < acetonitrile < methanol, suggesting that solvent strength increases in the reverse order: methanol (weakest) < acetonitrile < isopropanol (strongest). The narrowest, most symmetrical peaks are generally observed with acetonitrile as B-solvent. Solvent selectivity is different for these three B-solvents, as illustrated for a group of seven peaks marked by a bracket and an asterisk in Figure 13.9. With isopropanol as B-solvent, these peptides are poorly resolved, but their resolution increases progressively for methanol and acetonitrile. Other changes in selectivity with a change in B-solvent can also be seen in Figure 13.9.

Surfactants. Surfactants are sometimes added to the sample and/or mobile phase for the solubilization of hydrophobic, poorly soluble proteins and their improved recovery from the column [21]. For example, RPC methods for integral membrane proteins often employ ionic, zwitterionic, or nonionic surfactants as additives. Since adsorbed surfactant may be difficult to remove from the column, it is advisable to dedicate a column for the use of a given surfactant.

13.4.1.3 Temperature

A change in column temperature can improve the separation of peptides or proteins by RPC. First, operation at higher temperatures reduces mobile-phase viscosity and increases solute diffusion, each of which contribute to increased column efficiency and better resolution. Second, a change in column temperature can be used to optimize separation selectivity—as in the case of other ionic samples (Section 7.3.2.2). An example is provided in Figure 13.10*a–d*, for the gradient separation

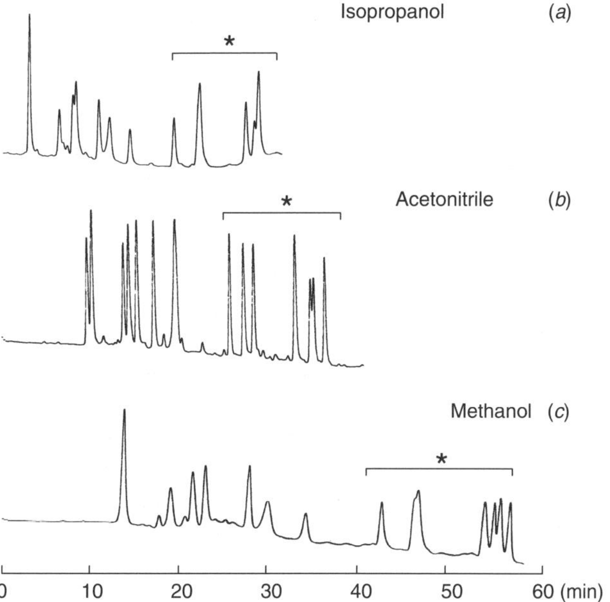

Figure 13.9 Effect of B-solvent on the separation of peptides by RPC. Sample described in text. Conditions: 250 × 4.1-mm C_8 column; linear gradients at 1%B/min; pH-2; 26°C; 1.0 mL/min. Adapted from [23].

of a tryptic digest at 30 and 50°C (only peaks 6–13 shown, out of a total of 18). Finally, an increase in temperature leads to increased protein denaturation, which generally favors narrower, more symmetrical peaks and increased recovery of the sample in RPC. On the downside, operation at high temperature can reduce column lifetime for some silica-based columns.

13.4.1.4 Gradient Elution

As noted above, RPC separations of biomolecules are characterized by rapid changes in retention for small changes in %B. As a consequence gradient elution is generally required for these samples. Isocratic retention k can be related approximately to mobile-phase composition (%B) by

$$\log k = \log k_w - S\phi \tag{13.1}$$

where ϕ (equal 0.01 × %B) is the volume-fraction of organic solvent, k_w is the (extrapolated) value of k for buffer as mobile phase ($\phi = 0$ or 0% B), and S is a constant for a specific solute and experimental conditions other than %B. The varying dependence of retention on %B for small molecules, peptides, and a small protein is illustrated in Figure 13.11; note that the slopes of these plots (values of S) increase with solute molecular weight M. The relationship between S and molecular weight M can be approximated by

$$S \approx 0.25\ M^{0.5} \tag{13.1a}$$

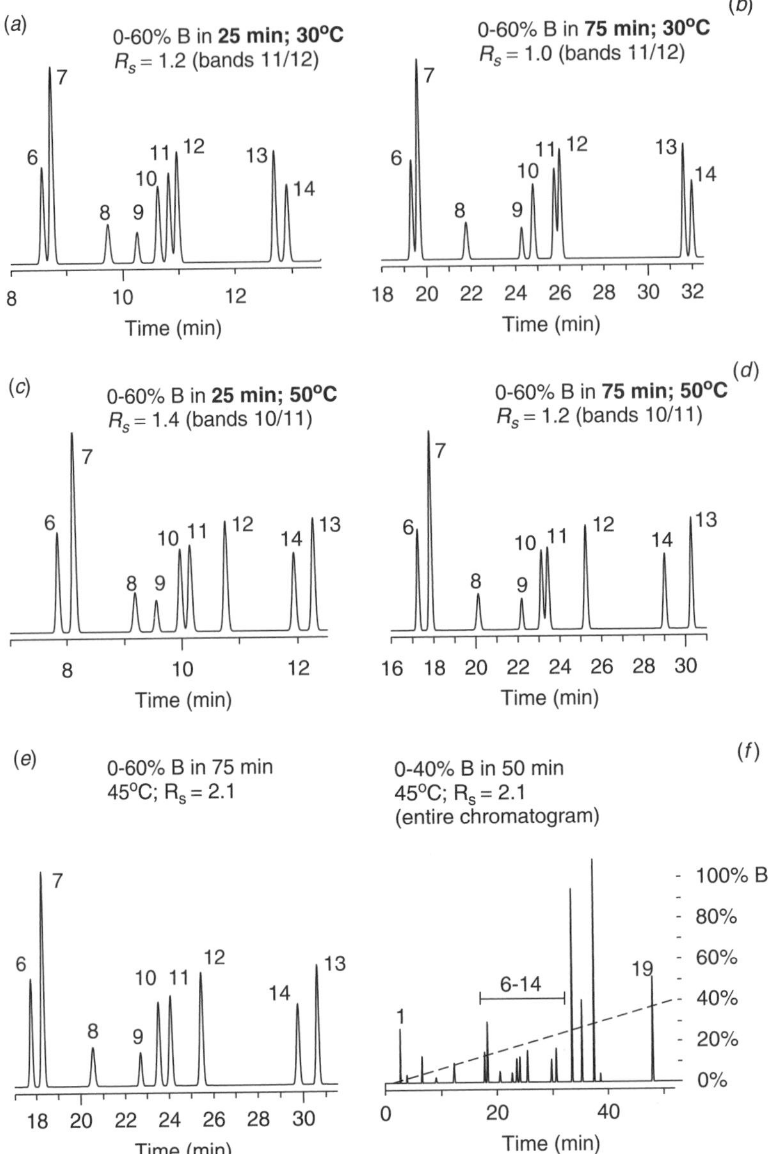

Figure 13.10 Separation of *rh*GH peptide digest, using different gradient times and temperatures in order to optimize selectivity and maximize resolution. Conditions: 150 × 4.6-mm C_{18} column; gradients of acetonitrile (B)–water + 0.1% TFA; 2.0 mL/min; other conditions indicated in figure. Simulations based on data of [24].

In gradient elution, retention is related to solvent strength (Section 9.2) by a relationship similar to Equation (13.1):

$$\log k^* = \log k_w - S\phi^* \qquad (13.2)$$

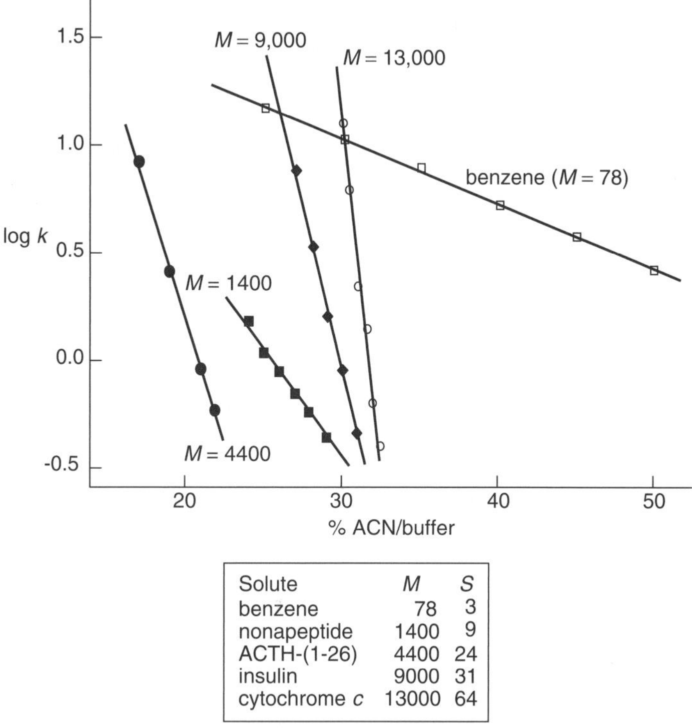

Solute	M	S
benzene	78	3
nonapeptide	1400	9
ACTH-(1-26)	4400	24
insulin	9000	31
cytochrome *c*	13000	64

Figure 13.11 Change in isocratic retention k with change in %B as a function of solute molecular weight. Adapted from [25].

where k^* is the median value of k during gradient elution and ϕ^* is the median value of ϕ (value when a band has moved halfway through the column). The value of k^* depends on gradient conditions

$$k^* = \frac{t_G F}{V_m \Delta\phi S} \tag{13.3}$$

where t_G is the gradient time, F is the flow rate, $\Delta\phi$ is the change in ϕ during the gradient (e.g., $A\phi = 0.55$ for a 5-60%B gradient), and V_m is the column dead-volume ($= t_0 F$). Equation (13.3) predicts that achieving satisfactory values of k^* ($1 \leq k^* \leq 10$) for molecules with large values of S (e.g., proteins) can be accomplished by using long gradient times with (if possible) a narrow gradient range (small value of $\Delta\phi$, i.e., a small difference in the initial and final values of %B for the gradient). Note that a change in gradient conditions that affects the value of k^* can change selectivity, similar to a change in %B and k in isocratic elution. Minor changes of this kind can be seen in Figure 13.10 for a change in gradient time. Similarly a change in flow rate while maintaining the same gradient time can cause changes in k^* and relative retention, as seen in the more dramatic example for a tryptic-peptide sample in Figure 13.12. A portion of each chromatogram in Figure 13.12*a* (marked by a bracket and arrow) is expanded in Figure 13.12*b*.

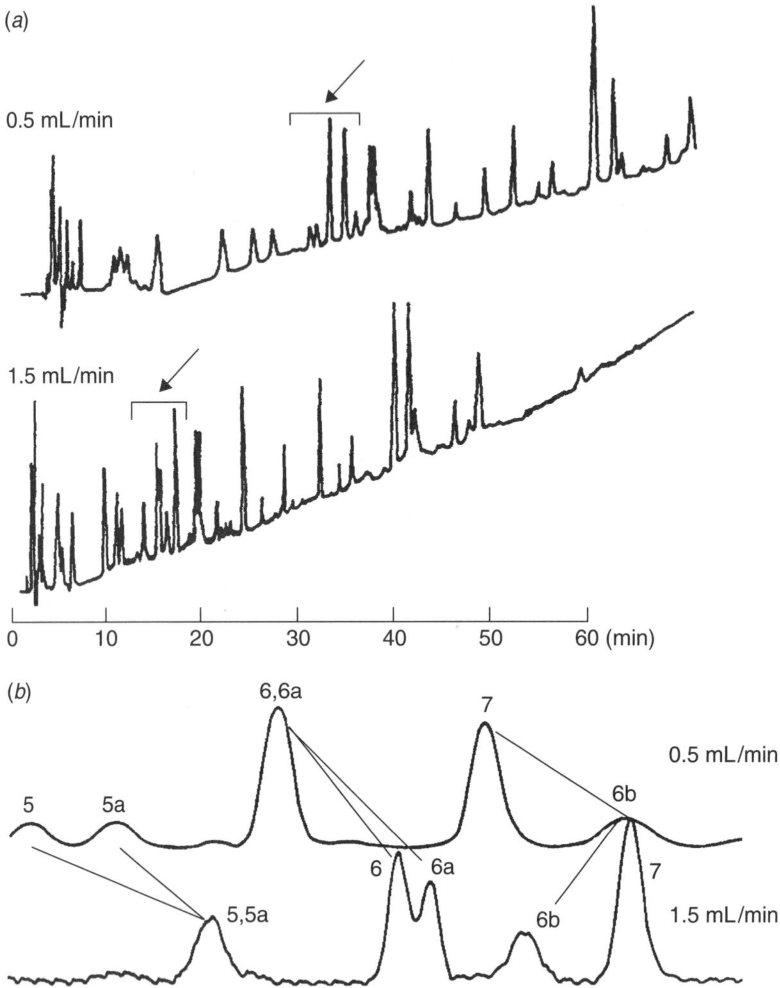

Figure 13.12 Effect of flow rate on selectivity in RPC gradient elution. Sample: peptides from the tryptic digest of myoglobin. Conditions: 80 × 6.2-mm C_8 column (5-μm particles); 10–70% ACN-buffer gradient in 60 minutes; (*a*) complete chromatograms; (*b*) expanded portions of chromatograms of (*a*) (indicated in [*a*] by brackets and arrows). Adapted from [26].

Plots of log *k* against %B as in Figure 13.11 adequately represent polypeptide retention data for B-solvent concentrations less than 50–80%. For > 50% B, a reversal of retention may occur occasionally (e.g., Fig. 13.13). Such behavior can be observed for both peptides and proteins, and has been interpreted as a transition from RPC for low %B to hydrophilic interaction chromatography (HILIC; Sections 8.6, 13.4.4) at high %B. That is, at sufficiently high %B retention increases with further increase in %B. This phenomenon can have two practical consequences for the chromatographer. First, the extension of a gradient to organic modifier concentrations above 60–70% may sometimes be unproductive for the complete elution of polypeptides from RPC columns. Second, attempts to strip contaminating

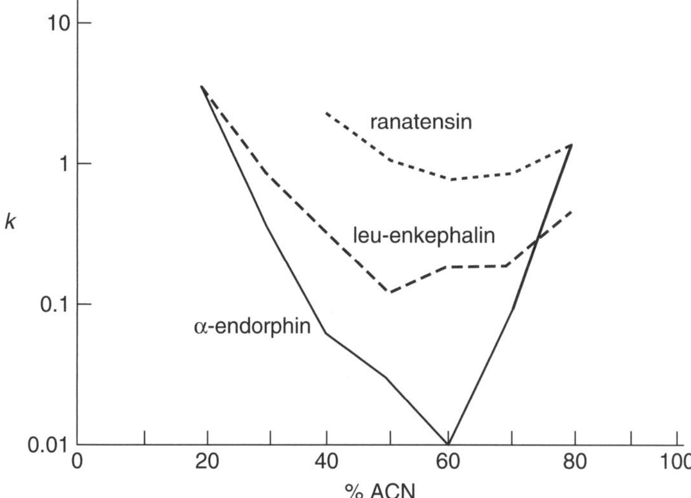

Figure 13.13 Mixed-mode retention in the RPC separation of different peptides. Conditions: C_{18} column, acetonitrile-buffer mobile phases (buffer is 20 mM ammonium acetate). Adapted from [27].

proteins bound to a RPC column by the use of a high %B mobile phase *may* not work—which is *not* to say that a gradient to 100% B will often be unsuccessful. The specific behavior of Figure 13.13 is more likely to be observed for older type-A columns (Section 5.2.2.2) and positively charged solutes. Our own experience with modern type-B RPC columns suggests that increased retention at higher % B is generally unlikely.

13.4.1.5 Effect of Polypeptide Conformation

The retention time of a small peptide can be estimated from its amino acid composition (Section 2.7.7 and Eq. 2.33). As polypeptide length increases beyond about 50 residues, however, such predictions become increasingly unreliable [23, 28, 29], suggesting that polypeptide conformation (which becomes more important for larger molecules) can play an important role in RPC retention and separation. For peptides that contain proline, the slow interconversion of *cis* and *trans* configurations of this amino acid can give rise to peak broadening [30] or even complete resolution of the two conformers [31]. The combination of low pH, the presence of organic solvent in the mobile phase, and hydrophobic RPC columns creates a denaturing environment for polypeptides, which is further enhanced at higher temperatures.

Conformational changes during migration of a protein through the column can have an adverse effect on protein separation. The injection of a protein into a hydrophobic RPC column will result in a total loss of quaternary structure, and partial or complete loss of tertiary structure. Denaturation of the protein exposes hydrophobic residues normally sequestered within the interior of the native protein, with a resulting increase in protein retention. If partial denaturation occurs *prior* to chromatographic migration—and if further denaturation is relatively slow—the resolution of native from fully denatured protein can result in the appearance of

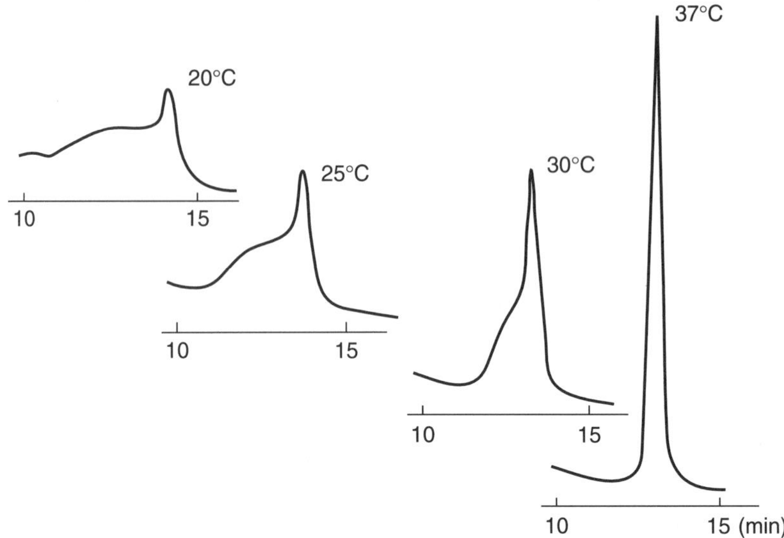

Figure 13.14 Effect of temperature on peak shape for ribonuclease. Conditions: 100 × 4.6-mm (10 μm) LiChrospher C4 column; solvent A, 10 mM H_3PO_4, pH-2.2; solvent B, 55/45 H_2O/1-propanol + 10 mM H_3PO_4; 5–85% B in 30 min; flow rate, 1 mL/min. Adapted from [32].

two distinct peaks. If relatively slow denaturation occurs *during* migration, the two forms may overlap, resulting in the appearance of a broad, misshapen peak. The slow interconversion of conformers can also cause increased peak broadening without the appearance of two peaks. Slow denaturation is associated with poor recovery of sample mass and "ghost peaks" (elution of the same sample in subsequent injections).

Partial denaturation during RPC separation as a function of temperature [32] was first demonstrated for the small protein ribonuclease (Fig. 13.14). At low temperatures, the chromatogram exhibits a sharp, late-eluting peak with a broad, early-eluting shoulder. Spectral measurements indicate that the sharp peak is the denatured molecule. The broad shoulder represents native protein generated by refolding of ribonuclease during elution. Chromatography at elevated temperatures favors full denaturation immediately upon injection, with conversion of the protein into a single, more hydrophobic (and more retained) species.

Separation problems in RPC due to conformational effects can be minimized by the use of conditions that favor the denatured state; for example, operation at elevated temperatures with a more hydrophobic column and low-pH mobile phase. The use of sample pretreatment with denaturing conditions and separation at 60°C has been used to obtain acceptable peak shapes and better recoveries for a variety of proteins differing in size and isoelectric points [33, 34]. See also the example of Figure 6.21 of [3], where recovery and peak shape for a recombinant protein was monitored from 50 to 90°C; peak shape continued to improve at higher temperatures, but recovery was a maximum at 70–80°C. In applications where denaturation is undesirable, selection of a gentler chromatographic mode such as ion exchange or hydrophobic interaction is recommended.

13.4.1.6 Capillary Columns and Nanospray Ionization Sources

RPC separation of protein digests coupled on-line to nanospray MS is the preferred method for protein identification and structural characterization in proteomics studies. In most cases sample amounts are limited and very high sensitivity is required. Capillary RPC columns are used for these applications, both to achieve decreased peak volumes and to improve nanospray performance. Capillary columns with internal diameters of 50 to 150 μm can provide a theoretical 2000- to 10,000-fold improvement in sensitivity relative to a conventional 4.6-mm i.d. column. These columns are operated at flow rates of 100 to 500 nL/min, which greatly improves the efficiency of desolvation and ionization in the nanospray process. Capillary columns must be used with an ultra-low dead-volume solvent-delivery system that is capable of providing precise and accurate flow at these low flow rates, so as to achieve repeatable and otherwise acceptable gradients. Dedicated nanoflow HPLC pumps are commercially available which meet these performance requirements (Section 3.5.4.1). Alternatively, a conventional HPLC pumping system can be configured with a splitting device to reduce the flow rate (the system dead-volume downstream from the splitter must be reduced to a minimum, through the use of capillary tubing and fittings).

Protein digests often contain salts which can contaminate the ionization source. Salts can be removed by installation of a peptide-trapping column and switching valve between the injector and the analytical column. Following injection, the trapping column is switched to waste to remove salts and contaminants, then switched to the analytical column for elution of peptides to the mass spectrometer. An alternative approach is the use of a vented column configuration [35]. In this system the trapping column is connected directly to the analytical column with a cross fitting. One arm of the cross serves as the high voltage connection for the nanospray source and another arm is connected to a valve that can be directed to waste (during injection and salt removal) or closed (during peptide elution).

13.4.1.7 RPC Method Development

Suggested starting conditions for the development of an RPC method for a peptide or protein sample are listed in Table 13.3. If the separation using these starting conditions is inadequate, peptide separations can be improved using the strategy outlined below. Especially for the case of proteins, issues of peak shape, carryover, and reproducibility must be addressed before moving forward with separation optimization. Poor protein peak shape often indicates the presence of conformers or multiple species under the starting conditions. Improving peak shape may require conditions that drive the protein into a single conformation, for example, denaturing conditions as a result of elevated temperature or the use of mobile phase additives, such as surfactants. Such conditions will usually eliminate the problem of carryover as well. Confirmation of injection-to-injection reproducibility, linearity, and elimination of carryover can be expedited by using short gradients for this purpose.

Once adequate peak shape and reproducibility have been achieved, the preferred strategy for optimizing the separation is to improve selectivity, for example, by changing temperature and gradient time—as described for small molecules in Section 9.3. Alternatively, computer simulation (Chapter 10) can be a more rapid and effective method for selecting the best separation conditions. In the example of

Table 13.3

Initial Conditions for RPC Method Development

Condition	Values for Different Samples		
	Peptides	Proteins	
Sample	$1 < M < 5$ kDa	$5 < M < 20$ kDa	$M > 20$ kDa
Sample treatment prior to injection	None	Add 8 M urea, store for 30 min	Add 8M urea, store for 30 min
Column[a]	150 × 4.6-mm, type-B C_{18} (8–12 nm pore-diameter), 3-μm particles	150 × 4.6-mm, type-B C_{18} (12–30 nm pore-diameter), 3-μm particles	50 × 4.6-mm, type-B C_4 (≥ 30 nm pore-diameter), 3-μm particles
Solvent A	0.1% TFA—water	0.1% TFA—water	0.1% TFA—water
Solvent B	0.10% TFA—ACN	0.10% TFA—ACN	0.10% TFA—ACN
Gradient range	0–60% B	5–100% B	5–100% B
Temperature	30–35°C	30–35°C[b]	30–35°C[b]
Flow rate (mL/min)	2.0	1.0	0.5
Gradient time (min)	25	50	50
k^*	2	1	1
%B/min	2.4	1.2	1.2
Value of S assumed	25	40	70

[a]Columns should be stable at low pH and temperatures ≤ 60°C; other column lengths, diameters and particle sizes can be used, in which case gradient time and flow rate should be adjusted to maintain similar values of k^* with acceptable pressure drop. The choice of ligand length (C_8, C_{18}) is less critical.
[b]Higher temperatures (e.g., 60–80°C) can be desirable for some protein samples, especially those with $M > 20$ kDa; column stability for these conditions should be verified before the use >50°C and pH < 2.5.

Figure 13.10, the initial four runs *a–d* can be used to predict the best combination of temperature and gradient time for optimal resolution (Fig. 13.10*e*). Once acceptable peak spacing is achieved, the gradient range can be trimmed to shorten overall separation time. For example, the gradient can be initiated at a %B-value just prior to elution of the first peak, and terminated at the %B-value just after elution of the last peak (Fig. 13.10*f*).

If no combination of gradient time and temperature yields acceptable resolution, the next step could be a change in the column or the composition of the A- or B-solvent; for example, an increase in TFA concentration, a change in pH, or the substitution of isopropanol for acetonitrile as B-solvent. After one or more of the latter changes in conditions, the four-run change in both gradient time and temperature (as in Fig. 13.10*a–d*) should be repeated, using the new conditions for other variables.

Finally, segmented gradients can be used to address particular separation problems. In the case of strongly adsorbed contaminants that must be removed from the column prior to the next sample injection, a final, steep gradient to 100% B can be used to clean the column. In the case of complex samples with clusters of poorly resolved components, a segment with a shallow gradient ramp can be inserted to improve their separation. This strategy is of limited value for small molecules; it is more likely to be successful for peptides, and especially for proteins [36].

13.4.2 Ion-Exchange Chromatography (IEC) and Related Techniques

Ion-exchange chromatography (IEC) can be used for analytical separations of peptides and proteins, but it is more frequently employed for the isolation and purification of proteins from laboratory to process scale [37]. The most important advantages of IEC for protein isolation include (1) the tendency of proteins to maintain their native conformation and biological activity during separation, (2) the relatively high binding capacity of IEC packings, and (3) high mass recoveries. Features (1) and (2) are favored by the use of mobile phases of moderate ionic strength and near-physiological pH. The most important feature of IEC for analytical applications is its unique selectivity relative to other modes of column chromatography. Three other chromatographic techniques (chromatofocusing, hydroxyapaptite chromatography, and immobilized-metal affinity chromatography; Sections 13.4.2.3–13.4.2.5) are related to IEC in that they also rely on ionic interactions between the column and sample.

Ion exchange is based on the reversible electrostatic interaction of charged groups on the packing with oppositely charged groups on the polypeptide (Section 7.4.1). The retention of a peptide or protein molecule P occurs as a result of the displacement of mobile-phase counterions X^+ by P^{+z} (or X^- by P^{-z}).

$$\text{(cation exchange)} \quad P^{+z}(m) + z(R^-)X^+(s) \Leftrightarrow (R^-)_z P^{+z}(s) + zX^+(m) \qquad (13.4)$$

$$\text{(anion exchange)} \quad P^{-z}(m) + z(R^+)X^-(s) \Leftrightarrow (R^+)_z P^{-z}(s) + zX^-(m) \qquad (13.5)$$

Here R^- or R^+ refers to a charged group (ligand) in the stationary phase, z is the charge on the protein molecule P^{+z} or P^{-z}, and (m) or (s) refers to a molecule in the mobile or stationary phase, respectively. A monovalent counter-ion X^+ or X^- is assumed in Equations (13.4) and (13.5). In cation-exchange chromatography, an anionic ligand (R^-) associates with cationic sites on the polypeptide. In anion-exchange chromatography, a positively charge ligand (R^+) binds to anionic groups on the polypeptide. Sample retention can be varied by altering the charge on the solute or—in some cases—the column ligand (Section 7.5.4) via a change in mobile-phase pH. A more common elution strategy is to vary the concentration of X^+ or X^- in the mobile phase, as discussed in Section 7.4.1, or to use gradient elution where the concentration of X^+ or X^- increases during the gradient (salt gradient). For reasons discussed below, the *apparent* charge $\pm z$ on the protein in Equations (13.4) and (13.5) can differ from the *net* charge.

Charged groups at the protein amino and carboxyl termini (as well as on amino-acid side-chains) strongly affect IEC retention. These groups have pK_a values between 2 and 13 (Table 13.4 and Fig. 13.1), so retention will be strongly dependent on mobile-phase pH. Note that the *local* environment of a charged amino-acid residue in a protein (i.e., surrounding mobile phase, and adjacent amino-acid groups within the molecule) can shift its apparent pK_a from the nominal value for the free amino acid. Charged post-translational modifications such as sialic acid, phosphate, and sulfate groups can also contribute to ionic retention.

Table 13.4

pK_a Values for Charged Amino Acids

Residue	pK_a in Amino Acid	pK_a in Protein
Terminal amino	8.8–10.8	6.8–7.9
Arginyl	12.5	≥12
Histidyl	6.0	6.4–7.4
Lysyl	10.8	5.9–10.4
Terminal carboxyl	1.8–2.6	3.5–4.3
Aspartyl	3.9	4.0–7.3
Glutamyl	4.3	4.0–7.3

Source: Reprinted from [37] with permission from Validated Biosystems.

The net charge $\pm z$ on a protein will depend on mobile-phase pH. At the pH where the sum of positive and negative charges are equal (the isoelectric point, or pI), no net IEC retention is expected. At pH values below its pI, a protein will have a net positive charge and should bind to a cation exchanger. At pH values above its pI, the protein will possess a net negative charge and should bind to an anion exchanger (Fig. 13.15*a*). This simple model can serve as a guide for selecting a column and mobile-phase pH, but in practice, a protein may exhibit anomalous binding behavior at or near its isoelectric point (Fig. 13.15*b*). The reason is that the charge on a protein may not be homogeneously distributed across its surface but instead clustered into different regions (contact areas) on the molecule (Section 13.3.2). As a result regions of excess charge can appear at different parts of the molecule, and these regions can interact with the column more or less independently of each other. Anomalous binding behavior can include binding at the isoelectric point, binding to an anion exchanger below the protein pI, or binding to a cation exchanger above the pI. Similarly a protein may fail to bind to an anion exchanger above its pI or to a cation exchanger below its pI. For example, β glucosidase (pI = 7.3) binds at pH-7.3 on an anion exchanger but fails to bind to a cation exchanger until the mobile-phase pH is two units below its pI (Fig. 13.15*b*). Chymotrypsin, with a pI of 9, binds at pH-9 on both an anion and a cation exchanger more than (Fig. 13.15*c*).

As a guideline, anion-exchange separations are often carried out at 1 to 1.5 pH units above a protein's pI, and cation-exchange separations at 1 to 1.5 pH units below the pI. Solubility and stability properties of the protein(s) of interest can limit the allowable ionic conditions for the separation. Virtually all protein purification schemes used in the biopharmaceutical industry contain one or multiple anion- and/or cation-exchange steps.

Since only a limited number of charged residues on the protein surface may interact with the stationary phase, small differences in the nature and positions of these charged residues can profoundly affect selectivity in ion-exchange chromatography [14]. In addition amino-acid substitutions within the interior of the protein may alter its conformation and affect ion-exchange selectivity indirectly by changing the positions of charged groups on the protein surface.

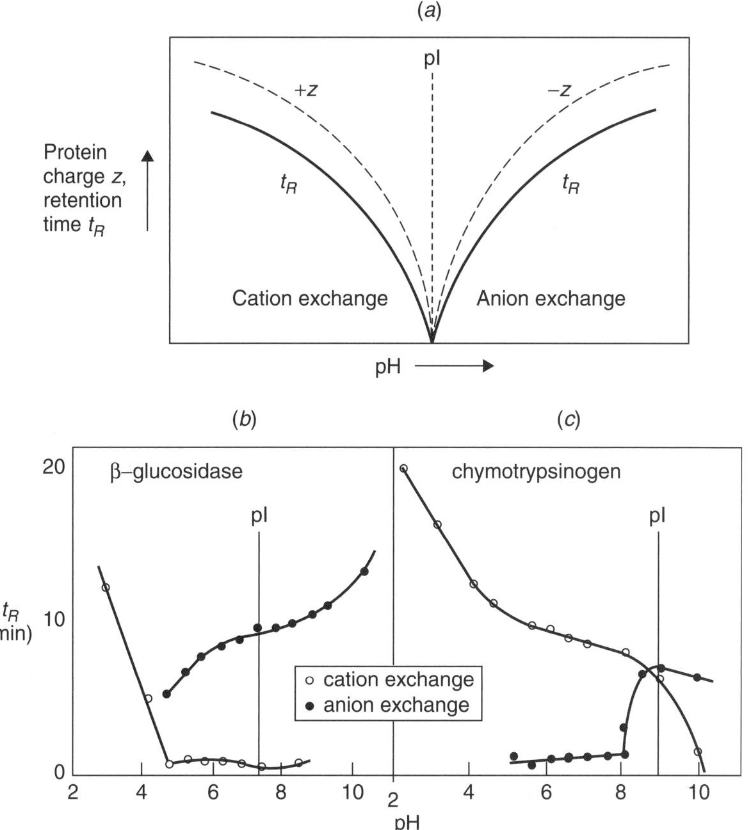

Figure 13.15 Protein retention on ion exchangers as a function of pH. Ideal behavior (*a*); actual behavior of β-glucosidase (*b*) and chymotrypsinogen (*c*). Adapted from [38].

13.4.2.1 Column Selection

Column-selection criteria include:

- particle size and pore diameter
- support composition
- ligand type
- ligand density

Particle size and pore diameter considerations are the same as described in Sections 13.3.1.1 and 13.3.1.2 for RPC.

Support Composition. The first supports for high-performance IEC were silica based, for the same reasons that silica was chosen for other modes of HPLC. However, early silica packings were unstable under preferred ion-exchange conditions (physiological pH, moderate salt concentration) and were gradually replaced by polymeric packings based on polystyrene-divinyl benzene or polymethacrylate. Although modern silica-based packings exhibit improved stability at neutral to alkaline pH, many labs continue to use polymer-based columns. For process chromatography, large-particle supports composed of semi-rigid gels such as cross-linked dextran, agarose, or polyacrylamide are preferred for their lower cost, and because they can withstand highly alkaline cleaning steps for the removal of endotoxins and other biological contaminants.

Ligand Type and Density. Within the respective categories of cation and anion exchange, IEC packings can be further divided into "strong" or "weak"—depending on the pK_a of the stationary-phase ionic ligand. Consequently the charge on the column and its binding capacity can vary with mobile-phase pH (Fig. 13.16). *Strong* ion-exchangers have pK_a values outside the normal pH-operating range of the column, and are therefore fully ionized—regardless of mobile-phase pH; see Table 13.5 for some common examples of IEC column ligands. Ionic groups in strong ion-exchangers include $-SO_3^-$ for cation exchange and $-N(CH_3)_3^+$ for anion exchange. *Weak* ion-exchangers have pK_a values *within* the operating range of the column, so their ion-exchange capacity varies with mobile-phase pH. Examples of

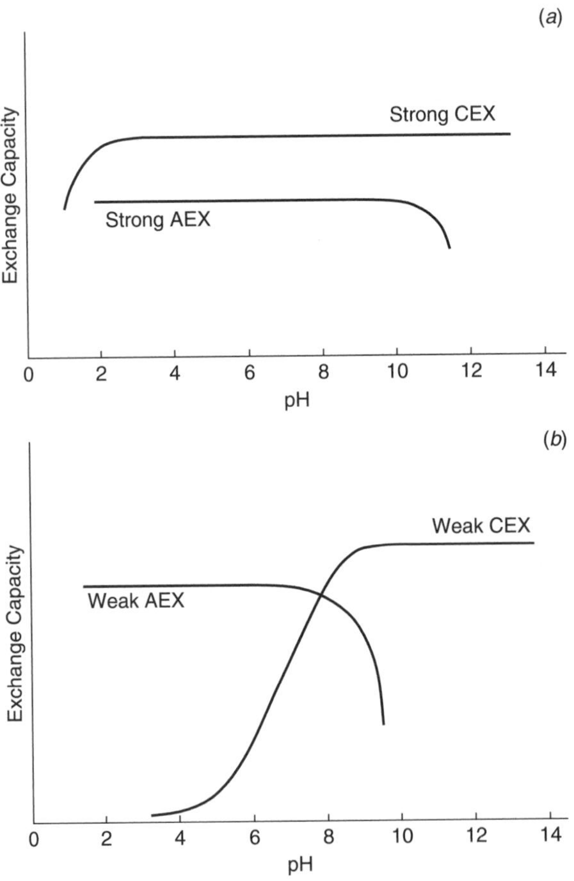

Figure 13.16 Capacities of ion-exchange groups. (*a*) Strong ion exchangers; (*b*) weak ion exchangers. Adapted from [39].

Table 13.5

Strong and Weak Ion-Exchange Ligands

Anion Exchange (AEX)	Cation Exchange (CEX)
Weak	Weak
DEAE (diethylaminoethyl) $-O-CH_2-CH_2-N^+H(CH_2CH_3)_2$	CM (Carboxymethyl) $-O-CH_2-COO^-$
PEI (polyethyleneimine) $(-NHCH_2CH_2)_n-N(CH_2CH_2-)_{n'}$ with $CH_2CH_2NH_2$ on N	
Strong	Strong
Q (quaternary ammonium) $-CHOH-CH_2-N^+(CH_3)_3$	S (sulfonate) $-CH_2-CH_2-CH_2-SO_3^-$

weak IEC groups include $-N(C_2H_5)_2H^+$ for weak anion exchange and $-COO^-$ for weak cation exchange. Weak anion-exchange columns of polyethyleneimine consist of a dense polymeric coating onto a silica support, yielding a column with high capacity and good stability under alkaline conditions. Strong ion-exchangers are often preferred, as their exchange capacity is independent of mobile-phase pH and their behavior is more predictable. The binding capacity of ion-exchangers depends on the surface area of the support and its charge density (μmoles/m^2). Typical ion-exchange capacities (i.e., for maximum uptake of sample by the column) for large-pore silica or polymer-based columns are in the range of 30 to 120 mg protein per milliliter of packing.

The linker group that joins the ion-exchange group to the support can contribute to the chromatographic properties of the column. For example, hydrophobic groups in the linker may participate in hydrophobic (reversed-phase) interactions with the solute. Such interactions can account for differences in column selectivity among different vendors who use the same ion-exchange functionality. *Tentacle* IEC stationary phases have a flexible hydrophilic linker (the "tentacle") that connects the charged group to the support [40]. These columns improve access of the protein to the charged group of the packing, thus enhancing binding capacity. In addition tentacle columns may exhibit reduced nonspecific interaction, improved binding kinetics, and reduced protein denaturation.

13.4.2.2 Mobile-Phase Selection

As noted above, control of retention (solvent strength) is usually achieved by varying the concentration of a displacing salt (counter-ion), rather than by changes in mobile-phase pH. Conditions that affect selectivity include:

- column (Section 13.4.2.1)
- mobile-phase buffer
- counter-ion salt type (as in Fig. 13.17)

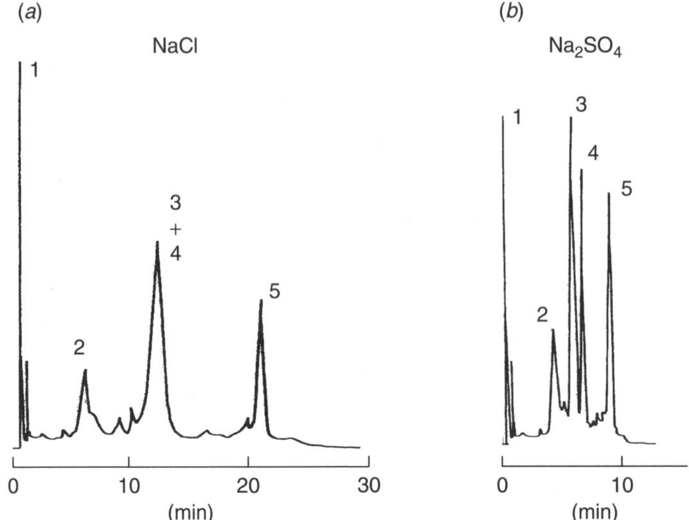

Figure 13.17 Effect of salt type on anion exchange separation of five proteins. Conditions: 50 × 4-mm Shim-pack WAX-2 column (Shimadzu); 0–0.5M of indicated salt in 20 min; pH-8 phosphate buffer; 1 mL/min. Adapted from [41].

- gradient steepness
- organic B-solvent (if used)
- other mobile-phase additives (especially surfactants)
- temperature

See also the discussion of Section 7.5.

Mobile-Phase Buffer. Achieving the desired retention and selectivity requires a careful selection and control of the mobile-phase pH. For a cation-exchange column, a mobile-phase pH near 6 is a good starting point, while a mobile-phase pH of 8 is appropriate for an anion exchanger. For good buffering capacity, the buffering agent should have a pK_a value within roughly 1.0 units of the target pH (Section 7.2.1.1), and a concentration of 0.02 to 0.1 M. Common buffers used for IEC are listed in Table 7.1. Note that some of these buffers absorb strongly at shorter UV wavelengths, especially if higher concentrations are used.

Counter-Ion. The most common elution strategy in IEC is the use of a gradient of increasing concentration of the counter-ion. The relative strength of different counter-ions follows their ranking in the Hofmeister series [37, 42]; see Table 13.6 or a similar series in Section 7.5.2. However, gradients that involve an increase in NaCl are most often used for both anion and cation exchange. Note that chloride is corrosive for stainless steel at low pH (<5) and should be removed from the HPLC system after use. However, special-purpose HPLC systems have been designed that enable the use of chloride under acidic conditions.

Organic Solvents and Surfactants. Organic solvents (e.g.. 1–10% methanol, propanol, or acetonitrile) can be added to the mobile phase to suppress hydrophobic

Table 13.6

Hofmeister Series of Lyotropic and Chaotropic Ions [36]

Increasing lyotropic (salting out) effect
SCN^- (least) < ClO_4^- < NO_3^- < Br^- < Cl^- < COO^- < SO_4^{2-} < PO_4^{3-} (most)
Increasing chaotropic (salting in) effect
Ba^{2+} (most) > Ca^{2+} > Mg^{2+} > Li^+ > Cs^+ > Na^+ > K^+ > Rb^+ > NH_4^+ (least)

Source: Data from [36].

interactions with the support or linker groups, and to decrease peak broadening or tailing (addition of as much as 50% organic solvent may be required in some cases, as in the example of Fig. 11.15 of [43]). Nonionic surfactants can also be used for the same reasons. Either of these mobile-phase additives can also maintain the solubility of very hydrophobic solutes such as membrane proteins. Ionic surfactants can not be used in ion-exchange chromatography.

13.4.2.3 Chromatofocusing

Chromatofocusing is a specialized form of IEC in which proteins are eluted from the column with a pH gradient [44–49]. Chromatofocusing is unique in that the pH gradient is formed within the column, by means of a single mobile phase that is a complex mixture of different buffering species. Although chromatofocusing can be performed with cation- or anion-exchangers, commercially available products are limited to anion exchange [48]. At the start of separation, proteins are retained by the anion exchanger, which has been pre-equilibrated at high pH for maximum retention of the sample. Then a low-pH buffer mixture is used as mobile phase, which, upon moving through the column, progressively titrates the charge on the column so that pH increases along the column, from inlet to outlet. Proteins migrate down the column in response to the changing pH and elute at or near their isoelectric points—a pH at which they can no longer bind to the exchanger. Elution is in order of descending protein pI values. Chromatofocusing is characterized by very high capacity, so it is useful for preparative separations. The technique is also capable of very high resolution, by virtue of focusing effects that generate sharp peaks 0.04 to 0.05 pH units in width. As is the case for conventional IEC (Figs. 13.15*b*,*c*), a protein can elute from a chromatofocusing column at a pH that is significantly different from its pI.

Successful and reproducible chromatofocusing separations depend on the use of buffers that contain multiple species, whose pK_a values span the range of the pH gradient, and that can achieve effective buffering across this range. Commercial chromatofocusing buffers are composed of a mixture of ampholytes (substances that may act as either an acid or a base). Alternatively, a combination of biological buffers such as Good's buffers [50] can be used. The ionic strength of the elution buffer must be kept low, in order to minimize salt-mediated elution (displacement of proteins by counter-ions). The improved resolution (or faster separation) of proteins whose pI values fall within a narrow range of values can be achieved by narrowing the pH range of the ampholyte or buffer blend (similar to a decrease in $\Delta\phi$ in gradient elution). Strong ion-exchange columns are preferred for chromatofocusing, since they are fully ionized—regardless of pH.

One shortcoming of chromatofocusing is the reduced solubility of proteins at their isoelectric point, a limitation which is exacerbated by the low ionic strength of the elution buffer. Protein solubility can be enhanced by an increase in salt concentration, but this will increase mobile-phase strength and compromise the separation. A preferred strategy for dealing with protein precipitation is the addition of zwitterions to the elution buffer. Additives such as taurine, glycine, and betaine promote protein solubility and can be used in concentrations up to 2M without affecting the ionic strength of the buffer. The addition of urea at concentrations of 1 to 2M also helps solubilize proteins; nonionic and zwitterionic surfactants may be used as well. Note, however, the tendency of urea to decompose to carbamates, which can covalently modify a protein.

Chromatofocusing is able to resolve isoforms of proteins that have different charge states, for example, post-translationally modified proteins that differ in the number of sialic acids or phosphate groups. The resolution of isoforms can be a limitation, if the goal is protein purification. The target protein is then resolved into multiple peaks, which dilutes the target protein and increases the risk of co-elution with sample contaminants. On the other hand, this characteristic of chromatofocusing can be an advantage, if only the characterization of isoforms is desired.

13.4.2.4 Hydroxyapatite Chromatography

This technique is frequently used in process chromatography for protein purification and the removal of contaminants [37]. Hydroxyapatite (HA) is a crystalline material composed of $Ca_{10}(PO_4)_6(OH)_2$ that serves both as the support and the stationary phase [51]. The multifunctional surface consists of positively charged pairs of calcium ions (C-sites) and clusters of six anionic oxygen atoms associated with triplets of phosphate ions (P-sites). The C- and P-sites and hydroxyls are distributed in a fixed pattern on the crystal surface [51–53], as illustrated in Figure 13.18. Early preparations of HA were unstable, but modern HA materials are sintered at high temperature to form ceramic hydroxyapatite (CHT), which is stable under chromatographic conditions. Columns packed with either 5- or 10-μm CHT particles are available for both analytical and preparative applications.

Protein interactions with CHT are complex (Fig. 13.18). Electrostatic interactions include attraction of protonated amino groups by P-sites and repulsion by C-sites (Fig. 13.18*a*). Similarly ionized carboxyl groups are attracted by C-sites and repelled by P-sites (Fig. 13.18*b*). Although the initial attraction of carboxyls to C-sites is electrostatic, the actual binding involves formation of much stronger coordination complexes between C-sites and clusters of protein carboxyl-groups [37]. Protein phosphate-groups bind C-sites even more strongly than protein carboxyl-groups. The selectivity of CHT for basic proteins is distinct from that of conventional cation exchange, due to the repulsion of amines by C-sites. Binding of weakly basic proteins can be enhanced by the addition of a low concentration of phosphate, which suppresses C-site repulsion of amines but does not block their interaction with P-sites [54]. Basic proteins can be eluted by gradients of sodium chloride or phosphate; a final salt concentration as high as 0.5M may be required. Although the

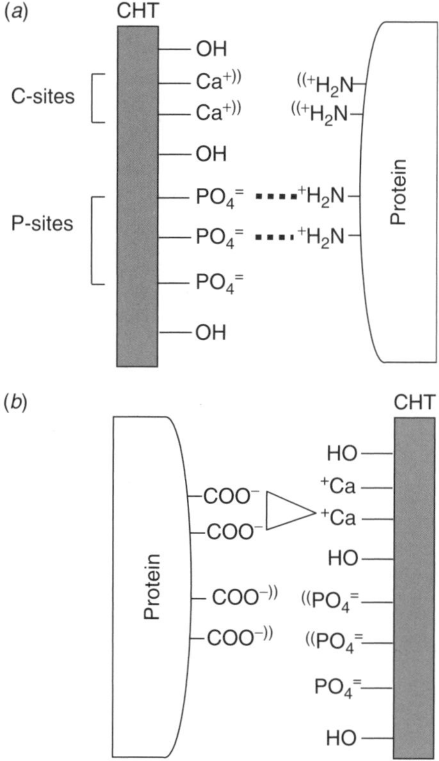

Figure 13.18 Binding to ceramic hydroxyapatite (CHT) of a basic protein (*a*) and an acidic protein (*b*). Double parenthesis indicate repulsion, dotted lines indicate ionic bonds, and triangular linkages indicate coordination bonds. Adapted from [37].

binding of basic proteins increases at lower pH, CHT is unstable below pH 5. Acidic proteins cannot be eluted with sodium chloride—even at concentrations > 0.3M; their elution requires the use of phosphate, citrate, or fluoride. This characteristic of CHT permits separation of basic proteins with an initial NaCl gradient, followed by elution of acidic proteins with a phosphate gradient.

CHT typically provides excellent recovery of protein mass and biological activity; it is used for protein purification from laboratory to process scale. The unique selectivity of CHT can enable the resolution of closely related species such as protein variants and glycoforms. It is used in the biopharmaceutical industry for the purification of antibodies and removal of contaminants such as endotoxins, nucleic acids, and viruses. The stability of CHT toward concentrated base, organic solvents, and chaotropes enables aggressive cleaning regimes to be applied after use.

13.4.2.5 Immobilized-Metal Affinity Chromatography (IMAC)

This separation mode, also known as metal-interaction chromatography (MIC), is based on the differential interaction of proteins with a metal ion [55–57]. The metal ion is immobilized by chelating groups that are attached to the support via a linker; see the example of Figure 13.19, which includes the various steps in its use. Several

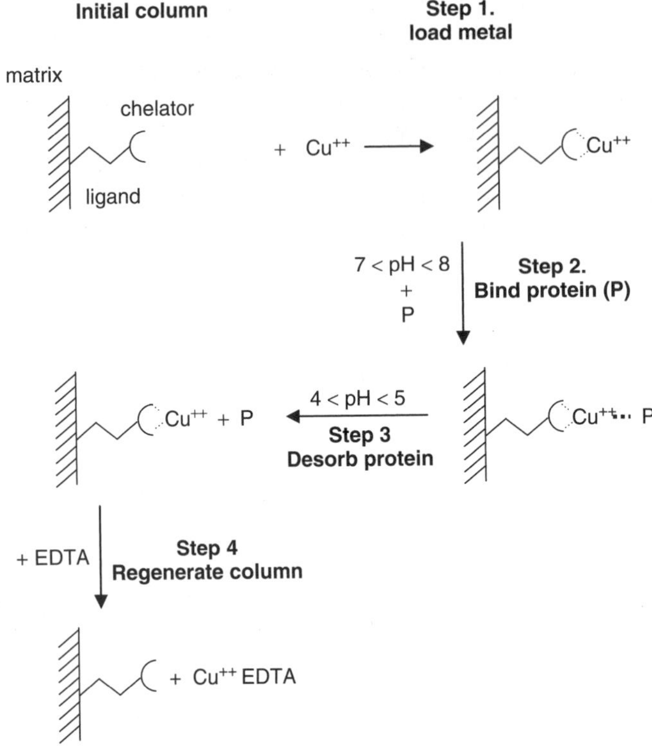

Figure 13.19 Steps in the use of IMAC. Adapted from [56].

amino-acid side chains in the protein can form coordination complexes with metals, so IMAC is a general method for protein separation. The primary interaction in IMAC is with the imidazole group of histidine in its unprotonated form [58]; the strength of metal binding by different amino-acid groups in the protein molecule decreases in the following order:

$$\text{his} > \text{trp} > \text{tyr} > \text{phe} > \text{arg} \sim \text{met} \sim \text{gly}$$

Cysteine residues can bind metals, but they may not be available on the protein surface in the reduced state, since they readily oxidize in the presence of metal ions [59]. Cysteine-containing proteins may therefore require a reducing environment (addition of 2-mercaptoethanol or dithiothreitol) in order to maintain the cysteine residues in their active (–SH) form. Aromatic residues can contribute indirectly to retention, by enhancing the binding of neighboring histidines [59].

The dominance of histidine binding to IMAC columns has been exploited by genetically engineering polyhistidyl sequences into target proteins for ease of purification. After preferential binding, elution, and recovery of the target protein, the polyhistidine sequence can be cleaved by means of carboxypeptidase A [59]. Phosphoproteins and phosphopeptides bind selectively to IMAC columns chelated with Fe^{+3} and Ga^{+3}, and IMAC has become a key tool in characterizing the phosphoproteome [60].

The high capacity of IMAC columns and the high recovery of protein mass and activity make this technique useful for preparative and process-scale chromatography. For protein purification, IMAC compares favorably with affinity chromatography in terms of binding strength and capacity and has the advantages of stability over a wide range of conditions and use-cycles, relatively mild elution conditions, and modest cost. In process chromatography, IMAC is best used as the initial step in a process sequence so that downstream steps such as IEC or hydrophobic interaction chromatography can eliminate any metals leached from the column during IMAC elution [37] (oxidation of protein residues can be catalyzed by metal ions).

Selectivity in IMAC can be controlled by the choice of:

- chelating ligand (i.e., the column)
- immobilized metal ion
- mobile-phase pH and ionic strength
- any mobile-phase additives used to enhance binding or elute proteins

Chelating Ligand. The chelating ligand–metal complex must be strong enough to be stable, but must also have metal-coordination sites available in order to bind the protein. The metal should be easily removed with a chelating agent such as EDTA, in order to allow column regeneration and conversion to another metallic form. The most common chelating groups used in IMAC are iminodiacetic acid (IDA) and tricarboxymethylethylenediamine (TED), which form tridentate and pentadentate metal complexes, respectively (Fig. 13.20). Although the pentadentate TED has stronger metal affinity, the tridentate IDA-metal complex leaves more metal coordination sites free for solute binding so that IDA-metal columns can exhibit higher protein affinity [55].The metal complex stability for IDA on an agarose support [56] is

$$Cu^{2+} > Ni^{2+} > Zn^{2+} \geq Co^{2+} > Fe^{2+} \gg Ca^{2+}$$

The corresponding affinity of TED for metals is

$$Fe^{3+} > Cu^{2+} > Ni^{2+} > Zn^{2+} \sim Co^{2+} > Fe^{2+} > Ca^{2+}$$

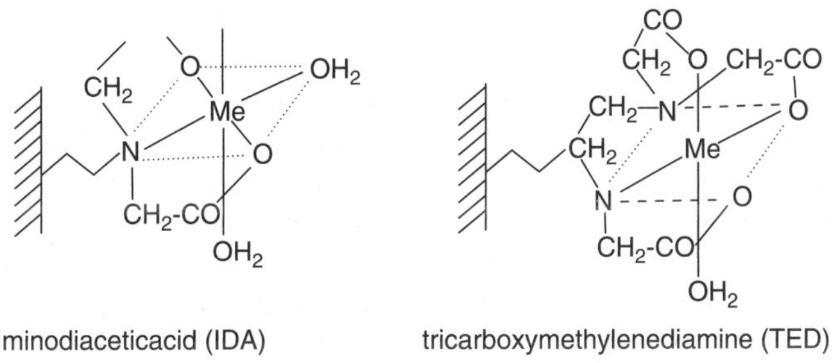

Iminodiaceticacid (IDA) tricarboxymethylenediamine (TED)

Figure 13.20 Structure of two common IMAC ligands. Adapted from [55].

Metal Ion. The selectivity of protein binding in IMAC depends primarily on the type of metal that is complexed with the chelating ligand. Protein–metal chelate interactions include coulombic and coordination bonding, while hydrophobic interactions may occur at high salt concentrations [55]. The most popular metals for IMAC are Cu^{2+}, Ni^{2+}, and Zn^{2+}. This popularity probably reflects the strong affinity of these metals for both the IMAC ligands and for proteins. The affinity of Cu^{2+} for imidazole is 15-fold greater than Ni^{2+}, which is 3-fold greater than Zn^{2+} or Co^{2+} [57]. The optimum pH and binding conditions are specific for a given protein–metal ion pair so they must be determined experimentally.

Protein Binding and Elution. In addition to metal ion complexation, IMAC columns have the potential for ion-exchange, ion-exclusion, and hydrophobic interactions. The initial mobile phase for protein binding and the final mobile phase for elution can be designed to minimize or exploit these effects. Metal-free chelating groups function as cation-exchange sites and can interact with cationic groups on proteins (for an ionic strength < 0.1M). High concentrations of salt (≥ 0.5 M) suppress ion-exchange interactions but promote hydrophobic interactions with the chelating group or its linker. An intermediate ionic strength can suppress ion-exchange interactions and reduce the risk of protein aggregation, particularly for antibodies [37]. A general observation is that the retention of acidic proteins tends to increase with salt concentration, while the retention of basic proteins initially decreases, then increases with increasing salt concentration [55]. Each of the latter separation conditions can be varied in order to optimize separation selectivity.

Binding and elution is strongly affected by mobile-phase pH. Proteins bind most strongly above the pK_a of the histidyl imidazole group ($pK_a \approx 7$), and binding strength is diminished as the pH drops below this value (and the histidine group becomes more ionized). Therefore a common elution strategy is to bind proteins at pH values between 7 and 8, followed by elution with a step or gradient to pH values between 4 and 5, in order to convert histidine residues to the ionized form.

An alternative elution strategy is the use of a displacing agent such as imidazole, histamine, histidine, glycine, or ammonia. The first three agents are equivalent in eluting strength, and generally stronger than the latter two. Displacing agents of increasing strength can be introduced in sequence to elute weakly retained proteins first, followed by strongly retained species [61]. Proteins can also be resolved by using a concentration gradient of a single displacing agent.

Like other separation techniques based on electrostatic interactions, IMAC ligands probe surface groups on the protein. Therefore conditions that alter or disrupt protein conformation can change selectivity in IMAC. The example in Figure 13.21 shows changes in retention and elution order caused by the addition of methanol to the mobile phase possibly reflecting changes in protein conformation). Two proteins differing by only a single, surface histine residue can be resolved by IMAC.

13.4.3 Hydrophobic Interaction Chromatography (HIC)

Hydrophobic interaction chromatography (HIC) was first described by Tiselius [63] in the late 1940s, and later characterized by Porath [64] and Gelotte [65]. Since the 1960s, HIC has been widely used for protein separations with carbohydrate-based

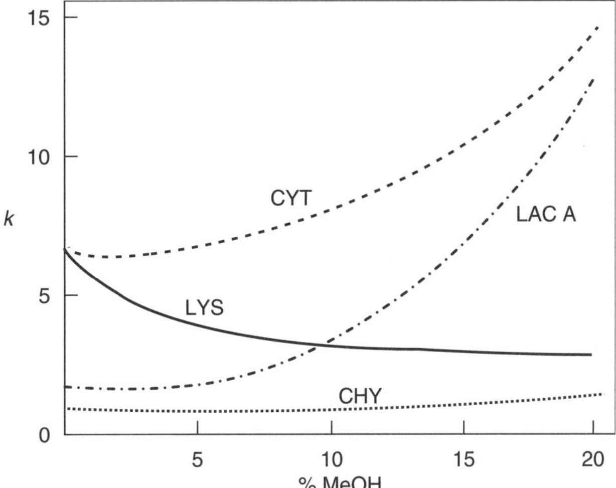

Figure 13.21 Effect of added methanol on IMAC retention. Column: Fe (III)–IDA silica. Mobile phase: varying %-methanol in 25 mM phosphate (pH 6) + 0.15M ammonium sulfate. CYT, cytochrome C; LYS, lysozyme; LAC A, β-lactoglobulin A; CHY, chymotrypsinogen A. Adapted from [62].

packings, and more recently with high-performance microparticulate supports [66, 67]. The principle of HIC is based on the interaction of proteins with mildly hydrophobic ligands in the presence of high concentrations of salt. Proteins largely maintain their conformation under these conditions, and retention results from the interaction of hydrophobic patches on the protein surface with the column ligand ("salting out"). HIC retention is unusual in that it is an entropy-driven process ([68] and see below), and uses a reverse gradient (from high- to low-salt concentration).

HIC is a gentle technique, with proteins eluting in their native conformation without loss of biological activity. It is therefore widely used for the preparative isolation of proteins in laboratory scale-up to process-scale applications. Although HIC and RPC share a retention mechanism based on hydrophobic interactions, the selectivity can be markedly different. The harsh conditions of RPC promote protein denaturation and exposure of internal hydrophobic residues, whereas retention in HIC only involves residues at the protein surface. A very different separation selectivity can therefore be expected.

13.4.3.1 Supports and Ligands for HIC

Both silica- and polymer-based supports are used for HIC, with pore sizes that are large enough to allow penetration of the protein. The support is typically covered with a polymeric, hydrophilic coating (or linker groups) in order to provide a wettable, noninteractive surface; hydrophobic ligands are then attached to the polymer or linker. The ligands are typically short-chain alkyl or phenyl groups, and retention increases with ligand length ([70, 71]; see the examples of Figure 13.22 for three different proteins). Ligands of 1 to 3 carbons in length promote retention at high salt and release at low salt. Longer ligands can cause excessive retention, as well as induce conformational changes.

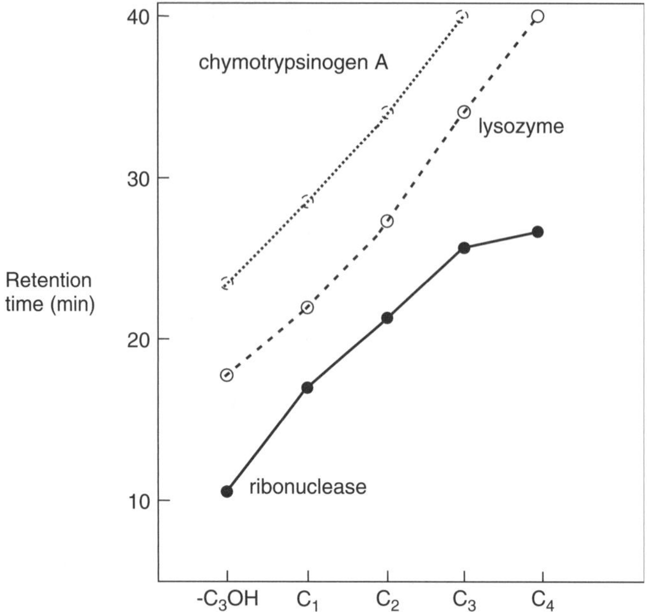

Figure 13.22 Effect of HIC ligand length on protein retention ($-C_3OH$, hydroxypropyl; C_1, methyl; etc.). Adapted from [69].

It should be noted that IEC and IMAC columns can each display HIC behavior under high-salt conditions, due to the hydrophobic contributions of spacer and cross-linking groups [68]. The latter columns can also exhibit multimodal retention behavior, depending on the operating conditions. When IEC and IMAC columns are used in a high-salt, HIC mode, the selectivity may be different from a conventional HIC column due to the contribution of non-HIC retention mechanisms.

13.4.3.2 Other Conditions

The selectivity and retention in HIC separation is influenced by:

- choice of antichaotropic salt and its concentration
- mobile-phase pH
- mobile-phase additives
- temperature

Antichaotropic Salt. The primary consideration in designing a HIC gradient is the selection of the salt and its concentrations at the beginning and end of the gradient. The ability of salts to promote retention in HIC parallels their effectiveness in protein precipitation as given by the Hofmeister salting-out series (Table 13.6); however, ammonium sulfate is the most widely used salt for HIC. Changing the salt type as well as its concentration can provide an opportunity for varying both retention and selectivity—but keep in mind the solubility and purity of the salt. Protein binding is achieved with an initial concentration of 1.5 to 3.0M salt; a reverse gradient to a lower salt concentration (or neat buffer) is then used for elution. For

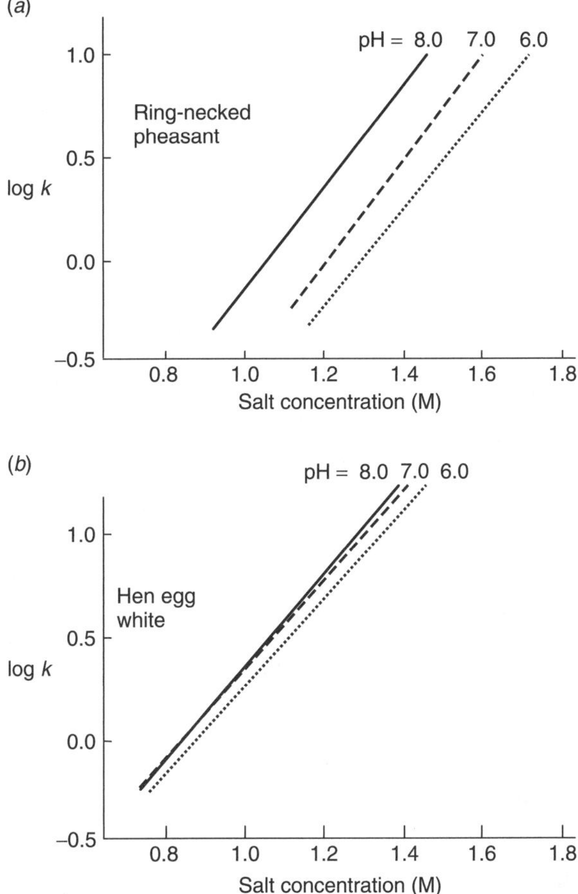

Figure 13.23 Effect of pH on HIC retention of two different avian lysozymes (ring-necked pheasant, hen egg white) with varying ammonium sulfate molality. Adapted from [72].

large-scale preparative applications, the use of ammonium sulfate under alkaline conditions must be approached with caution, as free ammonia can be liberated [37]. In such cases potassium sulfate is an acceptable substitute.

Mobile-Phase pH. This can affect selectivity if acidic or basic amino acids are located within the hydrophobic contact area. This is illustrated by the effect of histidine ionization state on the retention of avian lysozymes in HIC [72]. Ring-necked pheasant lysozyme has histidine residues in the contact region, and retention changes (Fig. 13.23*a*) as pH is varied across the pK_a of the imidazole group ($pK_a = 6$). Hen lysozyme, which has no histidines in the contact region, displays little pH-dependence of retention over the same pH range (Fig. 13.23*b*). A change in mobile-phase pH can therefore be used to separate these two protein variants.

Additives. Retention in HIC is based on hydrophobic interaction, and it therefore should be affected by the addition of surfactants to the mobile phase (which can bind to both the column and the protein, thereby reducing the hydrophobicity of each). Thus the addition of nonionic and zwitterionic surfactants reduces protein retention [68, 73]. Inclusion of surfactants is particularly useful in HIC separations of very hydrophobic species such as integral membrane proteins, which may require surfactants for solubilization. The addition of organic solvent to the mobile phase should also reduce protein retention, but this can be problematic for HIC. Organic solvents can induce conformational changes in the protein, and their use under conditions of high salt also introduces the risk of protein precipitation.

Temperature. HIC retention is entropy driven and therefore increases with temperature—the opposite of the usual effect of temperature on retention. This effect is enhanced by conformational changes at higher temperatures that make internal hydrophobic residues available for increased interaction. Conformational effects can be recognized by peak broadening at temperatures intermediate between native and denaturing conditions [74], for example, between 10 and 35°C in Figure 13.24.

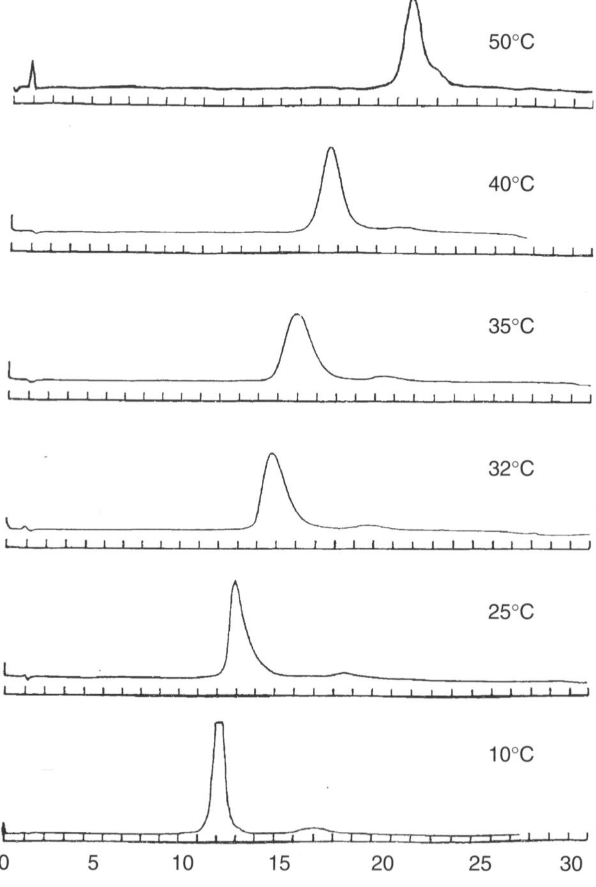

Figure 13.24 Effect of temperature on the HIC elution behavior of Ca $^{2+}$-depleted α-lactalbumin. See text for details. Reprinted from [72] with permission.

13.4.4 Hydrophilic Interaction Chromatography (HILIC)

Water-soluble solutes that are very hydrophilic and uncharged present the chromatographer with a challenge, as they are poorly retained in RPC and unretained by IEC. Hydrophilic interaction chromatography (HILIC; Section 8.6; [75, 76]) provides a solution to this problem. In this technique a polar stationary phase is used with an aqueous-organic mobile phase. In contrast to RPC, the aqueous component of the mobile phase (e.g., water or buffer) serves as the strong solvent and the organic component (usually acetonitrile) is now the weak solvent; that is, retention *increases* as %-organic increases (Fig. 13.25). Note also that retention increases for more polar solutes in Figure 13.25 (Arg [most polar] > p-Ser > Leu [least polar])—again the opposite of RPC. HILIC can be considered as a variant of normal-phase chromatography (NPC; Chapter 8).

In addition to achieving reasonable retention and separation of hydrophilic water-soluble analytes, HILIC has two other advantages. First, mobile phases with >50% acetonitrile are less viscous, which means lower pressures and higher plate numbers. Second, these organic-rich mobile phases are ideal for efficient desolvation in electrospray-ionization LC-MS.

13.4.4.1 Stationary Phases for HILIC

A variety of stationary phases have been used for HILIC [76], including underivatized silica [77], aminopropyl silica [78], amide silica [79], diol silica [80], sulfonated polystyrene-divinylbenzene [81], and poly(2-hydroxyethyl aspartamide) [75]. The use of bare silica as the stationary phase eliminates the problem of ligand bleed in LC-MS, which can occur when bonded phases are used. While the ionization

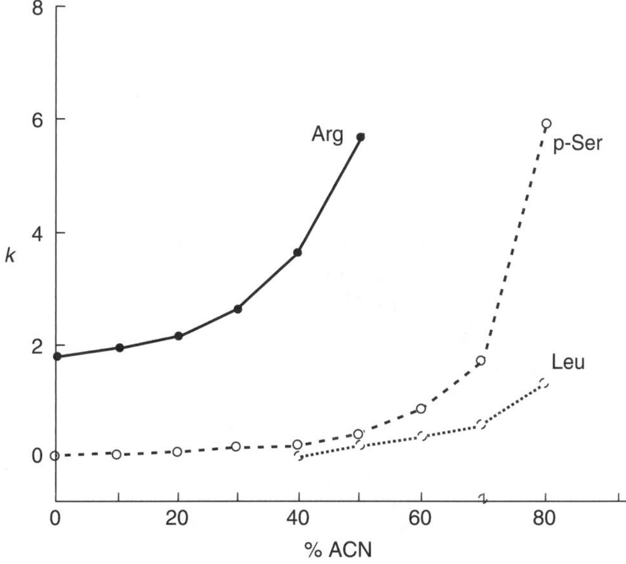

Figure 13.25 Retention behavior of amino acids in hydrophilic interaction chromatography (HILIC). Conditions: cation-exchange column (PolySulfoethyl A) used in HILIC mode; mobile phase, 25-mM TEAP (pH-5.0) with acetonitrile as indicated. Adapted from [75].

of silanols at pH ≥ 7 can introduce ion-exchange interactions and complicate the interpretation of the separation, the use of a mobile-phase pH < 7 avoids this problem. Alternatively, a bonded-phase HILIC column can be used; for example, silica-based amide columns have been used successfully for peptide separations [79]. Silica-based diol columns most closely approach the behavior of underivatized silica for HILIC separations, and minimize the problem of active silanols. Poly(2-hydroxyethyl aspartamide) is a stationary phase that was designed expressly for HILIC [77]; it is prepared by incorporating ethanolamine into a coating of polysuccinimide covalently bonded to silica. The resulting polyaspartamide coating is a derivative of asparagine, the most hydrophilic neutral amino acid. Highly sulfonated polystyrene-divinylbenzene columns can be operated in a HILIC mode using acetonitrile-water eluents [81], with the advantage of stability provided by the polymeric support. However, these columns also function as ion-exchangers. Bare-silica and amide-silica columns were most often used for HILIC at the time this book was written.

13.4.4.2 Mobile Phases for HILIC

Typical mobile phases for HILIC are mixtures of water (or aqueous buffer) and acetonitrile, with water content ranging from 5 to 40% [75]. Commonly used buffering agents are ammonium salts of formic or acetic acid, which are soluble in acetonitrile. These salts are also volatile for compatibility with electrospray ionization–mass spectrometry. The triethylamine salt of trifluoroacetic acid has also been recommended. Although the latter salt is volatile, however, TFA is notorious for causing suppression in electrospray ionization [17].

13.4.4.3 Application of HILIC to Peptides and Proteins

HILIC has been used successfully to separate peptides, and it exhibits a unique selectivity for peptides with hydrophilic post-translational modifications. Glycopeptide sequences with variations in glycan structure have been separated on an amide-silica column [82, 83], while a poly(2-hydroxyethyl aspartamide) column has been successful for the selective isolation of phosphopeptides [84]. The complementarity of HILIC and RPC was demonstrated in a study of the behavior of amphipathic α-helical peptides possessing hydrophobic and hydrophilic faces [85]. Substitutions in the hydrophilic face were shown to have little effect on RPC retention but to cause a change in relative retention in HILIC. Conversely, substitutions in the hydrophobic face affected RPC retention but not HILIC retention. This study also confirmed that different chromatographic contact areas are operative in the two modes.

HILIC has not been widely used for proteins, due to the limited solubility of proteins in high concentrations of organic solvents. The appearance of multiple peaks for single proteins during HILIC has been reported [86, 87] and attributed to on-column denaturation. However, HILIC has been used successfully for the separation of histones (a family of highly basic DNA-binding proteins), including acetylated [88], phosphorylated [89], and methylated [90] histone variants.

13.4.4.4 Electrostatic-Repulsion Hydrophilic-Interaction Chromatography (ERLIC)

This is a variation of HILIC that employs an ion-exchange column operated with a predominantly organic mobile phase [91]. With ERLIC, solutes can be retained via hydrophilic interaction (HILIC mode), even if they have the same charge as the stationary phase. As a result it is possible to simultaneously separate mixtures of acids and bases that might otherwise be difficult to separate by either HILIC or IEC. The ERLIC technique can be regarded as possessing both hydrophilic-interaction and ion-exchange selectivity. The important feature of this combination is the independence of hydrophilic and electrostatic effects, allowing their independent manipulation. This is illustrated in Figure 13.26 by a comparison of HILIC and ERLIC separations of a mixture of acidic, neutral and basic peptides. In the HILIC separation (Fig. 13.26*a*) with a neutral, hydrophilic column, a concentration of acetonitrile (%-ACN) that provides suitable retention of basic peptides yields inadequate retention for acidic and neutral peptides. In the ERLIC separation of Figure 13.26*b* (performed with a weak anion-exchange column at low pH), electrostatic repulsion of the basic peptides reduces their retention, allowing %-ACN to be increased so that all peptides are retained and resolved.

ERLIC has been used for the enrichment and separation of phosphopeptides from a tryptic digest of HeLa cell proteins [92]. Under conventional weak

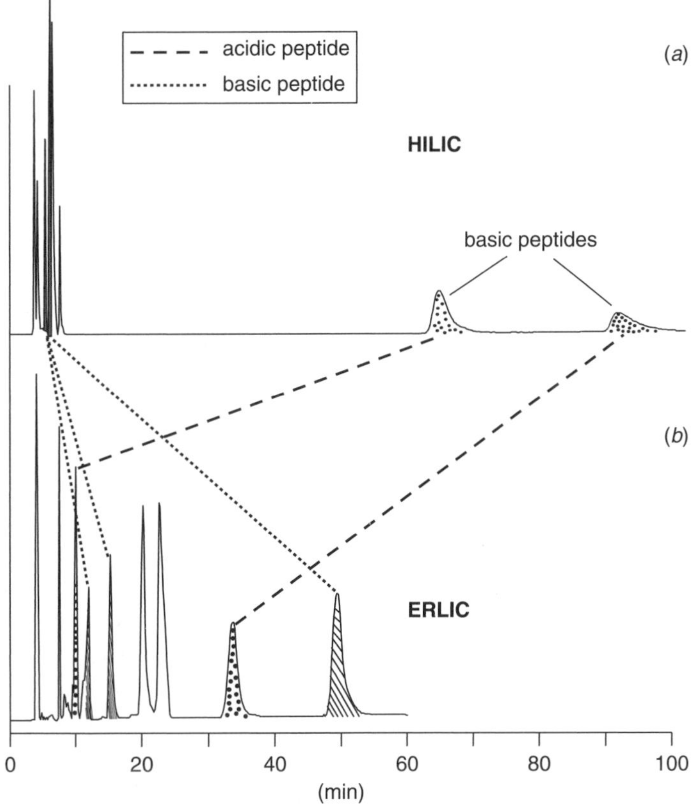

Figure 13.26 Separation of peptide standards by HILIC (*a*) compared with ERLIC (*b*). HILIC conditions: PolyHydroxyethyl A column (PolyLC, Columbia, Maryland); mobile phase, 20-mM Na-MePO$_4$ (pH-2.0) +63% acetonitrile. ERLIC conditions: column, PolyWAX LP; mobile phase, 20-mM Na-MePO$_4$ (pH-2.0) +70% acetonitrile. Adapted from [91].

anion-exchange conditions, the negative charge of a single phosphate group is insufficient to counteract the electrostatic repulsion of *N*-terminal and side-chain amines, and monophosphorylated peptides elute in or near the void volume. When a weak anion exchange column is operated in the ERLIC mode with higher %-ACN, hydrophilic interaction drives an increased interaction of phosphates, enabling the retention and resolution of mono- and multiple-phosphorylated peptides. ERLIC has also been applied to the enrichment and separation of sialylated glycopeptides [93].

13.4.5 Multidimensional Liquid Chromatography (MDLC) in Proteomics

Expression proteomics seeks to characterize the entire complement of proteins expressed in a cell or tissue under defined conditions, with the goal of identifying proteins that are differentially expressed under those conditions. Such proteins are candidates for biomarkers to be used in diagnosing and monitoring a disease state, or as targets for therapeutic intervention. In "bottom-up" proteomics the entire protein content of a cell or tissue lysate is cleaved with a proteolytic enzyme, and the resultant peptides are separated by chromatography and characterized by tandem mass spectrometry (LC-MS/MS). The challenge for bottom-up proteomics is the enormous complexity of the peptide mixture. The human proteome is estimated to contain from 100,000 to 2 million proteins, of which 10,000 to 20,000 are expressed at any given time. Proteolytic digestion (e.g., with trypsin) yields approximately 20 peptides for each protein, so the digests obtained from a lysate can be expected to contain 200,000 to 500,000 cleavage products. The unique feature of the proteomics experiment is that virtually every component of the sample is an analyte, so that an extraordinary resolving power is required. Separation of the components of such a complex mixture is beyond the capability of a single chromatographic mode. Proteomics studies require the coupling of two or more modes of chromatography (multidimensional liquid chromatography, MDLC) with MS detection (which adds an additional dimension to the analysis). See Section 9.3.10 for a general discussion of MDLC (or 2D-LC for the case of two chromatographic modes).

The key to achieving sufficient resolving power in MDLC is the sequential use of chromatographic modes that are orthogonal in selectivity, that is, that separate by totally different retention mechanisms. The model for orthogonality is two-dimensional gel electrophoresis (2D-GE [94]), which separates by charge (or isoelectric point) in the first dimension (isoelectric focusing), and by size in the second dimension (SDS-PAGE). In this case the resolving power or *peak capacity* (Section 9.3.9.1) of the combined separations is the product of the peak capacities for each separation. Thus, if 30 to 50 proteins can be distinguished in each separation, 2D-GE should be able to achieve the separation of 1000 to 3000 proteins. The primary constraint in MDLC is the use of RPC as the final dimension, as RPC conditions are compatible with on-line coupling to electrospray ionization systems. Selection of the initial separation(s) will be dictated by the MDLC approach taken. Size-exclusion chromatography (SEC) and IEC are compatible with a following RPC separation because sample fractions in a mainly aqueous mobile phase will be strongly retained on the RPC column (without excessive band broadening), allowing the injection of larger volumes. However, SEC is not an ideal choice because of its very low peak capacity (Section 13.8.5). Larger polypeptides also tend to elute later in RPC (as in SEC), so the two chromatographic modes are not completely independent or orthogonal (as desired for MDLC).

IEC is most often chosen as the first dimension in MDLC because the orthogonality of the two dimensions approximates that of 2D-GE. Cation exchange at low pH is preferred, since all peptides will be positively charged and retained on the ion exchanger. Cation-exchange columns have sufficient capacity to permit recovery and detection of both high- and low-abundance peptides (because larger samples can be injected). Both RPC under alkaline conditions (pH-10) and HILIC also have been shown to provide a high degree of orthogonal selectivity as the first dimension separation [95].

The three approaches used in proteomics for multidimensional chromatography are:

- discontinuous MDLC using fraction collection
- directly coupled MDLC
- MDLC using column switching

13.4.5.1 Use with Fraction Collection

The simplest and most straightforward approach to MDLC is collection of fractions from the first-dimension column, followed by injection of the individual fractions onto the second-dimension column [96]. The advantage of this off-line approach is that reagents used in the reduction and alkylation of proteins in the digestion step are eluted to waste, with no risk of their contaminating the ionization source. The disadvantage of off-line fraction collection is the necessity of operator intervention to collect and re-inject fractions. However, this approach can be automated by coupling the two dimensions using a multiple solvent-delivery HPLC system equipped with an automatic fraction collector and column-switching valve. Fractions are collected during the first-dimension separation, and then the column selection valve is switched to place the second-dimension column in line; the collected fractions become the sampling source for the second dimension and are analyzed in sequence. Several HPLC instrument manufacturers offer automated 2D systems using ion exchange (or chromatofocusing in one case) as the first dimension and reversed-phase chromatography as the second dimension (Beckman, Waters, Michrom Bioresources, Microtech Scientific).

13.4.5.2 Directly Coupled MDLC

Directly-coupled MDLC forms the basis for a multidimensional protein-identification technology (MuDPIT) developed by the Yates group [97, 98]. In this approach, a single capillary HPLC column consists of successive segments of two orthogonal stationary phases (cation exchange followed by RPC) for the separation of complex peptide mixtures; there are 12 to 15 elution cycles and resulting chromatograms. In a typical elution cycle, a part of the sample is displaced from the cation-exchange segment by aqueous buffer, following which the displaced fraction is eluted from the RPC segment by an acetonitrile-buffer gradient in 90 to 120 minutes. Successive elution cycles use increasing concentrations of ammonium acetate (500 mM for the last cycle). The column is directly interfaced to a nanospray-tandem MS. An evaluation of MuDPIT with yeast lysates as sample has demonstrated that (unlike 2D-GE) this approach has no inherent bias against low-abundance species, high- or low-mass proteins, strongly acidic or basic proteins, or hydrophobic proteins [99].

13.4.5.3 MDLC with Column Switching

Multidimensional liquid chromatography with column switching can provide the greatest flexibility and largest peak capacity, but at the expense of complexity and cost. Column switching can also eliminate the problem of introducing contaminants into the MS ionization source. Three systems of increasing complexity (the first for peptides, the others for proteins) illustrate different possible approaches to MDLC with column switching. The first example [100] consists of a first dimension strong cation-exchange column and an RPC second dimension. The sample is split into two fractions in the first dimension, unretained peptides and a retained fraction that is eluted with 0.5M salt. This simple binary fractionation is optimal for high-throughput application but is limited in terms of total peak capacity (only 2-fold greater than for a single separation).

A second example (comprehensive 2D-LC; Section 9.3.10) [101, 102] couples size-exclusion chromatography (SEC) as the first dimension with an RPC second dimension. The SEC separation is achieved using a set of six columns in sequence (1.8-m total length) so as to generate 90,000 theoretical plates for improved peak capacity. The effluent from the SEC columns is diverted in alternating fashion to one of two RPC columns plumbed in parallel. While peptides are being captured on one column, peptides from the preceding segment are resolved on the other column by means of a fast gradient separation. This arrangement reduces the overall time required by half. The fundamental limitation of this approach is the lack of full orthogonality between SEC and RPC for the two separation modes (as well as the limited peak capacity of all SEC separations). Hence only about 30% of the expected (relatively small) peak capacity was achieved. The system is of moderate complexity, requiring isocratic and gradient delivery systems with a switching valve (Fig. 9.25).

The third example, designed for MDLC separation of proteins, integrates sample preparation with the analytical separation [103]. An in-line sample-prefractionation column contains restricted-access media (RAM, Section 16.6.7.2) that excludes large proteins, while concentrating and separating small proteins by charge. The first dimension of the analytical MDLC system consists of a single ion-exchange column, with fractions from this column transferred to a suite of four RPC columns in alternation, two of which separate fractions from the first-dimension gradient, while the other two undergo sample injection and column regeneration. This highly complex MDLC system utilizes three gradient solvent-delivery systems, one isocratic pump, an autosampler, and four 10-port valves. See also the alternative procedure of Section 9.3.10.

13.5 SEPARATION OF NUCLEIC ACIDS

Nucleic acids carry ionized phosphates on the sugar-phosphate backbone, with hydrophobic nucleobases attached to the sugars. Therefore anion exchange, RPC, and HIC are each candidates for the separation of nucleic acids; all three modes have been used for analytical and preparative applications in nucleic acid research.

13.5.1 Anion-Exchange Chromatography

Anion-exchange chromatography (with either weak or strong anion-exchangers) is most often used for the separation of oligonucleotides. Ion-exchange interactions between solute phosphate-groups and the column increase with the number of phosphate groups, so retention increases with increasing chain length or solute molecular weight. Elution is generally accomplished with a gradient from low to high concentration of a neutral salt such as sodium chloride or sodium phosphate. Formamide can be added to the mobile phase to suppress self-association of the oligonucleotides. A typical application of anion-exchange chromatography is the purity determination of synthetic oligonucleotides prepared by solid-phase synthesis [104]. Following cleavage of the product from the synthesis support and removal of protecting groups, anion-exchange chromatography can resolve truncated failure sequences from the full-length product (Fig. 13.27).

Conventional anion-exchange chromatography is useful for separating oligonucleotides of up to about 30 bases in length. The separation of larger nucleic acids is limited by exclusion of the solute from the pore structure. In this case large-pore ($\geq$30 nm diameter) or nonporous supports are preferred. DNA restriction fragments in the range 50 to 1000 base-pairs have been separated on weak anion-exchange phases (Fig. 13.28) bonded either to (1) silica with 400-nm pores [105] or (2) nonporous polymeric resins [107]. An exception to this generalization is the anion-exchange separation of transfer RNAs (tRNAs); these molecules (which contain 75–90 nucleotides) have considerable secondary structure and a compact size which can penetrate 12 nm-pores (Fig. 13.29). The resolution seen in the separation of Figure 13.29 is surprisingly good, considering that most tRNAs are similar in size and would not be expected to be resolved based on charge differences alone. However, tRNAs vary in base composition and in methylation patterns, so this separation is likely based on mixed-mode effects arising from hydrophobic as well as electrostatic interactions [106, 107].

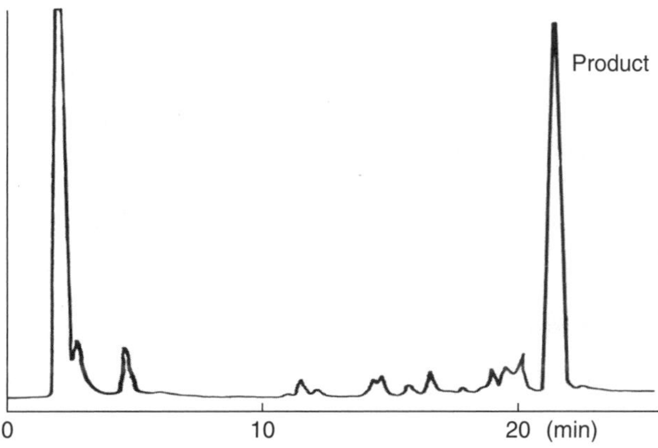

Figure 13.27 Analysis of the crude mixture of a chemically synthesized deprotected oligodeoxynucleotide with a strong anion-exchange column using a gradient from 1-mM KH_2 PO_4 + 60% formamide to 300-mM KH_2 PO_4 + 60% formamide. Adapted from [104].

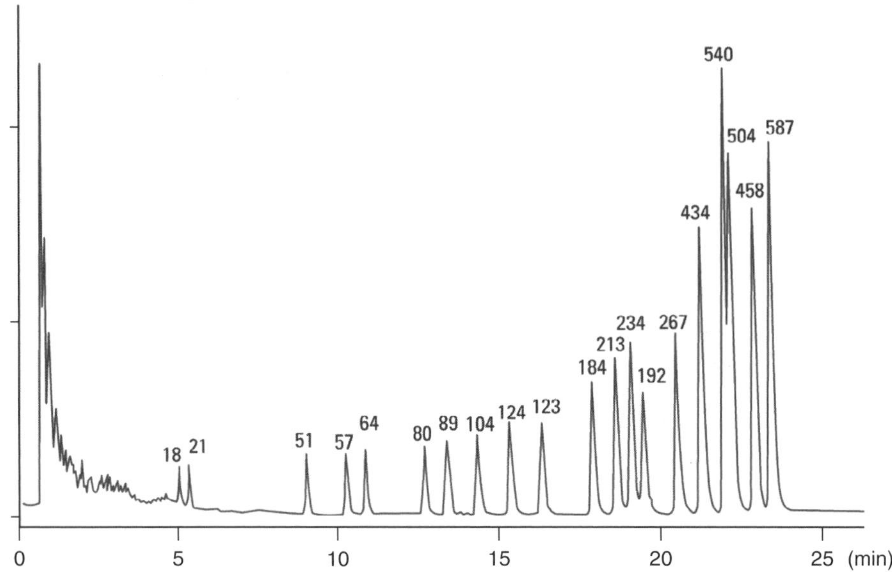

Figure 13.28 Separation of DNA restriction fragments on a TSK-DEAE-NPR nonporous-resin column using a gradient from 250-mM to 450-mM NaCl in 20-mM Tris-HCl (pH-9). Adapted from [106].

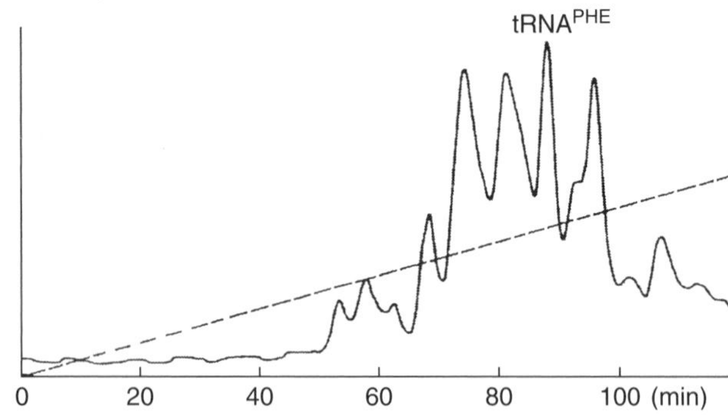

Figure 13.29 Fractionation of transfer RNA (tRNA) from *E. coli* on PEI-coated Hypersil (3-μm particles), 12-nm pore-diameter, 58.8% quaternized). Solvent A: 0.05 M potassium phosphate, 5% acetonitrile (pH-5.9). Solvent B: 0.05 M potassium phosphate, 5% acetonitrile (pH-5.9) + 1 M ammonium sulfate. Gradient: 0.25%/min from 50% solvent B. Flow rate: 0.5 mL/min. Adapted from [107].

13.5.2 Reversed-Phase Chromatography

RPC cannot be applied directly for the separation of nucleic acids, including oligonucleotides and larger species such as restriction fragments, as these highly charged, very polar molecules are unretained for typical separation conditions. As for other polar, ionized molecules, RPC retention can be increased by the addition of an ion-pair reagent to the mobile phase (ion-pair chromatography [IPC]; Section 7.4.1.2). For the negatively charged oligonucleotides, a positively charged ion-pair reagent such as tetrabutylammonium can provide increased retention and acceptable separation. The reagent is retained by the column, creating a positively charged stationary phase, which then acts as a hydrophobic anion-exchanger.

13.5.2.1 Oligonucleotides

The retention of oligonucleotides increases with increasing chain length and negative charge, as expected for anion exchange. As a result of hydrophobic interactions, differences in retention can occur for oligonucleotides of similar molecular weight but different base composition. Purification of synthetic oligonucleotides can be performed with IPC by taking advantage of the hydrophobic protective groups (dimethoxytrityl, DMT) used in synthesis [108]. In this approach chromatography is performed prior to deprotection; the protected product is strongly retained while failure sequences without DMT groups are poorly retained and well separated from the product (Fig. 13.30).

13.5.2.2 Restriction Fragments and PCR Products

Larger nucleic acids such as double-stranded DNA restriction fragments and products of the polymerase chain reaction (PCR) can be separated by IPC either on nonporous particles or porous polymer monoliths (Fig. 13.31). Both separations in Figure 13.31 demonstrate the excellent size-dependent separation efficiency of poly(styrene-divinylbenzene) when used for ion-pair separations of nucleic acids. In the case of double-stranded molecules, elution at moderate temperatures ($\leq 50^\circ C$) is in order of increasing chain length. The hydrophobic contribution of the bases is minimal because of the positioning of very hydrophilic phosphate and sugar groups on the outer surface of the helix [109]. At higher column temperatures ($>53^\circ C$), partial or complete strand separation occurs and strict size-dependent separation is lost due to contributions from base composition and sequence.

13.5.2.3 Denaturing HPLC

Denaturing HPLC (dHPLC) is a technique for the analysis of DNA sequence variations in individuals, in order to identify single-nucleotide polymorphisms (SNPs)

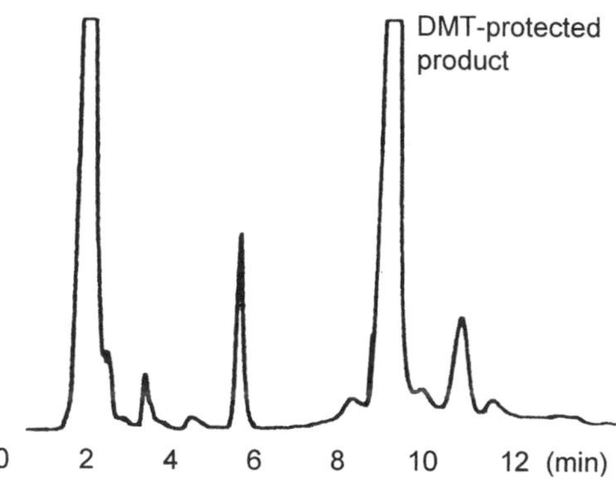

Figure 13.30 Gradient separation of a crude trityl-on 20-mer oligonucleotide by ion-pair chromatography. Conditions: 0.1M tetraethylammonium acetate (TEAA) as weak solvent and acetonitrile as strong solvent. Reprinted from [107] with permission.

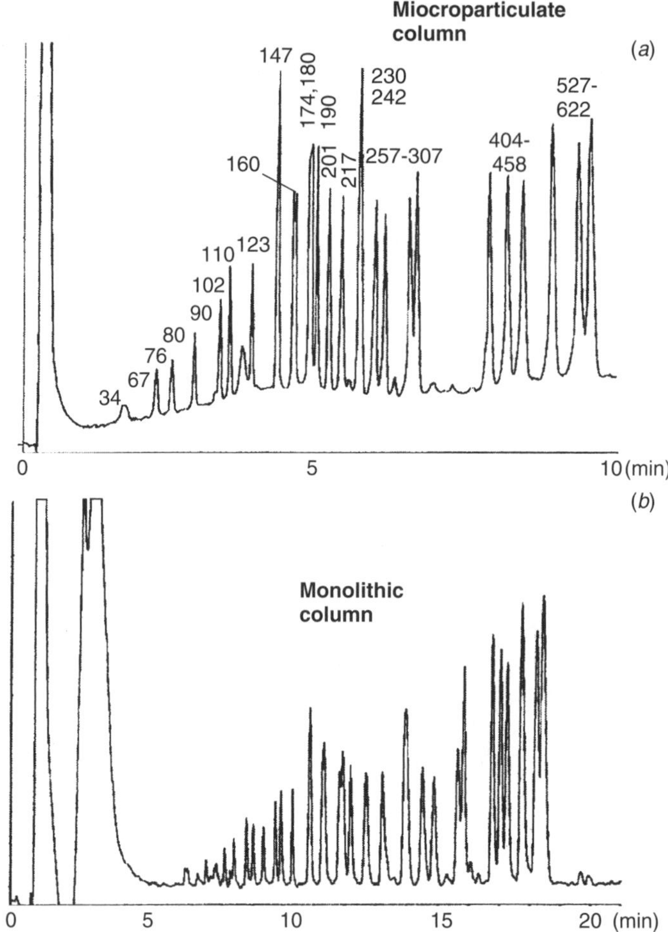

Figure 13.31 Comparison of ion-pair chromatography separation of double-stranded DNA restriction fragments by (*a*) a 2-μm nonporous PS-DVB-C18 columns compared with (*b*) a 0.2-mm i.d. P/S-DVB monolithic capillary column. Solvent-A: 100-mM TEAA + Na_4EDTA (pH-7.0). Solvent-B: 100-mM TEAA + Na_4EDTA (pH-7.0), 25% acetonitrile. Column temperature: (*a*) 50°C; (*b*) 49.7°C. The longer elution time for the monolith is a consequence of the gradient delay at the capillary column flow rate. Adapted from [109].

and disease-related mutations [109, 110]. The technique identifies SNP polymorphic sites in PCR-amplified sequences of normal (reference) DNA and the target DNA suspected of containing the variant locus. The co-amplified sequences are heated to separate individual DNA strands, re-annealed by gradual cooling, and then analyzed by dHPLC. If the SNP site in the target DNA matches that of the normal DNA, only two identical homoduplexes are observed. If the target DNA contains a variant base at the SNP site, mismatched heteroduplexes of normal and target strands are formed in addition to the homoduplexes.

Denaturing HPLC employs IPC with elevated temperatures to maintain amplified DNA sequences in their partially or fully denatured state, and this technique can be used to identify mutational sites by the chromatographic "signature" of the homo- and heteroduplexes.

The mismatched regions of heteroduplexes (containing unpaired "bubbles") have a reduced contact region with the ion-pairing surface. As a consequence they are less retained than the homoduplexes and are resolved from each other on the basis of the sequence difference in the bubble. IPC is performed at a temperature that maintains the heteroduplexes as stable entities. For a mixture of segments that are heterozygous for a single base change, a characteristic four-component chromatogram is observed (Fig. 13.32), with the last two peaks representing the matched homoduplexes and the earlier two peaks representing the reannealing of the sense strand of either homoduplex with the antisense strand of the other homoduplex.

13.5.2.4 RPC-5 Chromatography

RPC-5 was originally a low-pressure chromatographic technique that was developed in the 1960s and applied to the separation of tRNAs [111]. The technique uses a

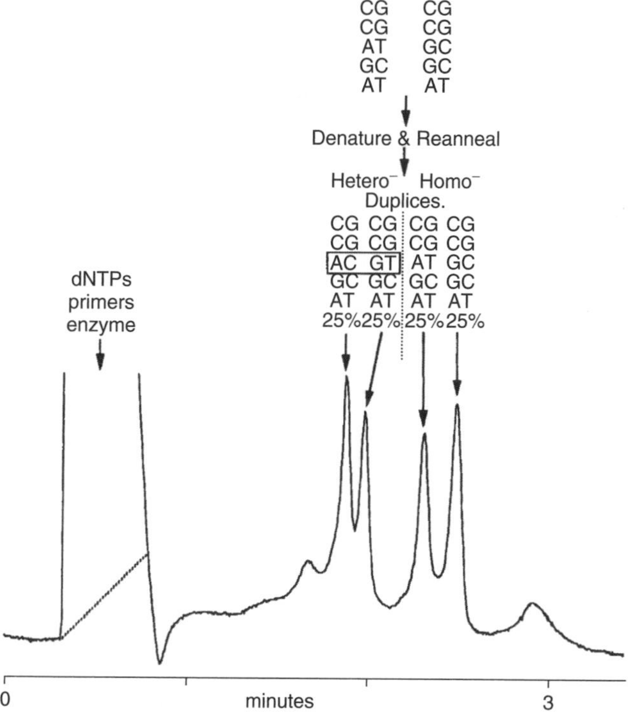

Figure 13.32 Denaturing-HPLC separation of two chromosomes as a mixture of PCR products denatured at 95°C, then re-annealed by gradual cooling to 65°C prior to analysis. In the presence of a mismatch, not only the two original homoduplexes are reformed, but simultaneously the sense and anti-sense strands of either homoduplex form two heteroduplices. The latter denature more extensively at the analysis temperature of 56°C and therefore are eluted earlier than the two homoduplices that undergo less pronounced denaturation. Adapted from [109].

polymeric polychlorotrifluoroethylene support with an adsorbed methyltrioctylammonium chloride liquid stationary phase. Although it was referred to as RPC, it is actually a mixed-mode separation that relies on both ionic and hydrophobic interactions. A later version of the RPC-5 mixed-mode approach uses HPLC technology based on 5-μm octadecyl silica coated with methyltrioctylammonium chloride [112].

13.5.3 Hydrophobic Interaction Chromatography

HIC has been applied to the separation of transfer RNAs using weakly hydrophobic columns with descending gradients of antichaotropic salts [104]. Large-pore silica-based supports with C_2 or C_4 stationary phases [113], and aminopropyl columns derivatized to introduce alkyl chains [114], have both been used with ammonium sulfate gradients as in the example of Figure 13.33. Selectivity can be varied by the addition of small amounts of 2-propanol to the mobile phase; transfer RNAs elute in order of increasing hydrophobicity. In the case of tRNAs aminoacylated with a hydrophobic amino acid such as valine, the aminoacylated tRNA eluted later than the non-aminoacylated form. Separation of tRNAs by HIC can also be accomplished using anion-exchange columns in the HIC mode [115, 116]. In this approach, tRNAs are bound under high-salt conditions and eluted with a reverse gradient to a salt concentration that is sufficiently low to elute tRNAs but still high enough to prevent ion-exchange interactions.

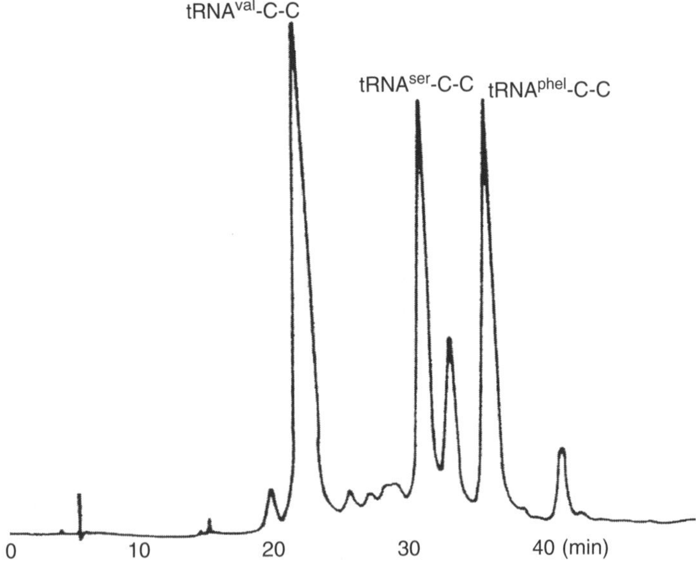

Figure 13.33 Resolution of specific tRNAs from yeast on an aminopropyl phase derivatized with hexanoic anhydride; elution was with a descending ammonium sulfate gradient. Reprinted from [114] with permission.

13.6 SEPARATION OF CARBOHYDRATES

The chromatographic analysis of neutral carbohydrates presents a formidable challenge, since they are poorly retained by RPC and (for a pH < 10) have no ionic groups that allows the use of ion-pair or ion-exchange chromatography. In addition most carbohydrates possess no chromophore or fluorophore, so optical detection methods, such as refractive index or light scattering, are usually employed (however, refractive-index detection cannot be used with gradient elution).

Three chromatographic modes have been used for the separation of carbohydrates: NPC or HILIC, ion-moderated partition, and high-pH anion-exchange chromatography. The suitability of each of these options for a particular application depends on the required sensitivity, resolution, and available instrumentation.

13.6.1 Hydrophilic Interaction Chromatography

In the early days of HPLC, amino columns (aminopropyl-bonded silica) with water-acetonitrile as mobile phase were widely used for the HILIC separation of carbohydrates. The major limitation of amino columns is marginal stability, while an important advantage is their increased rate of anomer mutarotation—which eliminates the problem of doublets formed by the resolution of anomers (diastereomers). The problem of short column lifetime has been partly addressed by the use of bare silica in conjunction with an amine-based mobile phase additive [117].

More practical HILIC separations of carbohydrates can be carried out with other columns: bare silica, amide-bonded silica, diol-bonded silica, hydrophilic polymer-clad silica, cyclodextrin-bonded silica, and polymeric ion-exchangers [118]. The use of these columns eliminates the reactivity of amino columns with reducing sugars [119]; they are also stable at elevated temperatures. While diol-bonded silica ("diol") columns have been used in the HILIC mode for carbohydrate separations, a disadvantage is peak broadening resulting from partial separation of anomers. The mobile phase can be supplemented with an amine additive to promote mutarotation of anomeric sugars [120], but these additives also hasten the dissolution of the silica support.

Silica-based polyaspartamide columns for HILIC can be used with water-acetonitrile mobile phases for carbohydrate separations [75]. In the oligoglycoside separation of Figure 13.34, note that retention decreases for lower %-acetonitrile (typical HILIC behavior). As with diol columns, the addition of an amine to the mobile phase can catalyze mutarotation to prevent peak broadening by anomers. Stationary phases bonded with α- or β-cyclodextrins (cyclodextrin columns) have been used with water-acetonitrile mobile phases for the separation of mono- and oligosaccharides [121]. Retention increases for a higher number of available hydroxyl groups on the molecule. These observations are consistent with a HILIC mechanism that involves interaction of the solute with hydroxyl groups located at the rim of the cyclodextrin—rather than formation of inclusion complexes within the cyclodextrin cavity (see related discussion of Section 14.6.4 and Fig. 14.18*a*). Some advantages of cyclodextrin columns include their greater stability and acceptable retention reproducibility.

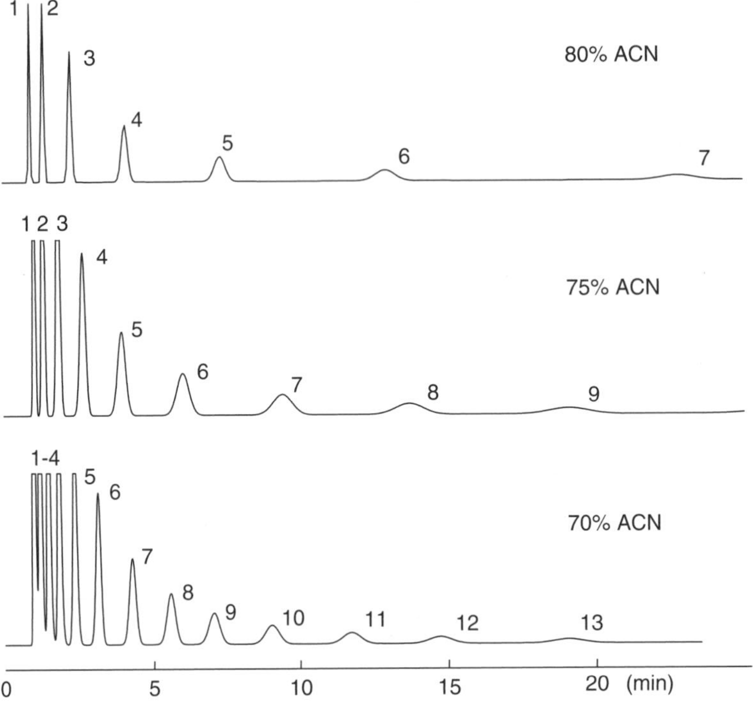

Figure 13.34 Isocratic HILIC separations of a homologous mixture of 3-hydroxy-2-nitropyridinyl-β-D-maltooligoglycosides. Conditions: 200 × 4.6-mm Poly-Hydroxyethyl A column; acetonitrile-water mobile phases; 2.0 mL/min. Numbers indicate degree of polymerization of each compound. Adapted from [75].

Polymer-based strong cation-exchange columns [122] and hydrophilic size-exclusion columns [123] have also been used for the separation of carbohydrates, with the usual HILIC mobile phases (acetonitrile-water mixtures). These separations are not based on either ion exchange or size exclusion, as evidenced by increased retention with increasing concentration of organic solvent.

13.6.2 Ion-Moderated Partition Chromatography

Sulfonated polystyrene-divinylbenzene resins are used for the separation of a wide variety of mono- and oligosaccharides, as well as mixtures of carbohydrates with alcohols and other small molecules [124, 125]. Separation is based on a combination of size exclusion and ligand exchange. Ligand exchange involves transition-metal ions that are tightly held by the resin sulfonic-acid groups; the metal ion then provides a positive charge that interacts with a very slight negative charge on the sugar molecule (the "ligand"). For oligosaccharide separations, the primary mechanism is size exclusion. Resins with a low percentage of cross-linking are preferred, in order to allow penetration of the oligosaccharides into the packing; Figure 13.35 shows the separation of oligosaccharides in a corn syrup. For monosaccharides, ligand exchange of the sugar hydroxyls with the fixed counter-ion on the resin is the primary mechanism.

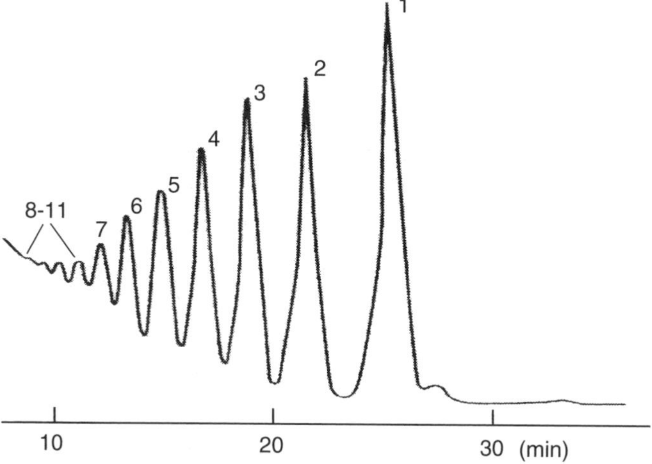

Figure 13.35 Analysis of oligosaccharides in corn syrup by ion-moderated partition chromatography on a 4% cross-linked strong cation exchange resin in the silver form. Sample: corn syrup (glucose, 1; 2–11, Dp 2–11). Conditions: 300 × 7.8-mm Aminex HPX 42-A column; water as mobile phase; 85°C; 0.4 mL/min. Adapted from [124].

A mechanism for ligand exchange has been proposed [126]: carbohydrate hydroxyls exchange with water molecules held in the hydration sphere of the fixed cation. The stability of the solute-cation complex increases with increased availability for coordination, and carbohydrate retention increases with the stability of the complex. Ligand-exchange selectivity is dependent on the nature of the counter-ion and the size, and on the structure and stereochemistry of the carbohydrate. Separations are isocratic with water or dilute solutions of sulfuric acid (5 – 10 mM) as mobile phase, usually at elevated temperatures (40–85°C). Recommended resin cross-linkages and ionic forms for particular applications are listed in Table 13.7.

Table 13.7

Strong Cation-Exchange columns for Ion-Moderated Partition Chromatography of Carbohydrates

Cross-Linkage Percentage	Ionic Form	Application
8	Calcium	Monosaccharides and class separation of di-, tri-, and tetrasaccharides
8	Lead	Pentoses and hexoses in wood products
8	Hydrogen	Carbohydrates in solution with fatty acids, alcohols, and ketones
8	Sodium	Sugars in samples with high salt
8	Potassium	Mono-, di-, and trisaccharides in corn syrup and brewing wort
4	Silver	Oligosaccharides
4	Calcium	mono- and disaccharides in starch hydrolysates

Note that these are fixed-ion resins; the column is converted to a specific form by the manufacturer before packing and is maintained in that form for the life of the column. In-column conversion from one form to another is not recommended, since resins can shrink and swell with changes in ion form, leading to likely column failure. Sodium-form columns are useful for samples containing high salt concentrations (e.g., molasses), potassium-form columns are useful for analysis of corn syrup, silver-form columns provide good selectivity for oligosaccharides, and calcium-form columns are used for analysis of starch hydrolysis products.

13.6.3 High-Performance Anion-Exchange Chromatography

Neutral carbohydrates are not retained on IEC columns under usual conditions, but they are weak acids that can partially ionize at $pH > 12$ (Table 13.8). Their separation can be achieved under alkaline conditions by means of polymer-based anion-exchange columns [127, 128]. Commercially available columns for high-performance anion exchange (HPAE) are based on polystyrene/divinylbenzene (for monosaccharide separations) or ethylvinylbenzene/divinylbenzene (for oligosaccharide separations). Both supports consist of nonporous particles covered with a fine layer of sulfonated latex microbeads (Fig. 13.36). Sugar alcohols are weaker acids than nonreduced sugars and require a high-capacity ion exchanger for their separation by HPAE. For the latter application a macroporous vinylbenzene-chloride/divinylbenzene functionalized with alkyl quaternary groups is used. Monosaccharides (including neutral and amino sugars) can be separated isocratically using a mobile phase of dilute aqueous sodium hydroxide. Separation of acidic carbohydrates (sialic acid, sialyated and phosphorylated oligosaccharides) can be achieved using sodium hydroxide/sodium acetate mobile phases, either isocratically or with a sodium acetate gradients. Oligo- and polysaccharides (including high-mannose, hybrid, and complex oligosaccharides) can be separated using sodium hydroxide or sodium hydroxide/sodium acetate gradients (Fig. 13.37). See also the carbohydrate separation of Figure 7.21.

HPAE is typically coupled with pulsed-amperometric detection (PAD). At high pH, carbohydrates are electrochemically oxidized at the surface of a gold electrode

Table 13.8

Dissociation Constants for Common Carbohydrates

Sugar	pK_a
Fructose	12.03
Mannose	12.08
Xylose	12.15
Glucose	12.28
Galactose	12.39
Dulcitol	13.43
Sorbitol	13.60
α-Methyl glucoside	13.71

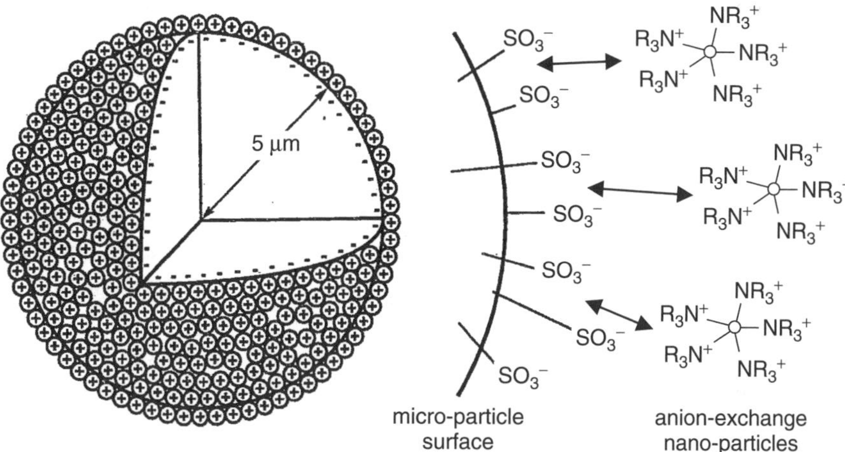

Figure 13.36 Pellicular anion-exchange-resin bead (Dionex). The bead structure consists of a nonporous sulfonated microparticle with surface-associated latex nanobeads that are functionalized with a strong anion-exchanger. The microparticles range in diameter from 5.5 to 10 μm, depending on the column type. The nanobeads range in diameter from 43 to 275 nm, also depending on column type. The nanobeads are immobilized on the microparticle surface by electrostatic interactions of the ion-exchange groups. Adapted from [127].

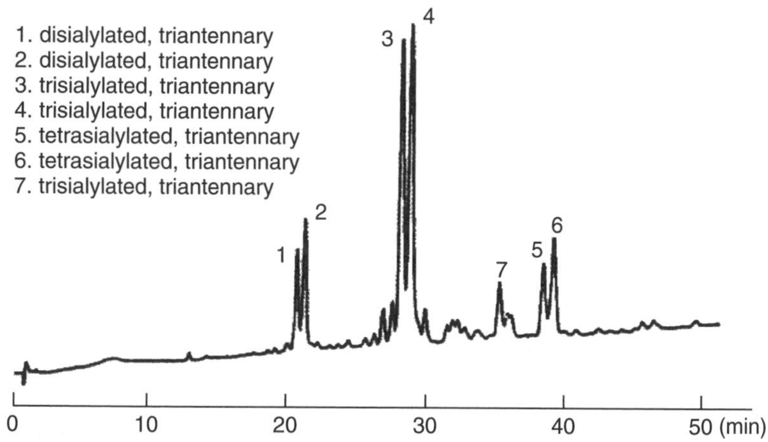

Figure 13.37 Separation of bovine fetuin *N*-linked oligosaccharide alditols by high-performance anion-exchange chromatography, using a gradient of 0.0–0.5M sodium acetate in 100-mM sodium hydroxide. Reprinted from [128] with permission from Dionex Corp.

by application of a positive potential [129]. The measured current from this reaction is proportional to carbohydrate concentration, with detection limits in the low picomole range. The reaction produces oxidation products that poison the electrode surface, so the detector is pulsed to a high voltage to clean the surface, then pulsed again to a low voltage to reduce gold oxide back to gold.

13.7 SEPARATION OF VIRUSES

The large size of the virus particle represents an important consideration in their chromatographic separation, one that has several consequences. First, viruses can be expected to display large values of S (reversed-phase) or m (ion exchange), such that small changes in mobile phase concentration will have large effects on their isocratic retention [3]. Therefore the isocratic separation of viruses is impractical for all modes of chromatography, except size-exclusion chromatography. In gradient elution, a virus will elute at a discreet point in the gradient, in most cases with $k^* \approx 0$. The second consequence of viral size is slow diffusion. For example, D_m for adenovirus is 5×10^{-8} cm^2/s, about 10-fold slower than for a large protein. Slow diffusion should result in low values of N^* and broad peaks. However, peak broadening due to slow diffusion is offset by the small value of k^* (Eq. 9.5, assuming a very large value of S), and observed peak widths for viruses are similar to those for proteins. The third consequence of viral size is their large hydrodynamic volume, which restricts their entry into the pore system of the packing. For this reason viral particles interact with only a small fraction of the total column surface area, and column capacities for viruses are 20- to 50-fold lower than for proteins.

Ion exchange, hydrophobic interaction, and metal-chelate chromatography have all been used for viral purification [130]. However, ion-exchange chromatography is most often employed as the capture step in virus purification because of its satisfactory yield and purity [3]. In the case of adenoviruses the isoelectric point of the surface is IP $\approx$ 6, so anion-exchange chromatography is the preferred procedure for separating viruses. Additives such as sucrose, magnesium, and glycerol that are used to stabilize the virus are compatible with anion-exchange conditions.

Both strong and weak anion-exchange columns have been employed for adenovirus purification, based on cross-linked polystyrene-divinyl benzene, hydrophilic polymer, polyacrylamide, or cross-linked agarose and dextran supports [5]. Monoliths have been used for viral separations [3], although these columns are not commercially available in the larger sizes that are required for large-scale purification. The mobile-phase pH should be $\approx$2 units above the isoelectric point of the virus, in order to avoid viral aggregation; pH values between 7.5 and 9 have been used for adenovirus purification. Sodium chloride gradients are generally used, with salt concentration of 0–0.3M for sample loading and $\geq$ 0.6M for elution. In a study of static binding capacity as a function of ionic strength, it was found that capacity passes through a maximum at about 0.3M. Loading of the virus at or slightly above 0.3M is recommended, in order to allow contaminating proteins to be washed from the column during the loading step [3]. Elevated column temperatures should be avoided, in order to minimize any loss of viral activity.

The advantage of IEC for adenovirus purification is its ability to distinguish aggregated and disrupted forms from the intact virus [130]. However, the removal of empty capsids from intact virus is less certain [5, 130]. The resolution of the early-eluting p53 adenovirus and late-eluting DNA is shown in Figure 13.38. For further details on virus separations, see [3].

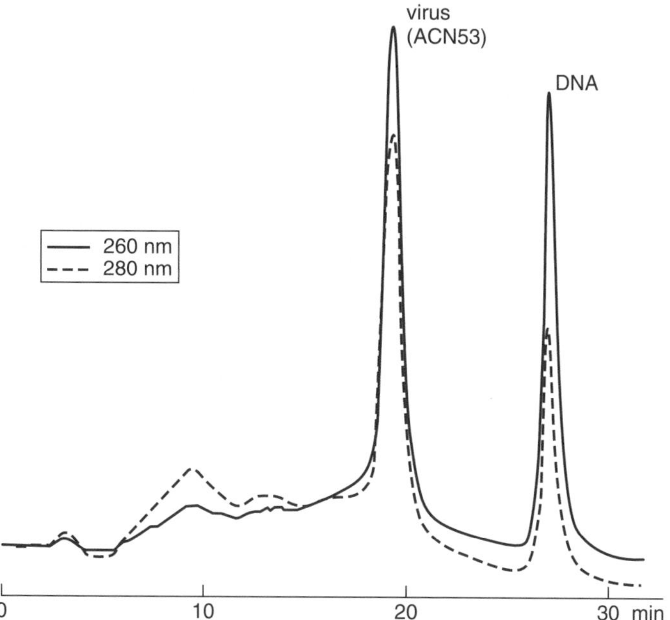

Figure 13.38 Separation of adenovirus by anion-exchange chromatography. Separation conditions: column, 50 × 6.6-mm Fractogel DEAE-650 M column; gradient, 300–600-mM NaCl (50-mM Tris, pH-8.0 plus 2-mM MgCl and 2% sucrose) in 10 min. Adapted from [130].

13.8 SIZE-EXCLUSION CHROMATOGRAPHY (SEC)

Size-exclusion chromatography (SEC) separates compounds according to their molecular size in solution, as a result of the exclusion of larger molecules from smaller pores in the column packing [131]. When applied to synthetic polymers with organic solvents as mobile phase and polymeric column packings (Section 13.10.3.1), the technique is referred to as *gel permeation chromatography* (GPC). When separating biopolymers such as proteins with aqueous buffers as mobile phases and hydrophilic column packings, the technique is termed *gel filtration*. Gel filtration can be used either as a preparative tool to isolate biologically active species (often in concert with other chromatographic techniques in a multi-stage purification process), or as an analytical tool to obtain information about solute molecular size or shape, aggregation state, or the kinetics of ligand-biopolymer binding.

Historically gel filtration has employed soft gels such as dextrans, agarose, or polyacrylamide [132–134]). These packings are compressible and therefore are only compatible with mobile-phase flow by means of gravity or low-pressure pumps. Soft gels may be stabilized by cross-linking, in which case they can be used with higher flow rates and pressures of a few hundred psi. Analytical gel filtration is most often carried out with rigid supports: a silica matrix modified with a hydrophilic stationary phase or cross-linked organic polymers. These materials are mechanically stable at pressures of a few thousand psi or higher, and can be used with HPLC systems.

13.8.1 SEC Retention Process

SEC is the simplest form of chromatography, in which retention depends only on the relative penetration (or "permeation") of solute molecules into and out of the pores of the stationary phase; molecules are separated on the basis of their size in solution (for polymers of the same chemistry and architecture, this size correlates with molecular weight). In contrast to other modes of chromatography such as RPC or IEC, in which solutes are retained by interacting with the stationary phase, SEC (under ideal conditions) involves no interaction of solute and stationary phase. Molecules that are too large to enter any of the pores elute in a volume of mobile phase that is equal to the interstitial volume between the stationary-phase particles (V_0). Molecules that are small enough to freely enter all of the pores elute in a volume equal to the interstitial volume plus the volume of the pore system (V_i). Molecules of intermediate sizes enter some fraction of the pore system, depending on their size or shape, and elute between V_0 and $V_0 + V_i$. The total mobile-phase volume or dead-volume V_m can be expressed as the sum of the interstitial volume and the pore volume:

$$V_m = V_0 + V_i \tag{13.6}$$

The extent to which a solute can penetrate the pore system is governed by its distribution coefficient K_D, which is related to its elution volume V_R by

$$K_D = \frac{V_{\mathrm{R}} - V_0}{V_i} \tag{13.7}$$

Equations (13.6) and (13.7) can be combined:

$$V_{\mathrm{R}} = V_0 + K_D V_i \tag{13.8}$$

From the expression above it can be seen that molecules too large to enter the pores will all have $K_D = 0$ and will co-elute at V_0. Similarly all molecules small enough to freely penetrate the entire pore system will have $K_D = 1$ and co-elute at V_m. Molecules of intermediate size will have K_D values between zero and one and will be separated according to size, with larger molecules eluting before smaller molecules.

The relationship between molecular size and retention volume can be used to estimate the molecular weight M of an analyte. A calibration plot of log M versus retention volume (or K_D) will exhibit an approximately linear segment between V_0 and V_i, as in Figure 13.39*a* (solid portion of curve). The range in sample molecular weights corresponding to this linear segment is referred to as the *fractionation* or *separation range*. If the plot is constructed using standard proteins whose shapes are similar to that of an analyte protein, the retention time of the analyte (or retention volume as in Fig. 13.39*b*) will correspond to a specific molecular weight on the plot, allowing an estimate of its molecular weight. The relationship between log M and K_D is nearly linear for K_D values between about 0.2 and 0.8, but with curvature at the ends of the plot (as in the SEC calibration curve of Fig. 13.39*a*). In theory, SEC column packings with a single, absolute pore size will provide a separation that spans 1.5 decades of molecular weight. In practice, however, this separation range

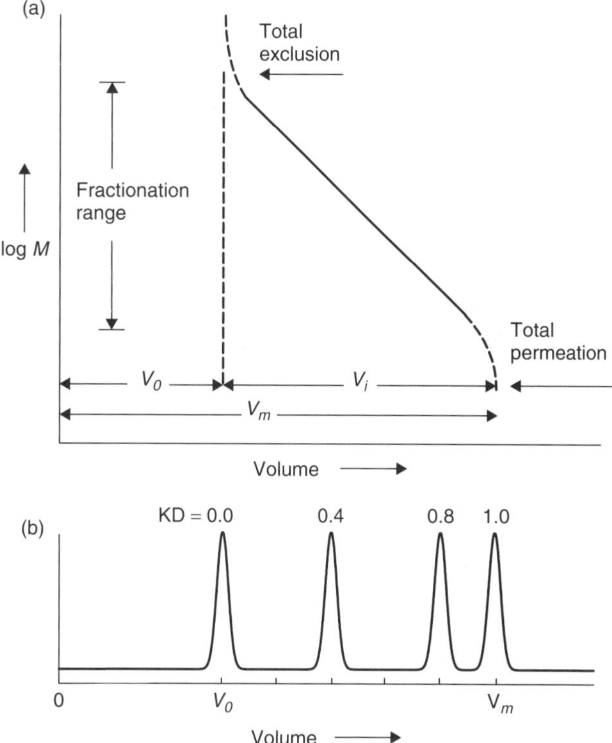

Figure 13.39 Hypothetical SEC calibration curve (*a*) and chromatogram (*b*). Adapted from [135].

will cover roughly two decades or a little more, since the size of the pores in a given packing material will vary somewhat.

Although SEC is often used to estimate protein molecular weight, it should be understood that retention is actually determined by the *hydrodynamic diameter* of the solute, which is only indirectly related to molecular weight. The hydrodynamic diameter of a protein (or other molecule) is related to its radius of gyration or Stokes radius, and this can vary with solute hydration and molecular shape. Two proteins with similar molecular weights but different shapes (e.g., spherical vs. oblate vs. rod-like, or native vs. denatured) will have different hydrodynamic diameters and therefore display significantly different retention volumes (Fig. 13.40). To obtain accurate molecular-weight estimates using gel filtration, it is necessary that the proteins used to construct the calibration plot and the analytes all have similar shapes. An alternative approach is to perform calibration and analysis under denaturing conditions, so that both calibrant and analyte proteins are converted to linear random-coil conformations, with retention times that better correlate with molecular weight.

13.8.2 Columns for Gel Filtration

The column packings used for SEC must be as inert as possible, so as to minimize any interactions with analytes (which would negate a relationship between solute

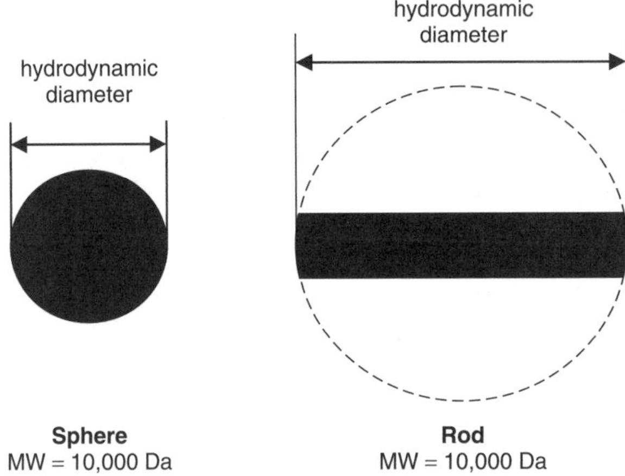

Figure 13.40 Molecular "size" and molecular shape compared. Adapted from [135].

molecular weight and SEC retention). For gel filtration, this is achieved by the use of hydrophilic packings whose interactions with the aqueous mobile phase are stronger than with protein solutes. The pore-volume of the packing should be as large as possible, and the support material should be mechanically stable for the required flow rates and pressures. Packings of different pore sizes may be needed for proteins of different molecular weight. Smaller pores are used for small molecules, larger pores for larger molecules. If a sample contains proteins of widely different molecular weights, two columns of different pore size can be connected in series (the order is not important). To obtain the greatest fractionation range, the connected columns should have packings with pores about 10-fold difference in size, and the two columns should be closely matched in terms of efficiency or values of H (Section 2.4.1).

13.8.2.1 Support Materials

Two types of materials are used for SEC columns: hydrophilic bonded silicas and hydrophilic organic polymers. Silica is the most widely used material for HPLC packings because of its mechanical stability, acceptable porosity, and availability in a range of pore sizes. However, the silica surface interacts strongly with proteins, so it must be derivatized. The most common approach is to react surface silanols with an organosilane reagent to form a diol-type or carbohydrate-like coating that is covalently attached to the silica. A limitation of silica-based SEC packings is their instability under alkaline conditions. Silica dissolves at pH values above 8, leading to reduced column lifetime. One manufacturer (the Zorbax GF columns from Agilent Technologies) uses a zirconyl cladding to stabilize the silica support for operation above pH-8.

Because of silica's limitations, several manufacturers offer gel-filtration columns based on hydrophilic organic polymers. These include polymethacrylate supports, proprietary hydrophilic polymers, and semi-rigid cross-linked agaroses and dextrans. These columns are more stable under high-pH operation but are less efficient and less able to tolerate high pressures.

13.8.2.2 Pore Size and Porosity

High-performance SEC packings are available in pore sizes ranging from 10 to 400 nm. A column should be selected with a pore size such that solutes of interest elute within $0.2 \leq K_D \leq 0.8$. Manufacturers provide calibration plots in their product literature for this purpose; see the example of Fig. 13.41 for two different gel-filtration columns (the x-axis in Fig. 13.41 is in units of retention time, corresponding to a specified flow rate). In this case the approximate fractionation range of the GF-250 column is $4000 \leq M \leq 200,000$ Da, while that of the GF-450 column is $10,000 \leq M \leq 1,000,000$ Da. By connecting the two columns in series, the fractionation range would be much wider: $4000 \leq M \leq 1,000,000$ Da.

Columns (of the same dimensions) with a narrow pore-size distribution will be characterized by higher resolution over a narrow range of analyte molecular weights, for example, $1000 \leq M \leq 30,000$. That is, such columns will exhibit a calibration plot with a shallow slope and a reduced fractionation range. Columns with a wide pore-size distribution will be characterized by a wider fractionation range, that is, a calibration plot with a steep slope. Sample resolution or the ability to separate two solutes of different molecular weight increases for a shallow slope of log M versus retention time or volume. Thus the choice of a particular column for a given sample represents a compromise between resolution and retention range.

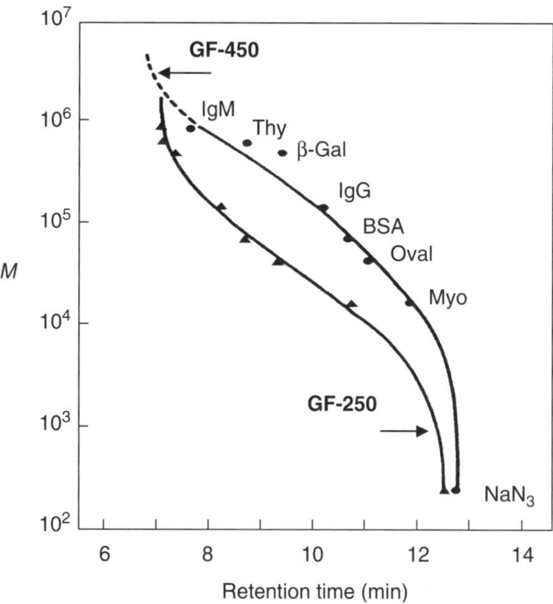

Figure 13.41 Calibration curves of proteins for GF-250 and GF-450 columns; 0.2M sodium phosphate (pH-7.5) mobile phase (nondenaturing conditions). Reprinted with permission of Agilent Technologies, Inc.

The porosity of an SEC column can be characterized by its phase ratio (V_i/V_0), as in Figure 13.39*a*. Other factors equal, resolution increases for a larger phase ratio. Soft-gel SEC packings have high porosities with phase ratios of 1.5 to 2.4 [136], while high-performance SEC packings have more modest phase ratios of 0.5 to 1.5 [137]. However, the disadvantage of lower phase ratios for high-performance SEC columns is more than offset by their higher efficiencies and faster analysis times. As particle pore-diameter and pore-volume increase, the mechanical strength of the support decreases.

13.8.2.3 Particle Diameter

As in interactive modes of chromatography, a reduction in particle diameter in SEC improves column efficiency. Column packings with particle diameters of 10 to 12 μm are available for less demanding applications, such as preparative separations, while SEC packings with particle diameters of 4 to 5 μm can be used for applications demanding higher resolution (as for molecular-weight analysis or analytical separations of protein mixtures).

13.8.2.4 Increasing Resolution

Resolution in SEC is controlled in two ways: (1) by selecting a column with a flatter slope of log M versus retention time (which effectively increases selectivity or values of α), and/or (2) by increasing the column plate number N. Higher values of N can be achieved by the use of smaller particle columns, lower flow rates, or an increase in column length. High-performance SEC columns are usually operated at flow rates of 1 mL/min or less for an 8-mm i.d. column. The small diffusion coefficients D_m for proteins and other biopolymers (or synthetic polymers) means that N usually increases significantly when flow rate is reduced (Section 2.3.1).

13.8.3 Mobile Phases for Gel Filtration

In contrast to interactive modes of chromatography, where the mobile phase is an active participant in the separation process, the mobile phase in SEC is simply a carrier that transports solute molecules through the column. The mobile phase is selected to maintain the solute in solution and in the appropriate conformation (e.g., native vs. denatured), to minimize column-solute interactions, and to maximize column lifetime. Thus the mobile phase may contain additives that suppress undesired interactions of the analyte with the support or the bonded stationary phase. These interactions may be electrostatic in nature, and in the case of silica-based columns, residual ionized silanols on the support often create a negative charge on the packing. For cationic solutes such as basic proteins, ionized silanols can result in a cation-exchange contribution to retention, so the solutes will elute later than predicted by a purely SEC mechanism. For anionic solutes, such as acidic proteins and nucleic acids, ion exclusion can result, with solutes eluting earlier than predicted. In severe cases, solutes may elute after V_M (ion exchange) or before V_0 (ion exclusion). A second type of non-ideal behavior in SEC is hydrophobic interaction, leading to increased solute retention.

Undesired analyte-column interactions can often be minimized by adjusting the salt concentration: increasing ionic strength reduces electrostatic interactions, and decreasing ionic strength reduces hydrophobic interactions. Thus an intermediate ionic strength will generally be required to avoid these non-ideal behaviors. A typical mobile phase for gel filtration is 100 mM potassium phosphate + 100 mM potassium chloride (pH-6.8). Hydrophobic interactions can also be reduced by adding a small amount (e.g., 5–10%) of an organic solvent such as methanol, ethanol, or glycerol.

13.8.4 Operational Considerations

Once the appropriate column, mobile phase, and flow rate have been selected, successful separations by SEC may require the adjustment of sample weight and concentration. Mobile-phase additives such as surfactants and salts can be used to maintain analyte solubility, or to suppress undesired analyte-stationary phase interactions. Alternatively, such interactions can be exploited in order to achieve a desired separation (Section 13.8.4.4).

13.8.4.1 Column Capacity

The loading capacity of SEC columns is relatively modest—compared to interactive modes of chromatograph—because high-molecular-weight sample solutions can be quite viscous. Viscous samples can result in undesired peak distortion and broadening for samples that are too concentrated—or for larger samples. Such samples therefore require either a smaller sample volume or a more dilute sample. A rule of thumb suggests that the sample-volume should be $\leq 2\%$ of the column-volume for a sample molecular weight of 10,000 Da, a value that will be greater for lower molecular-weight samples, and smaller for higher molecular-weight samples. A typical analytical SEC column with dimensions of 300 × 8.0-mm has $V_m = 10{-}11$ mL, providing a maximum sample-injection volume of about 200 μL. Because sample size in SEC is limited mainly by sample volume and viscosity (which increases with sample concentration), the weight limit for a protein sample and a 300 × 8.0-mm column is then about 1 to 2 mg. For larger sample weights or volumes, resolution may be compromised. Sample capacity will scale in proportion to column volumes, for different column lengths and diameters.

13.8.4.2 Use of Denaturing Conditions

For gel-filtration separation under nondenaturing conditions (as in Fig. 13.41), estimates of analyte molecular weight can be in error as a result of differences in molecular shape. Variations in molecular shape are less likely for denatured species, suggesting analysis and calibration with denaturing conditions whenever the shape of the native protein molecule is in question. The addition of a denaturant such as 4 to 6M guanidinium hydrochloride, 4 to 6M urea, or 0.1–1% sodium dodecylsulfate (SDS) to the mobile phase can be used to convert calibrants and analytes to random coil conformations. Denaturing conditions lower the fractionation range of a gel-filtration column, because of the increase in analyte hydrodynamic diameters [138]. Also surfactants, such as SDS, may bind strongly to the column and be difficult to remove; therefore it is advisable to dedicate the column to each application of this kind (or use an unretained denaturant such as urea or guanidine). High concentrations of chaotropic salts such as urea and guanidinium HCl in the mobile phase (for denaturing conditions) can compromise pump and injector seals, and should never be left standing in the HPLC system following use.

13.8.4.3 Column Calibration

When installing a new gel-filtration column, the values of V_0 and V_m should be determined using appropriate probes. The value of V_0 can be measured by injecting a large biopolymer whose molecular weight falls outside the exclusion limit of the column; high-M DNA (e.g., calf thymus DNA) is often used. The blue dextran used for measuring V_0 on soft gel columns can give erroneous values on some high-performance SEC columns because of hydrophobic binding with increased retention. The value of V_m is determined using a very hydrophilic small molecule that can be detected by UV. Popular choices are cyanocobalamin (vitamin B_{12}), glycyl tyrosine, or p-aminobenzoic acid [137].

Non-ideal interactions can be characterized by using small-molecule probes [139, 140] that elute at V_m. Cation-exchange interactions are indicated by retention times $t_R > V_m$ for arginine or lysine. Ion exclusion is shown by early elution of citrate or glutamic acid ($t_R > V_m$). Hydrophobic interactions can be detected by $t_R > V_m$ for phenylethyl alcohol or benzyl alcohol as solute. Inasmuch as proteins can denature at higher temperatures, SEC retention can be temperature dependent.

13.8.4.4 Exploiting Non-ideal Interactions

While non-ideal interactions can prevent accurate estimates of molecular weight, these interactions can also be used to advantage for the purpose of changing relative retention and improving resolution—especially for preparative separations. The above mentioned approaches for suppressing non-ideal interactions (Section 13.8.3; varying salt concentration, adding organic solvents, varying pH) also suggest means to enhance these interactions, depending on the interaction that is to be enhanced.

13.8.5 Advantages and Limitations of SEC

Size-exclusion chromatography offers several advantages that make it a desirable technique for both preparative and analytical applications. First, separations are relatively fast: with a 300 × 8.0-mm analytical column operated at 1 mL/min, all analytes should elute in about 10 minutes. Second, because the stationary phase is designed to eliminate interactions with the sample, SEC columns usually exhibit excellent recovery of mass and biological activity. Third, all separations are performed under isocratic conditions, which generally favors convenience.

There are also some limitations to SEC. First, the resolving power is quite modest, compared to interactive chromatography. The maximum number of baseline-resolved peaks in a separation is usually only 5 to 10, compared to several hundred for gradient RPC. For a gel-filtration column with a fractionation range from 10 to 500 kDa, this implies that two proteins can be resolved, if they differ in molecular weight by a factor of two. Thus analytical SEC is only useful for samples that contain a limited number of components with quite different molecular weights. The second limitation of SEC is its low volume- and mass-loading capacity. As a consequence of these two limitations, SEC is more likely to be used as a later step in a purification scheme. A third limitation of SEC is modest column lifetime, particularly for silica-based SEC columns. When operated with aqueous buffers at neutral pH, SEC column lifetime is typically shorter than that of a silica-based RPC column operated with aqueous-organic mobile phases. A final limitation of SEC is the accuracy of molecular-weight estimates, which are usually limited to rough estimates of molecular weight. While SDS-PAGE and mass spectrometry provide more accurate values, the first technique is laborious and the second is expensive. Coupling an SEC column to a static laser light-scattering detector (in conjunction with a concentration-sensitive detector), however, can provide accurate molecular-weight values [141].

13.8.6 Applications of SEC

SEC can be used for separating and characterizing analytes based on molecular size, and as a preparative tool for recovering purified material from a mixture.

13.8.6.1 Analytical Applications

Analytical applications of gel filtration include molecular-weight estimation, monitoring or characterizing protein folding and aggregation, and determining receptor-ligand interaction. As discussed above, the retention of biopolymers in SEC is governed by molecular size and shape. To obtain accurate estimates of molecular weight, the column must be calibrated with standards that possess the same shape as the analyte, or both analyte and calibrants must be converted to a random-coil (denatured) configuration—and maintained as such during the analysis. Gel filtration can also be used to characterize protein folding and aggregation. As an illustration of the former, Figure 13.42 summarizes several experiments that

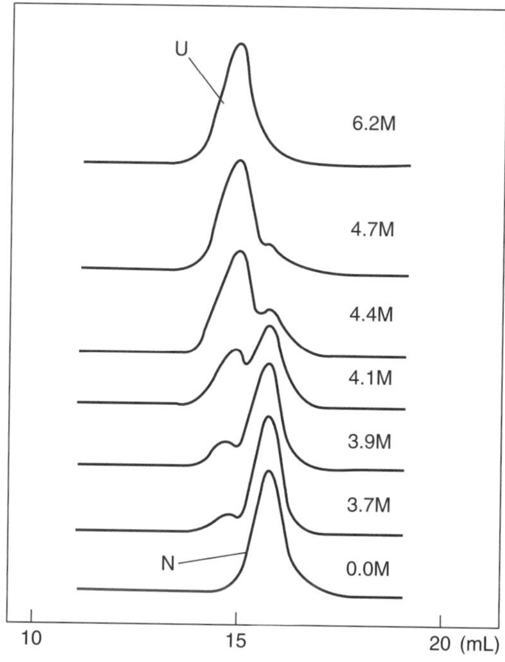

Figure 13.42 Use of SEC to monitor the folding states of lysozyme in the presence of increasing mobile phase concentrations of guanidine-HCl. *U*, unfolded (denatured) species; *N*, native species. Adapted from [142].

monitor the folding state of lysozyme in the presence of varying concentrations of guanidine-HCl [142] in the mobile phase. In the absence of guanidine (0.0M), the protein exists in the folded or native state (*N*) and has a retention volume of 15.5 mL. For a guanidine concentration of 6.2M, the protein is completely unfolded (*U*) or denatured; the unfolded (extended) protein molecule has a larger molecular size, and is therefore less retained (retention volume of 14.5 mL). For intermediate guanidine concentrations, the protein exists simultaneously in folded and unfolded states. An illustration of using SEC to monitor protein aggregation (Figure 13.43) is the determination of the aggregation state of human growth-hormone by the distribution of the protein among monomer, dimer, and oligomer [143]. Because the biological activity of a protein varies with its aggregation, measurements of aggregation are important for determining the quality of protein pharmaceuticals (as in this example).

Gel filtration is also able to determine receptor-ligand (e.g., protein–drug) interactions using zonal chromatography, Hummel–Dreyer methodology, or frontal analysis [144]. In zonal chromatography a mixture of protein and ligand is applied to the column. The protein-ligand complex elutes first and is separated from the free ligand. Quantitative analysis of the two species allows calculation of an affinity constant. Zonal chromatography can be used if dissociation of the protein-ligand complex is slow relative to the chromatographic process. In the Hummel–Dreyer method [145] the mobile phase contains the ligand, and a small volume of protein is injected onto the column. The elution profile exhibits a leading peak representing the protein-ligand complex, followed by a negative peak representing ligand-depleted mobile phase. The advantage of the Hummel–Dreyer approach is that protein is always in equilibrium with free ligand. It also requires only a small amount of protein. The method is applicable to protein-ligand complexes with rapid association-dissociation kinetics. In frontal analysis a large volume of protein and ligand is injected onto the column. The elution profile exhibits plateaus representing free protein, the complex in equilibrium with dissociated components, and free ligand. Frontal analysis enables determination of binding ratios under conditions where the concentration of species is known and constant, but requires large amounts of analytes.

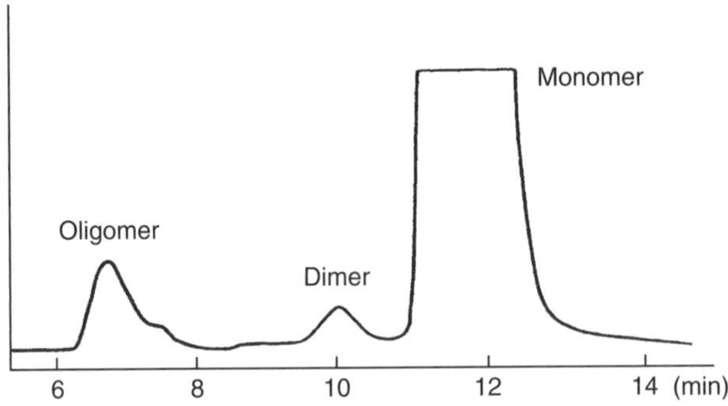

Figure 13.43 SEC separation of monomer, dimer, and oligomer forms of recombinant human growth hormone. Reprinted from [143] with permission.

13.8.6.2 Preparative Applications

Speed and gentle elution conditions make gel filtration a convenient method for the rapid isolation of biopolymers, with high recovery of mass and biological activity. Although resolving power is modest compared to other chromatographic modes, gel filtration is useful for purifying a target species from higher and lower molecular-weight components. The aqueous buffers used as mobile phases usually are compatible with subsequent purification steps such as IEC and RPC. The inert nature of gel-filtration columns allows the use of mobile phases supplemented with additives such as surfactants and organic modifiers that may be necessary to maintain the solubility of hydrophobic species, for example, membrane proteins.

Salts, buffering agents, and other small molecules will all elute in the total permeation volume in an SEC separation, well-resolved from biopolymers such as proteins and nucleic acids. Total run times can be limited to a few minutes, so gel filtration provides a quick means for desalting biological samples. For buffer exchange, the destination buffer is used as the mobile phase, and solutes elute in the new buffer resolved from the original sample buffer zone. Thus in a multi-step purification scheme, gel filtration can be used not only as a fractionation tool but as a link between otherwise incompatible chromatographic steps; for example, HIC and ion exchange.

13.9 LARGE-SCALE PURIFICATION OF LARGE BIOMOLECULES

The general principles of preparative liquid chromatography (prep-LC) are discussed in Chapter 15 and apply equally for the isolation or purification of large biomolecules. In this section, we will provide a brief background of the purification of biomacromolecules by prep-LC, as well as a representative example.

13.9.1 Background

The chromatographic purification of peptides and proteins on a laboratory scale has been underway for the past five decades. The principal chromatographic modes were originally limited to anion exchange, cation exchange, and gel filtration. The use of prep-LC for biomolecules has since expanded to other separation modes, including their purification on a commercial scale (downstream processing); no other procedure is competitive for the separation of closely related proteins and other biomolecules. The sales of biomolecules that have been purified by prep-LC today account for billions of dollars per year, and the number of products and their dollar value continue to climb. A wide variety of column packings or "resins" are now available, including different modes, supports (matrices), particle sizes, and porosities. Automated systems with column diameters of a meter or more allow large-scale prep-LC to be carried out at any desired scale. Both analytical and preparative chromatography are based on similar principles, as discussed in Chapter 15. In this section we highlight some aspects of prep-LC that are relevant for large-scale separations of biomacromolecules.

Several factors have contributed to the recent, rapid development of large-scale prep-LC for biomolecules. It is now recognized that protein molecules may undergo subtle modifications during processing, and these impurities must be removed from a final pharmaceutical product [146]. High-performance prep-LC is the only practical method for separating these closely related protein species on a large scale. An early example was the purification of insulin by Eli Lilly and Novo Nordisk, using large-scale gel filtration and ion exchange [147]. Subsequent improvements in both equipment and column packings have made this approach more versatile, more attractive, and more widely applicable [148]. Expanded-bed technology allowed the removal of particulates and soluble impurities in a single operation [149], and other separation modes also became available [150, 151]. Finally, the advent of recombinant DNA technology in the late 1970s made it possible to produce large amounts of human proteins in microbial or animal cells, resulting in high-value products—but with challenging purification needs.

Recombinant human insulin (rh-insulin; also called "bacterial-derived human insulin") was the first product based on recombinant DNA technology to be approved by the FDA. The manufacturing process resulted in several closely related insulin derivatives as impurities that could be resolved by analytical RPC, but not by any of the other prep-LC methods existing at that time [152, 153]. A commercial-scale HPLC-process for the purification of insulin was subsequently developed, as discussed in Section 13.9.2 below. In subsequent years, prep-LC in the RPC mode became the method of choice for purifying peptides and low-molecular-weight proteins [152, 154]. Other proteins (e.g., monoclonal antibodies) cannot be purified by RPC because they do not tolerate organic solvents. However, alternative chromatographic modes that use aqueous mobile phases have been more successful (e.g., affinity, metal chelation, hydroxyapatite, and hydrophic interaction chromatography [151]).

The purification of insulin by high-performance RPC served as an example for many other recombinant products, including growth hormone [155], erythropoietin [156], hirudin [157], cytokines [158, 159], insulin-like growth factor 1 (IGF-1) [160], and granuloma colony-stimulating factor (G-CSF) [161]. It soon became apparent that preparative high-performance RPC is a powerful and versatile method for purifying peptides and small proteins; the value of products purified by high-performance RPC has been estimated to account for approximately a third of total biotech product sales [162].

13.9.2 Production-Scale Purification of rh-Insulin

The purification process for recombinant human insulin is a rare example for which the details of the separation and its development have been published [162]. It is therefore instructive to review this separation, various aspects of which can be organized as follows:

- purification targets
- stationary phase
- packing the column
- stability of the product and column
- mobile phase
- separation

- column regeneration
- small-scale purification
- scale-up
- production scale purification
- product analysis

13.9.2.1 Purification Targets

The RPC purification goals were a purity >97.5%, with yields of $\geq$ 75%. Two different microbial production processes were originally considered: (1) a two-chain process in which insulin A- and B-chains were produced in separate fermentations then combined, and (2) a single-chain process in which proinsulin was formed by fermentation, then enzymatically converted to insulin by protease treatment. A single purification process was designed to handle either feedstock, as long as insulin-like impurities did not exceed 20%; in each case IEC, SEC, and high-performance RPC were used. The RPC separation will be discussed further. The mobile phase was required to be compatible with insulin stability and other steps in the process. The allowed cost of this purification step per pound of product was determined, based on expected product sales in the ton-per-year range.

13.9.2.2 Stationary Phases

The column packing was selected after screening products from five manufacturers. Different alkyl-chain lengths, particle diameters, and pore sizes were evaluated, using 150 × 9.4-mm columns. Best results were obtained with particle diameters $\leq$ 12 μm, pores of 12 to 15 nm, and C_8 or C_{18} ligands; these packings proved to be less fragile and easier to pack than particles with pores $\geq$ 30 nm. The final packing (Zorbax™ Process Grade C8) was selected on the basis of the latter preferred properties, as well as availability of the packing in the required quantities, and a demonstration by the manufacturer of batch-to-batch reproducibility.

13.9.2.3 Packing the Column

To achieve the required separation on a commercial scale, prep-LC columns were required with plate numbers N that were similar to values for analytical columns. Laboratory-scale columns were slurry-packed under high pressure. For larger columns containing 5 to 50 kg packing, axial-compression was used at a pressure of 750 psi. Values of N were measured by injecting small-molecule test samples; N was equal to 30,000 to 40,000 plates/m for the laboratory-scale columns, and 45,000 to 55,000 plates/m for the larger axial-compression columns. Axial-compression columns also maintained their performance for a longer time, by minimizing voids and channeling that normally occur as the packing deteriorates and/or settles.

13.9.2.4 Stability of the Product and Column

Insulin purification is best carried out with a mobile-phase pH of 3.0 to 4.0, as insulin solubility decreases to a minimum at its isoelectric point of 5.4 [163]. Under acidic conditions insulin deamidates to form monodesamido (A-21) insulin [164]. This undesirable reaction was avoided, however, because the RPC separation required only a few hours, and the insulin product could be rapidly recovered from the mobile phase by crystallization as the zinc salt [163]. At pH-7 or above, the monodesamide (B-3) derivative forms, which is undesirable. The column packing was stable over the pH range 2.0 to 8.0.

13.9.2.5 Mobile-Phase Composition

The separation of insulin and two impurities of interest are shown in Figure 13.44 for a mobile-phase pH that is either acidic (Fig.13.44*a*) or mildly alkaline (Fig.13.44*b*).

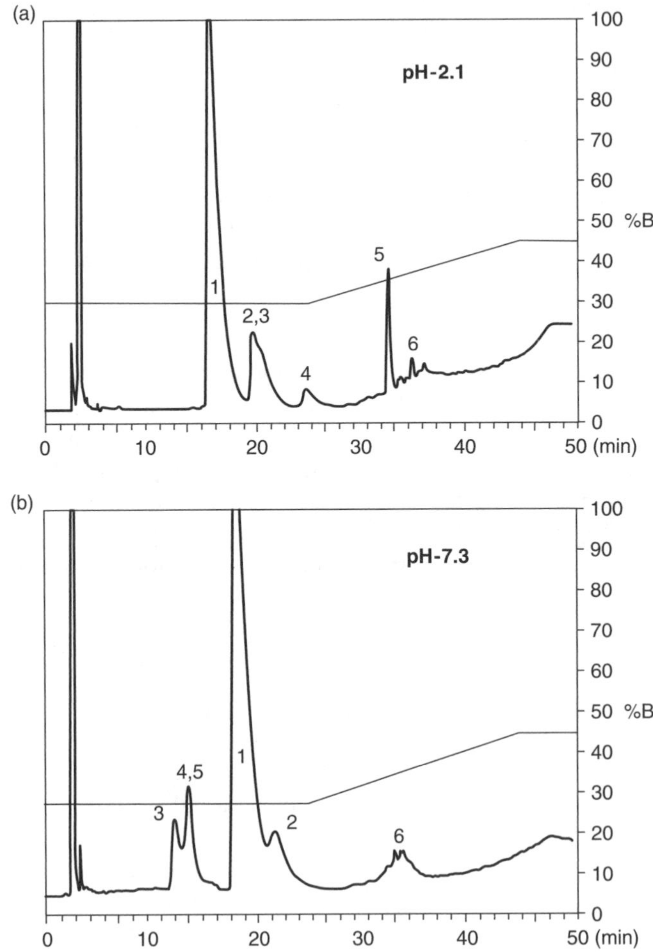

Figure 13.44 RPC separation of rh-insulin and insulin derivatives with (*a*) acid and (*b*) alkaline mobile phases. Sample: *1*, 7.5 μg rh-insulin and 1.3 μg of each insulin derivative: *2*, desamido A-21 insulin; *3*, *N*-carbamoyl-Gly insulin; *4*, *N*-formyl-Gly insulin; *5*, *N*-carbamoyl-Phe insulin; *6*, insulin dimers. Conditions: 250 × 3.5-mm Zorbax™ C8 column; 35°C; 1.0-ml/min; gradients: (*a*) solvent-A is pH-2.1 phosphate buffer; solvent-B is 50% acetonitrile/solvent-A; (*b*) solvent-A is pH-7.3 phosphate buffer; solvent-B is 50% acetonitrile/solvent-A Adapted from [166].

Better resolution was obtained under acidic conditions, and all of the impurities elute after insulin (very desirable!); for mildly alkaline conditions, the impurities elute both before and after insulin. Several acids were evaluated, including acetic, formic, propionic, and phosphoric acid. Acetic acid (0.25M) was selected, because high concentrations of insulin (>50 mg/mL) were found to dissolve in the monomeric form under these conditions. The tendency of insulin to aggregate restricted the choice of mobile-phase conditions because aggregation interfered with the separation.

Several B-solvents were evaluated, including ethanol, isopropanol, acetonitrile, and acetone. Ethanol and isopropanol gave poor resolution and a lower recovery (50–60%), while the solubility of insulin in isopropanol was poor. Acetone gave low yields (<50%) due to insulin precipitation. Acetonitrile provided the best separation, highest yields (75–85%), and did not interfere with the zinc precipitation step. Acetonitrile was also available in bulk and could be recovered by distillation; for these reasons it was selected as B-solvent.

13.9.2.6 Separation

Isocratic and gradient elution of insulin were compared. With gradient elution, the product could be eluted in less than one column volume, while isocratic elution required two or more column volumes. Gradient elution was selected because the smaller elution volume facilitated downstream processing. A gradient from 15 to 30% acetonitrile provided satisfactory purity and yield, with a minimal volume of mobile phase; step gradients with comparable performance could not be found. The saturation capacity of the column (Section 15.3.2.1) for insulin was approximately 85 mg insulin/mL packing. Insulin mass recovery exceeded 97%.

13.9.2.7 Column Regeneration

For the process to be economical, a column must be usable over many cycles. Samples of cellular origin tend to foul chromatographic columns rapidly because of the presence of lipids, nucleic acids, cell wall and cell membrane fragments, complex carbohydrates, and other cellular components. Column fouling can be reduced by proper design of steps that precede RPC, including filtration, precipitation, and IEC. Following each insulin separation, it was found necessary to clean the column by a wash with 60% acetonitrile/buffer (50 mM ammonium phosphate; pH-7.4).

13.9.2.8 Small-Scale Purification

A small-scale insulin purification procedure was developed on the basis of the experiments outlined above, including IEC before RPC and gel filtration after. The ion-exchange step removed most of the protein impurities while protecting the RPC column from potential column fouling. The role of the RPC separation column was as a "polishing" step for removal of species closely related to insulin. A series of scale-up runs were carried out next. Figure 13.45 illustrates the preparative RPC separation. A major product peak is seen, followed by impurities that elute after the main peak. The resolution of product from these impurities appears poor, but this is not the case. The analysis of fractions from the RPC separation showed that the center of the mainstream peak from 3.4 to 4.3 column volumes (C-Vs) was nearly pure insulin (98.7%) in 82% yield, while the side-stream fractions (3.3–3.4

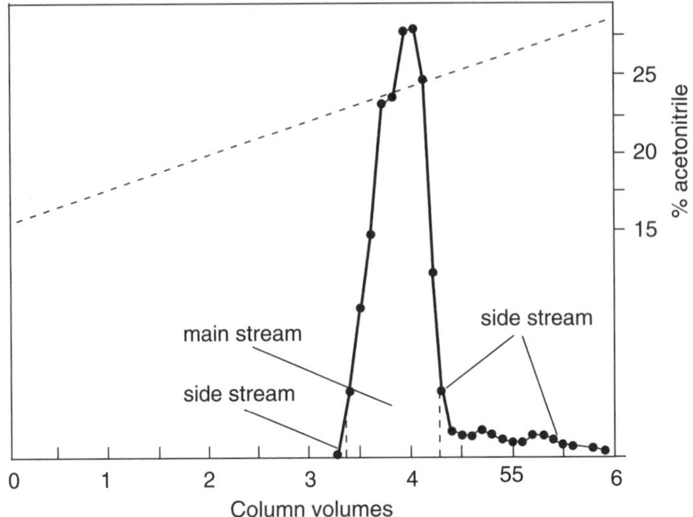

Figure 13.45 Preparative separation of rh-insulin. Column, 10 μm Zorbax™ Process Grade C8 (150 × 9.4-mm i.d.); load, 153 mg rh-insulin from proinsulin process; gradient, 17–29% acetonitrile in 0.25M acetic acid in six column volumes; flow rate, 0.3 mL/min. Fractions from 3.3 to 4.3 column volumes were pooled (mainstream). Fractions 3.2–3.3 and 4.4–5.4, plus the protein eluted during column regeneration (not shown), were combined as the side stream. Adapted from [162].

and 4.4–5.4 C-Vs plus recovered solvent from column regeneration) contained an additional 15% of the product (for re-separation)—for an overall insulin recovery of 97%. The initial purity of the sample for RPC separation was 91.5%. This example illustrates a common property of preparative separations: in contrast to analytical chromatography, prep-LC chromatograms may suggest poor separation of product from impurities, but when fractions are collected and analyzed, the results often are acceptable.

13.9.2.9 Scale-Up

Scale-up experiments were carried out next, using six, successively larger, axial-compression columns, with the bed-volume increased from 10 mL to 80L, as summarized in Tables 13.9 and 13.10. In each separation the weight of insulin applied to the column was 14 to 15 g per L of column-volume (C-V), the flow rate was 1.5 C-V/h, and the gradient slopes were 2%/C-V. Minor changes in gradient slope were necessary to maintain column performance, as measured by product purity and recovery. Flow rates were increased in proportion to the volume of the column as the process moved from lab to pilot plant to production. Purity was 98.5, 98.6, and 98.6% at lab scale, pilot scale, and production scale, respectively, while mainstream yields were 82, 79, and 83%. It may seem remarkable that mainstream purities, recoveries, and elution volumes remained consistent from lab scale to production scale over a 10,000-fold range of column volumes, but this should be true of scale-up if carried out properly (Section 15.1.2.1).

Table 13.9

Column Sizes Used for rh-Insulin Scale-up Studies

Column size (mm)	Internal volume (L)	Type	Insulin load (g)
150×9.4	0.01	Fixed	0.15
300×22	0.12	Fixed	1.7
500×50	0.59	Fixed	8.9
450×150	8	Axial	120
500×300	35	Axial	525
500×450	80	Axial	1200

Source: Data from [162].

Table 13.10

Summary of rh-Insulin Scale-up Studies

Column Size (Volume)	Scale	Operating Conditions			Product	
		Flow Rate (C-V/h)[a]	Gradient (%B/C-V)[b]	Load (mg/mL)	Purity	Yield
150 × 9.4-mm (10 mL)	Lab	1.6	2.2%	13	98.5%	82%
350 × 150-mm (6.2 L)	Pilot plant	1.5	2.0%	15	98.6	79%
570 × 300-mm (40 L)	Pilot plant	1.4	2.1%	15	98.6	83%

Source: Data from [162].
[a]C-V is empty column-volume.
[b]Change in %B per column-volume (C-V) of mobile phase; proportional to gradient steepness.

13.9.2.10 Production-Scale Purification

The purification of rh-insulin on a production scale was next carried out for both two-chain insulin and proinsulin. Separation was more challenging for the two-chain process because of a 20%-higher concentration of structurally related impurities. The production-scale conditions were: 48 × 30-cm column, 500 g insulin sample, and a gradient of 17–30% acetonitrile over 6 C-V at a flow rate of 1.4 C-V/h (0.8 L/min). The purity of the charge was 80%, and the mainstream purity was 98.5%. For insulin derived from the proinsulin process, the purity of the feedstock was higher (91%), which resulted in purified product of higher purity (99.1%). The corresponding purification of a 1-kg sample with a 48 × 45-cm column yielded comparable results.

The product fractions from high-performance RPC were subjected to an additional purification step by SEC. Two lots of rh-insulin from each process were then compared to insulin purified by conventional chromatography that did not use high-performance RPC. The results showed that purification by RPC results in higher purity levels, equivalent biopotency, and comparable low levels of contamination by endotoxin or host cell protein. No siloxanes (potential breakdown products from the stationary phase) were detected.

13.9.3 General Requirements for Prep-LC Separations of Proteins

Targets for yield and purity are needed, as well as methods (usually RPC) to measure yield and purity. Estimates of product weight (e.g., tons/year) and targets for the expected delivery time and purification cost are also required. A defined starting material is necessary, as well as an expected range of purity for the starting material. A standard for the purified product is helpful, but the standard can be produced as part of process development. Mass spectrometry and other physical methods can be useful for confirming the identity and purity of purified product. Additional development requirements may be imposed if the separation is intended to produce pharmaceutical products under current Good Manufacturing Practices (cGMP; [165, 166]) or ISO 9000 guidelines [167].

For protein products, a primary requirement is product stability during separation. While, in principle, proteins can be denatured during RPC and renatured afterward [168], this approach has generally not been favored for purifications by means of RPC. Not only must the product retain its biological activity, there should also be no detectable chemical changes such as oxidation, deamidation, or cleavage of peptide bonds. Mass spectrometry methods greatly simplify the task of detecting the latter modifications. Methods for assessing the stability of proteins under various conditions are well established. It is often preferable to measure the stability profile of a protein product before chromatography development, so that time is not wasted exploring modes or mobile phases that are incompatible with the product.

13.10 SYNTHETIC POLYMERS

Separations of synthetic polymers are usually carried out for one of two purposes: (1) determination of the molecular-weight distribution of a sample (Section 13.10.3.1), and/or (2) determination of different compound types or classes in the sample (Section 13.10.3.2). These applications differ fundamentally from other HPLC separations covered in this book. For this and other reasons the present section represents only an introduction to separations of synthetic polymers.

13.10.1 Background

Synthetic polymers are large, man-made molecules; in all cases they are formed from one or more different monomers, which occur in the molecule many times. If a single monomer is polymerized, the result is a *homopolymer* (Fig. 13.46*a*); short-chain members of such a sample are referred to as *oligomers*. An example is ethylene as monomer, with polyethylene as the resulting polymer ($C_2H_5{-}[C_2H_4{-}]_{p-2}{-}C_2H_5$, for the reaction of p ethylene molecules to form a polymer molecule). If two (or more) different monomers are used to create a synthetic polymer, we have a *copolymer* (Fig. 13.46*f*). Homopolymers can differ in length, as in Figure 13.46*a*, *c*, *e*. Molecular length is expressed either as the degree of polymerization p (i.e., where p is the number of monomeric units) or the molecular weight of the polymer.

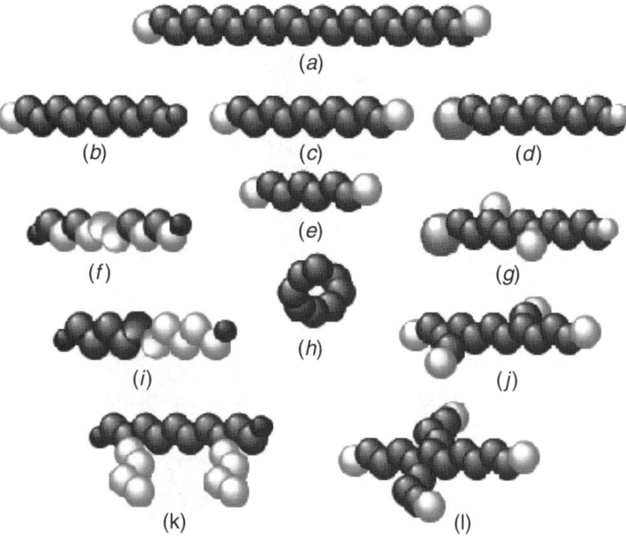

Figure 13.46 Different structures of synthetic polymers. (*a*), (*c*), and (*e*), linear homopolymers of varying length; (*b*), (*c*), and (*d*) are linear homopolymers with different end-groups; (*g*), functional groups have been introduced along the chain; (*h*), a cyclic homopolymer; (*f*), a random copolymer, (*i*), a block copolymer; (*k*), a graft copolymer; (*j*) and (*l*) are branched homopolymers, with short-chain branching and long-chain branching, respectively.

Finally, homopolymers can differ in molecular shape or topology. Linear molecules exist (Fig. 13.46*a*–*f*), as well as cyclic (small) oligomers (Fig. 13.46*h*) and related polymers. Depending on the synthetic process, branches may be deliberately or accidentally introduced (Fig. 13.46*g*, *j*–*l*). Individual molecules can differ in the number of branches, their length, and their position in the molecules. In the case of copolymers the sequence of the monomers is relevant. If this sequence is determined by a purely statistical process, *random copolymers* are formed. The opposite extreme is that of *block copolymers* (Fig. 13.46*i*), where long sequences of a single monomer occur within the molecule. Finally, there are *graft copolymers* (Fig. 13.46*k*), where chains formed from a different monomer are attached to the primary polymer backbone.

Polymer properties are affected by structural features, which are therefore important. A higher molecular weight generally leads to a stronger polymer. End-groups and functional groups are critically important for polymers used in reactive formulations such as adhesives, sealants, and coatings. Branching usually affects the processing properties of polymers. Block copolymers can have very different properties compared to random copolymers. One structural property that is not depicted in Figure 13.47 is the degree of stereoregularity of the chain, usually called *tacticity*. In an atactic polymer, the monomeric units are oriented in a random fashion; in an isotactic (or syndiotactic) polymer, all monomers are positioned in the same (or alternating) direction. Stereoregular polymers usually exhibit a much higher degree of crystallinity. As a result atactic polypropene (or "polypropylene," as it used to be called) is a soft plastic, whereas isotactic polypropene is hard and strong.

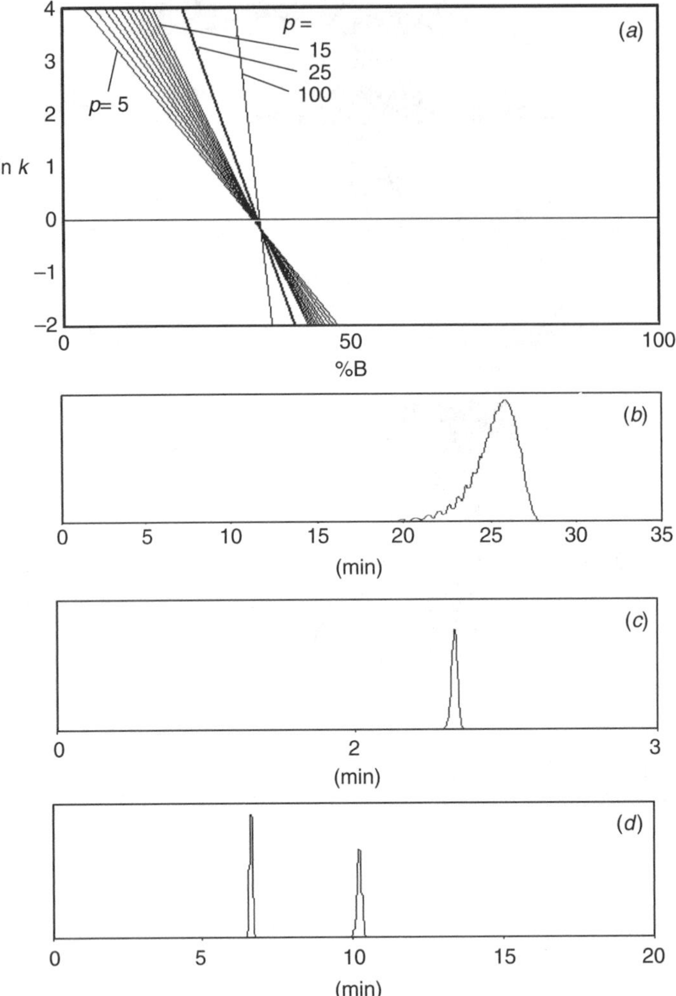

Figure 13.47 Polymer retention and separation in i-LC. (*a*) Illustration of the retention behavior of a homopolymeric series; (*b*) hypothetical illustration of the elution of a low-molecular-weight homopolymer with a 35-minute gradient; (*c*) similar separation as in (*b*) for a gradient time of 10 minutes; (*d*) hypothetical illustration of the separation of a polymeric blend by interactive liquid chromatography (i-LC).

Just as the properties of a synthetic polymer are affected by its molecular structure, so is polymer chromatographic behavior. This allows us to separate polymers based on molecular weight, chemical composition (functionality), stereoregularity, degree of branching, and so forth, as summarized in Table 13.11 for different separation modes.

Many physical properties of synthetic polymers are important in relation to their chromatographic behavior, none more so than their solubility. For any kind of material to be separated by liquid chromatography, it must be dissolved completely (any agglomerates or particles are detrimental to chromatographic separation).

Table 13.11

Effect of the Molecular Structure of Synthetic Polymers on their Chromatographic Behavior

	Molecular Weight	End Groups	Chemical Composition	Stereo-Regularity	Branching	Section
Size-exclusion chromatography	●	–[a]	○	–	○	13.8
Interactive liquid chromatography	○	○/●[b]	●	–	–[c]	13.10.3.2
Temperature-gradient interaction chromatography	●	●	○[d]	–	–	–

Note: ●, major effect; ○, minor effect; —, no significant effect.
[a]A significant adverse effect may be observed if end groups or functional groups show strong interactions with the stationary phase.
[b]Effect is strong in isocratic ("critical") chromatography of polymers. In gradient-elution LC it is usually overshadowed by the effects of molecular weight and, especially, chemical composition.
[c]A significant effect may be observed if branching introduces different or additional functional groups.
[d]Effect may be strong, but the technique is not usually applied for the separation of copolymers.

For many polymers, complete dissolution can be difficult; polyolefins (polyethene, polypropene), for example, require elevated temperatures and a high-boiling solvent such as trichlorobenzene. Many polar polymers (polyesters, polyamides, polyketones) require special solvents such as hexafluoroisopropanol. Polymers can also require a long time to dissolve because of the strong interactions between chains (including "entanglements") and the slow diffusion of large molecules.

Another important consideration is the HPLC detection of synthetic polymers. Many important types of polymers (poly-olefins, poly-acrylates, poly-alkoxides) lack UV chromophores. Consequently the RI detector is mainly used for isocratic separations of polymers (notably by SEC). Occasionally, infrared-absorption detectors (operating at a fixed wavelength) can be advantageous. In gradient separations, the evaporative light-scattering detector has been commonly used, but quantitation can be a problem; the charged-aerosol detector is a promising alternative. Although the characterization of (polar) polymers by mass spectrometry has greatly improved since the mid-1990s, the use of LC-MS for this purpose is still uncommon. Some common synthetic polymers, typical solvents used in size-exclusion chromatography, and the most commonly used detection methods are listed in Table 13.12. For additional information on possible detectors for polymer separation, see Chapter 3, Chapter 9 of [169], and [170–172].

13.10.2 Techniques for Polymer Analysis

Synthetic polymers are not amenable to the resolution of individual molecules, a difference that sets them apart from other samples for HPLC separation and analysis. Instead, polymer molecules come in a range of sizes or a *molecular-weight distribution* (MWD). While many techniques can be used to determine an average molecular weight for the sample, chromatographic methods such as size-exclusion

Table 13.12

Common Synthetic Polymers, Typical Solvents Used in Size-Exclusion Chromatography, and Commonly Used Detection Methods

Typical Synthetic Polymer	Typical Solvents					Commonly Used Detectors				
	Water	Methanol	Tetrahydrofuran (THF)	Hexafluoro Isopropanol (HFIP)	Trichloro Benzene (TCB)	UV	Refractive Index	Evaporative Light Scattering	Multi-angle Light Scattering	Viscometry
Polyethene oxide (polyethylene oxide, polyethyleneglycol)	○						●	○[c,d]	●	●
Polypropene oxide (polypropylene oxide, polypropyleneglycol)		●					○	●[c,d]	●	●
Polystyrene			●			●				
Poly (alkyl acrylate) and poly (alkyl methacrylate)[a]			●			○	●	○[c,d]	●	●
Aliphatic polyesters, aliphatic polyamides				●			●	○[c,d]	○	○
Aromatic polyesters, aromatic polyamides				○						
Polyethene (polyethylene), polypropene (polypropylene)					●[b]		○		●	●

Note: ● indicates applicability of a solvent or detector.
[a]For example, poly (methyl acrylate) and poly (methyl methacrylate).
[b]Requires operation at high temperatures (e.g., 150°C).
[c]More realistic option for interactive liquid chromatography than for size-exclusion chromatography.
[d]Charged-aerosol detection may be preferred alternative.

chromatography (SEC) are able to determine the MWD, as well as number-average or weight-average molecular weights [169]. In addition to a MWD, copolymers exhibit a chemical-composition distribution. Spectroscopic techniques such as Fourier-transform infrared (FTIR) spectroscopy and (especially) nuclear-magnetic resonance (NMR) spectroscopy can provide detailed information on chemical composition, while interactive liquid chromatography can separate different chemical types and provide a chemical-composition distribution (CCD). A nonexhaustive overview of a number of important techniques is provided in Table 13.13. It is seen that NMR is highly useful in the middle column (averages), whereas the right-side column (distributions) is dominated by chromatographic techniques.

13.10.3 Liquid-Chromatography Modes for Polymer Analysis

13.10.3.1 Size-Exclusion Chromatography

Size-exclusion chromatography (SEC) is reviewed in Section 13.8 and in [169]; its application for the determination of molecular weight or molecular-weight distribution (MWD) is similar for both synthetic polymers and biopolymers. There are two main differences between these two applications of SEC. For synthetic polymers, SEC is used to determine a molecular-weight distribution [169], whereas for biopolymers the goal is the estimation of molecular weight for individual compounds. Likewise the solvent used as mobile phase is often different; usually aqueous mobile phases are used for biopolymers (gel filtration), and organic solvents for synthetic polymers (gel permeation).

In the case of homopolymers, SEC can be coupled to other polymer-characterization methods, notably light-scattering and viscometry (for copolymers it is difficult to accurately correlate the resulting data [170]). Static light-scattering can be used to obtain accurate information on the (weight-average) molecular weight of polymer in the SEC effluent, provided that (1) we know how refractive index varies as a function of polymer concentration, and (2) the detector is properly calibrated. Also the concentration of the polymer in the effluent fraction must be accurately known, for example, by using a RI detector in conjunction with light-scattering.

13.10.3.2 Interactive Liquid Chromatography

In SEC, conditions are selected to suppress interactions between the analyte and the stationary phase as much as possible. In interactive liquid chromatography (i-LC), these interactions are used to separate molecules by chemical type or functionality. While i-LC separations of polymers are, in many ways, similar to the separation of small molecules by HPLC, there are two overriding differences: (1) the molecular-weight range of polymers (large number of individual species that differ in molecular weight), and (2) a systematic change in analyte retention as the size of the solute molecule increases. High-molecular-weight analytes typically exhibit larger changes in k for a given change in %B, as seem in the examples of Figure 13.11 for several peptides and proteins. A similar example for synthetic polymers is illustrated in Figure 13.47*a*, which illustrates schematically how retention varies with composition for oligomers and polymers that differ in their size or degree of polymerization p (number of monomers). For the oligomers of Figure 13.47*a*, p equals 5–15; for the polymers, p equals 25 and 100. The curves for larger molecules (larger p) are increasingly steep, to the extent that for large polymers there is only a

Table 13.13

Summary of Techniques for Determining Average Molecular Structures and Molecular Distributions of Synthetic Polymers

Polymer Property	Techniques for Determining Averages	Techniques for Determining Complete Distributions
Molecular weight	Osmometry light scattering	Size-exclusion chromatography Hydrodynamic chromatography Sedimentation Ultracentrigugation
Chemical composition	NMR FTIR pyrolysis GC-MS	Interactive liquid chromatography (mainly gradient elution)
Functionality (end groups or functional groups)	NMR titration	Interactive liquid chromatography (mainly isocratic)
Chain regularity	NMR	Temperature-rising elution Fractionation
Degree of branching	NMR[a]	Molecular-topology fractionation

[a]Appropriate for polymers with a relatively low molecular weight. Also NMR is generally considered appropriate only for determining short-chain branching, not long-chain branching.

very narrow range of mobile-phase composition (%B) for which the polymer can be eluted isocratically. For this reason gradient elution is usually the method of choice for separations by i-LC.

Gradient elution is usually carried out with linear gradients, corresponding to (roughly) constant values of gradient retention $k^*(\equiv k$; see Section 9.1.3) for different polymeric species. Under these conditions retention times for each peak in a polymer sample will correspond to the intersection of plots as in Figure 13.47*a* with a horizontal line that corresponds to a given value of k or k^*. For higher values of k^* (corresponding to a longer gradient; Eq. 9.5), there are larger differences in retention time for adjacent peaks (and therefore better resolution), compared to separations with a shorter gradient. Finally, for a sufficiently fast gradient (and small enough value of k^*), all solutes leave the column with the same retention time as a single peak. The latter behavior for the separation of a low-molecular-weight polymer with long and short gradients is illustrated in Figure 13.47*b*, *c*, respectively. In the long gradient of Figure 13.47*b*, the retention of individual oligomers differs enough so that there is a partial separation of the sample. In the short gradient of Figure 13.47*c*, this is no longer true, so a single peak is observed—corresponding to elution of all peaks at about the same time (e.g., $k \approx 1$ in Fig. 13.47*a*).

Separation as in Figure 13.47*c* or in Figure 13.47*a* for $k = 1$ is sometimes referred to as *pseudocritical chromatography*, as opposed to (isocratic) chromatography under critical conditions, as described in Section 13.10.3.4. Pseudocritical i-LC (i.e., with gradient-elution) is particularly useful for the separation of polymers according to chemical composition. This is illustrated in Figure 13.47*d*, which shows the separation of two different kinds of polymers. Retention is seen to be a function of chemical composition in this example, but not of molecular weight. Such pseudocritical conditions can be approached more closely for (1) higher molecular-weight polymers and (2) shorter gradient times. As a result gradient-elution i-LC is well suited for the determination of chemical-composition distributions. Figure 13.48*b*

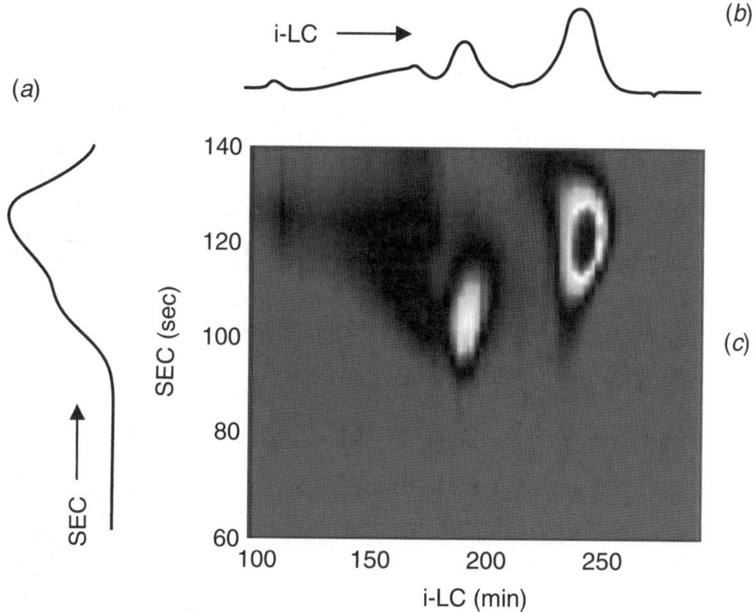

Figure 13.48 Comprehensive two-dimensional liquid chromatography (LC × SEC) of a copolymeric binder produced by two-stage emulsion polymerization of styrene and methyl methacrylate. (*a*) Second-dimension separation: SEC; (*b*) first-dimension separation, RPC linear gradient from 65 to 100% in 170 minutes; (*c*) 2D separation. UV detection at 214 nm. For further details, see [170].

provides a practical illustration where a high-molecular-weight copolymer of styrene and methyl methacrylate is very well separated by chemical type (note three peaks in the chromatogram, for three chemical types in the sample). For a further discussion of Figure 13.38, see Section 13.10.4.

13.10.3.3 Liquid Chromatography under Critical Conditions

Figure 13.47*a* suggests that there is a "critical" mobile-phase composition for which all oligomers co-elute. The potential benefit of working at or near critical conditions is that the effect of the homopolymeric chain on retention can be minimized, so that polymers with different structural elements (e.g., end-groups) can be separated as in Figure 13.47*d*—regardless of their MWD. This implies that critical conditions can be used to determine differences in polymer functionality [174]. NPC separations are especially suited for this kind of separation because of the large effect that a (polar) functional group can have on retention.

13.10.3.4 Other Techniques

Some other HPLC procedures for polymer separation are noted in Table 13.13. Very large polymers can be separated by field-flow fractionation (FFF) and by hydrodynamic chromatography (HDC), techniques that are outside the scope of the present book.

13.10.3.5 Chemical Composition as a Function of Molecular Size

A copolymer typically exhibits both molecular-weight and chemical-composition distributions. Depending on polymerization conditions, the chemical composition may or may not vary with polymer molecular weight. To investigate the presence of such chemical heterogeneity, we can couple SEC with a spectroscopic technique that yields chemical-composition information. Such a combined technique provides the average composition at each point in the SEC chromatogram, that is, for each molecular size. If only one of two monomers can be detected by UV, the combination of a UV detector and another concentration-sensitive detector (e.g., refractive index, RI) can in principle be used to follow the concentration of each monomer. Additional information can be obtained from combining SEC with either FTIR or NMR spectroscopy.

Although information about chemical composition as a function of molecular size can be very valuable, even the smallest SEC fractions can contain a variety of molecules that vary in both chemical composition and molecular weight. That is, differences in chemical composition can result in molecules with different molecular weights having the same molecular "size" in solution, as illustrated in Figure 13.49. A fraction obtained from a high-resolution SEC separation (rectangular box in Fig. 13.49) will contain molecules with the same molecular size (gyration radius R_g) in solution, but with different molecular weights. It is often important to know the chemical-composition distribution, rather than just the average chemical composition. Likewise the functionality-type distribution (FTD) may be more important than the average number of functional groups per molecule. This will be especially true if the chemical composition or the number of functional groups per molecule is known (or suspected) to vary. An example is reactive (pre-)polymers that are used in many formulations for sealants, adhesives, and coatings. Molecules without reactive (functional) groups will not react, molecules with one functional group will locally terminate the polymerization process, molecules with two functional groups will

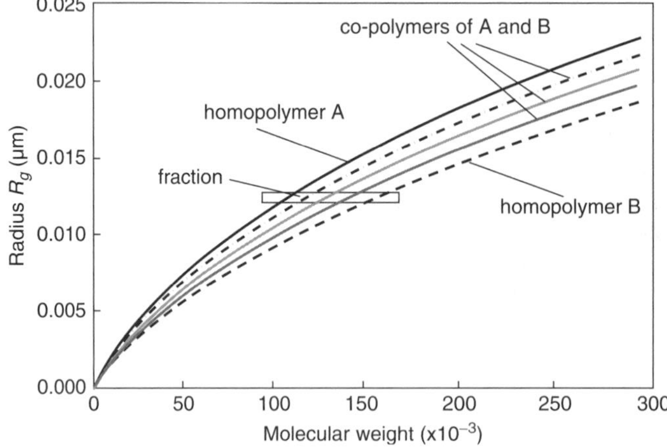

Figure 13.49 Schematic illustration of the relationship between molecular size and molecular weight for (co-)polymers of different composition. Lines represent (from top to bottom) homopolymer A, copolymer AB (75:25), AB (50:50), AB (25:75), and homopolymer B.

sustain the polymerization, and molecules with more than two functional groups promote the formation of resinous polymeric networks. Knowledge of only the average number of functional groups per molecule would be insufficient in this case.

13.10.4 Polymer Separations by Two-Dimensional Chromatography

In comprehensive two-dimensional liquid chromatography (LC × LC; Sections 9.3.10, 13.4.5), the entire sample is subjected to two different successive separations, while the separation obtained in the first dimension is preserved. To simultaneously determine two mutually dependent distributions, such as the combination of MWD and CCD (MWD × CCD), a technique that separates according to molecular weight (e.g., SEC) must be combined with one that separates (largely) according to composition, such as i-LC. Combination of the two separations (i-LC × SEC) then yields a two-dimensional chromatogram that represents an analysis of the sample according to both molecular weight and chemical composition; an example is shown in Figure 13.48. Corresponding one-dimensional separations are shown for SEC at the side, and for i-LC at the top of Figure 13.48. While neither of the latter one-dimensional separations provides an adequate separation of the total sample, the corresponding two-dimensional separation does. Another i-LC × SEC separation is shown in Figure 13.50, for a more complex sample: chain-end-functionalized poly(methyl methacrylates). The horizontal time-axis for the i-LC separation is indicative of the chemical composition of the copolymer (note labels at top of figure for the number of functional groups in the molecule); while the vertical time-axis for the SEC separation is related to its molecular weight.

Two-dimensional chromatograms such as those in Figures 13.48 and 13.50 can provide a useful qualitative picture of the composition of a copolymer. Different samples can be compared in great detail, and the results of such a comparison

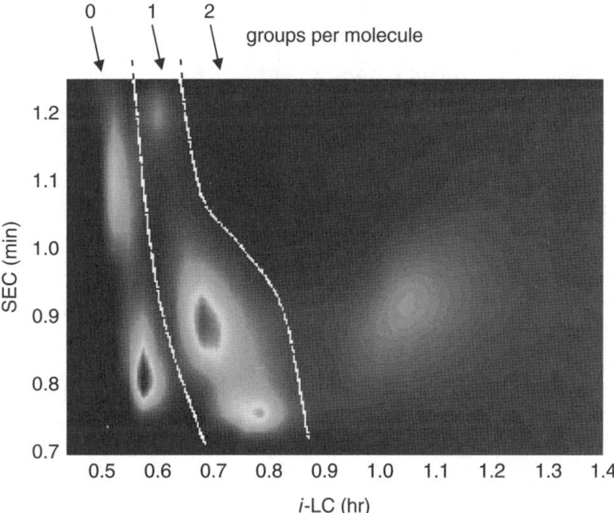

Figure 13.50 Two-dimensional separation of chain-end-functionalized poly(methyl methacrylates). The dashed lines indicate areas in the 2D-chromatogram that correspond to molecules with zero, one or two functional groups, as indicated at the top of the figure. Adapted from [172].

can be used to better understand the properties of polymeric materials or related polymerization processes [173]. Unfortunately, it is much more difficult to obtain quantitative information from such figures, as a number of complications arise. *First*, the relationship between SEC retention time and molecular weight depends also on polymer chemical composition and topology (e.g., degree of branching). *Second*, detector response also depends on these polymer properties.

To solve the first problem (retention not completely defined by molecular weight), we must know retention in SEC as a function of solute molecular weight *and* chemical composition; this can be accomplished by the use of appropriate copolymer standards. The second problem (varying response factor) is more of a challenge. When homopolymers are studied, the response factor may be nearly constant (i.e., independent of molecular weight) for UV detection. However, many polymers lack chromophors, which necessitates the use of refractive-index (RI) detection. Here the response factor (usually referred to as the refractive-index increment or d*n*/d*c*) tends to be nonconstant in the oligomeric region.

REFERENCES

1. G. T. Hermanson, *Bioconjugate Techniques*, Academic Press, San Diego, CA, 1996.
2. L. R. Snyder and M. A. Stadalius, *High-Performance Liquid Chromatography: Advances and Perspectives*, Vol. 4, C. Horváth, ed. Academic Press, San Diego, 1986, p. 195.
3. L. R. Snyder and J. W. Dolan, *High-Performance Gradient Elution*, Wiley-Interscience, Hoboken, NJ, 2007.
4. J. O. Konz, R. C. Livingood, A. J. Bett, A. R. Goerke, M. E. Laska, and S. L. Sagar, *Hum. Gene Ther.*, 16 (2005) 1346.
5. E. I. Trilisky and A. M. Lenhoff, *J.Chromatogr.*, 1142 (2007) 2.
6. M. A. Stadalius, B. F. D. Ghrist, and L. R. Snyder, *J. Chromatogr.*, 387 (1987) 21.
7. J. S. Richardson, *Adv. Protein Chem.*, 34 (1981) 167.
8. W. R. Melick-Adayan, V. V. Barynin, A. A. Vagin, V. V. Borisov, B. K. Vainshtein, B. K. Fita, M. R. N. Murthy, and M. G. Rossman, *J. Mol. Biol.*, 188 (1986) 63.
9. R. L. Cunico, K. M. Gooding, and T. Wehr, *Basic HPLC and CE of Biomolecules*, Bay Bioanalytical Laboratory, Richmond, CA, 1998.
10. M. T. W. Hearn and B. Grego, *J. Chromatogr.*, 282, (1983) 541.
11. W. Doerfler, in *Medical Microbiology*, 4th ed., S. Baron (ed.), Univ. TX Medical Branch, Galveston, 1996.
12. N. B. Afeyan, N. F. Gordon, I. Mazsaroff, L. Varady, S. P. Fulton, Y. B. Yang, and F. E. Regnier, *J. Chromatogr.*, 519, (1990) 1.
13. F. B. Rudolph, D. P. Wiesenborn, J. Greenhut, and M. L. Harrison, in *HPLC of Biological Macromolecules*, K. M. Gooding and F. E. Regnier, eds., Dekker, New York, 1990, p. 333.
14. F. E. Regnier, *Science*, 238 (1987).
15. M. A. Stadalius, H. S. Gold, and L. R. Snyder, *J. Chromatogr.*, 296 (1984) 31.
16. M. Kawakatsu, H. Kotaniguchi, H. Freiser, and K. M. Gooding, *J. Liq. Chromatogr.*, 18 (1995) 633.
17. A. Apfel, S. Fischer, G. Goldberg, P. C. Goodley, and F. E. Kuhlmann, *J. Chromatogr. A*, 712 (1995) 177.

18. D. V. McCalley, *LCGC*, 23 (2005) 162.
19. D. V. McCalley, *J. Chromatogr. A*, 1075 (2005) 57.
20. D. Guo, C. T. Mant, and R. S. Hodges, *J. Chromatogr.*, 386 (1987) 205.
21. M. T. W. Hearn, in *HPLC of Biological Macromolecules*, 2nd ed., K. M. Gooding and F. E. Regnier, eds., Dekker, New York, 2002, pp. 195–312.
22. J. E. Rivier, *J. Liq. Chromatogr.*, 1 (1978) 343.
23. D. Guo, C. T. Mant, A. K. Taneja, J. M. R. Parker, and R. S. Hodges, *J. Chromatogr.*, 359 (1986) 499.
24. W. Hancock, R. C. Chloupek, J. J. Kirkland, and L. R. Snyder, *J. Chromatogr. A*, 686 (1994) 31.
25. S. Terabe, S. Nishi, and T. Ando, *J. Chromatogr.*, 212 (1981) 295.
26. J. L. Glajch, M. A. Quarry, J. F. Vasta, and L. R. Snyder, *Anal. Chem.*, 58 (1986) 280.
27. C. T. Wehr and L. Correia, *LC at Work*, LC-121, Varian, Walnut Creek, CA 1980.
28. D. Guo, C. T. Mant, A. K. Taneja, and R. S. Hodges, *J. Chromatogr.*, 359 (1986) 519.
29. C. T. Mant, T. W. L. Burke, J. A. Black, and R. S. Hodges, *J. Chromatogr.*, 458 (1988) 193.
30. M. T. W. Hearn and B. Grego, *J. Chromatogr.*, 296 (1984) 61.
31. W. R. Melander, J. Jacobson, and C. Horváth, *J. Chromatogr.*, 234 (1982) 269.
32. S. Cohen, K. Benedek, Y. Tapuhi, J. C. Ford, and B. L Karger, *Anal. Biochem.*, 144 (1985) 275.
33. W. G. Burton, K. D. Nugent, T. K. Slattery, B. F. Johnson, and L. R. Snyder, *J.Chromatogr.*, 443 (1988) 363.
34. K. D. Nugent, W. G. Burton, T. K. Slattery, B. F. Johnson, and L. R. Snyder, *J.Chromatogr.*, 443 (1988) 381.
35. L. J. Licklider, C. C. Thoreen, J. Peng, and S. P. Gygi, *Anal. Chem.*, 74 (2002) 3076.
36. B. F. D. Ghrist and L. R. Snyder, *J. Chromatogr.*, 459 (1989) 43.
37. P. Gagnon, *Purification Tools for Monoclonal Antibodies*, Validated Biosystems, Tuscon, AZ, 1996.
38. W. Kopaciewicz, M. A. Rounds, J. Fausnaugh, and F. E. Regnier, *J. Chromatogr.*, 266 (1983) 3.
39. C. D. Scott in *Modern Practice of Liquid Chromatography*, J. J. Kirkland, ed., Wiley-Interscience, New York, 1971.
40. W. Muller, *J. Chromatogr.*, 510 (1990) 133.
41. M. T. Ueda and Y. Ishida, *J. Chromatogr.*, 386 (1987) 273.
42. R. Chicz and F. Regnier, *Met. Enzymol.*, 182 (1990) 392.
43. L. R. Snyder, J. J. Kirkland, and J. L. Glajch, *Practical HPLC Method Development*, 2nd ed., Wiley-Interscience, New York, 1997, p. 515.
44. L. Sluyterman and O. Elermsa, *J. Chromatogr.*, 150 (1978) 17.
45. L. Sluyterman and J. Wijdenes, *J. Chromatogr.*, 150 (1978) 31.
46. L. Sluyterman and J. Wijdenes, *J. Chromatogr.*, 206 (1981) 429.
47. L. Sluyterman and J. Wijdenes, *J. Chromatogr.*, 206 (1981) 441.
48. *Chromatofocusing with Polybuffer and PBE Handbook*, ed. AB, Publication 18-1009-07, Amersham Pharmacia Biotech, Uppsala, Sweden.
49. P. Gagnon, *Quarterly Resource Guide to Downstream Processing*, Validated Biosystems, Tuscon, AZ, 1999.

50. N. G. Good, G. D. Winget, W. Winter, T. N. Connally, S. Izawa, and R. M. M. Singh, *Biochemistry*, 5 (1966) 467.
51. T. Kawasaki and S. Takahashi, *Eur. J. Biochem.*, 152 (1985) 361.
52. T. Kawasaki, *J. Chromatogr.*, 151 (1978) 95.
53. T. Kawasaki, *J. Chromatogr.*, 157 (1978) 7.
54. M. J. Gorbunoff, *Anal. Biochem.*, 136 (1984) 425.
55. K. M. Gooding, Z. El Rassi, and C. Horváth, in *HPLC of Biologicial Macromolecules*, 2nd ed., K. M. Gooding and F. E. Regnier, eds., Dekker, New York, 2002, pp. 247–280.
56. L. Kagedal, in *Protein Purification*, J. C. Janson and L. Ryden, eds., VCH, New York, 1989, pp. 227–251.
57. F. H. Arnold, *Biotechnology*, 151 (1991) 9.
58. J. Porath and B. Olin, *Biochemistry*, 22 (1983) 162.
59. E. Hochuli, W. Bannwarth, H. Dobeli, R. Gentz, and D. Stuber, *Biotechnology*, 6 (1988) 1321.
60. B. Bodenmiller, L. N. Mueller, M. Mueller, B. Domon, and R. Aebersold, *Nature Methods*, 4 (2007) 231.
61. J. Porath, *J. Chromatogr.*, 443 (1988) 3.
62. Z. El Rassi and C. Horváth, *J. Chromatogr.*, 359 (1986) 241.
63. A. Tiselius, *Ark. Kem. Min. Geol.*, 26B (1948).
64. J. Porath, *Biochem. Biophys. Acta*, 39 (1960) 193.
65. B. Gelotte, *J. Chromatogr.*, 3 (1960) 330.
66. Y. Kato, T. Kitamura, and T. Hashimoto, *J. Chromatogr.*, 266 (1983) 49.
67. Y. Kato, T. Kitamura, and T. Hashimoto, *J. Chromatogr.*, 292 (1984) 418.
68. R. E. Shansky, S.-L. Wu, A. Figueroa, and B. L. Karger, in *HPLC of Biological Macromolecules*, K. M. Gooding and F. E. Regnier, eds., Dekker, New York, 1990, p. 95.
69. H. S. Frank and M. J. Evans, *J. Chem. Phys.*, 13 (1945) 507.
70. S. Shaltiel, Z. Er-el, *Proc. Natl. Acad. Sci. USA*, 52 (1973) 430.
71. D. L. Gooding, M. N. Schmuck, M. P. Nowlan, and K. M. Gooding, *J. Chromatogr.*, 359 (1986) 331.
72. J. L. Fausnaugh and F. E. Regnier, *J. Chromatogr.*, 359 (1986) 131.
73. D. B. Wetlaufer and M. R. Koenigbauer, *J. Chromatogr.*, 359 (1986) 55.
74. S. L. Wu, K. Benedek, and B. L. Karger, *J. Chromatogr.*, 359 (1986) 3.
75. A. J. Alpert, *J. Chromatogr.*, 499 (1990) 177.
76. M. Lafosse, B. Herbreteau, M. Dreux, and L. Morinallorym, *J. Chromatogr.*, 472 (1989) 209.
77. W. Naidong, *J. Chromatogr. B*, 796 (2003) 209.
78. B. A. Olsen, *J. Chromatogr. A*, 913 (2001) 113.
79. T. Yoshida, *Anal. Chem.*, 68 (1997) 3038.
80. H. Tanaka, X. Zhou, and O. Masayoshi, *J. Chromatogr. A*, 987 (2003) 119.
81. T. K. Chambers and J. S. Fritz, *J. Chromatogr. A*, 797 (1998) 139.
82. M. Wuhrer, C. A. M. Koeleman, A. M. Deelder, and C. N. Hokke, *Anal. Chem.*, 76 (2004) 833.
83. M. Wuhrer, C. A. M. Koeleman, C. H. Hokke, and A. M. Deelder, *Anal. Chem.*, 77 (2005) 886.

84. A. J. Ytterberg, R. R. Ogorzalek-Loo, P. Boontheung, J. Wohlschlegel, and J. A. Loo, abstract WP 523, 55th ASMS Conference on Mass Spectrometry and Allied Topics, Indianapolis, 2007.
85. C. T. Mant, J. R. Litowski, and R. S. Hodges, *J. Chromatogr. A*, 816 (1998) 65.
86. C. A. Mizzen, A. J. Alpert, L. Levesque. T. P. A. Kruck, and D. R. McLachlan, *J. Chromatogr. B*, 744 (2000) 33.
87. A. Jungbauer, C. Machold, and R. Hahn, *J. Chromatogr. A*, 1079 (2005) 221.
88. H. Lindner, B. Sarg, C. Meraner, and W. Helliger, *J. Chromatogr. A*, 743 (1996) 137.
89. H. Lindner, B. Sarg, C. Meraner, and W. Helliger, *J. Chromatogr. A*, 782 (1997) 55.
90. B. Sarg, W. Helliger, H. Talasz, E. Kooutzamani, and H. Lindner, *J. Biol. Chem.*, 279 (2004) 53–58.
91. A. J. Alpert, *Anal. Chem.*, 80 (2008) 62.
92. A. J. Alpert, G. Mitulovic, and M. Mechtler, poster P2412-W, 32nd Annual Symposium on High Performance Liquid Phase Separations and Related Techniques, Baltimore, 2008.
93. U. Lewandrowski, K. Lohrig, R. P. Zahedi, D. Wolters, and A. Sickmann, *Clin. Proteom.*, 4 (2008) 25.
94. P. H. O'Farrell, *J. Biol. Chem.*, 250 (1975) 4007.
95. M. Gilar, P. Olivova, A. E. Daly, and J. C. Gebler, *Anal. Chem.*, 77 (2005) 6426.
96. S. P. Gygi, B. Rist, S. A. Gerber, F. Turecek, M. H. Gelb, and R. Aebersold, *Nat. Biotechnol.*, 17 (1999) 994.
97. A. J. Link, J. Eng, D. M. Schieltz, E. Carmac, G. J. Mize, D. R. Morris, B. M. Garvik, and J. R. Yates, *Nat. Biotechnol.*, 17 (1999) 676.
98. D. A. Wolter, M. P. Washburn, and J. R. Yates, *Anal. Chem.*, 73 (2001) 5683.
99. M. P. Washburn, D. Wolters, and J. R. Yates, *Nat. Biotechnol.*, 19 (2001) 5683.
100. M. T. Davis, J. Beierle, E. T. Bures, M. D. McGinley, J. Mort, J. H. Robinson, C. S. Spahr, W. Yu, R. Luethy, and S. D. Patterson, *J. Chromatogr. B*, 752 (2001) 281.
101. G. J. Opiteck and J. W. Jorgenson, *Anal. Chem.*, 69 (1997) 2283.
102. G. J. Opiteck, S. M. Ramirez, J. W. Jorgenson, and M. A. Moseley *Anal. Biochem.*, 258 (1998) 349.
103. K. Wagner, T. Miliotis, G. Marko-Varga, R. Bischoff, and K. K. Unger, *Anal.Chem.*, 74 (2002) 809.
104. R. Bischoff and L. W. McLaughlin, in *HPLC of Biologicial Macromolecules*, K. M. Gooding and F. E. Regnier, eds., Dekker, New York, 1990, pp. 641–667.
105. R. Hecker, M. Colpan, and D. Riesner, *J. Chromatogr.*, 326 (1985) 251.
106. S. Nakatani, T. Tsuda, Y. Yamasaki, M. Moriyama, H. Watanabe, and Y. Kato, *Technical Report 78*, TosoHaas, Tokyo, 1995.
107. R. R. Drager and F. E. Regnier, *Anal. Biochem.*, 145 (1985) 47.
108. G. Zon, in *Characterization of Proteins: New Methods in Peptide Mapping*, W. S. Hancock, ed., CRC Press, Boca Raton, 1995, p. 301.
109. W. Xiao and P. J. Oefner, *Human Mutation*, 17 (2001) 439.
110. A. Premstaller and P. J. Oefner, in *Methods in Molecular Biology*,. 211, P.-Y. Kwok, ed., Humana Press, Totowa, NJ, 2002, p. 15.
111. R. L. Pearson, J. F. Weiss, and A. D. Kelmers, *Biochim. Biophys. Acta*, 228 (1971) 770.
112. R. Bischoff and L. W. McLaughlin, *Anal. Biochem.*, 151 (1985) 526.
113. J. D. Pearson, M. Mitchell, and F. E. Regnier, *J. Liq. Chromatogr.*, 6 (1983) 1441.

114. R. Bischoff and L. W. McLaughlin, *J. Chromatogr.*, 296 (1984) 329.
115. Z. el Rassi and C. Horváth, *J. Chromatogr.*, 326 (1985) 79.
116. Z. el Rassi and C. Horváth, *Chromatographia*, 19 (1984) 9.
117. S. C. Churms, *CRC Handbook of Chromatography: Carbohydrates*, Vol. 2, CRC Press, Boca Raton, 1991.
118. S. C. Churms, *J. Chromatogr. A*, 720 (1996) 75.
119. K. Koizumi, T. Utamura, Y. Kubota, and S. Hizukuri, *J. Chromatogr.*, 409 (1987) 396.
120. C. Brons and C. Olieman, *J. Chromatogr.*, 159 (1983) 79.
121. D. W. Armstrong and H. L. Jin, *J. Chromatogr.*, 462 (1989) 219.
122. S. Honda and S. Suzuki, *Anal. Biochem.*, 142 (1984).
123. A. S. Feste and I. Khan, *J. Chromtogr.*, 607 (1992) 7.
124. *Guide to Aminex® HPLC Columns*, Bulletin 1928, Bio-Rad Laboratories.
125. T. Jupille, *Amer. Lab.*, 13 (1981) 80.
126. R. W. Goulding, *J. Chromatogr.*, 103 (1975) 229.
127. *Analysis of Carbohydrates by High Performance Anion Exchange Chromatography with Pulsed Amperometric Detection (HPAE-PAD)*, Dionex Technical Note 20 (2000).
128. *Glycoprotein Oligosaccharide Analysis Using High-Performance Anion-Exchange Chromatography*, Dionex Technical Note 42 (1997).
129. *Optimal Settings for Pulsed Amperometric Detection of Carbohydrates Using the Dionex ED40 Electrochemical Detector*, Dionex Technical Note 21 (1998).
130. B. G. Huyghe, X. Liu, S. Sutjipto, B. J. Sugarman, M. T. Horn, H. M. Shepard, C. J. Scandella, and P. Shabram, *Hum. Gene Ther.*, 6 (1995) 1403.
131. W. W. Yau, J. J. Kirkland, and D. D. Bly, *Modern Size-Exclusion Liquid Chromatography*, Wiley-Interscience, New York, 1979.
132. J. Porath and P. Flodin, *Nature (London)*, 183, (1959) 1657.
133. S. Hjerten and R. Mosbach, *Anal. Biochem.*, 3, (1962) 109.
134. S. Hjerten, *Arch. Biochem. Biophys.*, 99, (1962) 466.
135. E. L. Johnson and R. L. Stevenson, in *Basic Liquid Chromatography*, Varian, Walnut Creek, CA, 1978, p. 150.
136. L. Hagel and J. C. Janson, in *Chromatography*, 5th ed., E. Heftmann, ed., Elsevier, Amsterdam, 1992, A267.
137. K. M. Gooding and F. E. Regnier, in *HPLC of Biological Macromolecules*, 2nd ed., K. M Gooding and F. E. Regnier, eds., Dekker, New York, 2002, p. 59.
138. B. F. D. Ghrist, M. A. Stadalius, and L. R. Snyder, *J. Chromatogr.*, 387 (1987) 1.
139. R. L. Cunico, K. M. Gooding, and T. Wehr, in *Basic HPLC and CE of Biomolecules*, Bay Bioanalytical Laboratory, Richmond, CA, 1999, p. 135.
140. E. Pfannkoch, K. C. Lu, F. E. Regnier, and H. G. Barth, *J. Chromatogr. Sci.*, 18, (1980) 430.
141. E. Folta-Stogniew and K. R. Williams, *J. Biomol. Techniques*, 10 (1999) 51.
142. V. N. Uversky, *Biochem.*, 32 (1993) 13288.
143. L. Hagel, *J. Chromatogr.*, 648, (1993) 19.
144. B. Sebille and N. Thuaud, In *Handbook of HPLC for the Separation of Amino Acids, Peptides, and Proteins*, Vol. 2, W. S. Hancock, ed., CRC Press, Boca Raton, 1984, pp. 379–391.
145. J. P. Hummel and W. J. Dreyer, *Biochim. Biophys. Acta*, 63 (1962) 530.
146. J. Curling., *Biopharm International*, (Feb. 2007) 10.

147. G. Walsh., *Appl. Microbiol Biotechnol.*, 67 (2005) 151.
148. L. Hagel, G. Jagschies, and G. K. Sofer, *Handbook of Process Chromatography: Development, Manufacturing, Validation and Economics*, 2nd ed., Elsevier, Amsterdam, 2007.
149. H. Chase, *Trends Biotechnol.*, 12 (1994) 296.
150. A. Jungbauer and E. Boschetti., *J. Chromatogr. B*, 662 (1994) 143.
151. A. Jungbauer, *J. Chromatogr. A*, 1065 (2005) 3.
152. P. Lu, C. D. Carr, P. Chadwick, M. Li, and K. Harrison, *BioPharm.*, (Sep. 2001) 19.
153. J. Rivier and R. McClintock, *J. Chromatogr.*, 268 (1983) 112.
154. J. Rivier, R. McClintock, R. Galyean, and H. Anderson. *J. Chromatogr.*, 288 (1983) 303.
155. E. I. Grimm and E. E. Logsdon, US Patent 4,612,367 (1986).
156. P. H. Lai and T. W. Strickland, US Patent 4,667,016 (1987).
157. R. Bischoff, D. Clesse, O. Whitechurch, P. Lepage, and C. Roitsch, *J. Chromatogr. A*, 476 (1989) 245.
158. D. I. Urdal, D. Mochizuki, P. J. Conlon, C. J. March, M. L. Remerowski, J. Eisenman, C. Ramthun, and S. Gillis, *J. Chromatogr. A*, 296 (1984) 171.
159. S. Hershenson, Z. Shaked, and J. Thomson, US Patent 4,961,969 (1990).
160. C. V. Olsen, D. H. Reifsnyder, E. Canova-Davis, V. T. Ling, and S. E. Builder, *J. Chromatogr. A* 675 (1994) 101.
161. V. Price, D. Mochizuki, C. J. March, D. Cosman, M. C. Deeley, R. Klinke, W. Clevenger, S. Gillis, P. Baker, and D. Urdal, *Gene*, 55 (1987) 28.
162. E. P. Kroeff, R. A. Owens, E. L. Campbell, R. D. Johnson, and H. I. Marks, *J. Chromatogr.*, 461 (1989) 45.
163. J. Brange, *The Physico-chemical and Pharmaceutical Aspects of Insulin*, Springer, Berlin, 1987.
164. E. P. Kroeff and R. E. Chance, *Proceedings of the FDA-USP Workshop on Drug and Reference Standards for Insulins, Somatotrophins and Thyroid-Axis Hormones*, United States Pharmacopeia Convention, Rockville, MD, 1982, pp. 148–162.
165. *Current Good Manufacturing Practice in Manufacturing, Processing, Packing, or Holding Of Drugs*, 21 CFR Part 210, http://www.fda.gov/cder/dmpq/cgmpregs.htm.
166. *Current Good Manufacturing Practice for Finished Pharmaceuticals*, 21 CFR Part 211, http://www.fda.gov/cder/dmpq/cgmpregs.htm.
167. International Organization for Standardization, http://www.iso.org/iso/home.htm.
168. M. T. W. Hearn, *Reversed-Phase High Performance Liquid Chromatography*, Academic Press, New York, 1984.
169. A. M Striegel, W. W Yau, J. J Kirkland, and D. D Bly, *Modern Size-Exclusion Liquid Chromatography*, 2nd ed., Wiley-Interscience, New York, 2009.
170. T. H. Mourey, *Int. J. Polym. Anal. Charact.*, 9 (2004) 97.
171. A. M Striegel, *Anal. Chem.*, 77 (2005) 104A.
172. W. F Reed, in *Multiple Detection in Size-Exclusion Chromatography*, A. M. Striegel, ed., ACS, New York, 2005, ch. 2.
173. W. M. C. Decrop et al., submitted for publication.
174. X. Jiang, P. J. Schoenmakers, X. Lou, V. Lima, J. L. J. van Dongen, and J. Brokken-Zijp, *J. Chromatogr.*, 1055 (2004) 123.
175. X. Jiang, A. van der Horst, V. Lima, and P. J. Schoenmakers, *J. Chromatogr*, 1076 (2005) 51.

CHAPTER FOURTEEN

ENANTIOMER SEPARATIONS

with Michael Lämmerhofer, Norbert M.Maier, and Wolfgang Lindner

Introduction to Modern Liquid Chromatography, Third Edition, by Lloyd R. Snyder, Joseph J. Kirkland, and John W. Dolan

14.1 INTRODUCTION

Previous chapters have described HPLC procedures of wide applicability; that is, separations that can be used for many different kinds of samples. Usually method development can be started with any of various columns or conditions, and resolution is then systematically improved by varying separation conditions. Consequently there are often many different ways of successfully separating a particular sample.

Such an approach cannot be used for enantiomers, however, which require highly specialized techniques and separation materials. Furthermore, in most cases, the selection of the column is the critical step; unless a suitable column is selected, subsequent changes in other conditions are unlikely to be successful. Nevertheless, enantiomer separations are today performed routinely in many research and routine laboratories, and are of great importance in the pharmaceutical industries. Various technologies and tools have been developed that provide a rich toolbox to separate virtually any sample. Among the available tools, *direct* separation with a chiral stationary phase (CSP; an enantioselective or "chiral" column) has become the predominant and most accepted procedure (but *not* the only procedure). However, the identification of the most suitable CSP/mobile-phase combination for a particular analyte can be challenging. Due to the specificity of molecular-recognition for each combination of CSP (*chiral selector*) and analyte—and its sensitivity to minor structural variations in either the solute or CSP—reliable predictions of an appropriate column from analyte molecular structure are as yet hardly possible.

Databases such as ChirBase (http://chirbase.u-3mrs.fr) can provide help in the selection of column and starting mobile phase, based on analyte structure [1, 2]. A database approach will be of greatest value when the enantiomers of interest are included in the database. This approach can also be useful for enantiomers of related structures—where presumably similar separation conditions will be successful, but it must fail for completely new structures. Alternatively, automated screening procedures are used in large pharmaceutical companies to solve this problem efficiently [3, 4]. This way the most promising CSP can be found quickly, followed by optimization of the mobile phase. Overall, direct HPLC enantiomer separation based on CSPs has become an extraordinarily powerful technology, and this will be the primary focus of the present chapter.

14.2 BACKGROUND AND DEFINITIONS

Non-enantiomeric separations can often be developed from only a general description of the sample components, for example, acidic, basic, or neutral solutes. In some cases the class of compounds within a sample (e.g., peptides, proteins, carbohydrates) suggest a range of suitable conditions, including column type (but not requiring a specific column). Consequently further details of solute molecular structure are unnecessary for non-enantioselective method development. This is not the case for enantiomeric separations, where solute molecular structure and related physicochemical properties play a critical role in method development. For this reason a basic understanding of the behavior of enantiomers is essential for their efficient separation.

14.2.1 Isomerism and Chirality

Isomers are molecules that possess identical atomic composition, yet are not superimposible upon each other [5, 6]. A classification of isomeric structures is given in Figure 14.1 [6]. As distinguished from the case of molecules which are actually identical ("homomers"), a structural isomer can be defined in various ways. Compounds whose atom-to-atom connections are different (e.g., 1-butanol, 2-butanol) are defined as *constitutional* isomers. *Stereoisomers*, by contrast, have identical atom-to-atom connections, but distinct orientation of atoms or groups in three-dimensional space. The latter can be further divided into *enantiomers* and *diastereomers* (or "diastereoisomers"). While the former are always chiral, the latter may also include nonchiral stereoisomers such as *cis/trans* isomers (e.g., *cis/trans*-1,2-dichloroethylene).

Chirality, the synonym for "handedness" (from the Greek word for hand), refers to the geometric property of an object which is nonsuperimposable on its mirror image (e.g., the left and right hand). Such an object has no symmetry elements of the second kind, such as a plane of symmetry, a center of inversion, or a rotation-reflection axis. Chirality may arise from various distinct chiral elements (Fig. 14.2): centers of chirality (stereogenic centers; Fig. 14.2*a*), chiral axes (axial

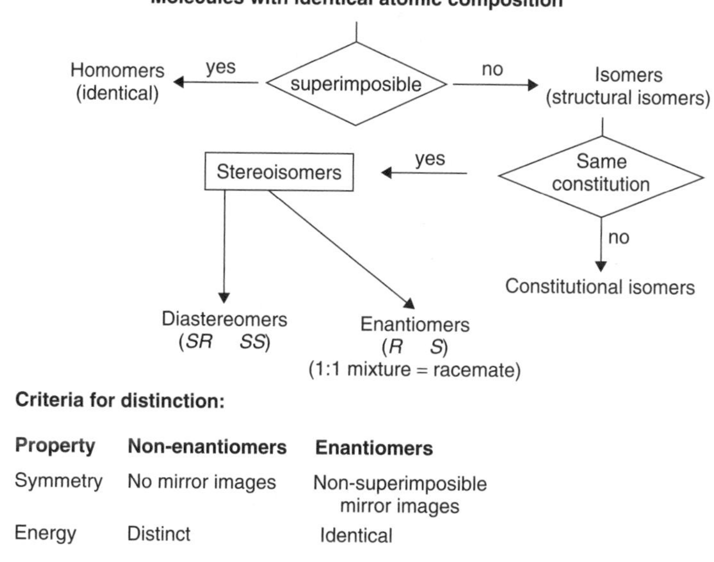

Property	Non-enantiomers	Enantiomers
Symmetry	No mirror images	Non-superimposible mirror images
Energy	Distinct	Identical

Figure 14.1 Classification of isomeric structures. Adapted from [6].

chirality; Fig. 14.2*b*), chiral planes (planar chirality; Fig. 14.2*c*), chiral helices (helical chirality; Fig. 14.2*d*), and topologically chiral elements (topological chirality). The most well known example is a center of chirality, where typically a carbon atom within the molecule is substituted by four different entities (see the example of carbons 1 and 2 in Figure 14.3*a*.

Enantiomers and diastereomers can be differentiated by either of two properties: symmetry or energy content (i.e., the free energy of the molecule). *Enantiomers* are molecules that are nonsuperimposable mirror images of each other having identical energy content (because of exactly identical atomic distances, angles, and torsions, as well as interatomic interactions). They are indistinguishable in an achiral environment and therefore cannot be separated by achiral chromatographic methods such as conventional reversed-phase chromatography (RPC). For example, (1*S*, 2*R*)-ephedrine and (1*R*, 2*S*)-ephedrine (Fig. 14.3*a*) are enantiomers. Moreover (1*R*, 2*R*)- and (1*S*, 2*S*)-pseudoephedrines are also enantiomeric to each other (Fig. 14.3*b*). Note that the corresponding enantiomers always exhibit opposite configurations at the two stereogenic centers (as in Fig. 14.3). An exactly equimolar mixture of enantiomers is called a *racemate*. All other mixtures of enantiomers with a composition deviating from 1:1 are defined as *nonracemic mixtures*.

Enantiomers are provided with stereochemical descriptors (e.g., *R* and *S*, or *L* and *D*, or + and −) that distinguish them (with the *R* and *S* system being

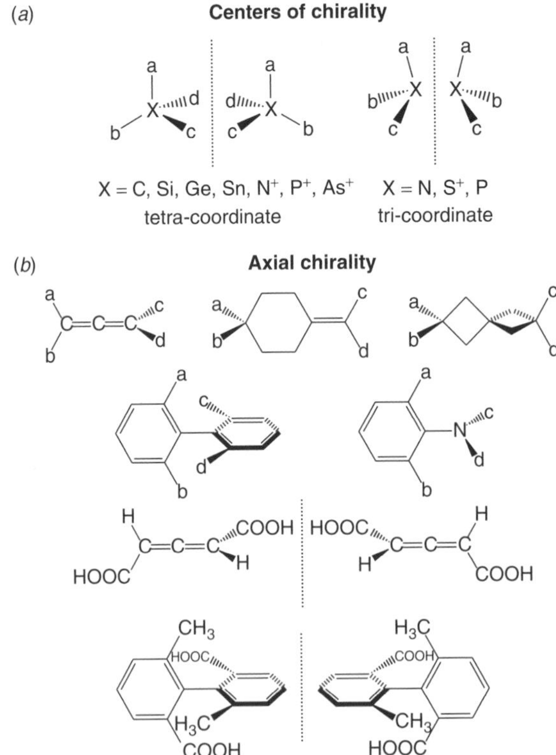

Figure 14.2 Structural features that contribute to chirality.

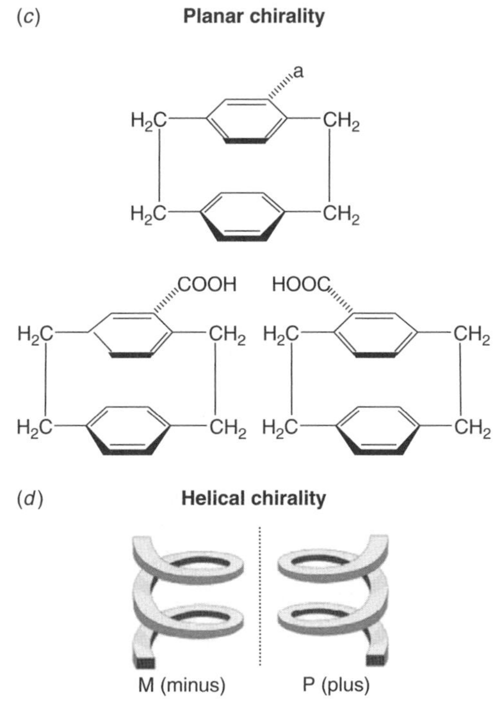

(c) Planar chirality

H_2C CH_2 a H_2C CH_2

COOH HOOC H_2C CH_2 H_2C CH_2 H_2C CH_2 H_2C CH_2

(d) Helical chirality

M (minus) P (plus)

Figure 14.2 (Continued).

highly preferred). *R* and *S* refer to configurations in which the substituents at the stereogenic center are in clockwise and counterclockwise arrangement regarding their Cahn–Ingold–Prelog priorities when the substituent with the lowest priority is oriented away from the observer. + and − refer to the optical rotation properties of a chiral molecule; that is, its ability to rotate the plane of linear polarized light to the right or left (dextrorotatory and levorotatory). L and D refer to the Fisher designations for amino acids and sugars that specify relative configurations chemically derived from D-(+)-glyceraldehyde [6].

Diastereomers are not mirror images of each other, are characterized by distinct physical and chemical properties, and can be separated by achiral chromatography. In the example of Figure 14.3, each stereoisomer of Figure 14.3*a* is a diastereomer of any of the stereoisomers in Figure 14.3*b* because they differ solely in the configuration of one stereogenic center rather than both (as for enantiomers). Thus ephedrines and pseudoephedrines are diastereomers because they are not mirror images; they can also be termed *epimers*, since only one of several stereogenic centers is inverted. This distinction between enantiomers and diastereomers is the fundamental basis for all chiral differentiation processes including all enantiomer separation concepts.

14.2.2 Chiral Recognition and Enantiomer Separation

Enantiomers have identical physicochemical properties, so their separation requires their conversion to either (1) diastereomers (the *indirect method*) or (2) diastereomeric complexes (the *direct method*) [7, 8]. Today the use of the indirect approach

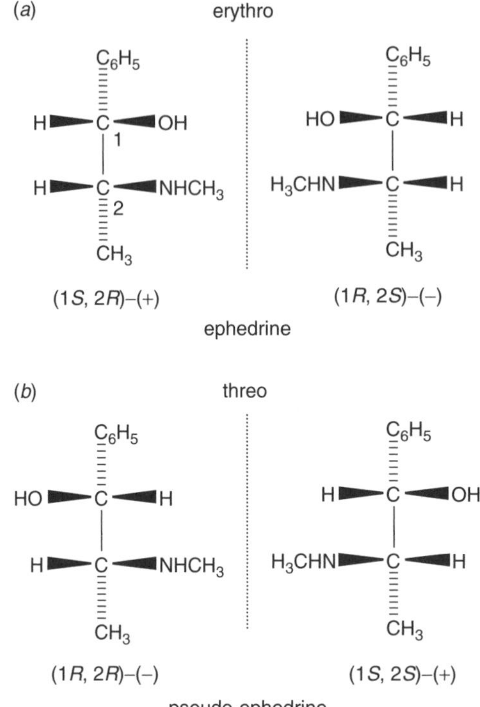

Figure 14.3 Structures of ephedrines and pseudoephedrines.

is decreasing because of certain problems discussed below [8]. Nevertheless, the indirect method is the procedure of choice for some applications, and its discussion in following Section 14.3 will cover issues that are also relevant to the later treatment of the direct method in Section 14.4. When planning an enantioselective separation by the direct method, different chromatographic modes can be used, as in the case of achiral chromatography. In this connection we will distinguish enantioselective separations by designating them as reversed phase (RP) or normal phase (NP); this contrasts with the previously used abbreviations RPC and NPC for achiral separations.

14.3 INDIRECT METHOD

The indirect method involves the *formation of diastereomers* by reaction of an analyte (*X*) in the *R* or *S* configuration with an enantiomerically pure compound (hereafter with *R*-configuration), which we will refer to as a chiral derivatizing reagent (CDR):

$$(R)\text{-}X + (R)\text{-}CDR \quad \rightarrow \quad (R,R)\text{-}X\text{-}CDR \tag{14.1}$$

and

$$(S)\text{-}X + (R)\text{-}CDR \rightarrow (S,R)\text{-}X\text{-}CDR \tag{14.1a}$$

The reactions above must go to completion, and the diastereomeric products must be stable (chemically and configurationally). The diastereomers (R,R)-X-CDR and (S,R)-X-CDR can then be separated by achiral chromatography (usually RPC).

One advantage of the indirect method is its use of conventional HPLC columns, which offer higher plate numbers compared to the enantioselective ("chiral") columns of Section 14.4. A higher column efficiency can be especially important for the measurement of impurities at the $<0.1\%$ level in complex mixtures, as required in pharmaceutial products. Even more important is the possible use of CDRs for enhanced detection by UV, fluorescent, electrochemical, or mass spectrometric means, for example, fluorescent tags for the sensitive detection of otherwise difficult-to-detect amino acids. The indirect method is also relatively economical, in contrast to the direct method with its requirement of a battery of different (and generally expensive) enantioselective columns.

A large number of CDRs have been developed that provide adequate diastereoselectivity and in some cases enhanced detection [8]. Chiral derivatizing reagents must be both chemically and stereochemically stable, and should be commercially available in both enantiomeric forms (e.g., *R* and *S*). The choice of *R* or *S* CDRs

Table 14.1

Commonly Employed Chiral Derivatizing Reagents (CDR) and Their Application

CDR	Analyte Class	Reference
1. (*R*) or (*S*)–α-methoxy-α-trifluoromethyl phenylacetic acid and corresponding acid chloride (Mosher's reagent)	Alcohols, amines	[9]
2. *O*,*O*′-dibenzoyl tartaric acid anhydride (DBTAAN)	Primary and secondary amines, alcohols, aminoalcohols	[10]
3. (*R*)- or (*S*)–1-(9-fluorenyl)ethyl chloroformate (FLEC)	Primary and secondary amines, amino acids	[11]
4. ortho-phthaldialdehyde (OPA) in combination with chiral thiols such as (*S*)-or (*R*)-enantiomers of *N*-acetyl-cysteine, *N*-t-Boc-cysteine, *N*-acetyl-penicillamine, 1-thio-β-glucose	Primary amines, primary amino acids	[12, 13]
5. 1-fluoro-2,4-dinitrophenyl-5-(*S*)-alanine amide (FDAA) (Marfey's reagent)	Primary and secondary amines, amino acids, thiols	[14, 15]
6. 2,3,4,6-tetra-*O*-acetyl-β-D-glucopyranosyl isothiocyanate (GITC)	Primary and secondary amines, amino acids, thiols	[16]

enables a reversal of elution order for two enantiomers, *which can be used to position a minor peak in front of a major peak, when one enantiomer is in considerable excess* (see Section 2.4.2 and compare Fig. 2.17*e*, *f*). Some popular examples for CDRs are given in Table 14.1.

A still popular, indirect method for the analysis of the enantiomer composition of amino acids is the use of the *o*-phthaldialdehyde (OPA) reagent with a chiral thiol such as *N*-acetyl-cysteine (CDR 4 in Table 14.1). The reaction scheme is shown in Figure 14.4. *OPA* derivatization is fast and amenable to automation, so derivative instability can be overcome by reacting just before injection (an excess of the non-fluorescent reagent will not interfere with detection). The chromatogram in Figure 14.5 shows the analysis of amino acids in a bacitracin sample, after its hydrolysis and oxidation of Cys to cysteic acid (Cya) by the OPA method [13]. This example illustrates the ability of the indirect approach to resolve several enantiomer pairs in a single separation, which is much less likely with a direct method (unless MS detection is used, which introduces additional chemoselectivity so that co-elution of species with distinct MS properties does not matter). Nevertheless, despite its apparent simplicity, the development of indirect enantiomer separation methods is far from a trivial task [8].

A crucial requirement of the indirect method is an analyte that possesses a selectively derivatizable functional group such as hydroxyl, amino, carboxylic, carbonyl, or thiol. Another important requirement is a derivatizing reagent that is chemically and (especially) enantiomerically pure, that is, an enantiomeric excess >99.9%. If the CDR contains a significant amount of enantiomeric impurity, erroneous quantification data will result. For example, the (*S*)-enantiomer impurity in (*R*)-CDR gives upon derivatization, besides the main diastereomeric pair of products, the formation of a second pair of diastereomers yielding all four stereoisomers (see Fig. 14.6; note that the impurity and its diastereomeric products are distinguished

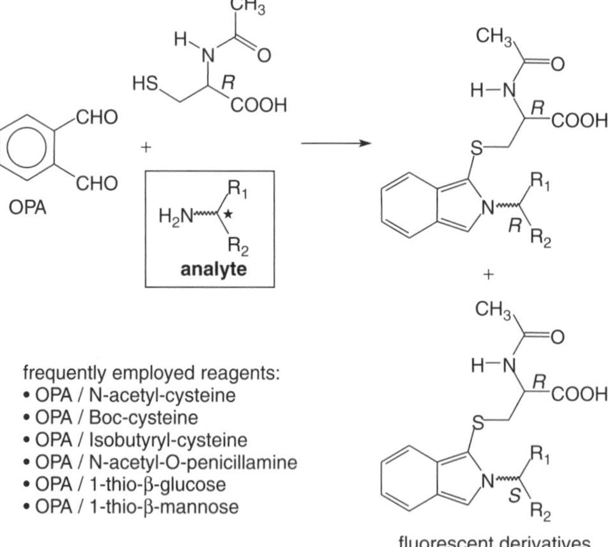

Figure 14.4 OPA-chiral thiol derivatization for the stereoselective analysis of primary amines and amino acids (indirect enantiomer separation).

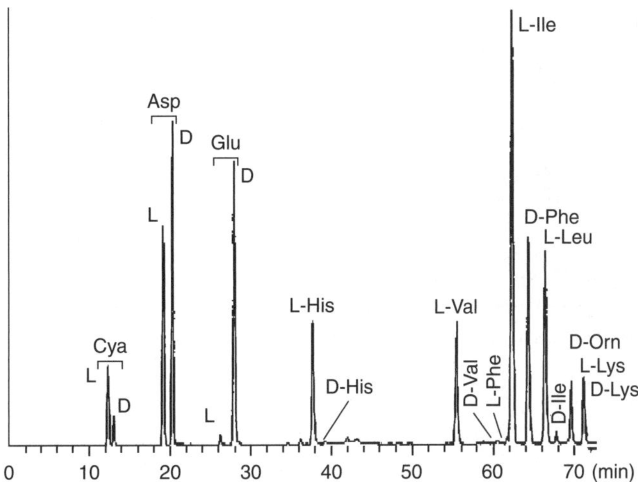

Figure 14.5 Enantiomeric separation of amino acids obtained from the hydrolysis of bacitracin A, followed by OPA-chiral thiol derivatization (indirect separation with fluorescence detection; Cya = cysteic acid). Adapted from [13].

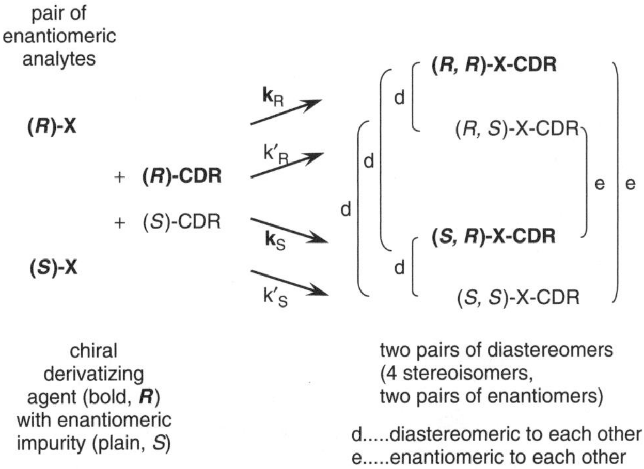

Figure 14.6 Reaction scheme for indirect HPLC enantiomer separation (in the presence of an *S*-CDR impurity in the chiral derivatization reagent *R*-CDR). All four stereoisomers are formed and two pairs of enantiomers, respectively (*d*, diastereomeric to each other; *e*, enantiomeric to each other).

by unbolded type). Achiral chromatography (e.g., RPC) will be unable to resolve the products that are enantiomeric to each other. Hence the stereoisomers arising from the enantiomeric contamination of the CDR will co-elute with the peaks of the opposite enantiomers (i.e., opposite configurations), and only two peaks will be observed. Needless to say, these co-elutions will prohibit accurate quantitation. Corrections are possible for CDR contamination, if the enantiomeric impurity level in the CDR is known; however, this adds considerable complexity to both method development and validation.

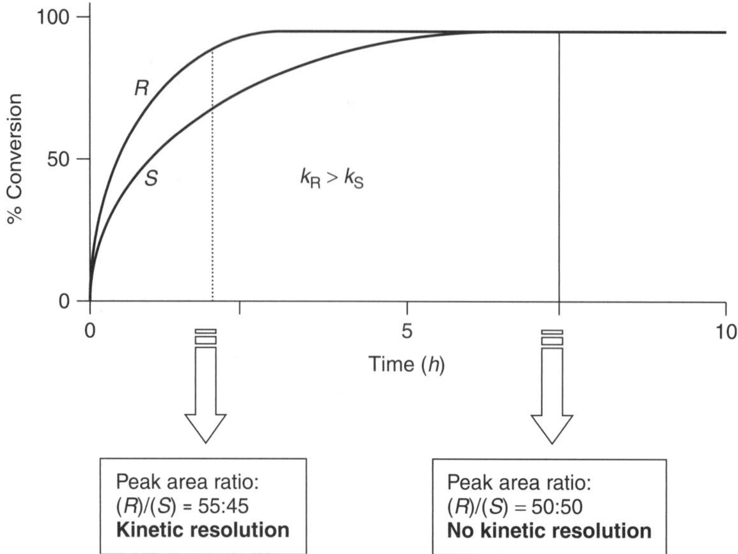

Figure 14.7 Illustrative kinetic profiles for the reaction of a sample with a chiral derivatizing reagent. The two enantiomers are assumed to have different rate constants and an identical detector response for the resulting diastereomers.

Two additional complications exist for the indirect method. *First*, for a given derivatization procedure, stereochemical integrity must be fully preserved; no racemization is allowed of the derivatizing reagent, analytes, or diastereomeric products [8]. The absence of racemization can be easily checked by derivatizing and analyzing a sample of known enantiomer composition. *Second*, it must be verified that the derivatization has reached completion, in order to avoid potential *kinetic resolution* problems (Fig. 14.7) [8]. Specifically, the reaction rates for derivatization of the *R*- and *S*-enantiomers of the analyte may differ; if the reaction is stopped before completion, this can give rise to a stereoisomer ratio that deviates from the actual ratio of enantiomers in the sample. For example, if the sample is a racemate and $k_R > k_S$ (with k being the rate constants of the derivatization reaction), the analyzed enantiomer composition after 2 minutes in Figure 14.7 would significantly deviate from the 1:1 ratio that is expected for a racemate. Systematic error due to kinetic resolution can be easily avoided by driving the derivatization reaction to completion. This is generally achieved by increasing the reaction temperature and time, and by the use of a large excess of the corresponding CDR (the CDR is typically employed in 10-fold molar excess relative to the enantiomers).

Another drawback of the indirect method is that enantiomeric ratios cannot be directly calculated from peak-area ratios measured in the final separation (as is possible with the direct method), since the two diastereomeric products can differ considerably in their detector response [8]. This holds true for most detection modes, including UV, fluorescence, and mass spectrometry, with response factors varying typically by a factor of 1.1 to 1.5. Consequently a correction for this difference in response factors will be necessary. External calibration with individual standards of the diastereomers circumvents this problem, but individual standards are not always available.

It might appear that these limitations make indirect methods questionable, which is certainly not the case. However, users should carefully check for all of the complications above, most important, the enantiomeric purity of the CDR [17] (it is unacceptable to rely only on specifications given by the supplier). Moreover the numerous method-validation issues that need to be addressed during the development of an indirect assay render this approach time-consuming, labor intensive, and more prone to systematic errors. Nevertheless, some enantioselective analysis assays are still carried out by indirect methods.

14.4 DIRECT METHOD

The direct approach relies on the reversible formation of (transient) *diastereomeric complexes* between the two enantiomers $[(R) - X$ and $(S) - X]$ and a chiral complexing agent termed *chiral selector* (CS). As in the case of the indirect method, the selector must be enantiomerically pure, but this requirement is less stringent for the direct method. If the difference in complex stabilities is sufficiently large for the diastereomeric associates, a less-pure selector can still be used in a direct method:

$$(R)\text{-}X + (R)\text{-}CS \quad \Leftrightarrow \quad (R)\text{-}X^{\cdots}(R)\text{-}CS \tag{14.2}$$

and

$$(S)\text{-}X + (R)\text{-}CS \quad \Leftrightarrow \quad (S)\text{-}X^{\cdots}(R)\text{-}CS \tag{14.2a}$$

The direct method circumvents the laborious derivatization with a CDR. Two distinct experimental modes for direct enantiomer separation exist: The *chiral mobile-phase-additive (CMPA) mode* (or simply, "additive mode") and the *chiral stationary-phase (CSP) mode*.

14.4.1 Chiral Mobile-Phase-Additive Mode (CMPA)

The additive mode makes use of an achiral stationary phase (e.g., a reversed-phase or normal-phase column) with a mobile phase that contains the chiral mobile-phase-additive (CMPA) at an appropriate concentration. The CMPA (selector, *CS*) may be present in the mobile phase and/or retained by the stationary phase as described by its distribution coefficient $K_{d,CS}$. Upon injection of the sample, various equilibria will be established that involve both the analyte *X* and the selector *CS* (Fig. 14.8; note that subscripts *m* and *s* refer to species in the mobile and stationary phases, respectively, and the *R*-form of the selector is assumed). These equilibria include:

- complex formation between selector (*R*)-*CS* and enantiomers (*R*)-*X* and (*S*)-*X* in the mobile phase, with association constants $K_{a,(R)\text{-}X}$ and $K_{a,(S)\text{-}X}$
- distribution of (*R*)-*X* and (*S*)-*X* between the mobile and stationary phases with distribution constants $K_{d,(R)\text{-}X}$ and $K_{d,(S)\text{-}X}$

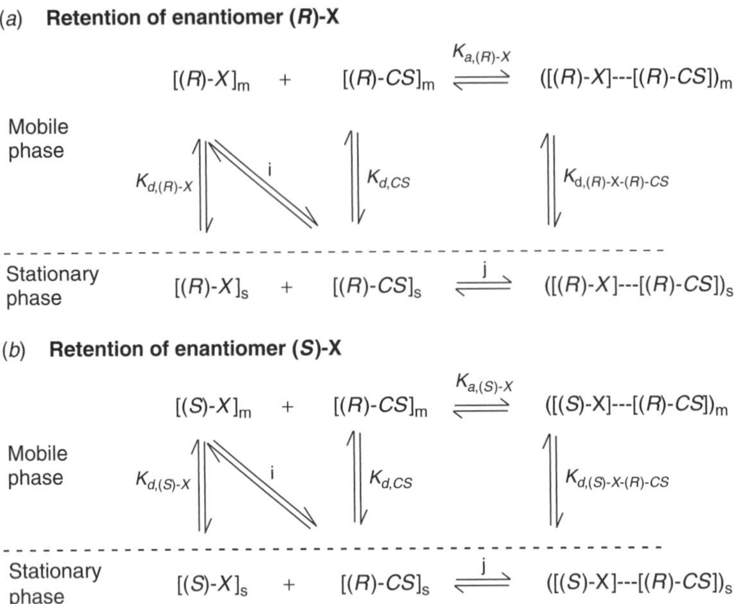

Figure 14.8 Equilibria for the retention of *R*- and *S*- enantiomers (chiral–mobile-phase additive, CMPA) mode. Subscripts *m* and *s* refer to corresponding species in mobile and stationary phases; K_a and K_d represent association and distribution constants, respectively.

- distribution of the complexed solutes $X \cdots CS$ between the mobile and stationary phases with distribution constants $K_{d,(R)\text{-}X\text{-}(R)\text{-}CS}$ and $K_{d,(S)\text{-}X\text{-}(R)\text{-}CS}$

Additionally the uncomplexed analyte $(R) - X$ can complex directly with the CMPA in the stationary phase (via processes *i* and *j* in Fig. 14.8). The same equilibria have to be considered for the *S*-enantiomer (*S*)-*X*. The observed solute retention factor *k* is then a weighted average of values of *k* for free and complexed *X*. If the CMPA is very strongly retained, it may saturate the stationary phase, leading to a situation similar to that of a dynamically coated CSP-column (note the similar situation for ion-pairing in Section 7.4.1 and Fig. 7.12).

The consequences of Figure 14.8 can be summarized as follows: Without the addition of the selector (*R*)-*CS*, no separation of the enantiomers (*R*)-*X* and (*S*)-*X* can occur, because $K_{d,(R)\text{-}X} = K_{d,(S)\text{-}X}$. When the CMPA is present in the mobile phase, and if its interaction with the analyte is significant and enantioselective, then the retention *k* of the two enantiomers should differ—hopefully leading to their chromatographic separation. In addition to the association of the two solute enantiomers with the CMPA (with different retention of free and complexed solute), stereoselectivity may originate from different retention of the diastereomeric associates, as well as nonequal adsorbate formation via pathway *i* or *j*. Stereoselectivity contributions from these individual processes may either enhance or attenuate each other. As the mobile-phase concentration of the CMPA is increased, there should be an increase in the separation of the two enantiomers. However, at sufficiently high concentrations of the CMPA, a decrease in separation is possible. The reason is that a high enough selector concentration can complex all of each enantiomers in the mobile phase, whereas differences in retention for the two enantiomers is favored when one is complexed to a greater extent than the other.

From the description of Figure 14.8, it should be clear that the choice of CMPA and its concentration in the mobile phase are primary determinants of enantioselectivity and a successful separation of the two enantiomers. However, other conditions can also play a role; for example, mobile-phase pH, ionic strength, different B-solvents (the organic solvent in RPC), and temperature can be important. CMPAs that have been utilized for chiral separation by HPLC include α-, β-, and γ-cyclodextrins and their derivatives [18], quinine and quinidine [19, 20], (+)- and (−)-10-camphorsulfonic acid [20], *N*-benzyloxycarbonyl-protected di- and tripeptides [19], chelating agents such as amino acids in combination with metal ions (adopting a chiral ligand-exchange chromatography approach) [21–23], and others.

The CMPA approach appears attractive because of its practical simplicity and relatively inexpensive columns. However, this approach suffers from a number of drawbacks, several of which are similar to problems encountered when an ion-pair reagent is added to the mobile phase (Section 7.4.3):

- close control of the temperature necessary because of its possible effects on the various equilibria of Figure 14.8, with a consequently less robust separation
- system peaks that result from differences in composition of the sample solvent and mobile phase (Section 7.4.3.1)
- incompatibility of some detectors with the presence of the CMPA in the mobile phase (e.g., UV-absorbing additives with UV detection; ion suppression with MS)
- different response factors for the two enantiomers because they can exist partly as the diastereomer-complexes in the mobile phase leaving the column and flowing through the detection cell; see the similar discussion for the indirect method (Section 14.3)
- expense and limited availability of CMPA reagents for all chiral compounds

Due to its numerous inherent drawbacks, today the CMPA mode has limited practical value for the HPLC separations of enantiomers. However, it should be noted that the CMPA mode is firmly established as the method of choice for enantiomer separations by *capillary electrophoresis*.

14.4.2 Chiral Stationary-Phase Mode (CSP)

The chiral stationary-phase mode (CSP) mode is generally the most straightforward and convenient means for chromatographic enantiomer separation; it is the method of choice for both analytical and preparative applications. The chiral selector is preferentially *covalently linked* or alternatively *strongly physically adsorbed* (e.g., by coating of a polymeric selector) to a chromatographic support (usually porous silica particles). The mobile phase is achiral, that is, devoid of any chiral constituents.

During migration of the sample through the column, the individual enantiomers are retained by association with the stationary-phase selector (similar to process *i*/*j* of Fig. 14.8). This way diastereomeric complexes (*R*)-*X*···(*R*)-*CS* and (*S*)-*X* ···(*R*)-*CS*

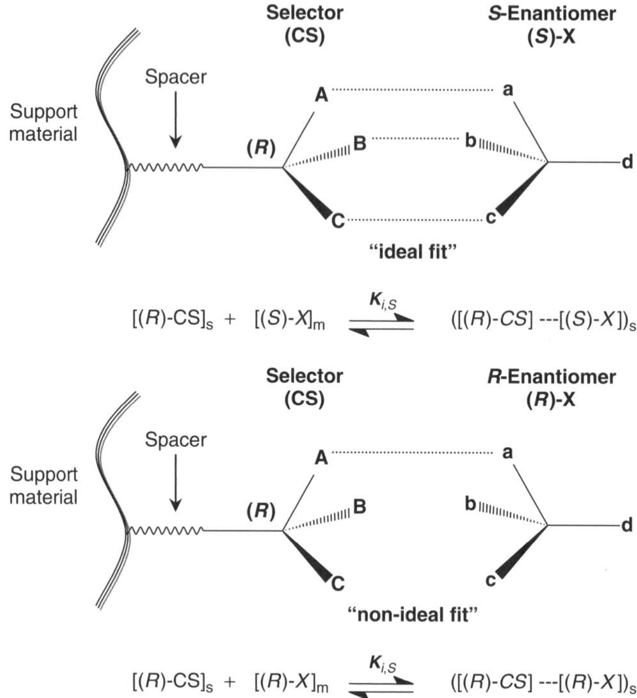

Figure 14.9 Three-point interaction model and associated interactions between *R*- or *S*- enantiomers and the CSP.

are formed in the stationary phase. If resulting values of k for the two enantiomers are sufficiently different, then their separation is possible. Consequently *successful separation requires a CSP that interacts more strongly with one enantiomer than the other*. The values of k are determined by the strengths of the resulting complexes, or the values of the equilibrium constants $K_{i,R}$ and $K_{i,S}$ in Figure 14.9—as discussed in the following Section 14.4.3.

CSPs and their corresponding enantioselective columns offer a number of striking advantages over both the indirect and additive approaches. Solutes that lack appropriate functional groups for derivatization can still be separated by the CSP approach. Minor enantiomeric impurities in the selector are only of concern insofar as these decrease the separation factor α for the two enantiomers [24, 25]. Other than this, none of the complications described above for the indirect method with an impure selector are present. Problems from the presence of the selector in the mobile phase leaving the column are also avoided with the CSP procedure. Thus the selector is absent, which avoids any interference with detection. The two enantiomers also possess an equal detector response—so enantiomeric composition can be directly derived from the area ratio for the monitored *R*- and *S*-enantiomer peaks—without any corrections. However, with chiroptical detectors (Section 4.10) a different response can be intentionally obtained for each enantiomers—with certain advantages (e.g., allowing assignment of the individual enantiomers as [+] and [−]).

Disadvantages of the CSP approach include relatively expensive columns in most cases (and often with shorter lifetimes than RPC columns), moderate column efficiencies (Section 14.5), and sometimes poor chemoselectivity for structural analogues (i.e., selectivity for enantiomers is often greater than for structural analogues that might also be present in the sample—leading to an overlap of the separated (*R*) and (*S*) enantiomer peaks by these analogues). It is important to note that enantiomer separations using CSPs have expanded the preparative isolation of enantiomers from laboratory scale (mg to g) up to production scale (kg to ton). The attractiveness of CSP separation lies in the short development times, ease of product recovery by the evaporation of volatile mobile phases, ready access to both enantiomers in high chemical and optical yield, and straightforward scalability (Section 15.1.2.1).

14.4.3 Principles of Chiral Recognition

14.4.3.1 "Three-Point Interaction Model"

Early attempts to rationalize chiral recognition at the molecular level have led to the formulation of geometric models, such as the *three-point attachment model* [26]. The latter model is still frequently utilized to visualize and explain the requirements for enantioselectivity when designing CSPs. In its original form this model states that at least three configuration-dependent attractive contact-points between a chiral receptor and a chiral substrate are required for chiral recognition. However, a fourth essential requirement is often neglected, namely the fact that the receptor is accessible only from one side and can therefore only be approached in one direction. The latter, simplistic model has been under debate since it was first formulated, and today it is known that not all three interactions need to be attractive (*three-point rule* by Pirkle [27]). Its adaptation for chiral recognition by CSPs is graphically illustrated in Figure 14.9.

Considerable confusion with the three-point interaction model has arisen from the question: "what does 'interaction' mean?" [27]. As noted above, both attractive and repulsive forces are included in "interaction," which can either stabilize or destabilize the formation of the analyte-CSP complex. Moreover many interactions are actually multipoint in nature, which minimizes the need for additional supportive interactions. For example, whereas hydrogen bonding and end-to-end dipole interactions are regarded as single-point interactions—therefore counting for only one interaction each, dipole-dipole stacking and $\pi-\pi$-interactions are effectively multipoint interactions and may count as at least two interaction points each [27]. Similarly molecules that contain chiral centers incorporated into rigid elements such as a cyclic ring require fewer interactions [28]. Two of the four bonds of the asymmetric centers are incorporated into a rigid ring, which enforces a molecular rigidity and makes the two stereoisomers more easily recognizable. It is a common perception in the field that a single interaction with a rigid plane or its surface can count for at least two interaction sites.

Consider next a chiral stationary phase that has a chiral selector CS, bonded to the surface of a suitable support via a spacer (Fig. 14.9). In this idealized model the analyte interacts solely with the selector, not with the spacer or support. The relative retention of the analyte enantiomers (*R* and *S*) are determined by their strength of interaction with the selector shown in Figure 14.9, as influenced by the three-dimensional orientation and spatial arrangement of complementary sites in the binding partners. Ideally there will be a perfect match between the chiral selector and the respective enantiomer of the analyte (*ideal fit*). The preferred enantiomer turns out to be the stronger-bound one [(*S*)-*X* in Fig. 14.9], which is characterized by a larger binding constant $K_{i,S}$ for the equilibrium reaction. It is therefore more strongly retained and elutes as the second enantiomer peak. For the other enantiomer (*R*)-*X*, there is a considerable spatial mismatch—at least for one of the interaction sites. Hence there is a *non-ideal* fit between (*R*)-*X* and *CS*, so that the corresponding binding constant $K_{i,R}$ will be significantly lower. The *R*-enantiomer therefore elutes earlier. Thermodynamic considerations for this selector-solute association and the adsorption process, respectively, are treated in more detail at the end of this chapter (see Section 14.7.).

An ideal fit and effective binding between analyte and selector can be achieved,

- if there is a size and shape complementarity, so that the analyte sterically fits the binding site of the selector—which is often arranged as a preformed pocket or cleft, as in the case of cyclodextrin selectors (*steric fit)*
- if the analyte and selector have complementary interaction sites (functional groups) arranged in a favorable geometric and spatial orientation, so that attractive *noncovalent intermolecular interactions* can become active (*functional fit* based on complementary interacting groups). These interactions drive the association between analyte and selector enantiomers and are basically *electrostatic* in nature; they comprise:
  - ionic interactions (electrostatic interactions between positively and negatively charged groups)
  - hydrogen bonding (between H-donor and H-acceptor groups)
  - ion–dipole, dipole–dipole (orientation forces), dipole–induced dipole (induction forces), and induced dipole–instantaneous dipole (dispersion forces) interactions
  - π–π-interactions (face-to-face or face-to-edge arrangement of electron-rich and electron-poor aromatic groups)
  - others such as quadrupolar, π–cation, π–anion interactions
- if hydrophobic regions of the selector and analyte are spatially matching so as to enable binding by hydrophobic interactions (*hydrophobic fit)*.
- if the flexibility of the analyte and selector allow an optimized binding in the stationary phase (*dynamic fit*). In particular cases, an *induced fit* through a conformational change of analyte and selector molecules upon complexation may further enhance the complex stability.

Intuitively, chiral recognition might be thought to increase for higher binding affinities between analyte and selector (i.e., large binding constants or low dissociation constants generally increase enantioselectivity). This concept has been known for a long time in pharmacology as Pfeiffer's rule [29], which states that the stereoselectivity of drugs will increase with their potency. The validity of this rule has been debated, and it is now accepted that for various reasons the principle is of limited applicability.

14.4.3.2 Mobile-Phase Effects

Chiral recognition is commonly regarded as a bimolecular process of analyte-CSP interaction, and the effect of the mobile phase is frequently ignored. The mobile phase can in fact play a significant role in enantioselectivity by determining the degree of solvation of interactive sites of the analyte and selector, and whether these sites are then available for intermolecular contact between the analyte and CSP. The choice of solvents, buffer salts, and pH are also major determinants of conformational preferences and of the ionization states of both analytes and CSPs. The mobile phase can also suppress detrimental, nonspecific interactions that deteriorate enantioselectivity (Section 14.7.3). Consequently a proper choice of mobile phase represents an important step in the development of both indirect, but especially direct enantiomer separations.

For the reasons above, the mobile phase must be considered as an important factor in enantiomer separations with CSPs, as it defines the interaction environment where chiral recognition takes place. Solvents can interfere or promote specific analyte–selector interactions and thus affect enantiorecognition. Solvents of high polarity attenuate the strength of electrostatic interactions, whereas hydrophobic interactions are present only in aqueous or hydro-organic mobile phases. Mobile-phase pH is especially significant in the case of ionizable analytes or CSPs, as ionic interactions require ionized entities. These ionic interactions can be weakened by an increase in mobile-phase ionic strength, similar to the case of ion-exchange retention (Section 7.5.2). Many other factors influence the selection of the mobile phase, but these are similar for all liquid chromatographic procedures: reversed-phase, normal-phase, ion-exchange, and so forth. For further details on mobile-phase selection, the reader is referred to Section 14.6 for individual CSPs, as well as relevant sections in earlier parts of the book.

14.5 PEAK DISPERSION AND TAILING

Compared to separations by RPC, the chromatographic efficiencies of enantioselective columns in following Section 14.6 are often fairly low. Plate numbers seldom exceed $N = 40{,}000$/m for columns of 5-μm particles, and peaks frequently tail to a greater extent than in other forms of HPLC. It can be assumed that these enantioselective column efficiencies are dependent on the same factors as for other HPLC columns (Section 2.4.1), so some additional factor must be involved, namely slow adsorption–desorption kinetics at the chiral sites. It is commonly accepted that the desorption process can be slow because of the formation of analyte-selector complexes that are stabilized by simultaneous, multiple interactions. Slow kinetics can be improved by the use of higher temperatures, but enantioselectivity often decreases with an increase in temperature. Consequently the lower efficiency of many CSP columns must be accepted as a fundamental problem, with no obvious means of its correction.

14.6 CHIRAL STATIONARY PHASES AND THEIR CHARACTERISTICS

Success in the use of enantioselective columns depends on the selection of a suitable CSP—one that is able to separate the target enantiomers. It is therefore important to know which chiral stationary phases are available, as well as their principal characteristics and typical operation conditions. It can also be helpful to understand how they distinguish between enantiomers and what analytes they are most useful for. These issues will be covered in the present section.

A broad variety of chiral molecules, of both natural and synthetic origin, have been considered as prospects for useful CSPs. A few hundred CSPs are today offered commercially, among which perhaps 20 to 30 CSPs are used most frequently—these columns are capable of separating most enantiomers likely to be presented for analysis. Selectors for the latter columns fall into the following CSP classes:

- macromolecular selectors of semisynthetic origin (polysaccharides; Section 14.6.1)
- macromolecular selectors of synthetic origin (poly(meth)acrylamides, (poly-tartramides; Section 14.6.2)
- macromolecular selectors of natural origin (proteins; Section 14.6.3)
- macrocyclic oligomeric or intermediate-sized selectors (cyclodextrins, macrocyclic antibiotics, chiral crown ethers; Section 14.6.4–14.6)
- synthetic, neutral entities of low molecular weight (Pirkle-type phases, brush-type CSPs; Section 14.6.7)
- synthetic, ionic entities of low molecular weight that provide for ion exchange (Section 14.6.8)
- chelating selectors for chiral ligand-exchange chromatography (Section 14.6.9)

The following sections summarize the most important characteristics of the most commonly employed, commercially available CSPs and enantioselective columns. These detailed insights into the way different CSPs achieve enantioselecitivity can be helpful in an initial selection of promising CSPs and other conditions for chiral separation, followed by trial-and-error experimentation to achieve an acceptable separation. *Specific recommendations for the use of certain, more favored CSPs and conditions are italicized and indented for easy access.*

14.6.1 Polysaccharide-Based CSPs

Polysaccharide selectors have a long tradition in enantioselective liquid chromatography. In 1973 Hesse and Hagel introduced microcrystalline cellulose triacetate (MCTA) as a polymeric selector material (without supporting matrix) for enantioselective liquid chromatography [30]. While MCTA exhibits widely applicable enantiorecognition and favorable loading capacities for preparative separations, it suffers from poor pressure stability, slow separations, and low chromatographic efficiency. A solution to the mechanical stability problem of MCTA was proposed by Okamoto and coworkers in 1984. The cellulose derivatives were coated at about 20 wt% onto the surface of macroporous silica beads (100 or 400 nm pore size) [31]. These materials exhibited considerably improved mechanical stability and much better efficiencies, and permitted HPLC enantiomer separations. Such coated polysaccharide-based CSPs were state-of-the-art for several decades.

In the following years a large variety of distinct polysaccharide derivatives, especially esters and carbamates of cellulose (consisting of 1,4-connected-β-D-glucose

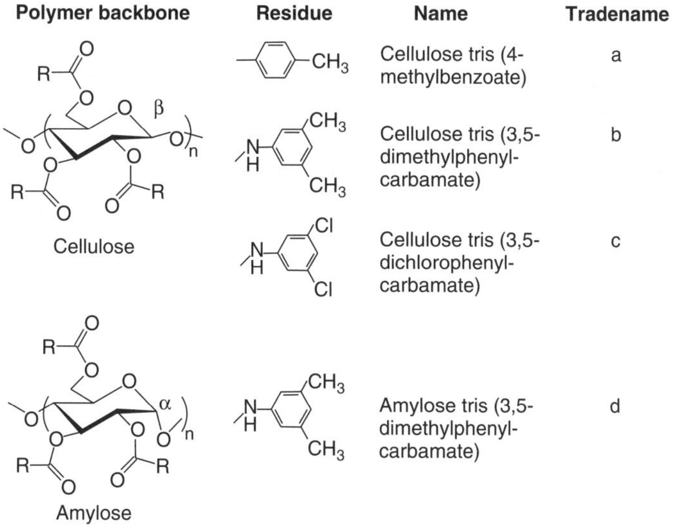

Polymer backbone	Residue	Name	Tradename
Cellulose	$-C_6H_4-CH_3$	Cellulose tris (4-methylbenzoate)	a
Cellulose	$-NH-C_6H_3(CH_3)_2$	Cellulose tris (3,5-dimethylphenyl-carbamate)	b
Cellulose	$-NH-C_6H_3Cl_2$	Cellulose tris (3,5-dichlorophenyl-carbamate)	c
Amylose	$-NH-C_6H_3(CH_3)_2$	Amylose tris (3,5-dimethylphenyl-carbamate)	d

Tradenames	Coated	Immobilized
a	Chiracel OJ	not available
b	Chiracel OD	Chiralpak IB
c	not available	Chiralpak IC
d	Chiralpak AD	Chiralpak IA
	Restricted solvent range	Full solvent range

Figure 14.10 Polysaccharide-based CSPs derived from cellulose and amylase, with corresponding column trade names.

units) and amylose (1,4-connected-α-D-glucose units) were prepared by Okomoto and coworkers and coated onto wide-pore silica (Fig. 14.10). The three most versatile coated polysaccharide CSPs are commercially available from Daicel Chemical Industries, Ltd. and Chiral Technologies:

- Chiralcel® OJ, based on cellulose tris(4-methylbenzoate)
- Chiralcel® OD, based on cellulose tris(3,5-dimethylphenylcarbamate)
- Chiralpak® AD, based on amylose tris(3,5-dimethylphenylcarbamate)

Since the expiration of the patents covering these CSPs, generic materials have become available from a number of suppliers (e.g., Eka Chemicals, Regis, Macherey-Nagel, Phenomenex) under different tradenames (e.g. Kromasil®, CelluCoat™, and ArmyCoat™, from Eka; Regis Cell from Regis; Lux Cellulose from Phenomenex). However, the overall performance of these generics often differ significantly from the original products in terms of chiral recognition capability, retention behavior, and column efficiency—because distinct supports and coating protocols are used for their fabrication (just as all C18 RPC columns are not equivalent).

Daicel and Chiral Technologies (as well as other suppliers) offer these CSPs in normal-phase (NP; to be operated with alkane-alcohol eluents or similar elution conditions) and reversed-phase versions (RP; for use with hydro-organic mobile phases; i.e., usually buffered aqueous-organic mixtures). These are presumed to differ in the type of support onto which the selectors are coated. The user manual recommends that columns with extension R be used with RP conditions, while other columns should avoid aqueous mobile phases. Whatever mobile phase is used, it is strongly recommended to dedicate the column to a specific mode and avoid switching between RP and NP conditions. The extension H indicates a high-performance version based on 5-μm particles. Most recently CSPs based on 3-μm supports became available as well (Chiralpak AD-3, Chiralcel OD- 3′ and the corresponding RP-versions, Chiralpak AD-3R, Chiralcel OD-3R) to provide higher column efficiency and faster separations for high-throughput chiral analysis. Because of slow adsorption-desorption kinetics (Section 14.5), however, the advantage of smaller particles can be less than for achiral separation.

The uniquely broad chiral-recognition capability of polysaccharide-type CSPs allows their use for a very broad range of sample types. This feature originates from various molecular and supramolecular structural features peculiar to these semi-synthetic macromolecules. The current understanding of chiral recognition principles for polysaccharide CSPs (largely derived from chromatographic experiments) is still at a rather immature state, despite their extensive use and importance. A limited number of studies have addressed the basis of polysaccharide molecular recognition: solution NMR of oligomeric surrogates [32–34], solid-state NMR [35], computational studies [34–37], ATR-IR [35], thermodynamics [36, 38, 39], and quantitative structure-property relationship studies [39–42]. However, these approaches have failed to provide much direction for the more effective use of polysaccharide CSPs.

The glucopyranose chains in cellulose and amylose derivatives have been shown to form helices, with the helical twist being less pronounced for the cellulose derivatives as compared to amylose derivatives (left-handed 4/3 helical structure for amylose tris(3,5-dimethylphenylcarbamate) [33]. Chiral recognition with polysaccharide-type CSPs may arise from three distinct features: (1) molecular chirality due to the presence of several stereogenic centers of the glucopyranose units, (2) conformational chirality due to the helical twist of the polymer backbone, and (3) supramolecular chirality from the alignment of adjacent polymer chains that form chiral cavities. These features, further enhanced by derivatization of the polysaccharide, provide the exceptional and versatile stereodiscriminating abilities of today's polysaccharide-type selectors.

From this general mechanistic picture it is clear that enantiorecognition for polysaccharide-type CSPs is defined not only by the type but also by the specific functionality of the respective cellulose and amylose derivative. The effect of the aromatic substituents on chiral recognition for cellulose tris(phenylcarbamate) derivatives has been studied in detail by Okamoto [43]. Chromatographic data were obtained for a set of 18 CSPs coated with different mono- or di-substituted phenylcarbamate derivatives; it was found that inductive and steric effects are crucial for the overall chiral recognition capacity. Ortho derivatives, regardless of the nature of the substituent, performed poorly; para-substituted derivatives carrying methyl-, ethyl-, chloro- and trifluromethyl-groups produced improved enantiorecognition. The best results were found with 3,4- and 3,5-dimethylphenyl- and dichlorophenylcarbamate derivatives, which have been subsequently selected as first choice for the preparation of commercial CSPs. For specific analytes, however, other derivatives may produce still higher enantioselectivities. For this reason, it is worthwhile to also include less common CSPs in the screening process—especially for preparative separations (see the extensive coated-polysaccharide-column lists of Daicel and Chiral Technologies).

As noted above, the supramolecular structure of a polysaccharide-type CSP has an enormous influence on its enantioselectivity. In the early studies of Hesse and Hagel [30] with microcrystalline MCTA material, it was demonstrated that this polymer showed major changes in chiral recognition performance after dissolution and re-precipitation. This observation was interpreted in terms of specific enantioselective binding sites (microcrystalline domains). In similar fashion Francotte and Zhang [44] found that the supramolecular organization of polysaccharide derivatives also has a major impact on the enantioselectivity of coated CSPs. Cellulose tris(4-methylbenzoate) coated from different solvents exhibited very different enantioselectivity, and—in some cases—inversion of enantiomer elution orders [44]. X-ray diffraction experiments support the hypothesis that solvent-induced alternations in the supramolecular structure account for this observation. These findings underline the fact that polysaccharide-type selectors respond very sensitively to external stimuli such as solvents, temperature, and additives that can lead to conformational changes of the CSP—with altered binding processes, memory effects, and so forth.

The complexity of the chiral recognition processes for polysaccharide-type CSPs renders a rational approach to method development difficult. Current strategies for chiral method development involve trial-and-error screening of various polysaccharide-type CSPs under multiple mobile-phase conditions—often using fully automated column- and solvent-switching. A number of studies have focused on identifying efficient screening routines to maximize the chance for success [45–48]: the most promising CSP in the NP mode is

Chiralpak AD > Chiralcel OD > Chiralcel OJ

If serial instead of parallel screening is utilized, columns should be tested in this order [45]. Interestingly, in the RP mode, Chiralcel OD-RH appears to be first choice rather than Chiralpak AD-RH. From these extended screening studies it is also clear that polysaccharide CSPs have extremely broad applicability. For example, Borman et al. reported the results of a comprehensive screening campaign for a set of over 100 chemically diverse racemates, employing HPLC, SFC, and CE separation techniques and systems [49]. With only three polysaccharide-type CSPs (Chiralpak AD, Chiralcel OD, and Chiralcel OJ), more than 70% of the racemic analytes could be resolved. It was also found that the individual polysaccharide derivatives often show complementary chiral recognition with respect to analyte structure; that is, different enantioselectivities (α) and different elution orders [50, 51]. This is particularly true for the cellulose and amylose tris(3,5-dimethylphenyl) derivatives (Chiralcel OD and Chiralpak AD), which for a given analyte frequently display reversed elution of the enantiomers.

Polysaccharide-based CSPs have traditionally been used as normal-phase packings; that is, their preferential mobile phase is hexane or heptane with 2-propanol or ethanol as strong solvents (B-solvents) [52]. Note that 2-propanol tends to show a different selectivity compared to ethanol for Chiralpak AD; elution can even be reversed by a change of these two solvents [53]. Recently an independent solid-state

NMR study confirmed that 2-propanol displaces hexane more efficiently from the polymeric selector, forming more ordered "solvent-CSP complexes" [54]. If the mobile phase is changed, sufficient equilibration time should be allowed to ensure the complete removal of the polar solvent. In order to achieve reproducible results, when switching from an additive-free to an additive-containing mobile phase (e.g., 0.1% trifluoroacetic acid or diethylamine), and vice versa, a prolonged equilibration time may be required [55].

In rare cases the use of particular additives may trigger transient or even persistent conformational changes in the polymeric structure of a polysaccharide derivative, either enhancing or attenuating its initial enantioselectivity. Persistent changes in selectivity were demonstrated to occur with amylose-based CSPs after operation with NP-type mobile phases that include diethylamine [56]; however, flushing with 2-propanol was shown to restore the original selectivity. Acid treatment of Chiralpak AD-H and Chiralcel OD-H CSPs also changed their performance, which was largely restored by washing with amine-containing mobile phases. Changes in enantioselectivity may also occur due to temperature-induced conformational changes of the selectors.

Although primarily used in the NP-mode, polysaccharide CSPs have been demonstrated to possess multimodal applicability. Besides the NP-mode and the already mentioned RP-mode [57], these columns can be used in the *polar-organic mode* (PO mode). The PO mode utilizes nonaqueous mobile phases that are made up of polar organic solvents, such as methanol or acetonitrile or a mixture of both, to which small amounts of organic acids (acetic acid, formic acid) and base (triethylamine, diethylamine, ammonia) are added as buffer constituents or competing agents. An important benefit of the PO mode, as opposed to normal-phase separation, is a better compatibility with (increasingly popular) electrospray MS detection, as well as better solubility of the analytes in the polar-organic mobile phase, and polar-organic sample matrix.

A change in enantioselectivity is often observed for the same compound upon a switch from normal-phase to reversed-phase mode [47]. As already noted, dedicated versions of Chiralpak AD-RH, Chiralcel OD-RH, and Chiralcel OJ-RH have been introduced for reversed-phase applications [57], coated onto a support that is compatible with RP separations. For neutral analytes, simple water-organic mixtures can be used (with acetonitrile being preferred to ethanol, 2-propanol, or methanol). While basic or acidic additives have little effect on the separation of neutral solutes, ionizable compounds, in contrast, require such additives to improve peak shapes. Phosphoric acid (pH-2) is recommended for acidic analytes. Enantiomer separation of basic analytes may be enhanced by addition of chaotropic counter-ions, such as perchlorate or hexafluorophosphate. Alternatively, basic buffer systems at pH-9 (preferably the relatively mild borate) can be employed to efficiently suppress dissociation of basic analytes [57]. For applications requiring mass-sensitive detection, volatile buffer systems must be used. For acidic analytes, phosphate can be replaced by formic acid, $pH > 3$; for basic analytes, borate can be replaced by ammonium bicarbonate buffer, pH-9.0. Nevertheless, it should be kept in mind that extended exposure to pH-9 will inevitably shorten column lifetime, as it is the case with most silica-based stationary phases.

Polysaccharide CSPs are also fully compatible with polar-organic mobile phases [58, 59], as mentioned above. A rather comprehensive study reported by Lynam [59]

	Coated CSPs	Immobilized CSPs
Standard mobile phases		
Normal phase conditions: Alkane/alcohol	✓	✓
Polar organic mode: Acetonitrile Alcohols (EtOH, MeOH)	✓	✓
Nonstandard mobile phases		
MTBE, Toluene, Choroform, Dichloromethane, Ethylacetate, THF, 1,4-dioxane, Acetone, DMSO (solvent strength increases in this order)	**Forbidden!** They will irreversibly destroy the coated CSP	✓

Figure 14.11 Allowable mobile phases for polysaccharide-based CSPs. A distinction is made between standard solvents to be used with all CSPs, and nonstandard solvents to be used only with immobilized CSPs (MTBE, methyl-t-butyl ether). Adapted from [60].

demonstrated that out of 80 test compounds, about 30% could be resolved with pure methanol, ethanol, or acetonitrile as mobile phases. Upon addition of hexane, analyte retention was increased as expected, while there was only a modest effect on enantioselectivity. The authors concluded that the chiral recognition processes operating with polar-organic mobile phases may be similar to those for normal-phase conditions.

The most critical limitation of the coated polysaccharide-type CSPs is their incompatibility with certain types of solvents, so-called nonstandard solvents, which include (among others) dichloromethane, chloroform, ethyl acetate, tetrahydrofuran, dioxane, toluene, and acetone (Fig. 14.11). The exposure of coated CSPs to the latter solvents induces swelling and/or dissolution of the physically adsorbed polymer layer, with destruction of the column. Such solvents must be strictly avoided, *even as constituents in the sample matrix*. To overcome this drawback, several research groups have developed fully solvent-resistant, immobilized polysaccharide-based CSPs by means of various procedures [61–63]. Since 2005 a set of three immobilized polysaccharide CSPs have become commercially available from Daicel and Chiral Technologies as:

- Chiralpak® IA (immobilized version of Chiralpak AD) [64]
- Chiralpak® IB (immobilized Chiralcel OD) [60]
- Chiralpak® IC [65] based on the cellulose tris(3,5-dichlorophenylcarbamate) selector that is not available in coated form (see Fig. 14.10)

These CSPs have been demonstrated to be fully compatible with any organic solvents, expanding the applicability to nonstandard solvents that cannot be used with the coated CSPs (Fig. 14.11). Moreover this new generation of immobilized polysaccharide-type CSPs also performs very well in PO, RP, and SFC mobile phases.

While the availability of additional solvents provides unique opportunities and considerable flexibility for method development, it necessarily involves more extended and elaborate screening—with additional labor. *To address this latter problem, systematic studies have created an efficient generic screening strategy based on the three commercially available, immobilized CSPs and a limited set of mobile phases. Using Chiralpak IA, IB, and IC and only five starting mobile phases for method development (alkane-2-propanol 80:20, alkane-ethanol 80:20, methyl tert-butylether-ethanol 98:2, alkane-tetrahydrofuran 70:30, and alkane-dichloromethane-ethanol 50:50:2), about 90% of a series of 70 randomly selected chiral analytes could be baseline separated [66].* Interestingly this screening strategy with nonstandard solvents and immobilized polysaccharide-type CSPs provided novel enantioselectivity that was not achievable with coated-version columns and standard-type mobile-phase conditions; for specific cases, greatly improved separations resulted [60].

Fully solvent-compatible, immobilized CSPs also offer unique advantages for preparative applications. Specifically, the poor solubility of chiral compounds in standard NP mobile phases is often the most serious bottleneck in the development of preparative enantiomer separation with coated polysaccharide-type CSPs—which, in general, have among the highest loading capacities of all commercially available CSPs [61]. This is particularly true for polar analytes, which are often sparingly soluble in alkane-based NP mobile phases. With immobilized CSPs, strong solvents such as tetrahydrofuran, dichloromethane, chloroform, and even dimethyl sulfoxide can be used for sample dissolution, without any concern for column stability. Residual, nonstandard solvents in the sample are also not an issue. Another beneficial feature of immobilized polysaccharide-type CSPs is that they can be regenerated with strong, nonstandard solvents such as tetrahydrofuran, dimethyl formamide, and even dimethyl sulfoxide, allowing the effective removal of strongly adsorbed sample components, and improving the reproducibility of subsequent separations.

A problem can arise if a method that was developed for a coated polysaccharide CSP is transferred to the corresponding immobilized CSP. The proprietary immobilization process appears to modify the enantiomer separation characteristics as compared to the coated versions [60, 67, 68]. For example, enantioselectivity may be lost upon exchange of the (coated) Chiralcel OD column by the (immobilized) Chiralpak IB column when using the standard alkane-based mobile phase (Fig. 14.12*b* vs. 14.12*c*), although both CSPs are based on the same polysaccharide derivative. Nevertheless, after re-optimization of the mobile-phase composition (including use of the tolerated nonstandard solvents), a greatly improved separation can be obtained (compare Fig. 14.12*a–b* or *c*). Comparable results were obtained with acidic (Figs. 14.12*a–c*), basic (Figs. 14.12*d–f*), and neutral compounds. These examples demonstrate a certain level of complementarity between coated and immobilized polysaccharide-type CSPs, which represents an advantage for both analytical and preparative method development. The combined use of both types of CSPs can provide chromatographers with an even more powerful tool for separation challenges of ever-increasing complexity. Certainly the introduction of the immobilized polysaccharide CSP technology represented an important milestone in the development of enantiomer separation.

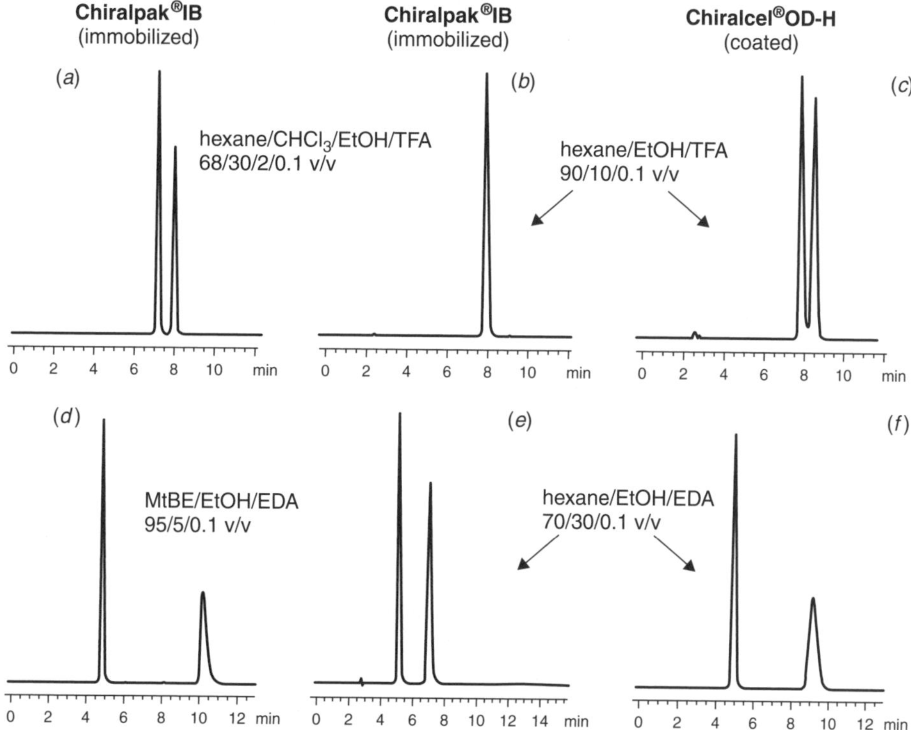

Figure 14.12 Effect of column type (immobilized vs. coated polysaccharide-based CSP) and mobile phase on enantioselectivity. Enantiomer separations of *N*-benzyloxycarbonyl-phenylalanine (*a–c*) and laudanosine (*d–f*) with Chiralpak IB (immobilized) and Chiralcel OD (coated). Flow rate, 1 mL/min; temperature, 25°C; UV detection at 230 nm. Note that the mobile phases used in (*a*) and (*d*) are forbidden mobile phases for the coated version Chiralcel OD. (EDA, ethylenediamine; MtBE, methyl tert-butylether). Reprinted with permission from ref. [60].

Finally, it should be noted that polysaccharide CSPs have now also been established as the first-choice of chiral phases for SFC enantiomer separation [69–72].

14.6.2 Synthetic-Polymer CSPs

A number of chiral synthetic polymers have been proposed as potential selectors, in an attempt to mimic the enantioselectivity of the semi-synthetic polysaccharides of Section 14.6.1. Like the polysaccharides, chiral synthetic polymers are constructed from identical chiral building blocks. However, the synthetic polymers do not achieve the same enantiorecognition as found for the polysaccharide CSPs. This may be the result of a much less ordered structure of the polymer chains, as well as the lack of pre-organized grooves and clefts for solute insertion.

Evidence that an ordered chiral hyperstructure is alone sufficient for effective chiral recognition has been furnished by Okamoto et al. [73]. A CSP was prepared from single-handed helical poly(triphenylmethacrylate), obtained by anionic polymerization from essentially achiral monomers (using a chiral catalyst); the polymer

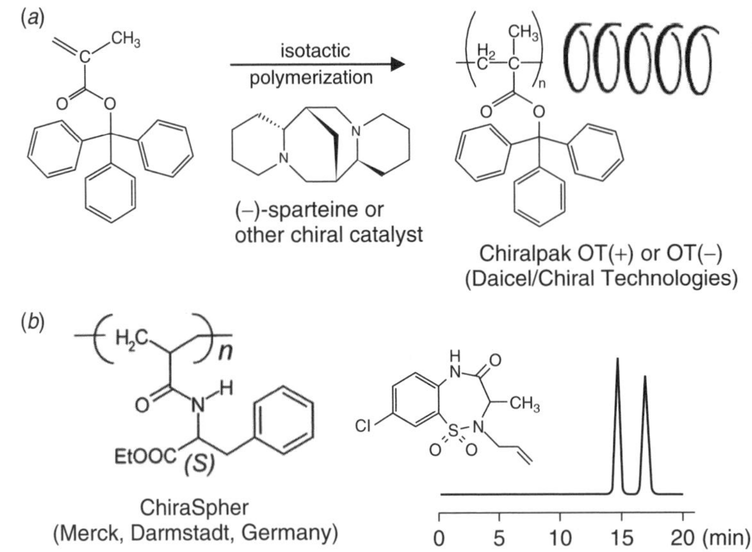

Figure 14.13 CSPs based on polymethacrylate-type chiral polymers. (*a*) Helically chiral poly(tritylmethacrylate) based CSP obtained by isotactic anionic polymerization in the presence of sparteine as catalyst (Chiralpak® OT(+) or OT(−) from Daicel and Chiral Technologies,). (*b*) poly[*N*-acryloyl-(*S*)-phenylalanine-ethyl ester] (ChiraSpher® from Merck KGaA, Darmstadt, Germany). Chromatographic conditions in (*b*): Column dimension, 250 × 4.6 mm column; mobile phase, 80:20 *n*-hexane/2-propanol; flow rate, 1 mL/min; temperature, 25°C. Adapted from [77] (*a*) and [78] (*b*).

chains were then coated onto macroporous silica (later commercialized by Daicel and Chiral Technologies under trade name Chiralpak OT+; Fig. 14.13*a*). The latter column successfully separated a number of chiral compounds that possessed multiple aromatic functionalities such as 1,1′-binaphthyl-2,2′-diol ($\alpha = 2.0$ with methanol as mobile phase at 5°C). Since this CSP suffers from rather poor chemical stability, it is now more of academic than practical interest.

In 1974 Blaschke introduced new polymeric materials in the shape of self-supporting cross-linked poly(meth)acrylamide polymer beads [74], prepared by suspension polymerization of acryl- or methacrylamides, with the chiral element residing as a stereogenic center in the amide side chain. To alleviate extensive swelling, and insufficient mechanical stability in high-pressure separations, silica-supported composite materials with more favorable chromatographic performances were later developed [75]. The CSP obtained by copolymerization of *N*-acryloyl-(*S*)-phenylalanine ethyl ester as monomer with vinylized-silica particles has been commercialized by Merck KGaA (Darmstadt, Germany) under the trade name ChiraSpher (Fig. 14.13*b*). This CSP, which is still used and has preparative capabilities [61], has been useful for the separation of a variety of polar pharmaceuticals with hydrogen donor-acceptor functionalities, employing normal phase conditions (usually n-hexane or n-heptane with a polar solvent such as alcohols, dioxane, or THF). A comprehensive review on this topic has been published by Kinkel [76].

Synthetic polymeric phases are commonly prepared by the classical "grafting-to" approach, that is, formation of the polymer from monomers in solution, with its subsequent anchoring to a support that has been modified by vinyl groups for copolymerization. Frequent problems are encountered with such polymer-type CSPs because of pore blockage and the inhomogeneous distribution of the polymer on the silica surface—leading to lower plate numbers for polymeric CSPs, compared to brush-type phases (Section 14.6.7).

Gasparrini and coworkers recently proposed an alternative procedure for the generation of silica-supported polymer-type composite CSPs. Instead of the "grafting-to" approach (Fig. 14.14*a*), a "grafting-from" technique (Fig 14.14*b*) was used to attach chiral selector moieties onto the surface of the silica particles [79]. In a first step, the radical initiator ("G") is immobilized onto the surface. After the addition of monomers and initiating polymerization by heating, the polymer chains begin to grow from the surface in a more regular way, resulting in a well-ordered surface-confined polymer layer. Detrimental polymer chains grown in solution can be removed during the washing steps. The procedure has been implemented successfully with (*trans*-1,2-diamino-1,2-diphenylethane)-N,N-diacrylamide as well as (*trans*-1,2-diaminocyclohexane)-N,N-diacrylamide as monomers (commercialized as P-CAP-DP and P-CAP in both enantiomeric forms by ASTEC). The synthesis of the latter is shown in Figure 14.14*c*. CSPs as in (Fig. 14.14*b*) exhibited a significantly higher column efficiency in terms of the control materials obtained by conventional grafting (i.e., a copolymerization or grafting-to procedure as in Fig. 14.14*a*). A broad range of chiral compounds, such as benzodiazepines, carboxylic acids and sulfoxides, esters, amides, lactones, and *N*-blocked amino acids can be separated primarily in the NP or PO mode.

A new class of CSPs based on network-type polymers has been proposed by Allenmark et al. [80, 81]. The preparation of these CSPs starts with O,O'-diaroyl-N,N'-diallyl-(R,R)-tartardiamide as chiral monomers, which are immobilized onto vinylized silica by cross-linking the monomers with multifunctional hydrosilanes—yielding a network polymer that has incorporated the bifunctional C_2-symmetric chiral selector as a thin film on the silica surface (Fig. 14.15*a*). Through network formation and multiple crosslinks of the polymer network to the vinylized silica, highly stable CSPs with low column bleed can be obtained (a general benefit of such CSPs). Different substitution patterns of the hydroxy groups can give rise to unique enantioselectivity. CSPs in which the tartramides are substituted with O,O'-bis(3,5-dimethylbenzoyl) and O,O'-bis[4-(*tert*-butyl)benzoyl] moieties are commercially available from Eka Chemicals (Bohus, Sweden) under the trade names Kromasil CHI-DMB and CHI-TBB. These CSPs exhibit useful enantiorecognition properties under NP conditions for a variety of pharmaceutically relevant entities, including acidic, neutral, and basic compounds that carry hydrogen donor-acceptor groups and $\pi-\pi$-interaction sites (as in the separation of Fig. 14.15*b*).

14.6.3 Protein Phases

The stereoselective binding of chiral compounds to proteins was an early discovery in biochemistry. Since nature offers a wide choice of stereoselective proteins, their exploitation as chiral selectors for enantioselective liquid chromatography seems obvious. Allenmark, Hermansson, Miwa, Haginaka, and others, have pioneered

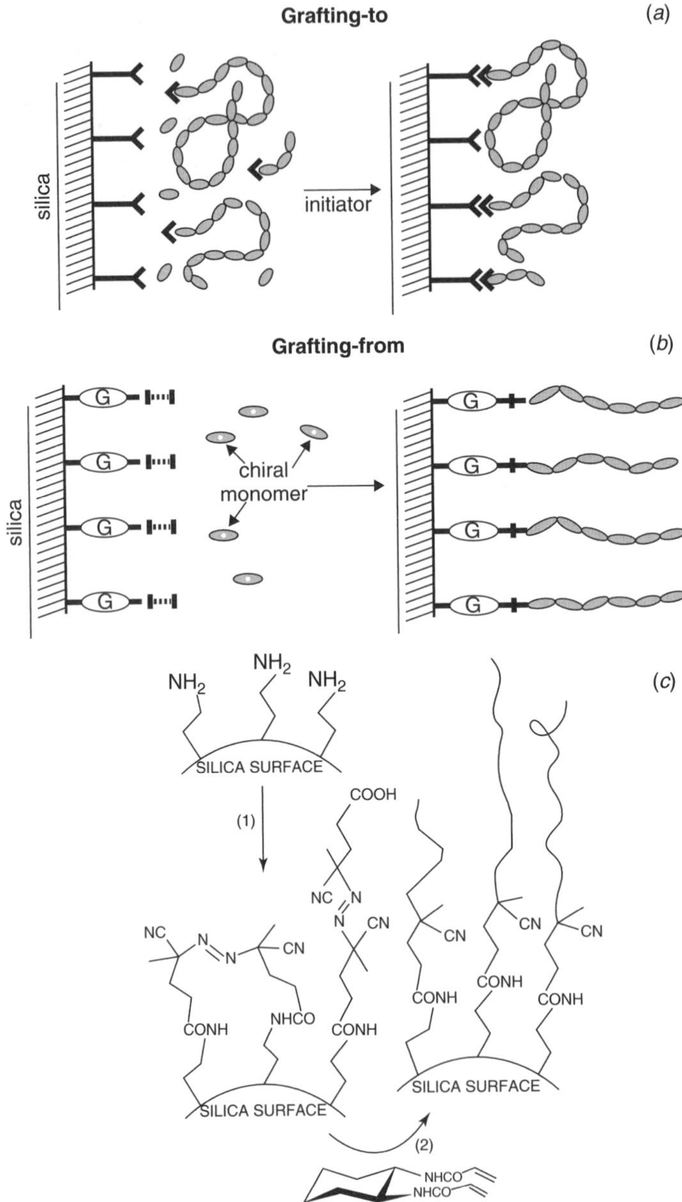

Figure 14.14 Preparation of synthetic-polymer CSPs (polyacrylate-type). (a) Grafting-to and (b) grafting-from approach; (c) reaction scheme for the synthesis of poly(diaminocyclohexane-*N*,*N*-diacrylamide)-based CSP prepared by the grafting-from concept (b). Adapted from [79].

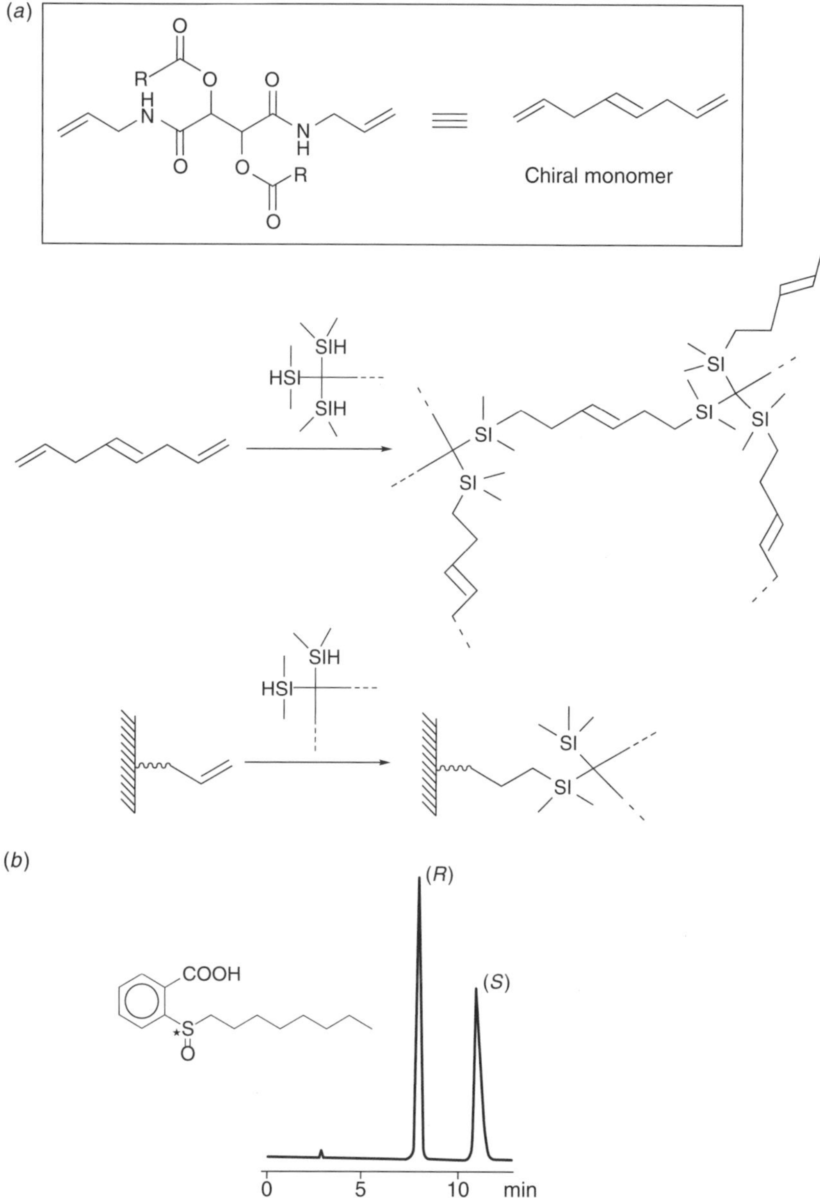

Figure 14.15 Synthetic tartardiamide-derived network-polymer based CSPs. (*a*) Preparation; (*b*) separation with 80:20 hexane/THF plus 0.05% TFA. Adapted from [80].

this field. Besides a wide variety of protein-based CSPs that are well documented in the literature [82–84], a number of such protein phases have become commercially available. Table 14.2 summarizes the most important protein phases, along with some of their characteristics and column trade names. Among them, α_1-acid glycoprotein (AGP) [85] and crude ovomucoid (OVM) [86] (commercially available as Chiral AGP and Ultron ES-OVM, respectively) exhibit the broadest enantiomer

Table 14.2

Important Protein-Type CSPs and Column Trade Names, with Some Characteristics of the Protein Selectors

Protein	Molecular Weight (kDa)	Carbohydrate (%)	Isoelectric Point	Column Trade Name (Supplier)
Serum albumin				
Human (HSA)	67	0	4.7	Chiral HSA (ChromTech)
Bovine (BSA)	68	0	4.7	Resolvosil BSA (Macherey Nagel)
α_1-Acid glycoprotein (AGP)	44	45	2.7	Chiral AGP (ChromTech)
Ovomucoid (OVM)	28	17-34	4.5	Ultron ES-OVM (Shinwa Chemical)
Cellobiohydrolase I (CBH)	60–70	6	3.6	Chiral CBH (ChromTech)
Avidin	66	20.5	9.5–10	Bioptic AV-1 (GL Sciences)
Pepsin	70–78	-	6.1–6.6	Ultron ES-Pepsin (Shinwa Chemical)

Source: Adapted from [83].

separation capabilities, covering a wide variety of neutral, acidic, and basic drugs. The range of applicable analytes is much narrower for other protein-type CSPs, and their use is correspondingly more limited. Cellobiohydrolase I (CBH I) [87] (Chiral CBH) is preferred for basic analytes (e.g. β-blockers), and human serum albumin (HSA) [88] (commercially available under trade name Chiral HSA) is used with acidic analytes. Apart from their use in assaying chiral compounds, HSA-based CSPs have received considerable attention for the study of drug–protein binding because the binding characteristics observed under physiological conditions in plasma are largely maintained when the protein is immobilized onto a silica support. This, unfortunately, does not apply for commercial AGP columns, which display considerably altered binding characteristics after immobilization that is due to the specific cross-linking procedure used by the manufacturer to improve column stability.

Structurally AGP consists of a single 181-residue peptide chain carrying five heteropolysaccharide moieties. The latter incorporate 14 sialic acid units that render AGP strongly acidic (pI = 2.7) [89]. AGP, covalently attached with cross-linking onto a chemically modified silica surface (by Hermansson [85]), tolerates a wide range of pH changes (3–7.5), high concentrations of organic solvents (up to 25%) and temperature up to 70°C. Crude chicken OVM-based CSPs, first described by Miwa et al., were later found to owe their chiral recognition ability to an impurity (ovoglycoprotein; 11%; w/w) in the crude OVM [90, 91]. In contrast, pure chicken ovomucoid turned out to have negligible enantiodiscrimination capabilities [92]. OVM-based CSP is stable in the pH range of 3 to 7.5 and tolerates organic solvents up to 50%; the separation temperature should be <40°C.

Protein phases are always used with aqueous or hydro-organic mobile phases (i.e., RP-mode). Mobile-phase pH, buffer type and strength, type and content of organic solvents, and additives—along with temperature—are the key variables for controlling retention and enantioselectivity. The appropriate conditions must be selected empirically, and depend on solute structure. For acidic and basic drugs, retention will vary with pH as a function of solute and selector ionization; ionic-strength variation can be a powerful tool to vary retention and improve selectivity. An increase in ionic strength can either decrease retention by shielding charges on the analyte or CSP or increase retention by salting out the analyte, thereby enforcing hydrophobic solute-sorbent interactions. The retention of neutral molecules can also be altered by a change in pH, as a result of changes in the protein structure/conformation (leading to altered binding properties). It is known, for example, that OVM can undergo reversible unfolding and refolding as a function of pH [93].

The addition of organic solvents (especially 1- or 2-propanol, or acetonitrile) reduces retention by weakening hydrophobic interactions. Enantioselectivity may thereby either decrease or increase, depending on whether these hydrophobic interactions occur at enantioselective or non-enantioselective sites. As an example of the potential effect of a change in organic solvent, verapamil was not separated on the AGP-type CSP with a mobile phase containing 1-propanol but was completely resolved when 1-propanol was replaced by acetonitrile [94]. It should be kept in mind that the addition of organic solvents may cause reversible changes of the secondary structure of the immobilized protein, thereby altering its enantioselectivity. *Organic solvent concentrations that are too high can cause irreversible changes and denaturation of the protein*, and must be avoided in order not to damage the column. As with any column, the manufacturers' instructions should be followed strictly. If on-line mixing is used, we strongly suggest ensuring that none of the reservoirs contains organic solvent at a higher concentration than that tolerated by the column, so as to avoid inadvertent destruction of the (expensive!) column

Cationic additives (alkylamines or quaternary-ammonium salts, such as *N*,*N*-dioctylamine and tetrabutylammonium bromide) and anionic additives (hydrophobic carboxylic acids, alkylsulfonates), which can compete with solutes for both ionic and hydrophobic interaction sites, have also been used to control retention and enantioselectivity. The addition of 50 μmol/L disodium 1,2-ethylenediamine-*N*,*N*,*N*′,*N*′-tetraacetic acid (EDTA) to the mobile phase is sometimes suggested for the purpose of shielding the protein phase from contamination by metals that may originate from the equipment. Temperature is another variable for the fine tuning of retention and enantioselectivity. Unlike many other columns, van't Hoff plots of protein phases frequently deviate from linearity (failure of Eq. 14.4). Such unusual temperature behavior can also be attributed to conformational changes of the protein. For example, the temperature dependence of enantioseparations of quinoline-substituted carboxylic acids on AGP suggests such a change in protein conformation between 30 and 35°C [95].

Because of the structural complexity of these macromolecular protein-selectors and their frequent conformational changes with conditions, not much is known about the fundamental retention process or how this relates to enantioselectivity. Beginning in the early 1980s, protein-based CSPs were popular—especially because they could be used with RP conditions and with aqueous samples, which was favorable for bioanalytical applications. Due to some major limitations, however, their use has since declined. Significant restrictions exist in terms of mobile-phase composition, as noted above. Improper use or storage, or elongated operation with more

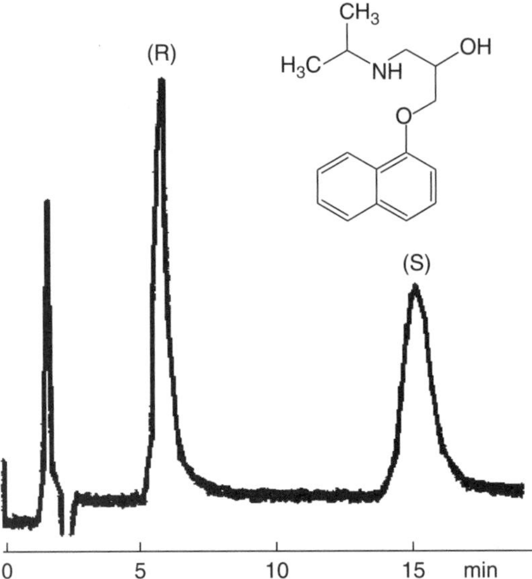

Figure 14.16 Separation of propranolol enantiomers on a protein column (cellobiohydrolase I; Chiral CBH I, ChromTech). Mobile phase: 0.01M acetate buffer, pH 5. Reprinted with permission from [83].

extreme conditions, can easily lead to chemical or biochemical changes of the protein selectors; this can adversely impact column longevity or enantioselectivity. Another disturbing limitation is a low to moderate column efficiency, as seen in the example separation of Figure 14.16 ($N = 300-700$). Even with large enantioselectivity values (as in Fig. 14.16), low plate numbers, combined with peak tailing, can make the analysis of 0.1% enantiomeric impurities impossible in some cases. This is especially important when the impurity is the second peak, as noted in Section 14.3; unfortunately, a simple switch to the CSP with opposite configuration and reversed elution order (to minimize the effects of peak tailing) is not possible for protein columns. For some samples, AGP and OVM display a (unpredictable) reversal of elution order. The complexity of the protein selectors limits insights into chiral recognition phenomena; yet several studies have been undertaken in this direction [96–103].

The value of protein phases is also considerably reduced by their very limited sample loadability [61]; the number of specific (enantioselective) interaction sites on the surface of the (large) protein molecule is usually small. The immobilization of smaller protein fragments that include the enantioselective sites might be expected to lead to higher sample loadability, but such experiments have so far proved unsuccessful [99, 100]. Consequently protein CSPs are of little value for preparative separations.

To summarize, the role of protein phases for enantiomer separations is today slowly declining; these columns are more and more being replaced by other CSPs (e.g., polysaccharides and macrocyclic antibiotics). While the use of protein columns for quality control is discouraged, they may still have some relevance in bioanalytical research, not least because of the compatibility of their mobile phases with electrospray interfaces (as opposed to typical normal-phase packings) [104].

14.6.4 Cyclodextrin-Based CSPs

Cyclodextrin (CD) bonded CSPs, introduced by Armstrong [105], are based on α-, β-, or γ-cyclodextrins: macrocyclic structures that are assembled from 6, 7, and 8 glucose units, respectively (Fig. 14.17). The glucose (glucopyranose) units are connected via α-1,4-linkages. These macrocyclic molecules adopt the shape of a truncated cone, and the number of glucose units determines the size of the cavity (0.57, 0.78, and 0.95 nm in diameter for α, β, and γ-CD, respectively) into which (preferentially hydrophobic) solutes or substituents of solutes can be inserted. The internal surface of the CD-cavity is hydrophobic, due to the carbon backbone of the sugar moieties. The upper and lower rim surfaces are hydrophilic, due to the presence of the primary hydroxyls (lower narrower rim) and secondary hydroxyls (upper wider rim). These CDs are usually bonded to silica gel at the narrower ring hydroxyls, either via ether linkage (*Cyclobond* columns, from ASTEC) or carbamate linkage (*ChiraDex* and *ChiraDex Gamma*, from Merck; *Ultron ES-CD* from Shinwa). Aside from CSPs with native CDs as selector, a few CD derivatives are also available (e.g., *Cyclobond I SP, RSP, RN, SN* or *Ultron ES-PhCD*) (see Fig. 14.17). Derivatization may enhance chiral recognition by providing additional interaction sites or by varying the size and access of the chiral inclusion cavity. Hence CSPs based on derivatized CDs can exhibit significantly different enantioselectivity, compared to their underivatized counterparts (native CDs).

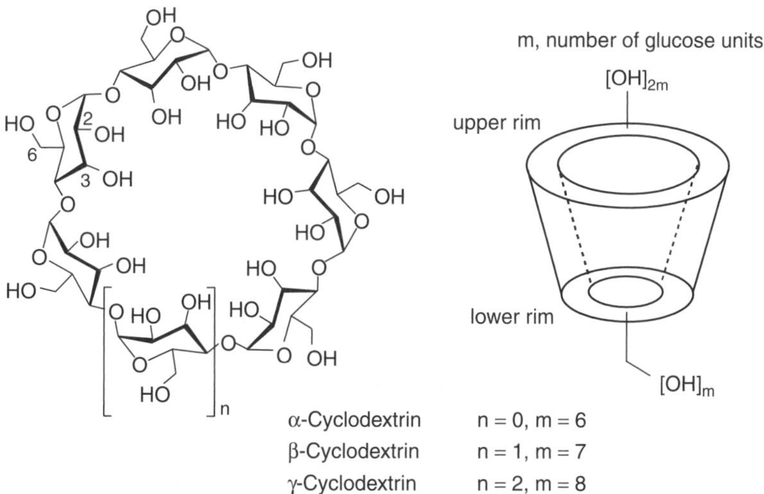

Commercial columns:

Cyclobond I (native β-CD), *II* (native γ-CD), and *III* (native α-CD) (from ASTEC)
Cyclobond I SP or *RSP* [(*S*)- or (*RS*)-2-hydroxypropylether-β-CD] (ASTEC)
Cyclobond I RN or *SN* [(*R*)-or (*S*)-1-(1-naphthyl)ethylcarbamate-β-CD] (ASTEC)
ChiraDex (native β-CD) and *ChiraDex Gamma* (native γ-CD) (from Merck)
Ultron ES-CD (native β-CD) and *Ultron ES-PhCD* (phenylcarbamoylated β-CD) (from Shinwa)

Figure 14.17 Structure of cyclodextrins and trade names of corresponding CSPs.

CD-based CSPs are truly multi-modal with regards to both elution conditions and chiral recognition mechanisms [18, 106]. They can be operated in RP, NP, and polar-organic modes. The molecular recognition mechanism was shown to differ for the different elution modes. It is commonly understood for the RP mode using aqueous or hydro-organic mobile phases (acetonitrile-buffer or methanol-buffer mixtures) that lipophilic solutes interact with CD selectors by inclusion complexation (Fig. 14.18*b*). A two-step (hydrophobic) mechanism has been proposed [107]: (1) penetration of the hydrophobic part of the analyte molecule into the CD cavity and (2) release of associated solvent (water) molecules from the analyte and CD molecules, followed by complex stabilization through hydrophobic interaction. Moreover hydrophilic interactions with hydroxyl groups at the upper and lower rim (hydrogen bonding, dipole–dipole interactions) may take place and positively contribute to complex stabilities. These combined effects plus conformational changes upon complexation (i.e., induced fit) can sometimes lead to extraordinary complex stabilities.

As in RPC, the concentration of organic solvent in the mobile phase can be used to control retention (for $1 \leq k < 10$). Solvent strength for CD-based CSPs increases in similar fashion as for achiral RPC:

$$\text{water} < \text{methanol} < \text{ethanol} < \text{propanol} \sim \text{acetonitrile} < \text{tetrahydrofuran}$$

The extent of analyte inclusion generally depends on the size of the CD cavity. According to this size-fit concept of inclusion complexation, higher affinity and greater enantioselectivity for the CSP-analyte pairs generally occur for the CD that gives the best match in terms of the size of hydrophobic portions of the solute with the CD cavity. Substituted phenyl, naphthyl, and heteroaromatic rings can conveniently be accommodated in a β-CD cavity, while larger analytes such as steroids fit preferentially into γ-CD, and smaller analytes prefer α-CD. This structure-binding relationship can be a helpful initial guide for column selection.

In the polar-organic mode (e.g., acetonitrile-methanol mixtures) the inner CD cavity is blocked by solvent molecules, so that inclusion complexation of lipophilic residues becomes thermodynamically unfavorable. Hence, solutes with

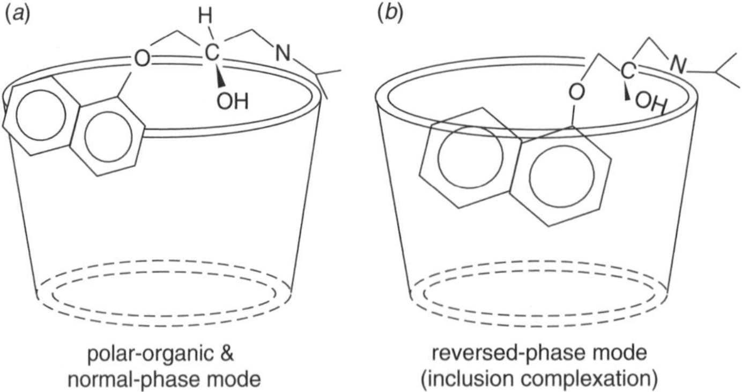

Figure 14.18 Molecular recognition mechanisms for cyclodextrin columns in polar-organic or normal-phase modes (*a*) and reversed-phase mode (*b*). Adapted from [108].

hydrophilic groups bind to the polar surface of the CD (either upper or lower rim; see Fig. 14.18*a*) [108]. The polar hydroxyls are surrounded by a chiral environment, with enantioselectivity resulting from differences in the strength of these polar interactions (hydrogen bonding, dipolar interactions) for the two enantiomers.

Solutes with more than one polar functional group, one of which is located at or close to the stereogenic center, are particularly amenable for cyclodextrin columns and the polar-organic mode—with great promise for a successful separation. Bulky groups near the stereogenic center facilitate the enantiorecognition process. Since many chiral drugs are polar, this elution mode can be quite useful—especially when RP and NP modes fail to resolve the enantiomers [109]. These polar-organic mobile phases for CD-based CSPs are typically composed of 0–15% methanol in acetonitrile plus 0.001–1.2% glacial acid plus 0.001–1.2% triethylamine and, for example, have been employed for enantiomer separations of β-blockers [110].

In the NP-mode (e.g., hexane or heptane mixtures with polar solvents such as alcohols), the internal CD cavity is also occupied by solvent molecules so that inclusion complexation does not occur. Interaction instead takes place at the polar surface of the CD or CD derivative. Only aromatic-derivatized CDs such as (*R*) or (*S*)–1-(1-naphthyl)ethyl carbamates have been shown to be effective CSPs for enantiomer separation in the normal-phase mode. In that case polar interactions such as hydrogen bonding and dipole–dipole interactions are supported by $\pi-\pi$ interactions, which seem to become crucial in this mode. Since inversions of elution orders have been observed when the configuration at the stereogenic center of the rim substituent was reversed, it can be concluded that enantioselective analyte binding is primarily controlled by the attached 1-(1-naphthyl)ethyl carbamate groups, rather than by the supporting CD cavity.

In recent years the importance of CD-based CSPs appears to be declining. Their applications overlap those of the more powerful macrocyclic antibiotics, which usually generate higher levels of enantioselectivity. Furthermore their preparative application is limited by a low sample-loading capacity. However, the compatibility of RP and polar-organic modes with the use of ESI-MS remains an important advantage for CD bonded CSPs.

14.6.5 Macrocyclic Antibiotic CSPs

Inspired by the stereoselective inclusion capabilities of macrocyclic cyclodextrins, Armstrong and coworkers have investigated other (more effective) macrocyclic natural compounds based on inclusion-complexation properties. From this research a new important CSP class was developed: the macrocyclic antibiotic CSPs. The first described CSP of this class was vancomycin-modified silica, introduced by Armstrong in 1994 [111]. Several structural analogues of this glycopeptide antibiotic were subsequently found to be powerful chiral selectors with complementary enantioselectivity (different enantioselectivities and elution orders), leading to CSPs based on vancomycin [111], teicoplanin [112], ristocetin A [113], and the aglycone of teicoplanin [114], all of which have been commercialized by Astec under the tradenames Chirobiotic V, Chirobiotic T, Chirobiotic R, and Chirobiotic TAG. Chirobiotic V and T are available with older (V1, T1) as well as newer (V2, T2) bonding chemistry and/or silica support. These CSPs have a broad range of application—which includes very polar compounds such as underivatized amino acids; these are today among the most powerful CSPs available. They have been described in several reviews [115–118].

Antibiotic selectors possess considerable structural complexity (Fig. 14.19). They share a common heptapeptide aglycone core, with aromatic residues that are bridged to each other to form a basket-like shape with shallow pockets for inclusion complexation and surface-confined carbohydrate moieties (Fig. 14.20). Inclusion complexation is often driven by polar interactions, notably for carboxylic acid-containing solutes: triple hydrogen-bonding of the carboxylic terminus (supported by other hydrogen bonding interactions), as well as $\pi-\pi$ and hydrophobic interactions (RP mode only). Together with the enantioselectivity that originates from multiple stereogenic centers, the heterogeneous multifunctionality and structural specificity of the glycopeptides provides a variety of potentially stereoselective binding options that appear to be the origin of enantioselectivity for a variety of pharmaceuticals.

Apart from their favorable structural features, the broad applicability of the macrocyclic antibiotics CSPs may also arise in part from their multimodal usage, which comprises RP, PO, and NP modes (Fig. 14.21) [111, 118, 119]. *The polar-organic mode is recommended as first choice, if the solute has more than one polar functional group (which applies to many chiral drugs), and if at least one of these groups is located at or close to the stereogenic center. Since hydrophobic interactions are disrupted by the PO mode, polar interactions such as ionic and dipole interactions as well as hydrogen bonding are of primary importance for enantioselectivity. A typical starting mobile phase is composed of methanol plus 0.1% acetic acid and 0.1% triethylamine as additives for the vancomycin CSP (additives should be increased to 1% for each for the teicoplanin CSPs). The ratio of acetic acid to amine is a key variable for adjusting enantioselectivity, while the total amount of the additives at given ratio mainly adjusts retention. If the solute cannot be eluted in the PO mode with about 1% of additives, the compound is too polar, and the RP mode is recommended. Typical starting conditions for the RP mode are mobile phases composed of THF-20 mM ammonium nitrate pH-5.5 (10:90; v/v) for vancomycin CSP, and methanol-0.1% triethylammonium acetate pH-4.1 (20:80; v/v) for the teicoplanin CSP. More detailed method-development procedures are provided in the* Chirobiotic Handbook *[119].*

In the RP mode, inclusion complexation may be driven or strongly supported by hydrophobic interactions, while multiple hydrogen bonds as well as ionic and dipole interaction contributions may be responsible for the stereoselective alignment of the solute in the binding cleft of the macrocyclic selector. If solutes cannot be sufficiently retained in the PO mode with as little as 0.01% of additives, or if solute solubility is limiting, the NP mode is a viable alternative. Typical starting conditions are a mixture of hexane-ethanol (80:20; v/v). Some other common elution conditions for different elution modes can be inferred from the caption of Figure 14.21. Overall, the use of different elution modes adds to the broad applicability of the macrocyclic antibiotic CSPs, and offers a tremendous flexibility of these CSPs for method development. It is further remarkable that these CSPs can be changed from one mode to the other without irreversible changes in performance.

From a statistical evaluation by the column supplier [120], which was later confirmed by researchers from industry [121], it was found that the highest success rate can be achieved with the polar-organic mode (~40%) or the reversed-phase

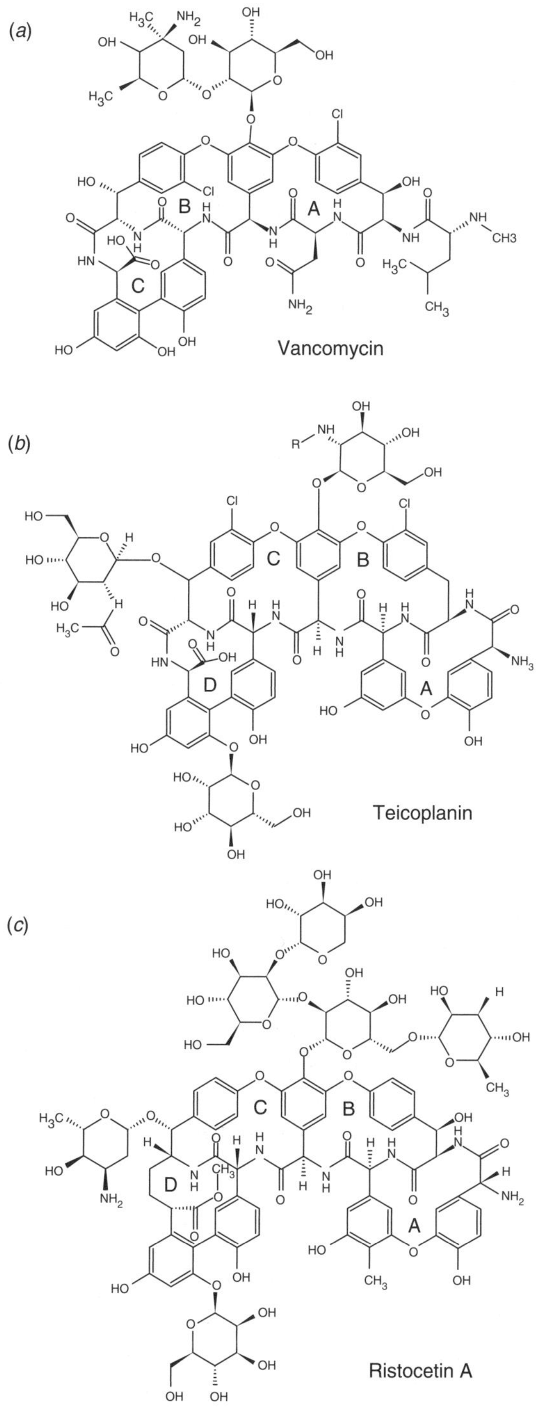

(*a*) Vancomycin

(*b*) Teicoplanin

(*c*) Ristocetin A

Figure 14.19 Structures of glycopeptide antibiotics: (*a*) Vancomycin (1 sugar moiety; 3 inclusion cavities A, B, C; molecular weight ≈ 1449 Da; pKa, 2.9, 7.2, 8.6, 9.6, 10.4, 11.7; pI, 7.2); (*b*) teicoplanin (molecular weight ≈ 1885 Da; 3 sugar moieties; R, decanoic acid residue; 4 inclusion cavities, A, B, C, D); (*c*) ristocetin A (molecular weight ≈ 2066 Da; 6 sugar moieties; 4 inclusion cavities A, B, C, D).

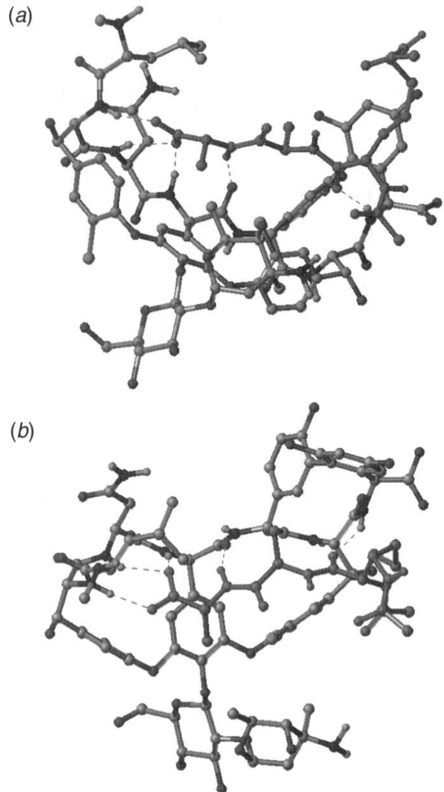

Figure 14.20 Basket-like shape and molecular recognition mechanism of vancomycin. X-ray crystal structure of a complex of vancomycin with N_α, N_ω-diacetyl-L-Lys-D-Ala-D-Ala. (*a*) Side view; (*b*) top view. X-ray crystal structure image was generated with SYBYL molecular modeling software (Tripos, St. Louis, MO) from fractional coordinates extracted from the Brookhaven protein databank (http://www.rcbs.org/pdb/).

mode (~40%), while normal-phase conditions appear to be less useful (~5%). In this context it should be noted, however, that many of the reported "RP" separations (with hydro-organic mobile phases) were actually HILIC-type (i.e., aqueous NP separations; Section 8.6), as pointed out by Wang et al. [122]. Actual RP separations can be obtained on the Chirobiotic phases with mobile phases that contain less than 20% organic solvent, while—for higher-% organic—polar interactions may be reinforced by the low dielectric constant media, and retention *increases* with increase in the organic solvent (suggesting HILIC behavior).

Although different macrocyclic antibiotic selectors have a close structural resemblance, they feature somewhat complementary enantioselectivity (Fig. 14.22), which is often helpful for method development. When a particular glycopeptide antibiotic CSP gives marginal or no separation, there is a fair chance that one of the other antibiotic CSPs can provide baseline separation. Subtle differences in the CSP structure and binding properties may be responsible for their complementary separation profiles; these have been attributed to distinct end-to-end distances of

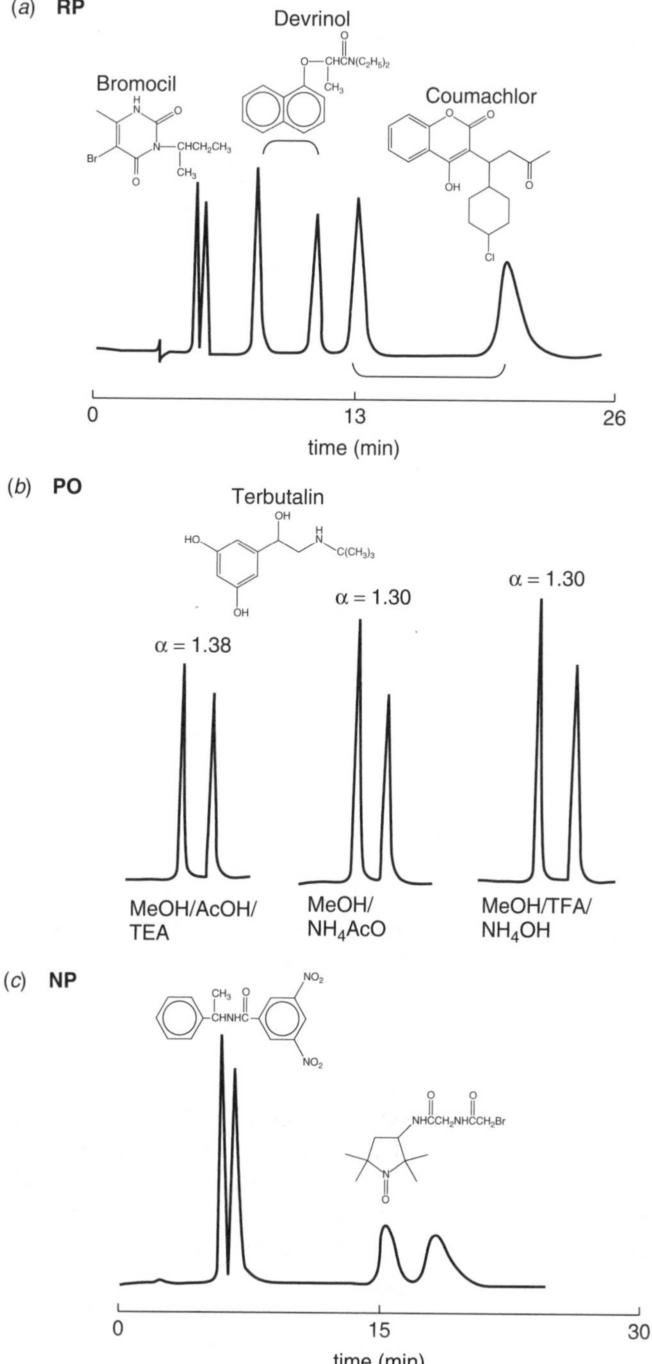

Figure 14.21 Separations by different elution modes with a macrocyclic antibiotic (vancomycin) column. (*a*) Reversed-phase (RP) mode; (*b*) polar-organic (PO) mode; (*c*) normal-phase (NP) mode. Experimental conditions: (*a*) 10:90 acetonitrile/1% triethylammonium acetate buffer, pH-7; (*b*) mobile phases from left to right: 100:0.3:0.2 MeOH/AcOH/TEA; 99:1 MeOH/20 mM NH_4AcO; 100:0.05:0.05 MeOH/TFA/NH_4OH; (*c*) 50:50 hexane/2-propanol. Adapted from [111] and [119].

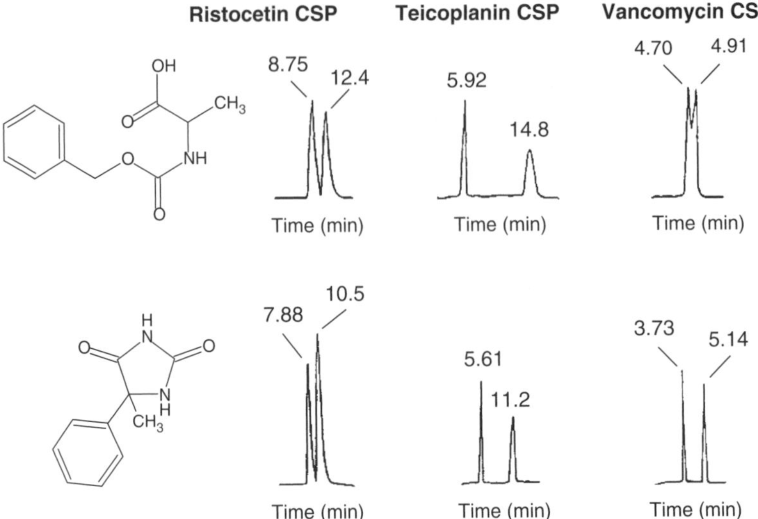

Figure 14.22 Complementarity of separation by different macrocyclic antibiotic columns, exemplified by enantiomer separations of Z-Ala (top) and 5-methyl-5-phenylhydantoin (bottom). Experimental conditions: Ristocetin A, 20:80 methanol/0.1% triethylammonium acetate buffer, pH-4.1; Teicoplanin, 20:80 methanol/1% triethylammonium acetate buffer, pH-4.1; vancomycin, 10:90 methanol/1% triethylammonium acetate buffer, pH-4.1. Adapted from [124].

the C-shaped aglycone (which becomes smaller in the order vancomycin, ristocetin, teicoplanin), and its helical twist (which increases in the same order), as well as the influence of various substituents and the sugar moieties [117].

In general, the broad applicability of the Chirobiotic CSP family includes chiral acids, bases, as well as amphoteric and neutral compounds—with the Chirobiotic T and V showing superior performance over Chirobiotic R [121]. Chirobiotic V especially shows enantiorecognition potential for neutral solutes, amides, esters, and amines (including aminoalcohols and cyclic amines). *Chirobiotic T can separate amino alcohols, underivatized and N-derivatized amino acids, and (di)peptides [116]; Chirobiotic R can be used for chiral acids such as hydroxy acids, substituted aliphatic acids, and other acids [117]. In this connection the enantiomer-separation ability of the teicoplanin CSP for underivatized natural and synthetic amino acids deserves particular attention [125]. With the teicoplanin CSP, α-values between 1.2 and 2.7 were reported for native amino acids, with baseline resolutions ($R_s > 1.5$) for all but His ($R_s = 0.8$). Amino-acid enantiomer separations on teicoplanin CSP are best carried out with plain ethanol-water (or methanol-water) mobile phases, but for acidic or basic amino acids a 0.1% triethylammonium buffer should be included in the mobile phase.*

An important study addressed the role of the carbohydrate moieties on the chiral-recognition capability of the teicoplanin-based CSP [114]. For a chromatographic comparison with the native teicoplanin-based CSP, a corresponding teicoplanin aglycone (TAG) analogue in which all the sugar moieties were chemically cleaved off was prepared and bonded to silica. While the overall retention of the

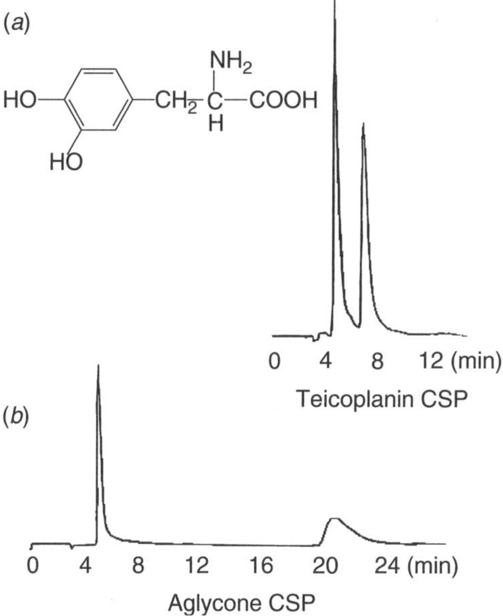

Figure 14.23 Enantiomer separation of D, L-DOPA on native teicoplanin CSP (*a*) and teicoplanin CSP after removal of sugars (aglycone) (*b*). Conditions: mobile phase, 60:40 methanol/triethylammonium acetate, pH-4.1; UV detection, 254 nm; temperature, 22°C; flow rate, 1 mL/min. Reprinted with permission from [114].

TAG CSP was quite similar as compared to the native teicoplanin CSP, the enantiorecognition profiles were significantly altered for most of the test compounds. Notably the sugar units considerably reduced enantioselectivity for amino acids (Fig. 14.23), which clearly indicates that the active chiral distinction site is located in the aglycone part of the CSP. For other solutes the opposite trend may be observed.

Chirobiotic phases are increasingly popular for the enantiomeric analysis of drugs and other xenobiotics in biofluids because of (1) their compatibility with these aqueous sample matrices and (2) their ideal ESI-MS compatibility in both reversed-phase and polar-organic modes [104, 123]. Drawbacks include the lack of CSPs that allow for a predictable reversal of elution order—when needed for the analysis of a trace-level enantiomer. Despite the claimed preparative applicability of these phases, their sample-loading capacity is limited, for the same reason as outlined for protein phases: a relatively low concentration of binding sites as compared to polysaccharide- and low-molecular-weight brush-type CSPs [61]. On the other hand, promising enantiomer separations of glycopeptide-type CSPs under SFC conditions have been demonstrated [126]. Comprehensive screens in this mode revealed that of a set of more than 100 chiral test solutes (including heterocycles, profens, β-blockers, sulfoxides, *N*-blocked amino acids and natural amino acids), more than 90% of these enantiomers could be resolved on commercial glycopeptide-type CSPs employing sub-/super-critical carbon dioxide–methanol as mobile phase, with low amounts of acidic and basic additives.

14.6.6 Chiral Crown-Ether CSPs

Stereoselective CSP-analyte complexation with chiral crown-ether CSPs and their first application as CSPs were pioneered by Cram and coworkers [127]. In this early work, two 1,1′-binaphthyl units were incorporated into a crown ether as replacement elements of two ethylene groups of the well-known 18-crown-6. Structural analogues of such 1,1′-binaphthyl-derived chiral crown-ether based CSPs were later developed by Shinbo's group using (3,3′-diphenyl-1,1′-binaphthyl)- or (6,6′-dioctyl-3,3′-diphenyl-1,1′-binaphthyl)-20-crown-6 dynamically coated onto octadecyl-silica [128]; a related CSP has been commercialized as Crownpak CR by Daicel and Chiral Technologies (Fig. 14.24*a*). Since they are of synthetic origin, both enantiomeric forms, with opposite elution orders, are available—denoted as Crownpak CR(+) and CR(−). Structural analogues are available, based on (18-crown-6)-2,3,11,12-tetracarboxylic acid (i.e., tartaric acid incorporated into crown ether) that is bivalently immobilized via two carboxylic functionalities onto 3-aminopropylated-silica [129]. These are commercially available as ChiroSil RCA(+) and SCA(−) (from Regis, Morton Grove, IL) and ChiralHyun-CR-1 (from K-MAC, Korea) (Fig. 14.24*b*).

The applications of such chiral crown-ether-based CSPs are essentially restricted to primary amines comprising mainly amino acids, amino acid esters, amino alcohols, and chiral drugs with free primary amino functionality. Typically aqueous mobile phases with pH between 1 and 3.5 are required to ensure full protonation of the solute. The resulting chiral ammonium ions can bind to the macrocyclic crown by inclusion complexation, driven by triple hydrogen-bond formation between the ammonium ion and the three oxygens of the crown (Fig. 14.25). Enantioselectivity may be governed by steric factors arising from the substituents of the chiral ammonium ions and the residues attached to the chiral moieties that are incorporated into the 18(20)-crown-6. Maintaining strongly acidic conditions appears also important to suppress silanol interactions that can be formed non-enantioselectively. This can be achieved by employing, for example, 5 mM perchloric acid in water or methanol-water mixtures (up to 15% methanol)

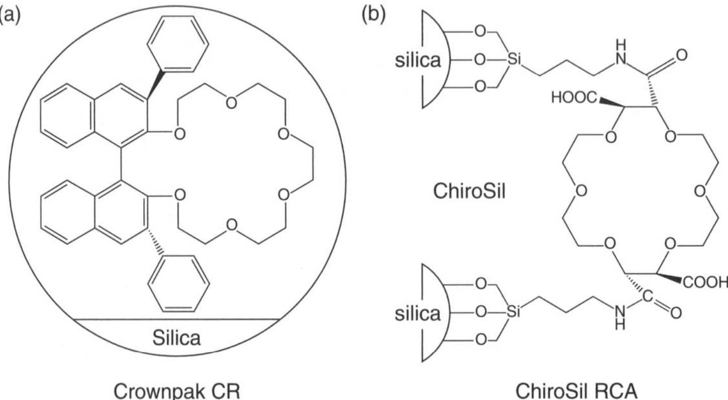

Figure 14.24 Commercially available chiral crown-ether based CSPs (adapted from online application notes provided by the suppliers).

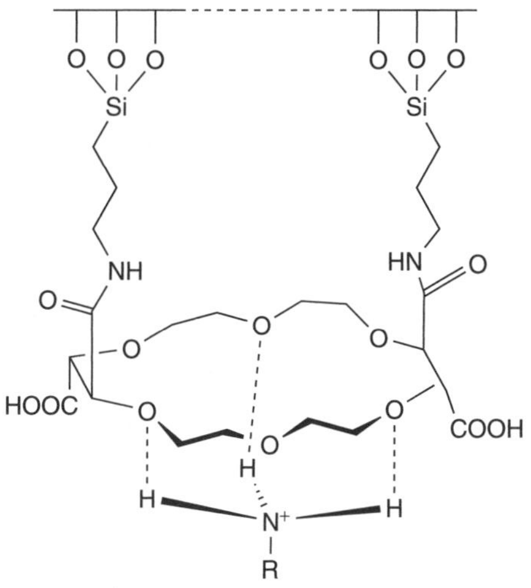

Figure 14.25 Molecular-recognition mechanism for chiral crown-ether CSPs: Schematic representation of solute-selector interaction driven by triple hydrogen bonding (adapted from the Regis webpage).

as mobile phase. Such harsh conditions can prove harmful for both the equipment and CSP, which has contributed to the limited popularity of these CSPs. Newer work on crown-ether-based CSPs can be found in [130–133].

14.6.7 Donor-Acceptor Phases

The first silica-bound CSPs with entirely synthetic selectors were developed in the late 1970s [134, 135]. Subsequent work by Pirkle and coworkers led in 1981 to the first commercialized CSP with a DNB-phenylglycine derivative immobilized ionicly onto silica. Later this synthetic chiral selector was grafted onto silica via a covalent amide linkage; this chiral packing material is still commercially available from Regis, Machery Nagel, and Merck as DNBPG (Fig. 14.26*a*). Such donor-acceptor-type CSPs (Brush-type CSPs) are based on chiral, low-molecular-weight selectors that are neutral, synthetic or semi-synthetic, and used in the NP mode. They are capable of generating enantioselectivity based on complementary, non-ionic attractive binding forces [27]. Hydrogen bonding, face-to-face or face-to-edge $\pi-\pi$ interaction (between electron-rich and electron-poor aromatic groups), and dipole–dipole stacking play important roles in stabilizing the selector-analyte complex and enantiorecognition. Enantioselectivity is often supported by steric interactions of bulky groups, which can represent effective steric barriers to a close selector-solute contact for one enantiomer. Due to the relative importance of hydrogen-bonding and other non-ionic electrostatic interactions, such CSPs are less effective in polar protic media, including the RP and PO modes. Because of the important contribution of Pirkle's group in this field, such donor-acceptor-type CSPs are now often referred to as Pirkle-type CSPs.

A number of powerful CSPs evolved early on from Pirkle's group as a result of systematic chromatographic [136] and spectroscopic [137–139] studies of chiral recognition phenomena, as well as the consistent exploitation of the reciprocity principle of chiral recognition [140, 141]. This reciprocity recognizes that the roles

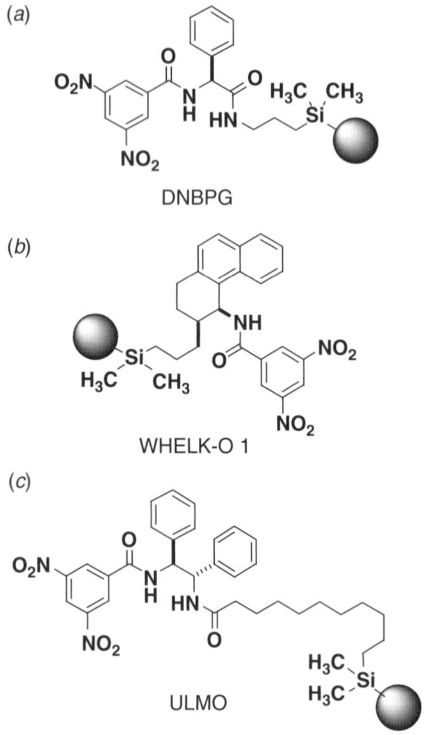

(a)

DNBPG

(b)

WHELK-O 1

(c)

ULMO

Figure 14.26 Structures of popular Pirkle-type donor-acceptor phases. (*a*) DNBPG; (*b*) WHELK-O 1; (*c*) ULMO.

of selector and analyte are interchangeable. Hence a single enantiomer of an analyte that is well resolved by a CSP with a given chiral selector will (after its immobilization at positions that are not involved in the chiral recognition process) be able to separate the racemate of this selector. Such concepts and tools have been used for the rational design of new advanced CSPs [136, 142].

As noted above, such donor-acceptor-type CSPs usually have been designed to exploit π–π-stacking interactions between electron-rich and electron-deficient aromatic systems as the primary attractions. Initially developed were either CSPs with π-acidic groups (with electron-deficient aromatic moieties, usually 3,5-dinitrobenzoyl) for π-basic solutes (with electron-rich aromatic groups) or CSPs with π-basic residues (e.g., naphthalene) for π-acidic solutes. The latter CSPs (e.g., *N*-2-naphthylalanine undecylester-derived CSP) [143] seemed to have less broad application and therefore disappeared form the market. Several of the early-invented π-electron acceptor phases from the Pirkle group, in contrast, are still available from Regis (e.g., DNBLeu, DNBPG, β-Gem 1, α-Burke 2, PIRKLE 1-J; see Table 14.3).

Eventually CSPs with both π-electron donor and acceptor moieties incorporated into a single selector turned out to be more powerful in terms of broader applicability. Along this line, the Whelk-O 1 phase was developed that has pre-organized clefts for solute insertion and allows for simultaneous face-to-face

Table 14.3

Commercially Available Donor-Acceptor (Pirkle-Type) CSPs

Chiral Selector	Column Trade Name	Supplier
π-electron acceptor/π-electron donor phases		
3-[1-(3,5-dinitro benzamido)-1,2,3,4-tetrahydrophenanthrene-2-yl]-propyl-silica	Whelk-O 1	Regis
11-[2-(3,5-dinitrobenzamido)-1,2-diphenylethylamino]-11-oxoundecyl-silica	ULMO	Regis
3-[*N*-(3,5-dinitrobenzoyl)-(*R*) – 1-naphthyl-glycine-amido]propyl-silica	Chirex 3005 (Sumichiral 2500)	Phenomenex (Sumitomo)
π-electron acceptor phases		
3-{3-{*N*-[2-(3,5-dinitrobenzamido-1-cyclohexyl)]-3,5-dinitrobenzamido}-2-hydroxy-propoxy}-propyl-silica	DACH-DNB	Regis
3-[3-(3,5-dinitrobenzamido)-2-oxo-4-phenyl-azetidine-1-yl]-propyl-silica	PIRKLE 1-J	Regis
5-(3,5-dinitrobenzamido)-4,4-dimethyl-5-dimethyl phosphonyl-pentanyl-silica	α-Burke 2	Regis
11-[*N*-(3,5-dinitrobenzoyl)-3-amino-3-phenyl-2-(1,1-dimethylethyl) propanoyl]-undecyl-silica	β-Gem 1	Regis
3-[*N*-(3,5-dinitrobenzoyl) leucine-amido]propyl-silica	Leucine (DNBLeu)	Regis
3-[*N*-(3,5-dinitrobenzoyl) phenylglycine-amido] propyl-silica	Phenylglycine (DNBPG)	Regis (Merck)
π-electron donor phases		
3-{*N*-[(*R*)-(α-naphthyl) ethylcarbamoyl]-(*S*)-indoline-2-carboxamido} propyl-silica (urea linkage)	Chirex 3022 (Sumichiral OA 4900)	Phenomenex (Sumitomo)
3-{*N*-[(*R*)-1-(α-naphthyl) ethylcarbamoyl]-(*S*)-*tert*-leucine-amido} propyl-silica (urea linkage)	Chirex 3020 (Sumichiral OA 4700)	Phenomenex (Sumitomo)

and face-to-edge π–π-interactions to facilitate chiral recognition [144, 145] (Fig. 14.26*b*). Inspired by the work of Pirkle, several other research groups followed this concept of CSPs based on synthetic, low-molecular-weight selectors. Among others, Oi and coworkers developed amide-type and urea-type CSPs, now commercialized as Sumichiral OA columns from Sumitomo or as Chirex columns from Phenomenex. A number of structural variants have been made accessible; the one denoted as Chirex 3005 (amide-type π-electron donor-acceptor phase) (see Table 14.3) appears to have the broadest applicability, followed by the Chirex 3022 and to minor degree Chirex 3020 [urea-type π-electron donor-phases derived from (*S*)-indoline-2-carboxylic acid and (*R*)-1-[α-naphthyl]ethylamine as well as (*S*)-*tert*-leucine and (*R*)-1-[α-naphthyl]ethylamine, respectively] (Table 14.3).

Other powerful π-donor-acceptor-type CSPs utilized C_2-symmetric diamine scaffolds such as bis-*N*, *N*′-(3,5-dinitrobenzoyl)-1,2-diamino cyclohexane from Gasparrini's group [146, 147] (DNB-DACH®, Regis) and *N*-3,5-dinitrobenzoyl-*N*′-undecanyl-1,2-diphenyl-1,2-diamine from Uray et al. (ULMO®, Regis; Fig. 14.26*c*) [148]. The latter CSP allows, for instance, the chromatographic separation of underivatized arylcarbinols as depicted in Figure 14.27. The evolution of CSPs in the Pirkle laboratory, as well as some design considerations and strategies that have lead to the modern donor-acceptor phases, have been comprehensively reviewed by Welch [142]. More recently some of the newer developments in this field were summarized and discussed by Gasparrini [149].

Since donor-acceptor phases are almost always used in the NP mode, method development is straightforward. It usually starts with a mixture of hexane or heptane/2-propanol (1–10% polar solvent). For basic solutes, 0.1% of a basic modifier such as diethylamine is added to the mobile phase; for acidic solutes, 0.1% of an acidic additive such as trifluoroacetic acid. After an initial separation, the polar-solvent content is adjusted to achieve a reasonable retention factor ($1 < k < 10$). If no baseline separation results, 2-propanol can be substituted by other polar solvents such as ethanol, dichloromethane, dioxane, methyl tert-butyl ether,

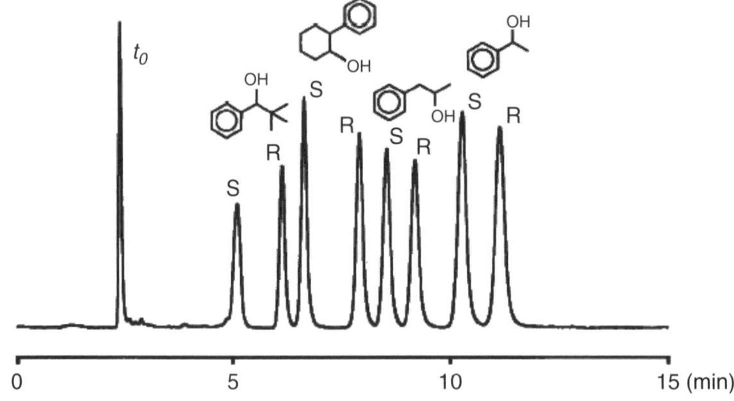

Figure 14.27 Separation of chiral alcohols on ULMO. Conditions: (*R*, *R*)-ULMO; 250 × 4-mm column; 99.5:0.5 *n*-heptane-isopropanol; 1 mL/min; 254 nm; 25°C. Reprinted with permission from [148].

or ethyl acetate. If RP conditions are required, enantioselectivity values usually drop significantly, since the retention- and selectivity-driving polar interactions are effectively nullified (or at least extremely weakened) by such strong, protic solvents. It should be noted that Regis offers a screening service for the Pirkle phases.

A number of characteristic benefits arise from the use of Pirkle-type CSPs. Since the building blocks of the selectors are available in both enantiomeric forms, CSPs can be developed in both configurations, allowing an opposite elution order for enantiomers (Section 14.3). The low molecular weight of these selectors, with their limited molecular dimensions, yields high surface concentrations of the CSP. As a result the sample loading capacities are much higher than for protein phases, macrocylic antibiotic CSPs, and cyclodextrin-based CSPs [61]. Synthetic donor-acceptor phases have also proved to be valuable tools for SFC enantiomer separation [150, 151].

14.6.8 Chiral Ion-Exchangers

Chiral ion-exchangers utilize ionizable selectors to exploit ionic interactions between oppositely charged selectors and analytes. Although a number of these CSPs are based on large molecules (e.g., protein CSPs, glycopeptide CSPs), we refer here to low-molecular-weight selectors that are similar to classical ion-exchangers yet have a chiral backbone. These CSPs can also be regarded as a subset of Pirkle phases, but carrying ionizable functional groups—thereby departing from the non-ionic interaction mode of the Pirkle-type CSPs. Several chiral ion-exchangers have been developed for the enantiomer separation of ionizable chiral compounds: chiral anion-exchangers based on cinchona alkaloid derivatives for chiral acids [152], chiral cation exchangers based on chiral amino sulfonic acids, and amino carboxylic acids for the separation of chiral bases [153], and zwitterionic ion-exchangers for the separation of both acids, bases, and zwitterionic solutes such as amino acids and peptides [154]. Only the chiral anion-exchangers with cinchonan carbamate selectors were commercially available at the time this book was published (under the tradename Chiralpak QN-AX and Chiralpak QD-AX; from Chiral Technologies) (Fig. 14.28*a*). The abbreviation AX refers to their anion-exchanger characteristics, while QN and QD denote the type of cinchona alkaloid employed as backbone of the selectors—quinine (QN) and quinidine (QD).

The selectors of these columns are highly enantioselective, as a result of five stereogenic centers. While configurations in position N_1, C_3, C_4 are fixed as 1*S*, 3*R*, 4*S*, those of carbon C_8 and C_9 are opposite in quinine (8*S*, 9*R*) and quinidine (8*R*, 9*S*) as well as separation materials derived therefrom. The experimental behavior of these cinchona-alkaloid derived CSPs is often under the stereocontrol of the stereogenic center of C_9; this gives them pseudoenantiomeric characteristics as a result of an opposite configuration of the two alkaloids at this chiral center. Aside from this peculiar configurational arrangement of the natural alkaloids, the exceptional enantiorecognition capability of the cinchonan carbamate-based chiral stationary phases arises also from several features: the bulky quinuclidine, the planar quinoline ring, and the semi-flexible carbamate group with a bulky *t*-butyl residue. These functionalities serve as potential binding sites, and they are structurally assembled to form a semi-rigid scaffold with predefined binding clefts for analyte insertion.

Much is known about the principal molecular recognition mechanisms of these semi-synthetic CSPs from various chromatographic [156–158], FTIR and

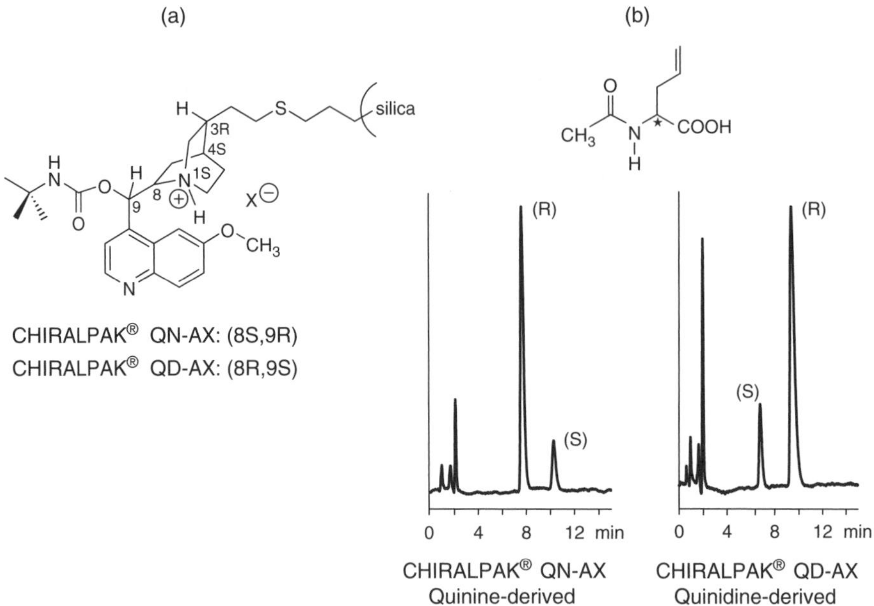

Figure 14.28 Commercially available cinchona alkaloid-derived chiral anion-exchangers. (*a*) Structure; (*b*) illustration of a reversal of elution order by change from the quinine-derived CSP to the corresponding pseudoquinidine-derived CSP. Experimental conditions: Column dimension, 150 × 4-mm column; mobile phase, 1% acetic acid in methanol; temperature, 25°C; flow rate, 1 mL/min; UV detection at 230 nm. Adapted from [155].

NMR spectroscopic [159–163], thermodynamic [163, 164], molecular modeling [161, 163], and X-ray diffraction studies [161–163,165]. If complementary H-donor-acceptor sites and aromatic moieties are incorporated into the guest molecule, favorable intermolecular H-bonding and π–π-interactions may result in stable complexes and exceptionally high enantioselectivities. Targeted optimization based on knowledge from the mechanistic studies mentioned above has led to a number of powerful CSPs [166], of which the commercially available ones provide broad applicability.

The cinchona alkaloid-based, anion-exchange columns offer excellent chiral resolving power for chiral carboxylic, sulfonic, phosphonic, and phosphoric acids [166], preferably by way of the PO or RP mode. Their applicability covers *N*-derivatized α-, β-, and γ-amino acids (Fig. 14.28*b*), their corresponding phosphonic, phosphinic, and sulfonic acid analogues, as well as many other pharmaceutically relevant chiral acids (e.g., arylcarboxylic acids, aryloxycarboxylic acids, hydroxy acids, pyrethroic acids, and a few underivatized amino acids).

If the cinchona-alkaloid based CSP is used with (weakly) acidic mobile phases, the quinuclidine nitrogen becomes protonated and acts as the fixed charge of the chiral anion-exchanger. Acidic analytes are then primarily retained by anion exchange, and retention can be explained by a stoichiometric displacement model [166]. Linear plots of log k versus the log of the counter-ion concentration [Z] (i.e., of the buffer anion) then result (Section 7.5.1 and Eq. 7.13). As discussed in Chapter 7, the slope of log k–log counter-ion concentration will, for a given column, vary with the charge on the analyte and the counter-ion, being steeper for a larger analyte charge and less steep for a larger counter-ion charge. A change in counter-ion concentration can be used to vary retention, often without much effect on enantioselectivity. The eluotropic strength (competitor effectiveness) increases in the order acetate $\leq$ formate $<$ phosphate $<$ citrate. A series of counter-ions (acid additives) in the PO mode have been tested confirming these trends for nonaqueous polar solvent-based mobile phases as well [167]. Other variables have a significant effect on enantioselectivity and allow for flexible method development: pH (RP mode), acid-base ratio (PO mode), and type and content of organic solvent(s) (RP and PO modes).

Preferred mobile phases are composed of methanol plus 0.5–2% glacial acetic acid, as well as 0.1–0.5% ammonium acetate (PO-mode), or methanol-ammonium acetate buffer (total buffer concentration in the mobile phase between 10 and 100 mM, pH 5–6) (RP mode). Methanol may be replaced by acetonitrile or methanol-acetonitrile mixtures, which are to some extent complementary (different enantioselectivities and elution orders) regarding their enantiorecognition capabilities.

As noted above, quinine and quinidine CSPs are actually diastereomers, but they behave like enantiomers. Therefore they usually (but not always) show opposite elution orders, as illustrated in Figure 14.28*b*. This complementarity in their chiral recognition profile can be systematically exploited in enantiomeric impurity-profiling applications and preparative enantiomer separations—since it is desirable to have the enantiomeric impurity elute first (Section 14.5). Cinchona carbamate-type CSPs also show great promise for preparative enantiomer separations, by virtue of their remarkable sample loadabilities. For example, adsorption isotherm measurements for FMOC-α-allylglycine on the *O*–9-*tert*-butylcarbamoylquinidine-CSP revealed a close to homogeneous adsorption mechanism with mass loading capacities of 20 mg/g CSP [168]. Although the primary application of these cinchonan carbamate CSPs are for chiral acids, recent studies showed that they can be used for neutral and basic compounds as well, via either RP [169] or NP mobile phases [170].

14.6.9 Chiral Ligand-Exchange CSPs (CLEC)

Chiral ligand-exchange CSPs allowed the first complete separation of a racemate by chromatography in the late 1960s. Davankov immobilized proline onto a polystyrene support and used this enantioselective matrix in combination with Cu(II)-ion containing mobile phases for the enantiomer separation of amino acids [171]. The basic principle of chiral ligand-exchange chromatography (CLEC) is the reversible coordination of immobilized selectors and analytes within the metal-ion coordination sphere that forms a mixed ternary metal-ion/selector/analyte complex (Fig. 14.29) [172]. Depending on the steric and functional properties of the analytes, these diastereomeric complexes result in enantioselectivity. During the chromatographic process the coordinated ligands are reversibly replaced by other ligands from the mobile phase such as ammonia and water. An important aspect of these separations is that the exchange of ligands at the metal center is fast; otherwise, column efficiency would be compromised.

An essential prerequisite for CLEC is the presence of metal-chelating functionalities in both the selector and analyte [172]. Suitable structures feature bidentate

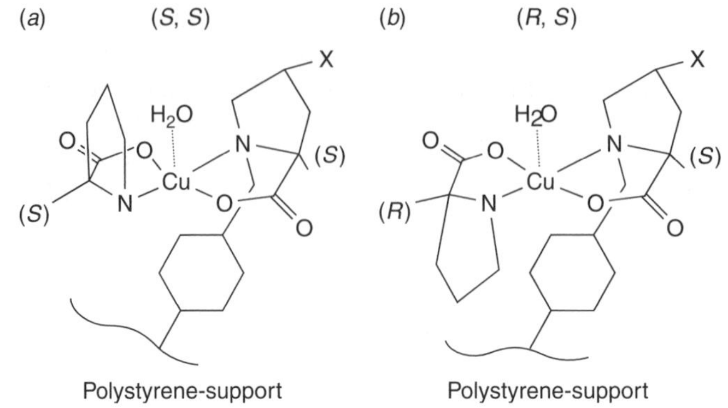

Figure 14.29 Principle of chiral ligand-exchange chromatography. Ternary diastereomeric Cu(II)-complexes of immobilized *S*-enantiomer of proline (X = H) (or hydroxyproline X = OH) ligand with *S*- and *R*-proline analytes, (*a*) and (*b*), respectively. Adapted from [173].

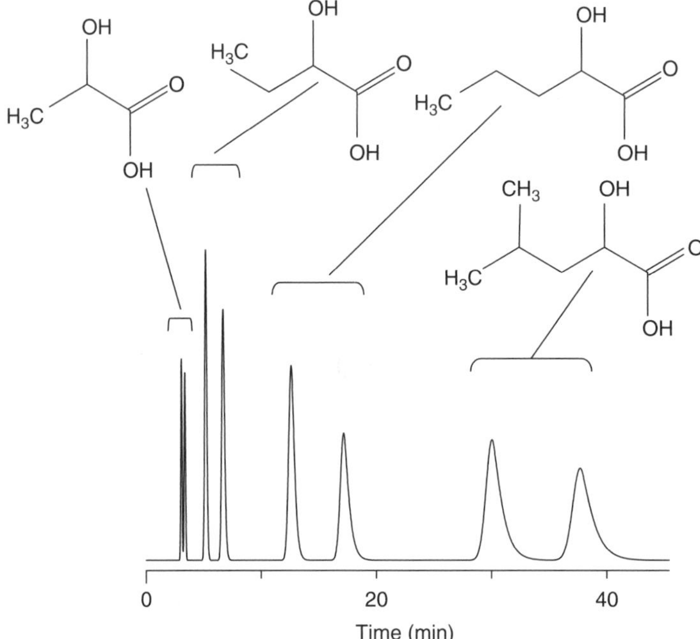

Figure 14.30 Enantiomer separation of hydroxy acids by chiral ligand-exchange chromatography (CLEC). Experimental conditions: column, CHIRALPAK MA; mobile phase, 10% ACN/H_2O plus 2-mM $CuSO_4$. Adapted from [8].

or tridentate ligands with two or three electron-donating functional groups, such as hydroxyl, amino, and carboxylic functionalities. Such structural prerequisites are typically found in α-amino acids, amino alcohols, and α-hydroxy acids (Fig. 14.30), representative compounds that have been separated by CLEC. Cu(II) is the preferred chelating metal ion, but Zn(II) and Ni(II) are suitable alternatives. As selectors for

CLEC-type CSPs, rigid cyclic amino acids, such as proline and hydroxyproline, have been shown to give the best results in combination with Cu(II). These chelating selectors are either (1) covalently anchored onto the surface of silica and organic polymer particles, respectively, or (2) dynamically coated onto reversed-phase materials (usually immobilized adsorptively by hydrophobic interactions based on the long alkyl chain substituents of the selectors; only a low %-organic in the mobile phase is tolerated). Because of the polar nature of the analytes for separation by CLEC, as discussed above, the mobile phase is aqueous or aqueous based. The mobile phase is usually doped with small quantities of metal ion, in order to compensate for loss of metal from the column packing during chromatography, thereby rendering the separation more stable.

The detection of nonchromophoric amino acids and hydroxy acids is possible as a result of their enhanced UV absorbance in the presence of Cu^{++}, while the presence of metal ions in the mobile phase may hamper mass spectrometric detection. Experimental conditions that can be varied in method development include mobile phase pH, type and concentration of buffer salts, nature and content of organic solvent, temperature, and the mobile-phase metal-ion concentration.

A number of covalently anchored and coated CSPs for CLEC are commercially available, including Chiralpak MA+ (based on *N*,*N*-dioctyl-L-alanine coated onto RP18) from Chiral Technologies, Nucleosil Chiral-1 (based on L-hydroxyproline chemically bonded to silica) from Macherey-Nagel, and Chirex 3126 (based on *N*,*S*-dioctyl-penicillamine coated onto RP18) from Phenomenex. In the past, CLEC was the only procedure that enabled the direct enantiomer separation of amino acids without derivatization. However, today the importance of chiral ligand-exchange chromatography is reduced, as a result of more favorable alternatives. More details can be found in a recent review [174].

14.7 THERMODYNAMIC CONSIDERATIONS

Analyte retention and enantioselectivity are, of course, based on the thermodynamics of the retention process—similar to the separation of achiral solutes, as discussed in preceding chapters. However, enantiomer separations are subject to some additional thermodynamic considerations.

14.7.1 Thermodynamics of Solute-Selector Association

The equilibrium constant K_i (see Fig. 14.9) of the solute-selector association can be related to thermodynamic parameters

$$\begin{aligned}\Delta G^0{}_i &= -RT \ln K_i \\ &= \Delta \mathrm{H}^0{}_i - \mathrm{T}\Delta \mathrm{S}_i{}^0 \end{aligned} \tag{14.3}$$

Here $\Delta G_i^0, \Delta H_i^0,$ *and* ΔS_i^0 refer to the standard free energy, enthalpy, and entropy changes upon the solute-selector complexation, *R* is the universal gas constant, *T* the absolute temperature (in *K*), and subscript *i* denotes the corresponding species (i.e., here R- or S-enantiomer); Ψ is the phase ratio (Section 2.3.1). Further manipulation of Equation (14.3) provides two additional relationships:

$$\ln K_i = -\frac{1}{T} \cdot \frac{\Delta H_i^\circ}{R} + \frac{\Delta S_i^\circ}{R} \tag{14.4}$$

and

$$\Delta\Delta G_{R,S}^\circ = \Delta G_R^\circ - \Delta G_S^\circ = -R \cdot T \cdot \ln \frac{K_{i,R}}{K_{i,S}} = -R \cdot T \cdot \ln \alpha \tag{14.5}$$

That is, plots of $\ln K_i$ against $1/T$ are predicted to be linear, with a slope that is proportional to ΔH_i^0. Likewise the separation factor α for two enantiomers R and S can be related to the difference in their standard free energies of solute-selector association $\Delta\Delta G^0{}_{R,S}$, as well as related differences in enthalpy change $\Delta\Delta H^0{}_{R,S}$ and entropy change $\Delta\Delta S^0{}_{R,S}$.

14.7.2 Thermodynamics of Direct Chromatographic Enantiomer Separation

If a single type of (enantioselective) solute-selector interaction is solely considered and other adsorption mechanisms do not exist for the solute, K_i in Equations (14.3) to (14.5) can be related to k and Ψ by

$$k = K_i \Psi \tag{14.6}$$

Values of $\Delta\Delta G^0{}_{R,S}$, $\Delta\Delta H^0{}_{R,S}$, and $\Delta\Delta S^0{}_{R,S}$ can be derived from values of α as a function of T, since the (usually unknown) phase ratio Ψ cancels in Equation (14.5) (but not in Eq. 14.4). Plots of $\ln k$ against $1/T$ are usually positive (k decreasing with T), implying a negative value of $\Delta H^0{}_i$ or an *enthalpically controlled* retention process. That is, attractive (mostly electrostatic type) noncovalent interactions between solute and selector result in values of $K_i \gg 1$. The latter contributions to retention are usually opposed by entropic effects, since the solute-selector complex is more ordered compared with the solute in the mobile phase. That is, $\Delta H^\circ > \Delta S^\circ$ and $\Delta\Delta H^\circ > \Delta\Delta S^\circ$, as observed for wide variety of different CSP-analyte mobile-phase systems. The usual result is a decrease in values of α for higher temperatures. The opposite behavior, an increase in enantioselectivity with T (called *entropically controlled* chiral recognition), has been observed in a few cases involving polysaccharide- and protein-type CSPs. The latter have been related to possible binding site-related (de)solvation phenomena [175] and/or conformational changes in backbones of the selector [176, 177]. Nonlinear plots of $\ln k$ against $1/T$ have also been observed occasionally [36]. Similar exceptions to a linear increase in $\ln k$ with $1/T$ have been observed for achiral separation as well (Section 2.3.2.2), possibly for similar reasons.

Unusual temperature-induced behaviors of another kind have been observed for the separation of chiral dihydropyrimidinones on polysaccharide CSPs [178]. Plots of $\ln k$ against $1/T$ were obtained by (1) heating the column from 10 to 50°C and (2) cooling from 50 to 10°C; the resulting plots for an ethanol-solvated Chiralpak AD-H column were not superimposable. That is, the system exhibited significant hysteresis, which was not the result of conformational changes of the polysaccharide column but rather a slow equilibration of the stationary phase when T is changed.

14.7.3 Site-Selective Thermodynamics

The discussion above overlooks the fact that enantioselective retention does not necessarily involve a single retention site [179]. While this observation is true also for achiral retention, there is an important difference for enantiomeric separation. That is, other sites are likely to be non-enantioselective; the latter (referred to as *type I* in distinction to enantioselective *type II* sites [179]) might consist of the supporting matrix (e.g., silica), linker groups, spacer units, residues stemming from silanol end-capping, and even non-enantioselective binding sites that involve the selector. The presence of type-I sites is well known to compromise enantioselectivity. While the binding affinity of type-I sites is usually much lower than for type-II sites, the concentration of type-I sites may exceed that of type-II sites by orders of magnitude, especially for the case of macromolecular selectors such as proteins (Section 14.6.3). Consequently the contribution of type-I sites to overall retention is usually not negligible, and experimental retention data represent *the sum of nonspecific (achiral) and specific (chiral) contributions to* k:

$$k_R = k_{I,R} + k_{II,R} \tag{14.7}$$

and

$$k_S = k_{I,S} + k_{II,S} \tag{14.7a}$$

Values of k in Equations (14.7) and (14.7a) are for the injection of a small sample (nonoverloaded separation), and subscripts I and II refer to type-I and type-II sites, respectively; subscripts R and S refer to values for the R- and S-enantiomers, respectively. The experimental enantioselective separation factor is given by $\alpha = k_R/k_S$ (for $k_R > k_S$), or

$$\alpha = \frac{k_{I,R} + k_{II,R}}{k_{I,S} + k_{II,S}} \tag{14.8}$$

Retention at site I is the same for both enantiomers (i.e., it is non-enantioselective), so $k_{I,R} = k_{I,S} = k_I$ *and*

$$\alpha = \frac{k_I + k_{II,R}}{k_I + k_{II,S}} \tag{14.8a}$$

If nonspecific retention is absent, $k_I = 0$ and $\alpha = k_{II,R}/k_{II,S}$. We assume that the R-enantiomer is more retained so that $k_{II,R}/k_{II,S} > 1$. For $k_I > 0$, the value of α in Equation (14.8a) decreases with increasing k_I and approaches 1 (no enantioselectivity) for $k_I \gg k_{II,R}$.

It is obvious that a maximization of enantiomer selectivity can be achieved either by maximizing the selectivity of the enantioselective type-II sites ($k_{II,R}/k_{II,S}$) or by minimizing the contribution to retention of the non-enantioselective type-I sites. When the goal is the interpretation of selector enantioselectivity (i.e., for type-II sites) as a function of the solute, selector, and experimental conditions, the intrinsic thermodynamic enantioselectivity ($k_{II,R}/k_{II,S}$) is the appropriate quantity, while the experimentally observed enantioselectivity (corresponding to α in Eq. 14.8a) can be misleading [179, 180].

From the preceding discussion it is clear that experimental values of α are only indirectly related to the various interactions that involve the solute and selector, as these values of α will reflect achiral as well as chiral interactions of solute with the stationary phase. The relative contributions of chiral and achiral sites to the observed enantioselectivity can be determined by fitting adsorption isotherm data for each enantiomers to a bi-Langmuir (two-site) model over a wide range in solute concentration. This procedure then provides values of $k_{I,R}, k_{II,R}, k_{I,S}$, and $k_{II,S}$ for small samples (linear-isotherm values). If isotherms are acquired at different temperatures, values of ΔH_i can be obtained for each enantiomer at each site (I and II) [181, 182]. Values of $\Delta\Delta G^0{}_{R,S}$, $\Delta\Delta H^0{}_{R,S}$, and $\Delta\Delta S^0{}_{R,S}$ can be derived and used to interpret the basis of enantioselectivity for a given system. By this methodology of adsorption isotherm measurements at variable temperatures, Guiochon and coworkers investigated, for example, the thermodynamics of 2,2,2-trifluoro-1-(9-anthryl)-ethanol (TFAE) [182] and 3-chloro-1-phenylpropanol (3CPP) [181] on *O*−9-*tert*-butylcarbamoylquinidine-modified silica under normal-phase conditions site-selectively.

REFERENCES

1. C. Roussel, J. Pierrot-Sanders, I. Heitmann, and P. Piras, in G. Subramanian, *Chiral Separation Techniques*, 2nd ed., Wiley-VCH, Weinheim, 2001, 95.
2. P. Piras and C. Roussel, *J. Pharm. Biomed. Anal.*, 46 (2008) 839.
3. T. Huybrechts, G. Török, T. Vennekens, R. Sneyers, S. Vrielynck, and I. Somers, *LCGC Europe*, 20 (2007) 320.
4. H. A. Wetli and E. Francotte, *J. Sep. Sci.*, 30 (2007) 1255.
5. V. A. Davankov, *Pure Appl. Chem.*, 69 (1997) 1469.
6. B. Testa, *Principles of Stereochemistry*, Wiley-VCH, Weinheim, 1998.
7. M. Lämmerhofer and W. Lindner, in *Separation Methods in Drug Synthesis and Purification*, K. Valkó, ed., Elsevier, Amsterdam, 2000, p. 337.
8. W. Lindner, in *Stereoselective Synthesis. Series: Methods of Organic Chemistry*, Vol. E21a, G. Helmchen, R. W. Hoffmann, J. Mulzer, E. Schaumann, eds., (Houben-Weyl), Thieme, Stuttgart, 1995, p. 225.
9. R. J. Bopp and J. H. Kennedy, *LCGC*, 6 (1988) 514.
10. W. Lindner, C. Leitner, and G. Uray, *J. Chromatogr.*, 316 (1984) 605.
11. F. Lai, A. Mayer and T. Sheehan, *J. Pharm. Biomed. Anal.*, 11 (1993) 117.
12. H. Brückner, R. Wittner, and H. Godel, *J. Chromatogr.*, 476 (1989) 3.
13. H. Brückner, T. Westhauser, and H. Godel, *J. Chromatogr. A*, 711 (1995) 201.
14. H. Brückner and C. Gah, *J. Chromatogr.*, 555 (1991) 81.
15. R. Bhushan and H. Brückner, *Amino Acids*, 27 (2004) 31.
16. N. Nimura, H. Ogura, and T. Kinoshita, *J. Chromatogr.*, 202 (1980) 75.
17. O. P. Kleidernigg, M. Lämmerhofer, and W. Lindner, *Enantiomer*, 1 (1996) 387.
18. F. Bressolle, M. Audran, T.-N. Pham, and J.-J. Vallon, *J. Chromatogr. B*, 687 (1996) 303.
19. A. Karlsson and C. Pettersson, *Chirality*, 4 (1992) 323.

20. C. Pettersson and E. Heldin, in *A Practical Approach to Chiral Separations by Liquid Chromatography*, G. Subramanian, ed., VCH, Weinheim, 1994, p. 279.
21. J. N. LePage, W. Lindner, G. Davies, D. E. Seitz, and B. L. Karger, *Anal. Chem.*, 51 (1979) 433.
22. R. Marchelli, R. Corradini, T. Bertuzzi, G. Galaverna, A. Dossena, F. Gasparrini, B. Galli, C. Villani, and D. Misiti, *Chirality*, 8 (1996) 452.
23. S. Zeng, J. Zhong, L. Pan, and Y. Li, *J. Chromatogr. B*, 728 (1999) 151.
24. V. A. Davankov, *Chromatograhia*, 27 (1989) 475.
25. P. Levkin, N. M. Maier, W. Lindner, and V. Schurig, submitted (2008).
26. C. E. Dalgliesh, *J. Chem. Soc.*, 132 (1952) 3940.
27. W. H. Pirkle and T. C. Pochapsky, *Chem. Rev.*, 89 (1989) 347.
28. V. A. Davankov, *Chirality*, 9 (1997) 99.
29. C. C. Pfeiffer, *Science*, 124 (1956) 29.
30. G. Hesse and R. Hagel, *Justus Liebigs Annalen der Chemie*, (1976) 996.
31. Y. Okamoto, M. Kawashima, and K. Hatada, *J. Am. Chem. Soc.*, 106 (1984) 5357.
32. E. Yashima, C. Yamamoto, and Y. Okamoto, *J. Am. Chem. Soc.*, 118 (1996) 4036.
33. C. Yamamoto, E. Yashima, and Y. Okamoto, *J. Am. Chem. Soc.*, 124 (2002) 12583.
34. Y. K. Ye, S. Bai, S. Vyas, and M. J. Wirth, *J. Phys. Chem. B.*, 111 (2007) 1189.
35. R. B. Kasat, N.-H. L. Wang, and E. I. Franses, *Biomacromolecules*, 8 (2007) 1676.
36. T. O'Brien, L. Crocker, R. Thompson, K. Thompson, P. H. Toma, D. A. Conlon, B. Feibush, C. Moeder, G. Bicker, and N. Grinberg, *Anal. Chem.*, 69 (1997) 1999.
37. E. Yashima, M. Yamada, Y. Kaida, and Y. Okamoto, *J. Chromatogr. A*, 694 (1995) 347.
38. R. W. Stringham and J. A. Blackwell, *Anal. Chem.*, 68 (1996) 2179.
39. T. D. Booth and I. W. Wainer, *J. Chromatogr. A*, 741 (1996) 205.
40. T. D. Booth and I. W. Wainer, *J. Chromatogr. A*, 737 (1996) 157.
41. T. D. Booth, W. J. Lough, M. Saeed, T. A. G. Noctor, and I. W. Wainer, *Chirality*, 9 (1997) 173.
42. T. D. Booth, K. Azzaoui, and I. W. Wainer, *Anal. Chem.*, 69 (1997) 3879.
43. Y. Okamoto, M. Kawashima, and K. Hatada, *J. Chromatogr.*, 363 (1986) 173.
44. E. Francotte and T. Zhang, *J. Chromatogr. A*, 718 (1995) 257.
45. C. Perrin, V. A. Vu, N. Matthijs, M. Maftouh, D. L. Massart, and Y. Vander Heyden, *J. Chromatogr. A*, 947 (2002) 69.
46. C. Perrin, N. Matthijs, D. Mangelings, C. Granier-Loyaux, M. Maftouh, D. L. Massart, and Y. Vander Heyden, *J. Chromatogr. A*, 966 (2002) 119.
47. N. Matthijs, C. Perrin, M. Maftouh, D. L. Massart, and Y. V. Heyden, *J. Chromatogr. A*, 1041 (2004) 119.
48. N. Matthijs, M. Maftouh, and Y. Vander Heyden, *J. Chromatogr. A*, 1111 (2006) 48.
49. P. Borman, B. Boughtelower, K. Cattanach, K. Crane, K. Freebairn, G. Jonas, I. Mutton, A. Patel, M. Sanders, and D. Thompson, *Chirality*, 15 (2003) S1
50. Y. Okamoto and E. Yashima, *Angew. Chem. Int. Ed.*, 37 (1998) 1021.
51. B. Chankvetadze, C. Yamamoto, and Y. Okamoto, *Chem. Lett.*, (2000) 1176.
52. E. Yashima, *J. Chromatogr. A*, 906 (2001) 105.
53. T. Wang, Y. W. Chen, and A. Vailaya, *J. Chromatogr. A*, 902 (2000) 345.
54. R. M. Wenslow, Jr. and T. Wang, *Anal. Chem.*, 73 (2001) 4190.
55. S. Caccamese, S. Bianca, and G. T. Carter, *Chirality*, 19 (2007) 647.

56. R. W. Stringham, K. G. Lynam, and B. S. Lord, Chirality, 16 (2004) 493.
57. K. Tachibana and A. Ohnishi, *J. Chromatogr. A*, 906 (2001) 127.
58. B. Chankvetadze, C. Yamamoto, and Y. Okamoto, *J. Chromatogr. A*, 922 (2001) 127.
59. K. G. Lynam and R. W. Stringham, *Chirality*, 18 (2005) 1.
60. T. Zhang, D. Nguyen, P. Franco, T. Murakami, A. Ohnishi, and H. Kurosawa, *Anal. Chim. Acta*, 557 (2006) 221.
61. E. R. Francotte, *J. Chromatogr. A*, 906 (2001) 379.
62. P. Franco, A. Senso, L. Oliveros, and C. Minguillón, *J. Chromatogr. A*, 906 (2001) 155.
63. T. Ikai, C. Yamamoto, M. Kamigaito, and Y. Okamoto, *J. Chromatogr. B*, 875 (2008) 2.
64. T. Zhang, C. Kientzy, P. Franco, A. Ohnishi, Y. Kagamihara, and H. Kurosawa, *J. Chromatogr. A*, 1075 (2005) 65.
65. T. Zhang, D. Nguyen, P. Franco, Y. Isobe, T. Michishita, and T. Murakami, *J. Pharm. Biomed. Anal.*, 46 (2008) 882.
66. P. Franco and T. Zhang, *J. Chromatogr. B*, 875 (2008) 48.
67. A. Ghanem and H. Aboul-Enein, *J. Liq. Chromatogr.*, 28 (2005) 2863.
68. A. Ghanem, H. Hoenen, and H. Y. Aboul-Enein, *Talanta*, 68 (2006) 602.
69. K. W. Phinney, *Anal. Bioanal. Chem.*, 382 (2005) 639.
70. J. L. Bernal, L. Toribio, M. J. D. Nozal, E. M. Nieto, and M. I. Montequi, *J. Biochem. Biophys. Meth.*, 54 (2002) 245.
71. M. Maftouh, C. Granier-Loyaux, E. Chavana, J. Marini, A. Pradines, Y. Vander Heyden, and C. Picard, *J. Chromatogr. A*, 1088 (2005) 67.
72. R. W. Stringham, *J. Chromatogr. A*, 1070 (2005) 163.
73. Y. Okamoto, S. Honda, I. Okamoto, and H. Yuki, *J. Am. Chem. Soc.*, 103 (1981) 6971.
74. G. Blaschke, *Chem. Ber.*, 107 (1974) 237.
75. G. Blaschke, W. Bröker, and W. Fraenkel, *Angew. Chem. Int. Ed.*, 25 (1986) 830.
76. J. N. Kinkel, in *A Practical Approach to Chiral Separations by Liquid Chromatography*, G. Subramanian, ed., VCH, Weinheim, 1994, p. 217.
77. T. Nakano, *J. Chromatogr. A*, 906 (2001) 205.
78. R. Cirilli, R. Costi, R. D. Santo, M. Artico, A. Roux, B. Gallinella, L. Zanitti, and F. L. Torre, *J. Chromatogr. A*, 993 (2003) 17.
79. F. Gasparrini, D. Misiti, R. Rompietti, and C. Villani, *J. Chromatogr. A*, 1064 (2005) 25.
80. S. G. Allenmark, S. Andersson, P. Möller, and D. Sanchez, *Chirality*, 7 (1995) 248.
81. S. Andersson, S. Allenmark, P. Moeller, B. Persson, and D. Sanchez, *J. Chromatogr. A*, 741 (1996) 23.
82. S. G. Allenmark and S. Andersson, *J. Chromatogr. A*, 666 (1994) 167.
83. M. C. Millot, *J. Chromatogr. B*, 797 (2003) 131.
84. J. Haginaka, *J. Chromatogr. B*, 875 (2008) 12.
85. J. Hermansson, *J. Chromatogr.*, 269 (1983) 71.
86. T. Miwa, M. Ichikawa, M. Tsuno, T. Hattori, T. Miyakawa, M. Kayano, and Y. Miyake, *Chem. Pharm. Bull.*, 35 (1987) 682.
87. P. Erlandsson, I. Marle, L. Hansson, R. Isaksson, C. Petterson, and G. Petterson, *J. Am. Chem. Soc.*, 112 (1990) 4573.

88. E. Domenici, C. Brett, P. Salvadori, G. Felix, I. Cahagne, S. Motellier, and I. W. Wainer, *Chromatographia*, 29 (1990) 170.
89. W. J. M. Kremer and L. H. Janssen, *Pharmacological Reviews*, 40 (1988) 1.
90. J. Haginaka, C. Seyama, and T. Murashima, *J. Chromatogr. A*, 704 (1995) 279.
91. J. Haginaka, C. Seyama, and N. Kanasugi, *Anal. Chem.*, 67 (1995) 2539.
92. J. Haginaka and H. Takehira, *J. Chromatogr. A*, 777 (1997) 241.
93. I. Kato, J. Schrode, W. J. Kohr, and M. Laskowski, *Biochemsitry*, 26 (1987) 193.
94. J. Hermansson, *Trends Anal. Chem.*, 8 (1989) 251.
95. M. S. Waters, D. R. Sidler, A. J. Simon, C. R. Middaugh, R. Thompson, L. J. August, G. Bicker, H. J. Perpall, and N. Grinberg, *Chirality*, 11 (1999) 224.
96. T. A. G. Noctor and I. W. Wainer, *Pharmaceut. Res.*, 9 (1992) 480.
97. H. Matsunaga and J. Haginaka, *J. Chromatogr. A*, 1106 (2006) 124.
98. F. Zsila, and Y. Iwao, *Biochem. Biophys. Acta*, 1770 (2007) 797.
99. S. Andersson, S. Allenmark, O. Erlandsson, and S. Nilsson, *J. Chromatogr.*, 498 (1990) 81.
100. J. Haginaka and N. Kanasugi, *J. Chromatogr. A*, 694 (1995) 71.
101. K. Tomoko, S. Akimasa, and M. Katsumi, *Pharmaceutical Research*, 23 (2006) 1038.
102. I. Petitpas, A. B. Bhattacharya, S. Twine, M. East, and S. Curry, *J. Biol. Chem.*, 276 (2001) 22804.
103. J. Stahlberg, H. Henriksson, C. Divne, R. Isaksson, G. Pettersson, G. Johansson, and T. A. Jones, *J. Mol. Biol.*, 305 (2001) 79.
104. J. Chen, W. A. Korfmacher, and Y. Hsieh, *J. Chromatogr. B*, 820 (2005) 1.
105. D. W. Armstrong and W. DeMond, *J. Chrom. Sci.*, 22 (1984) 411.
106. C. R. Mitchell and D. W. Armstrong, in *Chiral Separations*, G. Gübitz and M. G. Schmid, eds., Vol. 243, Humana Press, Totowa, NJ, 2004, p. 61.
107. M. V. Rekharsky and Y. Inoue, *Chem. Rev.*, 98 (1998) 1875.
108. D. W. Armstrong, L. W. Chang, S. C. Chang, X. Wang, H. Ibrahim, G. R. Reid, and T. Beesley, *J. Liq. Chromatogr.*, 20 (1997) 3279.
109. S. C. Chang, G. L. Reid, III, S. Chen, C. D. Chang, and D. W. Armstrong, *Trends Anal. Chem.*, 12 (1993) 144.
110. D. W. Armstrong, S. Chen, C. Chang, and S. Chang, *J. Liq. Chromatogr.*, 15 (1992) 545.
111. D. W. Armstrong, Y. Tang, S. Chen, Y. Zhou, C. Bagwill, and J.-R. Chen, *Anal. Chem.*, 66 (1994) 1473.
112. D. W. Armstrong, Y. Liu, and K. H. Ekborgott, *Chirality*, 7 (1995) 474.
113. K. Ekborg-Ott, Y. Liu, and D. W. Armstrong, *Chirality*, 10 (1998) 434.
114. A. Berthod, X. Chen, J. P. Kullman, D. W. Armstrong, F. Gasparrini, I. D'Acquarica, C. Villani, and A. Carotti, *Anal. Chem.*, 72 (2000) 1767.
115. T. J. Ward and A. B. Farris, III., *J. Chromatogr. A*, 906 (2001) 73.
116. I. Ilisz, R. Berkecz, and A. Peter, *J. Sep. Sci.*, 29 (2006) 1305.
117. T. L. Xiao and D. W. Armstrong, in *Chiral Separations*, G. Gübitz and M. G. Schmid, eds., Vol. 243, Humana Press, Totowa, NJ, 2004, 113.
118. I. D'Acquarica, F. Gasparrini, D. Misiti, M. Pierini, C. Villani, *Adv. Chromatogr.*, Vol. 46, CRC Press, Boca Raton, 2008, 108.
119. *ASTEC. Chirobiotic Handbook*, ASTEC, Whippany, NJ, 1997.

120. T. E. Beesley, J. T. Lee, and A. X. Wang, *Chiral Separation Techniques*, 2nd ed., G. Subramanian(ed.), Wiley-VCH, Weinheim, 2001, p. 25.
121. M. E. Andersson, D. Aslan, A. Clarke, J. Roeraade, and G. Hagman, *J. Chromatogr. A*, 1005 (2003) 83.
122. C. Wang, C. Jiang, and D. W. Armstrong, *J. Sep. Sci.*, 31 (2008) 1980.
123. H. Jiang, Y. Li, M. Pelzer, M. J. Cannon, C. Randlett, H. Junga, X. Jiang, and Q. C. Ji, *J. Chromatogr. A*, 1192 (2008) 230.
124. K. H. Ekborg-Ott, J. P. Kullman, X. Wang, K. Gahm, L. He, and D. W. Armstrong, *Chirality*, 10 (1998) 627.
125. A. Berthod, Y. Liu, C. Bagwill, and D. W. Armstrong, *J. Chromatogr. A*, 731 (1996) 123.
126. Y. Liu, A. Berthod, C. R. Mitchell, T. L. Xiao, B. Zhang, and D. W. Armstrong, *J. Chromatogr. A*, 978 (2002) 185.
127. L. R. Sousa, G. D. Y. Sogah, D. H. Hoffman, and D. J. Cram, *J. Am. Chem. Soc.*, 100 (1978) 4569.
128. T. Shinbo, T. Yamaguchi, K. Nishimura, and M. Sugiura, *J. Chromatogr.*, 405 (1987) 145.
129. M. H. Hyun, J. S. Jin, and W. Lee, *J. Chromatogr. A*, 822 (1998) 155.
130. M. H. Hyun, S. C. Han, B. H. Lipshutz, Y.-J. Shin, and C. J. Welch, *J. Chromatogr. A*, 910 (2001) 359.
131. R. J. Steffeck, Y. Zelechonok, and K. H. Gahm, *J. Chromatogr. A*, 947 (2002) 301.
132. M. H. Hyun, *J. Sep. Sci.*, 26 (2003) 242.
133. Y. Machida, H. Nishi, and K. Nakamura, *Chirality*, 11 (1999) 173.
134. W. Pirkle and D. House, *J. Org. Chem.*, 44 (1979) 1957.
135. F. Mikes and G. Boshart, *J. Chromatogr.*, 149 (1978) 455.
136. W. H. Pirkle, M. H. Hyun, A. Tsipouras, B. C. Hamper, and B. Banks, *J. Pharm. Biomed. Anal.*, 2 (1984) 173.
137. W. H. Pirkle and T. C. Pochapsky, *J. Am. Chem. Soc.*, 109 (1987) 5975.
138. W. H. Pirkle and S. R. Selness, *J. Org. Chem.*, 60 (1995) 3252.
139. W. H. Pirkle, P. G. Murray, and S. R. Wilson, *J. Org. Chem.*, 61 (1996) 4775.
140. W. H. Pirkle, D. W. House, and J. M. Finn, *J. Chromatogr.*, 192 (1980) 143.
141. W. H. Pirkle and R. Dappen, *J. Chromatogr.*, 404 (1987) 107.
142. C. J. Welch, *J. Chromatogr. A*, 666 (1994) 3.
143. W. H. Pirkle, T. C. Pochapsky, G. S. Mahler, D. E. Corey, D. S. Reno, and D. M. Alessi, *J. Org. Chem.*, 51 (1986) 4991.
144. W. H. Pirkle and C. J. Welch, *Tetrahedron: Asymmetry*, 5 (1994) 777.
145. M. E. Koscho, P. L. Spence, and W. H. Pirkle, *Tetrahedron Asymmetry*, 16 (2005) 3147.
146. F. Gasparrini, D. Misiti, and C. Villani, *Chirality*, 4 (1992) 447.
147. F. Gasparrini, D. Misiti, C. Villani, and F. La Torre, *J. Chromatogr.*, 539 (1991) 25.
148. N. M. Maier and G. Uray, *J. Chromatogr. A*, 732 (1996) 215.
149. F. Gasparrini, D. Misiti, and C. Villani, *J. Chromatogr. A*, 906 (2001) 35.
150. P. Macaudiere, A. Tambute, M. Caude, R. Rosset, M. A. Alembik, and I. W. Wainer, *J. Chromatogr.*, 371 (1986) 177.

151. A. M. Blum, K. G. Lynam, and E. C. Nicolas, *Chirality*, 6 (1994) 302.
152. M. Lämmerhofer and W. Lindner, *J. Chromatogr. A*, 741 (1996) 33.
153. C. V. Hoffmann, M. Lämmerhofer, and W. Lindner, *J. Chromatogr. A*, 1161 (2007) 242.
154. C. V. Hoffmann, R. Pell, M. Lämmerhofer, and W. Lindner, *Anal. Chem.*, submitted (2008).
155. M. Lämmerhofer, N. M. Maier, and W. Lindner, *Nachrichten aus der Chemie*, 50 (2002) 1037.
156. A. Mandl, L. Nicoletti, M. Lämmerhofer, and W. Lindner, *J. Chromatogr. A*, 858 (1999) 1.
157. N. M. Maier, L. Nicoletti, M. Lämmerhofer, and W. Lindner, *Chirality*, 11 (1999) 522.
158. M. Lämmerhofer, P. Franco, and W. Lindner, *J. Sep. Sci.*, 29 (2006) 1486.
159. J. Lesnik, M. Lämmerhofer, and W. Lindner, *Anal. Chim. Acta*, 401 (1999) 3.
160. R. Wirz, T. Buergi, W. Lindner, and A. Baiker, *Anal. Chem.*, 76 (2004) 5319.
161. N. M. Maier, S. Schefzick, G. M. Lombardo, M. Feliz, K. Rissanen, W. Lindner, and K. B. Lipkowitz, *J. Am. Chem. Soc.*, 124 (2002) 8611.
162. K. Akasaka, K. Gyimesi-Forras, M. Lämmerhofer, T. Fujita, M. Watanabe, N. Harada, and W. Lindner, *Chirality*, 17 (2005) 544.
163. W. Bicker, I. Chiorescu, V. B. Arion, M. Lämmerhofer, W. Lindner, *Tetrahedron: Asymmetry*, 19 (2008) 97.
164. W. R. Oberleitner, N. M. Maier, and W. Lindner, *J. Chromatogr. A*, 960 (2002) 97.
165. C. Czerwenka, M. Lämmerhofer, N. M. Maier, K. Rissanen, and W. Lindner, *Anal. Chem.*, 74 (2002) 5658.
166. M. Lämmerhofer and W. Lindner, *Adv. Chromatogr.*, Vol. 46, CRC Press, Boca Raton, 2008, 1.
167. K. Gyimesi-Forras, K. Akasaka, M. Lämmerhofer, N. M. Maier, T. Fujita, M. Watanabe, N. Harada, and W. Lindner, *Chirality*, 17 (2005) S134
168. R. Arnell, P. Forssén, T. Fornstedt, R. Sardella, M. Lämmerhofer, and W. Lindner, *J. Chromatogr. A*, 1216 (2009) 3480.
169. K. Gyimesi-Forras, J. Kökösi, G. Szasz, A. Gergely, and W. Lindner, *J. Chromatogr. A*, 1047 (2004) 59.
170. Gyimesi-Forras, N. M. Maier, J. Kökösi, A. Gergely, and W. Lindner, *Chirality*, 21 (2009) 199.
171. V. A. Davankov, *Enantiomer*, 5 (2000) 209.
172. V. A. Davankov, in *Chiral Separations*, G. Gübitz and M. G. Schmid, eds., Vol. 243, Humana Press, Totowa, NJ, 2004, 207.
173. V. A. Davankov, *J. Chromatogr. A*, 666 (1994) 55.
174. G. Gübitz and M. G. Schmid, in *Chiral Separation Techniques*, 3rd ed., G. Subramanian, ed., Wiley-VCH, Weinheim, Germany, 2007, p. 155.
175. W. H. Pirkle and P. G. Murray, *J. High Resol. Chromatogr.*, 16 (1993) 285.
176. R. W. Stringham and J. A. Blackwell, *Anal. Chem.*, 68 (1996) 2179.
177. O. Gyllenhaal and M. Stefansson, *Chirality*, 17 (2005).
178. F. Wang, D. Yeung, J. Han, D. Semin, J. S. McElvain, and J. Cheetham, *J. Sep. Sci.*, 31 (2008) 604.

179. G. Götmar, T. Fornstedt, and G. Guiochon, *Chirality*, 12 (2000) 558.
180. G. Götmar, T. Fornstedt, and G. Guiochon, *Anal. Chem.*, 72 (2000) 3908.
181. L. Asnin, K. Kaczmarski, A. Felinger, F. Gritti, and G. Guiochon, *J. Chromatogr. A*, 1101 (2006) 158.
182. G. Götmar, L. Asnin, and G. Guiochon, *J. Chromatogr. A*, 1059 (2004) 43.

CHAPTER FIFTEEN

PREPARATIVE SEPARATIONS

with Geoff Cox

Introduction to Modern Liquid Chromatography, Third Edition, by Lloyd R. Snyder, Joseph J. Kirkland, and John W. Dolan

15.1 INTRODUCTION

The aim of preparative liquid chromatography (prep-LC) is the collection of one or more purified compounds from a mixture. The scale of prep-LC can range from micrograms (for compound identification) to tens of metric tons (for production of a pharmaceutical product, as in Section 13.9.2), but usually the goal is the recovery of milligrams to grams (the main emphasis in this chapter). Larger scale separations will receive only brief attention (Section 15.6). A closely allied separation technique, supercritical fluid chromatography [1], is beyond the scope of the present chapter.

15.1.1 Column Overload and Its Consequences

Prep-LC involves *mass-overload* conditions, that is, sample weights that are large enough to affect peak widths and retention times. When developing an analytical separation, care is usually taken to ensure that the weight (or volume) of the sample does not exceed certain limits (Section 2.6), and that retention times and resolution will not vary with the amount of sample injected. As the weight of injected sample is increased, however, the detector will eventually become overloaded, retention times will decrease, and peaks will broaden and become distorted (compare Fig. 15.1*a*, *b*).

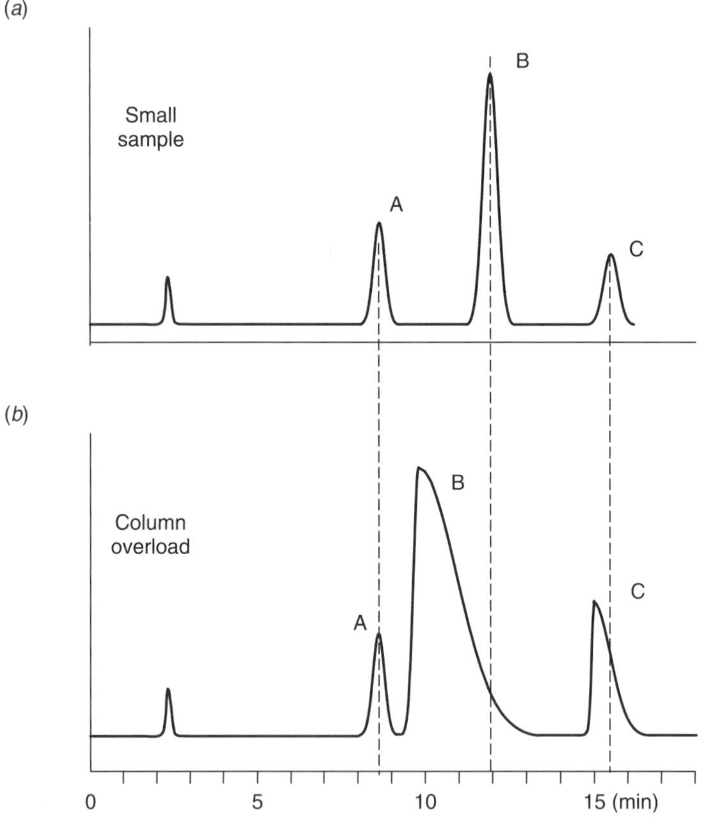

Figure 15.1 Hypothetical separations illustrating (*a*) an analytical separation and (*b*) a corresponding mass-overloaded touching-peak (T-P) separation.

Table 15.1

Requirements for Purified Samples

Objective	Product Weight Required
Tentative identification by instrumental methods	~ 1 mg
Positive identification and confirmation of structure	1–100 mg
Use as analytical standard (e.g., for calibrating an HPLC assay)	100 mg–2 g
Toxicology testing	10–100 g
Early phase 1 trials	200 g–2 kg

Detector overload or nonlinearity is often the first consequence of column overload, for example, when the UV absorbance of a peak exceeds 1 or 2 absorbance units (AU; Section 4.2.5). Detector nonlinearity can often be circumvented by using a short-path-length flow cell (a preparative cell), or by changing the detection wavelength so that the compound of interest (referred to hereafter as the *product*) absorbs less strongly. A large enough sample can also lead to changes in the separation, which we will refer to as *column overload* (as in Fig. 15.1*b*).

15.1.2 Separation Scale

The quantity of sample to be separated by prep-LC varies with the intended use of the purified product, as illustrated in Table 15.1 for small molecules (molecular weights <1000 Da) in pharmaceutical discovery and development. Note that column overload is determined by the weight of the largest peak of interest (usually the product peak), *not* the weight of the entire sample. For the purification of an impure product, however, where the initial purity is often >80%, there is little difference between sample and product weights. When we speak of sample weight in this chapter, we mean the weight of the product to be purified.

When the recovery of a few mg (or less) of a purified compound is required, the sample weight can be increased to the point where the space between the product peak and its nearest neighbor just disappears (i.e., corresponding to baseline resolution). The result is described as a *touching-peak* (or "touching-band") separation; touching-peak (T-P) separation corresponds to the largest quantity of sample that can be injected, while maintaining ≈100% recovery of the product with ≈100% purity. An illustrative T-P separation is shown in Figure 15.1*b*, where the sample weight has been increased (relative to the separation of Fig. 15.1*a*) until the major peak B expands to touch peak A. In this example the weight of peak A is not sufficient to affect either its retention time or peak width. Peak C is somewhat overloaded, but it has not broadened enough to touch peak B. It should be noted that the tail of both overloaded peaks elute near the analytical (small sample) retention times of Figure 15.1*a* (marked by the dashed lines in Fig. 15.1). A purified-product fraction (peak B in the example of Fig. 15.1*b*) can be collected as it leaves the column, using either manual or automated procedures (Section 15.2.4); solvent-free product can then be recovered by evaporation of the mobile phase (Section 15.2.5).

For the recovery of up to about 10 mg of purified product, more than a single sample injection may be required. When $\gg$10 mg of purified product are required, the large number of injection/collection cycles becomes inconvenient, especially if carried out manually. For larger sample weights, two different options exist: (1) use a larger diameter column (Section 15.1.2.1), or (2) optimize separation conditions for maximum resolution of the product peak (Section 15.1.2.2).

15.1.2.1 Larger Diameter Columns

For the use of a larger diameter ("semi-preparative") column, the column length is usually unchanged, and the identical column packing should be used. The flow rate and sample volume are increased in proportion to column cross-sectional area or d_c^2 (d_c is the column's inner diameter, i.d.); note that the equipment must be capable of this increase in flow rate. For example, when replacing a conventional analytical column (4.6-mm i.d.) with a 10-mm-i.d. column, both the flow rate and the sample volume should be increased by a factor of $10^2/4.6^2 = 4.73$. Under these conditions the same separation will be obtained for both small- and large-diameter columns (same retention times, peak widths, resolution, and column pressure P). The replacement of an analytical column by a larger diameter column in this way will be referred to as *scale-up*. Note that the column length can also be changed, in which case the sample size should be adjusted in proportion to column volume. While a longer column has a larger plate number N for analytical separations, this is less important in prep-LC, because N plays only a minor role in affecting separation (Section 15.3.1.2).

15.1.2.2 Optimized Conditions for Prep-LC

A second option is a change in selectivity that provides a better separation of the product. Most analytical separations are designed for the baseline separation of *all* peaks of interest, as in the optimized, small-sample separation of Figure 15.2*a*. In prep-LC, where usually a single product peak is to be recovered, only the resolution of the product peak from adjacent impurity peaks is important; the resolution of the product peak should therefore be as large as possible. This is illustrated in Figure 15.2*b*, where selectivity has been optimized for just the recovery of product peak 8—using the same general approach (Section 2.5.2) as for the development of the analytical separation of Figure 15.2*a* (i.e., a change in separation conditions that improves selectivity). Although some impurity peaks now overlap in Figure 15.2*b* (peaks 2–3, 5–6), T-P separation for peak 8 allows a much larger sample to be injected—as illustrated in Figure 15.2*c*. If the same sample weight as in Figure 15.2*c* is injected for the separation conditions of Figure 15.2*a* (see Fig. 15.2*d*), product peak 8 will no longer be well separated from impurity peak 7. Thus a much larger sample can be separated (with $\approx$100% recovery of pure peak 8), when the conditions of Figure 15.2*b* are used rather than those of Figure 15.2*a*.

15.1.2.3 Other Considerations

For the larger sample weights encountered in prep-LC, issues other than column dimensions or separation selectivity may also become important. One consequence of the chromatographic process is that components leaving the column are greatly

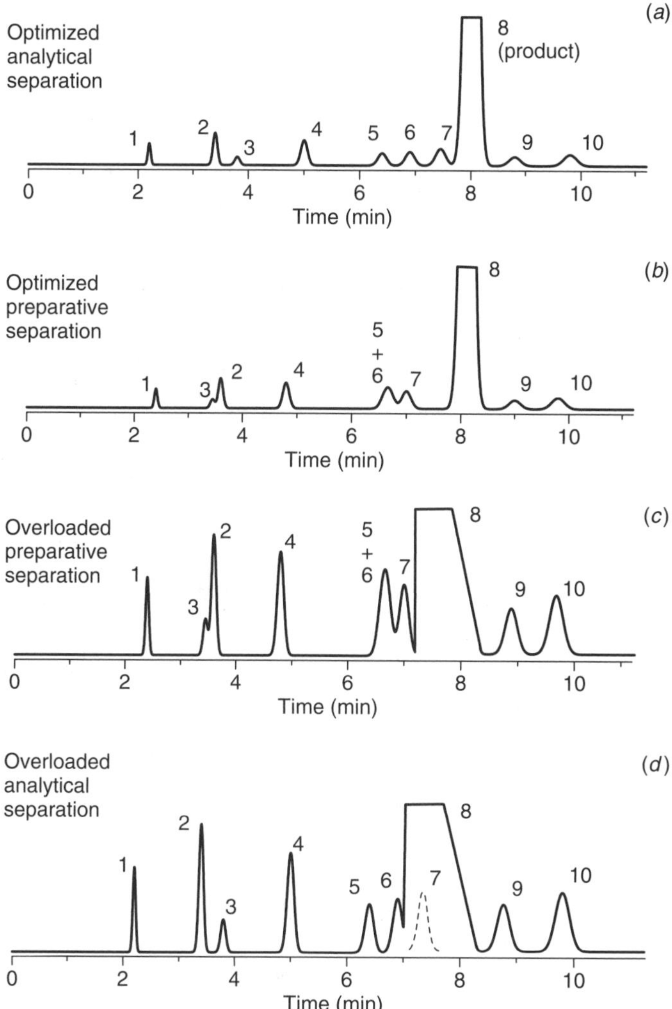

Figure 15.2 Analytical and preparative conditions compared for the optimum separation of a sample. Sample: a mixture of substituted anilines and benzoic acids. Conditions: 150 × 4.6-mm (5-μm) C_{18} column; acetonitrile-buffer mobile phases; flow rate 2.0 mL/min; other conditions noted in figure. (*a*) Conditions optimized for the separation and analysis of all compounds in the sample (small sample); (*b*) conditions optimized for the prep-LC purification of the product peak 8 (small sample); (*c*) T-P separation of sample with prep-LC conditions of (*b*) (large sample); (*d*) injection of large sample as in (*c*), but with analytical conditions of (*a*). Reprinted from [2] with permission of Wiley-Interscience.

diluted, and the ease of removing solvent from collected fractions is often a major issue. Solvent removal is generally easier for normal-phase chromatography (NPC) than for reversed-phase chromatography (RPC) because organic solvents are easier to evaporate than water. For the case of a few milligrams of product, dissolved in a few tens of milliliters of aqueous mobile phase from a RPC separation, solvent-free product can be recovered conveniently with a rotary evaporator. The removal of larger amounts of aqueous solvent, however, requires much more effort and cost; for this reason many (but not all) prep-LC separations tend to be carried out by NPC, rather than by RPC.

For sample weights >10 mg, analytical HPLC equipment is often too small, and its detectors too sensitive for prep-LC. For the separation of these larger samples, specialized equipment may be required that features high-flow pumps, automated sampling and fraction collection, and a detector fitted with a preparative flow cell.

15.2 EQUIPMENT FOR PREP-LC SEPARATION

As noted above, many small-scale prep-LC separations can be carried out with analytical chromatography systems (Chapter 3), possibly with minor modifications for increased injection volumes or decreased detector sensitivity. As sample weight increases beyond a few mg, however, it is more convenient to increase column size (scale-up) than to use (time-consuming and tedious) multiple injections with an analytical-scale column. This may require a corresponding change in equipment to a dedicated prep-LC system that allows higher flow rates and the processing of larger sample weights. Table 15.2 summarizes approximate guidelines for column size, equipment type, and flow rates for different scales of operation. For separations at the gram scale, a dedicated prep-LC system is usually necessary because of the required flow rates. Semi-preparative equipment is essentially similar to an analytical HPLC unit, but with a higher flow-rate pump and some arrangement for fraction collection. Small- and laboratory-scale prep-LC systems are typically used for the isolation of tens of grams to kilograms. In these systems columns with internal diameters as large as 11 cm are often used; larger diameter columns, which are appropriate for multi-kilogram scale projects, are better used within a kilo-lab or pilot plant to handle the large volumes of solvent required in an explosion-proof environment. These larger-diameter-column systems are outside the scope of the present chapter.

15.2.1 Columns

It is best to develop a prep-LC separation with an analytical column, using a column packing that is available in larger diameter columns. The use of small-diameter

Table 15.2

Approximate Sizes of Columns and Equipment Used for Prep-LC on a Laboratory Scale

Quantity Desired	Column Internal Diameter (mm)	Equipment	Scale-up Ratio for Flow Rate and Sample Size
<1 mg	4.6	Analytical (~1 mL/min)	(1)
1–100 mg	10	Analytical (~5 mL/min)	4.7
100 mg–5 g	20–30	Semi-preparative (20–50 mL/min)	19–42
5–100 g	30–50	Small-scale preparative (50–150 mL/min)	42–120
200g–2 kg	50–110	Lab-scale preparative (100–600 mL/min)	120–570

columns during method development minimizes any unnecessary consumption of sample and mobile phase, as well as the need for a more expensive prep-LC system. The columns used for semi-prep and lab-scale prep-LC are closely similar to those employed for analytical separations (Chapter 5). For sample weights of <10 mg, the analytical column itself can often be used since—depending on the separation—such columns may be compatible with injections of several milligrams of sample. A prep-LC column should be packed with the identical column packing that was used for the analytical column prior to scale-up. This ensures that there will be no change in relative retention (selectivity) between the two columns, which can be especially important for prep-LC (Section 15.3.2). Note that it is also important that the particle size be the same for both the analytical and prep-scale columns, so that the same (also optimized) column efficiency and resolution found for the smaller column will be duplicated for the larger column.

When moving to larger diameter columns (Table 5.2), with a corresponding increase in flow rate and sample volume, certain other considerations should be kept in mind. Be sure that the time lapse between the sample leaving the detector and entering the fraction collector is small, for both small- and large-diameter columns. Fractions are usually collected on the basis of the detector signal; if there is a significant volume of tubing between the detector cell and the fraction collector, there can be an appreciable time lag between detection of a peak and its collection; this can lead to mistakes in starting and ending fraction collection, with either a loss of product or its contamination by an adjacent impurity. Be aware of the tubing diameter that leads from the detector-cell outlet, which may be significantly larger than that used for the inlet. It is a simple matter to calculate the internal volume of the tubing, which with the flow rate determines the time lapse. With the higher flow rates used in prep-LC, the time lapse will be reduced proportionally—other factors equal.

The higher flow rates used with larger columns will increase the pressure drop across connecting tubing, if the same (analytical) equipment is used. Many HPLC systems are constructed with narrow capillary tubing (0.005–0.010-in. i.d.), in order to reduce extra-column peak broadening (Section 3.4). Narrow tubing will lead to a higher pressure drop, with a possible shut-down of the system, when the original analytical system is used with a flow rate of 5 to 10 mL/min with a 10-mm column (instead of the normal 1–2 mL/min used for 4.6-mm columns). This is especially true when more-viscous solvents are used, as in RPC. Special attention should be given to the tubing from the detector outlet, since small-i.d. tubing combined with higher flow rates can lead to a higher back-pressure and damage to the detector flow cell. It may be necessary to replace the original tubing with wider diameter tubing, (as large as 0.020-in. for 20-mm-i.d. columns). At the same time, keep in mind the effect of such a change on the time lapse between peak detection and collection (as discussed above). Instrument manufacturers can also advise the user on how to set up their equipment for prep-LC applications.

15.2.2 Sample Introduction

For small sample weights and the use of an analytical-scale HPLC system, sample introduction is usually carried out in one of two ways: (1) injection with the loop-injector that is part of the system, or (2) injection of the sample by means of a separate pump. In dedicated prep-LC systems, the sample is usually introduced with a sample pump that is different from the mobile-phase pump(s).

15.2.2.1 Loop Injectors

When using the standard injector that forms part of the analytical HPLC system, the maximum injection volume may be insufficient for prep-LC. The original sample loop can be replaced by a larger volume loop; these are available from a number of suppliers or can be made easily from bulk stainless-steel tubing. For systems fitted with autosamplers, large-sample-volume options are often available from the manufacturer. It is important to maintain the sample during injection as a cylindrical plug of approximately constant volume. Any dilution of the sample plug by mobile phase will increase the sample volume, which may compromise the separation (Section 15.3.2.2).

Care should be taken when choosing a larger loop, as peak spreading in an open tube is proportional to the sixth power of tube i.d. Thus, when the loop diameter is increased, the injected sample plug may exhibit increased tailing and broadening. The use of a longer (vs. wider) sample loop will also increase the width and tailing of the sample plug, but usually to a lesser degree. The trailing edge of the sample plug will be much more spread out than the front of the plug because the tail of the plug (but not the front) must traverse the length of the sample loop for properly designed sample injection (Section 3.6.1.2). The extent of the trailing edge can be determined by injecting a nonretained, UV-absorbing compound (i.e., with $k = 0$), then observing the resulting peak that leaves the column. If the sample plug entering the column does not tail significantly, the latter nonretained peak will be symmetrical.

One means of eliminating sample-tailing during injection is by partially emptying the sample loop. The filled sample loop is connected to the column for a time that is long enough to allow the required amount of sample to enter the column, but without introducing the end of the sample plug (that will be diluted with mobile phase). To achieve this result, the sample valve is switched from the inject position back to the load position before the sample loop is completely emptied. This ensures that the injected sample plug will not deteriorate the separation; however, the sample remaining in the loop may be lost.

15.2.2.2 Pump Injection

This technique is more convenient and applicable for large injection-volumes. It requires a 2-pump (i.e., gradient) system for isocratic prep-LC separations; the sample is introduced to the column by means of one pump, with subsequent elution of sample by the second (mobile-phase) pump. Best results are obtained with a high-pressure-mixing gradient system (Section 3.5.2.1), because of its minimal dead-volume after the gradient mixer. One of the two pumps is used to supply the sample, while the other pump delivers the (pre-mixed) mobile phase. The sample pump is first primed with the sample solution, after which injection is accomplished by simply switching the flow from the mobile-phase pump to the sample pump for a length of time (depending on flow rate) that will supply the required sample volume.

For pump injection with a low-pressure-mixing gradient system (Section 3.5.2.2), one of the solvent inlet lines to the mixer is used for delivering the sample. The sample is loaded by programming a step-gradient that switches from the mobile-phase line to the sample line and back again. Because of the larger dead-volume of low-pressure gradient systems (Section 3.5.2.2), it is advisable to measure the extent of peak broadening during injection, as described above (Section 15.2.2.1). Significant sample losses may occur with low-pressure gradient systems as a result of priming the pump and tubing; these sample losses can be both substantial and difficult to avoid.

Dedicated prep-LC units usually have a sample-injection pump that is separate from the mobile-phase pump(s). Where the sample volume is limited and the system volume is large, it is better to use manual injection—or aspirate the sample through a small tube directly into the sample pump. Sample injection is often operated in a stopped-flow mode; the mobile-phase pump is stopped, and the feed pump is actuated to pump the required sample volume directly to the column. This direct, on-column injection with a feed pump eliminates the tailing that may be seen in alternative systems where an injection valve with a large sample loop is used with injection of the entire contents of the loop.

15.2.3 Detectors

A general description of HPLC detectors is provided in Chapter 4. The present section will emphasize detectors and their characteristics that are most relevant for prep-LC.

15.2.3.1 UV Detectors

In most cases, the same UV detector can be used for both analytical and prep-LC applications. However, it is advisable to fit the detector with a short-path-length (≈1 mm) flow cell that allows operation at the optimum wavelength without detector overload. Some detector cells are available with variable path-lengths that can be selected for different separations. Despite the use of prep-LC flow cells, the sample absorbance can still exceed 1 to 2 AU, with peaks that are off scale and chromatograms that do not return to baseline—so that monitoring the separation becomes difficult or impossible. Excess detection sensitivity can be reduced by selecting a suitable (non-optimum, usually longer) wavelength, but note the possibility that impurities may absorb much more strongly than the product at the new wavelength, with resulting problems in recognizing the product peak for fraction collection. An example is shown in Figure 15.3, where the small-sample chromatogram (Fig. 15.3*a*, with detection at 280 nm) does not indicate any impurities with significant UV absorption at this wavelength (only two major peaks). When the sample load is increased (Fig. 15.3*b*), however, the detector signal is quickly overloaded at 280 nm. For detection at a longer detector wavelength (375 nm), in an attempt to bring the product peak on scale, the impurity peaks are relatively enhanced—to the point that the major components are no longer clearly identifiable. For such a sample another detector may be necessary.

15.2.3.2 Other Detectors

The refractive-index (RI) detector is not often sufficiently sensitive for analytical use, but it can be quite useful for prep-LC applications—precisely because of this

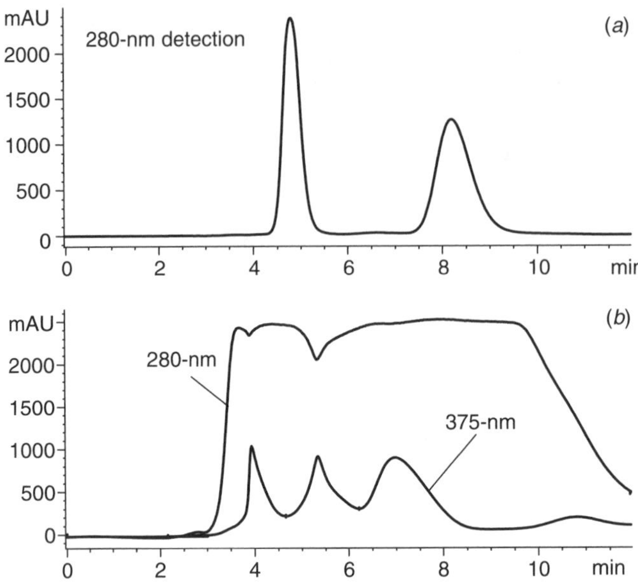

Figure 15.3 Difficulty in monitoring a prep-LC separation when the UV wavelength is changed to decrease detection sensitivity. (*a*) Analytical chromatogram, 280 nm; (*b*) preparative chromatogram, 280 and 375 nm.

insensitivity. Because the refractive index of the mobile phase can exceed that of some sample components, negative peaks are possible—a problem that need not prove serious if the fractions are collected manually, or as long as any automation software used for fraction collection can function correctly when negative peaks are present. For related components (which comprise the majority of samples) the RI detector provides similar detection sensitivity and a better representation of relative concentration than the UV detector—thus largely avoiding the problem of Figure 15.3*b*.

In principle, any detector can be used for prep-LC (Chapter 4). The size of the detector flow cell may preclude its use with the higher flow rates that are common in prep-LC, in which case a stream splitter can be used to bypass the flow cell. A low-dead-volume tee is inserted into the outlet line from the column and connected by short, small-diameter tubing to the detector. The flow through the cell is then controlled by the length and diameter of the tubing from the other branch of the tee to the fraction collector. Care should be taken in balancing the flow rates to ensure that the detector output is synchronized with the peaks entering the fraction collector. Ideally the volumes of the two tubes downstream of the splitter should be in the same ratio as the volumetric flow rates through them, in order to ensure that the peak arrives at the fraction collector and detector at the same time. Stream splitters should also be used for detectors that are destructive of the sample (evaporative light scattering or mass spectrometry [MS]). The use of MS detection for small-scale prep-LC is increasing, mainly for complex mixtures where compounds are collected based on their molecular weight.

15.2.4 Fraction Collection

Fraction collection can be carried out manually, by using the detector signal to determine when to begin and end fraction collection. For repetitive separations, however, operator fatigue rapidly ensues—with collection of the wrong fractions, or diversion of product to waste. When more than an occasional, small-scale prep-LC separation is contemplated, the use of a fraction collector is recommended. For some prep-LC systems the fraction collector will form an integral part of the system. It is also possible to purchase an add-on fraction collector, the most useful of which can be programmed to collect according to the detector signal. By means of a combination of fraction time window, peak threshold, and baseline slope, the correct assignment of fractions to the product container can be accomplished. These fraction collectors can function automatically and run continuously, or until the desired amount of sample has been processed.

Dedicated semi-preparative and preparative units have built-in fraction collectors. Small-scale prep-LC fraction collectors may use 96-well plates, while larger units generally have a fixed set of fraction-collection valves mounted in a manifold, to allow fractions to be selected and collected. Systems which use fixed-volume fraction-collection devices such as a 96-well plate or a multi-tube collector will generally collect on a time basis, to be sure that the fraction volumes are not exceeded. It is then the responsibility of the operator to combine those fractions that contain the purified product. Where collection occurs via a manifold of collection valves, the fractions can be collected in any suitably sized container, the required volume of which can be calculated from the peak width, the flow rate, and the number of injections to be made.

It is important to choose a fraction collector that meets the likely requirements of the prep-LC facility. When only a single product is recovered—as is often the case—only a few fraction-collection ports are needed. When more complex samples are separated into multiple fractions, more fraction-collection ports will be required. In the latter case a multi-tube fraction collector (similar to an autosampler) will usually be preferable—although care must be taken not to exceed the volume of the collection vessels. Fortunately, very complex samples are more likely to be encountered for small-scale separations in a research laboratory, while larger scale separations generally involve collection of only one or two components so that no more than 5 or 10 fraction ports are necessary. Because it is often found that small impurities may elute at the front or on the tail of the major peaks, it is common practice, at least during the early stages of a prep-LC separation, to collect narrow fractions at the front and tail of the peak to ensure that the desired purity is reached. Thus, as many as three fraction ports may be required for each compound to be purified. It is always better to collect too many fractions across a peak, and then combine the pure fractions, than to collect a single fraction that is too broad, and thus less pure.

15.2.5 Product Recovery (Removal of the Mobile Phase)

Product recovery can influence the entire plan for prep-LC method development, for example, the initial choice between RPC or NPC. When the separation results in several grams of product dissolved in several liters of an aqueous RPC mobile phase (possibly containing a nonvolatile buffer), separation of the product from the mobile phase is no longer a minor issue. The major difficulty is the elimination of water, which has a higher boiling point and higher heat of vaporization than typical organic solvents—resulting in long evaporation times. This not only limits the quantity of material that can be isolated in a given time but also exposes the product to a higher temperature during prolonged evaporation; this can lead to degradation of the product. For these reasons many chromatographers routinely select nonaqueous NPC (Section 8.4) for prep-LC. For NPC carried out with solvents that boil below 80°C (see Table I.3 of Appendix I), the recovery of purified product is relatively simple. NPC is also often favored by a potentially larger value of the separation factor α, which corresponds to larger allowable sample weights (as in the example of Figure 15.2; see also Section 15.3.2); this is especially true for closely related solutes such as isomers (Section 8.3.5).

RPC is not ruled out for prep-LC separation, as there are several options for the recovery of solvent-free product—depending on the nature of the product. Additionally any required buffers or additives can be selected from a list of volatile compounds (Section 7.2.1) to avoid the necessity of salt removal from the product. Many compounds are relatively insoluble in water, especially at a pH where they are not ionized: for example, a strongly acidic pH for an acid or an alkaline pH for a base (see Tables 7.2, 13.1, and 13.8 for solute pK_a values as a function of compound molecular structure). The appropriate adjustment of the pH of a product fraction, followed by partial evaporation of the fraction in order to eliminate most of the organic solvent, will often precipitate most of the product and allow its recovery by filtration. An alternative approach is the continuous extraction of the (partly aqueous) fraction with a water-immiscible organic solvent. The product is thus partitioned into an organic solvent which is then easily removed by evaporation. Either of these approaches also leaves any (nonvolatile) inorganic buffer behind in the aqueous phase.

In other cases dilution of the fraction with water, followed by passage through a solid-phase extraction (SPE) cartridge (Section 16.6), may allow the desired product to be retained by the cartridge. The cartridge can then be washed with water to remove nonvolatile buffer or salt, following which the product can be eluted from the cartridge with a water-miscible organic solvent that is more easily evaporated. Any small amount of water remaining in the product-fraction after SPE can be removed by azeotropic distillation in a rotary evaporator—following the addition of a suitable solvent that can form a volatile azeotrope with water. Chloroform is especially useful for this purpose, as any remaining water will be visually apparent as immiscible droplets. Because only 3% of a chloroform/water distillate is water, this process may need to be repeated once or twice to complete the removal of water. Other solvents (e.g., ethanol, dichloromethane) can also be used. Reference to tables of physical properties of solvent mixtures [3] can be helpful in finding a suitable azeotroping solvent.

The removal of water by lyophilization may be preferred for less stable, very water-soluble products. Prior elimination of nonvolatile buffers or additives may be required initially, as by ion-exchange chromatography (Section 16.6.2.3). Because the product is held at a temperature $<0°C$, lyophilization is commonly used for protein products, in order to prevent their denaturation or decomposition during drying.

15.3 ISOCRATIC ELUTION

Since 1980 there have been dramatic advances in our understanding and use of prep-LC. The group of Guiochon, in particular, has developed an extensive mathematical treatment of preparative chromatography, especially for large-scale, non-touching peak separations [4]. For separations on the laboratory scale (Table 15.2), the general principles of prep-LC are more useful than involved mathematical treatments that require computer calculations for their implementation. The present chapter will emphasize these general principles.

15.3.1 Sample-Weight and Separation

The present section provides a fundamental understanding of how column overload affects separation, but it has limited immediate application. For this reason the reader may prefer to skip to Section 15.3.2, *and return to the present section as appropriate.*

15.3.1.1 Sorption Isotherms

The sorption isotherm describes the distribution of the solute between the stationary and mobile phases at a given temperature, as a function of solute concentration in the mobile phase. Most HPLC separations obey the *Langmuir isotherm*, which describes the distribution of solute molecules between the mobile and stationary phases:

$$C_s = \frac{aC_m}{1 + b^* C_m} \tag{15.1}$$

Here C_s is the concentration of solute in the stationary phase, C_m is the solute concentration in the mobile phase, and a and b^* are constants for a given solute, mobile phase, stationary phase, and temperature. This model of sample uptake by the column assumes that solute retention takes place onto a planar surface with a defined maximum capacity—the filled *adsorbed monolayer* (Sections 8.2.1, 15.3.2.1; Fig. 15.6*a*).

For small solute concentrations C_m, $C_s = aC_m$, and from Equation (15.1),

$$k = \frac{C_s V_s}{C_m V_m} = \frac{C_s / C_m}{V_s / V_m} = K\Psi \tag{15.2}$$

where V_s and V_m are, respectively, the volumes of stationary and mobile phase within the column, $K = (C_s/C_m)$ is the solute distribution coefficient (also known as the Henry constant), and $\Psi = V_s/V_m$ is the phase ratio. From Equations (15.1) and (15.2) we see that K equals a for small values of C_m. For large values of $C_m, C_s = a/b^*$, corresponding to a filled solute monolayer. The quantity $(a/b^*)V_s$ equals the maximum uptake of the column by solute, which is defined as the *column saturation capacity* w_s (Section 15.3.2.1).

The fractional filling (θ) of the monolayer by solute is equal to C_s divided by the maximum value of C_s (equal a/b^*), or

$$\theta = \frac{b^* C_m}{1 + b^* C_m} \tag{15.3}$$

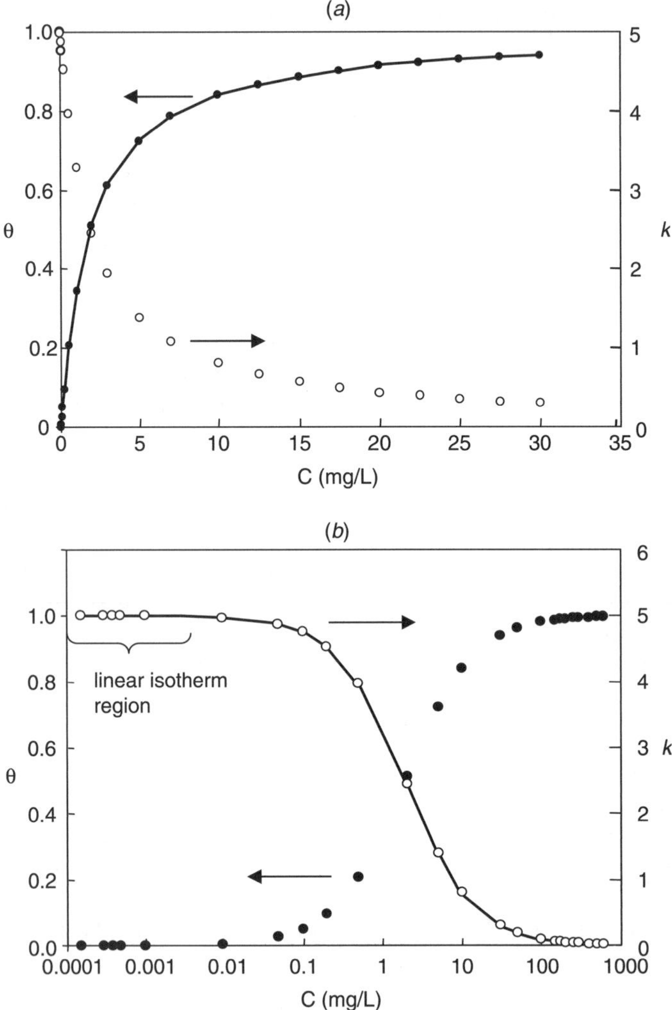

Figure 15.4 Illustration of a sorption isotherm for $K = 25$ and $\Psi = 0.2$. (*a*) Plot showing the influence of mobile-phase solute concentration on the retention factor k and surface coverage θ; (*b*) logarithmic plot illustrating the range of the linear isotherm.

Figure 15.4*a* shows a plot of θ vs. C_m for a Langmuir isotherm with $a/b^* = 25$ and $K = 13$, as well as corresponding values of k; values of k decrease as C_s increases. Figure 15.4*b* is a re-plot of Figure 15.4*a* with a logarithmic scale for C_m. The latter plot shows that k becomes constant (defined here as k_0) when C_m is sufficiently small (so-called linear-isotherm region; for $C_m > 0.0004$ in this example, $k_0 > k > 0.99k_0$).

15.3.1.2 Peak Width for Small versus Large Samples

The width W of a solute peak can be expressed as a function of W for a small sample weight (W_0) and an additional ("thermodynamic") peak broadening (W_{th})

due to an increase in sample weight [5, 6]:

$$W^2 = W_0^2 + W_{th}^2 \tag{15.4}$$

where

$$W_0^2 = \left(\frac{16}{N_0}\right) t_0^2(1 + k_0)^2 \tag{15.4a}$$

and

$$W_{th}^2 = 4t_0^2 k_0^2 \left(\frac{w_x}{w_s}\right) \tag{15.4b}$$

N_0 is the plate number for a small sample weight, t_0 is the column dead-time, k_0 is the retention factor for a small sample weight, w_x is the weight of solute injected, and w_s is the column saturation capacity (Section 15.3.2.1).

According to Equation (15.4) the effect of the column plate number N_0 on peak width becomes less important as sample size increases (and W_{th} becomes larger than W_0). This has important implications for prep-LC; for example, as the separation factor α is increased, and larger sample weights can be injected for T-P separation, the required plate number becomes smaller and peak width is less affected by those conditions that affect N (column length, particle size, flow rate). Inasmuch as larger values of N_0 require longer run times, this suggests that higher flow rates and or shorter columns (resulting in a decrease in N_0) will often be advantageous in prep-LC, in order to increase the amount of product that can be purified per hour, with little adverse effect on either product recovery or purity.

15.3.2 Touching-Peak Separation

Touching peak (T-P) separation was defined in Section 15.1.1. A weight of injected sample is selected such that the broadened product peak just touches one of the two surrounding peaks (as in Fig. 15.1*b*); this then allows a maximum weight of sample for ≈100% recovery of the separated product in ≈100% purity. T-P separation can be achieved by trial and error, guided by the following discussion.

If varying amounts of peak B in Figure 15.1 are injected, and the chromatograms superimposed, a series of so-called nesting right-triangles will result—as in the illustration of Figure 15.5. Here, the three overloaded product peaks (1, 2, 4) correspond to relative sample weights of 1, 2, and 4. The tail of each overloaded peak is located near the retention time for a small weight of injected sample, and for varying "small" weights of the sample there is no change in peak width (linear isotherm region; see Fig. 15.4*b*). The widths W of overloaded peaks increase approximately in proportion to the square root of the sample weight (Eq. 15.4*b*). At the bottom of Figure 15.5, a value of W is shown for the largest weight of injected sample (peak 4), measured from the start of the peak at 6 minutes to the retention time for a small sample (10 min). One way of determining the required weight of injected sample for T-P separation is as follows: After an initial injection of a small sample, an arbitrary increase in sample weight can be made for the next sample, for example, resulting in curve 1 of Figure 15.5. In this case, W must be increased

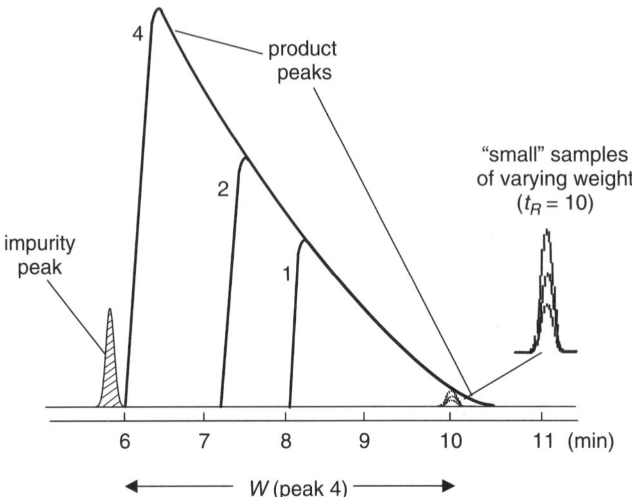

Figure 15.5 Effect of sample weight on peak shape for overloaded separation (superimposed chromatograms).

2-fold, to move the front of the product peak next to the adjacent impurity peak (shaded). Since peak width increases as the square root of the sample weight, the sample weight should therefore be increased 4-fold ($4^{0.5} = 2$).

The weight of injected product for T-P separation depends on (1) the value of the product separation factor α_0 for a small weight of injected sample, (2) the nature of the sample, and (3) the saturation capacity of the column (Section 15.3.2.1). If molecules of the product are not ionized, and a 10-nm-pore, 150 × 4.6-mm column is assumed, the weight of injected product for T-P separation will be about 3 mg for $\alpha_0 = 1.5$. For columns differing in length or i.d., the latter product weight for T-P separation will be proportional to the column internal volume. If α_0 equal 1.1 or 3.0, the corresponding sample weights will be about 0.2 and 10 mg, respectively. The extent of sample ionization and column characteristics together determine the column saturation capacity w_s (see Section 15.3.2.1). The weight of injected product w_x for T-P separation can then be approximated by

$$w_x = \left(\frac{1}{6}\right)\left(\frac{\alpha - 1}{\alpha}\right)^2 w_s \tag{15.5}$$

15.3.2.1 Column Saturation Capacity

An understanding of the column saturation capacity w_s (which we will refer to simply as "column capacity") is basic to any further discussion of prep-LC. The value of w_s corresponds to the maximum possible uptake of a solute molecule by the column, that is, the weight of solute that will fill an adsorbed monolayer completely (this corresponds to a large concentration of solute in the mobile phase); see the hypothetical example of Figure 15.6*a* for the adsorption of benzene onto a representative portion of the stationary-phase surface. Column capacity is specific to a particular column and can vary somewhat with both the solute and experimental

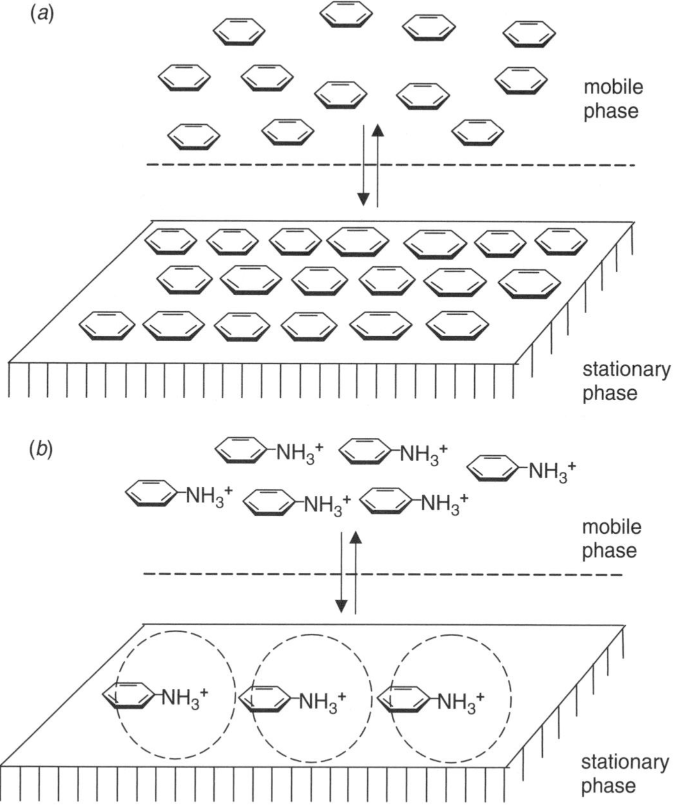

Figure 15.6 Illustration of column capacity as a function of solute ionization. Adsorption of benzene (*a*) and protonated aniline (*b*).

conditions. The column capacity will be proportional to the accessible surface area of the packing material, and for a non-ionized solute w_s can be approximated by

$$w_s(\text{mg}) = 0.4 \text{ (surface area in m}^2\text{)} \tag{15.6}$$

While this might suggest the use of a column packing with the largest possible surface area, higher surface packings have smaller diameter pores. If solute molecules are large relative to the pore diameter, they cannot enter the pores and take full advantage of the surface area (which is almost entirely contained within the pores) so that the effective column capacity will be reduced. For large-pore packings, conversely, the surface area and column capacity are both smaller. Thus intermediate pore-size packings provide the largest column capacities (e.g., pore diameters of $\approx$8 nm for solute molecular weights $<$500 Da); see the related discussion of Figure 13.7 for the retention of larger molecules as a function of column pore diameter.

Some characteristics of the solute molecule (other than size) are also important. Small solutes that adsorb perpendicular to the packing surface can result in a higher column capacity than solutes which lie flat on the surface (compare Fig. 8.3*b* with 8.3*a*). Large molecules, such as proteins, may assume a three-dimensional conformation when retained, resulting in a significantly greater thickness of the adsorbed monolayer; in this case the saturation capacity can be much greater than predicted by Equation (15.6)—as long as the pore diameter is large enough to admit the protein molecule. Charged (ionized) solute molecules of the same kind will repel each other when adsorbed, and this electrostatic repulsion can reduce column capacity by as much as two orders of magnitude. This is illustrated in Figure 15.6, where the uptake of a neutral solute (benzene) in Figure 15.6*a* is contrasted with that of an ionized solute (protonated aniline) in Figure 15.6*b*. Because of the much smaller column capacity for ionized solute molecules, prep-LC is preferably carried out under conditions that minimize sample ionization. However, because of the much greater retention of a non-ionized molecule vs. its ionized counterpart (e.g., a nonprotonated base vs. the protonated base), partially ionized solutes should have much larger column capacities than fully ionized species—if not as great as for completely neutral molecules.

15.3.2.2 Sample-Volume Overload

The volume of the injected sample for a T-P separation depends on the required product weight and the concentration of the product in the original sample solution (which may be limited by sample solubility). It has been estimated [5] that the sample volume will have little effect on prep-LC separation until the sample volume V_s exceeds half the volume of the peak being collected; V_p equals the peak width W times the flow rate F. Figure 15.7 shows simulated chromatograms for a 0.1-mL injection (solid line), a 1-mL injection (dotted line) and a 1.5-mL injection (dash-dotted line), while holding the sample-weight constant by varying its concentration (values of 36 g/L, 3.6 g/L and 2.4 g/L, respectively). For the 1.5-mL sample volume, significant peak overlap results, with only 85% recovery of pure peak A. It is clearly preferable to use the highest possible sample concentration. For the lowest sample concentration in Figure (15.7) (2.4 g/L), a T-P separation can only be achieved by reducing the injection volume to less than 1 mL, with a consequent reduction in the weight of pure A that can be recovered from each separation.

In order to avoid peak distortion and a deterioration of separation, as well as other problems, it is usually best to use the same solvent composition for both the sample and the mobile phase. However, sample solvents whose compositions differ from that of the mobile phase may be required in order to improve sample solubility (Section 15.3.2.3). Provided that the strength of the sample solvent and mobile phase are similar, there should be little adverse effect on peak width or shape from the use of a sample solvent and mobile phase that are not the same.

15.3.2.3 Sample Solubility

T-P separation requires a certain weight of the injected sample, preferably injected in a volume of mobile phase that is less than $^1/_2$ of the peak volume WF (as discussed above). Sample solvents that are weaker than the mobile phase are acceptable, and larger volumes of such sample solutions can be injected. However, sample solubility often decreases when %B is reduced, so the injection of larger volumes of sample dissolved in a weaker solvent may not provide a greater weight of injected sample. Means of dealing with the problem of limiting sample solubility include:

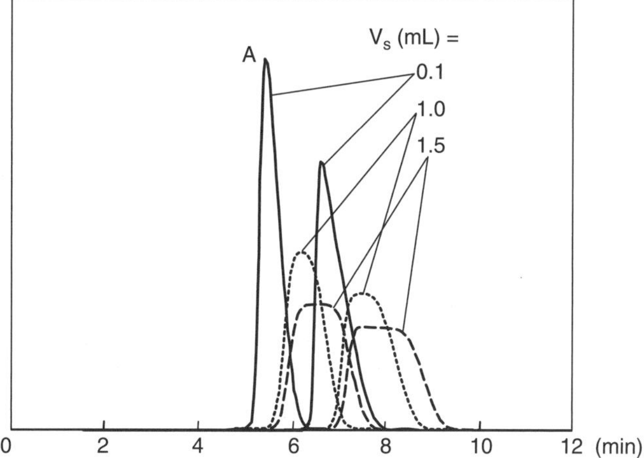

Figure 15.7 Chromatograms illustrating the effects of volume overload on the separation of two adjacent peaks. Sample volumes:____, 0.1 mL; ......, 1.0 mL; _ _ _, 1.5 mL. Simulation for a column 250 × 4.6 mm operated at 1 mL/min, two components at 1:1 composition, k(A) = 1.0, selectivity $\alpha = 1.5$, $N = 1000$. The two peaks are moderately mass-overloaded.

- a change in the mobile phase
- a change in the sample solvent
- an increase in column temperature

Change in Mobile Phase. The choice of prep-LC mode (RPC or NPC) should consider whether the sample is likely to be more soluble in aqueous solvents (RPC) or in organic solvents (NPC). The choice of B-solvent in each case can further influence sample solubility. On the other hand, the choice of mode and B-solvent also affects selectivity α, so a compromise between mobile-phase solubility and selectivity may be necessary when selecting the B-solvent.

Change in Sample Solvent. The sample solvent need not be the same as the mobile phase but, generally, should not be much stronger. Just as the use of stronger sample solvents adversely affects analytical separations (Sections 2.6.1, 17.4.5.3), peak broadening and distortion can also occur in prep-LC separations when the sample solvent is stronger than the mobile phase. *Usually a sample solvent that is similar in strength to the mobile phase will be the best choice.* A change in B-solvent for the sample solvent may improve solubility, especially for NPC where sample solubility can be varied independently of solvent strength. Thus, the same mobile-phase strength ε can be achieved with varying values of %B for different B-solvents (see example of Fig. 8.6), depending on the polarity (or ε^0 value) of the B-solvent. Since sample solubility is generally higher for a mobile phase with a higher value of %B, this suggests the use of a sample solvent with as large a value of %B as possible—while maintaining the same mobile-phase strength ε. For example, if the mobile phase consists of 4% ethyl acetate in hexane, Figure 8.6 suggests the use of 50% CH_2Cl_2 in hexane as a sample solvent that is likely to provide greater sample solubility without adverse effects on the separation.

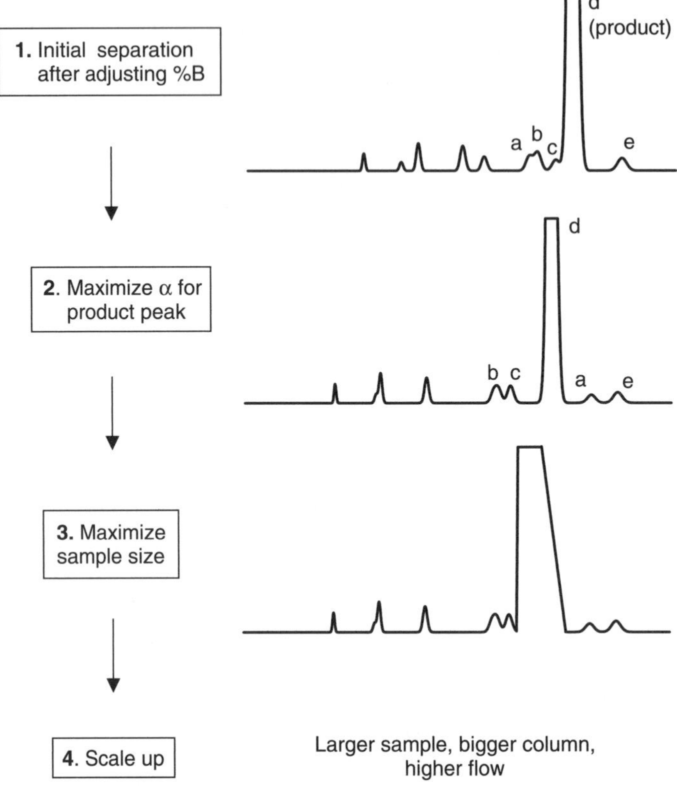

Figure 15.8 Prep-LC method development.

A change in pH of the sample solvent is another option for increasing sample solubility, when acidic or basic compounds are separated by RPC. In the latter case a mobile-phase pH that suppresses sample ionization is usually preferred for increased column capacity (as illustrated in Fig. 15.6), but sample solubility is often greater for ionized compounds. This suggests the use of a sample-solvent pH that provides increased sample ionization and solubility. Since an increase in solute ionization decreases its retention (undesirable for a sample solvent), the sample solvent can be lightly buffered, so that upon mixing with the mobile phase the resulting pH will be similar to that of the mobile phase. However, this procedure assumes a relatively small sample volume and a more heavily buffered mobile phase.

Whatever change in the sample solvent is considered, *the possibility of sample precipitation (with blockage of the injector, solvent tubing, or column) should be kept in mind when the sample mixes with the mobile phase.* In some cases precipitation may not occur instantaneously, allowing the sample to be taken up by the column before precipitation can occur. Experiments where aliquots of sample solution are mixed with the mobile phase and observed for a few minutes can indicate the likelihood of precipitation problems during the separation (see the related discussion of Section 7.2.1.2 for assessing buffer solubility). Alternatively, the likelihood of sample precipitation when using a sample solvent different than the mobile phase can be reduced by the technique of "at-column dilution" [7, 8]. The latter procedure introduces the sample via a separate pump to the head of the column so that sample and mobile phase are mixed together just prior to entering the column—with the likelihood that sample uptake by the stationary phase will occur before sample precipitation.

Increase in Column Temperature. A temperature increase will generally increase sample solubility, but the sample must also be heated during its passage from the sample container to the column inlet (a sample pump can be a significant heat sink!). A failure to heat any part of the system can lead to sample precipitation and blockage of the injector, solvent tubing, or column. An increase in column temperature will also reduce sample retention and may lead to smaller values of α—with an offsetting decrease in the weight of sample that can be injected.

15.3.2.4 Method Development

The development of a prep-LC separation is summarized in Figure 15.8; for steps 1 to 3 an analytical-scale column should be used (e.g., 150 × 4.6-mm). If a wider column is required for an increase in the weight of purified sample from each run (separation), this is selected in step 4.

Selection of Initial Conditions. Prior to Step-1 (initial separation), it is necessary to choose either RPC or NPC, as well as select the initial separation conditions (column type, mobile phase, temperature). Whereas RPC is generally favored for analytical separations, NPC is often preferred for prep-LC—especially when the sample is not water soluble and the required weight of purified product is more than 50 mg. We will assume NPC in the following discussion, but for the selection of initial RPC conditions, see Chapters 6 and 7.

If more than 10 mg of purified product will be required, be sure that the same column packing is available in larger-i.d. columns for scale-up (step 4). Next select a strong (B) and weak (A) solvent for the mobile phase (see Chapter 8 for NPC; e.g., ethyl acetate and hexane), and vary %B so that $1 \leq k \leq 10$ for the product and later-eluting peaks (if possible). Since peaks eluting before the product need not be separated from each other, their values of k can be <1. Similarly, if some impurities are strongly retained, they can be removed more quickly by washing the column with a stronger mobile phase after elution of the product peak; the column must then be equilibrated with the original mobile phase before injecting the next sample (gradient elution is an alternative for such samples).

Step 1 of Figure 15.8. An initial separation is carried out next, using a similar approach as for analytical method development, for example, using a strong mobile phase (e.g., 80% B) in order to achieve the elution of the entire sample within a reasonable time. The value of %B is next adjusted by trial and error to achieve $1 < k < 10$ for the product peak, with lower values of k favored for faster separation and the purification of a greater weight of product per hour (see Sections 2.5.1, 8.4). Alternatively, thin-layer chromatography (Section 8.2.3) can be used for this purpose, or an initial separation can be carried out using gradient elution—which in turn allows an estimate of a preferred %B value for isocratic separation

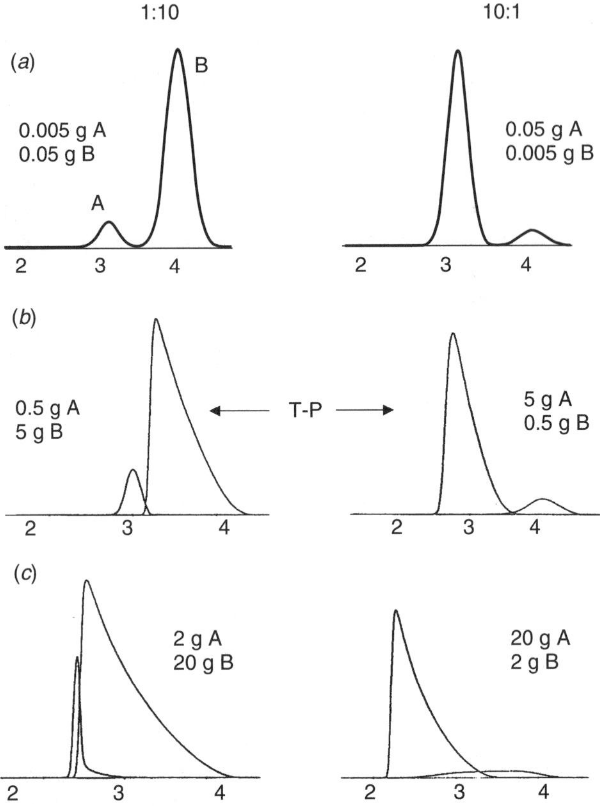

Figure 15.9 Isocratic separation of a two-component sample as a function of sample size. Computer simulations based on the Langmuir isotherm; Conditions: 250 × 50-mm column (7-μm), 210 mL/min flow rate, $N = 800$; $k = 1$ and 1.5, respectively. Sample weights indicated in figure. Adapted from [9].

(Section 9.3.1). If there is any doubt as to the identity of the product peak in these initial separations, this can be confirmed by a separate injection of pure product.

Step 2 of Figure 15.8. Following the adjustment of %B in step 1, separation conditions are varied for the best possible separation of the product peak from adjacent impurity peaks. Usually the product peak should be placed midway between the adjacent impurities on each side. Previous chapters provide a detailed discussion of how selectivity α can be optimized, depending on the kind of sample and whether NPC or RPC is used (see Table 2.2 for conditions that affect α). Because of the importance of maximizing α in prep-LC (Eq. 15.5), more work on step 2 may be warranted than for analogous analytical separations. Unlike the case of analytical separation, in prep-LC it is important—if possible—to avoid separation conditions that result in >50% ionization of the product molecule (see Section 15.3.2.1 and the discussion of Fig. 15.6). Large changes in α (without ionizing the product) are most likely to be achieved by a change in B-solvent or the column.

The same conditions used for this optimized separation can be used to assay fractions collected during prep-LC (but with the initial small-scale column). If the resolution of the product peak is $R_s \gg 2$ (desirable for prep-LC), the assay separations can be carried out with a shorter column and increased flow rate to speed up fraction analysis.

Step 3 of Figure 15.8. An initial estimate of the weight of injected sample is possible, based on (1) Equation (15.5), (2) a value of α, (3) the column capacity w_s, and (4) a rough estimate of sample purity (%-product in the sample). For a 150 × 4.6-mm column with 10-nm pores, and a product that does not ionize in the mobile phase, the column capacity can be estimated as $w_s \approx 150$ mg ($\approx$100 mg/g of column packing), from which the weight of sample for T-P separation is $w_s \approx (1/6) \times (150) \times ([\%\text{-product}]/100) \times ([\alpha - 1]/\alpha)^2$; if the product is partly or completely ionized, the allowed sample weight can be much lower than the latter estimate. Following a separation with this estimated sample weight, sample weight can be increased or decreased by trial and error to achieve T-P separation. Alternatively, the use of fully automated equipment allows a number of trial separations where sample size is varied; from such experiments the correct sample weight can be quickly determined. Once a promising separation is identified in this way, the product peak should be collected and assayed, in order to confirm $\approx$100% recovery and purity. It may also be worthwhile at this point to see if an increase in flow rate can maintain the latter separation, but with a reduced run time. The object of prep-LC is usually maximum production of purified product in minimum time, which favors short run times.

Step 4 of Figure 15.8. The final step in Figure 15.8 (scale-up) completes method development. The desired scale-up factor can be calculated from the results of step 3 (see Section 15.1.2.1), taking into account the availability of (1) columns of different i.d. and (2) equipment that can provide the required flow rate. A final separation with this larger column can then be carried out, allowing verification of the product recovery and purity obtained with the previous (smaller) column. Scale-up should result in essentially the same purity and recovery of product as found for the small-scale separation.

15.3.2.5 Fraction Collection

The usual goal of fraction collection—whether carried out manually or with an automated system—is to obtain a maximum yield of adequately pure product, with as little effort as possible. The initial step is to collect a number of fractions across the product peak, followed by their analysis for content and purity. These results can be used to determine the time during which the product peak should be collected (best "cut points") in the final separation(s), so as to achieve the purification goals (Section 15.4.1). Prior to finalizing the prep-LC procedure, a trial separation can be carried out to confirm the latter cut points. A few small fractions around each cut point can be collected for this purpose. Once the separation procedure and cut points are finalized, only a single product fraction need be collected. However, additional fractions can provide insurance against unanticipated changes in the separation.

15.4 SEVERELY OVERLOADED SEPARATION

A detailed study of severe column overload (i.e., sample sizes larger than those that correspond to T-P separation) is beyond the scope of the present book; however, it is useful to consider certain aspects of such separations. Such severely overloaded separations can result in a greater production of purified product per hour with reduced consumption of the mobile phase, as well as requiring smaller columns and smaller scale equipment—all of which can be of great practical importance. The disadvantage of such separations is that more effort is required for method development, and individual separations usually require the collection and analysis of several product fractions so that only adequately pure material is obtained. It may also be necessary to re-process product fractions that are insufficiently pure. The interested reader is referred to several texts [4, 9, 10] for further study.

15.4.1 Recovery versus Purity

As sample weight increases to the point of severe overload, the prediction of individual peak shapes becomes more uncertain. This is illustrated in Figure 15.9 for small-sample (Fig. 15.9*a*), T-P (Fig. 15.9*b*), and severely-overloaded (Fig. 15.9*c*) separations of a sample where the relative concentrations of the two components A and B vary from 1:10 to 10:1. Thus we can see what happens to a minor peak that elutes either before or after the (larger) product peak. For severe overload (Fig. 15.9*c*), when the impurity peak precedes the product peak, it is *displaced* and compressed so that peak height increases. There is also some overlap of the two peaks. When the impurity peak follows the product peak, it is dragged into the product peak (so-called tag-along effect). The relative importance of these two effects can be difficult to predict, so the optimum sample weight must be determined experimentally. This optimum weight will also vary with the relative concentrations of product and impurities.

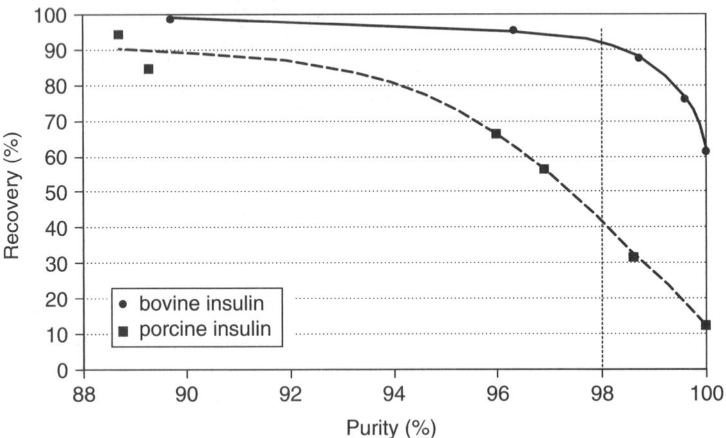

Figure 15.10 Plot of recovery against product purity for a given sample load (5 mg) for a 1:1 mixture of bovine and porcine insulins. Bovine insulin (solid line); porcine insulin (dashed line). Conditions: 250 × 4.6-mm (20-μm) C8 column; 10–29% acetonitrile–0.1% aqueous TFA in 10 minutes. Adapted from [13].

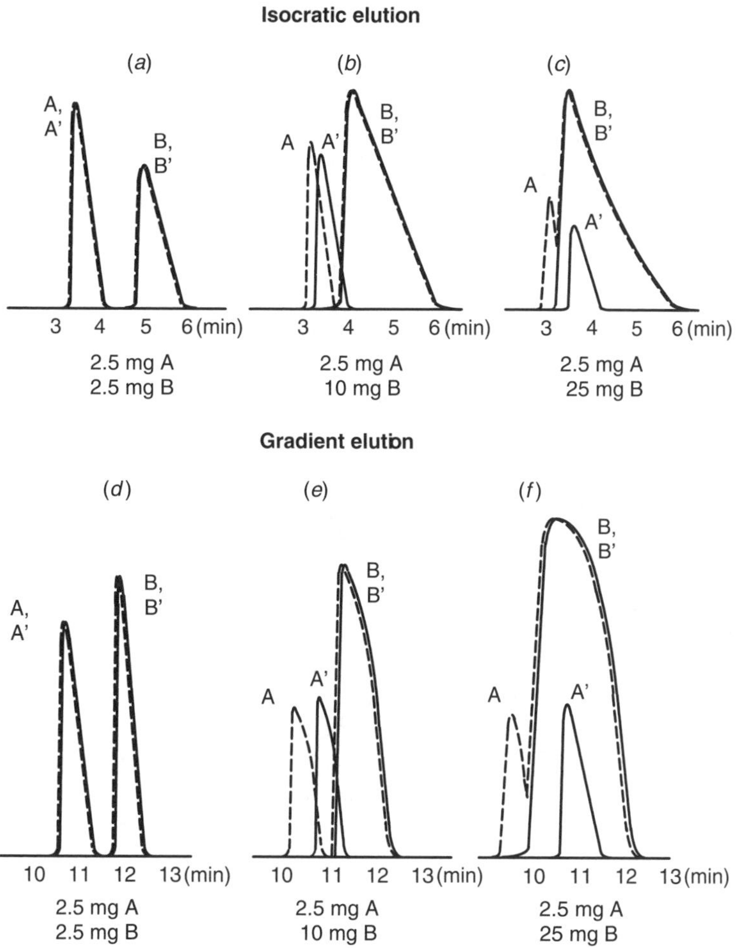

Figure 15.11 Similar effects of column overload in corresponding separations: (*a*) isocratic and (*b*) gradient elution. Separation of two xanthines (β-hydroxyethyltheophylline [A] and 7β-hydroxypropyltheophylline [B]) with k (isocratic) equal k^* (gradient). Sample weights shown in figure. Peaks labeled A′ and B′ are for the injection of samples of pure of A or B; peaks labeled A and B are for the separation of mixtures of A and B. Adapted from [2].

15.4.2 Method Development

The selection of an optimum sample size for severely overloaded prep-LC can start with an optimized T-P separation (as in Fig. 15.2*b*), followed by injecting successively larger sample weights. For each separation (or sample weight), a number of fractions that surround and include the product peak are collected and assayed, and the results are collected within a spreadsheet. Based on pooling the purest fractions, product recovery (or yield) can then be plotted against product purity, as in the examples of Figure 15.10 for two different products (bovine and porcine insulin). Similar plots will result for different sample weights, allowing selection of the most attractive sample weight. To a first approximation, the maximum weight of purified product with some predetermined purity (e.g., 98%) can be established in this way. For the examples of Figure 15.10 the recovery of 98%-pure material would be 92 and 42%, respectively, for bovine and porcine insulin. Similar plots for different sample sizes might result in a better compromise between the weight of purified product and its recovery.

The separation of Figure 15.10 was carried out with RPC using gradient elution. However, the same approach would be used for NPC or isocratic elution. The principle of estimating sample size is the same regardless of whether isocratic or gradient elution chromatography is being used. Optimum separation conditions—other than sample weight—may not be the same for T-P as compared with severely overloaded separation. In both cases a very considerable experimental effort can be required in order to simultaneously optimize both sample size and separation conditions.

15.4.2.1 Column Efficiency

As the (small-sample) separation factor α_0 increases for T-P separation, and a larger sample weight becomes possible, the effect of the small-sample column plate number N_0 on product resolution decreases (Eq. 15.4*a*). Smaller values of N_0 are therefore required, with little effect on the recovery or purity of the product. This is no longer the case for severely overloaded separations. Displacement effects as in Figure 15.9*c*, for a small peak that precedes a large peak, can improve separation. This is better shown in the isocratic separations of Figure 15.11*c*, where in the absence of sample displacement peak B would completely overlap peak A. Because of displacement, there is some separation of the two peaks (compare the similar situation of Fig. 15.9*c*). Sample displacement is highly advantageous in severely overloaded separation, but unlike the case of T-P separation, it appears to be favored by higher values of N_0 [11]. As a result large-scale separations are generally carried out with moderately efficient columns that use particle diameters of 10 to 15 μm.

15.4.2.2 "Crossing Isotherms"

This unusual behavior can arise for solutes with different saturation capacities. An example is seen in the separation of alcohols from phenols [13], where alcohols can have significantly higher saturation capacities than phenols. Figure 15.12*a* shows the RPC separation of benzyl alcohol and phenol for a small sample (10 μg), where benzyl alcohol elutes last. Figure 15.12*c* shows the same separation for a larger sample (1-mg phenol, 3-mg benzyl alcohol); the two peaks are almost completely separated. When the order of elution of a phenol and alcohol are reversed, while the weights of early- and late-eluting compound are held the same (as in Fig. 15.12*a*,*c*), a very different result is obtained; see Figure 15.12*b*,*d* for the separation of phenethyl alcohol and *p*-cresol. In the overloaded separation of Figure 15.12*d*, peak overlap is almost complete—contrasting strongly with the analogous separation of Figure 15.12*c*.

The reason for the contrasting separations of Figure 15.12 is somewhat complicated, but can be pictured in terms of "crossing isotherms"—as illustrated in Figures 15.12*e* (phenol and benzyl alcohol) and Figure 15.12*f* (phenethyl alcohol and *p*-cresol). For the separation of phenol and benzyl alcohol, the isotherms do not cross (Fig. 15.12*e*) because phenol is always more retained than benzyl alcohol, and

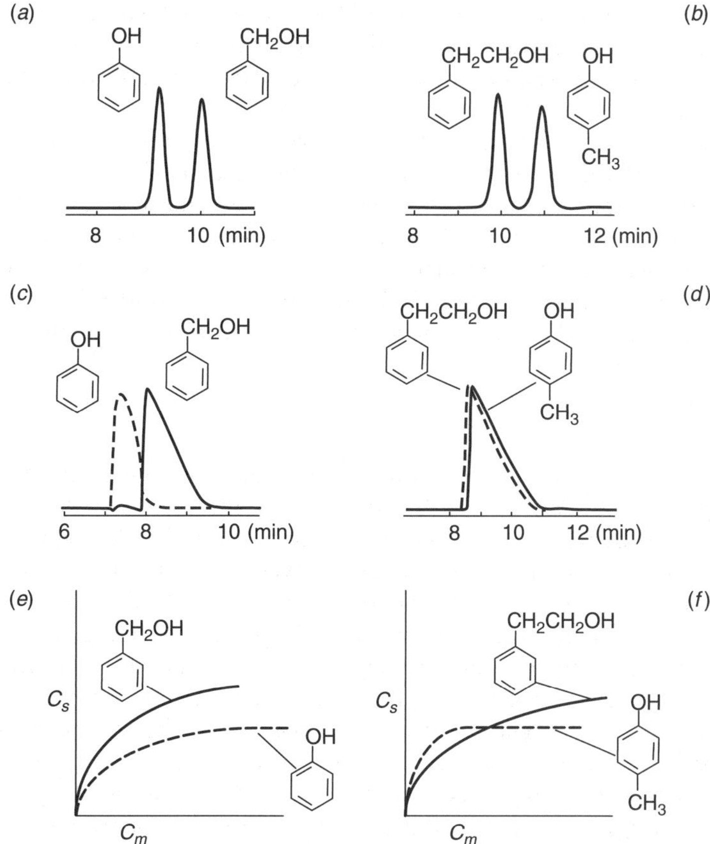

Figure 15.12 Example of crossing-isotherm behavior, with decrease in allowed sample weight for touching-peak separation. Conditions: 150 × 4.6-mm (5-μm) C_{18} column; methanol-water mobile phases; 1.0 mL/min. (*a*) 3 μg phenol and 7 μg benzyl alcohol (BA); (*b*) 3 μg 2-phenylethanol (PE) and 7 g *p*-cresol; (*c*) same as (*a*), except 4-mg sample weight; (*d*) same as (*b*), except 4-mg sample weight; (*e*, *f*) hypothetical isotherms corresponding to separations of phenol-benzyl alcohol and 2-phenylethanol/*p*-cresol, respectively. Adapted from [14].

the two compounds are well separated. For the separation of phenethyl alcohol and *p*-cresol (Fig. 15.12*f*), the greater retention of *p*-cresol for a small sample, combined with its smaller column capacity, leads to crossing of the isotherms for a sufficiently large sample. For the latter sample weight, the two compounds are equally retained, with no separation—as observed in the separation of Figure 15.12*d*. The latter explanation is intentionally oversimplified.

15.5 GRADIENT ELUTION

While gradient elution is often used for analytical separations and small-scale prep-LC, its use for large-scale separations can be less convenient and more costly.

An exception to this generalization occurs for the separation of large biomolecules because their isocratic retention can vary greatly for small changes in %B (Section 13.4.1.4), making isocratic elution impractical or impossible. An example of the industrial-scale purification of biosynthetic human insulin by gradient elution is discussed in Section 13.9.2. Method development for gradient separations closely parallels that for isocratic separation, as discussed in Chapter 9. Thus, when the gradient retention factor k^* is the same as k for isocratic elution, and other conditions are the same ("corresponding" separation; Section 9.13), the separation of a product peak from its impurities will be the same for both isocratic and gradient elution. Similarly any change in conditions that can improve isocratic selectivity can be used in the same way to improve gradient separation. Consequently virtually everything that applies for isocratic prep-LC in Section 15.3 applies equally for gradient elution. This will simplify our remaining discussion of gradient prep-LC in this section. For additional information about gradient prep-LC, see [2].

15.5.1 Isocratic and Gradient Prep-LC Compared

Figure 15.11*a–c* was used previously to compare the effect of sample size on an isocratic separation, where only the weights of two compounds in the sample are varied. A similar series of separations is shown in Figure 15.11*d–f* for the gradient separation of the same sample (compounds A and B) with the same conditions (except that gradient steepness replaces %B). In each case chromatograms are overlaid for (1) the separate injection of each compound (A′ and B′), and (2) the injection of the mixture (A plus B); see the related discussion of Section 2.6.2 and Fig. 2.24. The isocratic and gradient chromatograms for separations of equal sample weights (e.g., Fig. 15.11*a* vs. *d*, *b* vs. *e*, *c* vs. *f*) are seen to be virtually identical, with the exception of the more rounded ("shark-fin" shaped) peaks for overloaded gradient elution in Figure 15.11*d–f*.

This similarity of isocratic and gradient separations under comparable conditions was discussed in Section 9.1.3. For equivalent results as in Figure 15.11 for "corresponding" isocratic and gradient separations, the retention factor for each peak in the isocratic (k) and gradient (k^*) separations must be approximately equal, and all other separation conditions (column, A- and B-solvents, flow rate, temperature) must be the same. In the gradient separations of Figure 15.11*d–f*, separation conditions were adjusted so that (small-sample) values of k^* were equal to isocratic values of k in Figure 15.11*a–c*. As discussed in Section 9.2, values of k^* are determined by gradient conditions:

$$k^* = \frac{0.87 t_G F}{V_m \Delta\phi S} \tag{9.5}$$

Here t_G is the gradient time, F is flow rate, $\Delta\phi$ is the change in $\phi \equiv 0.01 \times$ (%B) during the gradient, S is related to the change in k for a given change in ϕ or %B (equal to $d[\log k]/d\phi$), and V_m is the column dead-volume (mL)—which can be determined from an experimental value of t_0 and the flow rate F (Section 2.3.1; $V_m = t_0 F$). Changes in isocratic separation as a result of a change in %B can be replicated in gradient elution, by a change in gradient time t_G (Eq. 9.5), so that the new values of both k and k^* are the same.

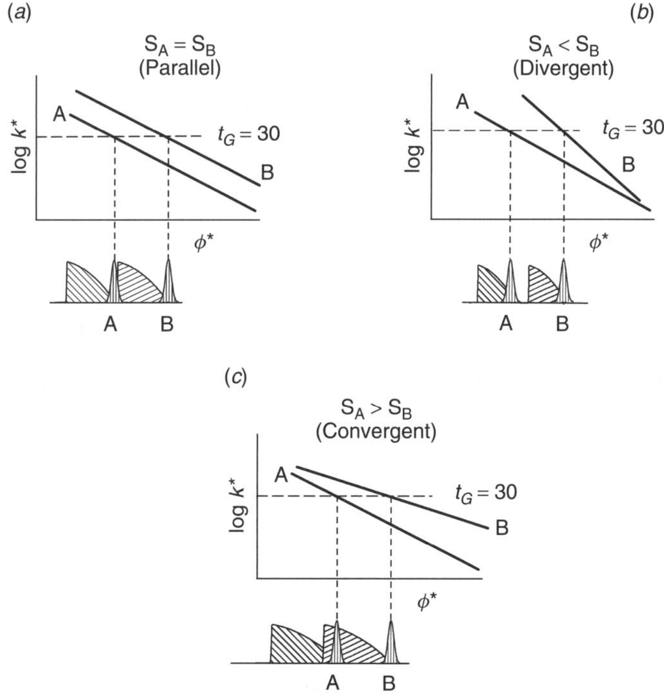

Figure 15.13 Effect of unequal values of *S* on the overload separation of two peaks by gradient elution. Adapted from [2].

15.5.2 Method Development for Gradient Prep-LC

The general plan of Figure 15.8 for isocratic method development can be followed for gradient elution also, with a few changes. The selection of initial conditions will be virtually the same, except than the initial separation will be carried out with a 0–100%B gradient, followed by narrowing the gradient range in most cases so as to shorten run time (Section 9.3.4). The optimization of the initial gradient separation for improved selectivity (step 2) can be carried out in the same way as for isocratic elution (Section 9.3.3), except that the goal is a maximum resolution for the product peak, rather than acceptable resolution of all peaks in the chromatogram. Aside from the choice of conditions for maximum α, the gradient program can be further modified so as to minimize separation time, while maintaining the resolution of the product peak from its adjacent impurity peaks (see Sections 9.3.4, 9.3.5). Maximizing sample weight (step 3) and scale-up (step 4) then proceed in exactly the same way as for isocratic prep-LC. For further details, see [2].

A complication not found for isocratic prep-LC is observed occasionally in gradient prep-LC. When two adjacent peaks have different values of *S*, this can affect the sample weight for T-P separation, as illustrated conceptually in Figure 15.13. Figure 15.13*a* gives the T-P gradient separation for the case of equal *S*-values for two compounds ("parallel" case), which is often a close approximation for most samples. At the top of Figure 15.13*a* is a plot of log k^* versus ϕ^* for each peak, where ϕ^* is the value of ϕ ($\equiv 0.01 \times$ %B) when the peak is at the column mid-point (see Eq. 9.5*a*); values of ϕ^* in Figure 15.13 track the time during the separations shown at the bottom of Figure 15.13*a*–*c* . The dotted lines connect the log k^*–ϕ^* plot for compounds A and B to (small-sample) peaks in the chromatogram below. The values of k^* and ϕ^* for each peak are determined by gradient conditions, with an assumed gradient time of 30 minutes for each separation in Figure 15.13.

Now assume that a large enough sample weight has been injected to allow peak *B* to cover the space between the two small-sample peaks (T-P separation), giving the wide cross-hatched peaks in the chromatogram of Figure 15.13*a*. We see that the vertical separation of the two log k^* versus ϕ plots is constant and equal to log α for each value of ϕ. Thus, at the beginning of elution of overloaded peak *B* (at a lower value of ϕ^*, corresponding to the elution of a small sample of A), α is the same ($= \alpha_o$) as at the end of elution; that is, the separation factor is not a function of sample weight. (Note that Eq. 15.5, which relates sample weight for T-P separation to values of α_o, assumes approximately equal values of *S* for the two adjacent peaks.)

Figure 15.13*b* is similar to that of 15.13*a* (same weight of injected sample for T-P separation in Fig. 15.13*a*), except that now the plots of log k^* versus ϕ^* are no longer parallel but diverge for lower values of ϕ^* ("divergent" case); that is, the value of *S* for compound B is greater than for compound A. For higher loading of the column (at lower values of ϕ^*), the vertical separation of the two log k^*–ϕ^* curves increases, corresponding to an increase in α with increasing sample weight. A larger value of α means a larger sample weight for T-P separation (Eq. 15.5), so the same injected weight of sample as in Figure 15.13*a* is no longer sufficient to cause the peaks to touch. That is, the divergent case allows a larger weight of injected sample (other factors equal), compared to the equal-*S* case of Equation (15.5) and Figure 15.13*a*.

Figure 15.13*c* illustrates the third possibility: log k^*–ϕ^* plots that converge for smaller ϕ^* ("convergent" case); namely *S* for compound B is less than for compound A. Now α decreases with increasing sample weight, and injection of the same weight of sample as in Figure 15.13*a* for T-P separation leads to a more rapid column overload with overlap of the two peaks. When convergent behavior is suspected (because of lower than expected sample weights for T-P separation), further changes in separation conditions should be considered—with the goal of reversing the elution order of the two peaks (product and nearest impurity bands). A similar approach can also be used to minimize the problem of crossing isotherms (Section 15.4.2.2). For a further discussion of the consequences of unequal *S*-values in gradient prep-LC, see [2].

15.6 PRODUCTION-SCALE SEPARATION

Production-scale separations are well beyond the scope of this book, but the simple theory and practice outlined here is still pertinent. Separations of this kind are usually highly optimized, so as to result in the highest possible production rate for the desired product, with the required purity and recovery. At this scale, process economics are of primary importance; the goal is a combination of purity, recovery, and production rate that yield the lowest cost per kg of the desired product, including the cost of removing mobile phase from the purified product. Separation conditions are usually developed empirically, using the approach of Figure 15.10, for samples much larger than correspond to T-P separation. For an example, see Section 13.9 for the production-scale separation of rh-insulin.

For production-scale separations, the use of simulated moving bed (SMB) techniques are increasingly important. This is a binary separation technique that relies on the simulation of a countercurrent separation system by the use of multiple columns and switching valves. The reader is referred to specialized texts on this topic [14]. Although this approach has been used for many decades in the petroleum industry (the Molex process for isolation of *p*-xylene) and the food industry (the Sorbex process for fructose-rich syrups), it has only been used by the pharmaceutical industry since the 1990s. The use of countercurrent separation with a continuous sample input and product output gives a more effective use of the chromatographic bed than the traditional procedures discussed in this chapter. SMB thus reduces both the size of the columns and the amount of solvent used (and therefore costs); it is widely used. The column efficiency required for countercurrent separations is relatively low, which allows the use of very short "pancake" columns. For example, successful enantiomer separations are carried out under SMB conditions with columns of 800- or 1000-mm i.d., but only 100 mm in length (e.g., packed with 20-μm particles).

REFERENCES

1. M. S Villeneuve and L. A. Miller in *Preparative Enantioselective Chromatography*, G. B. Cox, ed., Wiley-Blackwell, 2005.
2. L. R. Snyder and J. W. Dolan, *High Performance Gradient Elution*, Wiley, Hoboken, NJ, 2007, ch. 7.
3. Azeotropes: http://en.wikipedia.org/wiki/Azeotrope_(data).
4. G. Guiochon, A. Felinger, D. G. Shirazi, and A. M. Katti, *Fundamentals of Preparative and Non-linear Chromatography*, 2nd ed., Academic Press, Boston, 2006.
5. J. H. Knox and H. M. Pyper, *J. Chromatogr.*, 363 (1986) 1.
6. L. R. Snyder, G. B. Cox, and P. E. Antle, *Chromatographia*, 24 (1987) 82.
7. U. D. Neue, C. B. Mazza, J. Y. Cavanaugh, Z. Lu, and T. E. Wheat, *Chromatographia Suppl.*, 57 (2003) S-121.
8. U. D. Neue, T. E. Wheat, J. R. Mazzeo, C. B. Mazza, J. Y. Cavanaugh, F. Xia, and D. M. Diehl, *J. Chromatogr. A*, 1030 (2004) 123.
9. L. R. Snyder, J. J. Kirkland, and J. L. Glajch, *Practical HPLC Method Development*, 2nd ed., Wiley-Interscience, New York, 1997.
10. H. Schmidt-Traub, *Preparative Chromatography of Fine Chemicals and Pharmaceutical Agents*, Wiley-VCH, New York, 2005.
11. G. Guiochon and S. Ghodbane, *J. Phys. Chem.*, 92 (1988) 3682.
12. G. B. Cox and L. R. Snyder, *J. Chromatogr.*, 590 (1992) 17.
13. G. B. Cox and L. R. Snyder, *J. Chromatogr.*, 483 (1989) 95.
14. O. Dapremont in G. B. Cox, *Preparative Enantioselective Chromatography*, Wiley-Blackwell, New York, 2005.

CHAPTER SIXTEEN

SAMPLE PREPARATION

with Ronald Majors

Introduction to Modern Liquid Chromatography, Third Edition, by Lloyd R. Snyder, Joseph J. Kirkland, and John W. Dolan

16.1 INTRODUCTION

Sample preparation is an essential part of HPLC analysis, intended to provide a representative, reproducible, and homogenous solution that is suitable for injection into the column. The aim of sample preparation is to provide a sample aliquot that (1) is relatively free of interferences, (2) will not damage the column, and (3) is compatible with the intended HPLC separation and detection methods. The sample solvent should dissolve in the mobile phase without affecting sample retention or resolution, and without interfering with detection. It may also be necessary to concentrate the analytes and/or derivatize them for improved detection or better separation.

Sample preparation begins at the point of collection, extends to sample injection onto the HPLC column, and encompasses the various operations summarized in Table 16.1. Options 1 to 4 of Table 16.1—which include sample collection, transport, storage, preliminary processing, laboratory sampling, and subsequent weighing/dilution—all form an important part of sample preparation. Although these four steps in the HPLC assay can have a critical effect on the accuracy, precision, and convenience of the final method, only option 3 (preliminary sample processing) will be (briefly) discussed here. See [1–4] for a discussion of options 1, 2, and 4. This chapter will be devoted mainly to options 5 to 8 of Table 16.1, which encompass what is usually meant by *sample pretreatment* or *sample preparation* ("sample prep").

Whereas HPLC is predominantly an automated procedure, sample pretreatment often is performed manually. As a result sample pretreatment can require 60% or more of the total time devoted to routine analysis. Sample pretreatment includes a large number of methodologies, as well as multiple operational steps, and can therefore be a challenging part of HPLC method development. Finally, method precision and accuracy are often largely determined by the sample-pretreatment procedure [5–6], including operations such as weighing and dilution. For all these reasons the development of a sample-pretreatment procedure deserves careful planning.

A sample-pretreatment procedure should provide quantitative recovery of analytes, involve a minimum number of steps, and (if possible) be easily automated. Quantitative (99^{+}%) recovery of each analyte enhances sensitivity and assay precision, although this does not mean that all of the analyte present in the original sample must be included in the final injected sample. For example, in a given method that includes a series of sample-pretreatment steps, aliquots of intermediate fractions may be used for further sample preparation or for injection. If analyte recovery is significantly less than 100%, it must be reproducible. A smaller number

Table 16.1

Sample Preparation Options

Option Number	Option	Comment
1	Sample collection	Obtain representative sample using statistically valid procedures.
2	Sample storage and preservation	Use appropriate inert, tightly sealed containers and storage conditions; stabilize volatile, unstable, or reactive samples, if necessary; biological samples may require freezing.
3	Preliminary sample processing	Disperse or divide sample (drying, sieving, grinding, etc.) for more representative sample and to improve dissolution or extraction.
4	Weighing or volumetric dilution	Take necessary precautions for reactive, unstable, or biological materials; for dilution, use calibrated volumetric glassware.
5	Alternative sample processing methods	Consider, among these, solvent exchange, desalting, evaporation, or freeze drying.
6	Removal of particulates	Use filtration, centrifugation, solid-phase extraction.
7	Sample extraction	For methods for liquid samples, see Table 16.2; for solid samples, Tables 16.8 and 16.9.
8	Derivatization	Used to enhance analyte detection or improve separation; extra steps may add time, complexity and potential loss of sample (Section 16.12).

of sample-pretreatment steps—plus automation—reduces the overall time and effort required, improves assay precision, and decreases the opportunity for errors by the analyst.

Many sample-preparation techniques have been automated, and appropriate instrumentation is commercially available. Approaches to automation vary from using a robot to perform manual tasks, to dedicated instruments that perform a specific sample-preparation procedure. Although automation can be expensive and elaborate, it is often desirable when large numbers of samples must be analyzed, and the time or labor (per sample) required for manual sample preparation would be excessive. The decision to automate a sample-preparation procedure is often based on a cost justification or, in some cases, operator safety (e.g., to minimize exposure to toxic substances or other possible health hazards). A full coverage of sample-preparation automation is beyond the scope of this chapter; the reader is referred to recent textbooks on the subject [7–8].

16.2 TYPES OF SAMPLES

Sample matrices can be broadly classified as organic (including biological) or inorganic, and may be further subdivided into solids, semi-solids (including creams, gels, suspensions, colloids), liquids, and gases. For nearly every matrix some form of sample pretreatment will be required prior to HPLC analysis, even if only simple dilution.

Gaseous samples usually are analyzed by gas chromatography rather than by HPLC. Techniques such as canister collection, direct sampling via sample loops, headspace sampling, and purge-and-trap are used to collect and inject gases. However, volatile analytes that are labile, thermally unstable, or prone to adsorb onto metal surfaces in the vapor state are sometimes better handled by HPLC. Trapping is required to analyze gaseous samples by HPLC. The gas sample is either (1) passed through a solid support and subsequently eluted with a solubilizing liquid or (2) bubbled through a liquid that traps the analyte(s). An example of the HPLC analysis of a gaseous sample is the American Society for Testing Materials (ASTM) Method D5197-03 and United States Environmental Protection Agency (EPA) Method TO-11 for volatile aldehydes and ketones [9]. An air sample is passed through an adsorbent trap coated with 2,4-dinitrophenylhydrazine, which quantitatively converts aldehydes and ketones into 2,4-dinitrophenylhydrazones. The hydrazones are then eluted from the trap with acetonitrile and separated by reversed-phase HPLC (RPC).

Table 16.2 provides an overview of typical sample preparation procedures used for liquids and suspensions. The remainder of this chapter will be devoted to the pretreatment of samples of most concern: semi-volatile and non-volatile analytes in various liquid and solid matrices.

Sample preparation for solid samples can be more demanding than for liquid samples. In some cases, the sample is easily dissolved and is then ready for injection or further pretreatment. In other cases, the sample matrix may be insoluble in common solvents, and the analytes must be extracted from the solid matrix. There are also cases where the analytes are not easily removable from an insoluble matrix—because of inclusion or adsorption. Here more rigorous techniques such as Soxhlet extraction, pressurized fluid extraction (PFE), ultrasonication, or solid–liquid extraction may be necessary (Section 16.8.2). Table 16.8 lists some traditional methods for the recovery of analytes from solid samples, while Table 16.9 describes some more recent procedures. Once analytes have been quantitatively extracted from a solid sample, the resulting liquid fraction can either be injected directly into the HPLC instrument or subjected to further pretreatment.

Compared to gases or solids, liquid samples are much easier to prepare for HPLC. Many HPLC analyses are based on a "dilute and shoot" procedure, whereby the solubilized analyte concentration is reduced by dilution so as to not overload the column or saturate the detector, or to make the injection solvent more compatible with the mobile phase.

16.3 PRELIMINARY PROCESSING OF SOLID AND SEMI-SOLID SAMPLES

16.3.1 Sample Particle-Size Reduction

Solid samples should be reduced in particle size because finely divided samples (1) are more homogeneous, allowing more representative sampling with greater precision and accuracy, and (2) dissolve faster and are easier to extract because of their greater surface area. Methods for reducing the particle size of solid samples are outlined in Table 16.3.

Table 16.2

Typical Sample-Preparation Methods for Liquids and Suspensions

Sample Preparation Method	Principles of Technique	Comments
Solid phase extraction (SPE)	Similar process to HPLC. Sample is applied to, and liquid is passed through, a column packed with a solid phase that selectively removes analytes (or interferences) (Section 16.6).	Wide variety of stationary phases is available for the selective removal of desired inorganic, organic, and biological analytes.
Liquid–liquid extraction (LLE)	Sample is partitioned between two immiscible phases. Interference-free analytes are then recovered from one of the two phases (Section 16.5).	Beware of formation of emulsions. Values of K_D can be optimized by the use of different solvents or additives; continuous extraction or large volumes can be used for low K_D-values.
Dilution	Sample is diluted with a solvent that is compatible with the HPLC mobile phase.	To avoid excess peak broadening or distortion, dilution solvent should be miscible with, and preferably weaker than, the HPLC mobile phase.
Evaporation	Liquid is removed by gentle heating with flowing air or inert gas.	Do not evaporate too quickly; avoid sample loss on wall of container; don't overheat to dryness; best with inert gas like N_2.
Distillation	Sample is heated to boiling point of solvent and volatile analytes in the vapor phase are condensed and collected.	Mainly for samples that can be easily volatilized; some samples may decompose if heated too strongly. Vacuum distillation for high boilers.
Microdialysis	A semi-permeable membrane is placed between two aqueous liquid phases, and analytes transfer from one liquid to the other, based on their differential concentration.	Enrichment techniques such as SPE are required to concentrate dialysates; dialysis with molecular-weight-cutoff membranes can be used on-line to deproteinate samples prior to HPLC; ultrafiltration and reverse osmosis can also be used in a similar manner.
Lyophilization	Aqueous sample is frozen, and water is removed by sublimation under vacuum.	Good for nonvolatile organics; large sample-volume can be handled; possible loss of volatile analytes; good for recovery of thermally unstable analytes—especially biologicals.
Filtration	Liquid is passed through a paper or membrane filter or a SPE cartridge/disk to remove suspended particulates.	Highly recommended to prevent back-pressure problems and to preserve column life.
Centrifugation	Sample is placed in a tapered centrifuge tube and spun at high force (several times gravity, G); supernatant liquid is decanted.	Ultracentrifugation is not normally used for simple particulate removal.
Sedimentation	Sample is allowed to settle when left undisturbed; settling rate is dependent on Stoke's radius.	Extremely slow process; manual recovery of different size particulates at different levels, depending on settling rate.

Table 16.3

Methods for Reducing Sample Particle-Size

Particle-Size Reduction Method	Description of Procedure
Blending	Mechanical blender is used to chop a semi-soft substance into smaller parts or blend a nonhomogeneous sample into a more consistent form.
Chopping	Process of mechanically cutting a sample into smaller parts.
Crushing	Tungsten-carbide variable-jaw crushers can reduce the size of large, hard samples.
Cutting	Cutting mills can reduce soft-to-medium hard materials (<100-mm diameter).
Grinding	Manual or automated mortar-and-pestle are the most popular choice; both wet and dry grinding are used; particle sizes of ≈10 μm can be achieved.
Homogenizing	Any process used to make sample more uniform in texture and consistency by breaking down into smaller parts and blending.
Macerating	Process of breaking down a soft material into smaller parts by tearing, chopping, cutting, etc.
Milling	Various disk, rotor-speed, or ball mills can reduce soft-to-medium hard and fibrous materials to 80–100-μm size.
Mincing	Process of breaking down a meat or vegetable product into smaller parts by tearing, chopping, cutting, dicing, etc.
Pressing	Generally, the process of squeezing liquid from a semi-solid material (e.g., plants, fruit, meat).
Pulverizing	Electromechanically driven rod or vibrating base used to reduce particle size for either wet or dry samples; a freezer mill can be used with liquid N_2 to treat malleable samples.
Sieving	Process of passing a sample through a metal or plastic mesh of a uniform cross-sectional area (square openings of 3–123 μm) in order to separate particles into uniform sizes; both wet and dry sieving can be used.

16.3.2 Sample Drying

Solid samples are often received for analysis in a damp or wet state. Removal of water or drying the sample to constant weight is usually necessary for reliable assay. Inorganic samples, such as soil, should be heated to 100–110°C to ensure the removal of moisture. Hydrophobic organic samples seldom require heating, since water absorption is minimal. However, organic vapors can be adsorbed by solid organic samples, and a heating step can remove these contaminants. For hydroscopic or reactive samples (e.g., acid anhydrides) drying in a vacuum dessicator is recommended. Samples that can oxidize when heated in the presence of air should be dried under vacuum or nitrogen. Biological samples generally should not be heated to >100°C, and temperatures above ambient should be avoided to prevent sample decomposition. Sensitive biological compounds (e.g., enzymes) often are prepared in a cold room at <4°C to minimize decomposition. Such samples should be maintained at these low temperatures until the HPLC analysis step. Freeze-drying (lyophilization) often is used to preserve the integrity of heat-sensitive samples (especially biologicals). Lyophilization is performed by quick-freezing the sample, followed by removal of frozen water by sublimation under vacuum.

16.3.3 Filtration

Particulates should be removed from liquid samples prior to injection, because of their adverse effect on column lifetime as well as possible damage to tubing, injection valves, and frits. The most common methods for removing particulates from the sample are filtration, centrifugation, and sedimentation. Several approaches to filtration are described in Table 16.4. The lower the porosity of the filter medium, the cleaner is the filtrate, but the longer is the filtering time. Vacuum filtration speeds the process. Membrane filters in a disk format can be purchased for use with commercial filter holders/housings. However, most users prefer disposable filters

Table 16.4

Filtration in HPLC

Filtration Media	Typical Products	Recommended Use	Comments
Filter paper	Cellulose	Removal of larger particles (<40 μm)	Beware of filter-paper fibers getting into sample; ensure solvent compatibility of filter paper.
Membrane filters	Nylon, PTFE, polypropylene, polyester, polyethersulfone, polycarbonate, polyvinylpyrolidone	Removal of small particles (>10 μm)	Prefilter may be needed for dirty samples prior to filtration; avoid solvent incompatibility.
Functionalized membranes	Ion-exchange membranes, affinity membranes	Can remove both particulates and matrix interferences	Prefilter may be needed for dirty samples prior to filtration; avoid solvent incompatibility.
SPE cartridges	Silica- and polymer-based	Can remove both particulates and matrix interferences	Particles of silica-bonded phase can pass into filtrate; beware of plugging.
SPE disks	PTFE- and fiberglass-based	Can remove both particulates and matrix interferences	PTFE membranes are delicate, so handle with care; can pass a large volume at high flow rate; beware of plugging.

equipped with Luer fittings. The sample is placed in a syringe and filtered through the membrane using gentle pressure.

A variety of membrane materials, nominal porosities, and dimensions are available for filtration, and the manufacturers' literature provides specifications. The large cross-sectional areas of the membrane-disk-type filter allows for good flow characteristics and minimizes plugging. For most samples encountered in HPLC, filters in the range of 0.25- to 2-μm nominal porosity are recommended. The porosity values are approximate, and the type of membrane can have some influence on the filtration characteristics. The most popular sizes for sample filtration are 0.25- and 0.45-μm porosities. Membranes with 0.25-μm pores remove the tiniest particles (and large macromolecules). If the sample contains colloidal material or a large amount of fines, considerable pressure may be required to force the liquid sample through the filter. Sometimes a prefilter or depth filter (a thick filter with a large capacity for trapping larger particulates) is placed on top of the membrane to prevent plugging with samples containing these types of particulates.

An important consideration in filter selection is solvent compatibility with the membrane. If an inappropriate solvent is used, the filter may dissolve (or soften) and contaminate the filtrate. Manufacturers of membrane filters usually provide detailed information on the solvent compatibility of their products. More expensive functionalized membranes and SPE disks and cartridges are used not only for chemical interference removal but also to remove particulates.

16.4 SAMPLE PREPARATION FOR LIQUID SAMPLES

Table 16.2 provides an introduction to sample preparation methods for liquid samples. Most laboratories need only a few of these procedures. For example, distillation is limited to volatile compounds, although vacuum distillation for high-boiling compounds in environmental samples can extend the application of this technique [10]. Lyophilization is usually restricted to the purification and handling of biological samples. In the present discussion we will emphasize two methods used most often in most HPLC laboratories: liquid–liquid extraction (Section 16.5) and liquid–solid or solid–phase extraction (Section 16.6).

16.5 LIQUID–LIQUID EXTRACTION

Liquid–liquid extraction (LLE) is useful for separating analytes from interferences by partitioning the sample between two immiscible liquids or *phases*. One phase in LLE will usually be aqueous, and the second phase will be an organic solvent. More hydrophilic compounds prefer the polar aqueous phase, whereas more hydrophobic compounds will be found mainly in the organic solvent. Analytes extracted into the organic phase are easily recovered by evaporation of the solvent; analytes extracted into the aqueous phase often can be injected directly onto a RPC column. The following discussion assumes that an analyte is preferentially concentrated into the organic phase; similar approaches can be used when the analyte is extracted into the aqueous phase.

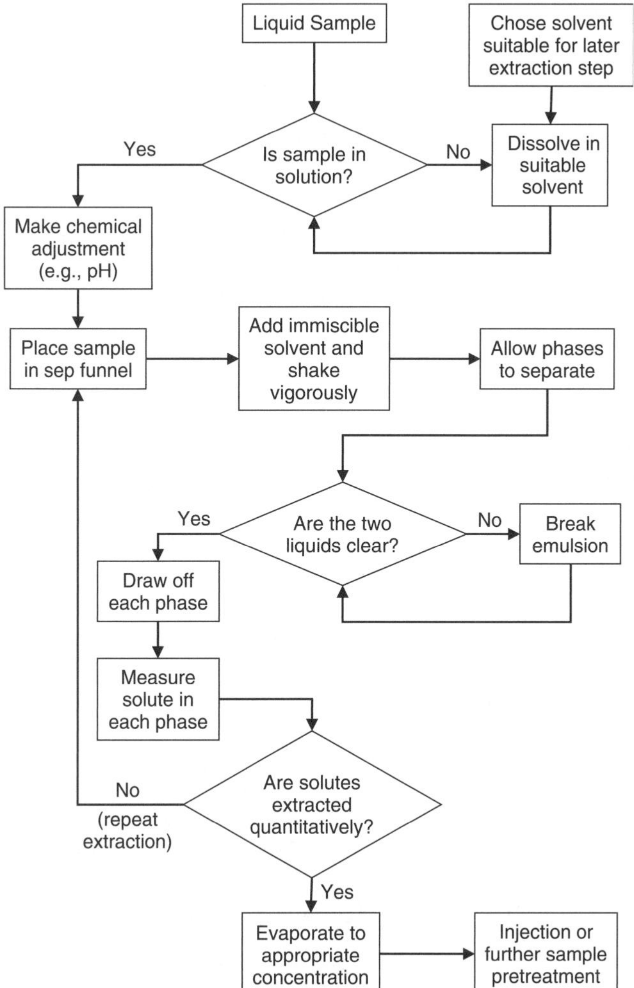

Figure 16.1 Summary of the steps involved in liquid-liquid extraction (LLE).

Figure 16.1 summarizes the steps involved in a LLE separation. Since extraction is an equilibrium process with limited efficiency (only one "theoretical plate"), significant amounts of the analyte can remain in both phases—even when K_D (or $1/K_D) \gg 1$. Chemical equilibria involving changes in pH, ion-pairing, and complexation, for example, can be used to enhance values of K_D and improve analyte recovery and/or eliminate interferences.

The following characteristics are desirable for a LLE organic solvent:

- low solubility in water (<10%)
- volatility for easy removal and concentration after extraction (Table I.3)
- compatibility with the HPLC detector (e.g., low UV absorbance) (Table I.3)
- polarity and hydrogen-bonding properties that enhance recovery of the analytes in the organic phase (Section 2.3.2.1) (Table I.4)
- high purity to minimize sample contamination

16.5.1 Theory

The Nernst distribution law states that any species will distribute between two immiscible solvents so that the ratio of the concentrations remains constant.

$$K_D = \frac{C_o}{C_{aq}} \tag{16.1}$$

where K_D is the distribution constant, C_o is the concentration of the analyte in the organic phase, and C_{aq} is the concentration of the analyte in the aqueous phase.

A more useful expression is the fraction E of analyte extracted, given by

$$E = \frac{C_o V_o}{C_o V_o + C_{aq} V_{aq}} = \frac{K_D \psi}{1 + K_D \psi} \tag{16.2}$$

where V_o is the volume of organic phase, V_{aq} the volume of aqueous phase, and ψ is the phase ratio V_o/V_{aq}.

Many LLE procedures are carried out in separatory funnels, and typically use tens or hundreds of milliliters of each phase. For one-step extractions K_D (or $1/K_D$) must be large (e.g., >10) for the quantitative recovery of analyte in one of the two phases, since the phase-ratio ψ must be maintained within a practical range of values, such as $0.1 < \psi < 10$ (see Eq. 16.2). In most separatory-funnel LLE procedures, quantitative recoveries (>99%) require two or more extractions. For successive multiple extractions, with pooling of the analyte phases from each extraction,

$$E = 1 - \left(\frac{1}{1 + K_D \psi}\right)^n \tag{16.3}$$

where n is the number of extractions. For example, if $K_D = 5$ for an analyte, and the volumes of the two phases are equal ($\psi = 1$), three extractions ($n = 3$) would be required for >99% recovery of the analyte. Several approaches can be used to increase the value of K_D:

- organic solvent can be changed to increase K_D.
- K_D can be increased if the analyte is ionic or ionizable by suppressing its ionization so as to make it more soluble in the organic phase
- the analyte can be extracted into the organic phase by ion pairing, provided that the analyte is ionized and an ion-pair reagent is added to the organic phase
- "salting out" can be used to decrease an analyte's concentration in the aqueous phase, by addition of an inert, neutral salt (e.g., sodium sulfate) to the aqueous phase.

16.5.2 Practice

Table 16.5 provides examples of typical extraction solvents, as well as some unsuitable (water-miscible) extraction solvents. Apart from miscibility considerations, the main selection criterion is the polarity P' of the solvent (Tables 2.3, I.4) in relation to that of the analyte. Maximum K_D occurs when the polarity of the extraction solvent matches that of the analyte. For example, the extraction of a polar analyte

Table 16.5

Extraction Solvents for Liquid—Extraction

Aqueous Solvents	Water-Immiscible Organic Solvents	Water-Miscible Organic Solvents (Unsuitable for LLE)
Pure water	Aliphatic hydrocarbons (hexane, isooctane, petroleum ether, etc.)	Alcohols (low molecular weight)
Acidic solution	Diethyl ether or other ethers	Ketones (low molecular weight)
Basic solution	Methylene chloride	Aldehydes (low molecular weight)
High salt (salting-out effect)	Chloroform	Carboxylic acids (low molecular weight)
Complexing agents (ion pairing, chelating, chiral, etc.)	Ethyl acetate and other esters	Acetonitrile
Combination of two or more above	Aliphatic ketones (C_6 and above)	Dimethyl sulfoxide
	Aliphatic alcohols (C_6 and above)	Dioxane
	Toluene, xylenes (UV absorbance!)	
	Combination of two or more above	

Note: Any solvent from the "aqueous solvents" column can be matched with any solvent of the "water-immiscible organic solvents" column; water-miscible organic solvents should not be used with aqueous solvents to perform LLE.

from an aqueous sample matrix would be best accomplished with a more polar (large P') organic solvent. An optimum-polarity organic solvent can be conveniently selected by blending two solvents of different polarity (e.g., hexane [$P' = 0.1$] and chloroform [$P' = 4.1$]), and measuring K_D vs. the composition of the organic phase [11]. A solvent mixture that gives the largest value of K_D is then used for the LLE procedure. Further changes in K_D can be achieved, with improvement in the separation of analytes from interferences, by varying organic-solvent *selectivity* in addition to polarity. Solvents from different regions of the solvent-selectivity triangle (Fig 2.9) are expected to provide differences in selectivity; see also the discussion of [12].

In solvent extraction, ionizable organic analytes often can be transferred into either phase, depending on the selected conditions. For example, consider the extraction of an organic acid from an aqueous solution. If the aqueous phase is buffered at least 1.5 pH units above its pK_a value, the analyte will be ionized and prefer the aqueous phase; less-polar interferences will be extracted into the organic phase. If the pH of the aqueous solution is lowered ($\ll pK_a$), so that the analyte is no longer ionized, the analyte will be extracted into the organic phase, leaving more-polar interferences in the aqueous phase. Successive extractions at high pH followed by low pH are able to separate an acid from both more- and less-polar interferences. Equilibria involving pH are discussed further in Section 7.2. Note that the principles of acid-base extraction as a function of pH are the same for LLE and RPC.

If the analyte K_D is unfavorable, additional extractions may be required for improved recovery (Eq. 16.3). For the case of an organic-soluble analyte, a fresh portion of organic solvent is added to the aqueous phase in order to extract additional solute; all extracts are then combined. For a given volume of final extraction-solvent, multiple extractions are generally more efficient in removing a solute quantitatively, as opposed to the use of a single extraction volume. *Back-extraction* can be used to further reduce interferences. For example, consider the example above of an organic-acid analyte. If the analyte is first extracted at low pH into the organic phase, polar interferences (e.g., hydrophilic neutrals, protonated bases) are left behind in the aqueous phase. If a fresh portion of high-pH aqueous buffer is used for the back-extraction of the organic phase, the ionized organic acid is transferred back into the aqueous phase, leaving less-polar interferences in the organic phase (the latter procedure is similar to successive extractions at *high* pH followed by *low* pH described above). Thus a two-step extraction with change of pH can allow the removal of both basic and neutral interferences, whereas a one-step extraction can eliminate one or the other of these interferences, but not both.

If K_D is not much greater than 1, or the required volume of sample is large, it may be impractical to carry out multiple extractions for quantitative recovery of the analyte—too many extractions are required, and the volume of total extract is too large (Eq. 16.3). If extraction is slow, a long time may also be required for the equilibrium to be established. In these cases continuous liquid–liquid extraction can be used, where fresh solvent is continually recycled through the aqueous sample. Continuous extractors that use heavier-than-water and lighter-than-water solvents have been described [13]. These extraction devices can run for extended periods (12–24 hr), and quantitative extractions (>99% recovery) can be achieved, even for less-favorable values of K_D.

For more efficient LLE, a countercurrent distribution apparatus can provide a thousand or more equilibration steps (but with more time and effort). This allows the recovery of analytes having K_D values near unity; countercurrent distribution also provides a better separation of analytes from interferences. Small-scale laboratory units are commercially available. For further information on these devices, see [14].

In some cases LLE can enhance analyte concentration in the extract fraction relative to its concentration in the initial sample. According to Equation (16.2), by choosing a smaller volume of organic solvent, the analyte concentration can be increased by the volumetric ratio of organic-to-aqueous phases (*assuming near-complete extraction into the organic phase or large K_D*). For example, assume 100 mL of aqueous sample, 10 mL of organic solvent, and a very large K_D (e.g., $K_D > 1000$). The concentration of the analyte in the organic phase will then increase by a factor of 10. For large ratios of aqueous-to-organic, a slight solubility of the organic solvent in the aqueous phase can reduce the volume of the recovered organic solvent significantly; this problem can be avoided by presaturating the aqueous solvent with organic solvent. Note that when the solvent ratio V_0/V_{aq} is small, the physical manipulation of two phases (including recovery of the organic phase) becomes more difficult.

16.5.3 Problems

Some practical problems associated with LLE include:

- emulsion formation
- analytes strongly adsorbed to particulates

- analytes bound to high-molecular-weight compounds (e.g., protein-drug interactions)
- mutual solubility of the two phases

16.5.3.1 Emulsion Formation

Emulsions are a problem that can occur with some samples (e.g., fatty matrices) under certain solvent conditions. If emulsions do not "break," with a sharp boundary between the aqueous and organic phases, analyte recovery can be adversely affected. Emulsions often can be broken by:

- addition of salt to the aqueous phase
- heating or cooling the extraction vessel
- filtration through a glass wool plug
- filtration through phase-separation filter paper
- addition of a small amount of different organic solvent
- centrifugation

16.5.3.2 Analyte Adsorption

If particulates are present in a sample, adsorption onto these particulates can result in a low recovery of the analyte. In such cases, washing the particulates after filtration with a stronger solvent often will recover the adsorbed analyte; this extract should be combined with the analyte phase from LLE. A "stronger" solvent for recovering adsorbed analyte may involve a change in pH, increase in ionic strength, or the use of a more polar organic solvent.

16.5.3.3 Solute Binding

Compounds that normally are recovered quantitatively in LLE may bind to proteins when plasma samples are processed, resulting in low recovery. Protein-binding is especially troublesome when measuring drugs and drug metabolites in physiological fluids. Techniques for disrupting protein binding in plasma samples include:

- addition of detergent
- addition of organic solvent, chaotropic agents, or strong acid
- dilution with water
- displacement with a more strongly binding compound

16.5.3.4 Mutual Phase-Solubility

"Immiscible" solvents have a small, but finite, mutual solubility, and the dissolved solvent can change the relative volumes of the two phases. Therefore it is a good practice to saturate each phase with the other, so that the volume of phase containing the analyte can be known accurately, allowing an optimum determination of analyte recovery. For values of the solubility of a solvent in water (or of water in the solvent), see [15].

16.5.4 Special Approaches to Liquid–Liquid Extraction

16.5.4.1 Microextraction

Extractions in this form of LLE are carried out with organic-aqueous ratios of 0.001 to 0.01. Analyte recovery may suffer, compared to conventional LLE, but the analyte concentration in the organic phase is significantly increased and solvent usage is greatly reduced. Such extractions are conveniently carried out in a volumetric flask. The organic extraction solvent is chosen to have a density less than that of water, so that the small volume of organic solvent accumulates in the neck of the flask for easy removal. For quantitative analysis, internal standards should be used and extractions of calibration standards carried out. Some modern autosamplers are capable of performing microextractions automatically on small volumes of aqueous samples in 2-mL vials provided that the position of the pickup needle is adjustable [16].

16.5.4.2 Single-Drop Microextraction (SDME)

This technique uses a 1- or 2-μL droplet of immiscible organic solvent, held at the end of a syringe needle, to extract and concentrate analytes from an aqueous (immiscible) sample [17]. Analytes diffuse into the droplet, resulting in a considerable increase in analyte concentration. When equilibrium is achieved, the microdrop is retracted into the syringe and then injected into an HPLC column or diluted in a microvial to achieve RPC mobile-phase compatibility. An example of the use of SDME is the RPC analysis of hypercins in plasma and urine [18]. Sometimes the use of a hollow-fiber membrane filled with a small volume of organic solvent is more useful in containing the solvent droplet. This procedure is often referred to as liquid-phase microextraction (LPME).

16.5.4.3 Solid-Supported Liquid–Liquid Extraction (SLE)

SLE replaces the separatory funnel in LLE with a small column that contains an inert support such as diatomaceous earth. An aqueous sample is first applied to the column, so as to coat the support with sample. A buffered immiscible solvent is then passed through the column, with extraction of any hydrophobic analytes. Samples that have been extracted in this way include diluted plasma, urine, and milk. The solvent moves through the column by means of gravity flow or a gentle vacuum. Because there is no vigorous shaking of the sample and extraction solvent, as in conventional LLE, there is no possibility of emulsion formation. The packed tubes are disposable, and the entire process is amenable to automation. Packed 96-well plates with several hundred milligrams of packing per well are suitable for the extraction of 150 to 200 μL of aqueous sample. Examples of commercial products that perform SLE are Varian's Hydromax (Palo Alto, CA), Biotage's Isolute HM-N (Charlottesville, VA), and Merck's Extrelut (Darmstadt, Germany).

16.5.4.4 Immobilized Liquid Extraction (ILE)

ILE involves extraction of hydrophobic analytes from an aqueous sample into a polymeric film comprising a phase similar to the bonded liquid phases used in capillary GC. The polymeric film can be applied to the cap of a vial, the inner walls of a 96-well plate, or inside a micropipette tip. The sample is first exposed to the film for extraction of analytes, followed by an aqueous wash to remove polar interferences, and a final wash with organic solvent to recover the analyte.

The sample can be directly injected into the chromatograph or evaporated and redissolved in a more HPLC-compatible solvent. Devices for ILE are supplied by ILE Inc. (Ferndale, CA) and other suppliers. ILE can be automated and compares favorably to SPE [19].

16.6 SOLID-PHASE EXTRACTION (SPE)

Solid-phase extraction is the most important technique used in sample pretreatment for HPLC. SPE can be used in similar fashion as LLE, but whereas LLE usually is a one-stage separation process, SPE is a chromatographic procedure that resembles HPLC and has a number of potential advantages compared to LLE:

- more complete extraction of the analyte
- more efficient separation of interferences from analytes
- reduced organic solvent consumption
- easier collection of the total analyte fraction
- more convenient manual procedures
- removal of particulates
- more easily automated

Because SPE is a more efficient separation process than LLE, it is easier to obtain a higher recovery of the analyte. LLE procedures that require several successive extractions to recover 99+% of the analyte often can be replaced by one-step SPE methods. With SPE it is also possible to obtain a more complete removal of interferences from the analyte fraction. Reversed-phase SPE procedures are the most popular because only small amounts of organic solvent are required while maintaining a higher concentration of analyte. There is no need for phase separation (as in LLE), so the total analyte fraction is easily collected in SPE, eliminating errors associated with variable or inaccurately measured extract volumes. In SPE there is no chance of emulsion formation. Finally, larger particulates are trapped by the SPE cartridge and do not pass through into the analyte fraction.

Some disadvantages of SPE compared with LLE include:

- potential variability of SPE packings
- irreversible adsorption of some analytes on SPE cartridges
- more-complex method development is required (up to 4 steps involved, Fig. 16.4)

The solvents used in LLE are usually pure and well defined, so that LLE separations are quite reproducible. Although the cartridges used in the past for SPE sometimes varied from lot to lot, initiatives to improve production quality have led to major improvements in cartridge reproducibility. The surface area of an LLE device (e.g., separatory funnel) is quite small (and less active) compared to an SPE cartridge (with its high-surface-area packing), so irreversible binding of analyte (with lower recoveries) is less likely with LLE vs. SPE.

16.6.1 SPE and HPLC Compared

In its simplest form, SPE employs a small, plastic, disposable column or cartridge, often the barrel of a medical syringe packed with 0.1 to 1.0 g sorbent. The sorbent is commonly a reversed-phase material (e.g., C_{18}-silica) that resembles RPC in its separation characteristics. In the following discussion we will assume reversed-phase SPE (RP-SPE) unless noted otherwise. Although silica-gel-based bonded-phase packings were introduced first, polymeric sorbents have become available in recent years and have been gaining in popularity. Compared to silica-based SPE packings, polymeric packings have several advantages: (1) higher surface area (thus higher capacity), (2) better wetability, (3) tolerance to partial drying after the conditioning step, without affecting recovery and reproducibility, (4) an absence of silanols (less chance of irreversible adsorption of highly basic compounds), and (5) a wide pH range (more flexibility in adjusting conditions).

In its most popular configuration, the SPE packing, is held in a syringe barrel by frits, similar to an HPLC column (Fig. 16.2*a*). The particle size (e.g., 40-μm average) typically is larger than that in HPLC (1.5–5-μm). Because shorter bed lengths, larger particles, and less well-packed beds are used, SPE cartridges are much less efficient than an HPLC column ($N < 100$). For cost reasons, irregularly shaped, type-A-silica packings (rather than spherical, type-B particles; see Section 5.2.2.2) usually are used in SPE. Recently spherical silicas for SPE have come on the market but have not impacted the sale of the most popular products. Polymeric sorbents, which generally are spherical, are more expensive than silica-based packings. Some SPE disks, however, use the more expensive spherical SPE packings with particle diameters in the 7-μm range. Overall, the principles of separation, selection of conditions, and method development are similar for both SPE and HPLC, except that SPE uses a series of isocratic steps during retention and elution of the analyte. One major difference between SPE and HPLC is that the SPE cartridge is usually used once and discarded, since potential interferences can remain on the cartridge, whereas HPLC columns are used many times.

16.6.2 Uses of SPE

SPE is used for six main purposes in sample preparation:

- removal of interferences and "column killers"
- concentration or trace enrichment of the analyte
- desalting
- solvent exchange
- in-situ derivatization
- sample storage and transport

16.6.2.1 Interference Removal

Interferences that overlap analyte peaks in the HPLC separation complicate method development and can adversely affect assay results. In some cases, especially for complex samples (e.g., natural products, protein digests), a large number of interferences in the original sample can make it almost impossible to separate these from one or more analyte peaks by means of a single HPLC separation. SPE can be used

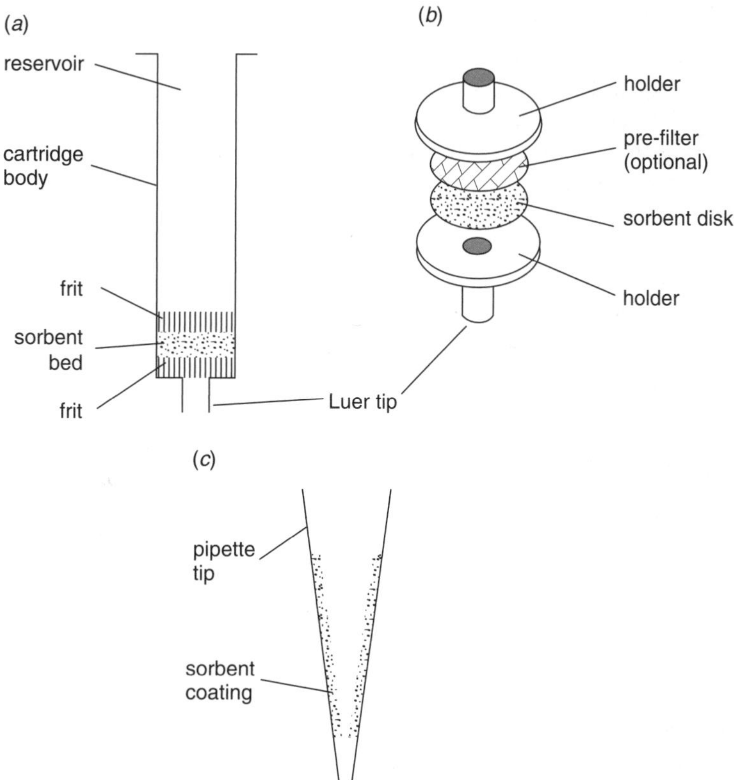

Figure 16.2 Different means for carrying out solid-phase extraction (SPE). (*a*) Disposable cartridge (syringe-barrel format); (*b*) disk; (*c*) micropipette tip (MPT).

to reduce or eliminate those interferences prior to HPLC. Some samples contain components, such as hydrophobic substances (e.g., fats, oils, greases), proteins, polymeric materials, or particulates that can plug or deactivate the HPLC column. These "column killers" often can be removed by RP-SPE.

16.6.2.2 Analyte Enrichment

SPE can be used to increase the concentration of a trace component. If an SPE cartridge can be selected so that $k \gg 1$ for the analyte, a relatively large volume of sample (e.g., several mL) can be applied before the analyte saturates the cartridge and begins to elute from the cartridge. An increase in analyte concentration (*trace enrichment*) can then be achieved, provided that the cartridge is eluted with a small volume of strong solvent ($k > 1$). An example of trace enrichment is the use of SPE to concentrate sub-ppb of polynuclear aromatic hydrocarbons [20] or pesticides [21] from environmental water samples using a RP-SPE cartridge. A strong solvent (e.g., ACN or MeOH) elutes these analytes from the cartridge in a small volume, which saves on evaporation time. The sample can then be redissolved in a solvent compatible with the subsequent HPLC separation. Alternatively, the eluted sample can be diluted directly into a suitable injection solvent.

16.6.2.3 Desalting

RP-SPE can be used to desalt samples, especially prior to ion-exchange chromatography (IEC) where a low-ionic strength sample is desirable. Conditions of pH and %-organic are selected to retain the analyte initially so that the inorganic salts can be washed from the cartridge with water. The analyte can then be eluted (salt free) with organic solvent [22].

16.6.2.4 Other Applications

The remaining applications of SPE—solvent exchange, in-situ derivatization, and sample storage/transport—are either seldom used or are less relevant for HPLC; for details, see [21, 23, 24].

16.6.3 SPE Devices

Several SPE configurations are used (Fig. 16.2):

- cartridge
- disk
- pipette tip
- 96-well plate
- coated fiber or stir bar

16.6.3.1 Cartridges

The most popular SPE configuration is the cartridge. A typical SPE disposable cartridge (syringe-barrel format) is depicted in Figure 16.2*a*. The syringe barrel is usually medical-grade polypropylene that is fitted with a Luer tip, so that a needle can be affixed to direct the effluent to a small container or vial. The frits that hold the particle bed in the cartridge are of made of PTFE, polypropylene, or stainless steel with a porosity of 10 to 20 μm, and thus offer little flow resistance. SPE cartridges may vary in design to fit an automated instrument or robotic system. SPE cartridges are relatively inexpensive, and they are discarded after a single use to avoid sample cross-contamination.

To accommodate a wide range of SPE applications, cartridges are available with packing weights of 35 mg to 2 g, as well as with reservoir volumes (the volume above the packing in the cartridge) of 0.5 to 10 mL. For very large samples, "mega" cartridges contain up to 10 g of packing with a 60-mL reservoir. Cartridges with a larger amount of packing should be used for dirty samples that may overload a low-capacity cartridge. However, cartridges containing 100 mg of packing or less are preferred for relatively clean liquid samples where cartridge capacity is not an issue, as well as for small sample volumes. Because of the higher surface areas of polymeric SPE packings, less packing is needed than for silica-based particles (2- to 60-mg). In most cases it is desirable to collect the analyte in the smallest possible volume, which means that the SPE cartridge generally should also be as small as possible.

16.6.3.2 Disks

The second most popular SPE configuration is the disk (Fig. 16.2*b*). SPE disks combine the advantages of membranes (see below) and solid-phase extraction. In their appearance, the disks closely resemble membrane filters: they are flat, usually ≤1-mm thick, and 4 to 96 mm in diameter. The physical construction of the SPE disks differs from membrane filters. SPE disks can be acquired in any of the following configurations:

- flexible or expanded PTFE networks filled with silica-based or resin packings
- rigid fiberglass disks with embedded packing material
- packing-impregnated polyvinylchloride
- derivatized membranes

Filled Disks. The packing in these disks generally comprise 60-90% of the total membrane weight. Some disks are sold individually and must be installed in a reusable filter holder. Others are sold preloaded in disposable holders or cartridges with Luer fittings for easy connection to syringes.

SPE disks and cartridges differ mainly in their length/diameter ratios (L/d): disks have $L/d < 1$ and cartridges have $L/d > 1$. Compared to SPE cartridges, this characteristic of the disk enables higher flow rates and faster extraction. "Dirty" water or water containing particulates, such as wastewater, can plug the porous disks, just as in the case of cartridges. So a prefilter is used prior to the SPE treatment. Some disk products come with a built-in prefilter. Channeling, which can cause uneven flow through poorly packed cartridges, is not a problem with disks. Due to the thinness of the disk (typically 0.5–2 mm), however, compounds with low k-values tend to have lower breakthrough volumes than for SPE cartridges.

SPE disks are especially useful for environmental applications, such as the analysis of trace organics in surface water, which often require a large sample volume to obtain the necessary sensitivity. The EPA has approved SPE technology as an alternative for large-volume LLE methods [25] in the preparation of water samples for HPLC analysis. Examples of approved methods include procedures for phenols [26], haloacetic acids in drinking water [27], and pesticides and polychlorinated biphenyls (PCBs) [28].

Embedded Disks. Low-bed-mass, rigid fiberglass disks with 1.5 to 30 mg of embedded packing material are useful for pretreating small clinical samples (e.g., plasma or serum; [29]). Their reduced sorbent mass and small volume reduces solvent consumption (and any related sample contamination by solvent impurities). An advantage of this type of disk is an absence of frits that are a possible further source of contamination.

Other Disks. Packing-impregnated polyvinylchloride (PVC) and derivatized membranes are used very little in SPE applications and are not discussed here.

16.6.3.3 Other SPE Formats

The move toward miniaturization in analytical chemistry has prompted the development of new formats for SPE:

- micropipette tip
- 96-well SPE plate
- coated fibers
- stir-bar sorbent extraction

The *micropipette tip* (MPT) format (Fig. 16.2*c*) permits the handling of submicroliter amounts of sample, such as biological fluids. Solid-phase extraction (SPE) has been performed with various packings that are placed in the pipette tip, or embedded in, or coated on the internal walls of the tip. With coatings on or embedded in the internal surface of the tip, liquid samples can be drawn up and expelled without undo pressure drop or plugging. Sample is drawn into the tip, where it interacts with the SPE packing. Next the tip may be rinsed to enhance cleanup, then is eluted with a strong solvent. Many popular SPE techniques have been adapted to MPTs, including reversed-phase-, ion exchange-, hydrophobic interaction-, hydrophilic interaction-, immobilized-metal affinity-, and affinity-chromatography. MPTs have mainly been used for purification, concentration, and selective isolation (e.g., affinity, metal chelation) of proteins and peptides and are an essential tool for MALDI and for other advanced MS techniques [30, 31]. One of the main advantages of micropipette tips is that they can be used with micropipettors or in liquid-handling automation.

The *96-well SPE plate* is another disk format, one that is well suited for automation and the SPE processing of a large number of small samples. In this format, 96 flow-through SPE "wells" of 0.5 to 2.0-mL volume contain small masses of packing (usually <100 mg) contained by small frits or embedded into individual disks. The plate is analogous to a 8×12 array of miniaturized SPE cartridges. The plates can be handled by robotic instrumentation to completely automate the SPE process. Dedicated 96-port evaporation stations can automatically evaporate the elution solvent from collection plates. The evaporated samples then can be reconstituted with a suitable injection solvent and injected directly from the 96-well plate.

Coated fibers are used for solid-phase microextraction (SPME). In this design a fused-silica fiber is coated with a polymeric stationary phase, such as a poly-dimethylsiloxane or a polyacrylate [32, 33]. The fiber is dipped into the solution to be analyzed, and analytes diffuse to and partition into the coating as a function of their distribution coefficients. Once equilibrium is achieved, the fiber is removed from solution and placed in the injection port of an HPLC valve or in an autosampler vial where analytes are displaced with a strong solvent.

Stir-bar sorbent extraction (SBSE) [34] is similar in concept to the use of coated fibers, but the greatly increased surface area allows for greater mass sensitivity. The stir bar, with a polymeric sorptive coating, is placed in an aqueous liquid, and the solution stirred while analyte/matrix partitioning takes place. After equilibrium the stir bar is removed, dried to remove traces of water, and then transferred to a special device where the analytes are displaced into the HPLC column. Both coated-fiber and stir-bar devices are more popular in gas chromatography than HPLC, where thermal desorption is more efficient in volatilizing sorbed analytes into the gas phase than solvent desorption in the liquid phase.

For purposes of brevity, a typical "SPE cartridge" will be assumed in the remainder of Section 16.6. In most cases the other SPE devices mentioned in Sections 16.6.3.2 and 16.6.3.3 will perform in a similar manner.

16.6.4 SPE Apparatus

The equipment needed for SPE can be very simple (Fig. 16.3). Gravity can be used as the driving force, but the flow through the cartridge with "real" samples can be quite slow. Although an interlocking syringe can be used to manually push solvent or sample through the cartridge (Fig. 16.3*a*), this approach can fail for samples that are viscous or which contain particulates, so vacuum-driven flow is preferred. For example, a vacuum flask can be used to handle one cartridge at a time (Fig. 16.3*b*). When several samples must be processed simultaneously, a vacuum-manifold system for processing multiple cartridges at a time is recommended (Fig. 16.3*c*). A removable

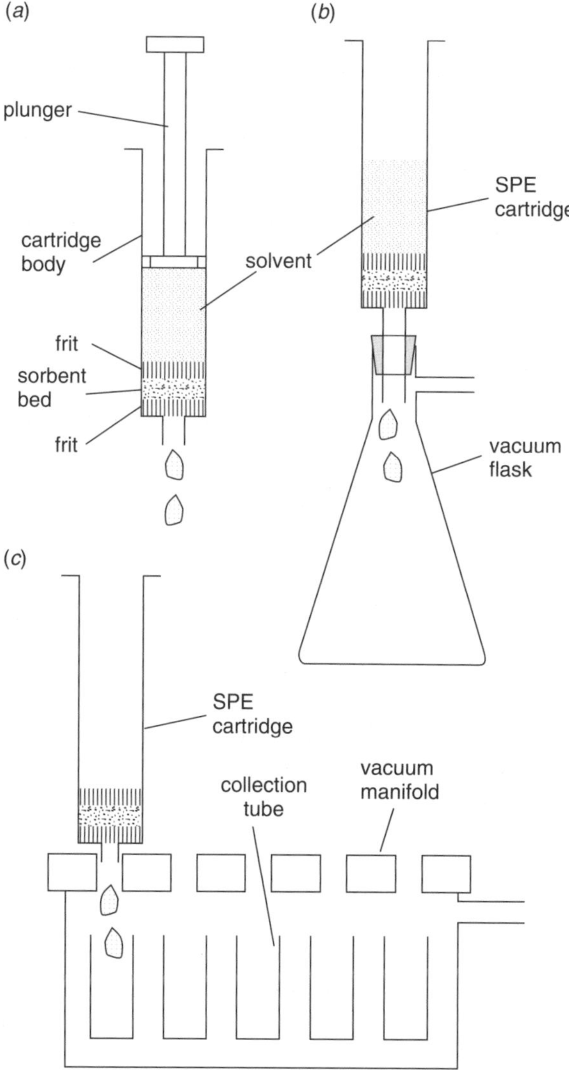

Figure 16.3 Equipment for carrying out solid-phase extraction (SPE). (*a*) Interlocking syringe; (*b*) vacuum bottle; (*c*) vacuum manifold.

rack is located inside the vacuum manifold to hold test tubes for eluant collection. In some units a vacuum bleed-valve, a flow-control valve for each cartridge, and a vacuum gauge are incorporated to allow better control of the solvent flow. Positive-pressure manifold-systems are also available that provide individual flow control for each of the cartridges. As the degree of sophistication increases, so does the price of the apparatus. Centrifugation is used less commonly to drive liquid through the cartridge.

Regardless of the method used to create flow through the SPE cartridge or other SPE device, the flow rate should allow sufficient time for the sample to contact the packing. Higher flow rates also decrease separation efficiency (i.e., the plate number N; Section 2.4). For typical SPE applications, flow rates $\leq$10 mL/min [35] for a cartridge and $\leq$50 mL/min for a 90 mm disk [36] are recommended.

When the number of samples increases to the point where SPE sample preparation becomes the "bottleneck," automation of the entire process becomes attractive. There are three basic approaches to SPE automation:

- dedicated SPE equipment
- modified *xyz* liquid-handling systems
- robotic workstations

The simplest and least expensive instrumentation is a *dedicated SPE device* that performs conditioning, loading, washing, and elution. Such systems may use standard syringe-barrel cartridges, special cartridges that are designed to fit the apparatus, SPE disks, or 96-well plates.

Modified liquid-handling systems are used mainly to perform liquid-handling functions such as dilution, mixing, and internal standard addition.

Robotic systems are the most versatile technique to perform sample preparation functions. Although a robot can be interfaced to devices that perform each of the steps of the SPE procedure, it is more time- and cost-effective to interface the robot to a dedicated SPE workstation. The robot serves to move sample containers to and from the SPE workstation, as well as to and from other sample preparation devices (e.g., balances, mixers, dilutors, autosamplers). Commercial robotics systems are available with different capabilities from many manufacturers (e.g., Beckman-Coulter, Gilson, Hamilton, Tomtec).

16.6.5 SPE Method Development

In its most popular form the application of SPE generally involves four steps (Fig. 16.4):

1. conditioning the packing
2. sample application (loading)
3. washing the packing (removal of interferences)
4. recovery of the analyte

Each of these four steps must be optimized.

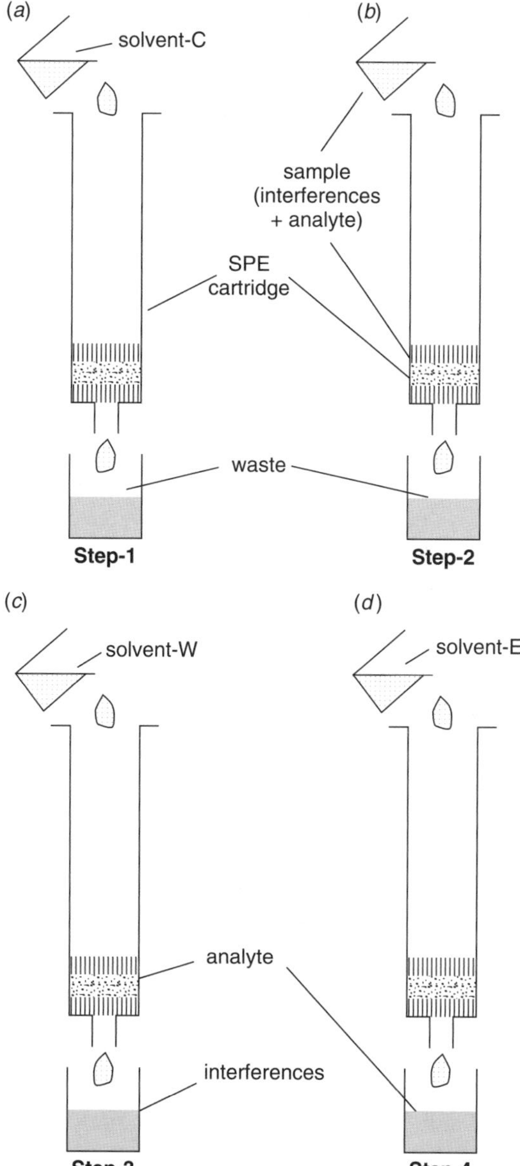

Figure 16.4 Successive steps in the application of solid-phase extraction (SPE). (*a*) Conditioning the packing; (*b*) sample application (loading); (*c*) washing the packing (removal of interferences); (*d*) recovery of the analyte.

16.6.5.1 SPE Steps

In this discussion we assume RP-SPE and initial retention of the analyte; for other phases and more detail for specific products, consult the manufacturer's literature. In *step 1* (Fig. 16.4*a*), performed prior to addition of sample, the packing is "conditioned" by passage through the cartridge of a few bed-volumes of solvent-C—typically methanol (MeOH) or acetonitrile (ACN). The role of the conditioning step is 2-fold: (1) it removes any impurities that may have collected while the cartridge was exposed to the laboratory environment, or are present in the cartridge supplied by the manufacturer, and (2) it allows the sorbent to be solvated. Solvation is important, as reversed-phase silica-based packings (especially C_8, C_{18}, or phenyl) that have been allowed to dry out often exhibit a considerable decrease in sample retention. In addition, varying states of packing dryness lead to nonreproducible analyte recoveries. On the other hand, polymeric packings with a balance of hydrophobic-hydrophilic surface character can dry out slightly and still maintain their performance.

After the packing is conditioned, the excess conditioning solvent should be removed by a flow of air through the cartridge until solvent no longer drips from the bottom of the cartridge (step 1a; not depicted in Fig. 16.4). However, the air flow should not be continued past this point, as this can dry the packing and adversely affect analysis reproducibility (especially with SPE disks; polymeric packings are more forgiving). A preconditioning water-wash is used next to ready the SPE cartridge for introduction of an aqueous sample (step 1b; not depicted in Fig. 16.4). Do not allow too much time (e.g., >5 min) between this water-conditioning step and the sample-loading step. If the packing sits in water too long, the solvating solvent may slowly partition into the water, thereby "de-wetting" the packing (Section 5.4.4.2).

Step 2 (Fig. 16.4*b*) in the SPE procedure involves sample application (loading); the sample, dissolved in a weak solvent (water or buffer with $\leq$10% organic), is added to the cartridge with strong retention of the analyte.

The sample for SPE can be applied with a pipette or syringe, or pumped into the cartridge. The latter method is more convenient for large sample volumes (e.g., >50 mL), such as environmental water samples. The sample and cartridge sizes must be matched so as not to overload the cartridge (see column capacity; Section 15.3.2.1). Remember that the capacity of the cartridge must be sufficient to handle the analytes, matrix, and interferences, all of which may be retained during the loading step.The sample solution should be passed through the cartridge without allowing the cartridge to dry out. The flow rate is not precisely controlled in SPE—as in HPLC—but it can be adjusted by varying the vacuum or the delivery rate from the syringe. Flow rates of 2 to 4 mL/min are usually acceptable.

Step 3 (Fig. 16.4*c*) provides for the removal of interferences by washing the cartridge with a solvent-W of intermediate strength. The wash-solvent strength and volume should be carefully chosen, as too large a volume and/or excessive solvent strength may result in partial elution of the analyte. Optimally, the wash step is discontinued just before analyte begins to leave the cartridge. In this way interferences that are more weakly retained than the analyte are washed from the cartridge, but no loss of analyte occurs. Water or buffer is often used for the wash solvent in RP-SPE, but this may not provide a maximum removal of interferences from the analyte fraction that is collected in step 4 (Fig. 16.4*d*). A small, controlled amount of organic solvent may be added to the wash solution (solvent-W) to aid in the removal of more hydrophobic interferences, but care must be taken that the analyte of interest is not removed at the same time. Because of the variability of the SPE separation from cartridge to cartridge, there must be some safety margin in the optimum wash-solvent strength and volume used to remove interferences from the cartridge. The primary goal is to collect 100% of the analyte in step 4; otherwise, poor and variable recoveries will result.

Step 4 (Fig. 16.4*d*) provides for elution and collection of the analyte fraction. If detection sensitivity is a major concern, then the goal should be the collection of the analyte in as small a volume as possible. This can be achieved with an elution solvent-E that is quite strong, so that $k \approx 0$ for the analyte band during elution. Alternatively, the use of a weaker solvent-E that still provides elution of the analyte (e.g., $k \approx 1$) will minimize the elution of more-strongly retained interferences that are preferentially left on the cartridge (highly desirable if isocratic elution is used during the HPLC separation). Evaporation to dryness is often required, since the elution solvent-E for SPE may be too strong a sample solvent for the HPLC injection. For this reason, choose a solvent-E that is relatively volatile; otherwise, an excessive time for evaporation may be required.

Adjusting the pH of the wash or elution solvent can be an effective way to moderate the retention and/or release of the analyte (e.g., acidic analytes will be more retained at low pH, and less retained at high pH). Fine-tuning the SPE cleanup can be further enhanced by use of "mixed-mode" SPE phases (Section 16.6.5.1). For example, a phase that includes both ion-exchange and RP characteristics can be used to advantage in the cleanup of ionizable analytes. Also, an SPE cleanup that is "orthogonal" to the analytical column (i.e., has different selectivity; Section 6.3.6.2) is likely to result in less overlap of analyte peaks by interferences.

SPE also can be used to retain impurities during the loading step, while allowing analyte(s) of interest to pass through the cartridge unretained. Here the SPE phase is chosen to retain the impurities and interferences, but not the analyte. This option does not provide for any concentration of the analyte in its SPE collected fraction. It is also not possible to separate the analyte from more weakly retained interferences. Therefore this SPE mode usually provides "dirtier" analyte fractions, whereas the procedure of Fig. 16.4 allows the separation of analyte from both weakly and strongly retained sample components. For this reason this procedure is used much less often and will not be discussed further.

16.6.5.2 SPE Packings

Because SPE represents a low-efficiency adaptation of HPLC, many packings used in HPLC are also available for SPE. Table 16.6 lists the more popular SPE packings and the analyte types for which they are suited. Bonded silicas are used most often, but other inorganic and polymeric materials are commercially available. In addition to the generic packings shown in Table 16.6, specialty packings are available for the isolation of drugs of abuse in urine [37–38], aldehydes and ketones from air [9], catecholamines from plasma [39], and many other popular assays. Florisil (activated magnesium silicate) and alumina are not used at present for HPLC, but can be useful for SPE; many published methods exist for the isolation of pesticides using Florisil [40]. The use of graphitized carbon for SPE has been increasing, especially for the removal of chlorophyll-containing plant extracts [41]. For instructions on the use of specific SPE products, consult the manufacturer's literature or one of the textbooks on the subject [7, 42–46].

Due to the nature of electrostatic interactions, ion exchange can be a powerful and selective SPE technique for ionic and ionizable compounds. Cation-exchange packings retain protonated bases and other cations, while anion-exchange packings retain ionized acids and other anions. Ion-exchange packings come in two forms:

Table 16.6

Various SPE Phases and Conditions

Mechanism of Separation	Typical Phases	Structure(s)	Analyte Type	Loading Solvent	Eluting Solvent[a]
Normal phase (adsorption)	Silica, alumina, florisil	–SiOH, AlOH, Mg_2SiO_3	Slightly to moderately polar	Small ε (e.g. hexane, $CHCl_3$/hexane; Fig. 8.6, Table 8.1)	Large ε (e.g. methanol, ethanol; Table 8.1, Fig. 8.6)
Normal phase (polar-bonded phase)	Cyano, amino, diol	–CN, $-NH_2$, –CH(OH)–CH(OH)–	Moderately to strongly polar	Small ε (e.g., hexane; Table 8.1)	Large ε (e.g. methanol, ethanol; Table 8.1)
Reversed phase (nonpolar bonded phase—strongly hydrophobic)	C_{18}, C_8 PS-DVB, DVB (polymeric)	C_{18}, C_8 PS-DVB, DVB	Hydrophobic (strongly nonpolar)	High P' (e.g. H_2O, dilute MeOH/H_2O, ACN/H_2O)	Intermediate P' (e.g., MeOH, ACN)
Reversed phase (nonpolar bonded phase—intermediate hydrophobicity)	Cyclohexyl, phenyl, diphenyl	[cyclohexyl, phenyl and diphenyl structures]	Moderately nonpolar	High P' (e.g., H_2O, dilute MeOH/H_2O, ACN/H_2O)	Intermediate P' (e.g., MeOH, ACN)
Reversed phase (nonpolar bonded phase—low hydrophobicity)	Butyl, ethyl, methyl	$(-CH_2-)_3CH_3$, $-C_2H_5$, $-CH_3$	Slightly polar to moderately nonpolar	High P' (e.g., H_2O)	Intermediate P' (e.g., MeOH, ACN)
Polymeric reversed phase (hydrophobic—hydrophilic balanced)	Polyamide, poly[*n*-vinylpyrrolidone-divinylbenzene(DVB)], methacrylate-DVB	Various polymers	Acidic, basic, neutral	Water or buffer	Intermediate P' (e.g., MeOH, ACN)
Anion exchange (weak)	Amino, 1°, 2°-amino	$(-CH_2-)_3$ NH_2, $(-CH2-)_3$ $NHCH_2CH_2NH_2$	Ionic (ionizable), acidic	Water or buffer ($pH = pK_a + 2$)	A. Buffer ($pH = pK_a - 2$)

					B. pH value where sorbent or analyte is neutral C. Buffer with high ionic strength
Anion exchange (strong)	Quaternary amine	$(-CH_2-)_3N^+(CH_3)_3$	Ionic (ionizable), acidic	Water or buffer ($pH = pK_a + 2$)	A. Buffer ($pH = pK_a - 2$) B. pH value where analyte is neutral C. Buffer with high ionic strength
Cation exchange (weak)	Carboxylic acid	$(-CH_2-)_3COOH$	Ionic (ionizable), basic	Water or buffer ($pH = pK_a - 2$)	A. Buffer ($pH = pK_a + 2$) B. pH where sorbent or analyte is neutral C. Buffer with high ionic strength[a]
Cation Exchange (Strong)	Alkyl sulfonic acid, aromatic sulfonic acid	$(-CH_2-)_3SO_3H$, $C_6H_5-SO_3H$	Ionic (ionizable), basic	Water or buffer ($pH = pK_a - 2$)	A. Buffer ($pH = pK_a + 2$) B. pH value where analyte is neutral C. Buffer with high ionic strength[a]

[a]For ion exchange, three possible elution conditions exist: A, buffer 2 units above (acids) or below (bases) pKa of analyte; B, pH where either analyte or sorbent (weak exchangers) is neutral; C, high ionic strength.

"strong" and "weak;" strong ion-exchangers are normally preferred if strong retention of the analyte is the main objective. Ionization (and thus retention) of weak ion-exchangers is a function of pH (Fig. 13.16). The choice of pH is a compromise between maintaining the ionic character of the stationary phase, and ensuring that the ionic analyte is remains in an ionic state. Thus pH becomes a powerful variable for both optimizing retention and releasing the analyte from a weak ion-exchanger.

SPE cartridge packings are generally of lower quality and cost than corresponding HPLC packings, and this contributes to the problem of batch-to-batch retention variability. Whereas high-purity type B column packings are preferred in RPC (Section 5.2.2.2), RP-SPE packings will generally be more "acidic" (type A); their silanol interactions will tend to be more pronounced and more variable from lot to lot. However, because SPE is usually practiced as an "on–off" technique, small differences in retention should be less important than in HPLC, where small differences in selectivity can be more important.

16.6.6 Example of SPE Method Development: Isolation of Albuterol from Human Plasma

The isolation of albuterol (I) will be used to illustrate a typical SPE application [47]. This drug is widely employed as a bronchodilator in the treatment of asthma:

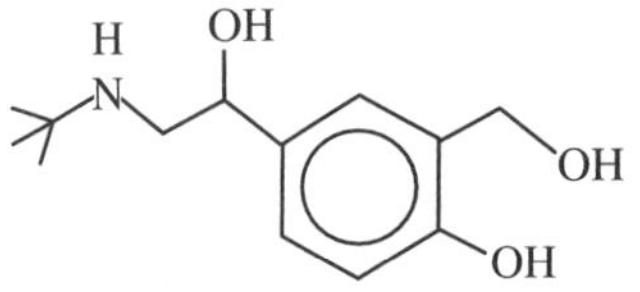

H OH
N
OH
OH

(I)

Albuterol (molecular weight 239 Da) is a polar, hydrophilic compound with two ionizable functional groups: a phenol (pK_a-9.4) and a secondary amine (pK_a-10.0). In aqueous solution it exists primarily in an ionic state at any pH. For these reasons albuterol partitions poorly into organic solvents from aqueous solutions. Albuterol possesses several polar and nonpolar functionalities that might be exploited for SPE retention. Any of five different modes (reversed-phase, cation exchange, anion exchange, normal-phase, or affinity) might be expected to retain the drug. A trial-and-error investigation was carried out with these five modes to find an SPE wash-solvent that would best remove interferences from the cartridge without releasing the analyte. A series of 17 different SPE phases from these five modes were scouted for best recovery with 23 solvent systems. Certain eluting solvents did not elute albuterol appreciably from some of the SPE cartridges, and these solvents were noted for possible use as wash solvents in step 3 (Fig. 16.4*c*).

Following the scouting experiments, four SPE cartridges (Table 16.7) were selected for further investigation. These phases appeared initially promising, with extracts showing low levels of endogenous plasma material, good HPLC system compatibility, and reasonable recoveries of albuterol from plasma. Two SPE phases (cyano and silica) proved acceptable, with the final method shown in Figure 16.5. In this method, albuterol was strongly retained during the rinse steps (Fig. 16.4*c*; steps

Table 16.7

SPE Results on Recovery of Albuterol from Plasma

SPE Cartridge Type	Elution Solvent	Percent Recovery	Comments
Cyano	19% 1M NH_4Oac + 90% MeOH	89	Clean extract; small volume; acceptable results
Silica	Same as above	94	Clean extract; small volume; acceptable results
Phenylboronate phase	0.1 M H_2SO_4	90	Clean extract; small volume but elution solvent too acidic for the HPLC system; unacceptable
C_{18}	Isopropanol	92	Extract not clean enough; trace enrichment not reliable; unacceptable

4 and 5 of Fig. 16.5); 0.5% of 1 M NH_4OAc in MeOH was required for elution (Fig. 16.4*d*; step 6 of Fig. 16.5).

16.6.7 Special Topics in SPE

16.6.7.1 Multimodal and Mixed-Phase Extractions

Most SPE procedures involve the use of a single separation mode (e.g., reversed phase) and a single SPE device (e.g., cartridge). However, when more than one type of analyte is of interest, or if additional selectivity is required for the removal of interferences, multimodal SPE can prove useful. Multimodal SPE refers to the intentional use of two (or more) sequential separation modes or cartridges (e.g., reversed phase and ion exchange). Experimentally, there are two approaches to multimodal SPE. In the serial approach, two (or more) SPE cartridges are connected in series. Thus, for the separate isolation of acids, strong bases, and neutrals, an anion- and cation-exchange cartridge could be connected in series. By adjusting the sample and wash-solvent to pH-7, both the acids and bases will be fully ionized. As a result the acids will be retained on the anion-exchange cartridge, the bases will be retained on the cation-exchange column, and the neutrals will pass through both columns (separated from acids and bases). The acids and bases can then be separately collected from each cartridge.

A second approach to multimodal SPE uses mixed phases. Here a single cartridge might possess two (or more) functional groups to retain multiple species, or to provide a unique selectivity. One popular application of multimodal SPE is the isolation of drugs of abuse and other pharmaceuticals from biological fluids [48]. Still another version of multimodal SPE is the use of layered packings [49], where two (or more) different packings are used to isolate differing molecular species.

16.6.7.2 Restricted Access Media (RAM)

RAM are a special class of SPE packings used for the direct injection of biological fluids such as plasma, serum or blood. Unlike SPE cartridges, RAM are actually HPLC

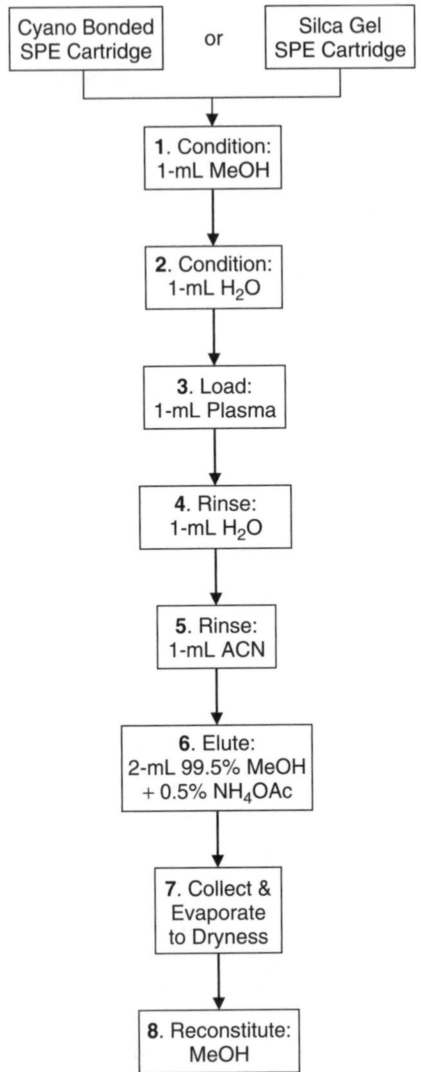

Figure 16.5 Isolation of albuterol from human plasma by means of solid-phase extraction (SPE).

columns that incorporate sample-preparation, and—unlike SPE cartridges—are characterized by high plate numbers and re-usability. RAM are most often selected for the analysis of low-molecular-weight drugs, their impurities, and metabolites [50–52]. Many variations of these packings have been described: (1) internal-surface reversed-phases, (2) shielded hydrophobic phases, (3) semi-permeable surfaces, (4) dual-zone phases, and (5) mixed functional phases. See [50] for a description and tabulation of commercial products.

Dual-mode porous packings are used in the most popular RAM columns. These packings are used typically for the analysis of drugs in blood, because proteins present in these samples will accumulate on a RPC column—leading to its failure after a few injections. The packing consists of small-pore particles with a C_8 or C_{18} layer covering the inside of the pores, and a nonretentive hydrophilic layer covering the exterior of the particle. Proteins are unable to access the pores because of their large size, and are unretained by the particle exterior; consequently they pass through the column with $k \approx 0$. Small-molecule analytes can enter the pores and are retained sufficiently to elute after the proteins (often using gradient elution). Because proteins are not retained on these columns, a column can be used for a considerable number of samples. However, some care must be exercised in the choice of mobile-phase pH and organic solvent; otherwise protein precipitation can occur—with fouling of the column.

Alternatively, a RAM column can be connected via a switching valve to a conventional RPC column (two-dimensional separation; column-switching, Section 16.9). The valve is initially positioned for elution of the RAM column to waste, and the sample is injected; after the proteins leave the column, the valve is switched to connect the two columns. Analytes are then eluted from the RAM column and enter the RPC column for further processing, usually by means of gradient elution. The RPC column also has a longer lifetime, as plasma proteins never contact this column (e.g., [53]).

16.6.7.3 Molecular-Imprinted Polymers (MIPs)

MIPs are among the most selective phases used in SPE, being designed for enhanced retention of a specific analyte. A MIP is a stable polymer with recognition sites that are adapted to the three-dimensional shape and functionalities of an analyte of interest (much like antibody binding). The most common approach involves noncovalent imprinting; this MIP synthesis is shown schematically in Figure 16.6. An analyte is used as a template, and is chemically coupled with a monomer (most often methacrylic acid or methacrylate; Fig. 16.6*a*,*b*). After polymerization (Fig. 16.6*c*), the bound analyte is cleaved to yield a selective binding site (receptor; Fig. 16.6*d*). The selective interactions between the analyte and the MIP include hydrogen bonding, ionic, and/or hydrophobic interactions. The action of a MIP is based on a "lock-and-key" fit, where a selective receptor or cavity on the surface of a polymer perfectly fits the analyte that was used to prepare the MIP. The concept is similar to immunoaffinity (IA) SPE phases (Section 16.6.7.4), but obtaining a suitable antibody for these IA sorbents can be very time-consuming. An introductory article [54] outlines the basics of MIP technology, while review articles [55–58] and a book [59] provide detailed information on the use and potential of MIPs in SPE.

Incomplete removal of analyte template from the MIP during its preparation is one of the main problems. This residual analyte frequently bleeds, resulting in baseline drift and interference with the assay of the desired analyte—especially for low analyte concentrations. There may be some swelling or shrinkage of the MIP with a change in solvent, which can modify the size of the receptor and reduce the retention of the target analyte. A major disadvantage of the MIP approach is that each sorbent must be custom made, either by the user or an outside supplier. Because of the high cost of synthesizing a MIP, their use is restricted to high-volume assays or when there is no other way to perform sample cleanup. Recently several off-the-shelf MIPs have been commercialized (by MTP Technologies, Lund, Sweden) for:

- clenbuterol in biological fluids

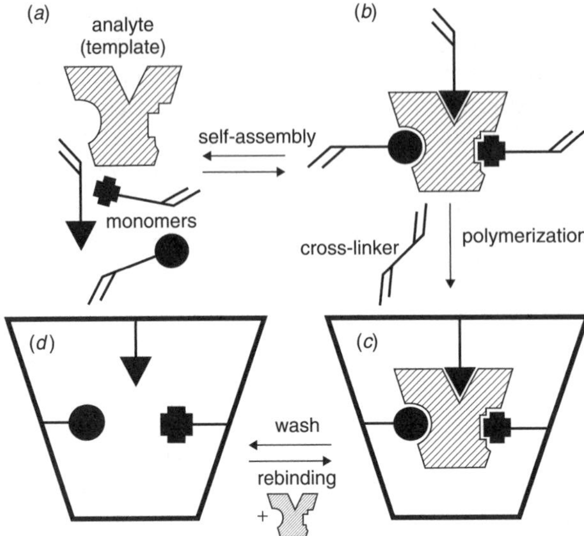

Figure 16.6 Synthesis of molecular-imprinted polymer (MIP). (*a*) Analyte plus monomers; (*b*) formation of analyte-monomer complex; (*c*) analyte-polymer complex; (*d*) selective-binding site.

- beta agonists, multi-residue extractions in urine and tissue samples
- NNAL (4-methylnitrosamino-1-(3-pyridyl)-1-butanol), tobacco-specific nitrosamine in biological matrices
- riboflavin (vitamin B2) in aqueous samples
- triazines, multi-residue extraction in water, soil, and food products
- chloramphenicol, antibiotic in biological matrices
- beta blockers, multi-residue extractions in water and biological samples

16.6.7.4 Immunoaffinity Extraction of Small Molecules

Immunoaffinity packings are based on antibodies that are attached to a particle. As in the case of MIPs, the analyte is retained by a receptor that is highly complementary, so as to provide a "lock-and-key" fit. As a result immunoaffinity packings are quite specific for an individual analyte, and are used for the selective extraction and concentration of individual compounds or classes of compounds from the sample—often in a one-step process. Antibodies for large biomolecules are readily available and have been used for many years in immunology and medical research (affinity chromatography). Because antibodies for small molecules are more difficult to obtain, the development of small-molecule immunoaffinity extraction is more recent and less developed. Some excellent review articles describe immunoaffinity extractions in more detail [60–64].

As long as an antibody can be prepared, the numbers of immunoaffinity packings can be almost unlimited. However, a great deal of time and effort is required in their production, so their use is restricted for the same reasons as for MIP packings (Section 16.6.7.3). Nevertheless, several commercial immunoaffinity packings have become available. Class-specific packings are available for a variety of pharmaceutical, food, and environmental applications [65]. As an example of the use of an immunoaffinity packing, a procedure for the separation of four aflatoxins as a group has been reported [66].

16.6.7.5 QuEChERS and Dispersive SPE

QuEChERS (Quick, Easy, Cheap, Effective, Rugged, and Safe) is an extraction technique for the sample preparation of pesticides in high-moisture samples such as vegetables [67]. The technique uses simple glassware and a minimal amount of organic solvent, followed by the successive addition of salt plus buffer and a SPE packing (Fig. 16.7). The initial addition of a hydrophilic solvent such as acetonitrile or acetone to a homogenized portion of a vegetable sample allows the extraction of pesticides into the solvent (step 2 of Fig. 16.7). Subsequent addition of salt (step 3) leads to separation of the organic solvent from water associated with the sample, and promotes extraction of pesticides into the organic solvent. The internal standard is added next (step 4), and after shaking and centrifugation, an aliquot of the organic phase is subjected to further cleanup using dispersive SPE: the addition of small amounts of bulk SPE packing (e.g., C_{18}, graphitized carbon, amino) to the extract for the purpose of removing interferences from the extract (step 5). After sample cleanup the supernatant is sampled and analyzed (step 7). Step 6 is an optional step for pesticides that are unstable at intermediate pH values.

QuEChERS has been found particularly useful for screening the food supply for multiple pesticides. Official QuEChERS methods from American Association of Official Analytical Chemists (AOAC). Method 2007.01 and the DIN-adopted European Standard Method EN 15662 are now available. QuEChERS has been investigated for over 500 pesticides in a variety of fruit and vegetable matrices [63, 65, 68]; analyte recoveries (for concentrations of 100 ng/g) generally range from 70 to 110%, with a variability of less than 10%. The technique has been extended to new matrices such as meat and fish products, as well as analytes such as antibiotics and other drugs [69–70].

16.6.7.6 Class-Specific SPE Cartridges

Over the years, specialty phases have been introduced that are compound- or class-specific. While MIP and immunoaffinity packings (Sections 16.6.7.3, 16.6.7.4) can provide extreme specificity or selectivity, specialty packings with special functional groups can selectively interact with certain compound classes. Under basic conditions immobilized phenylboronic acid (PBA) selectively binds analytes that possess vicinol diols (e.g., sugars and catechols). Other compounds that are also selectively retained include alpha-hydroxy acids, aromatic *o*-hydroxy acids and amides, and aminoalcohol-containing compounds. Covalent bonds between packing and analyte are formed, allowing interfering compounds to be washed from the packing with a variety of different solvents. Once washed, the covalent bonds can be broken by washing the phase with an acidic buffer/solvent that hydrolyzes the covalent bonds. A popular application of the PBA phases is the isolation of catecholamines in biological fluids [41].

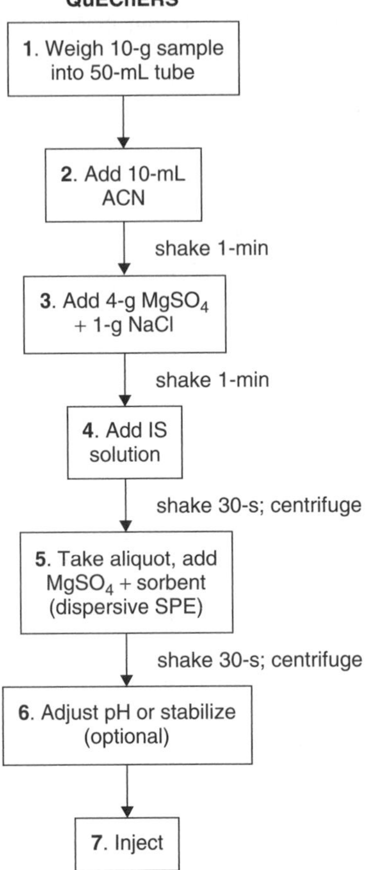

Figure 16.7 Application of QuEChERS (Quick, Easy, Cheap, Effective, Rugged, and Safe) extraction of pesticides from high-moisture vegetable samples.

16.7 MEMBRANE TECHNIQUES IN SAMPLE PREPARATION

With the exception of filtration- and SPE-membranes, membrane separation techniques have not been widely used for HPLC sample preparation. Microporous, semi-permeable membranes permit selective filtration because of the size of their micropores. Compared to SPE or LLE, membrane separations are slower and less likely to increase the analyte concentration by orders of magnitude. A successful application of membrane separation requires either removal of analyte from the acceptor side by trapping (trace enrichment) or by a change in the chemical state of the analyte (e.g., change from a charged to an uncharged species). An advantage of membrane separation techniques for RP-HPLC analysis is that both the donor and acceptor liquids are usually water or buffer. Analytes can also be moved across a membrane by chemical or electrochemical gradients. Membrane separations can be carried out in a static system or in a flowing system, with the latter more amenable to automation. Ultrafiltration, reverse osmosis, dialysis, microdialysis, and electrodialysis are examples of techniques that use membranes for concentration, purification, and separation of analytes. For more information on membrane separations, see [71].

16.8 SAMPLE PREPARATION METHODS FOR SOLID SAMPLES

A sample must be in a liquid state prior to HPLC analysis. Some insoluble solids contain soluble analytes, such as additives in a solid polymer, fats in food, and polyaromatic hydrocarbons in soil. Contacting the sample with solvent allows the extraction of analytes into the solvent. The solvent is separated from the solid residue by decanting, filtration, or centrifugation, and the filtrate is further treated, if necessary, prior to HPLC analysis. Tables 16.8 and 16.9 summarize some techniques used for the extraction ("leaching") of soluble analytes from an insoluble solid matrix.

Table 16.8

Traditional Methods for Sample Preparation of Solid Samples

Sample Preparation Method	Principles of Technique	Comments
Solid–liquid extraction	Sample and solvent are placed in a stoppered container and agitated; the resulting solution is separated from the solid by filtration (sometimes called "shake-flask" method).	Solvent is sometimes boiled or refluxed to improve solubility; sample should be in a finely divided state to aid the leaching process.
Soxhlet extraction	Sample is placed in an extraction thimble; refluxing solvent flows through the thimble and dissolves analytes that are continuously collected in a boiling flask (Sections 16.8.1, 16.8.2.1).	Extraction occurs by contact with the pure solvent; sample must be stable at the boiling point of the solvent; a slow process, but extraction is carried out unattended until complete; excellent recoveries (used as standard to which other solid extraction methods are compared).
Homogenization[a]	Sample and solvent are homogenized in a blender or mechanical homogenizer to a finely divided state; solvent is removed for further workup.	Used for plant and animal tissue, food, environmental samples; organic or aqueous solvent can be used; dry ice or diatomaceous earth can be added to make sample flow more freely.
Sonication[a]	Use of ultrasound to create vigorous agitation at the surface of a finely divided solid material; either a sonotrode probe or ultrasonic bath can be used.	Heat can be added to increase rate of extraction; safe; rapid; best for coarse, granular materials.
Dissolution[a]	Sample is treated with dissolving solvent and taken directly into solution with or without chemical change.	Some samples may not dissolve directly (e.g., digestion or other pretreatment needed); filtration may be required after dissolution.

[a]Not discussed in text.

Table 16.9

Modern Extraction Methods for Solid Samples

Method of Sample Pretreatment	Principles of Technique	Comments
Supercritical fluid extraction (SFE)	Sample is placed in a flow-through container, and supercritical fluid (e.g., CO_2) is passed through the sample; after depressurization, the extracted analyte is collected (Section 16.8.2.2).	To affect the "polarity" of SFE fluid, density can be varied and solvent modifiers added.
Pressurized fluid extraction (PFE)/accelerated solvent extraction (ASE)	Sample and extraction solvent are placed in a sealed container and heated above the solvent's boiling point, causing pressure in the vessel to rise; extracted sample is removed and transferred to a vial for further treatment (Section 16.8.2.3).	Greatly increases speed of liquid–solid extraction process; may be automated; extracted sample is diluted and requires further concentration.
Automated Soxhlet extraction (ASE)	Combines hot solvent leaching and Soxhlet extraction; sample thimble is first immersed in boiling solvent, then thimble is raised for conventional Soxhlet extraction/rinsing (Section 16.8.2.1).	Uses less solvent than traditional Soxhlet; decreased extraction time due to the two-step process.
Microwave-assisted extraction (MAE)	Sample is placed in an open or closed container and heated by microwave energy (Section 16.8.2.4).	Closed container MAE allows use of pressure to improve extraction.
Matrix solid-phase dispersion (MSPD)	Bonded-phase support is used as an abrasive to produce disruption of sample-matrix architecture and as a "bound" solvent to aid complete sample disruption during the sample blending process.	Solid or viscous sample ($\approx$0.5 g) is homogenized in a mortar, or other suitable container, with about 2 g of SPE sorbent (e.g., C_{18}). Blend is transferred to an empty column and analytes are eluted with solvent.

16.8.1 Traditional Extraction Methods

No one solvent-extraction technique can be used for all samples. Table 16.8 lists several traditional methods for the pretreatment of solid samples. *Soxhlet extraction* has been used for more than a hundred years, is time tested, and accepted by most scientists. Regulatory agencies such as the EPA, the Food and Drug Administration (FDA), and their equivalents in other countries readily approve these classical methods of extracting solid samples. However, older methods often use large amounts of organic solvent, which has led to miniaturization in recent years. As the oldest form of efficient extraction, Soxhlet extraction is the accepted standard for comparison with newer extraction technologies for solid samples.

Solvent extraction can assume many forms. The *shake-flask* method, whereby a solvent is added to the sample followed by agitation, works well when the analyte is highly soluble in the extraction solvent and the sample is quite porous. For fast extraction, the sample should be finely divided. Heating or refluxing the sample-plus-solvent can speed extraction. For faster and more complete extraction, ultrasonic agitation (sonication) often allows more effective solid–liquid contact, while providing gentle heating that further aids extraction. *Sonication* is a recommended procedure for the pretreatment of many solid environmental samples, such as US EPA Method 3550 [72] for extracting non-volatile and semi-volatile organic compounds from solids such as soils, sludges, and wastes. For this method different extraction solvents and sonication conditions are recommended, depending on the type of analytes and their concentration in the solid matrix.

Soxhlet extraction has been the most widely used method for the extraction of solids. In this procedure the solid sample is placed in a Soxhlet thimble (a disposable porous container made of stiffened filter paper), and the thimble is placed in the Soxhlet apparatus (Fig. 16.8). The extraction solvent is refluxed; it subsequently condenses into the thimble and extracts the soluble analytes. The apparatus is designed to siphon the extract each time the chamber holding the thimble fills with solution. The siphoned solution containing the dissolved analytes returns to the boiling flask, and the process is repeated until the analyte has been transferred from the solid sample to the flask. Because the sample is contained in the boiling extraction solvent, it must be stable at elevated temperature. Only clean (distilled) warm solvent is used to extract the solid in the thimble, which increases the efficiency of the extraction vs. the simple shake-flask method—a key advantage of classical Soxhlet extraction. Method development consists of finding a volatile solvent (e.g., boiling point $<100^{\circ}C$) that has a high solubility for the analyte and a low solubility for the solid sample matrix.

Soxhlet extractions are usually slow (12–24 hr, or more), but the process takes place unattended. Modern Soxhlet extractors can speed up extraction 8- to 10-fold as described below (Section 16.8.2.1). The most common extractors use hundreds of milliliters of very pure (and expensive!) solvent, but small-volume extractors and thimbles are available for mg-size samples. The analyte concentration, the necessary mass to obtain a representative sample, and the chromatographic detector sensitivity, together determine the required sample size.

16.8.2 Modern Methods for Extracting Solids

Table 16.9 lists several modern methods used for extracting solid samples.

16.8.2.1 Modern Soxhlet Extraction

Classical Soxhlet extraction has been improved to reduce extraction time 8- to 10-fold [73]. An automated apparatus lowers and raises the sample thimble, so as to either place the thimble in the boiling solvent or raise the thimble for conventional Soxhlet extraction. The sample is first totally immersed in the boiling solvent, where the higher temperature of the boiling solvent speeds up extraction of the analyte by

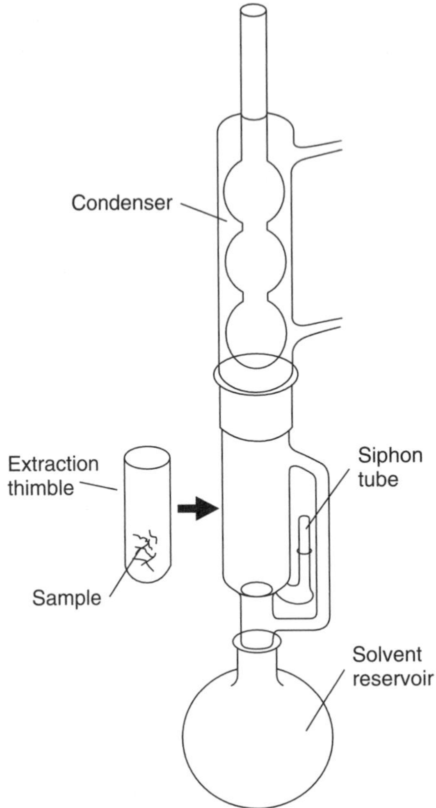

Figure 16.8 Apparatus for Soxhlet extraction.

increasing both its solubility and diffusion rate. Then, to flush residual extract from the sample, the thimble is raised above the boiling solvent, and conventional Soxhlet extraction proceeds. Finally, the solvent is boiled off to concentrate the analyte. Automated Soxhlet extraction is approved by the EPA for the extraction of organic analytes from soil, sediment, sludges, and waste solids (Method 3541 [74]).

Further improvements in the above procedure for faster Soxhlet extraction have been proposed. Focused microwave-assisted Soxhlet extraction can reduce extraction time further for environmental solid samples, by using microwave absorbing solvents [75, 76]. The use of water as an extraction solvent makes the process much more environmentally friendly [77].

16.8.2.2 Supercritical Fluid Extraction (SFE)

When first introduced in the early 1990s, supercritical fluid extraction (SFE) was thought to be preferred for all solid-sample extraction problems. SFE was viewed as a safe, solvent-free technique that allowed easy removal of the carbon dioxide used for analyte solubilization. Moreover SFE was expected to allow faster, more complete extraction of diverse samples such as soil, sand, sludge, fly ash, foods, and polymers. New applications of SFE are still being reported, but no new instruments or major accessories for existing instruments have been commercialized since 2000. While SFE is an important process for industrial-scale purification, its analytical applications (including sample pretreatment) are today relatively unimportant. Numerous reasons can be cited for the continuing decrease in SFE for sample pretreatment, but these are outside the scope of the present book. For details, on the use of SFE, see [78–82].

16.8.2.3 Pressurized Fluid-Extraction (PFE)/Accelerated Solvent Extraction (ASE)

Originally introduced by Dionex (Sunnyvale, CA) as ASE, the technique was generalized to PFE by the EPA. PFE is now well accepted as a modern alternative to Soxhlet extraction; it achieves analyte recoveries equivalent to those from Soxhlet extraction, but with less solvent and in only 10 to 20 minutes. Sample weights of 1 to 30 g are possible, with extraction cells that range in size from 1 to 100 mL. PFE uses the same solvents as classical Soxhlet and sonication methods, and method development is therefore easy. The finely divided sample is placed in an extraction cell located in an oven, and a pump transfers solvent from one or more reservoirs into the extraction cell. Only one sample at a time can be extracted, but an autosampler that can handle up to 24 samples for unattended operation is available.

Extractions can be performed in a static and/or dynamic mode. Once the extraction is completed in the static mode, a nitrogen purge is used to transfer the extract to a collection vial. In the PFE process the sample is contained in a large volume of solvent, so an additional step is needed to concentrate the analyte of interest (e.g., RPC, evaporation). The EPA method 3545A [83], entitled "Pressurized Fluid Extraction" provides for the extraction of water-insoluble or slightly water-soluble organic compounds from soils, clays, sediments, sludges, and waste solids. This method is applicable to the extraction of semi-volatile organic compounds such as pesticides, herbicides, polychlorinated biphenyls (PCBs), and dioxins. PFE, sometimes referred to as pressurized solvent extraction (PSE), uses elevated temperatures (100–180°C) and pressures (1500–2000 psi).

An example of PFE is the extraction of active ingredients from natural-product pharmaceuticals [84] at 100°C and 1500 psi. Samples, in capsule form, are opened, the ground material is collected and 1 to 3 g are loaded into the extraction cells. Only $\approx$15 mL of the extraction solvent, ACN, is used for each sample, with a 14-minute cycle time per sample. Analysis is performed by RPC, with external calibration. Results (3–4% CV) are comparable to those obtained by Soxhlet and sonication, but were faster, used less solvent and were fully automated.

16.8.2.4 Microwave-Assisted Solvent Extraction (MAE)

Microwave-assisted extraction (sometimes referred to an microwave-accelerated solvent extraction) of organic analytes [85] is an alternative to the traditional liquid-solid extractions described earlier. Microwave heating is advantageous in that the extraction solvent is heated internally rather than by convective heating. The temperature in the extraction container is determined by the solvent used for extraction (not by the set point of an external oven or heater), and as the temperature of the solvent in the closed container rises, the pressure also increases. As a result extraction takes place faster than with convection heating.

Water has a high dielectric constant and readily absorbs microwave energy, but organic solvents, such as hydrocarbons, do not absorb microwave energy and therefore are not heated. Microwave extractions with these non–microwave-absorbing organic solvents requires indirect heating via the placement of a microwave-absorbing device in the extraction vessel. These devices are usually rods or a powder of a microwave-absorbing material such as carbon black coated with an inert polymeric substance (e.g., PTFE) that will not contaminate the sample. Multiple samples can be simultaneously processed in a microwave oven, so the sample throughput can be very high compared to some of the other liquid–solid extraction techniques. Commercial systems have the capacity to treat up to 24 extraction vessels.

Microwave extraction is a widely accepted alternative to Soxhlet, PFE/ASE, and other liquid–solid extraction techniques. For example, EPA Method 3546 [86] is a MAE procedure for extracting water-insoluble or slightly water-soluble organic compounds (pesticides, herbicides, PCBs, etc.) from soils, clays, sediments, sludges, and solid wastes. Extraction takes place at 100 to 115°C and 50 to 175 psi in a closed vessel containing the sample and organic solvent(s). Analyte recoveries are equivalent to those from Soxhlet extraction (Method 3540 [87]), but use less solvent and take significantly less time.

16.9 COLUMN-SWITCHING

Column-switching was discussed in Sections 2.7.6, 3.6.4.1, 9.3.10, and 13.4.5.3. It is a powerful technique for sample preparation (discussed here), as well as for two-dimensional separation (Section 9.3.10). For sample preparation, a portion of the chromatogram from an initial column (column 1; e.g., the enrichment column of Fig. 3.23) is selectively transferred to a second column (column 2; e.g., the analytical column of Fig. 3.23) for further separation. Column-switching for sample preparation is used for:

- removal of "column killers" prior to column 2
- removal of late-eluters prior to column 2
- removal of interferences that can overlap analyte bands in column 2
- trace enrichment

The achievement of one or more of these goals often results in increased sample throughput due to the presentation of a cleaner sample to the column. Unwanted interferences can be directed to waste or backflushed to prevent their transfer to the HPLC column. The goal of column-switching is separation of the analyte from interfering compounds by the initial column; that is, the same goal as for SPE.

While column-switching is similar to the HPLC analysis of fractions provided by SPE, several advantages exist:

- SPE cartridges are used only once and discarded; the initial column in column-switching is used repeatedly
- the initial column has a higher efficiency (e.g., 5-μm d_p) compared to an SPE cartridge (e.g., 40-μm d_p)
- less sample loss occurs with column-switching
- many valve configurations are possible for heart cutting, backflushing, diverting contaminants directly to waste, and so forth.

See Section 3.6.4.1 and Figure 3.23 for further discussion of column-switching and an example for trace enrichment.

16.10 SAMPLE PREPARATION FOR BIOCHROMATOGRAPHY

In the separation of biomolecules, sample preparation almost always involves the use of one or more pretreatment techniques. As is the case for HPLC of other sample types, no one sample-preparation technique can be applied to all biological samples. The sample preparation approaches used in modern biochromatography are often the same techniques that were used in classical biochemistry, such as dialysis, chemical precipitation, column chromatography, and centrifugation. Currently there is a growing interest in not only the application of these classical approaches but also newer sample preparation technologies to the fields of molecular biology, biotechnology, and the various "omics" (e.g., proteomics, genomics, metabolomics). In these areas the samples are often complex, are available only in small quantities, and require the utmost care in handling. The requirements for the recovery of biopolymers with structural and functional integrity often demand that the sample preparation be rapid and gentle.

The complex nature of most biological samples necessitates some form of preliminary sample manipulation to achieve better separation results and to prolong the life of the HPLC column. The actual recommended sample preparation technique(s) will depend on the nature of the sample (e.g., molecular weight, presence of additives, endogenous interferences, particulates, or other unwanted components).

In addition to filtration for particulate removal, chromatographic principles (including affinity) can be used to clean up biological samples. Table 16.10 provides a listing of sample preparation techniques that may be used in the flow-through mode using cartridge, disk, or column format. Some of these techniques can be performed in a batch mode, where the media is poured into the sample in a liquid form, allowed to stay in contact usually with agitation, and then is removed by filtration or by pouring off the liquid phase leaving behind the compound(s) of interest sorbed onto the stationary phase or contained in the liquid phase. Although slower than the column approach, batch sorption is easier to perform.

The flow-through format is more widely used. Although dilution is a possible consequence, the flow-through column approach is more useful for removing the last traces of the analyte of interest. Convenient, pre-packed cartridges and membrane disks, that offer less flow resistance due to their large cross-sectional areas, are readily available from manufacturers. Sometimes kits are purchased that contain all the media, chemicals and accessories necessary to perform a cleanup process. Liquids can be transported through the flow-through devices with applied pressure, vacuum, or centrifugation. Many of the sample preparation techniques of Table 16.10 use retardation of ionic species by means of ion exchange or ion retardation. Other procedures use hydrophobic interaction and adsorption to retain macromolecules, while letting ionic compounds and smaller molecules pass through.

Besides the chromatographic principles covered in Table 16.10, cleanup of biological samples while maintaining biological activity can be accomplished using other approaches. For example, dialysis is a time-tested process using membranes to clean up and desalt biological samples. Miniaturized dialysis kits permit the

Table 16.10

Sample-Preparation Techniques in Biochromatography

Requirement	Most Frequently Used Approaches	Species Retained	Typical Applications
Antibody purification	Affinity chromatography; hydroxylapatite chromatography	IgG and subclasses	IgG concentration in serum, ascites, and tissue culture media; fluorescent labeled antibodies with unreacted fluorescent tag.
Buffer and reagent ultrapurification	Ion exchange	Trace cations and anions	Removal of ions that cause band broadening or high background.
	Adsorption	Trace organics	Neutral PS-DVB, alumina, and silica will remove polar organics from buffers; water can be removed from organic solvents.
Deionization	Mixed-bed ion exchange	Ions	Deionization of carbohydrates before HPLC; separation of ionic contaminants from proteins; reagent preparation; separation of anions from carbohydrates, dextrans, and polyhydric alcohols.
		Proteins	Deionization of proteins containing hydrophobic molecules.
Desalting and buffer exchange	Ion exchange	Cations, anions	Desalting amino acids for better TLC and HPLC analysis.
	Gel filtration	Large molecules are eluted before salts and small molecules	Desalting proteins and nucleic acids with masses > 6000 Da.
	Ion retardation	Cations, anions	Removal of salts and ionic detergent from protein and amino-acid samples.
	Reversed phase	Hydrophobic analytes	Desalting of polypeptide solutions.
Detergent removal	Ion exchange	Cationic detergents; anionic detergents	From proteins, enzyme reactivation.
	Adsorption	Nonionic detergents	Triton X-100 from protein solutions.
	Ion retardation	Anionic detergents	Excess SDS from samples.
Metal concentration or removal	Ion exchange	Cations	Removal of metals and salts from aqueous medium.

Table 16.10

(*Continued*)

Requirement	Most Frequently Used Approaches	Species Retained	Typical Applications
	Chelating resins	Polyvalent cations	Removal of copper, iron, heavy metals, calcium, and magnesium.
	Adsorption	Metal-organic complexes	Metals complexed with polar or hydrophobic complexing agents.
Particulate removal	Filtration	Particulate matter	Pretreatment to protect HPLC frits and valves; filtration of culture medium.
Plasmid purification, probe cleanup	Gel filtration	Low-molecular-weight contaminants	Removal of unincorporated radioactive nucleotides from labeling reaction mixture.
	Adsorption	Large DNA	Removal of RNA, protein, and other cellular compounds.
	Ion exchange	Ethidium bromide or propidium iodide	Removal from plasmid visualization experiments.
Protein concentration	Ion exchange	Water	Improve sensitivity of electrophoretic and HPLC analysis.
		Proteins	Separation of proteins and low molecular weight substances.
	Adsorption	Hydrophobic proteins	C_{18} solid-phase extraction to remove hydrophobic proteins from hydrophilic proteins.
Removal or concentration of anions and cations	Ion exchange	Cations, anions	Removal of ions from aqueous solutions; concentration of large proteins; removal of mineral acids.
Removal or concentration of organics	Adsorption	Polar organic	Removal of nonionic detergents and lipids; separation of ethidium bromide from nucleic acid preparations.
	Gel filtration (organic)	High-molecular-weight compounds are eluted before small molecules	Separation of soluble organic compounds with masses $<150{,}000$ Da from complex sample matrices.

efficient dialysis of small sample volumes. Electrodialysis is a more rapid approach for desalting or buffer exchange that is a gentle technique and provides excellent desalting without loss of biological activity of proteins.

Ultrafiltration (UF) uses centrifugation as the driving force for membrane filtration. Membrane filters with molecular weigh cutoffs in the tens of KDa are held in a centrifuge cartridge that allows solvent, salts, and small molecules to pass through the membrane while macromolecules larger than the cutoff value are retained and concentrated above the membrane. Because the membranes are selected to show low nonspecific adsorption, UF results in good recovery and little loss of biological activity.

Many of the sample-preparation techniques already covered, such as liquid–liquid extraction (Section 16.5), are directly applicable to the fractionation of biological samples. Often the combination of two sample preparation procedures results in an overall improvement in cleanup efficiency. For example, the use of chromatographic media combined with dialysis can provide excellent concentration of protein solutions.

With the advent of proteomics, biological researchers are searching for new drug targets and biomarkers of disease at very low concentrations in human serum/plasma. Unfortunately, the predominant, less interesting proteins such as human serum albumin (HSA) and immunoglobulin (IgG) account for a high percentage of protein in these body fluids and are orders of magnitude more concentrated than the low abundance proteins that need to be identified and quantified. Classical techniques for removal of the high abundance proteins usually deplete only one or a few of them. New affinity-based products can deplete from 6 to 20 high-to-medium-abundance proteins leaving the thousands of lower-abundance proteins for further pretreatment. Affinity-based products contain antibodies that are specific for high-abundance proteins, and these products are available both as spin tubes and flow-through columns [65]. High-abundance proteins are retained on the column while low-abundance proteins pass through the column for collection and subsequent concentration by trace-enrichment techniques. The high-abundance proteins are then released by a change of buffer and can be either discarded or saved. Finally, the column is regenerated and can be re-used for several hundred injections. A combination of affinity depletion and multidimensional LC-MS/MS has been used to investigate trace levels of up- and down-regulated proteins in biological fluids [88].

16.11 SAMPLE PREPARATION FOR LC-MS

To obtain the maximum performance from LC-MS and LC-MS/MS, sample pretreatment usually is required. When MS/MS was first introduced, it was widely believed that its specificity of would eliminate the need for sample preparation. Thus HPLC was regarded as little more than a sample-introduction tool. However, it was naïve to assume that the MS alone would be able to solve all separation problems, and in subsequent years it was realized that *both* chromatographic separation prior to the detector and sample pretreatment prior to injection were required to avoid certain problems, such as previously unrecognized "matrix effects" in LC-MS and LC-MS/MS.

Sample pretreatment can be used to minimize problems associated with the sample and/or sample matrix, such as:

- spectral interference
- system compromise
- adduct formation
- ion suppression

Ions that appear at the same or nearly the same m/z value as the component of interest can cause *spectral interference*. Various MS detector designs exist (Section 4.14), with differing ability to distinguish between ions of similar m/z. The MS detectors used for routine HPLC analysis are of sufficiently low mass resolution that sample pretreatment may be required to remove co-eluting compounds that have similar m/z values to the analyte(s) of interest.

System compromise can occur when non-volatile sample or mobile-phase components precipitate in the LC-MS interface and degrade detector performance. The mobile-phase components that are most likely to precipitate in the atmospheric-pressure- or electrospray-ionization source are buffer salts and ion-pair reagents, so volatile buffers and reagents are used to solve this problem. Sample matrix elements, such as proteins, can also precipitate in the interface. Proteins are most commonly removed by precipitation, LLE (Section 16.5), or SPE (Section 16.6). Additional protein removal techniques are summarized in Table 16.11.

Adduct formation between another ion and the component of interest shifts the m/z value at which the component of interest appears in the spectrum. Adducts can be beneficial or detrimental, but in either case the amount of adduct formation needs to be controlled. Adduct ions such as sodium, potassium, and ammonium can originate from the sample itself, from reagents, or even from the container holding the sample. Adduct formation also can be used as a way to improve signals for macromolecules. However, uncontrolled adduct formation generally is undesirable and may require specific sample preparation procedures to reduce or eliminate it. Techniques for the removal of ions are listed in Table 16.10.

Ion suppression results when interferences are present that suppress (or compete with) the ionization of the analyte. Ion suppression is the most critical MS interference because it can be caused by components that do not appear in the mass spectrum. Phospholipids are one class of matrix components that are especially potent in causing ion suppression; titania sorbents can specifically remove these compounds [101]. In biological samples the natural variation in endogenous compound concentrations from one sample to another can cause varying levels of ion suppression. This variation in turn contributes to unacceptable variability in the signal response for the compounds of interest. Another type of ion suppression occurs when very strong ion-pairs are formed that are not broken apart in the API interface. Ion-pairing agents, such as trifluoroacetic acid, have been shown to contribute to ion suppression [102], and therefore their use in LC-MS should be avoided where possible. LLE (Section 16.5) and SPE (Section 16.6) are two techniques commonly used to remove ion-suppressing materials from the sample.

Table 16.11

Techniques for Removal of Protein from Biological Fluids

Protein Removal Technique	Principle	Reference(s)
Precipitation	Organic solvent (e.g., ACN), acid solution (e.g., perchloric), or salt solution (e.g., sodium sulfate) is added with agitation to a solution of plasma. The protein precipitates and forms a bead upon centrifugation. Supernatant can be analyzed by HPLC.	89–90
Restricted access media (RAM)	Solution containing protein is injected into a RAM column to separate protein from small molecules. See Section 16.6.7.2.	50–54
Turbulent flow chromatography	A large particle (~50–um) small diameter bonded silica RP column (1-mm i.d.) is run at high linear velocities (up to 8-mL/min). Although not truly turbulent flow, these high linear velocities do not allow the slower diffusing proteins to penetrate the packing pores, and they are flushed to waste. Smaller molecules are retained within the pores by RPC and are eluted to an analytical column or the MS interface.	91–92
Ion-exchange chromatography	Proteins are retained on an ion-exchange column at the proper pH; uncharged small molecules may pass through unretained onto an analytical column.	93–94
Size-exclusion chromatography	By selection of the appropriate pore size, proteins can be excluded and elute from the column first while the small molecular weight compounds elute later.	95–96
Reversed phase chromatography	The use of a C3 or C4 phase on wide-pore silica will retain proteins and have less retention of polar drugs, which can be eluted first.	97–98
High-abundance protein depletion	High-abundance proteins (up to 20) from human and other plasma/serum samples are depleted by antibody affinity phases. See Section 16.6.7.4.	99–100

16.12 DERIVATIZATION IN HPLC

Derivatization involves a chemical reaction between an analyte and a reagent to change the chemical and physical properties of the analyte. The four main uses of derivatization in HPLC are to:

- improve detectability
- change the molecular structure or polarity of analyte for better chromatography
- change the matrix for better separation
- stabilize an analyte

Ideally a derivatization reaction should be rapid, quantitative, and produce minimal by-products. Excess reagent should not interfere with the analysis or be easily removed from the reaction matrix.

With the increased popularity of LC-MS and LC-MS/MS, especially in the field of bioanalysis, many laboratories prefer this approach to high sensitivity and selective detection, rather than contend with the relatively time-consuming, labor-intensive approach of compound derivatization. Derivatization often is a last

Table 16.12

Functional Group and Derivatization Reagents

Functional Group	UV Derivatives[a,b]	Fluorescent Derivatives[a,c]
Carboxylic acids	PNBDI	BrMaC
Fatty acids Phosphonic acids	DNBDI	BrMmC
	PBPB	
Alcohols	DNBC	
	Dabsyl-Cl	
	NIC-1	
Aldehydes	PNBA	Dansyl hydrazine
Ketones	DNBA	
Amines, 1°		Fluorescamine OPA
Amines, 1° and 2°	DNBC	NBD-Cl
	SNPA	NBD-F
	SDNPA	Dansyl-Cl
	Dabsyl-Cl	
	NIC-1	
Amino acids (peptides)	SBOA	Fluorescamine
	SDOBA	OPA
	Dabsyl-Cl	NBD-Cl
		NBD-F
		Dansyl-Cl
Isocyanates	PNBPA	
	DNBPA	
Phenols	DNBC	NBD-Cl
	Dabsyl-Cl	NBD-F
	NIC-1	Dansyl-Cl
Thiols	Dabsyl-Cl	NBD-Cl
		NBD-F
		OPA

[a]See Table 16.13 for list of derivatives.
[b]Typically aromatic derivatives enhancing UV detection at 254 nm.
[c]Typically aromatic derivatives for enhanced fluorescence detection.

resort when developing a method. The introduction of pre- or post-column reaction that provides sample derivatization adds complexity, other sources of error to the analysis, and increases the total analysis time. While these procedures can be automated, the analyst must ensure that the derivatization step is quantitative (if necessary) and that there are no additional impurities introduced in the analysis. Although derivatization has its drawbacks, it may still be required to solve a specific separation or detection problem—as when mass spectral detection is not available. Reagents are available that react selectively with specific functional groups to form derivatives with enhanced UV- or fluorescence-detection characteristics. Some of the more common functional groups that can be reacted are listed in Table 16.12; the reagents are listed in Table 16.13. Figures 4.13, 4.36, and 4.37 show examples of derivatization to enhance fluorescence detection. In addition to derivatization to enhance detection, derivatization is used to enable separation of enantiomers, such as by the use of the reagents listed in Table 14.1. For more information on the use of derivatization in HPLC, see Sections 4.16 and 14.3, or consult one of the books dedicated to derivatization [103–108].

Table 16.13

Derivatization Reagents

UV Derivatives		Fluorescent Derivatives	
Dabsyl-Cl	4-Dimethylaminiazobenzene-4-sulphinyl	NBD-Cl	7-Chloro-4-nitrobenzo-2-oxa-1,3-diazole
DNBA	3,5-Dinitrobenzyloxyamine hydrochloride	NBD-F	7-Fluoro-4-nitrobenzo-2-oxa-1,3-diazole
NIC-1	1-Naphthylisocyanate	Fluorescamine	4-Phenylsprio(furan-2(3H),1′-phthalan-3,3-dione
PBPB	*p*-bromophenacyl bromide		
PNBA	*p*-Nitrobenzyloxyamine hydrochloride	OPA	*o*-Phthaldehyde
		Dansyl-Cl	5-Dimethylaminonaphthalene-1-sulfonyl chloride
PNBDI	*p*-Nitrobenzyl-*N*,*N*′-diisopropylisourea	BrMmC	4-Bromomethyl-7-methoxycoumarin
DNBDI	3,5-Dinitrobenzyl-*N*,*N*′-diisopropylisourea	BrMaC	4-Bromomethyl-7-acetoxycoumarin
PNBPA	*P*-Nitrobenzyl-*N*-*n*-propylamine hydrochloride		
DNBPA	3,5-Dinitrobenzyl-*N*-*n*-propylamine hydrochloride		
SNPA	*N*-Succinimidyl-*p*-nitrophenylacetate		
SDNPA	*N*-Succinimidyl-3,5-dinitrophenylacetate		
DNBC	3,5-Dinitrobenzyl chloride		

Note: Functional group reactivity is listed in Table 16.12.

REFERENCES

1. P. Gy, *Sampling for Analytical Purposes*, Wiley, New York, 1998.
2. D. C. Harris, *Quantitative Chemical Analysis*, 7th ed., Freeman, New York, 2003, pp. 701–795.
3. S. Mitra, ed., *Sample Preparation Techniques in Analytical Chemistry*, Wiley-Interscience, Hoboken, NJ, 2003.
4. J. R. Dean, *Methods for Environmental Trace Analysis*, Wiley, New York, 2003, ch. 3.
5. L. R. Snyder and S. van der Wal, *Anal. Chem.*, 53 (1981) 877.
6. V. Meyer, *LCGC*, 20 (2002) 106.
7. D. A. Wells, *High Throughput Bioanalytical Sample Preparation: Methods and Automation*, Elsevier, Amsterdam, 2003.
8. F. Settle, Ed., *Handbook of Instrumental Techniques for Analytical Chemistry*, Prentice-Hall, Upper Saddle River, NJ, 1997.
9. C. M. Druzik, D. Grosjean, A. Van Neste, and S. Parmar, *Int. J. Environ. Anal. Chem.*, 38 (1990) 495.
10. M. Hiatt, *Anal. Chem.*, 67 (1995) 4044.
11. L. R. Snyder, *Chemtech*, 9 (1979) 750.
12. L. R. Snyder, *Chemtech*, 10 (1980) 188.
13. T. S. Ma and V. Horak, *Microscale Manipulations in Chemistry*, Wiley, New York, 1976.
14. N. B. Mandava and Y. Ito, eds., *Countercurrent Chromatography: Theory and Practice*, Dekker, New York, 1988.
15. J. A. Riddick and W. B. Bunger, *Organic Solvents*, Wiley-Interscience, New York, 1970.
16. R. E. Majors and K. D. Fogelman, *Amer. Lab.*, 25(2) (1993) 40W.
17. M. A. Jeannot and F. F. Cantwell, *Anal. Chem.*, 68 (1996) 2236.
18. E. M. Gioti, D. C. Skalkos, Y. C. Fiamegos, and C. D. Stalikas, *J. Chromatogr. A*, 1093 (2005) 1.
19. J. M. Stevens, M. Crawford, M. Halvorson, and R. Woleb, Pittsburgh Conference, paper 1550-4, 2008.
20. W. F. Lane and R. C. Loehr, *Environ. Sci. Technol.*, 256(5) (1992) 983.
21. D. Barcelo, *Analyst*, 116 (1991) 681.
22. P. D. MacDonald and E. S. P. Bouvier, eds., *Solid Phase Extraction Applications Guide and Bibliography*, Waters, Milford, MA, 1995.
23. F. X. Zhou, J. M. Thorne, and I. S. Krull, *Trends Anal. Chem.*, 11 (1992) 80.
24. D. R. Green and D. Le Pape, *Anal. Chem.*, 59 (1987) 699.
25. A. Alfred-Stevens and J. W. Eichelberger, in *Methods for Determination of Organic Compounds in Drinking Water (Supplement 1)*, Environmental Monitoring Systems Laboraotroy, Office of R&D, US EPA, Cincinnati, OH, 1990, pp. 33–63.
26. *3M Empore Extraction Disks Method Summary: Phenols*, Pub. 78-6900-3714-4 (113.05) R1, 3M Corp., St. Paul, MN, 1994.
27. *3M Empore Extraction Disks Method Summary, EPA Method 552.1: Haloacetic Acids and Dalapon in Drinking Water*, Pub. 78-6900-373-4 (113.05) R1, 3M Corp., St. Paul, MN, 1994.
28. *3M Empore Extraction Disks Method Summary: Pesticides and Polychlorinated Biphenyls*, Pub. 78-6900-3715-1 (113.05) R1, 3M Corp., St. Paul, MN, 1994.

29. G. M. Hearne and D. O. Hall, *Amer. Lab.*, 25(1) (1993) 28H.
30. R. Majors, *LC/GC*, 23 (2005) 358.
31. R. E. Majors and A. Shukla, *LC/GC*, 23 (2005) 646.
32. A. A. Boyd Boland and J. B. Pawliszyn, *Anal. Chem.*, 68 (1996) 1521.
33. R. Shirey, *Advances and Applications of Solid Phase Microextractios (SPME)*, paper 23602 at Analytica, Munich, 1996, pp. 23–26.
34. E. Balrussen, P. Sandra, F. David, and C. A. Cramers, *J. Microcol. Sep.*, 11 (1999), 737.
35. B. A. Bidlingmeyer, *Liq. Chromatogr.*, 2 (1984) 578.
36. T. A. Dirksen, S. M. Price, and S. J. St. Mary, *Amer. Lab.*, 25(18) (1993) 24.
37. G. E. Platoff and J. A. Gere, *Forens. Sci. Rev.*, 3 (1991) 117.
38. X.-H. Chen, J.-P. Franke, K. Ensing, J. Wijsbeek, and R. A. De Zeeuw, *J. Chromatogr.* 613 (1993) 289.
39. V. Dixit and V. M. Dixit, *J. Liq. Chromatogr.*, 14 (1991) 2779.
40. R.-C. Hsu, I. Biggs, and N. K. Saini, *J. Agric. Food Chem.*, 39 (1991) 1658.
41. T. Tanaka, T. Hori, T. Asada, K. Oikawa, and K. Kawata, *J. Chromatogr. A*, 1175 (2007) 181.
42. N. J. K. Simpson, ed., *Solid-Phase Extraction: Principles, Techniques, and Applications*, Dekker, New York, 2000.
43. E. M. Thurman and M. S. Mills, *Solid-Phase Extraction: Principles and Practice*, Wiley, New York, 1998.
44. J. S. Fritz, *Analytical Solid-Phase Extraction*, Wiley-VCH, New York, 1999.
45. N. Simpson and K. C. van Horne, *Sorbent Extraction Technology Handbook*, Varian, Harbor City, CA, 1993.
46. M. J. Telepchak, *Forensic and Clinical Applications of Solid Phase Extraction*, Humana Press, Totowa, NJ, 2004.
47. R. Bland, *Proceedings of the 3rd Annual International Symposium on Sample Preparation and Isolation using Bonded Silicas*, Analytichem International, Harbor City, CA, 1986, pp. 93–116.
48. E. J. Rook, M. J. X. Hillebrand, H. Rosing, J. M. van Ree, and J. H. Beijnen, *J. Chromatogr. B*, 824 (2005) 213.
49. M. Raisglid and M. F. Burke, abstract 653, 48th Pittsburgh Conference on Analytical Chemistry and Applied Spectroscopy, Atlanta, GA, 1997.
50. K.-S. Boos and A. Rudolphi, *LCGC*, 15 (1997) 602.
51. K.-S. Boos and A. Rudolphi, *LCGC*, 15 (1997) 814.
52. N. M. Cassiano, V. V. Lima, R. V. Oliveira, A. C. de Pietra, and Q. B. Cass, *Anal. Bioanal.Chem.*, 384 (2006) 1462.
53. D. R. Doerge, M. L. Churchwell, and K. Berry Delclos, *Rapid Commun. Mass Spectrometry*, 14 (2000) 673.
54. K. Ensing, C. Berggren, and R. E. Majors, *LCGC*, 19 (2001) 942.
55. S. G. Dmitrienko, V. V. Irkha, A. Yu. Kuznetsova, and Yu. A. Zolotov, *J. Anal. Chem.*, 59 (2005) 808.
56. C. Baggiani, L. Anfossi, and C. Giovannoli, *Current Pharma. Anal.*, 2 (2006) 219.
57. J. O. Mahony, K. Nolan, M. R. Smyth, and B. Mizaikoff, *Anal.Chim.Acta*, 534 (2005) 31.
58. P. A. G. Cormack and A. Z. Elorza, *J. Chromatogr. B*, 804 (2004) 173.
59. S. Piletsky and A. Turner, *Molecular Imprinting of Polymers*, Landes Bioscience, Austin, TX, 2006.

60. M.-C. Hennion and V. Pichon, *J. Chromatogr. A*, 1000 (2003) 29.
61. V. Pichon, N. Delaunary-Bertoncini, and M.-C. Hennion, in *Comprehensive Analytical Chemistry*, Vol. 37 J. Pawliszyn, ed., Elsevier, Amsterdam, 2002, p. 1081.
62. N. Delaunay, V. Pichon, and M.-C. Hennion, *J. Chromatogr. B*, 745 (2000) 15.
63. D. Stevenson, *J. Chromatogr. B*, 745 (2000) 39.
64. N. Zolotarjova, P. Mrozinski, and R. E. Majors, *LCGC*, 25 (2007) 118.
65. R. E. Majors, *LCGC*, 25 (2007) 16.
66. J. Stroka, E. Anklam, U. Jörissen, and J. Gilbert, *JAOAC Int.*, 83 (2000) 320.
67. M. Anastassiades, S. J. Lehotay, D. Stajnbaher, and F. J. Schenck, *JAOAC Int.*, 86 (2003) 412.
68. M. Anastassiades, http://www.quechers.com/docs/quechers_recov.pdf (2004).
69. S. J. Lehotay, A. de Kok, M. Hiemstra, and P. Van Bodegraven, *JAOAC Int.*, 88 (2005) 595–614.
70. C. F. Fagerquist, A. R. Lightfield, and S. J. Lehotay, *Anal.Chem.*, 77 (2005) 1473.
71. N. Jakubowska, Z. Polkowska, J. Namiesnik, and A. Przyjazny, *Crit. Rev. Anal. Chem.* 35 (2005) 217.
72. *Method 3550C, Ultrasonic Extraction*, US EPA, http://www.epa.gov/epawaste/hazard/testmethods/sw846/online/3_series.htm.
73. E. L. Randall, *JAOAC Int.*, 57 (1974) 1165.
74. *Method 3541, Automated Soxhlet Extraction*, US EPA, http://www.epa.gov/epawaste/hazard/testmethods/sw846/online/3_series.htm.
75. J. L. Luque-Garcia and L. de Castro. *J. Chromatogr A.*, 998 (2003) 21.
76. J. L. Luque-García, M. J. Ramos, M. J. Martínez-Bueno, and M. D. Luque de Castro, *Chromatographia*, 62 (2005) 69.
77. J. L. Luque-Garcia and M. D. Luque de Castro., *Anal. Chem.*, 73 (2001) 5903.
78. J. M Levy, *LCGC*, 17(6S) (1999) S14.
79. R. Smith, *J. Chromatogr. A*, 1000 (2003) 3.
80. L. T. Taylor, *Techniques in Analytical Chemistry: Supercritical Fluid Extraction*, Wiley, New York, 1996.
81. M. D. Luque De Castro, M. Valcarcel, and M. T. Tena, *Analytical Supercritical Fluid Extraction*, Springer, New York, 1994.
82. E. D. Ramsey, ed., *Analytical Supercritical Fluid Extraction Techniques*, Kluwer Academic, Dordrecht, 1998.
83. *Method 3545A, Pressurized Fluid Extraction (PFE)*, US EPA, http://www.epa.gov/epawaste/hazard/testmethods/sw846/online/3_series.htm.
84. *Accelerated Solvent Extraction (ASE) of Active Ingredients from Natural Products*, Application note 335, Dionex Corp., Sunnyvale, CA.
85. G. LeBlanc, *LCGC*, 17(6S) (1999) S30.
86. *Method 3546, Microwave Extraction*, US EPA, http://www.epa.gov/epawaste/hazard/testmethods/sw846/online/3_series.htm.
87. *Method 3540C, Soxhlet Extraction*, US EPA, http://www.epa.gov/epawaste/hazard/testmethods/sw846/online/3_series.htm.
88. N. Tang and C. Miller, *An Integrated Approach to Improve Sequence Coverage by Combining LC-MALDI MS/MS and NanoLC-LC-MS/MS*, Agilent Technologies, Santa Clara, CA, Publication Number 5989-3500EN (2005).
89. S. R. Souverain and J.-L. Veuthey, *J. Pharm. Biomed. Anal.*, 35 (2004) 913.

90. R. Herráez-Hernández, P. Campíns-Falcó, and A. Sevillano-Cabeza, *Chromatographia*, 33 (1992) 177.

91. L. Ynddal and S. H. Hansen, *J. Chromatogr. A*, 1020 (2003) 59.

92. J. Ayrton, G. J. Dear, W. J. Leavens, D. N. Mallett, and R. S. Plumb, *Rapid Comm. Mass Spec.*, 11 (1997) 1953.

93. K. N. Frayn and P. F. Maycock, *Clin. Chem.* 29 (1983) 1426.

94. Y.-J. Xue, J. B. Akinsanya, J. Liu, S. E. Unger, *Rapid Comm. Mass Spec.*, 20 (2006) 2660.

95. J. Porath, *J. Protein Chem.*, 16 (1997) 463.

96. F. Laborda, M. V. Vicente, J. M. Mir, and J. R. Castillo, *Anal. Bioanal. Chem.*, 357 (1997) 1618.

97. C. Shaw, *Meth. Mol. Biol.*, 64 (1996) 101.

98. Y. Wang and J. Seneviratne, *Curr. Proteomics*, 5 (2008) 104.

99. J. Martosella, N. Zolotarjova, H. Liu, G. Nicol, and B. E. Boyes, *Proteomics*, 13 (2005) 3304.

100. N. Zolotarjova, J. Martosella, G. Nicol, J. Bailey, B. E. Boyes, and W. C. Barrett, *Electrophoresis*, 25 (2004) 2402.

101. Y. Ikeguchi and H. Nakamura, *Anal. Sci.*, 16 (2000) 541.

102. A. Apffel, S. Fishcher, G. Goldberg, P. C. Goodley, and F. E. Kuhlman, *J. Chromatogr. A.*, 712 (1995) 177.

103. S. Ahuja, *Selectivity and Detectability Optimization in HPLC*, Wiley, New York, 1989.

104. B. King and G. S. Graham, *Handbook of Derivatives for Chromatography*, Heyden, Philadelphia, 1979.

105. H. Lingeman, *Detection-Oriented Derivatization Techniques in Liquid Chromatography*, Dekker, New York, 1990.

106. J. F. Lawrence and R. W. Frei, *Chemical Derivatization in Liquid Chromatography*, 3rd ed., Elsevier, Amsterdam, 1985.

107. T. Toyo'oka, *Modern Derivatization Methods for Separation Science*, Wiley, West Sussex, UK, 1999.

108. K. Blau and J. M. Halket, eds., *Handbook of Derivatives for Chromatography*, 2nd ed, Wiley, West Sussex, UK, 1993.

CHAPTER SEVENTEEN

TROUBLESHOOTING

QUICK FIX

If your goal is to solve a problem quickly and not to read about troubleshooting, one of two approaches is recommended. Look through the detailed table of contents at

Introduction to Modern Liquid Chromatography, Third Edition, by Lloyd R. Snyder, Joseph J. Kirkland, and John W. Dolan

the front of the book and find the section of this chapter that relates to your problem. Alternatively, go to Section 17.5 at the end of this chapter and use Tables 17.2 through 17.11 to guide you through the troubleshooting process and give you a cross-reference to the text for more information. Each chapter in this book contains troubleshooting information relative to the topic of discussion; these are easily located by consulting the appropriate topic in the index at the end of the book.

17.1 INTRODUCTION

When the second edition of this book was written, troubleshooting was a major part of the job of the HPLC operator. Problems related to excessive pump-pressure pulsation, pressure shocks from manual injectors, short lifetimes for pump seals and detector lamps, and general instrument problems abounded. Columns were subject to failure as a result of shipping damage or column-bed collapse, and column inlet-frit replacement was so common that most manufacturers shipped several extra frits with each column. A detailed understanding of gradient elution was in its infancy, and this lack of understanding resulted in unexpected changes in the chromatogram with changes in flow rate or other gradient adjustments, and the transfer of gradient methods was quite difficult.

Fast-forward 30 years, and the situation has changed dramatically. HPLC hardware is much more reliable, with routine maintenance intervals of 6 to 12 months or longer, instead of on a weekly or monthly basis. With the advent of higher purity, type-B column packings, improved particles, and better column-packing procedures, column problems are a small fraction of those of their ancestors. Gradient elution is well understood, and with a little care, it can perform as well as isocratic techniques, even in the hands of relatively inexperienced users.

This chapter focuses on HPLC problems and how to correct them. At the core of troubleshooting is the prevention of problems (Section 17.2)—a major strategy for reliable HPLC system operation. Problems are most easily isolated by use of a disciplined approach (Section 17.3) to identify specific symptoms (Section 17.4).

Troubleshooting and problem prevention are greatly aided by a good understanding of how the HPLC system operates; readers are encouraged to get this information by reading Chapter 3 (equipment) and the appropriate part(s) of Chapter 4 (detectors). At a minimum, Section 3.10 (maintenance) should be reviewed; maintenance practices tie in closely to troubleshooting and will be cross-referenced regularly in this chapter. Of course, do not forget to check the troubleshooting and preventive maintenance sections of the instrument manuals for specific instructions that apply to your HPLC system.

Another aspect of troubleshooting and preventive maintenance is the question of who will make the necessary repairs. A discussion of this topic is covered in Section 3.10.3.1; Table 3.7 lists recommended repair activities for different personnel. The final determination of who is qualified to make repairs is made by a combination of laboratory policy, regulatory requirements, training, and personal mechanical aptitude. In any event, a good understanding of instrument operating principles and the troubleshooting process will benefit workers at all levels of competence and responsibility.

Finally, it is important to recognize that HPLC troubleshooting is difficult to condense into a single chapter, as in the present case. Other resources exist, and you are encouraged to explore these for additional help. There are numerous books on the subject. References [1, 2] are a good place to start, but new material is constantly being written—a Google of "HPLC troubleshooting books" at the time of this writing yielded 13,600 hits! (but *not* 13,600 books). Another excellent source of regular advice on HPLC troubleshooting is the "LC Troubleshooting" column, found each month in LCGC magazine and written by one of the authors [3]. On-line discussion groups also provide troubleshooting support. One of the best of these is Chromatography Forum [4], a discussion of troubleshooting and related problems that is monitored and contributed to by experts in all aspects of HPLC. There also are on-line expert-system tools, such as the HPLC Wizard [5] that can help you isolate and identify HPLC problems.

Each laboratory should establish its own preventive maintenance program. A list of possible items to include can be found under the "Best practices" entry in this book's index.

17.2 PREVENTION OF PROBLEMS

The best kind of HPLC problem is the one that does not occur. For this to be the general pattern for an HPLC system, a structured preventive-maintenance program, such as that described in Section 3.10.2, should be established. Four elements are important in this process. First, a *system performance test* (Section 17.2.1) is used to establish that the HPLC system can operate properly under ideal conditions. Second, *periodic maintenance* (Section 17.2.2) needs to be performed to repair or replace parts that have a limited lifetime due to normal wear. Third, a *system suitability test* (Section 17.2.3) should be run prior to each batch of samples to ensure that the method and equipment are working at a level that will produce acceptable data. Finally, a *repair and maintenance record* (Section 17.2.4) should be kept as proof of maintenance and to establish failure patterns for each HPLC system.

17.2.1 System Performance Tests

An HPLC system performance test is described in Section 3.10.1; a brief summary is included here—consult Section 3.10.1 for details. There are three types of tests that can be used to test system performance: Installation Qualification, Operational Qualification, and Performance Qualification (Section 17.2.1.1). All three groups of tests should be carried out when the HPLC system is new. The gradient tests (Section 17.2.1.2) and additional tests (Section 17.2.1.3) should be performed every 6 to 12 months to ensure continued reliable HPLC operation.

Throughout this book it is assumed that the HPLC system is operating in reversed-phase mode with a UV detector. Many of the tests that are described work best with UV detection, including the tests listed and cross-referenced in this section. It may be possible to devise ways to accomplish the same testing goals with non-UV detectors, but some tests—such as the various gradient performance tests—will be very difficult to perform with certain detectors (e.g., LC-MS or LC-MS/MS). For this reason we recommend that every laboratory have at least one UV detector available to facilitate performance testing and troubleshooting.

17.2.1.1 Installation Qualification (IQ), Operational Qualification (OQ), and Performance Qualification (PQ) Tests

A combination of tests designed by the instrument manufacturer (IQ, OQ) and the user (PQ) show that the HPLC system works as it was designed to and performs according to the published instrument specifications. These tests are performed when the instrument is new, and the PQ test is repeated periodically afterward. The results of these tests play an important role in the divide-and-conquer problem-isolation strategy (Section 17.3.1), by allowing the user to check the current performance of the instrument against its original performance when it was known to be working properly. This helps to answer the "system or method?" question that often arises when an HPLC problem occurs.

17.2.1.2 Gradient Performance Test

If the HPLC system has gradient capabilities, the gradient performance test (Section 3.10.1.2) should be run, even if the system is used only for isocratic applications with on-line mixing of the mobile phase. The gradient performance test checks the linearity and accuracy of mobile-phase preparation as well as measures a value of the gradient dwell volume. If the system is used only for isocratic methods with on-line mixing, an alternative test is to inject a standard solution under isocratic conditions. Then exchange the A- and B-solvent reservoirs and adjust the program to deliver the same mobile phase (e.g., 30% A = water + 70% B = MeOH becomes 30% B = water and 70% A = MeOH) and inject again. The retention times in both cases should be the same. These tests ensure that when a mobile-phase mixture is programmed into the system controller, the desired mixture is delivered to the column. See Section 17.4.6.1 for specific examples of gradient performance test failures.

17.2.1.3 Additional System Tests

A few more tests will round out the overall testing of the performance of the HPLC system (Section 3.10.1.3). A flow-rate check will determine if the pump is operating properly, and the pressure bleed-down test checks the outlet check-valves and/or pump-seal integrity. A retention-reproducibility test double-checks the flow rate and verifies that on-line mobile-phase preparation is consistent. The peak-area reproducibility test ensures that the autosampler is working as it should. See Section 17.4.6.2 for specific examples of failure of the additional system tests.

17.2.2 Periodic Maintenance

A regular maintenance program will allow the HPLC system to work more reliably and encounter fewer breakdowns. A list of recommended preventive maintenance items is contained in Table 3.6 and discussed in Section 3.10.2.1. As a rule, time spent in preventive maintenance, especially to ensure that the system is cleaned out regularly and normal-wear parts (e.g., pump seals) are replaced before failure, will be time well spent and will reduce overall operating costs. Laboratories that always wait for a failure to occur before performing HPLC maintenance usually spend more time and money on troubleshooting and repair than those who regularly maintain their HPLC systems.

17.2.3 System-Suitability Testing

System suitability refers to a series of injections, pre-defined for a specific method, that are made prior to running samples to ensure that the HPLC system is working properly and that the system is able to generate data that will meet the acceptance criteria of the method. To accomplish this task, the system-suitability sample must be able to adequately assess method performance—usually a synthesized sample made from standards mixed in a real or formulated sample matrix. These tests may include a subset of the performance qualification test (Sections 17.2.1, 3.10.1.1) as well as other tests, such as retention-time and/or peak-area reproducibility, that demonstrate that the HPLC is working suitably as a system. System-suitability testing should assess the most meaningful parameters of the separation (e.g., resolution, accuracy, reproducibility) to ensure the quality of the chromatographic data. In other words, system-suitability testing should answer the question, "Does the HPLC system currently work acceptably for this method?" Whether or not system-suitability tests are required by a regulatory agency or local operating procedures, it is wise to make some kind of a test prior to running samples to minimize the risk of collecting meaningless data and wasting precious samples—as well as time and resources. A historical record (Section 17.2.4) of system-suitability tests can also be useful to track down the source of a problem. See Section 12.3 for further information on system-suitability testing.

17.2.4 Historical Records

The value of historical records in establishing instrument failure patterns and preventive maintenance programs cannot be underestimated. Section 3.10.3.2 described a suggested set of records, comprising at least three elements that should be maintained for each HPLC system. A record of the *system configuration* will include sufficient information to identify each system component (pump, autosampler, detector, etc.) that makes up a specific HPLC system. This, combined with sample batch-records (e.g., sample identification associated with injection number), should allow correlation of each sample run with a specific HPLC configuration. *Maintenance records* will provide a written record of all system maintenance activities. The results of *system checks*, such as performance qualification and gradient performance tests (Section 17.2.1) also should be included in the records for each HPLC system.

Historical records serve two purposes. The first is to provide documentation to a regulatory agency that proper procedures for instrument qualification, use, and maintenance were in place and were followed. The second is to provide data to help design a preventive maintenance program. After a sufficient period of time, such as one to two years, depending on the intensity of HPLC system use, enough data should be available to determine failure patterns. If several HPLC systems in the laboratory are nominally identical (i.e., same brand and model), data from several systems may be pooled. For example, to determine a pump-seal replacement cycle, the pump-seal replacement interval records could be pooled for all instruments. Usually it is best to replace the pump seals prior to failure, so that particles generated from seal wear do not block tubing or frits downstream from the pump. If seal replacement was performed at 7, 6, 8, 12, and 11 months during 5 different repair incidents, the data suggest that it would be prudent to proactively replace the seals every 6 months so as to avoid the consequences of seal failure. A similar procedure can be used to identify other replacements or repairs that should be added to a preventive maintenance list. As a rule, if the lifetime of a part can be anticipated, replacement of the part at 70–80% of its anticipated lifetime is a good target for preventive service activities—this gets most of the useful service from the part while reducing risk of lost data when the part fails.

17.2.5 Tips and Techniques

This section contains a collection of tips and techniques that can be used to isolate and correct HPLC problems, as well as prevent them from happening in the future. The topics highlight some of the most common problem areas in HPLC operation—by paying special attention to these, you should be able to reduce problem frequency, as well as the amount of time spent to correct problems when they occur. These are individually cross-referenced throughout this chapter.

17.2.5.1 Removing Air from the Pump

The internal parts of the HPLC pump and associated hardware have many small, often angular, passages that can trap air bubbles. Sometimes a sharp tap on a pump head with a wooden or plastic object, such as a screwdriver handle, will dislodge bubbles. A system flush with a thoroughly degassed, low-viscosity, low-surface-tension solvent such as methanol will sometimes dissolve bubbles that resist displacement using other techniques. The use of degassed solvents on a routine basis will prevent the accumulation of bubbles in the system, since the solvent will have an additional capacity for dissolved gas that will solubilize tiny bubbles before they become a problem. Every HPLC system will work more reliably if the mobile phase is degassed.

17.2.5.2 Solvent Siphon Test

All HPLC systems will perform more reliably if the reservoirs are elevated relative to the pump, so that a slight siphon head-pressure helps deliver mobile phase to the pump. To ensure a free flow of solvent to the mixer, it is important to check the solvent inlet-line frits occasionally. Because many pumps have an "asymmetric" duty cycle in which they spend more time delivering solvent than refilling, the flow rate during the intake stroke can be much higher than the overall average flow rate. Therefore one should expect the reservoir to be able to deliver several times more solvent by siphon action than will be required by the pump. To test this feature, disconnect the solvent inlet-line at the mixer (low-pressure mixing) or the pump inlet (high-pressure mixing) and allow the solvent to siphon through the tubing. A 10-fold excess of solvent is a good rule of thumb for adequate delivery. For example, if the typical operation of the system is 1 mL/min, at least 10 mL/min of solvent through the siphon should be expected. This will supply enough solvent so that starvation of the pump will never be an issue. If the flow is lower than expected, check for a blocked solvent inlet-line frit, a pinched inlet-line, or a poorly vented reservoir. A restricted solvent supply in low-pressure-mixing systems also can cause mobile-phase proportioning errors (Section 17.4.6.1).

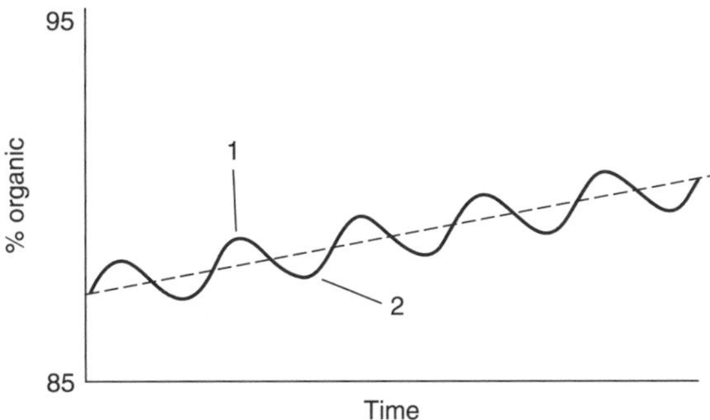

Figure 17.1 Programmed shallow-gradient profile (---), profile actually observed (—). Mobile phase is stronger at point 1 than 2, even though point 1 occurs earlier in the gradient.

17.2.5.3 Pre-mixing to Improve Retention Reproducibility in Shallow Gradients

Complex mixtures often require high-resolution separations that utilize shallow gradients. Also gradient slopes of <1%/min often are required for high-molecular-weight compounds, such as peptides and proteins so that reasonable k^*-values can be obtained for these samples with large S-values (Sections 9.4, 13.4.1.4). In some cases the HPLC equipment cannot generate these shallow gradients with sufficient accuracy to obtain an acceptable separation. Low-pressure mixers generate the gradient by mixing alternate pulses of the A- and B-solvents; because on-line mixing is never complete, a small residual variation in mobile phase composition (and strength) remains when the mobile phase reaches the column. This is visualized in Figure 17.1 as the solid trace overlaid on the programmed gradient (dashed line). Thus there is an oscillation of values of %B around the programmed gradient throughout the separation. For this very shallow gradient, the value of %B at point 1 is greater than at point 2, even though point 1 appears earlier in the gradient. To correct this problem, the mobile phases can be pre-mixed.

For example, when pure solvents are used for the A- and B-solvents, a gradient of 40–50% B over 50 minutes requires that the system accurately deliver the gradient at a slope of 0.2%/min, which may challenge the precision of some equipment. When A and B are pre-mixed to 40% B (in the A-reservoir) and 50% B (in the B-reservoir), the system can be programmed to deliver a solvent gradient of 0–100% in 50 minutes, or 2%/min, a much easier job for most gradient equipment. Thus pre-mixing the mobile phase in this way can overcome the limitations of the equipment.

The same pre-mixing technique can be used to reduce baseline noise in both isocratic and gradient separations when the detector is operated at maximum sensitivity. For example, isocratic methods with refractive index detection (Section 4.11) will have less baseline noise if the mobile phase is pre-mixed instead of using on-line mixing. Gradient baselines generally are improved if 5% of the B-solvent is mixed in the A-reservoir and 5%A in the B-reservoir.

17.2.5.4 Cleaning and Handling Check Valves

As an alternative to the replacement of pump check valves with new parts, faulty check valves often can be rejuvenated by sonication in alcohol. Note that the inlet and outlet check valves seldom are interchangeable, and if they are not clearly marked, an identifying mark should be scribed on the check-valve body to indicate position and flow direction. Some check valves will come apart when inverted, so it is wise to check for this prior to sonication, or you may be surprised to find small parts in the sonicator after a cleaning attempt. We recommend placing each check valve in a separate beaker with enough methanol or isopropanol to cover the check valve; then sonicate for 5 to 10 minutes. In our experience, this will fix leaky check valves most of the time, presumably by removing unwanted contaminants from the ball and/or seat of the seal. If a check-valve comes apart, clean the parts in alcohol, then carefully reassemble them using forceps; avoid contacting the internal valve parts with paper, cloth, or fingers, because a small bit of fiber or a fingerprint can cause the valve to leak. See Section 17.4.2.2 for additional discussion on the possible causes of and corrections for check-valve sticking.

17.2.5.5 Leak Detection

Mobile-phase leaks may be obvious—or not. Drips, puddles, and leak alarms usually make location of the leak simple. When buffers are used, a white, crystalline deposit may show up on a slowly leaking fitting—even where no liquid is obvious. A simple leak detector for hard-to-find leaks can be made from a piece of thermal-printer paper (e.g., a credit-card receipt). Cut a narrow, pointed strip of thermal paper (e.g., 1×5 cm, pointed at one end) and use the pointed end to probe suspected fittings, seals, or other possible leak sources. The paper will turn black when it contacts organic solvent; this can be useful for locating leaks that are hard to detect by other means.

17.2.5.6 Repairing Fitting Leaks

Correcting a leaky fitting may be as simple as tightening the fitting one-quarter turn to see if a leak can be stopped. If this does not fix the problem, disassemble the fitting, rinse it, and reassemble it or replace the ferrule with a new one. *Do not* overtighten a stainless steel fitting, because the internal parts can distort enough that the ferrule will "mushroom" out beyond the fitting threads, making the connection impossible to disassemble. For stubborn fitting-leaks, PEEK (poly-ether-ether-ketone) ferrules often are superior to stainless steel because they deform sufficiently to seal with otherwise imperfect surfaces. When a leak is encountered with PEEK fittings and tubing, it is best to shut off the mobile phase flow, loosen the nut, reseat the tubing in the fitting body, and re-tighten the nut. Sometimes when a PEEK fitting is tightened with the flow on, the tube end can slip in the fitting, creating a small cavity at the tip of the tube, which in turn can cause unwanted peak broadening (extra-column effects, Section 17.4.5.3).

17.2.5.7 Cleaning Glassware

Organic residues on "clean" glassware can be the source of ghost peaks or baseline drift in blank gradients, so a thorough cleaning of the glassware is essential. Various techniques to clean glassware have been recommended [6]. Extra rinsing (10 rinses with tap water followed by 10 rinses with deionized water) of glassware that had been washed with laboratory dishwashing detergent was found to be satisfactory in one study [7]. Other workers [8] avoid detergents altogether, preferring to rinse glassware (including reservoirs, pipettes, and graduated cylinders) used only for HPLC mobile phases with water and then a clean organic solvent [6]. To avoid inadvertent contamination of the mobile phase by glassware, use only the cleanest possible glassware and take care that the cleaning process does not add contaminants to the glass surfaces.

17.2.5.8 For Best Results with TFA

Trifluoroacetic acid (TFA) is a widely used additive for gradient-elution mobile phases. TFA is readily miscible with both water and organic solvents, provides a low-pH mobile phase (0.1% TFA $\approx$ pH-1.9), acts as an ion-pairing reagent with biomolecules (Section 13.4.1.2), can be used at wavelengths of $<$220 nm, and is sufficiently volatile to use with mass spectrometric or evaporative light-scattering detectors. TFA is available in a highly purified form suitable for HPLC use, but it degrades rapidly upon exposure to air. For best results purchase HPLC-grade TFA (or equivalent spectral grade) in 1-mL ampoules and use the entire ampoule at one time. TFA is available in larger containers (e.g., 25 mL) at a much lower cost per mL, but—in the experience of one of the authors—it is impossible to prevent rapid degradation of the reagent once the bottle is opened, even when working in an inert atmosphere and carefully resealing the bottle. However, once mixed with water, resulting TFA solutions are fairly stable (e.g., 1 week). With UV detection at $<$220 nm and gradient elution, some drift with TFA-acetonitrile mobile phases may be observed, because the absorbance of TFA depends on the %-ACN in the mobile phase [9]. To minimize baseline drift, add the same amount of TFA (e.g., 0.1%) to both the A- and B-solvents and work at 215 nm, where the absorbance is fairly constant for TFA/ACN mixtures [9]. An alternative to preparing TFA-containing solvents in the laboratory is to buy HPLC-grade solvents with TFA (or other additives) already added from one of the HPLC-grade solvent suppliers (e.g., Burdick and Jackson, J.T. Baker).

17.2.5.9 Improved Water Purity

In the examples of Section 17.4.5.2 and accompanying Figures 17.11 and 17.12, the problem of mobile-phase contamination by impure water or additives is discussed. It is important to use the highest quality reagents in order to avoid unnecessary background peaks. This usually means purchasing HPLC-grade reagents for all salts, buffers, and organic solvents. Most laboratories prepare their own HPLC-grade water with a water purification system, such as the Milli-Q system (Millipore). Such water purifiers combine physical filtration ($\leq$0.2 μm), ion exchange, and adsorption on carbon to remove organic contaminants. Ultraviolet photo oxidation may also be carried out in order to kill bacteria and oxidize organic contaminants [6].

A further cleanup of the water may be required for maximal removal of background peaks, for example, for stability-indicating pharmaceutical assays—where peaks with areas $\geq$0.05% of the parent-peak area must be quantified. Passing the water through a C_{18} HPLC column works [10], but this is inconvenient. For high-pressure-mixing systems, another option is to install a C_{18} guard column between the A-pump and the mixer [10–12]. This "scrubber" column will trap organic materials before they reach the mixer and prevent them from entering the analytical column and producing background peaks. Other devices have been suggested for a final polishing of water prior to use. These include passing the water through a C_{18} SPE cartridge prior to use, or using a low-back-pressure in-line filter before the mixer (low-pressure mixing) or A-pump (high-pressure mixing) [6, 13].

17.2.5.10 Isolating Carryover Problems

"Carryover" describes the repeated appearance of a peak in later chromatograms, when the sample is only injected in an initial run. That is, remnants of the sample remain after the first run and are somehow introduced into one or more subsequent blank runs. There are three main types of carryover:

- volumetric carryover
- adsorptive carryover
- incomplete elution

Each of these carryover problems is described below, with some tips for distinguishing among them and correcting the problem.

In *volumetric carryover* a small amount of sample is trapped in the sample-injection system and is unintentionally injected the next time the injector cycles. It is characterized by a constant percent-area of the carryover peak relative to the previous peak. For example, if an injection gave a peak of 100,000 area-counts and 1% carryover was seen, a blank injection following a normal injection would have a peak 1% as large as the original (1000 area counts) with the same retention time. An additional blank injection would have 1% of the 1% peak, or 10 area counts. This constant serial dilution in subsequent blank injections characterizes volumetric carryover. Because of the dilution effect it is rare to have any carryover peak after 2 or 3 blank injections.

If volumetric carryover is suspected, look for small unintentional volumes where sample might get trapped and then diluted out, such as poorly assembled fittings that only contact the mobile phase during sample injection. Sometimes if a system overpressure occurs, tubing held by PEEK ferrules may slip slightly and create small gaps within the fitting, loosen the fitting and reseat the tube end as described in Section 17.2.5.6. Ineffective autosampler flushing between samples also can be a problem source, make sure that the autosampler wash mechanism is working properly. A worn injection valve rotor can contribute to volumetric carryover; rotor replacement should correct this (see Section 17.4.1.4 for additional autosampler-related discussion). When column-inlet frits become blocked, sample may pool in a stagnant channel and become purged only when an injection occurs. Back-flushing or replacing the column (see split or distorted peaks in Section 17.4.5.3) should solve this problem.

Adsorptive carryover may appear to resemble volumetric carryover, but it does not disappear as rapidly in subsequent injections. For example, the first blank may have 1% carryover from the original peak, but the second blank may also have 1% of the original peak or maybe a bit less, not 1% of the 1% peak. The source of such carryover is sample adsorption on surfaces within the system or column. Sample adsorption on the internal polymeric surfaces of the sample injector, the autosampler loop, and the inside of the injector needle are common sources of adsorption. Injection of a very hydrophobic sample dissolved in a polar solvent (e.g., water) is one common cause of adsorptive carryover. Addition of a few percent of an organic solvent to the injection solvent often will correct the problem. Also it may be useful to increase the strength and volume of the autosampler wash solvent or change the nature of the surfaces (e.g., replace a stainless-steel loop with a PEEK one). For a detailed discussion of sample adsorption in the injection valve, see [14]

Sometimes volumetric carryover and adsorptive carryover can occur together. In such cases a constant-fraction drop-off in peak size would be seen in the first and perhaps second blank injection, as the true carryover peak disappeared, but the adsorptive peak would persist for later injections. When all the solutions mentioned above have been tried, and carryover still is unacceptable, you may need to adjust the sample injection sequence so that all high-concentration samples are followed by a blank injection (i.e., ignored) or that low-concentration samples are never injected directly after high-concentration samples. When approximate sample concentrations are not known in advance, the method instructions can be written to ignore a sample with a small peak following a large one; the sample then is re-injected following a blank to get an accurate peak size.

Incomplete elution is highly unlikely in gradient elution. With isocratic separation, if an analyte does not leave the column during the run, it may elute in a following injection as an unexpectedly wide peak. See Section 17.4.4.1 for a further discussion of late-eluted peaks.

17.3 PROBLEM-ISOLATION STRATEGIES

The ability to isolate problems is a skill that is honed with practice and depends, in part, on a personal aptitude for such activities. In this section we describe 6 rules of thumb that are recommended components of a problem-isolation strategy. Experienced users will likely have at least some of these rules incorporated into their personal approach to isolating problems.

Remember to keep safety in mind whenever you are working with an HPLC system. Eye protection should be worn at all times—a few drops of mobile phase seldom will cause a problem on your skin, but in your eyes it may result in irritation or a more serious injury. The high pressure of an HPLC system might at first seem dangerous, but the mobile-phase flow rate is low, so a broken piece of tubing or a leak may cause a mess, but that rarely is a physical danger. However, be careful to avoid trying to stop a leak by pressing against the leak with your finger or thumb—pressures are sufficient to inject mobile phase through your skin, which can cause serious tissue damage. Normal laboratory safety precautions usually are sufficient for HPLC troubleshooting and maintenance (e.g., eye protection, a lab coat, and in some cases solvent-resistant gloves).

17.3.1 Divide and Conquer

This is an essential part of troubleshooting. Make changes that allow you to eliminate potential problems—the more sources eliminated with each change, the better. A typical example is to run a *new-column test* to determine if a problem is related to the analytical method (including the original column) or the hardware. Just install a new column and repeat the manufacturer's column-performance test (Section 3.10.1.3, Table 3.5). If you get the same results (e.g., plate number and retention time within ≈10%) as the column manufacturer, you know the HPLC system is working satisfactorily and the method (and/or the original column) is more likely the problem source. The column-performance test checks isocratic performance; you may want to supplement this with a gradient linearity or gradient-step test (Section 3.10.1.2).

17.3.2 Easy versus Powerful

It is important to balance which tests are done first, so as to make the best use of time. For example, if the problem takes longer than normal retention time, and there is more than one peak in the chromatogram, determine the retention-time ratio for each peak in the original and current chromatograms. If this ratio is constant for each peak, a decrease in flow rate is likely. This can be confirmed by a flow-rate check, which is easy and fast. Although it may not be as likely an answer to the problem to make up a new batch of mobile phase, a flow-rate check can be chosen first for convenience and speed. Of course, common sense should lead you to focus on the more common problem areas, even if they are not as easy to troubleshoot.

17.3.3 Change One Thing at a Time

Also called *The Rule of One*, this reminds us to use the scientific method during troubleshooting. Make a change and evaluate the result. Sometimes it is faster to make several changes at a time, but this offers little insight into the real source of the problem, a knowledge of which can be used to solve the problem and design preventive maintenance procedures for the future.

17.3.4 Address Reproducible Problems

Also called *The Rule of Two*, make sure the problem happens at least twice. Chromatographic problems that are not reproducible are difficult to troubleshoot, and it is even more difficult to know that they have been corrected. Make sure that the problem you are trying to solve is sufficiently reproducible that you can be confident you have corrected it. For example, if you have an extra peak or "spike" in a chromatogram, but do not see it with a reinjection of the same sample or in other samples, how will you know if you fixed the problem by making some change in the system?

17.3.5 Module Substitution

Replacing a suspect part with a known good part, whether it is a column, check valve, circuit board, detector, or other part, is one of the easiest and most powerful ways to isolate a problem. This strategy constitutes a good argument to have multiple installations of a given brand and model of HPLC system in a laboratory—so that there are more equivalent parts to interchange. Always keep plenty of consumable items on hand, such as filters, frits, guard columns, columns, tubing, and fittings, so that they are available for substitution.

17.3.6 Put It Back

This goes hand in hand with module substitution and reminds us that if we have substituted a known good part for a suspect part, and it does not correct the problem, we should re-install the original part. This avoids accumulating used parts of questionable quality. Of course, use common sense—it does not make sense to put the old seal back if replacing a pump seal did not solve the problem.

17.4 COMMON SYMPTOMS OF HPLC PROBLEMS

One of the most effective ways to isolate and correct an HPLC problem is always to be on the lookout for the common problem symptoms listed below. A good habit that will help identify problems early—before they cause real trouble—is to carefully examine the system every day. For example, during system equilibration at the beginning of the day, trace the flow path through the system, looking for unusual conditions. Start with the reservoir (enough solvent, no visible contamination), trace the transfer tubing to the pump (no leaks, steady pressure), then the autosampler (no puddles, enough wash solvent), column oven (no leaks), detector (no leaks), and waste bottle (sufficient capacity). By following this procedure regularly, you may find problems that can be corrected prior to running samples; you will also get to know how the system works when it is working properly, so that you can recognize when something is wrong. For example, it is unlikely that you will record the sound of the system, but you will get to know the normal clicks and hums that occur when the system is working well. You are then more likely to recognize that "something doesn't sound right," so you can investigate further.

The list of common symptoms in this section is not exhaustive, but it should cover the most common HPLC problems. For additional help, consult one of the references [1–5] listed in Section 17.1. The text in the present section is conveniently used with the data of Tables 17.2 through 17.11, where each symptom is listed with potential sources of the problem and a cross-reference to the discussion below. *Note that these tables are grouped for convenient cross-referencing in Section* 17.5 *at the end of the chapter*. If you get beyond your level of expertise (see Section 3.10.3.1), do not hesitate to ask for help from a more experienced worker, or request a repair visit from a service technician. The first time you perform a mechanical procedure, such as pump-seal replacement, it may be useful to supplement the manufacturer's diagram(s) with digital photographs or a sketch of your own so that you can confidently reassemble all the parts properly.

Problems can exhibit more than one symptom. For example, a loose fitting is likely to exhibit both a leak and low pressure. To avoid repetition, a detailed discussion of a problem will not be repeated for each symptom linked to that problem. Finally, the symptoms and problems in this section are based on the assumption that the system and/or method was working properly prior to observing the symptom. For this reason it is a good idea to consider what changes have been made since the system last worked acceptably—this may narrow down the possible problem sources.

17.4.1 Leaks

Mobile-phase leaks are one of the most common sources of HPLC problems, and one of the easiest to locate and identify. Many HPLC systems contain leak sensors in various places throughout the system where leaks are likely to occur—as in the column oven. Usually the sensor comprises a pair of electrical contacts that, when bridged by a small amount of liquid, will activate an alarm to alert the user to the presence of a leak. These sensors are located at a low point in the module, so that any solvent will collect at this point and trigger the sensor.

Some leaks are obvious, such as those that are clearly visible or identified by a leak sensor. Slow leaks may be more difficult to locate because the liquid evaporates before a drip is apparent (however, slow leaks may result in a visible deposit of buffer crystals at the point of leakage). To locate such micro leaks, a homemade leak detector can be useful (Section 17.2.5.5).

A short summary of the causes of leaks in the HPLC system is contained in Table 17.3 and cross-referenced to the following sections.

17.4.1.1 Pre-pump Leaks

Leaks before the pump are in the low-pressure portion of the HPLC system (Fig. 3.1), and most commonly are due to a loose low-pressure fitting (Section 3.4.2.1). Whenever parts that are made of two different materials are threaded together, they may tend to loosen over time because of differences in thermal expansion coefficients. Loosening is also accelerated by vibrations, as occur on the mobile-phase proportioning manifold for low-pressure mixing systems. If vibration-loosening is a recurring problem, several manufacturers sell low-pressure fittings with lock nuts to hold the low-pressure nuts more securely. If a loose fitting is found, tighten it (e.g., 1/4–1/2 turn). If the fitting still leaks after additional tightening, take the fitting apart and examine it for damage or contamination. If in doubt, replace the fitting. Nearly all low-pressure fittings are interchangeable (as long as the thread-pitch matches, i.e., English or metric threads with the same diameter and number of threads per unit length), so it is common to switch to another brand of fittings when factory-supplied low-pressure fittings fail.

With high-pressure mixing systems (Fig. 3.14), the only low-pressure connections are at the pump where the mobile-phase inlet tubing is connected to an inlet check-valve, solvent-selection device, or a splitter to supply solvent to two pump heads. The source of the leak should be obvious in this case.

With low-pressure-mixing systems (Fig 3.15), in addition to the low-pressure connections at the pump, a proportioning manifold is used to blend the proper proportions of solvents to form the mobile phase. The fittings at this manifold are another possible source of leaks. The proportioning valves can also leak, although this occurs rarely. If fluid is present at the base of a proportioning valve, gently tighten the screws holding the proportioning valve(s) in place. These may be metal screws threaded into a plastic block, so be careful—they are easy to overtighten and strip the threads. If the proportioning valves continue to leak, it is best to exchange the entire proportioning manifold with a new one—the parts are carefully matched at the factory for best performance.

It is also possible to have an air leak on the low-pressure side of the pump. Air can leak in through a gap too small to allow liquid to leak out. Air leaking in will cause the pump pressure to be lower than normal (see Section 17.4.2.2 for more details).

17.4.1.2 Pump Leaks

There are many parts of the pump that can leak, and leaks often are accompanied by low- or variable-pressure symptoms (Sections 17.4.2.2, 17.4.2.3). Low-pressure fittings generally are used on the pump inlet (Section 17.4.1.1) and high-pressure fittings on the outlet (Section 17.4.1.3); consult these Sections for more details. The remaining possible sources of pump leakage are the check valves, the pump seals, and auxiliary components, such as pulse dampeners.

External leaks of fluid from *check valves* result from a check valve that is loose, cross-threaded, or contains a damaged seat. A loose check valve is the most common source of leakage. Tighten the check valve (e.g., 1/8–1/4 turn) with a wrench to correct this problem. Be sure to hold the pump firmly so that the whole pump is not twisted when you tighten the check valve. If this does not correct the problem, make sure that the check valve is properly threaded into the pump head. Turn off the pump and remove the check valve. Check valves should turn freely with your fingers once they are initially loosened. If a check valve requires a wrench to remove it or insert it most of the way, the threads may be damaged. Check the threads for damage and replace a damaged check valve—in some cases the threads in the pump head may be damaged, requiring a pump-head replacement. When the check valve is removed, examine the seat where the check valve contacts the pump head at the bottom of the threaded portion. This seat often is made of a hard plastic and can crack—if the seat is damaged or cracked, replace it. Consult the pump manual for details about your specific pump.

Pump seals are a common source of pump leaks, if they are not replaced regularly according to a preventive maintenance program (Section 17.2.2). The function of the pump seals is described in Section 3.5.1. Because the pump seal continuously rubs on the moving piston during pump operation, it is the most wear-prone part of the HPLC system. As the seal wears, it will become less effective, with resulting leaks and/or pressure fluctuations. If pump seals are replaced every 6 to 12 months, it is rare that pump-seal failure will be encountered; for this reason we strongly recommend periodic pump-seal replacement as part of preventive maintenance. In addition to leaks, pump-seal failure usually results in the generation of small particles that work themselves downstream and block tubing, frits, or columns. If the seal leaks, usually liquid will appear below the pump head between the inlet check valve and the pump body—most pumps have a small drain hole in this position.

Before replacing the pump seal, consult the pump operator's manual for specific recommendations for your pump. A general description of seal replacement follows:

- Turn off the pump and disconnect the inlet and outlet tubing at the inlet and outlet check valves.
- Remove the pump-head retaining screws; this usually requires a hexagonal ("Allen") wrench. Successively loosen each screw by a small amount until the pump head is free (so that the pump head is not twisted in the process). Pull the pump head straight off the pump so that no sideways pressure is

placed on the (fragile) piston; alternatively, with some pumps, the head and piston come out as a unit.

- Remove the pump seal. If the pump came with a seal-removal tool, use it. Alternatively, a brass wood-screw can be used like a cork screw to remove the seal. If you elect to pry the seal out of the pump head, use a plastic tool, not a screwdriver or other metal tool, to avoid damaging the pump head. Do not use the piston to pry out the seal because the piston is likely to break. As you remove the seal, be sure to note how the seal is installed—the open side of the seal, where the spring is visible, should be facing the high-pressure section of the pump.
- Clean the piston. Usually rinsing with alcohol and wiping the piston with a disposable laboratory wipe is sufficient to clean the piston. If a residue remains, sometimes polishing the piston with a small amount of toothpaste will remove the residue (be careful to avoid fluoride-containing toothpaste if the system is used for ion chromatography). The toothpaste is sufficiently abrasive to remove buffer or other deposits, without scratching the sapphire piston.
- Inspect the piston for any scratches or defects. A simple way to highlight any problems is to hold a laser pointer up to the end of the piston, which will cause it to act as a light pipe and glow brightly. Any imperfections in the piston surface are either scratches or residues. Examine the end of the piston to be sure it is not broken or rough. If further cleaning does not remove these imperfections, replace the piston.
- Install the new seal. This usually can be done by pressing the new seal into the pump head with a fingertip. Be sure to install it in the correct orientation. Most seals come with a flange that prevents improper installation, but some seal designs can be installed backward.
- Lubricate the seal and piston with a squirt of alcohol from a wash bottle so that it slides together easily.
- Slide the pump head back on the piston, taking care not to twist the head and break the piston.
- Successively tighten each retaining screw a small amount so that the pump head is not twisted, possibly breaking the piston. If a torque specification is given in the manual, tighten the screws as directed; otherwise, tighten the screws firmly.
- Reconnect any tubing connections, and purge the pump (without a column connected, to allow particles to be flushed out); the pump is now ready for use.

Leakage can occur at *auxiliary pump components* such as pulse dampeners, mixers, or pressure transducers. If the leak is at a fitting, follow the normal procedure for correcting leaks at high-pressure fittings (Section 17.4.1.3)—first tighten the fitting, and if this does not fix the leak, clean or replace the fitting. If the leak is in a pulse dampener or mixer, and the cause of the leak is not obvious or cannot be corrected easily, the component may need to be replaced.

17.4.1.3 High-Pressure Leaks

High-pressure fittings are the most common location of leaks in the HPLC system. Such leaks become obvious when a leak sensor triggers an alarm or a puddle is discovered. More subtly, the pressure may be low or fluctuating (Sections 17.4.2.2, 17.4.2.3), or a white deposit of buffer crystals may appear where the nut threads into the fitting body. Sometimes an elusive leak can be identified using a piece of thermal paper as a probe (Section 17.2.5.5). (If you are not familiar with the parts and proper assembly of high-pressure fittings, read Section 3.4.2.2 and consult Figs. 3.8 and 3.9.)

In most cases a leaky fitting can be fixed by tightening the fitting 1/4 to 1/2-turn. Note that with PEEK fittings the pump should be turned off, the nut loosened, the tubing pushed firmly into the fitting, and then the nut should be re-tightened; otherwise, the tube end can slip in the fitting, causing a void volume in the fitting (see Fig. 3.09*b*). If this additional tightening does not stop the leak, disassemble the connection, rinse it out, and reseal it. If it still leaks, replace the ferrule. PEEK ferrules can be used to overcome minor imperfections in the surface of the seat. While PEEK fittings are suitable for traditional HPLC system pressures (up to 6000 psi or 400 bar), note that higher pressure systems usually require stainless-steel fittings for secure, leak-free connections.

17.4.1.4 Autosampler Leaks

Nearly all HPLC systems are operated with autosamplers today. But, if your system uses a manual injector, many of the same corrective measures apply, because the injection valves are very similar for both manual injectors and autosamplers. Injection-valve and autosampler design and operation are discussed in Sections 3.6.1 and 3.6.2, respectively. Leaks at the low- or high-pressure fittings in the autosampler can be addressed in the same way, respectively, as leaks before the pump (Section 17.4.1.1) or with other high-pressure fittings (Section 17.4.1.3). Other leaks can be divided into two categories: those associated with the injection valve, and those external to the valve.

Leaks associated with the *injection valve* occur either at one of the fittings or connections, or as a result of injector rotor-seal leakage. Push-to-fill autosamplers (Section 3.6.2.2, Fig. 3.21) and most manual injectors make the connection between the needle and the injector valve with a low-pressure seal, shown in a schematic in Figure 17.2. A nut (n) and ferrule (f) are used to hold a polymeric sleeve (s) in place in the valve body (b). At the ferrule the sleeve is crimped slightly (c) so that it seals against the needle. This seal can wear or deform over time so that fluid leaks out at the top of the sleeve when the sample is transferred into the loop. A slight tightening of the nut, often will tighten the crimp slightly and correct the leakage problem. In other cases the sleeve may need to be replaced. Consult the autosampler manual for specific instructions for your autosampler.

Needle-in-loop autosamplers (Section 3.6.2.3, Fig. 3.22) require a high-pressure seal between the needle and the injection valve. This seal is made of a hard plastic, such as PEEK or other hard material, that can wear, become distorted, or crack over time. Sometimes tightening the nut that holds the high-pressure seal in place will stop a leak, but usually a leak at the high-pressure seal will require installation of a new seal.

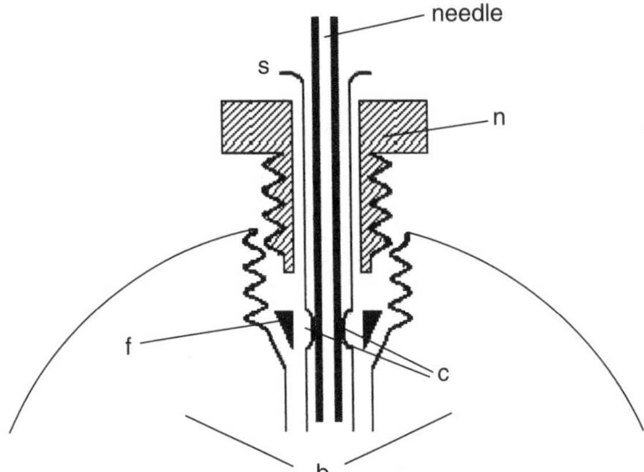

Figure 17.2 Schematic of low-pressure needle-seal. Valve body (b), nut (n), ferrule (f), polymeric sleeve/seal (s) and distorted portion of sleeve (crimp) that forms a seal around the needle (c).

Failure of the *injection rotor-seal* can result in leaks at the injection valve. The standard 6-port injection valve uses a rotating seal, the *rotor*, which turns against a stationary *stator*, to which the tube connections are made. The rotor typically is made of a hard polymer, such as PEEK, Vespel, or one of the fluoropolymers, and contains tiny passages that connect the fluid passages in the stator. The stator may consist of the stainless-steel valve body, or it may be an insert in the valve made of ceramic or another hard, smooth surface. In the schematic of Figure 3.17, the rotor comprises the portion of the valve inside the circle, whereas the stator is the part immediately outside the circle. In the most popular injection valves, the rotor is a flat disk with three kidney-shaped grooves in the surface, similar to that shown in the sketch of Figure 17.3*a*. These grooves line up with the ends of the tube connections on the stator, shown as the dashed circles in Figure 17.3*b*. In one position (e.g., load), the flow channels that result connect ports 1–2, 3–4, and 5–6, as seen in Figure 17.3*b*. When the rotor is rotated 60° (e.g., to the inject position), a different set of ports is connected, in this example, 2–3, 4–5, and 6–1.

If a small piece of hard material, such as a particle of column-packing or a bit of stainless steel from a poorly cut tube, gets caught in one of the passages of the injector, it can scratch the rotor. This can form a connecting passage between two of the grooves, as shown between ports 4 and 5 in Figure 17.3*c* (arrow). The result is *cross-port leakage*, where fluid from one hydraulic portion of the system leaks into another. This can show up as fluid leaking out the injection or waste port, as a problem of precision due to liquid leaking into or out of the sample loop, or sometimes as a carryover problem (Section 17.2.5.10). Rotor-seal (and possibly stator) replacement will be required to correct this problem. Also be sure to clean the remainder of the valve thoroughly to remove any particulate matter.

More commonly the rotor seal will fail as a result of normal frictional wear. The rotor seals are designed with a service lifetime of >100,000 cycles. For many laboratories, this will mean several years of operation before failure, so routine replacement of the rotor seal does not make much sense—unless the system has a

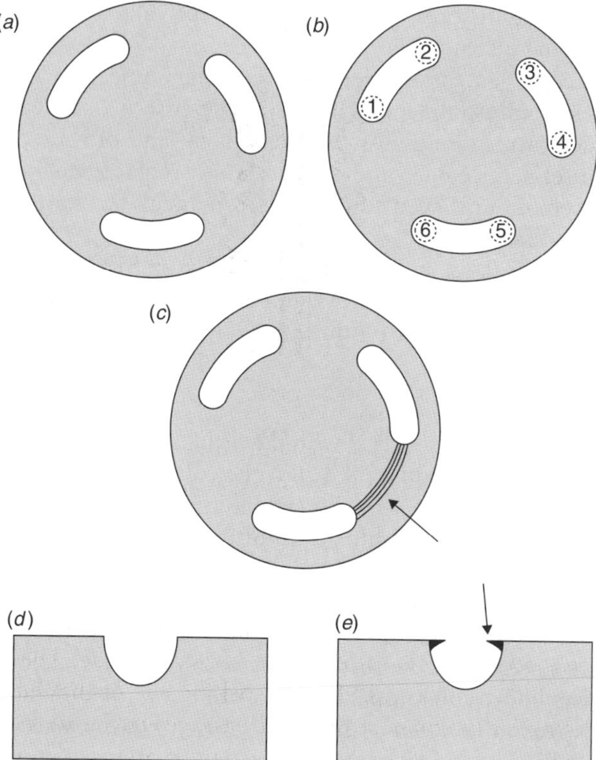

Figure 17.3 Schematic of injector rotor-seal. (*a*) View showing kidney-shaped grooves in surface of polymeric seal; (*b*) as in (*a*), showing connecting ports (numbered, dashed circles) for tubing; (*c*) scratch between ports 5 and 6 of *b*, causing cross-port leakage. (*d*) Cross section of kidney-shaped groove (normal condition); (*e*) worn seal resulting in burr of seal material at edge of groove (arrow).

counter to automatically record the number of injection cycles, so as to allow the remaining rotor seal life to be estimated. When viewed in cross section, the grooves of the rotor shown in Figure 17.3*a* are U-shaped as in Figure 17.3*d*. As the surface is worn during normal operation, the rotor will become slightly thinner and often a small burr of rotor material will form at the open edges of the groove, as seen at the arrow in Figure 17.3*e*. These burrs can break off and shed particles that can block tubing or frits. As the surface or the rotor wears, the contact pressure between the rotor and the stator is reduced, and eventually the seal will leak. Replacement of the rotor should solve this problem; thoroughly clean the valve before reassembly.

The pressure limit of the seal between the valve rotor and stator is related to how tightly the two surfaces are held together. Higher pressure-limits require that the surfaces be held together more tightly, but this also increases the friction between the surfaces and the effort to rotate the valve. Normally the injection valves are adjusted to withstand 6000 psi (400 bar) for traditional HPLC applications. For U-HPLC use (>6000 psi, Section 3.5.4.3) the two surfaces must be held together more tightly, so rotor lifetimes are expected to be shorter; alternative injector designs may overcome this problem. When an injection valve is disassembled for servicing, the rotor-to-stator sealing pressure may need to be adjusted; consult the service manual for additional instructions.

A major exception to the general use of rotary injection valves is found in some Waters-brand autosamplers, which typically use a "seal-pack" design that incorporates slider valves and high-pressure seals instead of rotary valves. Some of these parts are user-serviceable, and some require replacement of a subassembly with a new or rebuilt unit. Consult the user manual for more information on troubleshooting and repair.

Other points of leakage in autosamplers will vary from one design to another and often are unique to one model. If tightening or replacing a connecting fitting does not correct the problem, consult the autosampler manual for more information.

17.4.1.5 Column Leaks

Leaks at the column will be associated with the fittings. Leaks at the tube connections are treated as outlined in Section 17.4.1.3. If the column end-fitting itself is leaking, it may be possible to stop the leak by tightening the nut 1/4-turn. These larger fittings will take more effort to tighten than fittings for the 1/16-in. o.d. connecting tubing. If the fitting continues to leak, it may be best to discard the column, because disassembly of the column end-fitting can result in permanent damage to the column. In the past, it was common to remove the column-inlet fitting to replace the frit if a blockage was suspected, but with today's column-packing techniques, removing the end-fitting may allow column packing to ooze out and permanently damage the column. For this reason *removal of the end-fitting for examination or repair is no longer recommended.*

Cartridge-type columns comprise a disposable column that is held in a reusable holder. If the end-fitting on a cartridge column leaks, try tightening it to correct the problem. If it still leaks, disassemble the holder, rinse the fitting, and reassemble. In some cases the polymeric seal between the column and the holder may need to be replaced to stop a leak.

17.4.1.6 Detector Leaks

The pressure at the column outlet is lower than at the inlet by an order-of-magnitude or more, so detectors are subject to much lower pressures than the preceding high-pressure components. Because leak-related problems correlate strongly with the local system pressure, leaks at the detector are much less common than in other parts of the HPLC system.

Fittings at the detector inlet usually are the same type of high-pressure compression fittings used in other high-pressure parts of the system; leaks at these tubing connections should be treated as described in Section 17.4.1.3. Some detectors use 1/32-in. o.d. tubing instead of the standard 1/16-in. tubing used elsewhere in the system. This smaller diameter tubing is easily twisted and kinked, so take extra care when working with it. Some detectors operate at sufficiently low pressure on the inlet side that low-pressure plastic fittings can be used; many detectors use low-pressure fittings on the detector outlet because the pressure is quite low. Correction of leaks in low-pressure fittings is described in Section 17.4.1.1.

UV detectors (Section 4.4) are the most popular HPLC detectors. A generic version of the detector cell is shown in Figure 17.4. The cell typically comprises a

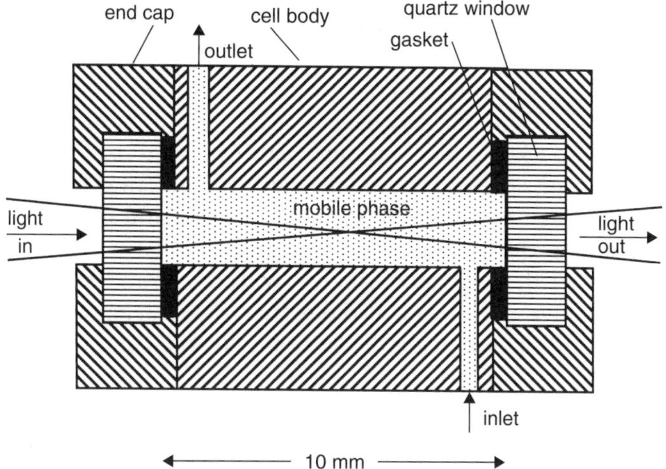

Figure 17.4 Schematic of typical UV-detector flow cell.

stainless-steel block (e.g., 10-mm long) with a hole (e.g., 1-mm diameter) drilled through it; quartz windows are held on the ends by an end-cap with O-ring seals. The tube connections to the cell body at the inlet and outlet may be compression-type fittings, which can be adjusted in the normal way (Section 17.4.1.3) if leaks occur. In other cases the tubing is soldered or welded to the flow-cell body, in which case factory repair or replacement will be required. The inlet tubing connected to the cell usually is thin-walled, narrow-diameter tubing (e.g., 1/32-in. o.d.) that also functions as a heat-exchanger to stabilize the temperature in the flow cell. Be careful not to twist or kink this thin tubing when tightening fittings.

Because air bubbles create noise in the cell, the UV detector often is operated with an after-market back-pressure restrictor on the cell outlet (Section 4.2.1). This creates enough pressure (e.g., 50–100 psi) to keep the bubbles in solution, but not so much pressure as to cause the window seals to leak. Spring-loaded back-pressure restrictors work well to accomplish this. An alternative is to use a narrow-bore waste tube (e.g., $\leq$0.010-in. i.d. = $\leq$0.25 mm i.d.), but as the flow rate is increased, the pressure also is increased. With this type of restrictor, a high flow rate may create sufficient back-pressure to exceed the upper pressure-limit of the detector cell (e.g., 150 psi), causing leakage at the window seals. Once a detector cell leaks, it may continue to leak if the window seals become distorted. Some detector cells are user-serviceable, whereas others will require service by a factory-trained technician, or complete replacement—consult the detector manual for more details on your model.

Other types of detectors have different flow-path and detector-cell designs. For any flow-through detector (fluorescence, electrochemical, conductivity, etc.), leaks can occur at the detector flow cell. Evaporative detectors, such as the evaporative light scattering, corona discharge, or mass spectrometric detector will have fluid leaks only at the interface, so leak isolation may be a little easier. Consult the appropriate section of Chapter 4 for a general discussion and generic design of specific detectors. This information, plus details from the detector manual, will help you locate and correct leaks in specific detector models.

17.4.2 Abnormal Pressure

Abnormal pressure is always a symptom of some other problem in the HPLC system. Normal system pressure will be different for different HPLC systems and applications, so it is a good idea to record the system pressure on a regular basis. For example, if you use a batch record or run sheet, record the pressure as part of the system suitability records; then you will be able to compare a questionable high- or low-pressure symptom to the historical record (Section 17.2.4). Conventional HPLC systems are capable of operation up to 6000 psi (400 bar), but most applications operate at pressures of 1500 to 3000 psi (100–200 bar). U-HPLC systems are designed to operate at pressures >6000 psi—as much as 15,000 psi (1000 bar) or more. However, as of this writing, these systems typically are operated in the 8000 to 10,000 psi (550–650 bar) range.

Ideally the pressure will be constant during an isocratic separation; however, small pressure fluctuations in the range of 1–2% of the operating pressure (e.g., 10–20 psi, 1–2 bar) are normal for many applications. Pressure also will vary with mobile-phase composition, as illustrated in Table 17.1, *so pressure changes during gradient elution are normal.* Methanol (MeOH) is more viscous than acetonitrile (ACN), and blends of MeOH and water are considerably more viscous than either MeOH or water alone. On the other hand, the pressure generated by ACN/water mixtures decreases in a fairly linear fashion as the mixture is changed from 100% water to 80% ACN. More detailed information on solvent viscosity can be found in Table I.3 of Appendix I.

Table 17.1

Example Pressures for Various Solvent Blends

	B-Solvent[a]	
% B-Solvent[a]	Acetonitrile	Methanol
0	1650 psi[b]	1650 psi[b]
20	1840	2520
40	1700	3265
60	1390	2920
80	1010	2150
100	710	1080

[a]A-solvent is water.
[b]Approximate (calculated) pressure for 250 × 4.6-mm, 5-μm particle column operated at 35°C and 2 mL/min (Eq. 2.13a).

Most HPLC pumps have pressure-limit settings that serve to shut off the pump if the limits are exceeded. The upper pressure-limit protects the system against damage or leaks if a blockage occurs causing excessive pressure. Most workers will set the upper limit well above the normal maximum operating pressure for all methods. For example, if normal method pressures are 2000 to 3000 psi, the upper limit might be set in the 4000 to 5000 psi region. This will provide some protection, yet is not low enough to result in frequent pump shutoff. The lower pressure-limit shuts the pump off if the pressure drops below the set point, such as when a mobile-phase reservoir runs dry. There is not much danger of pump damage if the solvent supply is exhausted, and the pump will not pump air, so the column will not dry out. Nevertheless, it is a good idea to set a lower pressure-limit in the 20 to 50 psi region so that the pump stops if the solvent supply is interrupted. Pressure problems are discussed below and summarized in Table 17.4.

17.4.2.1 Pressure Too High

Higher-than-normal pressure is a symptom of a blockage (assuming that the system settings are correct). Most commonly the pressure will rise gradually with successive batches of samples as debris collects on the in-line filter, guard column, or frit at the head of the column. This is normal with methods for the analysis of samples that may not be completely free of particulate matter. Remember, the pressure can increase by as much as 60% over the starting pressure during a gradient (Section 17.4.2, Table 17.1), so set the upper pressure-limit to accommodate this normal pressure fluctuation. It is also important to remember that column pressure is related to particle size, column length, and column diameter (Eq. 2.13*a*). Nominally equivalent columns can also differ in their pressure drop (e.g., two different brands of 150×4.6-mm i.d., 5-μm particle C_{18} columns). Finally, HPLC systems designed for higher pressure operation (e.g., U-HPLC) often have very small orifices and reduced tubing diameters (e.g., $\leq$0.005-in. or $\leq$0.12-mm i.d.), resulting in $>$1000 psi of system pressure without a column installed, so allowances for normal background pressure must be taken into account during troubleshooting.

Occasionally a sudden pressure increase will occur, and likely trigger the upper pressure-limit with a resulting pump shutdown. This can happen upon injection of a very dirty sample, such as untreated plasma, if sufficient particulate matter is present to completely block a frit or a piece of connecting tubing. Blockage can also take place when buffer and organic solvent are mixed on-line under conditions where the buffer solubility is poor and precipitation occurs, such as blending phosphate buffer and acetonitrile in some HPLC equipment.

Problem isolation for excessive-pressure problems is quite simple. Just work your way upstream from the column outlet, loosening the tubing connections as you go, with the pump operating. (If you strongly suspect a blockage at a particular point in the system, such as the in-line filter, start there and save time.) When a fitting is loosened and a sudden pressure drop results, the blockage is immediately downstream from that fitting. Remember, there normally is $>$1000 psi pressure drop across the column (depending on the flow rate), so much of the pressure drop observed when loosening the column-inlet connection is normal. With conventional HPLC systems designed for operation $\leq$6000 psi, once the column is removed, the pressure should be very low (e.g., $\leq$100 psi). As noted above, systems designed for higher pressures (Section 3.5.4.3) and sub–3-μm particles may have a system pressure of 1000 psi or more with no column attached. For reference purposes, it may be useful to go through the blockage isolation procedure with a normally operating system, to see what the normal system pressure is at each connection (for some fixed flow rate).

Once the problem location is isolated, appropriate corrective action should be taken. For example, if a piece of tubing is blocked, the tubing should be replaced. If the in-line filter is blocked, replace its frit. If the column is blocked, column reversal may help (the procedure is included in the discussion of split and distorted peaks in Section 17.4.5.3); otherwise, it may be necessary to replace the column.

Before putting the system back into service, consider if steps need to be taken to prevent the problem from recurring—or reduce its impact the next time. For example, if the problem is frit blockage from sample particulates, you may want to institute a filtration or centrifugation step during sample preparation, and use an in-line filter (Section 3.4.2.3). If buffer precipitation blocked a tube in the pump or mixer, consider reducing the buffer concentration or pre-mixing the buffer and organic solvent.

17.4.2.2 Pressure Too Low

System pressure that is too low is a sign of a leak or a pump problem (assuming all the system settings are correct). If the pressure is cycling, the problem is more likely at the pump (see the further discussion in Section 17.4.2.3), whereas if the pressure is steady and low, a leak is more likely. Because low-pressure and cycling-pressure problems are closely related and sometimes hard to tell apart, consult Section 17.4.2.3 for additional information.

If the pump has shut off due to the lower pressure-limit, and the cause is not obvious (e.g., an empty mobile-phase reservoir), restart the system and see what happens. Sometimes the lower pressure-limit sensor is too sensitive and will stop the pump if a momentary drop in pressure occurs, such as with the passage of a bubble through the pump. If this is a regular occurrence, it may be best to disable the lower pressure-limit.

Mobile-phase leaks can also cause low pressure; identify and correct the source of the leak using the instructions of Section 17.4.1 and Table 17.3. Although mobile-phase leaks are the most common type of leak, it is possible for air to leak into the system through a loose low-pressure fitting (Section 17.4.1.1). If efforts to find a liquid leak are fruitless, tighten each of the low-pressure fittings to see if this corrects the problem. If a proportioning valve is not sealed properly, it may be possible to pull air into the system through an unused solvent supply tube.

A sticking inlet check valve can prevent the pump from building sufficient pressure. Check-valve sticking is particularly problematic with ball-type check valves (Fig. 3.12*a*) when used with acetonitrile (ACN). When ACN is used as a solvent, the machined surface of the sapphire valve seat can catalyze the polymerization of minor contaminants in the ACN (aliphatic amines) [15, 16]. This polymer then results in a smoothing of the contact surface, so the ruby ball sticks to the seat via increased surface tension. Sonicating the check valve in methanol seems to correct this problem, at least temporarily. There is also speculation that sonication in dilute nitric acid might serve to remove this polymeric buildup [17], but this had not been confirmed at the time this book was completed. HPLC pumps that use active check valves (Fig. 3.12*b*,*c*) are not subject to this sticking problem; unfortunately, pumps with ball-type check valves cannot be retrofitted with active check valves.

A pump that is starved for mobile phase will not be able to generate the expected pressure. Check that sufficient mobile phase is available at the pump, by carrying out the siphon test described in Section 3.2.1. Impediments to free-flowing mobile phase include blocked inlet-line frits, faulty proportioning valves, and pinched or blocked tubing.

Insufficient mobile-phase degassing or insufficient pump purging can leave enough air in the mobile phase that the pump loses prime, thereby lowering the pressure. Ensure that the mobile phase is properly degassed (Section 3.3), and purge the pump by opening the purge valve, then pumping 10 to 20 mL of degassed mobile phase to waste (a high flow rate will sometimes displace bubbles in the pump). Frequent bubble problems can point to a faulty degassing module. If this is suspected, try an alternate degassing method (Section 3.3) to isolate the problem.

A worn pump seal can cause leaks (Section 17.4.1.2), but before leaks are apparent, the seal problem may prevent the pump from being able to provide the expected pressure. Carefully check for seal leaks at the drain hole on the bottom of the pump head (between the inlet check valve and the pump body), and check the system logbook (Section 17.2.4) to see if the system is due for a scheduled seal replacement. Seal replacement (see Section 17.4.1.2) should correct any problems. When the pump head is removed, check to see if there is any piston damage—a scratched or broken piston also can cause the pump to underperform.

17.4.2.3 Pressure Too Variable

As mentioned above, the HPLC system pressure in the normally fluctuates. Typically this fluctuation is 1–2% (e.g., 10–20 psi, 1–2 bar) of the operating pressure, but this will vary between systems and applications. It is therefore a good idea to make a note of the normal pressure variation as part of the records kept for each batch of samples (Section 17.2.4). Also keep in mind the normal pressure cycle during each run for gradient elution.

A summary of variable-pressure symptoms and solutions is given in Table 17.4. The most common sources of these problems are bubbles in the pump, sticky check valves, worn pump seals, broken pump pistons, and an inadequate mobile-phase supply to the pump. The identification and correction of these problems is almost exactly the same as for low-pressure problems (Section 17.4.2.2), so further instructions are not needed. The main difference between low-pressure and variable-pressure problems is that the latter may be limited to one pump head of a dual-piston pump, or one pump of a two-pump system. Thus part of the system may be working normally (higher pressure) while part of the system is not delivering enough mobile phase (lower pressure). The two most likely problem sources are a bubble in a pump head or a sticky inlet check valve. The simplest initial approach to correcting the problem is first to purge the pump to see if this fixes the problem. If it does not, sonicate the check valves (Section 17.4.2.2) in methanol. Additional information is found in Table 17.4.

17.4.3 Variation in Retention Time

Retention times for analytes should be constant within a sample batch (e.g., same day, same batch of mobile phase) if all chromatographic conditions are held constant. If a change in retention is observed, it indicates that at least one condition has changed. Because it is impossible to hold all variables exactly constant, there is a normal variation in retention time for every method. Typically this is in the range of

±0.5% of the retention time or ±0.0.02 to 0.05 minute; this normal variation can be determined from historic sample-batch records. When the retention-time variation exceeds the normal variation, steps should be taken to assess the cause and take corrective action.

A tool to implement the initial divide-and-conquer approach (Section 17.3.1) is to calculate the retention factor k (Eq. 2.6), and then compare changes in k and retention time t_R with the help of Table 17.5. Table 17.5 will guide you to one of the following sections for more information about possible changes in the mobile phase, column, or column temperature. As discussed in more detail below, mobile-phase changes tend to occur in a stepwise fashion when some intentional change is made, column changes usually take place over a period of weeks or months, and temperature changes tend to cycle during the day. These general patterns can greatly aid the identification of the problem source.

An alternate way to classify retention-time problems is by the observed change in t_R—increased, decreased, or variable. If you want to approach the problem in this manner, consult Table 17.6 and Section 17.4.3.6 first. The following sections (17.4.3.1–17.4.3.5) cover each of these symptoms. It may be appropriate at this point to run the system reference test (Sections 3.10.1.3, 17.2.1) to determine if the problem is related primarily to the equipment or the method. Finally, there is much overlap among the various causes of variable retention, so it is a good idea to read all of Sections 17.4.3.1 through 17.4.3.5 in order to gather as many ideas as possible, if the solution to the problem is not quickly reached.

17.4.3.1 Flow-Rate Problems

A change in the flow rate will change t_R but not k, because k is independent of flow rate, while t_R varies in inverse proportion to flow rate (an exception can occur for pressures >5000 psi, because of a slight dependence of k on pressure). Never underestimate the power of operator error—it is a good idea to verify that the proper flow rate is selected. Once a setting error has been eliminated, the only possible cause of a higher than normal flow rate is a problem with the system controller, which will require the skills of a trained technician to fix.

Flow rates that are lower than normal cause retention times to be too long, a problem that can be caused by bubbles in the pump, pump starvation, faulty check valves or pump seals, or leaks. Sections 17.4.1 and 17.4.2.2 describe corrective actions for these problems. As discussed in Section 17.4.2.3 for pressure, the causes of variable pressures or flow rates often are the same as low-pressure or low-flow problems

17.4.3.2 Column-Size Problems

A rare, but possible cause of a change in t_R but not k is installation of the wrong column size, but correct type (e.g., Symmetry C18). Of course, the column size cannot change without operator intervention. The obvious fix is to look at the column label and then install the proper column.

17.4.3.3 Mobile-Phase Problems

A change in the mobile phase (e.g., %B) can result in changes in both t_R and k. When a new mobile-phase batch is prepared incorrectly, any changes in retention will be noticed at that time—whether the mobile phase is pre-mixed ("off-line") or prepared by on-line mixing. When on-line mixing systems are used, small variations in mobile-phase composition may occur due to problems with the proportioning system. Alternatively, small, continuous changes in the mobile phase (and sample retention) can occur over time, although this is much less common. For example, a volatile buffer, such as ammonium carbonate, may evaporate so as to change the pH of the mobile phase. If continuous helium sparging is used for degassing, a volatile organic component of the mobile phase could be selectively evaporated.

Errors in formulating the mobile phase are a likely cause of shifts in retention (with increased retention for a reduction in %B, and vice versa). The rule of 2.5 (Section 2.5.1) indicates that a 10% change in %B will change the retention factor by approximately a factor of 2.5 times; a 1% error in mobile phase %B can account for $\approx$10% change in k.

An error in mobile-phase pH can have a much larger effect on the retention of acidic or basic solutes (Section 7.3.4.1) than neutral analytes. The concentration of mobile-phase additives, such as ion-pairing reagents (Section 7.4.1.2), also can affect retention. During method development the robustness of the separation to small changes in mobile-phase composition should have been examined (Section 12.2.6). The results from robustness testing can be useful in determining the specific mobile-phase error that was made. From a practical standpoint, however, the most direct solution is to make up a new batch of mobile phase and determine whether the problem has been corrected.

17.4.3.4 Stationary-Phase Problems

With continued use of the column, changes in retention and selectivity are common, but t_0 is unaffected; consequently values of both k and t_R will change. Retention shifts due to a change in the stationary phase rarely are the only symptom observed. Usually the plate number N will also drop, peak tailing will increase, and the column pressure will rise. Past records of method use (Section 17.2.4) in combination with recent data on the performance of a column (values of N, pressure, etc.) can be used to avoid its (highly undesirable!) failure during the assay of a series of samples. As a rule, a column lifetime of 500 to 2000 sample injections should be expected for most applications (and will account for $<$1% of the total cost of analysis). Some methods can degrade columns more quickly, while other methods may allow a longer use of the column. Expect shorter column lifetimes when the column is operated outside the $2 < \text{pH} < 8$ region or at temperatures $>50^\circ$C. The use of in-line filters (Section 3.4.2.3) and guard columns generally will extend column life. In any event, the column should be considered a consumable item that will wear out (hopefully gradually) over time.

The easiest way to check for stationary-phase related problems is to replace the column (module substitution, Section 17.3.5). If a guard column is in use, first replace or remove the guard column to see if the problem is resolved. If the column repeatedly fails prematurely, check to be sure that it is operated within its recommended limits (consult the column care-and-use instructions for specific guidelines). In some cases it may be appropriate to find and use an equivalent column (Section 5.4.2) that is more stable. If the failure is due to the injection of dirty samples, additional sample-preparation steps (Chapter 16) may be necessary.

17.4.3.5 Temperature Problems

Changes in column temperature affect values of t_R and k. A 1°C increase in column temperature will normally decrease retention by 1–2% (Section 2.3.2.2), so a method that is operated without column-temperature control will be subject to changes in retention as the temperature of the laboratory changes during the day. Temperature changes also can influence selectivity (Section 6.3.2), so shifts in relative retention may also be observed. Many laboratories have stable daytime temperatures, but for energy conservation do not provide the same quality of temperature control at night. Also, even though the laboratory temperature is relatively constant (as measured at a wall-mounted thermostat), the local temperature can fluctuate significantly, especially if a heating duct directs air at or near the HPLC system. For this reason problems related to temperature tend to be exhibited as cyclic changes in retention throughout the day. Temperature-related retention problems can be corrected by using a column oven operated in a range where it has stable temperature control (Section 3.7). Inadequate column temperature control also can cause peak shape problems, as described in Section 17.4.5.3.

17.4.3.6 Retention-Problem Symptoms

This section discusses retention-time problems in terms of symptoms; see the related items in Table 17.6.

Abrupt changes in retention are usually easy to isolate. If these occur when the column is changed, the column itself is the most likely cause. Re-installation of the previous column should confirm this. Column-to-column variation is much less common with today's high-purity, type-B silica columns, but was commonplace with the lower-purity, type-A columns that may still be in use for some legacy methods. Legacy methods may require adjustment of the mobile phase with each new column in order to meet system suitability; an alternative is to order several columns from the same batch of packing material. Redevelopment of the method for a more robust separation is another solution, but it may not be economically feasible. Substitution of an equivalent column (Sections 5.4.2, 6.3.6.1) that is more reproducible is another option. Also, don't overlook the possibility that the wrong column was inadvertently installed.

If the change in retention occurred when a new mobile phase was formulated, the simplest solution is to make another batch of the mobile phase. Be sure that the correct mobile-phase pH is used (Section 7.2.1), and that the pH is adjusted prior to the addition of organic solvent.

Abrupt changes in retention are fairly common when a gradient method is transferred from one HPLC system to another. This usually is due to differences in the system dwell-volume between different equipment (Section 9.2.2.4). Sometimes these differences can be compensated by a change in mobile-phase conditions, the injection timing, or modification of the system plumbing (Section 9.3.8.2; also Section 5.2.1 of [18]).

If retention changes abruptly when none of the above conditions exist, and there is no obvious change in the system operating conditions, it is likely that there is an equipment problem (e.g., check-valve failure), a leak (Table 17.3), a bubble (Table 17.4), or a column-temperature problem (Section 17.4.3.5).

Drifting retention times are a symptom of some instability in the system. When a method is set up, it is not uncommon for retention times to drift for the first few injections; this may be even more pronounced when a new column is installed. The most likely cause of retention-time drift for RPC is incomplete equilibration of the mobile phase and column. Incomplete equilibration can be especially pronounced for ion-pair separations, where 20 to 50 column volumes may be required for equilibration (Section 7.4). For most isocratic methods, however, retention times should stabilize after the first two or three injections. For gradient elution, an increase of the equilibration time between runs may be required to stabilize retention times, especially if the first few peaks in the run are eluted close to t_0 (Section 9.3.7).

A less-common cause of retention-time drift is the presence of slowly equilibrating active sites on the column that become saturated after several injections. When this is the case, several "priming" injections to deactivate the column (Sections 3.10.2.2, 13.3.1.4) may solve the problem. Make several large-mass injections of the sample in a row (it usually is not necessary to make a complete run for each injection, just inject several times with perhaps a 30-second delay between injections), then allow the normal method cycle to run. Sometimes priming injections are required just once for a column, whereas other samples may require priming injections each time the method is started.

If retention time drifts in a continuous fashion over an entire sample batch, it suggests that something is continuously changing in the method; for example, the mobile phase may be unstable. The use of a volatile buffer (e.g., ammonium carbonate) coupled with helium sparging can result in evaporation of the buffer with a change in mobile-phase pH. Similarly loss of the organic component of the mobile phase can occur, but this is uncommon during the course of a day. Re-formulation of the mobile phase on a daily basis may be necessary for some methods. If helium sparging is used (Section 3.3.2), note that it takes only one volume of helium to degas an equal volume of mobile phase (e.g., 1-L of He for 1-L of mobile phase), so a few minutes of vigorous sparging is all that is needed. If continuous sparging is necessary for pump or detector stability, turn down the helium supply to a trickle rather than allow vigorous sparging to continue. If the presence of a small amount of dissolved air is not a problem, in-line vacuum degassing (Section 3.3.3) usually is more convenient and is adequately effective in most cases—without causing mobile-phase evaporation.

Variable retention times for some or all peaks between chromatograms are symptoms that some variable is not adequately controlled. In one example where retention-time variation was observed only in the middle of gradient runs, the cause was related to a mobile-phase proportioning problem (see Section 5.5.4.1 of [18]). An intermittent check-valve failure will cause intermittent flow-rate, and thus retention changes. Temperature fluctuations in the laboratory can change retention on a run-to-run basis. Usually the causes and fixes for variable retention times are similar to those for drifting retention.

When *retention times have decreased*, several possible causes exist. If retention-time loss correlates with larger injected sample-mass and right-triangle peak shapes (e.g., Fig. 17.15*a*), mass overload of the column is likely. Reduction of the injected sample weight should correct this problem. See the discussion of tailing and distorted peaks in Section 17.4.5.3 for more information on mass overload.

When all peaks in the chromatogram show reduced retention, the problem is associated with the column, mobile-phase, temperature, or flow rate. Consult Table 17.5 and the appropriate discussion in Sections 17.4.3.1 through 17.4.3.5 for more information.

When only some peaks in the run have shorter-than-normal retention times, an unexpected change in the system chemistry is suggested; for example, a change in ionization of acidic or basic solutes. Check the mobile-phase pH (prior to addition of organic). Usually a change in the %B will affect all peaks in the run (though not necessarily in an identical way); if this is suspected, make a new batch of mobile phase. Note also that the accuracy of on-line mixing of the mobile phase can vary among different HPLC systems. An aging column can also affect the retention of just some peaks in the chromatogram; installation of a new column will serve to identify the column as the problem source.

Inadequate retention of polar samples is sometimes a problem during RPC method development. If the sample is ionic, it may be possible to change the mobile-phase pH so that the sample is converted to its non-ionized form, which will be less polar and better retained (Section 7.3). An alternative is to use ion pairing to improve the retention of ionic samples (Section 7.4). If the sample is neutral, use of a more polar mobile phase (less strong solvent) should increase retention. However, if the %-organic is $\leq$5%, column de-wetting may occur for alkyl-silica columns (Section 5.4.4.2), with resultant loss of retention. Use of a column containing embedded polar groups or "AQ" type columns may be useful. If other attempts to retain polar compounds by RPC are not successful, a change to normal phase (Chapter 8) and especially hydrophilic interaction chromatography (HILIC, Section 8.6) may provide the desired results. See the additional discussion regarding poor retention of polar solutes in Section 6.6.1.

Retention times that are too long usually have similar causes as those that are too short. Refer to Table 17.5, Sections 17.4.3.1 through 17.4.3.5, and the discussion of smaller than expected retention.

17.4.4 Peak Area

With today's data systems, quantification by peak area is much more common than by peak height (Section 11.2.3), so we will assume peak-area measurements for the current discussion; however, the same troubleshooting process can be used for either peak-height or area problems. If a change in retention accompanies a peak-area problem, first correct the retention problem before addressing the peak-size problem.

Peak-area response for most methods will be very consistent over time. For example, repetitive injections of the same, well-retained sample (e.g., $k>2$) with UV detection and a signal-to-noise ratio of $S/N > 100$, peak area should vary <1% between runs (Section 3.10.1.3). However, smaller peaks, shorter retention times, and/or the use of some other detectors may generate less reproducible results. The following discussion of peak-area related problems is organized by (1) peaks that are larger than expected (Section 17.4.4.1), including peaks in blanks and carryover, (2) smaller than expected peaks (Section 17.4.4.2), and (3) peak areas that are variable from run to run (Section 17.4.4.3). A summary of symptoms and solutions is listed in Table 17.7. In this section, we will assume that the method had been working properly for previous sample batches.

17.4.4.1 Peak Area Too Large

For peak areas that are too large, the first step is to determine if the problem is reproducible, and if it is related to just one sample or solute, or all samples. Answers to these questions usually will require re-injecting one or more samples and/or examining several chromatograms from a batch of samples. If the area is not reproducible between several injections of the same sample, see Section 17.4.4.3 (variable areas). If the sizes of all peaks vary in the same proportion, check to be sure that the correct injection volume is selected. Another possible cause is faulty sample preparation—check to be sure that the dilution or concentration steps were done properly. If the areas for different peaks in the chromatogram have changed by different proportions, the detector settings may be at fault. Check the detector wavelength (UV detector, Section 4.4), interface adjustments (evaporative detectors, Sections 4.12–4.14), time constant, and so forth.

Peaks that appear in a blank injection generally come from one of two sources: late elution or carryover. A peak that is not fully eluted in one run can appear in the next (or later) run; if the sample contains other components, the extra peak will be much broader than the neighboring peaks. This is illustrated in Figure 17.5, where in *a* a broad peak X (arrow) appears at approximately 2 minutes in the chromatogram. In Figure 17.5*b*, the run of Figure 17.5*a* is extended, showing peak X both in the previous run (at ≈2 min) and at its normal place in the chromatogram (≈7 min). If peak X must be quantified in the run, the run can be extended as in Figure 17.5*b* to include the peak in the correct run. If the peak is not of interest, several options are available. The run can be extended as in Figure 17.5*b*, the run time can be adjusted so that the peak appears in the following chromatogram in a region where no other peaks are present, a step-gradient can be used to flush the peak from the column, or sample cleanup can be modified to remove the peak from the sample prior to injection. Carryover results when a small portion of the sample is trapped in or adsorbed on the surfaces of the autosampler and shows up when a blank is injected. Check for carryover as described in Section 17.2.5.10.

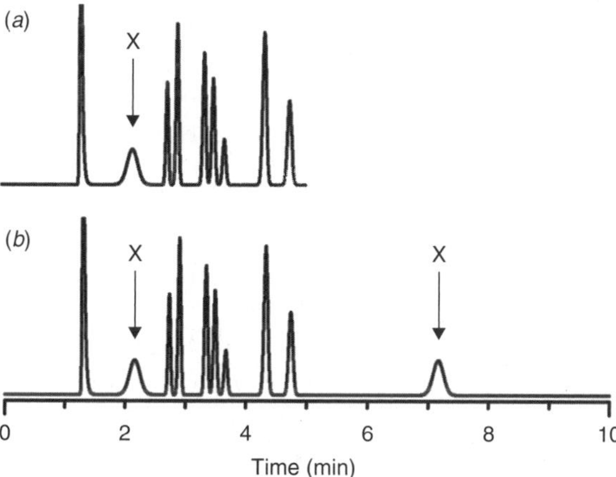

Figure 17.5 Example of late elution. (*a*) Broad peak (X) appears out of place in chromatogram; (*b*) entire chromatogram; extended run time allows peak to elute in proper position in chromatogram (≈7 min).

17.4.4.2 Peak Area Too Small

Peak areas that are smaller than expected can have the same root cause as peak areas that are too large, and the process discussed above (Section 17.4.4.1) can be followed to isolate and identify problems due to small peaks. Of course, carryover and late-elution problems are less applicable for peaks that are too small. Other less common causes of small peaks are a detector time constant that is too large (Section 4.2.3.1), a data sampling rate that is too slow (Section 11.2.1.1), peaks that are off scale (underintegrated), or peaks that are improperly integrated (Section 11.2.1.4).

17.4.4.3 Peak Area Too Variable

If the precision of a method is worse than it has been historically, this will appear as peak areas (or heights) that are more variable than expected. If there also is a retention-time problem, it is best to correct it first (Section 17.4.3). There are many possible causes of variability in peak areas, some of which are also discussed in Section 11.2.4. Nearly any step in sample preparation and analysis can contribute to peak-area variation. Some of the more likely sources are discussed below.

The first step is to determine if the results from a single sample are consistent. If replicate injections of the same sample give consistent peak areas, all the processes from sample injection onward are working properly. The source of the problem then has to be something prior to placing the sample in its vial. Possible problems of this kind include sampling, equipment, and sample preparation errors. *Sampling* is the process of selecting a representative (in this case, equivalent) sample (Section 16.3)—if the master sample is not homogeneous, subsamples may not be equivalent. If volumetric or gravimetric laboratory *equipment* is not accurate or operating properly, error can be introduced, a common source of such error is a pneumatic pipette that is worn beyond acceptable tolerances. The typical sample-preparation process (Chapter 16) has multiple steps in each of which small errors are possible that can affect analyte recovery (e.g., filtration, evaporation, dilution). In a stepwise manner modify the sample preparation process or circumvent specific steps to isolate the source of the problem.

If replicate injections of the same sample give inconsistent peak areas, the problem is likely due to the processes that take place from sample injection onward. The most likely sources are the autosampler, pump, detector, or data-processing steps. First check the *autosampler* by rerunning the reproducibility test of Section 3.10.1 to see how it compares to past tests (Section 17.2.4); make any necessary repairs. *Pump* malfunction can lead to a change in mobile-phase flow rate, another possible source of peak-area variation (check this by running a flow-rate test, Section 3.10.1.3). *Detection* problems, such as detector overload or poor wavelength selection might affect one peak and not another. If detector overload is suspected (very large peaks, e.g., >1 AU for a UV detector), dilute the sample or inject a smaller volume to see if smaller peaks give more consistent areas. For LC-MS detectors with an electrospray interface (Section 4.14.1.1), a poorly performing spray tip can result in different amounts of sample getting into the MS at different times in the chromatogram. The *integration* and data workup process might have problems, such as if a peak had a start or stop time improperly set, or the data sampling rate was too slow (Section 11.2.1). Another occasional case of variable peak area can occur if a frozen sample is not properly thawed and/or mixed prior to injection. A gradient of analyte concentrations may then occur from the top to the bottom of a vial. In this case replicate injections from such a sample may show a descending or ascending (depending on the nature of analyte and matrix) series of peak areas.

17.4.5 Other Problems Associated with the Chromatogram

In addition to the symptoms discussed in the preceding sections, chromatograms often exhibit obvious defects in appearance which can be used to isolate the cause of the problem. This section covers three of these:

- baseline drift
- baseline noise
- peak shape

17.4.5.1 Baseline Drift Problems

Baseline drift is defined as a continuous rise and/or fall of the chromatographic baseline extending over a period of tens of minutes to hours (Section 4.2.3.1). Drift can occur in a rising, falling, or cycling pattern, as well as exhibit other characteristics. Some of the symptoms and causes of drift are summarized in Table 17.8. It should be noted that some drift is expected;, for example, one UV detector specifies drift of $\leq 2 \times 10^{-4}$ AU/hr at 250 nm at constant room temperature and with air in the cell and $\leq 3 \times 10^{-4}$ AU/hr with a room temperature fluctuation of $\leq 2°$C [19].

Periodic drift is characterized by a cyclic pattern, with the baseline rising and then falling (or vice versa) over one or more runs. This is most common with gradient elution within a single run, as a result of a mismatch of the detector response to the mobile phase A- and B-solvents. This is illustrated in the baselines of Figure 17.6 [20]. Baseline (Fig.17.6*a*) is for a gradient run from 5–80% water/MeOH at 215 nm, with drift of ≈0.9 AU (because MeOH has much stronger absorbance than water at this wavelength; see data of Table I.2, Appendix I). Such drift is normal and

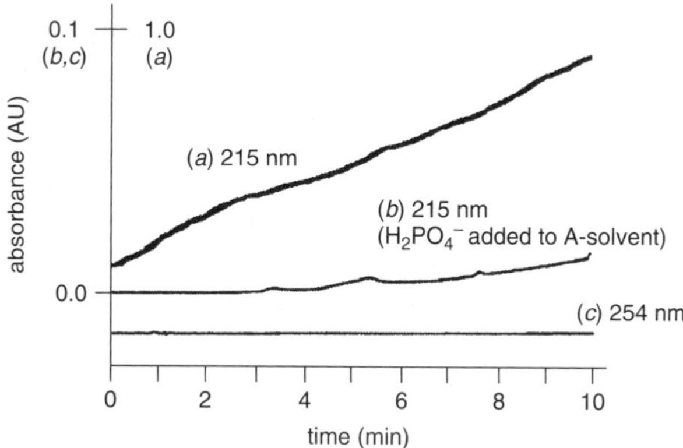

Figure 17.6 Baselines obtained using water-methanol or phosphate-methanol gradients, 5–80% B in 10 minutes. (*a*) Gradient at 215 nm and 1.0 AU full-scale; solvent A: water; solvent B: methanol; (*b*) same as (*a*), except solvent A: 10 mM potassium phosphate (pH-2.8) and 0.1 AU full-scale; (*c*) same as (*a*), except 254 nm and 0.1 AU full-scale. Adapted from [20].

is a problem only if it precludes accurate integration of the chromatogram. If the drift is unacceptable, there are three general approaches for addressing the problem. One option is to add a UV-absorbing reagent to the A-solvent. In the example of Figure 17.6*b*, the use of 10-mM phosphate buffer (pH-2.8) instead of water reduced the drift of Figure 17.6*a* by nearly 30-fold. Because drift will be less severe at longer wavelengths, another option is to increase the detection wavelength, provided that the sample response is acceptable at the new wavelength (UV detection is assumed; other detectors may offer other options). The effect of a wavelength change is seen by comparing Figure 17.6*a* (215 nm) with Figure 17.6*c* (254 nm). Alternatively, a less-absorbing organic solvent might be chosen. In this case ACN could be used instead of MeOH (not shown); ACN has negligible drift at 215 nm and may be used successfully for gradients at 200 nm or above. Of course, a change in mobile-phase A or B can change the chromatographic selectivity, so further adjustments in the method may be necessary (only applicable for method development).

Negative baseline drift can be a greater problem because data systems typically stop integrating when the detector reads less than −0.1 AU (−10% drift). Thus, if the gradient-elution baseline of Figure 17.7*a* [20] is encountered, it is likely that the baseline will drop off scale in a negative direction, with loss of the data (it was possible to collect this baseline only by turning off the auto-zero function and manually setting the baseline start at +1 AU). As in Figure 17.6*a*, *c*, the drift of Figure 17.7 is much less at 254 nm (Fig. 17.6*c*) than 215 nm (Fig. 17.6*a*). The negative drift of Figure 17.7*a* could be converted into a (more acceptable) positive drift by adding a UV-absorbing buffer to the B-solvent (Fig. 17.8*a* [20]). Another possible fix with some data systems is to adjust the scale of the data channel to a range of 0.0 to −1.0 AU.

In some cases, however, the use of mobile-phase additives as in Figure 17.8*a* cannot correct severe, negative drift. In the example of Figure 17.9*a*, the baseline for this ammonium bicarbonate-methanol gradient exhibits a negative dip in the

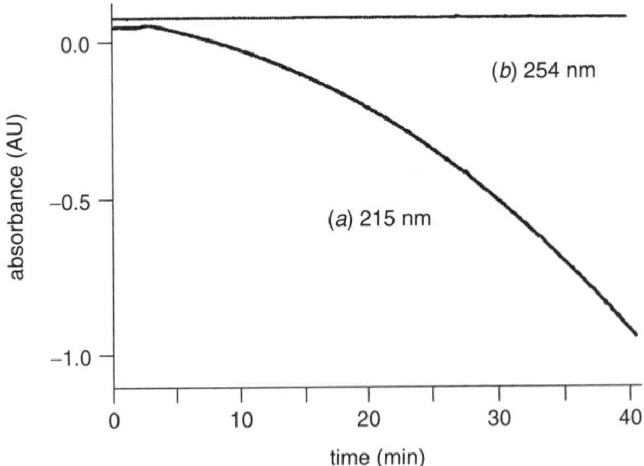

Figure 17.7 Baselines obtained using ammonium acetate-methanol gradients. Solvent A: 25-mM ammonium acetate (pH-4); solvent B: 80% methanol in water; gradient: 5–100% B in 40 minutes. (*a*)215-nm detection; (*b*)254 nm. Adapted from [20].

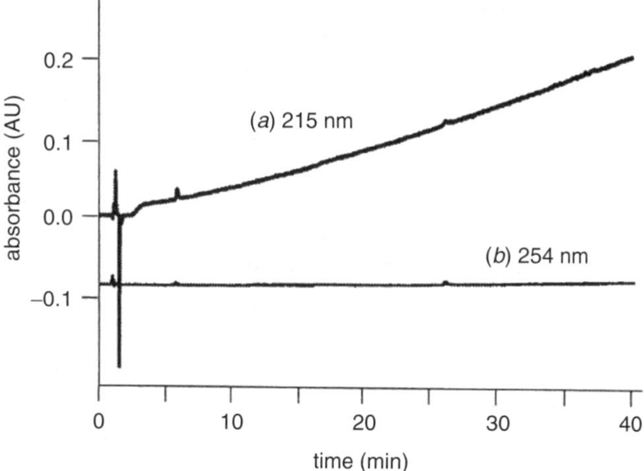

Figure 17.8 Baselines obtained using equimolar ammonium acetate-methanol gradients as in Figure 17.8, but with buffer added to B-solvent. Solvent A: 25-mM ammonium acetate (pH-4) in 5% methanol; solvent B: 25-mM ammonium acetate in 80% methanol; gradient 0–100% B in 40 minutes. (*a*) 215-nm detection; (*b*)254 nm. Adapted from [20].

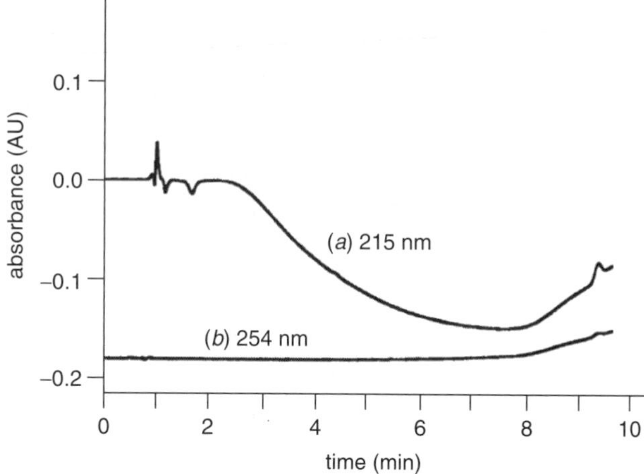

Figure 17.9 Baselines obtained using ammonium bicarbonate-methanol gradients. Solvent A: 50-mM ammonium bicarbonate (pH-9); solvent B: methanol; gradient: 5–60% B in 10 minutes (*a*) 215-nm detection; (*b*)254 nm. Adapted from [20].

middle at 215 nm. Adjustment of the absorbance of either the A- or B-solvent cannot solve this problem. Although this mobile phase is unacceptable for detection at 215 nm (Fig. 17.9*a*), detection at 254 nm (Fig. 17.9*b*) poses no problem. An alternative detector might also be used; for example, bicarbonate mobile phases are commonly used with LC-MS, without creating baseline problems. Fluorescence detection is another option used to obtain flat baselines for gradient elution of fluorescent analytes.

A *change in temperature of the column (and mobile phase)* is another major cause of periodic baseline drift. A change in mobile-phase temperature changes the refractive index of the mobile phase and the transmission of light through the UV-detector cell. If the column is operated without adequate temperature control (Section 3.7.1), the baseline is likely to drift as the laboratory temperature changes. Temperature-related baseline drift can be confirmed by related changes in retention times with temperature. See Section 17.4.3.5 for further discussion of temperature-related problems.

Other types of isocratic baseline drift are not cyclic, and these may arise from different causes. *Slow system equilibration* after a change of conditions (mobile phase, column, column temperature, flow rate, etc.) will result in initial baseline drift that usually subsides within 30 to 60 minutes. Baseline drift associated with equilibration may be accompanied by retention-time drift. Similarly, when a detector is first turned on, the detector response may drift for a few minutes or even hours as the lamp, electrodes, or other detector elements warm up and stabilize.

17.4.5.2 Baseline Noise Problems

Disturbances in the baseline are referred to as baseline noise. The characteristics of baseline noise can help identify its source. Baseline disturbances can be periodic or random, and the duration of the disturbances can be shorter (short–term noise) or longer (long-term noise) than the width of a chromatographic peak. Moreover, baseline noise is superimposed upon any baseline drift. In addition to the discussion below, consult Table 17.9 as well as Sections 3.3.1 (degassing), 3.8.3 (data rates), 4.2.3 (noise), 11.2.1.1 (data sampling contributions), 11.2.4.2 (chromatographic sources), and 11.2.4.3 (detection sources).

High-frequency short-term noise shows up as the "buzz" on the baseline (e.g., Fig. 4.5) resulting from electronic noise on the electrical circuits. This has a period of 60 Hz (North America) or 50 Hz (most of the rest of the world), depending on the frequency of the alternating-current electrical supply. High-frequency noise usually can be significantly reduced as discussed in Section 4.2.3.1 by the use of a cleaner electrical supply (e.g., use an uninterruptable power supply, UPS) and/or selection of a larger detector time-constant. Figure 4.5 shows the reduction of noise by approximately 300-fold by the use of a simple noise filter.

Random and *low-frequency* short-term noise can result from several different sources. Insufficient *degassing* can lead to the introduction of air bubbles into the HPLC system. Bubbles trapped in the pump head(s) can also cause baseline disturbances as the pressure fluctuates from one piston stroke to the next, giving a regular pattern to the baseline noise. Bubbles in the pump should be accompanied by pressure fluctuations as described in Section 17.4.2.3. Bubbles that make it through the pump, or that are formed after the pump by mixing inadequately degassed mobile phase in high-pressure-mixing systems, often will be kept in solution due to the system pressure. However, when the dissolved air leaves the column, the pressure is greatly reduced and the bubbles may reform. As the bubbles pass through the detector, random, sharp spikes may appear, especially with optical detectors (e.g., UV-visible, Section 4.4; fluorescence, Section 4.5; refractive index, Section 4.11). Detectors that evaporate the mobile phase (e.g., Sections 4.12–4.14) are, of course, not susceptible to mobile-phase bubble problems. If the bubble is trapped in the flow cell, a large shift in baseline may result. Adding a back-pressure restrictor after the detector (Section 4.2.1) may solve bubble problems in optical detectors.

Electrical spikes are similar to bubbles. But to distinguish their presence from bubbles, turn off the pump flow and monitor the baseline. If the spiking continues, the problem is electronic; if the spiking stops and the baseline remains steady, the problem is due to a bubble. The use of better degassing procedures (Section 3.3.1) is the first line of defense against bubbles. A back-pressure restrictor (Section 4.2.1) will keep bubbles in solution until after they leave the detector.

The selection of a *data collection rate* that is too fast can result in excessive short-term baseline noise. As described in Section 3.8.3, the data rate should be set to collect ≈20 points across the peak. Higher data rates will increase the baseline noise while having little benefit on the amount of signal collected, so the signal-to-noise ratio (Section 4.2.3) will worsen. Lower data rates may reduce baseline noise, but this risks reducing the signal as well, so the signal-to-noise ratio may suffer.

Long-term noise shows up as baseline disturbances that are comparable in size (or wider) to normal peaks. One common source of long-term noise is the presence of *late-eluted materials* in the sample (see the discussion of Fig. 17.5 in Section 17.4.4.1). As retention time increases for solutes or background interferences in the sample, the band width increases and the peak height decreases. Late-eluting peaks from prior separations can accumulate over time, resulting in a drifting and erratic baseline. A strong-solvent flush of the column (e.g., 25 mL of methanol or acetonitrile) often will remove strongly retained material from the column. For this reason a strong-solvent flush is recommended following each batch of samples (isocratic separation assumed). For some methods a column flush may be needed more often. Gradient methods usually are less susceptible to late-eluted interferences because they have a strong-solvent column-wash built into every run. Heroic efforts to remove strongly retained materials (e.g., flushing with acid, base, chaotropes, or methylene chloride) can be effective but can also damage the column. A better approach is to use improved sample pretreatment (Chapter 16) to reduce the sample burden of late-eluted materials. Remember, the column is a consumable item. Once 500 or so samples are analyzed, the cost per sample for the column becomes a trivial portion of the overall analysis cost, so column replacement often is a better choice than extensive column cleaning or sample pre-treatment.

Sometimes long-term noise shows up as regular baseline fluctuations, as in Figure 17.10 (note that the y-axis is 1 mAU full scale). Usually cyclic baseline disturbances are caused by pump problems and will be accompanied by pressure

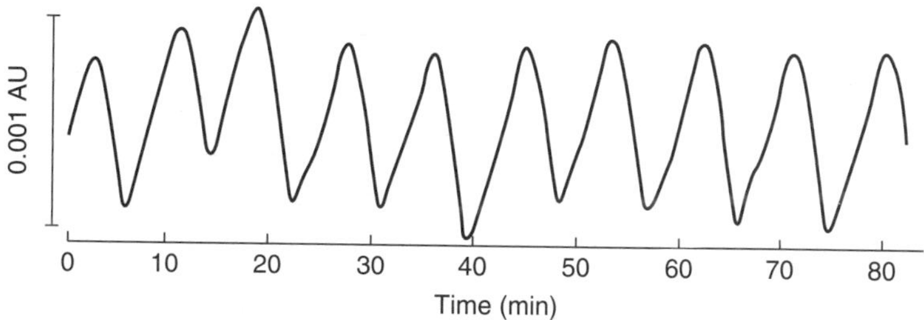

Figure 17.10 Cyclic baseline noise that was attributed to interference from an electronic air filter in the laboratory. Adapted from [21].

pulsations (Section 17.4.2.3) or leaks (Section 17.4.1). In the case of Figure 17.10 isolation of the HPLC system from the electrical circuit, as well as pump and detector troubleshooting, could not solve the problem. Rather, movement of the system away from its original location reduced the frequency of the noise, and with sufficient distance eliminated it. Although the source of noise was never definitively identified, the noise was attributed to interference from an electronic air filter [21].

Artifact or "ghost" peaks in blank gradients represent a special type of long-term noise. An example of this is shown in Figure 17.11*a* for a gradient of 5–80% ACN-phosphate buffer (pH-7) in 15 minutes, with a hold at 80%B [7]. Ideally the baseline should be free of peaks in this blank gradient (the baseline drift is caused by differences in absorbance of the A- and B-solvents as discussed above). The most likely source of peaks in blank gradients is contamination of the A-solvent, since these contaminants tend to concentrate at the head of the column during equilibration between runs—followed by their elution during the gradient. A simple way to confirm A-solvent contamination is to increase the equilibration time between runs (flowing A-solvent). If the contaminants arise from the A-solvent, all the peaks should increase roughly in proportion to the increased equilibration time (i.e., a larger volume of A-solvent, with an increase of collected contaminants). In the present example, the 10-minute equilibration of Figure 17.11*a* was extended to 30 minutes (Fig. 17.11*b*) and the gradient was repeated. It can be seen that the peaks are each about three times larger, so contamination of the A-solvent is confirmed. Further isolation of the problem identified the pH-meter probe as the source of contamination in this example [7]. Figure 17.12 compares results for a blank run made with buffer prepared by dipping the pH probe in the buffer to adjust the pH (Fig. 17.12*a*) with results from the use of buffer made without contact with the pH probe (Fig. 17.12*b*). Additional examples of gradient ghost peaks originating

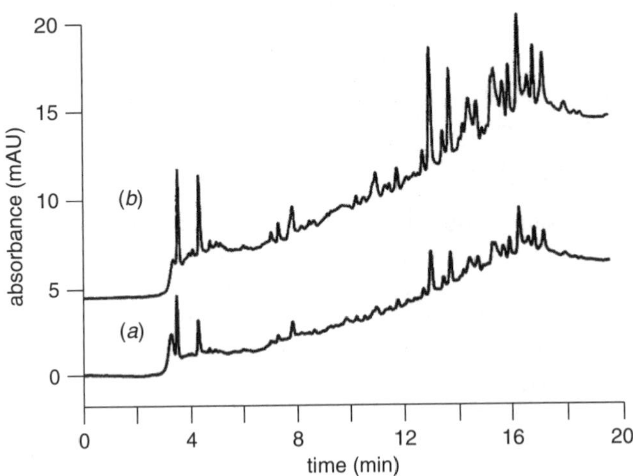

Figure 17.11 Effect of impurities in the A-solvent on a gradient chromatogram: blank gradient runs after (*a*) 10-minute and (*b*) 30-minute equilibration. C_{18} column; gradient 5–80% ACN–10-mM phosphate buffer (pH-7) in 15-minute plus 5-minute hold at 80%; UV detection at 215 nm. Adapted from [7].

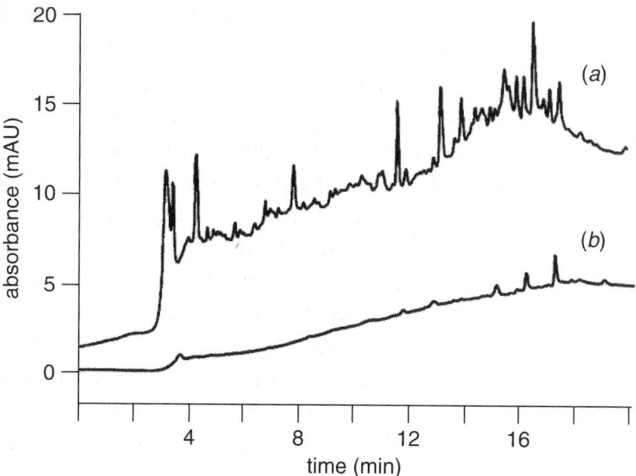

Figure 17.12 Comparison of (*a*) contaminated buffer (same conditions as Fig. 17.12*a*), and (*b*) buffer prepared with extra-clean glassware and no exposure to the pH probe. All other conditions as in Figure 17.12. Adapted from [7].

from water and other reagents can be found in Section 5.5.4 of [18]. An excellent discussion of artifact peaks in gradients also can be found in [6].

17.4.5.3 Peak Shape Problems

The ideal chromatographic peak is a symmetrical, Gaussian curve. Deviations from an ideal peak shape can be quantified by the peak tailing factor or peak asymmetry (Section 2.4.2), as illustrated in Figure 2.16*a*. Peak shape problems may or may not be accompanied by abnormal retention times. Deviations from symmetry can be classified as:

- tailing peaks
- fronting peaks
- broad peaks
- split or distorted peaks

These irregular peak shapes are discussed below, with a corresponding summary of symptoms, causes, and solutions in Table 17.10. Section 2.4.2 also contains a detailed discussion of some of the causes of peak shape irregularities.

Peak tailing is the most common peak-shape problem. New-column specifications often allow peaks with tailing factors $TF \leq 1.2$, so a small amount of peak tailing should be considered as normal. Although regulatory agency guidelines [22] allow TF ≤ 2 for pharmaceutical methods, peaks with $TF< 1.5$ are preferred. Peak tailing tends to increase over time, due to deterioration of the column; when $TF > 2$ is observed, action should be taken to reduce peak tailing (e.g., replace the column). If all peaks are severely tailing, see the discussion below for split or distorted peaks. If the early peaks tail more than later peaks, extra-column peak broadening may be the source of the problem. This is illustrated in Figure 17.13, where the tailing factor ranges from $TF \approx 2.5$ for the first peak to $TF \approx 1.2$ for the last peak. If

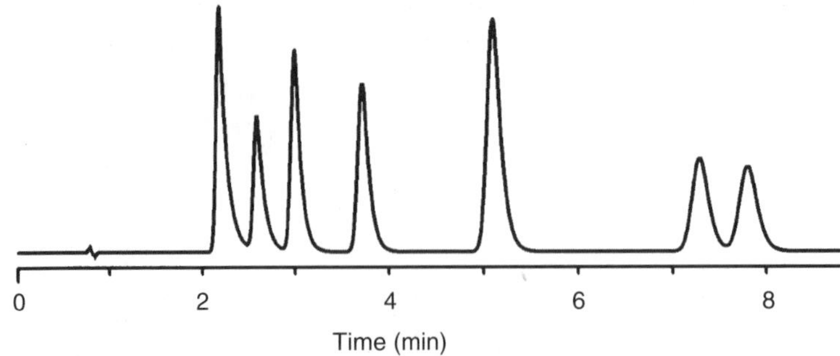

Figure 17.13 Peak tailing from extra-column effects. $TF \approx 2.5$ for first peak, $TF \approx 1.5$ for last peak.

extra-column peak broadening is suspected, reduce extra-column volume, such as by using shorter lengths of smaller i.d. tubing and ensuring that all connections are made properly (Sections 2.4.1, 3.9).

In the past a common cause of peak tailing was the *strong interaction of ionized basic compounds* BH^+ with ionized silanols–SiO^- on the column (Section 7.3.4.2):

$$BH^+ + \mathrm{SiO}^- K^+ \Leftrightarrow BH^+\mathrm{SiO}^- + K^+$$

However, this problem is becoming less frequent with the use of less-acidic columns made from type-B silica (Section 5.2.2.2). Apart from a change to a type-B column, peak tailing of this kind can also be reduced by changes in the mobile phase. Silanol ionization and tailing decrease as the pH is lowered, while solute ionization and tailing decrease for a pH $\gg$ pK_a for the solute. Increased ion-pairing of the solute BH^+ or addition of the competing ion triethylamine (in the protonated form) to the mobile phase can be effective in reducing peak tailing. An increase in ionic strength can also reduce peak tailing, although this is usually less effective than are other changes in the mobile phase. However, these changes in column or mobile phase need to be addressed during method development; otherwise, the method may require re-validation.

An example of the effect of a change in the mobile phase on peak tailing is shown in Figure 17.14 for the analysis of 4 proteins by gradient elution with a TFA/ACN mobile phase on 3 columns of different purity silica (column A, high purity [i.e., type-B]; B, intermediate purity; C, low purity [type-A]). When 0.1% TFA is used in the mobile phase, there is little difference between the chromatograms observed on the three columns (Fig. 17.14*a–c*). However, when the TFA concentration is reduced to 0.01% TFA, the proteins show strong interaction with the lower purity columns, as exhibited by greater peak tailing (Fig. 17.14*d–e*), and longer retention times for the low-purity column (*f*). The effect of insufficient TFA becomes more pronounced as the column is changed from high-purity (*d*) to intermediate-purity (*e*) to low-purity (*f*) silica, because of greater silanol ionization. TFA is used as an ion-pairing reagent for proteins (Sections 7.4, 13.4.1.2), so the advantage of higher TFA concentration may be due to increased ion-pairing, although a corresponding decrease of pH and increase of ionic strength may also contribute to better peak shape.

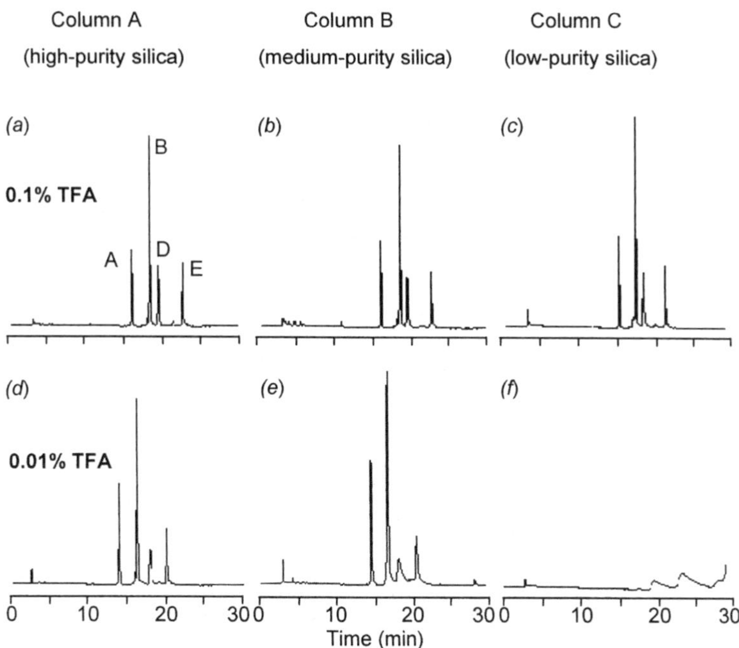

Figure 17.14 Influence of trifluoroactic acid concentration on peak tailing for columns of varying silica purity. (*a*, *d*) High-purity silica (column A); (*b*, *e*) intermediate purity silica (column B); (*c*, *f*) low-purity silica (column C). (*a-c*) 0.1% Trifluoroacetic acid (TFA); (*d-f*) 0.01% TFA. C_{18}, 30-nm pore silica particles, 5–70% ACN/TFA gradients. Sample: A, ribonuclease A; B, cytochrome C; C, holo-transferin, D, apomyoglobin. Data courtesy of Advanced Chromatography Technologies (ACT).

Injection of too large a mass of sample can result in *mass overload* of the column. Mass overload can occur for one or more peaks in the chromatogram, and peak tailing then takes on a right-triangle appearance with a concurrent reduction in retention time as the mass on column is increased, as seen in Figure 17.15*a* (right to left, 0.01 to 5 μg [23]). *Mass* overload is confirmed if dilution of the sample or injection of a smaller sample volume gives a longer retention time and a reduction in peak tailing. Fully ionized compounds exhibit mass overload for sample weights about 50-fold smaller than for other compounds (Section 15.3.2.1); for a 4.6-mm-i.d. column, mass overload and peak tailing occur for 1 > μg of an ionized solute, as opposed to about 50 μg of a neutral solute.

Peak fronting is much less common in RPC than is peak tailing. As with peak tailing, a small amount of peak fronting can be tolerated; many column manufacturers' specifications allow for some peak fronting: TF $\geq$ 0.9. Peak fronting has been attributed to temperature problems with ion pairing [24, 25], but these reports are for older type-A, low-purity silica columns and do not seem to be prevalent with type-B, higher purity columns. Usually fronting is attributed to a void in the column (column collapse), and the result can be quite dramatic, as shown in Figure 17.16*b* [26]. In this case the C_{18} column was operated at pH-9, above its recommended operating pH. Sample analysis proceeded with normal peak shape (Fig. 17.16*a*) for $\approx$500 injections, then suddenly, from one injection to the next,

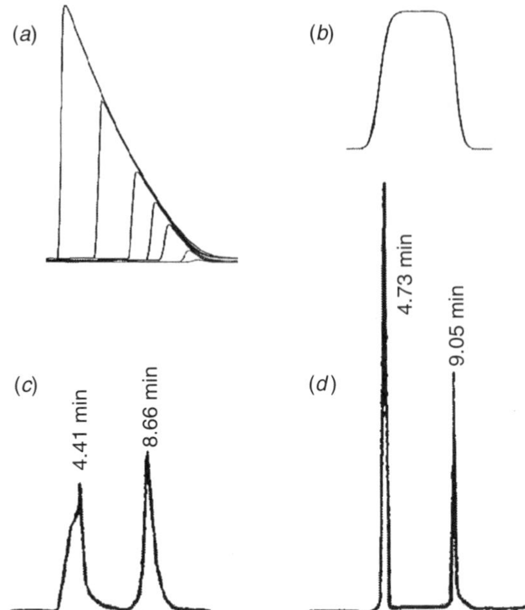

Figure 17.15 Examples of peak distortion. (*a*) Right-triangle peak shape as a result of a too-large sample weight; right to left 0.01–5 μg nortryptiline; 150 × 4.6-mm i.d. (3.5-μm) XTerra MS C18 column; 28% ACN pH-2.7 (20-mM formic acid); adapted from [23]. (*b*) Flat-topped peak characteristic of either injection volume overload or detector overload. (*c*) Distortion due to injection of too large a volume of too strong an injection solvent (30 μL of 100% ACN injected in a 18% ACN mobile phase); (*d*) same conditions as (*c*), except mobile phase used as injection solvent. Conditions for (*c, d*): 250 × 4-mm Lichrosorb RP-18, 18:81:1 ACN-water-acetic acid mobile phase; adapted from [29].

the peak began to front (Fig. 17.16*b*). No change in peak area was observed, so data collection for the portion of the run with the fronting peak could be used for quantitative purposes. Column flushing and other restorative measures could not regenerate the column, so the column was discarded. Such a pattern of failure is typical for this method and is attributed to a collapse of the column bed due to dissolution of the silica at high pH. For this method, it was deemed more prudent to replace the column each time it failed rather than go to the time and expense to redevelop and revalidate the method for a more stable column. Other potential causes of fronting include limited analyte solubility in the mobile phase, a tendency of analyte molecules to aggregate, and conformational changes in the analyte molecules.

Broad peaks may be a precursor of split or distorted peaks, so the discussion of these peak types (later in this section) should be consulted if the suggestions below do not solve the problem. Broad peaks have a calculated plate number N that is significantly less than the specified value for a new column. Because columns are initially tested by the manufacturer with "ideal" samples and conditions, the resulting plate number N may be significantly larger than for "real" samples and conditions. It is best to measure the plate number (or peak width for gradient elution) on a new column and one or more sample solutes for each method; this

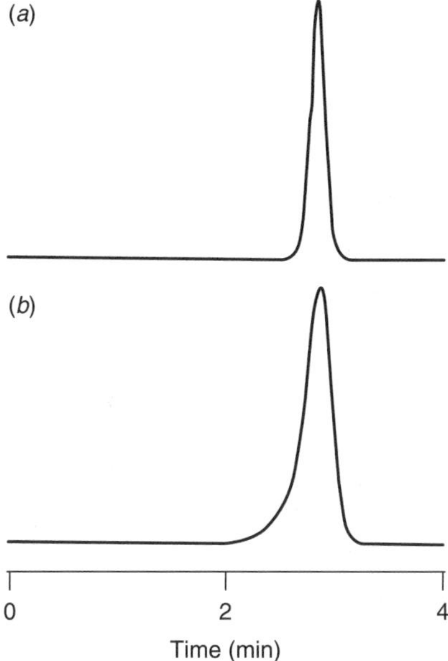

Figure 17.16 Peak fronting due to column collapse. (*a*) Normal chromatogram; (*b*) peak fronting after ≈500 samples run at pH-9 with a C_{18} column. Adapted from [26].

can serve as a reference value for the method. Acceptable peak broadening will vary from one application to the next, and a reference value should be determined as part of the system suitability test. As a rule of thumb, if N for a routine assay with a new column is less than about 75% of N under the manufacturer's test conditions, the cause for this lower value of N should be investigated.

As the column ages, a reduction in plate number will occur, so it is normal for peaks to become increasingly broad over the lifetime of the column. If peak broadening is observed after 500 or more injections, this is the likely cause. Flushing the column with strong solvent (e.g., 20–30 mL ACN or MeOH) may improve peak shape, but column replacement is often the most expedient corrective action.

If a chromatogram exhibits a single broad peak and the surrounding peaks are narrower, the single broad peak is likely a late-eluting peak from the prior injection. Late elution can be confirmed by extending the run to allow sufficient time for the peak to elute in the correct run (Section 17.4.4.1, Fig. 17.5).

Extra-column effects may generate tailing peaks as in Figure 17.13, with some peak broadening. Reduce extra-column effects by using shorter lengths of narrower diameter tubing to connect the column to the autosampler and detector.

Signal processing problems also can result in broad peaks. If a detector's time constant (noise filter) is used, it should be no larger than 1/10 the width of the narrowest peak of interest. For example, a 1-second time constant is suitable for a 10-second wide peak (measured at the baseline), but 5-second time constant may broaden the peak excessively. In a similar manner the data rate of the data system should be sufficiently fast to collect a minimum of 20 data points across the peak (Section 11.2.1.1). With some detectors, such as MS, peak smoothing can be used to improve the appearance of the peak. However, excessive smoothing can broaden peaks.

Too large an injection volume can result in broader peaks that may even appear to be flat-topped—as in Figure 17.15*b*. For injection in mobile phase as the injection solvent, the injection volume should be no more than ≈15% of the volume of the first peak of interest (Sections 2.6.1, 3.6.3, 15.3.2.2, Table 3.3) for a 5% increase in peak width (5% loss in resolution). If sample solvents are used that are more than ≈10% weaker than the mobile phase, larger injection volumes may be possible (Sections 2.6.1, 3.6.3). Flat-topped peaks are also characteristic of detector overload. At low concentrations, the detector response will increase in the normal manner for an increase in analyte concentration (e.g., the front and back of a peak), but when the detector is overloaded, no increase in response is seen for an increase in concentration—the peak appears with a flat top.

Split or distorted peaks can appear for just one peak, several, or all the peaks in a chromatogram. Examples of split peaks throughout the chromatogram are shown in Figure 17.17. Split or distorted peaks for all peaks in the chromatogram is a classic symptom of a *blocked frit* or (less commonly) a column void, and often this is accompanied by an increase in pressure. The effects of a blocked frit or column void are illustrated in Figure 17.18. Figure 17.18*a* represents the inlet of a normal column with a frit in place; when the sample is injected, all portions of the sample stream (arrows) arrive at the top of the column at the same time. The chromatographic separation thus starts for the entire front edge of the sample at the same time. When the frit is partially blocked (Fig. 17.18*b*), the sample stream is distorted such that some portions of the sample reach the head of the column late—with peak tailing. A void at the head of the column (Fig. 17.18*c*) may also cause peak distortion, but usually a strongly fronting peak is the result (as in Figure 17.16*b*). Because these distortions happens before any chromatographic separation has taken place, each peak is distorted in the same way as the peaks migrate through the column.

Reversal of the column is the best way to flush particulate matter from the top of the inlet frit and restore normal peak shapes. After reversing the column, flow 20 to 30 mL of solvent through the column to waste (not the detector), then reconnect the column and leave it in the reverse-flow direction (the problem may recur if the column is returned to its original direction). For example, the chromatogram of Figure 17.19*a* was observed with all peaks doubled [27]. Injection of a reference

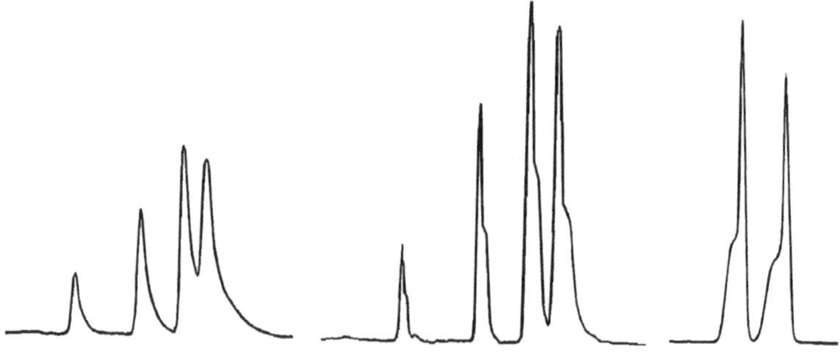

Figure 17.17 Examples of similar peak distortion for all peaks in the chromatogram, attributed in each case to a partially blocked column-inlet frit.

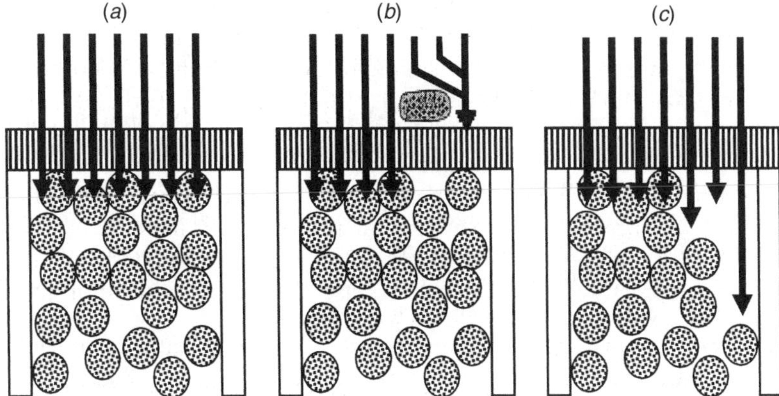

Figure 17.18 Peak distortion at the column inlet. Arrows represent flow streams for sample components as they enter the column. (*a*) Normal flow of sample through column-inlet frit during injection. (*b*) Distortion of sample stream due to partially blocked column-inlet frit. (*c*) Distortion of sample stream due to void at head of column.

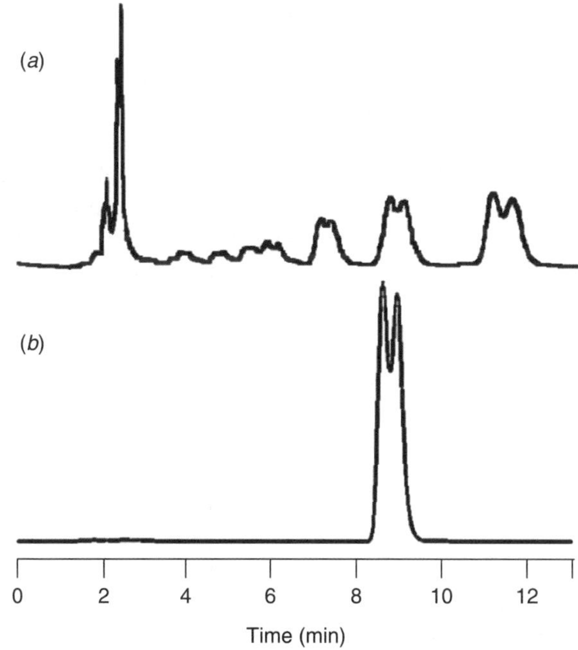

Figure 17.19 Split peaks attributed to a partially blocked column-inlet frit. (*a*) Splitting observed for all peaks in a normal sample; (*b*) splitting for reference standard. Adapted from [27].

standard also resulted in a split peak (Fig. 17.19*b*), but reversal of the column corrected the problem. Permanent column-reversal is not recommended for some columns. For example, some 3-μm particle columns use a 0.5-μm porosity frit on the outlet end and a 2-μm frit on the inlet end. The 2 μm is not fine enough to hold the 3-μm particles in the column, but a brief back-flushing usually will cause no harm if the column is returned to the normal flow direction. If in doubt, consult the column care-and-use instructions for flow-direction limitations. In the past it was common to replace the column inlet frit when blockage was suspected, but today's columns may be permanently damaged when the column end-fitting is removed, so this technique is no longer recommended. Use of a 0.5-μm porosity in-line filter (Section 3.4.2.3), sample filtration or centrifugation, and/or better sample cleanup may prevent the recurrence of blocked column frits.

Inadequate control of the column temperature can cause distorted peaks, especially for peaks later in the chromatogram. In the example of Figure 17.20*a*, the mobile phase is preheated to the column temperature (56°C), and acceptable peak shapes are observed. In Figure 17.20*b* the temperature of the mobile phase entering the column was 39°C (i.e., 17°C lower than the column temperature because of inadequate pre-heating of the mobile phase). As a result peaks are broader than those in Figure 17.20*a*, and later peaks are visibly distorted. Peak broadening and distortion in Figure 17.20*b* are caused by a radial temperature-gradient along and across the column, as illustrated in Figure 17.21. In Figure 17.21*a*, the incoming mobile phase and the column are in thermal equilibrium, so the mobile-phase temperature is the same throughout the column, and the sample band (shown as dots) travels at a uniform velocity down the column. When cooler mobile phase is introduced into the column (Fig. 17.21*b*), the mobile phase at the center of the column is cooler (and

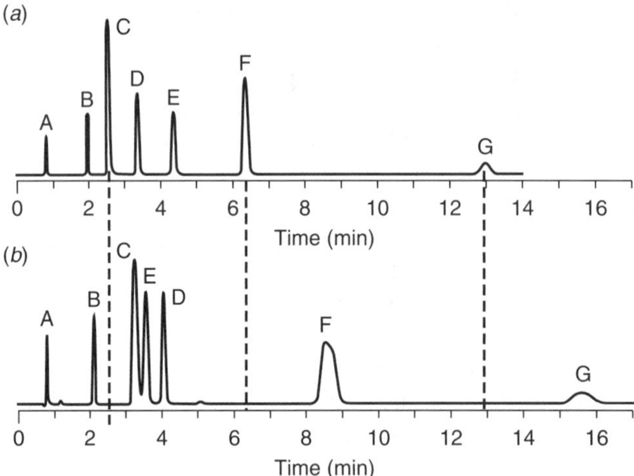

Figure 17.20 Effect of temperature mismatch of incoming mobile-phase and column. (*a*) Inlet solvent 56°C; oven 56°C (matched temperatures). (*b*) Inlet solvent 39°C; oven 56°C (unmatched temperatures). Sample: A, uracil; B, nitroethane; C, phthalic acid; D, 3,5-dimethylaniline; E, 4-chloroaniline; F, 3-cyanobenzoic acid; and G, 1-nitrobutane. 150 × 4.6-mm (5 μm) Zorbax SB-C_{18} column; 90/10 50 mM potassium phosphate (pH-2.6)/ACN at 2.0 mL/min. Adapted from [28].

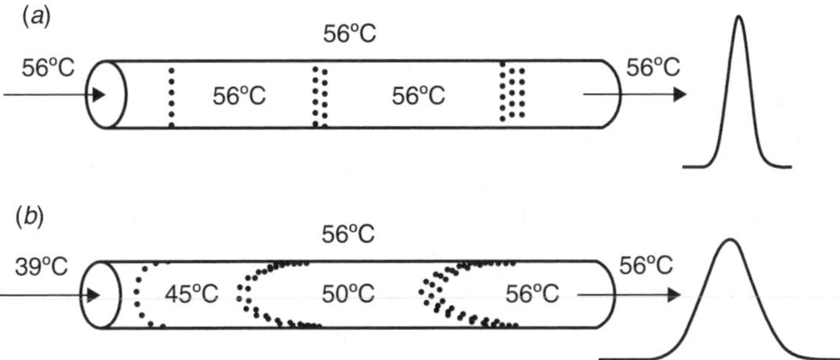

Figure 17.21 Band broadening due to thermal effects. (*a*) Ideal case, no thermal effects; (*b*) effect of incoming mobile phase that is at a lower temperature than the column. Adapted from [28].

more viscous) than that at the walls. Then molecules of the sample near the center flow more slowly through the column than at the column wall, so the sample band widens and can become distorted. Because the average temperature in the column is lower when cooler mobile phase is introduced, retention times also increase—with some changes in selectivity (note that peaks D and E appear in reversed order in Fig. 17.21*a*, *b*). The solution to the problem of Figure 17.20 is better control of mobile-phase temperature, specifically by pre-heating the incoming mobile phase (Section 3.7.1). When the temperature of the incoming mobile phase is within ±6°C of the column, peak distortion is not likely to occur [28].

Three causes of misshapen peaks are related to *sample injection*: distortion from too strong a sample solvent, broadening from too large an injection volume (see discussion under broad peaks), and tailing from too large an injection mass (see the discussion above of mass overload). If the sample solvent used for the injection is too strong, the solvent will not be diluted quickly enough by the mobile phase, so part of the sample may travel too quickly through the column before it is fully diluted. This can cause peak distortion, especially of early peaks in the chromatogram, as illustrated in Figure 17.15*c*. Use of a sample solvent that is too strong can reduce retention time. In the chromatogram of Figure 17.15*c* [29], a 30-μL of sample diluted in 100% ACN was injected with a mobile phase of 18% ACN—which is much too strong an injection solvent for this injection volume. The first peak is badly distorted, and the second peak is broadened. The retention times are also shorter than normal for both peaks. This problem can be corrected either by (1) diluting the sample so that the solvent is no stronger than the mobile phase, and/or (2) reducing the injection volume to ≤10 μL (for a 150 × 4.6-mm column; smaller volumes for smaller volume columns). When the sample of Figure 17.15*c* was diluted in mobile phase, a 30-μL injection gave normal peak shape and retention times (Fig. 17.15*d*) [29]. Sample solvent effects are discussed further in Sections 2.6 and 3.6.3.2.

Another cause of distorted peaks is *degradation or chemical change* of an analyte as it passes through the column, when a compound is not stable under the chromatographic conditions. If degradation takes place rapidly at the head of the column, possibly catalyzed by the metal frit of the column, the decomposed sample is chromatographed without further change in sample composition, yielding original and reacted peaks of normal appearance. However, if the rate of decomposition is

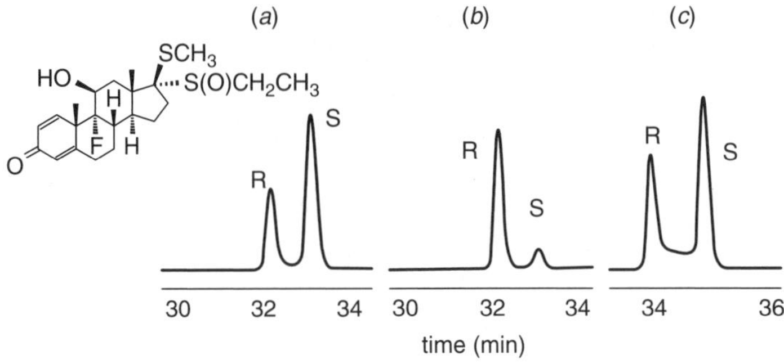

Figure 17.22 Separation of tipredane epimers. Conditions: (*a*, *b*) 100 × 4.6-mm Hypersil ODS column; 29/32/62% acetonitrile/pH-7.2 buffer in 0/10/20 min; 1.5 mL/min; 26°C. (*c*) 150 × 3.9-mm Resolve C_{18} column and similar, but not identical gradient conditions. (*a*) Injection of S-epimer; (*b*) injection of R-epimer; (*c*) injection of S-epimer. Adapted from [31, 32].

slow, the sample may degrade while the sample transits through the column, resulting in a distorted peak—the result of two (or more) distinct molecular structures passing through the column, with the ratio of their concentrations changing during the separation [30, 31].

An example of both fast and slow sample reactions is provided in Figure 17.22 [31, 32] for gradient separations of two tipredane epimers (structure shown in the figure) under different conditions. In Figure 17.22*a*, the pure S-epimer was injected, and peaks for both the *R*- and S-epimers are observed in the chromatogram (i..e, reaction of S-epimer to *R*). Because the two peaks are sharp and well-separated, the reaction of *R* to S must have occurred prior to significant elution through the column. The injection of the pure R-epimer (Fig. 17.22*b*) shows a similar, but reduced conversion to the alternate epimer. The two separations of (Fig. 7.22*a*) and (Fig. 7.22*b*) were each carried out on a Hypersil ODS column. When the column was changed to Resolve C_{18}, the separation of an injection of the pure S-epimer in Figure 17.22*c* was obtained. In this case a characteristic "saddle" is observed between the two peaks, indicating that the sample reaction occurred more slowly—*during* the separation, rather than primarily during sample injection.

If degradation is suspected, this can often be confirmed by changing the chromatographic conditions (temperature, pH, etc.) to speed or slow the rate of degradation. For example, increasing or decreasing column temperature will usually speed or slow the rate of sample reaction, with a predictable effect on peak shape. For the sample of Figure 17.22 it was found that a higher temperature accelerates sample reaction, while a higher mobile-phase pH slows the reaction.

17.4.6 Interpretation of System Performance Tests

HPLC system performance tests were described in detail in Section 3.10.1 and summarized briefly in Section 17.2.1. Of particular importance for identifying hardware problems are the gradient performance test (Section 3.10.1.2) and the additional system tests (Section 3.10.1.3). *We recommend that these tests be run every 6 to 12 months on each HPLC system to ensure that optimal equipment performance is obtained.* A summary of failed performance test symptoms and solutions is given in Table 17.11. The following discussion uses case-study examples to illustrate some of the problems that can be identified using these tests.

17.4.6.1 Interpretation of Gradient Performance Tests

The gradient performance test described in Section 3.10.1.2 contained the following elements:

- gradient linearity
- dwell-volume determination
- gradient step-test
- gradient proportioning valve (GPV) test

These tests often are run as a set, and the results of one test are related to the results of other tests. At other times the results of a test run suggest running another of the tests. The examples below illustrate the interrelationships of the tests, as well as the results of failed tests. The discussion is organized as a set of five case studies of problems that were highlighted as a result of the gradient performance tests. Each example is followed to completion so as to show how the various tests apply to real problems.

Case 1. The *gradient linearity* test was described in Section 3.10.1.2. This test comprises replacing the column with a piece of capillary tubing and running a linear gradient from 100% water to 0.1% acetone-water, monitored at 265 nm. The typical result in Figure 3.26 shows a delay at the beginning, corresponding to the dwell-volume, followed by a linear transition to 100% B, with slight rounding at the ends of the gradient. Visual inspection of the plot usually is sufficient to determine linearity; the plot can also be printed and a line can be drawn next to the curve for reference, as shown in Figure 17.23 (dashed line). In this case [33] the overall plot was linear, but at about 25%, 50%, and 75%B there were slight offsets in the plot. These results suggest that the mobile-phase proportioning process generated an error at each of these points in the curve.

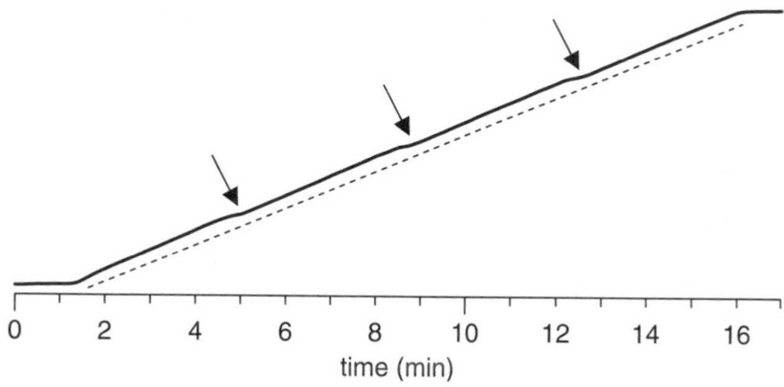

Figure 17.23 Plot for a linear gradient, using a system with faulty proportioning valves. Arrows show deviations from linearity; dashed line drawn below plot for reference. Gradient 0–100%B in 15 min at 1 mL/min; Solvents: A = water, B = 0.1% acetone in water; detection UV 265 nm. Adapted from [33].

The next step was to examine the *gradient step-test*. The step-test is also described in Section 3.10.1.2 and a portion of a successful test is shown in Figure 3.27. The step-test comprises a series of steps of varying mobile-phase composition 0–100% B, using the same acetone-water mobile phase described above. The recommended steps are 10%B increments between 0 and 100%B, plus two additional steps at 45 and 55%B. For the present example, the latter series of steps (Fig. 17.24*a*) passed the acceptance criteria of ±1% B from the target values [33]; nevertheless, deviations from linearity were obvious in the linearity test (Fig. 17.23). Next, the step-test was repeated in the vicinity of the questionable results but re-programmed for 1% steps. The results for the 45–55%B test region are shown in Figure 17.24*b*, where all steps look normal except the 50–51% step (arrow), where the step size is obviously too small. This same type of deviation was observed at 25–26% B and 75–76% B.

The HPLC system in this case study relied on low-pressure mixing (Section 3.5.2.2), so a *gradient-proportioning valve* (GPV) test (Section 3.10.1.2) was the next step in problem isolation. This test relies on alternate steps of water compared to acetone-water using various combinations of the 4 solvent supply lines (lines A and B are water, C and D are acetone-water), with a normal test result appearing as in Figure 3.28. For the present case, the results of Figure 17.25 [33] were observed. The maximum acceptable deviation between the highest and lowest plateaus is 5%, whereas Figure 17.25 has a 12% deviation between the first two steps. Because the A-solvent is common between the two steps, it is a likely source of the problem. The suspected problem's source was first a partially blocked solvent inlet frit or solvent transfer tube between the reservoir and the proportioning valve; however, a siphon test (Section 3.21) showed that there were no significant restrictions in the inlet frit

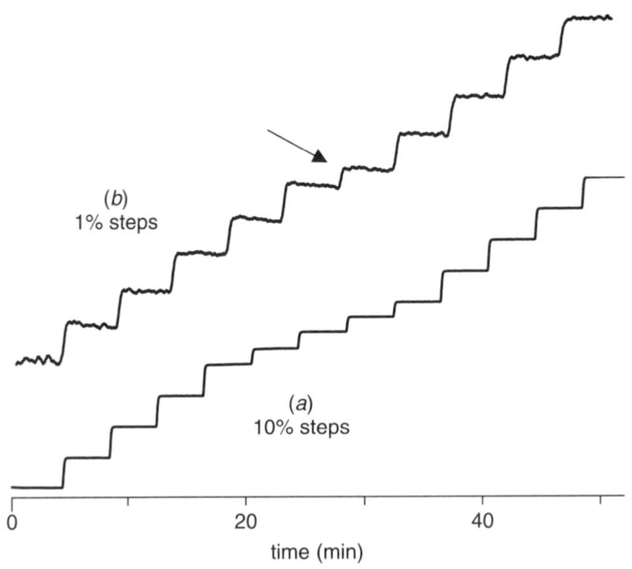

Figure 17.24 Gradient step-test results for HPLC system of Figure. 17.23. (*a*) Steps of 0, 10, 20, 30, 40, 45, 50, 55, 60, 70, 80, 90, and 100% B; (*b*) upper trace is 45–55% in 1% steps. Arrow showing "short" step between 50 and 51%B. Adapted from [33].

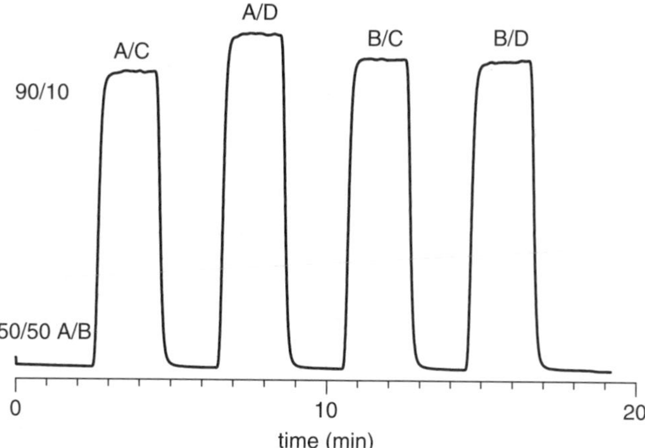

Figure 17.25 Gradient proportioning valve test for HPLC system of Figure 17.23. Baseline is generated by 50:50 A:B; the remaining plateaus are 90:10 A:C, A:D, B:C, and B:D from left to right. Solvents: A = B = water; C = D = 0.1% acetone in water. Adapted from [33].

or connecting tubing. Other potential corrective actions were to degas the mobile phase (Section 3.3), sonicate the check valves (17.2.5.4), and replace the pump seals (Section 3.5.1), but these did not correct the problem. Several attempts were made to adjust the proportioning algorithm in the control software, and although improvements were observed, the problem persisted. Finally, the proportioning valve assembly was replaced, with a step-test result of 0.9% maximum deviation—well within the 5% limit [33].

A less severe, but more common *gradient-linearity*-test failure occurs when a linear 0–100%B gradient is programmed but appears as a segmented 0/50/100%B gradient, with a slightly different slope for the 0–50% segment rather than the 50–100% segment. Usually the controlling software is at fault, and it can be adjusted for some HPLC systems; consult the pump service manual or the manufacturer for specific recommendations.

Case 2. An example of a more dramatic failure of the *gradient step-test* and *gradient linearity* test is shown in Figure 17.26. In this case the operator was unable to obtain reproducible retention times [34]. A gradient step-test (Fig. 17.26*a*) and a linearity test (Fig. 17.26*b*) were run, with obviously unacceptable results. The cause was suspected to be trapped air bubbles in the pumping system because occasional pressure fluctuations were observed. Thorough purging of the system with degassed solvent accompanied by tapping on each component of the system with a screwdriver handle (to dislodge bubbles adhering to internal surfaces) resulted in a series of bubbles in the waste stream. The two tests were rerun, with acceptable results. (See Section 17.2.5.1 for further hints on removing entrapped air.)

Case 3. In the final example of a failed *gradient step-test*, the method worked well, with acceptable retention time, precision, accuracy, and resolution. However, when the gradient step-test was run, the results of Figure 17.27 were obtained [34]. It can be seen that a small secondary step is located between each major

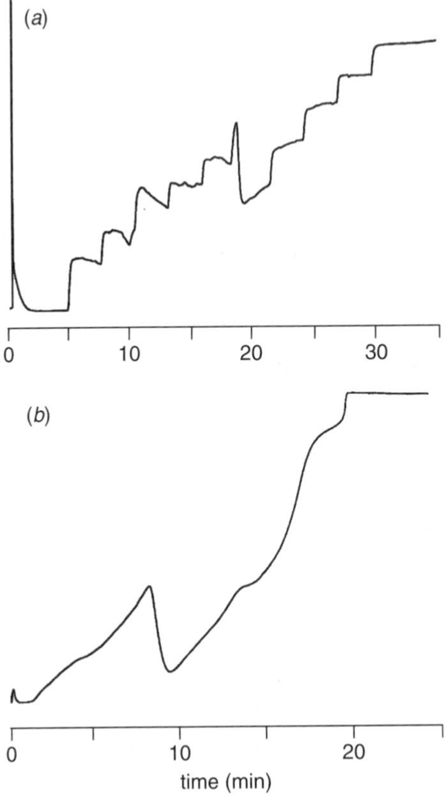

Figure 17.26 Results for an unacceptable (*a*) gradient step-test, and (*b*) gradient linearity test. Adapted from [34].

step (note the small step marked by the arrow between the 10 and 20% B major steps). In the process of eliminating likely causes, the autosampler was replaced with a manual injector (module substitution, Section 17.3.5), at which time the problem disappeared. Replacement of two stainless-steel frits within the autosampler corrected the problem. It was not clear why these blocked frits generated the secondary steps of Figure 17.27. In retrospect, it may be that the frits controlled flow through a flow-bypass channel that is used in some autosampler designs to minimize pressure pulses to the column [35–37]. In such designs part of the mobile phase flow bypasses the injection valve so that flow is not shut off when the injection valve is rotated (for additional information, see p. 238 of [38]). One of the authors has observed retention time and peak width problems when the flow through such a passage was disturbed; a disturbance in the gradient can also result from such partial blockage.

Case 4. In some cases a failed *gradient linearity* test can reflect the inappropriate use of the HPLC instrument rather than an instrument failure per se. It was noted in Section 3.10.1.2 that rounding of the gradient occurs at its beginning and end (Fig. 3.26). This rounding is normally minor and unlikely to affect the separation. However, when the *gradient-volume* ($= t_G F$) is comparable to or

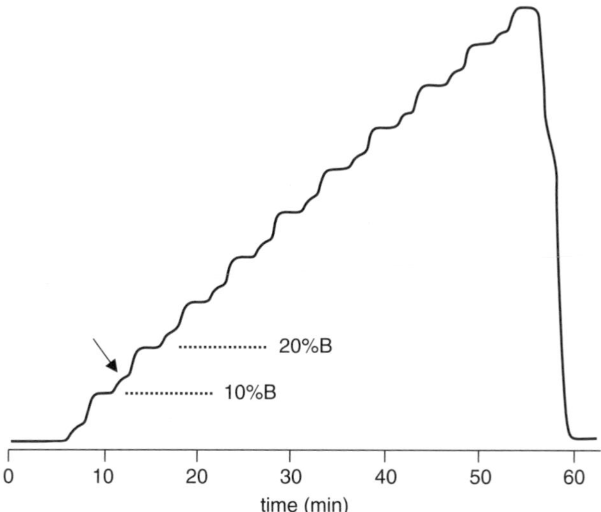

Figure 17.27 Unacceptable step test showing small secondary steps between each major step. Adapted from [34].

smaller than the dwell-volume, this rounding becomes more pronounced, so as to create serious distortion of the gradient. An example is shown in Figure 17.28 for two different gradients [39]. A low-pressure-mixing system ($V_D \approx 1$ mL) was used with a 100 × 2.1-mm i.d. column at 0.2 mL/min, and the gradient reached the detector at ≈5 minutes (see dashed traces of expected gradient profiles at the column outlet in Fig. 17.28*a*, *b*). A 9.9-minute gradient ($t_G F = 9.9 \times 0.2 \approx 2$ mL) generated the expected linear profile with little apparent distortion, as seen in the solid line of Figure 17.28*a*. This gradient was followed by a 1-minute isocratic hold and a 0.1-minute step gradient back to the initial conditions. In this case the gradient-volume (2 mL) was significantly larger than the dwell-volume of 1.0 mL.

For Figure 17.28*b*, a steep, 1.4-minute gradient was run, followed by a 3.5-minute isocratic hold and a 0.1-minute step back to the initial conditions. The gradient volume $t_G F = 1.4 \times 0.2 \approx 0.3$ mL, which is much smaller than the dwell-volume; severe distortion of the gradient can be predicted in this case, as observed in Figure 17.28*b*. Although it may be possible to generate reproducible gradients under the conditions of Figure 17.28*b* on one instrument, it is unlikely that such a steep-gradient method will transfer to a second instrument without problems. Further evidence of severe distortion can be seen in both Figure 17.28*a*, *b* during the re-equilibration phase (slow return of gradient to starting %B). For more details on the effects of gradient rounding on separation, see pp. 393–396 of [18].

Case 5. The *gradient dwell-volume* can be determined from the same experiment used to check gradient linearity (Section 3.10.1.2 and Fig. 3.26). The effect of the dwell-volume on the separation is discussed in Section 9.2.2.4. Differences in dwell-volume among different gradient HPLC systems are one of the primary reasons that gradient methods are difficult to transfer from one system to another. There are two common effects that are observed when a method is run on systems with

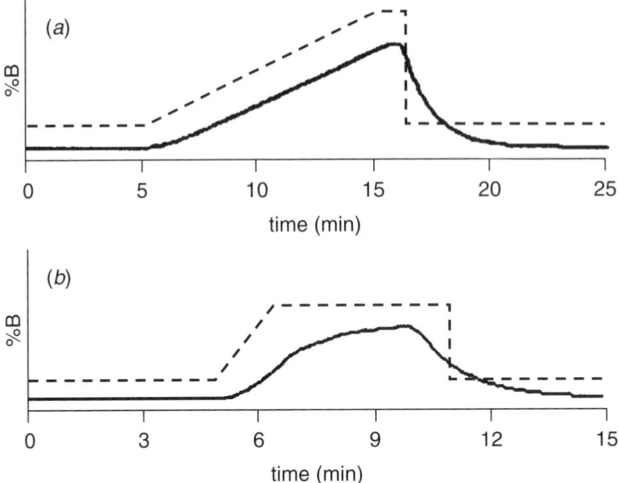

Figure 17.28 Gradient distortion due to excessive gradient rounding. (*a*) Gradient of 20/20/50/50/20%B at times 0.0/0.1/10/11/11.1 min (little rounding); (*b*) 20/20/40/40/20%B at 0.0/0.1/1.5/5/5.1 min (excessive rounding). Solid lines show water/water–acetone gradient trace; dashed lines show gradient program (vertically offset for clarity). Data of [39].

different dwell-volumes, as illustrated in Figure 17.29. In this case a 10–40% B gradient was run over 12 minutes at a flow rate of 1 mL/min. The system of Figure 17.29*a* had a dwell-volume of 1 mL, whereas the system in 17.29*b* had a dwell-volume of 3 mL. These differences in dwell-volume translate into the effective gradients shown by the dashed lines in Figure 17.29, with a 1- and 3-minute delay before the gradient reaches the column in Figure 17.29*a* and *b*, respectively. With later-eluting solutes for which the *k*-value at the start of the gradient (k_0) is sufficiently large, the primary effect of a dwell-volume difference is a shift in retention times equivalent to the difference in dwell-time. This is seen as the increase in retention of peaks 3 to 9 by 2 minutes [(3 mL—1 mL)/1 mL/min = 2 min] in Figure 17.29*b*. However, for peaks that elute early in the chromatogram, and especially for "irregular" solutes (Section 2.5.2.2), a change in relative peak spacing (and α) may also occur. This is illustrated by peaks 1 and 2 (note the loss in resolution for these peaks in Fig. 17.29*b*, *despite* an expected increase in resolution for early peaks and a greater dwell-volume (Section 9.2.2.3). One way of looking at this is as follows: For Figure 17.29*a*, early peaks migrate for 1 minute under isocratic conditions and 2 to 3 minutes under gradient conditions, whereas in Figure 17.29*b*, there is 3 minutes of isocratic migration plus 2 to 3 minutes of gradient elution. This change in the ratio of isocratic/gradient migration results in a difference in effective values of *k* (k^*) and peak spacing. With later peaks the initial isocratic migration is insignificant, so only a retention-time shift is observed due to the delay of the gradient arriving at the head of the column.

There are several ways to compensate for differences in dwell-volume for two HPLC systems. For the example of Figure 17.29, there are three possible solutions, assuming that the method developed on system of Figure 17.29*a* is transferred to system 17.29*b*. If the HPLC equipment is capable, programming an *injection delay* is the simplest solution. The gradient in Figure 17.29*b* would be started and the

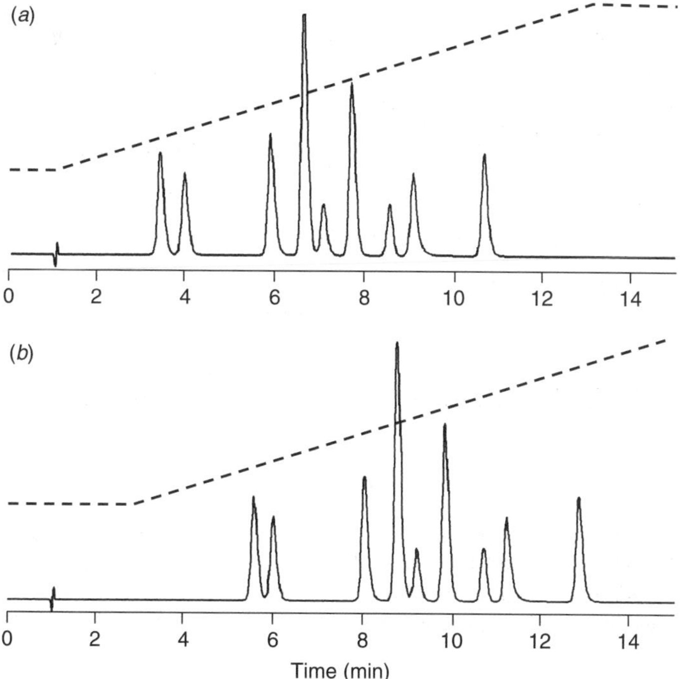

Figure 17.29 Effect of varying gradient dwell-volume on retention and selectivity. (*a*) 1-mL dwell volume system; (*b*) 3-mL dwell volume system. 100 × 4.6-mm column; 10–40%B gradient in 12 minutes at 1 mL/min.

injection would be programmed to occur 2 minutes after the gradient started. This would have the effect of shifting the chromatogram 2 minutes to the right relative to the dashed gradient overlay, so the sample would see exactly the same gradient as for Figure 17.29*a*. Unfortunately, not all HPLC systems have the capability of making a delayed injection. A second approach is to develop a *maximum dwell-volume* method. This requires advanced knowledge of the dwell-volume of system 17.29*b*. The dwell-volume of the initial system is adjusted to equal the largest dwell volume of any system to which the method will be transferred. In the present case, system 17.29*a* would be adjusted by adding 2 min (2 min at 1 mL/min = 2 mL) of isocratic hold at the beginning of each run so the total effective dwell-volume would be 3 mL. When the method is transferred, this additional hold would be dropped and the resulting gradients would be identical for both systems. If several systems with different dwell-volumes are to be used, the method would be developed with a dwell-volume equivalent to the largest dwell volume and the combination of isocratic hold and true dwell-volume would then be adjusted so that the effective dwell-volume is the same in all cases. A third possibility is to *adjust the initial %B* for system 17.29*b*. This involves a combination of starting the gradient at a higher %B and changing the times of the isocratic and gradient segments. See Section 5.2.1.3 of [18] for an example of this technique. A final technique for compensating for dwell-volume differences applies when the initial system has a larger dwell-volume than the new system, for example if the present method was developed on Figure 17.29*b* and transferred to Figure 17.29*a*. In this case an isocratic hold of 2 minutes can be added to the gradient of Figure 17.29*a*, giving an equivalent delay time for the start of each gradient. This approach is closely related to the maximum-dwell-volume technique discussed above. A more detailed discussion of adjusting for gradient dwell-volume differences can be found in Section 5.2.1.3 of [18].

17.4.6.2 Interpretation of Additional System Tests

Section 3.10.1.3 described several additional tests that should be run on a regular basis:

- flow-rate check
- pressure bleed-down test
- retention reproducibility test
- peak-area reproducibility test

If the results of the *flow-rate check* exceed ±2% of the flow-rate setting, the source of the problem should be investigated. Lower than normal flow is most common. First check for leaks (Section 17.4.1, Table 17.3). Other likely causes of subnormal flow rate are bubbles in the pump, faulty check valves, and worn pump seals. Sometimes the flow rate for organic solvents will be different than that for pure water with the same pump and settings, because of the greater compressibility of organic solvents. Solvent compressibility is usually ignored by the end user, but many pumps have adjustments to compensate for solvent compressibility (see the pump operator's manual). It is rare that the flow rate is higher than set (with the exception of extreme compressibility adjustments). First check that the settings were made correctly and repeat the test. Consult the pump operator's manual for instructions on how to adjust the flow-rate calibration if the problem persists. For an additional discussion of flow-rate problems, see Section 17.4.3.1 and the general discussion of pump operation in Section 3.5.

The *pressure bleed-down* test is a check of the pump's ability to hold pressure under static conditions. If the bleed-down test (Section 3.10.1.3) results in a pressure loss of >15% in 10 minutes, the cause should be identified and corrected. Because it is based on blocking the pump outlet tubing, the bleed-down test reflects the integrity of the pressure-limiting pump component farthest downstream (closest to the blocked outlet tubing). First check for leaks between the pump and the location of the plugged outlet tubing. If the pump uses an outlet check valve(s), this is the most likely point of failure. If the pump does not use an outlet check valve(s), a component further upstream is responsible for the problem—failure may be due to the pump seal, inlet check valve (dual-piston pumps, Section 3.5.1.1), or intermediate check valve (accumulator-piston pumps, Section 3.5.1.2). Replace any questionable pump seals and sonicate (Section 17.2.5.4) or replace questionable check valves.

Retention-time reproducibility generally should be better than ±0.05 minutes (1 S.D.), but some methods may exhibit poorer reproducibility. If the questionable retention-precision values are for a method that has been run before, check the results against historic values (Section 17.2.4), and run the retention reproducibility test (Section 3.10.1.3, Table 3.5). If no leaks are present (Section 17.4.1, Table 17.3), the problem is most likely related to the pump and/or mobile-phase mixing process. Run the gradient performance test (Section 3.10.1.2) and isolate the problem (Section 17.4.6.1).

Peak-area reproducibility will vary according to the application, with acceptable values ranging from ≤ 1–2% (content assays) to 15–20% (trace analysis), so it is best to compare test results with the past performance of the method (Section 17.2.4). When the standard peak-area reproducibility test (Section 3.10.1.3) exceeds 0.5–1%, check for autosampler problems. Make sure that any vents or vent-needles are not blocked. Check the sample needle for partial blockage or damage. Check any seals or syringes in the sampling system. Check the needle-to-injection-valve seal. Ensure that the detector time-constant is not set too slow. See Section 3.6 and the autosampler operator's manual for more ideas.

17.5 TROUBLESHOOTING TABLES

Most of the troubleshooting tables from this chapter are gathered in this section for convenience in simultaneously referring to more than one table. Various approaches can be used to identify a problem. First, consult the outline at the beginning of the chapter for the appropriate section containing topics of interest. Alternatively, use Table 17.2 as a guide to the present section. Table 17.2 lists the major symptoms likely to be encountered and cross-references other tables in this section as well as the appropriate section of text that will be useful. The remaining tables contain more specific information to help isolate specific problems. For the most part, these are organized like an outline, with the left-hand column giving a high-level symptom (e.g., the location of a leak in Table 17.3); as you move to the right across the table, each column gives more detailed information about isolation and possible sources of the problem. The right-hand column usually contains suggested solutions and often provides additional cross-references to discussions of the problem. The tables alone may be sufficient to isolate and solve a problem, or you may need to refer to the associated text material—depending on the specific problem and your level of experience.

Table 17.2

HPLC Problem Symptoms

Problem Symptom	More Information
Leaks	Table 17.3, Section 17.4.1
Pressure	Table 17.4, Section 17.4.2
Retention time	Tables 17.5, 17.6, Section 17.4.3
Peak area	Tables 17.7, Section 17.4.4
Baseline drift	Table 17.8, Section 17.4.5.1
Baseline noise	Table 17.9; Section 17.4.5.2
Peak shape	Table 17.10 Sections 17.4.5.3, 2.4.2
Failed system performance tests	Table 17.11, Section 17.4.6

Table 17.3

Leaks (Section 17.4.1): Symptoms and Sources

Location	Sub-location/Symptom	Possible Source	Solution
Pre-pump (17.4.1.1)	Fitting	Loose fitting	Tighten or replace
	Proportioning manifold (low-pressure mixing)	Loose or damaged valve	Tighten valve or replace manifold
Pump (17.4.1.2)	Fittings	Loose or damaged	Tighten or replace
	Check valve	Loose	Tighten
		Cross-threaded	Replace; may require new pump head
		Seal damaged	Replace
	Pump seal	Worn seal	Replace
	Auxiliary component	Loose fitting; damaged component	Tighten or replace
High-pressure fitting (17.4.1.3)	Stainless steel	Loose or contaminated	Tighten 1/4-turn; clean or replace
	PEEK	Loose or contaminated	Turn off pump, reseat tube, retighten; not for use >6000 psi
Autosampler or manual injector (17.4.1.4)	Fittings	Loose or damaged	See fittings adjustments, Sections 17.4.1.1, 17.4.1.3
	Low-pressure needle seal	Worn or loose seal	Adjust or replace
	High-pressure needle seal	Worn or damaged seal	Adjust or replace
	Cross-port leakage	Damaged rotor seal	Clean, replace rotor seal; may require stator replacement.
	External leaks not at fittings	Worn rotor seal	Clean, replace rotor seal
		Insufficient rotor-to-seal pressure	Adjust
	Seal-pack failure (Waters only)	Worn or damaged seals	Service or replace

Table 17.3

(*Continued*)

Location	Sub-location/Symptom	Possible Source	Solution
Column (17.4.1.5)	Tube connections	Loose or damaged	See fittings adjustments, Section 17.4.1.3
	End-fitting	Loose or damaged	Tighten; may require new column
	End-fitting on cartridge column	Loose or damaged fitting; damaged seal	Tighten; replace seal; may require new assembly
Detector (17.4.1.6)	Connecting fittings	Loose or damaged	See fittings adjustments, Sections 17.4.1.1, 17.4.1.3
	Cell leaks	Loose fitting	Tighten
		Over-pressure damage	Remove source of excessive pressure
		Failed window seal	Adjust or replace
	Non-UV detector	See Chapter 4 and operator's manuals	

Table 17.4

Pressure Problems (Section 17.4.2): Symptoms and Sources

Symptom	Location/Symptom	Possible Source	Solution
Pressure too high (17.4.2.1)	General	Wrong mobile phase; temperature too low; flow rate too high; wrong column; wrong particle size	Use correct conditions
	Blocked tubing	Buffer precipitation; sample particulates	Replace tubing; remove problem source
	Blocked or partially blocked frit	In-line filter	Replace frit; consider additional sample cleanup
		Guard column	Replace guard column
		Analytical column	Reverse column (if permitted); replace column; consider using in-line filter
	Upper pressure limit mid-gradient	Normal gradient pressure changes	Adjust upper pressure limit to accommodate

(*continued overleaf*)

Table 17.4

(*Continued*)

Symptom	Location/Symptom	Possible Source	Solution
Pressure too low (17.4.2.2)	General	Wrong mobile phase; temperature too high; flow rate too low; wrong column	Use correct conditions
	Lower pressure limit	Insufficient mobile phase	Refill reservoir
		Overly sensitive limit switch	Disable lower pressure limit
		Massive leak	Isolate and fix (Section 17.3.1)
	High-pressure fitting	Loose or contaminated fitting	Tighten, clean, or replace
	Low-pressure fitting	Loose or contaminated fitting	Tighten, clean, or replace
	Low-pressure mixing manifold	Air leak from unused solvent line; leaky proportioning valve	Prime unused lines with organic solvent; replace faulty proportioning valve
	Inlet check valve	Sticking with ACN mobile phases	Sonicate in MeOH; replace
	Pump starvation (failed siphon test)	Blocked inlet-line frit	Replace
		Faulty proportioning valve	Replace
		Pinched or blocked tubing	Clear or replace
	Bubbles in pump	Inadequate degassing or purging; degasser failure	Degas mobile phase and purge pump; replace degasser
	Pump seal	Worn pump seal	Replace
	Pump piston	Broken or scratched piston	Replace
Pressure too variable (17.4.2.3)	Bubbles in pump	Inadequate degassing or purging	Degas mobile phase and purge pump
	Inlet check valve	Sticking with ACN mobile phases	Sonicate in MeOH; replace
	High-pressure fitting	Loose or contaminated fitting	Tighten, clean, or replace
	Low-pressure fitting	Loose or contaminated fitting	Tighten, clean, or replace

Table 17.4

(Continued)

Symptom	Location/Symptom	Possible Source	Solution
	Low-pressure mixing manifold	Air leak from unused solvent line; leaky proportioning valve	Prime unused lines with organic solvent; replace faulty proportioning valve
	Pump starvation (failed siphon test)	Blocked inlet-line frit	Replace
		Faulty proportioning valve	Replace
		Pinched or blocked tubing	Clear or replace
	Pump seal	Worn pump seal	Replace
	Pump piston	Broken or scratched piston	Replace

Table 17.5

Interpreting Retention-Factor k and Retention-Time t_R Changes (Section 17.4.3)

k Changes	t_R Changes	Possible Cause	More Information
No	Yes	Flow rate (settings, leaks, bubbles, pump problems)	Section 17.4.3.1
No	Yes	Wrong column size	Section 17.4.3.2
Yes	Yes	Mobile-phase error (%B, pH, additives)	Section 17.4.3.3
Yes	Yes	Stationary-phase problem (aging, dirty samples)	Section 17.4.3.4
Yes	Yes	Temperature problem (poor control)	Section 17.4.3.5

Table 17.6

Retention-Time Problems (Section 17.4.3.6): Symptoms and Sources

Symptom	Location/Symptom	Possible Source	Solution
Abrupt change in t_R	When column changed	Wrong column; poor column-to-column reproducibility (especially type-A columns)	Use correct column; use columns from same batch; adjust method for more robust separation; use more reproducible columns (e.g., type-B); Section 2.5.4.6, Table 17.5
	When mobile phase changed	Improperly formulated mobile phase	Make new batch of mobile phase; Table 17.5

(*continued overleaf*)

Table 17.6

(*Continued*)

Symptom	Location/Symptom	Possible Source	Solution
	In gradient elution, with change in instrument	Different dwell volume	Adjust or compensate dwell volume; Section 9.3.8.2
	No apparent change in conditions	Equipment, bubble, or temperature problem	Check for leaks (Table 17.3, Section 17.4.1); check for pressure problems (Table 17.4, Section 17.4.2); check *k* changes (Table 17.5, Section 17.4.3); Section 2.5.4.6
Retention drifts	During first few injections only	Slow column equilibration; column "loading" for sample	Normal for some columns and samples; allow longer equilibration; ignore first few injections (Section 3.10.2.2); Sections 2.7.1 (general), 7.4.3.2 (ion-pair), 8.5 (normal-phase), 9.3.7 (gradient elution)
	Over entire sample batch	Unstable mobile phase; mobile-phase evaporation; excessive helium sparging	Use stable mobile phase; cover reservoir to reduce evaporation; reduce helium sparging time or change to different degassing technique (Section 3.3)
Retention too small	Only for large-mass injections for some or all peaks, with right-triangle peak shape	Mass overload	Reduce mass of sample injected; Table 17.11, Sections 17.4.5.3, 2.6.2
	All peaks	Mobile phase error; flow rate, column, or temperature change	See Table 17.5
	Some peaks	pH problem or other mobile-phase error; column aging	Make appropriate adjustments; Section 7.3 (pH); replace column
	Only for polar solutes	Ionic samples	Change mobile-phase pH (Section 7.3); use ion pairing (Section 7.4)

Table 17.6

(Continued)

Symptom	Location/Symptom	Possible Source	Solution
		Neutral or non-ionic samples	Try EPG or AQ column; use normal phase (Chapter 8) or HILIC (Section 8.6)
Retention too large	All peaks	Mobile-phase error; flow rate, column, or temperature change; leak	See Tables 17.5, 17.3
	Some peaks	pH problem or other mobile-phase error; column aging	Make appropriate adjustments; Section 7.3 (pH); replace column

Table 17.7

Peak-Area Problems (Section 17.4.4): Symptoms and Sources

Symptom	Location/Symptom	Possible Source	Solution
Peak area to large (17.4.4.1)	All peaks larger by same proportion	Injection volume too large	Adjust
		Sample preparation errors	Use correct procedure
		Detector settings wrong	Adjust
	Different peaks in sample larger by different amounts	Detector settings wrong	Adjust
	Peak(s) present in blank	Late elution from prior injection	Extend run; add strong-solvent flush; modify sample pretreatment
		Carryover	Add autosampler wash steps; change wash solvent; adjust connections; change sample loop material; rearrange injection order
Peak area too small (17.4.4.2)	All peaks smaller by same proportion	Injection volume too small	Adjust
		Sample preparation errors	Use correct procedure

(*continued overleaf*)

Table 17.7

(*Continued*)

Symptom	Location/Symptom	Possible Source	Solution
		Detector settings wrong	Adjust
	Different peaks in sample smaller by different amounts	Detector settings wrong	Adjust
Peak area too variable (17.4.4.3)	Replicate injections give constant areas	Problem prior to injection	Check sampling and sample pretreatment steps
	Replicate injections give inconsistent areas	Problem from injector onward	Check autosampler, pump, detector, and integration

Table 17.8

Baseline-Drift Problems (Section 17.4.5.1): Symptoms and Sources

Symptom	Location/Symptom	Possible Source	Solution
Periodic drift	1 cycle/run	Normal with gradient elution	Ignore; use higher wavelength; change solvent; add UV absorber
	Several hours or 1 cycle/day	Laboratory temperature cycle	Thermostat oven; move system from drafts; adjust lab HVAC
Other drift	Baseline drifts, then stable	Normal equilibration; detector warm-up	Wait until baseline stabilizes

Table 17.9

Baseline-Noise Problems (Section 17.4.5.2): Symptoms and Sources

Symptom	Location/Symptom	Possible Source	Solution
Short-term noise	High-frequency (50 or 60 Hz)	"Dirty" electrical supply; too small of detection time constant; too high of data rate	Use UPS to cleanup supply; use larger time constant or RC filter; use slower data collection rate
	50–60 Hz < noise < peak width	Inadequate mobile-phase degassing	Use better degassing technique

Table 17.9

(Continued)

Symptom	Location/Symptom	Possible Source	Solution
Long-term noise	Cyclic	Bubble in pump; piston seal or piston damage; sticking or leaking check valve	Degas mobile phase; service pump; clean or replace check valve
	Appears as chromatographic peak	Late-eluted solute	Extend run; add strong-solvent flush to method; change sample cleanup
	Random, irregular baseline disturbances	Accumulated late-eluted non-polar materials from sample	Add strong-solvent flush to method; replace column; change sample cleanup
	Peaks appear in blank gradients ("ghost" peaks, Figs. 17.11, 17.12)	Contaminated mobile phase	Replace mobile-phase reagents with fresh and/or higher purity reagents

Table 17.10

Peak-Shape Problems (Section 17.4.5.3): Symptoms and Sources

Symptom	Symptom	Symptom/Possible Source	Solution
Peak tailing	$TF \leq 1.2$	Normal	None required
	$1.2 < TF \leq 2$	If increased over time: column aging or contamination; mobile phase error	Flush or replace column; prepare new batch of mobile phase
		If constant, may be normal	No action or explore solutions for $TF > 2$
	$TF > 2$	For all peaks	Also see split or distorted peaks
		Early peaks tail more than later peaks (Fig. 17.13): extra-column effects	Reduce extra-column volume (Sections 2.4.1, 3.9)

(*continued overleaf*)

Table 17.10

(*Continued*)

Symptom	Symptom	Symptom/Possible Source	Solution
		Worse for bases and ionic compounds (Fig. 5.9*a*): silanol effects; trace metal contamination	Adjust pH; use higher purity or less-active column; see Sections 5.4.4.1, 7.3.4.2
		Inadequate buffering or mobile-phase additive concentration	Increase buffer or additive concentration; see Section 7.2.1.1
	Right-triangle peak tailing accompanied by earlier retention (Fig. 17.15*a*)	Mass overload of the column	Reduce injection volume or dilute sample for smaller mass on column (Section 2.6.2)
Peak fronting	$TF \geq 0.9$	Normal	None required
	$TF < 0.9$	For ion pairing, especially with type-A columns	Change temperature ±5–10°C may help; change to type-B column
		Column void or bed collapse	Replace column; operate column below high pH-limit
Broad peaks	$N > 75\%$ of manufacturer's test	May be normal	No action required; compare to new column test with sample solutes
	High-molecular-weight compounds (proteins, polymers, etc.)	Some broadening normal	No action required
	Gradual broadening over >500 injections	Normal column aging	Flush column with strong solvent; replace column
	In the presence of narrower peaks (Fig. 17.5)	Late elution	Extend run; adjust run length; use gradient flush; improve sample cleanup
	Accompanied by longer retention times	Column temperature too low	Increase temperature

Table 17.10

(*Continued*)

Symptom	Symptom	Symptom/Possible Source	Solution
	Narrower peaks broadened more than normally wider peaks	Detector time constant too large; integrator data-rate too low; excessive smoothing (especially MS detection)	Use time constant $\leq$1/10 peak width; collect a minimum of 20 points across a peak; use less smoothing
	early peaks broad or flat-topped (Fig. 17.16*b*)	Too large an injection volume	Reduce injection volume; dilute injection solvent (Sections 2.6.1, 3.6.3)
Split or distorted peaks	All peaks distorted in same manner (Figs. 17.17, 17.19)	Partially blocked column inlet-frit; column void	Reverse-flush column; use in-line filter or better cleanup; replace column (Section 5.8)
	Broad or distorted peaks, especially at end of run (Fig. 17.20*b*), accompanied by change in retention time	Mismatch of mobile phase and column temperature	Use better temperature control; use mobile phase pre-heater (Section 3.7.1)
	Early peaks distorted (Fig. 17.16*c*), usually accompanied by earlier retention	Too large a volume of too strong a sample-injection solvent	Use more dilute injection solvent; reduce injection volume (Sections 2.6.1, 3.6.2.2)

Table 17.11

Failed System Performance Tests (Section 17.4.6): Symptoms and Sources

Failed Test	Symptom	Symptom/Possible Source	Solution
Gradient linearity	Steps in gradient	Blocked reservoir frit; bad proportioning valve	Check/replace frit; replace proportioning manifold
	Segmented appearance in linear gradient	Software control error	Adjust control parameters (see service manual)

(*continued overleaf*)

Table 17.11

(*Continued*)

Failed Test	Symptom	Symptom/Possible Source	Solution
	Gradient distortion (Fig. 17.28*b*)	Gradient mixing volume too large relative to gradient volume	Use shallower gradient; reduce mixing volume
Gradient step-test failure	Steps in gradient step-test are of uneven height and/or are distorted	Bubble, check-valve failure, or leak;	Degas mobile phase, clean or replace check valve, fix leak; blocked reservoir frit; bad proportioning valve
Gradient proportioning valve (GPV) test failure	GPV test steps >5%	Blocked reservoir frit; restricted solvent supply tubing; bad proportioning valve	Replace frit; clear or replace tubing; replace proportioning manifold
Dwell-volume differences between systems	Changes in retention and/or selectivity between systems	Normal for differences in dwell-volume	Adjust method to compensate for dwell-volume differences
Flow-rate check failure	Flow rate > ±2% from set value	Bubble; bad check valve or pump seal; leak; wrong compressibility set	Degas mobile phase; clean or replace check valve, replace pump seal; adjust compressibility (or ignore)
Pressure bleed-down failure	>15% pressure loss in 10 min	Check valve or pump-seal failure; leak	Clean or replace check valve, replace pump seal; fix leak
Retention time reproducibility failure	> ±0.05 min for standard test; more than normal method performance	Bubbles, leaks, check-valve or pump-seal failure; pump or mixing failure	Degas solvents, fix leaks, clean or replace check valve, replace pump seal; run gradient performance test to isolate further
Peak area reproducibility failure	>1% imprecision in standard test; more than normal method performance	Autosampler problem	Clean or replace needle; replace seals; see autosampler manual

REFERENCES

1. J. W. Dolan and L. R. Snyder, *Troubleshooting LC Systems*, Humana Press/Springer, Clifton, NJ, 1989.
2. V. R. Meyer, *Pitfalls and Errors of HPLC in Pictures*, Wiley-VCH, Weinheim, 2006.
3. J. W. Dolan, "LC Troubleshooting," in *LCGC*, a monthly column (1983-present).
4. http://www.chromforum.org.

5. http://www.lcresources.com/resources/TSWiz/.
6. S. J. Williams, *J. Chromatogr. A*, 1052 (2004) 1.
7. M. D. Nelson and J. W. Dolan, *LCGC*, 16 (1998) 992.
8. C. K. Cheung and R. Swaminathan, *Clin. Chem.*, 33 (1987) 202.
9. C. T. Mant and R. S. Hodges, eds., *High-Performance Liquid Chromatography of Peptides and Proteins: Separation, Analysis, and Conformation*, CRC Press, Boca Raton, 1991.
10. J. W. Dolan, J. R. Kern, and T. Culley, *LCGC*, 14 (1996) 202.
11. D. W. Bristol, *J. Chromatogr.*, 188 (1980) 193.
12. P.-L. Zhu, L. R. Snyder, and J. W. Dolan, *J. Chromatogr. A*, 718 (1995) 429.
13. J. W. Dolan, *LCGC*, 11 (1993) 640.
14. J. W. Dolan, *LCGC*, 9 (1991) 22.
15. M. L. Ledtje and D. Long, Jr., US Patent 5,002,662 (March 26, 1991).
16. J. W. Dolan, *LCGC*, 26 (2008) 532.
17. J. W. Dolan, personal communication.
18. L. R. Snyder and J. W. Dolan, *High-Performance Gradient Elution*, Wiley, Hoboken, NJ, 2007.
19. *SPD-10Avp UV-Vis Detector Instruction Manual*, Shimadzu, Kyoto, 1997.
20. N. S. Wilson, R. Morrison, and J. W. Dolan, *LCGC*, 19 (2001) 590.
21. J. W. Dolan, *LCGC*, 23 (2005) 370.
22. *Reviewer Guidance: Validation of Chromatographic Methods*, USFDA-CDER, Nov. 1994, http://www.fda.gov/cder/guidance/index.htm.
23. D. V. McCalley, *Anal. Chem.*, 78 (2006) 2532.
24. M. T. W. Hearn, ed., *Ion-pair Chromatography. Theory and Biological and Pharmaceutical Applications*, Dekker, New York, 1985.
25. E. Rajakylä, *J. Chromatogr.*, 218 (1981) 695.
26. R. D. Morrison and J. W. Dolan, *LCGC*, 23 (2005) 566.
27. J. W. Dolan, *LCGC*, 25 (2008) 610.
28. R. G. Wolcott, J. W. Dolan, L. R. Snyder, S. R. Bakalyar, M. A. Arnold, and J. A. Nichols, *J. Chromatogr. A*, 869 (2000) 211.
29. T.-L. Ng and S. Ng, *J. Chromatogr.*, 389 (1985) 13.
30. W. R. Melander, H.-J. Lin, and C. Horváth, *J. Phys. Chem.*, 88 (1984) 4527.
31. M. R. Euerby, C. M. Johnson, I. D. Rushin, and D. A. S. Sakunthala Tennekoon, *J. Chromatogr. A*, 705 (1995) 229.
32. M. R. Euerby, C. M. Johnson, I. D. Rushin, and D. A. S. Sakunthala Tennekoon, *J. Chromatogr. A*, 705 (1995) 219.
33. J. J. Gilroy and J. W. Dolan, *LCGC*, 22 (2004) 982.
34. T. Culley and J. W. Dolan, *LCGC*, 13 (1995) 940.
35. J. W. Dolan, *LCGC*, 13 (1995) 940.
36. U. D. Neue, personal communication (1995).
37. T. Eidenberger, personal communication (1995).
38. J. W. Dolan and L. R. Snyder, *Troubleshooting LC Systems*, Humana Press, Totowa, NJ, 1989.
39. G. Hendriks, J. P. Franke, and D. R. A. Uges, *J. Chromatogr. A*, 1089 (2005) 193.

APPENDIX I

PROPERTIES OF HPLC SOLVENTS

Solvents are used in HPLC for formulating mobile phases, for dissolving the sample, and for carrying out sample preparation. Mobile-phase solvents are of primary concern, because their properties must often fall within narrow limits for acceptable performance. However, these same properties also influence the choice of the sample-injection solvent and solvents used for sample preparation. Table I.1 lists several solvent properties that can be important when selecting solvents for an HPLC application. Some of these properties have been discussed previously in one or more sections of this book (second column of Table I.1). The present appendix contains several tables that list values of one or more solvent properties (third column of Table I.1). A brief comment on each solvent property is given in the last column of Table I.1; this serves as an introduction to following sections that deal with individual solvent properties.

I.1 SOLVENT-DETECTOR COMPATIBILITY

I.1.1 UV Detection

The mobile phase will preferably have an absorbance $A < 0.2$ AU at the wavelength used for detection of the sample; a lower absorbance may mean improved assay precision and better results with gradient elution, but higher absorbances may be acceptable for some isocratic separations. Table 1.2 summarizes values of solvent absorbance at different wavelengths (200–260 nm) for solvents that are used for RPC (exclusive of NARP). Very rarely, there may be a reason to use UV detection at a wavelength <200 nm, for the detection of solutes with low absorptivity at higher wavelengths.

Because water does not absorb at 200 nm or above, the absorbance of aqueous mobile phases that contain these solvents will equal the pure-solvent absorbance times the volume-fraction ϕ of the B-solvent in the mobile phase. For example, a mobile phase of 25% B would have the following absorbance values A for different B-solvents at 215 nm: ACN, 0.00 AU; MeOH, 0.09 AU; degassed MeOH, 0.05 AU; THF, 0.22 AU; IPA, 0.07 AU. Note that degassing methanol lowers the

Introduction to Modern Liquid Chromatography, Third Edition, by Lloyd R. Snyder, Joseph J. Kirkland, and John W. Dolan

Table I.1

Table I. 1 Solvent Properties of Interest in HPLC

Property	Section Reference	Table of Values	Comment
UV cutoff	4.4	I.2	For UV detection; useful solvents depend on wavelength required for sample detection
Refractive index	4.11	1.3	For RI detection; low values generally preferred
Polarity	2.3.2.1, 6.2.1, 8.2.1	I.4	Determines solvent strength for $1 \leq k \leq 10$
Selectivity	6.3, 8.3.2	I.4	Determines differences in solvent-type selectivity
Sample solubility	15.3.2.3		Can be important for injection of large samples in prep-LC or trace analysis
Viscosity	2.4.1	I.3, I.5	Determines column pressure drop; low values of viscosity desirable
Boiling point		I.3	Affects pump performance and safety; higher boiling solvents preferred
Miscibility			Important for choice of RPC organic solvent and sample solvent
Density		I.6	Required for the (more accurate) formulation of mobile phases by weight
Stability			Not usually an issue
Safety		I.7	Generally important, but not usually critical

concentration of oxygen in the pure solvent, with a resulting decrease in solvent absorbance by about 1/3 over the range 200 to 240 nm. When the mobile phase is degassed, as by helium sparging, the absorbance of other B-solvents will also be lowered, in proportion to the amount of oxygen that is normally present in the solvent. Oxygen is more soluble in less polar solvents such as THF and IPA; the absorbance values of Table 1.2 for these solvents may therefore be higher than found in practice.

While water should not absorb light at wavelengths ≥200 nm, this assumes that the water has been properly purified. Specifications for HPLC-grade water are described in ASTM D1193, but water for use with low-wavelength detection may require total organic carbon (TOC) levels below 50 ppb, as well as high resistivity values. Table 1.2 also lists UV absorbance data for some commonly used buffers, and Table 1.3 provides UV cutoff wavelengths for several additional solvents.

Table I.2

UV Absorbance as a Function of Wavelength of Various Solvents and Buffers Used for RPC

	Absorbance at Indicated Wavelength (nm)								
	200	205	210	215	220	230	240	250	260
Solvents									
Acetonitrile	0.06	0.02	0.02	0.01	0.00	0.00	0.00	0.00	0.00
Methanol	1.0+	1.0	0.53	0.35	0.23	0.10	0.04	0.02	0.01
Methanol (degassed)	1.0+	0.76	0.35	0.21	0.15	0.06	0.02	0.00	0.00
Tetrahydrofuran	1.0+	1.0+	1.0+	0.85	0.70	0.49	0.30	0.17	0.09
Isopropanol	1.0+	0.98	0.46	0.29	0.21	0.11	0.05	0.03	0.02
Buffers									
Acetate									
Acetic acid, 1%	1.0+	1.0+	1.0+	1.0+	1.0+	0.87	0.14	0.01	0.00
Ammonium salt 10 nM	1.0+	0.94	0.53	0.29	0.15	0.02	0.00	0.00	0.00
Carbonate									
$(NH_4)HCO_3$, 10 mM	0.41	0.10	0.01	0.00	0.00	0.00	0.00	0.00	0.00
Formate									
Sodium salt, 10 mM	1.00	0.73	0.53	0.33	0.20	0.03	0.01	0.01	0.01
Phosphate									
H_3PO_4, 1%	0.00	0.00	0.00	0.00	0.00	0.00	0.00	0.00	0.00
KH_2PO_4, 10 mM	0.03	0.00	0.00	0.00	0.00	0.00	0.00	0.00	0.00
K_2HPO_4, 10 mM	0.53	0.16	0.05	0.01	0.00	0.00	0.00	0.00	0.00
$(NH_4)_2HPO_4$, 10 mM	0.37	0.13	0.03	0.00	0.00	0.00	0.00	0.00	0.00
sodium salt, pH-6.8, 10 mM	0.20	0.08	0.02	0.01	0.00	0.00	0.00	0.00	0.00
Trifluoroacetic acid									
0.1% in water	1.0+	0.78	0.54	0.34	0.20	0.06	0.02	0.00	0.00
0.1% in ACN	0.29	0.33	0.37	0.38	0.37	0.25	0.12	0.04	0.01

Source: Data of [1, 2].

It is preferable that the detector response of the mobile phase remains constant during gradient elution. This requires that the A-solvent (water) and the B-solvent each respond similarly. When UV detection is used at low wavelengths, this may not be the case, especially for B-solvents other than acetonitrile. A related problem is a variation in the absorbance of the buffer or other mobile-phase additives as %B changes. Each of these effects is discussed in Section 17.4.5.1.

I.1.2 RI Detection

For isocratic separation, the choice of mobile phase is usually not limited for RI detection. Detection sensitivity can be increased by selecting a mobile phase whose RI-value is more different than that of sample components (Table 1.3). Detectors based on differential measurement (e.g., RI) cannot be used for gradient elution because of the usual large difference in response for the A- and B-solvents.

Table I.3

Miscellaneous Solvent Properties

Solvent	UV Cutoff (nm)[a] [2]	RI[b] [3]	Viscosity (cP) [3]	Boiling Point (°C)[c] [3]	ε (silica)[d]
Acetone	330	1.359	0.36	56	0.53
Acetonitrile	190	1.344	0.38	82	0.52
1-Butanol	215	1.399	2.98	118	0.40
1-Chlorobutane	220	1.402	0.45	78	0.20
Chloroform	245	1.446	0.57	61	0.26
Cycohexane	200	1.424	1.00	81	0.00
Dimethyl formamide	268	1.430	0.92	153	—
Dimethylsulfoxide	268	1.478	2.24	189	0.50
1,4-Dioxane	215	1.422	1.37	101	0.51
Ethyl acetate	256	1.372	0.45	77	0.48
Heptane	200	1.388	0.40	98	0.00
Hexane	195	1.375	0.31	69	0.00
Isooctane	215	1.391	0.50	99	0.00
Methanol	205	1.328	0.55	65	0.70
Methyl-*t*-butyl ether	210	1.369	0.27	55	0.48
Methylethyl ketone	329	1.379	0.43	80	0.40
Methylene chloride	233	1.424	0.44	40	0.30
i-Propanol	205	1.377	2.40	82	0.60
n-Propanol	210	1.386	2.30	97	0.60
Tetrahydrofuran	212	1.407	0.55	66	0.53
Toluene	284	1.497	0.59	111	0.22
Water	190	1.333	1.00	100	

[a]Wavelength at which solvent absorbs 1.0 AU in a 10-mm cell.
[b]Refractive index.
[c]Boiling point.
[d]Solvent strength parameter [4].

I.1.3 MS Detection

The MS interface evaporates the mobile phase, so mobile phases comprising water, organic solvent, and volatile additives are used (i.e., no nonvolatile buffers or salts). Because the mobile phase is removed, UV absorbance is of no concern for LC-MS, but the solvents must be free of particulates.

I.2 SOLVENT POLARITY AND SELECTIVITY

Table 1.4 lists solvent properties that affect solvent strength, selectivity, and solubility. These "normalized selectivity" properties recognize three contributions of the solvent to solute–solvent interaction: solvent hydrogen-bond (H-B) acidity α_2^H/Σ H-B basicity β_2/Σ, and dipolarity π^*/Σ. The latter parameters are the basis of

Table I.4

Solvent Selectivity Characteristics

Solvent	Normalized Selectivity[a]				
	H-B Acidity α_2^H/Σ	H-B Basicity β_2/Σ	Dipolarity π^*/Σ	P'[b]	ε[c]
Acetic acid	0.54	0.15	0.31	6.0	6.2
Acetone	0.06	0.38	0.56	5.1	20.7
Acetonitrile	0.15	0.25	0.60	5.8	37.5
Benzene	0.14	0.86	0.00	2.7	2.3
Chloroform	0.43	0.00	0.57	4.1	4.8
Diethyl ether	0.00	0.64	0.36	2.8	4.3
Dimethylsulfoxide	0.00	0.43	0.57	7.2	4.7
Ethanol	0.39	0.36	0.25	4.3	24.6
Ethylacetate	0.00	0.45	0.55	4.4	6.0
Ethylene chloride	0.00	0.00	1.00	3.5	10.4
Formamide	0.33	0.21	0.46	9.6	182
Glycol	0.38	0.23	0.39	6.9	37.7
Hexane	0.00	0.00	0.00	0.1	1.9
Isopropanol	0.22	0.35	0.43	3.9	19.9
Methanol	0.43	0.29	0.28	5.1	32.7
Methylacetate	0.05	0.40	0.55	≈ 5	6.7
Methylene chloride	0.27	0.00	0.73	3.1	8.9
Methylethyl ketone	0.05	0.40	0.55	4.7	18.5
Methyl-*t*-butylether	0.00	≈ 0.6	≈ 0.4	≈ 2.4	≈ 4
N,N-dimethylformamide	0.00	0.44	0.56	6.4	36.7
Nitromethane	0.17	0.19	0.64	6.0	35.9
Pyridine	0.42	0.58	0.00	5.3	12.4
Sulfolane	0.00	0.17	0.83		43.3
Tetrahydrofuran	0.00	0.49	0.51	4.0	7.6
Toluene	0.17	0.83	0.00	2.4	2.4
Triethylamine	0.00	0.84	0.16	1.9	2.4
Trifluoroethanol	0.68	0.00	0.32		
Water	0.43	0.18	0.45	10.2	80

[a]Values from [4].
[b]Polarity index; values from [5].
[c]Dielectric constant; values from [3].

the solvent-selectivity triangle (Fig. 2.9). Solvent polarity P' is a measure of overall solvent polarity. Sample solubility tends to correlate with values of P'—"like dissolves like," so samples tend to be more soluble in solvents of similar P'. Less polar compounds, such as hydrocarbons, will be preferentially dissolved by solvents with low values of P', and the reverse will be true for solvents with high values of P'. The dielectric constant ε similarly correlates with the ability of the solvent to

Table I.5

Viscosity of RPC Mobile Phases as a Function of Composition (%B) and Temperature (*T*)[a]

T	%B										
°C	0	10	20	30	40	50	60	70	80	90	100
15	1.10	1.43	1.72	1.92	2.00	2.02	1.91	1.69	1.40	1.05	0.63
	1.10	1.18	1.23	1.30	1.09	0.98	0.89	0.81	0.70	0.54	0.40
20	1.00	1.32	1.57	1.75	1.83	1.83	1.72	1.52	1.25	0.93	0.60
	1.00	1.14	1.10	1.13	0.99	0.90	0.81	0.69	0.56	0.50	0.37
25	0.89	1.18	1.40	1.56	1.62	1.62	1.54	1.36	1.12	0.84	0.56
	0.89	1.01	0.98	0.98	0.89	0.82	0.72	0.59	0.52	0.46	0.35
30	0.79	1.04	1.23	1.36	1.43	1.43	1.36	1.21	1.01	0.76	0.51
	0.79	0.90	0.87	0.86	0.80	0.74	0.65	0.52	0.45	0.43	0.32
35	0.70	0.92	1.07	1.19	1.24	1.26	1.21	1.09	0.91	0.69	0.46
	0.70	0.73	0.78	0.76	0.72	0.68	0.59	0.47	0.43	0.39	0.30
40	0.64	0.82	0.96	1.05	1.11	1.12	1.08	0.98	0.83	0.64	0.42
	0.64	0.72	0.70	0.68	0.65	0.62	0.54	0.44	0.41	0.36	0.27
45	0.58	0.75	0.87	0.96	1.00	1.02	0.98	0.89	0.76	0.58	0.39
	0.58	0.61	0.64	0.61	0.59	0.58	0.50	0.43	0.38	0.33	0.25
50	0.54	0.71	0.82	0.89	0.93	0.94	0.90	0.82	0.70	0.54	0.37
	0.54	0.60	0.60	0.57	0.55	0.53	0.46	0.41	0.36	0.31	0.24
55	0.51	0.67	0.77	0.84	0.88	0.88	0.84	0.76	0.65	0.50	0.36
	0.51	0.53	0.56	0.53	0.51	0.49	0.43	0.38	0.34	0.29	0.23
60	0.47	0.61	0.70	0.77	0.81	0.81	0.79	0.72	0.61	0.47	0.33
	0.47	0.52	0.53	0.50	0.49	0.46	0.41	0.35	0.37	0.27	0.22

Source: Data from [6, 7].
[a]The composition is given as %B (v/v), where B is either methanol (top) or acetonitrile (bottom); for example, the viscosity of 30% methanol water at 30°C is 1.36. See [8, 9] for viscosity compared to composition, temperature, pressure, as well as compressibility data.

Table I.6

Density of Solvents for RPC Mobile Phases [3]

Solvent	Density at Temperature (g/mL)		
	20°C	22°C	25°C
Acetonitrile	0.7822	0.7800	0.7766
Methanol	0.7913	0.7894	0.7866
2-Propanol	0.7855	0.7838	0.7813
Tetrahydrofuran	0.8892	0.8874	0.8847
Water	0.9982	0.9977	0.9970

Note: Error of 1°C = 0.1%, error in weight = ≪0.1%, error in %v.

dissolve ionized solutes or buffers; high values of ε favor increased solubility for ionized compounds. Table I.5 provides viscosity values for mixtures of MeOH/water and ACN/water as a function of temperature. These data are useful in estimating column pressure drop (Eq. 2.13). Table I.6 provides densities for some common solvents, to facilitate the more accurate formulation of reversed-phase mobile phases by weighing each solvent in the mixture.

I.3 SOLVENT SAFETY

The solvents commonly used for HPLC are often flammable and moderately toxic. Consequently most of these solvents should be stored in a secure, metal cabinet. Solvent flammability can be roughly assessed by the flash point, values of which are listed in Table 1.7. Common experience suggests that methanol is only moderately flammable, so that solvents with flash points above 12°C should not normally present a problem in terms of fire safety. However, solvents with lower flash points present a greater danger and should be treated accordingly.

Many factors can contribute to solvent toxicity, and solvents other than water should be manipulated in a hood. A very rough measure of immediate toxicity is the solvent LD_{50} value (Table 1.7), the administered amount in mg/kg body weight that causes mortality in 50% of the population. However, solvents in the laboratory

Table I.7

Flammability and Toxicity Data for Various Solvents

Solvent	Flash Point (°C)[a]	Threshold Limit[b] (ppm)
Acetone	−18	1000
Acetonitrile	42	40
Carbon tetrachloride	None	10
Chloroform	None	25
Ethyl acetate	13	400
Ethyl ether	−45	400
Heptane	−4	85
Hexane	−26	100
Methanol	12	200
Methyl-*t*-butyl ether	−28	—
Methylene chloride	None	500
n-Propanol	27	200
i-Propanol	12	400
Tetrahydrofuran	−20	200
Water	None	none

[a]Data from [2].

[b]Maximum allowable concentration of solvent vapor in the work place air as established by governmental regulation [2].

are rarely ingested; rather, their primary effect is by contact or inhalation. Rubbing alcohol (*i*-propanol) with an LD_{50} of 400 is clearly not a problem in terms of either contact or inhalation. Material Safety Data Sheets (MSDS) should be consulted before handling any solvent or reagent.

REFERENCES

1. J. B. Li, *LCGC*, 10 (1992) 856.
2. *High-Purity Solvent Guide*, Burdick & Jackson Laboratories, Muskegon, MI, 1980.
3. J. A. Riddick and W. B. Bunger, *Organic Solvents*, Wiley-Interscience, New York, 1970.
4. L. R. Snyder, in *High-performance Liquid Chromatography: Advances and Perspectives*, Vol. 3, C. Horväth, ed., Academic Press, New York, 1983, p. 157.
5. K. Valko, L. R. Snyder and J. L. Glajch, *J. Chromatogr.*, 656 (1993) 501.
6. L. R. Snyder, *J. Chromatogr. Sci.*, 16 (1978) 223.
7. H. Colin, J. C. Diez-Masa, G. Guiochon, T. Czajkowska, and I. Miedziak, *J. Chromatogr.*, 167 (1978) 41.
8. M. A. Quarry, R. L. Grob, and L. R. Snyder, *J. Chromatogr.*, 285 (1984) 1.
9. J. Billen, K. Broeckhoven, A Liekens, K. Choikhet, G. Rozing, and G. Desmet, *J. Chromatogr. A*, 1210 (2008) 30.

APPENDIX II

PREPARING BUFFERED MOBILE PHASES

II.1 SEQUENCE OF OPERATIONS

Buffered mobile phases can be prepared by the following sequence of operations:

1. combine the buffer ingredients with water to obtain the aqueous buffer (solution A)
2. confirm or adjust the pH of solution A with a pH meter
3. combine a given volume (e.g., 200 mL) of organic buffer (solution B) with a given volume (e.g., 800 mL) of solution A from step 2 to obtain the final mobile phase (20% organic buffer in this example)
4. check the pH of the final mobile phase (optional)

Because a pH measurement for a mobile phase that contains organic buffer is unreliable due to drift of the pH meter, step 4 is only useful for detecting major errors in the formulation or comparing two solutions with the same organic content. Most laboratories elect to skip step 4.

The usual approach in step 1 is to formulate aqueous buffers of differing pH (A1 and A2), and then combine these two solutions in the correct proportions to obtain final solution A with the desired pH. If the pH is adjusted in step 2, the same two starting solutions can be used to titrate the final buffer to the desired pH as measured by the pH meter. The precision of a pH measurement (step 2) in most laboratories is usually no better than ±0.05 to 0.10 pH unit, which can cause significant changes in resolution for some samples (Section 7.3.4.1). When an HPLC method is pH sensitive, step 2 should be used only for an approximate confirmation of pH. By combining accurate weights of the buffer ingredients with accurate volumes of distilled and degassed water (without further adjusting pH), the pH of the buffer solution can be controlled within narrow limits (±0.02 unit). Buffer concentrations whose pH is known quite accurately are also commercially available.

Introduction to Modern Liquid Chromatography, Third Edition, by Lloyd R. Snyder, Joseph J. Kirkland, and John W. Dolan

Table II.1

Preparation of Low-pH Phosphate Buffers of Defined pH

Required pH	Volume (mL) of A1[a]	Volume (mL) of A[b]
2.0	565	435
2.2	455	545
2.4	345	655
2.6	250	750
2.8	175	825
3.0	110	890
3.2	55	945

[a]Solution of 0.1 M phosphoric acid; the phosphoric acid used to prepare this stock solution must be titrated to confirm the amount of phosphoric acid present.
[b]Solution of 0.1 M sodium monophosphate; combine 13.8 g of NaH_2PO_4 monohydrate with water in a 1-L flask.

It is common practice to adjust the buffer pH with a concentrated acid. For example, solution A2 of Table II.1 might be prepared and titrated to the desired pH with concentrated phosphoric acid. This still produces a buffer at the desired pH, but the ionic strength of the buffer will be higher than if equimolar solutions of A1 and A2 are blended. While it is unlikely to make much difference in the chromatographic results obtained by the two techniques (titrating with concentrated acid vs. equimolar blending) for RPC, some separations can be sensitive to differences in ionic strength (especially ion exchange). It is best to describe in the method documentation exactly how a buffer is to be prepared, and to follow these directions—consistency in mobile-phase preparation is generally important and will give more reliable results.

Acidic or basic additives are sometimes added to the mobile phase for various purposes. When such additives are not used as the primary buffer, they should be added to the desired quantity (concentration) of the buffer first; the mixture should then be adjusted to the desired pH by titrating with acid or base.

II.2 RECIPES FOR SOME COMMONLY USED BUFFERS

The pH of a buffered solution remains approximately constant as the buffer is diluted or concentrated, or when one ionized cation (e.g., Na^+, K^+) or anion (e.g., Cl^-, Br^-) is replaced by another. Tables II.1 to II.3 describe the preparation of some buffers that are commonly used in RPC (adapted from [1])—using the mixing of two solutions A1 and A2, each of which have an equal concentration of the buffering species. The specified volumes of solutions A1 and A2 are combined and mixed to give the final buffer solution of a required pH. The formulations of Tables II.1 to II.3 are based on a final buffer concentration of 0.1M and sodium as cation. Formulations for other buffer concentrations and/or the use of different cations (K^+ is usually preferred) can be inferred from these data. The pH of buffers that are more dilute or more concentrated, or that contain different cations, may differ slightly

Table II.2

Preparation of Acetate Buffers of Defined pH

Required pH	Volume (mL) of A1[a]	Volume (mL) of A2[b]
3.6	926	74
3.8	880	120
4.0	820	180
4.2	736	264
4.4	610	390
4.6	510	490
4.8	400	600
5.0	296	704
5.2	210	790
5.4	176	824
5.6	96	904

[a]Solution of 0.1 M acetic acid; combine 6.0 g (5.8 mL) of glacial acetic acid with water in a 1-L flask.
[b]Solution of 0.1 M sodium acetate; combine 8.2 g of sodium acetate (or 13.6 g sodium acetate trihydrate) with water in a 1-L flask.

Table II.3

Preparation of Intermediate-pH Phosphate Buffers of Defined pH

Required pH	Volume (mL) of A1[a]	Volume (mL) of A2[b]
5.6	948	52
5.8	920	80
6.0	877	123
6.2	815	185
6.4	735	265
6.6	685	315
6.8	510	490
7.0	390	610
7.2	280	720
7.4	190	810
7.6	130	870
7.8	85	915
8.0	53	947

[a]solution of 0.1 M monobasic sodium monophosphate; combine 13.8 g of monobasic sodium monophosphate monohydrate with water in a 1-L flask.
[b]solution of 0.1 M dibasic sodium phosphate; combine 26.8 g of $Na_2HPO_4 \cdot 7H_2O$ with water in a 1-L flask.

from these values. The exact pH value of the mobile phase is usually unimportant in method development. What is important is that the final pH of the mobile phase can be reproduced (preferably within ±0.02 unit) each time a new batch of mobile phase is prepared. Note that solutions only buffer effectively ±1 pH unit from the pK_a value of the ionizable constituent (Section 7.2.1). Although the mobile phase may be used at a temperature other than ambient, the pH at ambient is assumed for the buffers of Tables II.1 to II.3 and should be used to describe the final mobile phase.

As an alternative to Tables II.1 to II.3 as guides for buffer preparation, many on-line buffer calculators are available (search for "HPLC buffer calculator") that provide for the use of several additional buffers. For example, one such calculator ("The Buffer Wizard," Zirchrom, Anoka, MN, www.zirchrom.com) provides buffer preparation instructions. Input the acid, base, desired buffer concentration, and pH, and the calculator provides instructions for preparation, along with warnings about buffer capacity, column stability, and so forth.

REFERENCE

1. G. Gomori, in Meth. Enzymology, S. P. Colowicxk and N. O. Kaplan, eds., Academic Press, New York, 1955, p. 145.

INDEX

Introduction to Modern Liquid Chromatography, Third Edition, by Lloyd R. Snyder, Joseph J. Kirkland, and John W. Dolan